AF734327

Volkswagen EuroVan

Repair Manual
1992, 1993, 1994, 1995, 1996, 1997, 1998, 1999

Gasoline, Diesel, TDI, 5-cylinder and VR6, including MultiVan and CV Camper

Selected Books and Repair Information From Bentley Publishers

Motorsports

Alex Zanardi: My Sweetest Victory
Alex Zanardi and Gianluca Gasparini
ISBN 978-0-8376-1249-2

The Unfair Advantage
Mark Donohue and Paul van Valkenburgh
ISBN 978-0-8376-0069-7

Equations of Motion - Adventure, Risk and Innovation *William F. Milliken*
ISBN 978-0-8376-1570-7

Engineering

Mechanical Ignition Handbook: The Hack Mechanic Guide to Vintage Ignition Systems
Rob Siegel ISBN 978-0-8376-1767-1

Hack Mechanic Guide to European Automotive Electrical Systems
Rob Siegel ISBN 978-0-8376-1751-0

Bosch Automotive Handbook *Robert Bosch GmbH* ISBN 978-0-8376-1732-9

Bosch Fuel Injection and Engine Management *Charles O. Probst, SAE*
ISBN 978-0-8376-0300-1

Maximum Boost: Designing, Testing, and Installing Turbocharger Systems
Corky Bell ISBN 978-0-8376-0160-1

Supercharged! Design, Testing and Installation of Supercharger Systems
Corky Bell ISBN 978-0-8376-0168-7

Scientific Design of Exhaust and Intake Systems *Phillip H. Smith & John C. Morrison* ISBN 978-0-8376-0309-4

Physics for Gearheads
Randy Beikmann ISBN 978-0-8376-1615-5

Audi

Audi A4 Service Manual: 2002-2008, 1.8L Turbo, 2.0L Turbo, 3.0L, 3.2L
Bentley Publishers
ISBN 978-0-8376-1574-5

Audi TT Service Manual: 2000-2006, 1.8L turbo, 3.2 L *Bentley Publishers*
ISBN 978-0-8376-1625-4

Audi A6 (C5 platform) Service Manual: 1998-2004 *Bentley Publishers*
ISBN 978-0-8376-1670-4

BMW

Memoirs of a Hack Mechanic
Rob Siegel ISBN 978-0-8376-1720-6

BMW 3 Series (F30, F31, F34) Service Manual: 2012-2015 *Bentley Publishers*
ISBN 978-0-8376-1752-7

BMW X3 (E83) Service Manual: 2004-2010 *Bentley Publishers*
ISBN 978-0-8376-1731-2

BMW X5 (53) Service Manual: 2000-2006 *Bentley Publishers*
ISBN 978-0-8376-1643-8

BMW 3 Series (E90, E91, E92, E93) Service Manual: 2006-2011
Bentley Publishers ISBN 978-0-8376-1723-7

BMW 3 Series (E46) Service Manual: 1999-2005 *Bentley Publishers*
ISBN 978-0-8376-1657-5

BMW 3 Series (E36) Service Manual: 1992-1998 *Bentley Publishers*
ISBN 978-0-8376-1709-1

BMW Z3 (E36/7) Service Manual: 1996-2002 *Bentley Publishers*
ISBN 978-0-8376-1617-9

BMW 5 Series (E60, E61) Service Manual: 2004-2010 *Bentley Publishers*
ISBN 978-0-8376-1689-6

BMW 5 Series (E39) Service Manual: 1997-2003 *Bentley Publishers*
ISBN 978-0-8376-1672-8

Corvette

Corvette: America's Star-Spangled Sports Car *Karl Ludvigsen*
ISBN 978-0-8376-1659-9

Corvette by the Numbers: The Essential Corvette Parts Reference 1955-1982 *Alan Covin*
ISBN 978-0-8376-0228-2

Zora Arkus-Duntov: The Legend Behind Corvette *Jerry Burton*
ISBN 978-08376-0858-7

Porsche

Porsche 911 (996) Service Manual: 1999-2005 *Bentley Publishers*
ISBN 978-0-8376-1710-7

Porsche 911 (993) Service Manual: 1995-1998 *Bentley Publishers*
ISBN 978-0-8376-1719-0

Porsche Boxster Service Manual: 1997-2004 *Bentley Publishers*
ISBN 978-0-8376-1645-2

Porsche 911 Carrera Service Manual: 1984-1989 *Bentley Publishers*
ISBN 978-0-8376-1696-4

Porsche 911 SC Service Manual: 1978-1983 *Bentley Publishers*
ISBN 978-0-8376-1705-3

Porsche: Excellence Was Expected
Karl Ludvigsen
ISBN 978-0-8376-0235-6

Porsche — Origin of the Species
Karl Ludvigsen
ISBN 978-0-8376-1331-4

Volkswagen

Volkswagen Rabbit, GTI Service Manual: 2006-2009 *Bentley Publishers*
ISBN 978-0-8376-1664-3

Volkswagen Jetta Service Manual: 2005-2010 *Bentley Publishers*
ISBN 978-0-8376-1616-2

Volkswagen Jetta, Golf, GTI Service Manual: 1999-2005 *Bentley Publishers*
ISBN 978-0-8376-1678-0

Volkswagen Jetta, Golf, GTI: 1993-1999, Cabrio: 1995-2002 Service Manual *Bentley Publishers*
ISBN 978-0-8376-1660-5

Volkswagen GTI, Golf, Jetta Service Manual: 1985-1992 *Bentley Publishers*
ISBN 978-0-8376-1637-7

Volkswagen Corrado Repair Manual: 1990-1994 *Bentley Publishers*
ISBN 978-0-8376-1699-5

Volkswagen Passat, Passat Wagon Service Manual: 1998-2005
Bentley Publishers
ISBN 978-0-8376-1669-8

MINI Repair Manuals

MINI Cooper Service Manual: 2007-2013 *Bentley Publishers*
ISBN 978-0-8376-1730-5

MINI Cooper Service Manual: 2002-2006 *Bentley Publishers*
ISBN 978-0-8376-1639-1

Mercedes-Benz

Mercedes-Benz C-Class (W202) Service Manual 1994-2000
Bentley Publishers
ISBN 978-0-8376-1692-6

Mercedes Benz E-Class (W124) Owner's Bible: 1986-1995
Bentley Publishers
ISBN 978-0-8376-0230-1

Mercedes-Benz Technical Companion
Staff of The Star and members of Mercedes-Benz Club of America
ISBN 978-0-8376-1033-7

B BentleyPublishers®
Automotive Reference

Bentley Publishers has published service manuals and automobile books since 1950. For more information, please contact Bentley Publishers at 1734 Massachusetts Avenue, Cambridge, MA 02138 USA, or visit our web site at
BentleyPublishers.com

Volkswagen EuroVan

Repair Manual

1992, 1993, 1994, 1995, 1996, 1997, 1998, 1999

Gasoline, Diesel, TDI, 5-cylinder and VR6, including MultiVan and CV Camper

BentleyPublishers.com

BENTLEY PUBLISHERS® | Automotive Reference™

Bentley Publishers, a division of Robert Bentley, Inc.
1734 Massachusetts Avenue
Cambridge, MA 02138 USA
800-423-4595 / 617-547-4170

Information that makes the difference®

BentleyPublishers®.com

Technical contact information

We welcome your feedback. Please submit corrections and additions to our Volkswagen technical discussion forum at:

`http://www.BentleyPublishers.com`

Updates and Corrections

We will evaluate submissions and post appropriate editorial changes online as updates or tech discussion. Appropriate updates and corrections will be added to the book in future printings. Check for updates and corrections for this book before beginning work on your vehicle. See the following web address for additional information:

`http://www.BentleyPublishers.com/updates/`

WARNING—Important Safety Notice

Do not use this manual unless you are familiar with automotive repair procedures and safe workshop practices. This manual illustrates workshop procedures required for service work; it is not a substitute for full and up-to-date information from the vehicle manufacturer or for proper training as an automotive technician. Note that it is not possible for us to anticipate all of the ways or conditions under which vehicles may be serviced or to provide cautions as to all of the possible hazards that may result.

We have endeavored to ensure the accuracy of the information in this manual. Please note, however, that considering the vast quantity and the complexity of the service information involved, we cannot warrant the accuracy or completeness of the information contained in this manual.

FOR THESE REASONS, NEITHER THE PUBLISHER NOR THE AUTHOR MAKES ANY WARRANTIES, EXPRESS OR IMPLIED, THAT THE INFORMATION IN THIS BOOK IS FREE OF ERRORS OR OMISSIONS AND WE EXPRESSLY DISCLAIM THE IMPLIED WARRANTIES OF MERCHANTABILITY AND OF FITNESS FOR A PARTICULAR PURPOSE, EVEN IF THE PUBLISHER OR AUTHOR HAVE BEEN ADVISED OF A PARTICULAR PURPOSE, AND EVEN IF A PARTICULAR PURPOSE IS INDICATED IN THE MANUAL. THE PUBLISHER AND AUTHOR ALSO DISCLAIM ALL LIABILITY FOR DIRECT, INDIRECT, INCIDENTAL OR CONSEQUENTIAL DAMAGES THAT RESULT FROM ANY USE OF THE EXAMPLES, INSTRUCTIONS OR OTHER INFORMATION IN THIS BOOK. IN NO EVENT SHALL OUR LIABILITY WHETHER IN TORT, CONTRACT OR OTHERWISE EXCEED THE COST OF THIS MANUAL.

Before attempting any work on your Volkswagen, read the **WARNINGS** and **CAUTIONS** on page vii and viii, and any **WARNING** and **CAUTION** that accompanies a procedure in the service manual. Review the **WARNINGS** and **CAUTIONS** on page vii and viii each time you prepare to work on your Volkswagen.

Your common sense and good judgment are crucial to safe and successful service work. Read procedures through before starting them. Think about whether the condition of your car, your level of mechanical skill, or your level of reading comprehension might result in or contribute in some way to an occurrence which might cause you injury, damage your car, or result in an unsafe repair. If you have doubts for these or other reasons about your ability to perform safe repair work on your car, have the work done at an authorized Volkswagen dealer or other qualified shop.

Part numbers listed in this manual are for identification purposes only, not for ordering. Always check with your authorized Volkswagen dealer to verify part numbers and availability before beginning service work that may require new parts.

Special tools required to perform certain service operations are identified in the manual and are recommended for use. Use of tools other than those recommended in this service manual may be detrimental to the car's safe operation as well as the safety of the person servicing the car.

The vehicle manufacturer will continue to issue service information updates and parts retrofits after the editorial closing of this manual. Some of these updates and retrofits will apply to procedures and specifications in this manual. We regret that we cannot supply updates to purchasers of this manual.

Copies of this manual may be purchased from authorized Volkswagen dealers, most automotive accessories and parts dealers specializing in Volkswagen automobiles, from selected booksellers, or directly from the publisher.

ISBN 978-0-8376-1681-0 (3-Volume Set)
ISBN 978-0-8376-1682-7 (Volume 1)
ISBN 978-0-8376-1683-4 (Volume 2)
ISBN 978-0-8376-1684-1 (Volume 3)

Bentley Stock No. VV99 **Job code: VV99-10** **VWoA, Inc. Part No. LPV 800 149**

Library of Congress Cataloging-in-Publication Data
Volkswagen EuroVan: official factory repair manual 1992,1993, 1994, 1995, 1996, 1997, 1998, 1999: gasoline, diesel, TDI, 5-cylinder, and VR6, including multivan and CV camper
p. cm.
ISBN 978-8376-0335-3 (alk. paper)
1. EuroVan van --Maintenance and repair--Handbooks, manuals, etc.
I. Robert Bentley, inc.
TL230.5.E92V55 1997
629.28'722--DC21
97-31787
CIP

 The paper used in this publication is acid free and meets the requirements of the National Standard for Information Sciences-Permanence of Paper for Printed Library Materials. ∞

Manufactured in the United States of America.

202510R003

Foreword

Service to Volkswagen owners is a top priority of the Volkswagen organization that has always included the continuing development and introduction of new and expanded services. In line with this purpose, Bentley Publishers in cooperation with Volkswagen of America, Inc. has introduced this Volkswagen Official Factory Repair Manual.

This Volkswagen EuroVan manual covers the model years 1992, 1993, 1994, 1995, 1996, 1997, 1998, and 1999. It includes Volkswagen EuroVan models equipped with gasoline and diesel engines. This manual was created specifically to cover only those models built for sale in the United States and Canada.

This manual has been prepared from the information that the Volkswagen organization provides to its own factory-trained technicians. It has been developed primarily with the Volkswagen dealership technician in mind. The aim throughout has been clarity and completeness, with step-by-step procedures and factory specifications.

For the automotive service professional, this manual includes the specifications and procedures that were available in an authorized Volkswagen dealer service department as this manual went to press. The Volkswagen owner who has no intention of working on his or her car will find that owning and referring to this manual will make it possible to be better informed and to more knowledgeably discuss repairs with a professional automotive service technician.

Disclaimer

We have endeavored to ensure the accuracy of the information in this manual. When the vast array of data presented in the manual is taken into account, however, no claim to infallibility can be made. We therefore cannot be responsible for the result of any errors that may have crept into the text. Please also read **WARNING–Important Safety Notice** on the copyright page.

Prior to starting a repair procedure, read the procedure, **Warnings and Cautions** on page vii and viii and the **warnings** and **cautions** that accompany the procedure. Reading a procedure before beginning work helps you determine in advance the need for specific skills, identify hazards, prepare for appropriate capture and handling of hazardous materials, and the need for particular tools and replacement parts such as gaskets.

For those intending to do maintenance and repair on their Eurovan, it is essential that safety equipment be used and safety precautions observed when working on your vehicle. A minimum safety equipment list includes hand protection, eye protection, and a fire extinguisher. A selection of good quality hand tools is also needed. This includes a torque wrench to ensure that fasteners are tightened in accordance with specifications.

Special tools that may be required for a particular repair procedure are usually identified by their Volkswagen special tool number and illustrated. A thorough review of each procedure is recommended to determine in advance the complexity of the repair work and the need for particular tools and replacement parts such as gaskets.

Some of the information in this manual applies only to cars of a particular model year or range of years. For example, “m.y. 1993” refers to the 1993 model year. The model year does not necessarily match the calender year in which the car was manufactured or sold. To be sure of the model year of a particular car, check the vehicle identification number (VIN) on the car. Technical changes made in production within a model year are identified in this manual by listing the VIN for the first car produced with this change.

The VIN is a unique sequence of 17 characters assigned by Volkswagen to identify each individual car. WV2MC0709PH016886 is an example. Each of the 17 letters and numbers indicates certain facts about the car and its manufacture. The characters in the 7th and 8th spaces are Volkswagen's model designation. The letter in the 10th space indicates model year. The last six digits are the unique sequential production number. References to VIN in this manual will often include only the last several characters, such as 709PH016886 or 70_P_016886 from the sample VIN above.

The Volkswagen VIN can be found on a plate mounted on the top of the instrument panel, on the driver's side where the number can be seen through the windshield, and on an identification label on the left door jamb. This manual covers Volkswagen EuroVan from model years 1992 through 1999, with these VINs:

position	1	2	3	4	5	6	7	8	9	10	11	12	13	14	15	16	17		1	2	3	4	5	6	7	8	9	10	11	12	13	14	15	16	17
1992:							7	0		N		0	0	0	0	0	1	to							7	0		N		9	9	9	9	9	9
1993:							7	0		P		0	0	0	0	0	1	to							7	0		P		9	9	9	9	9	9
1994:							7	0		R		0	0	0	0	0	1	to							7	0		V		9	9	9	9	9	9
1995:							7	0		S		0	0	0	0	0	1	to							7	0		S		9	9	9	9	9	9
1996:							7	0		T		0	0	0	0	0	1	to							7	0		T		9	9	9	9	9	9
1997:							7	0		V		0	0	0	0	0	1	to							7	0		V		9	9	9	9	9	9
1998:							7	0		W		0	0	0	0	0	1	to							7	0		W		9	9	9	9	9	9
1999:							7	0		X		0	0	0	0	0	1	to							7	0		X		9	9	9	9	9	9

Parts of this manual apply only to certain models or components. Notice especially that the Fuel Injection and Ignition, Engine, and Transmission Repair Group numbers are repeated, each Group covering a particular engine, fuel and ignition system, or transmission type. Most pages are clearly labeled to indicate the systems or components to which the information applies.

Bentley Publishers encourages comments from the readers of this manual with regard to errors, and/or suggestions for improvement of our product. These communications have been and will be carefully considered in the preparation of this and other manuals. If you identify inconsistencies in the manual, you may have found an error. Please contact the publisher and we will endeavor to post applicable corrections on our website. Posted updates and corrections should be reviewed before beginning work. Please see the following web address:

`http://www.BentleyPublishers.com/updates/`

Volkswagen continues to issue service information and parts retrofits after the editorial closing of this manual. Some of this updated information may apply to procedures and specifications in this manual. For the latest information, please see the following web address:

`http://www.ebahn.com/vw`

Volkswagen offers extensive warranties, especially on components of the fuel delivery and emission control systems. Before deciding to repair a car that may be covered wholly or in part by any warranties issued by Volkswagen of America, Inc., consult an authorized Volkswagen dealer. You may find that the dealer can make the repair either without charge or at minimal cost. Regardless of its age, or whether it is covered by warranty, a Volkswagen is an easy car to get serviced. If at any time a repair is too difficult, or requires special equipment that is not available, a trained Volkswagen dealer service technician is ready to do the job for you.

Bentley Publishers

Please read these WARNINGS and CAUTIONS before proceeding with maintenance and repair work.

WARNING—

• Some repairs may be beyond your capability. If you lack the skills, tools and equipment, or a suitable workplace for any procedure described in this manual, we suggest you leave such repairs to an authorized Volkswagen dealer service department, or other qualified shop.

• Volkswagen is constantly improving its cars. Sometimes these changes, both in parts and specifications, are made applicable to earlier models. Therefore, before starting any major jobs or repairs to components on which passenger safety may depend, consult an authorized Volkswagen dealer about Technical Bulletins that may have been issued since the editorial closing of this manual.

• Do not re-use any fasteners that are worn or deformed in normal use. Some fasteners are designed to be used only once, and are unreliable and may fail if used a second time. This includes, but is not limited to, nuts, bolts, washers, circlip and cotter pins. Always replace these fasteners with new parts.

• Never work under a lifted car unless it is solidly supported on stands designed for the purpose. Do not support a car on cinder blocks, hollow tiles or other props that may crumble under continuous load. Never work under a car that is supported solely by a jack. Never work under the car while the engine is running.

• If you are going to work under a car on the ground, make sure that the ground is level. Block the wheels to keep the car from rolling. Disconnect the battery negative terminal (ground strap) to prevent others from starting the car while you are under it.

• Never run the engine unless the work area is well ventilated. Carbon monoxide (CO) kills.

• Do not attempt to work on a vehicle if you do not feel well. You increase the danger of injury to yourself and others if you are tired, upset or have taken medication or any other substance that may impair you or keep you from being fully alert.

• Disconnect the battery negative (–) terminal (ground strap) whenever you work on the fuel system or the electrical system. Do not smoke or work near heaters or other fire hazards. Keep an approved fire extinguisher handy.

• Batteries produce explosive hydrogen gas. Keep sparks, lighted matches and open flame away from the top of the battery. If hydrogen gas escaping from the cap vents is ignited, it will ignite gas trapped in the cells and cause the battery to explode.

• Connect and disconnect battery cables, jumper cables or a battery charger only with the ignition switched off, to prevent sparks. Do not quick-charge the battery (for boost starting) for longer than one minute and do not allow charging voltage to exceed 16.5 volts. Wait at least one minute before boosting the battery a second time.

• Illuminate your work area adequately but safely. Use a portable safety light for working inside or under the car. Make sure the bulb is enclosed by a wire cage. The hot filament of an accidentally broken bulb can ignite spilled fuel or oil.

• Tie long hair behind your head. Do not wear a necktie, a scarf, loose clothing, or a necklace when you work near machine tools or running engines. If hair, clothing, or jewelry were to get caught in the machinery, severe injury could result.

• Finger rings, bracelets and other jewelry should be removed so that they cannot cause electrical shorts, get caught in running machinery, or be crushed by heavy parts.

• Catch draining fuel, oil, or brake fluid in suitable containers. Do not use food or beverage containers that might mislead someone into drinking from them. Store flammable fluids away from fire hazards. Wipe up spills at once, but do not store the oily rags, which can ignite and burn spontaneously.

• Always observe good workshop practices. Wear goggles when you operate machine tools or work with battery acid. Gloves or other protective clothing should be worn whenever the job requires working with harmful substances.

• The Air-Conditioning (A/C) system is filled with chemical refrigerant, which is hazardous. The A/C system should be serviced only by trained technicians using approved refrigerant recovery/recycling equipment, trained in related safety precautions, and familiar with regulations governing the discharging and disposal of automotive chemical refrigerants.

• Do not expose any part of the A/C system to high temperatures such as open flame. Excessive heat will increase system pressure and may cause the system to burst.

• Some aerosol tire inflators are highly flammable. Be extremely cautious when repairing a tire that may have been inflated using an aerosol tire inflator. Keep sparks, open flame or other sources of ignition away from the tire repair area. Inflate and deflate the tire at least four times before breaking the bead from the rim. Completely remove the tire from the rim before attempting any repair.

• Friction materials such as brake or clutch discs may contain asbestos fibers. Do not create dust by grinding, sanding, or by cleaning with compressed air. Avoid breathing asbestos fibers and asbestos dust. Breathing asbestos can cause serious diseases such as asbestosis or cancer, and may result in death.

• Greases, lubricants and other automotive chemicals contain toxic substances, many of which are absorbed directly through the skin. Read manufacturer's instructions and warnings carefully. Use hand and eye protection. Avoid direct skin contact.

• An airbag is an explosive device. Handled improperly or without adequate safeguards, it can be accidently activated and cause serious personal injury. To prevent personal injury or airbag system failure, **only trained Volkswagen dealer service technicians** should test, disassemble or service airbag system.

WARNINGS and CAUTIONS continued on next page ▶

Please read these WARNINGS and CAUTIONS before proceeding with maintenance and repair work.

WARNING (continued) —

• ALWAYS disconnect the power supply before working on the airbag system, or when doing repairs that require removing airbag system components. Disconnect the battery negative (–) terminal and cover the battery.

• The airbag unit should not be removed, nor should any work be performed on the airbag system, without first reviewing the warnings and cautions that accompany the airbag system repair information in this manual.

• Never apply stickers or any other coverings on the steering wheel. Do not let chemical cleaners, oil or grease come into contact with vinyl covering of the airbag unit.

• Never open or otherwise attempt to repair airbag system parts. Always use new parts. Never leave airbag parts or the partially disassembled airbag system unattended.

• Never use a test light or volt meter to conduct electrical tests on the airbag system. The airbag system must only be tested by trained Volkswagen Service technicians using special airbag tester equipment. The airbag system must only be tested while it is installed in the car.

• Do not expose an airbag unit to temperatures above 194°F (90°C), even for a brief period, while handling it during the repair process. Keep the unit clear of heat sources such as hot plates, soldering irons, heat lamps, welding equipment, or the like.

• Always replace airbag system components that have been mechanically damaged (bubbles, cracks, etc.).

• Always set down a removed airbag unit so that the horn pad is facing upward.

• The storage, transportation and disposal of airbag units are subject to the laws for flammable solids.

• When driving or riding in an airbag-equipped vehicle, never hold test equipment in your hands or lap while the vehicle is in motion. Objects between you and the airbag can increase the risk of injury in an accident.

CAUTION—

• If you lack the skills, tools and equipment, or a suitable workshop for any procedure described in this manual, we suggest you leave such repairs to an authorized Volkswagen dealer or other qualified shop. We especially urge you to consult an authorized Volkswagen dealer before beginning repairs on any car that may still be covered wholly or in part by any of the extensive warranties issued by Volkswagen of America, Inc.

• Volkswagen is constantly improving its cars and sometimes these changes, both in parts and specifications, are made applicable to earlier models. Therefore part numbers in this manual are listed for reference only. Always check with an authorized Volkswagen dealer parts department for the latest information.

• Before starting a job, make certain that you have all the necessary tools and parts on hand. Read all the instructions thoroughly, do not attempt shortcuts. Use tools that are appropriate to the work and use only replacement parts meeting Volkswagen specifications. Makeshift tools, parts and procedures will not make good repairs.

• Use pneumatic and electric tools only to loosen threaded parts and fasteners. Never use these tools to tighten fasteners, especially on light alloy parts. Always use a torque wrench to tighten fasteners to the tightening torque specification listed.

• Be mindful of the environment and ecology. Before you drain the crankcase, find out the proper way to dispose of the oil. Do not pour oil onto the ground, down a drain, or into a stream, pond or lake. Consult local ordinances that govern the disposal of wastes.

• When connecting cables or ground straps, all connections must be cleaned to bare metal. Be sure mounting bolts are tight.

• Do not replace fuse links with fuse links of other gauge or with standard wire. Always replace the entire length of the fuse link.

• Before doing any electrical welding on vehicles equipped with Anti-lock Brakes (ABS), disconnect the battery negative terminal (ground strap) and the ABS Control Module connector.

• On cars equipped with ABS, the ABS control unit cannot withstand temperatures from a paint-drying booth or a heat lamp in excess of 203°F (95°C) and should not be subjected to temperatures in excess of 185°F (85°C) for more than two hours.

• On cars equipped with anti-theft radios, make sure you know the correct radio activation code before disconnecting the battery or removing the radio. If the wrong code is entered into the radio when power is restored, that radio may lock up and be rendered inoperable, even if the correct code is then entered.

• When boost-charging the battery, first remove the fuses for the Engine Control Module (ECM), the Transmission Control Module (TCM), the ABS Control Module, and the On-Board computer. In cases where one or more of these components is not separately fused, disconnect the Control Module connector(s).

• Do not quick-charge the battery (for boost starting) for longer than one minute. Wait at least one minute before boosting the battery a second time.

◀ **WARNINGS and CAUTIONS continued from previous page**

You are now in ▶ Volume 2

Automatic Transmission 098

Section H

WARNING!
Before proceeding with any work, please read the warnings and cautions immediately following this index. Also read any warning and caution that accompanies a procedure in the manual.

39 Final drive, Differential

Please read these WARNINGS and CAUTIONS before proceeding with maintenance and repair work.

WARNING—

- Some repairs may be beyond your capability. If you lack the skills, tools and equipment, or a suitable workplace for any procedure described in this manual, we suggest you leave such repairs to an authorized Volkswagen dealer service department, or other qualified shop.

- Volkswagen is constantly improving its cars. Sometimes these changes, both in parts and specifications, are made applicable to earlier models. Therefore, before starting any major jobs or repairs to components on which passenger safety may depend, consult your authorized Volkswagen dealer about the Technical Bulletins that may have been issued since the editorial closing of this manual.

- Do not re-use any fasteners that are worn or deformed in normal use. Many fasteners are designed to be used only once and become unreliable and may fail when used a second time. This includes, but is not limited to, nuts, bolts, washers, self–locking nuts or bolts, circlips and cotter pins. Always replace these fasteners with new parts.

- Never work under a lifted car unless it is solidly supported on stands designed for the purpose. Do not support a car on cinder blocks, hollow tiles or other props that may crumble under continuous load. Never work under a vehicle that is supported solely by a jack. Never work under the car while the engine is running.

- If you are going to work under a car on the ground, make sure that the ground is level. Block the wheels to keep the car from rolling. Disconnect the battery negative terminal (ground strap) to prevent others from starting the vehicle while you are under it.

- Never run the engine unless the work area is well ventilated. Carbon monoxide (CO) kills.

- Do not attempt to work on your vehicle if you do not feel well. You increase the danger of injury to yourself and others if you are tired, upset or have taken medicine or any other substance that may impair you or keep you from being fully alert.

- Disconnect the battery negative (–) terminal (ground strap) whenever you work on the fuel system or the electrical system. Do not smoke or work near heaters or other fire hazards. Keep an approved fire extinguisher handy.

- Batteries produce explosive hydrogen gas. Keep sparks, lighted matches and open flame away from the top of the battery. If hydrogen gas escaping from the cap vents is ignited, it will ignite gas trapped in the cells and cause the battery to explode.

- Connect and disconnect battery cables, jumper cables or a battery charger only with the ignition switched off, to prevent sparks. Do not quick–charge the battery (for boost starting) for longer than one minute, and do not allow charging voltage to exceed 16.5 volts. Wait at least one minute before boosting the battery a second time.

- Illuminate your work area adequately but safely. Use a portable safety light for working inside or under the car. Make sure the bulb is enclosed by a wire cage. The hot filament of an accidentally broken bulb can ignite spilled fuel or oil.

- Tie long hair behind your head. Do not wear a necktie, a scarf, loose clothing, or a necklace when you work near machine tools or running engines. If your hair, clothing, or jewelry were to get caught in the machinery, severe injury could result.

- Finger rings, bracelets and other jewelry should be removed so that they cannot cause electrical shorts, get caught in running machinery, or be crushed by heavy parts.

- Catch draining fuel, oil, or brake fluid in suitable containers. Do not use food or beverage containers that might mislead someone into drinking from them. Store flammable fluids away from fire hazards. Wipe up spills at once, but do not store the oily rags, which can ignite and burn spontaneously.

- Always observe good workshop practices. Wear goggles when you operate machine tools or work with battery acid. Gloves or other protective clothing should be worn whenever the job requires working with harmful substances.

- The Air-Conditioning (A/C) system is filled with a chemical refrigerant, which is hazardous. The A/C system should be serviced only by trained technicians using approved refrigerant recovery/recycling equipment, trained in related safety precautions, and familiar with regulations governing the discharging and disposal of automotive chemical refrigerants.

- Do not expose any part of the A/C system to high temperatures such as an open flame. Excessive heat will increase system pressure and may cause the system to burst.

- Some aerosol tire inflators are highly flammable. Be extremely cautious when repairing a tire that may have been inflated using an aerosol tire inflator. Keep sparks, open flame or other sources of ignition away from the tire repair area. Inflate and deflate the tire at least four times before breaking the bead from the rim. Completely remove the tire from the rim before attempting any repair.

- Friction materials such as brake or clutch discs may contain asbestos fibers. Do not create dust by grinding, sanding, or by cleaning with compressed air. Avoid breathing asbestos fibers and asbestos dust. Breathing asbestos can cause serious diseases such as asbestosis or cancer, and may result in death.

- Greases, lubricants and other automotive chemicals contain toxic substances, many of which are absorbed directly through the skin. Read manufacturer's instructions and warnings carefully. Use hand and eye protection. Avoid direct skin contact.

- An airbag is operated by an explosive device. Handled improperly or without adequate safeguards, it can be accidently activated and cause serious personal injury. To prevent personal injury or airbag system failure, **only trained Volkswagen dealer service technicians** should test, disassemble or service the airbag system.

WARNINGS and CAUTIONS continued on next page ▶

Please read these WARNINGS and CAUTIONS before proceeding with maintenance and repair work.

WARNING (continued) —

- ALWAYS disconnect the airbag power supply before using a computer memory saver. Failure to disconnect the airbag power supply may result in accidental activation of the airbag.
- The airbag unit should not be removed, nor should any work be performed on the airbag system, without first reviewing the warnings and cautions that accompany the airbag system repair information in this manual.
- Never apply stickers or any other coverings on the steering wheel. Do not let chemical cleaners, oil or grease come into contact with the vinyl covering of the airbag unit.
- Never open or otherwise attempt to repair airbag system parts. Always use new parts. Never leave airbag parts or the partially disassembled airbag system unattended.
- Never use a test light or volt meter to conduct electrical tests on the airbag system. The airbag system must only be tested by trained Volkswagen Service technicians using special airbag tester equipment. The airbag system must only be tested while it is installed in the car.
- Do not expose an airbag unit to temperatures above 194°F (90°C) even for a brief period, while handling it during the repair process. Keep the unit clear of heat sources such as hot plates, soldering irons, heat lamps, welding equipment, or the like.
- Always replace airbag system components that have been mechanically damaged (bubbles, cracks, etc.).
- Always set down a removed airbag unit so that the horn pad is facing upward.
- When driving or riding in an airbag-equipped vehicle, never hold test equipment in your hands or lap while the vehicle is in motion. Objects between you and the airbag can increase the risk of injury in an accident.
- The storage, transportation and disposal of airbag units are subject to the laws for flammable solids.

CAUTION—

- If you lack the skills, tools and equipment, or a suitable workshop for any procedure described in this manual, we suggest you leave such repairs to an authorized Volkswagen dealer or other qualified shop. We especially urge you to consult an authorized Volkswagen dealer before beginning repairs on any car that may still be covered wholly or in part by any of the extensive warranties issued by Volkswagen of America, Inc.
- Volkswagen is constantly improving its cars and sometimes these changes, both in parts and specifications, are made applicable to earlier models. Therefore, part numbers listed in this manual are for reference only. Always check with your authorized Volkswagen dealer parts department for the latest information.
- Before starting a job, make certain that you have all the necessary tools and parts on hand. Read all the instructions thoroughly, do not attempt shortcuts. Use tools that are appropriate to the work and use only replacement parts meeting Volkswagen specifications. Makeshift tools, parts and procedures will not make good repairs.
- Use pneumatic and electric tools only to loosen threaded parts and fasteners. Never use these tools to tighten fasteners, especially on light alloy parts. Always use a torque wrench to tighten fasteners to the tightening torque specification listed.
- Be mindful of the environment and ecology. Before you drain the crankcase, find out the proper way to dispose of the oil. Do not pour oil onto the ground, down a drain, or into a stream, pond, or lake. Consult local ordinances that govern the disposal of wastes.
- When connecting cables or ground straps, all connections must be cleaned to bare metal. Be sure mounting bolts are tight.
- Do not replace fuse links with fuse links of other gauge or with standard wire. Always replace the entire length of the fuse link.
- Before doing any electrical welding on vehicles equipped with Anti-lock Brakes (ABS), disconnect the battery negative terminal (ground strap) and the ABS Control Module connector.
- On cars equipped with ABS, the ABS control unit cannot withstand temperatures from a paint–drying booth or a heat lamp in excess of 203°F (95°C) and should not be subjected to temperatures in excess of 185°F (85°C) for more than two hours.
- On cars equipped with anti-theft radios, make sure you know the correct radio activation code before disconnecting the battery or removing the radio. If the wrong code is entered when power is restored, the radio may lock up and be rendered inoperable, even if the correct code is then entered.
- When boost-charging the battery, first remove the fuses for the Engine Control Module (ECM), the Transmission Control Module (TCM), the ABS Control Module, and the On-Board computer. In cases where one or more of these components is not separately fused, disconnect the Control Module connector(s).
- Do not quick-charge the battery (for boost starting) for longer than one minute, and do not exceed 16.5 volts at the battery with the boosting cables attached. Wait at least one minute before boosting the battery a second time.

◀ **WARNINGS and CAUTIONS** continued from previous page

Transmission identification

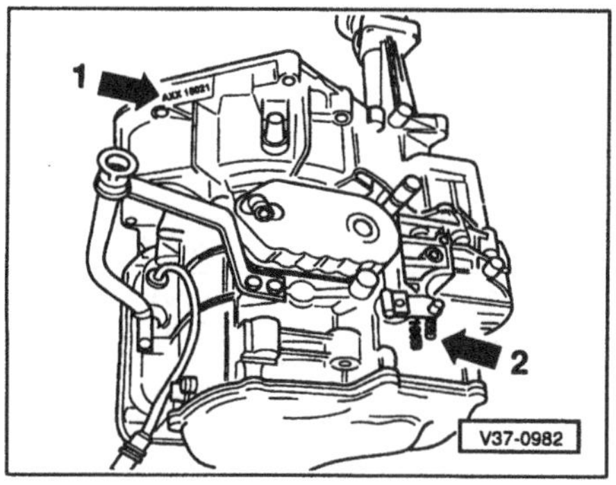

The 4-speed automatic transmission 098 is installed in the EuroVan 1992 ➤ in combination with a 5-cylinder engine. Applications ⇒ page H 00-3.

Location on transmission

◄ Code letters -arrow 1-

Automatic transmission 098 -arrow 2-

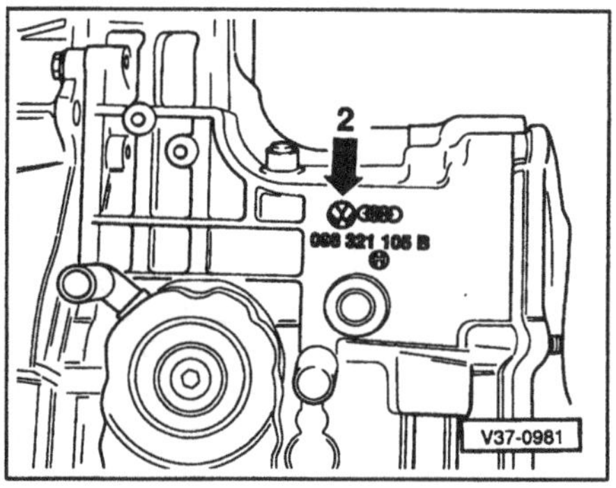

◄ **Automatic transmission 098 -arrow 2-**

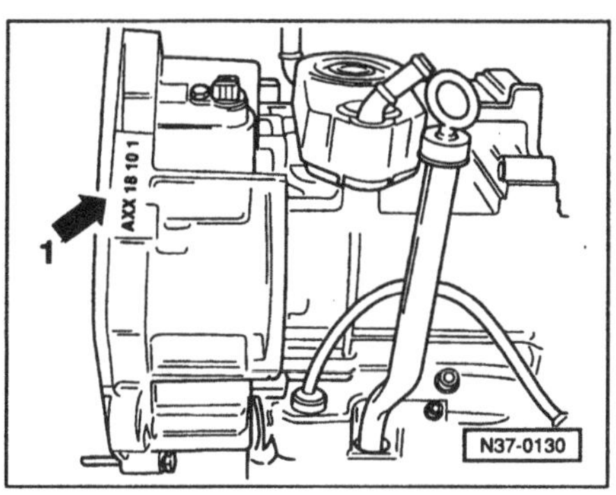

◄ **Code letters and date of manufacture of transmission -arrow 1-**

Example:	AXX	18	10	1
	Code letters	Day	Month	Year (1991) of manufacture

The code letters of the transmission are also given on the vehicle data plates.

Code letters, transmission application, ratios, equipment

Transmissions for vehicles with Transmission Range (TR) Program Switch -E122-

Automatic Transmission			**098**					
Transmission	**Code letters**		**AXX**		**AXW**		**AXY**[1]	
	Manufactured	from to	10.91 08.92		10.91 08.92		10.91 12.91	
Torque converter	**Code letters**		**QBCA**		**QBCA**		**QBCA**	
Valve body	**Code letters**		**QAC**		**QAC**		**QAC**	
	Manufactured	from to	10.91 08.92		10.91 08.92		10.91 12.91	
Number of plates			Inner	Outer	Inner	Outer	Inner	Outer
	Clutch -K1-		5	4	5	4	5	4
	Clutch -K2-		5	5	5	5	5	5
	Clutch -K3-		5	4	5	4	5	4
	Brake -B1-		6	6	6	6	6	6
	Brake -B2-		6	7	6	7	6	7
Application	Model		EuroVan 1993 ➤		EuroVan 1993 ➤		EuroVan 1993 ➤	
	Engine				2.4 L - 57 kW (76 hp) Diesel		2.5 L – 81 kW (109 hp)	

[1] Transmission CEU can be used to replace AXY. If CEU is used, the drive plate with modified TDC marking must be installed ⇒ Removing and installing drive plate; Repair Group 13.

Transmission	**Code letters**		**AXX**	**AXW**	**AXY**
Ratios	1st gear		2.714	2.714	2.714
	2nd gear		1.551	1.551	1.551
	3rd gear		1.000	1.000	1.000
	4th gear		0.679	0.679	0.679
	Reverse gear		2.111	2.111	2.111
Intermediate drive	No. of teeth	Input gear	44	44	44
		Output gear	51	51	51
	Ratio		1.159	1.159	1.159
Final drive	No. of teeth	Drive pinion	16	15	16
		Final drive gear	73	74	73
	Ratio		4.563	4.933	4.563
Drive shaft	Flange diameter mm (in.)		108 (4.25)	108 (4.25)	108 (4.25)
ATF cooler	Equipment		Pressure-side 6-row	Pressure-side 6-row	Pressure-side 6-row

Automatic transmission			098					
Transmission	**Code letters**		**ARW** [1]		**CEU**		**CEV**	
	Manufactured	from to	10.91 12.91		12.91 08.92		01.92 08.92	
Torque converter	**Code letters**		**QBCA**		**QBCA**		**QBCA**	
Valve body	**Code letters**		**QAC**		**QAC**		**QAC**	
	Manufactured	from to	10.91 12.91		12.91 08.92		01.92 08.92	
Number of plates			Inner	Outer	Inner	Outer	Inner	Outer
	Clutch -K1-		5	4	5	4	5	4
	Clutch -K2-		5	5	5	5	5	5
	Clutch -K3-		5	4	5	4	5	4
	Brake -B1-		6	6	6	6	6	6
	Brake -B2-		6	7	6	7	6	7
Application	Model		EuroVan 1993 ➤		EuroVan 1993 ➤		EuroVan 1993 ➤	
	Engine		2.5 L – 81 kW (109 hp)		2.5 L - 81 kW (109 hp)		2.5 L – 81 kW (109 hp)	

[1] Transmission CEV can be used to replace ARW. If CEV is used, the drive plate with modified TDC marking must be installed ⇒ Removing and installing drive plate; Repair Group 13.

Transmission	**Code letters**		**ARW**	**CEU**	**CEV**
Ratios	1st gear		2.714	2.714	2.714
	2nd gear		1.551	1.551	1.551
	3rd gear		1.000	1.000	1.000
	4th gear		0.679	0.679	0.679
	Reverse gear		2.111	2.111	2.111
Intermediate drive	No. of teeth	Input gear	44	44	44
		Output gear	51	51	51
	Ratio		1.159	1.159	1.159
Final drive	No. of teeth	Drive pinion	15	16	15
		Final drive gear	74	73	74
	Ratio		4.933	4.563	4.933
Drive shaft	Flange diameter mm (in.)		108 (4.25)	108 (4.25)	108 (4.25)
ATF cooler	Equipment		Pressure-side 6-row	Pressure-side 6-row	Pressure-side 6-row

Automatic transmission			098					
Transmission	**Code letters**		**CKB**		**CHZ**		**CKA**	
	Manufactured	from to	08.92 12.92		08.92 12.92		08.92 12.92	
Torque converter	**Code letters**		**QBCA**		**QBCA**		**QBCA**	
Valve body	**Code letters**		**QAC**	**QAE[1)]**	**QAC**	**QAE[1)]**	**QAC**	**QAE[1)]**
	Manufactured	from to	08.92 09.92	10.92 12.92	08.92 09.92	10.92 12.92	08.92 09.92	10.92 12.92
Number of plates			Inner	Outer	Inner	Outer	Inner	Outer
	Clutch -K1-		5	4	5	4	5	4
	Clutch -K2-		5	5	5	5	5	5
	Clutch -K3-		5	4	5	4	5	4
	Brake -B1-		6	6	6	6	6	6
	Brake -B2-		6	7	6	7	6	7
Application	Model		EuroVan 1993 ➤		EuroVan 1993 ➤		EuroVan 1993 ➤	
	Engine		2.4 L – 57 kW (76 hp) Diesel		2.4 L - 57 kW (76 hp) Diesel		2.5 L – 81 kW (109 hp)	

1) From transmission build date 01 10 2, the valve body QAE can be installed if wiring is effected by a conductor strip.

Transmission	**Code letters**		**CKB**	**CHZ**	**CKA**
Ratios	1st gear		2.714	2.714	2.714
	2nd gear		1.551	1.551	1.551
	3rd gear		1.000	1.000	1.000
	4th gear		0.679	0.679	0.679
	Reverse gear		2.111	2.111	2.111
Intermediate drive	No. of teeth	Input gear	44	44	44
		Output gear	51	51	51
	Ratio		1.159	1.159	1.159
Final drive	No. of teeth	Drive pinion	15	16	15
		Final drive gear	74	73	74
	Ratio		4.933	4.563	4.933
Drive shaft	Flange diameter mm (in.)		108 (4.25)	108 (4.25)	108 (4.25)
ATF cooler	Equipment		Pressure-side 6-row	Pressure-side 6-row	Pressure-side 6-row

Automatic transmission			098					
Transmission	**Code letters**		**CHY**					
	Manufactured	from to	08.92 12.92					
Torque converter	**Code letters**		**QBCA**					
Valve body	**Code letters**		**QAC**	**QAE**[1]				
	Manufactured	from to	08.92 09.92	10.92 12.92				
Number of plates			Inner	Outer	Inner	Outer	Inner	Outer
	Clutch -K1-		5	4				
	Clutch -K2-		5	5				
	Clutch -K3-		5	4				
	Brake -B1-		6	6				
	Brake -B2-		6	7				
Application	Model		EuroVan 1993 ➤					
	Engine		2.5 L – 81 kW (109 hp)					

[1] From transmission build date 01 10 2 the valve body QAE can be installed if wiring is effected by a conductor strip.

Transmission	**Code letters**		**CHY**		
Ratios	1st gear		2.714		
	2nd gear		1.551		
	3rd gear		1.000		
	4th gear		0.679		
	Reverse gear		2.111		
Intermediate drive	No. of teeth	Input gear	44		
		Output gear	51		
	Ratio		1.159		
Final drive	No. of teeth	Drive pinion	16		
		Final drive gear	73		
	Ratio		4.563		
Drive shaft	Flange diameter mm (in.)		108 (4.25)		
ATF cooler	Equipment		Pressure-side 6-row		

Transmissions for vehicles with electronic program switch in Transmission Control Module (TCM) -J217-

Automatic transmission			098					
Transmission	**Code letters**		**CFZ**[1]		**CGA**[1]		**CGK**[1]	
	Manufactured	from to	01.93		01.93		01.93	
Torque converter	**Code letters**		**QBCA**		**QBCA**		**QBCA**	
Valve body	**Code letters**		**QBB**		**QBB**		**QBB**	
	Manufactured	from to	01.93		01.93		01.93	
Number of plates			Inner	Outer	Inner	Outer	Inner	Outer
	Clutch -K1-		5	5	5	5	5	5
	Clutch -K2-		5	5	5	5	5	5
	Clutch -K3-		5	4	5	4	5	4
	Brake -B1-		5	5	5	5	5	5
	Brake -B2-		6	7	6	7	6	7
Application	Model		EuroVan 1993 ➤		EuroVan 1993 ➤		EuroVan 1993 ➤	
	Engine		2.4 L – 57 kW (76 hp) Diesel		2.4 L - 57 kW (76 hp) Diesel		2.5 L – 81 kW (109 hp)	

[1] On vehicles with modified shift elements, planetary gearbox, gear program switch, and electronic shift program in Transmission Control Module (TCM) -J217-, the Transmission Range (TR) Program Switch -E122- on the instrument panel has been discontinued.

Transmission	**Code letters**		**CFZ**	**CGA**	**CGK**
Ratios	1st gear		2.714	2.714	2.714
	2nd gear		1.441	1.441	1.441
	3rd gear		1.000	1.000	1.000
	4th gear		0.743	0.743	0.743
	Reverse gear		2.884	2.884	2.884
Intermediate drive	No. of teeth	Input gear	44	44	44
		Output gear	51	51	51
	Ratio		1.159	1.159	1.159
Final drive	No. of teeth	Drive pinion	17	16	17
		Final drive gear	72	73	72
	Ratio		4.235	4.563	4.235
Drive shaft	Flange diameter mm (in.)		108 (4.25)	108 (4.25)	108 (4.25)
ATF cooler	Equipment		Pressure-side 6-row	Pressure-side 6-row	Pressure-side 6-row

Automatic transmission			098					
Transmission	**Code letters**		**CGH**[1]		**CTL**[1]			
	Manufactured	from to	01.93		01.94			
Torque converter	**Code letters**		**QBCA**		**QBCA**			
Valve body	**Code letters**		**QBB**		**QBB**			
	Manufactured	from to	01.93		01.94			
Number of plates			Inner	Outer	Inner	Outer	Inner	Outer
	Clutch -K1-		5	5	5	5		
	Clutch -K2-		5	5	5	5		
	Clutch -K3-		5	4	5	4		
	Brake -B1-		5	5	5	5		
	Brake -B2-		6	7	6	7		
Application	Model		EuroVan 1993 ➤		EuroVan 1993 ➤			
	Engine		2.5 L – 81 kW (109 hp)		2.5 L - 81 kW (109 hp)			

[1] On vehicles with modified shift elements, planetary gearbox, gear shift program, and electronic shift program in Transmission Control Module (TCM) -J217-, the Transmission Range (TR) Program Switch -E122- on the instrument-panel has been discontinued.

Transmission	**Code letters**		**CGH**	**CTL**	
Ratios	1st gear		2.714	2.714	
	2nd gear		1.441	1.441	
	3rd gear		1.000	1.000	
	4th gear		0.743	0.743	
	Reverse gear		2.884	2.884	
Intermediate drive	No. of teeth	Input gear	44	44	
		Output gear	51	51	
	Ratio		1.159	1.159	
Final drive	No. of teeth	Drive pinion	16	15	
		Final drive gear	73	74	
	Ratio		4.563	4.933	
Drive shaft	Flange diameter mm (in.)		108 (4.25)	108 (4.25)	
ATF cooler	Equipment		Pressure-side 6-row	Pressure-side 6-row	

Capacities

Planetary gearbox

Capacities	Planetary gearbox	Automatic transmission
Initial filling	6.1 L (6.5 qt.)	098
Oil change	approx. 3.5 L (3.7 qt.)	
Lubricant	ATF Dexron or VW ATF	

Final drive

Capacities	Final drive	Automatic transmission
Initial filling	1.15 L (1.22 qt.)	098
Oil change	Filled for life No change	
Lubricant	Gear oil (synthetic) SAE75 W90	

Note:

- *Gear oil SAE 75 W90 (synthetic oil) is available as a replacement part:*

  Container 0.5 L (0.53 qt.) – Part No. G 052 145 A1

  Container 1.0 L (1.06 qt.) – Part No. G 052 145 A2
- *VW ATF 1.0 L (1.06 qt.) – Part No. G 052 162 A2*

Repair instructions

Proper tools and the maximum possible care and cleanliness are essential for satisfactory and successful transmission repairs. Basic safety precautions apply when doing vehicle repairs.

Standard procedures used throughout the Repair Manual are summarized here.

Special tools

A summary of the special tools used in this Repair Manual can be found in the Special Tools/Catalog microfiche.

Transmission

When an automatic transmission is replaced, the ATF in the planetary gearbox, ⇒ page H 37-46 and the gear oil in the final drive, ⇒ page H 39-1 is to be checked and filled as necessary. Capacities and specifications ⇒ page H 00-15.

- When installing the transmission, make sure that the dowel sleeves for engine/transmission alignment are installed correctly.

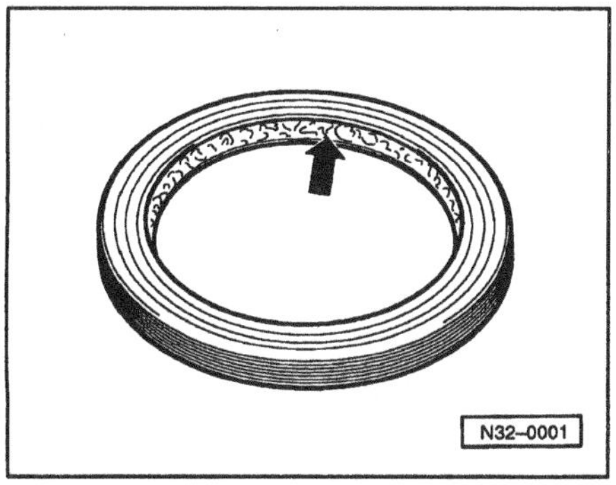

Gaskets, oil seals

- Always install new O-rings.
- Radial shaft seals: Pack space between sealing lips with grease before installing.
- After installing, check gear oil level or ATF level; fill to correct level, if necessary.

Locking elements

- Do not overstretch circlips; replace if necessary.
- Circlips must be properly seated in the grooves.

Nuts, bolts

- Nuts and bolts for attaching covers and housings must be loosened and tightened evenly in an alternating pattern.
- Do not distort particularly sensitive parts, e.g., valve body, and loosen or tighten diagonally in stages.
- The torques given apply to dry nuts and bolts.
- Threads of bolts that are secured by locking fluid are to be cleaned with a wire brush. Then, insert bolts with Part No. AMV 185 100 A1.
- Always replace self-locking nuts.

Bearings

- Install needle bearings with the inscribed side (thicker sheet metal) facing the installation drift.
- Lubricate all bearings in transmission with gear oil before installing. Oil with special care for measuring friction torque.
- Heat the inner races of tapered roller bearings to approximately 100°C (212°F) before installing.
- Do not interchange the inner and outer races of same size bearings.
- Always replace tapered roller bearings which are mounted on a shaft together and use bearings from the same manufacturer.

Shims

- Measure shims at several points with a micrometer. Different tolerances in the same size shim make it possible to obtain the exact shim thickness required.
- Check for signs of burrs and damage. Only install shims that are in perfect condition.

Inner plates

- Place new inner plates in ATF for 15 minutes before installing.

Valve body

- Replace the valve body if any of the selector elements are scorched.

Clutches

- Clutches K1, K2, and K3 should initially be disassembled only for cleaning. If parts are faulty, replace clutch.

On Board Diagnostic (OBD) Program

- Before performing repairs to the automatic transmission, determine the cause of the malfunction as precisely as possible with the aid of the On Board Diagnostic capabilities of the Transmission Control Module (TCM) ⇒ Repair Group 01.

On Board Diagnostic (OBD)

Function

The Transmission Control Module (TCM) -J217- for the automatic transmission is equipped with a Diagnostic Trouble Code (DTC) memory.

If malfunctions occur in the monitored sensors or components, they are stored in the memory together with an indication of the type of malfunction.

Malfunctions which occur only occasionally are known as sporadic malfunctions and are coded as such.

The Transmission Control Module (TCM) -J217- for the automatic transmission analyses the information and distinguishes different malfunctions ⇒ Diagnostic Trouble Code (DTC) tables, starting on page H 01-20, and stores them.

If malfunctions do not reoccur within:

- min. 5 km (3 miles) or 6 minutes or
- max. 20 km (12 miles) or 24 minutes

they are stored as sporadic malfunctions.

Electrical malfunctions that affect vehicle performance can be determined with the VAG 1551 Scan Tool (ST).

The possibilities offered by OBD can only be fully exploited with operating mode 1, "Rapid data transfer" of VAG 1551 Scan Tool (ST).

Functions which the VAG 1551 Scan Tool (ST) can register ⇒ page H 01-17, List of selectable functions.

Safety functions of Transmission Control Module (TCM)

If critical malfunctions occur when driving, the transmission will continue to operate in an emergency running mode.

If the malfunction occurs in "D", "3" or "2", 3rd gear emergency running mode is activated.

If the malfunction occurs in "1","P", "N" or "R", the emergency running mode for that stage is activated.

If the engine is started again in the emergency running mode and the malfunction occurs with the selector lever in "D", "3" or "2", Third Gear is activated hydraulically, until the malfunction is repaired.

If malfunctions exist which result in emergency running, the transmission remains in the emergency running mode until the ignition is turned off.

Malfunctions which may activate the emergency running mode:

Open circuit in wiring, short-circuit, malfunctions in electrical or hydraulic components.

Transmission Control Module (TCM) malfunction recognition

If a malfunction exists for a certain time, it is stored as a static malfunction. If the malfunction no longer exists after a pre-specified time or distance, it is stored as a sporadic malfunction.

These malfunctions which are stored in the Diagnostic Trouble Code (DTC) memory, will be displayed as "sporadic malfunctions" when retrieved by VAG 1551 Scan Tool (ST). "SP" appears on the right of the display in such cases. If the printer is turned on, "sporadic malfunctions" are printed out after the malfunction is addressed.

Malfunctions which are stored in the Diagnostic Trouble Code (DTC) memory as sporadic malfunctions are automatically erased after 1,000 km (600 miles) or 20 hours.

Technical data of On Board Diagnostic

Memory ◆ Permanent memory ◆ Volatile memory	 yes no
Data output ◆ Rapid data transfer ◆ Blink code output	 yes no
Output Diagnostic Test Mode	no
Basic setting	yes
Transmission Control Module (TCM) coding	no
Reading measured value block (VAG 1551 Scan Tool function 08)	yes
Location of components	⇒ page H 01-5

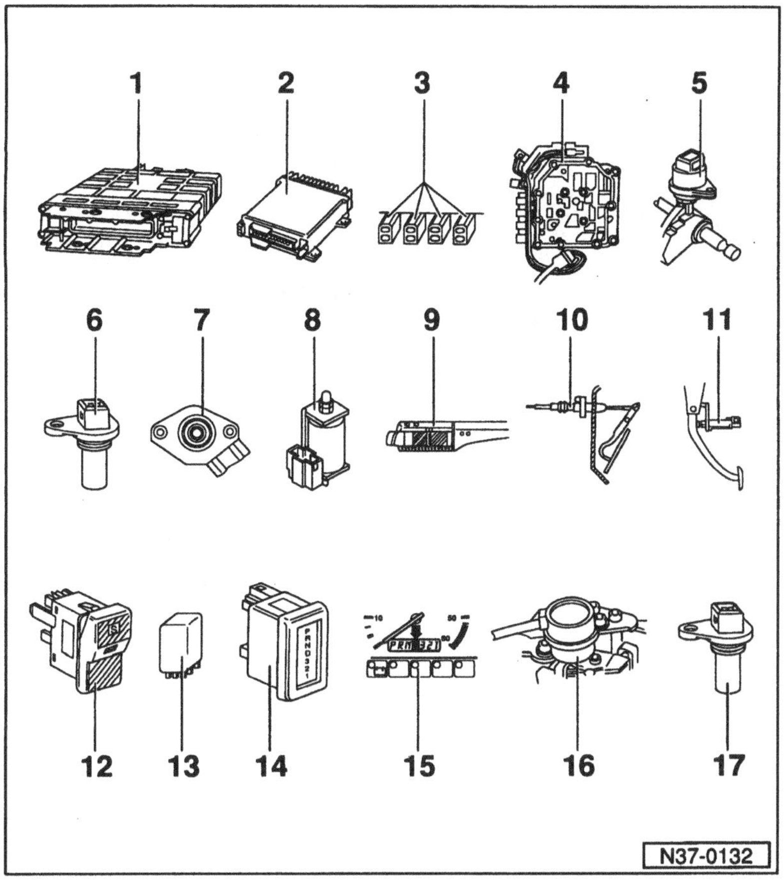

Electronic component locations

Note:

If Engine Control Module (ECM) -J257- or Transmission Control Module (TCM) -J217- are replaced, or if repairs are performed on the Throttle Position (TP) Sensor -G69-, the system must be restored to the basic setting ⇒ page H 01-33, Initiating basic setting.

1 - **Transmission Control Module (TCM) -J217-**
- ♦ Location: Transmission Control Module (TCM) -J217- is located in right footwell on A pillar
- ♦ Checked by On Board Diagnostic (OBD)

2 - **Engine Control Module (ECM) -J257-**
- ♦ Location: Depends on engine equipment ⇒Repair Group 24

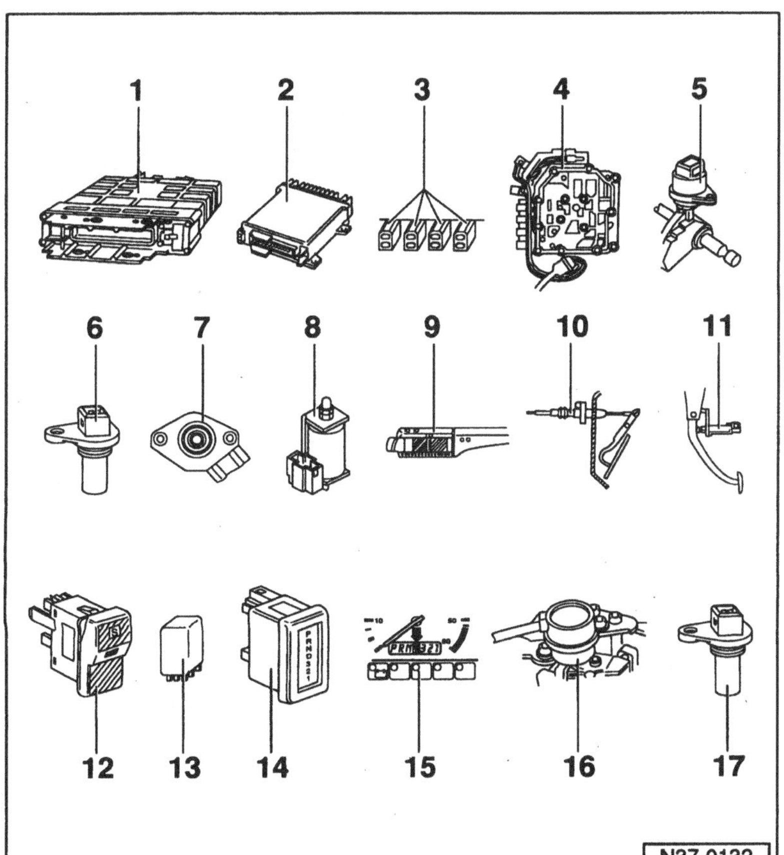

3 - **Data Link Connector (DLC)**
- ♦ Location: The Data Link Connectors (DLC) are located above the relay plate ⇒ Connecting VAG 1551 Scan Tool (ST) and selecting functions, page H 01-12

4 - **Valve body**
- ♦ Location: In transmission "098" behind oil pan ⇒ page H 38-36
- ♦ The Solenoid Valves -N88-, -N89-, -N90-, -N91-, -N92-, -N93-, -N94- and the transmission temperature sensor -G93- are attached to the valve body
- ♦ Components are checked by OBD

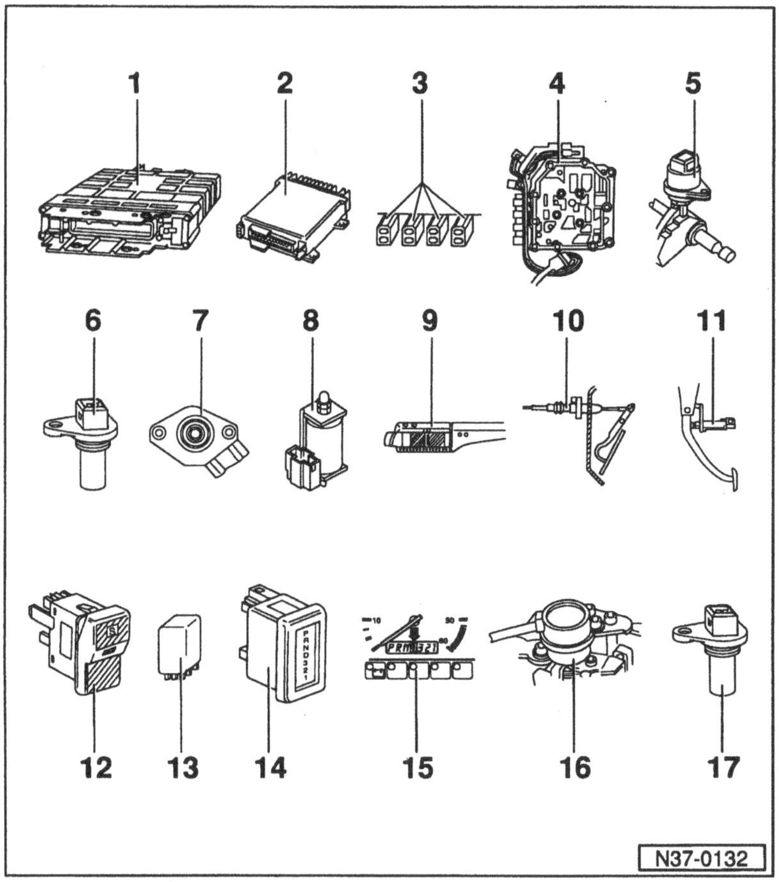

5 - Multi-Function Transmission Range (TR) Switch -F125-

- Location: At rear of transmission housing ⇒ page H 38-45
- Checked by OBD

6 - Vehicle Speed Sensor (VSS) -G68-

- Location: On top of transmission housing ⇒ page H 38-45
- Checked by OBD

7 - Throttle Position (TP) Sensor -G69-

- Location: The Throttle Position (TP) Sensor is located on the engine throttle valve housing
- Checked by OBD
- Various types according to engine version
- Removing and installing ⇒ Repair Group 24

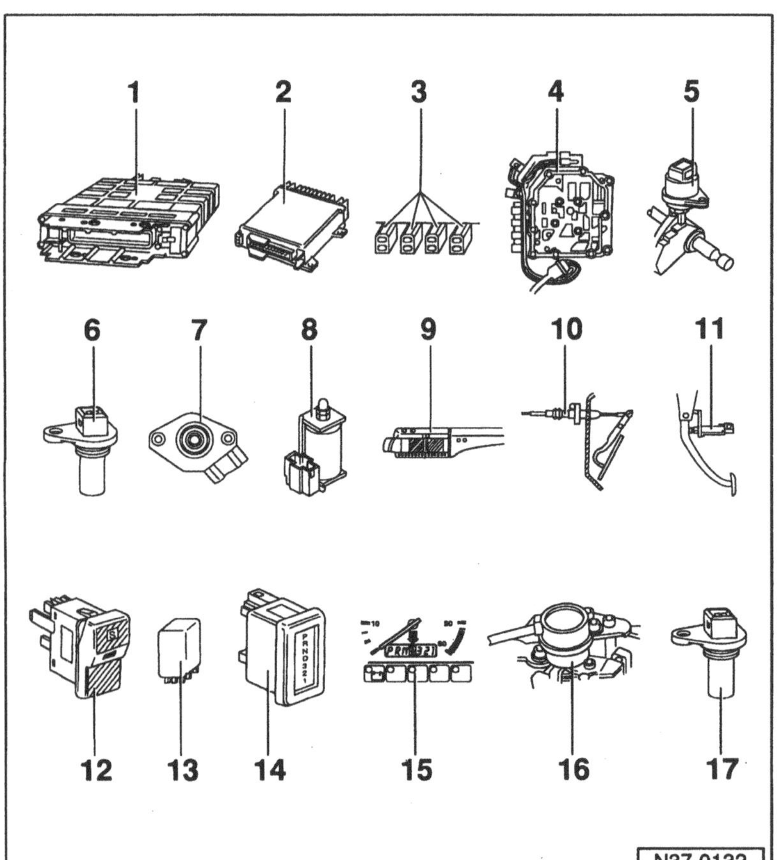

8 - Shift Lock Solenoid -N110-

- Location: Solenoid is located on selector lever mount ⇒ page H 37-1
- Can be checked electrically ⇒ page H 01-55 and by reading measured value block ⇒ page H 01-35

9 - Cruise Control Switch -E45-

- Location: Steering column switch
- Can be checked by reading measured value block ⇒ page H 01-35

10 - Kickdown Switch -F8-

- Location: The Kickdown Switch -F8- is integrated in the throttle cable in front of engine compartment
- Can be checked electrically ⇒ page H 01-55 and by reading measured value block ⇒ page H 01-35
- Adjusting throttle cable ⇒ Repair Group 20

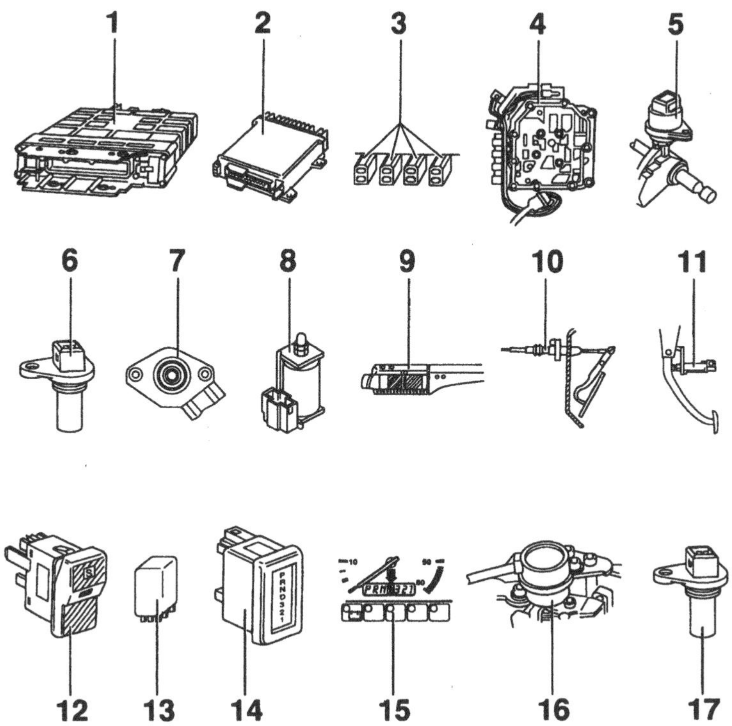

N37-0132

11 - Brake Light Switch -F-

- Location: The Brake Light Switch is located on the pedal cluster
- Can be checked electrically ⇒ page H 01-55 and by reading measured value block ⇒ page H 01-35

12 - Transmission Range (TR) Program Switch -E122-

- Location: The Transmission Range (TR) Program Switch -E122- is located in the center of the instrument panel ⇒ page H 37-1
- Is no longer installed on transmissions for vehicles with electronic shift program ⇒ page H 00-11
- Can be checked electrically ⇒ page H 01-55 and by reading measured value block ⇒ page H 01-35

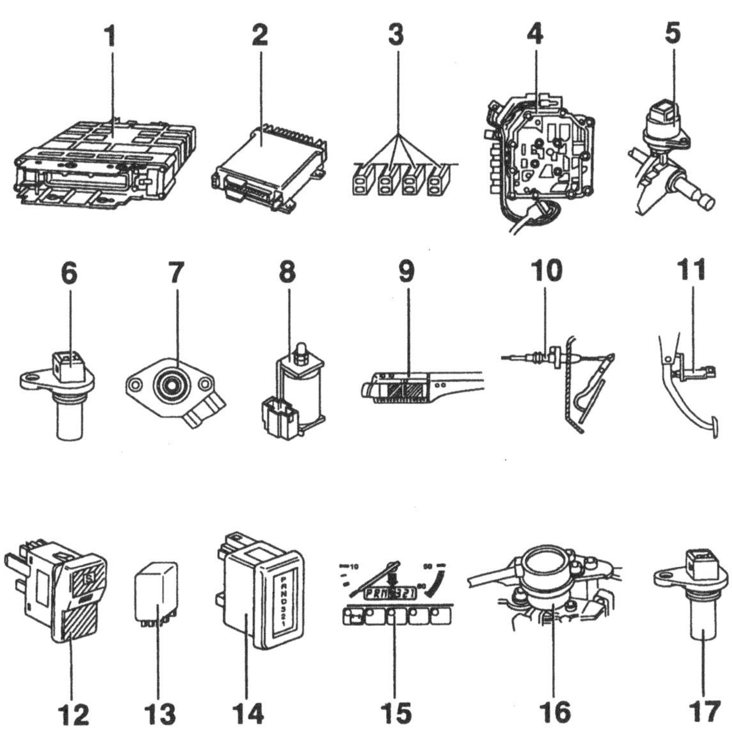

N37-0132

13 - Park/Neutral Position (PNP) Relay -J226-

- Location: Relay plate
⇒"Wiring diagrams"

14 - Transmission Range (TR) Selector Lever Display -Y5-

- Location: The Transmission Range (TR) Selector Lever Display -Y5- is located in the center of instrument panel
- Discontinued after moving display to instrument cluster ⇒ item 15

15 - Transmission Range (TR) Selector Lever Display -Y5-

- Location: The Transmission Range (TR) Selector Lever Display is located in the instrument-cluster

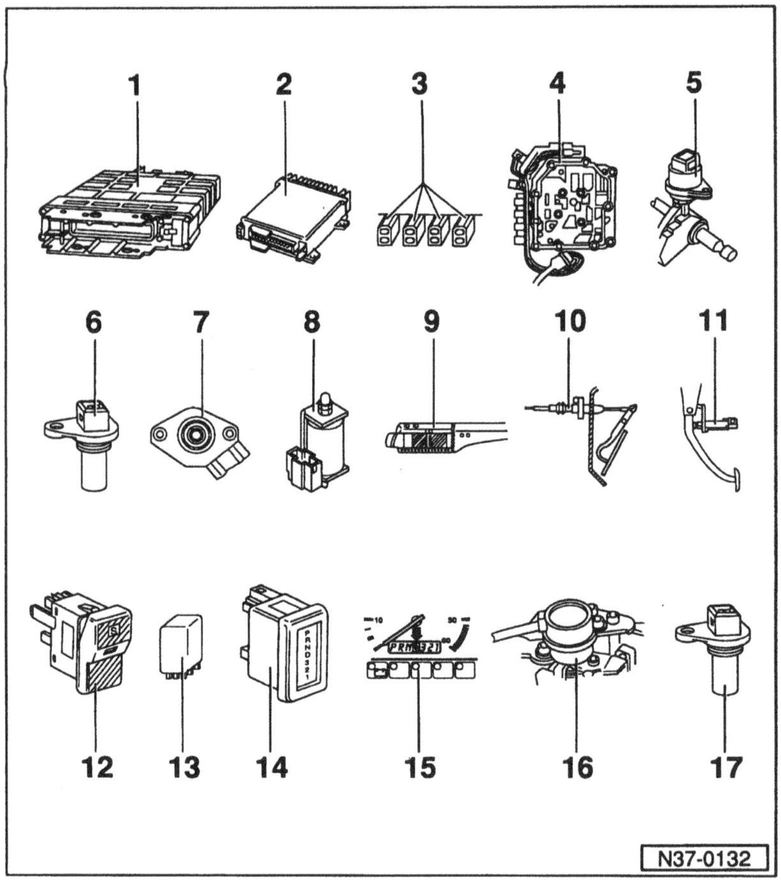

16 - **Throttle Position (TP) Sensor -G69- (for Diesel engine)**
- Location: The Throttle Position (TP) Sensor is located on injection pump throttle lever
- Checked by OBD
- Can be checked and adjusted in "reading measured value block" mode
  ⇒ page H 01-35
- Removing and installing
  ⇒ Repair Group 23

17 - **Engine Speed (RPM) Sensor -G28- (for Diesel engine)**
- The Engine Speed (RPM) Sensor is located on top of torque converter
  ⇒ Removing and installing differential, page H 39-18

VAG 1551 Scan Tool (ST), connecting and selecting functions

Test conditions:

- Battery Positive Voltage (B+) OK.
- Fuses 14 (S14) and 21 (S21) OK.
- Check Ground (GND) connections for transmission:

– Check Ground (GND) connections for corrosion and poor contact; repair if necessary.

– Ground (GND) connecting point is located on right of relay plate.

– Check battery Ground (GND) strap and Ground (GND) strap between battery and transmission.

- Selector lever in position "P" and parking brake applied.

– Remove stowage compartment in front of relay plate.

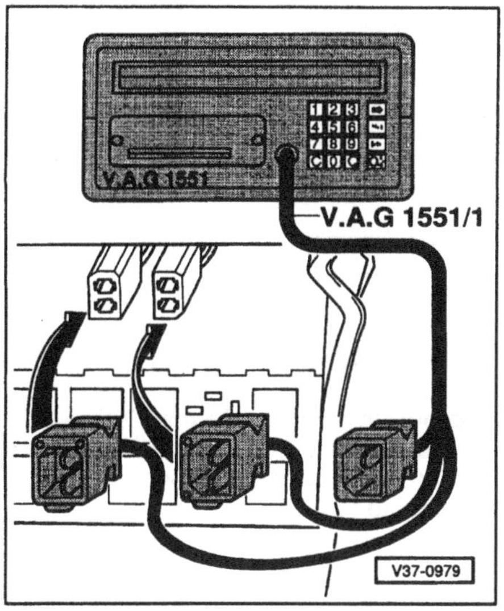

– Connecting VAG 1551 Scan Tool (ST):

– With ignition turned off, connect VAG 1551 Scan Tool (ST) with VAG 1551/1 adaptor cable as follows to Data Link Connectors (DLC) above relay plate.

– Connect black plug for VAG 1551 Scan Tool (ST) voltage supply to black DLC.

– Connect white plug to white DLC. (The blue plug is not required.)

VAG – On Board Diagnostic	HELP
1 – Rapid data transfer[1)]	
2 – Blink code output[1)]	

Indicated on display:

[1)] operating modes -1- and -2- are displayed alternately

Notes:

- *Additional operating instructions can be printed out by pressing the HELP key of VAG 1551 Scan Tool (ST).*
- *The → key is used to advance the Scan Tool program.*
- *An automatic check can be carried out in the "Rapid data transfer" operating mode. All vehicle control modules will then be interrogated automatically.*

⇒ VAG 1551 Scan Tool (ST) operating instructions.

– Turn on ignition.

– Turn on printer with PRINT key (indicator lamp in key lights up).

– Press key -1- for "Rapid data transfer" mode.

Rapid data transfer	HELP
Enter address word XX	

Indicated on display:

– Press keys -0- and -2-. (To enter address word 02, for "transmission electronics".)

Rapid data transfer Q 02 transmission electronics

◄ Indicated on display:

– Press -Q- key to confirm input.

098927731A AG4 transmission 098 0843 Coding 00000 WSC 131071

◄ Indicated on display:

The Transmission Control Module (TCM) identification, TCM coding, and the dealership number of the VAG 1551 Scan Tool (ST) are displayed:

Transmission Control Module (TCM) identification

Depending upon build level (program level) the Transmission Control Module (TCM) can indicate a different Transmission Control Module (TCM) identification, than that shown in the example. Location of Transmission Control Module (TCM)
⇒ Parts microfiche.

- ♦ 098 927 731 A: Part No. (example)
- ♦ AG4 Gearbox 098 (Digimat): 4-speed automatic transmission 098
- ♦ 0843: EPROM (program level)
- ♦ Coding 00000: not required for current version.
- ♦ WSC 131071: dealership number (example) of the VAG 1551 Scan Tool (ST) with which the most recent coding was performed.

Control module does not answer! HELP

◄ Indicated on display:

By pressing the HELP key, a list of possible malfunctions is printed out.

– After eliminating possible causes of malfunctions, enter address word 02, for "transmission electronics" again.

– Press -Q- key to confirm input.

If "Control module does not answer!" appears again:

Control module does not answer! HELP

◄ Indicated on display:

Check power supply, Battery Positive Voltage (B+), to Transmission Control Module (TCM).

– Perform electrical test, test step 1
⇒ page H 01-55.

– Check wiring connections to Data Link Connector (DLC) sockets.
⇒ *Electrical Wiring Diagrams*

– See Diagnostic Trouble Code (DTC) table, under DTC 65535 for Control Module Malfunctioning
⇒ page H 01-20.

– Press → key.

Rapid data transfer HELP Select function XX

◄ Indicated on display:

By pressing the HELP key, a list of the possible functions is printed out.

List of selectable functions

		Page
01 -	Check Transmission Control Module (TCM) version ⇒ Connecting VAG 1551 Scan Tool (ST) and selecting functions	H 01-12
02 -	Check DTC Memory	H 01-18
04 -	Initiate basic setting	H 01-33
05 -	Erase DTC Memory	H 01-31
06 -	End output	
08 -	Read measured value block	H 01-35

Further functions, which can be printed out by pressing the HELP key, need not be considered.

– After checking and completing a function VAG 1551 Scan Tool (ST) returns to following start position:

Rapid data transfer	HELP
Select function XX	

◄ Indicated on display:

Diagnostic Trouble Code (DTC) Memory, checking

– Connect VAG 1551 Scan Tool (ST) and enter address word 02, "Transmission Electronics" and advance program until "Select function XX" appears in display
⇒ from page H 01-12.

Rapid data transfer	HELP
Select function XX	

◄ Indicated on display:

– Press keys -0- and -2-. (The function "Check DTC Memory" is selected with 02).

Rapid data transfer	Q
02 – Check DTC memory	

◄ Indicated on display:

– Press -Q- key to confirm input.

X DTC recognized!

◄ The number of stored Diagnostic Trouble Codes (DTC) or "No DTC recognized" appears in the display.

The stored malfunctions are displayed in turn and printed out.

– Press → key.

After the last malfunction has been displayed and printed out, the malfunctions must be corrected as described in the Diagnostic Trouble Code (DTC) table
⇒ from page H 01-20.

– Press → key.

Rapid data transfer	HELP
Select function XX	

◄ Indicated on display:

Note:

After interrogating the DTC memory and correcting the malfunctions:

– *Erase DTC memory*
  ⇒ page H 01-31.

Diagnostic Trouble Code (DTC) table

Notes:

- *All the possible malfunctions which are recognized by the Transmission Control Module (TCM) -J217- and displayed and printed out by the VAG 1551 Scan Tool (ST) are listed below grouped according to the Diagnostic Trouble Code (DTC) number.*
- *If malfunction occurs only occasionally or if the DTC Memory was not erased after correcting the malfunctions, these malfunctions will be displayed as "sporadic malfunctions" for a stipulated period of time*
  ⇒ Transmission Control Module (TCM) recognition of malfunction, page H 01-3.
- *If faulty components are found, test the wiring to the components for short circuits and open circuits according to the wiring diagram.*
- *For "Open circuit" malfunctions, check first for contact corrosion or water contamination and if necessary, replace. With solenoid malfunctions, check the 10 pin connector at the transmission between the valve body conductor strip and the wiring harness.*
- *The malfunction code is printed out in the "rapid data transfer" mode only when the printer of VAG 1551 Scan Tool (ST) is turned on.*

  Example: Diagnostic Trouble Code (5-position) 65535

Output on printer VAG 1551 Scan Tool (ST)	Possible cause of malfunction	Corrective action
No DTC recognized!	If "No DTC recognized" appears after making repairs, the On Board Diagnostic program is completed. In the event that the automatic transmission does not operate properly despite running the On Board Diagnostic program, make repairs according to the Repair Manual.	

Output on printer VAG 1551 Scan Tool (ST)	Possible cause of malfunction	Corrective action
00258 Solenoid Valve 1 -N88- Open circuit[1)] Short to Ground (GND)[1)]	Open circuit or short to Ground (GND) Solenoid Valve 1 -N88- faulty	– Check wiring, connections, and solenoid coil – Read measured value block ⇒ page H 01-35; display group number 04 – Perform test step 6[2)] – Replace valve body ⇒ page H 38-36
00260 Solenoid Valve 2 -N89- Open circuit[1)] Short to Ground (GND)[1)]	Open circuit or short to Ground (GND) Solenoid Valve 2 -N89- faulty	– Check wiring, connections, and solenoid coil – Read measured value block ⇒ page H 01-35; display group number 04 – Perform test step 7[2)] – Replace valve body ⇒ page H 38-36
00262 Solenoid Valve 3 -N90- Open circuit[1)] Short to Ground (GND)[1)]	Open circuit or short to Ground (GND) Solenoid Valve 3 -N90- faulty	– Check wiring, connections, and solenoid coil – Read measured value block ⇒ page H 01-35; display group number 04 – Perform test step 8[2)] – Replace valve body ⇒ page H 38-36

[1)] One of these displays appears in addition to relevant component.

[2)] ⇒ page H 01-55, Electrical test

Output on printer VAG 1551 Scan Tool (ST)	Possible cause of malfunction	Corrective action
00264 Solenoid Valve 4 -N91- Open circuit[1)] Short to Ground (GND)[1)]	Open circuit or short to Ground (GND) Solenoid Valve 4 -N91- faulty	– Check wiring, connections, and solenoid coil – Read measured value block ⇒ page H 01-35; display group number 04 – Perform test step 9[2)] – Replace valve body ⇒ page H 38-36
00266 Solenoid Valve 5 -N92- Open circuit[1)] Short to Ground (GND)[1)]	Open circuit or short to Ground (GND) Solenoid Valve 5 -N92- faulty	– Check wiring, connections, and solenoid coil – Read measured value block ⇒ page H 01-35; display group number 04 – Perform test step 10[2)] – Replace valve body ⇒ page H 38-36
00268 Solenoid Valve 6 -N93- Open circuit[1)] Short to Ground (GND)[1)]	Open circuit or short to Ground (GND) Solenoid Valve 6 -N93- faulty	– Check wiring, connections, and solenoid coil – Read measured value block ⇒ page H 01-35; display group number 02 – Perform test step 11[2)] – Replace valve body ⇒ page H 38-36

[1)] One of these displays appears in addition to relevant component.

[2)] ⇒ page H 01-55, Electrical test

Output on printer VAG 1551 Scan Tool (ST)	Possible cause of malfunction	Corrective action
00270	Open circuit or short to Ground (GND)	– Check wiring, connections, and solenoid coil
Solenoid Valve 7 -N94-	Solenoid Valve 7 -N94- faulty	– Read measured value block ⇒ page H 01-35; display group number 04
Open circuit[1] Short to Ground (GND)[1]		– Perform test step12[2] – Replace valve body ⇒ page H 38-36
00281	Open circuit in wiring	– Check wiring, connections, and solenoid coil
Vehicle Speed Sensor (VSS) -G68-	Vehicle Speed Sensor (VSS) -G68- faulty	– Read measured value block ⇒ page H 01-35; display group number 02 – Perform test step 15[2]
No signal		– Replace Vehicle Speed Sensor (VSS) -G68- ⇒ Disassembling and assembling parking lock, page 38-26 – Replacing input gear ⇒ Removing and installing input gear, page H 39-23

[1] One of these displays appears in addition to relevant component.

[2] ⇒ page H 01-55, Electrical test

Output on printer VAG 1551 Scan Tool (ST)	Possible cause of malfunction	Corrective action
00293	Open circuit in wiring	– Check wiring, connections, and solenoid coil
Multi-Function Transmission Range (TR) Switch -F125-	Multi-Function Transmission Range (TR) Switch -F125- faulty	– Read measured value block ⇒ page H 01-35; display group number 01
Undefined switch state		– Perform test step 5[2] – Replace Multi-Function Transmission Range (TR) Switch -F125- ⇒ Disassembling and assembling parking lock, page H 38-45
00300	Open circuit in wiring	– Check wiring and connections
Transmission Fluid (ATF) Temperature Sensor -G93-[1]	Transmission Fluid (ATF) Temperature Sensor -G93- faulty	– Read measured value block ⇒ page H 01-35; display group number 05 – Perform test step 16[2]
No malfunction identified		– Replace valve body ⇒ page H 38-36

[1] A faulty Transmission Fluid (ATF) Temperature Sensor -G93- is indicated.

[2] ⇒ page H 01-55, Electrical test

Output on printer VAG 1551 Scan Tool (ST)	Possible cause of malfunction	Corrective action
00518	Open or short circuit	– If in addition malfunction 00638 is shown, then correct this first – Check wiring and connections against the wiring diagram
Throttle Position (TP) Sensor -G69- Signal outside tolerance	Throttle Position (TP) Sensor -G69- faulty	– Read measured value block ⇒ page H 01-35; display group numbers 01 and 03 – Perform test step 2[1)] – Replace Throttle Position (TP) Sensor -G69- ⇒ Repair Group 24 – Bring system to basic setting ⇒ page H 01-33
	Throttle Position (TP) Sensor -G69- on Diesel engines faulty	– Replace Throttle Position (TP) Sensor -G69- on Diesel engine ⇒ Repair Group 23 – Adjust Throttle Position (TP) Sensor -G69- on Diesel engine – Read measured value block ⇒ page H 01-35; display group number 01 and 03 – Perform test step 2[1)] – Bring system to basic setting ⇒ page H 01-33

[1)] ⇒ page H 01-55, Electrical test

Output on printer VAG 1551 Scan Tool (ST)	Possible cause of malfunction	Corrective action
00529 RPM information missing	Open circuit in wiring	– Check wiring and connections against the wiring diagram – Read measured value block ⇒ page H 01-35; display group numbers 03 and 05 – Check Engine Control Module ⇒ Repair Group 24 – Check Engine Speed (RPM) Sensor -G28- on Diesel engines – Perform test step 18[1)]
00532 Battery Positive (B+) Voltage	Battery faulty Insufficient system voltage	– Test Battery Positive Voltage (B+) – Read measured value block ⇒ page H 01-35; display group number 02 – Test system voltage to Transmission Control Module (TCM) -J217- perform test step1[1)]

[1)] ⇒ page H 01-55, Electrical test

Output on printer VAG 1551 Scan Tool (ST)	Possible cause of malfunction	Corrective action
00545	Open circuit or short to Ground (GND)	– Check wiring and connections against the wiring diagram
Engine/transmission electrical connection	Open connection between Engine Control Module (ECM) and Transmission Control Module (TCM) -J217-	– Read measured value block ⇒ page H 01-35; display group number 05 ⇒ Repair Group 24
Open circuit[1] Short to Ground (GND)[1]		– Bring system to basic setting ⇒ page H 01-33
00596	10 pin connector between valve body conductor strip and wiring harness	– Check wiring and connections against the wiring diagram
Short circuit between solenoid valve lines	Conductor strip to valve body faulty	– Perform test steps 6 to 12 and 16[1] – Replace valve body ⇒ page H 38-36

[1] ⇒ page H 01-55, Electrical test

Output on printer VAG 1551 Scan Tool (ST)	Possible cause of malfunction	Corrective action
00638	Open circuit or short to Ground (GND)	– Check wiring and connections against the wiring diagram
Engine/transmission electrical connection 2	Broken connection between Engine Control Module (ECM) and Transmission Control Module (TCM) -J217-	– Read measured value block ⇒ page H 01-35; display group number 05 ⇒ Repair Group 24
No signal		– Bring system to basic setting ⇒ page H 01-33
00641	Transmission becomes too hot, max. 148° C (298° F). If ATF temperature is too high, transmission shifts down into next lower gear.	– Check ATF level ⇒ page H 37-46 – Read measured value block ⇒ page H 01-35; display group number 05; read off ATF temperature
ATF temperature	Trailer load too heavy	
Signal too large	ATF level low	

[1] ⇒ page H 01-55, Electrical test

Output on printer VAG 1551 Scan Tool (ST)	Possible cause of malfunction	Corrective action
00652	Electrical/hydraulic malfunction	– Read measured value block ⇒ page H 01-35; display group number 04 and determine the gear in which malfunction occurs
Transmission range controller	Clutch or valve body faulty	– Replace valve body ⇒ page H 38-36
Incorrect signal		– Replace clutch ⇒ Disassembling and assembling planetary gearbox, page H 37-57
00660	Open circuit in wiring	– Check wiring and connections – Read measured value block ⇒ page H 01-35; display group number 01
Kick Down Switch/Throttle Position Sensor	Throttle Position (TP) Sensor -G69- faulty	– Perform test step 14 and Throttle Position (TP) Sensor -G93- test step 2[1]
Incorrect signal	Kick Down Switch -F8- faulty	– Adjust accelerator cable or replace ⇒ Repair Group 20

[1] ⇒ page H 01-55, Electrical test

Output on printer VAG 1551 Scan Tool (ST)	Possible cause of malfunction	Corrective action
65535	Electrical interference from outside sources or poor Ground (GND) connections	– Check wiring, connections, and Ground (GND) cables to Transmission Control Module (TCM) -J217-
Transmission Control Module (TCM) -J217- faulty	Transmission Control Module (TCM) -J217- faulty	– Bring system into basic setting ⇒ page H 01-33

Notes:

*The Transmission Control Module (TCM) -J217- should **not** be replaced ⇒ page H 01-5, and the system should not be restored to the basic setting, ⇒ page H 01-33, until the cause of the malfunction has been determined and the following malfunctions have been corrected:*

- *Mechanical malfunctions*
- *Hydraulic malfunctions*
- *All affected electrical components and cable connections.*

Diagnostic Trouble Code (DTC) Memory, erasing

Requirement:

- DTC Memory checked ⇒ page H 01-18.

After DTC Memory has been checked:

Rapid data transfer HELP	
Select function XX	

Indicated on display:

– Press keys -0- and -5-. (To select function 05, "Erase DTC memory".)

Rapid data transfer Q	
05 Erase DTC memory	

Indicated on display:

– Press -Q- key to confirm input.

Attention!
DTC memory was not interrogated.

Indicated on display:

Note:

- *If the ignition was turned off between checking and erasing the DTC memory, the DTC memory will not be erased.*
- *Follow the sequence of operations, i.e. always check the DTC memory before attempting to erase it.*

Rapid data transfer	→
DTC Memory is erased!	

Indicated on display:

(DTC Memory will be erased approximately 5 seconds after the display appears.)

The DTC Memory is now erased.

Note:

Wait about 1 minute before checking the DTC Memory again.

System cannot be interrogated!

Indicated on display:

1 DTC recognized!	
00811	3333
System cannot be interrogated	

Print out with printer turned on:

System (Transmission Control Module (TCM) -J217-) given insufficient time to recognize malfunctions.

– Wait about 1 minute before checking DTC Memory again.

– After checking and erasing DTC Memory, test drive vehicle and check DTC Memory again.

When the DTC Memory is checked, the following display should appear:

"No DTC recognized!"

Basic setting, initiating

Note:

The basic setting must be initiated after performing any of the following repairs:

- *Replacing engine*
- *Replacing Engine Control Module (ECM)*
- *Replacing/altering throttle valve*
- *Adjusting throttle valve (setting idling speed).*
- *Replacing Throttle Position (TP) Sensor -G69-*
- *Altering setting of Throttle Position (TP) Sensor -G69- (engine)*
- *Replacing Diesel engine injection pump*
- *Changes to control lever, altering setting of residual quantity or maximum speed adjustment screw on Diesel engines*
- *Replacing Transmission Control Module (TCM) -J217-*

– Connect VAG 1551 Scan Tool (ST) and enter address word 02, for "Transmission Electronics" and advance VAG 1551 Scan Tool (ST) program until "Select function XX" appears in display, ⇒ from page H 01-12.

Rapid data transfer	HELP
Select function XX	

◄ Indicated on display:

– Press keys -0- and -4-. (To select function 04, "Initiating basic setting").

Note:

Accelerator Pedal must remain in closed throttle position.

Rapid data transfer	Q
04 Basic setting	

◄ Indicated on display:

– Press -Q- key to confirm input.

Basic setting	HELP
Enter display group number XX	

◄ Indicated on display:

– Press keys -0- twice (to input display group 00).

– Press -Q- key to confirm input.

System in basic setting	→

◄ Indicated on display:

System is now in the basic setting.

– Depress Accelerator Pedal as far as kickdown and hold in this position for 3 seconds.

– Press → key.

Rapid data transfer	HELP
Select function XX	

◄ Indicated on display:

Measured value block, reading

– Connect VAG 1551 Scan Tool (ST) and enter address word 02 for "Transmission Electronics". Advance VAG 1551 Scan Tool (ST) program until "Select function XX" is indicated on display, ⇒ from page H 01-12.

Rapid data transfer	HELP
Select function XX	

◄ Indicated on display:

– Press keys -0- and -8-. (To select function 08, "Reading measured value block".)

Rapid data transfer	Q
08 – Reading measured value block	

◄ Indicated on display:

– Press -Q- key to confirm entry.

Reading measured value block
Enter display group number XX

◄ Indicated on display:

List of selectable display group numbers

Display group No.	Display field	Designation
01	1	Selector lever position
	2	Throttle Position (TP) Sensor voltage
	3	Accelerator Pedal value
	4	Switch positions
02	1	Actual current of Solenoid Valve 6 -N93-
	2	Specified current of Solenoid Valve 6 -N93-
	3	Battery Positive Voltage (B+)
	4	Voltage at Vehicle Speed Sensor (VSS) -G68-
03	1	Vehicle speed
	2	Engine speed
	3	Gear selected
	4	Accelerator Pedal value
04	1	Solenoid valves
	2	Gear selected
	3	Selector lever position
	4	Vehicle speed
05	1	Transmission Fluid (ATF) Temperature
	2	Switch output
	3	Gear to be selected
	4	Engine speed

Reading measured value block 1 →
1 2 3 4

◄ Indicated on display:

(Specification
⇒ page H 01-38)

– There are always 4 display fields in the measured value block (displayed in physical units, if necessary.)

Notes:

- *Explanation of values in the individual display fields ⇒ Test table, page H 01-38.*
- *If the printer is turned ON, the current display is printed out on the log.*
- *If the specified values of the Throttle Position (TP) Sensor are not reached in display field -2-, the system should first be restored to the basic setting ⇒ page H 01-33.*
- *If the values in all display fields are as specified:*

– Press → key.

Rapid data transfer HELP
Select function XX

◄ Indicated on display:

Test table

Group No.	Display field	Designation	Test conditions		Specified readout on VAG 1551	Corrective action
01	1	Selector lever position	Station-ary	P	P	– Check Multi-Function Transmission Range (TR) Switch -F125-: Perform test step 5[1]
			Selector	R	R	
			lever	N	N	
			in	D	D	
				3	3	
				2	2	
cont.				1	1	

[1] ⇒ page H 01-55, Electrical test

Group No.	Display field	Designation	Test conditions		Specified readout on VAG 1551	Corrective action
01	2	Voltage of Throttle Position (TP) Sensor -G69-	Stationary gasoline engine	Idling min. Idling max.	0.156 V 0.8 V	– When accelerating from idling to Wide Open Throttle, voltage value increases constantly – Perform test step 2[1)]
				Wide Open Throttle - min. Wide Open Throttle - max.	3.5 V 4.680 V	– Bring system into basic setting ⇒ page H 01-33
		Voltage of Throttle Position (TP) Sensor -G69- for Diesel engines	Stationary Diesel engine	Idling min. Idling max.	0.6 0.8	– Specified value not reached at idling: Set Throttle Position (TP) Sensor to specified value by adjusting on injection pump control lever – Perform test step 2[1)]
cont.				Wide Open Throttle - min. Wide Open Throttle - max.	3.5 V 4.680 V	– Bring system into basic setting ⇒ page H 01-33

[1)] ⇒ page H 01-55, Electrical test

Group No.	Display field	Designation	Test conditions		Specified readout on VAG 1551	Corrective action
01	3	Accelerator Pedal value	Stationary	Idling	0...1 %	– Check Throttle Position (TP) Sensor -G69- ⇒ Repair Group 23 or 24
				Wide open throttle	99...100 %	– Bring system into basic setting ⇒ page H 01-33
	4	Switch position Display 1	Brake	Operated	1	– Perform test step 4[1)]
				Not operated	0	
		Traction control system 2	Traction control	Activated	1	not applicable
cont.				Not activated	0	

[1)] ⇒ page H 01-55, Electrical test

Group No.	Display field	Designation		Test conditions	Specified readout on VAG 1551	Corrective action
01	4	Transmission Range (TR) Program Switch -E122- (up to 12.92)	Display 3	"S" selected	1	– Perform test step 17[1]
				"E" selected	0	
		Electronic Shift Program (from 1.93)	3	"S" selected	1	Electronic Shift Program is integrated in control module and activates the "E" or "S" program automatically
cont.				"E" selected	0	⇒ Transmissions for vehicles with electronic shift program, ⇒ page H 00-11

[1] ⇒ page H 01-55, Electrical test

Group No.	Display field	Designation		Test conditions		Specified readout on VAG 1551	Corrective action
01	4	Kickdown Switch -F8-	4		Operated	1	– Perform test step 14[1]
					Not operated	0	
		Multi-Function Transmission Range (TR) Switch -F125-	5	Selector lever in	R, N, D, 3, 2	1	– Check Multi-Function Transmission Range (TR) Switch -F125-: Perform test step 5[1]
					P, 1	0	
			6	Selector lever in	P, R, 2, 1	1	
					N, D, 3	0	
			7	Selector lever in	P, R, N, D	1	
					3, 2, 1	0	
			8	Selector lever in	P, R, N,	1	
					D, 3, 2, 1	0	

[1] ⇒ page H 01-55, Electrical test

Group No.	Display field	Designation	Test conditions		Specified readout on VAG 1551	Corrective action
02	1	Actual current of solenoid valve 6 -N93-	Stationary	Wide Open Throttle	0.0 A	– Perform test step 11[1]
				Idling max.	1.1 A	
	2	Specified current of solenoid valve 6 -N93-	Stationary	Wide Open Throttle	0.0 A	
				Idling max.	1.1 A	
	3	Battery Positive Voltage (B+)	Stationary	min.	10.8 V	– Check battery, replace if necessary
				max.	16.0 V	
	4	Vehicle Speed Sensor (VSS) -G68-	Stationary	min.	2.20 V	– Perform test step 15[1]
				max.	2.52 V	
03 cont.	1	Vehicle Speed Sensor (VSS) -G68-	Driving[2]		...km/h (mph)	Speedometer reading and readout on VAG 1551 Scan Tool (ST) may vary slightly.

[1] ⇒ page H 01-55, Electrical test

[2] While driving, a second technician is needed to read the specified values

Group No.	Display field	Designation	Test conditions		Specified readout on VAG 1551	Corrective action
03	2	Engine speed (RPM)	Engine running		rpm	– Tune engine if necessary ⇒ Repair Group 23 or 24
	3	Gear selected	Driving[2]	Neutral	0	– Check solenoid valves: Perform test steps 6 to 12[1]
				Reverse	R	
				1 hydraulic	1	
				2 hydraulic	2	
				3 hydraulic	3 H	
				3 mechanical[3]	3 M	
				4 mechanical	4	
	4	Accelerator Pedal value	Driving[2]	Idling	0...1 %	– Check Throttle Position (TP) Sensor -G69- ⇒ Repair Group 24
				Wide Open Throttle	99...100 %	– Check Kick Down Switch -F8- ⇒ Repair Group 20 – Bring system into basic setting ⇒ page H 01-33

[1] ⇒ page H 01-55, Electrical test

[2] While driving, a second technician is needed to read the specified values

[3] "3M" not applicable to transmissions with modified shift program from 01.93 ⇒ page H 00-11.

Group No.	Display field	Designation		Test conditions		Specified readout on VAG 1551	Corrective action
04	1	Solenoid Valves: -N88-, -N89-, -N90-, -N91-, -N92-, -N94-		Driving[2]			The solenoid valves are switched according to the driving conditions.
			Display 1	-N88-	switched	1	– Perform test steps 6 to 10 and 12[1]
					not switched	0	
			2	-N89-	switched	1	
					not switched	0	
			3	-N90-	switched	1	
					not switched	0	
			4	-N91-	switched	1	
					not switched	0	
			5	-N92-	switched	1	
					not switched	0	
			6	-N94-	switched	1	
cont.					not switched	0	

[1] ⇒ page H 01-55, Electrical test

[2] While driving, a second technician is needed to read the specified values

Group No.	Display field	Designation	Test conditions		Specified readout on VAG 1551	Corrective action
04	2	Gear engaged	Driving[2]			– Check solenoid valves: Perform test steps 6 to 12[1]
				Neutral	0	
				Reverse	R	
				1 hydraulic	1	
				2 hydraulic	2	
				3 hydraulic	3 H	
				3 mechanical[3]	3 M	
cont.				4 mechanical	4	

[1] ⇒ page H 01-55, Electrical test

[2] While driving, a second technician is needed to read the specified values

[3] "3M" not applicable to transmissions with modified shift program from 01.93 ⇒ page H 00-11.

Group No.	Display field	Designation	Test conditions	Specified readout on VAG 1551	Corrective action
04	3	Selector lever position	Driving[2)] P	P	– Check Multi-Function Transmission Range (TR) Switch -F125-: Perform test step 5[1)]
			R	R	
			N	N	
			D	D	
			3	3	
			2	2	
cont.			1	1	

1) ⇒ page H 01-55, Electrical test

2) While driving, a second technician is needed to read the specified values

Group No.=	Display field	Designation	Test conditions	Specified readout on VAG 1551	Corrective action
04	4	Vehicle speed	Speed at which vehicle driven[2)]	...km/h (mph)	Speedometer reading and readout on VAG 1551 Scan Tool (ST) may vary slightly
05	1	Transmission Fluid (ATF) Temperature	Stationary with engine running. Exact temperature is displayed from approx. 50°C (112°F) or higher	...°C (°F)	The ATF level is checked at a temperature of approx. 50°C - 70°C (112°F - 158°F) – Perform test step 16[1)]
	2	Selector outputs Display 1	Driving[2)] Engine management is turned on	1	– Check routing of wiring against the wiring diagram – Replace Engine Control Module (ECM) ⇒ Repair Group 24
			is turned off	0	– Replace Transmission Control Module (TCM) -J217- ⇒ page H 01-5
		2	turned on	1	– Bring system into basic setting ⇒ page H 01-33
cont.			turned off	0	

1) ⇒ page H 01-55, Electrical test

2) While driving, a second technician is needed to read the specified values

Group No.	Display field	Designation	Test conditions		Specified read out VAG 1551	Corrective action
05	2	Selector outputs	Shift Lock Solenoid -N110-			– Check wiring diagram
		Display 3		is turned on	1	– Check Shift Lock Solenoid -N110-: Perform test step 3[1]
				is turned off	0	
		4		turned on	1	
				turned off	0	
		5	Cruise control			– Check wiring diagram
				turned on	1	– Check cruise control ⇒ “Electrical Wiring Diagrams”
cont.				turned off	0	

[1] ⇒ page H 01-55, Electrical test

Group No.	Display field	Designation	Test conditions		Specified readout on VAG 1551	Corrective action
05	2	Selector outputs Display 6	Air conditioner	was turned off	1	– Check wiring diagram
				was not turned off	0	– Check air conditioner ⇒ Repair Group 87
		7	Park/Neutral signal Selector lever in			– Check wiring diagram
				P, N	1	
cont.				1, 2, 3, D	0	

Group No.	Display field	Designation	Test conditions		Specified readout on VAG 1551	Corrective action
05	3	Gear engaged	Driving[2]	Neutral	0	– Check solenoid valves: Perform test steps 6 to 12[1]
				Reverse	R	
				1 hydraulic	1	– If no gearshifts are performed, a clutch or brake may be faulty
				2 hydraulic	2	
				3 hydraulic	3 H	– Replace Transmission Control Module (TCM) -J217- ⇒ page H 01-5
				3 mechanical[3]	3 M	
				4 mechanical	4	
	4	Engine speed (RPM)	Driving[2] with engine running		RPM	– Tune engine if necessary ⇒ Repair Group 23 or 24

[1] ⇒ page H 01-55, Electrical test

[2] While driving, a second technician is needed to read the specified values

[3] "3M" not applicable to transmissions with modified shift program from 01.93 ⇒ page H 00-11.

Solenoid valves, checking while driving

Check DTC Memory before checking solenoid valves while driving ⇒ page H 01-18 and correct malfunctions.

Check wiring to solenoid valves according to wiring diagram.

The solenoid valves can be checked while driving by accessing "read measured value block" function 08; "display group number 04" and display field -1- using VAG 1551 Scan Tool (ST).

VAG 1551 Scan Tool (ST) display zone 1 is 6 figures (0000 00) and is read as follows:

	Display field 1:					
	Display 1	Display 2	Display 3	Display 4	Display 5	Display 6
	-N88-	-N89-	-N90-	-N91-	-N92-	-N94-
not activated	0	0	0	0	0	0
activated	1	1	1	1	1	1

The table shows Solenoid Valves -N88-, -N89-, -N90- and -N91- controlled in each selector lever position. The solenoid valves control the switching valves to the relevant gears.

During gear changes, control of the solenoid valves may vary from the following table for a short period.

Solenoid Valves -N92- and -N94- are supplementary valves that affect gear selection changes and are only controlled during gear changes. They are displayed in positions 5 and 6.

On transmissions up to 12.92, the activation of the solenoid valves will be briefly different to that described in the table when shifting from 3H - 3M and from 4 - 3M.

Transmissions up to 12.92

Group No.	Display field	Designation	Test conditions	Specified readout on VAG 1551	Corrective action if not to specification
04	1	Solenoid Valves: -N88-, -N89-, -N90-,	Selector lever in: P	1 0 0 1 00	Solenoid valves are selected according to driving conditions.
		-N91-, -N92-, -N94-	R	0 0 0 0 00	
			N	1 0 0 1 00	
			:D[1] 1	0 0 0 1 00	
			2	0 1 0 1 00	– Test steps 6 to 10 and 12[2]
			3H	0 0 0 0 00	
			3M	0 0 1 0 00	
			4	1 1 1 1 00	
			:3[1] 1	0 0 0 1 00	
			2	0 1 0 1 00	
			3H	0 0 0 0 00	
			:2[1] 1	0 0 0 1 00	
			2	0 1 0 1 00	
			:1[1] 1	0 0 0 1 00	

[1] While driving, a second technician is needed to read the specified values.

[2] ⇒ page H 01-55, Electrical test

Transmissions from 01.93

Group No.	Display field	Designation	Test conditions	Specified readout on VAG 1551	Corrective action if not to specification
04	1	Solenoid Valves: -N88-, -N89-, -N90-,	Selector lever in: P	1 0 0 1 00	Solenoid valves are selected according to driving conditions.
		-N91-, -N92-, -N94-	R	0 0 0 0 00	
			N	1 0 0 1 00	
			:D[1] 1	0 0 0 1 00	
			2	0 1 0 1 00	
			3H	0 0 0 0 00	
			4	1 1 1 1 00	
			:3[1] 1	0 0 0 1 00	– Perform test steps 6 to 10 and 12[2]
			2	0 1 0 1 00	
			3H	0 0 0 0 00	
			:2[1] 1	0 0 0 1 00	
			2	0 1 0 1 00	
			:1[1] 1	0 0 0 1 00	

[1] While driving, a second technician is needed to read the specified values.

[2] ⇒ page H 01-55, Electrical test

Transmission: Electrical testing

Perform only the recommended test steps listed in the Diagnostic Trouble Code (DTC) table and measured value block test table (targeted entry).

Perform all the measures stated in the column "Corrective action".

Notes:

- *Use the digital multimeter Fluke 83 (US1119) or equivalent for testing.*
- *The specified values are valid for ambient temperatures ranging from 0°C – 40°C (32°F – 104°F).*
- *If the readings obtained differ from the specified values, determine the cause by using the wiring diagrams.*
- *If the readings obtained differ only slightly from the specified values, clean the sockets and connectors of the test equipment and test leads. Repeat the test. Before replacing any components, test wiring and connections and, particularly if specified values are below 10 Ω, repeat the resistance measurement on component.*
- *Use VAG 1598 test box with VAG 1598/9 adapter cable and auxiliary cables from VAG 1594 for connecting the test equipment.*
- *The contact numbers of the connector and the socket numbers of the VAG 1598 test box correspond* *⇒ page H 01-58.*

> ***CAUTION!***
>
> ***To avoid damage to the electronic components, select the appropriate function and measurement range on the test equipment before connecting the test leads.***

Test requirements:

- Battery Positive Voltage (B+) OK.
- Fuses 14 (S14) and 21 (S21) OK.
- Transmission Ground (GND) connections OK:

  Ground (GND) connection is located on the left next to the relay panel.

– Check battery Ground (GND) strap and Ground (GND) strap between battery and transmission.

– Turn off ignition for all test steps, disconnect multi-pin connector from Transmission Control Module (TCM) -J217- and connect VAG 1598 test box to wiring harness connector using VAG 1598/9 adapter cables.

- The Transmission Control Module (TCM) -J217- is located in right-hand footwell on A pillar.

Voltage measurement:

VAG 1598/9 adapter cables connected to Transmission Control Module (TCM) -J217-

Resistance measurement:

VAG 1598/9 adapter cables disconnected from Transmission Control Module (TCM) -J217-

– If readings obtained differ from specified values, perform measures for corrective actions in right-hand column of test table ⇒ from page H 01-59.

Terminal assignment for 38-pin Transmission Control Module (TCM) -J217- connector (also sockets on VAG 1598 Test Box)

1 – Ground (GND) (terminal 31)
2 – Solenoid Valve 4 -N91-
3 – Solenoid Valve 3 -N90-
4 – not assigned
5 – Park/Neutral Position (PNP) signal
6 – K-wire for diagnosis
7 – not assigned
8 – Kickdown for air conditioner
9 – Throttle Position (TP) Sensor -G69- signal
10 – Throttle Position (TP) Sensor -G69- + 5 Volt
11 – not assigned
12 – Transmission Range (TR) Selector Lever -Y5- display
13 – Vehicle Speed Sensor (VSS) -G68- screening
14 – not assigned
15 – Multi-Function Transmission Range (TR) Switch -F125-
16 – Multi-Function Transmission Range (TR) Switch -F125-
17 – Kick Down Switch -F8-
18 – Supply voltage for solenoid valves
19 – Supply voltage (terminal 15)
20 – Shift Lock Solenoid -N110-
21 – Solenoid Valve 7 -N94-
22 – Solenoid Valve 1 -N88-
23 – Solenoid Valve 2 -N89-
24 – Solenoid Valve 5 -N92-
25 – Solenoid Valve 6 -N93-
26 – Brake Light Switch -F- signal voltage
27 – Gasoline engine = TD-(RPM) signal
Diesel engine = Engine Speed (RPM) Sensor -G28-
28 – Gasoline engine = Ignition timing influence
Diesel engine = Engine Speed (RPM) Sensor-G28-
29 – Throttle Position (TP) Sensor -G69- Ground (GND)
30 – Transmission Fluid (ATF) Temperature Sensor -G93-
31 – Selector Lever Display -Y5-
32 – Vehicle Speed Sensor (VSS) -G68-
33 – Vehicle Speed Sensor (VSS) -G68-
34 – Multi-Function Transmission Range (TR) Switch -F125-
35 – Multi-Function Transmission Range (TR) Switch -F125-
36 – Transmission Range (TR) Program Switch -E122- or electronic shift program
37 – Idling switch
38 – Cruise control system

Test table

Switch to voltage measuring range 20 V					
Test step	**VAG 1598 sockets**	**Test of**	**• Test conditions – Additional operations**	**Specified value**	**Corrective action**
1	19 + 1	Supply voltage from Transmission Control Module (TCM) -J217-	• Ignition switched on	approx. Battery Positive Voltage (B+)	– Check wiring from contact 1 to Ground (GND) – Check wiring from pin 19 to terminal 15
2 cont.	10 + 29	Throttle Position (TP) Sensor -G69-	• Ignition switched on – Disconnect connector from Throttle Position (TP) Sensor -G69-	4.6 to 5 V	– Check routing of wiring per wiring diagram – Replace Transmission Control Module (TCM) -J217- ⇒ page H 01-5

Switch to voltage measuring range 20 V						
Test step	**VAG 1598 sockets**	**Test of**	**• Test conditions – Additional operations**		**Specified value**	**Corrective action**
2	9 + 29	Throttle Position (TP) Sensor -G69-	• Ignition switched on			– When accelerating from idling to Wide Open Throttle, voltage value increases constantly
			Gasoline engine	Idling - min. Idling - max.	0.156 V 0.8 V	– Adjust Throttle Position (TP) Sensor -G69-, replace if necessary ⇒ Repair Group 24
				Wide Open Throttle - min. Wide Open Throttle - max.	3.5 V 4.68 V	
			Diesel engine	Idling - min. Idling - max.	0.6 V 0.8 V	– Adjust Diesel engine Throttle Position (TP) Sensor, replace if necessary ⇒ page H 01-35; Read measured value block, display group number 01
				Wide Open Throttle - min. Wide Open Throttle - max.	3.5 V 4.68 V	– Restore system to basic setting ⇒ page H 01-33

Switch to voltage measuring range 20 V					
Test step	**VAG 1598 sockets**	**Test of**	**• Test conditions – Additional operations**	**Specified value**	**Corrective action**
3	19 + 20	Shift Lock Solenoid -N110-	• Ignition switched on • Selector lever in P or N	approx. Battery Positive Voltage (B+)	– Check wiring diagram – Replace Transmission Control Module (TCM) -J217- ⇒ page H 01-5 – Restore system to basic setting ⇒ page H 01-33
			• With brake pedal depressed	0 V	– Perform test step 4, if necessary replace Transmission Control Module (TCM) -J217- ⇒ page H 01-5 – Restore system to basic setting ⇒ page H 01-33
4	26 + 1	Brake Light Switch -F-	• Ignition switched on • Brake pedal not depressed	0 V	– Check routing of wiring per wiring diagram – Replace Brake Light Switch -F-
			– Brake pedal depressed	approx. Battery Positive Voltage (B+)	

Switch to voltage measuring range 20 V					
Test step	**VAG 1598 sockets**	**Test of**	**• Test conditions – Additional operations**	**Specified value**	**Corrective action**
5	34 + 1	Multi-Function Transmission Range (TR) Switch -F125-	• Ignition switched on	4.5 to 5 V	– Check routing of wiring per wiring diagram
			– Selector lever position R, N, D, 3, and 2		– Replace Multi-Function Transmission Range (TR) Switch -F125- ⇒ Disassembling and assembling parking lock, page H 38-45
			– Selector lever position P and 1	0 to 0.8 V	
	15 + 1		– Selector lever position P, R 2, and 1	4.5 to 5 V	
			– Selector lever position N, D, and 3	0 to 0.8 V	
	35 + 1		– Selector lever position P, R, N, and D	4.5 to 5 V	
			– Selector lever position 3, 2, and 1	0 to 0.8 V	
	16 + 1		– Selector lever position P, R, and N	approx. Battery Positive Voltage (B+)	
			– Selector lever position D, 3, 2, and 1	0 to 0.8 V	

Switch to resistance measuring range 200 Ω					
Test step	**VAG 1598 sockets**	**Test of**	**• Test conditions – Additional operations**	**Specified value**	**Corrective action**
6	22 + 18	Solenoid Valve 1 -N88-	• Ignition switched off	55 to 65 Ω	– Check routing of wiring per wiring diagram
	22 + 1		• Transmission Control Module (TCM) -J217- disconnected	infinite Ω[1]	– Replace valve body ⇒ page H 38-36
7	23 + 18	Solenoid Valve 2 -N89-	• Ignition switched off	55 to 65 Ω	– Check routing of wiring per wiring diagram
	23 + 1		• Transmission Control Module (TCM) -J217- disconnected	infinite Ω[1]	– Replace valve body ⇒ page H 38-36
8	3 + 18	Solenoid Valve 3 -N90-	• Ignition switched off	55 to 65 Ω	– Check routing of wiring per wiring diagram
	3 + 1		• Transmission Control Module (TCM) -J217- disconnected	infinite Ω[1]	– Replace valve body ⇒ page H 38-36
9	2 + 18	Solenoid Valve 4 -N91-	• Ignition switched off	55 to 65 Ω	– Check routing of wiring per wiring diagram
	2 + 1		• Transmission Control Module (TCM) -J217- disconnected	infinite Ω[1]	– Replace valve body ⇒ page H 38-36

[1] Switch Digital Multimeter Fluke 83 to largest Ω range

Switch to resistance measuring range 200 Ω					
Test step	**VAG 1598 sockets**	**Test of**	**• Test conditions – Additional operations**	**Specified value**	**Corrective action**
10	24 + 18	Solenoid Valve 5 -N92-	• Ignition switched off	55 to 65 Ω	– Check routing of wiring per wiring diagram
	24 + 1		• Transmission Control Module (TCM) -J217- disconnected	infinite Ω[1]	– Replace valve body ⇒ page H 38-36
11	25 + 18	Solenoid Valve 6 -N93-	• Ignition switched off	4.5 to 6.5 Ω	– Check routing of wiring per wiring diagram
	25 + 1		• Transmission Control Module (TCM) -J217- disconnected	infinite Ω[1]	– Replace valve body ⇒ page H 38-36
12	21 + 18	Solenoid Valve 7 -N94-	• Ignition switched off	55 to 65 Ω	– Check routing of wiring per wiring diagram
	21 + 1		• Transmission Control Module (TCM) -J217- disconnected	infinite Ω[1]	– Replace valve body ⇒ page H 38-36
13	19 + 20	Shift Lock Solenoid -N110-	• Ignition switched off • Transmission Control Module (TCM) -J217- disconnected	14 to 25 Ω	– Check routing of wiring per wiring diagram – Replace Shift Lock Solenoid -N110- ⇒ Servicing selector mechanism, page H 37-1

[1] Switch Digital Multimeter Fluke 83 to largest Ω range

Switch to resistance measuring range 200 Ω					
Test step	**VAG 1598 sockets**	**Test of**	**• Test conditions – Additional operations**	**Specified value**	**Corrective action**
14	1 + 17	Kick Down Switch -F8-	• Ignition switched off • Transmission Control Module (TCM) -J217- disconnected • Accelerator Pedal not depressed	infinite Ω[1]	– Check routing of wiring per wiring diagram – Adjust accelerator cable or replace ⇒ Repair Group 20
			– Depress Accelerator Pedal as far as kick down	less than 1.5 Ω	

[1] Switch Digital Multimeter Fluke 83 to largest Ω range

Switch to resistance measuring range 20 K Ω					
Test step	**VAG 1598 sockets**	**Test of**	**• Test conditions – Additional operations**	**Specified value**	**Corrective action**
15	32 + 33	Vehicle Speed Sensor (VSS) -G68-	• Ignition switched off • Transmission Control Module (TCM) -J217- disconnected		– Check routing of wiring per wiring diagram – Replace Vehicle Speed Sensor (VSS) -G68- ⇒ Disassembling and assembling parking lock, page H 38-45
			min.	0.8 K Ω	
			max.	0.9 K Ω	

Switch to resistance measuring range 2 M Ω					
Test step	**VAG 1598 sockets**	**Test of**	**• Test conditions – Additional operations**	**Specified value**	**Corrective action**
16	30 + 18	Transmission Fluid (ATF) Temperature Sensor -G93-	• Ignition switched off • Transmission Control Module (TCM) -J217- disconnected • Transmission Fluid (ATF) Temperature		– Check routing of wiring per wiring diagram – Replace valve body ⇒ page H 38-36
			approx. 20°C (68°F)	0.247 M Ω	
			– Switch digital multimeter Fluke 83 to 200 K Ω approx. 60°C (140°F) approx. 120°C (248°F)	48.8 K Ω 7.4 K Ω	

Switch to resistance measuring range 200 Ω					
Test step	**VAG 1598 sockets**	**Test of**	**• Test conditions – Additional operations**	**Specified value**	**Corrective action**
17	36 + 1	Transmission Range (TR) Program Switch -E122-	• Ignition switched off • Transmission Control Module (TCM) -J217- disconnected • Transmission Range (TR) Program Switch -E122- not operated	infinite Ω[1]	– Check wiring diagram – Replace Transmission Range (TR) Program Switch -E122- ⇒ Disassembling and assembling shift mechanism, page H 37-6
			• Transmission Range (TR) Program Switch -E122- operated	less than 1.5 Ω	

[1] Switch Digital Multimeter Fluke 83 to largest Ω range

Switch to resistance measuring range 20 K Ω					
Test step	**VAG 1598 sockets**	**Test of**	**• Test conditions – Additional operations**	**Specified value**	**Corrective action**
18	27 + 28	Engine Speed Sensor (ESS) -G28- (for Diesel engine)	• Ignition switched off • Transmission Control Module (TCM) -J217- unplugged	0.8 K Ω	– Check wiring diagram – Distance to pressure plate max. 1.20 mm (0.047 in.) – Replace Engine Speed (RPM) Sensor (ESS) -G28- (for Diesel engine): Engine Speed (RPM) Sensor (ESS) -G28- is located on upper torque converter ⇒ Removing and installing differential, page H 39-18,

Torque converter

The ATF pump is driven by the torque converter.

When installing the torque converter, ensure that both drive pins engage in the recesses in the ATF pump inner wheel.

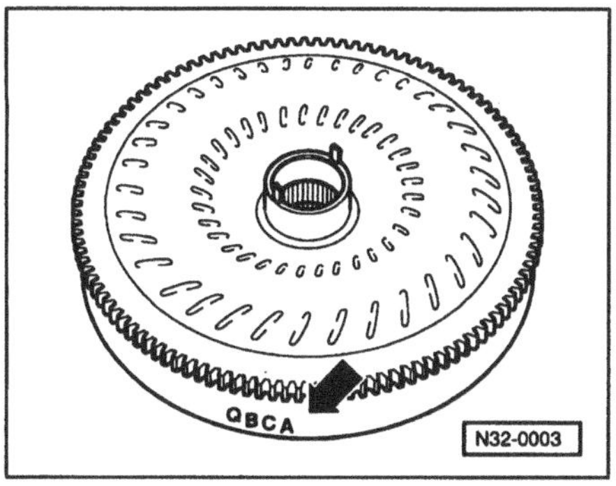

Torque converter, identification

◄ Various torque converters are available. They are identified by code letters -arrow-.

Transmission/torque converter application
⇒ from page H 00-3

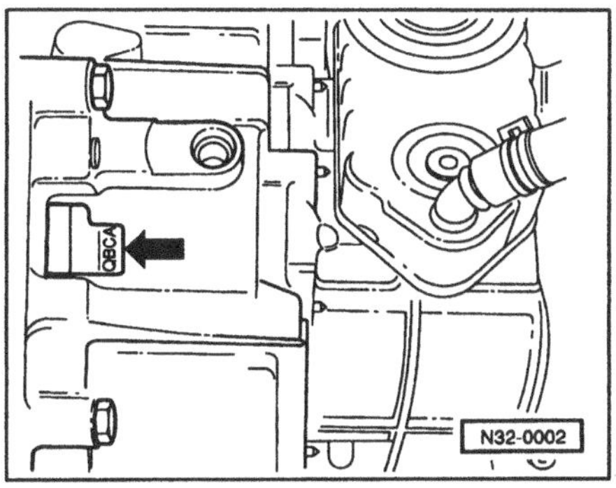

◄ **Torque converter code letters**

(Transmission installed)

Torque converter, draining

If the torque converter has become contaminated by particles resulting from wear or abrasion, or when performing a major overhaul of the transmission, drain the torque converter as follows:

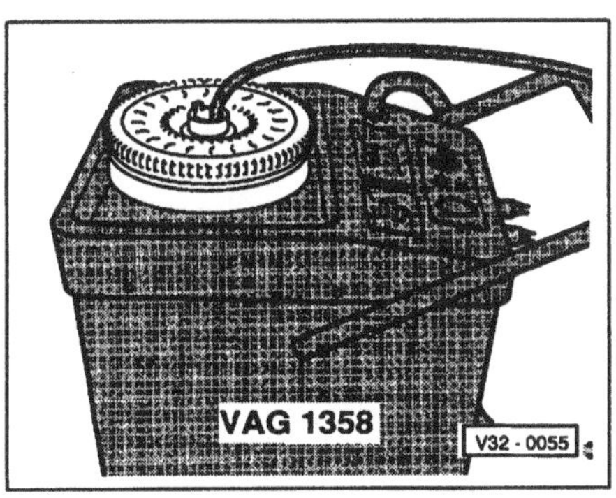

◄ – Remove transmission fluid (ATF) from torque converter using VAG 1358 A extraction system or EZ1 fluid extractor or equivalent.

Torque converter oil seal, removing and installing

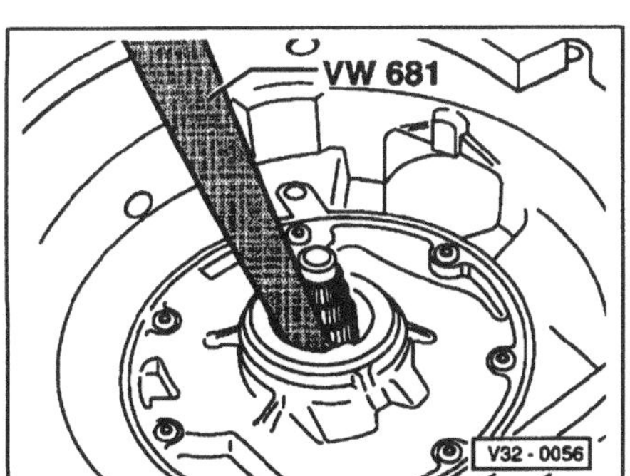

◄ **Torque converter oil seal, removing**

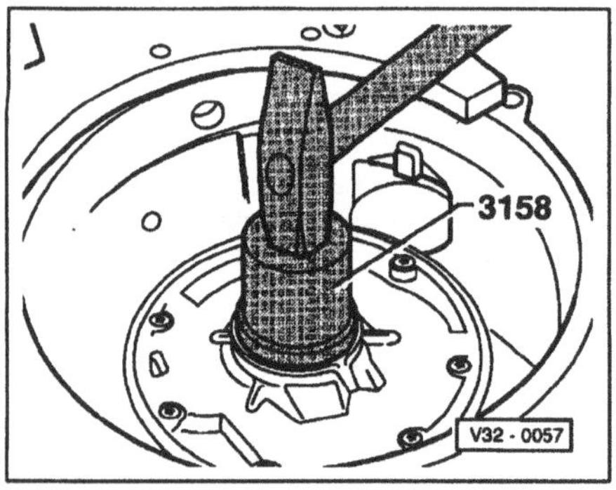

◄ Torque converter oil seal, driving in until flush

Selector lever mechanism, servicing

CAUTION!

Before working on the vehicle with the engine running, move the selector lever into the "P" position and apply the hand-brake.

Selector lever mechanism, checking

Selector lever in "P" position and ignition switched on:

- Brake pedal not depressed:

  Selector lever is locked and cannot be shifted out of "P" position. Shift Lock Solenoid -N110- blocks selector lever.

- Brake pedal depressed:

  Shift Lock Solenoid -N110- releases selector lever. It is now possible to shift into a driving gear.

Selector lever in "N" position and ignition switched on:

- **Brake pedal not depressed:**

  Selector lever is locked and cannot be shifted out of "N" position. Shift Lock Solenoid -N110- blocks selector lever.

- **Brake pedal depressed:**

  Shift Lock Solenoid -N110- releases selector lever. It is now possible to shift into a driving gear.

Note:

- *The starter must not be operative with the selector lever in "1", "2", "3", "D", and "R" positions.*

- *When travelling at speeds above 5 km/h (3 mph)and shifting into "N" position, Shift Lock Solenoid -N110- must not engage and block selector lever. Selector lever can be shifted into a driving gear.*
- *When travelling at speeds below 5 km/h (3 mph) (almost stopped) and shifting into "N" position, Shift Lock Solenoid -N110- must not engage until after approx. 1 sec. Selector lever cannot be shifted out of "N" position until the brake pedal is depressed.*

Selector lever cable, checking and adjusting

Checking

- Remove selector lever cable locking washer from lever/selector shaft.
- Remove selector lever cable from lever.
- Move selector lever from "P" to "1".

- Shift mechanism and selector lever cable must move freely. If necessary, replace selector lever cable or service shift mechanism.

Always replace selector lever cable locking element on transmission support bracket after disassembling.

Adjusting

– Selector lever in "P" position.

– Loosen adjustment screw on selector lever in selector lever housing.

– Bring selector shaft lever in position "P" on transmission. Locking lever must engage in park lock wheel, both front wheels blocked.

– Tighten adjustment screw on selector lever in selector lever housing to 20 Nm (15 ft lb).

Ignition key removal lock, checking (vehicles with ignition key removal lock)

– Set selector lever to "P" position and switch ignition off. Ensure that ignition key can be removed from vehicle with selector lever in "P" position only.

– Remove ignition key. Selector lever cannot be shifted out of "P" position.

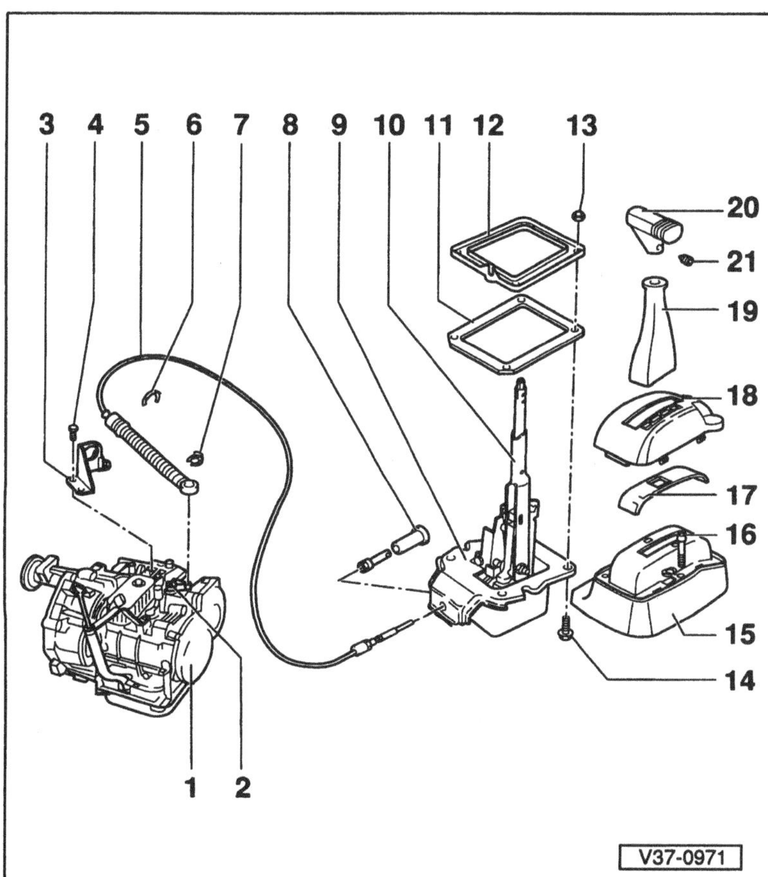

Shift mechanism, disassembling and assembling

1 - Transmission

2 - Lever
- ♦ For selector shaft

3 - Support bracket
- ♦ For selector lever cable

4 - Bolt
- ♦ Tightening torque: 10 Nm (7 ft lb)

5 - Selector lever cable with guide tube
- ♦ Do not bend or kink the cable. Lightly grease cable eyelet and cable ends before installing
- ♦ Boot and protective cover must be installed correctly otherwise moisture can enter the selector lever cable

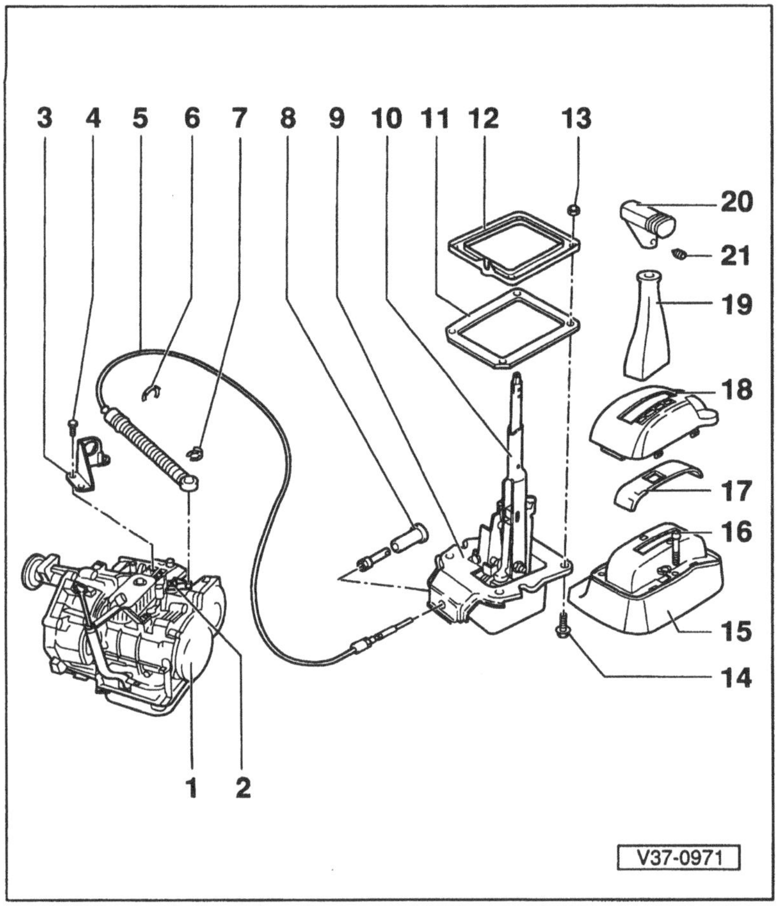

- Check that boot and protective cover are undamaged and seated correctly. Replace if necessary
- Check guide tube for tight fit
- Tighten selector lever cable on anchorage/selector lever housing to 20 Nm (15 ft lb). Make sure recessed side of selector lever cable is towards adjustment screw
- Secure to support bracket with locking element
- Secure to lever/selector shaft with locking washer

6 - Locking element

- For selector lever cable
- Always replace

7 - Locking washer

- For selector lever cable

8 - Protective sleeve

- Insert into selector lever housing before attaching selector lever cable

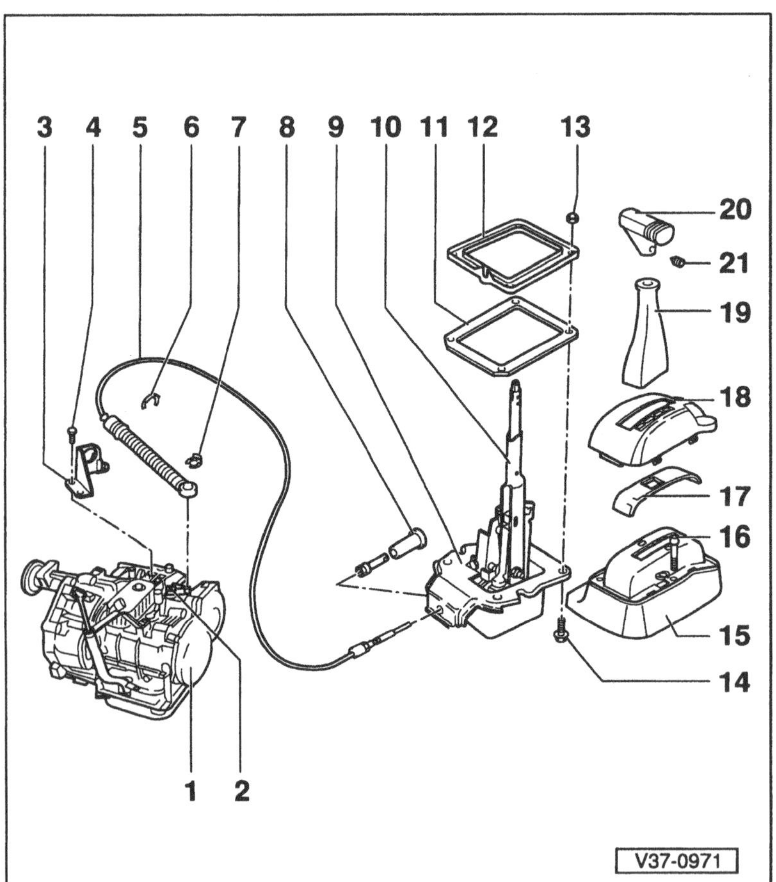

9 - Selector lever housing

- With installed selector lever
- Need not be removed to replace selector lever cable or individual parts of shift mechanism

10 - Selector lever

- Removing and installing ⇒ page H 37-10
- Disassembling and assembling ⇒ page H 37-13

11 - Gasket

- Attach to selector lever housing with gasket cement

12 - Frame

- For securing the console and the ignition key removal lock cable

13 - Nut

- Tightening torque: 12 Nm (9 ft lb)

14 - Bolt

- Tightening torque: 12 Nm (9 ft lb)

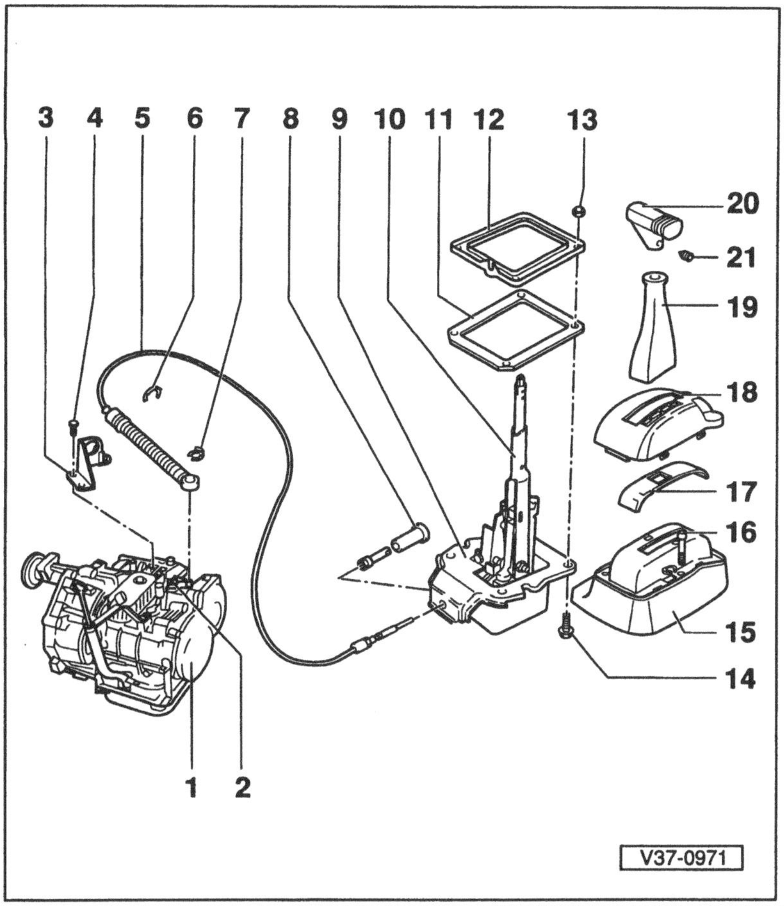

15 - Console

16 - Bolt
- ♦ Tightening torque: 2 Nm (18 in. lb)

17 - Cover strip

18 - Cover
- ♦ With selector indicator

19 - Selector lever trim

20 - Selector lever handle
- ♦ Lightly grease push rod before installing
- ♦ Depending on the application, secure with set screw or grub screw

21 - Set screw
- ♦ Tightening torque: 1.5 Nm (13 in. lb)
- ♦ Insert grub screw with D 000 600

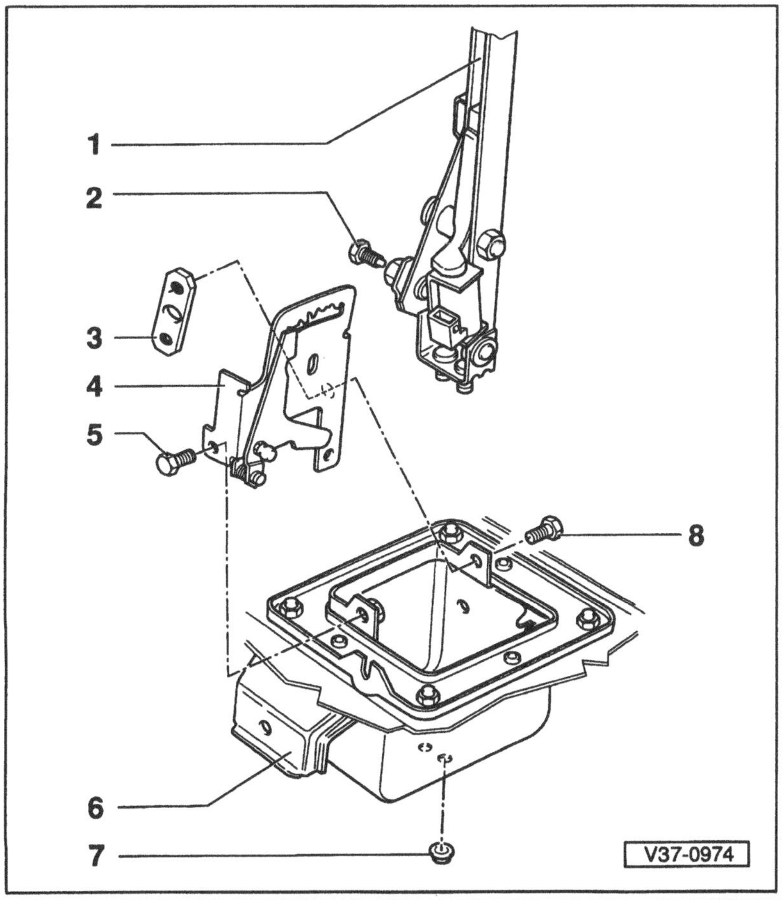

Selector lever, removing and installing

1 - Selector lever
- ♦ Assemble before installing in selector lever housing
- ♦ Disassembling and assembling ⇒ page H 37-13
- ♦ Wiring harness ⇒ Fig. 1

2 - Bolt
- ♦ Tightening torque: 20 Nm (15 ft lb)
- ♦ For securing the selector lever cable

3 - Plate with threaded holes

4 - Detent plate
- ♦ Insert into selector lever housing before installing selector lever
- ♦ Disassembling and assembling ⇒ page H 37-13

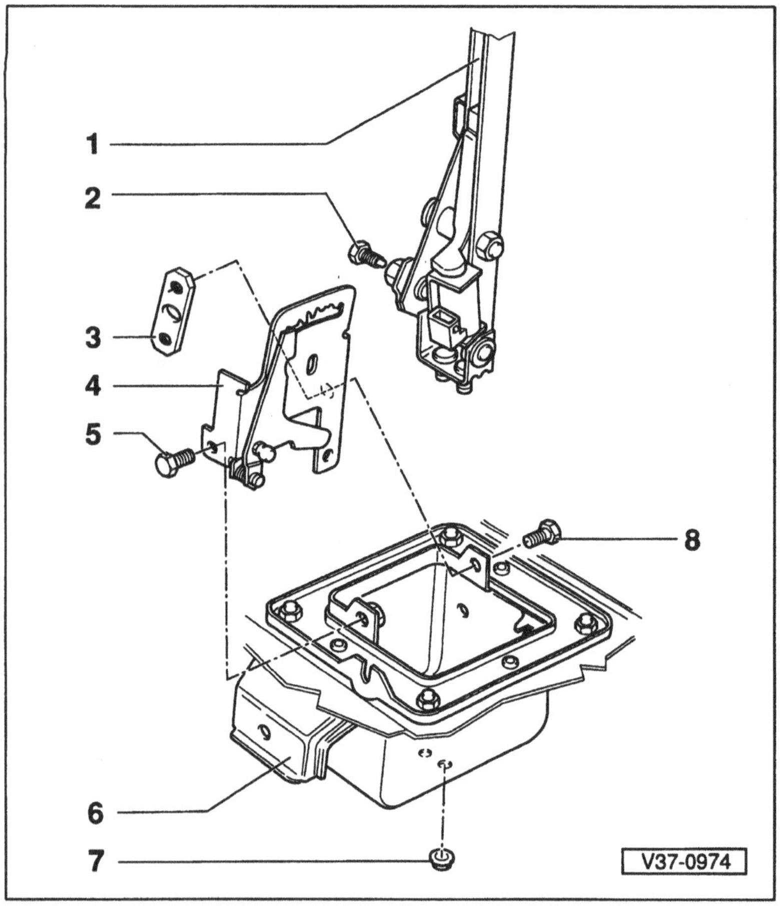

5 - **Bolt**
- Tightening torque: 25 Nm (18 ft lb)

6 - **Selector lever housing**
- Need not be removed to replace selector lever cable or individual parts of shift mechanism

7 - **Nut**
- Tightening torque: 15 Nm (11 ft lb)

8 - **Bolt**
- Tightening torque: 25 Nm (18 ft lb)

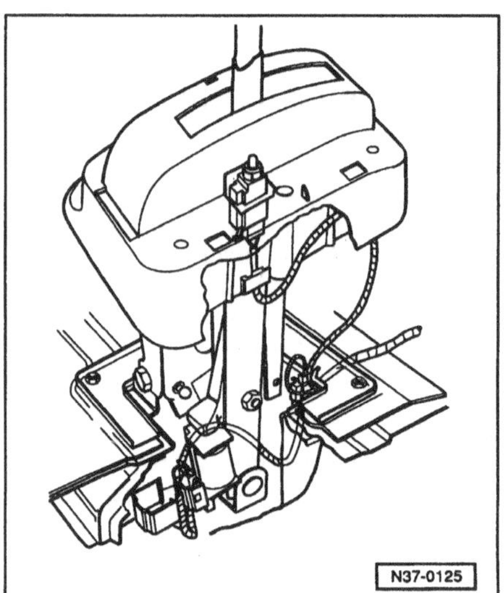

◄ **Fig. 1 Wiring harness for indicator lighting and Shift Lock Solenoid -N110-**

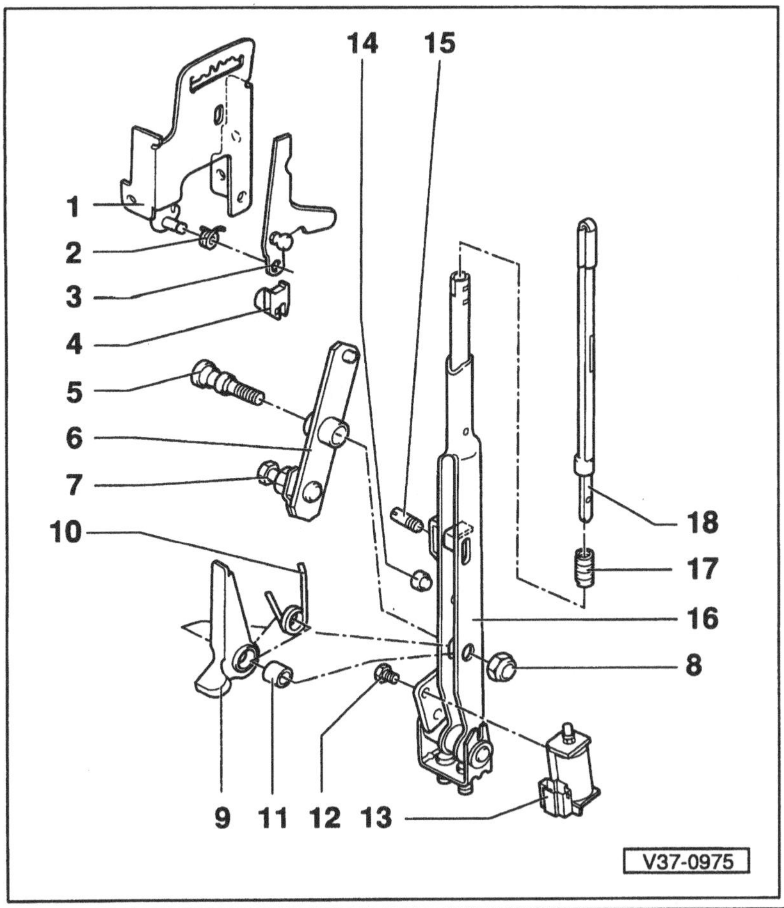

Selector lever, disassembling and assembling

Note:

Lubricate all joints and sliding surfaces with multi purpose grease.

1 - Detent plate
- ♦ Before installing the selector lever in selector housing, assemble selector lever, detent plate, spring, locking lever, and pivot piece.

2 - Spring
- ♦ For locking lever
- ♦ Hook into hole of detent plate

3 - Locking lever
- ♦ For vehicles with ignition key removal lock

4 - Pivot piece

5 - Combi-bolt

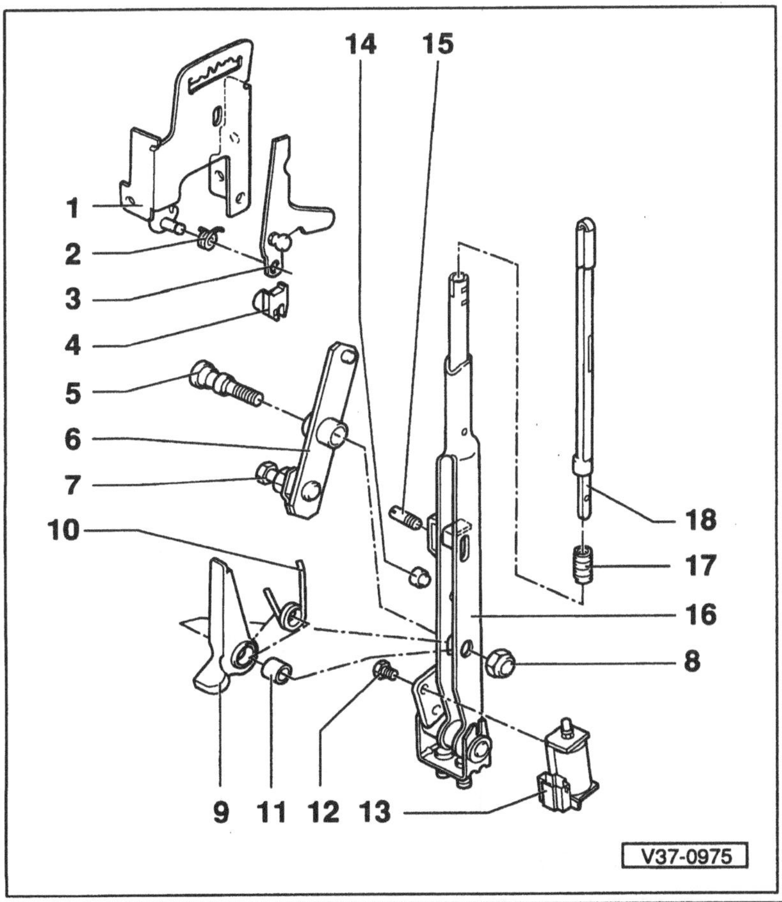

6 - Relay lever

7 - Bolt
- ♦ For securing the selector lever cable

8 - Hex nut
- ♦ Tightening torque: 25 Nm (18 ft lb)
- ♦ Always replace

9 - Locking lever
- ♦ Insert with spring and bushing into the selector lever

10 - Spring
- ♦ Insert with locking lever and bushing into the selector lever

11 - Bush
- ♦ Insert bushing in locking lever

12 - Hex bolt
- ♦ Tightening torque: 3 Nm (27 in lb)
- ♦ Install with thread–locking fluid, Part No. AMV 185 101 A1

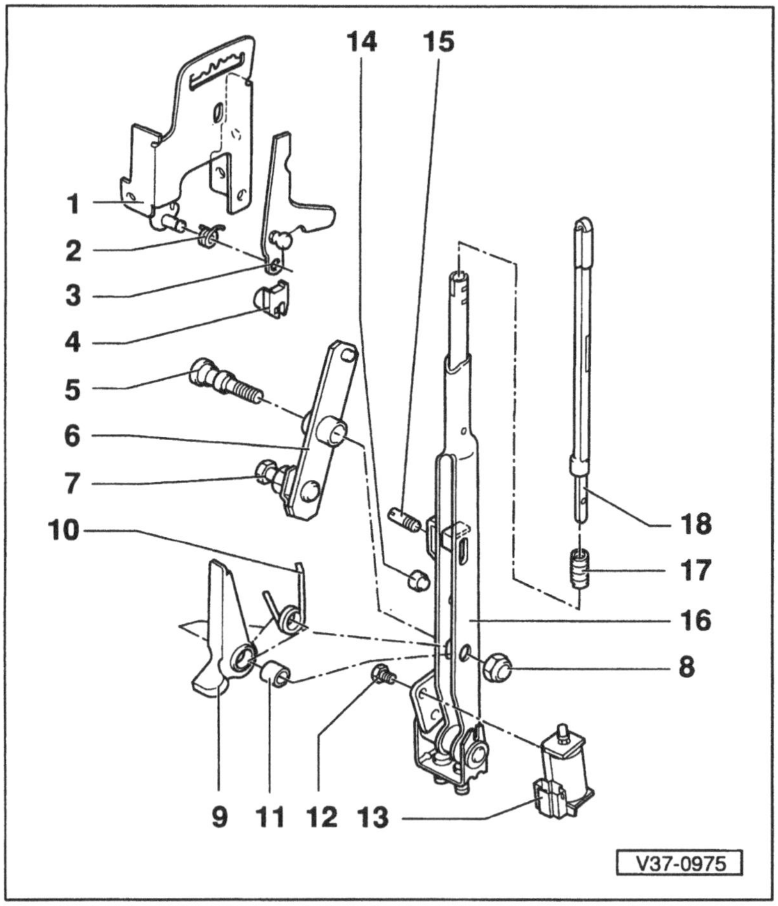

13 - **Shift Lock Solenoid -N110-**
- Can be checked electrically ⇒ page H 01-55 and by measured value block ⇒ page H 01-35

14 - **Slide**
- Press into the selector lever hole

15 - **Set screw**
- Install with thread–locking fluid, Part No. AMV 185 101 A1

16 - **Selector lever**
- Assemble before installing in selector lever housing

17 - **Spring**

18 - **Push rod**
- Grease in area of sliding surfaces

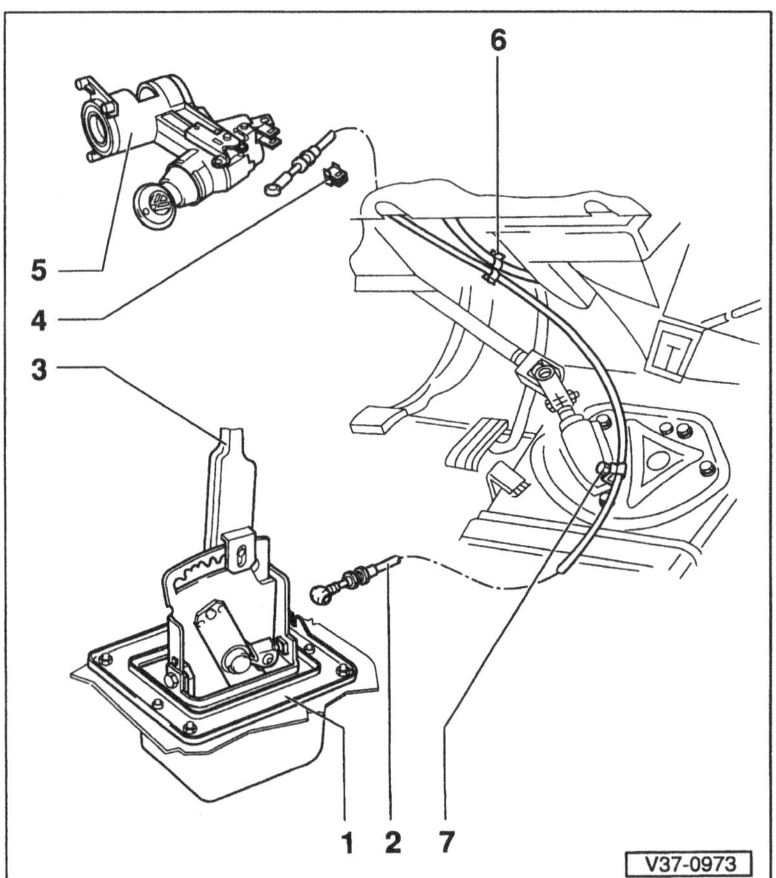

Locking cable, removing and installing

(vehicles with ignition key removal lock)

Note:

After installing adjust locking cable ⇒ page H 37-19.

1 - **Frame**

2 - **Locking cable**
- Must not be kinked
- Adjusting ⇒ page H 37-19
- Checking ignition key removal lock ⇒ page H 37-5

3 - **Selector lever**
- Removing and installing ⇒ page H 37-10
- Disassembling and assembling ⇒ page H 37-13

4 - **Clip**
- To secure locking cable
- Always replace

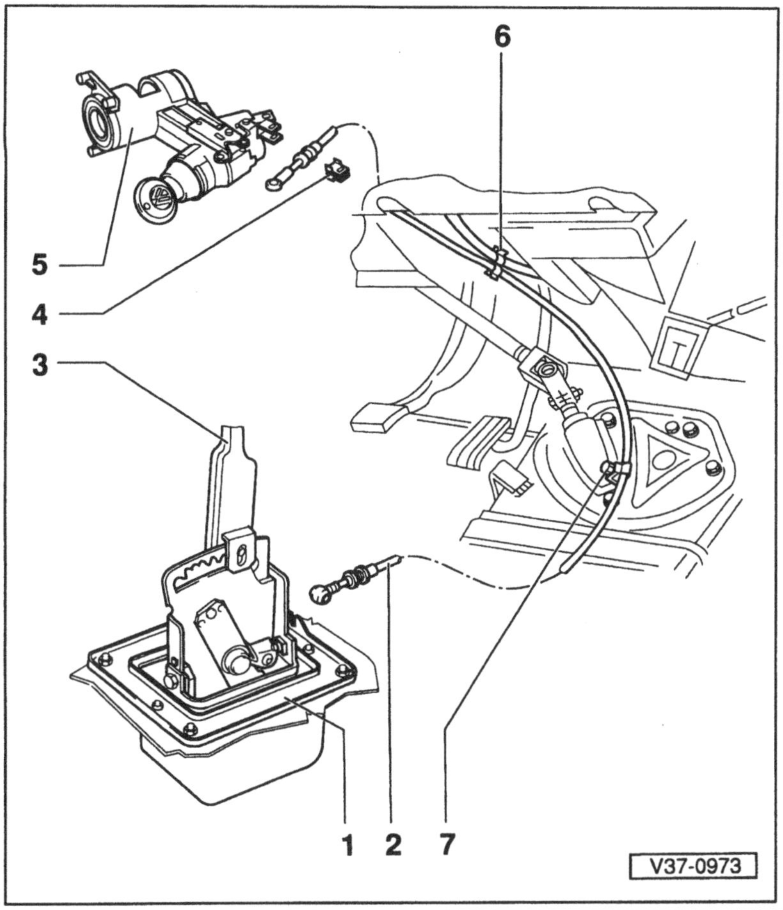

5 - Ignition/starter switch

6 - Mounting
- For locking cable
- Secure to heater hose

7 - Mounting
- For locking cable
- Secure to base plate and pull locking cable through mounting eye

Removing

– Remove selector lever handle, cover, cover strips, and selector lever console.

– Remove locking cable from support bracket and remove from locking lever ball head.

– Remove ignition/starter switch trim.

⇒ Electrical system; Repair Group 90; Removing and installing instrument cluster.

– Press locking cable clip out of ignition/starter switch support bracket and remove locking cable from ball head.

– Remove footwell vent console.

⇒ Heating, air conditioning; Repair Group 80, Servicing heating

– Remove locking cable.

Installing

Installation is performed in the reverse order.

Notes:

- *Ensure proper routing of locking cable when installing.*
- *Adjust locking cable after installing*
  ⇒ page H 37-19.

Locking cable, adjusting (vehicles with ignition key removal lock)

Note:

Locking cable must only be adjusted in its installation position
⇒ page H 37-16.

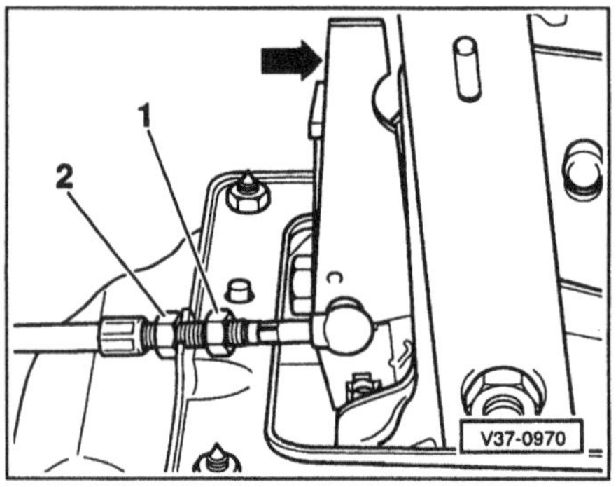

– Selector lever in "P" position.

– Switch ignition/starter switch "OFF".

◄ – Loosen locking cable nuts -1- and -2- at frame.

– Push locking lever -arrow- until it touches selector lever.

– Secure locking cable in this position on frame.

– Perform functional check
⇒ page H 37-5.

Shift Points, checking

Test Requirements

- Activate On Board Diagnostic (OBD) Program
  - Interrogate Diagnostic Trouble Code (DTC) memory
  - Begin basic adjustment
- Check Automatic Transmission Fluid (ATF) level
⇒ page H 37-46
- Check engine settings
⇒ Repair Group 24

Transmission shifts too early or too late into the next gear

When testing shift points be sure to note that the speedometer can deviate from the specified figure within the permissible tolerance.

The table lists the gear shifts by transmission code letters.

H = Hydraulic shift

M = Mechanical shift

For vehicles with a shift program switch, the 3 H – 3 M shift can occur at higher speeds through torque converter slip, as specified in the charts.

For vehicles equipped with an electronic shift program, the "ECO" or "Sport" program is shifted automatically by the Transmission Control Module (TCM) depending on the operating speed of the Accelerator Pedal.

For the electronic Transmission Range (TR) Program Switch -E122- only the kickdown shift points are given.

To check the shift points, connect VAG 1551 Scan Tool.
⇒ On Board Diagnostic (OBD) program, Repair Group 01; Reading test value block.

CAUTION!

When driving or riding in airbag equipped vehicles, never hold test equipment in your hands or lap when the vehicle is in motion. Objects between you and the airbag can increase the risk of injury in an accident.

Shift Points in mph

Transmission Code	Shift	Wide Open Throttle		Kick Down	
		ECO mph	Sport mph	ECO mph	Sport mph
ARW, CEV, CKA	1 H – 2 H	13 ... 16	18 ... 22	18 ... 22	
	2 H – 3 H	24 ... 28	35 ... 39	35 ... 39	
	3 H – 3 M	41 ... 44	52 ... 55	52 ... 55	
	3 M – 4 M	49 ... 53	58 ... 61	64 ... 68	
	4 M – 3 M	44 ... 40	58 ... 54	66 ... 63	
	3 M – 3 H	25 ... 22	35 ... 32	35 ... 32	
	3 H – 2 H	24 ... 20	31 ... 27	35 ... 32	
	2 H – 1 H	6 ... 2	16 ... 12	19 ... 15	
AXY, CEU, CHY	1 H – 2 H	14 ... 18	20 ... 23	20 ... 23	
	2 H – 3 H	26 ... 29	38 ... 42	38 ... 42	
	3 H – 3 M	44 ... 47	56 ... 59	56 ... 59	
	3 M – 4 M	53 ... 56	62 ... 66	69 ... 73	
	4 M – 3 M	47 ... 44	62 ... 59	71 ... 68	
	3 M – 3 H	28 ... 24	38 ... 35	38 ... 35	
	3 H – 2 H	25 ... 22	31 ... 27	38 ... 35	
	2 H – 1 H	7 ... 3	17 ... 13	20 ... 16	
AXX, CHZ	1 H – 2 H	10 ... 13	17 ... 21	17 ... 21	
	2 H – 3 H	25 ... 29	33 ... 36	33 ... 36	
	3 H – 3 M	25 ... 29	37 ... 41	37 ... 44	
	3 M – 4 M	47 ... 50	55 ... 59	55 ... 59	
	4 M – 3 M	47 ... 43	55 ... 51	55 ... 51	
	3 M – 3 H	27 ... 23	33 ... 29	35 ... 33	
	3 H – 2 H	23 ... 20	33 ... 29	32 ... 29	
	2 H – 1 H	6 ... 2	15 ... 12	19 ... 15	

Transmission Code	Shift	Full Throttle		Kick Down	
		ECO mph	Sport mph	ECO mph	Sport mph
AXW,CKB	1 H – 2 H	9 ... 13	16 ... 20	16 ... 20	
	2 H – 3 H	23 ... 27	30 ... 34	30 ... 34	
	3 H – 3 M	23 ... 27	34 ... 38	34 ... 38	
	3 M – 4 M	43 ... 47	51 ... 55	51 ... 55	
	4 M – 3 M	43 ... 41	51 ... 47	51 ... 47	
	3 M – 3 H	25 ... 22	30 ... 27	34 ... 30	
	3 H – 2 H	24 ... 20	30 ... 27	30 ... 27	
	2 H – 1 H	6 ... 2	14 ... 11	18 ... 14	

Shift Points in mph (with electronic program switch in control module)

Transmission Code	Shift	Kick Down
		ECO/Sport mph
CFZ	1 H – 2 H	19 ... 23
	2 H – 3 H	38 ... 42
	3 H – 4 M	58 ... 62
	4 M – 3 H	55 ... 51
	3 H – 2 H	36 ... 32
	2 H – 1 H	20 ... 16
CGA	1 H – 2 H	18 ... 22
	2 H – 3 H	35 ... 39
	3 H – 4 M	54 ... 58
	4 M – 3 H	51 ... 47
	3 H – 2 H	33 ... 30
	2 H – 1 H	19 ... 15
CGK	1 H – 2 H	22 ... 26
	2 H – 3 H	46 ... 50
	3 H – 4 M	71 ... 74
	4 M – 3 H	70 ... 66
	3 H – 2 H	46 ... 42
	2 H – 1 H	24 ... 20

Transmission Code	Shift	Kick Down ECO/Sport mph
CGH	1 H – 2 H 2 H – 3 H 3 H – 4 M 4 M – 3 H 3 H – 2 H 2 H – 1 H	20 ... 24 43 ... 46 66 ... 69 65 ... 61 43 ... 39 22 ... 18
CTL	1 H – 2 H 2 H – 3 H 3 H – 4 M 4 M – 3 H 3 H – 2 H 2 H – 1 H	19 ... 22 39 ... 43 61 ... 64 60 ... 56 40 ... 36 20 ... 17

Stall Speed, checking

- Connect tachometer (e.g., VAG 1367 engine-tester ignition system), or read engine speed (RPM) in measured value block with VAG 1551 Scan Tool.
- Secure vehicle with parking and foot brake.
- Start engine and place selector lever in "D" position.
- Increase engine speed (RPM) briefly (max. 5 sec.), if necessary repeat test after 20 sec. The stall speed (RPM) listed in the chart must be attained.
- If the stall speed is not attained, the following malfunctions are probable causes:

Stall speed too high:	Forward clutch slipping, freewheel slipping
Stall speed up to 200 RPM too low:	Check engine settings
Stall speed over 200 RPM too low:	Torque converter malfunction

In cases of doubt, always check the engine settings first and correct if necessary.

The correct torque converter must always be installed with the transmission.

Stall Speed Chart

Transmission Code	Torque Converter Code Letter	Stall Speed (RPM)
AXY, ARW, CEU, CEV, CKA, CHY	QBCA	2400 ... 2700
AXX, AXW, CKB, CHZ,	QBCA	2250 ... 2550
CFZ, CGA	QBCA	2250 ... 2550
CGK, CGH, CTL	QBCA	2450 ... 2750

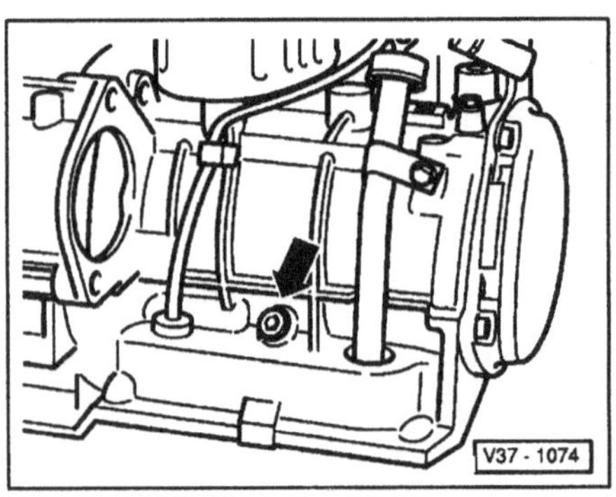

Main Pressure, measuring

- Before checking the main pressure, check the Diagnostic Trouble Code (DTC) memory and, if necessary, correct any electrical malfunctions.
- Activate Diagnostic Trouble Code (DTC) memory,
  ⇒ Repair Group 01, On Board Diagnostic (OBD) program

◄

- Remove bolt from main pressure bore -arrow-.
- Connect pressure manometer

● Always replace bolt for main pressure bore.

- Connect VAG 1702 pressure tester or equivalent and tighten knurled screw.
- Bleed pressure tester line by loosening knurled screw with engine running.

Note:

Retighten knurled screw after bleeding.

– Bring Automatic Transmission Fluid (ATF) temperature to 60°C (140°F). Connect VAG 1561 Scan Tool and read temperature in measured value block.
⇒ Repair Group 01, On Board Diagnostic (OBD) program]

Main Pressure Specified Values:

Automatic Transmission	Selector Lever Positions and Test Conditions			
096			Connector for solenoid valves removed [1]	
	D Idle	R Idle	D Engine Speed approx. 2000 RPM	R Engine Speed approx. 2000 RPM
Main pressure in bar	3.4 ... 3.8	6.5 ... 7.5	10.1 ... 11.3	23.0 ... 24.0

[1] After reconnecting the solenoid valve connector, check and erase Diagnostic Trouble Code (DTC) memory.
⇒ Repair Group 01, On Board Diagnostic (OBD) program.

– If possible, perform main pressure test on a chassis dynamometer.

If deviations occur from the specified values the following could be the cause:

Possible Causes	**Corrective action**
Idle Speed too high	Adjust engine
ATF pump faulty	Check ATF Pump; replace if necessary
Control valve sticking in valve body	Replace valve body

Transmission, removing and installing

Removing

– Fit support frames 3250 before lifting vehicle.

⇒ Suspension, Wheels, Brakes, Steering; Repair Group 40

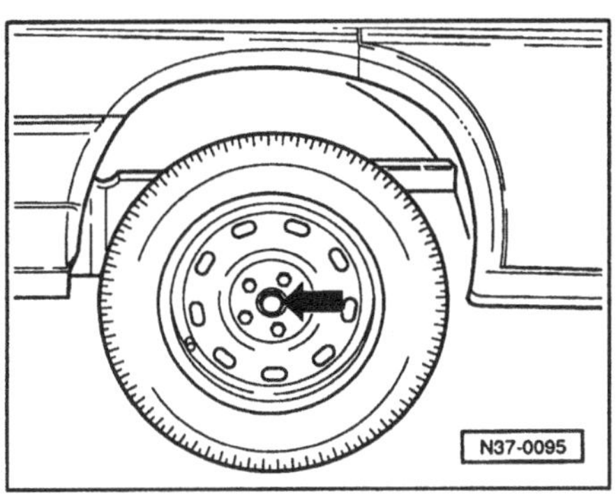

◄ – Remove left-hand drive shaft/wheel hub bolt -arrow-.

– Obtain radio code on vehicles with coded radio.

– Disconnect Battery Ground (GND) strap.

– Remove Battery covering.

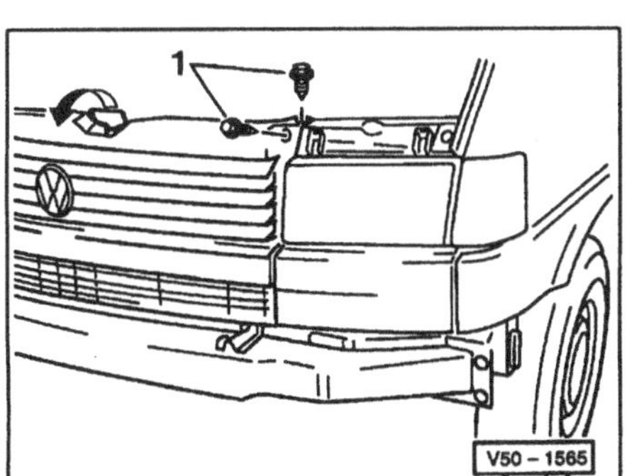

◄ – Remove bolts -1- and lift radiator out in direction of arrow.

– Remove upper engine/transmission connecting bolts.

– Remove transmission dipstick and plug transmission filler tube.

– Remove noise insulation tray.

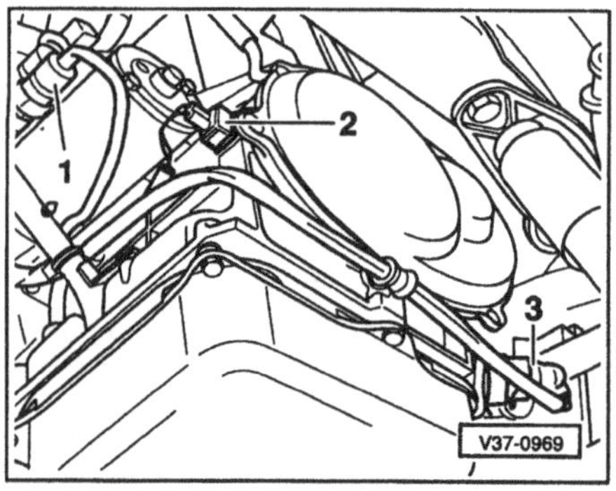

◄ – Disconnect electrical connections -1-, -2-, and -3- from transmission.

1 - Valve body connection

2 - Vehicle Speed Sensor (VSS) -G68-

3 - Multi-Function Transmission Range (TR) Switch -F125-

– Remove Multi-Function Transmission Range (TR) Switch, wiring, with bracket, and set to side.

– Remove Starter.

– Disconnect speedometer drive shaft from bearing housing, or

– Pull speedometer connector off bearing housing.

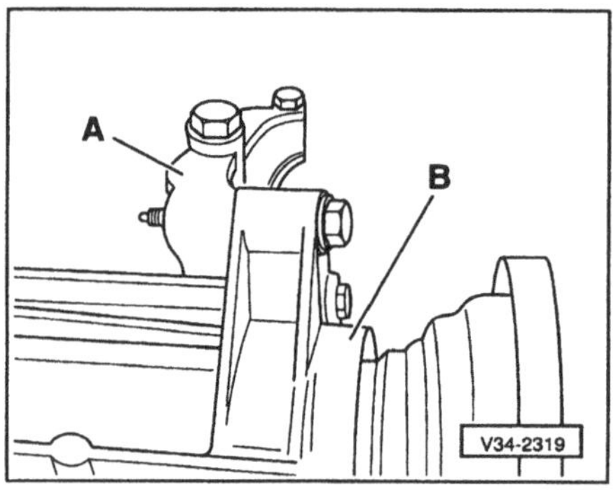

◄ – Remove transmission carrier -A- from engine and bearing housing -B-.

– Disconnect front exhaust pipe at exhaust manifold.

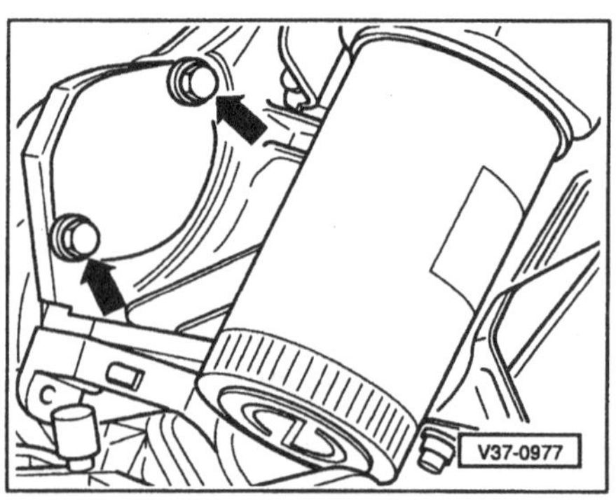

◄ – Remove bolts -arrows- from cover plate.

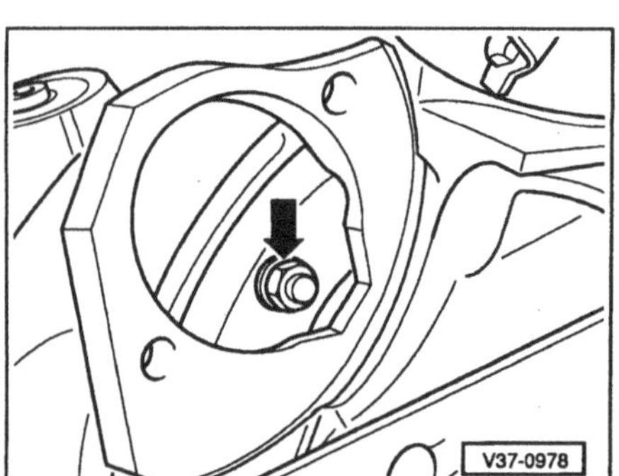

◄ – Remove nut -arrow- from torque converter (Qty. 3).

– Disconnect both drive shafts at flanges.

– Raise right-hand drive shaft and secure.

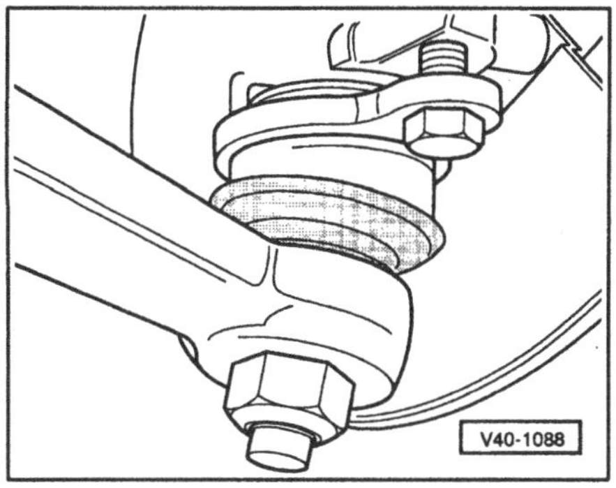

- Separate left-hand wheel bearing housing/lower axle link bolted connection.
- Remove bolt for left-hand shock absorber/anti-roll bar coupling rod.
- Compress left-hand shock absorber.
- Install 3184 removal device to support engine/transmission:
- Install engine bracket 3227 on engine block.

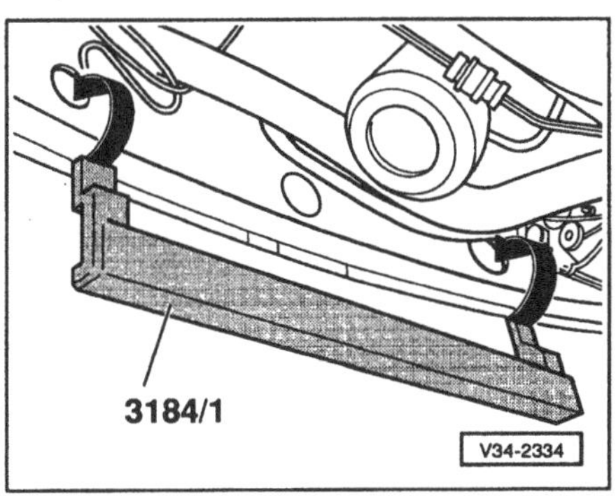

- Attach 3184/1 auxiliary support for 3184 removal device into cross panel -arrow-. If necessary, remove wiring harness and lines from cross panel.

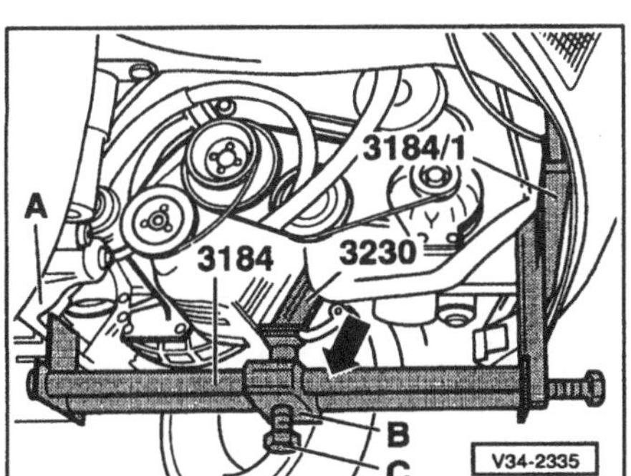

- Install 3184 removal device into 3184/1 auxiliary support and into subframe -A-. Thread spindle -C- into hole on engine mounting.
- Align 3184 removal device so that slide -B- is resting against right-hand tube -arrow-.
- Secure 3184 removal device to subframe -A- and to 3184/1 auxiliary support by tightening angle plates.
- Release weight from subframe using spindle -C- to raise engine/ transmission.

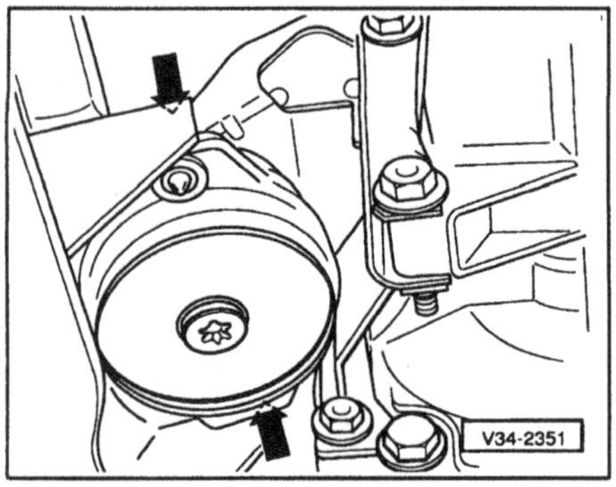

◄ – Remove bracket with rubber bonded mounting from rear of subframe -arrows-.

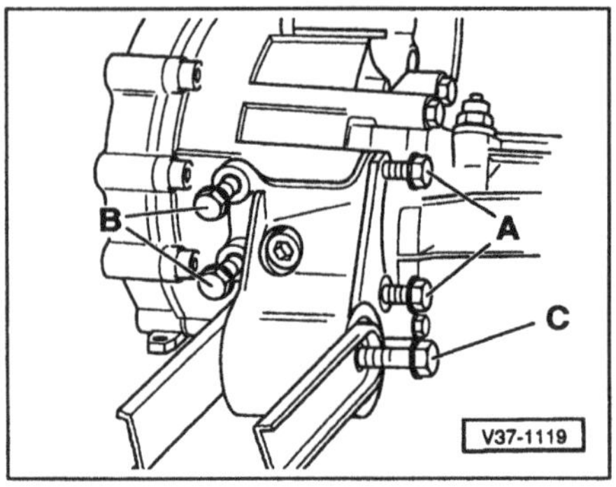

Vehicles from 08.92

◄ – Remove all bolts (A, B, C) and remove bracket with bonded rubber mounting.

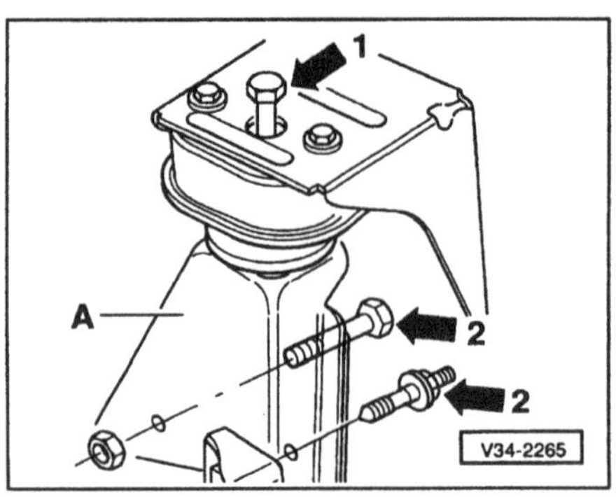

◄ – Unbolt mounting bracket -A- from bonded rubber mount at body -arrow 1-.

– Lower engine using removal device spindle.

– Remove bolts for mounting bracket -A- from transmission -arrow 2-.

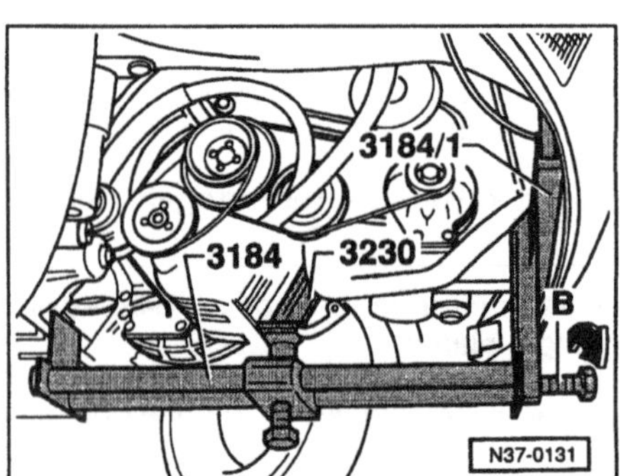

◄ – Ensuring that selector lever cable, Multi-Function TR Switch, and drive shafts are not damaged, carefully swivel engine/transmission forward and lower.

- To swivel forward: turn spindle -B- to left
- To swivel back: turn spindle -B- to right

This prevents the selector lever cable from being over-stressed.

– Unclip selector lever cable securing clips and remove selector lever cable.

- Clamp-off ATF cooler hoses with -3094- and remove hoses from ATF cooler.
- Tilt engine/transmission forward and lower.

● Provide sufficient clearance between engine and vehicle longitudinal member. If necessary, loosen Generator and remove tensioning arm from engine.

● To ensure Multi-Function Transmission Range (TR) Switch is not damaged, remove left-hand drive shaft.

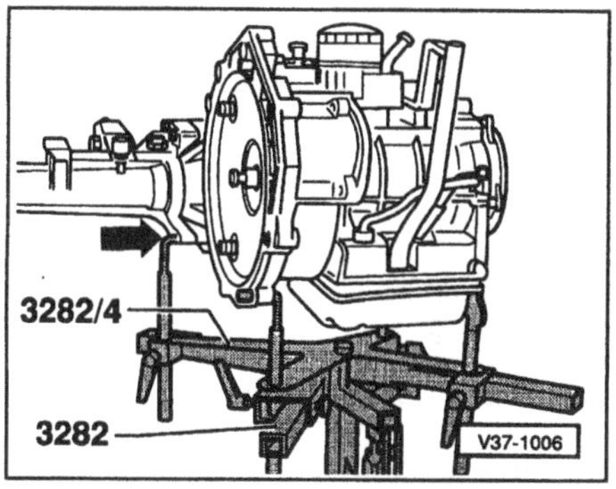

◄ – Set up 3282 transmission support.

- Remove bolt -arrow- securing bearing housing to allow transmission mounting hook to be installed.

Note:

♦ *The 3282 transmission support is set up with the 3282/4 adjusting plate for removing the automatic transmission 098.*

♦ *The symbols on the adjusting plate indicate the mounts required.*

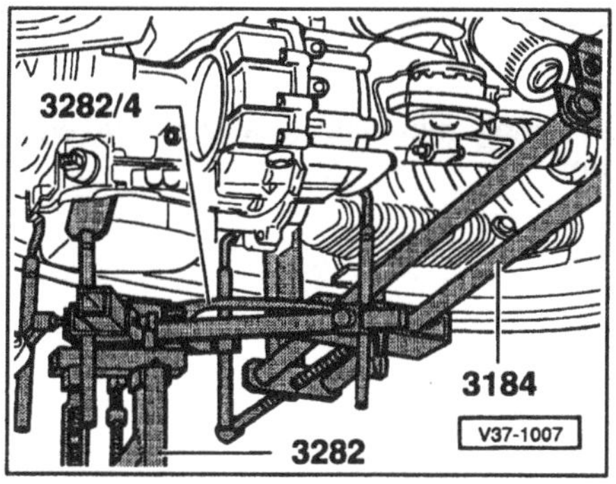

◄ – Move VAG 1383A jack with 3282 transmission support under transmission and support.

- Remove lower engine/transmission connecting bolts.
- Press transmission off engine. When doing this, press torque converter out of drive plate.
- Press torque converter against ATF pump.
- Lower transmission slightly.
- Tilt transmission forward and guide right-hand drive flange past 3184 removal device.
- Swivel transmission and carefully lower.
- Secure torque converter to prevent it from falling out.

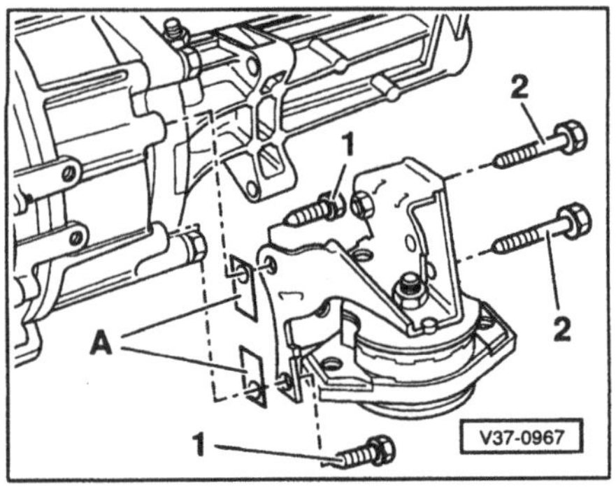

◄ If the transmission, the bearing housing, or the mount is replaced, the air gap between transmission housing and mount must be measured with a feeler gauge when the transmission is removed. The necessary spacer plates can then be reinstalled. To make this measurement, remove hex bolts -1-, and tighten hex bolts -2- to specified torque.

Air gap	Spacer plate	Part No.
to 0.5 mm (0.0197 in.)	none	–
to 1.0 mm (0.0393 in.)	x	701 399 219
above 1.0 mm (0.0393 in.)	x	701 399 219 A

The following modifications are to be made to transmissions with a M 12 hole in torque converter housing:

– Enlarge holes for hex bolts -1- in mount to 13 mm.

– Install new bolts M12 x 30 with washers.

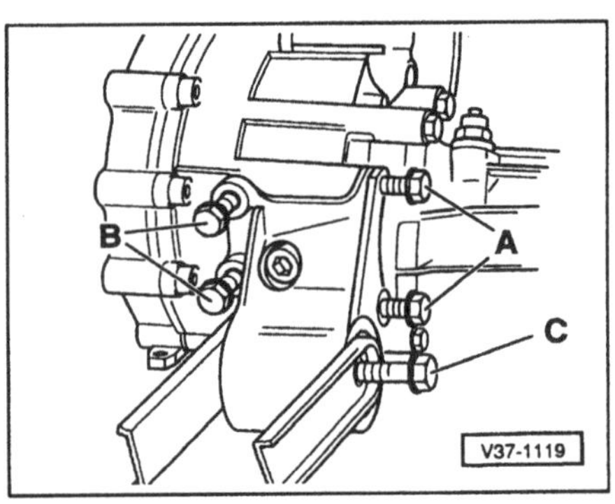

Vehicles with modified assembly carrier and rear assembly mounting from 08.92

◄ New method of securing assembly mounting to torque converter housing and assembly carrier.

A - Hex bolt M12 x 35

B - Hex bolt M10 x 40

C - Hex bolt M 16 x 116

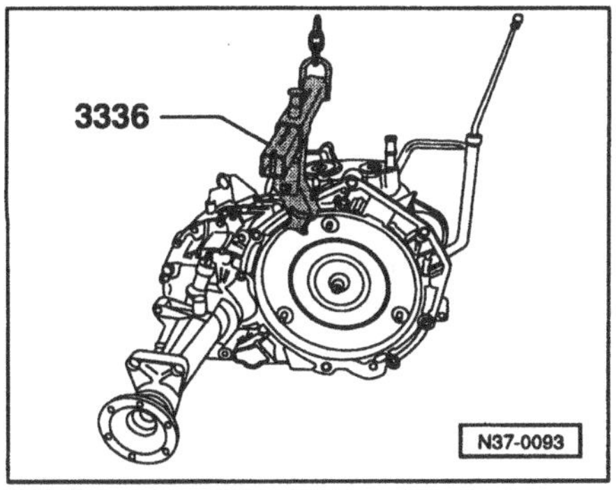

Transporting the automatic transmission

3336 special tool may be used to transport the automatic transmission and when setting up 3282 transmission support.

◄ – Secure 3336 special tool to transmission housing flange.

– Position pins so that 8 holes are visible in chain direction.

Installing

Installation is performed in the reverse order.

When installing the torque converter, ensure that both drive pins engage in the ATF pump inner wheel recesses.

– Before installing, ensure that dowel sleeves for aligning engine/transmission are correctly located.

– For vehicles from 08.92, place bracket with bonded rubber mounting (⇒ page H 37-35) in assembly carrier before installing transmission.

– When installing transmission, fit both drive shaft flanges in flange shafts.

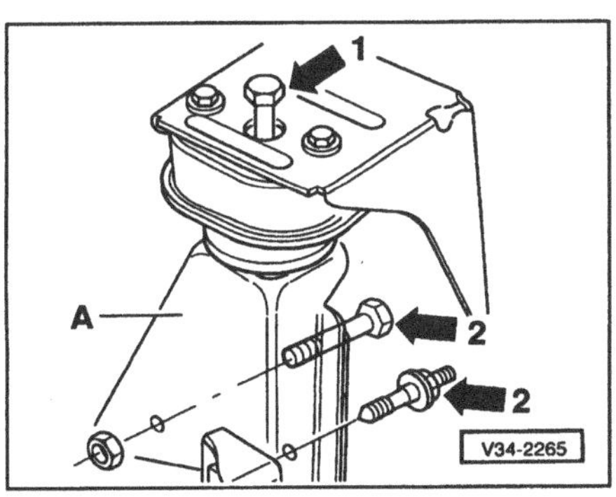

◄ – Before lifting engine/transmission securely tighten mounting bracket -A- bolts -arrows 2-.

– After lifting engine/transmission, securely tighten bolt -arrow 1-.

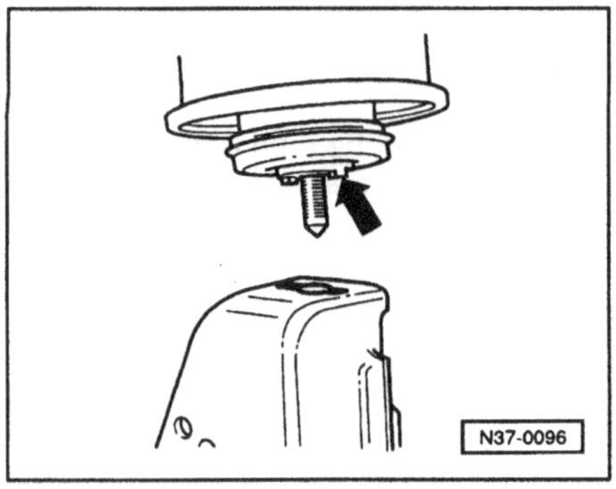

◄ – Ensure that bonded rubber mounting lugs -arrow- are seated in mounting bracket.

> ***CAUTION!***
>
> ***Do not remove 3184 removal device until the bolts securing the left-hand bearing bracket have been securely tightened.***

– Replace selector cable anchorage locking washer.

– Check selector lever cable adjustment. Adjust if necessary ⇒ page H 37-4.

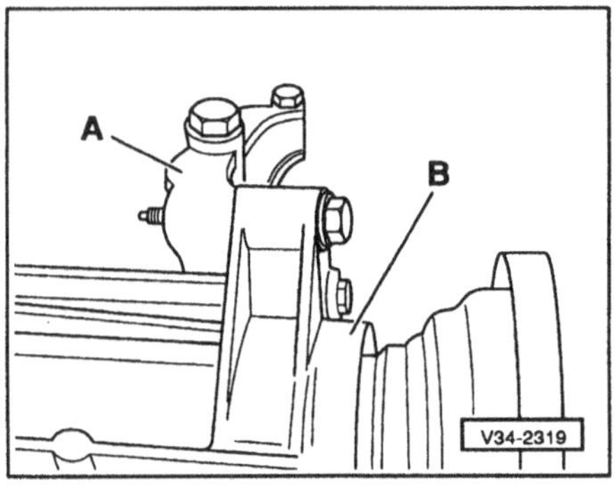

◄ – First tighten transmission carrier -A- to bearing housing -B- and then to engine.

– Left-hand drive shaft: Replace hex socket-head bolt on transmissions built up to 01 06 2 with a multiple-toothed socket-head bolt and shorten multiple-toothed socket-head bolt, M 10 x 48, by 3 mm (0.118 in.).

Tightening torques

Drive shaft to flange Multiple toothed socket-head bolts [1]	80 Nm (59 ft lb)	Hex bolt on drive shaft (Always replace bolt)	150 Nm (111 ft lb) plus 1/4 turn
Torque converter to drive plate	60 Nm (44 ft lb)	Bonded mounting to assembly carrier	45 Nm (33 ft lb)
Transmission to engine M 8	20 Nm (15 ft lb)	Bonded rubber mounting console to:	
Transmission to engine M 10	60 Nm (44 ft lb)	Torque converter housing M10 [2]	45 Nm (33 ft lb)
Transmission to engine M 12	80 Nm (59 ft lb)	Bearing housing	65 Nm (48 ft lb)
Left bearing bracket to bonded mounting	65 Nm (48 ft lb)		
Left-hand bearing bracket to transmission	65 Nm (48 ft lb)	Vehicles with modified assembly carrier	from 08.92
Cover plate on transmission	20 Nm (15 ft lb)	Assembly mounting to assembly carrier	200 Nm (148 ft lb)

Continued on next page

Tightening torques (continued)

Front exhaust pipe to exhaust manifold			Assembly mounting to torque converter		
	M 6	10 Nm (84 in. lb)		M 10	40 Nm (30 ft lb)
	M 8	25 Nm (18 ft lb)		M 12	100 Nm (74 ft lb)
	M 10	40 Nm (30 ft lb)	Wheel bearing housing to lower ball joint (Always replace bolts)		55 Nm (41 ft lb) plus 1/4 turn
Transmission carrier to bearing housing	M 8	25 Nm (18 ft lb)	Shock absorber to axle link		160 Nm (118 ft lb)
Transmission carrier to engine M 10		60 Nm (44 ft lb)			

1) Left-hand drive shaft: Replace hex socket-head bolt on transmissions built up to 01 06 2 with a multiple toothed socket-head bolt and shorten multiple toothed socket-head bolt, M 10 x 48, by 3 mm (0.118 in.).

2) For transmissions with modified console and M 12 threaded hole in torque converter housing: 100 Nm (74 ft lb)

ATF level, checking and topping up

ATF level, checking

Conditions for performing check

- When checking ATF level the transmission must not be in emergency running mode.
- Vehicle parked on level surface, parking brake applied
- Selector lever at "P" and engine running at idling speed

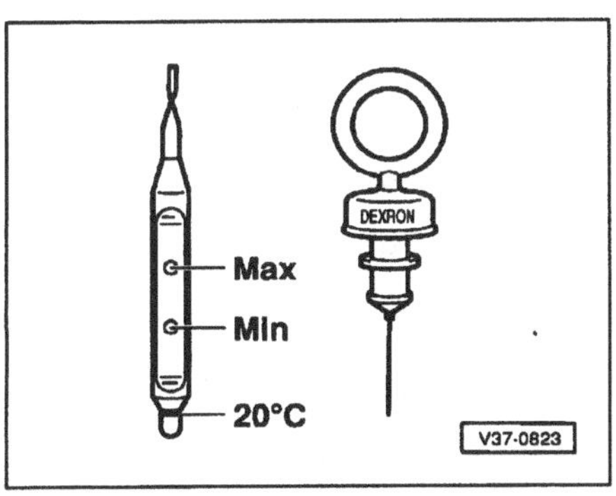

- When cold, the ATF should extend up to the 20°C (68°F) marking
- ATF must be up to normal operating temperature of approx. 60°C (140°F).

  Operating temperature is measured with VAG 1551 Scan Tool ⇒ page H 01-35, reading measured value block. Check ATF level and fill or extract as necessary.

  It is essential that the ATF level is between the Min. and Max. markings.

> ***CAUTION!***
>
> ***Too much or too little ATF will affect the operation of the transmission. The ATF level must therefore be checked at regular intervals.***

ATF, topping up

> ***CAUTION!***
>
> - ***Too much ATF will cause transmission malfunctions and must therefore be extracted. This can be done with VAG 1358 A/1 extraction system or EZ1 fluid extractor.***
> - ***Only ATF with the designation Dexron (or VW ATF) may be used. Do not use any lubricant additives.***

– Pour in ATF through a funnel. Approx. 3 liters (3.2 qts) for an initial filling.

– Start engine and shift through all selector lever positions with vehicle stationary.

– Check ATF level
⇒ page H 37-46.

Note:

Use a lint–free cloth to wipe off the dipstick.

– Refill as necessary

The ATF level changes with the temperature of the fluid. The Min. and Max. markings on the dipstick relate to an ATF temperature of 60°C (140°F).

Transmission Gear Selector Elements in transmissions up to 12.92

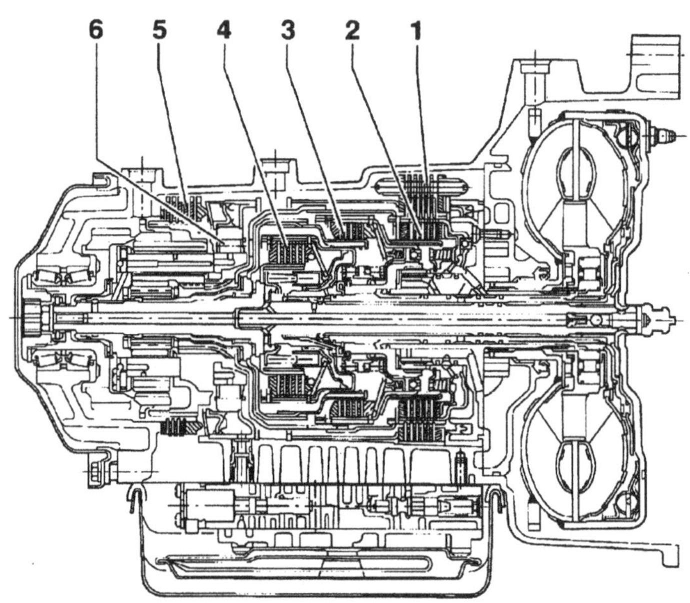

V37-0833

Note:

The illustration V37-0833 is from automatic transmission 096. The application of automatic transmission 098 selector elements is identical.

Position of selector elements

1 = 2nd and 4th gear brake -B2-

2 = Reverse gear clutch -K2-

3 = 1st to 3rd gear clutch -K1-

4 = 3rd and 4th gear clutch -K3-

5 = Reverse gear brake -B1-

6 = Free wheel clutch

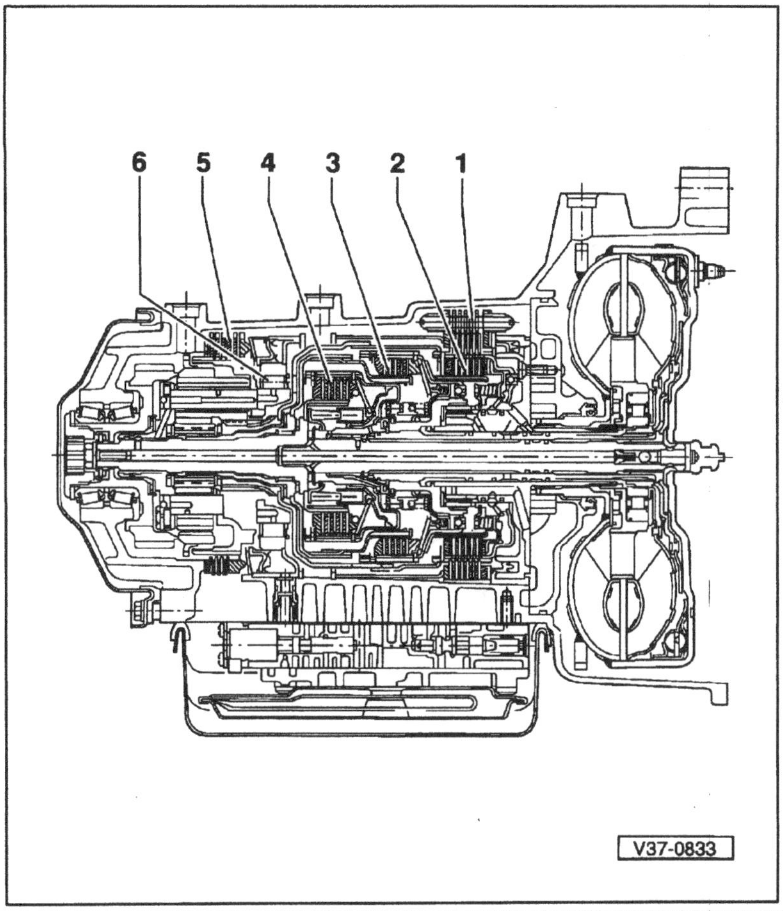

The following table shows which gear selector elements are operated in the individual gears to help you deal with complaints regarding poor performance or general drive train malfunctions. The table makes it possible to determine which gear selector elements are not operating properly.

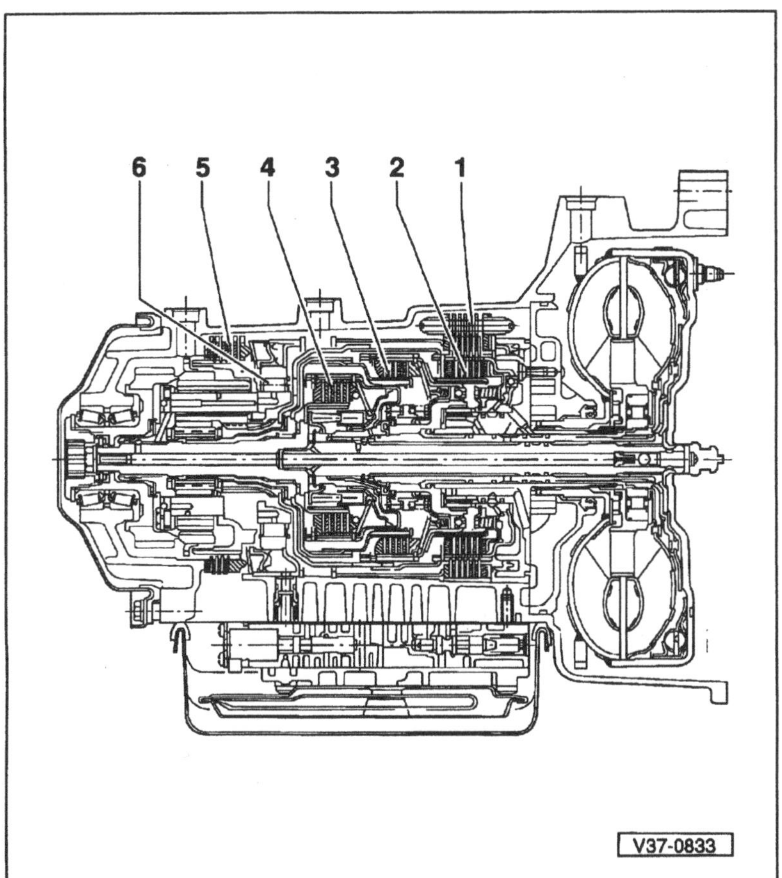

Position of selector elements on transmissions up to 12.92

Vehicles with Transmission Range (TR) Program Switch -E122-

Gear	B1	B2	K1	K2	K3	F
R	x			x		
1H			x			x
2H		x	x			
3H			x	x		
3LT1			x		x	
3M			x	x	x	
3LT2				x	x	
4M		x			x	

X = clutches, brakes or free wheel engaged

H = hydraulic

M = mechanical

LT1 = Load splitting

LT2 = Load splitting

Before starting repairs to the transmission, first perform the On Board Diagnostic (OBD) Program ⇒ Repair Group 01.

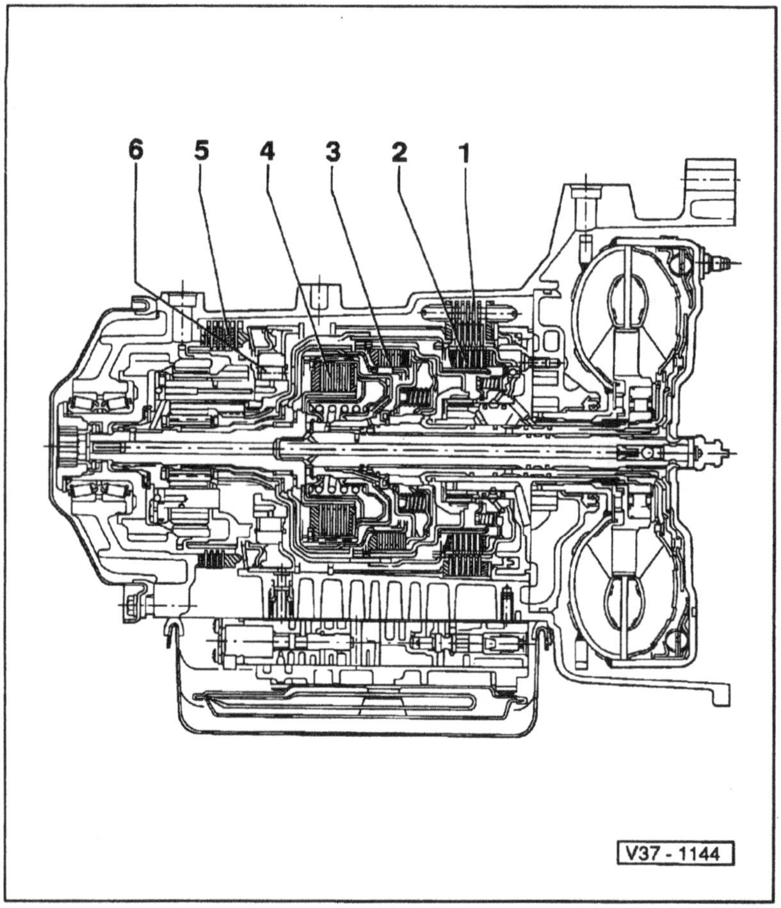

Transmission Gear Selector Elements in transmissions from 01.93

Note:

The illustration V37-1144 is from automatic transmission 096. The application of automatic transmission 098 selector elements is identical.

Position of gear selector elements

1 = 2nd and 4th Gear Brake -B2-

2 = Reverse Gear Clutch -K2-

3 = 1st to 3rd Gear Clutch -K1-

4 = 4th Gear Clutch -K3-

5 = Reverse Gear Brake -B1-

6 = Free wheel clutch

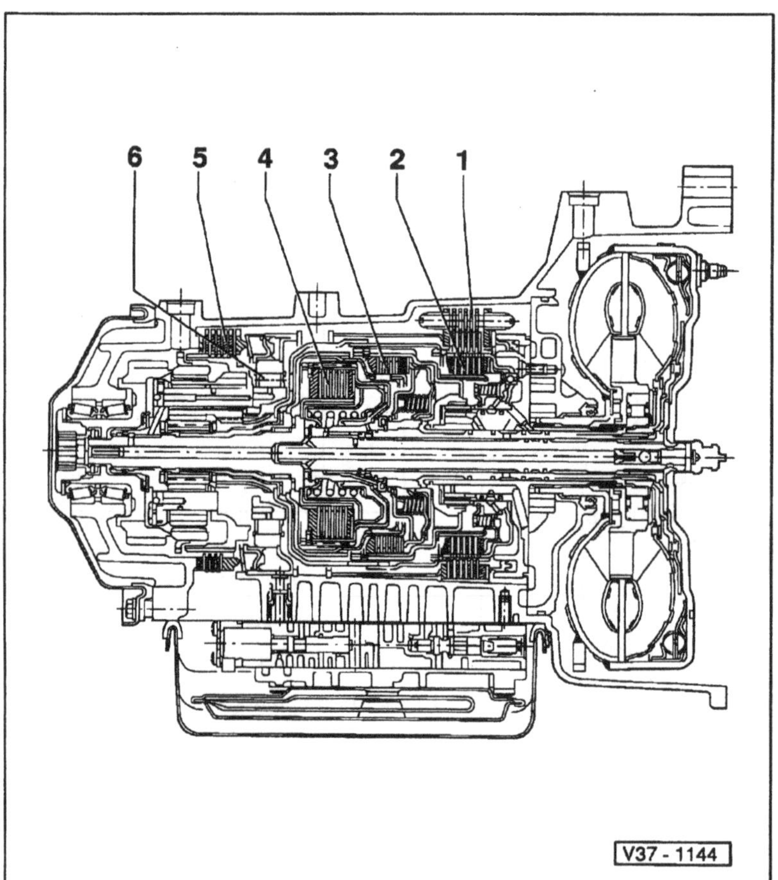

The following table shows which gear selector elements are operated in the individual gears to help you deal with complaints regarding poor performance or general drive train malfunctions. The table makes it possible to determine which gear selector elements are not operating properly.

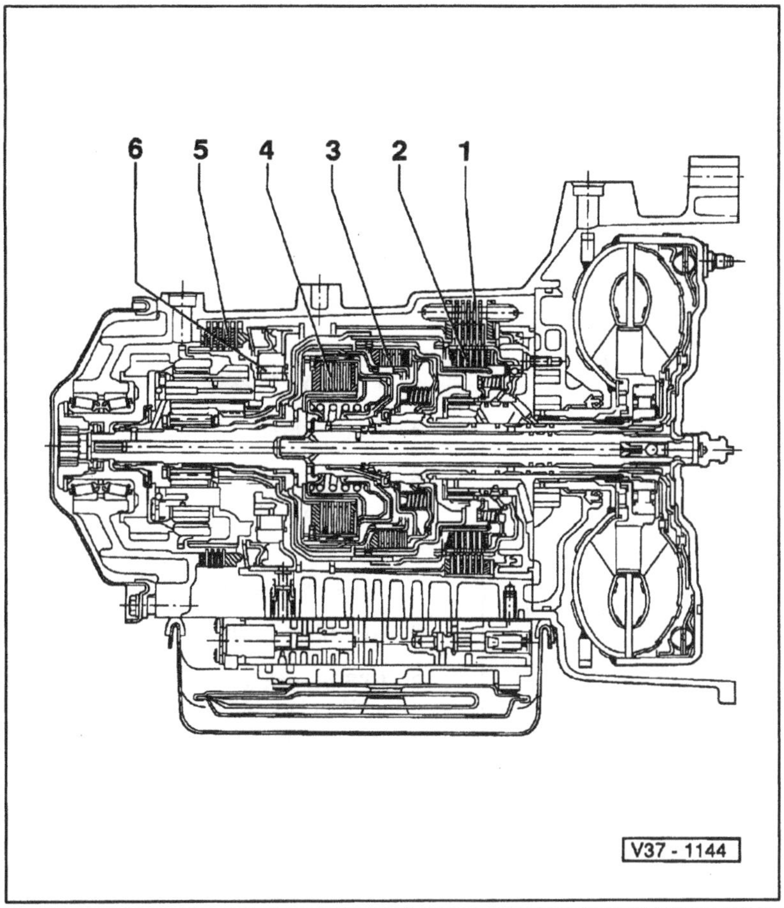

Position of selector elements on transmissions from 01.93

Vehicles with electronic program switch in Transmission Control Module (TCM)

Gear	B1	B2	K1	K2	K3	F
R	x			x		
1H			x			x
2H		x	x			
3H			x	x		
4M		x			x	

X = clutches, brakes or free wheel engaged

H = hydraulic

M = mechanical

Before starting repairs to the transmission, first perform the On Board Diagnostic (OBD) program
⇒ Repair Group 01.

Transmission, disassembling and assembling

Removing and installing ATF cooler and ATF filler tube
⇒ page H 37-55.

Disassembling and assembling planetary gearbox -Overview-
⇒ page H 37-57.

Disassembling and assembling planetary gearbox
⇒ page H 37-72.

Disassembling
⇒ page H 37-72

Assembling
⇒ page H 37-78

Overview of all planetary gearbox adjustments
⇒ page H 37-91.

Removing and installing valve body
⇒ page H 38-36.

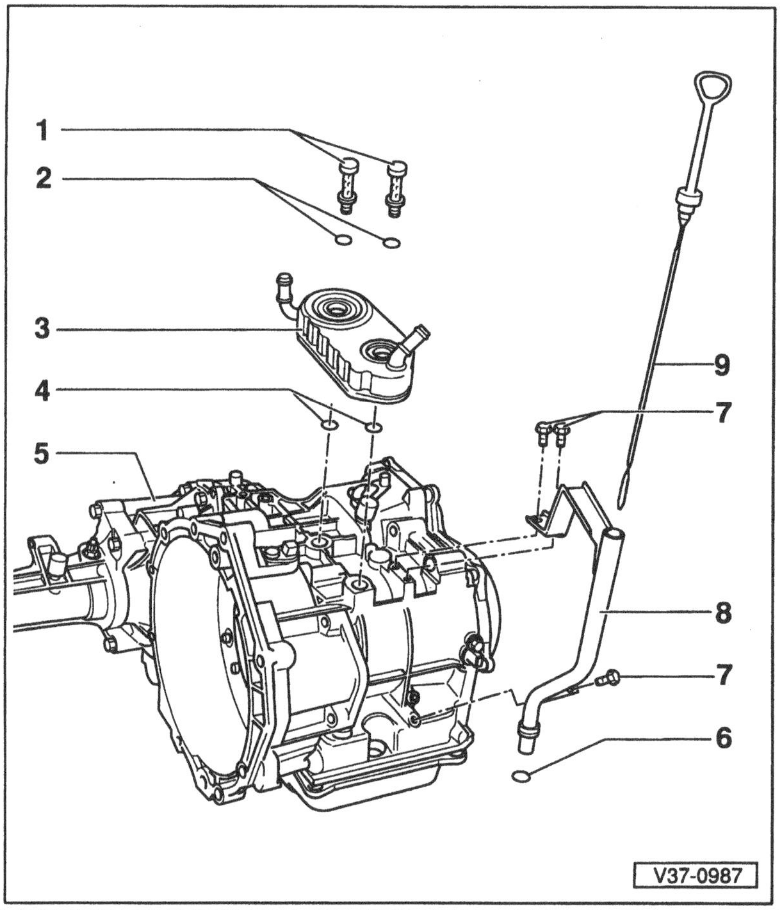

ATF cooler and ATF filler tube, removing and installing

1 - Banjo bolt
- ♦ Tightening torque: 35 Nm (26 ft lb)

2 - O-ring
- ♦ Always replace

3 - ATF cooler

4 - O-ring
- ♦ Always replace

5 - Transmission housing

6 - O-ring
- ♦ Always replace

7 - Bolt
- ♦ Tightening torque: 10 Nm (7 ft lb)

8 - ATF filler tube

9 - ATF dipstick
- ♦ Checking ATF level ⇒ page H 37-46

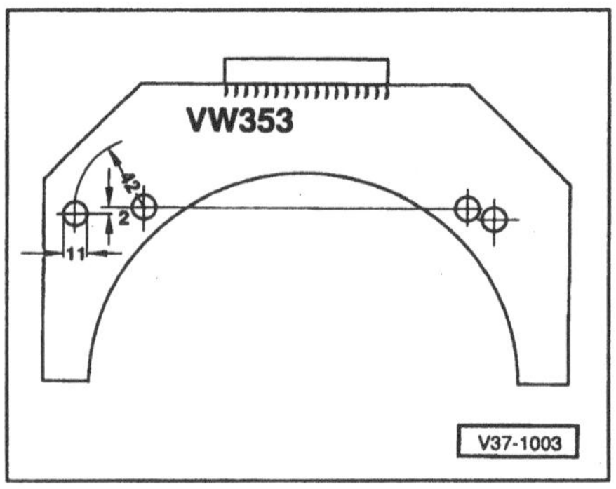

VW 353 Transmission Support, modifying

◄ To secure the automatic transmission 098 to assembly bracket, a new hole is required in VW 353 transmission support.

Dimensions given are in mm (in.).

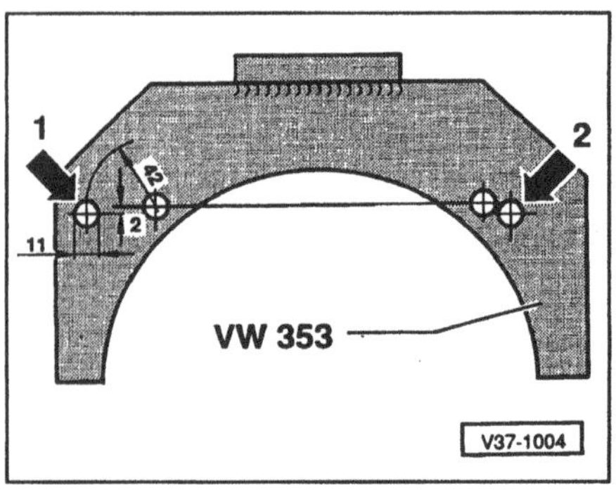

◄
- To mark required hole, attach transmission housing to VW 353 transmission support with 10 mm bolt -arrow 2-.
- Mark VW 353 transmission support through appropriate hole in transmission housing -arrow 1-.
- Drill 11 mm (0.433 in.) hole -arrow 1- in VW 353 transmission support.

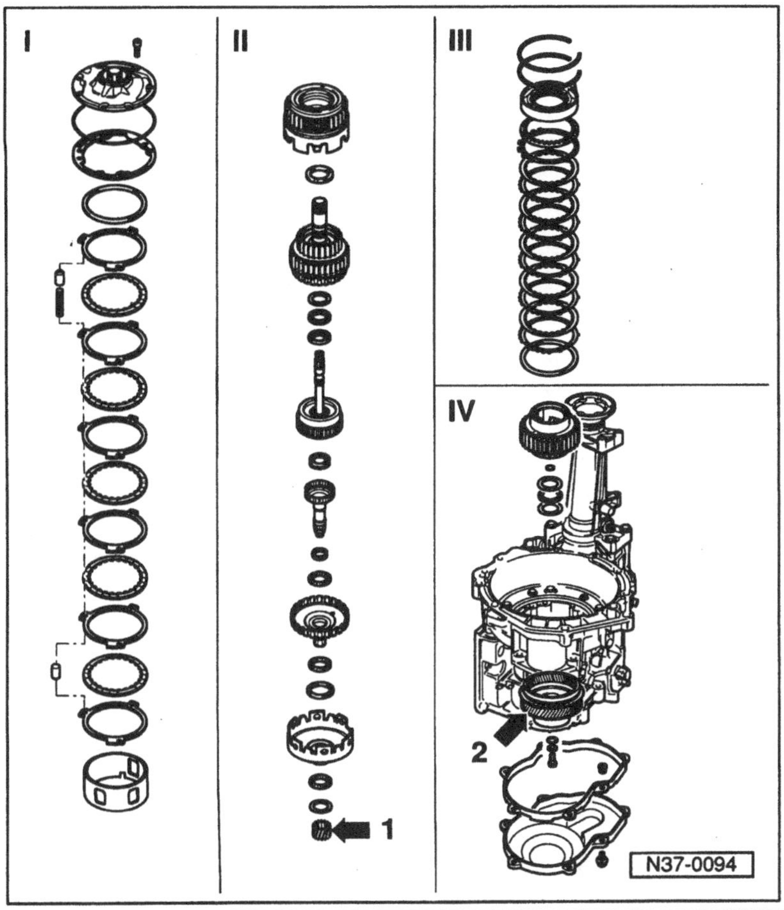

Planetary gearbox, disassembling and assembling, overview

Note:

Disassembling planetary gearbox ⇒ page H 37-72,

Assembling ⇒ page H 37-78.

I - Removing and installing ATF pump to support tube ⇒ page H 37-59

II - Removing and installing reverse clutch -K2- to small sun gear ⇒ page H 37-62

Note:

The small sun wheel -arrow 1- cannot be removed from the planet carrier in transmissions from 01.93 ⇒ page H 00-11.

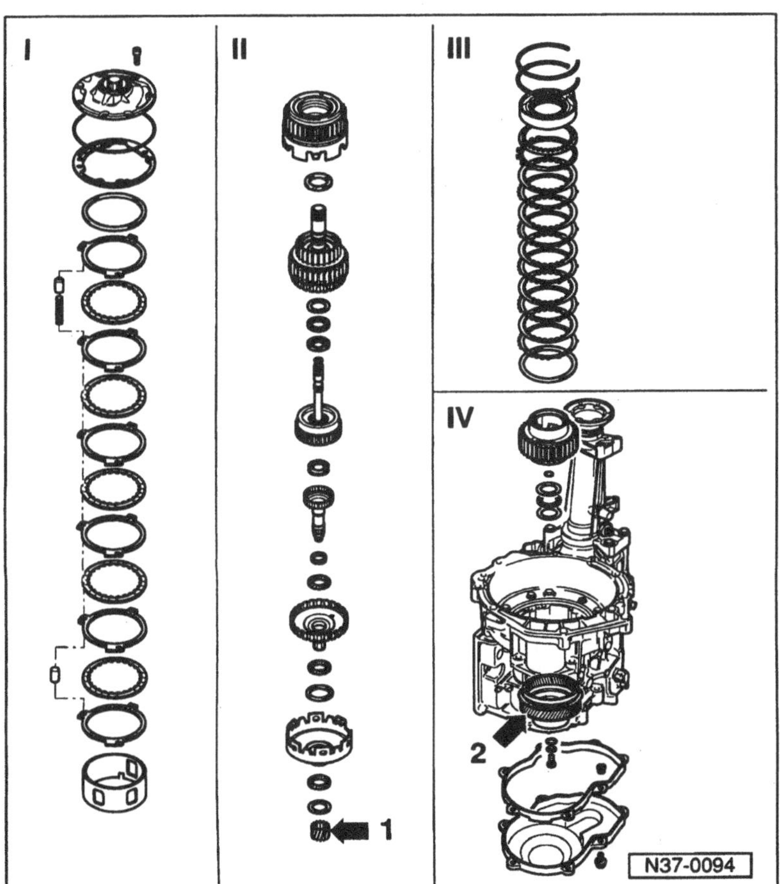

III - Free wheel and reverse gear brake -B1-, removing and installing ⇒ page H 37-66

Note:

To remove and install free wheel, first remove valve body together with plugs ⇒ page H 38-36.

IV - Planet carrier, removing and installing ⇒ page H 37-69

Note:

It is not necessary to remove input gear -arrow 2-.

Removing and installing input gear ⇒ page H 39-23.

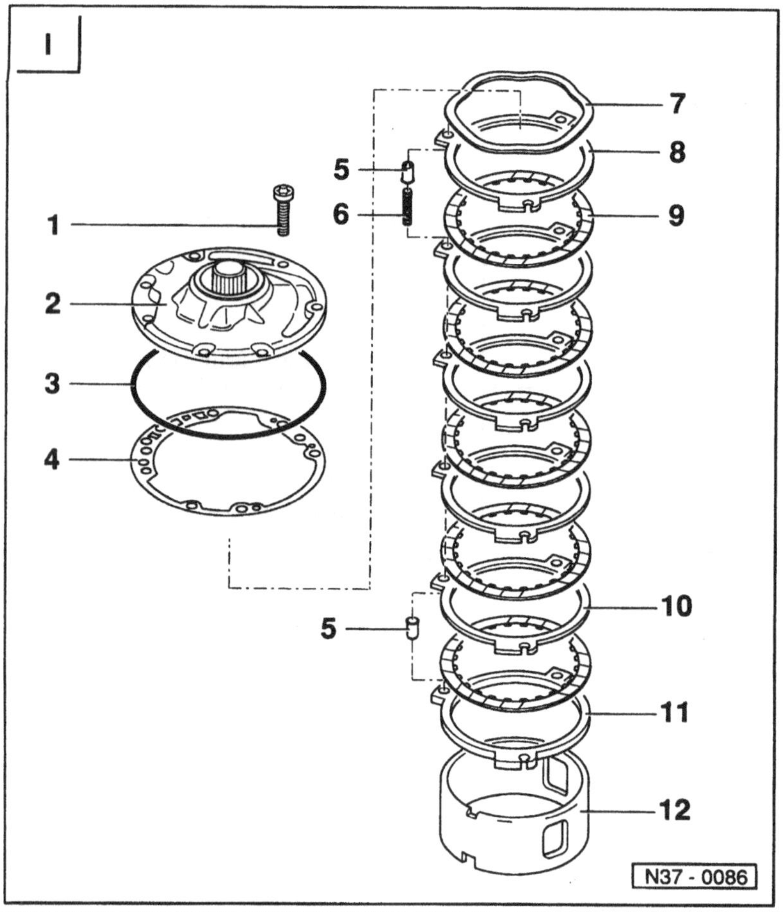

I - ATF pump to supporting tube, removing and installing

⇒ Overview, page H 37-57

Note:

Disassembling planetary gearbox ⇒ page H 37-72,

Assembling ⇒ page H 37-78.

1 - Bolt
- ♦ Qty. 7
- ♦ Tightening torque: 8 Nm (71 in lb) plus an additional 1/4 turn (90)
- ♦ Further turns of 90° (1/4 turn) may be done in several stages

2 - ATF pump with -B2- piston
- ♦ Disassembling and assembling ⇒ page H 38-1

3 - O-ring
- ♦ Always replace
- ♦ Place on ATF pump

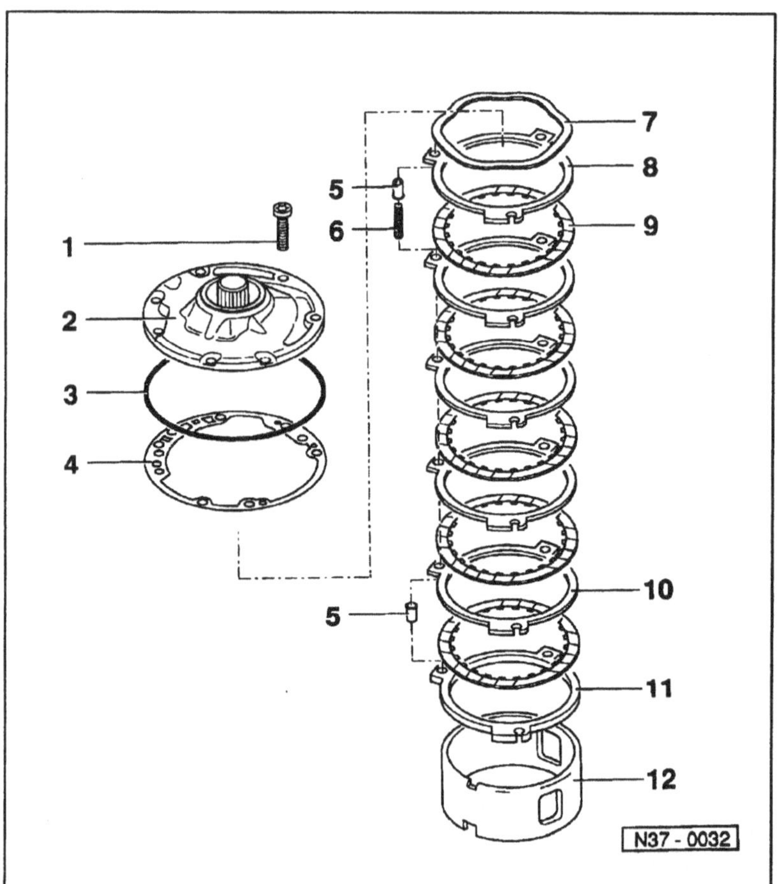

4 - Gasket
- ♦ Always replace

5 - Spring cap, Qty. 6
- ♦ Install 3 spring caps after installing first outer plate
- ♦ Install 3 spring caps before installing last outer plate

6 - Spring
- ♦ Quantity 3

7 - Corrugated spring washer

8 - Outer plate -B2-
- ♦ Quantity ⇒ from page H 00-3
- ♦ Determining thickness of outer plate ⇒ Adjusting 2nd and 4th gear brake -B2-, page H 37-115

9 - Inner plate -B2-
- ♦ Quantity ⇒ from page H 00-3
- ♦ Before installing place new inner plates in ATF for 15 minutes

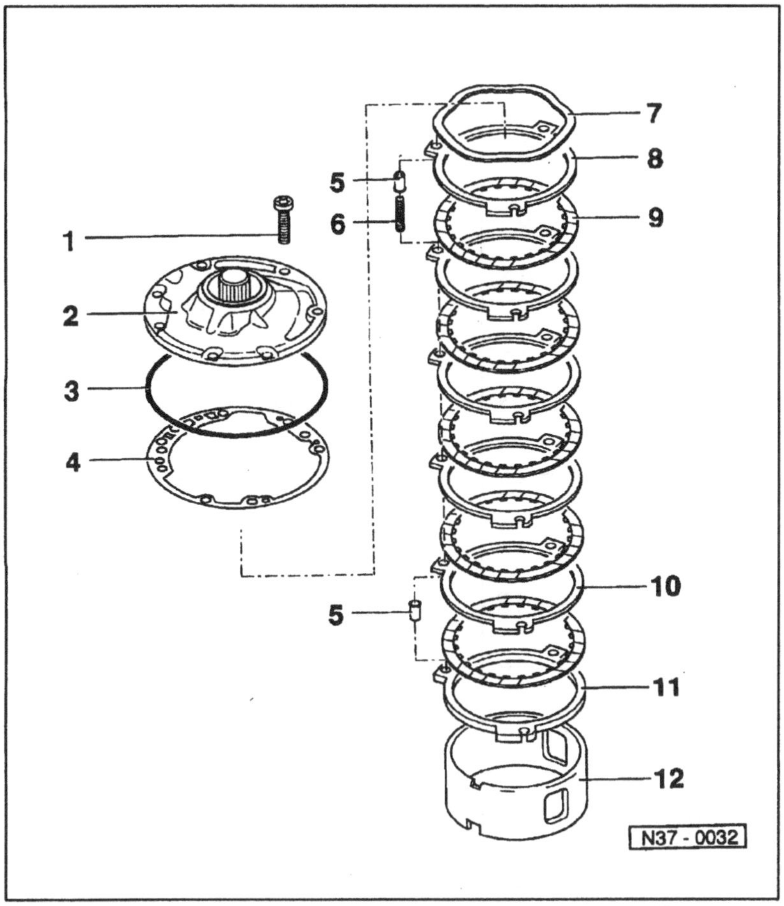

10 - **Outer plate -B2-**
- Always install 2 mm (0.079 in.) thick outer plates
- Quantity
  ⇒ from page H 00-3

11 - **Outer plate -B2-**
- Install 3 mm (0.118 in.) thick outer plate onto supporting tube

12 - **Supporting tube -B2-**
- For the -B2- plate package
- Length of -B2- with

4 inner plates	72.3 mm (2.847 in.)
5 inner plates	68.6 mm (2.701 in.)
6 inner plates	64.9 mm (2.555 in.)

- Quantity of -B2- inner plates ⇒ from page H 00-3
- Install so that groove engages in wedge of free wheel

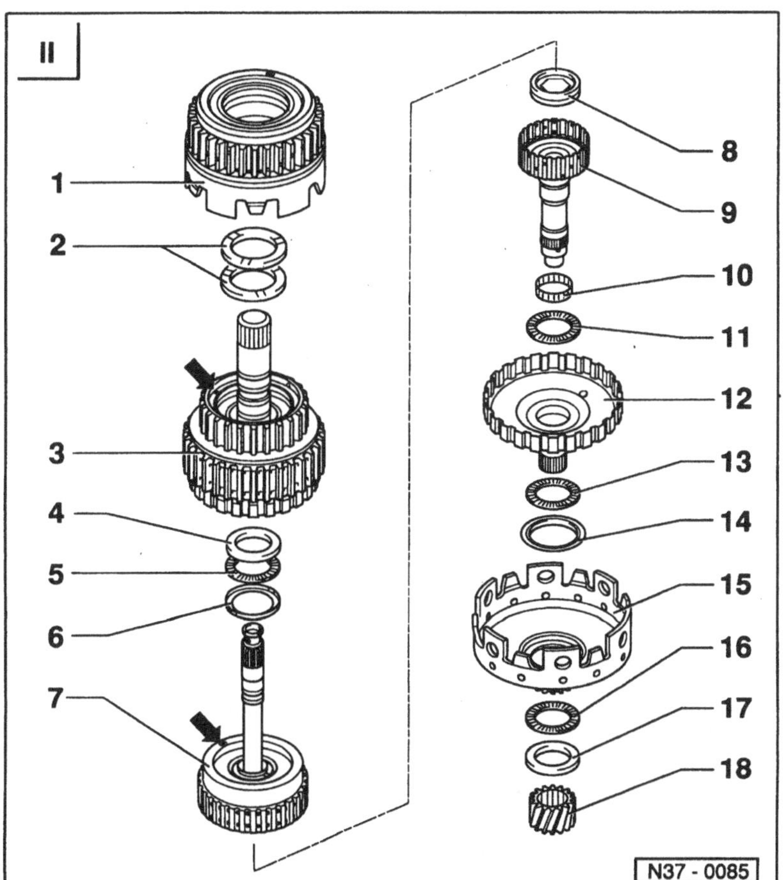

II - Reverse gear clutch -K2- to small sun gear, removing and installing

⇒ Overview, page H 37-57

Note:

Disassembling planetary gearbox ⇒ page H 37-72,

Assembling ⇒ page H 37-78.

1 - **Reverse gear clutch -K2-**
- Disassembling and assembling ⇒ page H 38-23

2 - **Shim**
- Determining thickness ⇒ Adjusting clutch play between -K1- and -K2-, page H 37-108
- 1 or 2 shims can be installed

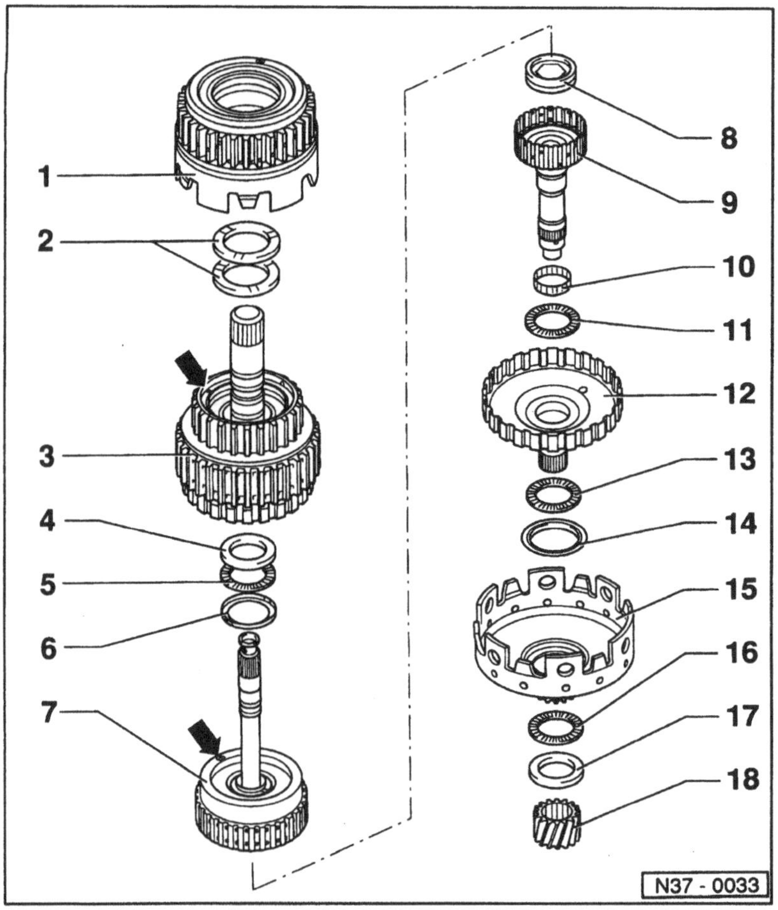

3 - 1st to 3rd gear clutch -K1- with turbine shaft

- With ball valve -arrow- disassembling and assembling ⇒ page H 38-7
- Without ball valve (for transmissions from 01.93, ⇒ page H 00-11) disassembling and assembling ⇒ page H 38-16

4 - Axial needle bearing washer

5 - Axial needle bearing

6 - Axial needle bearing washer

- Lugs face towards axial needle bearing

7 - 3rd and 4th gear clutch -K3- with pump shaft

- With ball valve -arrow- disassembling and assembling ⇒ page H 38-26
- 4th gear clutch -K3- without ball valve (for transmissions from 01.93, ⇒ page H 00-11) disassembling and assembling ⇒ page H 38-31

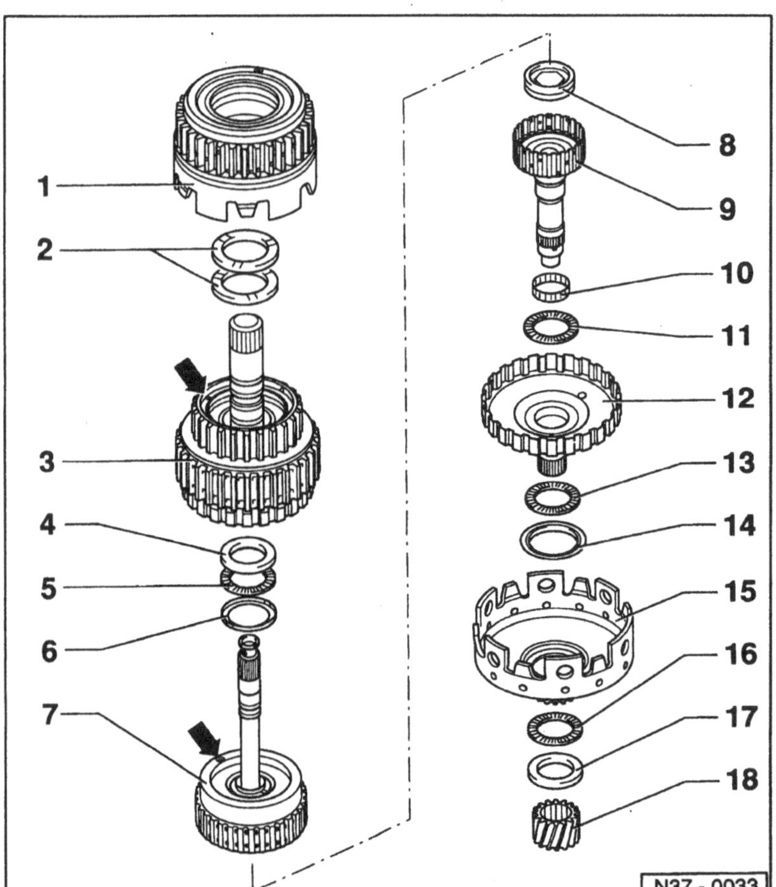

8 - Axial needle bearing with washer

- Axial needle bearing face towards small drive shaft

9 - Small drive shaft

- Modified -K3- clutch from 01.93

10 - Needle bearing

11 - Axial needle bearing

12 - Large drive shaft

- Modified -K1- clutch from 01.93

13 - Axial needle bearing

14 - Axial needle bearing washer

- With shoulder

15 - Large sun gear

- On transmissions up to 12.92, with 27 teeth ⇒ page H 00-3
- On transmissions from 01.93, with 24 teeth ⇒ page H 00-11

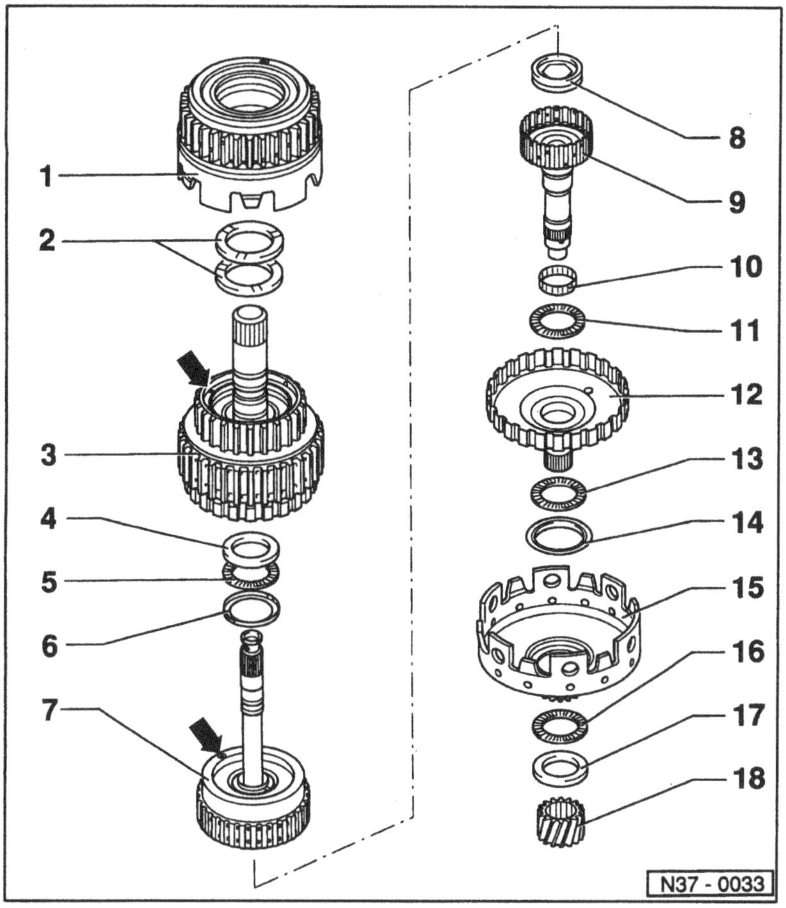

16 - Axial needle bearing
- On transmissions up to 12.92, ⇒ page H 00-3, outer diameter: 41 mm (1.614 in.)
- On transmissions from 01.93, ⇒ page H 00-11, outer diameter 39.1 mm (1.539 in.)

17 - Axial needle bearing washer

18 - Small sun gear
- Cannot be removed from the planet carrier in transmissions from 01.93, ⇒ page H 00-11

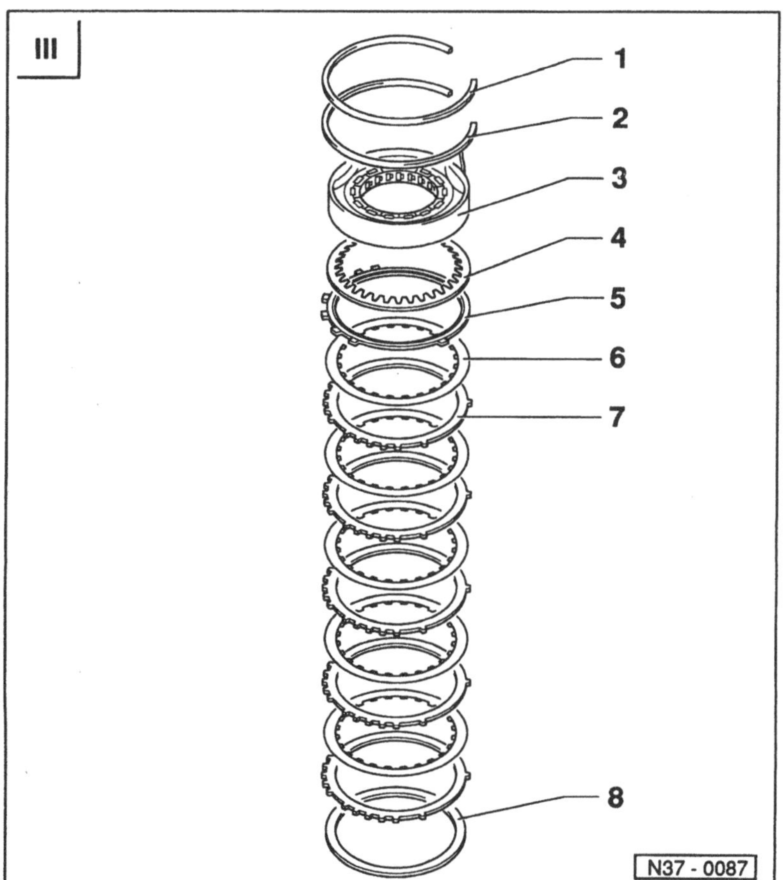

III - Free wheel and reverse gear brake -B1-, removing and installing

⇒ Overview, page H 37-57

Note:

Disassembling planetary gearbox ⇒ page H 37-72,

Assembling ⇒ page H 37-78.

1 - Circlip
- For supporting tube -B2-

2 - Circlip
- For free wheel

3 - Free wheel with -B1- piston
- Before removing free wheel, remove valve body and sealing plugs ⇒ page H 38-36
- Disassembling and assembling ⇒ page H 38-2

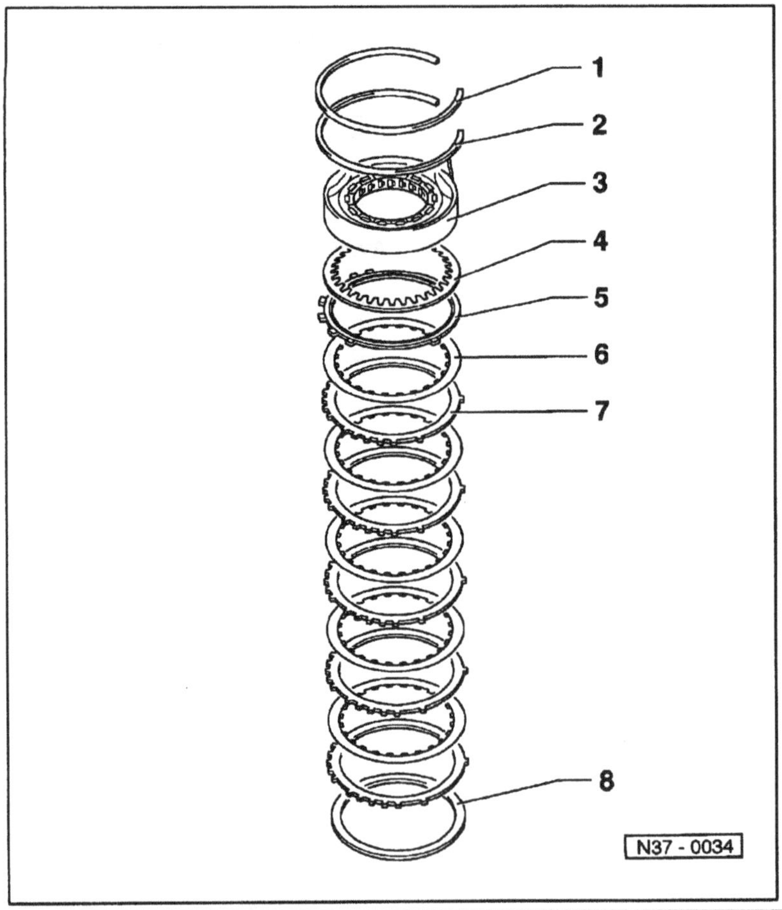

4 - Dished spring

- Install with convex side facing free wheel

5 - Pressure plate -B1-

- Install with flat side facing plates
- On transmissions up to 12.92, 7.5 mm (0.2953 in.) thick and a 2 mm (0.0787 in.) thick outer plate
- On transmissions from 01.93: 13.5 mm (0.5315 in.) thick and a 1.5 (0.0591 in.) mm thick outer plate

6 - Inner plate, -B1-

- Quantity of inner plates -B1- ⇒ transmissions up to 12.92, from page H 00-3 ⇒ transmissions from 01.93, page H 00-11
- Before installing, place plates in ATF for 15 minutes

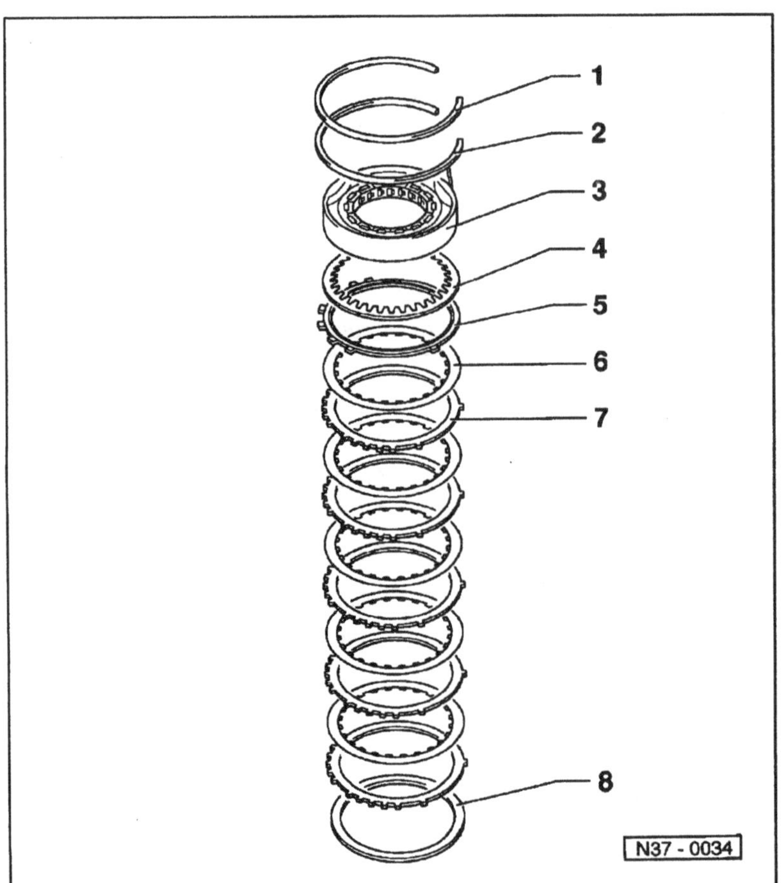

7 - Outer plate, -B1-

- Quantity of outer plates -B1- ⇒ transmissions up to 12.92, from page H 00-3 ⇒ transmissions from 01.93, page H 00-11

8 - Shim

- Determining thickness ⇒ page H 37-101, adjusting reverse gear brake -B1-

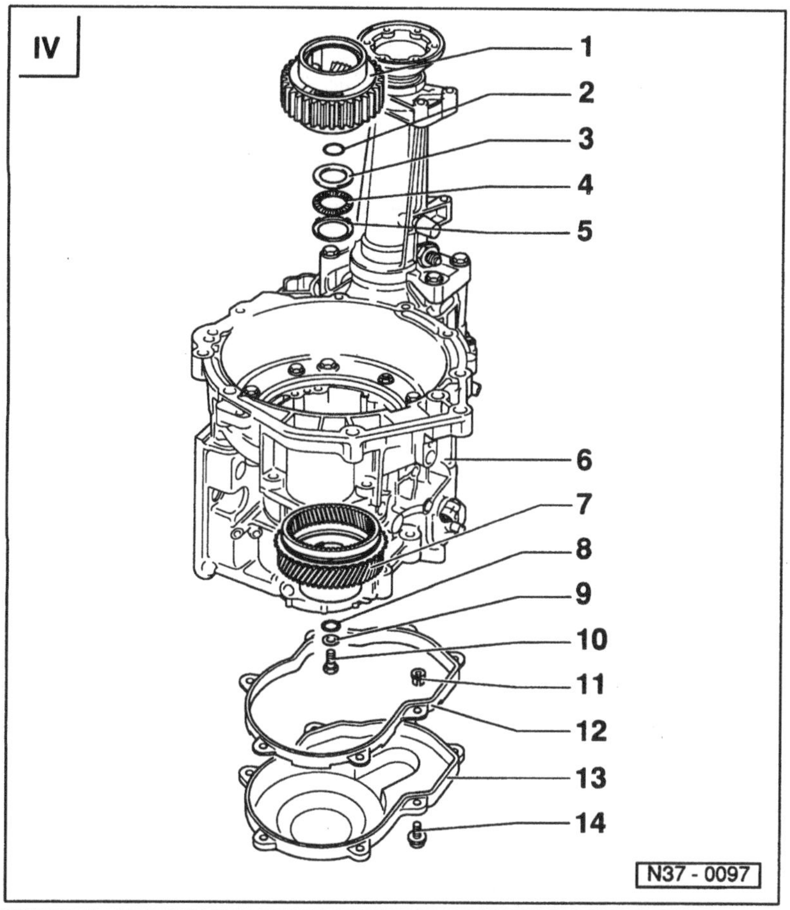

IV - Planet carrier and transmission housing with input gear and cover

⇒ Overview, page H 37-57

Note:

Disassembling planetary gearbox ⇒ page H 37-72,

Assembling ⇒ page H 37-78.

1 - Planet carrier
- Modified ratios for transmissions from 01.93, ⇒ page H 00-11
- The small sun gear cannot be removed on transmissions from 01.93,
- Adjusting ⇒ page H 37-93

2 - O-ring
- Always replace
- Install in planet carrier

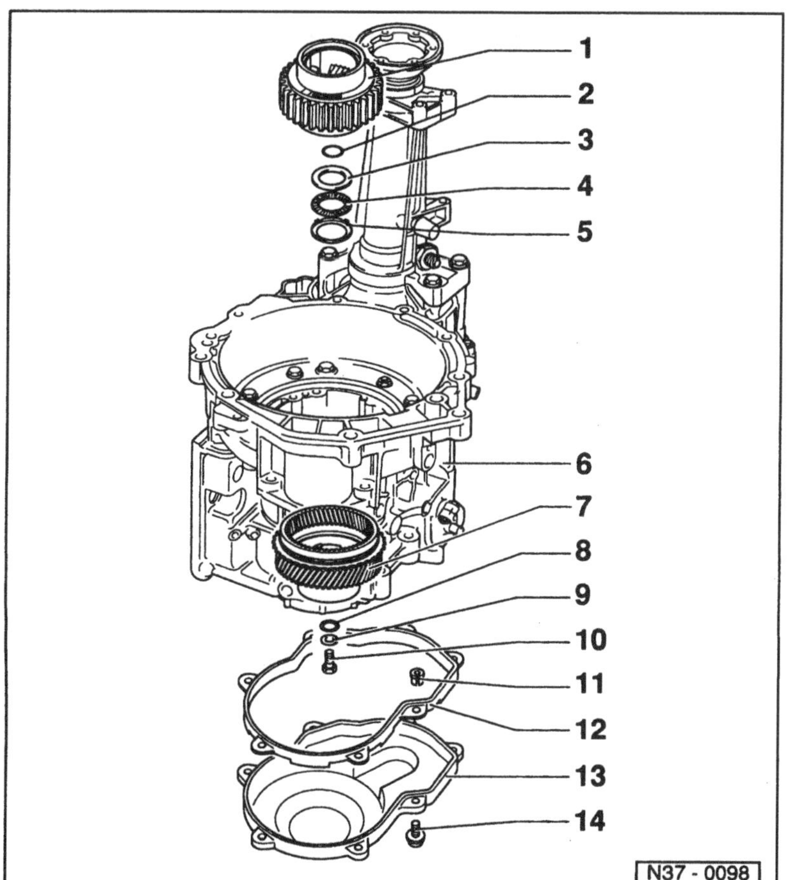

3 - Axial needle bearing washer

4 - Axial needle bearing

5 - Axial needle bearing washer
- Install smooth side in input gear

6 - Transmission housing

7 - Input gear
- Do not remove to disassemble planetary gearbox
- Removing and installing input gear or axial needle bearing for planet carrier ⇒ page H 39-23

8 - Planet carrier shim
- Determining thickness, ⇒ Adjusting planet carrier, page H 37-93

9 - Washer

10 - Bolt
- For small input shaft
- Tightening torque: 30 Nm (22 ft lb)

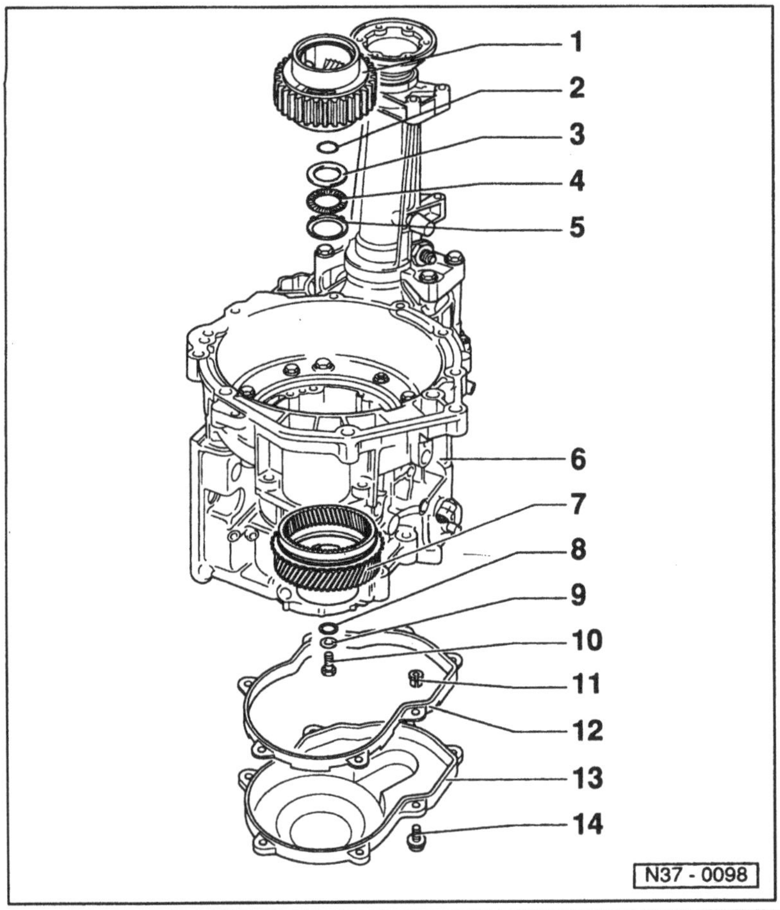

11 - Spacer bushing
- ♦ Clip quantity 7 into gasket

12 - Gasket
- ♦ Always replace

13 - Cover

14 - Bolt
- ♦ Tightening torque: 8 Nm (71 in lb)

Planetary gearbox, disassembling and assembling

Disassembling

– Seal ATF cooler connections.

– Drain ATF.

- ATF can be removed using VAG 1358 A/1 extraction system or EZ1 fluid extractor.

– Remove torque converter.

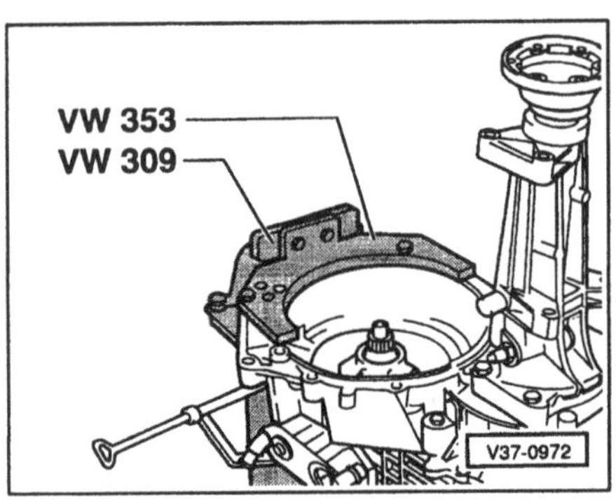

◄ – Secure transmission to assembly support with bolts -1- and -2-.

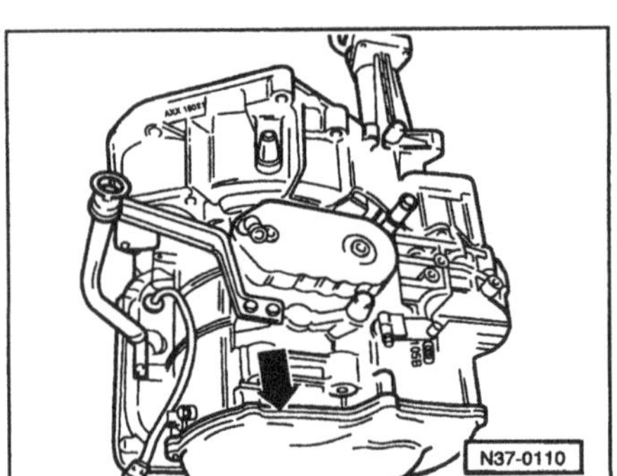

◄ – Remove transmission housing cover with gasket -arrow-.

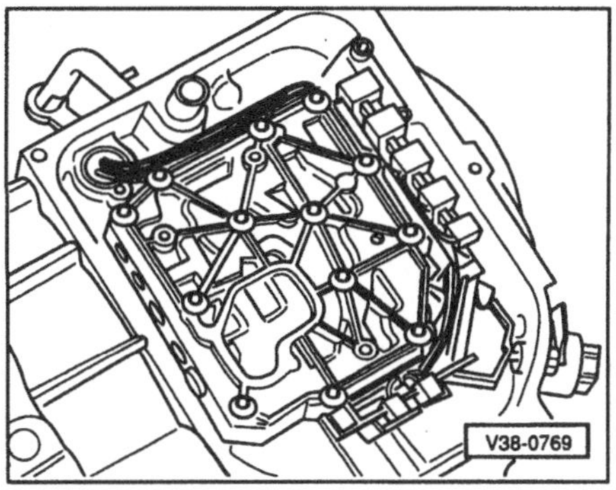

- Remove oil pan.
- Remove ATF screen.
- ◄ Remove valve body ⇒ page H 38-36.

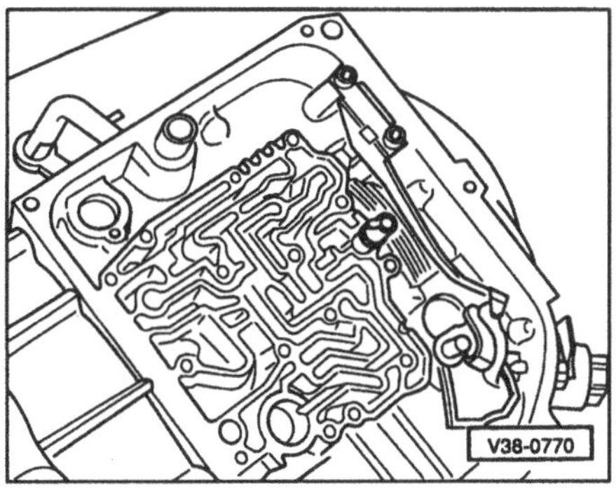

- ◄ Remove sealing plugs for -B1-.

To remove ATF pump,

- Remove ATF pump bolts,
- ◄ Screw bolts -A- (M8) into tapped holes of ATF pump.
- Press ATF pump off transmission by turning bolts -A- evenly.

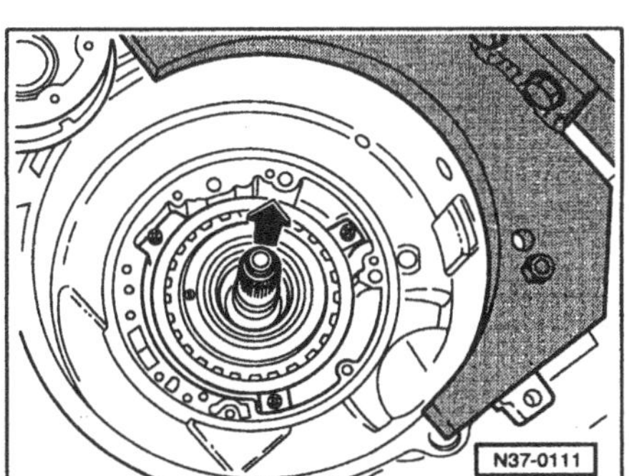

- ◄ Remove all clutches with supporting tube, -B2- plates, springs and spring caps together.

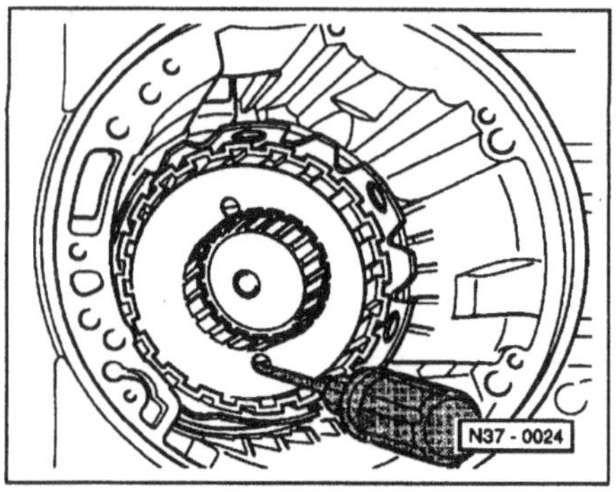

◄ – Install screwdriver through hole of large drive shaft and large sun gear to loosen or tighten bolt for small drive shaft.

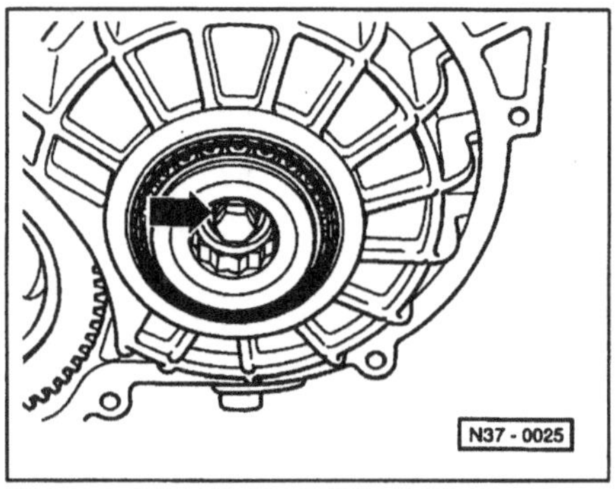

◄ – Loosen small drive shaft bolt -arrow-.

– Remove small drive shaft bolt, washer, and shim.

- Planet carrier axial needle bearing remains in transmission/input gear.

– Pull out small drive shaft.

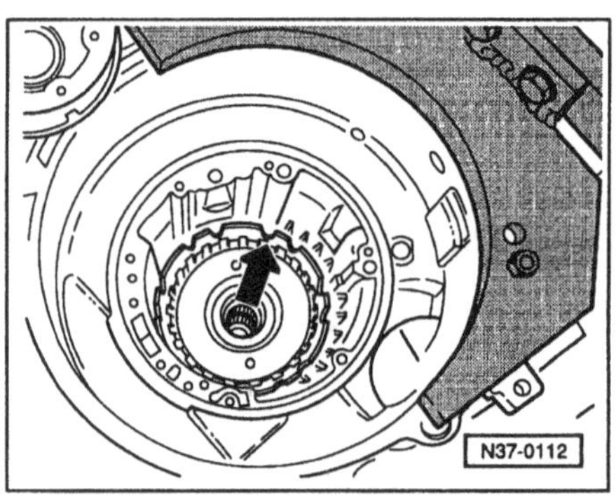

◄ – Pull out large drive shaft -arrow-.

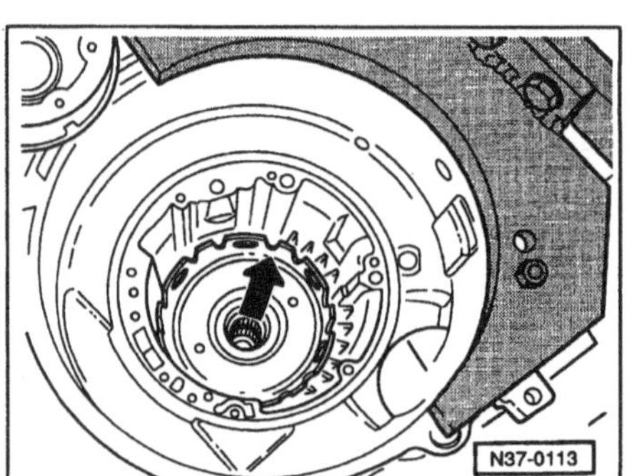

◄ – Pull out large sun gear.

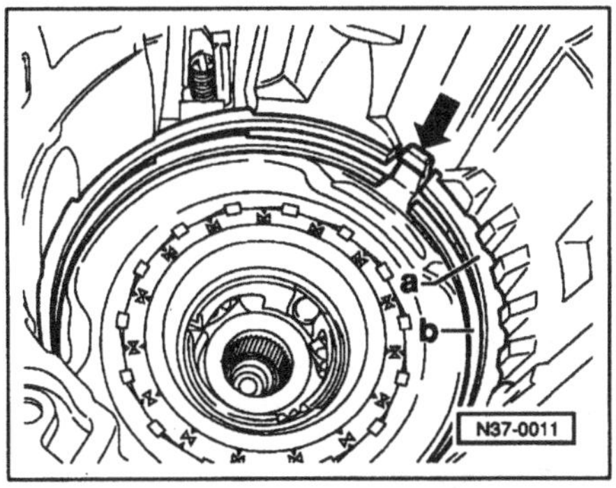

- Remove supporting tube circlip -a- and free wheel circlip -b-.
- Remove free wheel from transmission housing by pulling on free wheel retaining wedge -arrow- with pliers.

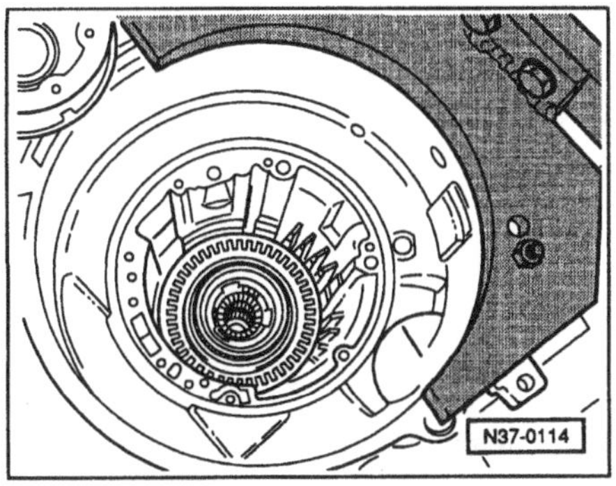

- Pull out planet carrier with dished spring.
- Remove reverse gear brake -B1- plates.

Note:

- *It is not necessary to remove input gear to disassemble planetary gearbox. Removing and installing input gear ⇒ page H 39-23.*
- *Assembling planetary gearbox ⇒ page H 37-78.*

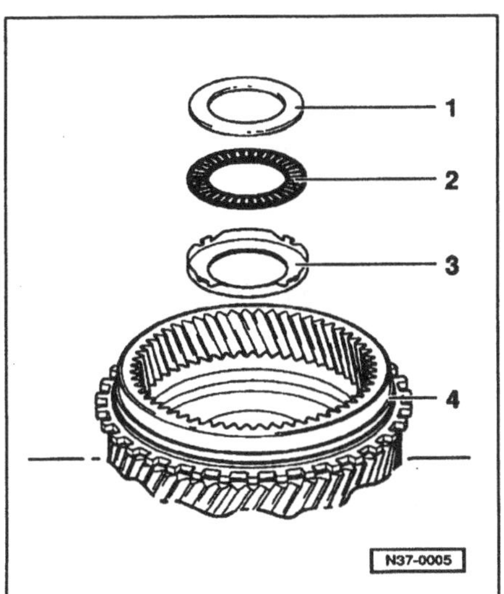

Assembling

- Install axial needle bearing with washer onto input gear.

1 - Axial needle bearing washer

2 - Axial needle bearing

3 - Axial needle bearing washer, install smooth side into input gear.

4 - Input gear, installed in transmission housing ⇒ page H 37-57.

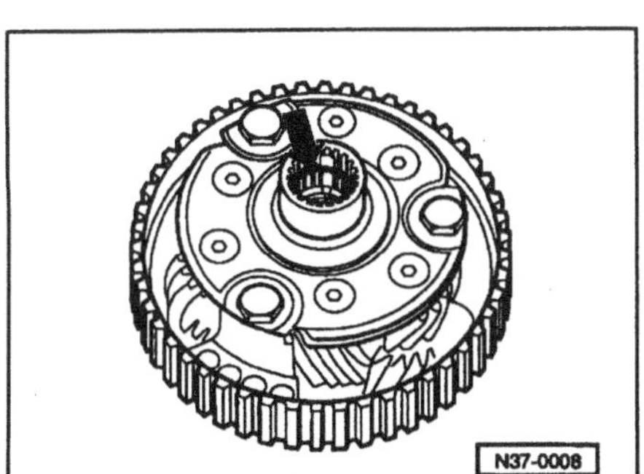

- Install O-ring into planet carrier.

Note:

After replacing planet carrier, adjust ⇒ page H 37-93.

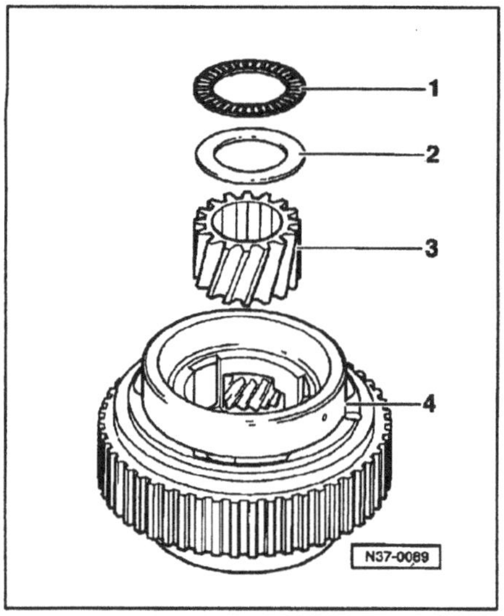

– Install small sun gear with washer and axial needle bearing into planet carrier.

1 - Axial needle bearing

♦ Transmissions up to 12.92,
⇒ page H 00-3
Outer diameter: 41 mm (1.614 in.)

♦ On transmissions from 01.93,
⇒ page H 00-11,
Outer diameter: 39.1 mm (1.540 in.)

2 - Axial needle bearing washer

3 - The small sun gear cannot be removed from the planet carrier on transmissions from 01.93,
⇒ page H 00-11.

4 - Planet carrier

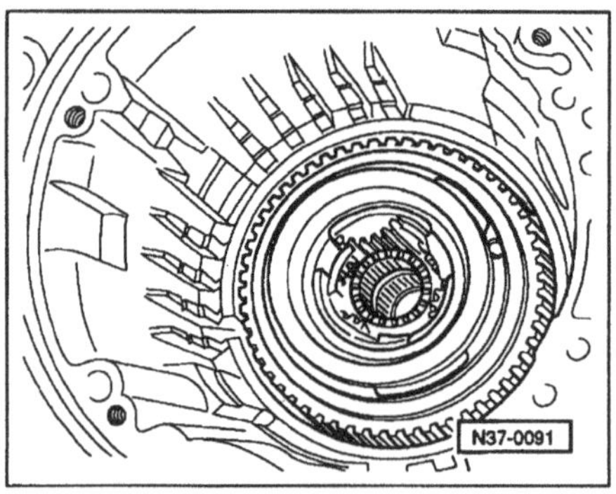

– Install planet carrier into input gear.

– Centralize washer and axial needle bearing in small sun gear.

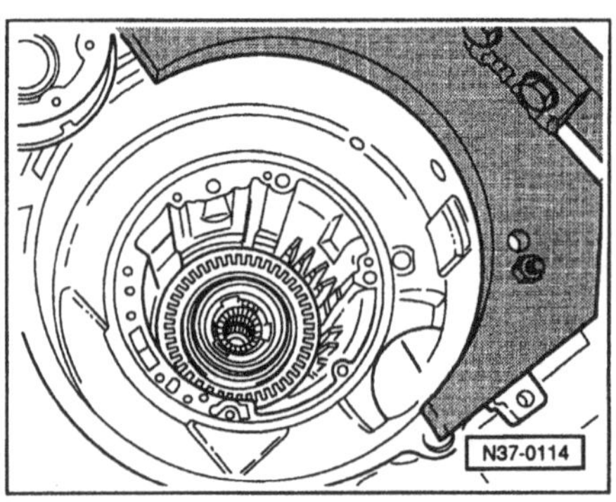

– Install inner and outer -B1- plates.

– Install pressure plate with flat side facing plates.

Thickness of pressure plate varies
⇒ page H 37-67, pressure plate (Item 5).

– Install dished washer with convex side facing toward free wheel.

Note:

When replacing

- *Transmission housing*
- *Free wheel*
- *Piston for reverse gear brake -B1-*
- *Plates*

adjust -B1-
⇒ page H 37-101.

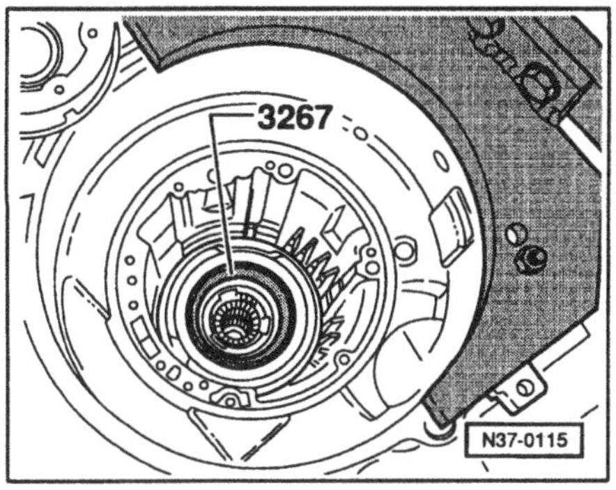

- Preload free wheel rollers with 3267 assembly ring and install free wheel.

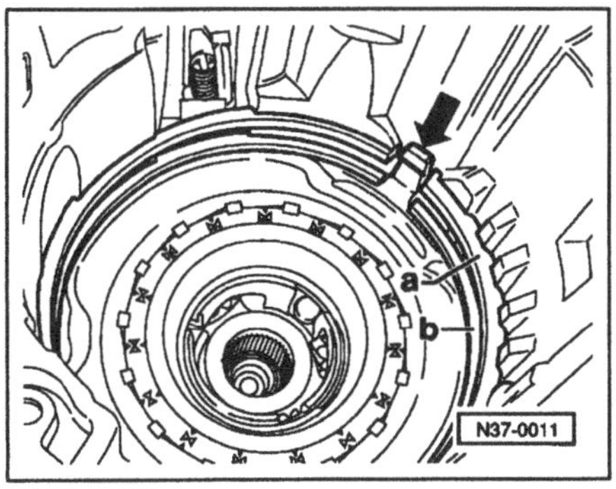

- Install circlip for free wheel -b- and circlip for supporting tube -a-.
- Install circlip gaps over free wheel retaining wedge -arrow-.
- Perform check measurement of -B1-
  ⇒ page H 37-107

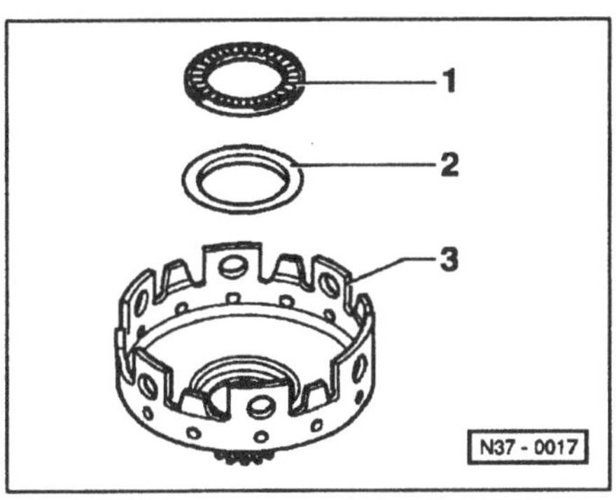

- Install washer with axial needle bearing in large sun gear.

1 - Axial needle bearing

2 - Washer with shoulder (install with smooth side facing axial needle bearing)

3 - Large sun gear

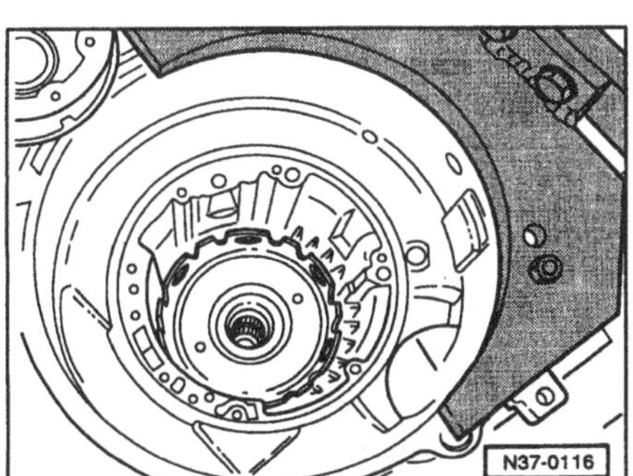

- Install large sun gear with bearing.
- Centralize both washers with axial needle bearing between sun gears.

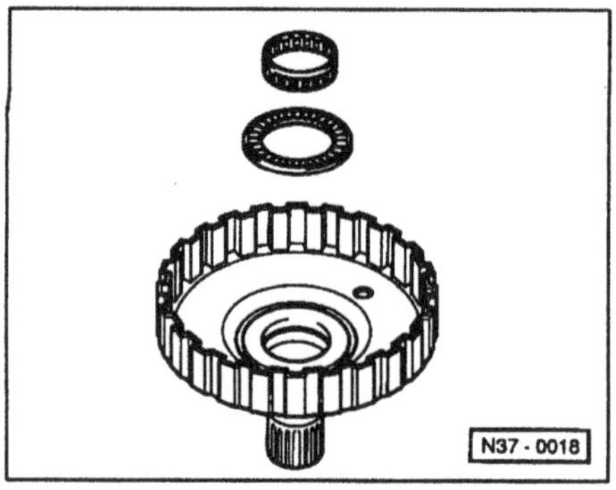

- Install axial needle bearing and needle bearing in large drive shaft.

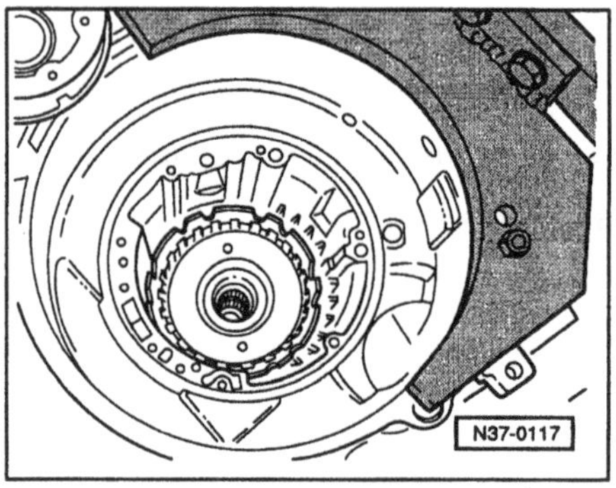

- Install large drive shaft with bearing.

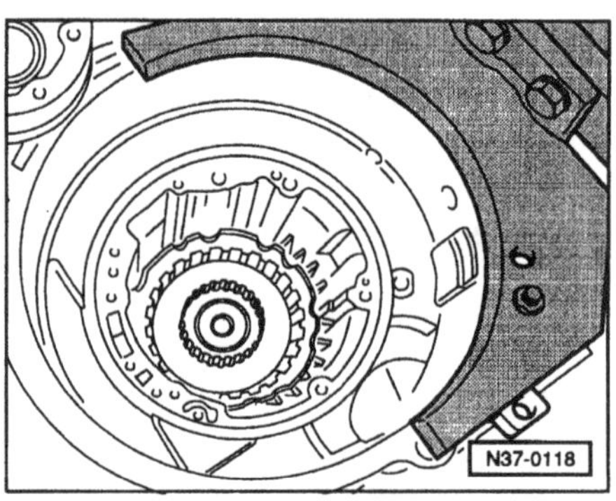

- Install small drive shaft.

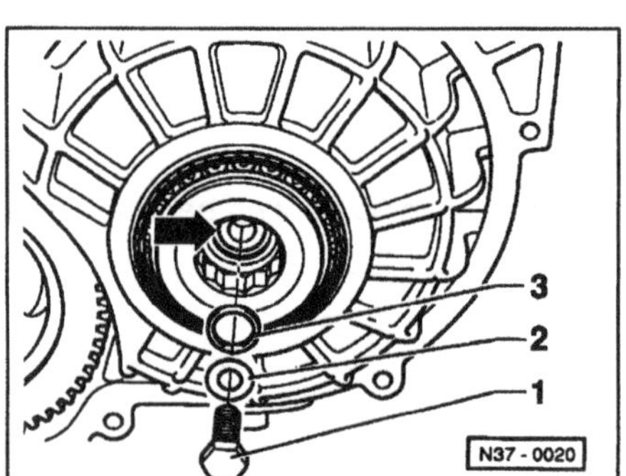

- Install small drive shaft bolt -1- with washer -2- and shim -3-.
  Tightening torque: 30 Nm (22 ft lb).

Notes:

- *Install shim -3- onto small drive shaft shoulder -arrow-.*
- *Determining shim thickness*
  ⇒ planet carrier, adjusting, page H 37-93.

- Perform planet carrier check measurement
  ⇒ page H 37-100.

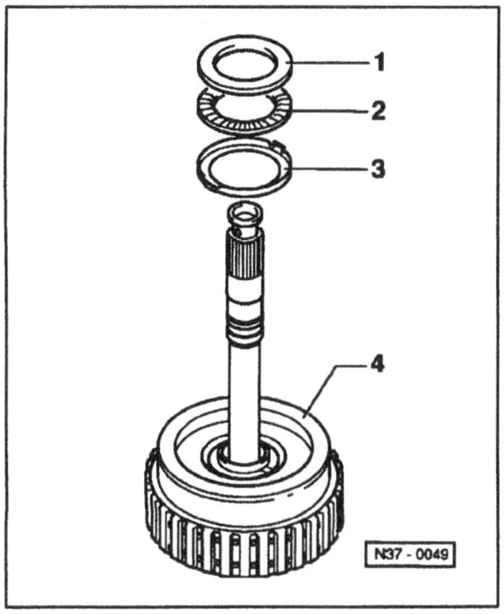

◄ – Install axial needle bearing with washers in -K3-.

1 - Axial needle bearing washer

2 - Axial needle bearing

3 - Axial needle bearing washer (install with smooth side facing -K3-)

4 - 3rd and 4th gear clutch -K3-

On transmissions from 01.93, the -K3- clutch only operates Fourth Gear
⇒ page H 00-11.

Note:

Make sure piston rings seat correctly.

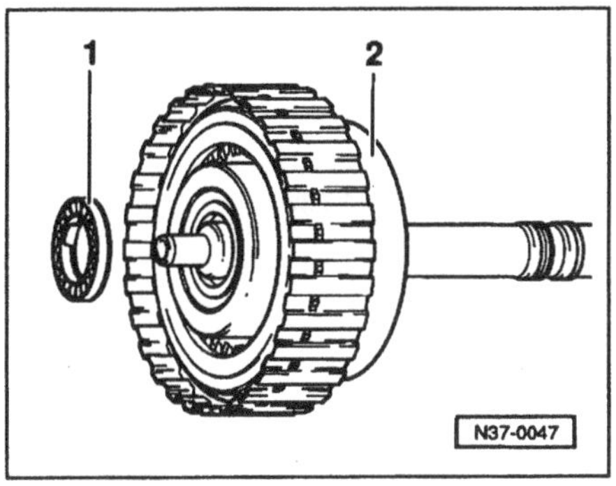

◄ – Install axial needle bearing with washer on 3rd and 4th gear clutch -K3-.

1 - Axial needle bearing with washer

– Moisten axial needle bearing washer with ATF so that the bearing, when installing sticks to -K3-.

– Install bearing washer on -K3-.

2 - 3rd and 4th gear clutch -K3-

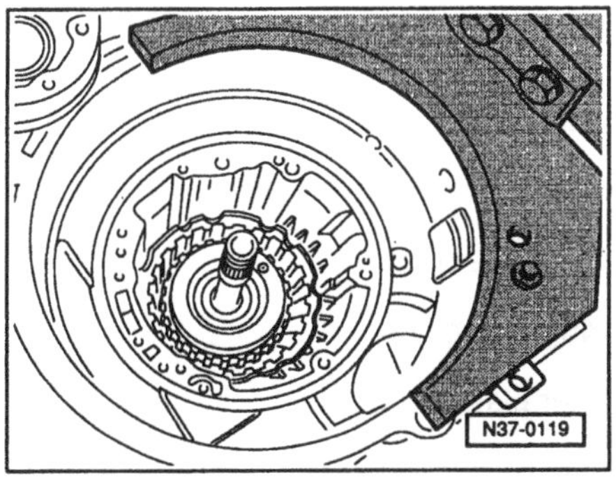

◄ – Install 3rd and 4th gear clutch -K3-.

Transmissions from 01.93, the -K3- clutch only operates Fourth Gear
⇒ page H 00-11.

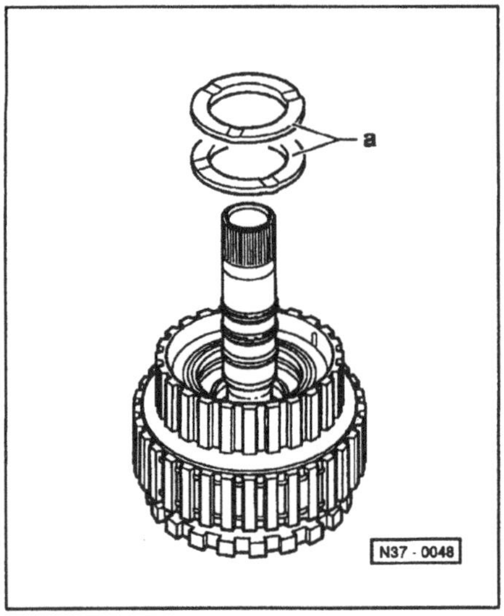

◄ – Install shims -a- in -K1-.

Notes:

- *Remeasure shim when replacing -K1-, -K2- or ATF pump ⇒ clutch play between -K1- and -K2-, adjusting, page H 37-108.*
- *1 or 2 shims can be installed.*

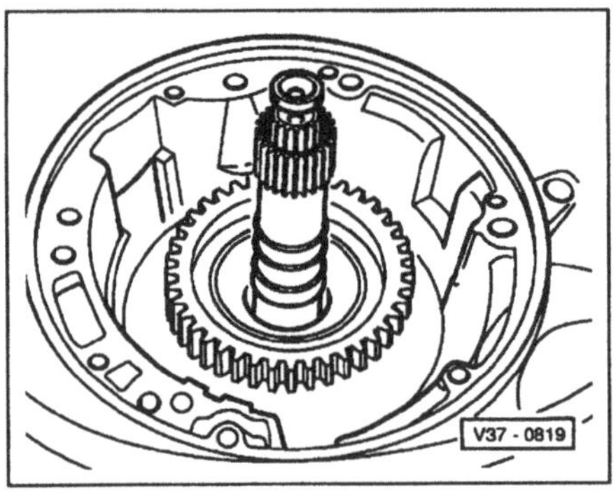

◄ – Install 1st to 3rd gear clutch -K1-.

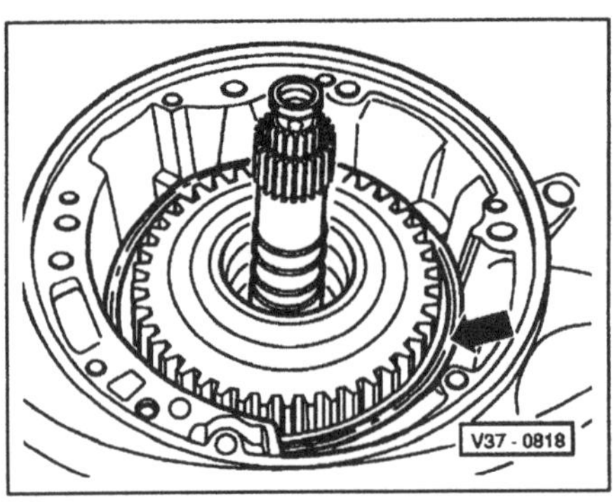

– Install reverse gear clutch -K2-.

◄ – Install supporting tube for plate set -B2- -arrow- so that supporting tube groove engages in free wheel wedge.

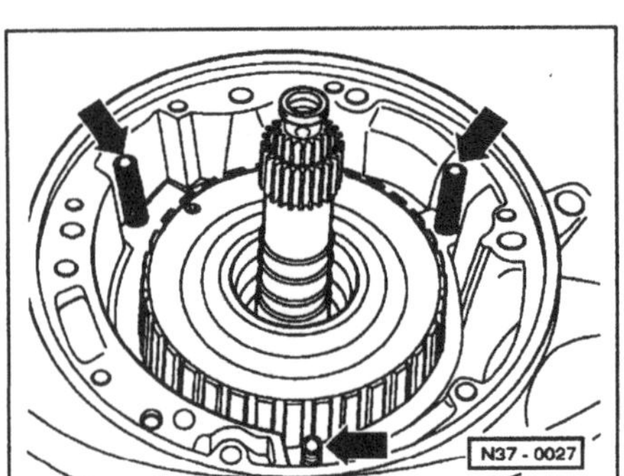

◄ – Install -B2- plates as follows:

  – First install a 3 mm (0.118 in.) thick outer plate

  – Install three spring caps in outer plate

  – Install compression springs -arrows-

  – Install all but the last outer plate

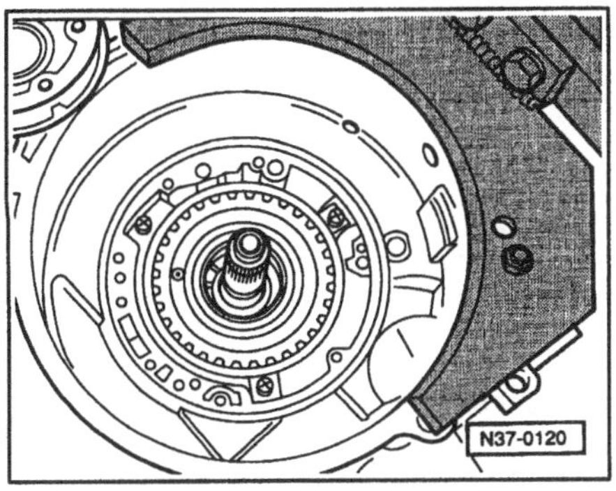

- Before installing last outer plate, install three spring caps onto compression springs.

Note:

When replacing

- *Supporting tube*
- *the ATF pump*
- *or the plates*

the 2nd and 4th gear brakes -B2- are to be adjusted, *⇒ page H 37-115.*

- Replace corrugated washer.
- Replace ATF pump seal.
- Install O-ring on ATF pump.

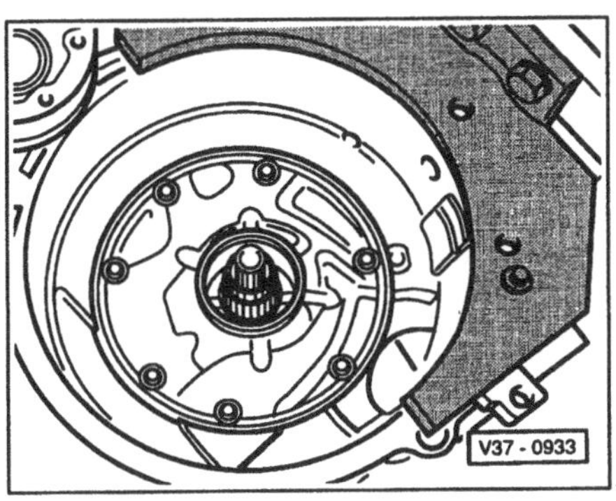

- Install ATF pump.
- Tighten bolts evenly, alternating crosswise.

- Tightening torque: 8 Nm (71 in lb) plus an additional 1/4 turn (90)

Note:

- *Ensure that the O-ring is not damaged.*
- *Additional turn of 90° (1/4 turn) can be done in several stages.*

- Perform check measurement of clutch play ⇒ page H 37-114.
- Install sealing plugs, valve body and oil pan ⇒ page H 38-36.
- Install cover with gasket and spacer bushings.

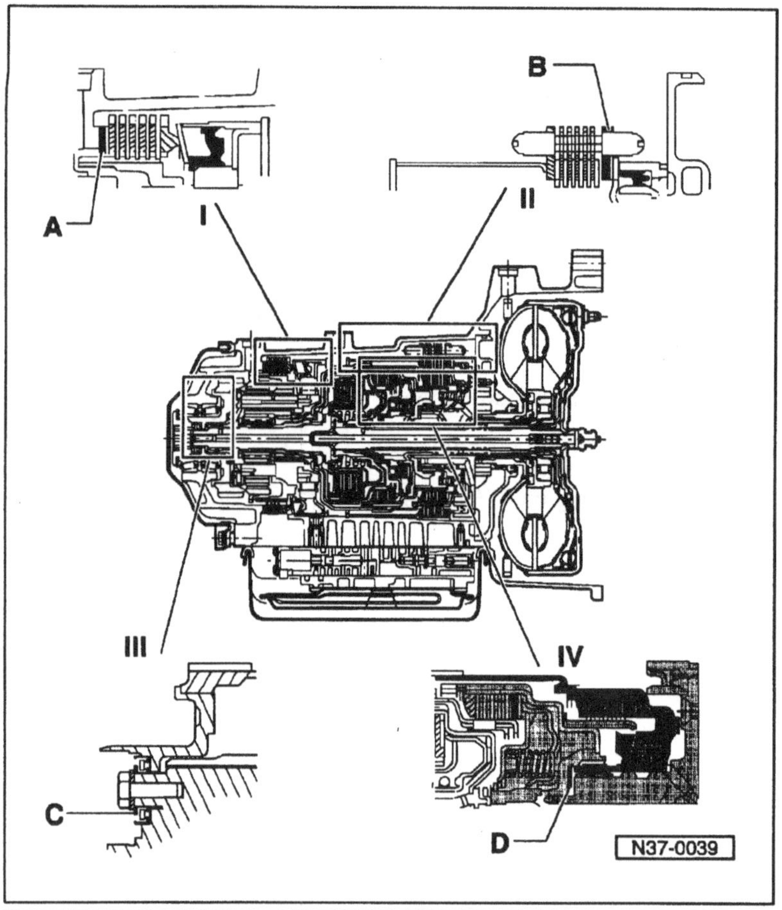

Planetary gearbox adjustments, overview

Note:

The illustration N37-0039 is from automatic transmission 096. Adjustment work to planetary gearbox on automatic transmission 098 is identical.

I = Reverse gear brake -B1-

- A - Shim
- Determining thickness ⇒ adjusting reverse gear brake -B1-, page H 37-101

II = 2nd and 4th gear brake -B2-

- B - Outer plate
- Determining thickness ⇒ adjusting 2nd and 4th gear brake -B2-, page H 37-115

III = Planet carrier

- C - Shim
- Determining thickness ⇒ adjusting planet carrier, page H 37-93

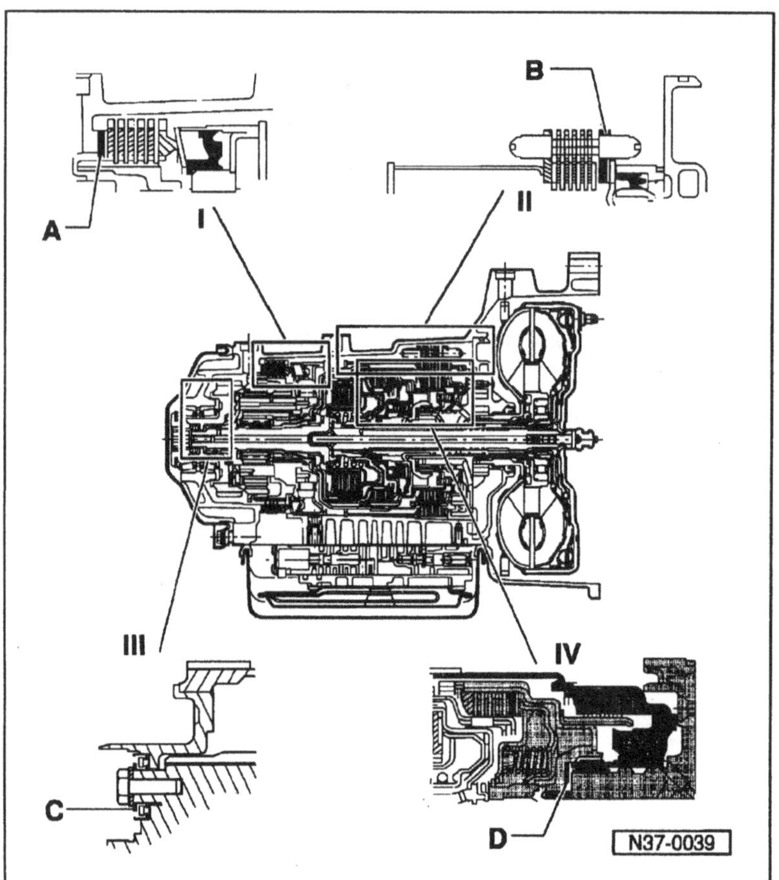

IV = Clutch play

- D - Shim
- Determining thickness ⇒ adjusting clutch play between -K1- and -K2-, page H 37-108

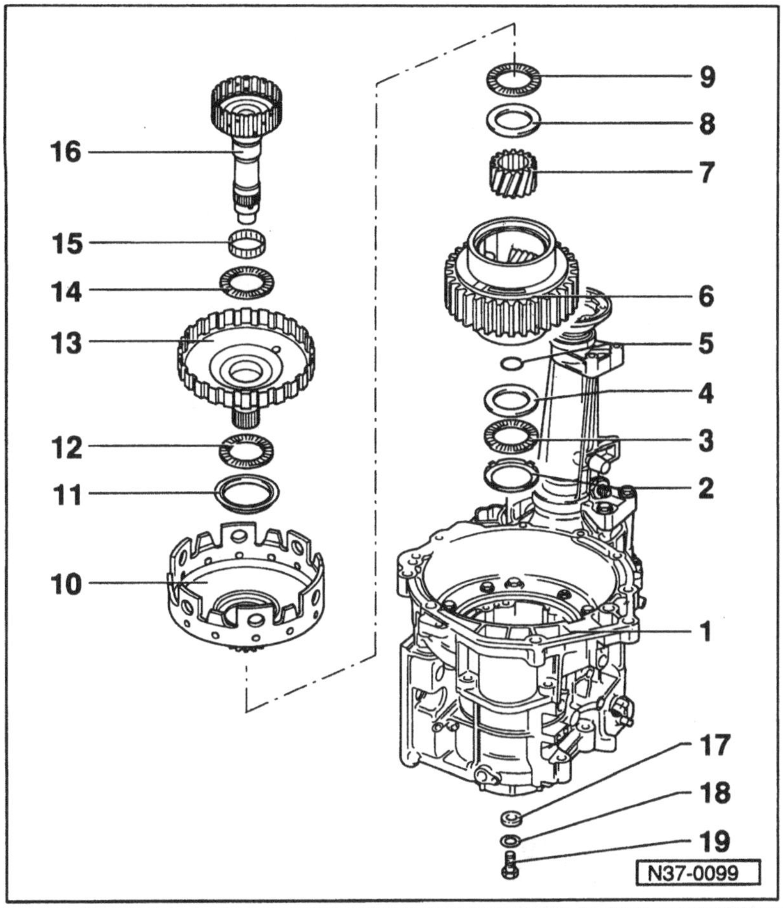

Planet carrier, adjusting

Overview of components to adjust planet carrier

– When adjusting planet carrier install components in transmission housing without shim -17-.

1 - Transmission housing
- ♦ With input gear and axial needle bearing
- ♦ Axial needle bearing remains in input gear
- ♦ If axial needle bearing is damaged ⇒ Removing and installing input gear, page H 39-23

2 - Axial needle bearing washer
- ♦ Install smooth side in input gear

3 - Axial needle bearing

4 - Axial needle bearing washer

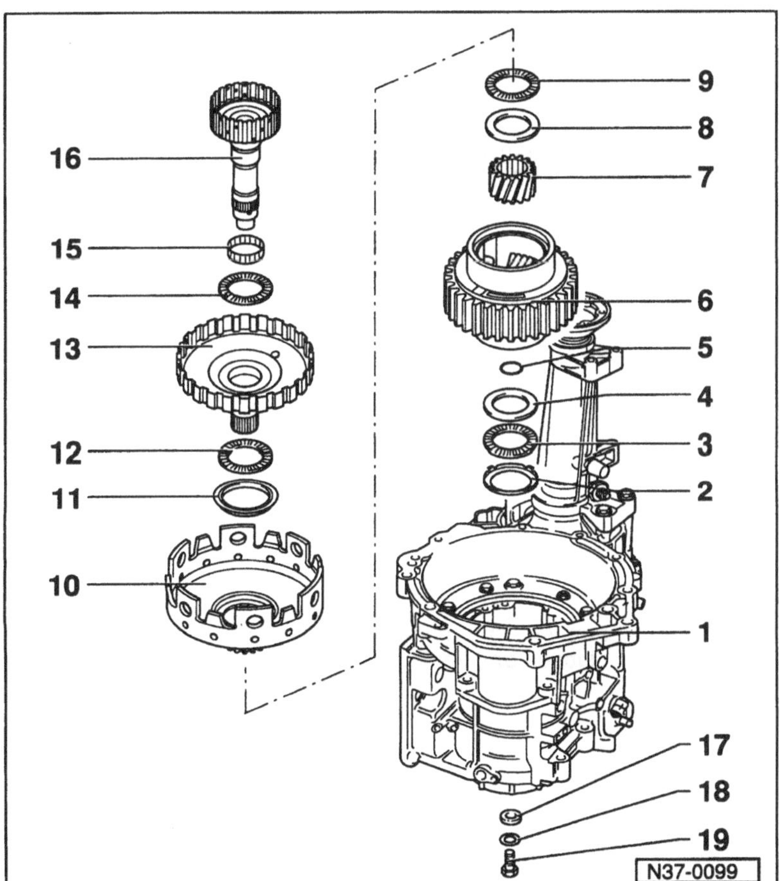

5 - O-ring
- ♦ Install into planet carrier ⇒ Assembling planet carrier, page H 37-78

6 - Planet carrier

7 - Small sun gear
- ♦ Cannot be removed from the planet carrier in transmissions from 01.93 ⇒ page H 00-11

8 - Washer

9 - Axial needle bearing

10 - Large sun gear

11 - Washer
- ♦ Install into large sun gear ⇒ Assembling planetary gearbox, page H 37-78

12 - Axial needle bearing

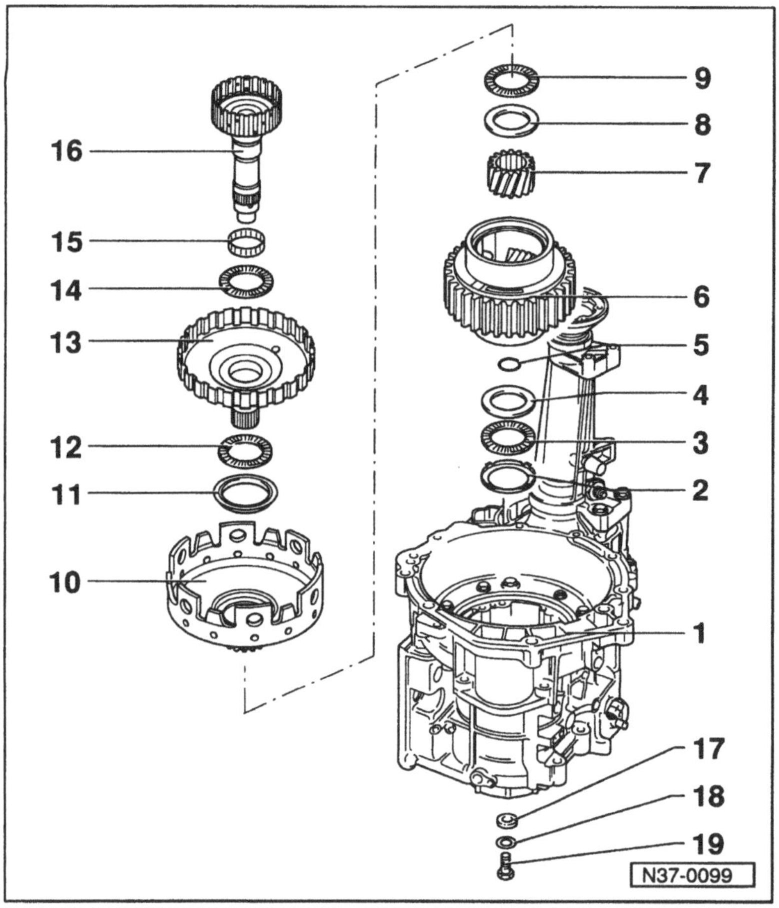

13 - Large drive shaft

14 - Axial needle bearing

15 - Needle bearing

16 - Small drive shaft

17 - Shim
- ♦ Do not install when adjusting planet carrier

18 - Washer

19 - Small drive shaft bolt
- ♦ Tightening torque: 30 Nm (22 ft lb)

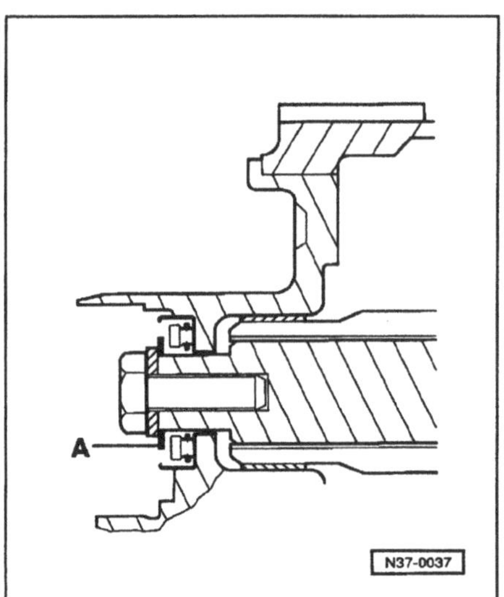

◄ **Determining shim -A-**

– Install all components -2- through -16- to adjust planet carrier into transmission housing ⇒ page H 37-93.

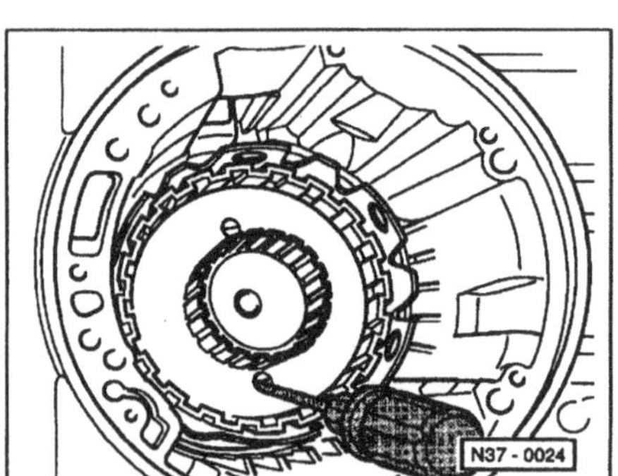

◄ – Insert screwdriver through hole of large drive shaft and large sun gear to loosen and tighten bolt for small drive shaft.

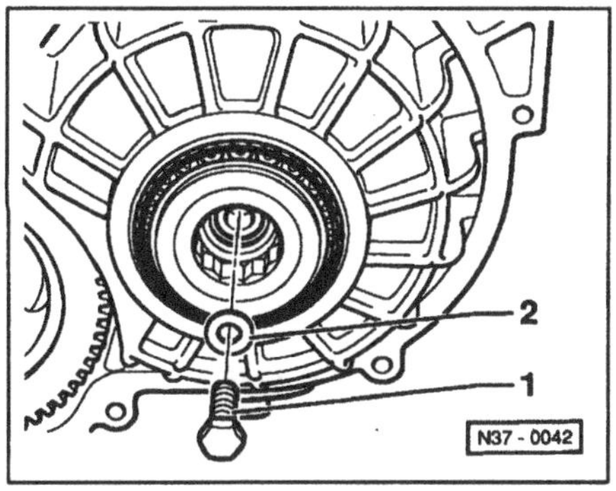

◄ – Install small input shaft bolt -1- with washer -2- but without shim.

- Tightening torque: 30 Nm (22 ft lb)

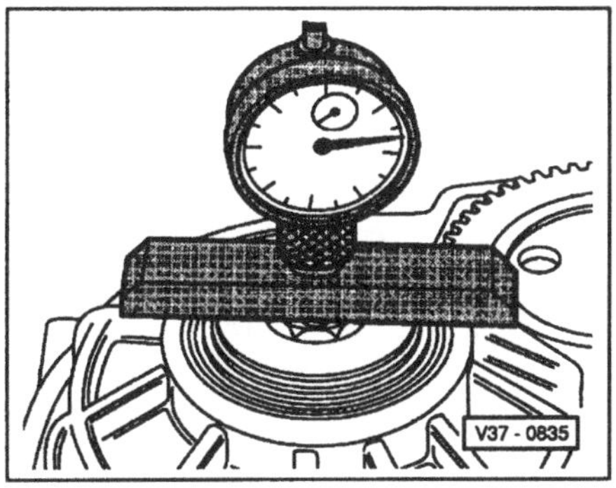

◄ – Set up dial indicator so that top of gauge is positioned on bolt head center with 1 mm (0.039 in.) preload.

– Zero dial indicator.

– Move small input shaft up and down and record reading.

Example: 2 mm (0.079 in.)

– Determine thickness of shim according to table and obtain Part Number from parts catalog microfiche.

Table of shims:

Indicator reading mm (in.)	Shims mm (in.)
1.26 (0.0496) ... 1.35 (0.0531)	1.0 (0.0393)
1.36 (0.0535) ... 1.45 (0.0571)	1.1 (0.0433)
1.46 (0.0575) ... 1.55 (0.0610)	1.2 (0.0472)
1.56 (0.0614) ... 1.65 (0.0650)	1.3 (0.0512)
1.66 (0.0653) ... 1.75 (0.0689)	1.4 (0.0551)
1.76 (0.0693) ... 1.85 (0.0728)	1.5 (0.0591)
1.86 (0.0732) ... 1.95 (0.0768)	1.6 (0.0630)
1.96 (0.0772) ... 2.05 (0.0807)	1.7 (0.0669)
2.06 (0.0811) ... 2.15 (0.0846)	1.8 (0.0709)
2.16 (0.0850) ... 2.25 (0.0886)	1.9 (0.0748)
2.26 (0.0890) ... 2.35 (0.0925)	2.0 (0.0787)
2.36 (0.0929) ... 2.45 (0.0965)	2.1 (0.0827)
2.46 (0.0969) ... 2.55 (0.1004)	2.2 (0.0866)
2.56 (0.1008) ... 2.65 (0.1043)	2.3 (0.0906)
2.66 (0.1047) ... 2.75 (0.1083)	2.4 (0.0945)
2.76 (0.1087) ... 2.85 (0.1122)	2.5 (0.0984)
2.86 (0.1126) ... 2.95 (0.1161)	2.6 (0.1024)
2.96 (0.1165) ... 3.05 (0.1201)	2.7 (0.1063)
3.06 (0.1205) ... 3.15 (0.1240)	2.8 (0.1102)
3.16 (0.1244) ... 3.25 (0.1280)	2.9 (0.1142)

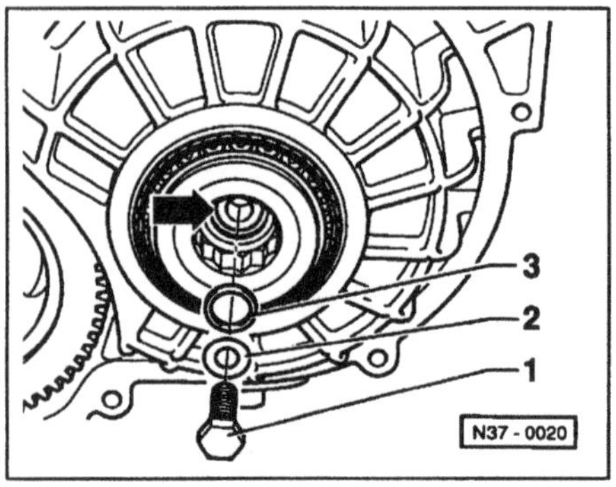

Example:

Measurement: 2.00 mm (0.079 in.)

Install 1.7 mm (0.067 in.) thick shim.

– Remove small input shaft bolt.

◄ – Install correct shim -3- on small input shaft -arrow-.

– Tighten small input shaft bolt -1- with washer -2-

● Tightening torque: 30 Nm (22 ft lb)

– Perform planet carrier check measurement.

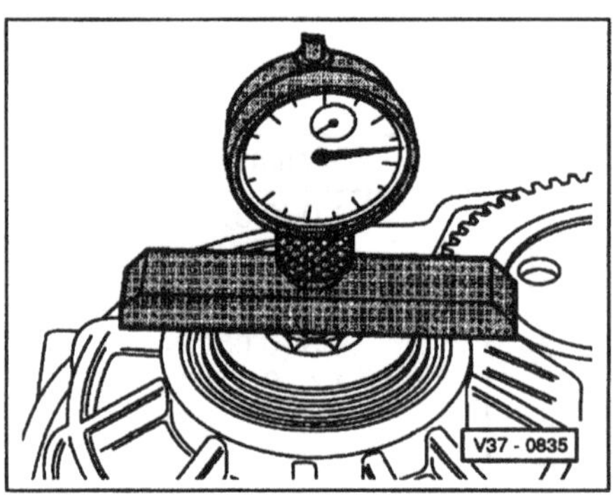

Planet carrier check measurement

◄ – Install dial indicator in VW 382/7 measuring bar and place the tip on the bolt of the small input shaft.

– Move small input shaft up and down and take reading of play on dial indicator.

Play:

♦ **min. = 0.23 mm (0.0091 in.)**

♦ **max. = 0.37 mm (0.0146 in.)**

Note:

If the planet carrier has been adjusted with reverse gear brake -B1- and free wheel removed, install reverse gear brake -B1-, ⇒ Planetary gearbox, assembling, page H 37-78, before installing planet carrier.

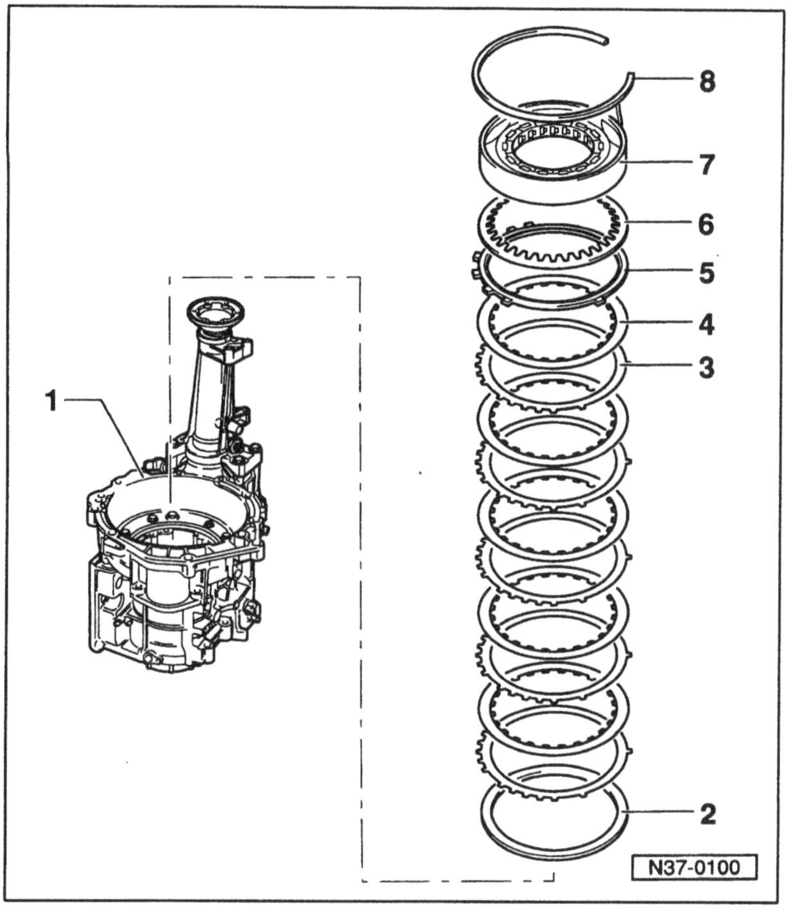

Reverse gear brake -B1-, adjusting

Overview of components to adjust reverse gear brake -B1-

1 - Transmission housing

2 - Shim

3 - Outer plate -B1-
- Quantity
  ⇒ from page H 00-3

4 - Inner plate -B1-
- Quantity
  ⇒ from page H 00-3

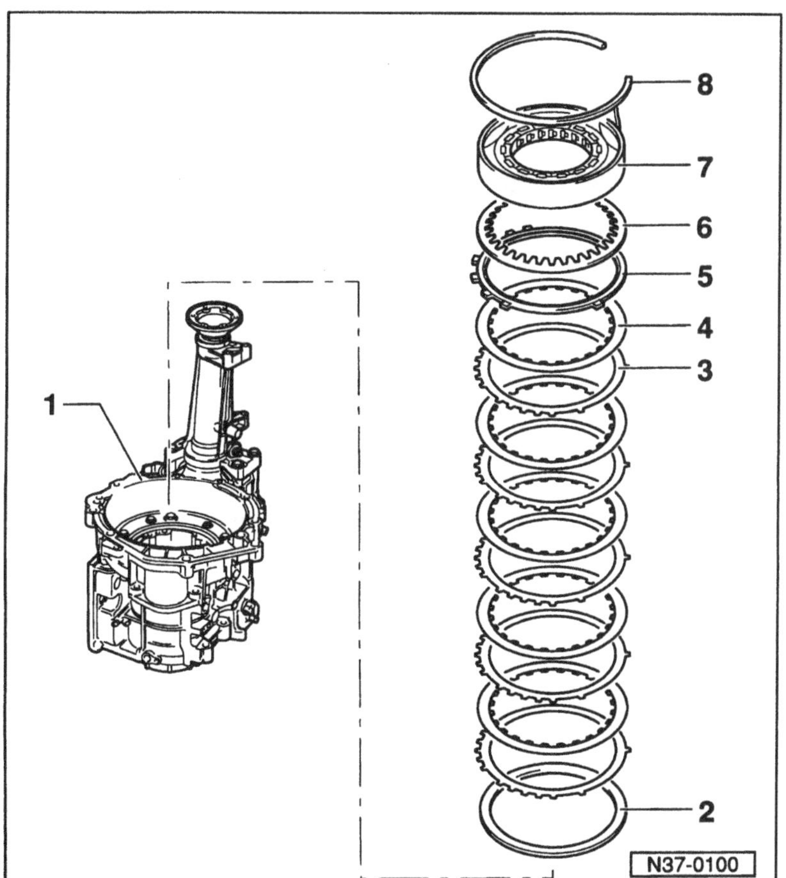

5 - Pressure plate
- Install with flat side facing plates
- 7.5 mm (0.2935 in.) thick with a 2 mm (0.0787 in.) thick outer plate
- 13.5 mm (0.5315 in.) thick with a 1.5 mm (0.0591 in.) thick outer plate

6 - Dished washer
- Convex side faces toward free wheel

7 - Free wheel with -B1- piston

Note:

Before removing or installing free wheel remove valve body and take out sealing plugs
⇒ page H 38-36.

8 - Circlip
- For free wheel

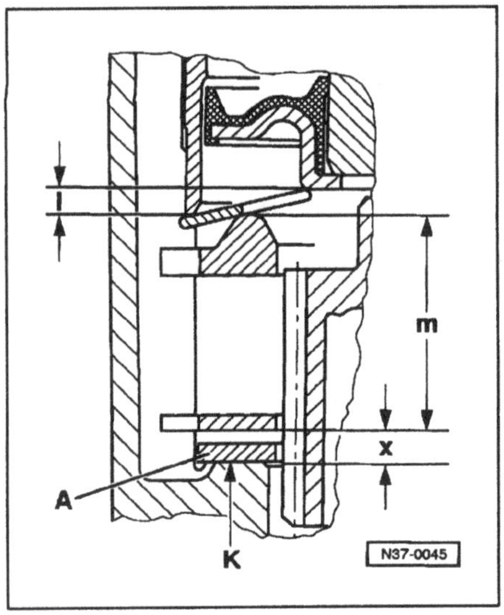

Determining shim -A- thickness

Shim thickness is determined by gap "x" and selected from table, ⇒ page H 37-106.

Gap $x \quad k \quad \frac{l}{2} - m$

- A = Shim
- x = Gap
- l = Position of piston in free wheel
- m = Height of plate set including pressure plate
- K = Constant = 30.5 mm (1.2008 in.)

  "K" is determined by installation height in transmission housing and is not adjustable.

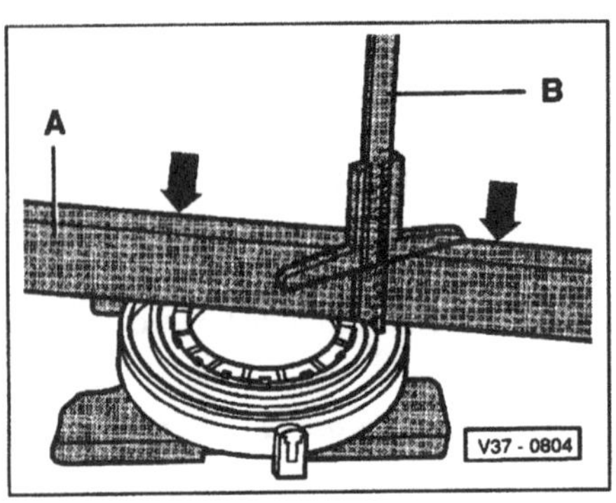

Calculating "l":

- Press piston in direction of arrow as far as stop.
- Place straightedge -A- onto outer race of free wheel.
- Measure to inner edge of piston with depth gauge -B-.

Example:

Reading		=	51.8 mm (2.0393 in.)
– Straightedge height		=	– 48.2 mm (1.8976 in.)
Calculated dimension	"l"	=	3.6 mm (0.1417 in.)

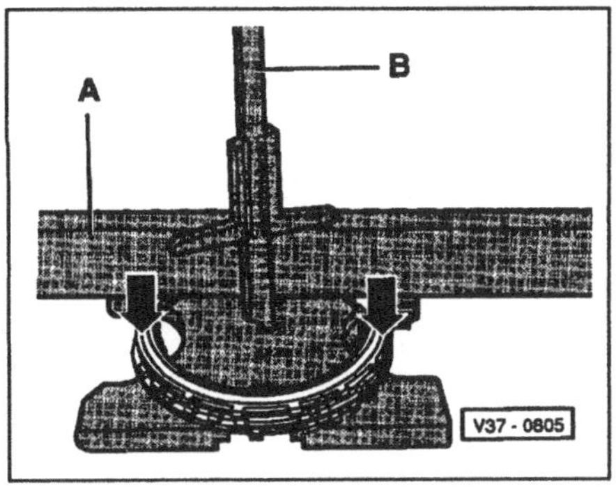

Calculating "m":

– Place straightedge -A- on thrust plate.

– Compress set of plates including pressure plate in direction of arrow and measure thickness of plate set with depth gauge -B-.

Example:

Reading		=	77.2 mm (3.0393 in.)
– Straightedge height		=	– 48.2 mm (1.8976 in.)
Calculated dimension	"m"	=	29.0 mm (1.1417 in.)

Gap x   k $\frac{l}{2}$–m

30.5 $\frac{3.6}{2}$–29.0   3.3 mm

(1.2008 $\frac{0.1417}{2}$–1.1417   0.1299 in.)

– Determine size of shims according to table:

Table of shims:

Gap "x" mm (in.)	Shims mm (in.)
2.36 (0.0929) ... 2.45 (0.0965)	1.0 (0.0393)
2.46 (0.0969) ... 2.55 (0.1004)	1.1 (0.0433)
2.56 (0.1008) ... 2.65 (0.1043)	1.2 (0.0472)
2.66 (0.1047) ... 2.75 (0.1083)	1.3 (0.0512)
2.76 (0.1087) ... 2.85 (0.1122)	1.4 (0.0551)
2.86 (0.1126) ... 2.95 (0.1161)	1.5 (0.0591)
2.96 (0.1165) ... 3.05 (0.1201)	1.6 (0.0630)
3.06 (0.1205) ... 3.15 (0.1240)	1.7 (0.0669)
3.16 (0.1244) ... 3.25 (0.1280)	1.8 (0.0709)
3.26 (0.1283) ... 3.35 (0.1319)	1.9 (0.0748)
3.36 (0.1323) ... 3.45 (0.1358)	1.0 + 1.0 (0.0393 + 0.0393)
3.46 (0.1362) ... 3.55 (0.1398)	1.0 + 1.1 (0.0393 + 0.0433)
3.56 (0.1402) ... 3.65 (0.1437)	1.1 + 1.1 (0.0433 + 0.0433)
3.66 (0.1441) ... 3.75 (0.1476)	1.1 + 1.2 (0.0433 + 0.0472)
3.76 (0.1480) ... 3.85 (0.1516)	1.2 + 1.2 (0.0472 + 0.0472)
3.86 (0.1520) ... 3.95 (0.1555)	1.2 + 1.3 (0.0472 + 0.0512)
3.96 (0.1559) ... 4.05 (0.1594)	1.3 + 1.3 (0.0512 + 0.0512)
4.06 (0.1598) ... 4.15 (0.1634)	1.3 + 1.4 (0.0512 + 0.0551)
4.16 (0.1638) ... 4.25 (0.1673)	1.4 + 1.4 (0.0551 + 0.0551)

– Determine thickness of shims according to table and obtain Part No. from parts microfiche.

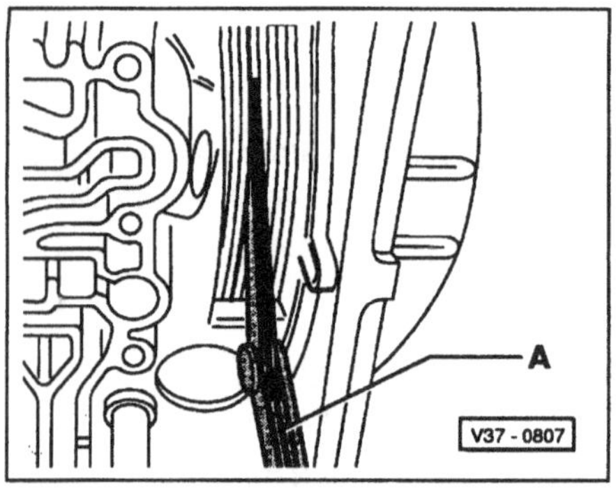

– After determining shim for -B1- perform check measurement.

Reverse gear brake -B1- check measurement

– Install parts up to free wheel and secure with circlip
⇒ Assembling planetary gearbox, page H 37-78.

◄ – Measure play between plates with feeler gauge -A-.

Play:

min. = 1.20 mm (0.0472 in.)

max. = 1.80 mm (0.07087 in.)

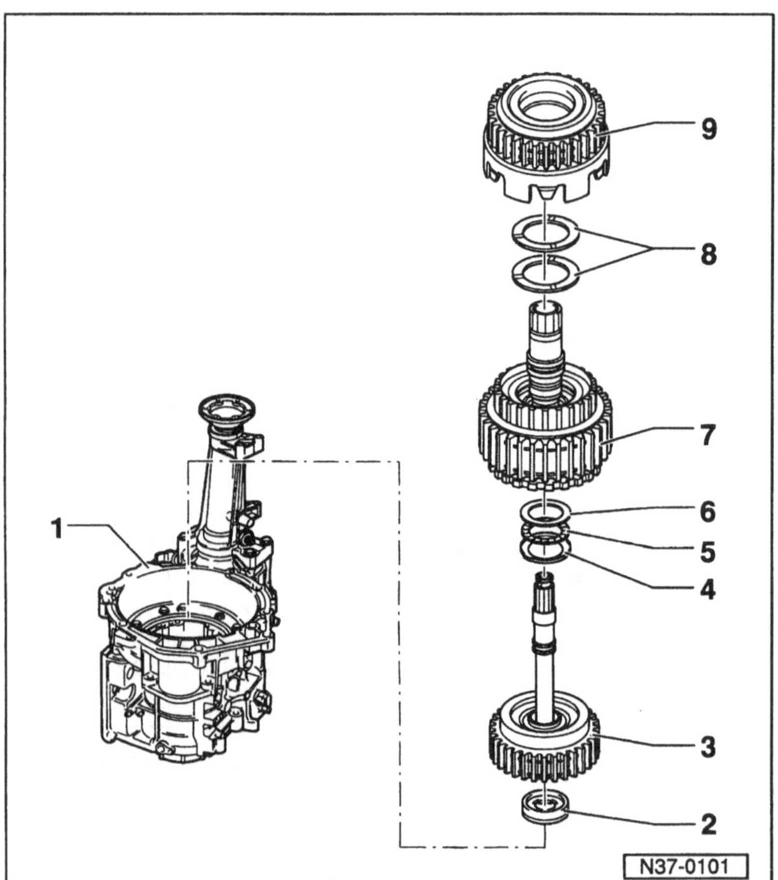

Clutch play between -K1- and -K2-, adjusting

– To adjust clutch play install components without shim -8-
⇒ Assembling planetary gearbox, page H 37-78.

1 - Transmission housing

2 - Axial needle bearing with washer
- ♦ Axial needle bearing faces toward small input shaft
- ♦ Washer faces toward -K3-

3 - 3rd and 4th gear clutch -K3- with pump shaft
- ♦ On transmissions from 01.93, the -K3- clutch only operates Fourth Gear ⇒ page H 00-11

4 - Axial needle bearing washer
- ♦ Lugs face towards axial needle bearing

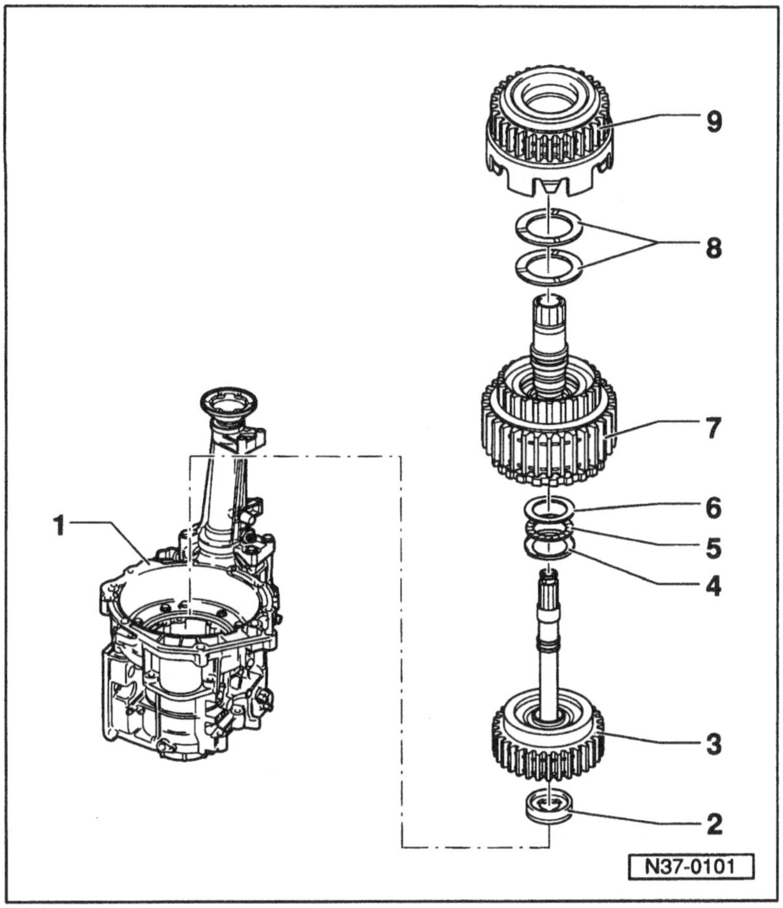

5 - Axial needle bearing

6 - Axial needle bearing washer

7 - 1st to 3rd gear clutch -K1- with turbine shaft

8 - Shim
- Do not install when adjusting clutch play
- 1 or 2 shims can be installed

9 - Reverse gear clutch -K2-

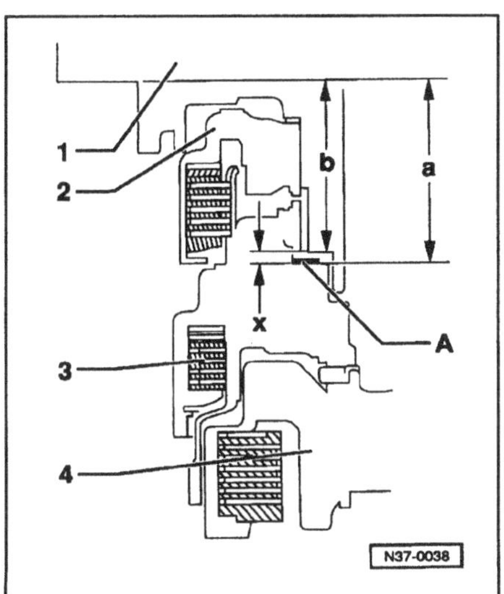

◄ **Determining shim -A-**

A - Shim

1 - ATF pump

2 - Reverse gear clutch -K2-

3 - 1st to 3rd gear clutch -K1-

4 - 4th gear clutch -K3-

Note:

On transmissions from 01.93 the -K3- clutch only operates Fourth Gear
⇒ page H 00-11.

– Determine gap “x” and calculate shim thickness.

Gap x = a – b

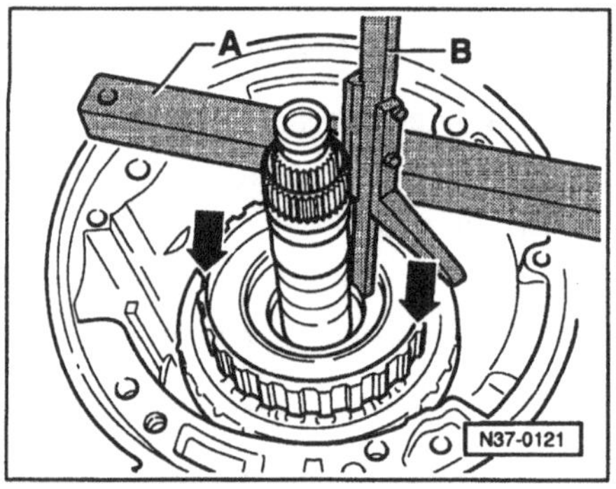

Calculating "a":

– Place straightedge -A- onto transmission housing.

– Press -K1- down in direction of arrow and measure distance into -K1- with depth gauge -B-.

Example:

Reading 1 = 88.5 mm (3.4842 in.)

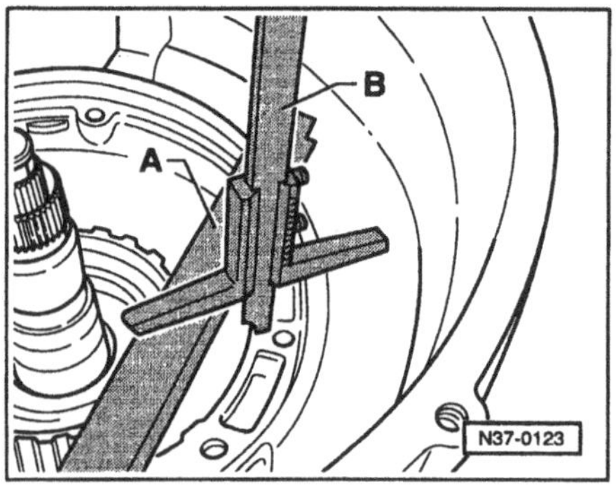

– Measure with depth gauge -B- to pump flange on transmission housing.

Example:

Reading 2 = 34.3 mm (1.3504 in.)

Reading 1		=	88.5 mm (3.4842 in.)
– Reading 2		=	– 34.3 mm (1.3504 in.)
Calculated dimension	"a"	=	54.2 mm (2.1338 in.)

= Size measured from pump flange/transmission housing into -K1-.

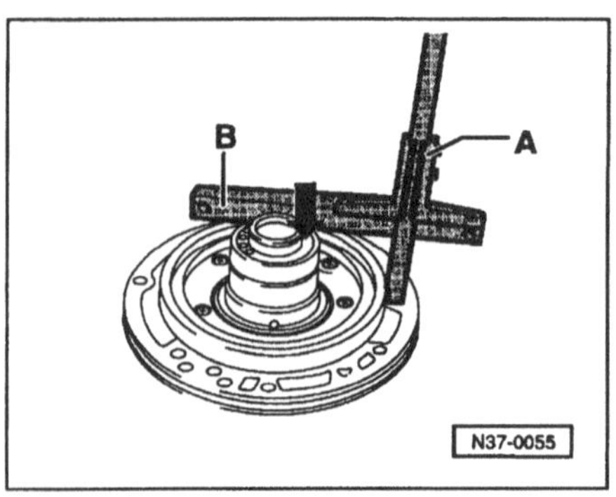

Calculating "b":

– Place straightedge -B- onto stator support (-arrow-) and measure with depth gauge -A- to gasket of pump flange.

Example:

Reading		=	70.5 mm (2.7756 in.)
– Straightedge height		=	– 19.5 mm (0.7677 in.)
Calculated dimension	"b"	=	51.0 mm (2.0079 in.)

Gap x = a – b

= 54.2 mm – 51.0 mm = 3.2 mm

= (2.1339 in. – 2.0079 in. = 0.01260 in.)

– Determine size of shim according to table:

Table of shims:

Gap "x" mm (in.)	Shims mm (in.)
... 2.54 (0.1000)	1.4 (0.0551)
2.55 (0.1004) ... 3.09 (0.1217)	1.0 (0.0393) + 1.0 (0.0393)
3.10 (0.1220) ... 3.49 (0.1374)	1.2 (0.0472)+ 1.2 (0.0472)
3.50 (0.1378) ... 3.89 (0.1531)	1.4 (0.0551)+ 1.4 (0.0551)
3.90 (0.1535) ... 4.29 (0.1689)	1.6 (0.0630)+ 1.6 (0.0630)
4.30 (0.1693) ... 4.69 (0.1846)	1.8 (0.0709)+ 1.8 (0.0709)
4.70 (0.1850) ... 5.04 (0.1984)	1.2 (0.0472)+ 1.2 (0.0472)+ 1.6 (0.0630)
5.05 (0.1988) ... 5.25 (0.2067)	1.2 (0.0472)+ 1.2 (0.0472) + 1.8 (0.0709)

– Determine thickness of shims according to table and obtain Part No. from parts microfiche.

– Perform clutch play check measurement after installing ATF-pump.

Check measurement (clutch play)

The clutch play cannot be measured until the ATF pump has been installed.

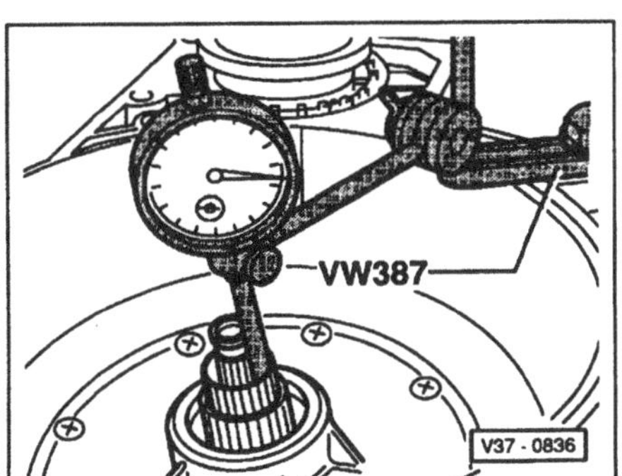

– Attach dial indicator holder to transmission housing and place dial indicator on turbine shaft with a preload of 1 mm (0.0393 in.).

– Move turbine shaft up and down and read axial play on dial indicator.

Play:

- **min. = 0.5 mm (0.0197 in.)**
- **max. = 1.2 mm (0.0472 in.)**

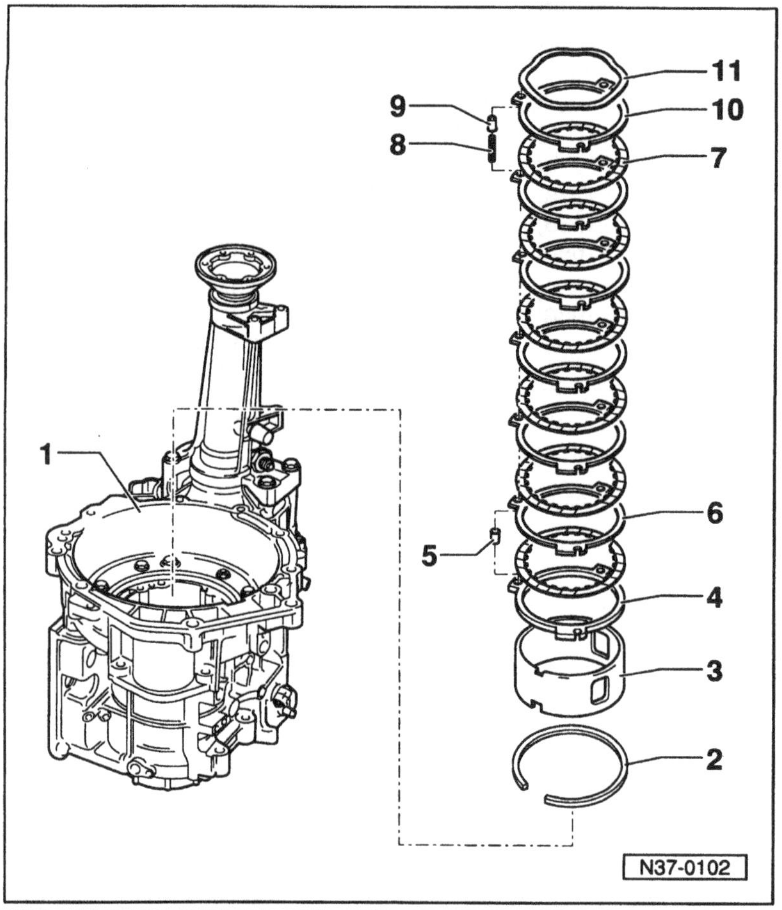

2nd and 4th gear brake -B2-, adjusting

– Install components to adjust 2nd and 4th gear brake -B2- but without corrugated washer -11- and last outer plate -10-
⇒ Assembling planetary gearbox, page H 37-78.

1 - Transmission housing

2 - Circlip
- ♦ For supporting tube

3 - Supporting tube -B2-
- ♦ Install so that notch locates on free wheel wedge

4 - Outer plate -B2-
- ♦ 3 mm thick (0.118 in.)

5 - Spring cap
- ♦ Install after installing outer plate

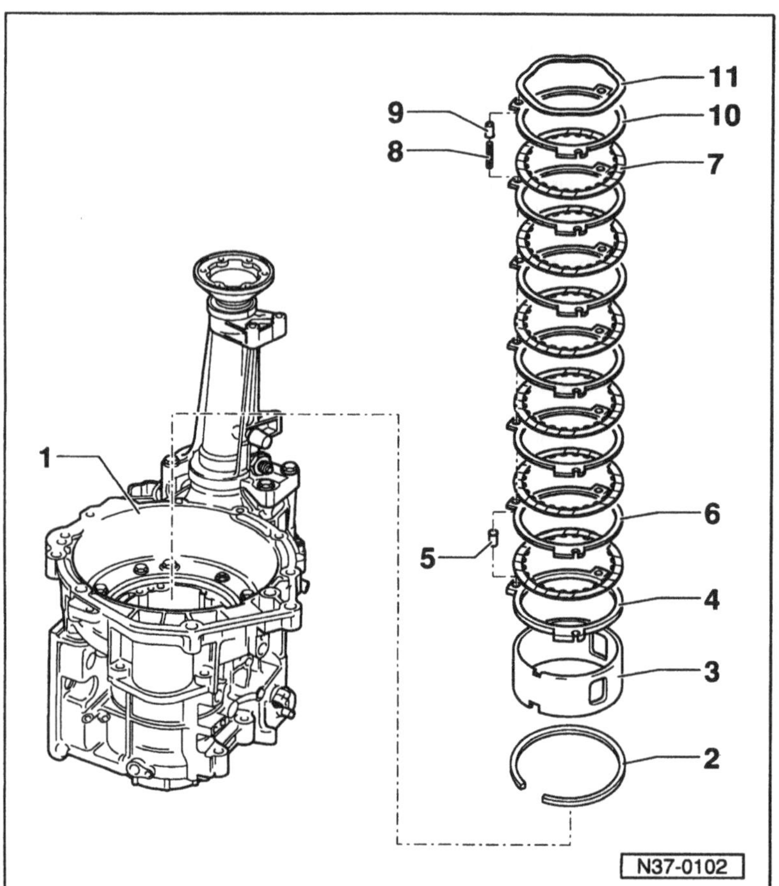

6 - Outer plate -B2-
- ♦ Always install 2 mm (0.0797 in.) thick outer plates
- ♦ Quantity
⇒ from page H 00-3

7 - Inner plate -B2-
- ♦ Quantity ⇒ from page H 00-3

8 - Spring

9 - Spring cap
- ♦ Install before installing last outer plate

10 - Outer plate -B2-
- ♦ Do not install when adjusting -B2-
- ♦ 2 outer plates can be installed.

11 - Corrugated washer
- ♦ Do not install when adjusting -B2-

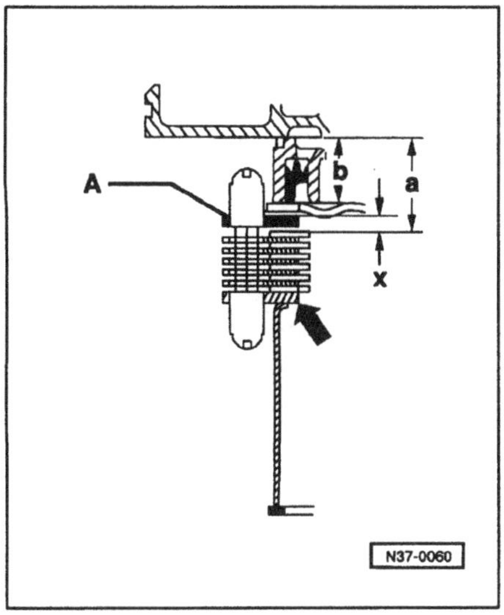

◄ **Calculating thickness of last outer plate -A-**

The thickness of the outer plate is determined by gap "x" and is selected from table, ⇒ page H 37-120.

Gap "x" = a – b – 3.6 mm (0.1417 in.)

A - Outer plate

Note:

First outer plate -arrow- is always 3 mm (0.118 in.) thick.

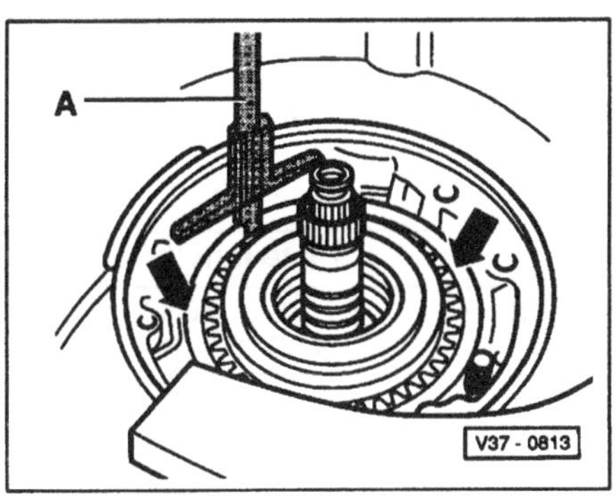

Calculating "a":

◄ – Measure from pump flange/transmission housing to the last inner plate with depth gauge -A-. Press inner plate down (in direction of arrow) for this step.

Example:

Reading for	"a"	=	30.2 mm (1.1890 in.)

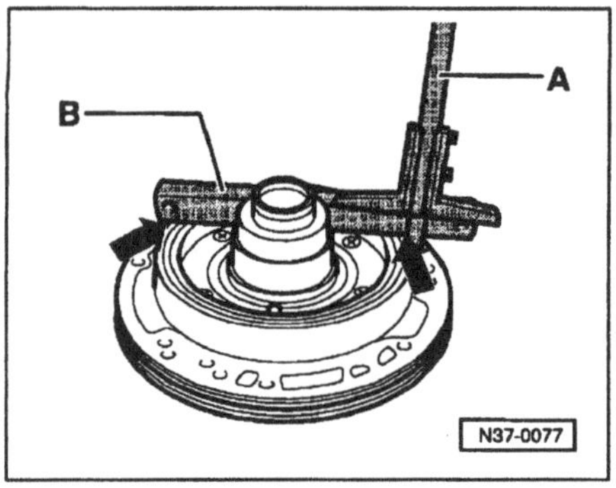

Calculating "b":

◄ – Place straightedge -B- onto stator support (-arrow-) and measure with depth gauge -A- to gasket of pump flange.

Example:

Reading		=	40.1 mm (1.5787 in.)
– Straightedge height		=	– 19.5 mm (0.7677 in.)
Calculated dimension	"b"	=	20.6 mm (0.8110 in.)

Gap x = a – b – 3.6 mm (0.1417 in.)

= 30.2 – 20.6 – 3.6 = 6.00 mm

= (1.1889 in. – 0.8110 in.– 0.1417 in. = 0.2362 in.)

– Determine thickness of outer plate according to table:

Table of plates:

Gap "x" mm (in.)	Plate mm (in.)
4.25 (0.1673) ... 4.49 (0.1768)	2.75 (0.1083)
4.50 (0.1772) ... 4.74 (0.1866)	3.00 (0.1180)
4.75 (0.1870) ... 4.99 (0.1965)	3.25 (0.1280)
5.00 (0.1969) ... 5.24 (0.2063)	3.50 (0.1378)
5.25 (0.2067) ... 5.49 (0.2161)	3.75 (0.1476)
5.50 (0.2165) ... 5.74 (0.2260)	2.00 (0.0787) + 2.00 (0.0787)
5.75 (0.2264) ... 5.99 (0.2358)	2.00 (0.0787) + 2.25 (0.0886)
6.00 (0.2362) ... 6.24 (0.2457)	2.25 (0.0886) + 2.25 (0.0886)
6.25 (0.2461) ... 6.49 (0.2555)	2.25 (0.0886) + 2.50 (0.0984)
6.50 (0.2559) ... 6.74 (0.2654)	2.50 (0.0984) + 2.50 (0.0984)
6.75 (0.2657) ... 7.00 (0.2756)	2.50 (0.0984) + 2.75 (0.1082)

– Determine thickness of outer plate according to table and obtain Part Number from parts microfiche.

- Two outer plates can be installed.

AG4 Automatic Transmission, troubleshooting

Troubleshooting Chart

Complaint	Possible Cause	Corrective Action
No.1 Drive plate torn	Centering bushing missing engine/ transmission	– Replace centering bushing engine/transmission and driven plate ⇒ Repair Manual Engine Repair Group 13 of installed engine (allocation per code letter)
No. 2 Transmission covered with oil	Transmission leaking	– Clean engine/transmission and localize where transmission or engine is leaking. If necessary use lead detector spray Part No. G001 900 01 – If transmission leaking, seal off faulty components or replace. – Replace sealing rings, seals, bolt connections or transmission housing (torque housing) ⇒ Repair Manual Automatic Transmission 096 or 098
	Too much ATF oil added	– Clean transmission, check ATF level suction off if necessary ⇒ Automatic Transmission 096 or 098; check ATF level and replace
	Too much differential oil added	– Clean transmission, check final drive oil level and suction off if necessary ⇒ Repair Manual 096 or 098; check final drive oil.

Complaint	Possible Cause	Corrective Action
No. 3 ATF leaks out in area of torque converter	Clearance of driving plate not OK	– Adjust clearance of driving plate ⇒ Repair Manual Engine Repair Group 13 of installed engine (allocation according to code letter)
	Torque converter bushing faulty	– Replace torque converter ⇒ Repair Manual Automatic Transmission 096 or 098; Identification torque converter
	Sealing ring for torque converter faulty	– Replace sealing ring for torque converter; if surface on torque converter is damaged, replace torque converter. ⇒ Repair Manual Automatic Transmission 096 or 098; sealing ring for torque converter removing and installing.
	ATF pump leaks	– Disassemble ATF Pump and assemble and replace faulty components.
No. 4 ATF in coolant or coolant fluid in ATF	ATF Cooler Faulty	– Replace ATF cooler as well as coolant and exchange ATF twice if necessary – If too much fluid in ATF, planetary gear as well as clutch must be disassembled, cleaned and reassembled. ⇒ Repair Manual Automatic Transmission 096 or 098, Planetary gear disassembled and reassembled.

Complaint	Possible Cause	Corrective Action
No. 5 ATF and differential oil have been interchanged	ATF fluid or differential oil level too low, no visible leakage on transmission from outside	– Correct ATF as well as differential oil level – Drive vehicle a longer driving stretch and check replaced ATF oil and differential oil level; if fill height has changed, sealing ring for drive shaft and O-ring for bearing support must be replaced. ⇒ Repair Manual Automatic Transmission 096 or 098; drive shaft removing and installing
No. 6 Shift operation is difficult	Shift operation is difficult outside of transmission Shift operation in transmission is difficult	– Remove selector lever cable at lever/selector shaft – If shift operation is difficult outside of transmission, Repair shift operation. ⇒ Repair Manual Automatic Transmission 096 or 098; Repairing shift operation – If shift operation in transmission is difficult, disassemble park lock mechanism and assemble. ⇒ Repair Manual Automatic Transmission 096 or 098; Park lock mechanism assembling and disassembling
No. 7 Selector lever position: 1	1st to 3rd gear engagement K1 or reverse gear brake B1 faulty	– Repair 1st to 3rd gear engagement - K1- or reverse rear brake -B1- ⇒ Repair Manual Automatic Transmission 096 or 098; Planetary gear disassembling and assembling

Complaint	Possible Cause	Corrective Action
No. 8 Selector lever position: D, 3, 2 Automatic Transmission does not shift into 1st gear (no transmission of power)	1st to 3rd gear engagement K1 or one–way clutch is faulty	– Repair 1st to 3rd gear engagement K1 one–way clutch ⇒ Repair Manual Automatic Transmission 096 or 098; Planetary gear disassembling and assembling
No. 9 Selector lever position: D, 3, 2 Automatic Transmission does not shift into 2nd gear	2nd and 4th gear – brake B2 faulty	– Repair 2nd and 4th gear brake B2 ⇒ Repair Manual Automatic Transmission 096 or 098; Planetary gear disassembling and assembling
No. 10 Selector lever position: D or 3 Automatic Transmission does not shift into Third Gear	Reverse gear – engagement K2 is faulty	– Repair reverse gear engagement K2 ⇒ Repair Manual Automatic Transmission 096 or 098; Planetary gear disassembling and assembling
No. 11 Selector lever position: D Automatic Transmission does not shift into Fourth Gear	4th gear (3rd and 4th gear [1]) – engagement K3 or 2nd and 4th gear – brake is faulty	– Repair 4th gear (3rd and 4th gear [1]) – engagement K3 or 2nd and 4th gear–brake ⇒ Repair Manual Automatic Transmission 096 or 098; Planetary gear disassembling and assembling

[1] for transmissions up to December 1992 with Program switch

Complaint	Possible Cause	Corrective Action
No. 12 Selector Lever Position: R Automatic Transmission does not shift into reverse gear	Reverse gear – engagement K2 faulty Reverse gear – brake B1 faulty	– Repair reverse engagement K2 or reverse gear brake B1 ⇒ Repair Manual Automatic Transmission 096 or 098; planetary gear disassembling and assembling
No. 13 Selector lever position: D, 3, 2 or 1 No drive in all gears.	1st to 3rd gear – engagement K1 faulty One-way clutch or reverse gear brake B1 faulty	– Repair 1st to 3rd gear engagement K1 or one-way clutch or reverse gear brake B1 ⇒ Repair Manual Automatic Transmission 096 or 098; planetary gear disassembling and assembling
No. 14 Shifting not possible	Valve or solenoid stuck in valve body	– Replace valve body ⇒ Repair Manual Automatic Transmission 096 or 098; planetary gear disassembling and assembling
No. 15 Uncontrolled or hard shifting	Short between the valve lines or cable guide rails	– Replace valve body ⇒ Repair Manual Automatic Transmission 096 or 098; Removing and installing valve body
No. 16 Shifting: During gear change is hard	Check in which gear change the hard shifting occurs Affected shift elements faulty	– Check triggering of shift elements ⇒ Repair Manual Automatic Transmission 096 or 098; transmission with shift elements – Replace shift elements ⇒ Repair Manual Automatic Transmission 096 or 098; planetary gear disassembling and assembling

Complaint	Possible Cause	Corrective Action
No. 17 Transmission shifts in emergency mode	Wrong control module installed.	– Select control module from parts microfiche and replace if necessary
	Circuit board (cable guide rail) faulty. Wires to solenoid valve faulty.	– Conduct On Board Diagnostic (OBD) program and electrical check then replace wires or components as required. ⇒ Repair Manual Automatic Transmission 096 or 098; On Board Diagnostic (OBD) or ⇒ "Electrical Wiring Diagrams" for respective vehicle.
	Valve in valve body stuck	– Replace valve body ⇒ Repair Manual Automatic Transmission 096 or 098; removing and installing valve body
No. 18 Park lock mechanism	Selector lever cable faulty or installed wrong	– Replace and adjust selector lever cable ⇒ Repair Manual Automatic Transmission 096 or 098; checking and adjusting selector lever cable
	Pawl, park lock gear, or operation of pawl faulty	– Repair pawl, park lock gear, and operation of pawl ⇒ Repair Manual Automatic Transmission 096 or 098; Pawl, disassembly and assembly

Complaint	Possible Cause	Corrective Action
No. 19 Noises in final drive	Tapered roller bearing loud	– Replace taper roller bearing ⇒ Repair Manual Automatic Transmission 096 or 098; Disassemble and assemble final drive
	Drive shaft loud	– Replace drive shaft ⇒ Repair Manual Automatic Transmission 096 or 098; removing and installing drive shaft
	Pinion gear loud	– Replace pinion gear ⇒ Repair Manual Automatic Transmission 096 or 098; removing and installing drive shaft
	Drive gear loud	– Replace drive gear ⇒ Repair Manual Automatic Transmission 096 or 098; removing and installing drive gear
	Differential gear loud	– Replace differential gear ⇒ Repair Manual Automatic Transmission 096 or 098; removing and installing differential gear

Complaint	Possible Cause	Corrective Action
No. 20 Operation not OK (shaking in idle)	Throttle body not tight (sometimes only during engine/transmission movement)	– Check throttle body and air duct for leaks ⇒ Repair Manual Engine Repair Group 24 of installed engine (application according to code letter)
No. 21 Park lock mechanism	Selector lever cable faulty or installed wrong Pawl, park lock gear, or operation of pawl faulty	– Replace control module ⇒ Repair Manual Engine Repair Group 24 of installed engine (application according to code letter)
No. 22 Vehicle does not start or in wrong selector lever position	Relay for starter lock and backup lights faulty	– Replace relay ⇒ "Electrical Wiring Diagrams" for respective vehicle.
No. 23 Selector lever cannot be shifted from "N" or "P" with brake applied "N" and "P" are not locked	Shiftlock Solenoid -N110- faulty Faulty Transmission Control Module -J217-	– Replace Shift–lock Solenoid -N110- ⇒ Repair Manual Automatic Transmission 096 or 098; disassembly and assembly of shift operation – Replace Transmission Control Module -J217- ⇒ Repair Manual Automatic Transmission 096 or 098; On Board Diagnostics (OBD)

Complaint	Possible Cause	Corrective Action
No. 24 Multi-Function Transmission Range (TR) Switch -F125- contacts burned	Multi-Function Transmission Range (TR) Switch -F125- faulty Wiring and connectors burned Wires as well as seal of wiring faulty Multi-Function Transmission Range (TR) Switch -F125- contact corrosion	– Replace Multi–Function Transmission Range (TR) Switch -F125- ⇒ Repair Manual Automatic Transmission 096 or 098; disassembly and assembly of park lock mechanism – Replace wires and connectors – For 096 and 098 transmissions until 02.94 the connector routing for Multi-Function Transmission Range (TR) Switch -F125- will be changed ⇒ "Electrical Wiring Diagrams" for respective vehicle.

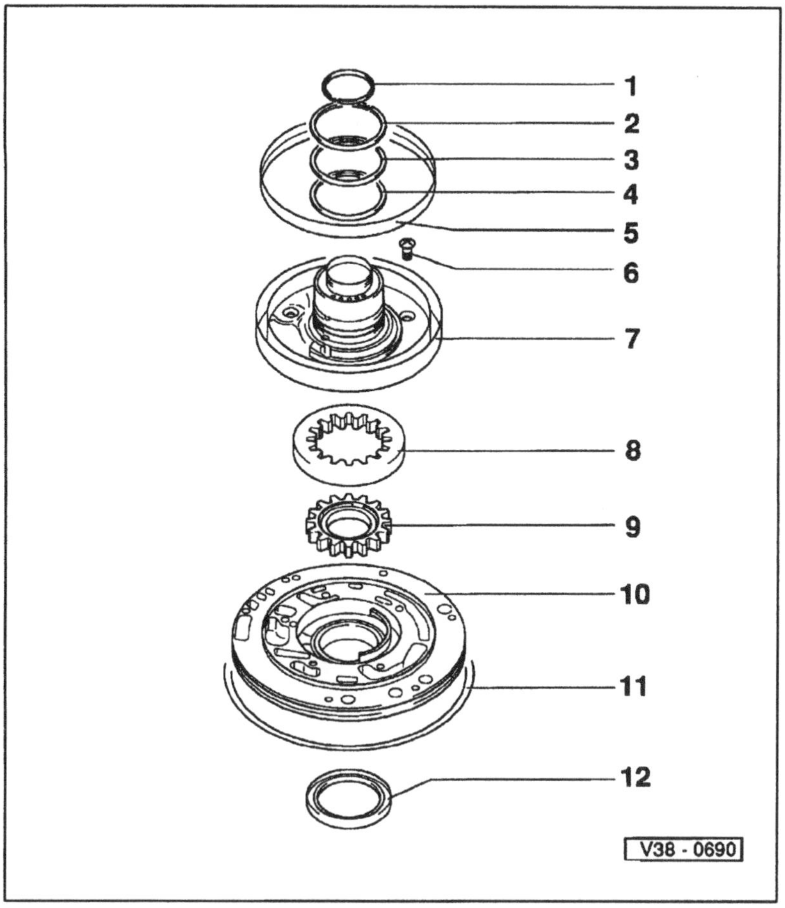

ATF pump, disassembling and assembling

1 - Piston ring
- Check that ring is correctly located
- Installing
  ⇒ Fig. 2

2 - Piston ring
- Check that ring is correctly located
- Installing
  ⇒ Fig. 2

3 - Piston ring
- Check that ring is correctly located
- Installing
  ⇒ Fig. 2

4 - Thrust washer

5 - Piston
- Sealing lips are vulcanized to the piston
- Moisten sealing lips with ATF before installing
- Turn piston slightly when installing

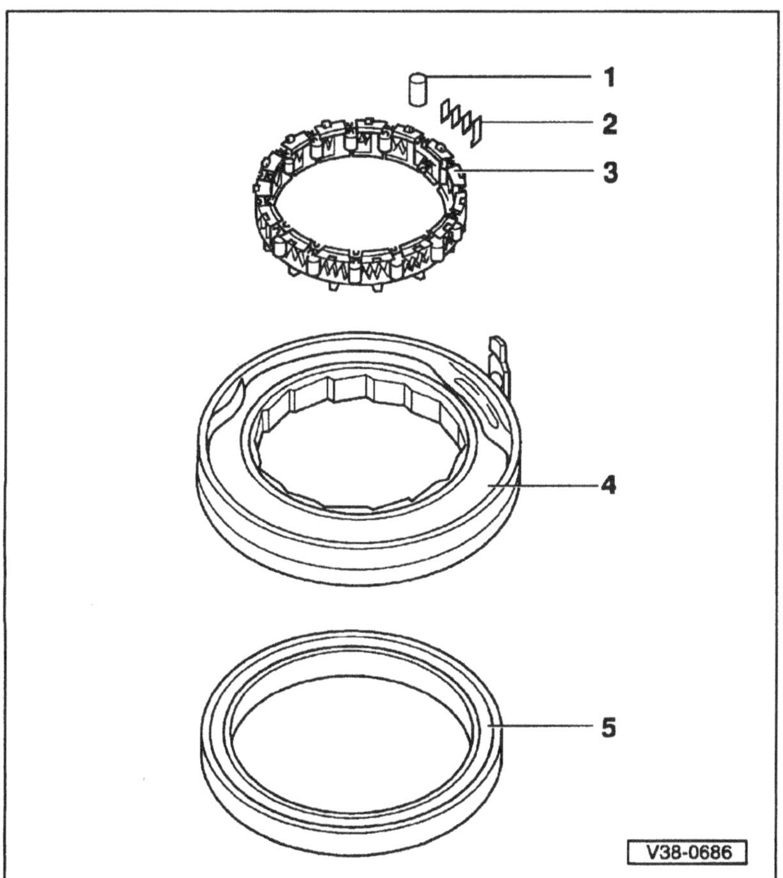

Free wheel with -B1- piston, disassembling and assembling

1 - Rollers
- Installing
  ⇒ Fig. 1

2 - Springs
- Installing
  ⇒ Fig. 1

3 - Cage
- Installing
  ⇒ Fig. 2
- Securing
  ⇒ Fig. 3

4 - Outer ring

5 - Piston
- Sealing lips are vulcanized to piston
- Moisten sealing lips with ATF before installing
- Turn piston slightly while installing
- Installed position
  ⇒ Fig. 4

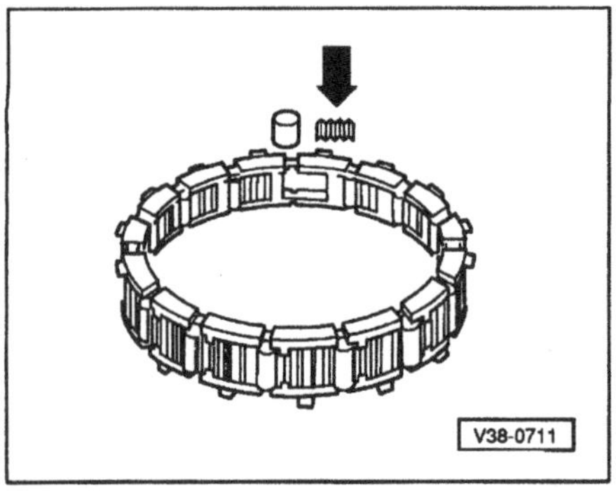

◄ **Fig. 1 Rollers and springs, installing**

Install springs so that spring -arrow- engages in cage.

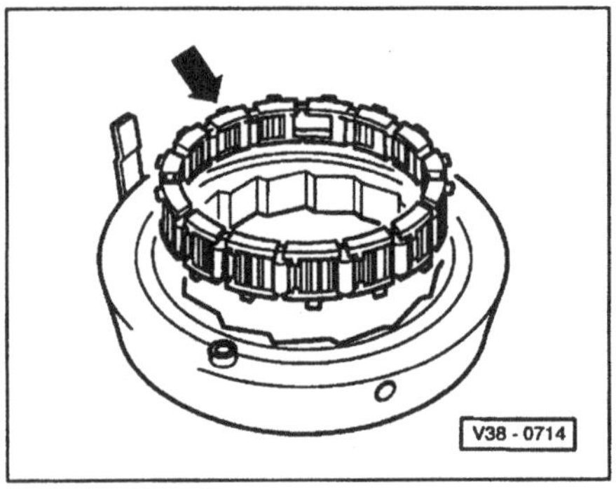

◄ **Fig. 2 Cage with springs and rollers, installing**

Install with large lugs -arrow- facing up.

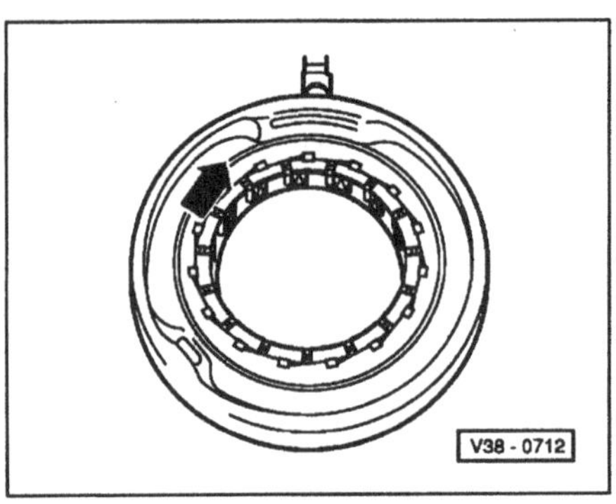

◄ **Fig. 3 Cage, securing**

Secure cage by turning in direction of arrow as far as stop.

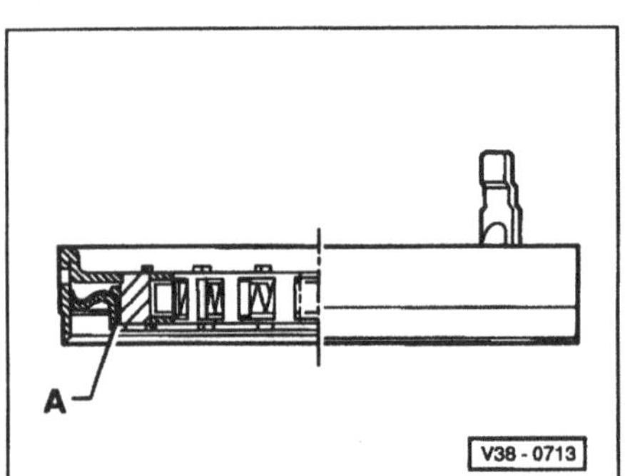

◄ **Fig. 4 Piston, installed position**

Install piston -A- as shown.

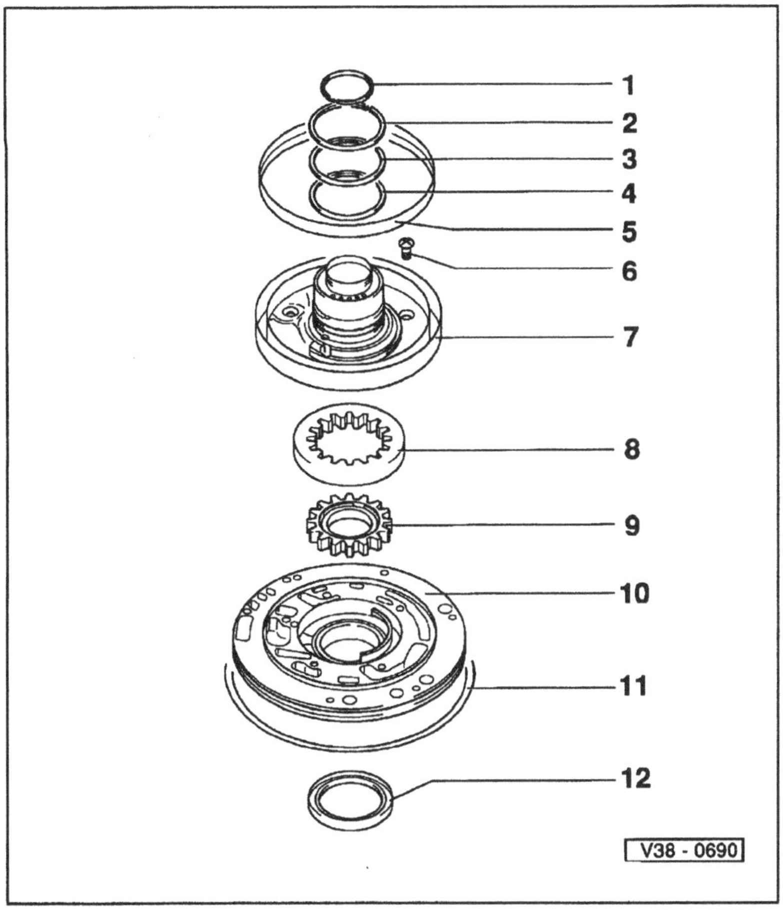

6 - Bolt

- Tightening torque: 10 Nm (7 ft lb) plus an additional 1/8 turn (45°)

7 - Stator support

8 - Outer gear

- Manufacturer's marking faces stator support
- If outer gear is incorrectly installed, the ATF pump may not operate freely

9 - Inner gear

- Larger groove faces stator support

10 - Housing for ATF pump

11 - O-ring

- Always replace

12 - Torque converter oil seal

- Removing and installing ⇒ page H 32-2

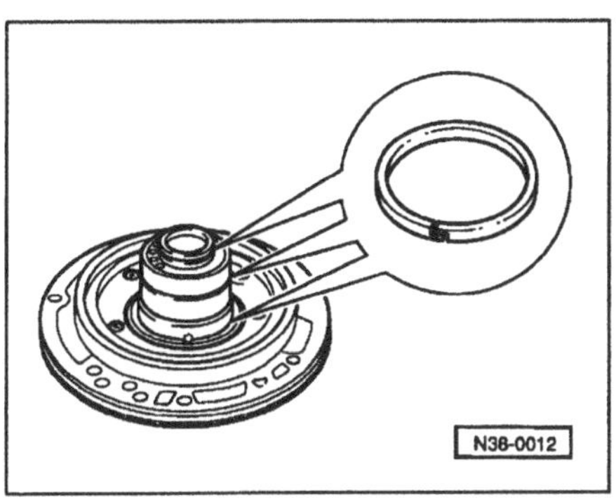

◄ **Fig. 1 Piston rings, checking for correct location**

Ensure that ends of piston rings are hooked together.

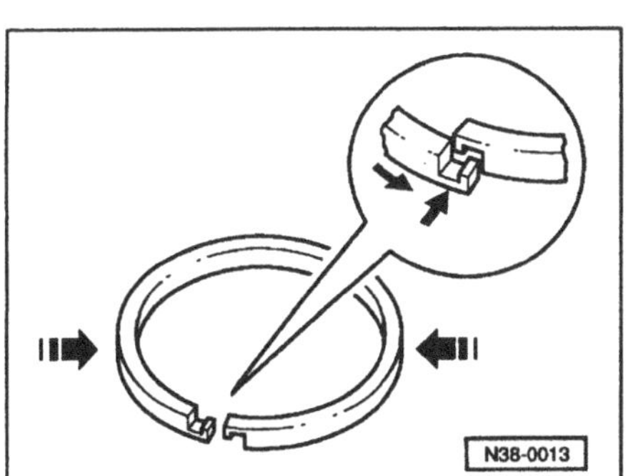

◄ **Fig. 2 Piston rings, installing and hooking ends together**

– Place piston ring in groove.

– Hook ends together by pressing sides of ring in and guiding ends over one another.

Note:

Do not twist piston ring out of groove on one side.

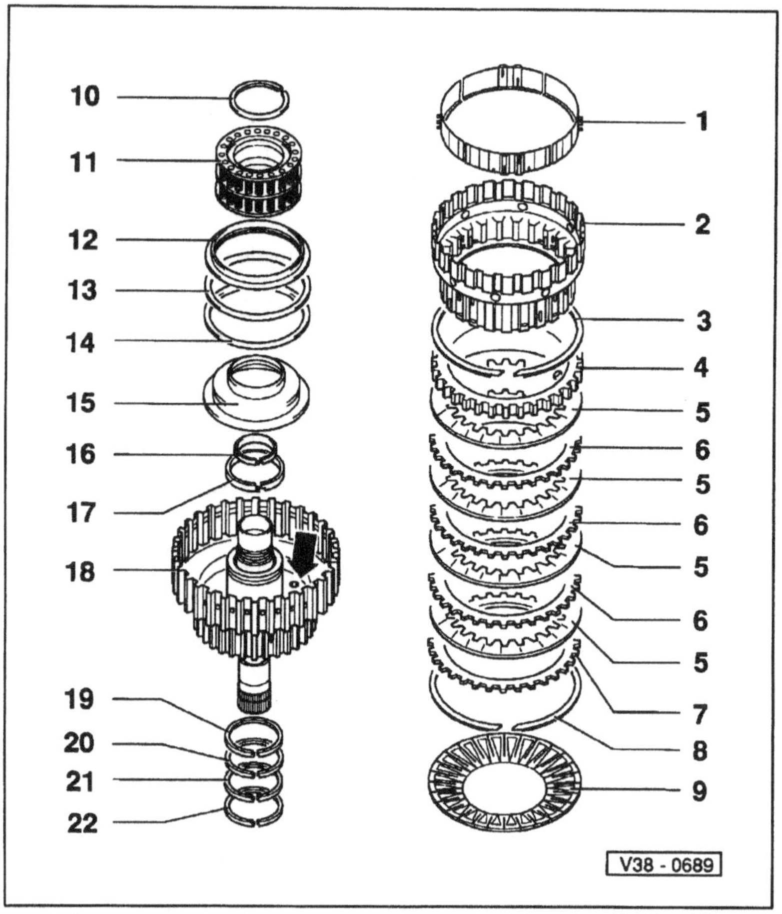

1st to 3rd gear clutch -K1- with turbine shaft on transmissions up to 12.92, disassembling and assembling

Transmissions up to 12.92
⇒ page H 00-3

Note:

When performing repairs to the clutch, make sure that the ball valve -arrow- is not damaged.

1 - Supporting ring
- 4-piece, unclip from inner plate carrier
- Installing
  ⇒ Fig. 4

2 - Inner plate carrier
- Installing
  ⇒ Fig. 6
- Assembling
  ⇒ Fig. 4

3 - Circlip
- Different thicknesses for different clutch assemblies
- Mark after removing and install in same position

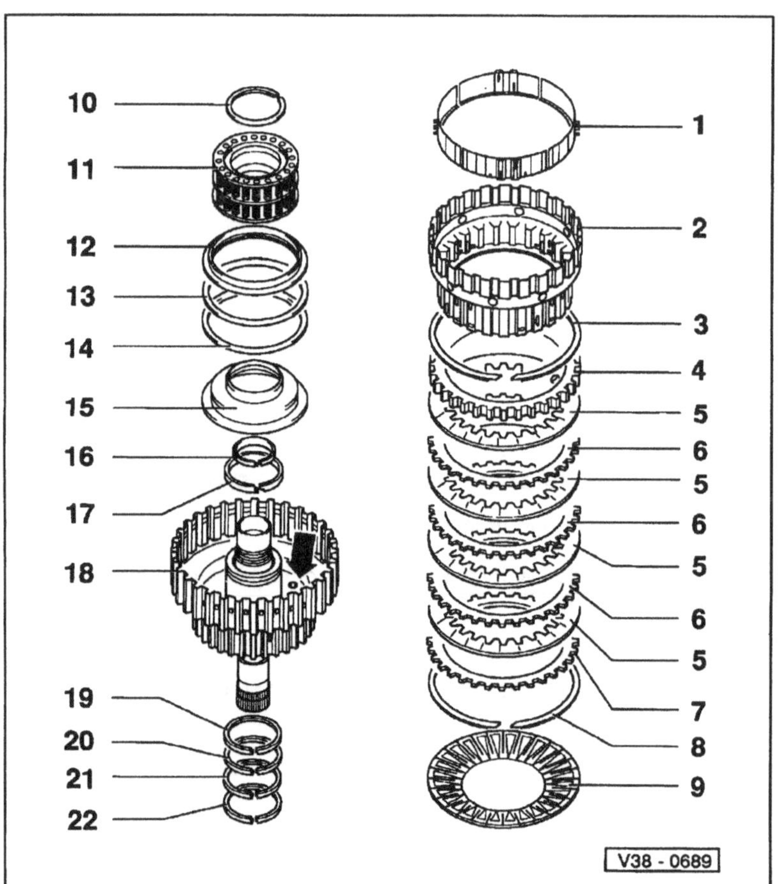

4 - Thrust plate
- Smooth side faces inner plate
- Install together with inner plate carrier
  ⇒ Fig. 4

5 - Inner plate
- Quantity
  ⇒ page H 00-3
- Install together with inner plate carrier
  ⇒ Fig. 4

6 - Outer plate
- Quantity
  ⇒ page H 00-3
- Install together with inner plate carrier
  ⇒ Fig. 4

7 - Thrust plate
- Curved side to diaphragm spring
- Installing
  ⇒ Fig. 5

8 - Circlip
- Installing
  ⇒ Fig. 3

9 - Diaphragm spring
- Install with curved side to piston

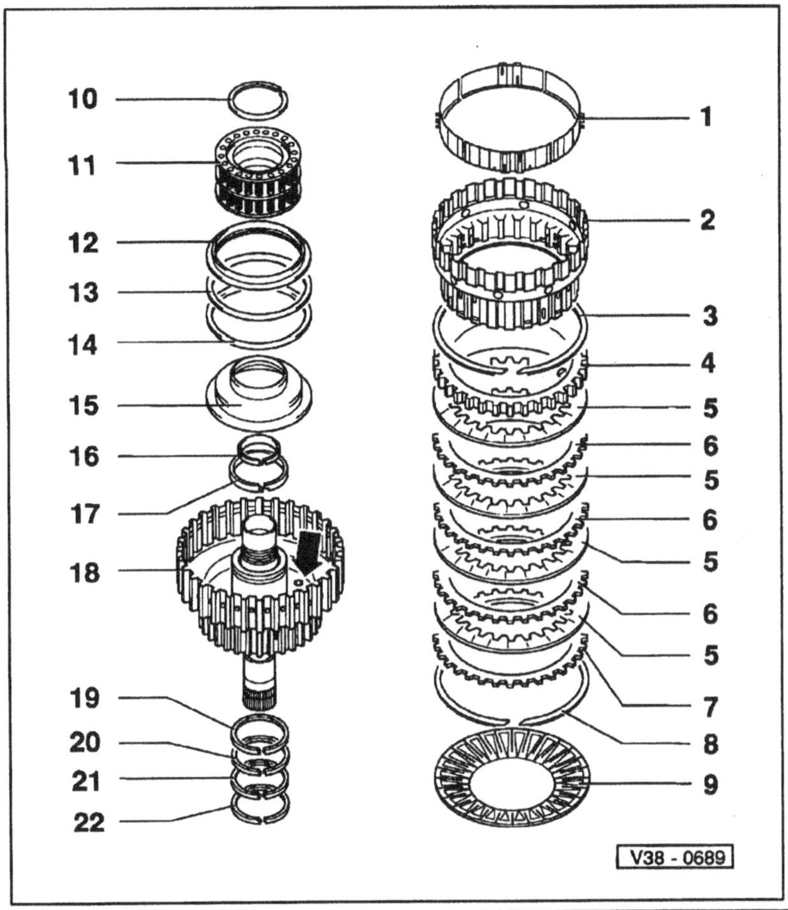

10 - Circlip
- Removing and installing ⇒ Fig. 2

11 - Intermediate spring ring
- With spring supporting shell and compression springs

12 - Operating ring
- Install with rounded surface facing intermediate spring ring

13 - Disc spring
- Curved side faces operating ring
- Installation position ⇒ Fig. 1

14 - Disc spring
- Curved side faces operating ring
- Installation position ⇒ Fig. 1

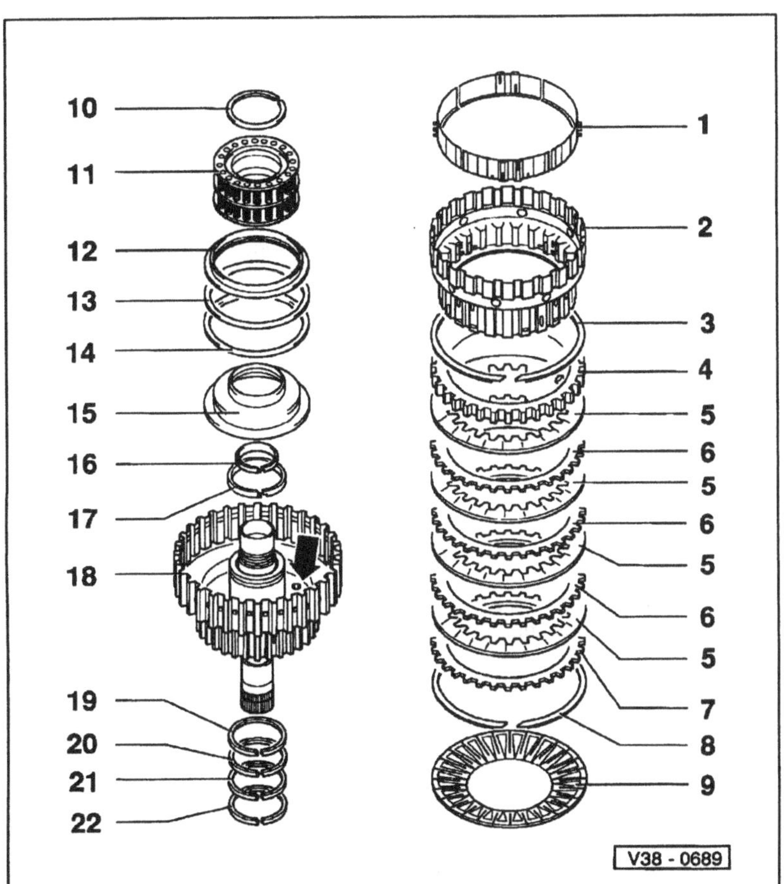

15 - Piston
- Sealing lips are vulcanized to piston
- Moisten sealing lips with ATF before installing
- Turn piston slightly while installing

16 - Piston ring (small diameter)
- Check that ring is correctly located ⇒ Fig. 7
- Installing ⇒ Fig. 8

17 - Piston ring
- Check that ring is correctly located ⇒ Fig. 7
- Installing ⇒ Fig. 8

18 - Clutch housing with turbine shaft
- Check bearing bushings. Replace housing if bushings are damaged.
- Before installing inner plate carrier install diaphragm spring -9-, circlip -3- and one inner and outer plate in ring clutch housing ⇒ Fig. 5

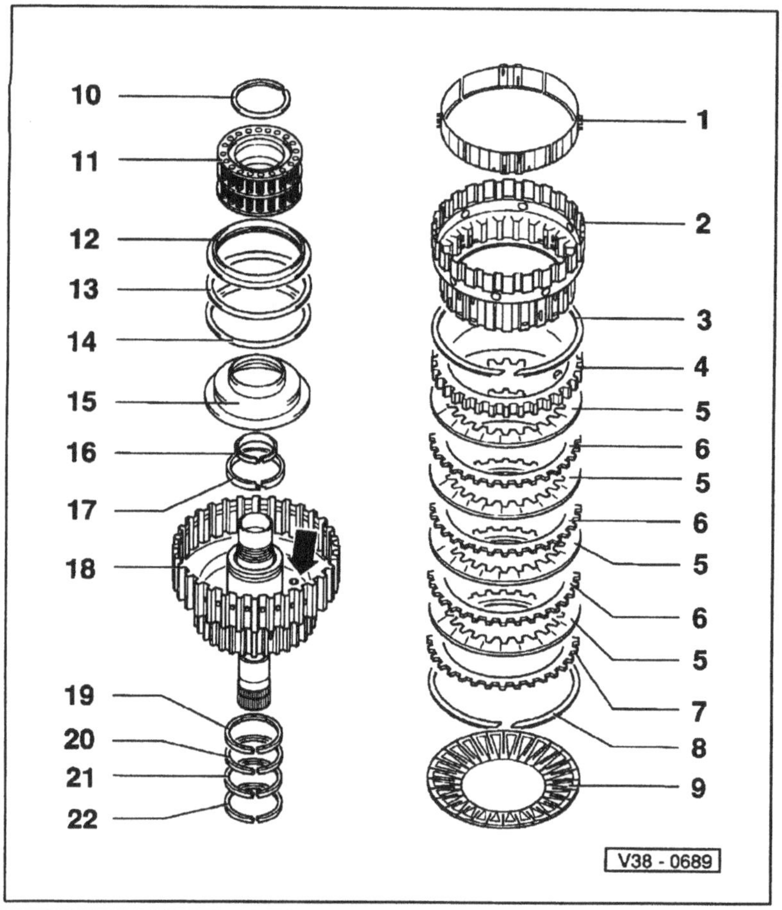

19 - Piston ring

- Check that ring is correctly located ⇒ Fig. 7
- Installing ⇒ Fig. 8

20 - Piston ring

- Check that ring is correctly located ⇒ Fig. 7
- Installing ⇒ Fig. 8

21 - Piston ring

- Check that ring is correctly located ⇒ Fig. 7
- Installing ⇒ Fig. 8

22 - Piston ring

- Check that ring is correctly located ⇒ Fig. 7
- Installing ⇒ Fig. 8

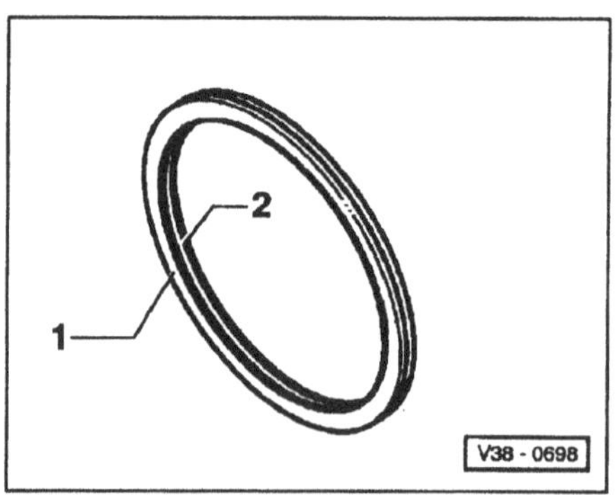

◄ **Fig. 1 Disc springs, installation position**

– Install disc springs -1- and -2- so that when placed together outer edges touch.

Note:

There is a clearance between the two springs at the inner diameter.

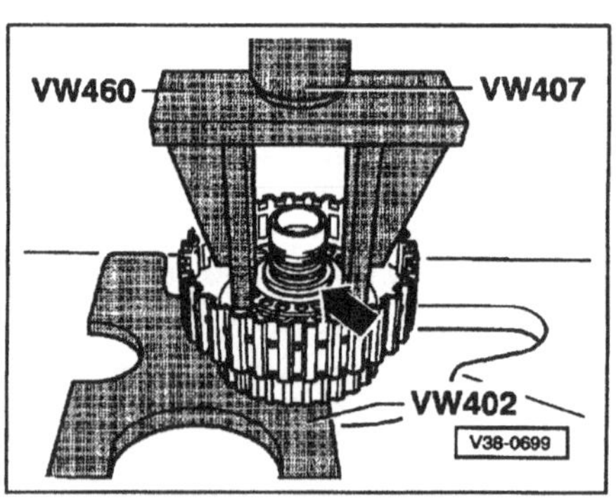

◄ **Fig. 2 Circlip, removing and installing**

– Press intermediate spring ring down far enough for circlip -arrow- to be installed in groove.

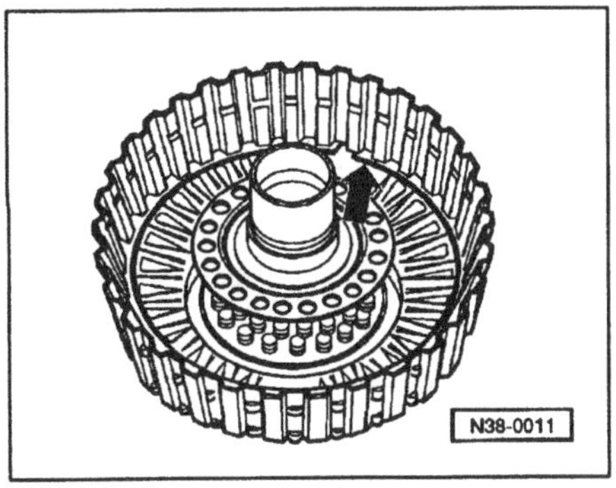

◄ **Fig. 3 Circlip, installing**

– Install circlip -arrow- in groove.

Note:

The diaphragm spring is under slight tension.

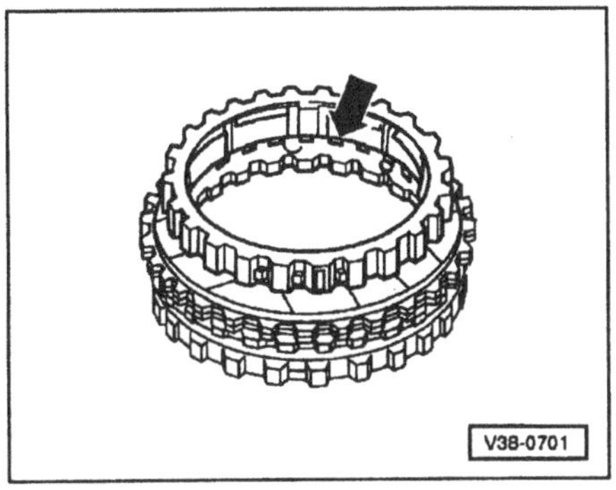

◄ **Fig. 4 Inner and outer plates, placing on inner plate carrier and installing support ring**

– Install thrust plate and inner and outer plates.

– Install supporting ring -arrow-.

Note:

An inner and an outer plate must be placed in the clutch housing before installing the inner plate carrier.

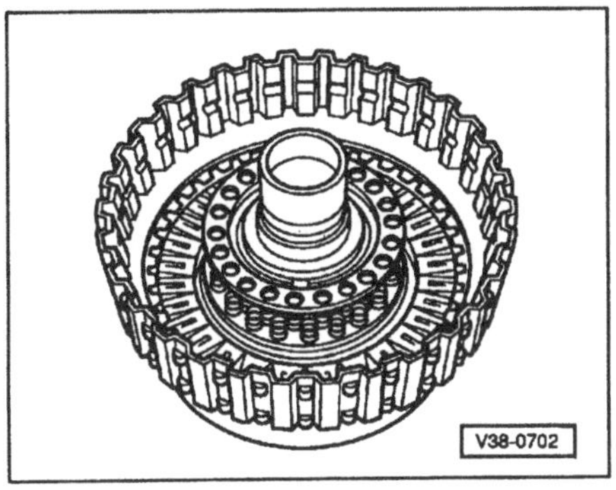

◄ **Fig. 5 Pressure plate, first inner and first outer plates, installing into clutch housing**

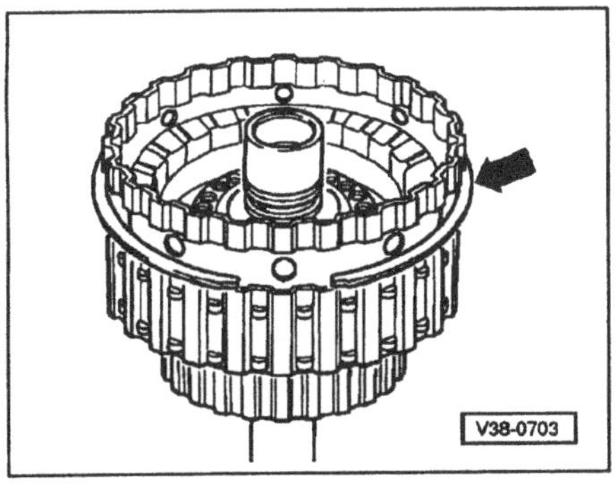

◄ **Fig. 6 Inner plate carrier, installing and installing circlip -arrow-**

Lift inner plate carrier slightly to install circlip.

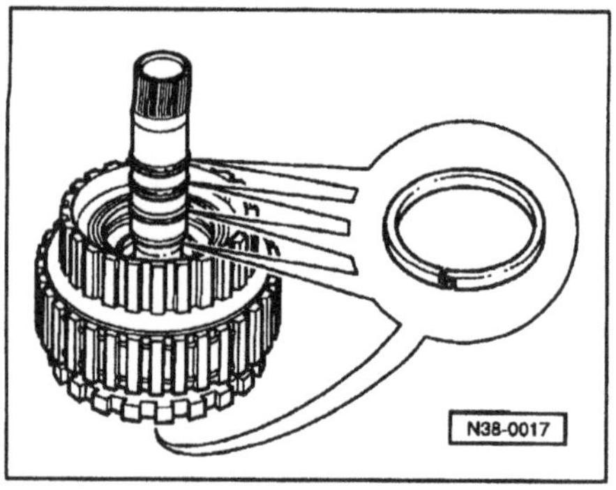

Fig. 7 Piston rings, checking for correct installation

There are two piston rings installed on the turbine shaft inside the clutch housing.

Ensure that ends of piston rings are hooked together.

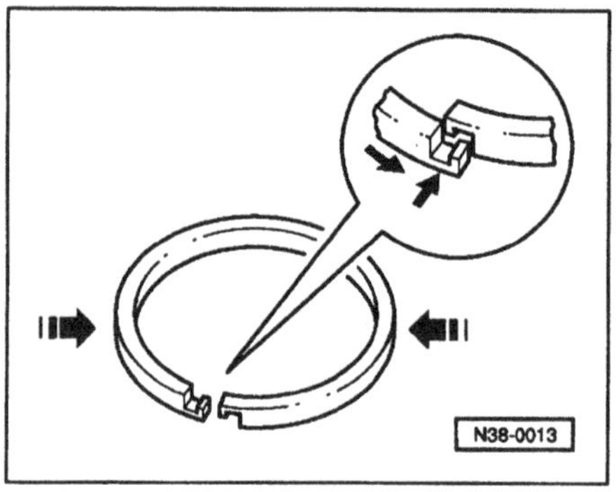

Fig. 8 Piston ring, installing and hooking ends together

- Place piston ring in groove.
- Hook ends together by pressing sides of ring in and guiding ends over one another.

Note:

Do not twist piston ring out of groove on one side.

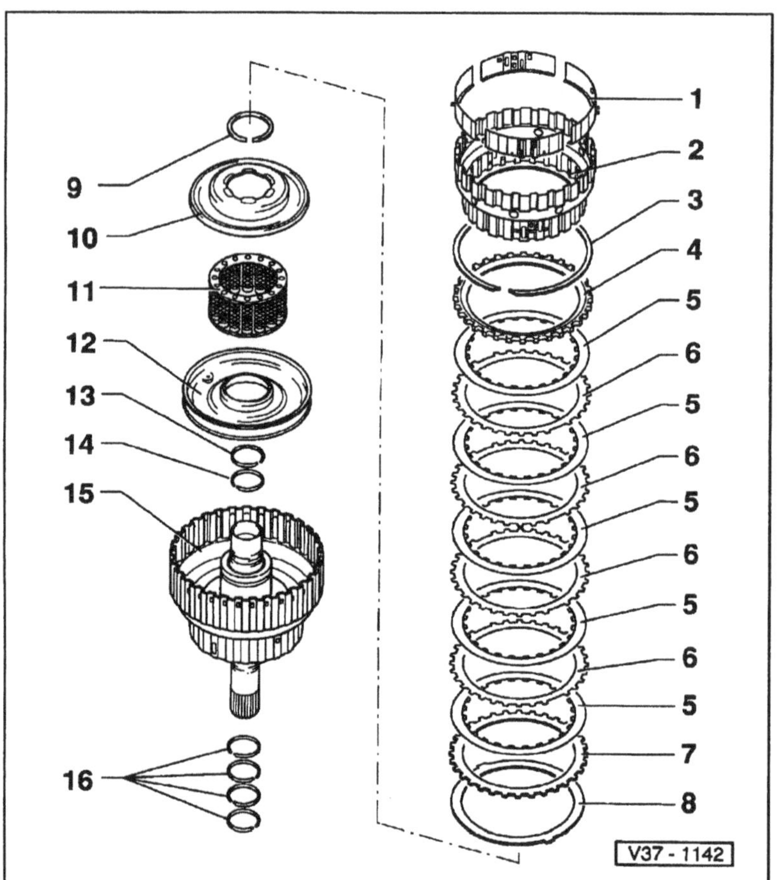

1st to 3rd gear clutch -K1- with turbine shaft on transmissions from 01.93, disassembling and assembling

Transmissions from 01.93,
⇒ page H 00-11

1 - Supporting ring
- ♦ 4-piece, unclip from inner plate carrier
- ♦ Installing ⇒ Fig. 2

2 - Inner plate carrier
- ♦ Installing ⇒ Fig. 4
- ♦ Assembling ⇒ Fig. 2

3 - Circlip
- ♦ Different thicknesses for different clutch assemblies
- ♦ Mark after removing and install in same position

4 - Thrust plate
- ♦ Smooth side faces inner plate
- ♦ Install with inner plate carrier ⇒ Fig. 2

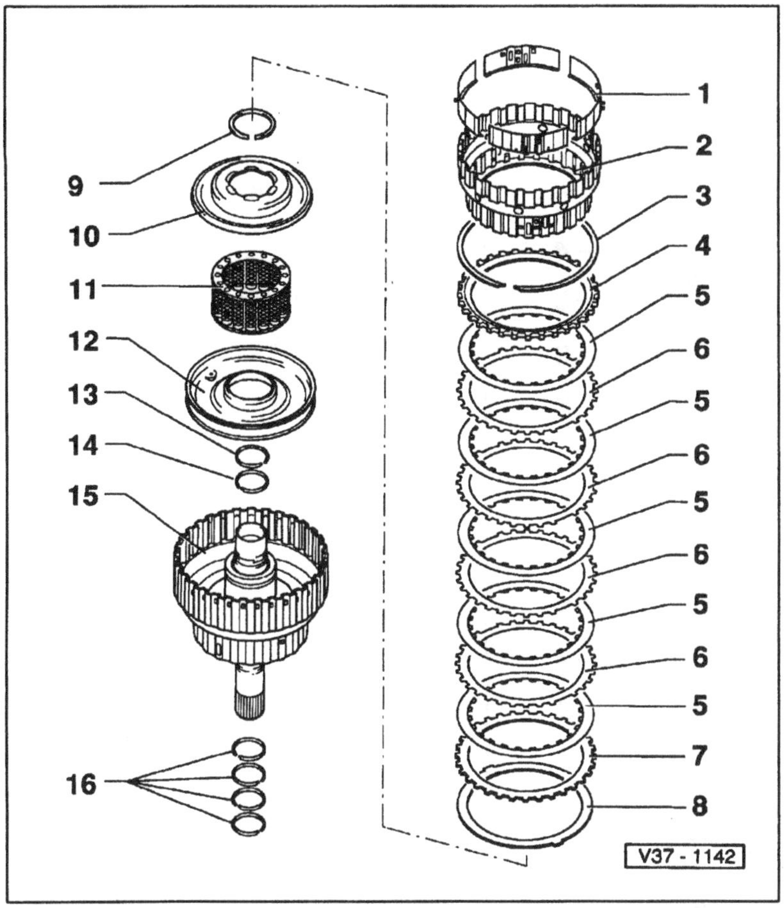

5 - Inner plate
- Quantity
  ⇒ page H 00-3
- Installing
  ⇒ Fig. 2

6 - Outer plate
- Quantity
  ⇒ page H 00-3
- Installing
  ⇒ Fig. 2
- Always 1.5 mm (0.0591 in.) thick

7 - Outer plate
- Always 2 mm (0.0787 in.) thick
- Installing
  ⇒ Fig. 2

8 - Corrugated washer

9 - Circlip
- Removing and installing
  ⇒ Fig. 1

10 - Piston cap
- Sealing lips are vulcanized to cap
- Moisten sealing lips with ATF before installing
- Turn cap slightly while installing

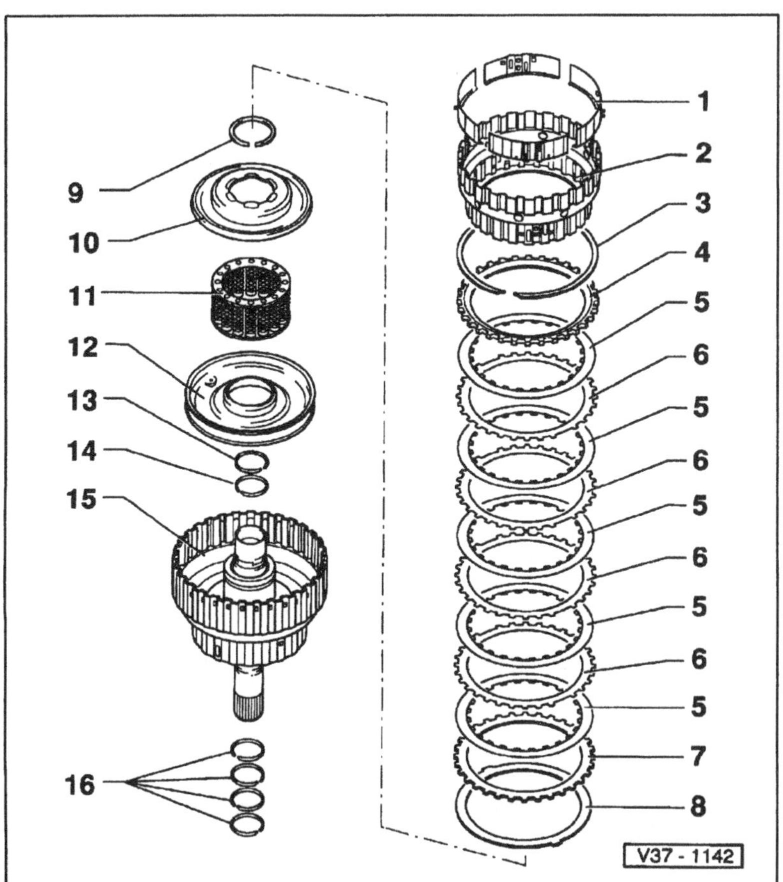

11 - Intermediate spring ring
- With spring supporting shell and compression springs

12 - Piston
- Sealing lips are vulcanized to piston
- Moisten sealing lips with ATF before installing
- Turn piston slightly when installing

13 - Piston ring (small diameter)
- Check that ring is correctly located
  ⇒ Fig. 5
- Installing
  ⇒ Fig. 6

14 - Piston ring
- Check that ring is correctly located
  ⇒ Fig. 5
- Installing
  ⇒ Fig. 6

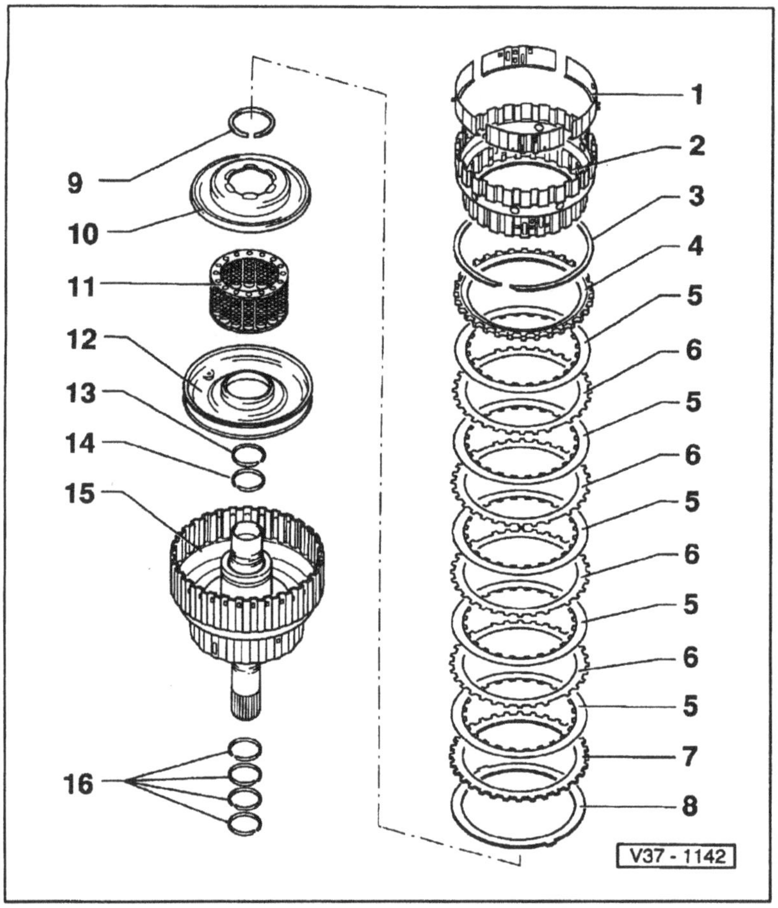

15 - **Clutch housing with turbine shaft**

- Check bearing bushings. Replace housing if bushings are damaged.
- Before installing the inner plate carrier corrugated washer -8-, install inner and outer plates in clutch
  ⇒ Fig. 3

16 - **Piston ring**

- Check that ring is correctly located
  ⇒ Fig. 5
- Installing
  ⇒ Fig. 6

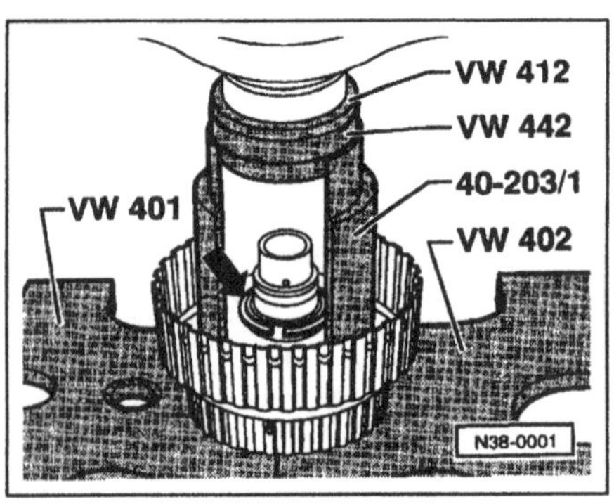

◄ **Fig. 1 Circlip. removing and installing**

Press piston cap with intermediate spring ring down far enough for circlip -arrow- to be installed in groove.

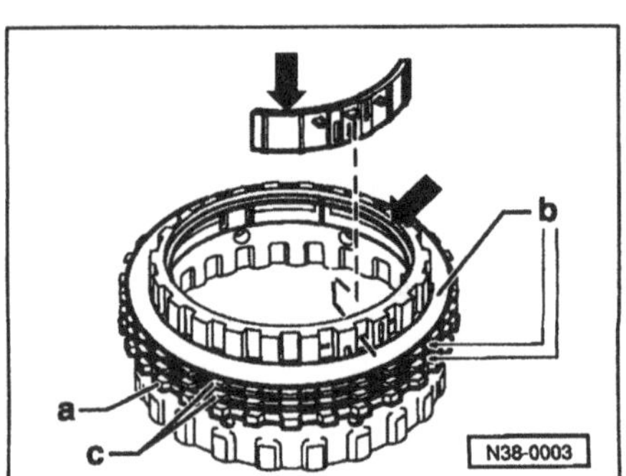

◄ **Fig. 2 Thrust plate and plates, installing onto inner plate carrier**

– Install thrust plate -a- with smooth side facing inner plate and stepped side towards inner plate carrier.

– Install 3 inner plates -b- and 2 outer plates -c-.

– Install supporting ring with step -arrow- after installing thrust plate as well as inner and outer plates.

Notes:

- *Locate lugs correctly.*
- *Installing corrugated washer and further plates in clutch housing ⇒ Fig. 3.*

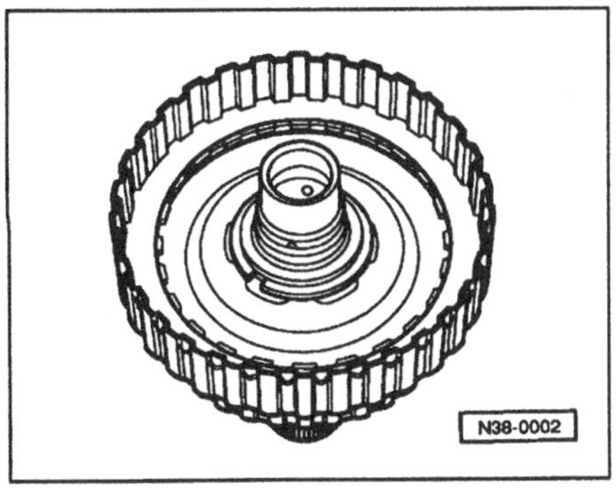

Fig. 3 Corrugated washer and inner and outer plates, installing in clutch housing

- Install corrugated washer.
- Install 2 mm (0.0787 in.) thick outer plate.
- Install remaining inner and outer plates.

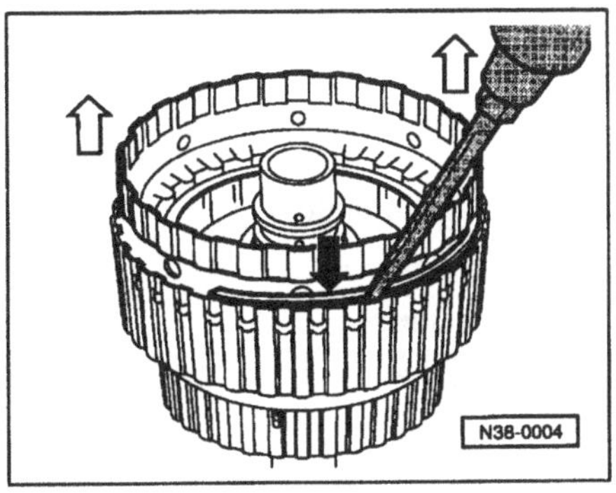

Fig. 4 Inner plate carrier, installing and installing circlip -arrow-.

Lift inner plate carrier slightly to install circlip.

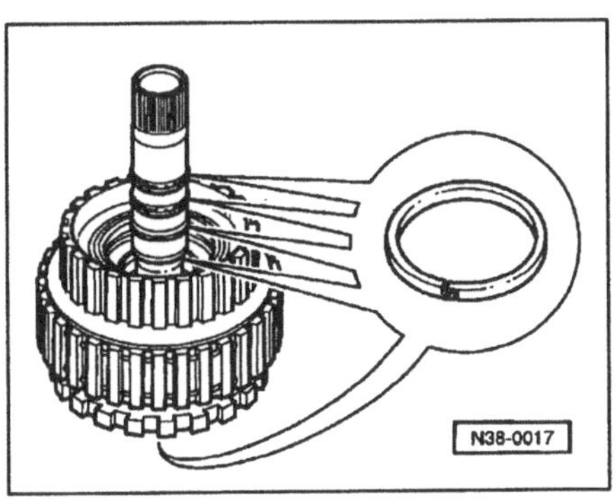

Fig. 5 Piston rings, checking for correct seating

There are two piston rings installed on the turbine shaft inside the clutch housing.

Ensure the ends of the piston rings are hooked together.

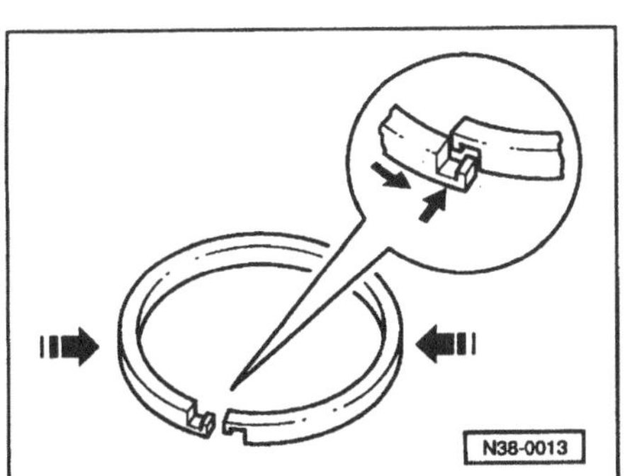

Fig. 6 Piston ring, installing and hooking ends together

- Place piston ring in groove.
- Hook ends together by pressing sides of ring in and guiding ends over one another.

Note:

Do not twist piston ring out of groove on one side.

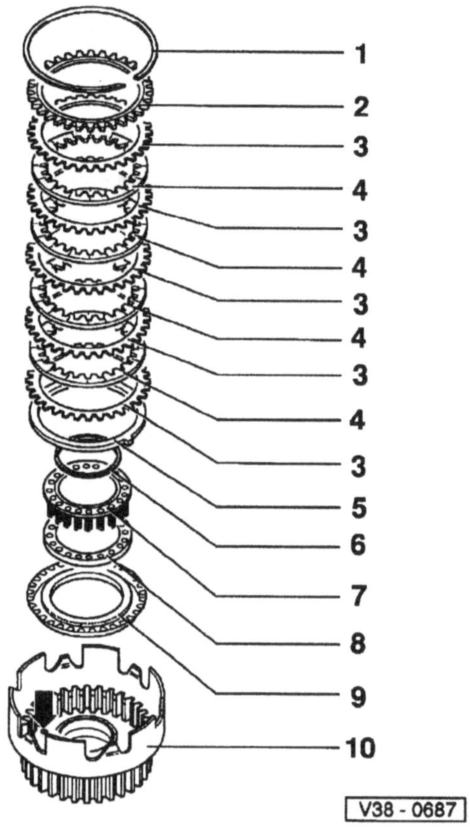

Reverse gear clutch -K2-, disassembling and assembling

Note:

When performing repairs to the clutch, make sure that the ball valve -arrow- is not damaged.

1 - Circlip

- Mark after removing and install in same position

2 - Thrust plate

- Smooth side facing inner plates
- Thickness with clutches of

4 inner plates	6.5 mm (0.256 in.)
5 inner plates	4.9 mm (0.193 in.)

3 - Outer plate

- Quantity for transmissions up to 12.92,
  ⇒ page H 00-3
- Quantity for transmissions from 01.93,
  ⇒ page H 00-11

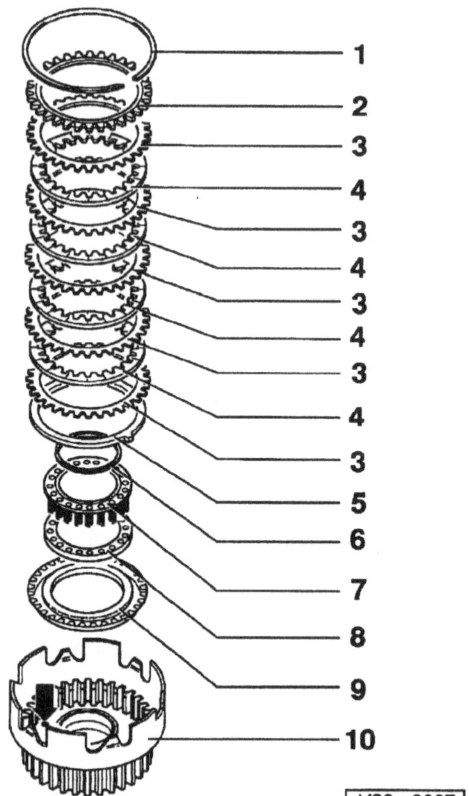

4 - Inner plate

- Quantity for transmissions up to 12.92,
  ⇒ page H 00-3
- Quantity for transmissions from 01.93,
  ⇒ page H 00-11

5 - Corrugated washer

6 - Circlip

- Removing and installing
  ⇒ Fig. 1
- Installation position
  ⇒ Fig. 2

7 - Spring supporting plate

- With springs
- Installation position
  ⇒ Fig. 2

8 - Spring supporting ring

9 - Piston

- Sealing lips are vulcanized to piston
- Moisten sealing lips with ATF before installing
- Turn piston slightly while installing

10 - Clutch housing with clutch tube

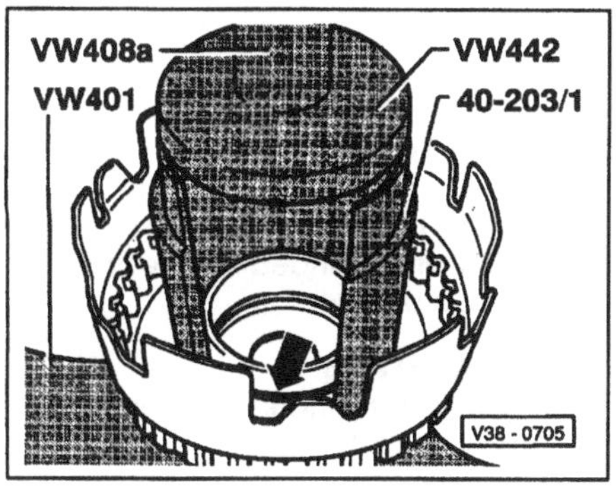

◄ **Fig. 1 Circlip -arrow-, removing and installing**

Removing

– Carefully press down spring supporting plate (do not bend) and spread circlip.

Installing

– Install new circlip and carefully press down with spring supporting plate.

Note:

The plate must be centered relative to the hub or the spring supporting plate will be bent.

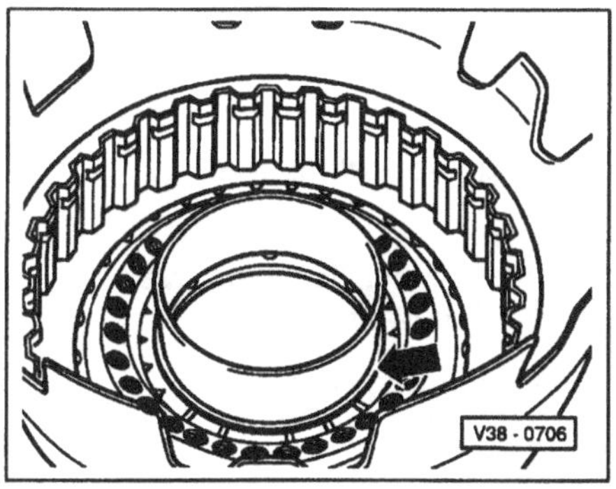

◄ **Fig. 2 Spring supporting plate, installation position / checking circlip**

Spring supporting plate must be centered relative to hub.

Circlip -arrow- must be seated in groove of clutch housing.

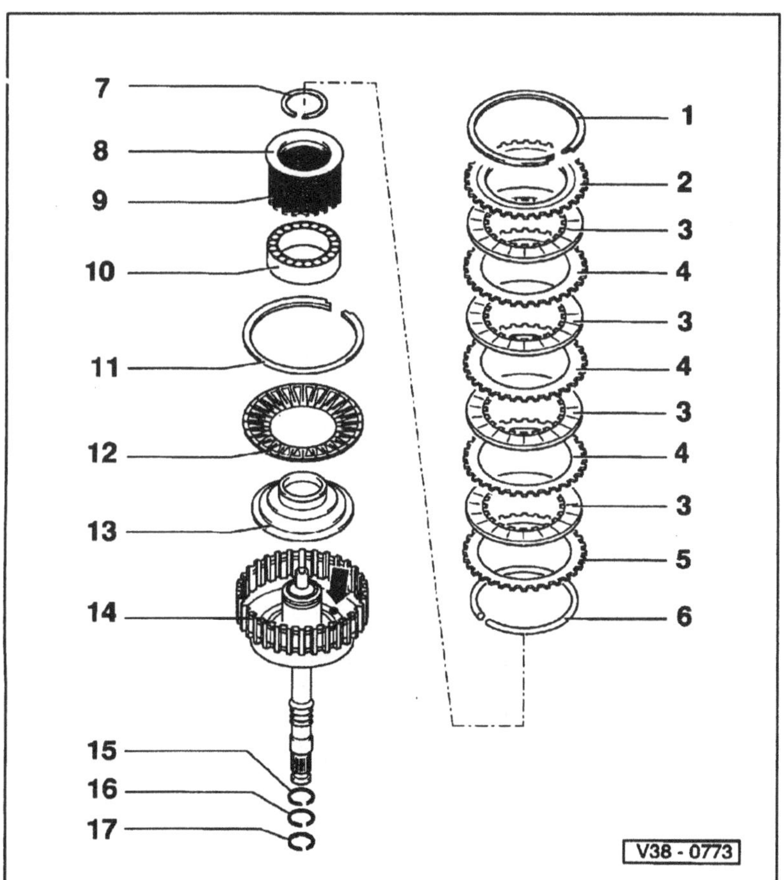

3rd and 4th gear clutch -K3- with pump shaft on transmissions up to 12.92, disassembling and assembling

Transmissions up to 12.92,
⇒ page H 00-3

Note:

When performing repairs to the clutch, make sure that the ball valve -arrow- is not damaged.

1 - **Circlip**
- ♦ Different thickness for different clutch assemblies
- ♦ Mark after removing and install in same position

2 - **Thrust plate**
- ♦ Install with stepped side facing circlip

3 - **Inner plate**
- ♦ Quantity
⇒ page H 00-3

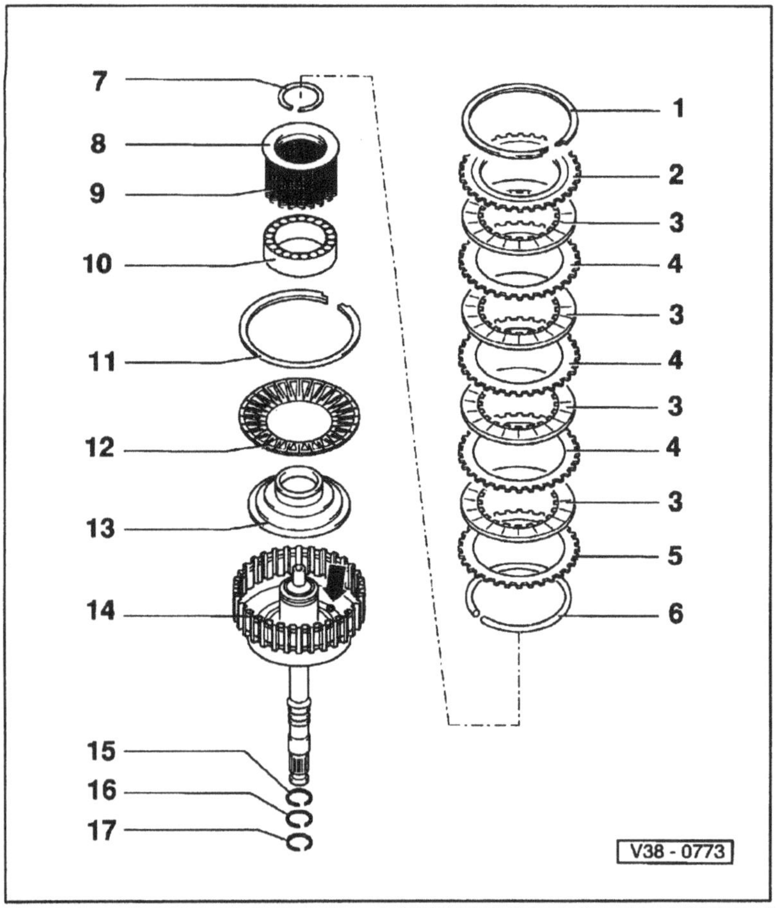

4 - Outer plate
- Quantity ⇒ page H 00-3

5 - Pressure plate
- Install with rounded side facing corrugated ring

6 - Corrugated ring
- Install with rounded side facing pressure plate

7 - Circlip
- Removing and installing ⇒ Fig. 1

8 - Spring supporting ring

9 - Springs
- Place onto spring supporting ring

10 - Spring supporting shell

11 - Circlip
- Must be installed in groove of clutch housing

12 - Diaphragm spring
- Install with curved side facing piston

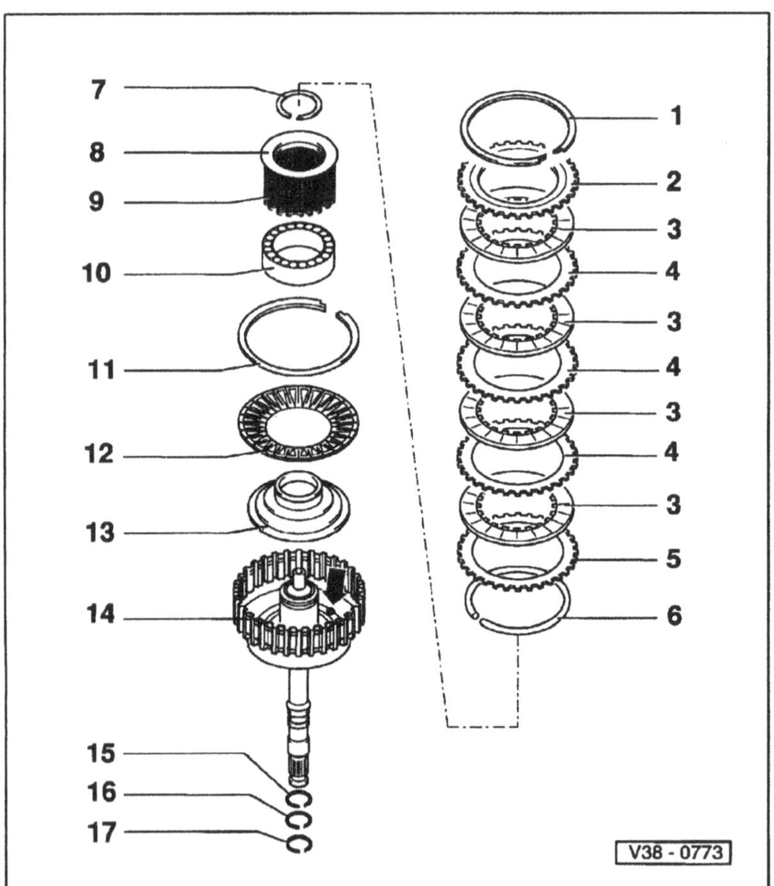

13 - Piston
- Sealing lips are vulcanized to piston
- Moisten sealing lips with ATF before installing
- Turn piston slightly while installing

14 - Clutch housing with pump shaft

15 - Piston ring
- Replace only faulty piston rings
- Piston ring is destroyed when removed
- Check for correct location ⇒ Fig. 2
- Installing ⇒ Fig. 3

16 - Piston ring
- Check for correct location ⇒ Fig. 2
- Installing ⇒ Fig. 3

17 - Piston ring
- Check for correct location ⇒ Fig. 2
- Installing ⇒ Fig. 3

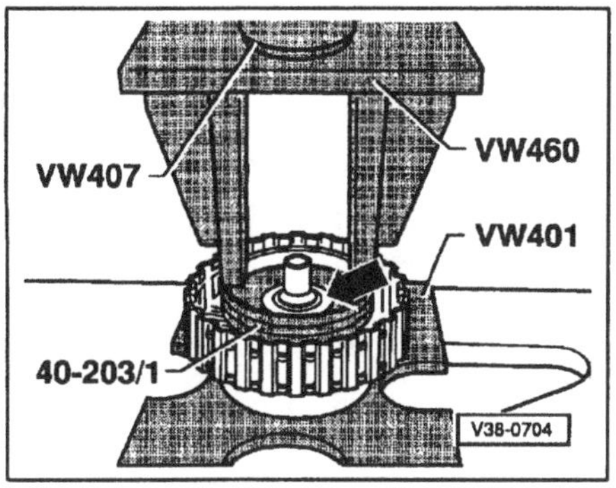

Fig. 1 **Circlip -arrow-, removing and installing**

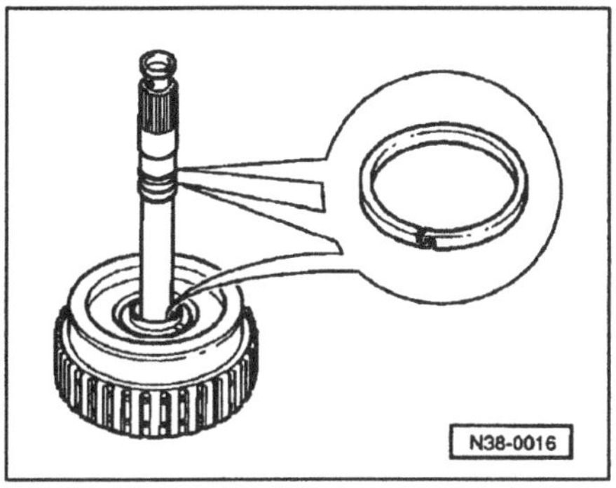

Fig. 2 **Piston rings, checking for correct seating**

Ensure that ends of piston rings are hooked together.

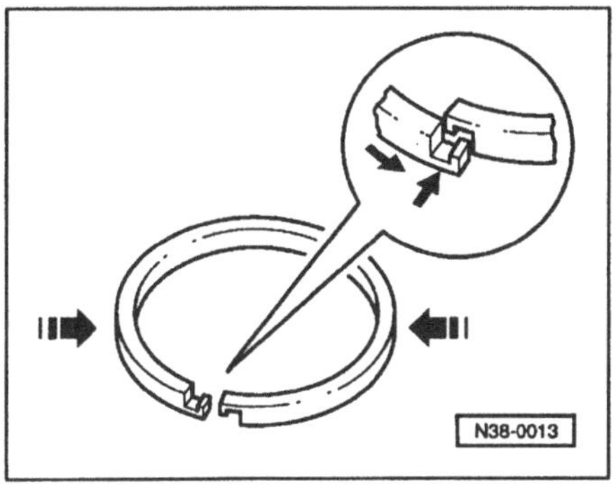

Fig. 3 **Piston ring, installing and hooking ends together**

– Place piston ring in groove.

– Hook ends together by pressing sides of ring in and guiding ends over one another.

Note:

Do not twist piston ring out of groove on one side.

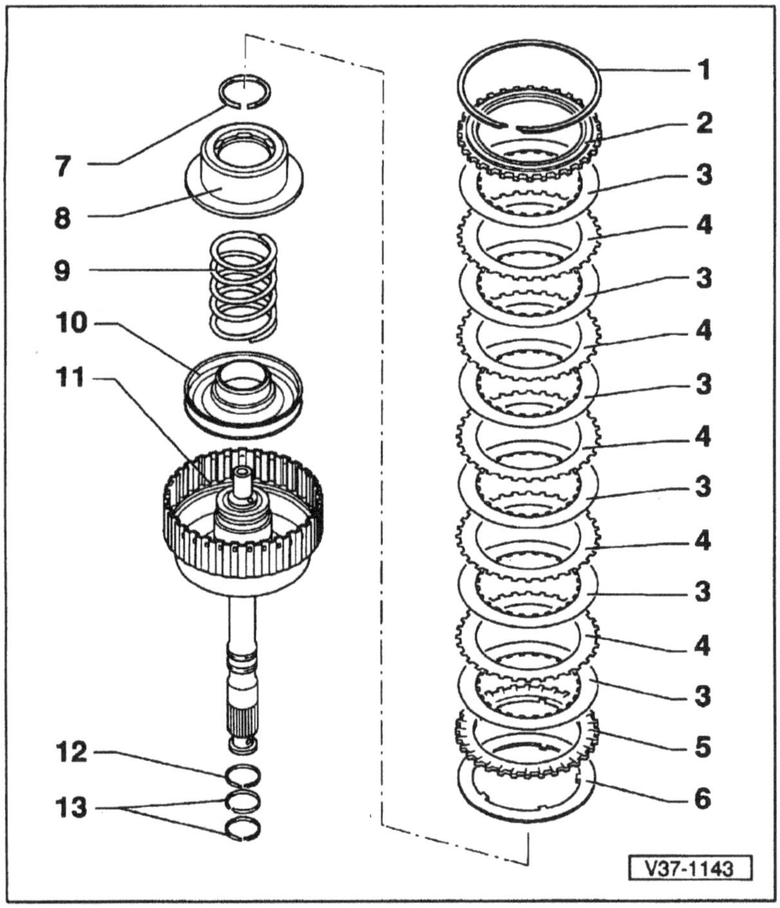

4th gear clutch -K3- with pump shaft on transmissions from 01.93, disassembling and assembling

Transmissions from 01.93,
⇒ page H 00-11

1 - Circlip
- ♦ Different thickness for different clutch assemblies
- ♦ Mark after removing and insert at same position

2 - Thrust plate
- ♦ Install with stepped side facing circlip

3 - Inner plate
- ♦ Quantity
  ⇒ page H 00-11

4 - Outer plate
- ♦ Quantity
  ⇒ page H 00-11

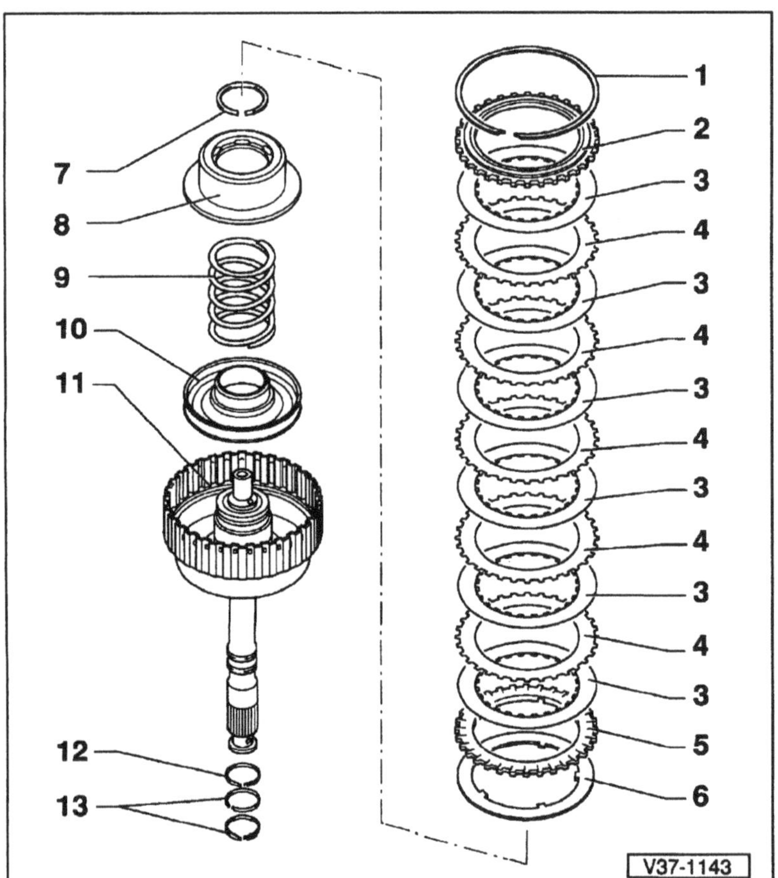

5 - Pressure plate
- ♦ Install with peened corrugated washer facing piston

6 - Corrugated washer
- ♦ Peened to thrust washer

7 - Circlip
- ♦ Removing and installing
  ⇒ Fig. 1

8 - Piston cap
- ♦ Sealing lips are vulcanized to piston cap
- ♦ Moisten with ATF before installing
- ♦ Turn piston cap slightly when installing
- ♦ When removing and installing, press down and release circlip -7-.

9 - Spring
- ♦ Seat between piston and piston cap

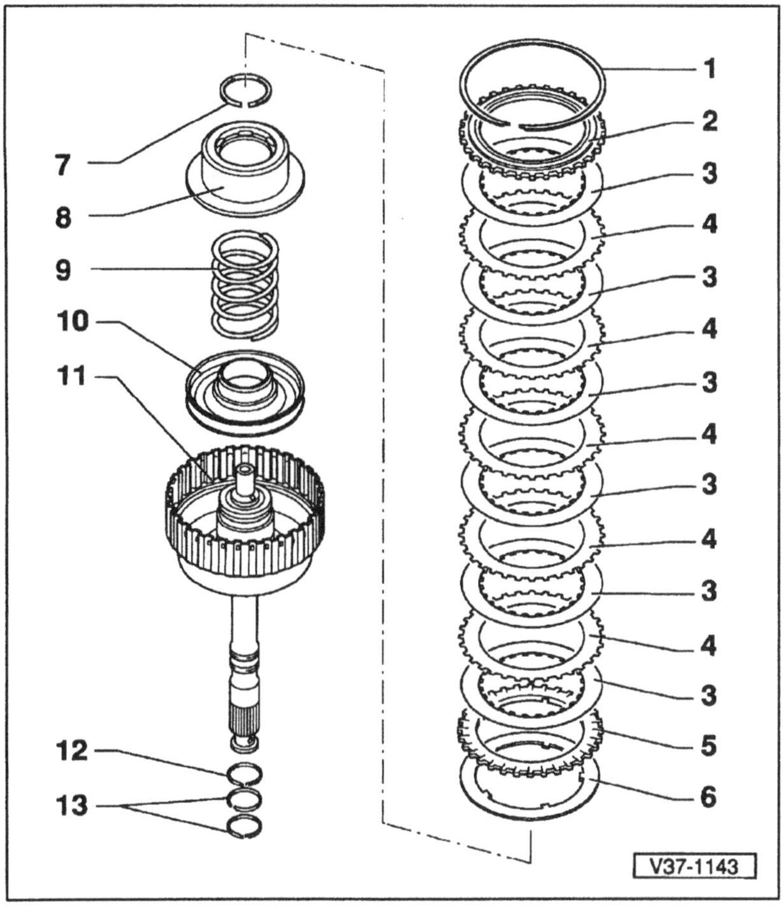

10 - Piston
- Sealing lips are vulcanized to piston
- Moisten with ATF before installing
- Turn piston slightly while installing

11 - Clutch housing with pump shaft
- Height varies according to number of inner and outer plates

12 - Piston ring
- Replace only faulty piston rings
- Piston ring is destroyed when removed
- Check that ring is correctly positioned ⇒ Fig. 2
- Installing ⇒ Fig. 3

13 - Piston ring
- Check that ring is correctly positioned ⇒ Fig. 2
- Installing ⇒ Fig. 3

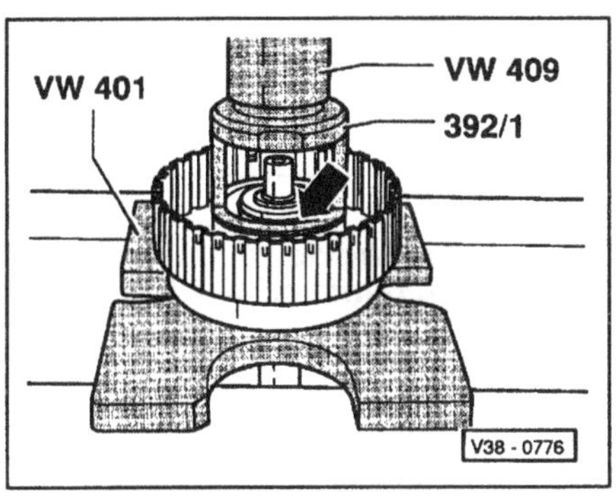

◄ **Fig. 1 Circlip -arrow-, removing and installing**

Press down carefully on piston cap until the circlip can be installed.

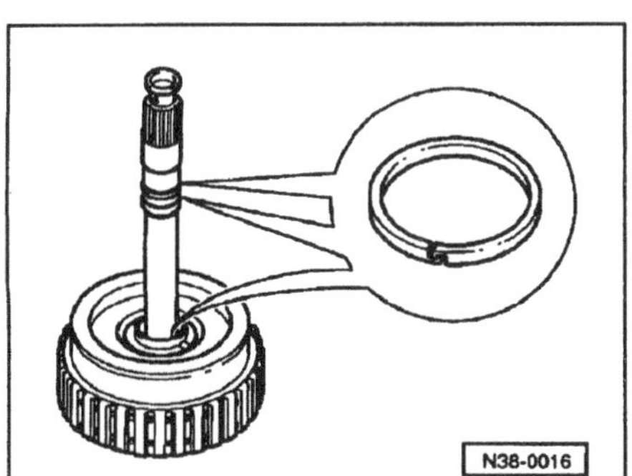

◄ **Fig. 2 Piston rings, checking for correct seating**

There is a piston ring installed on the turbine shaft inside the clutch housing -arrow-.

Ensure that ends of piston rings are hooked together.

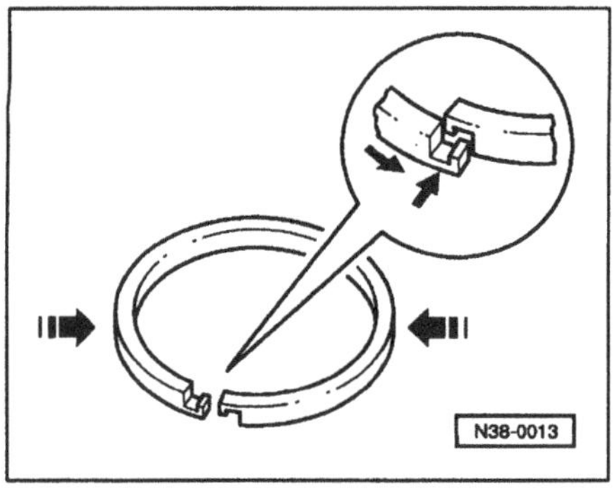

◄ **Fig. 3 Piston ring, installing and hooking ends together**

– Place piston ring in groove.

– Hook ends together by pressing sides of ring in and guiding ends over one another.

Note:

Do not twist piston ring out of groove on one side.

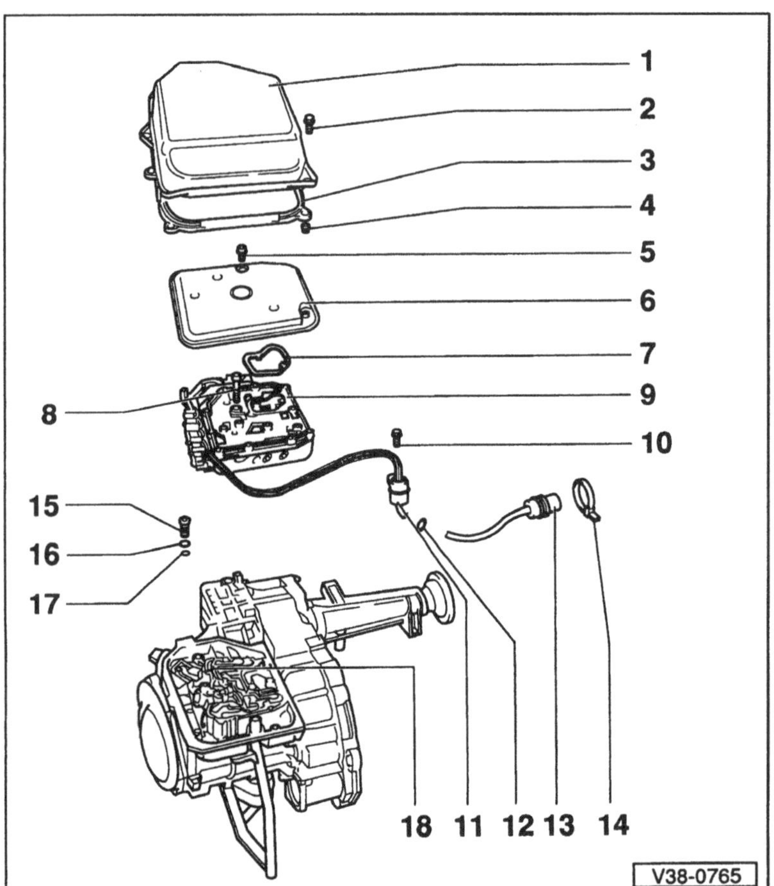

Valve body, removing and installing

CAUTION!

Do not run engine or tow vehicle with oil pan removed or when there is no ATF in transmission.

Notes:

- *The valve body can also be removed with the transmission installed in the vehicle.*
- *After re-installing the oil pan, fill with ATF* *⇒ page H 37-46.*
- *Always replace a valve body that is fouled or faulty.*
- *Removing and installing conductor strip with Transmission Fluid Temperature Sensor -G93-,* *⇒ page 38–43.*

1 - Oil pan

2 - Bolt
- Tightening torque: 12 Nm (9 ft lb)

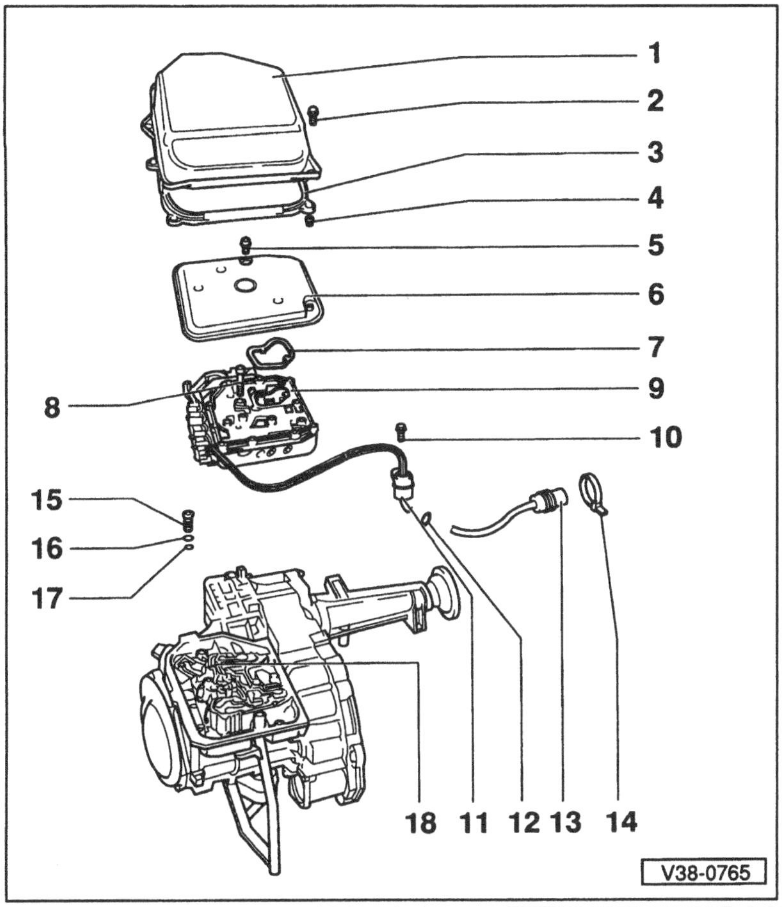

3 - Gasket
- Always replace
- Press spacer bushings into gasket

4 - Spacer bushing
- Press into gasket

5 - Bolt
- Tightening torque: 8 Nm (71 in lb)

6 - ATF screen
- Press into gasket
- Installing ⇒ Fig. 5

7 - Gasket
- Always replace
- Press into ATF screen

8 - Bolt
- Tightening torque: 5 Nm (44 in lb)

9 - Valve body
- Marking ⇒ Fig. 6
- Installing with wiring harness ⇒ Fig. 2
- Installing with conductor strip ⇒ Fig. 3

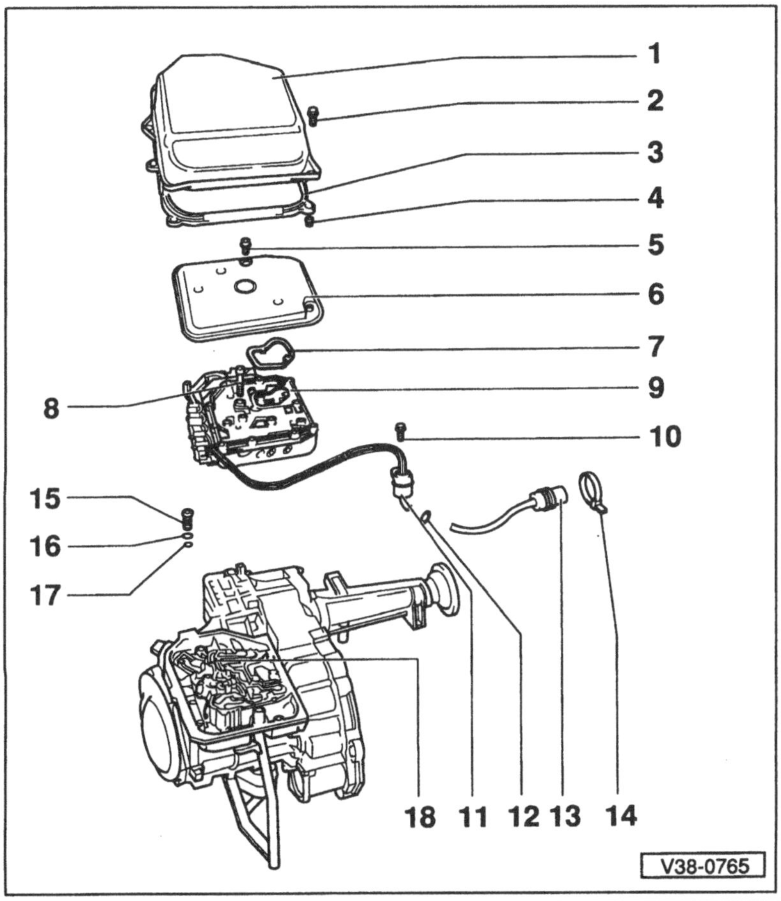

10 - Bolt
- Tightening torque: 10 Nm (7 ft lb)

11 - Cable gland
- Install with O-ring

12 - O-ring
- Always replace

13 - Cable connector
- Attach to starter

14 - Clip
- For securing cable connector

15 - Sealing plug
- Take out before removing and installing free wheel
- Install so that lug is located in groove on housing ⇒ Fig. 1
- Place O-rings onto sealing plug

16 - O-ring
- Always replace

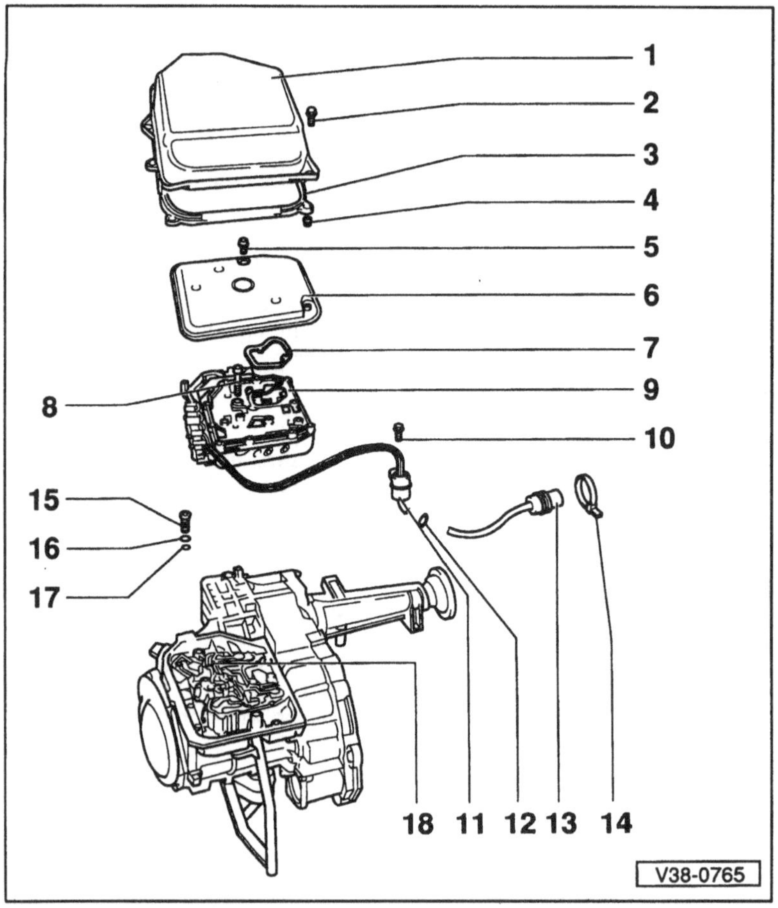

17 - **O-ring**
- Always replace

18 - **Operating rod for manual selector valve**
- Adjusting ⇒ Fig. 4
- Install with stepped side facing manual selector valve

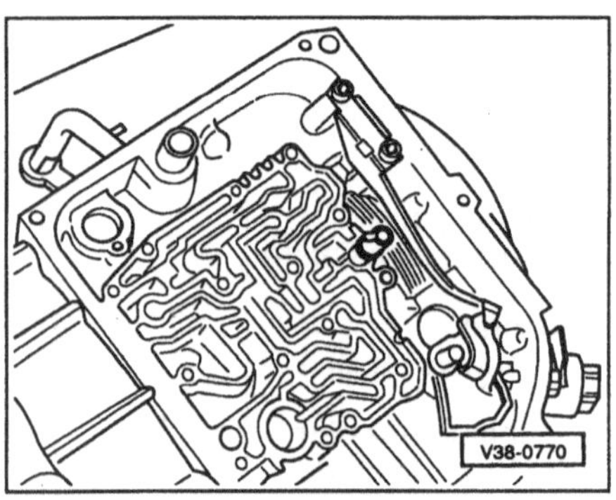

◄ **Fig. 1 Sealing plug, installing**

Remove sealing plug from housing before removing or installing free wheel otherwise sealing plug and O-ring will be damaged.

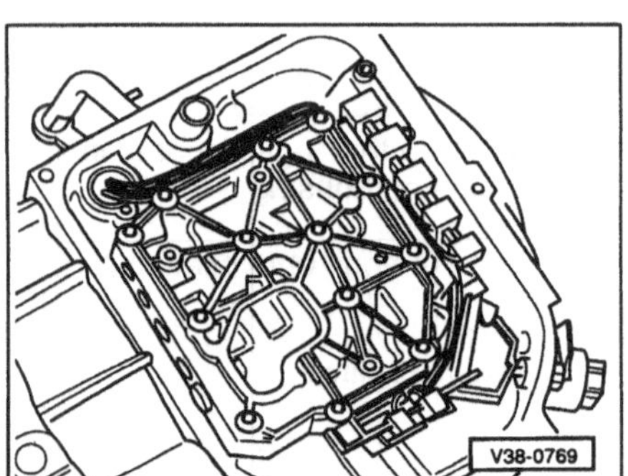

◄ **Fig. 2 Valve body with wiring harness, installing**

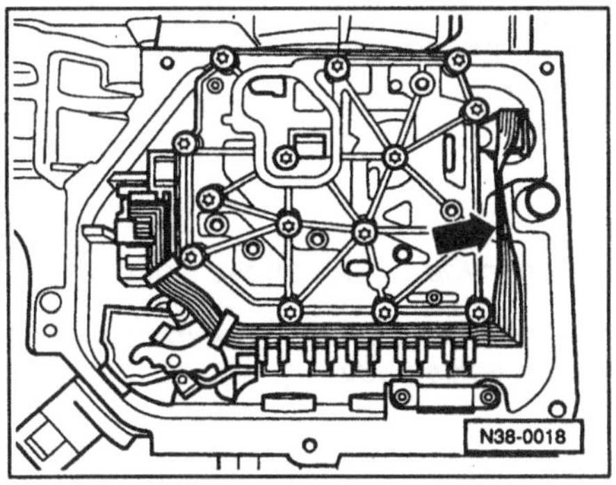

◄ **Fig. 3 Valve body with conductor strip, installing**

Route conductor strip as illustrated. When installing, do not kink or twist.

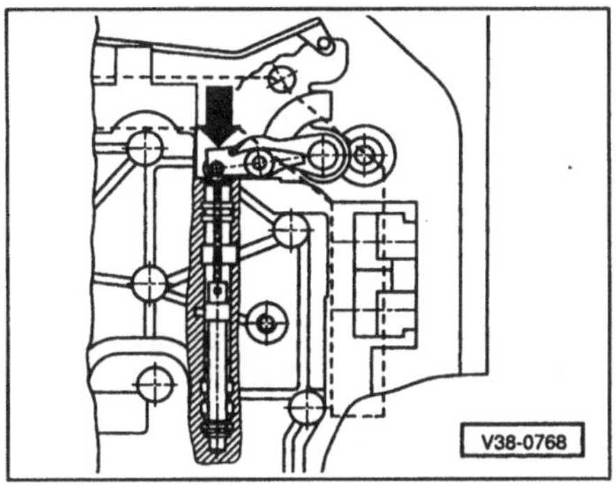

◄ **Fig. 4 Operating rod for manual selector valve, adjusting**

– Move selector shaft to position "P".

– Push operating rod with manual selector valve into valve body as far as stop (direction of arrow) and tighten bolt to 3.5 Nm (31 in lb).

Note:

- *Manual selector valve must always be resting against the stop. Counter-hold in direction of arrow while tightening bolt.*
- *Always replace manual selector valve bolt.*

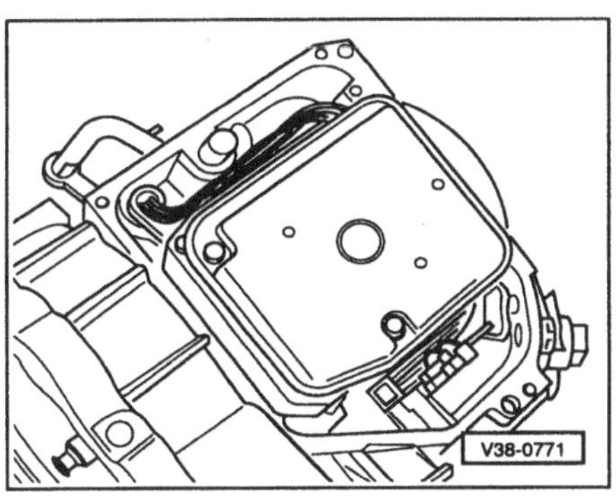

◄ **Fig. 5 ATF screen, installing**

Press gasket into ATF screen before installing screen.

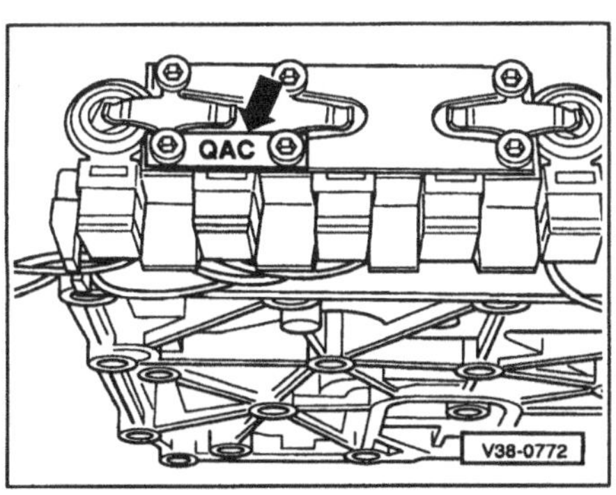

◄ **Fig. 6 Valve body, identification**

Code letters are stamped on a metal tab. Metal tab must remain on the valve body.

Valve body/transmission application
⇒ from page H 00-3

Note:

If the valve body does not have an identification tab, it can remain installed in the transmission. When replacing the valve body, make sure of correct application
⇒ from page H 00-3.

Conductor strip with Transmission Fluid Temperature Sensor -G93-, removing and installing

Note:

- *The conductor strip is located in the oil pan on the valve body.*
- *The conductor strip can be changed with the transmission installed without removing the valve body.*

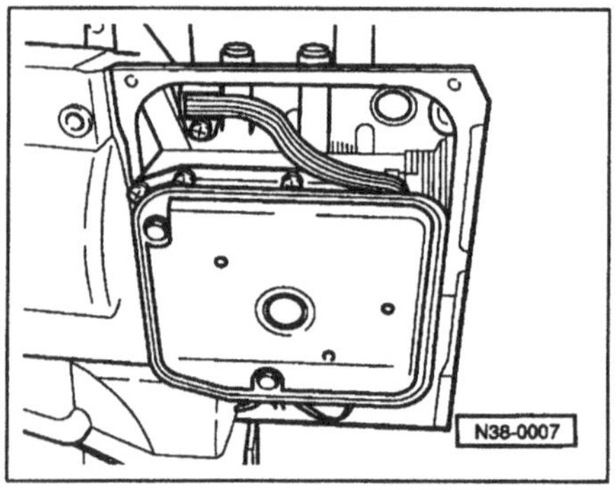

Removing

– Drain Automatic Transmission Fluid (ATF).

– Raise vehicle and remove protective plate beneath ATF pan.

– Disconnect conductor strip harness connector.

◄ – Remove ATF pan and ATF screen.

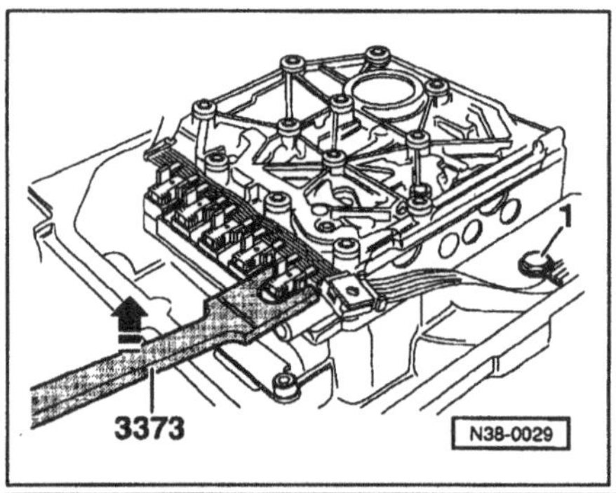

◄ – Place special tool 3373 into position on stop under solenoid valve connector, and pry conductor strip off in direction shown (arrow).

– Remove bolt -1- and remove conductor strip.

Note:

If the locating points are damaged, the valve body must be replaced.

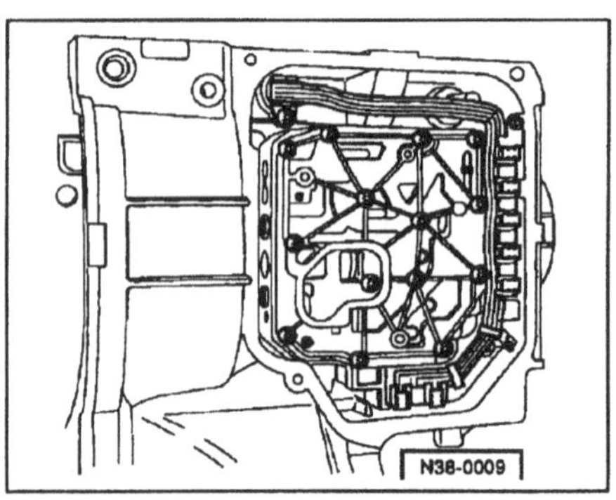

Installing

◄ – Route conductor strip as shown; do not kink or twist.

– Press gasket into ATF screen before installing screen.

– Replace ATF pan gasket; clip bushing into gasket.

Tightening torques

Conductor strip to transmission housing	10 Nm (7 ft lb)
ATF screen to valve body	8 Nm (71 in lb)
ATF pan to transmission housing	12 Nm (9 ft lb)

– Add 3 liters (3.2 US qt) ATF, check ATF level and top up as necessary
⇒ page H 37-47

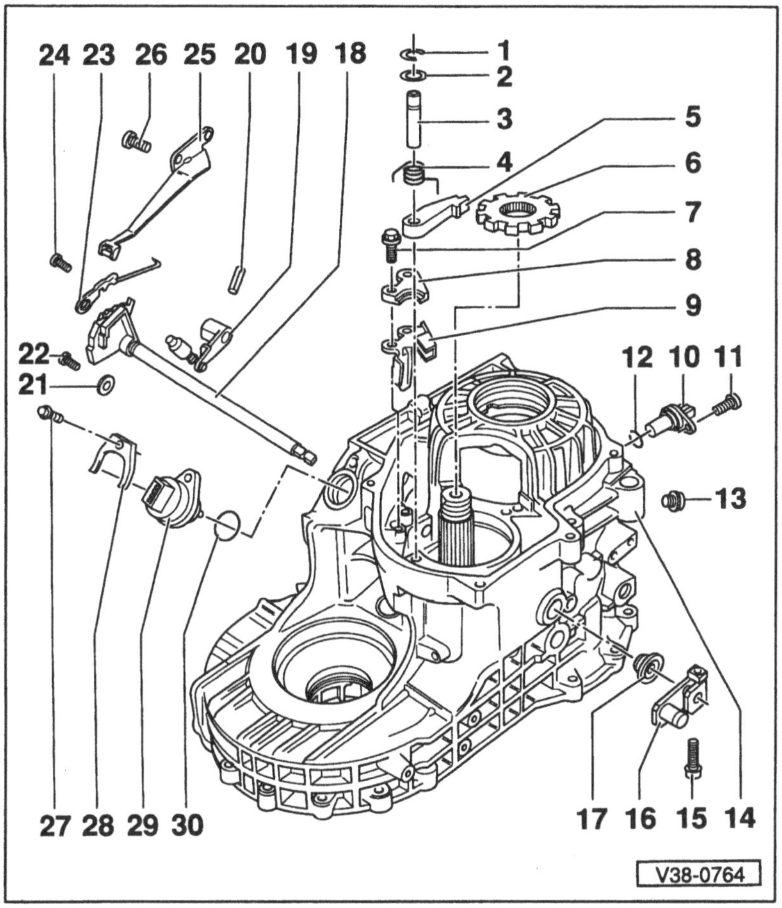

Parking lock, disassembling and assembling

Note:

- *Multi-Function Transmission Range (TR) Switch -F125- and Vehicle Speed Sensor (VSS) -G68- can be tested in the vehicle by the On Board Diagnostic (OBD) program* *⇒ page H 01-1, On Board Diagnostic (OBD).*
- *The output gear must be removed to disassemble and assemble the parking lock* *⇒ Removing and installing drive pinion, page H 39-29.*

1 - Circlip
- Always replace

2 - Washer

3 - Shaft for detent lever
- Is secured by circlip and bearing cap of drive pinion
- According to version with groove for circlip

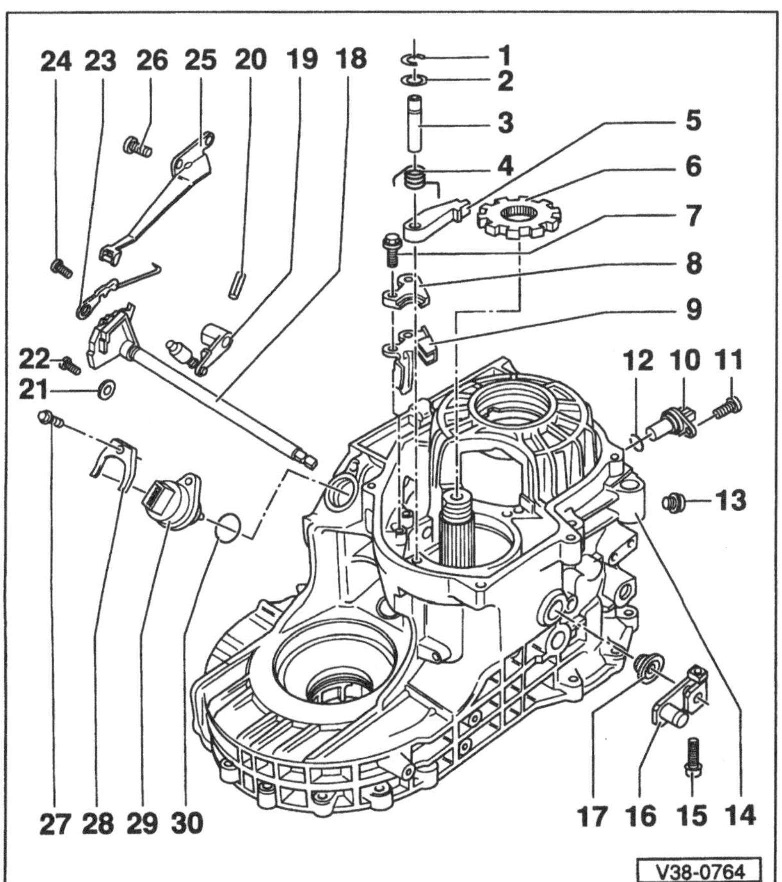

4 - Return spring
- Installing
  ⇒ Fig. 4

5 - Detent lever
- Installing with return spring
  ⇒ Fig. 4

6 - Parking lock gear
- Rounded side faces teeth of drive pinion

7 - Bolt
- Tightening torque: 14 Nm (10 ft lb)

8 - Supporting plate
- Installing
  ⇒ Fig. 3

9 - Guide plate
- Install before supporting plate
  ⇒ Fig. 3

10 - Vehicle Speed Sensor (VSS) -G68-
- Can be tested in the vehicle by On Board Diagnostic (OBD)
  ⇒ page H 01-1

11 - Bolt
- Tightening torque: 10 Nm (7 ft lb)

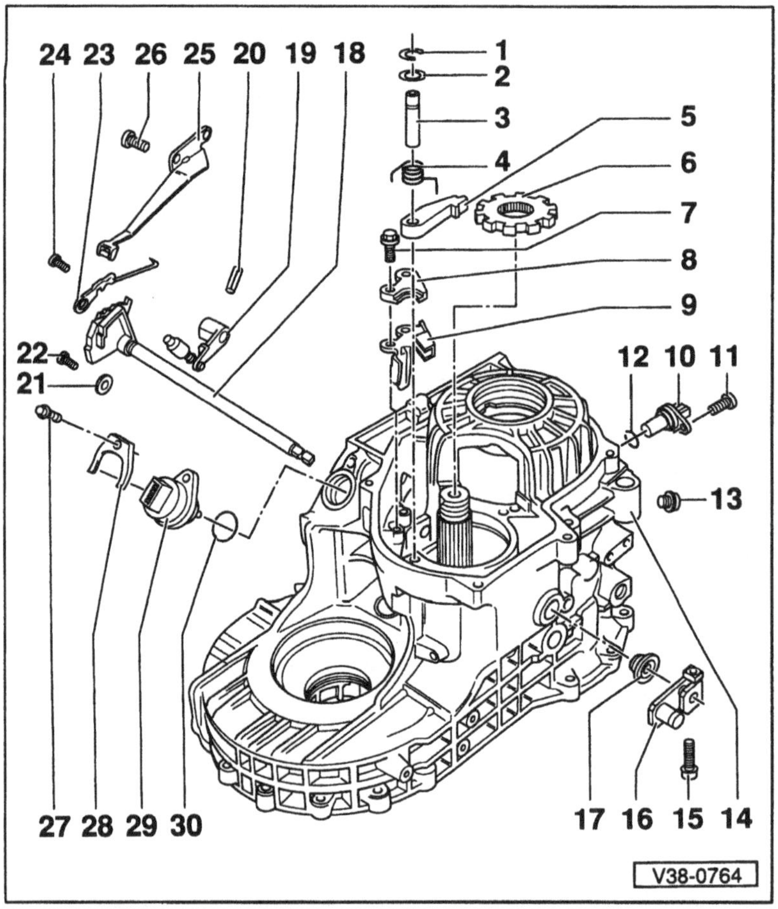

12 - O-ring
- Always replace

13 - Bolt
- Tightening torque: 15 Nm (11 ft lb)
- With sealing ring
- Always replace

14 - Transmission housing

15 - Bolt
- Tightening torque: 10 Nm (7 ft lb)

16 - Lever
- For selector shaft

17 - Seal
- Pry out with screwdriver
- Drive in flush with VW 423 sleeve

18 - Selector shaft with selector segment and detent segment
- Install with engaging lever
- Removing and installing ⇒ Fig. 2

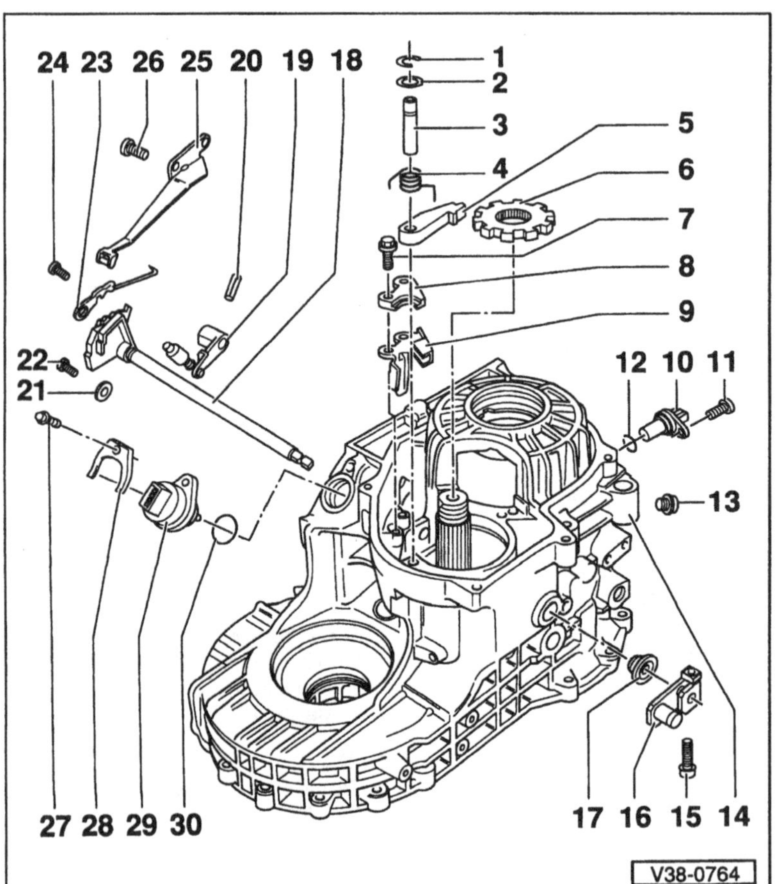

19 - Engaging lever
- Install together with selector shaft and detent segment

20 - Spring pin
- Drive in after installing engaging lever ⇒ Fig. 1
- Drive out with drift

21 - Locking washer
- Removing and installing ⇒ Fig. 2

22 - Bolt
- Tightening torque: 10 Nm (7 ft lb)

23 - Manual selector valve control
- Adjusting ⇒ page 38–41,

24 - Bolt
- Tightening torque: 10 Nm (7 ft lb)

25 - Spring for selector segment

26 - Bolt
- Tightening torque: 10 Nm (7 ft lb)

27 - Bolt
- Tightening torque: 10 Nm (7 ft lb)

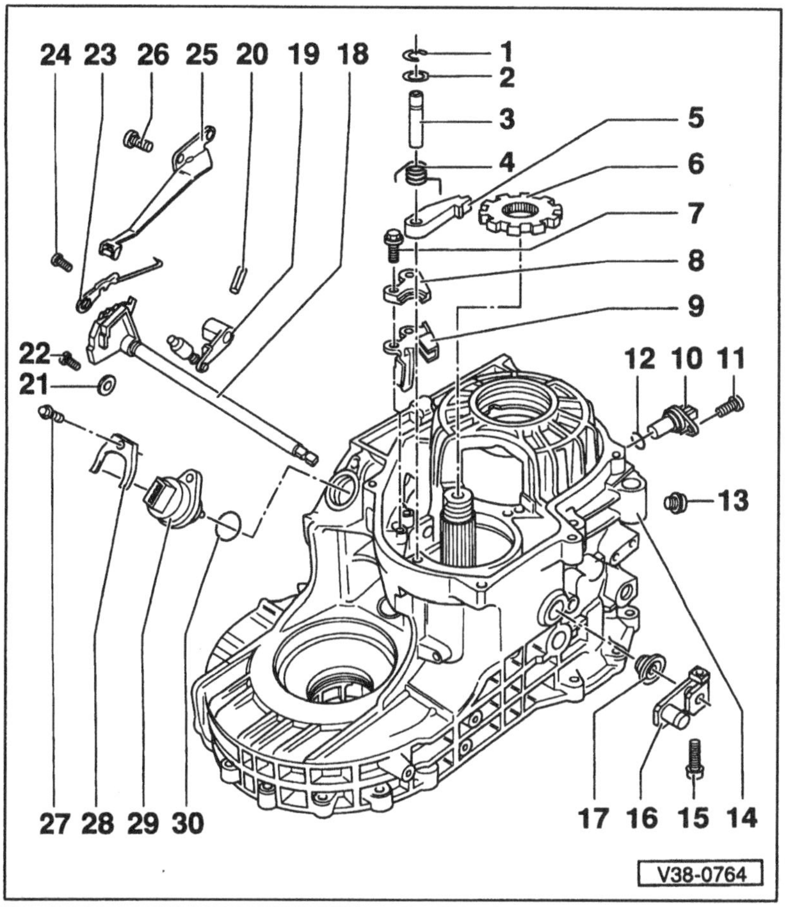

28 - Bracket
- ♦ For Multi-Function Transmission Range (TR) Switch -F125-

29 - Multi-Function Transmission Range (TR) Switch -F125-
- ♦ To remove
- – Extract ATF
- – Remove insulation tray
- – Remove connector and place drain pan under transmission
- – While installing, make sure connector is correctly sealed
- – After installing top off ATF ⇒ page H 37-47
- ♦ Install so that 1 pin each is resting on a slope of the detent segment
- ♦ Can be tested in the vehicle by On Board Diagnostic (OBD) ⇒ page H 01-1

30 - O-ring
- ♦ Always replace

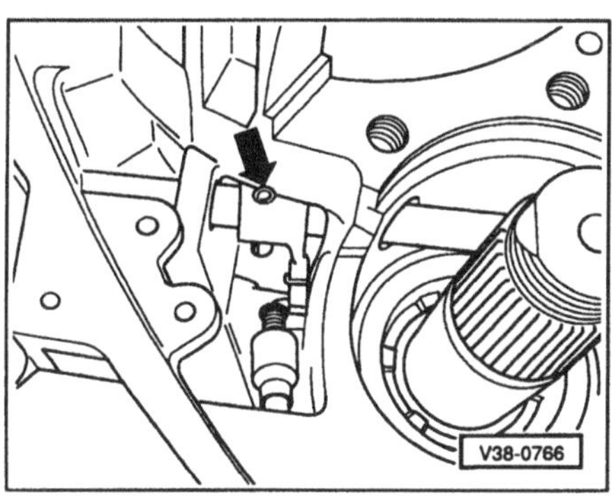

◄ **Fig. 1 Spring pin -arrow-, driving out and in**

Driving out

– Drive spring pin for engaging lever out only so far as to remove the selector shaft.

Note:

Spring pin of engaging lever may drop into planetary transmission.

Driving in

– Drive in spring pin flush after installing selector shaft and engaging lever.

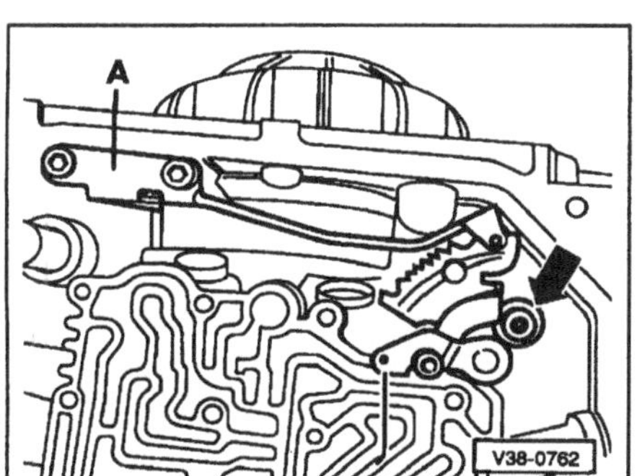

◄ **Fig. 2 Removing and installing selector shaft with detent and selector segments**

– Remove bolt -arrow- and take out locking washer.

Notes:

- ♦ *The locking washer is placed into the groove of the selector shaft and secures the shaft from dropping out.*
- ♦ *The selector shaft can be removed or installed after taking out the spring -A- for the selector segment.*

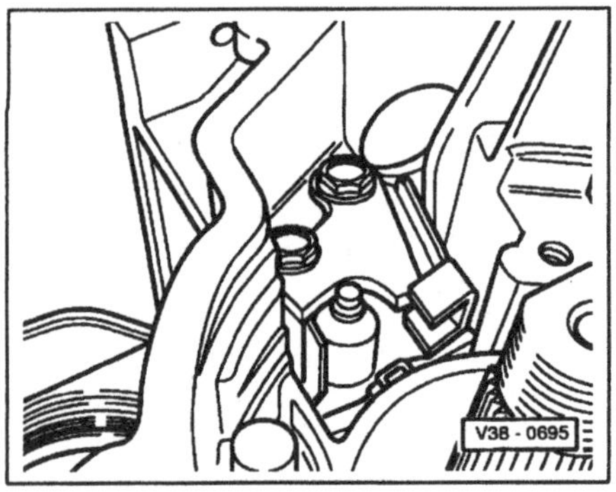

◄ **Fig. 3** **Installing guide plate and supporting plate**

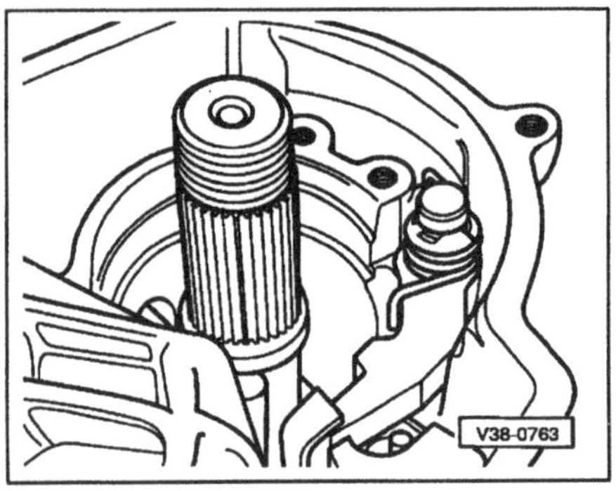

◄ **Fig. 4** **Installing pawl with return spring**

Oil in final drive, checking level

Check oil level with the transmission installed.

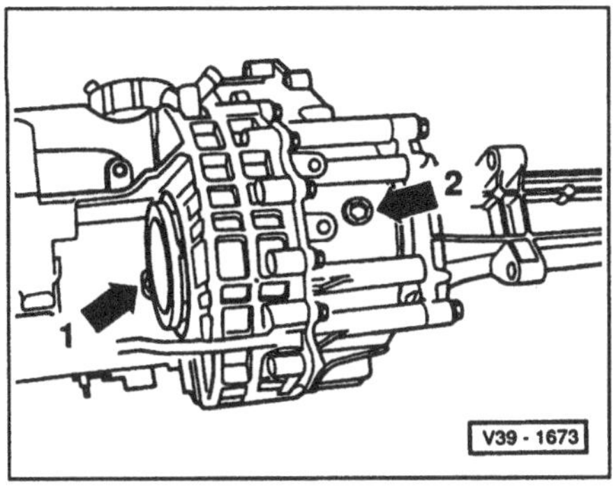

Checking

– Remove plug to check gear oil level and top gear oil up to lower edge of drilling.

Specification
⇒ page H 00-15

Note:

◄ *The plug next to the flange shaft -arrow 1- or the plug in the torque converter housing -arrow 2- can be removed to check the gear oil level.*

Vehicles from 08.92 with modified assembly carrier:

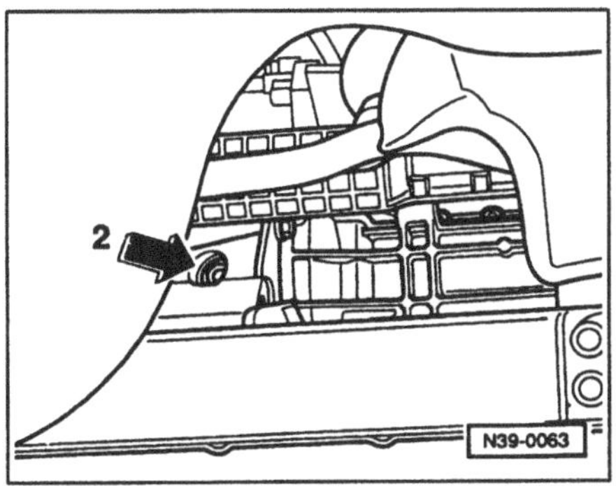

◄ – Remove plug -arrow 2- through hole in assembly carrier.

– Check gear oil level and top gear oil up to lower edge of drilling.

– Install plug.

- Tightening torque: 20 Nm (15 ft lb)

Oil seal for flange shaft or drive flange, removing and installing

(Transmission installed)

– Install support frame 3250 before lifting vehicle.

⇒ Suspension, Wheels, Brakes, Steering; Repair Group 40

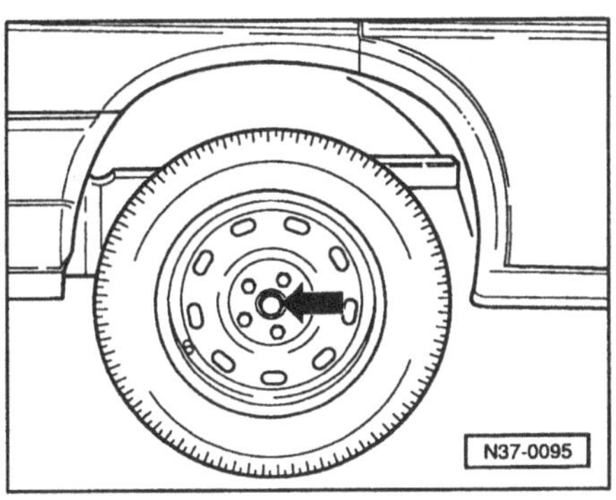

◄ – Remove bolt -arrow- for drive shaft/wheel hub.

– Remove noise insulation tray.

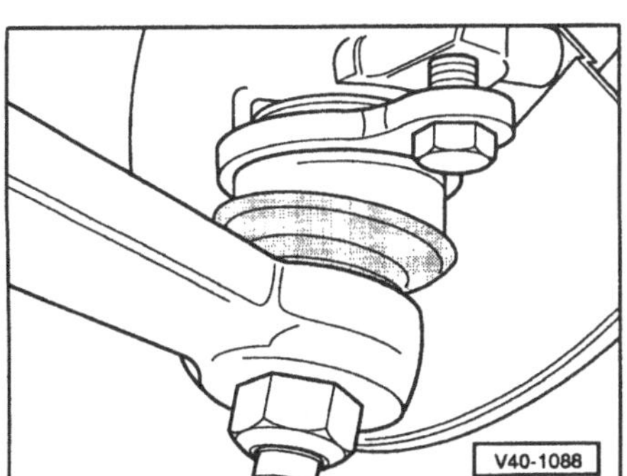

◄ – Disconnect wheel bearing housing/lower ball joint bolted connection.

– Remove bolt for left-hand shock absorber/stabilizer bar coupling rod.

– Press drive shaft out of wheel bearing housing.

– Compress shock absorber.

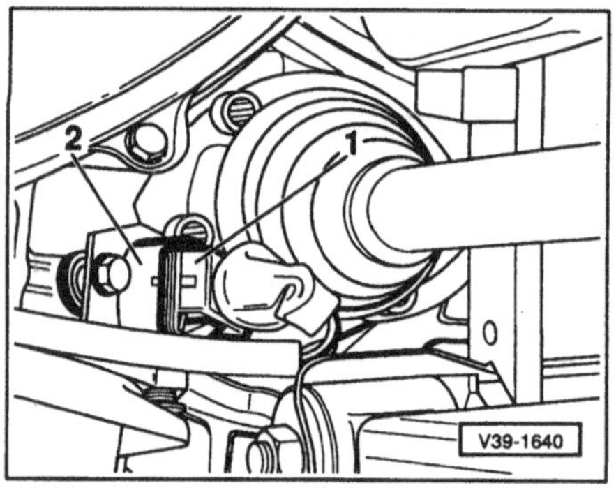

– Remove bolts and disconnect drive shaft from drive flange.

Replacing right-hand oil seal
⇒ page H 39-6

When replacing left-hand oil seal:

◄ – Pull connector -1- from Multi-Function Transmission Range (TR) Switch -2-.

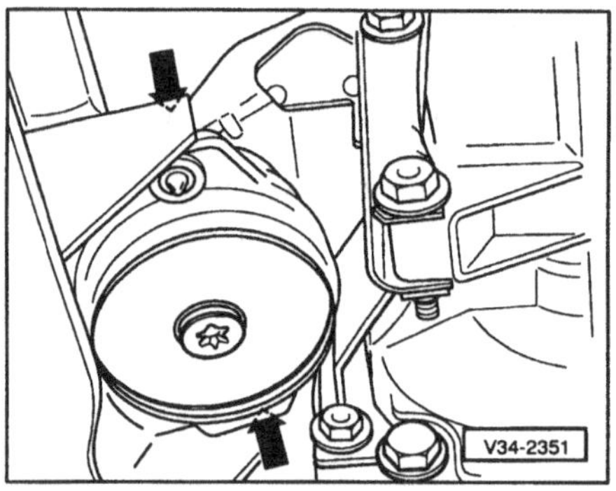

◄ – Remove bolts and disconnect rear mounting from assembly carrier -arrows-.

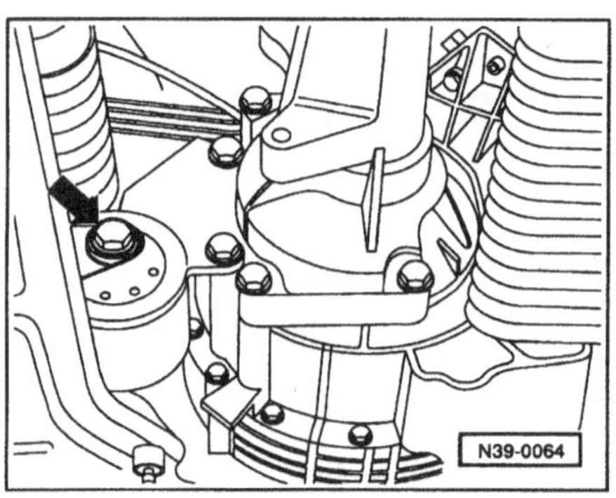

◄ – Remove bolt -arrow- on vehicles with modified assembly carrier.

– Remove left-hand drive shaft. Press engine/transmission slightly forward to to allow removal.

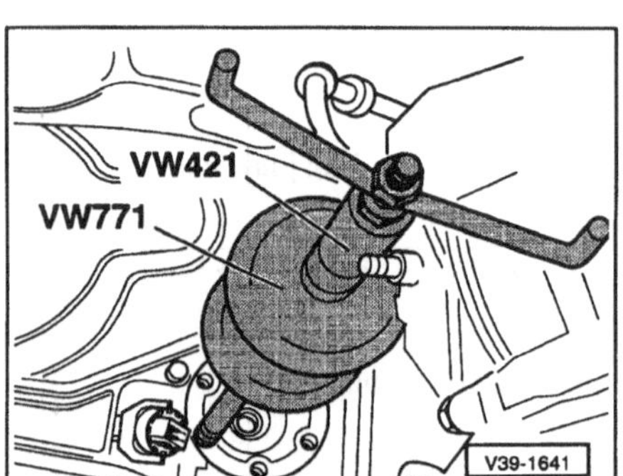

◄ – Remove flange shaft using VW 771 slide hammer and VW 421 tube.

Note:

Always replace flange shaft circlip.

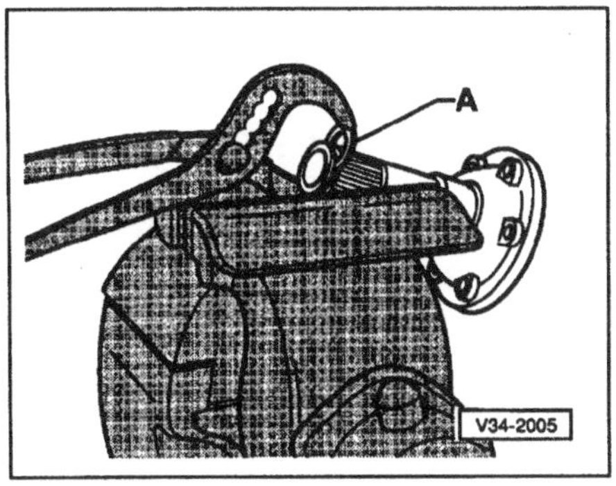

◄ – Clamp flange shaft in vice using protective vice clamps.

– Press circlip out of flange shaft groove using a new circlip -A-.

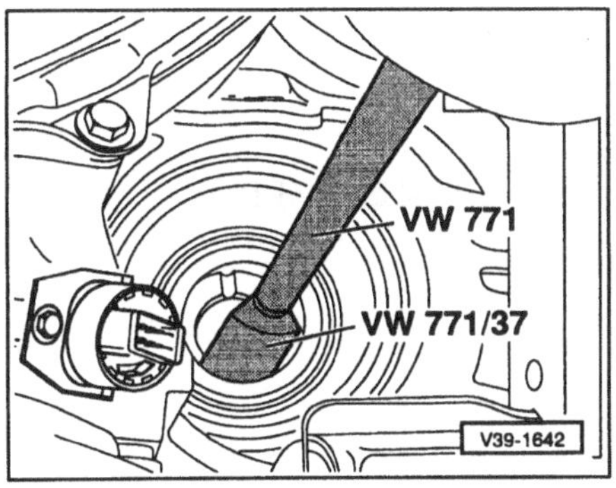

◄ – Pull flange shaft oil seal out using VW 771 slide hammer and VW 771/37.

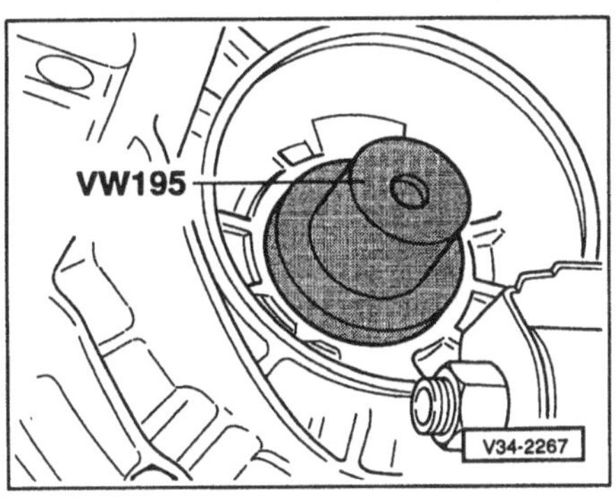

◄ – Press oil seal onto stop with VW 195 arbor.

– Fill space between seal and dust lip with multi-purpose grease.

– Install flange shaft.

– Install drive shaft.

– With transmission installed, check final drive oil level ⇒ page H 39-1.

When replacing right-hand oil seal:

– Remove drive shaft.

– Pierce sealing cap in drive flange with a screwdriver and pry out.

– Remove drive flange circlip.

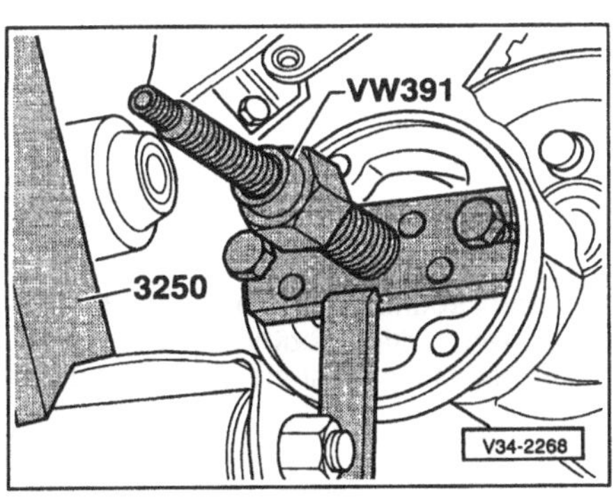

◄ – Remove drive flange with VW 391 puller.

– Remove drive flange oil seal with VW 681 extractor lever.

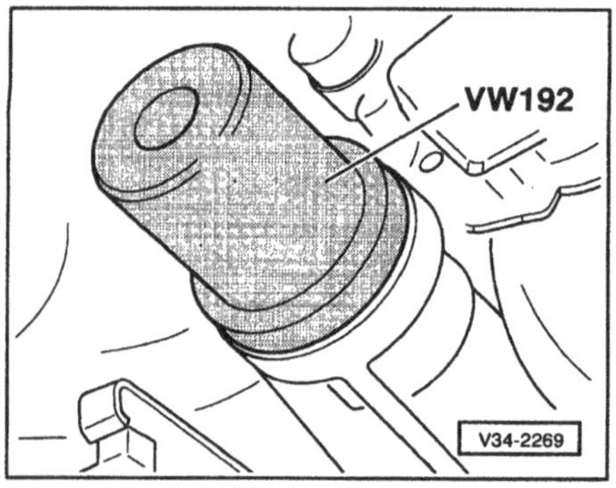

– Press in oil seal up to stop with VW 192 arbor.

– Fill space between seal and dust lip with multi-purpose grease.

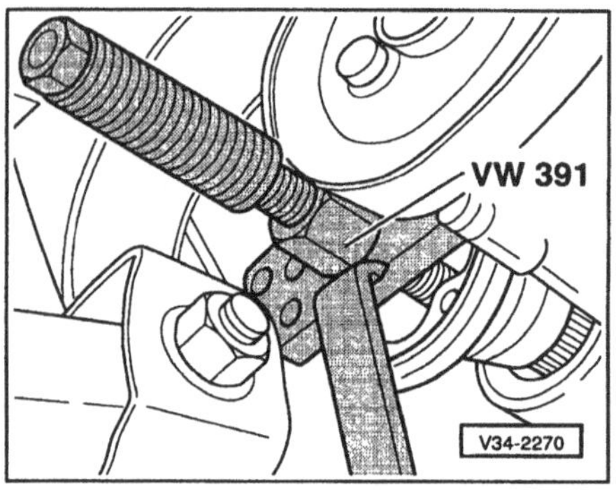

– Remove drive flange with VW 391 puller.

– Install new circlip and new sealing cap.

– Install drive shaft.

– With transmission installed, check final drive oil level ⇒ page H 39-1.

Tightening torque:

Wheel bearing housing to lower ball joint	55 Nm (41 ft lb) plus 1/4 turn
Shock absorber to axle link	160 Nm (118 ft lb)
Hex bolt to drive shaft (Always replace bolt)	150 Nm (111 ft lb) plus 1/4 turn
Bonded mounting to assembly carrier	45 Nm (33 ft lb)
Vehicles from 08.92 with modified assembly carrier	200 Nm (148 ft lb)
Wheel bolts to wheel hub	160 Nm (118 ft lb)
Drive shaft to drive flange [1] Multiple-toothed socket head bolt[1]	80 Nm (59 ft lb)

[1] Left-hand drive shaft: Replace hex socket head bolt with multiple toothed socket head bolt and shorten multiple toothed socket head bolt M10 x 48, by 3 mm (0.118 in.) on transmissions with build date prior to 01 06 2.

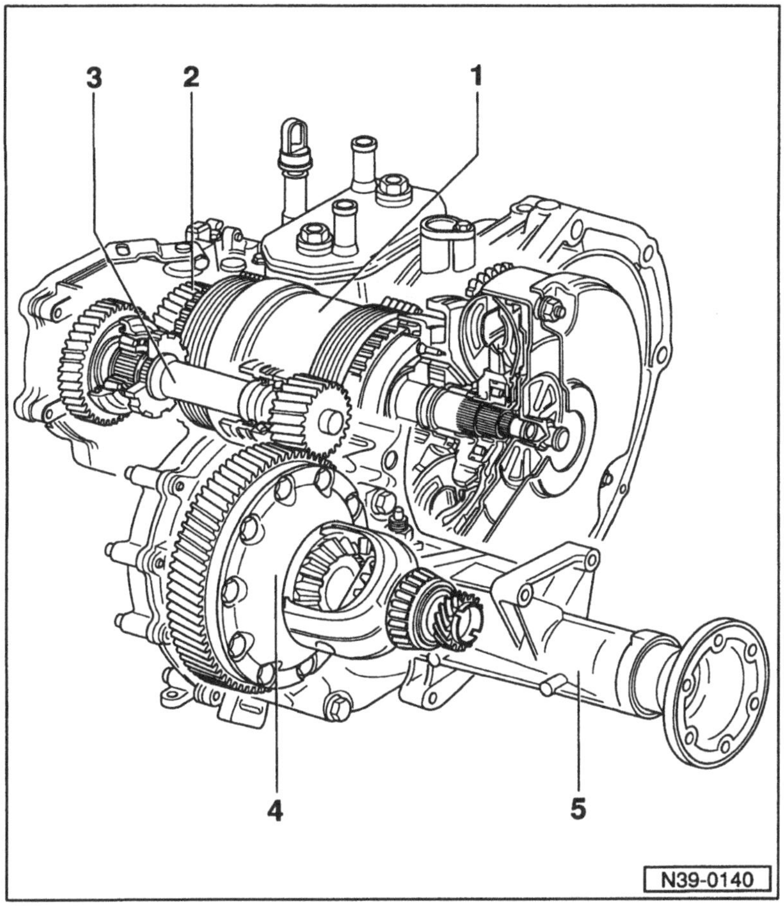

Final drive, disassembling and assembling

Repair overview of final drive

1 - Planetary gear
- Disassembling and assembling ⇒ page H 37-57

2 - Input gear
- Removing and installing ⇒ page H 39-23
- Adjusting ⇒ page H 39-59
- Adjusting final drive (overview) ⇒ page H 39-49

Note:

Repairs on the input gear can only be performed after the component parts of the planetary gear have been removed.

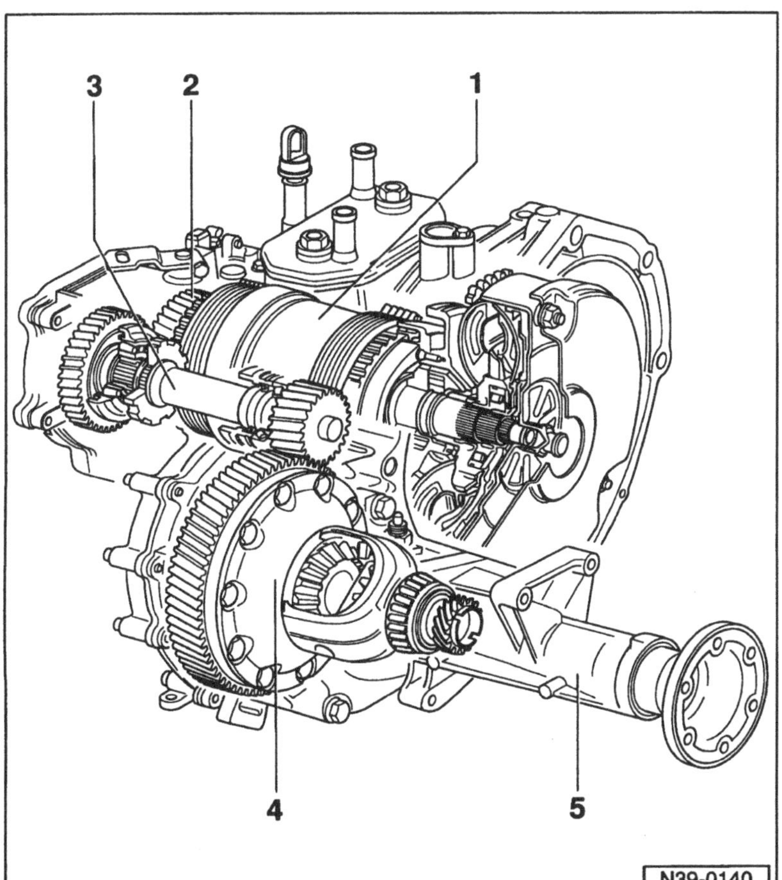

3 - Drive pinion
- Removing and installing ⇒ page H 39-29
- Adjusting ⇒ page H 39-53
- Adjusting final drive (overview) ⇒ page H 39-49

4 - Differential
- Removing and installing ⇒ page H 39-18
- Disassembling and assembling ⇒ page H 39-39
- Adjusting ⇒ page H 39-62
- Adjusting final drive (overview) ⇒ page H 39-49

5 - Bearing housing
- Removing and installing ⇒ page H 39-11

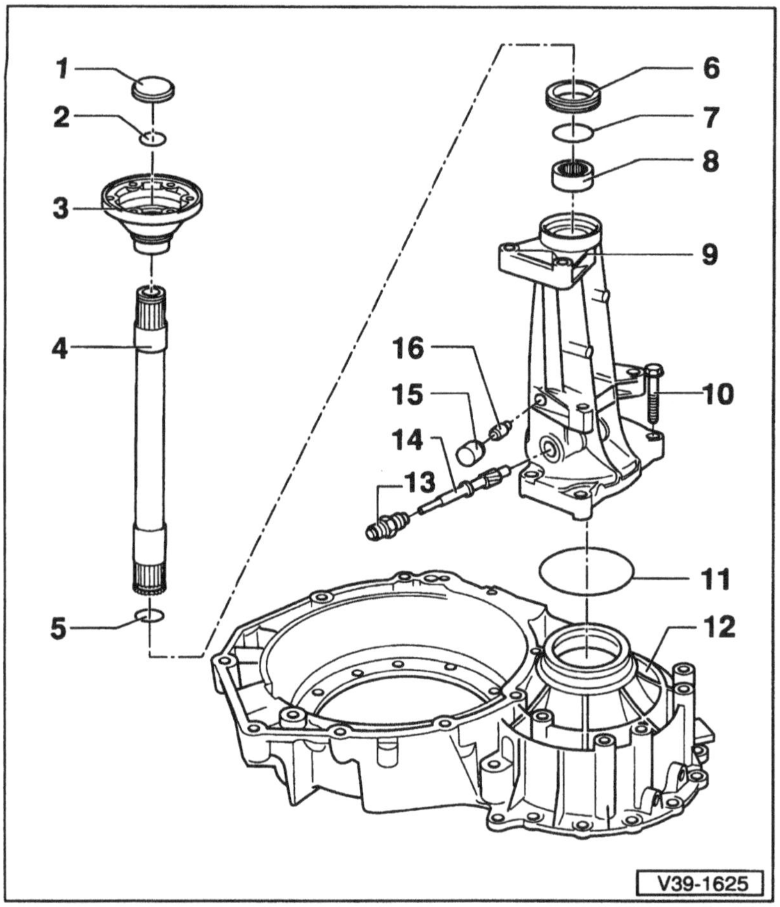

Bearing housing, removing and installing

1 - Sealing cap
- Always replace
- Pry out with screwdriver

2 - Circlip

3 - Drive flange
- Pressing off ⇒ Fig. 2
- Pressing on ⇒ Fig. 3

4 - Output shaft
- Drive out with drive flange ⇒ Fig. 1

5 - Circlip
- Always replace
- Removing and installing ⇒ Fig. 4

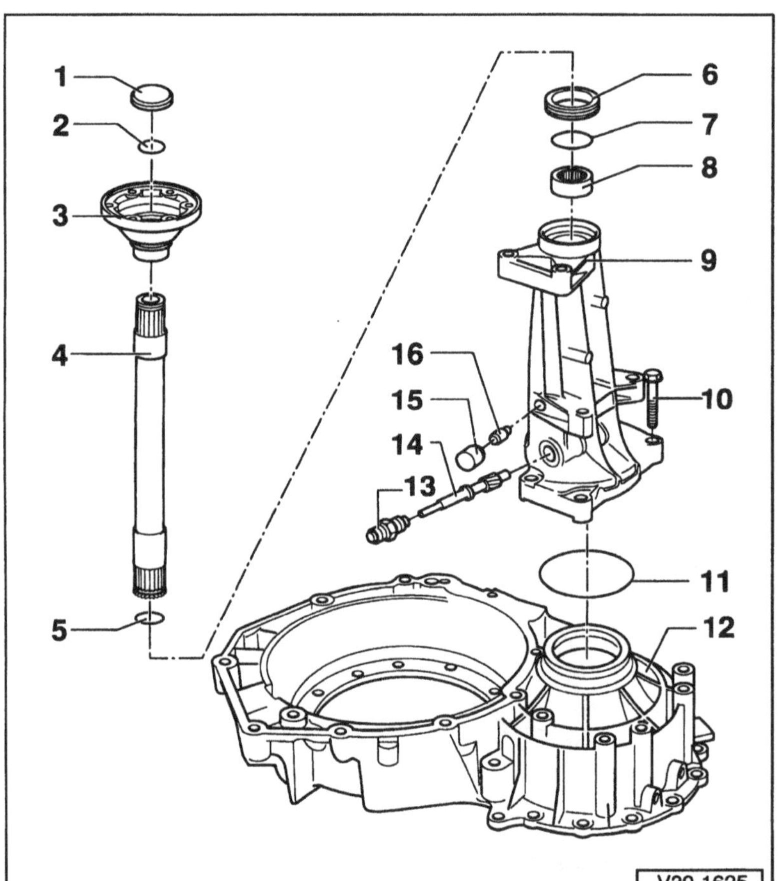

6 - Oil seal for drive flange
- Before installing, pack space between sealing lips with multi-purpose grease
- Can be replaced with transmission installed ⇒ page H 39-2
- Remove with VW 681 extractor-lever
- Driving in ⇒ Fig. 7

7 - Circlip
- For needle roller bearing
- Always replace

8 - Needle roller bearing
- Pulling out ⇒ Fig. 5
- Pressing in ⇒ Fig. 6

9 - Bearing housing

10 - Bolt
- Tightening torque: 40 Nm (30 ft lb) plus an additional 1/4 turn (90°) further
- Additional 1/4 (90°) turn can be made in several stages.

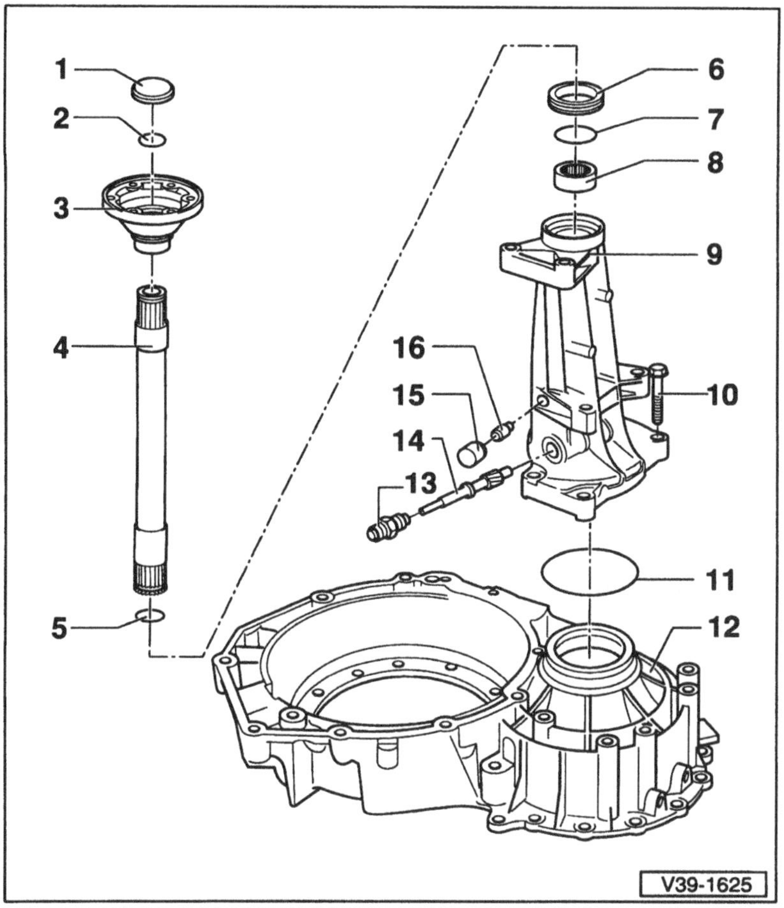

11 - O-ring
- ♦ Always replace
- ♦ Coat with grease

12 - Torque converter housing
- ♦ Removing
  ⇒ page H 39-18 Removing and installing differential

13 - Guide bushing
- ♦ For speedometer drive

14 - Speedometer drive
- ♦ Remove before removing bearing housing

15 - Breather cap
- ♦ For final drive

16 - Breather pipe
- ♦ Remove using pliers
- ♦ Drive onto stop with VW 460/2 drift

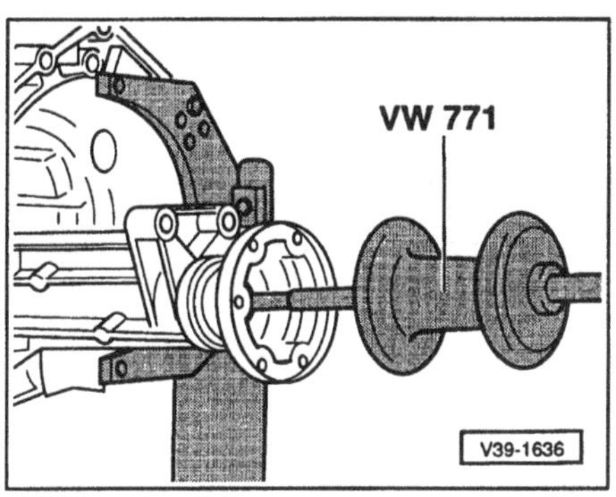

◄ **Fig. 1 Output shaft with drive flange, driving out**

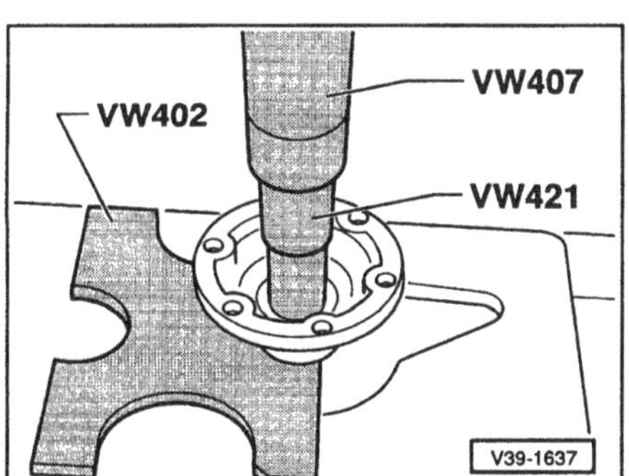

◄ **Fig. 2 Drive flange, pressing off**

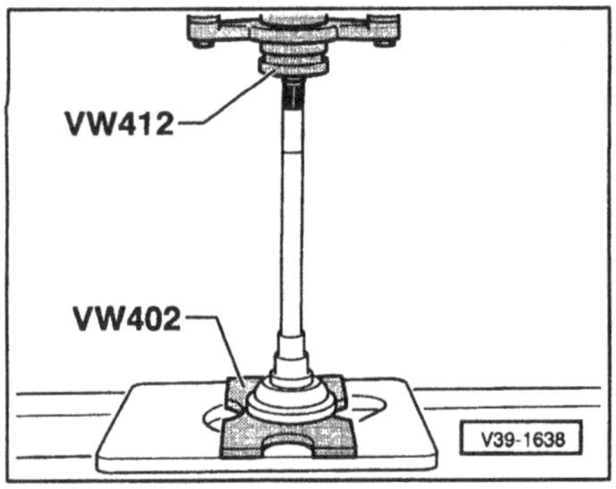

◄ **Fig. 3 Drive flange, pressing on**

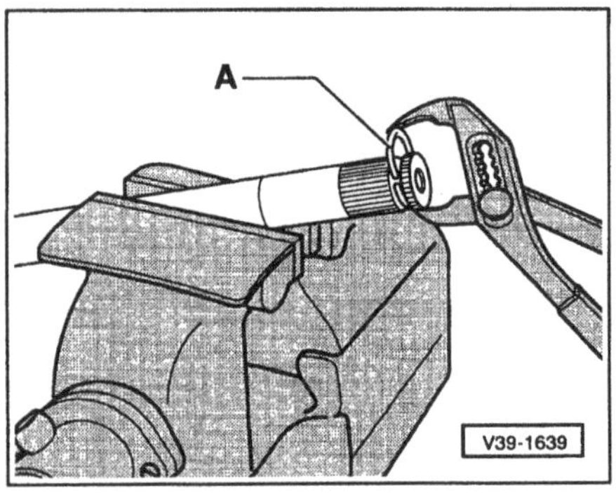

◄ **Fig. 4 Circlip, removing and installing**

- Clamp output shaft in vise using protective vise clamps.
- Press circlip out of output shaft groove using new circlip -A-.

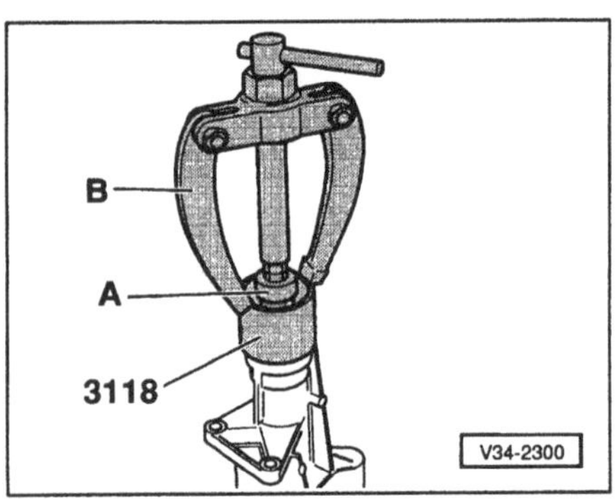

◄ **Fig. 5 Needle roller bearing, pulling out**

A - Internal puller, Kukko 21/5 or equivalent

B - Counter support, Kukko 22-2 or equivalent

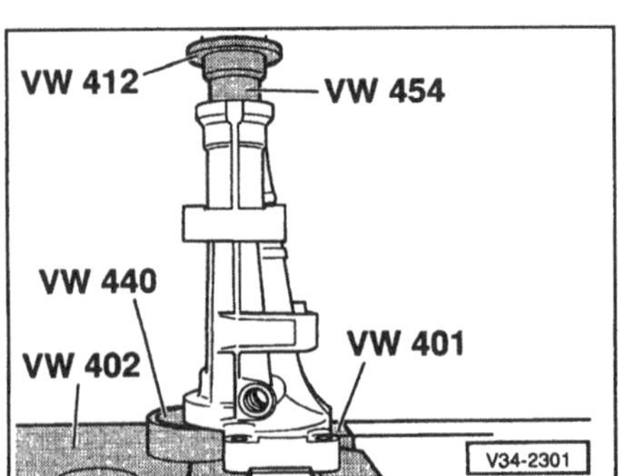

◄ **Fig. 6 Needle roller bearing, installing**

- Press needle roller bearing into stop.

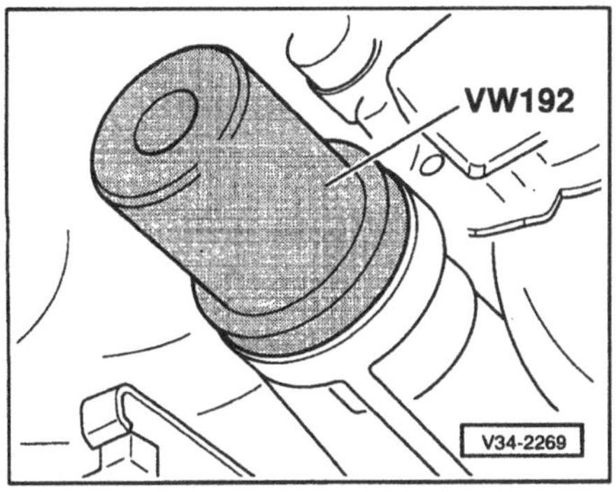

Fig. 7 Seal, driving into stop using VW 192

Pack space between seal and dust lip with multi-purpose grease.

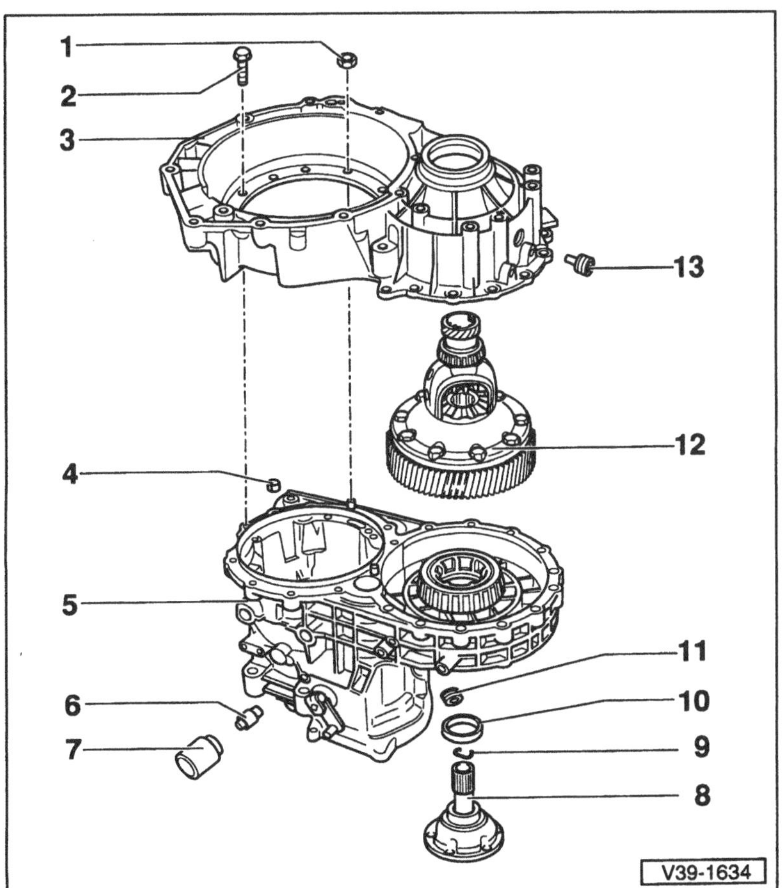

Differential, removing and installing

Note:

The drive pinion does not need to be removed for removing and installing the differential.

1 - Nut
- ♦ Qty. 2
- ♦ Tightening torque: 25 Nm (18 ft lb)

2 - Screw
- ♦ Tightening torque: 25 Nm (18 ft lb)

3 - Torque converter housing
- ♦ Replacement part is applicable according to transmission code letters

⇒ Replacement parts catalog microfiche
- ♦ On vehicles with Diesel engine, the Engine Speed (RPM) Sensor -G28-, is installed in the top of the torque converter housing ⇒ Fig. 3

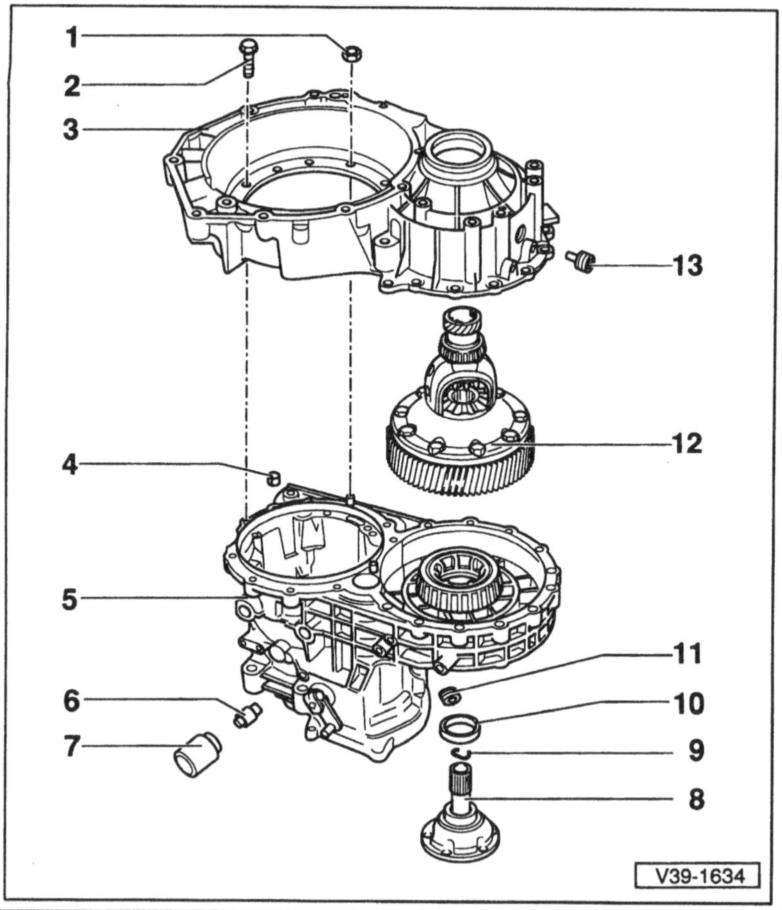

4 - Sleeve
- To center the transmission and torque converter housing

5 - Transmission housing
- Before assembling transmission and torque converter housings together, coat sealing surfaces with AMV188 200

6 - Breather tube
- Remove using pliers
- Drive into stop using VW 460/2 drift

7 - Breather cap
- For planetary transmission

8 - Flange shaft
- Remove using VW 771 slide hammer
- Drive in using plastic hammer

9 - Circlip
- Always replace
- Removing and installing ⇒ Fig. 1

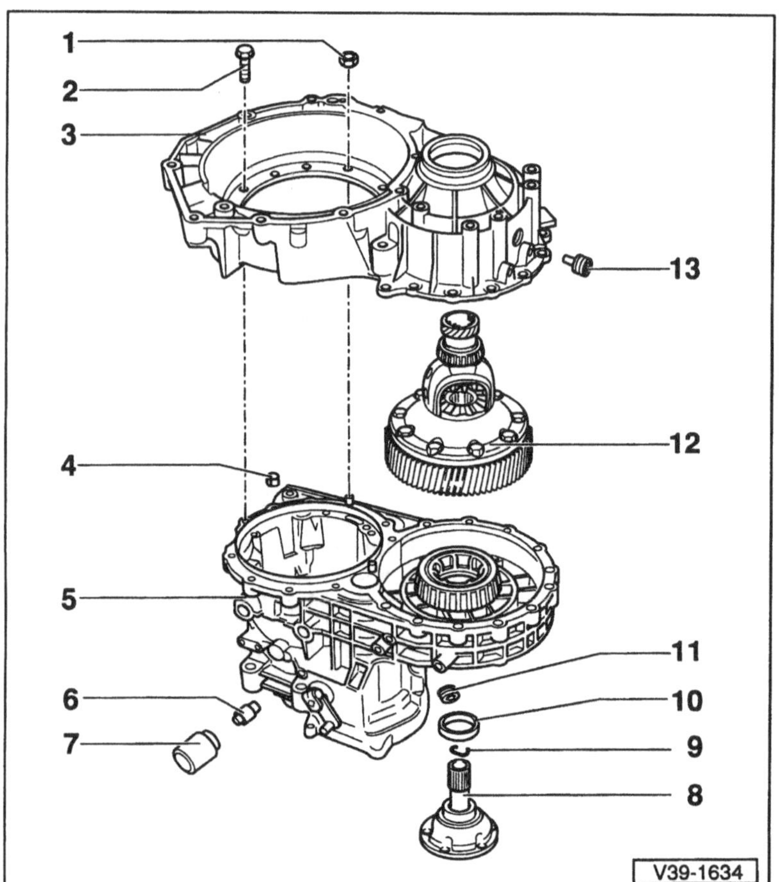

10 - Oil seal for flange shaft
- Before installing, pack space between sealing lips with multi-purpose grease
- Can be replaced with transmission installed ⇒ page H 39-2
- Remove with VW 681 extractor lever
- Driving in ⇒ Fig. 2

11 - Plug
- Tightening torque: 20 Nm (15 ft lb)
- Checking gear oil with transmission installed ⇒ page H 39-1

12 - Differential
- Disassembling and assembling ⇒ page H 39-39

13 - Plug
- Tightening torque: 20 Nm (15 ft lb)
- Checking gear oil with transmission installed ⇒ page H 39-1

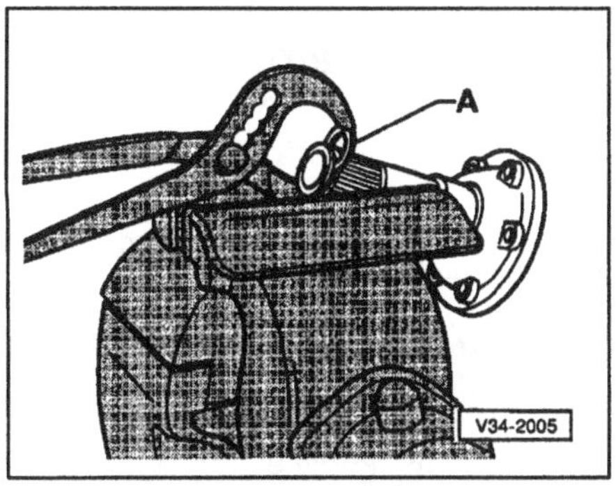

Fig. 1 Circlip, removing and installing

– Clamp flange shaft in vise using protective vise clamps.

– Press circlip out of flange shaft groove using new circlip -A-.

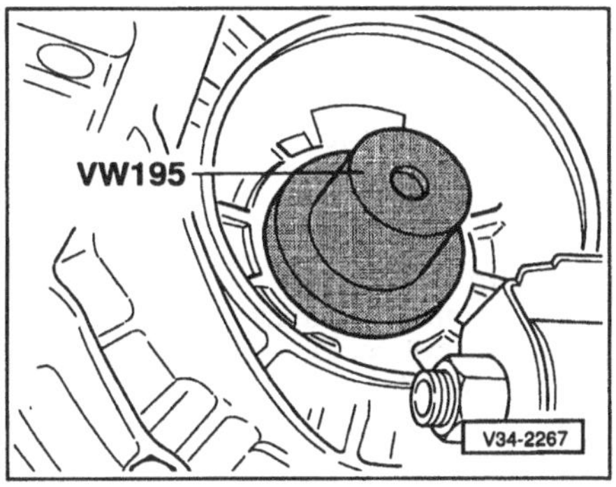

Fig. 2 Seal, driving in onto stop using VW 195

Pack space between seal and dust lip with multi-purpose grease.

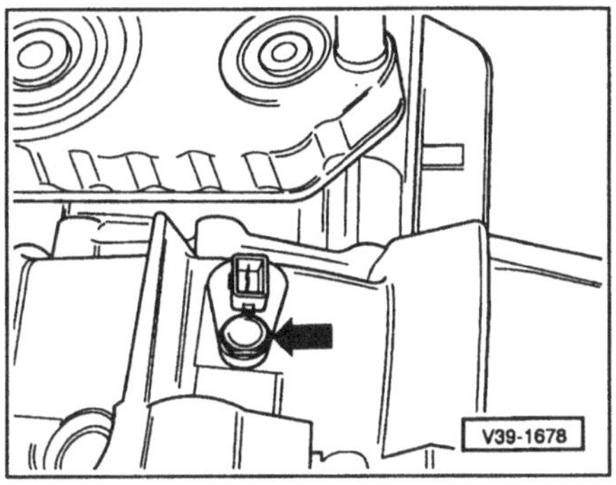

Fig. 3 Engine Speed (RPM) sensor

(only for Diesel engines)

– Remove bolt -arrow- and pull Engine Speed (RPM) Sensor -G28- out of torque converter housing.

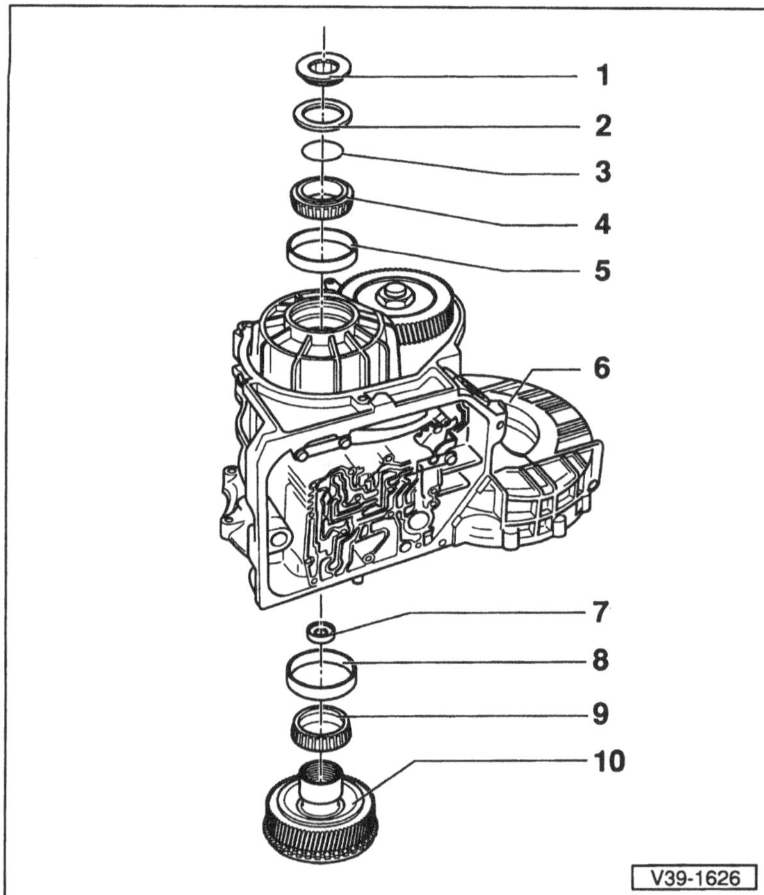

Input gear, removing and installing

Note:

Drive pinion and differential do not need to be removed.

1 - Socket-head fastener
- Tightening torque: 250 Nm (184 ft lb)
- Engage parking lock to remove and install
- Use 22 mm wrench socket
- Before installing socket-head fastener, install axial needle bearing -7-

2 - Dished washer
- Curved side faces socket-head fastener

3 - Shim
- Determining thickness ⇒ Adjusting input gear, page H 39-59

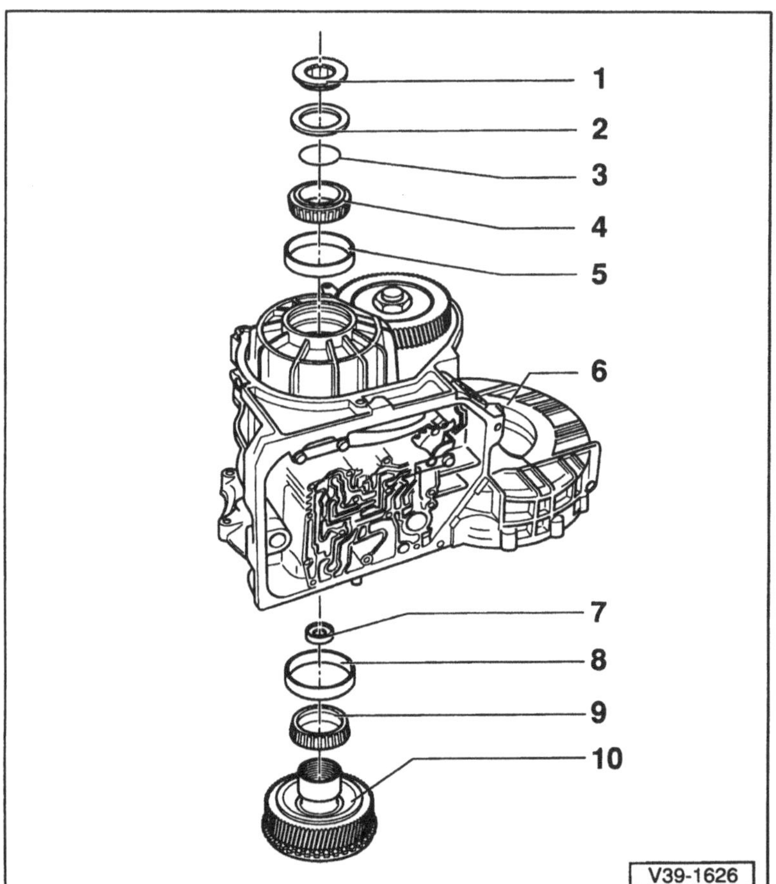

4 - Tapered roller bearing inner race
- Install so that the lugs engage in the groove of the tapered roller bearing inner race -9- ⇒ Fig. 3
- Installation position ⇒ Fig. 4
- After measuring thickness of shim, fit onto input gear with AMV 185 101 A1 thread locking fluid

5 - Tapered roller bearing outer race
- Drive out with drift
- Drive in with 30-205 thrust pad
- Install with AMV 185 101 A1 thread locking fluid

6 - Transmission housing

7 - Axial needle bearing
- Install with flat side facing input gear
- Install into input gear before installing socket-head fastener -1-

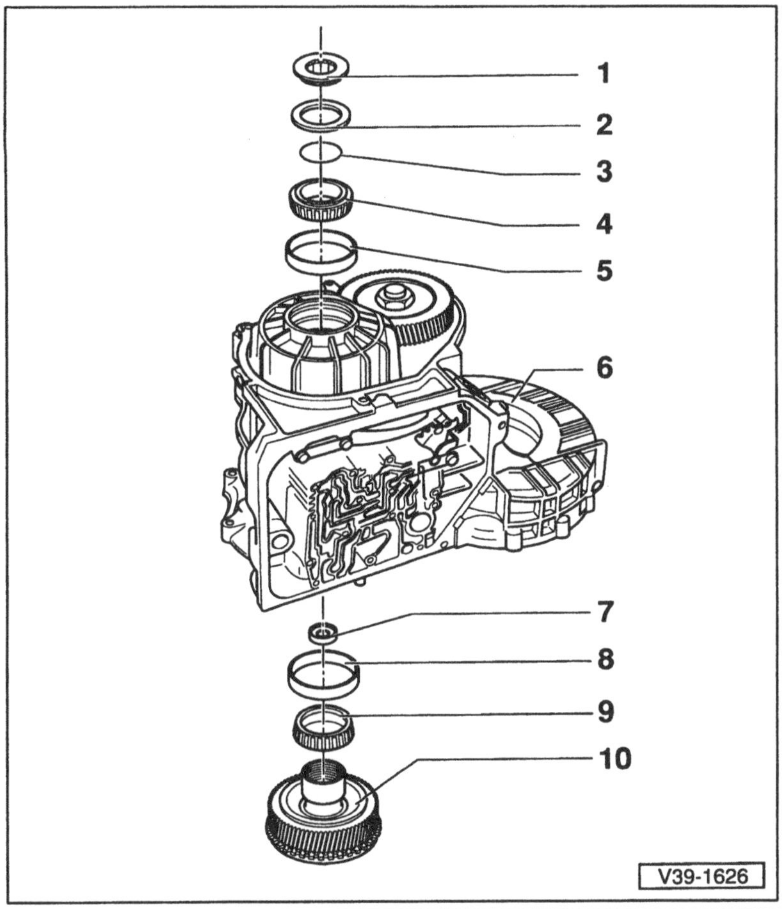

8 - Tapered roller bearing outer race

- Drive out with drift
- Drive in with 30-205 thrust pad and appropriately long drift
- Install with AMV 185 101 A1 thread locking fluid

9 - Tapered roller bearing inner race

- Pulling off ⇒ Fig. 1
- Pressing on ⇒ Fig. 2
- Installation position ⇒ Fig. 4
- Install with AMV 185 101 A1 thread locking fluid

10 - Input gear

- With impulse wheel for Vehicle Speed Sensor (VSS) -G68-
- Number of teeth ⇒ page H 00-3
- Adjusting ⇒ page H 39-59
- If signs of damage exist, always replace input gear and output gear together
- Removing ⇒ Fig. 5

Note:

If input gear is replaced, adjust planet carrier *⇒ page H 37-93.*

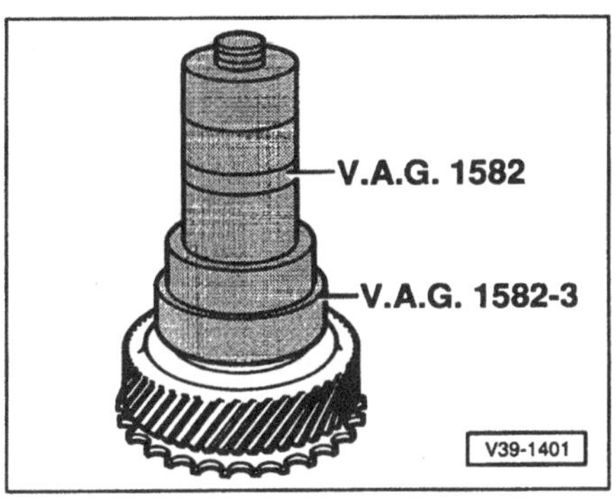

◄ **Fig. 1 Tapered roller bearing inner race, pulling off**

Place press piece on input gear.

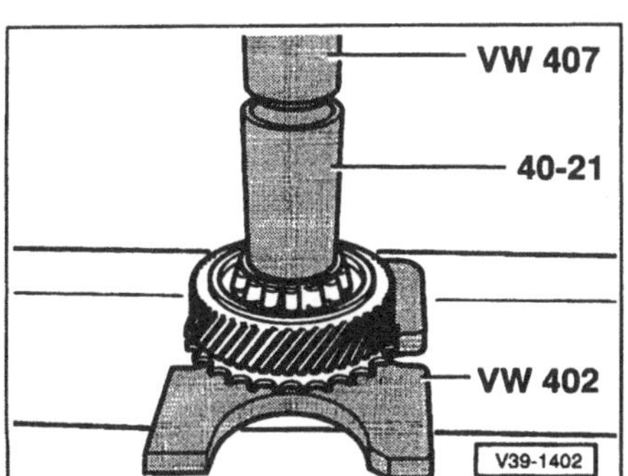

◄ **Fig. 2 Tapered roller bearing inner race, pressing on**

Note:

A press piece must be placed below the hub of the input gear to avoid damaging it.

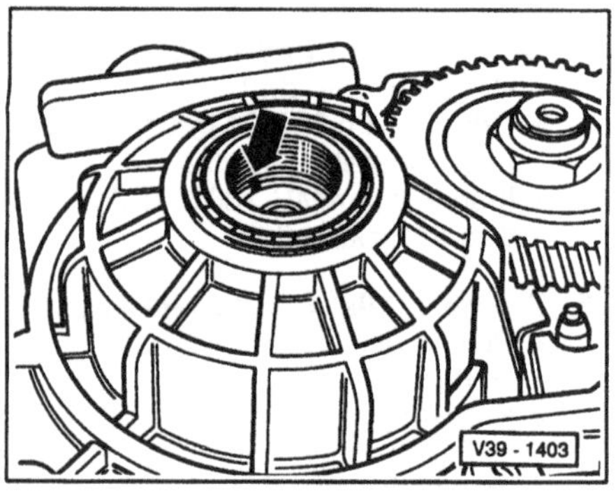

◄ **Fig. 3 Tapered roller bearing inner race, installing**

Install tapered roller bearing inner race so that the lugs of the inner race engage in the grooves of the opposite inner race -arrow-.
⇒ Fig. 4

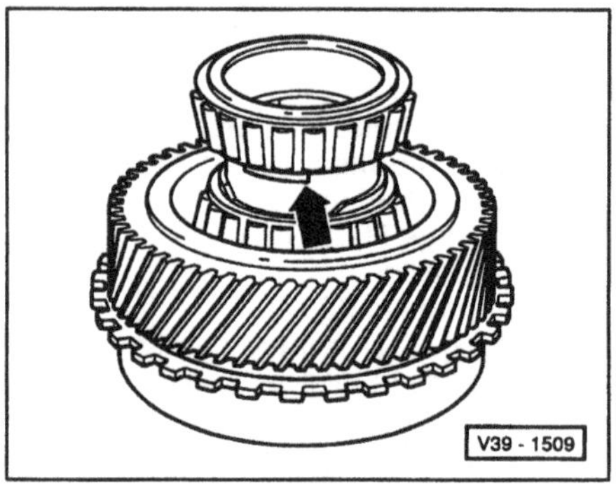

◄ **Fig. 4 Tapered roller bearing inner race, installation position**

Lug -arrow- must engage in groove of the opposite inner race.

Note:

Tapered roller bearings without lugs can also be installed.

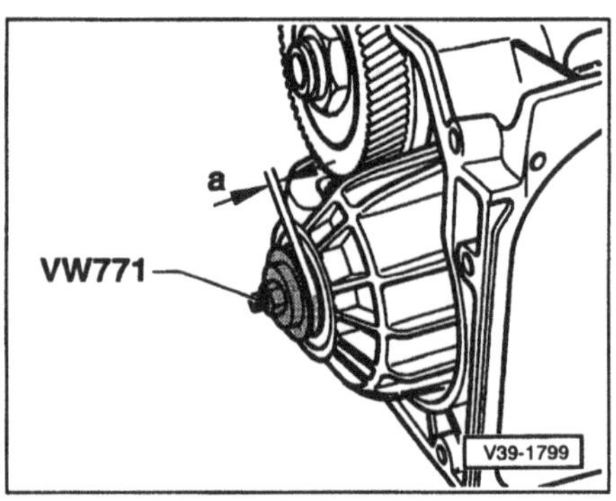

◄ **Fig. 5 Input gear, removing**

- Install socket-head fastener without dished washer, shim and axial needle bearing far enough for gap size "a" = 3 mm (0.118 in.) to exist between socket-head fastener and tapered roller bearing inner race.
- Drive out input gear with VW 771 slide hammer as far as the stop of the tapered roller bearing inner race.
- Remove VW 771 slide hammer and remove input gear.

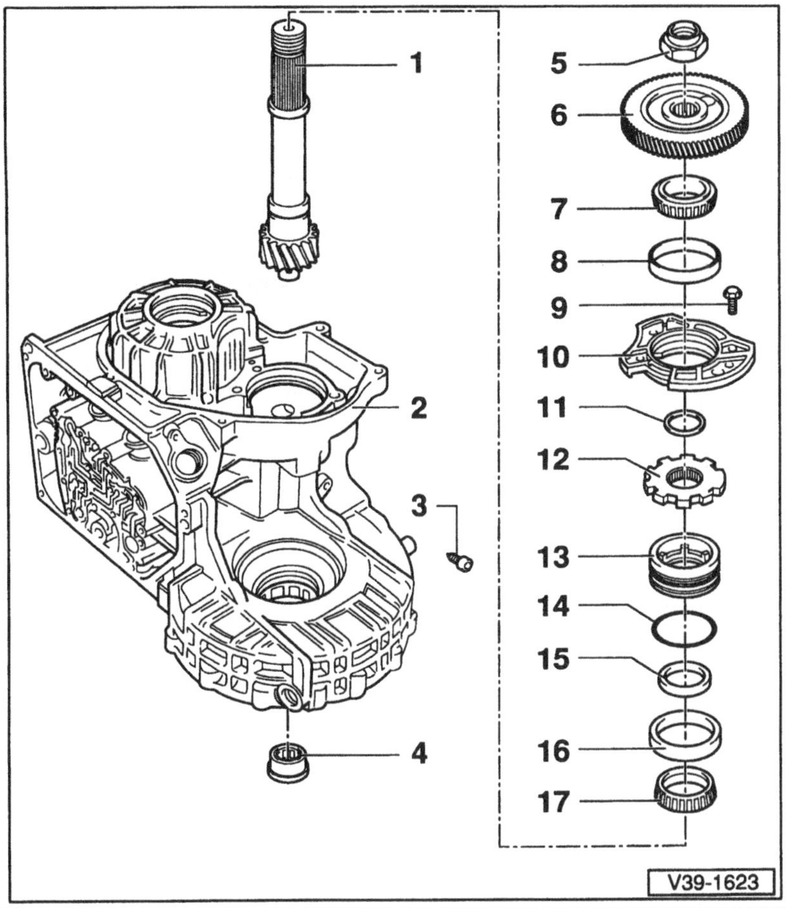

Drive pinion, removing and installing

Note:

To remove the drive pinion, the selector shaft must be removed after taking out the parking lock
⇒ page H 38-45, disassembling and assembling parking lock.

1 - Drive pinion
- Number of teeth ⇒ from page H 00-3
- Adjusting ⇒ page H 39-53
- Removing ⇒ Fig. 8

2 - Transmission housing

3 - Screw
- Tightening torque: 12 Nm (9 ft lb)
- Remove screw before removing or installing bearing supporting ring -13-.
- Install with AKD 456 000 01 thread locking fluid
- Remove screw before removing or installing bearing supporting ring -13-.
- Install with AKD 456 000 01 thread locking fluid

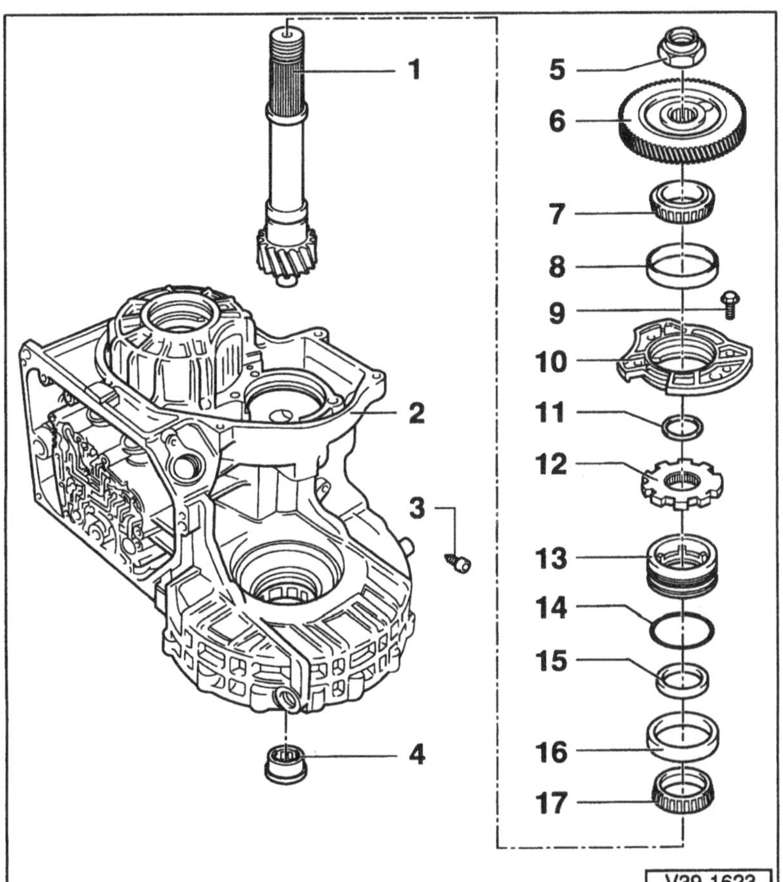

4 - Needle roller bearing
- Pressing out ⇒ Fig. 10
- Pressing in ⇒ Fig. 11

5 - Hex nut
- Tightening torque: 250 Nm (184 ft lb)
- Engage parking lock to remove or install
- Secure with drift

6 - Output gear
- Number of teeth ⇒ from page H 00-3
- Pulling off ⇒ Fig. 1
- If signs of damage exist, always replace input gear and output gear together

7 - Tapered roller bearing inner race
- Pulling off ⇒ Fig. 2
- Pressing on ⇒ Fig. 3

8 - Tapered roller bearing outer race
- Drive out with drift
- Pressing in ⇒ Fig. 4

9 - Bolt
- M 6 - 14 Nm (10 ft lb)
- M 8 - 25 Nm (18 ft lb)

10 - Bearing cap

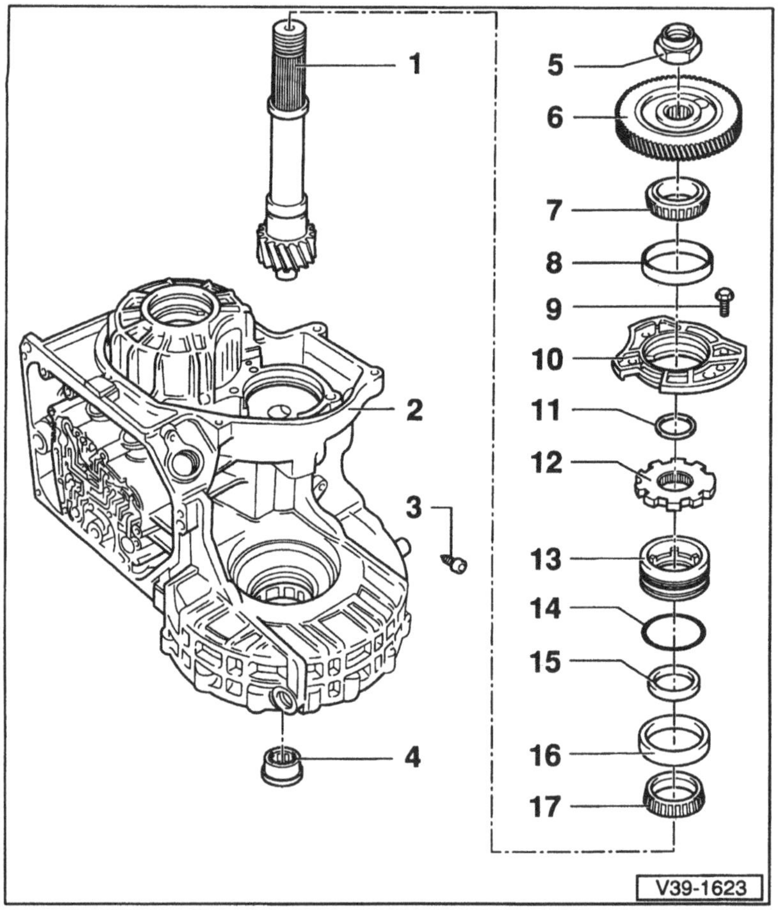

11 - Shim

- Determining thickness
  ⇒ page H 39-53, Adjusting drive pinion

12 - Parking lock gear

- Rounded side faces teeth of drive pinion

13 - Bearing supporting ring

- Tightening torque: 200 Nm (148 ft lb)
- Remove screw -3- before removing or installing bearing supporting ring.
- Removing or installing
  ⇒ Fig. 5
- Bearing supporting ring cannot be unscrewed until the selector shaft has been removed
  ⇒ page H 38-45, Disassembling and assembling parking lock
- Install bearing supporting ring carefully. Sealing lip and spring must not be damaged.
- Supporting ring with seal
  ⇒ Fig. 6

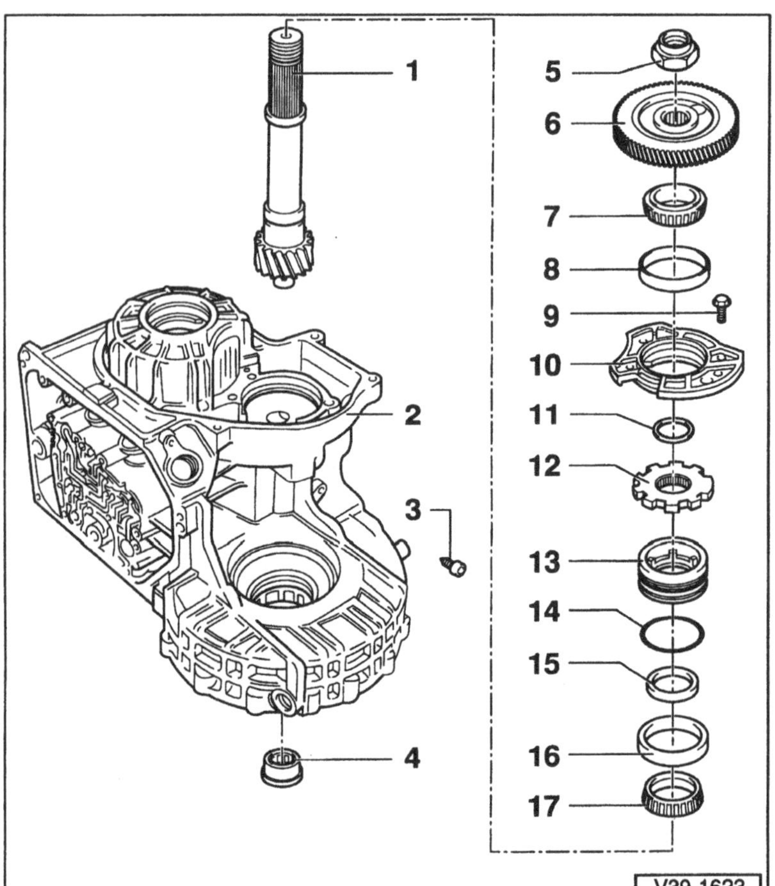

14 - O-ring

- Always replace
- Place onto bearing supporting ring –13-

15 - Oil seal for drive pinion

- Remove with screwdriver
- Installation position
  ⇒ Fig. 6
- Driving in
  ⇒ Fig. 7

16 - Tapered roller bearing outer race

- Driving in
  ⇒ Fig. 9

17 - Tapered roller bearing inner race

- Drive off with drift
- Pressing on
  ⇒ Fig. 12

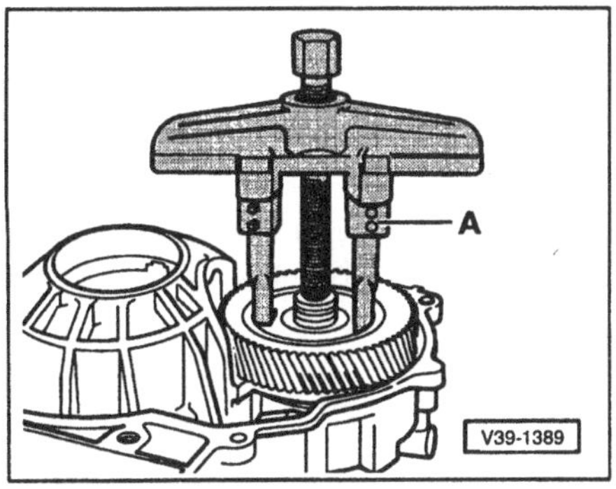

◄ **Fig. 1 Output gear, pulling off**

A – Kukko 20-10 and Matra V-172 pulling hooks.

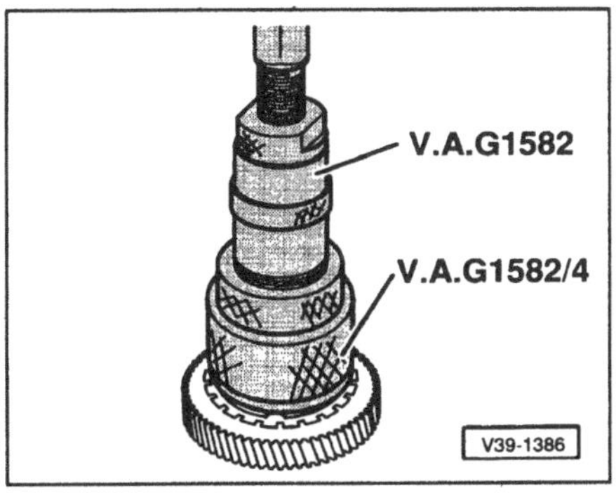

◄ **Fig. 2 Tapered roller bearing inner race, pulling off**

Place press piece onto collar of output gear.

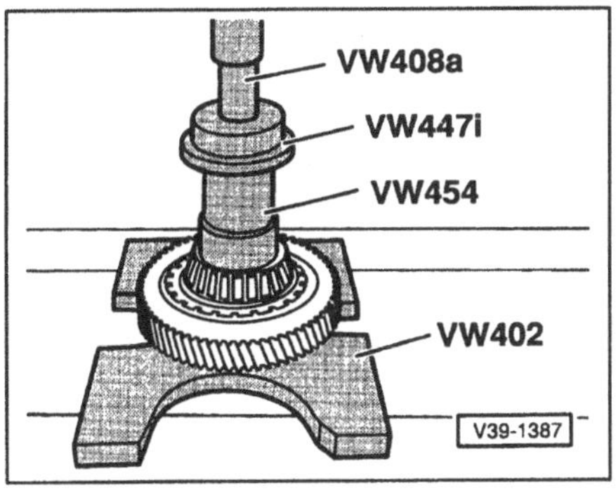

◄ **Fig. 3 Tapered roller bearing inner race, pressing on**

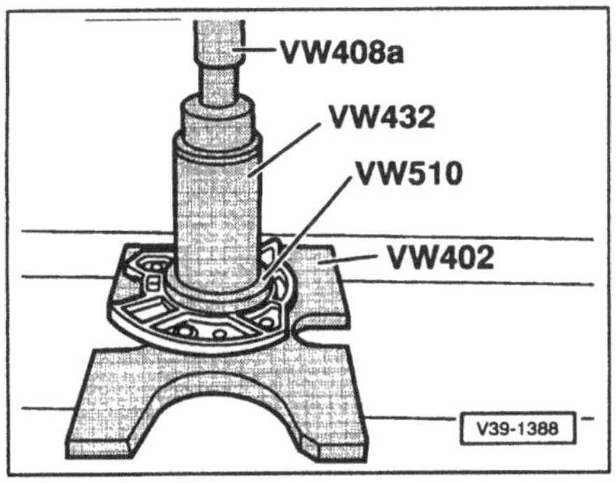

◄ **Fig. 4 Tapered roller bearing outer race, pressing in**

- Press tapered roller bearing outer race in onto stop in bearing cap.

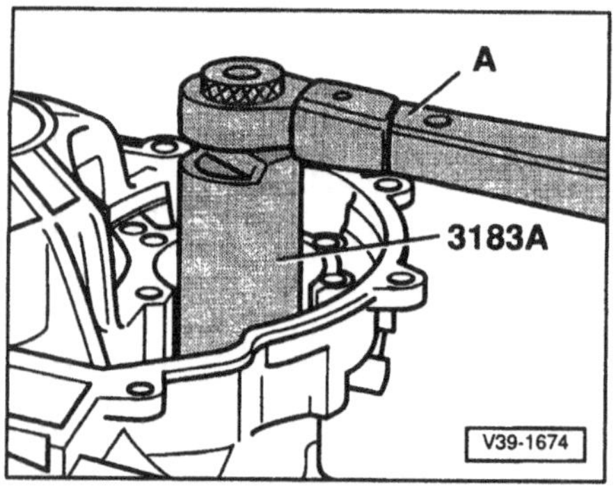

◄ **Fig. 5 Bearing supporting ring, removing and installing**

A – Torque wrench

– Tighten bearing supporting ring to 200 Nm (148 ft lb) and secure with locking screw. Install screw with AKD 456 000 01 thread locking fluid.

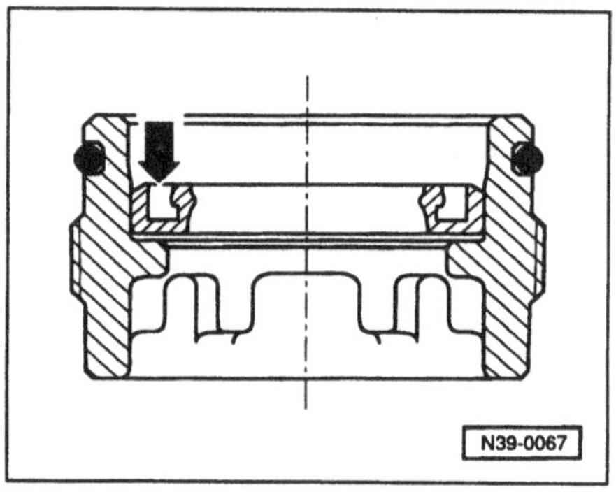

◄ **Fig. 6 Bearing supporting ring with seal for drive pinion**

Installation position of seal:

Open side -arrow- faces towards tapered roller bearing outer race

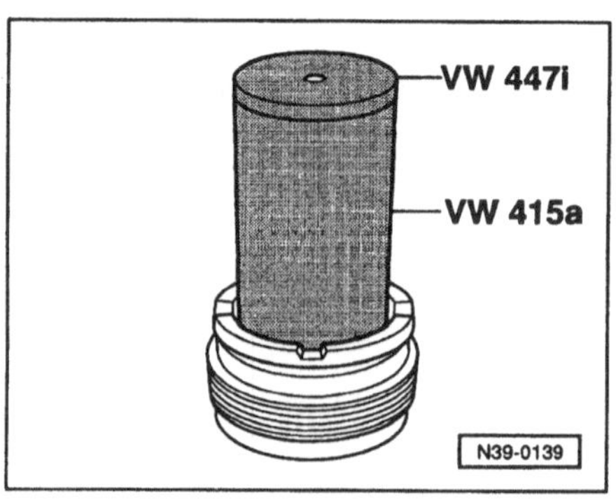

◄ **Fig. 7 Drive pinion oil seal, driving in**

– Install oil seal so that sealing lip faces special tool.

Note:

Install oil seal carefully. Sealing lip and spring must not be damaged.

– Drive oil seal in onto stop.

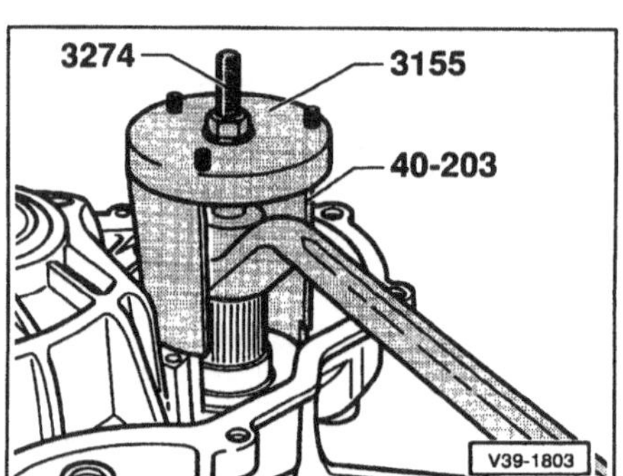

◄ **Fig. 8 Drive pinion, removing**

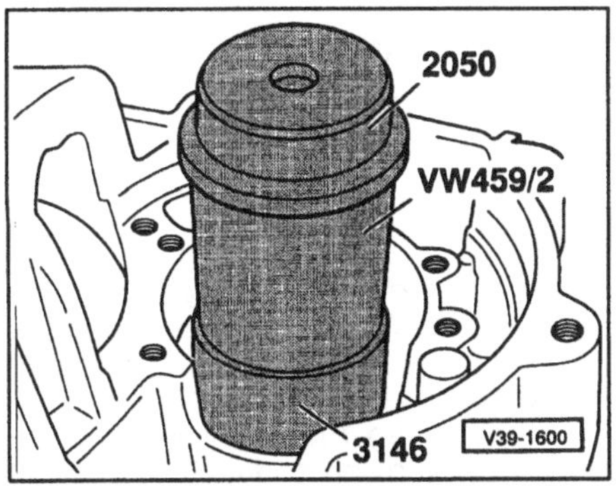

Fig. 9 Tapered roller bearing outer race, driving in

– Install drive pinion with tapered roller bearing inner race, then drive in outer race.

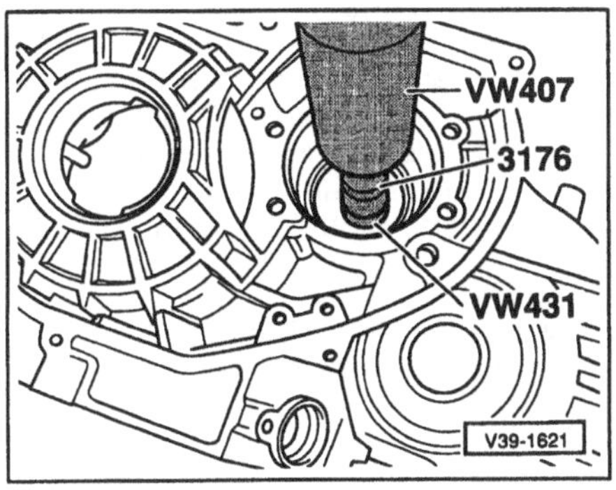

Fig. 10 Needle roller bearing, pressing out

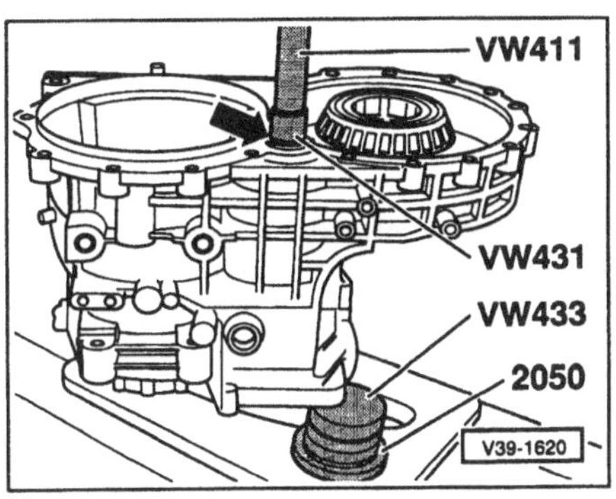

Fig. 11 Needle roller bearing, pressing in

– Position reduced side -arrow- to outside and press needle roller bearing onto stop.

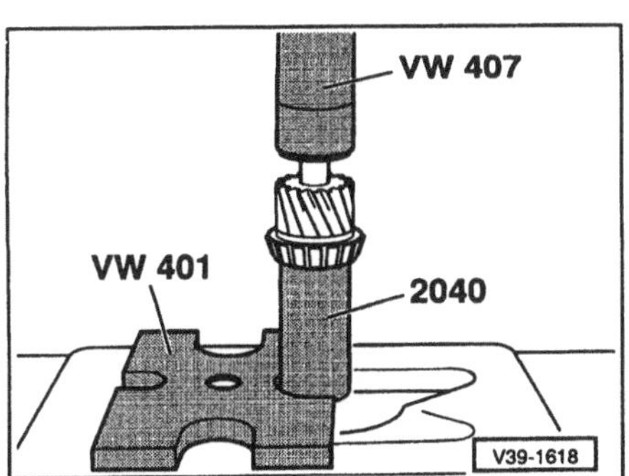

Fig. 12 Tapered roller bearing inner race, pressing on

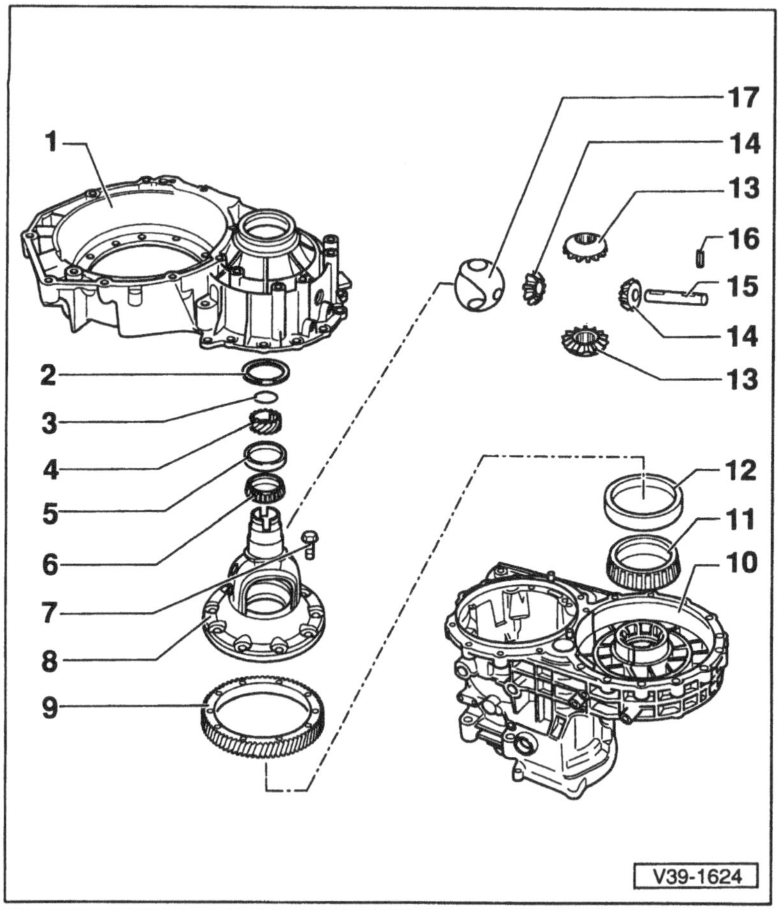

Differential, disassembling and assembling

Notes:

- *Heat tapered roller bearing inner race to 100 °C (212 °F) before pressing on.*

1 - Torque converter housing

- Removing and installing
  ⇒ Removing and installing differential, page H 39-18
- As a replacement part it is allocated according to transmission code letters

⇒ Replacement parts catalog microfiche

- On vehicles with Diesel engine the Engine Speed (RPM) Sensor -G28- is installed in the top of the torque converter housing
  ⇒ page H 39-22, Fig. 3

2 - Shim

- Determining thickness
  ⇒ page H 39-62, Adjusting differential

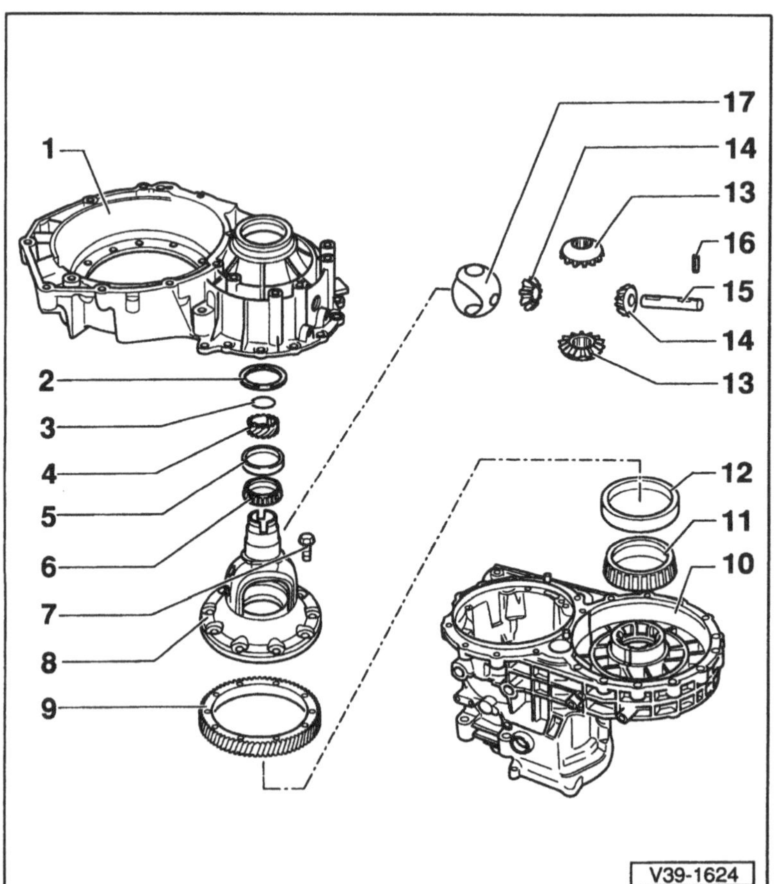

3 - Circlip

- For securing speedometer drive gear

4 - Speedometer drive gear

- Seat drive gear lugs into differential housing grooves, straight face outwards

5 - Tapered roller bearing outer race

- Remove and install only when differential housing heated
- Pressing out
  ⇒ Fig. 1
- Pressing in
  ⇒ Fig. 2

6 - Tapered roller bearing inner race

- Pulling off
  ⇒ Fig. 3
- Pressing on
  ⇒ Fig. 4

7 - Bolt

- Tightening torque: 80 Nm (59 ft lb) plus an additional 1/4 turn (90°) further

8 - Differential housing

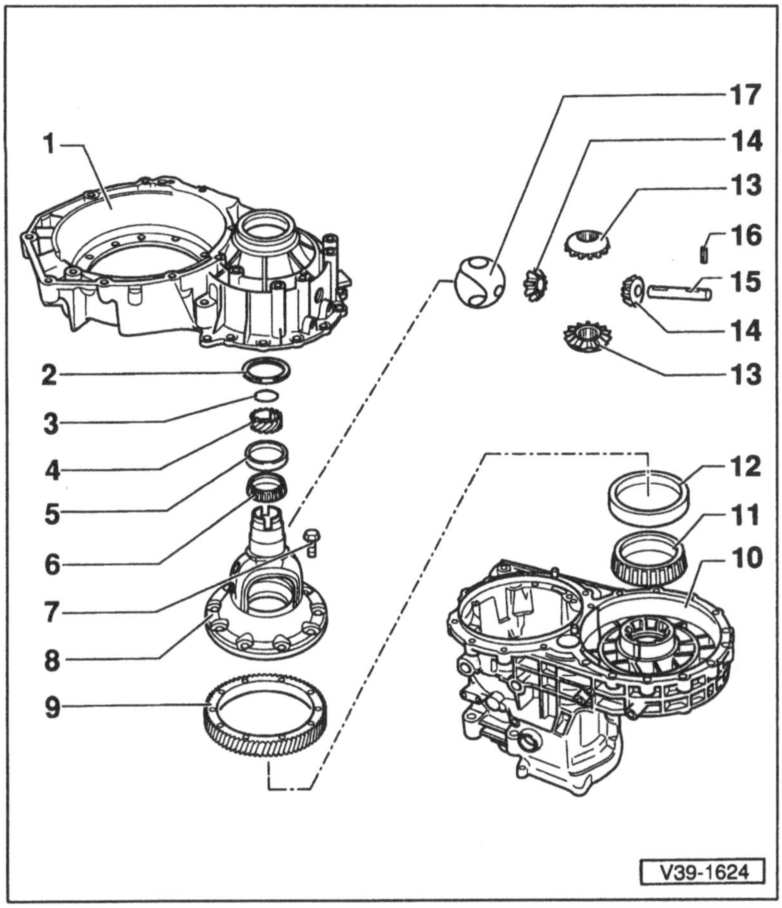

9 - Final drive gear wheel
- Number of teeth ⇒ page H 00-3
- Install heated ⇒ Fig. 5

10 - Transmission housing
- Removing and installing ⇒ Removing and installing differential, page H 39-18
- As a replacement part it is identified by transmission code letters

⇒ Replacement parts catalog microfiche
- Before assembling transmission and torque converter housings together, coat sealing surfaces with AMV188 200

11 - Tapered roller bearing inner race
- Pulling off ⇒ Fig. 6
- Pressing into transmission housing ⇒ Fig. 7

12 - Tapered roller bearing outer race
- Pulling off ⇒ Fig. 8
- Pressing into differential housing ⇒ Fig. 9

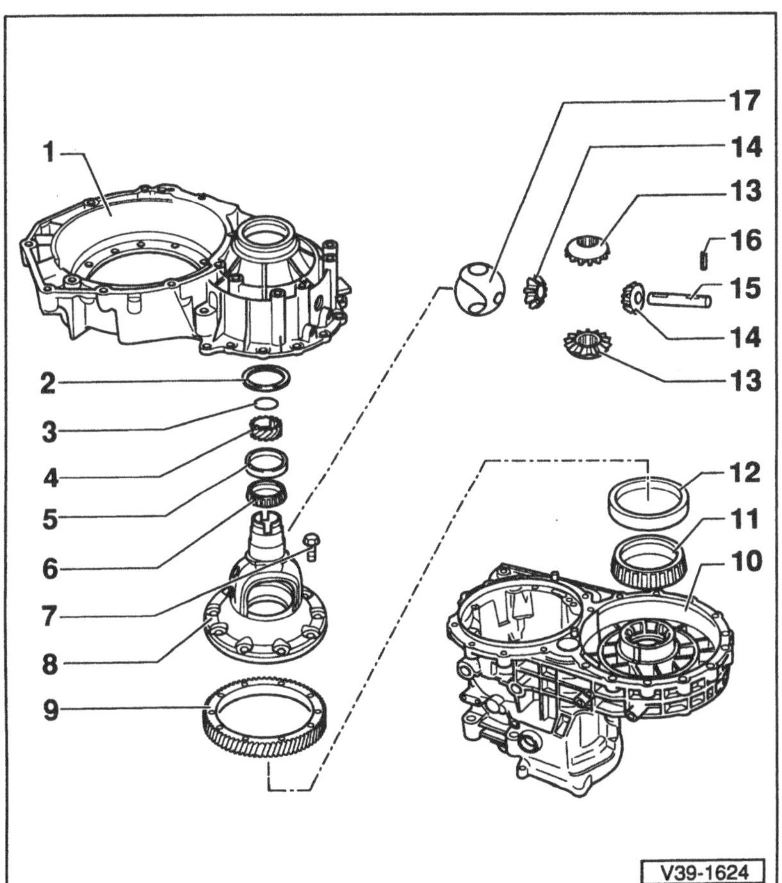

13 - Large bevel gears
- Installing ⇒ Fig. 10

14 - Small bevel gears
- Installing ⇒ Fig. 10

15 - Shaft for bevel gears
- Drive out with drift
- When driving in, do not damage one-piece thrust washer

16 - Spring pin
- For securing shaft for bevel gears
- Removing and installing spring pin with circumferential groove ⇒ Fig. 11

17 - One-piece thrust washer
- Place into differential housing before installing bevel gears

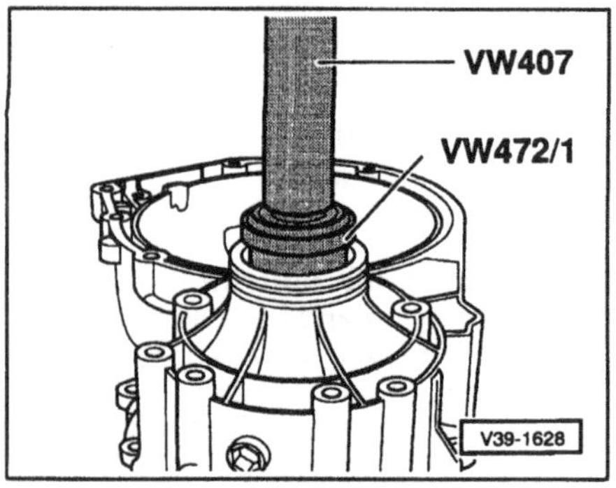

◄ **Fig. 1 Tapered roller bearing outer race, pressing out**

Note:

Only press tapered roller bearing outer race in when torque converter housing is heated.

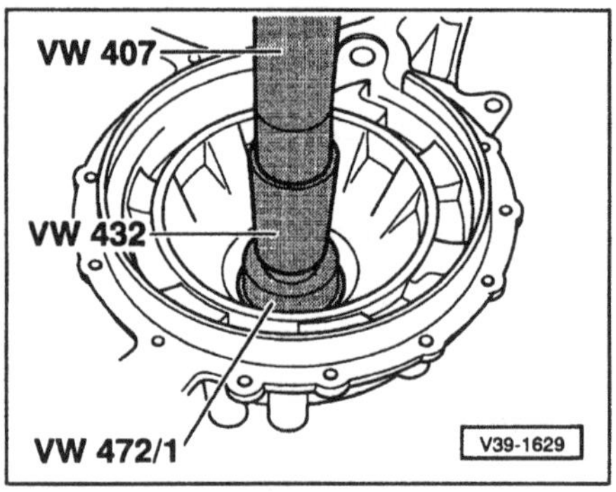

◄ **Fig. 2 Tapered roller bearing outer race, pressing into torque converter housing**

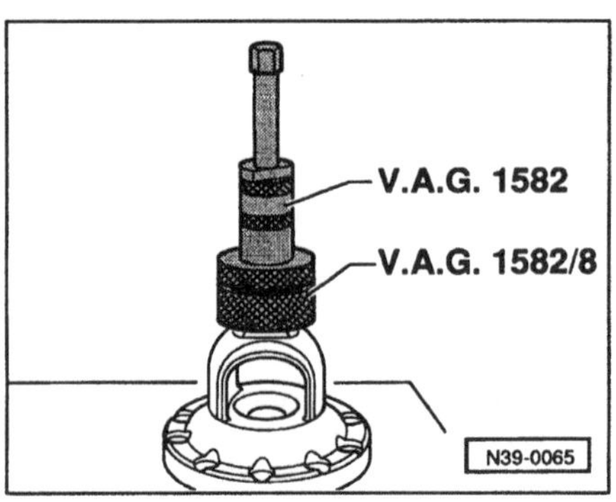

◄ **Fig. 3 Tapered roller bearing inner race, pulling off**

Place 40-105 thrust piece on differential housing.

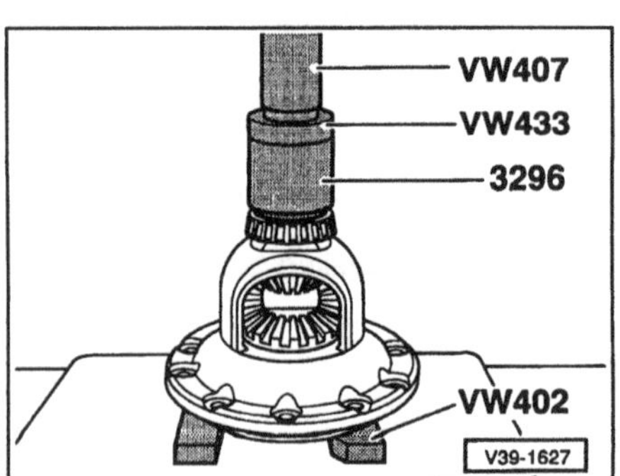

◄ **Fig. 4 Tapered roller bearing inner race, pressing on**

- Press tapered roller bearing inner race onto stop on differential housing.

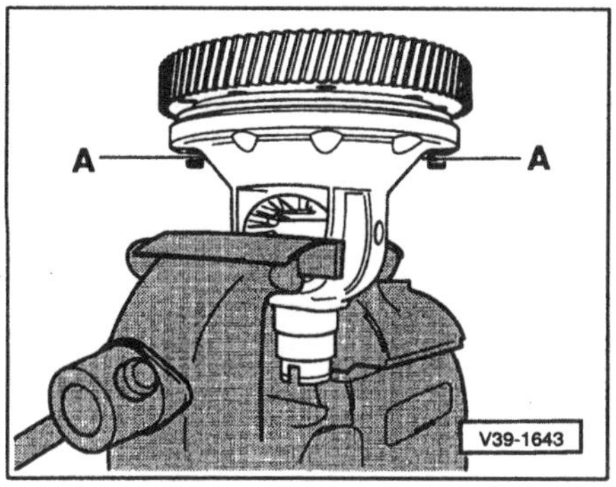

◄ **Fig. 5 Heated final drive gear wheel, installing**

– Heat final drive gear wheel to approx. 100°C (212°F).

– Install final drive gear wheel over centering pins -A- (local manufacturer) on differential housing.

Note:

Gear wheel recess faces toward differential housing.

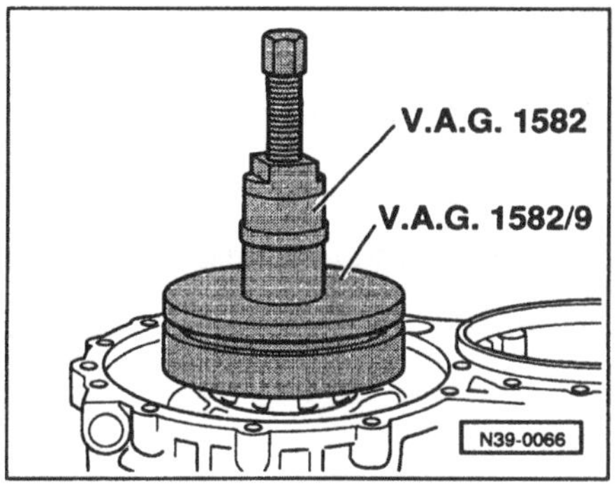

◄ **Fig. 6 Tapered roller bearing inner race, pulling off**

– Install VW 412 thrust disc on transmission housing.

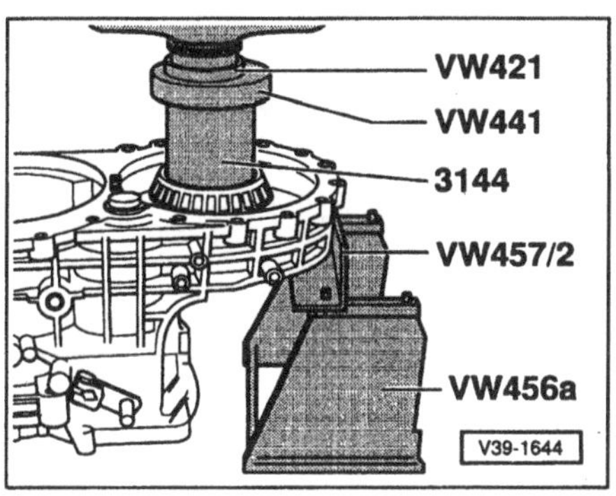

◄ **Fig. 7 Tapered roller bearing inner race, pressing into transmission housing**

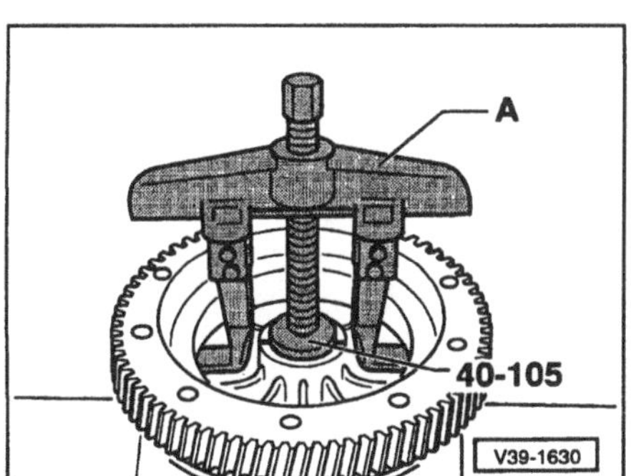

◄ **Fig. 8 Tapered roller bearing outer race, pulling off**

A - Kukko 20-10

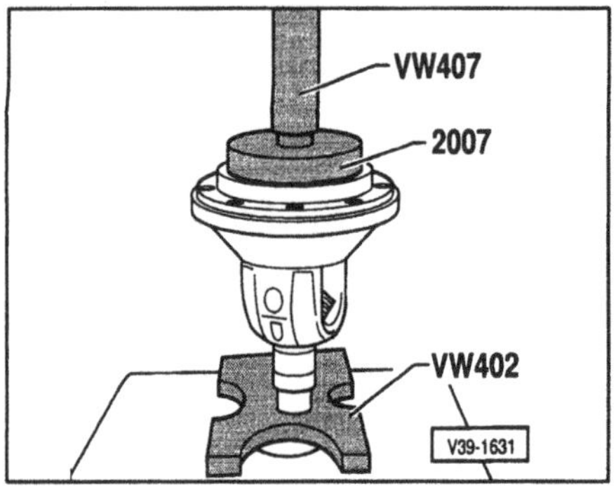

Fig. 9 **Tapered roller bearing outer race, pressing onto stop in differential housing**

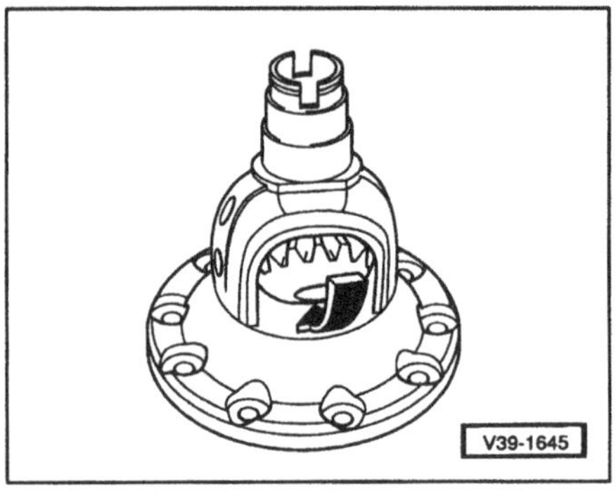

Fig. 10 **Differential bevel gears, installing**

- Install one-piece thrust washer with gear oil.
- Install small bevel gears, drive in shaft and secure with spring pin.
- Install large bevel gears offset 180° and swivel in in direction of arrow.

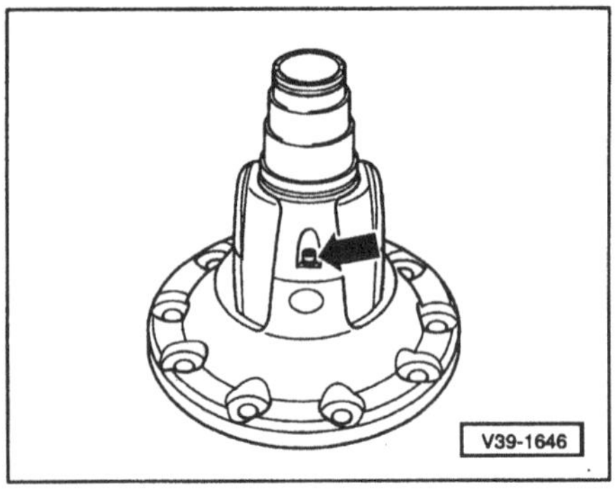

Fig. 11 **Spring pin with circumferential groove, removing and installing**

Removing

- Drive out spring pin using a chisel.

Installing

- Drive in spring pin with groove facing outward.
- Drive in spring pin up to stop.

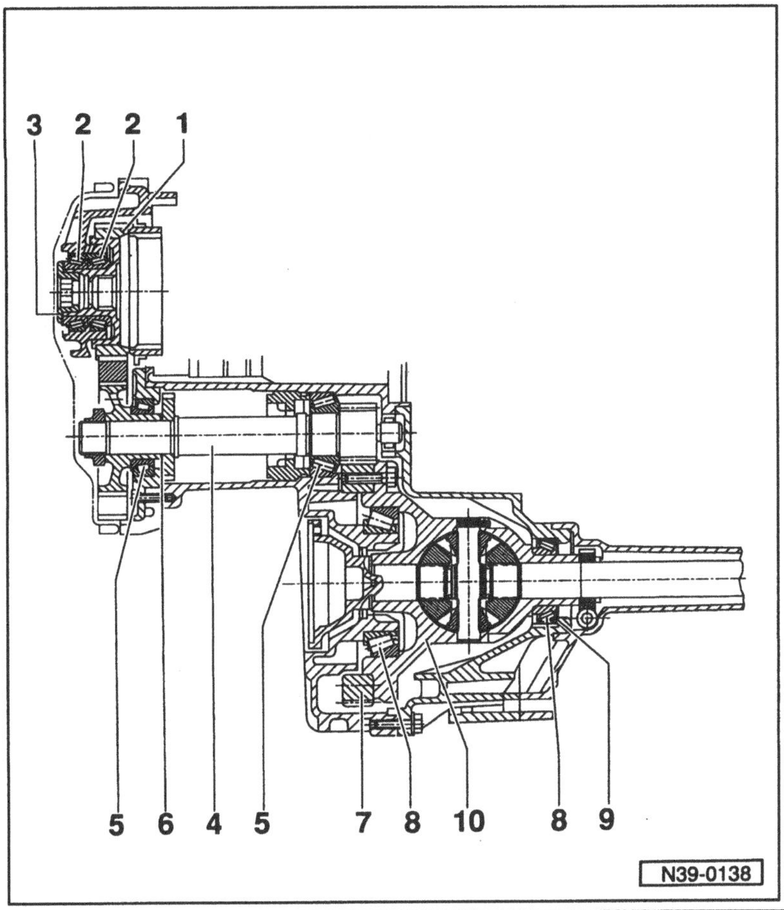

Final drive, adjusting

Overview of adjustments

1 - Input gear
- Adjusting
  ⇒ page H 39-59

2 - Tapered roller bearings for input gear
- Always replace together
- Measure friction torque by drive pinion. Friction torque of new tapered roller bearings must be 180 – 220 Ncm (16 – 20 in. lb) greater than friction torque of pinion shaft

3 - Shim for input gear
- Determine thickness on basis of table
  ⇒ page H 39-56
- Determine part number from parts catalog microfiche

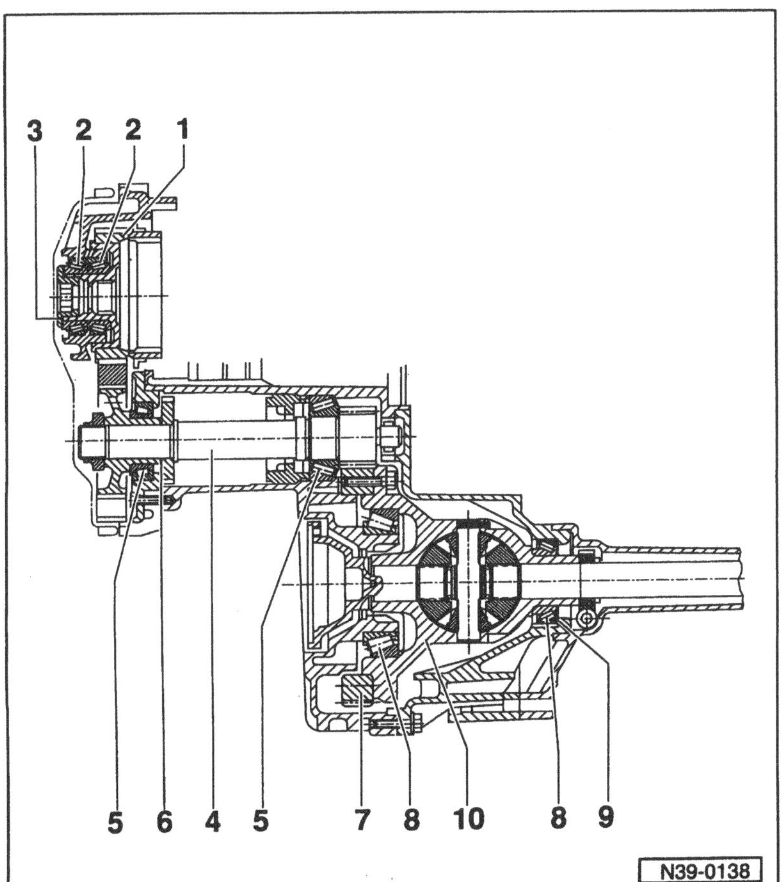

4 - Drive pinion
- Adjusting
  ⇒ page H 39-53
- If input gear, differential and drive pinion tapered roller bearings are replaced then the overall friction torque must be 220 – 420 Ncm (19 – 37 in. lb)

5 - Tapered roller bearings for drive pinion
- Always replace together
- Re-determine shim when replacing tapered roller bearings
- Friction torque of new tapered roller bearings must be 80 – 120 Ncm (7 – 11 in. lb)

6 - Shim for drive pinion
- Determine thickness on basis of table ⇒ page H 39-56
- Determine part number from parts catalog microfiche

7 - Final drive gear wheel

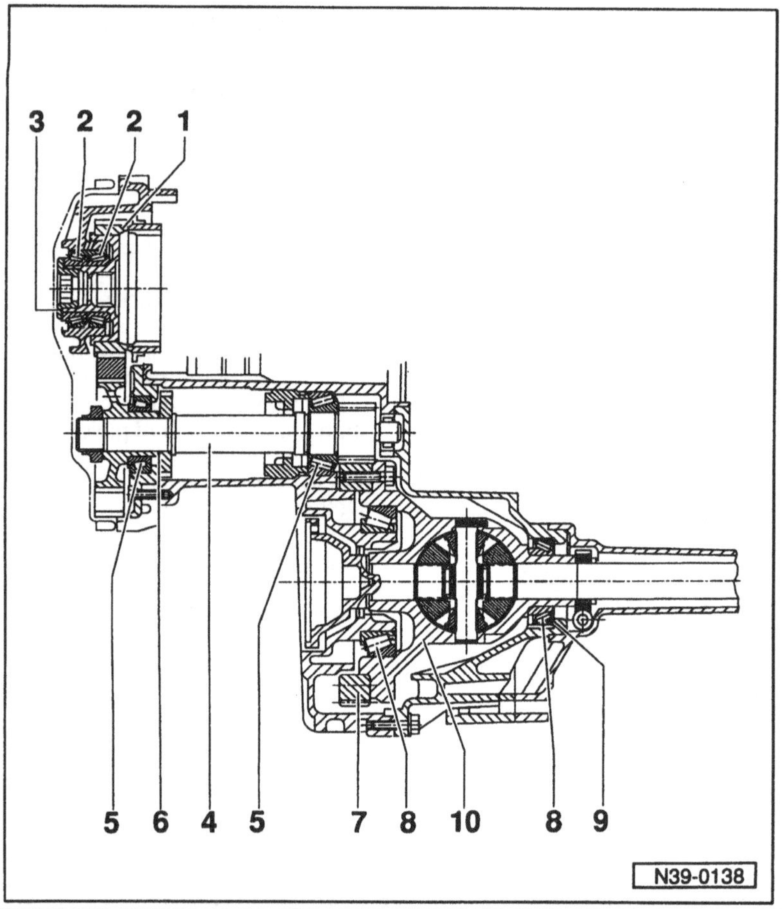

8 - **Tapered roller bearings for differential**
- Always replace together
- Measure friction torque through the drive pinion. Friction torque of new tapered roller bearings must be 50 – 90 Ncm (4 – 8 in. lb) greater than drive pinion friction torque

9 - **Shim for differential**
- Determine thickness from table ⇒ page H 39-66

10 - **Differential**
- Adjusting ⇒ page H 39-62

Adjustments to be completed after replacing individual components

When working on transmission, only the following components require adjustment if individual components have been replaced that directly affect the final drive adjustment. To eliminate unnecessary adjustments, observe following table:

	To adjust:		
Components being replaced:	**Differential** ⇒ **page H 39-62**	**Adjusting input gear**[1] ⇒ **page H 39-59**	**Adjusting drive pinion with output gear**[1] ⇒ **page H 39-53**
Transmission housing	X	X	X
Torque converter housing	X		
Input gear[1]		X	X
Input gear tapered roller bearing		X	
Output gear[1]		X	X
Drive pinion			X
Drive pinion tapered roller bearing			X
Differential housing with gear wheel	X		
Tapered roller bearing for differential, bearing body or shim	X		
Supporting ring, park lock gear or bearing cover			X

[1] Always replace input gear and output gear together

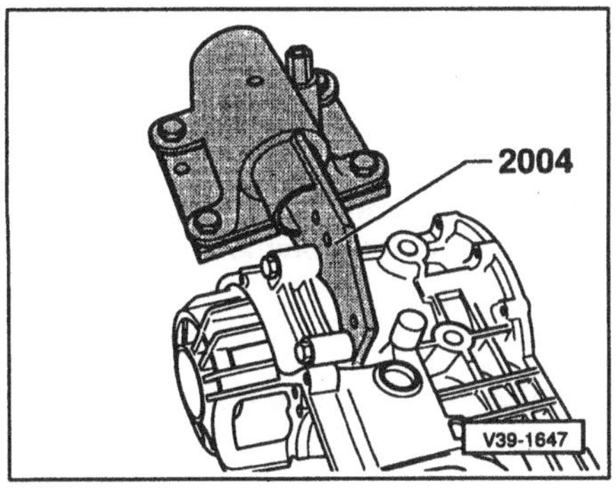

Drive pinion, adjusting

◄ – Secure transmission housing to 2004 transmission holder

– Place washers between transmission housing and transmission holder 2004 and secure transmission holder with 8 mm bolts.

Conditions:

- Supporting ring for drive pinion bearing is installed and secured.
- Parking lock mechanism and parking lock gear are installed.

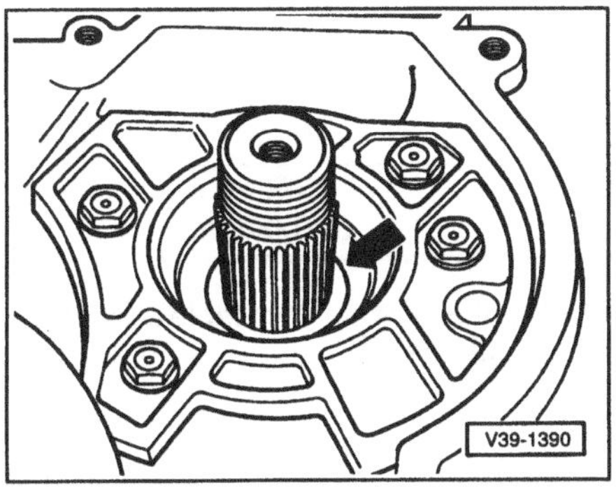

Adjusting:

◄ – Place two shims, each 1.5 mm (0.591 in.) thick -arrow- onto drive pinion.

– Place output gear with tapered roller bearing onto drive pinion and tighten hex nut (41 mm AF) to 250 Nm (184 in lb); engage parking lock when performing this step.

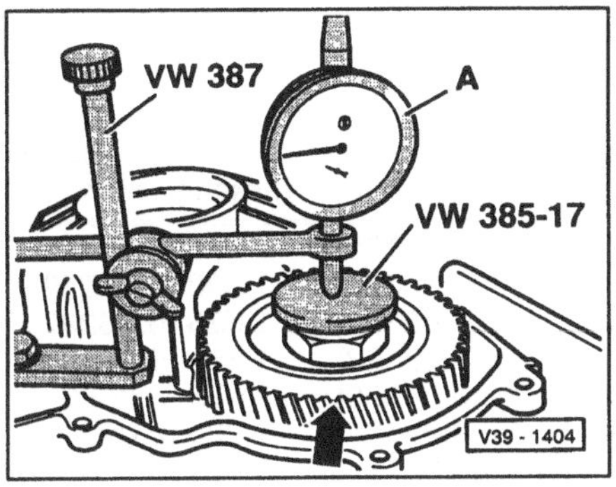

◄ – Position dial indicator -A- and zero indicator with a preload of 1 mm (0.0393in.).

– Move drive pinion up and down. Read off and record axial play at dial indicator and record (example: 0.93 mm (0.0366 in.)).

Note:

Do not turn the drive pinion while performing measurement because the bearings will settle and give a false measurement.

Determining shims

The specified bearing preload is achieved when the reading, 0.93 mm (0.0366 in.), the preload on the bearings, 0.12 mm (0.0047 in.), and the amount for compression, 0.10 mm (0.0039 in.) are deducted from the 2 x 1.5 mm (0.0590 in.) thick shims.

Example:

Shims inserted, 2 x 1.5 mm (0.0590 in.)	3.00 mm (0.1181 in.)
- Reading	0.93 mm (0.0366 in.)
- Preload	0.12 mm (0.0047 in.)
- Amount for compression	0.10 mm (0.0039 in.)
= Thickness of shims required	1.85 mm (0.0728 in.)

– Remove output gear. Determine size of shim from table and install shim onto drive pinion. Table of shims ⇒ page H 39-56.

– Re-install output gear and tighten hex nut to 250 Nm (184 ft lb) and secure with drift.

Table of shims for drive pinion and input gear

Gauged thickness mm (in.)	Shim to be installed mm (in.)	Gauged thickness mm (in.)	Shim to be installed mm (in.)
0.975 (0.03838) ... 0.999 (0.03933)	1.000 (0.03937)	1.550 (0.06102) ... 1.574 (0.06196)	1.575 (0.06200)
1.000 (0.03937) ... 1.024 (0.04031)	1.025 (0.04035)	1.575 (0.06200) ... 1.599 (0.06295)	1.600 (0.06299)
1.025 (0.04035) ... 1.049 (0.04129)	1.050 (0.04133)	1.600 (0.06299) ... 1.624 (0.06393)	1.625 (0.06397)
1.050 (0.04133) ... 1.074 (0.04228)	1.075 (0.04232)	1.625 (0.06397) ... 1.649 (0.06492)	1.650 (0.06496)
1.075 (0.04232) ... 1.099 (0.04326)	1.100 (0.04330)	1.650 (0.06496) ... 1.674 (0.06590)	1.675 (0.06594)
1.100 (0.04330) ... 1.124 (0.04425)	1.125 (0.04429)	1.675 (0.06594) ... 1.699 (0.06688)	1.700 (0.06692)
1.125 (0.04429) ... 1.149 (0.04523)	1.150 (0.04527)	1.700 (0.06692) ... 1.724 (0.06787)	1.725 (0.06791)
1.150 (0.04527) ... 1.174 (0.04622)	1.175 (0.04625)	1.725 (0.06791) ... 1.749 (0.06885)	1.750 (0.06889)
1.175 (0.04625) ... 1.199 (0.04720)	1.200 (0.04724)	1.750 (0.06889) ... 1.774 (0.06984)	1.775 (0.06988)
1.200 (0.04724) ... 1.224 (0.04818)	1.225 (0.04822)	1.775 (0.06988) ... 1.799 (0.07082)	1.800 (0.07086)
1.225 (0.04822) ... 1.249 (0.04917)	1.250 (0.04921)	1.800 (0.07086) ... 1.824 (0.07181)	1.825 (0.07185)
1.250 (0.04921) ... 1.274 (0.05015)	1.275 (0.05019)	1.825 (0.07185) ... 1.849 (0.07279)	1.850 (0.07283)
1.275 (0.05019) ... 1.299 (0.05114)	1.300 (0.05118)	1.850 (0.07283) ... 1.874 (0.07377)	1.875 (0.07381)
1.300 (0.05118) ... 1.324 (0.05212)	1.325 (0.05216)	1.875 (0.07381) ... 1.899 (0.07476)	1.900 (0.07480)
1.325 (0.05216) ... 1.349 (0.05311)	1.350 (0.05314)	1.900 (0.07480) ... 1.924 (0.07574)	1.925 (0.07578)
1.350 (0.05314) ... 1.374 (0.05409)	1.375 (0.05413)	1.925 (0.07578) ... 1.949 (0.07673)	1.950 (0.07677)
1.375 (0.05413) ... 1.399 (0.05507)	1.400 (0.05511)	1.950 (0.07677) ... 1.974 (0.07771)	1.975 (0.07775)
1.400 (0.05511) ... 1.424 (0.05606)	1.425 (0.05610)	1.975 (0.07775) ... 1.999 0.07870)	2.000 (0.07874)
1.425 (0.05610) ... 1.449 (0.05704)	1.450 (0.05708)	2.000 (0.07874) ... 2.024 (0.07968)	2.025 (0.07972)
1.450 (0.05708) ... 1.474 (0.05803)	1.475 (0.05807)	2.025 (0.07972) ... 2.049 (0.08066)	2.050 (0.08070)
1.475 (0.05807) ... 1.499 (0.05901)	1.500 (0.05905)	2.050 (0.08070) ... 2.074 (0.08165)	2.075 (0.08169)
1.500 (0.05905) ... 1.524 (0.05999)	1.525 (0.06003)	2.075 (0.08169) ... 2.099 (0.08263)	2.100 (0.08267)
1.525 (0.06003) ... 1.549 (0.06098)	1.550 (0.06102)	2.100 (0.08267) ... 2.124 (0.08362)	2.125 (0.08366)

Different tolerances enable the required shim thickness to be measured precisely.

Part numbers for shims:
⇒ Parts catalog microfiche

Gauged thickness mm (in.)	Shim to be installed mm (in.)
2.125 (0.08366) ... 2.149 (0.08460)	2.150 (0.08464)
2.150 (0.08464) ... 2.174 (0.08559)	2.175 (0.08562)
2.175 (0.08562) ... 2.199 (0.08657)	2.200 (0.08661)
2.200 (0.08661) ... 2.224 (0.08755)	2.225 (0.08759)
2.225 (0.08759) ... 2.249 (0.08854)	2.250 (0.08858)
2.250 (0.08858) ... 2.274 (0.08952)	2.275 (0.08956)
2.275 (0.08956) ... 2.299 (0.09051)	2.300 (0.09055)
2.230 (0.09055) ... 2.324 (0.09149)	2.325 (0.09153)
2.325 (0.09153) ... 2.349 (0.09248)	2.350 (0.09251)
2.350 (0.09251) ... 2.374 (0.09346)	2.375 (0.09350)
2.375 (0.09350) ... 2.399 (0.09444)	2.400 (0.09448)
2.400 (0.09448) ... 2.424 (0.09543)	2.425 (0.09547)
2.425 (0.09547) ... 2.449 (0.09641)	2.450 (0.09645)
2.450 (0.09645) ... 2.474 (0.09740)	2.475 (0.09744)
2.475 (0.09744) ... 2.499 (0.09838)	2.500 (0.09842)
2.500 (0.09842) ... 2.524 (0.09936)	2.525 (0.09940)
2.525 (0.09940) ... 2.549 (0.10035)	2.550 (0.10039)
2.550 (0.10039) ... 2.574 (0.10133)	2.575 (0.10137)
2.575 (0.10137) ... 2.599 (0.10232)	2.600 (0.10236)
2.600 (0.10236) ... 2.624 (0.10330)	2.625 (0.10334)
2.625 (0.10334) ... 2.649 (0.10429)	2.650 (0.10433)
2.650 (0.10433) ... 2.674 (0.10527)	2.675 (0.10531)
2.675 (0.10531) ... 2.700 (0.10629)	2.700 (0.10629)

Different tolerances enable the required shim thickness to be measured precisely.

Part numbers for shims:

⇒ Parts catalog microfiche

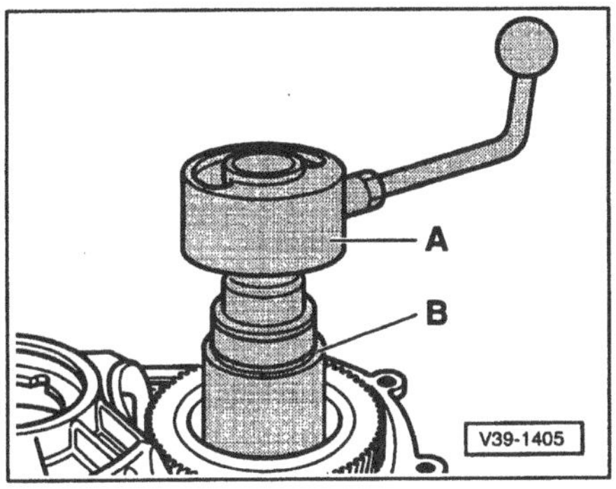

◄ **Checking friction torque**

A – Torque wrench, commercially available, 0 – 600 Ncm (0 – 53 in lb)

B – Wrench socket (41 mm AF)

– Oil bearing with gear oil.

- Friction torque of new tapered roller bearings must be 80 – 120 Ncm. (7 – 11 in lb)

Input gear, adjusting

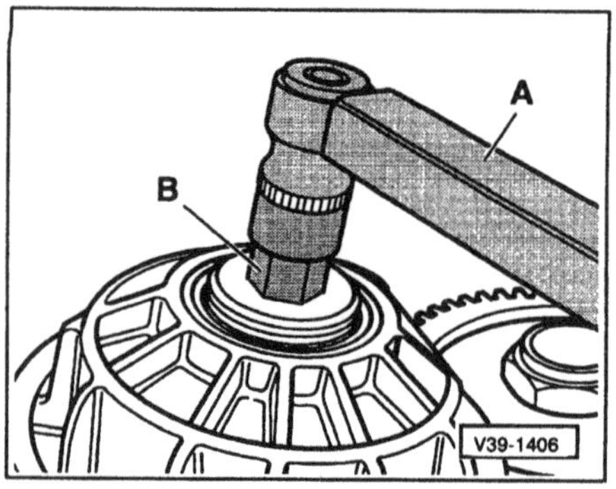

◄ – Tighten socket-head fastener to 100 Nm (74 ft lb) without dished washer and without shim.

A – Torque wrench
B – 22 mm socket

The parking lock must be engaged.

– Remove socket-head fastener.

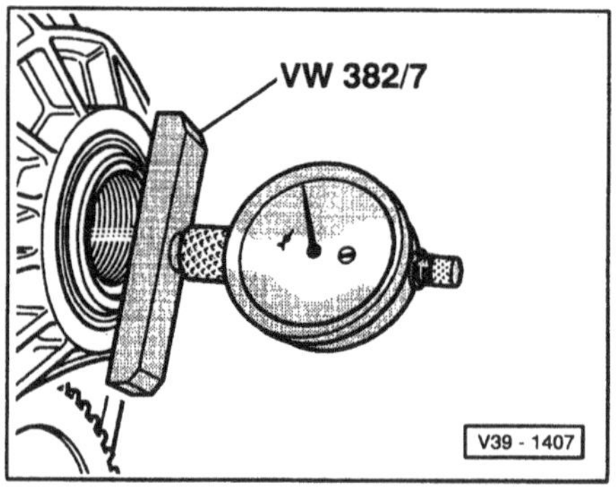

◄ – Measure distance between input gear and tapered roller bearing inner race.

– Insert dial indicator and zero with a 3 mm (0.1181 in.) preload. Record reading Example: 1.00 mm (0.0393 in.).

The thickness of the dished washer, 1.50 mm (0.0590 in.), must be added to the reading: 1.00 mm (0.0393 in.) + 1.50 mm (0.0590 in.) = 2.50 mm (0.0984 in.).

Determining shim thickness

The specified bearing preload is achieved when the bearing preload of 0.18 mm (0.0070 in.) is subtracted from the measured value – (dished washer + reading) of 2.50 mm (0.9842 in.).

Example:

Dished washer + reading	2.50 mm (0.0984 in.)
- Bearing preload	0.18 mm (0.0070 in.)
= Thickness of shim required	2.32 mm (0.0913 in.)

– Determine size of shim from table and install into drive shaft. Table of shims
⇒ page H 39-56.

– Insert axial needle bearing and tighten socket-head fastener including dished washer to 250 Nm (184 ft lb).

Note:

Insert axial needle bearing for input shaft before tightening socket-head fastener. Installation position
⇒ Removing and installing input gear, page H 39-23.

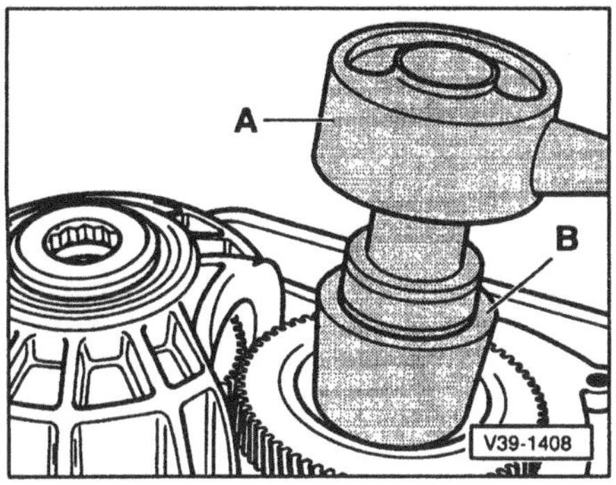

◄ **Checking friction torque**

A – Torque wrench, commercially available 0 – 600 Ncm (0 – 53 in lb)

B – Wrench socket (41 mm)

The friction torque of new tapered roller bearings must be 180 – 220 Ncm (16 – 20 in lb) greater than friction torque at the drive pinion.

Example:

	Friction torque
Drive pinion (from measurement)	= 100 Ncm[1)] (9 in lb)
Input gear (average)	= 200 Ncm (18 in lb)
Friction torque of input gear + drive pinion	= 300 Ncm (27 in lb)

– Moisten bearings with ATF before installing

[1)] Insert the friction torque measured for the drive pinion if used bearings are reinstalled.

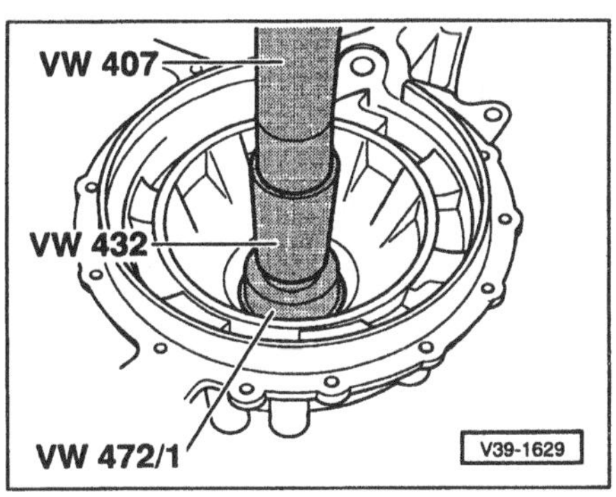

Differential, adjusting

◄ – Press tapered roller bearing outer race, without shim, into torque converter housing.

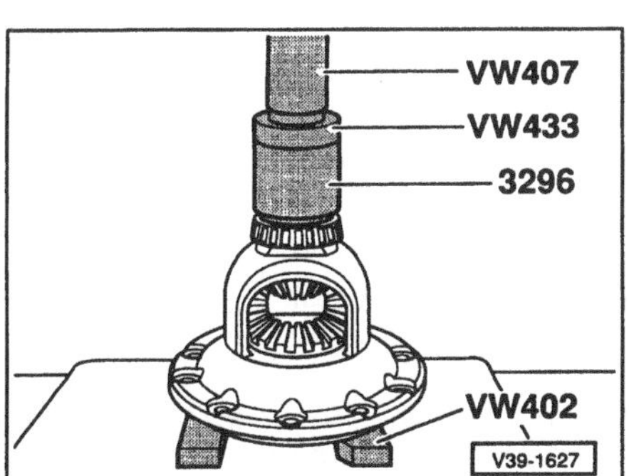

◄ – Press tapered roller bearing inner race onto differential housing.

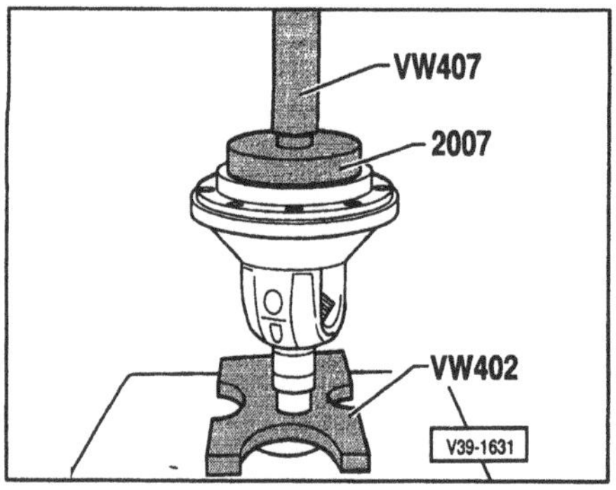

◄ – Press tapered roller bearing outer race onto stop in differential housing.

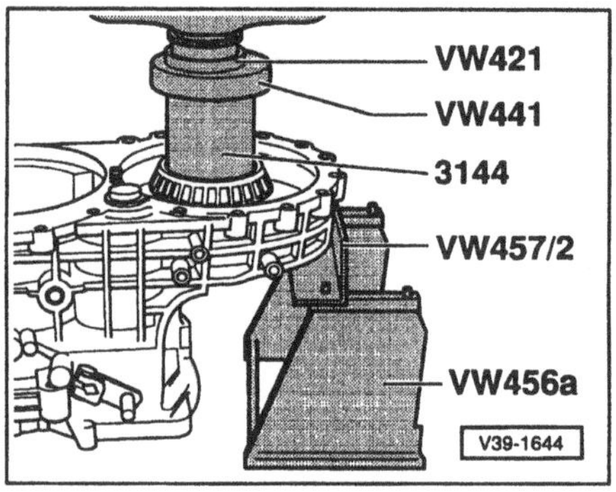

◄ – Press tapered roller bearing inner race into differential housing.

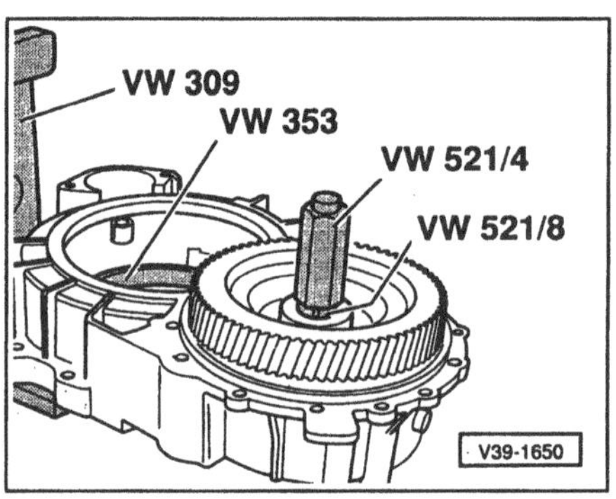

◄ – Bolt torque converter housing onto VW 353 transmission support and install differential.

– Install special tool into differential housing.

– Install transmission housing on differential and install every second bolt. Tighten to 25 Nm (18 ft lb).

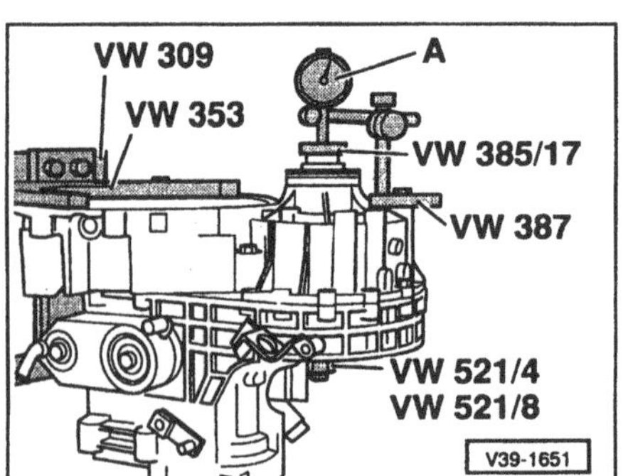

◄ – Assemble special tool to measure play.

– Zero dial indicator with 1 mm (0.0393 in.) preload.

– Lift and lower differential.

– Read and record axial play from dial indicator.

Note:

Do not turn differential while performing measurement because the bearings will settle and give a false measurement.

Determining shim thickness

The specified bearing preload is achieved if the bearing preload of 0.12 mm (0.0047 in.) and the settling amount of 0.10 mm (0.0039 in.) is added to the figure obtained from the measurement.

Example:

Measured value		0.60 mm (0.0236 in.)
+	Preload	0.12 mm (0.0047 in.)
+	Settling allowance	0.10 mm (0.0039 in.)
=	Thickness of shim to be installed	0.82 mm (0.0322 in.)

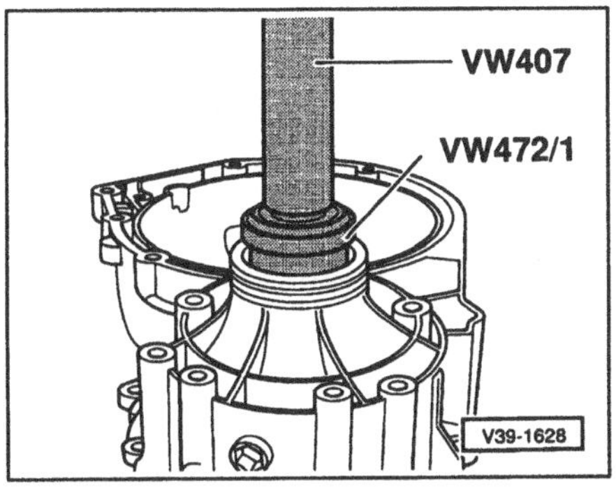

– Unbolt transmission housing and remove differential from torque converter housing.

– Press tapered roller bearing outer race out of torque converter housing.

Table of shims for differential

Thickness mm (in.)	Part number	Thickness mm (in.)	Part number
0.65 (0.0255)	02B 409 210	1.05 (0.0413)	02B 409 210 H
0.70 (0.0275	02B 409 210 A	1.10 (0.0433)	02B 409 210 J
0.75 (0.0295)	02B 409 210 B	1.15 (0.0452)	02B 409 210 K
0.80 (0.0314)	02B 409 210 C	1.20 (0.0472)	02B 409 210 L
0.85 (0.0334)	02B 409 210 D	1.25 (0.0492)	02B 409 210 M
0.90 (0.0354)	02B 409 210 E	1.30 (0.0511)	02B 409 210 N
0.95 (0.0374)	02B 409 210 F	1.35 (0.0531)	02B 409 210 P
1.00 (0.0393)	02B 409 210 G	1.40 (0.0551)	02B 409 210 Q

If the thickness measured is greatest than the sizes in the table then 2 shims equal to the thickness of the shim to be installed may be used.

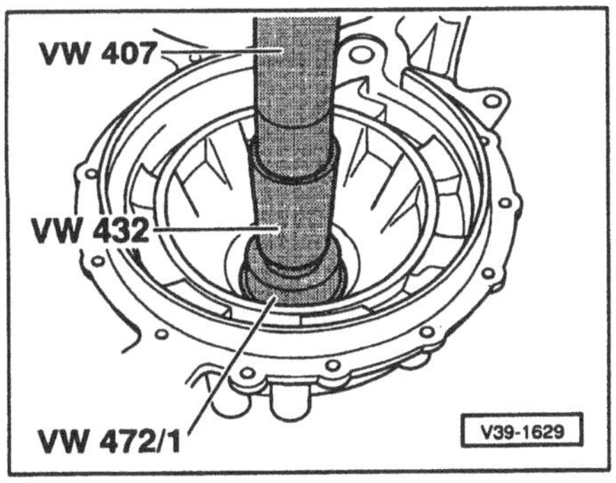

– Press tapered roller bearing outer ring, with shim, into torque converter housing.

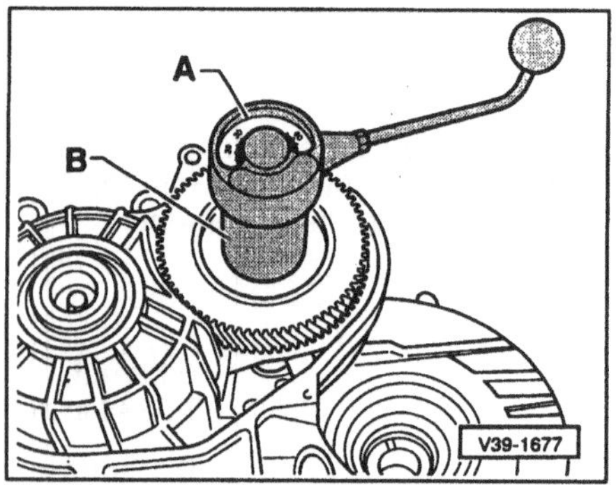

Checking friction torque

A – Torque wrench, commercially available 0 – 600 Ncm (0 – 53 in lb)

B – Wrench socket (41 mm AF)

The friction torque of new tapered roller bearings must be 50 – 90 Ncm (4 – 8 in. lb) greater than at the drive pinion.

Example:

	Friction torque
Drive pinion (measured value)	= 300 Ncm[1)] (27 in lb)
Differential (average)	= 70 Ncm (6 in. lb)
Friction torque of differential + drive pinion	= 370 Ncm (33 in lb)

– Oil bearing beforehand with gear oil.

[1)] Install the friction torque measured for the drive pinion in the case of used bearings.

Automatic Transmission 01P

Section J

WARNING!
Before proceeding with any work, please read the warnings and cautions immediately following this index. Also read any warning and caution that accompanies a procedure in the manual.

37 Automatic transmission—Controls, Housing

38 Automatic transmission—Gears, Shafts

39 Final drive, Differential—Front

Please read these WARNINGS and CAUTIONS before proceeding with maintenance and repair work.

WARNING—

- Some repairs may be beyond your capability. If you lack the skills, tools and equipment, or a suitable workplace for any procedure described in this manual, we suggest you leave such repairs to an authorized Volkswagen dealer service department, or other qualified shop.

- Volkswagen is constantly improving its cars. Sometimes these changes, both in parts and specifications, are made applicable to earlier models. Therefore, before starting any major jobs or repairs to components on which passenger safety may depend, consult your authorized Volkswagen dealer about the Technical Bulletins that may have been issued since the editorial closing of this manual.

- Do not re-use any fasteners that are worn or deformed in normal use. Many fasteners are designed to be used only once and become unreliable and may fail when used a second time. This includes, but is not limited to, nuts, bolts, washers, self–locking nuts or bolts, circlips and cotter pins. Always replace these fasteners with new parts.

- Never work under a lifted car unless it is solidly supported on stands designed for the purpose. Do not support a car on cinder blocks, hollow tiles or other props that may crumble under continuous load. Never work under a vehicle that is supported solely by a jack. Never work under the car while the engine is running.

- If you are going to work under a car on the ground, make sure that the ground is level. Block the wheels to keep the car from rolling. Disconnect the battery negative terminal (ground strap) to prevent others from starting the vehicle while you are under it.

- Never run the engine unless the work area is well ventilated. Carbon monoxide (CO) kills.

- Do not attempt to work on your vehicle if you do not feel well. You increase the danger of injury to yourself and others if you are tired, upset or have taken medicine or any other substance that may impair you or keep you from being fully alert.

- Disconnect the battery negative (–) terminal (ground strap) whenever you work on the fuel system or the electrical system. Do not smoke or work near heaters or other fire hazards. Keep an approved fire extinguisher handy.

- Batteries produce explosive hydrogen gas. Keep sparks, lighted matches and open flame away from the top of the battery. If hydrogen gas escaping from the cap vents is ignited, it will ignite gas trapped in the cells and cause the battery to explode.

- Connect and disconnect battery cables, jumper cables or a battery charger only with the ignition switched off, to prevent sparks. Do not quick–charge the battery (for boost starting) for longer than one minute, and do not allow charging voltage to exceed 16.5 volts. Wait at least one minute before boosting the battery a second time.

- Illuminate your work area adequately but safely. Use a portable safety light for working inside or under the car. Make sure the bulb is enclosed by a wire cage. The hot filament of an accidentally broken bulb can ignite spilled fuel or oil.

- Tie long hair behind your head. Do not wear a necktie, a scarf, loose clothing, or a necklace when you work near machine tools or running engines. If your hair, clothing, or jewelry were to get caught in the machinery, severe injury could result.

- Finger rings, bracelets and other jewelry should be removed so that they cannot cause electrical shorts, get caught in running machinery, or be crushed by heavy parts.

- Catch draining fuel, oil, or brake fluid in suitable containers. Do not use food or beverage containers that might mislead someone into drinking from them. Store flammable fluids away from fire hazards. Wipe up spills at once, but do not store the oily rags, which can ignite and burn spontaneously.

- Always observe good workshop practices. Wear goggles when you operate machine tools or work with battery acid. Gloves or other protective clothing should be worn whenever the job requires working with harmful substances.

- The Air-Conditioning (A/C) system is filled with a chemical refrigerant, which is hazardous. The A/C system should be serviced only by trained technicians using approved refrigerant recovery/recycling equipment, trained in related safety precautions, and familiar with regulations governing the discharging and disposal of automotive chemical refrigerants.

- Do not expose any part of the A/C system to high temperatures such as an open flame. Excessive heat will increase system pressure and may cause the system to burst.

- Some aerosol tire inflators are highly flammable. Be extremely cautious when repairing a tire that may have been inflated using an aerosol tire inflator. Keep sparks, open flame or other sources of ignition away from the tire repair area. Inflate and deflate the tire at least four times before breaking the bead from the rim. Completely remove the tire from the rim before attempting any repair.

- Friction materials such as brake or clutch discs may contain asbestos fibers. Do not create dust by grinding, sanding, or by cleaning with compressed air. Avoid breathing asbestos fibers and asbestos dust. Breathing asbestos can cause serious diseases such as asbestosis or cancer, and may result in death.

- Greases, lubricants and other automotive chemicals contain toxic substances, many of which are absorbed directly through the skin. Read manufacturer's instructions and warnings carefully. Use hand and eye protection. Avoid direct skin contact.

- An airbag is operated by an explosive device. Handled improperly or without adequate safeguards, it can be accidently activated and cause serious personal injury. To prevent personal injury or airbag system failure, **only trained Volkswagen dealer service technicians** should test, disassemble or service the airbag system.

WARNINGS and CAUTIONS continued on next page ▶

Please read these WARNINGS and CAUTIONS before proceeding with maintenance and repair work.

WARNING (continued) —

- ALWAYS disconnect the airbag power supply before using a computer memory saver. Failure to disconnect the airbag power supply may result in accidental activation of the airbag.
- The airbag unit should not be removed, nor should any work be performed on the airbag system, without first reviewing the warnings and cautions that accompany the airbag system repair information in this manual.
- Never apply stickers or any other coverings on the steering wheel. Do not let chemical cleaners, oil or grease come into contact with the vinyl covering of the airbag unit.
- Never open or otherwise attempt to repair airbag system parts. Always use new parts. Never leave airbag parts or the partially disassembled airbag system unattended.
- Never use a test light or volt meter to conduct electrical tests on the airbag system. The airbag system must only be tested by trained Volkswagen Service technicians using special airbag tester equipment. The airbag system must only be tested while it is installed in the car.
- Do not expose an airbag unit to temperatures above 194°F (90°C) even for a brief period, while handling it during the repair process. Keep the unit clear of heat sources such as hot plates, soldering irons, heat lamps, welding equipment, or the like.
- Always replace airbag system components that have been mechanically damaged (bubbles, cracks, etc.).
- Always set down a removed airbag unit so that the horn pad is facing upward.
- When driving or riding in an airbag-equipped vehicle, never hold test equipment in your hands or lap while the vehicle is in motion. Objects between you and the airbag can increase the risk of injury in an accident.
- The storage, transportation and disposal of airbag units are subject to the laws for flammable solids.

CAUTION—

- If you lack the skills, tools and equipment, or a suitable workshop for any procedure described in this manual, we suggest you leave such repairs to an authorized Volkswagen dealer or other qualified shop. We especially urge you to consult an authorized Volkswagen dealer before beginning repairs on any car that may still be covered wholly or in part by any of the extensive warranties issued by Volkswagen of America, Inc.
- Volkswagen is constantly improving its cars and sometimes these changes, both in parts and specifications, are made applicable to earlier models. Therefore, part numbers listed in this manual are for reference only. Always check with your authorized Volkswagen dealer parts department for the latest information.
- Before starting a job, make certain that you have all the necessary tools and parts on hand. Read all the instructions thoroughly, do not attempt shortcuts. Use tools that are appropriate to the work and use only replacement parts meeting Volkswagen specifications. Makeshift tools, parts and procedures will not make good repairs.
- Use pneumatic and electric tools only to loosen threaded parts and fasteners. Never use these tools to tighten fasteners, especially on light alloy parts. Always use a torque wrench to tighten fasteners to the tightening torque specification listed.
- Be mindful of the environment and ecology. Before you drain the crankcase, find out the proper way to dispose of the oil. Do not pour oil onto the ground, down a drain, or into a stream, pond, or lake. Consult local ordinances that govern the disposal of wastes.
- When connecting cables or ground straps, all connections must be cleaned to bare metal. Be sure mounting bolts are tight.
- Do not replace fuse links with fuse links of other gauge or with standard wire. Always replace the entire length of the fuse link.
- Before doing any electrical welding on vehicles equipped with Anti-lock Brakes (ABS), disconnect the battery negative terminal (ground strap) and the ABS Control Module connector.
- On cars equipped with ABS, the ABS control unit cannot withstand temperatures from a paint–drying booth or a heat lamp in excess of 203°F (95°C) and should not be subjected to temperatures in excess of 185°F (85°C) for more than two hours.
- On cars equipped with anti-theft radios, make sure you know the correct radio activation code before disconnecting the battery or removing the radio. If the wrong code is entered when power is restored, the radio may lock up and be rendered inoperable, even if the correct code is then entered.
- When boost-charging the battery, first remove the fuses for the Engine Control Module (ECM), the Transmission Control Module (TCM), the ABS Control Module, and the On-Board computer. In cases where one or more of these components is not separately fused, disconnect the Control Module connector(s).
- Do not quick-charge the battery (for boost starting) for longer than one minute, and do not exceed 16.5 volts at the battery with the boosting cables attached. Wait at least one minute before boosting the battery a second time.

◀ **WARNINGS and CAUTIONS** continued from previous page

Transmission, identification

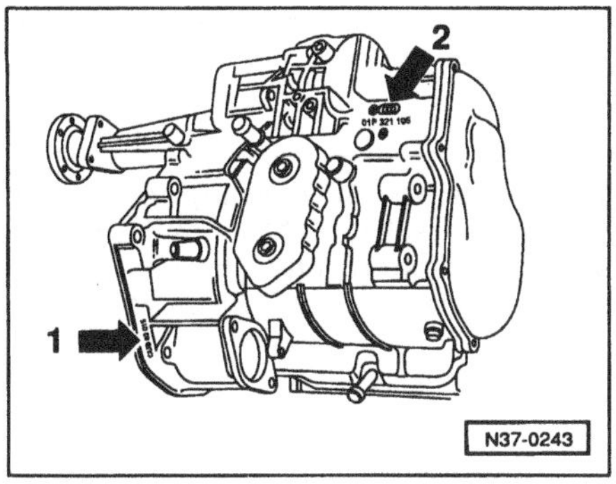

The 4–speed automatic transmission 01P is installed in the EuroVan m.y. 1996 ➤ in combination with the 6–cylinder engine. Applications ⇒ Page J 00-6.

Location on transmission

◄ Code letters (arrow 1)

Automatic transmission 01P (arrow 2)

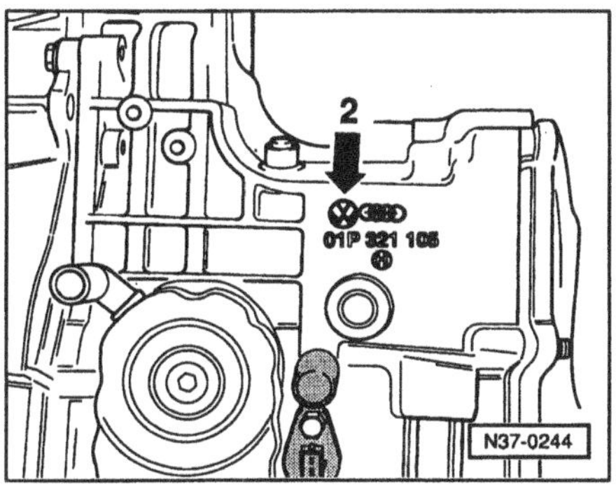

◄ Automatic transmission 01P (arrow 2)

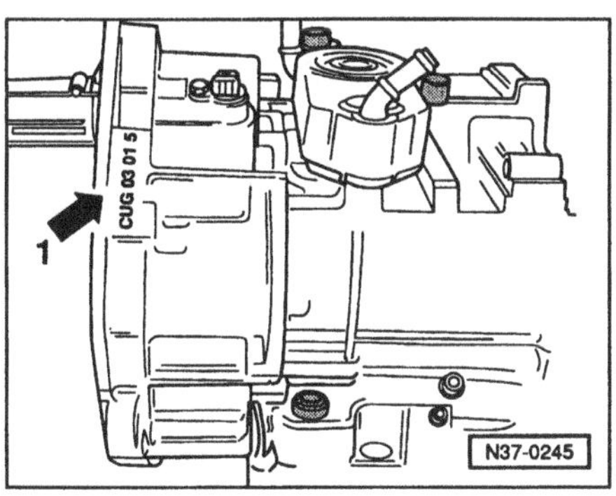

◄ **Code letters and date of manufacture -arrow 1-**

Example:	**CUG**	**03**	**01**	**5**
	I	I	I	I
	I	I	I	I
	Code letters	Day	Month	Year (1995) of manufacture

The code letters of the transmission are also given on the vehicle data plates.

Automatic transmission 01P, overview

There are several fundamental differences between the 4–speed automatic transmission 01P and the Automatic transmission 098. These changes are briefly mentioned below.

Transmission

The 4–speed automatic Transmission 01P is equipped with 4 hydraulically activated forward gears. When the lock–up clutch in the torque converter is closed these forward gears become mechanically driven gears by eliminating the torque converter slip.

Torque converter

The torque converter is equipped with a lock–up clutch. The lock–up clutch is closed, dependent on load and speed, and drives 1st, 2nd, 3rd and 4th gears mechanically, independent of slip.

Control module –J217– with driving situation gear changes (Fuzzy Logic)

Driving situation adaptation takes place automatically.

Advantages:

- Gear changes will be consistent with better fuel economy.
- Maximum engine output is always available
- Individual adaptation of gear change points in all driving situations
- Gear change points are infinitely variable

Gear change point variations for hills

An additional gear change map automatically selects gear changes for hills, dependent upon Accelerator Pedal position and driving speed.

- Gear change map for extreme uphill stretches is matched to engine output
- Gear change map for extreme downhill stretches is matched to the braking effect of the engine

New ATF for Volkswagen (VW ATF)

The new VW ATF has a yellowish color.

The planetary gearbox is filled with VW ATF.

- Due to a modified checking procedure for the ATF, the ATF dipstick has been discontinued.
- Checking the ATF level or changing the ATF is now done during the inspection service.
- Only VW ATF must be used.

Container size: 1.0 L – Part No. G 052162 A2

> ***CAUTION!***
>
> ***Only the VW ATF available as a replacement part can be used in the automatic transmission 01P.***

Code letters, transmission applications, ratios, equipment

Automatic transmission			01P					
Transmission	Code letters		**CUF**		**CUH**		**CUG**	
	Manufactured	from to	01.95		01.95		01.95	
Torque converter	Code letters		**QBDB/QBDC**		**QBDB/QBDC**		**QBDB**	
Valve body	Code letters		**QCB**		**QCB**		**QCB**	
	Manufactured	from to	01.95		01.95		01.95	
Number of plates			Inner	Outer	Inner	Outer	Inner	Outer
	Clutch –K1–		5	5	5	5	5	5
	Clutch –K2–		5	5	5	5	5	5
	Clutch –K3–		5	4	5	4	5	4
	Brake –B1–		5	5	5	5	5	5
	Brake –B2–		6	7	6	7	6	7
Application	Model		EuroVan 1992➤		EuroVan 1992➤		EuroVan 1992➤	
	Engine		2.4L – 57 kW (75 hp) Diesel		2.4L – 57 kW (75 hp) Diesel		2.5L – 81 kW (109 hp)	

Automatic transmission			01P		
Transmission	Code letters		**CUF**	**CUH**	**CUG**
Ratios	1st gear		2.714	2.714	2.714
	2nd gear		1.441	1.441	1.441
	3rd gear		1.000	1.000	1.000
	4th gear		0.743	0.743	0.743
	Reverse gear		2.884	2.884	2.884
Intermediate drive	Number of teeth	Input gear	44	44	44
		Output gear	51	51	51
	Ratio		1.159	1.159	1.159
Final drive	Number of teeth	Drive pinion	17	16	17
		Final drive gear	72	73	72
	Ratio		4.235	4.563	4.235
Drive shaft	Flange diameter		108mm	108mm	108mm
ATF cooler	Equipment		Pressure side 6-row	Pressure side 6-row	Pressure side 6-row

Automatic transmission			01P					
Transmission	Code letters		**CUJ**		**CUV**		**DKT**	
	Manufactured	from to	01.95		01.95		12.95	
Torque converter	Code letters		**QBDB**		**QBDB**		**QBDB**	
Valve body	Code letters		**QCB**		**QCB**		**QCB**	
	Manufactured	from to	01.95		01.95		12.95	
Number of plates			Inner	Outer	Inner	Outer	Inner	Outer
	Clutch –K1–		5	5	5	5	5	5
	Clutch –K2–		5	5	5	5	5	5
	Clutch –K3–		5	4	5	4	5	4
	Brake –B1–		5	5	5	5	5	5
	Brake –B2–		6	7	6	7	6	7
Application	Model		EuroVan 1992➤		EuroVan 1992➤		EuroVan 1992➤	
	Engine		2.5L – 81 kW (109 hp)		2.5L – 81 kW (109 hp)		2.5L – 81 kW (109 hp)	

Automatic transmission			01P		
Transmission	Code letters		**CUJ**	**CUV**	**DKT**
Ratios	1st gear		2.714	2.714	2.714
	2nd gear		1.441	1.441	1.441
	3rd gear		1.000	1.000	1.000
	4th gear		0.743	0.743	0.743
	Reverse gear		2.884	2.884	2.884
Intermediate drive	Number of teeth	Input gear	44	44	44
		Output gear	51	51	51
	Ratio		1.159	1.159	1.159
Final drive	Number of teeth	Drive pinion	16	15	15
		Final drive gear	73	74	74
	Ratio		4.056	4.933	4.933
Drive shaft	Flange diameter		108mm	108mm	108mm
ATF cooler	Equipment		Pressure side 6-row	Pressure side 6-row	Pressure side 6-row

Automatic transmission			01P					
Transmission	Code letters		**CWC**		**CWE**		**CWD, DLM**	
	Manufactured	from to	01.96		01.96		01.96	
Torque converter	Code letters		**QBDB**		**QBDB**		**QBDB**	
Valve body	Code letters		**QCB**		**QCB**		**QCB**	
	Manufactured	from to	01.96		01.96		01.96	
Number of plates			Inner	Outer	Inner	Outer	Inner	Outer
	Clutch –K1–		5	5	5	5	5	5
	Clutch –K2–		5	5	5	5	5	5
	Clutch –K3–		5	4	5	4	5	4
	Brake –B1–		5	5	5	5	5	5
	Brake –B2–		6	7	6	7	6	7
Application	Model		EuroVan 1992➤		EuroVan 1992➤		EuroVan 1992➤	
	Engine		2.4L – 57 kW (75 hp) Diesel		2.4L – 57 kW (75 hp) Diesel		2.5L – 81 kW (109 hp)	

Automatic transmission			01P		
Transmission	Code letters		**CWC**	**CWE**	**CWD, DLM**
Ratios	1st gear		2.714	2.714	2.714
	2nd gear		1.441	1.441	1.441
	3rd gear		1.000	1.000	1.000
	4th gear		0.743	0.743	0.743
	Reverse gear		2.884	2.884	2.884
Intermediate drive	No. of teeth	Input gear	44	44	44
		Output gear	51	51	51
	Ratio		1.159	1.159	1.159
Final drive	No. of teeth	Drive pinion	17	16	17
		Final drive gear	72	73	72
	Ratio		4.235	4.563	4.235
Drive shaft	Flange diameter		Tripod	Tripod	108mm
ATF cooler	Equipment		Pressure side 6-row	Pressure side 6-row	Pressure side 6-row

Automatic transmission			01P					
Transmission	Code letters		**CWF**		**CWG**		**CWH**	
	Manufactured	from to	01.96		01.96		01.96	
Torque converter	Code letters		**QBDC**		**QBDC**		**QBDB**	
Valve body	Code letters		**QCB**		**QCB**		**QCB**	
	Manufactured	from to	01.96		01.96		01.96	
Number of plates			Inner	Outer	Inner	Outer	Inner	Outer
	Clutch –K1–		5	5	5	5	5	5
	Clutch –K2–		5	5	5	5	5	5
	Clutch –K3–		5	4	6	5	6	5
	Brake –B1–		5	6	5	6	6	6
	Brake –B2–		6	7	6	7	6	7
Application	Model		EuroVan 1992➤		EuroVan 1992➤		EuroVan 1992➤	
	Engine		2.5L – 81 kW (109 hp)		2.8L – 103kW (139 hp)		2.8L –103kW (139 hp)	

Automatic transmission			01P		
Transmission	Code letters		**CWF**	**CWG**	**CWH**
Ratios	1st gear		2.714	2.714	2.714
	2nd gear		1.441	1.441	1.441
	3rd gear		1.000	1.000	1.000
	4th gear		0.743	0.743	0.743
	Reverse gear		2.884	2.884	2.884
Intermediate drive	Number of teeth	Input gear	44	44	44
		Output gear	51	51	51
	Ratio		1.159	1.159	1.159
Final drive	Number of teeth	Drive pinion	16	19	18
		Final drive gear	73	70	71
	Ratio		4.563	3.684	3.944

Capacities

Planetary transmission

Capacities	Planetary transmission	Automatic transmission type
Initial filling	5.35 ltr.	01P
Oil change	approx. 3.5L	
Lubricant	VW ATF	

VW ATF is obtainable as a replacement part.

Container size 1.0 ltr. – Part No. G 052 162 A2

CAUTION!

Only the VW ATF available as a replacement part can be used in the automatic transmission 01P.

Final drive

Capacities	Planetary transmission	Automatic transmission type
Initial filling	1.15L	01P
Oil change	Filled for life No change	
Lubricant	Gear oil SAE75 W90 (synthetic oil)	

Note:

VW gear oil: 1.0L container – Part No. G 052 145 A2

Repair instructions

The maximum possible care, cleanliness and proper tools are essential to ensure satisfactory and successful transmission repairs. The usual basic safety precautions for servicing vehicles also apply when working on the transmission.

A number of generally applicable instructions for individual repair operations, which are otherwise mentioned at various points in the Repair Manual, are summarized here. They apply to this Repair Manual.

Special tools

A summary of the special tools used in this Repair Manual can be found in Special tools/Workshop equipment catalogs.

Transmission

When an automatic transmission is exchanged, the VW ATF in the planetary gearbox, ⇒ page J 37-42, and the gear oil in the final drive, ⇒ page J 39-1, is to be checked and topped up as necessary. Capacities and specifications, ⇒ page J 00-14.

- ♦ When installing the transmission, ensure that the dowel sleeves are correctly installed.

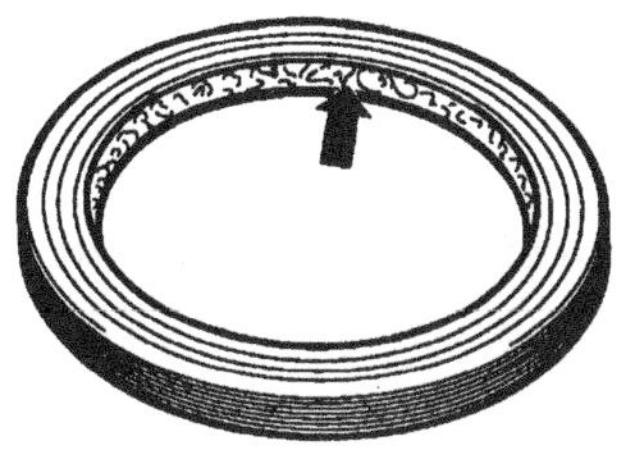
N32-0001

Gaskets, oil seals

◄ ♦ Radial shaft seals: Pack space between sealing lips with grease –arrow– before installing.

♦ Always install new O-rings.

♦ After installing check ATF level in final drive or planetary gearbox, adjust to correct level, if necessary.

Locking elements

♦ Do not overstretch circlips; replace if necessary.

♦ Circlips must be properly seated in grooves.

Nuts, bolts

♦ Nuts and bolts for attaching covers and housings must be loosened and tightened in a crosswise pattern.

♦ Do not distort particularly sensitive parts, e.g., valve body; and loosen or tighten diagonally, in stages.

♦ The tightening torques stated apply to unoiled nuts and bolts.

♦ Threads of bolts that are secured by locking fluid are to be cleaned with a wire brush. Then, install bolts with AMV 185 100 A1.

♦ Always replace self-locking nuts.

Bearings

♦ Install needle bearings with the inscribed side (thicker sheet metal) facing the installation drift.

♦ Lubricate all bearings in the transmission with gear oil before installing. Oil with special care for measuring friction torque.

♦ Heat the inner races of taper roller bearings to approx. 100°C (212°F) before installing.

♦ Do not interchange the inner and outer races of same size bearings.

♦ Always replace tapered roller bearings, which are mounted on a shaft, together and only use bearings from a single manufacturer.

Shims

♦ Measure shims at several points with a micrometer. Different tolerances make it possible to obtain the exact shim thickness required.

♦ Check for signs of burrs and damage. Only install shims which are in perfect condition.

Inner plates

- Place new inner plates in ATF fluid for 15 minutes before installing.

Valve body

- Replace the valve body if any of the selector elements are scorched.

Clutches

- Clutches K1, K2 and K3 should initially be disassembled only for cleaning. If parts are faulty, replace clutch. Application of the clutches is by the transmission code letters.

On Board Diagnostic (OBD)

- Before performing repairs to the automatic transmission, determine the cause of the malfunction as precisely as possible with the aid of Repair Group 01, On Board Diagnostic (OBD).

On Board Diagnostic (OBD), function

The automatic transmission is controlled electro-hydraulically.

The term "On Board Diagnostic (OBD)" is specifically for the electronic/electric control.

The transmission control module -J217- receives information from components that effect the selection of gears and passes the information on to solenoid valves that control the valves in the valve body.

The control module is equipped with a Diagnostic Trouble Code (DTC) Memory so that in the event of an electronic/electrical component failure, or an open circuit, the malfunction can be determined quickly. Malfunctions are recognized by electrical signals and are stored in the DTC Memory.

Safety functions of transmission control module

If malfunctions occur in the monitored sensors or components, these are stored in the DTC Memory together with an indication of the type of malfunction.

Malfunctions which only occur once are known as sporadic malfunctions (temporary). Sporadic malfunctions are additionally identified as such.

After analyzing the information, the automatic transmission control module distinguishes between sporadic and static (permanent) malfunctions ⇒ DTC table, from page J 01-33, and stores these.

If malfunctions do not reoccur within a minimum of 5 km (3 miles) or 6 minutes, a maximum of 20 km (12 miles) or 24 minutes, they are stored as sporadic malfunctions.

Electrical malfunctions that affect vehicle performance can be determined with the Scan Tool VAG 1551.

The possibilities offered by On Board Diagnostic (OBD) can only be fully exploited with operating mode 1, "Rapid data transfer" of Scan Tool VAG 1551.

Functions which the Scan Tool VAG 1551 can register ⇒ page J 01-30, List of selectable functions.

Safety functions of transmission control module

If critical malfunctions occur when driving, the transmission will continue to operate in an emergency (auxiliary) running mode. If the malfunction occurs in "D", "3" or "2" then 3rd–gear emergency mode is activated.

If the malfunction occurs in "1", "P", "N" or "R" then the emergency running mode for that stage is activated.

If the engine is started again in the emergency running mode and the malfunction occurs when selector lever is in position "D", "3" or "2" then 3rd gear is activated hydraulically, until the malfunction is corrected.

For malfunctions which can lead to emergency running:

The transmission will remain in emergency running mode until the malfunction is no longer recognized by the control module over a predetermined period.

Malfunctions which may lead to emergency running:

Open circuit in wiring, short-circuit, electrical or hydraulic components faulty.

Transmission control module malfunction recognition

If a malfunction exists, it is stored as a static malfunction. If the malfunction no longer exists after a predetermined period or distance traveled, it is considered a sporadic malfunction.

Malfunctions which are stored in the memory as sporadically occurring malfunctions, will be displayed as "sporadic malfunctions" when retrieved by Scan Tool VAG 1551. "SP" appears on the right of the display in such cases. If the printer is switched on, the phrase "sporadic malfunctions" is printed out after the malfunction is addressed.

Malfunctions which are stored in the memory as sporadically occurring malfunctions are automatically erased after 1,000 km (600 miles) or 20 hours.

Technical data, On Board Diagnostic (OBD)

Memory	
♦ Permanent memory	yes
♦ Volatile memory	no
Data output	
♦ Rapid data transfer	yes
♦ Blink code output	no
Output Diagnostic Test Mode	no
Basic setting	yes
Control module coding	no
Reading measuring value block	yes
Locations of electrical/electronic components	⇒ page J 01-7

On Board Diagnostic (OBD)

Troubleshooting with Scan Tool VAG 1551

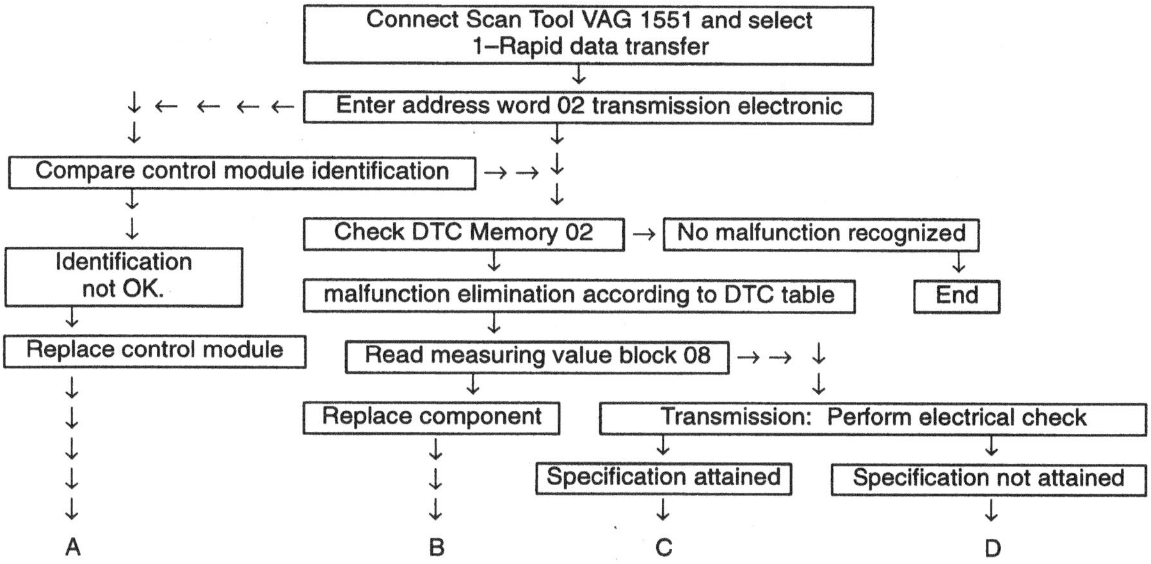

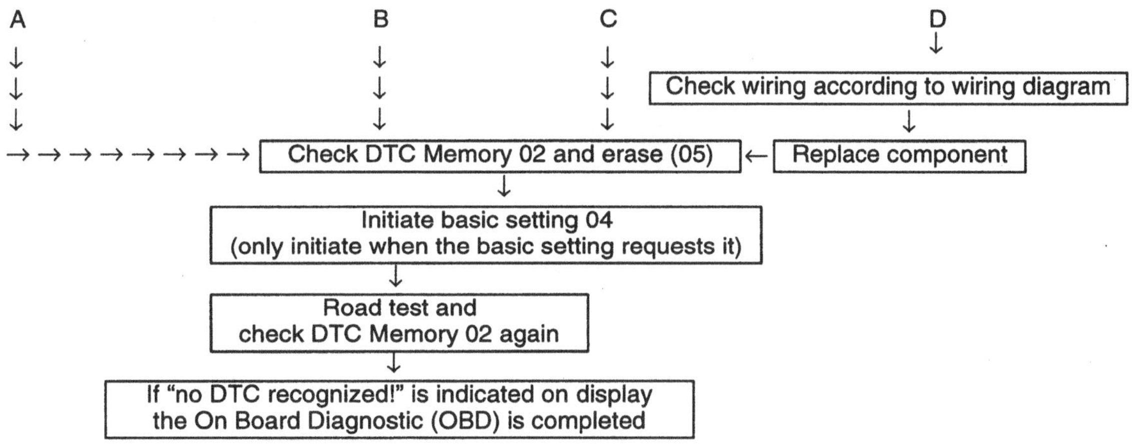

If after completing the On Board Diagnostic (OBD) the automatic transmission continues to operate with malfunctions, then continue troubleshooting using program.

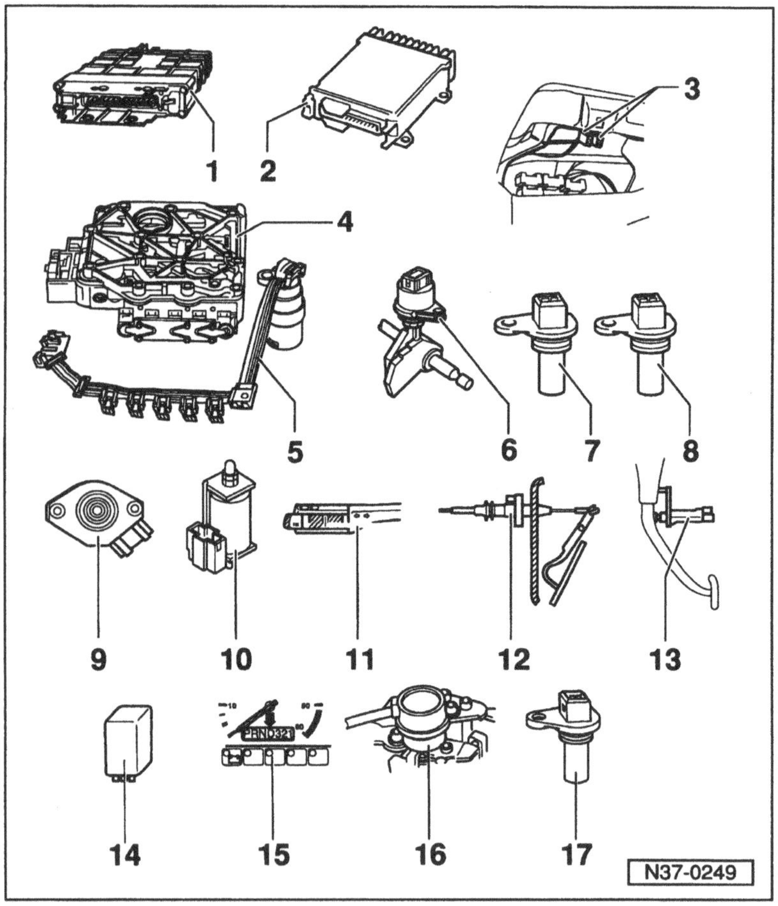

Component locations

1 - Transmission Control Module -J217-
- Location ⇒ Fig. 1 and Fig. 2
- Removing ⇒ Fig. 2 and Fig. 3
- Installing ⇒ Fig. 2 and Fig. 4
- Checked by On Board Diagnostic (OBD)

2 - Engine control module
- Location ⇒ Fig. 5
- Removing and installing

⇒ Repair Group 24

If engine or transmission control modules are replaced, the system must be returned to basic setting ⇒ page J 01-47, Initiating basic setting.

3 - Data Link Connector (DLC)
- Location ⇒ Fig. 6 and Fig. 7

4 - Valve body
- Location ⇒ Fig. 8
- Removing and installing ⇒ Page J 38-20
- The Solenoid Vaves –N88-,-N89-, -N90-, -N91-, -N92-, -N93- and -N94- are attached to valve body
- Components are checked by On Board Diagnostic (OBD)

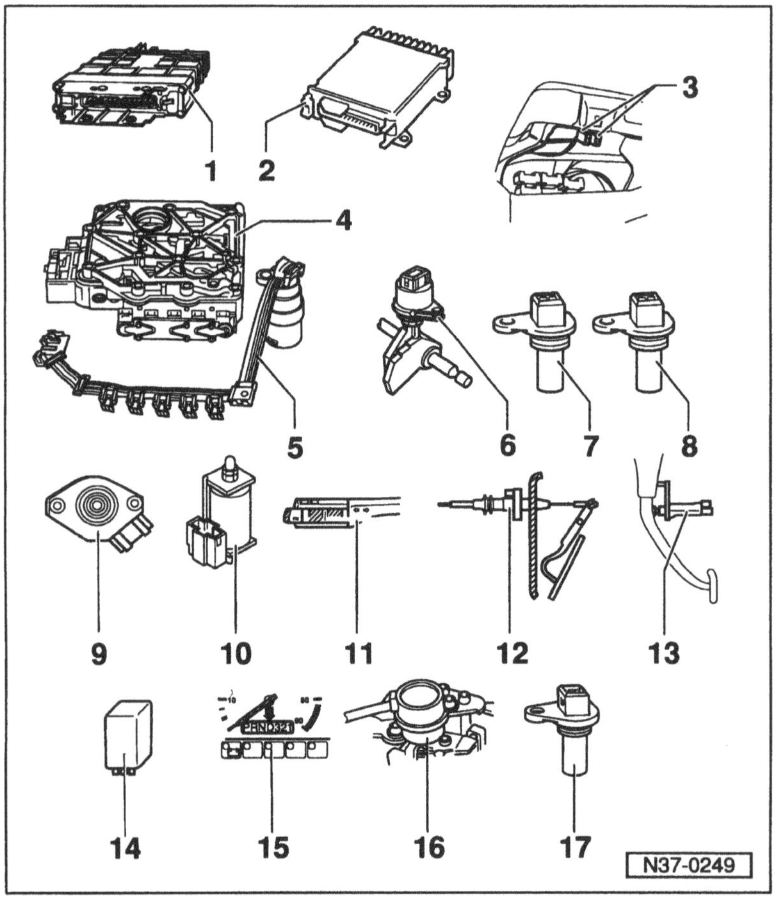

5 - Conductor strip with integrated Transmission Fluid Temperature Sensor -G93-
- Location ⇒ Fig. 9
- Removing and installing ⇒ Fig. 10
- -G93- is checked by On Board Diagnostic (OBD)

6 - Multi-Function Transmission Range (TR) Switch -F125-
- Location and removing and installing ⇒ Fig. 11
- Checked by On Board Diagnostic (OBD)

7 - Transmission Vehicle Speed Sensor -G38-
- Location and removing and installing ⇒ Fig. 12
- The connector T2 to wiring harness is white
- Checked by On Board Diagnostic (OBD)

> ***CAUTION!***
>
> ***Do not interchange connector T2 for sensors -G38- and -G68-.***

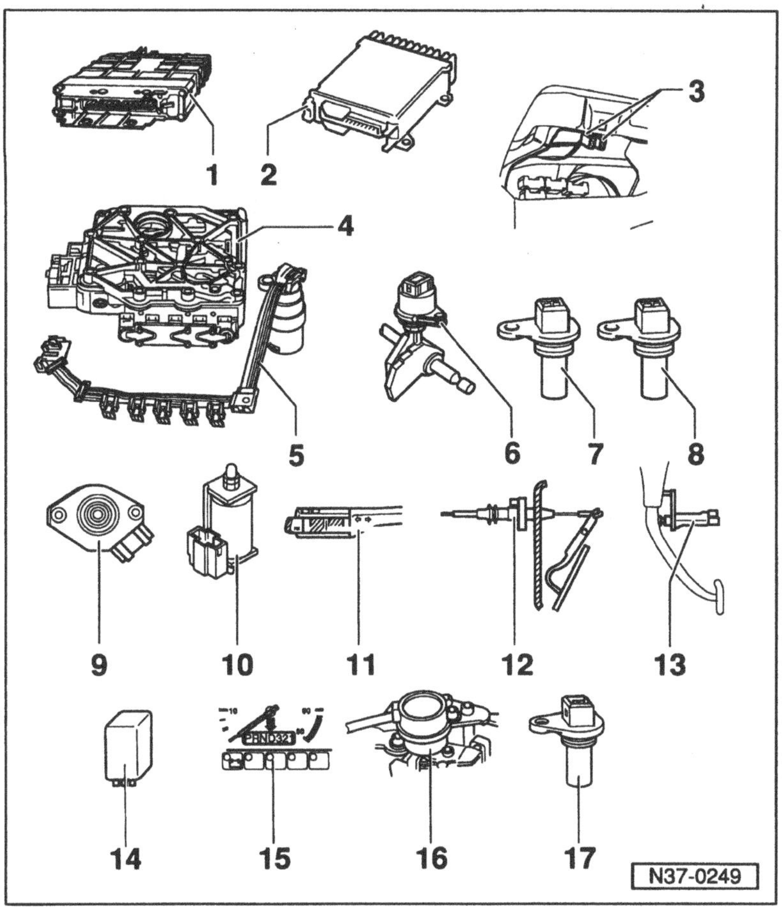

8 - Vehicle Speed Sensor -G68-

- Location and removing and installing ⇒ Fig. 13
- The connector T2 to wiring harness is black
- Checked by On Board Diagnostic (OBD)

9 - Throttle Position (TP) Position -G69-

- Location ⇒ Fig. 14
- Removing and installing

⇒ Repair Group 24 or 25

- After repairs initiate basic setting ⇒ Page J 01-47
- The signal of TP Sensor can be checked by On Board Diagnostic (OBD).

10 - Shift Lock Solenoid -N110-

- Location ⇒ Fig. 15
- Removing and installing ⇒ Page J 37-1, servicing gear change lever
- Can be checked electrically ⇒ page J 01-73 and via reading measuring value block ⇒ page J 01-49

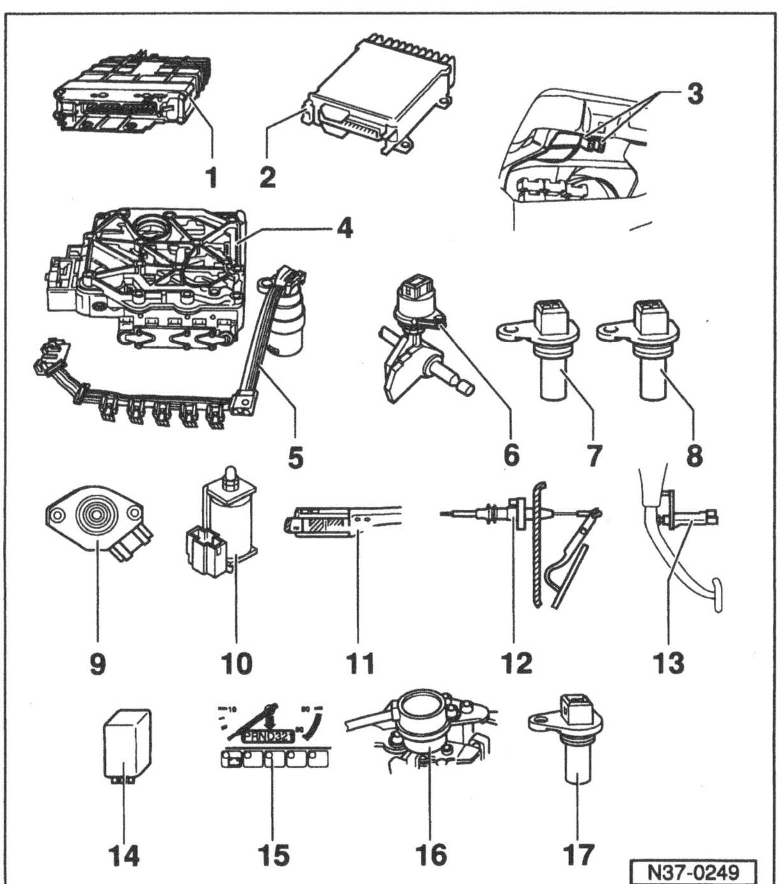

11 - Cruise Control Switch -E45-

- Location ⇒ Fig. 16

⇒ Removing and installing

- Repair Group 94; Servicing steering column switch
- Can be checked via reading measuring value block ⇒ page J 01-49

12 - Kick Down Switch -F8-

- Location ⇒ Fig. 17
- When removing and installing kick down switch, remove and install throttle cable, then adjust afterward Adjusting throttle cable

⇒ Repair Group 20

- Can be checked electrically ⇒ page J 01-73 and via reading measuring value block ⇒ page J 01-49

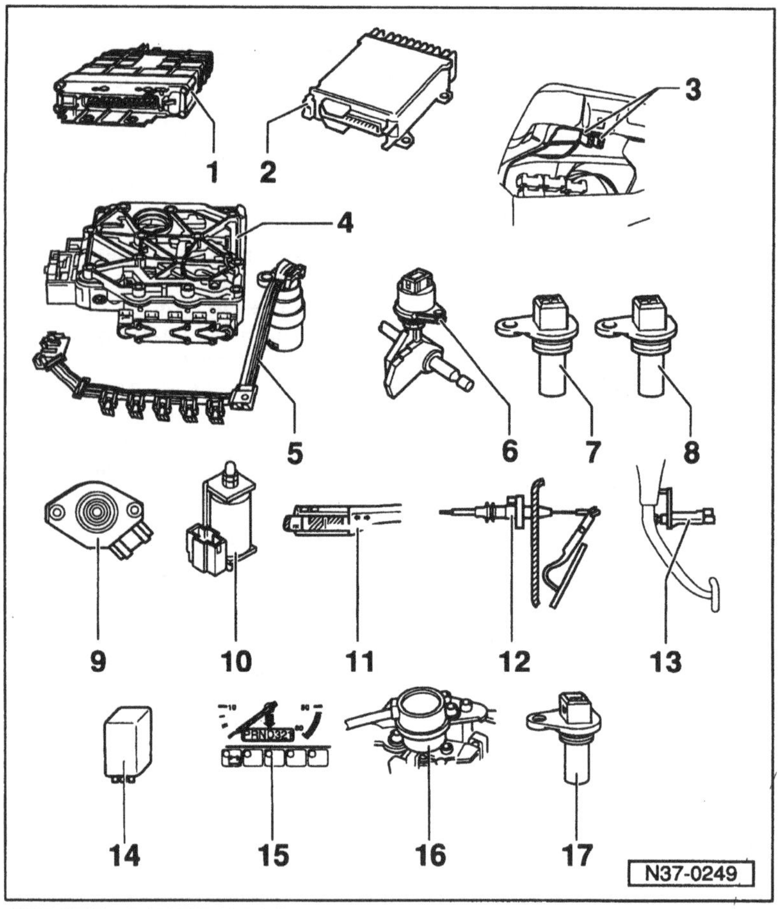

13 - Brake Light Switch -F-

- Location: ⇒ Fig. 18
- Removing and installing

⇒ Repair group 47; Assembly overview: Pedal cluster, brake pedal

- Can be checked electrically ⇒ page J 01-73 and via reading measuring value block ⇒ page J 01-49

14 - Park/Neutral Position Relay -J226-

- Location ⇒ Fig. 19

15 - Transmission Range Selector Lever Display -Y5-

- Location ⇒ Fig. 20

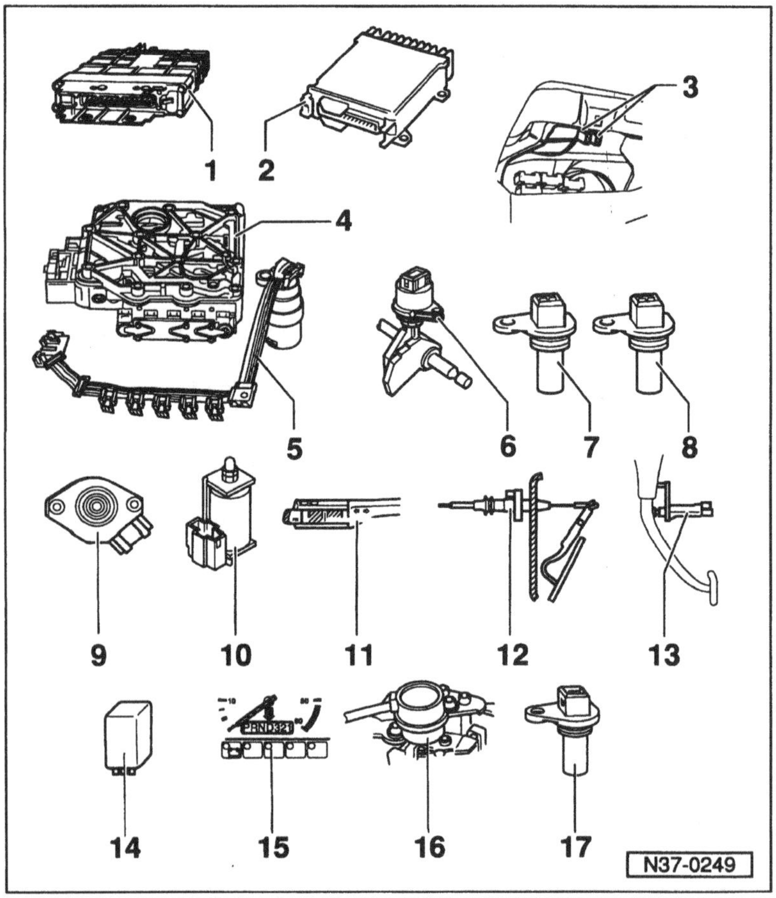

16 - Throttle Position (TP) Sensor -G69- (for Diesel engine)

- Location ⇒ Fig. 21
- Can be checked and adjusted in measuring value block ⇒ page J 01-49
- Removing and installing

⇒ Repair Group 23

- Checked by On Board Diagnostic (OBD)

17 - Engine Speed (RPM) Sensor -G28- (for Diesel engine)

- Location ⇒ Fig. 22
- Checked by On Board Diagnostic (OBD)

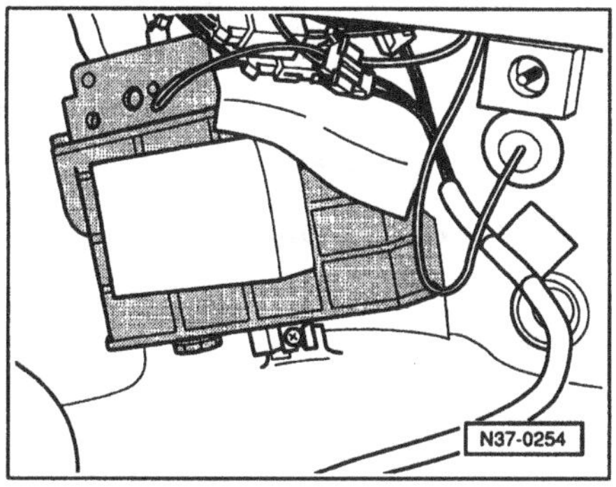

Fig. 1 Transmission Control Module -J217-, location in vehicles up to 12.95

The control module is located in the footwell on the "A" pillar.

– Remove screw to remove.

Fig. 2 Transmission Control Module -J217-, location / removing / installing in vehicles from 01.96

The control module is located in the left footwell at the "A" pillar.

– Remove cover on relay plate and unhook plate.
⇒ Electrical system, Repair Group 97, Relay plate, removing and installing

– Disconnect multi-pin connector and remove from control module.

– Remove control module.

Installation is the reverse of removal.

Fig. 3 Transmission Control Module -J217-, removing (vehicles up to 12.95)

– Release multi-pin connector, then pull connector off control module.

– Remove control module.

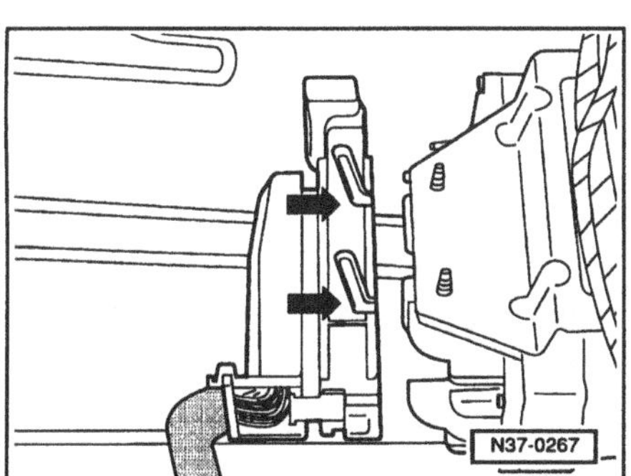

Fig. 4 Transmission Control Module -J217-, installing (vehicles up to 12.95)

– Locate multi-pin connector on control module pins -arrows- then lock multi-pin connector.

– Install screw.

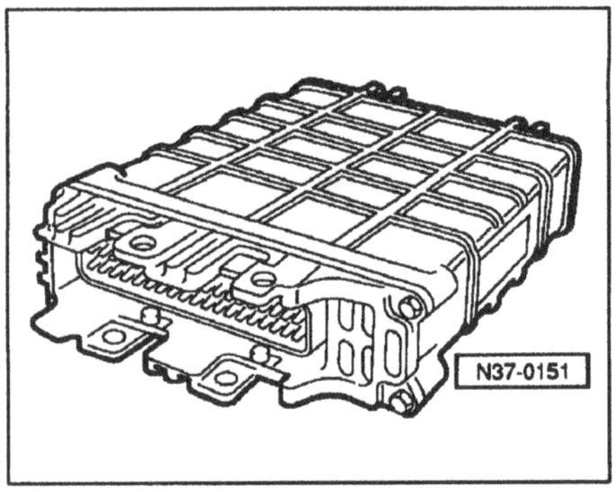

◄ **Fig. 5 Engine Control Module, location**

The control module is located in plenum chamber, right.

Removing and installing Engine Control Module
⇒ Repair group 24

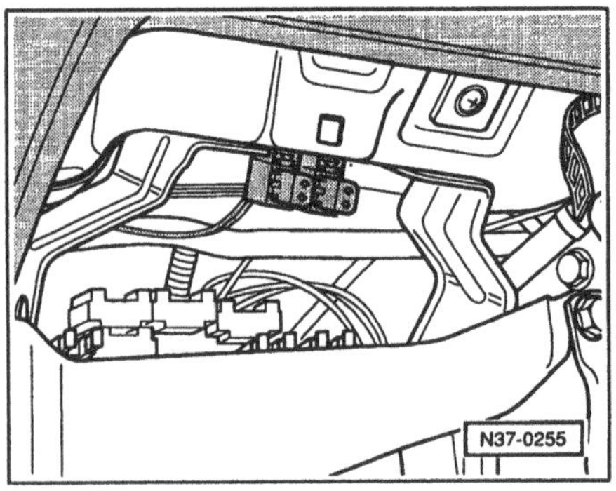

◄ **Fig. 6 Data Link Connector (DLC), location in vehicles up to 12.95**

The DLC is located above the relay station.

– Remove stowage tray in front of relay station.

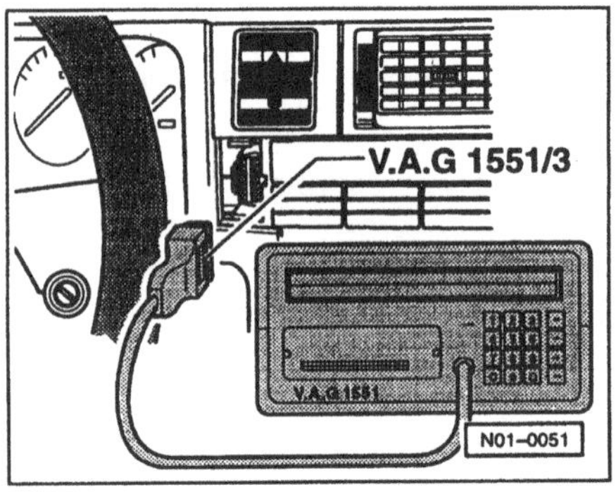

◄ **Fig. 7 Data Link Connector (DLC), location in vehicles from 01.96**

The DLC is located behind a cover to the right of the instrument cluster.

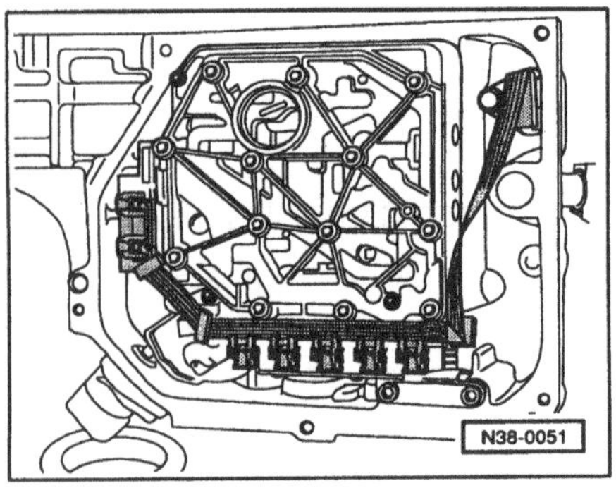

◄ **Fig. 8 Valve body, location**

The valve body is bolted to underside of transmission housing inside oil pan.

The Solenoid Valves -N88-, -N89-, -N90-, -N91-, -N92-, -N93- and -N94- are attached to the valve body.

Removing and installing valve body ⇒ Page J 38-20.

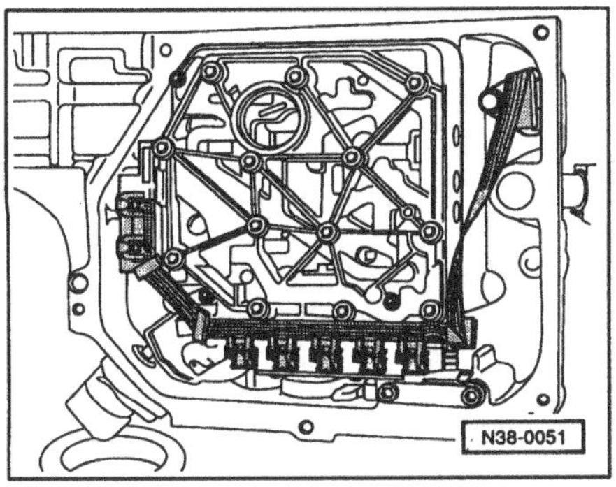

◄ **Fig. 9 Conductor strip with integrated Transmission Fluid Temperature Sensor -G93-, location**

The conductor strip is located in the oil pan on the valve body.

The conductor strip can be changed with the transmission installed without removing the valve body.

Do not kink or damage the conductor strip.

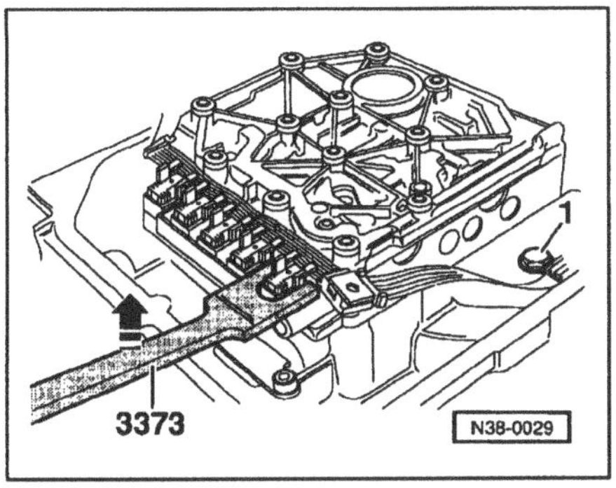

◄ **Fig. 10 Conductor strip, removing and installing**

- Separate wiring connector at transmission and remove retaining clamp.
- Remove oil pan
- Remove guide -1-.
- Pry conductor strip off solenoid valve in direction of arrow with special tool 3373.

Note:
If the locating points are damaged, the valve body must be replaced.

Installation is performed in the reverse order.

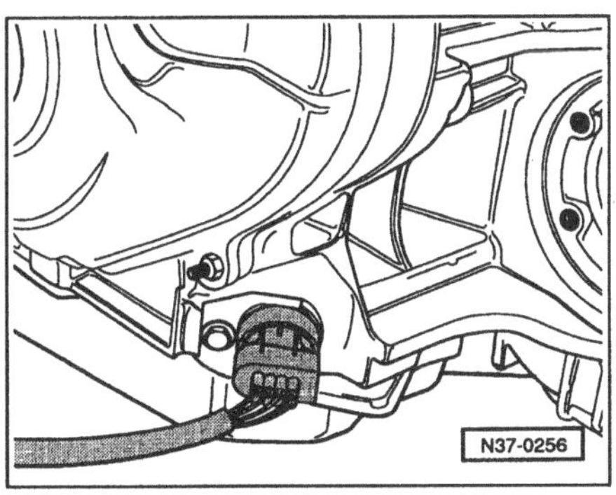

◄ **Fig. 11 Multi-Function Transmission Range (TR) Switch -F125-, location**

The multi-function switch is located on transmission lower section.

Multi-Function TR Switch -F125-, removing and installing

- Disconnect connector from switch.
- Remove bolt and retainer.
- Place oil tray under transmission to catch ATF.
- Remove multi-function switch.
- Replace seal.

Installation is performed in the reverse order.

- Tighten bolt to 10 Nm (7 ft lb).
- Check ATF level and top-up ⇒ Page J 37-42.

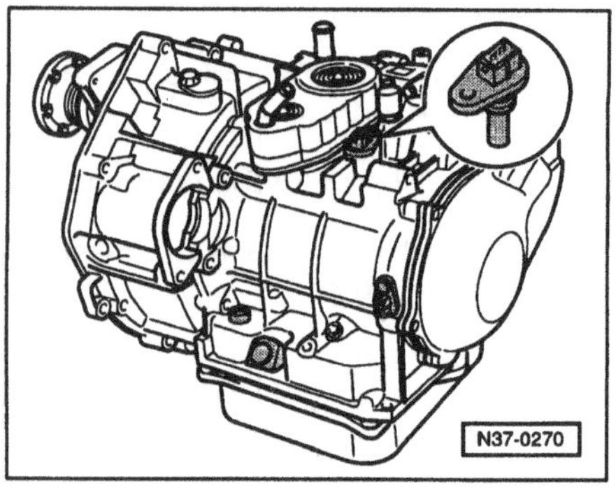

Fig. 12 Transmission Vehicle Speed Sensor (VSS) -G38-, location

The VSS -G38- -arrow- is located on top of the transmission.

VSS -G38-, removing and installing

– Disconnect connector from sensor.

– Remove bolt and pull sensor out.

– Replace seal.

Installation is performed in the reverse order.

– Tighten bolt to 10 Nm (7 ft lb).

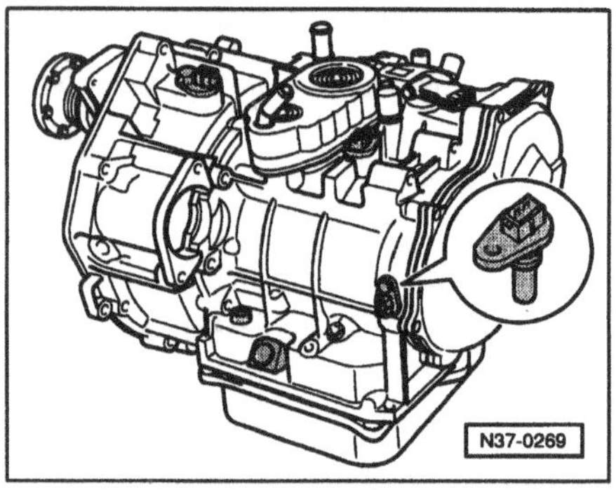

Fig. 13 Vehicle Speed Sensor (VSS) -G68-, location

The VSS -G68- -arrow- is located on front of transmission.

VSS -G68-, removing and installing

– Disconnect black connector from sensor.

– Remove bolt and pull sensor out.

– Replace seal.

Installation is performed in the reverse order.

– Tighten bolt to 10 Nm (7 ft lb).

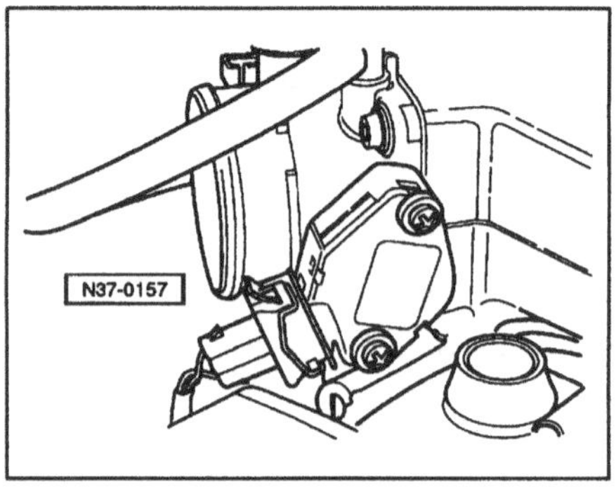

Fig. 14 Throttle Position (TP) Sensor-G69-, location

The TP Sensor -G69- is located on the throttle valve housing (engine). Construction can vary, according to engine.

TP Sensor -G69-, removing and installing

Removing and installing is described in relevant injection and ignition assembly group, compare engine code.
⇒ Repair Group 24

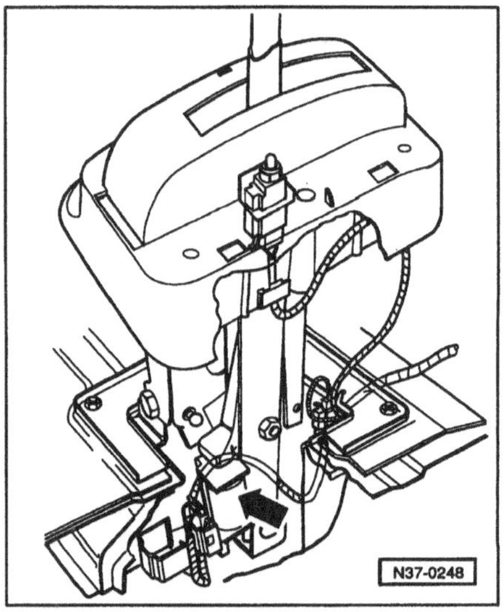

Fig. 15 Shift Lock Solenoid -N110-, location

The Shift Lock Solenoid -N110- is located on selector lever -arrow-.

Shift Lock Solenoid -N110-, removing and installing
⇒ Servicing shift mechanism, Page J 37-1.

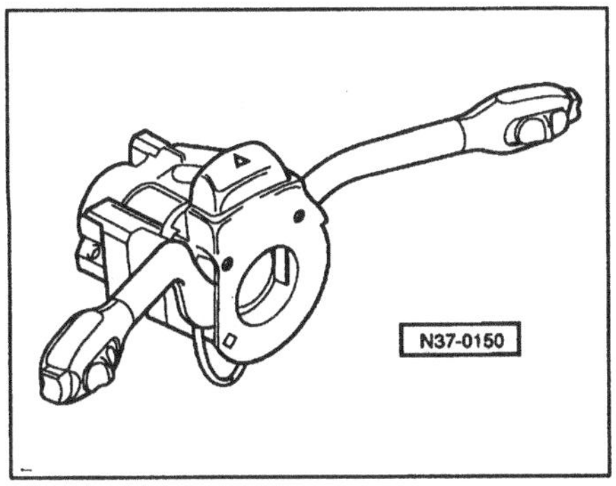

Fig. 16 Cruise Control Switch -E45-, location

The Cruise Control Switch -E45- is located on the steering column.

Cruise Control Switch -E45-, removing and installing
⇒ Servicing steering column switch; Repair group 94

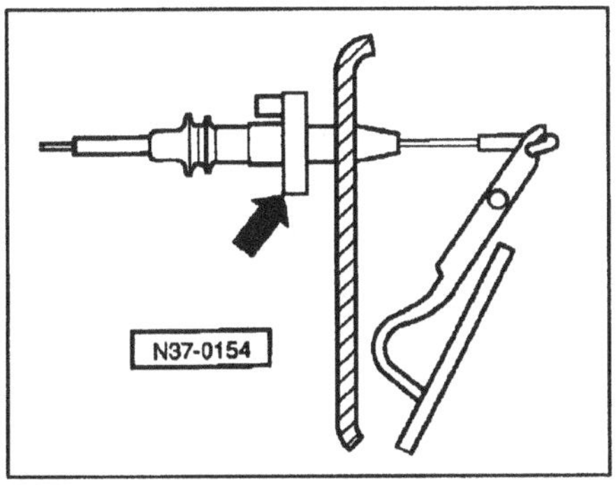

Fig. 17 Kick Down Switch -F8-, location

The Kick Down Switch -E45- -arrow- is integrated into the accelerator cable and is located in the engine compartment.

Kick Down Switch -F8-, removing and installing

- Remove accelerator cable to remove and install the kickdown switch, then reinstall cable and adjust

⇒ Repair group 20; Adjusting accelerator cable

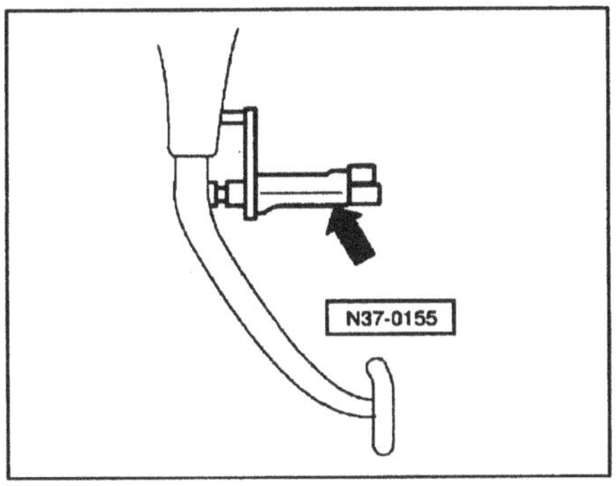

Fig. 18 Brake Light Switch -F-, location

The Brake Light Switch -E45- -arrow- is located on the pedal cluster.

Brake Light Switch -F-, removing and installing
⇒ Repair group 47; Assembly overview: Pedal cluster, brake pedal

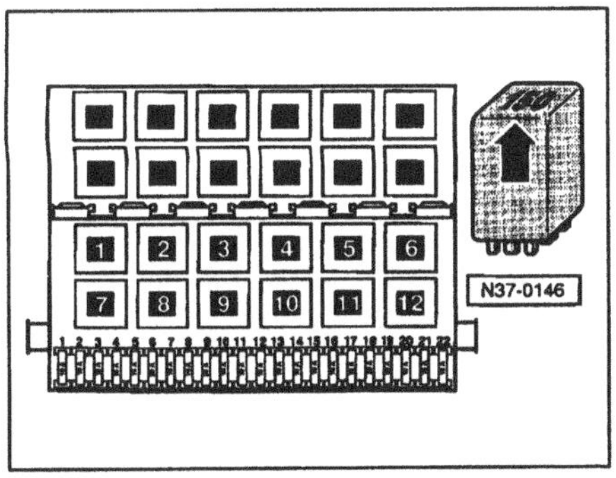

Fig. 19 Park/Neutral Position (PNP) Relay - J226-, location

The PNP Relay -J226- is located on the additional relay carrier behind the instrument panel.

Relay is marked with number "150" -arrow-.

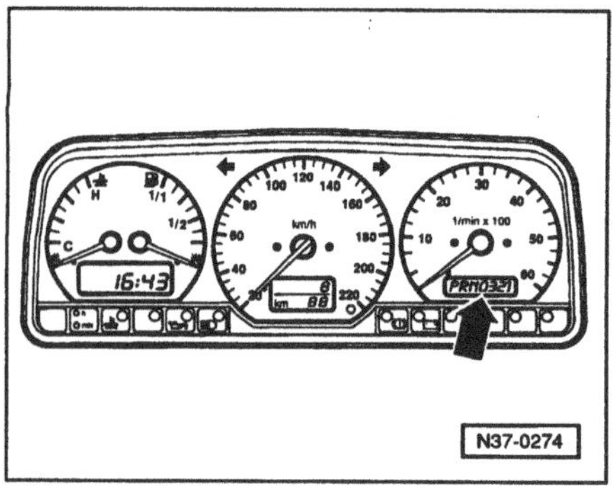

Fig. 20 Transmission Range (TR) Selector Lever Display -Y5-, location

The TR Selector Lever Display -Y5- -arrow- is located in the instrument cluster.

TR Selector Lever Display -Y5-, removing and installing
⇒ Repair group 90; Removing and installing instrument cluster

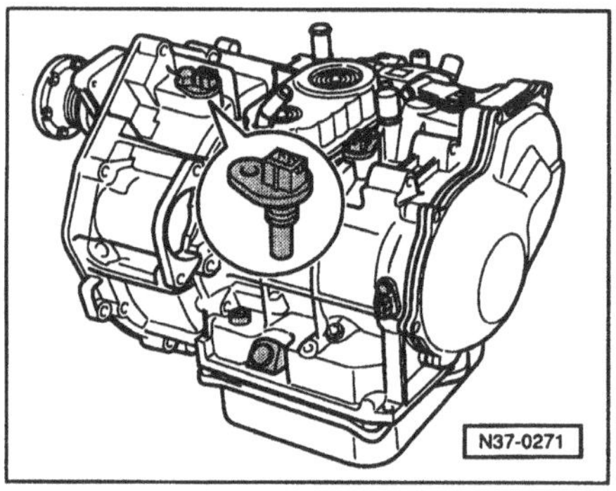

Fig. 21 Diesel engine Throttle Position (TP) Sensor -G69-, location

The TP Sensor is located on injection pump control lever.

TP Sensor-G69- (Diesel engine), removing and installing
⇒ Repair group 23

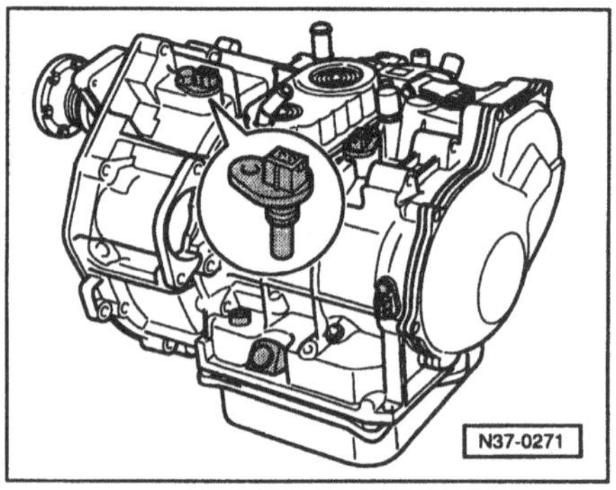

Fig. 22 Diesel engine Engine Speed (RPM) Sensor - G28-, location

The RPM Sensor-arrow- is located on top of transmission.

Engine Speed (RPM) Sensor (Diesel engine), removing and installing

- Disconnect connector from sensor.
- Remove bolt and pull sensor out.
- Replace seal.

Installation is performed in the reverse order.

- Tighten bolt to 10 Nm (7 ft lb).

On Board Diagnostic (OBD), connecting Scan Tool VAG 1551 and selecting functions

Test conditions:

- Vehicle voltage supply OK.
- Fuses 14 and 21 OK.
- Check Ground (GND) connections for transmission:
- Selector lever in position "P" and parking brake applied.

– Check Ground connections for corrosion and poor contact; repair if necessary.

- Ground connecting point is located on right of relay plate.

– Check Battery Ground strap and Ground strap between Battery and transmission.

– Remove storage compartment in front of relay plate.

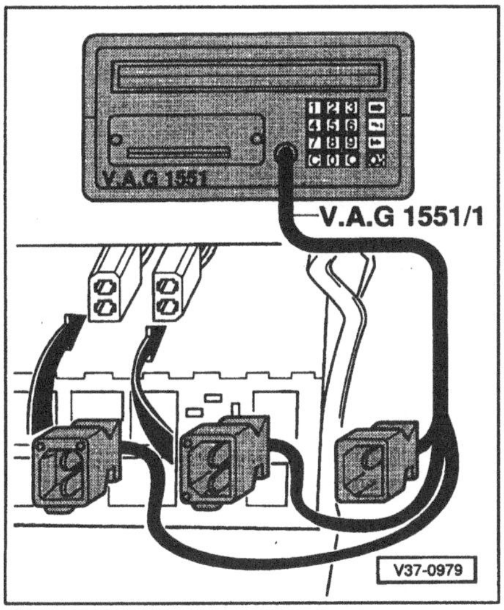

◄ – With ignition switched off, connect Scan Tool VAG 1551 with harness VAG 1551/1 to Data Link Connector (DLC) above relay plate as follows:

– Connect black plug for Scan Tool voltage supply to black DLC.

– Connect white plug to white DLC. (The blue plug is not required.)

VAG–On Board Diagnostic (OBD) HELP
1 – Rapid data transfer[1)]
2 – Blink code output[1)]

Indicated on display:

[1)] appears alternately

Notes:

- *Additional operating instructions can be retrieved by pressing the HELP key of VAG 1551.*
- *The → key is used for advancing within the program sequence.*
- *An automatic check (keys 00) can be carried out in the operating mode 1, "Rapid data transfer". Then all vehicle control modules will be interrogated automatically.*

⇒ Scan Tool VAG 1551 operating instructions.

- Switch on ignition.
- Switch on printer with the Print key (indicator lamp in key lights up).
- Press key 1 for "Rapid data transfer" mode.

Rapid data transfer HELP
Enter address word XX

Indicated on display:

- Press keys 0 and 2. (Enter the address word "transmission electronics" with 02.)

Rapid data transfer Q
02 transmission electronics

Indicated on display:

- Confirm entry with key Q.

01P927733A AG4 transmission 01P 2029
Coding 00000 WSC 00000

Indicated on display:

The control module identification, the coding and the dealership number (WSC) of the VAG 1551 are displayed:

Control module identification

Depending upon build level (program level) the control module can indicate a different control module identification to that shown in the example. Allocation of control module ⇒ Parts catalog.

- 01P 927 733 A: Part No.
- AG4 transmission 01P: 4-speed automatic transmission 01P
- 2209:EPROM (Program level)
- Coding 00000: is not required at present.
- WSC 00000: VAG 1551 dealership number with which the last coding was performed.

Control module does not answer! HELP

Indicated on display:

- By pressing the HELP key, a list of possible malfunction causes is printed out.

- After eliminating the possible causes of malfunctions, again enter the address word 02, for "transmission electronics" and confirm.

If "Control module does not answer!" appears:

Control module does not answer!	HELP

◄ Indicated on display:

Check control module voltage supply.

- Perform test step 1 ⇒ Page J 01-73, Electrical check.

- Check wiring connections to DLCs.
⇒ Electrical Wiring Diagrams, EuroVan 1993

- ⇒ DTC table, page J 01-33 under DTC 65535 – control module malfunctioning

- Press → key.

Rapid data transfer	HELP
Select function XX	

◄ Indicated on display:

- After the HELP key is pressed, a list of the possible functions is printed out.

List of selectable functions

		Page
01–	Check control module version ⇒ Performing On Board Diagnostic (OBD)	J 01-28
02–	Check DTC Memory	J 01-31
04–	Initiate basic setting	J 01-47
05–	Erase DTC Memory	J 01-45
06–	End ouptut	
08–	Read measuring value block	J 01-49

Further functions, which can be printed out by pressing the HELP key, need not be considered.

After checking a function the VAG 1551 returns to the following start position:

Rapid data trasfer	HELP
Select function XX	

◄ Indicated on display:

Diagnostic Trouble Code (DTC) Memory, checking

- Connect Scan Tool (ST) VAG 1551 and enter address word, "02 Transmission electronics" and advance until "Select function XX" appears in the display.

Rapid data transfer	HELP
Select function XX	

◄ Indicated on display:

- Press keys 0 and 2.
  (The function "Checking DTC Memory" is selected with 02).

Rapid data transfer	Q
02 – Checking DTC memory	

◄ Indicated on display:

- Confirm entry with key Q.

X DTCs recognized!

◄ The number of stored malfunctions or "No DTC recognized" appears in the display.

The stored malfunctions are displayed in turn and printed out.

- Press → key.
- After the last DTC has been displayed and printed out, the malfunctions should be corrected as described in the DTC table ⇒ from page J 01-33.
- Press → key.

Rapid data transfer	HELP
Select function XX	

◄ Indicated on display:

Note:

After checking the DTC Memory and correcting the malfunctions:

- Erase DTC Memory ⇒ Page J 01-45.

Diagnostic Trouble Code (DTC) table

Notes:

- *All the possible malfunctions which are recognized by the Transmission Control Module -J217- and displayed on the VAG 1551 when the DTC Memory is checked, are listed below grouped according to the DTC number.*
- *If malfunctions occur only occasionally or if the DTC Memory was not erased after correcting the malfunctions, these faults will be displayed as "sporadic faults" for a stipulated period of time* *⇒ transmission control module malfunction recognition, Page J 01-3.*
- *If components are indicated as being faulty when checking the DTC Memory, additionally test wiring to the components for short and open circuits according to wiring diagram Electrical Wiring Diagrams, EuroVan 1993 ➤.*
- *The DTC is printed out in the "rapid data transfer" mode only when the printer of VAG 1551 is switched on. Example: DTC (5-position) 65535*

Output on printer VAG 1551	
No DTC recognized!	If "No DTC recognized" appears after performing repairs, the On Board Diagnostic (OBD) is completed. If the transmission does not operate properly despite OBD, repair as per troubleshooting program.

Output on printer VAG 1551	Possible cause	Corrective steps
00258 Solenoid Valve 1 -N88- Open circuit[1)] Short to Ground[1)]	Open circuit or short to Ground Solenoid Valve 1 -N88- faulty	– Check wiring and connections according to wiring diagram[2)] – Read measuring value block ⇒ page J 01-49; display group number 04 – Perform electrical test ⇒ from page J 01-73
00260 Solenoid Valve 2 -N89- Open circuit[1)] Short to Ground[1)]	Open circuit or short to Ground Solenoid Valve 2 -N89- faulty	– Check wiring and connections according to wiring diagram[2)] – Read measuring value block ⇒ page J 01-49; display group number 04 – Perform electrical test ⇒ from page J 01-73
00262 Solenoid Valve 3 -N90- Open circuit[1)] Short to Ground[1)]	Open circuit or short to Ground Solenoid Valve 3 -N90- faulty	– Check wiring and connections according to wiring diagram[2)] – Read measuring value block ⇒ page 01–49; display group number 04 – Perform electrical test ⇒ from page J 01-73

[1)] One of these displays appears in addition to relevant component.

[2)] First check connections for contact corrosion or water ingress and replace if necessary. If solenoid DTCs are displayed then especially check the 10-pin connector on transmission between valve body conductor strip and wiring harness.

Output on printer VAG 1551	Possible cause	Corrective steps
00264 Solenoid Valve 4 -N91- Open circuit[1] Short to Ground[1]	Open circuit or short to Ground Solenoid Valve 4 -N91- faulty	– Check wiring and connections according to wiring diagram[2] – Read measuring value block ⇒ page J 01-49; display group number 07 – Perform electrical test ⇒ from page J 01-73
00266 Solenoid Valve 5 -N92- Open circuit[1] Short to Ground[1]	Open circuit or short to Ground Solenoid Valve 5 -N92- faulty	– Check wiring and connections according to wiring diagram[2] – Read measuring value block ⇒ page J 01-49; display group number 04 – Perform electrical test ⇒ from page J 01-73
00268 Solenoid Valve 6 -N93- Open circuit[1] Short to Ground[1]	Open circuit or short to Ground Solenoid Valve 6 -N93- faulty	– Check wiring and connections according to wiring diagram[2] – Read measuring value block ⇒ page J 01-49; display group number 02 – Perform electrical test ⇒ from page J 01-73

[1] One of these displays appears in addition to relevant component.

[2] First check connections for contact corrosion or water ingress and replace if necessary. If solenoid DTCs are displayed then especially check the 10-pin connector on transmission between valve body conductor strip and wiring harness.

Output on printer VAG 1551	Possible cause	Corrective steps
00270 Solenoid Valve 7 -N94- Open circuit[1] Short to Ground[1]	Open circuit or short to Ground Solenoid Valve 7 -N94- faulty	– Check wiring and connections according to wiring diagram[2] – Read measuring value block ⇒ page J 01-49; display group number 07 – Perform electrical test ⇒ from page J 01-73
00281 Vehicle Speed Sensor -G68-	Open circuit in wiring Vehicle Speed Sensor -G68- faulty	– Check wiring and connections according to wiring diagram[2] – Read measuring value block ⇒ page J 01-49; display group number 02 – Perform electrical tests ⇒ from page J 01-73 – Replace Vehicle Speed Sensor -G68- ⇒ from page J 01-19, Fig. 11
No signal	Input gear, impulse wheel loose	– Replacing input gear ⇒ Removing and installing input gear, page NO TAG

[1] One of these displays appears in addition to relevant component.

[2] First check connections for contact corrosion or water ingress and replace if necessary. If solenoid DTCs are displayed then especially check the 10-pin connector on transmission between valve body conductor strip and wiring harness.

Output on printer VAG 1551	Possible cause	Corrective steps
00293 Multi–Function Transmission Range (TR) Switch -F125- Undefined switch state	Open circuit in wiring Multi–Function TR Switch -F125- faulty	– Check wiring and connections according to wiring diagram[1)] – Read measuring value block ⇒ page J 01-49; display group number 01 – Vehicles produced up to October, 1995, perform electrical test ⇒ from page J 01-73 – Vehicles produced from October, 1995, can only be checked using the "Measuring Value Block". – Replace Multi–Function TR Switch -F125- ⇒ page J 01-18, Fig. 9

[1)] First check connectors for contact corrosion or water ingress and replace if necessary.

Output on printer VAG 1551	Possible cause	Corrective steps
00297 Transmission Vehicle Speed Sensor -G38- No signal	Wiring open circuit Transmission Vehicle Speed Sensor -G38- faulty	– Check wiring and connections according to wiring diagram[2)] – Perform electrical check ⇒ from page J 01-73 – Replace Transmission Vehicle Speed Sensor -G38- ⇒ page J 01-19, Fig. 10
00300 Transmission Fluid Temperature Sensor -G93-[1)] No DTC type recognized	Open circuit in wiring Transmission Fluid Temperature Sensor -G93- faulty	– Check wiring and connections according to wiring diagram[2)] – Read measuring value block ⇒ page J 01-49; display group number 05 – Perform electrical test ⇒ from page J 01-73

[1)] A faulty Transmission Fluid Temperature Sensor -G93- is indicated.

[2)] First check connections for contact corrosion or water ingress and replace if necessary. If solenoid DTCs are displayed then especially check the 10–pin connector on transmission between valve body conductor strip and wiring harness.

Output on printer VAG 1551	Possible cause	Corrective steps
00518	Open or short circuit	– If in addition DTC 00638 is displayed, then correct this first – Check wiring and connections according to wiring diagram
Throttle Position (TP) Sensor -G69-	Throttle Position (TP) Sensor -G69- faulty	– Read measuring value block ⇒ page J 01-49; display group numbers 01 and 03 – Perform electrical tests ⇒ from page J 01-73 – Replace Throttle Position (TP) Sensor -G69- ⇒ Repair Group 24 – Bring system to basic setting ⇒ page J 01-47
Signal outside tolerance	Diesel engine Throttle Position (TP) Sensor -G69- faulty	– Replace Diesel engine Throttle Position (TP) Sensor -G69- ⇒ Repair Group 23 – Adjust Diesel Engine Throttle Position (TP) Sensor -G69- – Read measuring value block ⇒ page J 01-49; display group number 01 and 03 – Return system to basic setting ⇒ page J 01-47

Output on printer VAG 1551	Possible cause	Corrective steps
00529	Open circuit in wiring	– Check wiring and connections according to wiring diagram
Speed information missing		– Read measuring value block ⇒ page J 01-49; display group number 03 – Check engine control module ⇒ Repair Group 01 for relevant engine code – Diesel engines: Check Engine Speed (RPM) Sensor –G28– – Perform electrical tests ⇒ from page J 01-73
00532	Battery faulty	– Test Battery voltage ⇒ Repair Group 27
Supply voltage	Insufficient system voltage for valves	– Read measuring value block ⇒ page J 01-49; display group number 02 – Test supply voltage to Transmission Control Module -J217- – Perform electrical tests ⇒ from page J 01-73

Output on printer VAG 1551	Possible cause	Corrective steps
00545	Open circuit or short Ground	– Check wiring and connections according to wiring diagram
Engine/transmission electrical connection	No connection engine/transmission control module	– Read measuring value block ⇒ page J 01-49; display group number 05 – Check engine control module ⇒ Repair Group 01 for relevant engine code
Open circuit[1)] Short to Ground[1)]		– Return system to basic setting ⇒ page J 01-47
00596	10-pin connector between valve body conductor strip and wiring harness	– Check wiring and connections according to wiring diagram
Short circuit between valve wiring	Conductor strip to valve body faulty	– Perform electrical tests ⇒ from page J 01-73 – Replace conductor strip ⇒ page J 38-20

[1)] One of these displays appears in addition to relevant component.

Output on printer VAG 1551	Possible cause	Corrective steps
00638	Open circuit or short to Ground	– Check wiring and connections according to wiring diagram
Engine/transmission electrical connection 2	No connection engine/transmission control module	– Read measuring value block ⇒ page J 01-49; display group number 05 – Check engine control module, replace if necessary ⇒ Repair Group 01 for relevant engine code
No signal		– Return system to basic setting ⇒ page J 01-47
00641	Transmission becomes too hot, max. 148°C. If ATF temperature is too high, transmission shifts down into next lower gear.	– Check ATF level ⇒ page 37–40 – Read measuring value block ⇒ page J 01-49; display group number 05; read off ATF temperature
ATF temperature	Trailer load of vehicle too high	
Signal too large	ATF level not in order Transmission Fluid Temperature Sensor -G93- faulty	– Replace conductor strip ⇒ page J 38-20

Output on printer VAG 1551	Possible cause	Corrective steps
00652	Electrical/hydraulic malfunction	– Read measuring value block ⇒ page J 01-49; display group 04 and establish by driving in gear in which the malfunction occurs
Gear monitoring	Clutch or valve body faulty	
Implausible signal		
00660	Open circuit in wiring	– Check wiring and connections according to wiring diagram
Kick Down Switch -F8-/Throttle Position (TP) Sensor -G69-	Throttle Position (TP) Sensor -G69- faulty	– Perform repairs as described in "Corrective steps" 00518 –Throttle Position Sensor -G69-
Implausible signal	Kick Down Switch -F8- faulty	– Read measuring value block ⇒ page J 01-49; display group number 01 – Perform electrical tests ⇒ from page J 01-73 – Adjust accelerator cable or replace ⇒ Repair Group 24

Output on printer VAG 1551	Possible cause	Corrective steps
01192	Torque converter clutch slips	– Check slip of torque converter clutch
Torque converter clutch	Valve body malfunction	– Read measuring value block ⇒ page J 01-49; display group 07
Mechanical malfunction		
01236	Wiring break or short to Ground	– Check routing of wiring according to wiring diagram
Shift Lock Solenoid -N110-	Solenoid for Shift Lock Solenoid -N110- malfuntion	– Perform test step 13 ⇒ page J 01-86 – Replace Shift Lock Solenoid -N110- ⇒ J 37-10
Open circuit[1)] Short to Ground[1)]		
65535		– Replace control module ⇒ page J 01-7
Control module faulty	Transmission Control Module -J217- faulty	– Return system into basic setting ⇒ page J 01-47

[1)] One of these displays appears in addition to relevant component.

Notes:

The Transmission Control Module -J217- should not be replaced page J 01-7, and the system be restored to the basic setting pageJ 01-47 until the possible cause of the malfunction has been determined and the following malfunctions have been corrected:

- *mechanical malfunctions*
- *hydraulic malfunctions*
- *all affected electrical/electronic components and wiring connections.*

Diagnostic Trouble Code (DTC) Memory, erasing

Requirement:

- DTC Memory checked ⇒ page J 01-31.

After DTC Memory has been interrogated:

Rapid data transfer HELP
Select function XX

◄ Readout in display:

- Press keys 0 and 5. (The function "Erase DTC Memory" is entered with 05.)

Rapid data transfer Q
05 Erase DTC Memory

◄ Readout in display:

- Confirm entry with key Q.

Attention!
DTC Memory was not checked.

◄ Readout in display:

Note:

- *If the ignition was switched off, e.g., between checking the DTC Memory and erasing DTC Memory, the DTC Memory is then not erased.*
- *Adhere strictly to the sequence of operations, i.e., first of all check DTC Memory.*

Rapid data transfer →
DTC Memory is erased!

◄ Indicated on display:

(The DTC Memory will be erased approximately 5 seconds after the display appears.)

The DTC Memory is now erased.

Note:

Wait about 1 minute before again checking the DTC Memory.

System cannot be checked!

◄ Indicated on display:

1 DTC recognized!
00811 3333
System cannot be checked

◄ Print out with printer switched on:

Transmission Control Module -J217- given no time to recognize DTCs.

- Wait about 1 minute before again checking DTC Memory.
- After checking and erasing DTC Memory, road test vehicle and again check DTC Memory.

When the DTC Memory is checked, the following display should appear:

"No DTC recognized!"

Basic setting, initiating

Note:

The basic setting should be initiated after performing the following lowing repairs:

- *Exchanging engine*
- *Replacing engine control module*
- *Replacing/altering throttle valve*
- *Adjusting throttle valve (setting idling speed).*
- *Replacing Throttle Position (TP) Sensor -G69-*
- *Altering TP Sensor -G69- setting, e.g., when adjusting closed throttle position switch (engine)*
- *Replacing Diesel engine injection pump*
- *Changes to control lever, altering setting of residual quantity or maximum speed adjustment screw on diesel engines*
- *Replacing Transmission Control module -J217-*

– Connect Scan Tool VAG 1551 and enter address word "02 Transmission electronics" and advance until "Select function XX" appears in the display

Rapid data transfer	HELP
Select function XX	

◄ Indicated on display:

– Press keys 0 and 4. (The function "Initiating basic setting" is selected with 04).

Note:

Accelerator Pedal must remain in closed throttle position.

Rapid data transfer	Q
04 Basic setting	

◄ Indicated on display:

– Confirm entry with key Q.

Basic setting	HELP
Enter display group number XX	

◄ Indicated on display:

– Press keys 0 and 0.

– Confirm entry with key Q.

System in basic setting	0→

◄ Indicated on display:

The system is now in the basic setting.

– Depress Accelerator Pedal as far as kick down and hold in this position for 3 seconds.

– Press key →.

Rapid data transfer	HELP
Select function XX	

◄ Indicated on display:

Measuring value block, reading

– Connect Scan Tool VAG 1551 and enter address word "02 Transmission electronics" and advance until "Select function XX" indicated on display ⇒ from page J 01-25.

Rapid data transfer	HELP
Select function XX	

◄ Indicated on display:

– Press keys 0 and 8. (The function "Read measuring value block" is selected with 08.)

Rapid data transfer	Q
08 – Reading measuring value block	

◄ Indicated on display:

– Confirm entry with key Q.

Reading measuring value block
Enter display group number XX

◄ Indicated on display:

– Enter display group number ⇒ List of selectable display group numbers, Page J 01-50.

– Confirm entry with key Q.

Reading measuring value block		1	→
→1	→2	→3	→4

◄ – There are always 4 display zones in the measuring value block. Decoding the individual values in display zones 1 to 4 ⇒ Test table page J 01-52.

List of selectable display group numbers

Reading measuring value block		1	→
P	0.8V	0%	00000111

◄

Reading measuring value block		2	→
0.983	0.985 A	12.76 V	2.50 V

◄

Reading measuring value block		3	→
0 km/h	900 rpm	0	0%

◄

Reading measuring value block		4	→
1001 00	0	P	0 km/h

◄

Display group No.	Display field	Designation
01	1	Selector lever position
	2	Throttle position sensor voltage
	3	Accelerator Pedal value
	4	Switch positions
02	1	Actual Solenoid Valve 6 -N93- voltage
	2	Specified Solenoid Valve 6 -N93- voltage
	3	Battery Positive (B+) voltage
	4	Voltage at Vehicle Speed Sensor -G68-
03	1	Road speed
	2	Engine speed
	3	Selected gear
	4	Accelerator Pedal value
04	1	Solenoid valves
	2	Selected gear
	3	Selector lever position
	4	Road speed

Reading measuring value block	5	→		
45°C	0011011	0	900 rpm	

Reading measuring value block	6	→

Reading measuring value block	7	→		
1H	200 rpm	900 rpm	0%	

Reading measuring value block	8	→

Display group No.	Display field	Designation
05	1	ATF temperature
	2	Switch output
	3	Gear to select
	4	Engine speed
06	1	not applicable
	2	
	3	
	4	
07	1	Selected gear
	2	Lock-up clutch slip
	3	Engine speed
	4	Accelerator Pedal value
08	1	not applicable
	2	
	3	
	4	

Notes:

- *If the printer is switched on, the current display is printed out on the log.*
- *If the specified values are reached in all the display fields:*

– Press key →

Rapid data transfer	HELP
Select function XX	

Indicated on display:

CAUTION!

When driving or riding in airbag equipped vehicles, never hold test equipment in your hands or lap when the vehicle is in motion. Objects between you and the airbag can increase the the risk of injury in an accident

Test table

Group No.	Display field	Designation	Test conditions		Specified readout on VAG 1551	Corrective steps
01	1	Multi-Function Transmission Range (TR) Switch -F125-	Stationary	P	P	– Check Multi-Function TR Switch -F125- ⇒ Perform electrical tests, from page J 01-73
			Selector	R	R	
			lever	N	N	
			in	D	D	
				3	3	
				2	2	
				1	1	

> ***CAUTION!***
> ***When driving or riding in airbag equipped vehicles, never hold test equipment in your hands or lap when the vehicle is in motion. Objects between you and the airbag can increase the risk of injury in an accident.***

Group No.	Display field	Designation	Test conditions		Specified readout on VAG 1551	Corrective steps
01	2	Voltage of Throttle Position (TP) Sensor -G69-	Stationary	Idling min. Idling max.	0.156 V 0.8 V	When accelerating from idling to Wide Open Throttle (WOT), voltage figure increases constantly – Check TP Sensor ⇒ Perform electrical tests, from page J 01-73
				Wide Open Throttle (WOT) min.	3.5 V	– Adjusting/replacing TP Sensor ⇒ Repair Group 24
				WOT max.	4.680 V	– Return system to basic setting ⇒ page J 01-47

Group No.	Displ ay field	Designation	Test conditions		Specified readout on VAG 1551	Corrective steps
01	2	Voltage of Diesel engine Throttle Position (TP) Sensor -G69-	Station-ary Diesel engine	Idling min. Idling max.	0.6 V 0.8 V	– Prescribed idling speed values not achieved: adjust injection pump to specification by turning TP Sensor on control lever – Check TP Sensor ⇒ Perform electrical tests, from page J 01-73
				Wide Open Throttle (WOT) min.	3.5 V	– Return system to basic setting ⇒ page J 01-47
Cont'd				WOT max.	4.680 V	

Group No.	Displ ay field	Designation	Test conditions		Specified readout on VAG 1551	Corrective steps
01	3	Accelerator Pedal value	Station-ary	Idling[1]	0...1%	When accelerating from idling to WOT, % figure constantly in-creases
				Wide Open Throttle (WOT)	99...100%	– Return system to basic setting ⇒ page J 01-47
	4	Switch positions				
		Brake Light Switch -F- Display 1	Brake	Operated	1	– Check Brake Light Switch -F- ⇒ Perform electrical tests, from page J 01-73
				Not operated	0	
		Traction control system 2		Activated	1	not applicable
Cont'd				Not activated	0	

[1] Engine with Mono Motronic coolant temperature min. 80°C

Group No.	Display field	Designation	Test conditions	Specified readout on VAG 1551	Corrective steps
01	4	3	Can be disregarded		
		Kick Down Switch -F8- Display 4	Kick down: Operated	1	– Check kick down switch ⇒ – Perform electrical tests, from page J 01-73
			Not operated	0	
		Multi-Function TR Switch -F125- 5	Selector lever in: R, N, D, 3, 2	1	– Check Multi-Function TR Switch -F125- ⇒ Perform electrical tests, from page J 01-73
			P, 1	0	
		6	Selector lever in: P, R, 2, 1	1	
			N, D, 3	0	

CAUTION!

When driving or riding in airbag equipped vehicles, never hold test equipment in your hands or lap when the vehicle is in motion. Objects between you and the airbag can increase the risk of injury in an accident.

Group No.	Display field	Designation	Test conditions	Specified readout on VAG 1551	Corrective steps
01	4	Multi-Function TR Switch -F125- Display 7	Selector lever in: P, R, N, D	1	– Check Multi–Function TR Switch -F125- ⇒ Perform electrical tests, from page J 01-73
			3, 2, 1	0	
		8	Selector lever in: P, R, N	1	
			D, 3, 2, 1	0	

CAUTION!

When driving or riding in airbag equipped vehicles, never hold test equipment in your hands or lap when the vehicle is in motion. Objects between you and the airbag can increase the risk of injury in an accident.

Group No.	Display field	Designation	Test conditions		Specified readout on VAG 1551	Corrective steps
02	1	Actual current of Solenoid Valve 6 -N93-	Stationary	Wide Open Throttle (WOT)	0.0 A	– Check Solenoid Valve 6 -N93- ⇒ Perform electrical tests, from page J 01-73
				Idling max.	1.1 A	
	2	Specified current of Solenoid Valve 6 -N93-	Stationary	WOT	0.0 A	
				Idling max.	1.1 A	
	3	Battery voltage	Stationary	min.	10.8 V	– Check Battery, replace if necessary – Check voltage supply to Transmission Control Module -J217- ⇒ Perform electrical tests, from page J 01-73
Cont'd				max.	16.0 V	– Replace Transmission Control Module -J217- ⇒ page J 01-7 – Return system to basic setting ⇒ page J 01-47

Group No.	Display field	Designation	Test conditions		Specified readout on VAG 1551	Corrective steps
02	4	Vehicle Speed Sensor (VSS) -G68-	Stationary	min.	2.20 V	– Check VSS -G68- ⇒ Perform electrical tests, from page J 01-47
				max.	2.52 V	
03	1	Vehicle speed	Driving[1]		...km/h	Speedometer reading and readout on VAG 1551 may differ slightly
	2	Engine speed	With engine running		rpm	– Tune engine if necessary ⇒ Repair Group 24 or 25

[1] When driving, a second mechanic is needed for reading the specified values.

CAUTION!

When driving or riding in airbag equipped vehicles, never hold test equipment in your hands or lap when the vehicle is in motion. Objects between you and the airbag can increase the risk of injury in an accident.

Group No.	Display field	Designation	Test conditions		Specified readout on VAG 1551	Corrective steps
03	3	Gear selected	Driving[1]	Neutral	0	– Check solenoid valves ⇒ Perform electrical tests, from page J 01-73
				Reverse	R	
				1 hydraulic	1	
				2 hydraulic	2H	
				2 mechanical	2M	
				3 hydraulic	3H	
				3 mechanical	3M	
				4 hydraulic	4H	
				4 mechanical	4	
	4	Accelerator Pedal value	Driving[2]	Idling	0...1%	When accelerating from idling to WOT, % figure constantly increases
				Wide Open Throttle (WOT)	99...100%	– Return system to basic setting ⇒ page J 01-47

[1] When driving, a second mechanic is needed for reading the specified values.

CAUTION!

When driving or riding in airbag equipped vehicles, never hold test equipment in your hands or lap when the vehicle is in motion. Objects between you and the airbag can increase the risk of injury in an accident.

Measuring value block, reading: Display group number 04; "Checking solenoid valves while driving"

- The solenoid valves can be checked while driving via "read measuring value block 08"; "display group number 04"
- The table shows the Solenoid Valves –N88–, and –N90– as activated in each selector lever position. These solenoid valves control the switching valves to the relevant gears.
- The solenoid valve -N91- activates the lock–up clutch modulation valve and is not covered here.
- The Solenoid Valves -N92- and -N94- are supplementary valves which affect selection changes and are only controlled during gear changes and displayed in zones 5 and 6.
- VAG 1551 display zone 1 has 6 characters (0000 00) and is read as follows:

Display on VAG 1551	Display zone 1: Display 1	Display 2	Display 3	Display 4	Display 5	Display 6
1551	Solenoid Valve 1 –N88–	Solenoid Valve 2 –N89–	Solenoid Valve 3 –N90–	Solenoid Valve 4 (can be disregarded)	Solenoid Valve 5 -N92-	Solenoid Valve 7 -N94-

- Non–activated solenoid valves display "0" and activated solenoid valves display "1".
- All solenoid valves are checked via On Board Diagnostic (OBD) ⇒ checking DTC Memory, page J 01-31

CAUTION!

When driving or riding in airbag equipped vehicles, never hold test equipment in your hands or lap when the vehicle is in motion. Objects between you and the airbag can increase the risk of injury in an accident.

Checking solenoid valves while driving

Group No.	Display box	Designation	Test conditions Selector lever in:	Specified readout on VAG 1551	Corrective steps
04	1	Solenoid valves/indicated on	P	1 0 1 0 00	Solenoid valves are selected according to driving condition
		VAG 1551 display	R[1]	0 0 1 0 00	– Perform electrical check ⇒ from page J 01-73
			N	1 0 1 0 00	
		–N88– display 1	D[1] 1H	0 0 1 0 00	
		–N89– display 2	1M	0 0 1 0 00	– Continued troubleshooting according to OBD guide, ⇒ page J 01-5
		–N90– display 3	2H	0 1 1 0 00	
		-N91- display 4	2M	0 1 1 0 00	
		-N92- display 5	3H	0 0 0 0 01	
		-N94- display 6	3M	0 0 0 0 01	
			4H	1 1 0 0 01	
Cont'd			4M	1 1 0 0 01	

[1] When driving with gears selected a second mechanic is required for reading the specified values.

Group No.	Display box	Designation	Test conditions Selector lever in:	Specified readout on VAG 1551	Corrective steps
04	1	Solenoid valves/indicated on	3[1] 1H (1M)	0 0 1 0 00	Solenoid valves are selected according to driving condition
		VAG 1551:	1M	0 0 1 0 00	– Perform electrical check ⇒ from page J 01-73
		–N88– display 1	2H	0 1 1 0 00	
		–N89– display 2	2M	0 1 1 0 00	
		–N90– display 3	3H	0 0 0 0 01	– Continue troubleshooting according to OBD guide, ⇒ page J 01-5
		-N92- display 5	3M	0 0 0 0 01	
		–N88– display 1	2[1] 1H	0 0 1 0 00	
		–N89– display 2	1M	0 0 1 0 00	
		–N89– display 2	2H	0 1 1 0 00	
		–N90– display 3	2M	0 1 1 0 00	
		–N88– display 1	1[1] 1H	0 0 1 0 00	
		–N89– display 2	1M	0 0 1 0 00	

[1] When driving with gears selected a second mechanic is required for reading the specified values.

Group No.	Displ ay field	Designation	Test conditions		Specified readout on VAG 1551	Corrective steps
04	2	Gear selected	Driving[1]	Neutral	0	– Check solenoid valves ⇒ Perform electrical tests, from page J 01-73
				Reverse	R	
				1 hydraulic	1	
				2 hydraulic	2H	
				2 mechanical	2M	
				3 hydraulic	3H	
				3 mechanical	3M	
				4 hydraulic	4H	
				4 mechanical	4M	

[1] When driving with gears selected a second mechanic is required for reading the specified values.

CAUTION!

When driving or riding in airbag equipped vehicles, never hold test equipment in your hands or lap when the vehicle is in motion. Objects between you and the airbag can increase the risk of injury in an accident.

Group No.	Displ ay field	Designation	Test conditions		Specified readout on VAG 1551	Corrective steps
04	3	Selector lever position	Driving[1]	P	P	– Check Multi–Function Transmission Range (TR) Switch ⇒ Perform electrical tests, from page J 01-73
				R	R	
				N	N	
				D	D	
				3	3	
				2	2	
Cont'd				1	1	

[1] When driving with gears selected a second mechanic is required for reading the specified values.

CAUTION!

When driving or riding in airbag equipped vehicles, never hold test equipment in your hands or lap when the vehicle is in motion. Objects between you and the airbag can increase the risk of injury in an accident.

Group No.	Displ ay field	Designation	Test conditions	Specified readout on VAG 1551	Corrective steps
04	4	Vehicle speed	Speed at which vehicle is driven[1]	...km/h	The speedometer reading and readout on VAG 1551 may differ slightly
05	1	ATF temperature, ATF level is checked at approx. 35°C–45°C	Stationary with engine running. The exact temperature is displayed from approx. 30°C	...°C	– Check Transmission Fluid Temperature Sensor -G93- ⇒ Perform electrical check, from page J 01-73
	2	Selector outputs	Driving[1] Engine management (ignition timing influence)		– Check wiring according to current flow diagram
		Display 1[2]	is switched on	1	– Replace engine control module ⇒ Repair Group 24
			is switched off	0	– Replace Transmission Control Module -J217- ⇒ page J 01-7
		2[2]	switched on	1	– Return system to basic setting ⇒ page J 01-47
Cont'd			switched off	0	

[1] When driving, a second mechanic is needed for reading the specified values.

[2] VAG 1551 must always show the same number "1" or "0" in displays 1 and 2.

CAUTION!

When driving or riding in airbag equipped vehicles, never hold test equipment in your hands or lap when the vehicle is in motion. Objects between you and the airbag can increase the risk of injury in an accident.

Group No.	Displ ay field	Designation	Test conditions	Specified readout on VAG 1551	Corrective steps
05	2	Selector outputs	Shift Lock Solenoid -N110-		– Check routing of wiring on basis of wiring diagram
		Display 3	is switched on	1	– Check Shift Lock Solenoid -N110- ⇒ Perform electrical check, from page J 01-73
			is switched off	0	
		4	switched on	1	
			switched off	0	
		5	Cruise control system		– Check routing of wiring according to wiring diagram – Check cruise control system
			switched off	0	⇒ Electrical Wiring Diagrams, EuroVan 1993➤
Cont'd			switched off	0	

Group No.	Displ ay field	Designation	Test conditions		Specified readout on VAG 1551	Corrective steps
05	2	Selector outputs				
		Display 6	Air con-ditioner	was switched off	1	– Check routing of wiring according to wiring diagram
				was not switched off	0	– Check air conditioner ⇒ Repair Group 87
		7	Park/Neutral signal Selector lever in			– Check routing of wiring according to wiring diagram
				P, N	1	
Cont'd				1, 2, 3, D	0	

Group No.	Displ ay field	Designation	Test conditions		Specified readout on VAG 1551	Corrective steps
05	3	Gear to select	Driving[1]	Neutral	0	– Check Multi-Function Transmission Range (TR) Switch ⇒ Perform electrical tests, from page J 01-73
				Reverse	R	
				1 hydraulic	1H	– If gearshifts do not occur, a clutch or brake may be faulty
				2 hydraulic	2H	
				2 mechanical	2M	
				3 hydraulic	3H	– Replace Transmission Control Module -J217- page ⇒ J 01-7
				3 mechanical	3M	
				4 hydraulic	4H	
				4 mechanical	4	
	4	Engine speed	Driving[1] with engine running		rpm	– Tune engine if necessary ⇒ Repair Group 24 or 25

[1] When driving, a second mechanic is needed for reading the specified values.

CAUTION!

When driving or riding in airbag equipped vehicles, never hold test equipment in your hands or lap when the vehicle is in motion. Objects between you and the airbag can increase the risk of injury in an accident.

Group No.	Display field	Designation	Test conditions	Specified readout on VAG 1551	Corrective steps
06	Can be disregarded				
07	1	Gear selected[1]	Neutral	0	– Check solenoid valves ⇒ Perform electrical tests, from page J 01-73
			Reverse	R	– If gearshifts do not occur, a clutch or brake may be faulty
			1 hydraulic	1H	
			1 mechanical	1M	
			2 hydraulic	2H	
			2 mechanical	2M	
			3 hydraulic	3H	
			3 mechanical	3M	
			4 hydraulic	4H	
Cont'd			4 mechanical	4M	– Replace Transmission Control Module -J217- ⇒ page J 01-7

[1] When driving, a second mechanic is needed for reading the specified values.

CAUTION!

When driving or riding in airbag equipped vehicles, never hold test equipment in your hands or lap when the vehicle is in motion. Objects between you and the airbag can increase the risk of injury in an accident.

Group No.	Display field	Designation	Test conditions		Specified readout on VAG 1551	Corrective steps
07	2	Torque converter lock-up clutch slip	While driving[1] with engine running	In the hydraulic gears[2]	0 ... stall speed	– Check wiring according to wiring diagram
		Solenoid Valve 4 -N91- is activated	Torque converter lock-up clutch closed	Engine speed: 2000...3000 rpm in the mechanical gears[3]	0 ... 130 rpm	– Check solenoid Valve 4 -N91- ⇒ perform electrical checks from page J 01-73 – Check transmissions, Repair Group 37 – Replace torque converter
	3	Engine speed	With engine running		...rpm	– If necessary tune engine ⇒ Repair group 24 or 25
	4	Accelerator Pedal		Idling	0 ... 1%	When accelerating from idling to WOT the percentage value constantly increases
				Wide Open Throttle (WOT)	99 ... 100%	– Return system to basic setting ⇒ page J 01-47
08	Can be disregarded					

[1] When driving, a second mechanic is needed for reading the specified values.

[2] The torque converter lock-up clutch must be open (gears will be hydraulically activated "H").

[3] The gear shift must be completed. The torque converter lock-up clutch must be closed and the Accelerator Pedal value held constant (gears will be mechanically activated "M").

Electrical test

Test requirements:

- Battery Positive Voltage (B+) OK.
- Fuses 14 and 21 OK.
- Ground (GND) connections OK:

  The Ground connection point is located on the left next to the relay plate.

  Check Battery Ground strap and Ground strap between Battery and transmission.

Notes:

- *Use the digital multimeter Fluke 83 (US1119) with auxiliary harness from VAG 1594 for testing.*

> ***CAUTION!***
>
> ***Select the appropriate measuring range at the tester before connecting the test leads to avoid damaging the electronic components.***

- *The specified values are valid for an ambient temperature from 0°C...40°C (32°F.. 104°F)*
- *If the readings obtained differ from the specified values, determine the malfunction according to the wiring diagram.*

- *If the readings obtained differ only slightly from the specified values, clean terminals and connectors of the testers and test leads and repeat test. Before replacing the particular components, check the wiring and connections and, particularly if the specified values are below 10 Ω, repeat the resistance measurement on the component.*

Electrical check

Test requirements, notes ⇒ Page 01–73

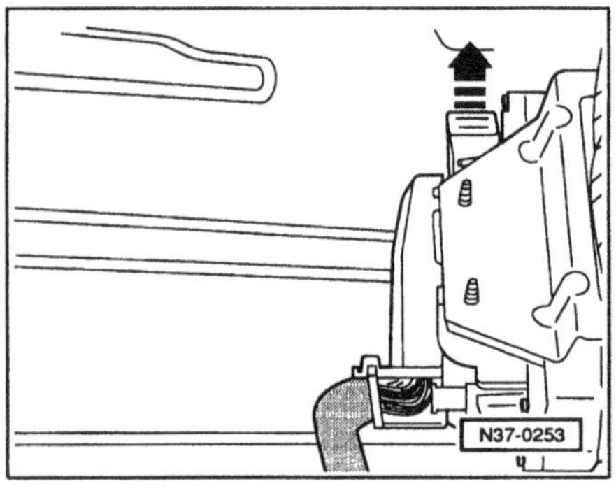

The control module -arrow- is located in footwell on "A" pillar

– Switch ignition off.

◄ – Release multi-pin connector in direction of arrow and remove control module.

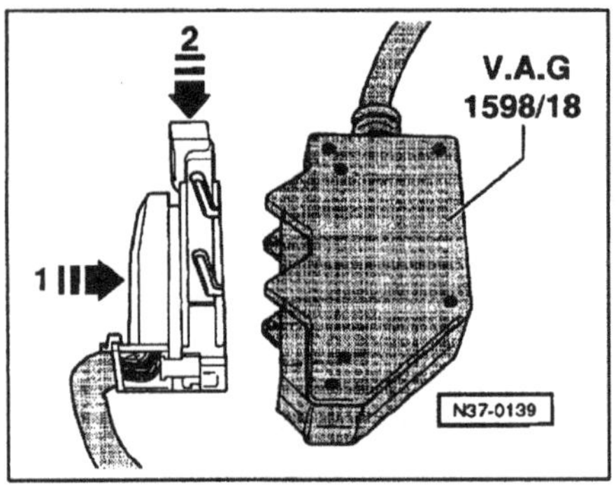

◄ – Connect test box VAG 1598/18 to multi-pin connector -1- and lock, in direction of arrow -2-.

Using test box VAG 1598/18 the wiring can be checked according to wiring diagram.

After electrical checks:

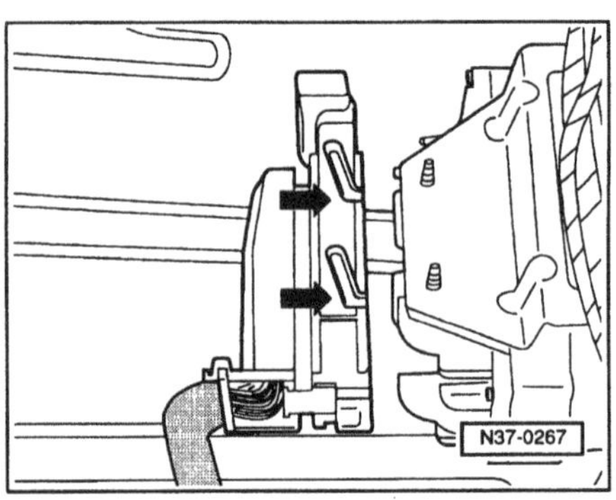

◄ – Connect multi-pin connector to Transmission Control Module -J217- and lock.

Note:

When connecting the control module ensure that the guides -arrows- are engaged on the control module pins.

Transmission Control Module -J217- multi-pin connector (68 pin) assignment (sockets on VAG 1598/18)

1– Ground (terminal 31)	20– Vehicle Speed Sensor -G68-
2– Vacant	21– Transmission Vehicle Speed Sensor -G38-
3– Vacant	22– Solenoid Valve 6 -N93- supply voltage
4– Vacant	23– Supply voltage (terminal 15)
5– Throttle Position Sensor -G69- signal	24– Diagnosis K–wire
6– Transmission Fluid Temperature Sensor -G93-	25– Vacant
7– Vacant	26– Vacant
8– Vacant	27– Vacant
9– Solenoid Valve 3 –N90–	28– Throttle Position Sensor -G69- Ground
10– Solenoid Valve 7 -N94-	29– Shift Lock Solenoid -N110-
11– Park/neutral signal	30– Vacant
12– Kickdown for air conditioner	31– Vacant
13– Ignition timing influence	32– Vacant
14– Vacant	33– Vacant
15– Brake Light Switch -F- signal voltage	34– Vacant
16– Kick Down Switch -F8-	35– Vacant
17– Vacant	36– Diagnosis L–wire
18– Multi-Function TR Switch -F125-	37– Vacant
19– TD (revolution) signal	38– Vacant

39– Vacant	54– Solenoid Valve 2 -N89-
40– Multi–Function TR Switch -F125-	55– Solenoid Valve 1 -N88-
41– Load signal from engine control module	56– Solenoid Valve 5 -N92-
42– Diesel engine rpm sensor (screening)	57– Selector lever display
43– Vehicle Speed Sensor -G68- (screening)	58– Solenoid Valve 6 -N93-
44– Transmission Vehicle Speed Sensor -G38- (screening)	59– Vacant
45– Supply voltage (terminal 30)	60– Cruise control system (input terminal 15)
46– Vacant	61– Cruise control system (output)
47– Solenoid Valve 4 -N91-	62– Multi-Function TR Switch -F125-
48– Vacant	63– Multi-Function TR Switch -F125-
49– Vacant	64– Engine Speed Sensor -G28- (Diesel engine)
50– Throttle Position Sensor -G69- (5 volt)	65– Vehicle Speed Sensor -G68-
51– Vacant	66– Transmission Vehicle Speed Sensor -G38-
52– Vacant	67– Solenoid valves voltage supply
53– Vacant	68– Terminal 30 (output)

List of test steps (multi-pin connector with 68 pins)

♦ Only perform test steps as listed in DTC table and measuring value block.

Component checked		Component checked	
Supply voltage from Transmission Control Module -J217-	– Perform test step 1	Solenoid Valve 5 -N92-	– Perform test step 10
Shift Lock Solenoid -N110-	– Perform test steps 2 and 13	Solenoid Valve 6 -N93-	– Perform test step 11
Brake Light Switch -F-	– Perform test step 3	Solenoid Valve 7 -N94-	– Perform test step 12
Throttle Position Sensor -G69-	– Perform test step 4	Kick Down Switch -F8-	– Perform test step 14
Multi-Function TR Switch -F125-	– Perform test step 5	Transmission Fluid Temperature Sensor -G93-	– Perform test step 15
Solenoid Valve 1 -N88-	– Perform test step 6	Vehicle Speed Sensor -G68-	– Perform test step 16
Solenoid Valve 2 -N89-	– Perform test step 7	Transmission Vehicle Speed Sensor -G38-	– Perform test step 17
Solenoid Valve 3 -N90-	– Perform test step 8	Engine Speed Sensor -G28- (for Diesel engine)	– Perform test step 18
Solenoid Valve 4 -N91-	– Perform test step 9		

Switch to voltage measuring range 20 V					
Test step	VAG 1598/18 sockets	Test of	• Test conditions – Additional operations	Specified value	Corrective steps
1	23 + 1	Supply voltage form Transmission Control Module -J217-	• Ignition switched on	approx. Battery voltage	– Check wiring according to wiring diagram – Check wiring from contact 1 to Ground – Check wiring from contact 23 to terminal 15
2	29 + 15	Shift Lock Solenoid -N110-	• Ignition switched on • Brake pedal not depressed	approx. Battery voltage	– Check wiring according to wiring diagram
			– Brake pedal depressed	0.2 V	– Replace Shift Lock Solenoid -N110- ⇒ Servicing selector mechanism, page J 37-1
3	15 + 1	Brake Light Switch -F-	• Ignition switched on • Brake pedal not depressed	0 V	– Check wiring according to wiring diagram – Replace Brake Light Switch -F-
			– Brake pedal depressed	approx. Battery voltage	⇒ Repair Group 47; Assembly overview: Pedal cluster, brake pedal

Switch to voltage measuring range 20 Ω

Test step	VAG 1598/18 sockets	Test of	• Test conditions – Additional operations	Specified value	Corrective steps
4		Throttle Position (TP) Sensor -G69-	• Ignition switched off		– Check wiring according to wiring diagram
			Accelerator Pedal position:		– The resistance measurement will constantly change when operating Accelerator Pedal from idling to WOT
	5 + 28		Idling min. Idling max.	0.7 kΩ 1.8 kΩ	– Adjust TP Sensor, replace if necessary ⇒ Repair Group 24
			Wide Open Throttle (WOT) min. WOT max.	2.1 kΩ 3.9 kΩ	– Return system to basic setting ⇒ page J 01-47
	5 + 50	Throttle Position (TP) Sensor -G69-[1)]	Idling min. Idling max.	2.1 kΩ 3.9 kΩ	
			WOT min. WOT max.	0.7 kΩ 1.8 kΩ	

Switch to voltage measuring range 20 Ω

Test step	VAG 1598/18 sockets	Test of	• Test conditions – Additional operations	Specified value	Corrective steps
4		Throttle Position (TP) Sensor -G69- Diesel engine	• Ignition switched off		– Check wiring according to wiring diagram
			Accelerator Pedal position:		– The resistance measurement will constantly change when operating Accelerator Pedal from idling to WOT
	5 + 28		Idling min. Idling max.	0.8 kΩ 1.2 kΩ	– Adjust TP Sensor for Diesel engine, replace if necessary Read measuring value block page 01–49; display group number 01
			Wide Open Throttle (WOT) min. WOT max.	1.4 kΩ 2.0 kΩ	– Return system to basic setting ⇒ page J 01-47
	5 + 50		Idling min. Idling max.	1.5 kΩ 2.3 kΩ	
			WOT min. WOT max.	1.0 kΩ 1.4 kΩ	

Switch to voltage measuring range 200 Ω					
Test step	VAG 1598/18 sockets	Test of	• Test conditions – Additional operations	Specified value	Corrective steps
5	63 + 1	Multi-Function Transmission Range (TR) Switch -F125-	– Selector lever position R, N, D, 3 and 2	infinite Ω[1]	– Check wiring according to wiring diagram – Check connector on multi-function switch for contact corrosion, Replace if necessary
			– Selector lever position P and 1	0.8 to 1 Ω	– Replace Multi-Function TR Switch -F125- ⇒ page J 01-18, Fig. 9
	40 + 1		– Selector lever position P, R, 2 and 1	infinite Ω[1]	
			– Selector lever position N, D and 3	0 to 0.8 V	
	62 + 1		– Selector lever position P, R, N and D	infinite Ω[1]	
			– Selector lever position 3, 2 and 1	0 to 0.8 V	

[1] Switch multimeter to largest Ω range

Switch to voltage measuring range 200 Ω					
Test step	VAG 1598/18 sockets	Test of	• Test conditions – Additional operations	Specified value	Corrective steps
5	18 + 1		Switch to voltage measuring range 20 V		
			– Selector lever position P, R and N	approx. Battery voltage	
			– Selector lever position D, 3, 2 and 1	0 V	
6	55 + 67	Solenoid Valve 1 -N88-	• Ignition switched off	55 to 65 Ω	– Check routing of wiring according to wiring diagram
	55 + 1			infinite Ω[1]	– Replace conductor strip or valve body ⇒ page J 38-20
7	54 + 67	Solenoid Valve 2 -N89-	• Ignition switched off	55 to 65 Ω	– Check routing of wiring according to wiring diagram
	54 + 1			infinite Ω[1]	– Replace conductor strip or valve body ⇒ page J 38-20
8	9 + 67	Solenoid Valve 3 -N90-	• Ignition switched off	55 to 65 Ω	– Check routing of wiring according to wiring diagram
	9 + 1			infinite Ω[1]	– Replace conductor strip or valve body ⇒ page J 38-20

[1] Switch multimeter to largest Ω range

Switch to voltage measuring range 200 Ω					
Test step	VAG 1598/18 sockets	Test of	• Test conditions – Additional operations	Specified value	Corrective steps
9	47 + 67	Solenoid Valve 4 -N91-	• Ignition switched off	4.5 to 6.5 Ω	– Check routing of wiring according to wiring diagram
	47 + 1			infinite Ω[1]	– Replace conductor strip or valve body ⇒ page J 38-20
10	56 + 67	Solenoid Valve 5 -N92-	• Ignition switched off	55 to 65 Ω	– Check routing of wiring according to wiring diagram
	56 + 1			infinite Ω[1]	– Replace conductor strip or valve body ⇒ page J 38-20
11	58 + 22	Solenoid Valve 6 -N93-	• Ignition switched off	4.5 to 6.5 Ω	– Check routing of wiring according to wiring diagram
	58 + 1			infinite Ω[1]	– Replace conductor strip or valve body ⇒ page J 38-20
	22 + 1				– Replace conductor strip or valve body ⇒ page J 38-20
12	10 + 67	Solenoid Valve 7 -N94-	• Ignition switched off	55 to 65 Ω	– Check routing of wiring according to wiring diagram
	10 + 1			infinite Ω[1]	– Replace conductor strip or valve body ⇒ page J 38-20

[1] Switch multimeter to largest Ω range

Switch to voltage measuring range 200 Ω					
Test step	VAG 1598/18 sockets	Test of	• Test conditions – Additional operations	Specified value	Corrective steps
13	23 + 29	Shift Lock Solenoid -N110-	• Ignition switched off	14 to 25 Ω	– Check routing of wiring according to wiring diagram – Replace Shift Lock Solenoid -N110- ⇒ Servicing selector mechanism, page J 37-1
14	1 + 16	Kick Down Switch -F8-	• Ignition switched off • Accelerator Pedal not depressed	infinite Ω[1]	– Check routing of wiring according to wiring diagram – Adjust accelerator cable or replace ⇒ Repair Group 20
			– Depress Accelerator Pedal as far as kick down	less than 1.5 Ω	

[1] Switch multimeter to largest Ω range

Switch to voltage measuring range 2 M Ω					
Test step	VAG 1598/18 sockets	Test of	• Test conditions – Additional operations	Specified value	Corrective steps
15	6 + 67	Transmission Fluid Temperature Sensor -G93-	• Ignition switched off • ATF temperature approx. 20°C – Switch multimeter to 200 K Ω approx. 60°C approx. 120°C	 0.247 M Ω 48.8 K Ω 7.4 K Ω	– Check wiring according to wiring diagram – Replace conductor strip or valve body ⇒ page J 38-20

Switch to voltage measuring range 20 K Ω					
Test step	VAG 1598/18 sockets	Test of	• Test conditions – Additional operations	Specified value	Corrective steps
16	20 + 65	Vehicle Speed Sensor -G68-	• Ignition switched off min. max.	 0.8 K Ω 0.9 K Ω	– Check wiring according to wiring diagram – Replace Vehicle Speed Sensor -G68- ⇒ page J 01-19, Fig. 11
17	21 + 66	Transmission Vehicle Speed Sensor -G38-	• Ignition switched off min. max.	 0.8 K Ω 0.9 K Ω	– Check wiring according to wiring diagram – Replace Transmission Speed Sensor -G38- ⇒ page J 01-19, Fig. 10

Switch to voltage measuring range 20 K Ω					
Test step	VAG 1598/18 sockets	Test of	• Test conditions – Additional operations	Specified value	Corrective steps
18	19 + 64	Engine Speed (RPM) Sensor -G28- (for Diesel engine)	• Ignition switched off • Transmission Control Module -J217- disconnected	0.8 K Ω	– Check wiring according to wiring diagram – Distance to carrier plate max. 1.20 mm – Replace Engine Speed (RPM) Sensor -G28- ⇒ page J 01-24, Fig. 20

Torque converter

ATF pump is driven via the torque converter.

When installing the torque converter, ensure that both drive pins engage in the recesses of the ATF pump inner wheel.

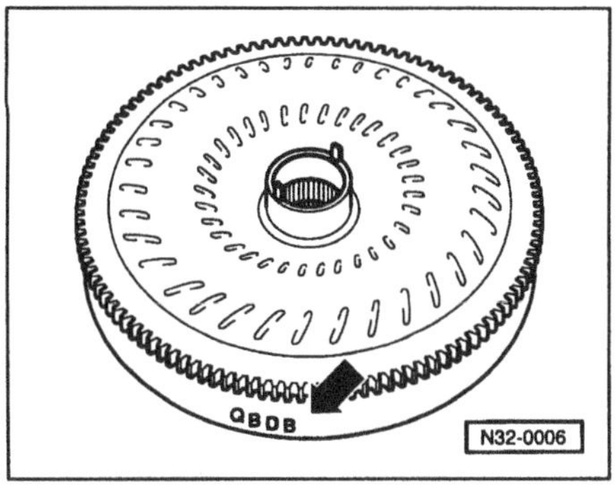

Torque converter, identification

◄ There are various torque converters. They are identified by code letters (arrow).

Transmission/torque converter applications
⇒ from Page J 00-6

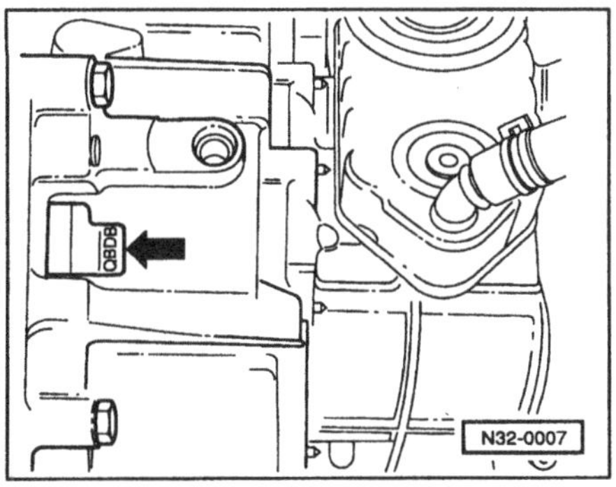

◄ **Torque converter code letters**

(Transmission installed)

Torque converter, draining

If the torque converter has been fouled by friction or, when performing a major overhaul of the transmission, drain the torque converter as follows:

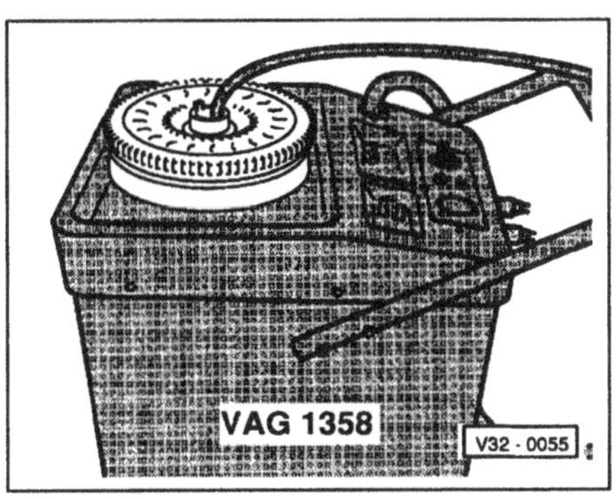

◄ – Extract ATF from torque converter using VAG 1 358A with VAG 1358A/1 probe, or EZ1 fluid extractor.

Torque converter oil seal, removing and installing

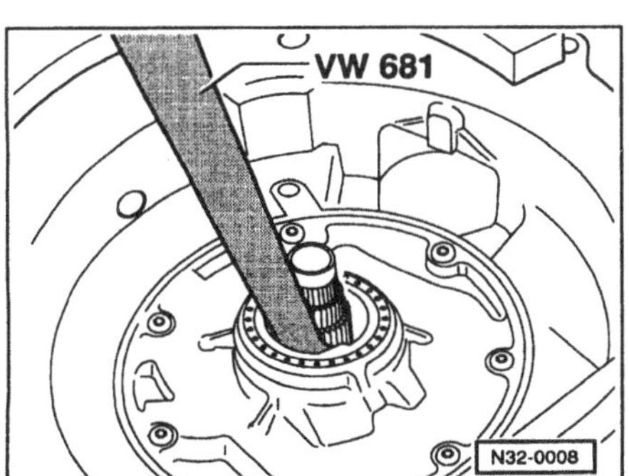

◄ **Torque converter oil seal, removing**

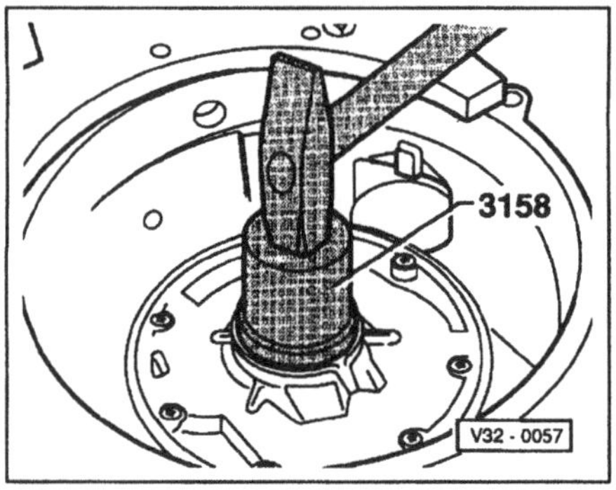

◄ **Torque converter oil seal, driving in until flush**

Shift mechanism, servicing

> ***CAUTION!***
> ***Before working on the engine when it is running, move the transmission selector lever into the "P" position and apply the parking brake.***

Selector mechanism, checking

Selector lever in "P" position; ignition switched on

- Brake pedal not depressed:

  Selector lever is locked and cannot be shifted out of "P" position. Shift Lock Solenoid -N110- blocks selector lever.

- Brake pedal depressed:

  Solenoid for selector lever lock releases selector lever. It is now possible to shift into a driving gear.

Selector lever in "N" position; ignition switched on

- Brake pedal not depressed:

  Selector lever is locked and cannot be shifted out of "N" position. Shift Lock Solenoid -N110- blocks selector lever.

- Brake pedal depressed:

  Shift Lock Solenoid -N110- releases selector lever. It is possible to shift into a driving gear.

Note:

- *The starter must not be able to operate with the selector lever in positions "1" "2" "3" "D" and "R"*
- *When traveling at speeds above 5 km/h (3 mph) and shifting into selector lever position "N", the Shift Lock Solenoid -N110- must not engage and block the selector lever. The selector lever can be shifted into a driving gear.*
- *When traveling at speeds below 5 km/h (3 mph – almost stopped) and shifting into selector lever position "N", the Shift Lock Solenoid -N110- must not engage until after approximately 1 second. The selector lever cannot be shifted out of "N" position until the brake pedal is depressed.*

Selector lever cable, checking and adjusting

Checking

– Remove selector lever cable locking washer from lever/selector shaft.

– Take selector lever cable off lever.

– Move selector lever from "P" to "1".

● Shift mechanism and selector lever cable must move freely. If necessary replace selector lever cable or service shift mechanism.

Note:

Always replace the selector lever cable locking element on the transmission support bracket after disassembling.

Adjusting

● Selector lever in "P" position.

– Loosen adjustment screw on selector lever in selector lever housing.

– Bring selector shaft lever into position "P" on transmission. Locking lever must engage in park lock wheel, both front wheels blocked.

– Tighten adjustment screw on selector lever, in selector lever housing, to 20 Nm (15 ft lb).

Ignition lock, checking

– Shift selector lever to "P" position and turn ignition key to end ("OFF") position.

● It must only be possible to remove ignition key with selector lever in "P" position.

– Remove ignition key.

● Selector lever cannot be shifted out of "P" position.

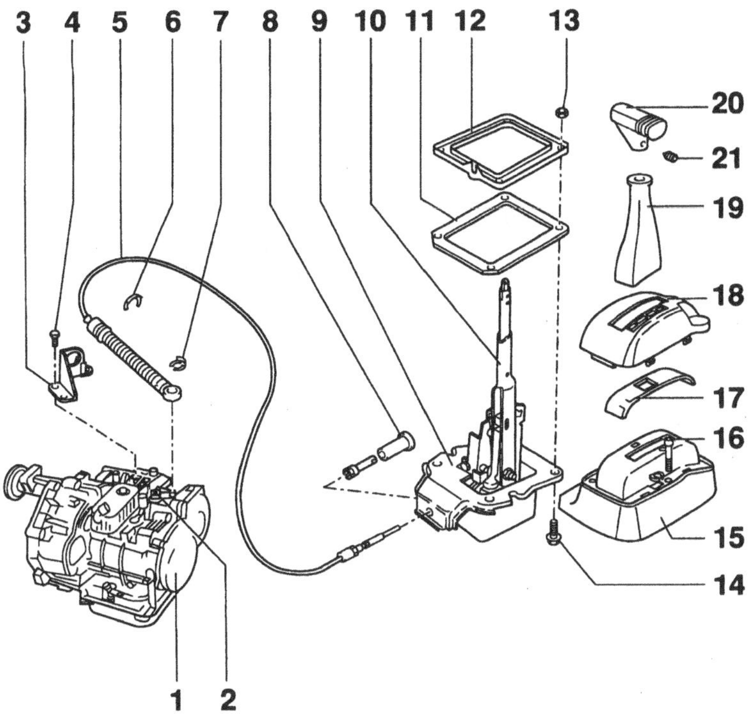

N37-0251

Shift mechanism, disassembling and assembling

1 - Transmission

2 - Lever
- For selector shaft

3 - Support bracket
- For selector lever cable

4 - Bolt, 10 Nm (7 ft lb)

5 - Selector lever cable with guidetube
- Do not bend or kink
- Lightly grease cable eyelet and cable ends before installing
- Boot and protective cover must be undamaged and installed correctly or moisture can enter the selector lever cable;
- Check guide tube for tight fit
- Tighten selector lever cable on anchorage/selector lever housing to 20 Nm (15 ft lb) ensuring recessed side of selector lever cable is towards adj. screw
- Secure to support bracket with locking element
- Secure to lever/selector shaft with locking washer

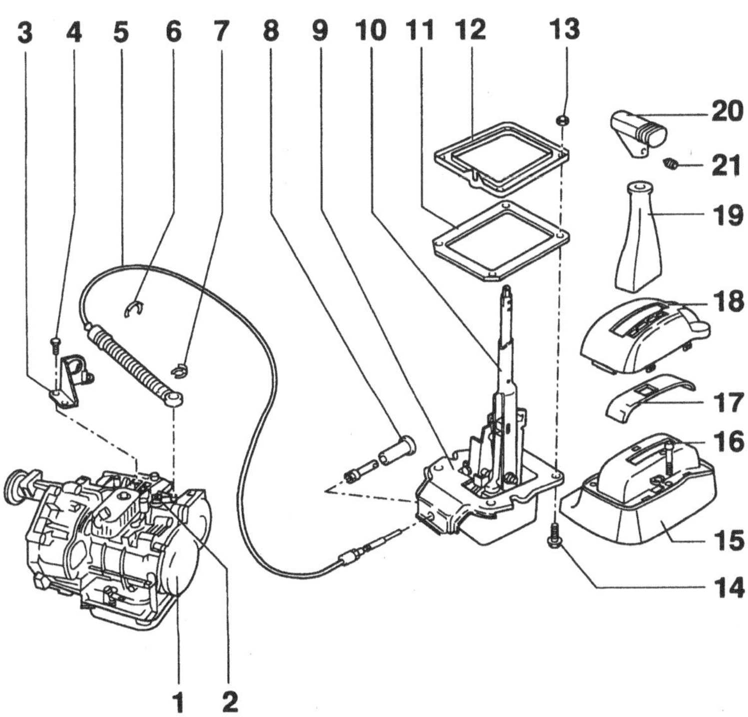

N37-0251

6 - Corrugated locking element
- For selector lever cable
- Always replace

7 - Locking washer
- For selector lever cable

8 - Protective sleeve
- Install in housing before attaching selector lever cable

9 - Selector lever housing
- With installed selector lever
- Need not be removed to replace selector lever cable or individual parts of shift mechanism

10 - Selector lever
- Removing and installing ⇒ Page J 37-9
- Disassembling and assembling ⇒ Page J 37-12

11 - Gasket
- Glue to selector lever housing

12 - Frame
- For securing console and ignition

13 - Nut, 12 Nm (9 ft lb)

14 - Bolt, 12 Nm (9 ft lb)

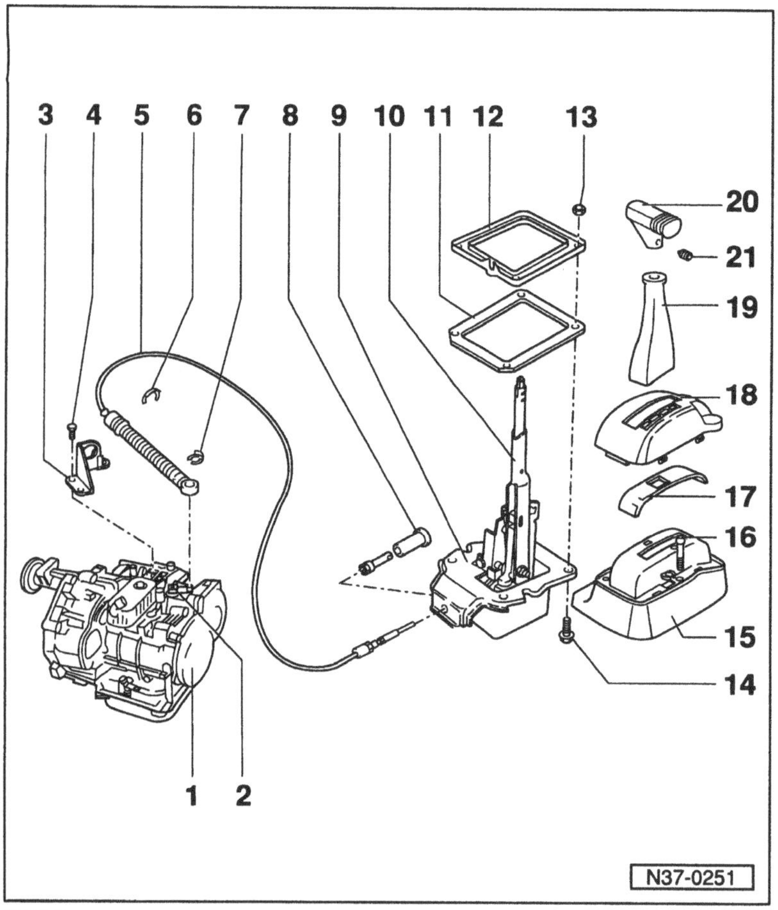

15 - Console

16 - Bolt, 2 Nm (18in.1b)

17 - Cover strip

18 - Cover
- With selector indicator

19 - Selector lever trim

20 - Selector lever handle
- Lightly grease push rod when installing
- Secure with set screw or grub screw

21 - Set screw, 1.5 Nm (14 in. lb)
- Install with D 000 600

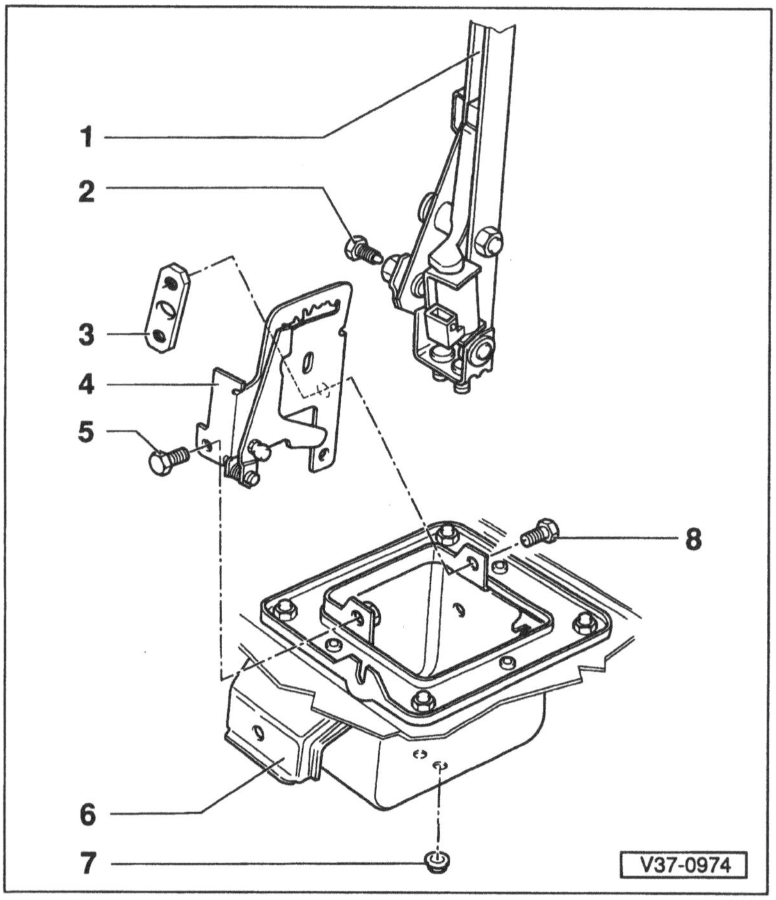

Selector lever, removing and installing

1 - Selector lever
- Assemble before installing in selector lever housing
- Disassembling and assembling ⇒ Page J 37-12
- Wiring routing ⇒ Fig. 1

2 - Bolt, 20 Nm (15 ft lb)
- For securing the selector lever cable

3 - Plate with threaded holes

4 - Detent plate
- Insert into selector lever housing before installing selector lever
- Disassembling and assembling ⇒ Page J 37-12

5 - Bolt, 25 Nm (18 ft lb)

6 - Selector lever housing
- Need not be removed to replace selector lever cable or individual parts of shift mechanism

7 - Nut, 15 Nm (11 ft lb)

8 - Bolt, 25 Nm (18 ft lb)

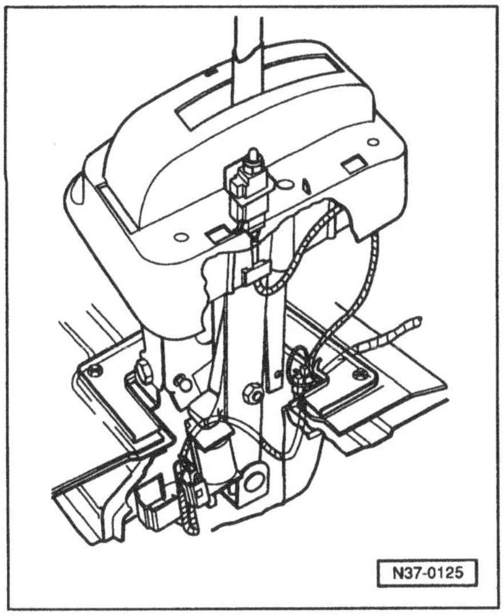

◄ **Fig. 1 Indicator lighting and Shift Lock Solenoid -N110- wiring, routing**

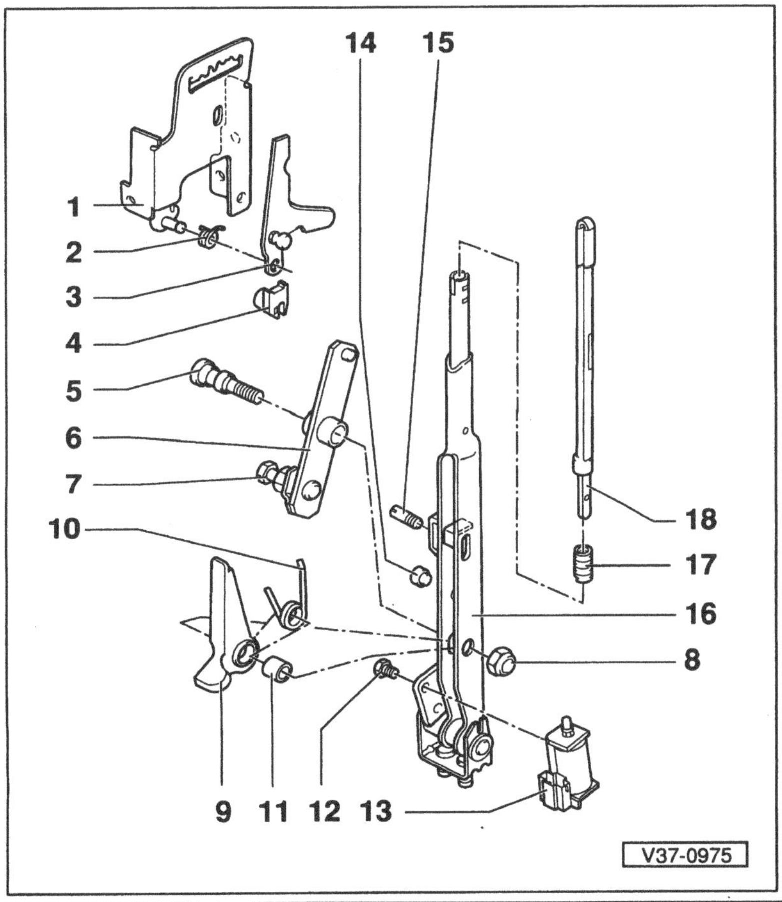

Selector lever, disassembling and assembling

Note:

Lubricate all joints and sliding surfaces with multi-purpose grease.

1 - Detent plate
- Before installing selector lever in housing, assemble selector lever, detent plate, spring, locking lever and pivot piece.

2 - Spring
- For locking lever
- Hook into hole of detent plate

3 - Locking lever
- For vehicles with ignition key

4 - Pivot piece

5 - Combi-bolt

6 - Relay lever

7 - Bolt
- For securing selector lever cable

8 - Hex nut, 25 Nm (18 ft lb)
- Always replace

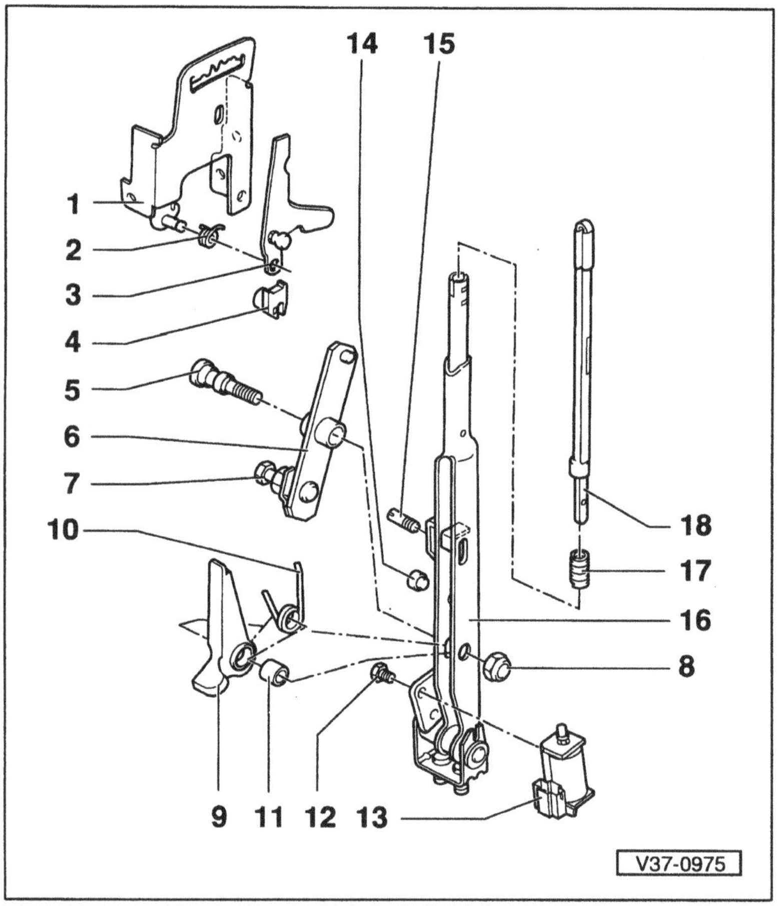

9 - **Locking lever**
- Install with spring and bushing

10 - **Spring**
- Install with locking lever and bushing into selector lever

11 - **Bushing**
- Install in locking lever

12 - **Hex bolt, 3 Nm (27 in. lb)**
- Install with locking fluid AMV 1 85101 A1

13 - **Shift Lock Solenoid -N110-**
- Can be checked electrically ⇒ Page J 01-73, and by measuring value block ⇒ Page J 01-49

14 - **Slide**
- Press into the selector lever

15 - **Set screw**
- Install with locking fluid AMV 185 101 A1

16 - **Selector lever**
- Assemble prior to installation

17 - **Spring**

18 - **Push rod**
- Grease in area of sliding surfaces

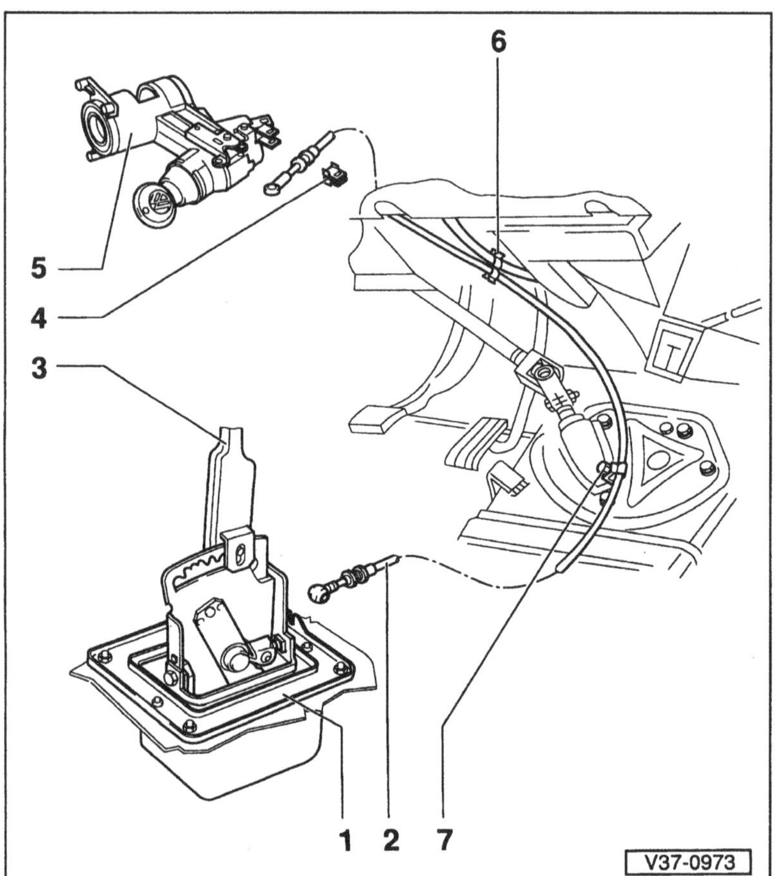

Ignition locking cable, removing and installing

Note:

After installing adjust locking cable ⇒ Page J 37-14.

1 - **Frame**

2 - **Locking cable**
- Must not be kinked
- Adjusting ⇒ Page J 37-14
- Checking ignition key lock ⇒ Page J 37-4

3 - **Selector lever**
- Removing and installing ⇒ Page J 37-8
- Disassembling ⇒ Page J 37-10

4 - **Clip**
- To secure locking cable
- Always replace

5 - **Ignition/starter switch**

6 - **Mounting**
- For locking cable
- Secure to heater hose

7 - **Mounting**
- Secure to base plate and pull locking cable through mounting eye

Removing

– Remove selector lever grip, cover, cover strips and selector lever console.

– Remove locking cable from frame and remove from locking lever ball head.

– Remove ignition/starter switch trim.
⇒ Electrical system; Repair Group 90; Removing and installing instrument cluster

– Press locking cable clip out of ignition/starter switch support and remove locking cable from ball head.

– Remove footwell vent console.
⇒ Heating, air conditioning; Repair Group 80

– Remove locking cable.

Installing

Installation is performed in the reverse order.

Notes:

♦ *Ensure proper routing of locking cable when installing.*

♦ *Adjust locking cable after installing ⇒ Page J 37-14*

Ignition locking cable, adjusting

Note:

Locking cable must only be adjusted in its installation position.

– Shift selector lever into "P" position and turn ignition key to end ("OFF") position.

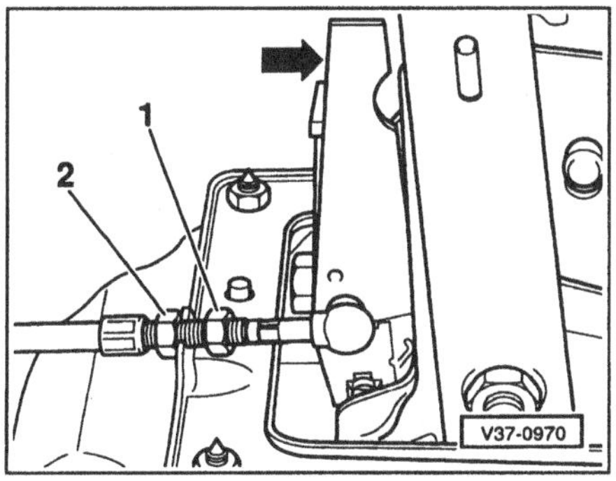

◄ – Loosen locking cable nuts -1- and -2- at frame.

– Push locking lever -arrow- until it touches selector lever.

– Secure locking cable in this position on frame.

– Perform functional check ⇒ Page J 37-4.

Transmission, checking

Note:

The automatic transmission is only to be checked as described after having first completed the On Board Diagnostic (OBD), ⇒ Page J 01-25, and the cause of damage cannot be located and corrected.

Shift points, checking

⇒ Page J 37-16

Stall speed, checking

⇒ Page J 37-20

Measuring main pressure

⇒ Page J 37-22

Shift points, checking

Test Requirements

- Activate On Board Diagnostic Program

– Check Diagnostic Trouble Code (DTC) Memory

- Check Automatic Transmission Fluid (ATF) level
  ⇒ page J 37-42

- Check engine settings
⇒ Repair Group 24

Transmission shifts too early or too late into the next gear

When testing shift points be sure to note that the speedometer can deviate from the specified figure within the permissible tolerance.

The table lists the gear shifts by transmission code letters.

H = Hydraulic shift

M = Mechanical shift

To check the shift points, connect VAG 1551 Scan Tool ⇒ On Board Diagnostic (OBD) program, Repair Group 01

Shift Points in km/h (mph)

Note

During the warm-up phase, with warm ATF and uphill or downhill driving, the shift points are automatically changed by the control module depending on the driving situation . These changes can mean deviations from the chart.

Transmission code	Shift	Kickdown km/h (mph)
CUF	1 H – 1 M	31...37 (19...23)
	1 M – 2 H	31...37 (19...23)
	2 H – 2 M	61...67 (38...42)
	2 M – 3 H	61...67 (38...42)
	3 H – 3 M	95...101 (59...63)
	3 M – 4 H	95...101 (59...63)
	4 H – 4 M	95...101 (59...63)
	4 M – 4 H	99...93 (62...58)
	4 H – 3 M	99...93 (62...58)
	3 M – 3 H	99...93 (62...58)
	3 H – 2 M	62...56 (39...34)
	2 M – 2 H	62...56 (39...34)
	2 H – 1 M	26...20 (16...12)
	1 M – 1 H	26...20 (16...12)

Transmission code	Shift	Kickdown km/h (mph)
CUH	1 H – 1 M	29...35 (18...22)
	1 M – 2 H	29...35 (18...22)
	2 H – 2 M	56...62 (35...39)
	2 M – 3 H	56...62 (35...39)
	3 H – 3 M	88...94 (55...58)
	3 M – 4 H	88...94 (55...58)
	4 H – 4 M	88...94 (55...58)
	4 M – 4 H	92...86 (57...53)
	4 H – 3 M	92...86 (57...53)
	3 M – 3 H	92...86 (57...53)
	3 H – 2 M	58...52 (36...32)
	2 M – 2 H	58...52 (36...32)
	2 H – 1 M	24...18 (15...11)
	1 M – 1 H	24...18 (15...11)
CUG	1 H – 1 M	39...45 (24...28)
	1 M – 2 H	39...45 (24...28)
	2 H – 2 M	78...84 (48...52)
	2 M – 3 H	78...84 (48...52)
	3 H – 3 M	87...93 (54...58)
	3 M – 4 H	115...121 (71...75)
	4 H – 4 M	115...121 (71...75)
	4 M – 4 H	118...112 (73...70)
	4 H – 3 M	118...112 (73...70)
	3 M – 3 H	75...69 (47...43)
	3 H – 2 M	75...69 (47...43)
	2 M – 2 H	75...69 (47...43)
	2 H – 1 M	29...23 (18...14)
	1 M – 1 H	29...23 (18...14)

Transmission code	Shift	Kickdown km/h (mph)
CUV	1 H – 1 M 1 M – 2 H 2 H – 2 M 2 M – 3 H 3 H – 3 M 3 M – 4 H 4 H – 4 M 4 M – 4 H 4 H – 3 M 3 M – 3 H 3 H – 2 M 2 M – 2 H 2 H – 1 M 1 M – 1 H	33...39 (21...24) 33...39 (21...24) 67...73 (42...45) 67...73 (42...45) 74...80 (46...50) 98...104 (61...65) 98...104 (61...65) 102...96 (63...60) 102...96 (63...60) 65...59 (40...37) 65...59 (40...37) 65...59 (40...37) 25...19 (16...12) 25...19 (16...12)
CUJ	1 H – 1 M 1 M – 2 H 2 H – 2 M 2 M – 3 H 3 H – 3 M 3 M – 4 H 4 H – 4 M 4 M – 4 H 4 H – 3 M 3 M – 3 H 3 H – 2 M 2 M – 2 H 2 H – 1 M 1 M – 1 H	36...42 (22...26) 36...42 (22...26) 72...78 (45...48) 72...78 (45...48) 81...87 (50...54) 107...113 (66...70) 101...113 (66...70) 110...104 (68...65) 110...104 (68...65) 70...64 (43...40) 70...64 (43...40) 70...64 (43...40) 27...21 (17...13) 27...21 (17...13)

Stall speed, checking

– Connect tachometer (e.g. VAG 1367 engine tester ignition system) or read engine speed (RPM) in measuring value block with VAG 1551 Scan Tool.

– Secure vehicle with parking and foot brake.

– Start engine and place selector lever in "D" position.

– Increase engine speed (RPM) briefly.

If the stall speed is not attained, the following malfunctions are probable causes:

Stall speed too high:	Forward clutch slipping, freewheel slipping
Stall speed up to 200 RPM too low:	Check engine settings
Stall speed over 200 RPM too low:	Torque converter malfunction

In case of doubt, always check the engine settings first and correct if necessary.

The correct torque converter must always be installed in the transmission.

Stall Speed Chart

Transmission code	Torque converter Code letters	Stall speed (RPM)
CUF	QBDB	2050...2350
CUH	QBDB	2050...2350
CUG	QBDB	2150...2450
CUJ	QBDB	2150...2450
CUV	QBDB	2150...2450

Main pressure, measuring

Before checking main pressure, check Diagnostic Trouble Code (DTC) Memory and, if necessary, correct any electrical malfunctions.

– Activate DTC Memory
⇒ Repair Group 01, On Board Diagnostic (OBD) program

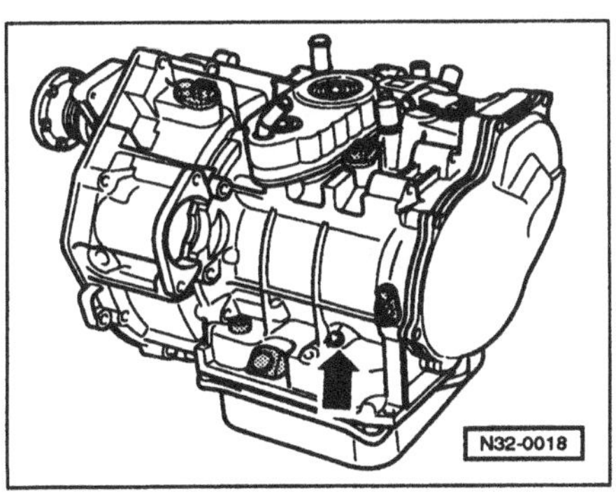

◄ – Remove bolt for main pressure hole -arrow-.

● Always replace bolt at main pressure hole.

– Connect VAG 1702 pressure gauge and tighten knurled screw.

– Bleed VAG 1702 pressure gauge line by loosening knurled screw with engine running.

Note:

Tighten knurled screw after bleeding.

– Bring automatic transmission fluid (ATF) temperature to 60°C (140°F).
– Connect VAG 1551 Scan Tool and read temperature in measuring value block.
⇒ Repair Group 01, On Board Diagnostic (OBD) program

Main pressure specified values

Automatic transmission	Selector lever positions and test conditions			
01P			Connector for solenoid valves removed [1]	
	D Idle	R Idle	D Engine speed approx. 2000 RPM	R Engine speed approx. 2000 RPM
Main pressure in bar	3.4 ... 3.8	6.5 ... 7.5	10.1 ... 11.3	23.0 ... 24.0

[1] After connecting the solenoid valve connector, check and erase Diagnostic Trouble Code (DTC) Memory.
⇒ Repair Group 01, OBD program
If possible, perform main pressure test on a chassis dynamometer.

Possible causes	**Corrective action**
Idle speed too high	Adjust engine
ATF pump faulty	Check ATF pump, replace if necessary
Control valve sticking – valve body	Replace valve body

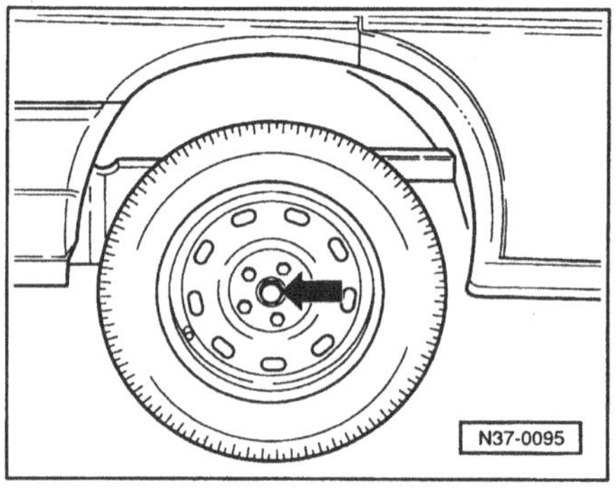

Transmission, removing and installing (vehicles up to 12.95)

Note:

Removing and installing transmissions in vehicles produced from 01.96, ⇒ page J 37-41.

Removing

– Install support frames 3250 before lifting vehicle.
⇒ Suspension, Wheels, Brakes, Steering (5-cylinder engines) Manual, Repair Group 40

– Remove left-hand drive shaft/wheel hub bolt -arrow-.

– Obtain radio code on vehicles with coded radio.

– Disconnect battery ground strap.

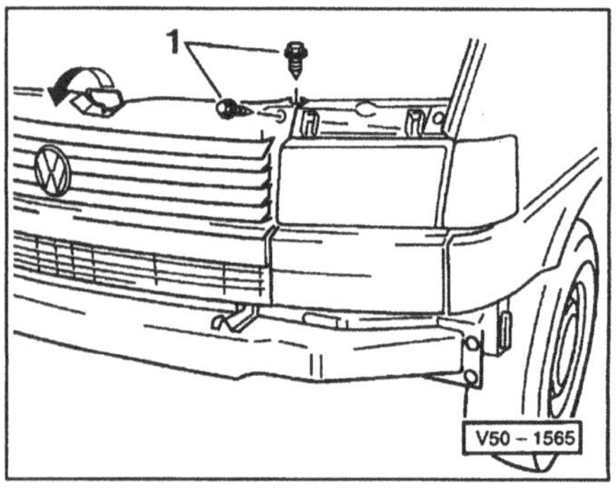

– Remove battery covering.

– Remove bolts -1- and fold radiator out in direction of arrow.

– Disconnect upper white connector from transmission (Transmission Vehicle Speed Sensor -G38-).

– Remove upper engine/transmission connecting bolts.

– Remove noise insulation tray.

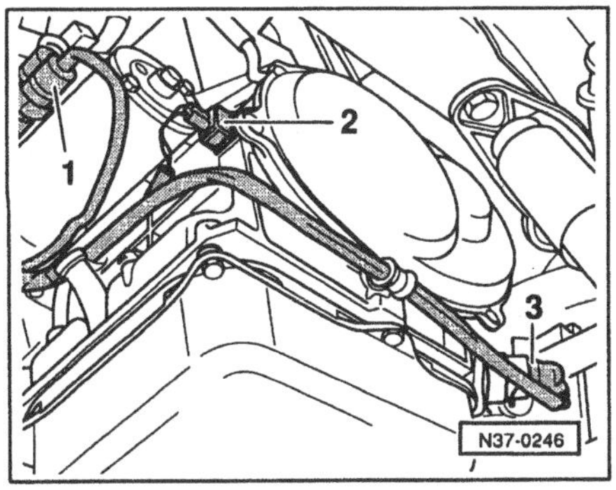

- Disconnect electrical connections -1- to -3- from transmission.

  1 – Valve body connection

  2 – Vehicle Speed Sensor -G68- (black connector)

  3 – Multi-Function Transmission Range (TR) Switch -F125-

  - Remove multi-function switch wiring, with bracket, from transmission and place to side.

- Remove starter.

- Disconnect speedometer drive shaft from bearing housing, or

- Pull speedometer connector off bearing housing.

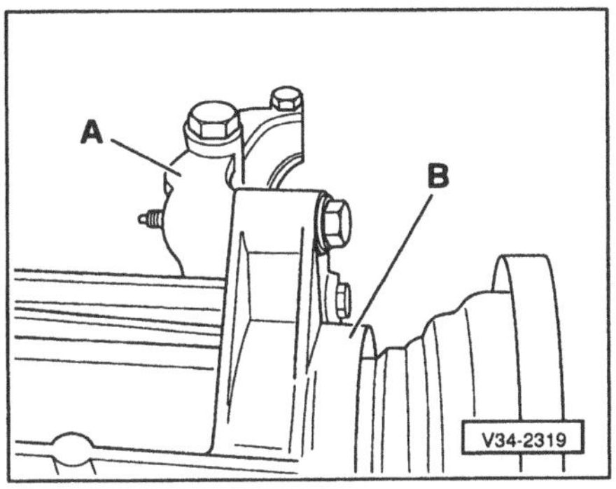

- Remove transmission carrier -A- from engine and bearing housing -B-.

- Disconnect front exhaust pipe at exhaust manifold.

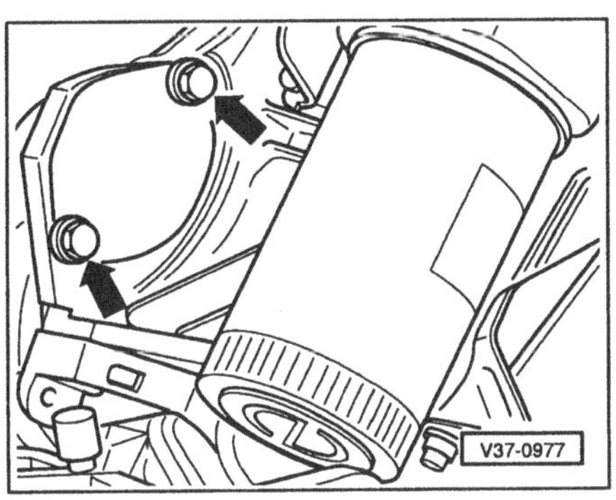

- Remove bolts -arrows- from cover plate.

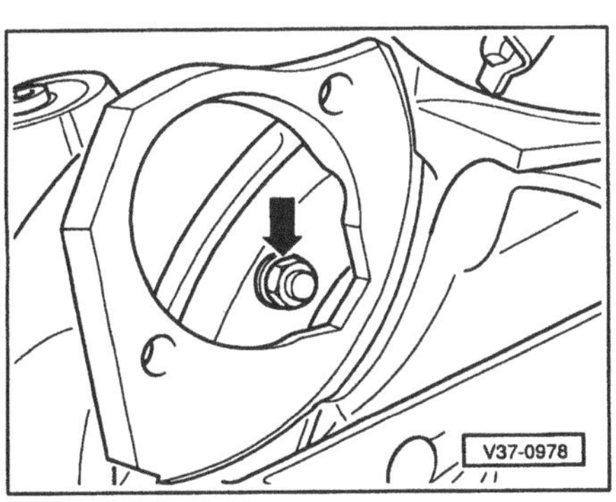

- Remove nut -arrow- from torque converter (Qty. 3).

- Disconnect both drive shafts at flanges.

- Raise right-hand drive shaft and secure.

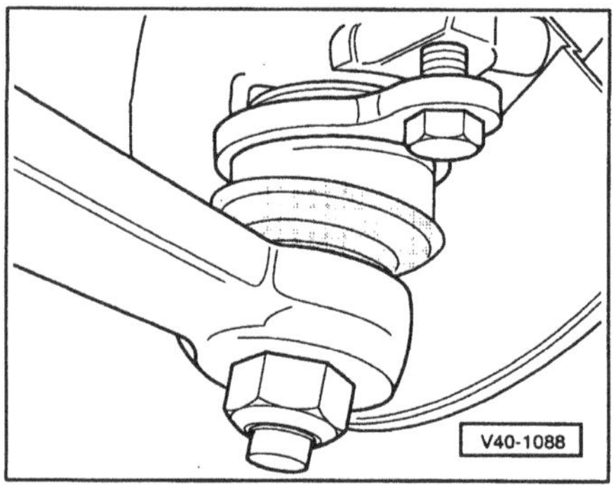

- Separate left-hand wheel bearing housing/lower axle link bolted connection.
- Remove bolt for left-hand shock absorber/stabilizer bar coupling rod.
- Compress left-hand shock absorber.

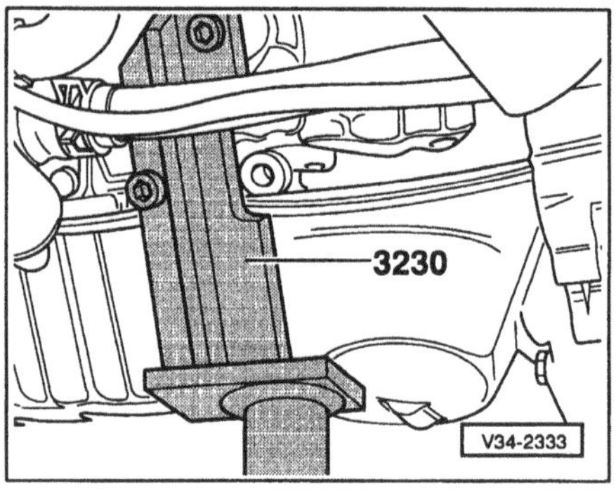

- Install removal device 3184 to support engine/transmission:
- Attach engine support to engine block.
  - ♦ Vehicles with 5-Cyl. engine – support 3227

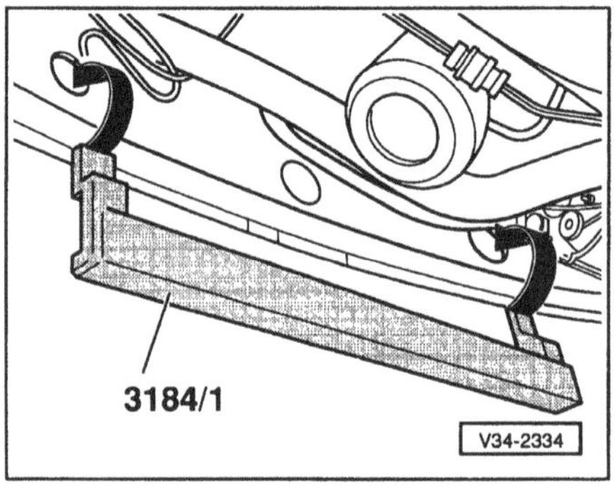

- Attach auxiliary support 3184/1 for removal device 3184 into cross panel -arrows-. If necessary, remove wiring harness and lines from cross panel.

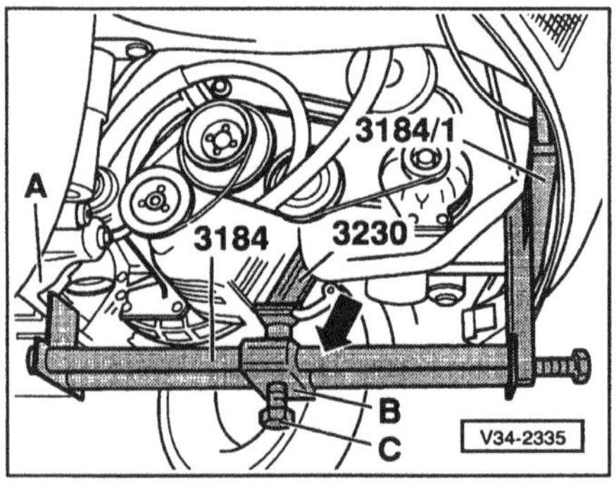

- Attach removal device 3184 into auxiliary support 3184/1 and into subframe -A-.
- Thread spindle -C- into hole on engine mounting.
- Align 3184 removal device so that slide -B- is resting against right-hand tube -arrow-.
- Secure 3184 removal device to subframe -A- and to auxiliary support 3184/1 by tightening angle plates.
- Release weight from subframe using spindle -C- to raise engine/transmission.

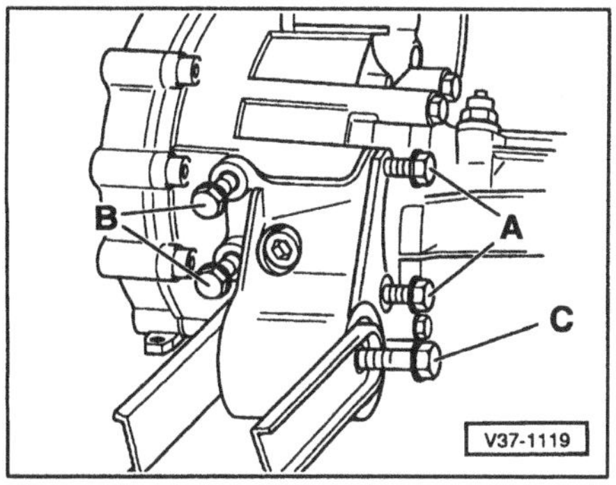

- Remove bonded rubber mounting from rear of assembly carrier -arrows-.
- Remove all bolts (A, B, C) and remove bonded rubber mounting.

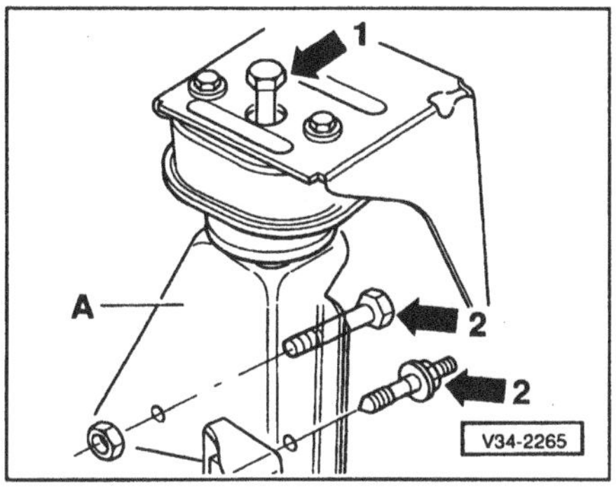

- Unbolt bearing bracket -A- from bonded rubber mounting at body -arrow 1-.
- Lower engine using removal device spindle.
- Unbolt bearing block -A- from transmission -arrow 2

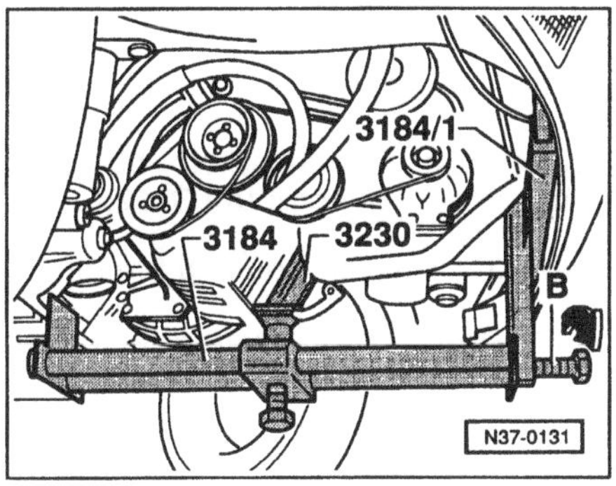

- Carefully swivel engine/transmission forward and lower, ensuring that selector lever cable, multi-function switch and drive shafts are not damaged.
- To swivel forward: turn spindle -B- to left
- To swivel back: turn spindle -B- to right

Note:

Ensure that selector lever cable is not overstressed.

- Unclip selector lever cable securing clips and remove selector lever cable.
- Clamp-off ATF cooler hoses with 3094 clamps or equivalent and remove hoses from ATF cooler.
- Tilt engine/transmission further forward and lower.
- Ensure sufficient clearance between engine and vehicle longitudinal member. If necessary, loosen Generator and remove tensioning bracket from engine.
- Remove left-hand drive shaft to ensure multi-function switch is not damaged.

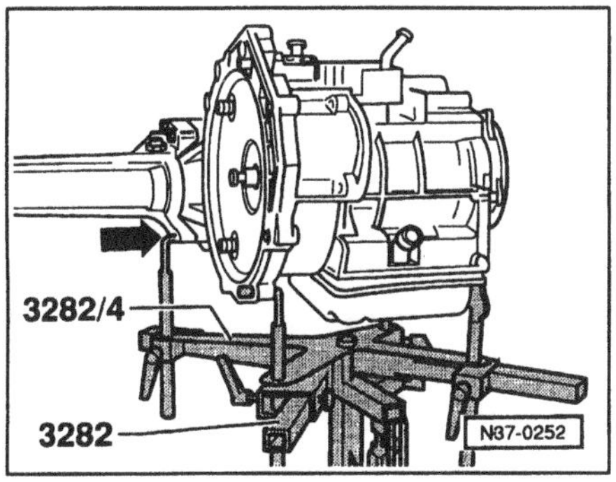

- Set up transmission support 3282.
- Remove bearing housing securing bolt -arrow- to enable transmission mounting hook to be inserted.

Use the transmission mount 3282 with the adjusting plate 3282/4 for removing the automatic transmission 01P.

Note:

The symbols on the adjusting plate indicate the adapters required.

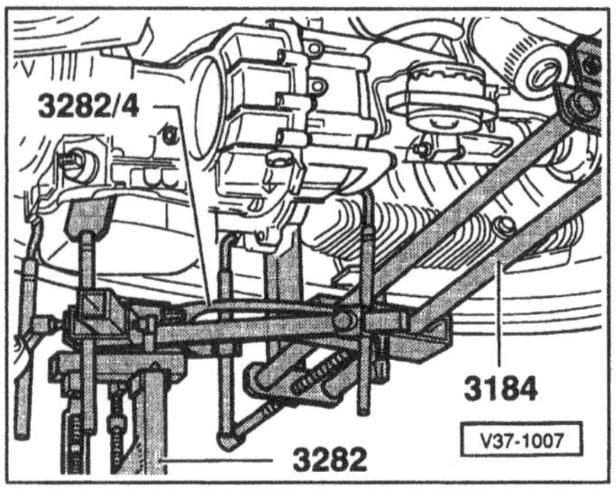

◄ – Move VAG 1383 A jack with transmission mount 3282 under transmission and support it.

– Remove lower engine/transmission connecting bolts.

– Press transmission off engine and press torque converter out of drive plate.

– Press torque converter against ATF pump.

– Lower transmission slightly.

– Tilt transmission forward and guide right-hand drive flange past removal device 3184.

– Swivel transmission and carefully lower.

– Secure torque converter to prevent it from falling out.

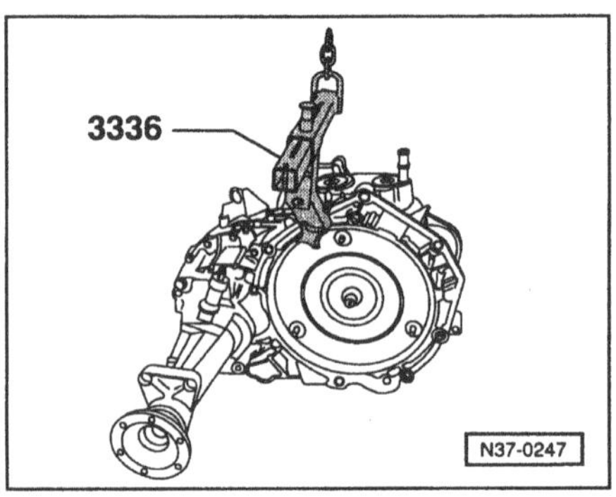

Transporting the automatic transmission

Special tool 3336 may be used to transport the automatic transmission and when setting up transmission mount 3282.

◄ – Secure special tool 3336 to transmission housing flange.

– Position pins so that 8 holes are visible in chain direction.

Installing

Installation is performed in the reverse order.

When installing the torque converter, ensure that both drive pins engage in the ATF pump inner wheel recesses.

- Before installing, ensure that dowel sleeves for aligning engine/transmission are correctly located.
- When swinging transmission in, install both drive shaft flanges in flange shafts.

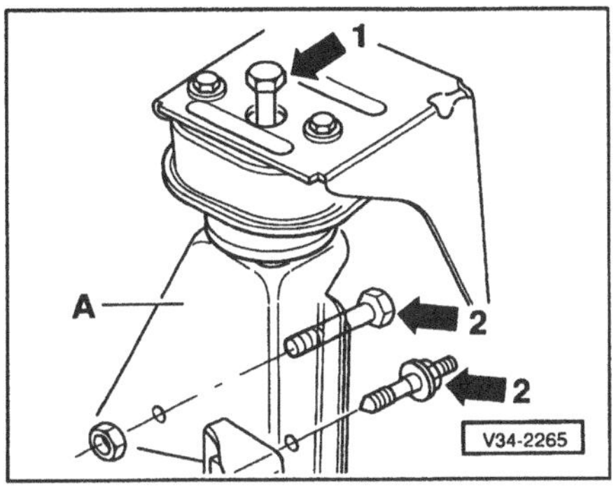

◄ – Before lifting engine/transmission, securely tighten bearing block -A- bolts -arrows 2-.

- After lifting engine/transmission, securely tighten bolt -arrow 1-.

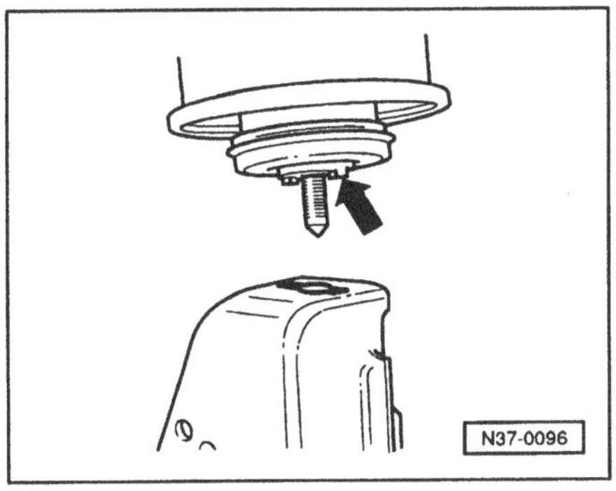

◄ – Ensure that bonded rubber mounting lugs -arrow- are seated correctly in bracket.

> ***CAUTION!***
>
> ***Do not remove 3184 removal device until the bolts securing the left-hand bearing bracket have been securely tightened.***

- Replace selector cable anchorage locking washer.
- Check selector lever cable adjustment, adjust if necessary ⇒ Page J 37-3.

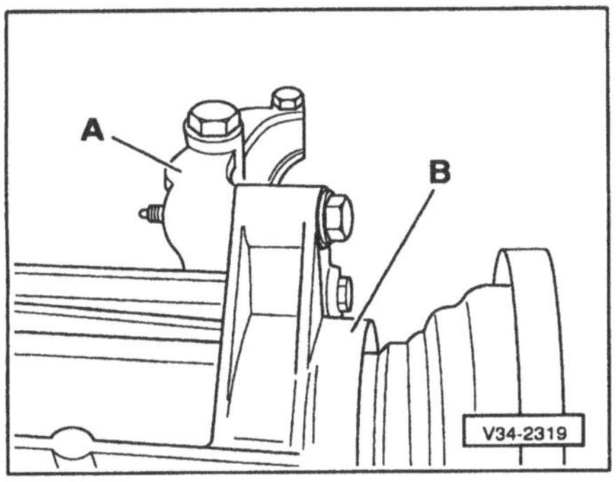

◄ – First tighten transmission carrier -A- to bearing housing -Band, then to engine.

Transmission, removing and installing (vehicles from 01.96)

Beginning in January 1996, the EuroVan is produced with a longer front end. If the transmission is to be removed or installed, note the following:

- Special tools are required – supports 10–222 A/6 and additional hook 10–222 A/7
- Subframe and transmission mount have been modified (self-aligning mount).

⇒ List of transmission code letters/specifications, Repair Group 00.

- Output to right drive shaft operates through a free-wheel drive flange.
- Support frames 3250 can no longer be used.

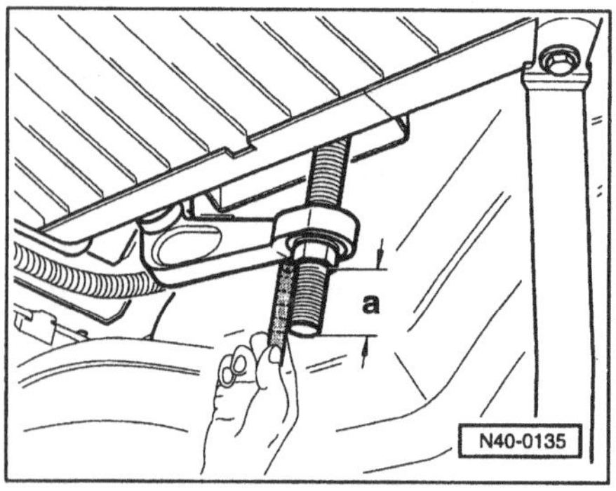

◄ **Removing**

– Raise vehicle and measure projection of nut thread to end of threaded rod -A- and record measurement.

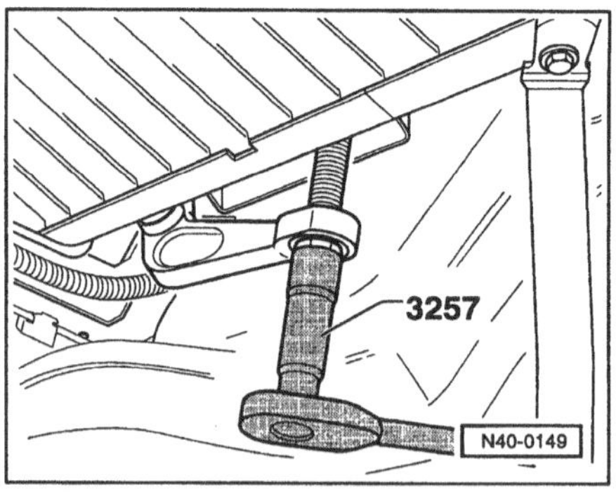

◄ – Remove tension on torsion bar with special tool 3257.
⇒ Suspension, Wheels, Brakes, Steering, Repair Group 40.

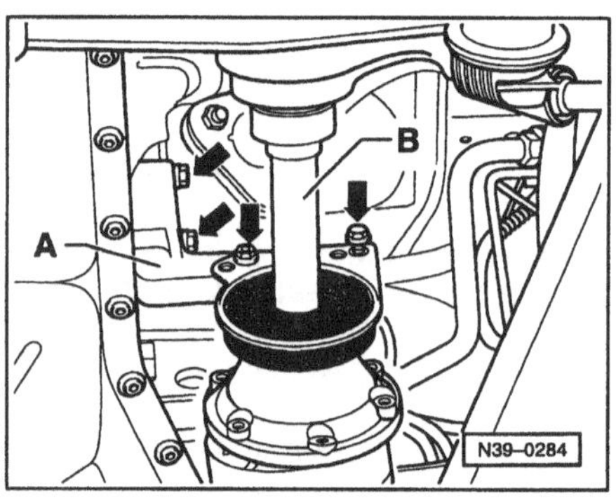

◄ – Remove bracket –A– of bearing for right drive flange –arrows–.

– Turn steering to right side and remove drive flange –B–.

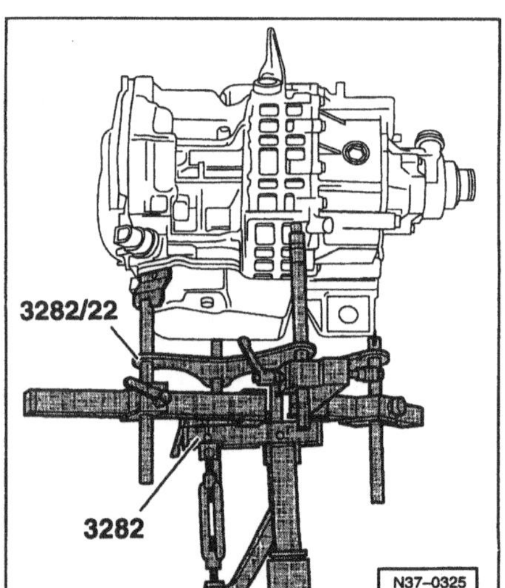

◄ – The transmision support 3282 is used with adjustment plate 3282/22, for removing the automatic transmission 01P.

– Move transmission jack VAG 1383 A, or equivalent, with transmission support 3282 and adjustment plate 3282/22 into place and support transmission.

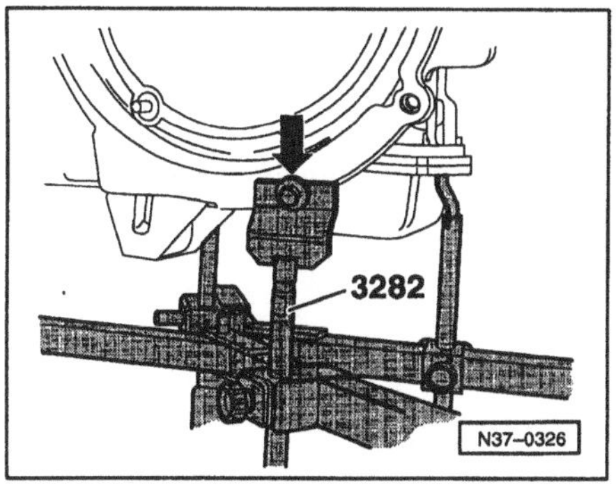

◄ – Attach transmission support and adapter to hole in torque converter cover plate -arrow-

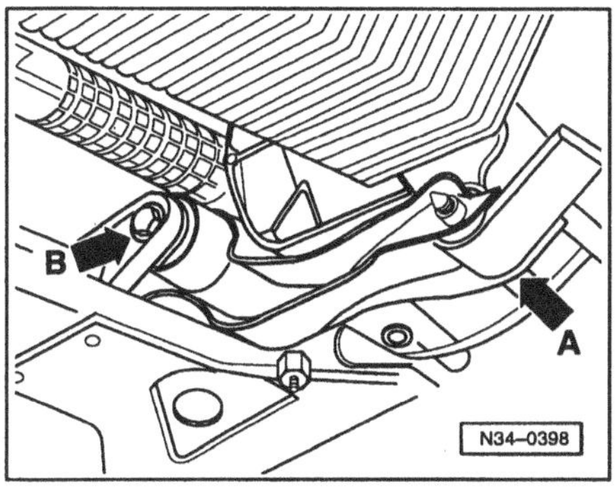

Modified rear transmission mount

The rear transmission mount is a self-aligning mount.

◄ **Tightening torque**

Bolt -A- = 80 Nm (59 ft lb) + 90° (1/4 turn)

Bolt -B- = 200 Nm (148 ft lb)

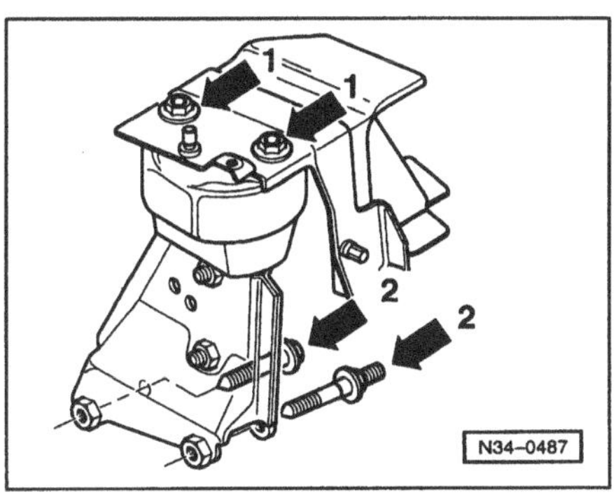

Modified left transmission mount

◄ – Remove complete transmission mount by removing nuts -1- and bolts -2- after removing transmission.

Tightening torque

Nuts -1- = 55 Nm (41 ft lb)

Bolts -2- = 50 Nm (37 ft lb) + 90° (1/4 turn)

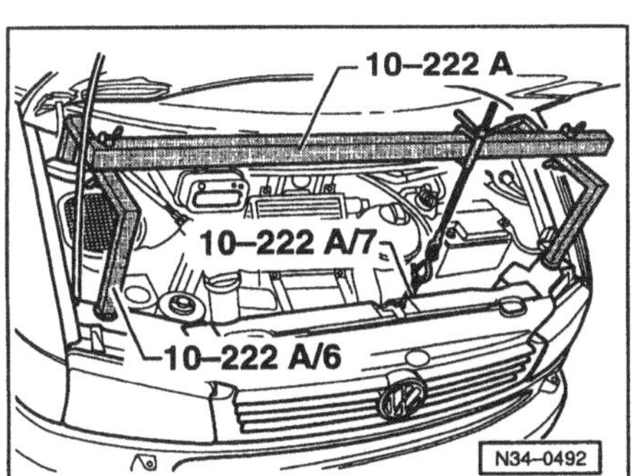

The radiator and radiator support are not folded forward as with earlier versions (with 5-cylinder engine).

◄ The engine is held with 10–222A in connection with supports 10–222A/6 and additional hook 10–222A/7 if necessary.

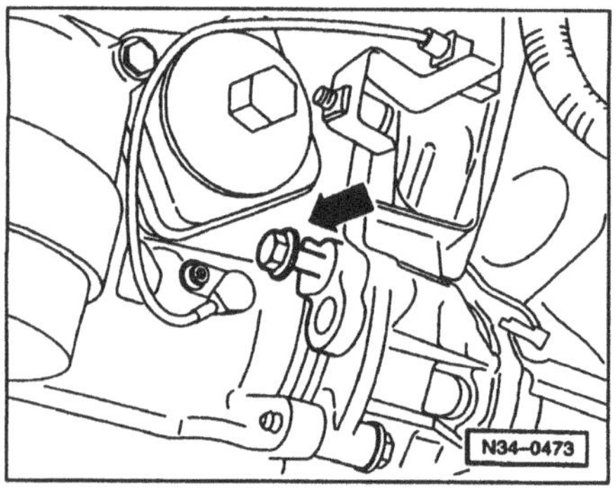

Vehicles with VR6 engine

◄ – Remove bolt connecting engine/transmission -arrow- .

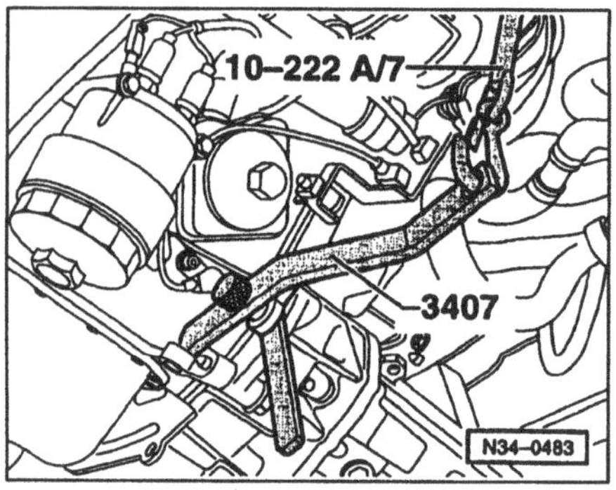

◄ – Attach bracket 3407 to engine.

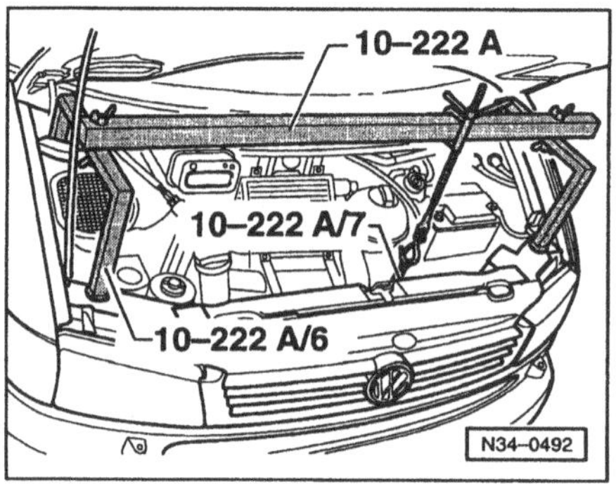

◄ The engine is held with 10–222A along with braces 10–222A/6. Additional hook 10–222A/7 connects to bracket 3407 (⇒ illustration N34–0483).

Tightening torques

Drive shaft to drive flange	80 Nm (59 ft lb)	Hex bolt on drive shaft ♦ Always replace bolt	150 Nm (111 ft lb) plus 1/4 turn (90°)
Torque converter to drive plate	60 Nm (44 ft lb)	Bonded mounting to subframe	45 Nm (33 ft lb)
Transmission to engine M 8	20 Nm (15 ft lb)	Assembly mounting to subframe	200 Nm (148 ft lb)
Transmission to engine M 10	60 Nm (44 ft lb)	Assembly mounting to torque convert-er housing	
Transmission to engine M 12	80 Nm (59 ft lb)	M10	40 Nm (40 ft lb)
Left-hand mounting bracket to bonded rubber mount	65 Nm (48 ft lb)	M 12	100 Nm (74 ft lb)
Left-hand mounting bracket to transmis-sion	65 Nm (48 ft lb)	Wheel bearing housing to lower ball joint ♦ Always replace bolts	55 Nm (41 ft lb) + 1/4 turn (90°)
Cover plate on transmission	20 Nm (15 ft lb)	Shock absorber to control arm	160 Nm (118 ft lb)
Front exhaust pipe to exhaust manifold		Bracket for drive flange bearing (from 01.96) to engine to bearing	50 Nm (41 ft lb) 27 Nm (20 ft lb)
M 6	10 Nm (7 ft lb)	Rear transmission mount (from 01.96)	see page J 37-37
M 8	25 Nm (18 ft lb)	Left transmission mount (from 01.96)	see page J 37-38
M 10	40 Nm (30 ft lb)		
Transmission carrier to bearing housing		Transmission carrier to engine	
M 8	25 Nm (18 ft lb)	M10	60 Nm (44 ft lb)

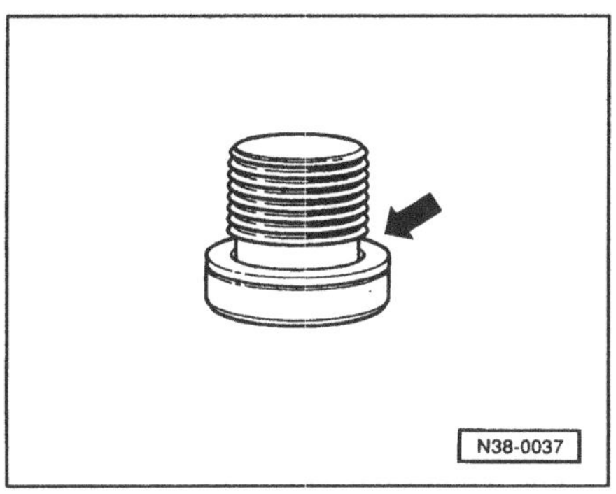

ATF level, checking and topping up

Only use VW ATF. VW ATF is available as a replacement part:

♦ Container size 1.0 Liter – Part No. G 052 162 A2

Note:

If the ATF level is checked, the seal -arrow- on the ATF level plug must always be replaced.

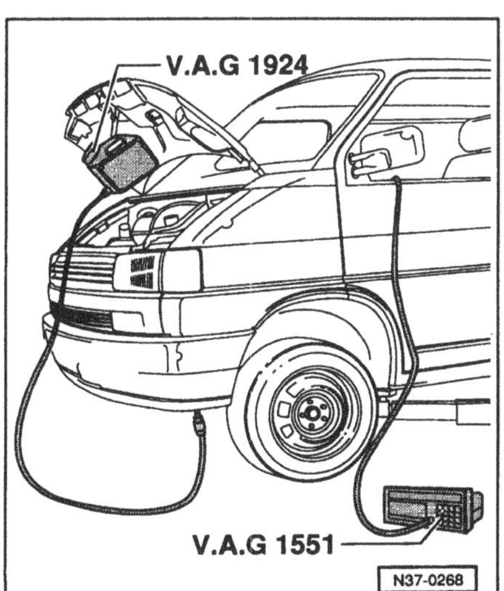

ATF level, checking

Conditions for performing check

- Transmission not in emergency running mode,
- ATF temperature not above 30°C (86°F)
- Vehicle level
- Selector lever at "P"

– Secure reservoir of ATF filling system VAG 1924 to vehicle.

– Connect VAG 1551 Scan Tool and enter address word 02, for "transmission electronics" and advance until "Select function XX" indicated on display ⇒ from page J 01-25.

- Start engine.
- Raise vehicle.

Rapid data transfer	HELP
Select function XX	

◄ Indicated on display:

- Press keys 0 and 8. (The function "Read measuring value block" is selected with 08.)

Rapid data transfer	Q
08–Reading measuring value block	

◄ Indicated on display:

- Confirm entry with key Q.

Reading measuring value block
Enter display group number XX

◄ Indicated on display:

- Press keys 0 and 5. (05 selects display group number 05)
- Confirm entry with key Q.

Reading measuring value block		5	→
40°C	0011011	0	900 rpm

◄ Indicated on display:

The first display zone shows the ATF temperature.

Note:

The ATF temperature must not be more than approximately 30°C (86° F).

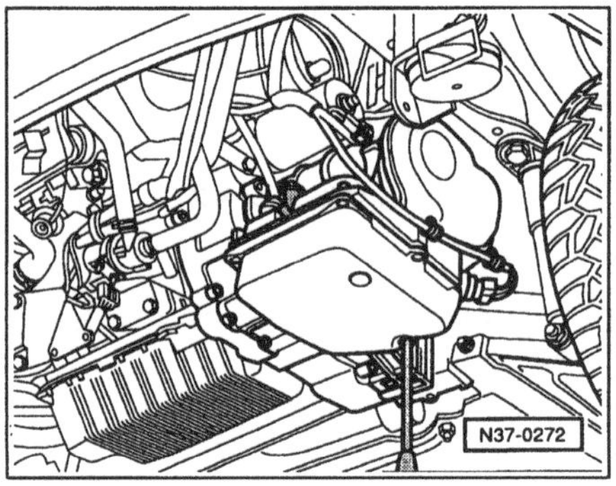

◄ - Remove ATF level plug from oil pan -arrow-.

- ● ATF present in the overflow pipe runs out.
- Bring ATF to test temperature.
- ● Test temperature: 35°C to 45°C (95 to 113°F).

If ATF continues to drip out of plug hole:

ATF need not be topped up.

- Install new seal to plug and tighten to 15 Nm (11 ft lb). The ATF level check is completed.

If only the ATF present in the overflow pipe runs out of the hole (no further dripping):

Top up ATF level.

ATF level, topping up

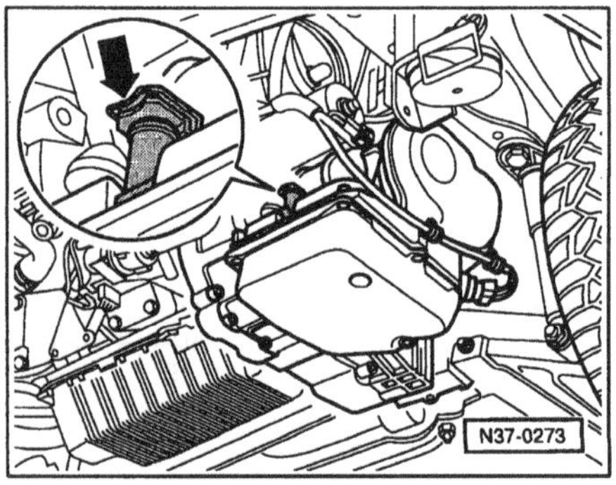

◄ - Remove cap -arrow- from filler pipe.

Note:

Some transmissions have a plug with a cap. The cap is destroyed during removal and must always be replaced.

- Pry cap out with a screwdriver and remove plug.

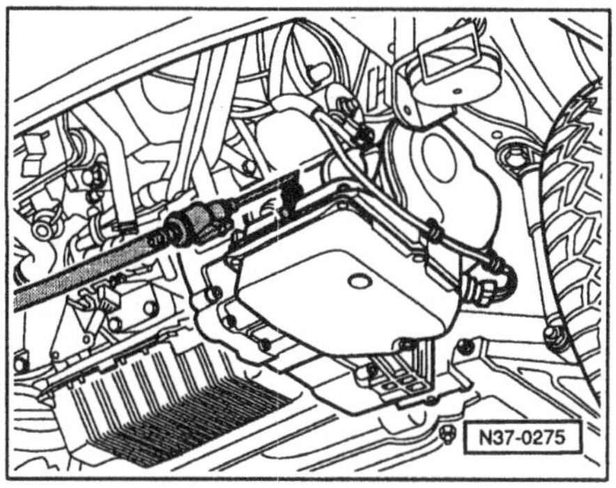

- Fill with ATF using VAG 1924 until ATF runs out of level hole -arrow-.

> ***Warning!***
>
> ***If the ATF level is too high or too low it will adversely affect the operation of the transmission.***

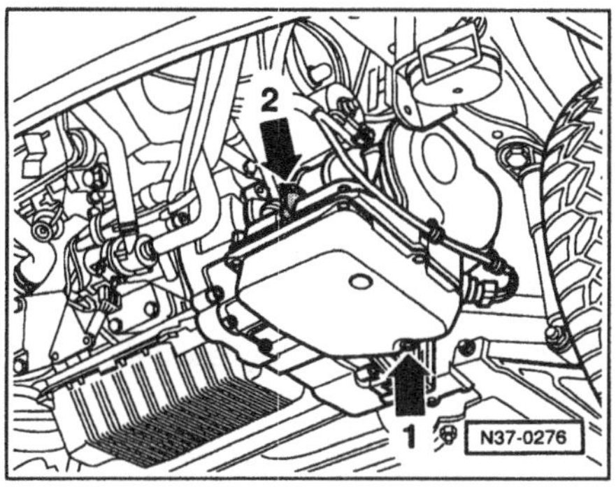

- Install new seal to level plug -arrow 1- and tighten to 15 Nm (11 ft lb).
- Push cap -arrow 2- on filler pipe until spring retainer locates correctly.

Note:

Some transmissions have a plug with a cap. The cap is destroyed during removal and must always be replaced.

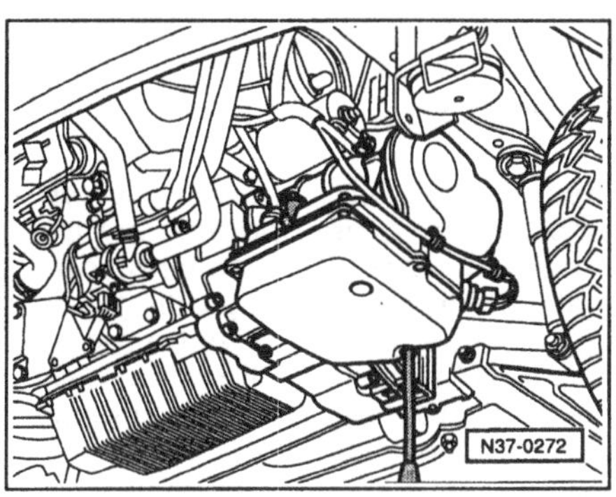

ATF, replacing

- Remove ATF level plug from oil pan -arrow-.
- Remove overflow pipe through level hole.
- Drain ATF.
- Install overflow pipe.
- Screw level plug in hand-tight.
- Fill with 3 Liters (3.2 qt) of ATF through filler pipe.
- Then start engine and shift through all selector lever positions with vehicle stationary.
- Check and top up ATF level ⇒ Page J 37-42.

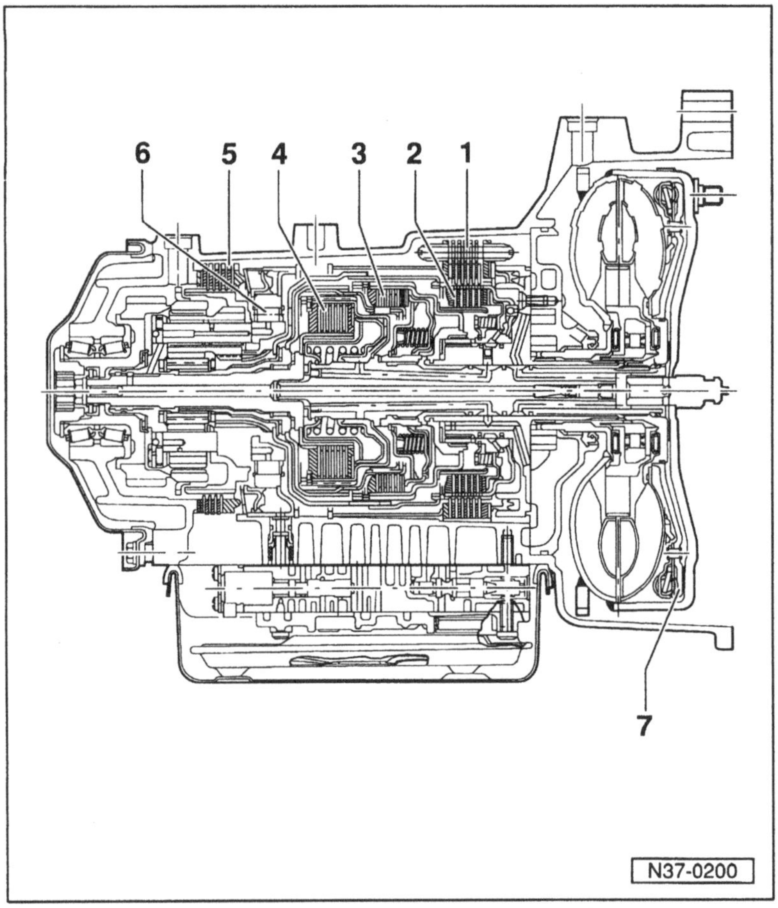

Transmission with selector elements

Note:

The illustration N37-0200 is from automatic transmission 01 M. The application of transmission 01P selector elements is identical.

Position of selector elements

1 = 2nd and 4th gear brake -B2-

2 = Reverse gear clutch -K2-

3 = 1st to 3rd gear clutch -K1-

4 = 4th gear clutch -K3-

5 = Reverse gear brake -B1-

6 = Free wheel clutch

7 = Lock-up clutch integrated in torque converter -LC-

Before starting repairs to the transmission, first perform the On Board Diagnostic (OBD) program ⇒ Repair Group 01.

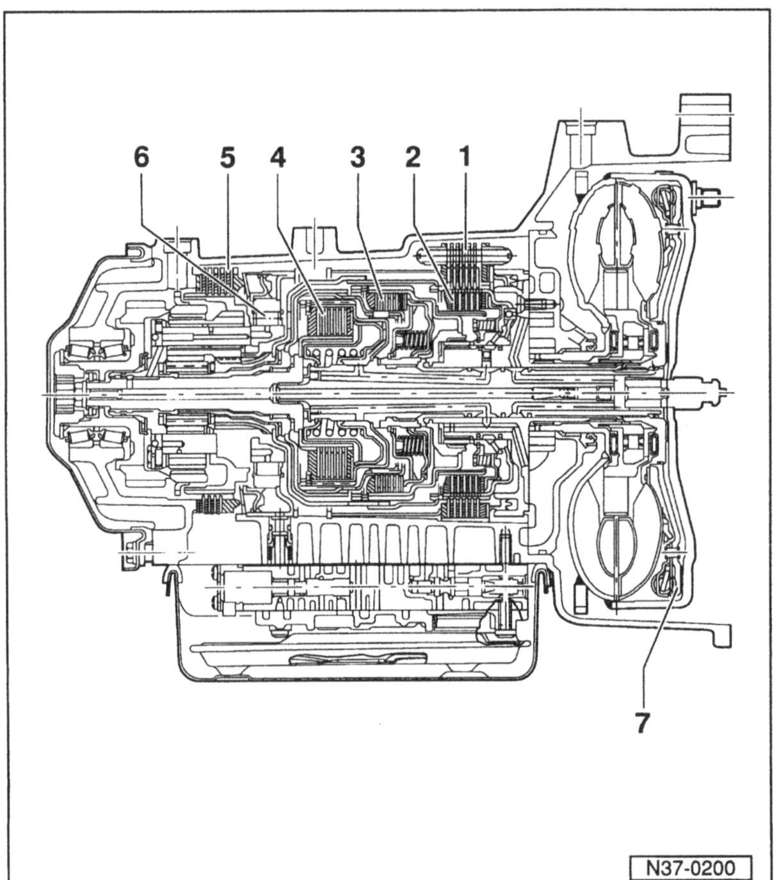

Position of selector elements

The following table shows which selector elements are operated in the individual gears. This can help you deal with complaints regarding poor performance or with drive train malfunctions in general. The table makes it possible to determine which gear selector elements are not operating properly.

All gears will be shifted hydraulically.

All forward gears are driven mechanically at a predetermined load and speed via the torque converter lock-up clutch.

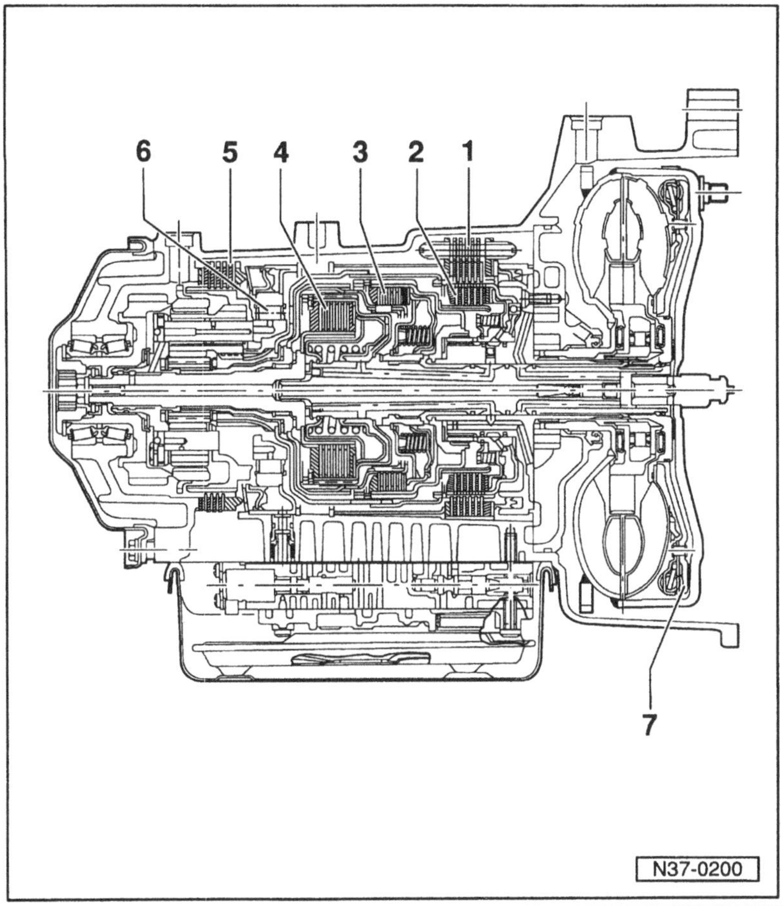

Gear	B1	B2	K1	K2	K3	F	LC
R	X			X			
1 H			X			X	
1 M			X			X	X
2 H		X	X				
2 M		X	X				X
3 H			X		X		
3 M			X		X		X
4 H		X			X		
4 M		X			X		X

X = clutches, brakes or free wheel engaged

H = hydraulic

M = mechanical

Transmission, disassembling and assembling

Removing and installing ATF cooler and ATF filler tube ⇒ Page J 37-51.

Disassembling and assembling planetary gearbox – Overview– ⇒ Page J 37-52.

Disassembling and assembling planetary gearbox ⇒ Page J 37-64.

Disassembling ⇒ Page J 37-64

Assembling ⇒ Page J 37-71

Overview of all planetary gearbox adjustments ⇒ Page J 37-80.

Removing and installing valve body ⇒ Page J 37-20

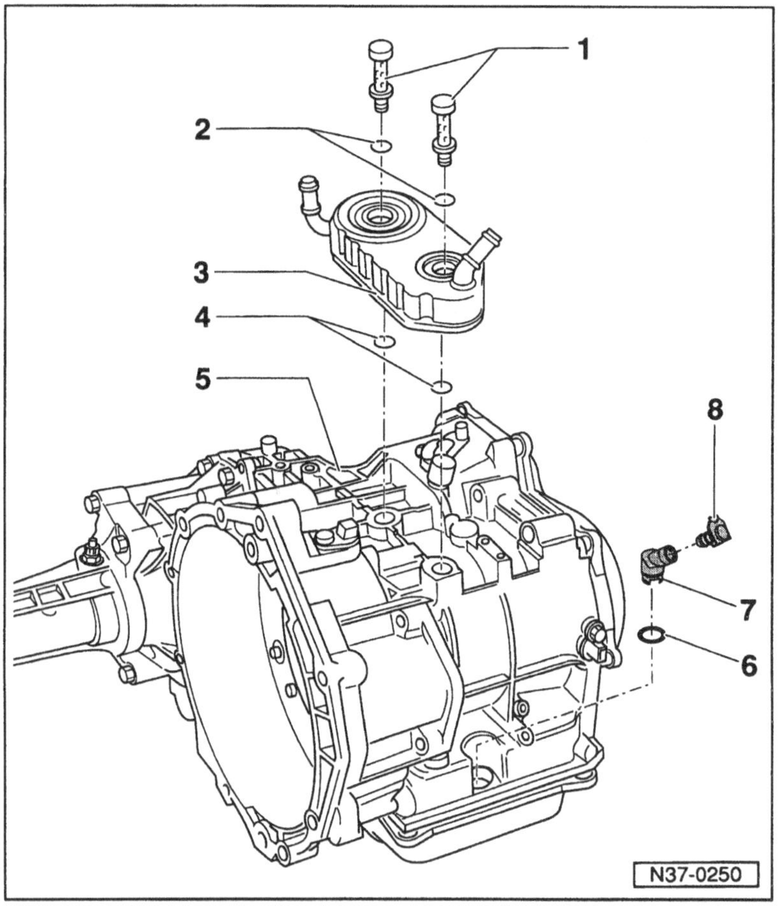

ATF cooler and ATF filler tube, removing and installing

1 - **Banjo bolt**
- ♦ Tightening torque: 35 Nm (26 ft lb)

2 - **O-ring**
- ♦ Always replace

3 - **ATF cooler**

4 - **O-ring**
- ♦ Always replace

5 - **Transmission housing**

6 - **O-ring**
- ♦ Always replace

7 - **ATF filler tube**

8 - **Cap**
- ♦ Checking ATF level ⇒ Page J 37-42

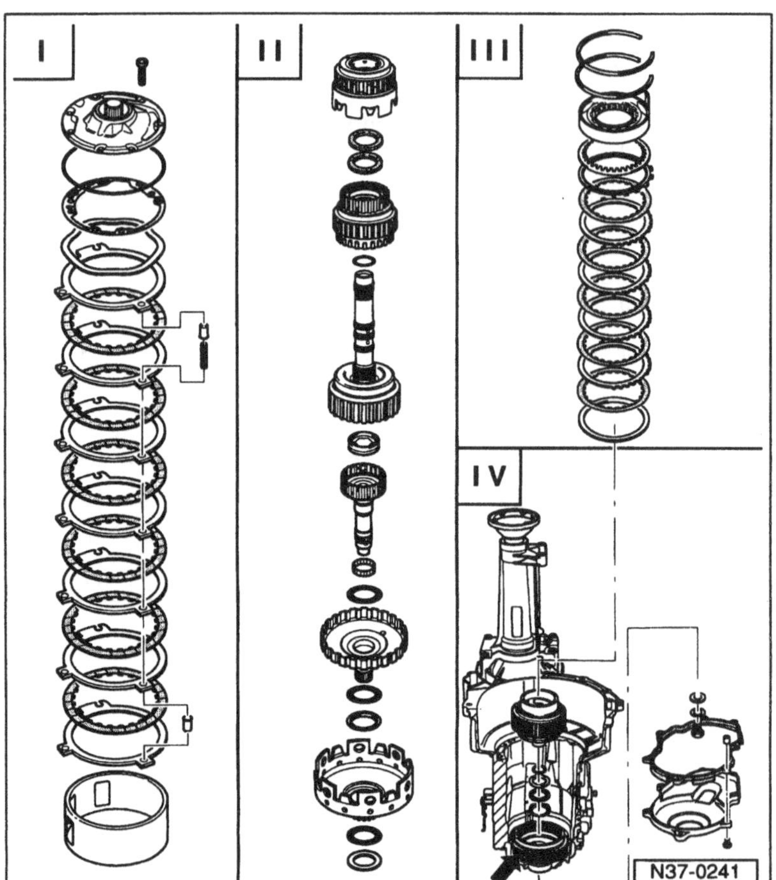

Planetary gearbox, disassembling and assembling, overview

Note:

Disassembling planetary gearbox ⇒ Page J 37-64,

Assembling ⇒ Page J 37-71.

I – **Removing and installing ATF pump to support tube ⇒ Page J 37-54**

II – **Removing and installing reverse clutch -K2- to small sun gear ⇒ Page J 37-57**

III – **Removing and installing free wheel and reverse gear brake -B1- ⇒ Page J 37-60**

Note:

To remove and install the free wheel, first remove the valve body together with plugs ⇒ Page J 38-20.

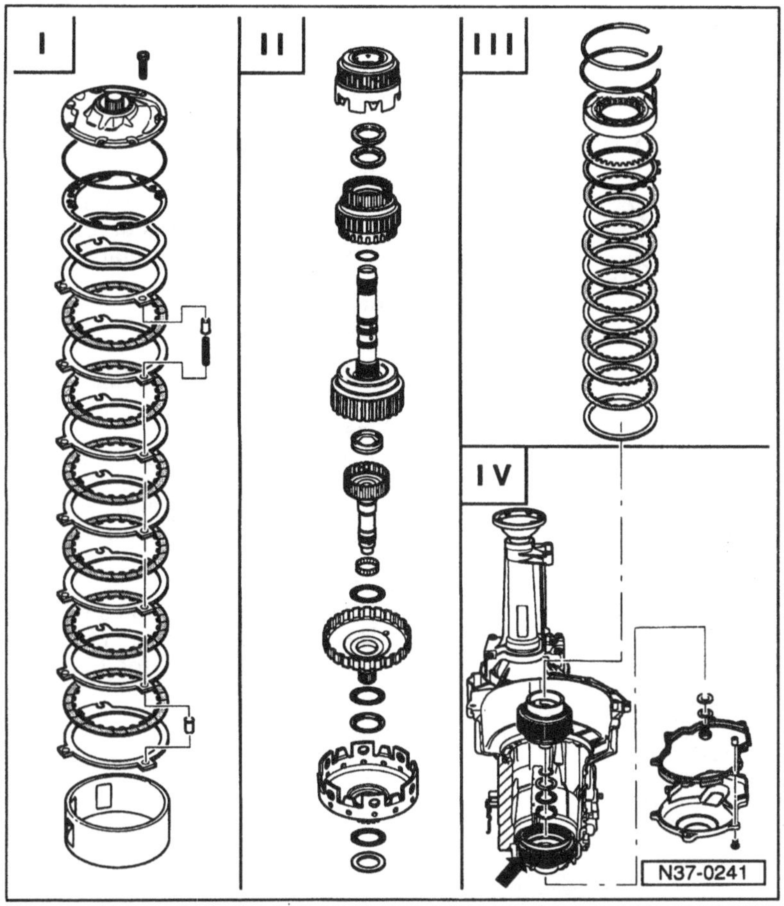

IV – Removing and installing plan et carrier ⇒ **Page J 37-62**

Note:

It is not necessary to remove the in-put gear -arrow-.

Input gear, removing and installing ⇒ Page J 39-22.

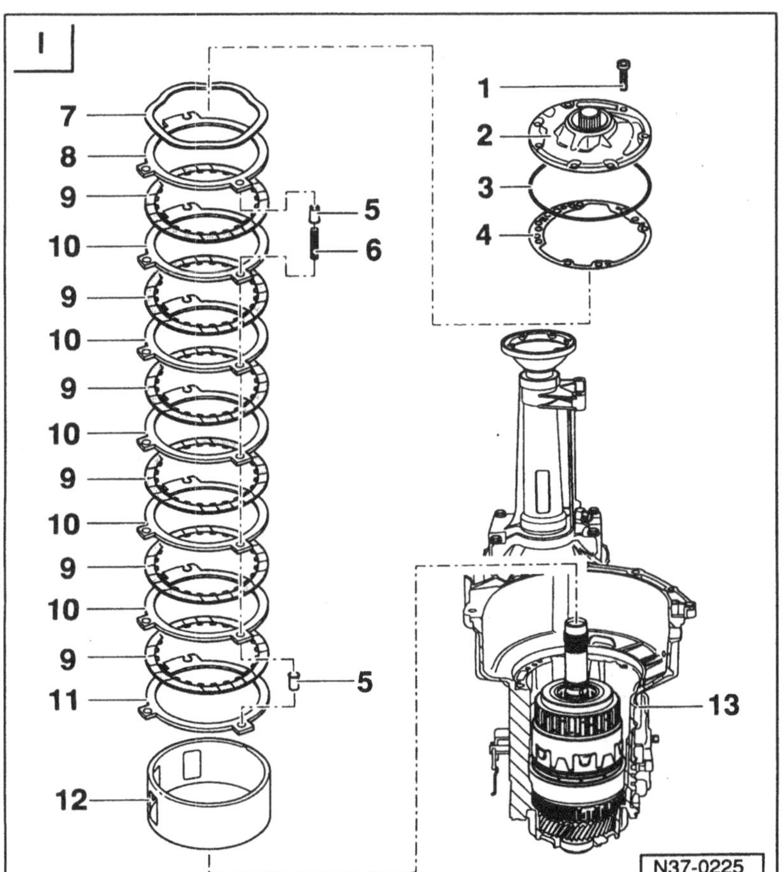

I – ATF pump to supporting tube, removing and installing

⇒ Overview, Page J 37-52

Note:

Planetary gearbox, disassembling ⇒ Page J 37-64,

Assembling ⇒ Page J 37-71.

1 - Bolt
- Qty. 7
- Tighten to 8 Nm (71 in. lb) and turn 90° further (1/4 turn)
- Further turn of 90° may be done in several stages

2 - ATF pump with -B2- piston
- Disassembling and assembling ⇒ Page J 38-1

3 - O-ring
- Always replace
- Place on ATF pump

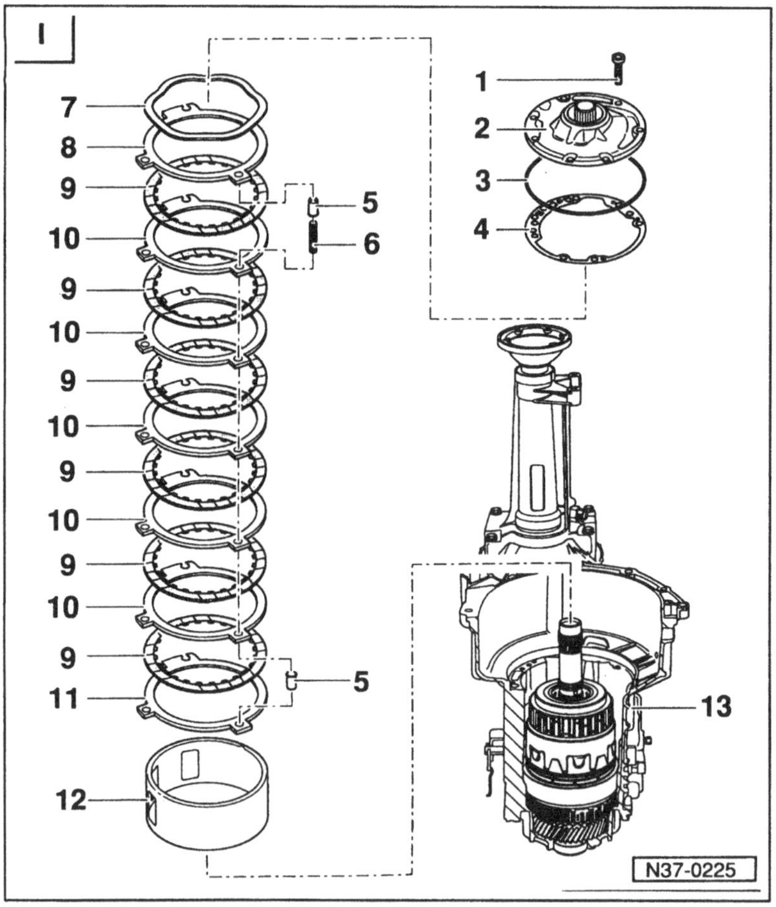

4 - Gasket
- Always replace

5 - Spring cap, Qty. 6
- Install 3 spring caps after installing first outer plate
- Install 3 spring caps before installing last outer plate

6 - Spring
- Qty. 3

7 - Corrugated spring washer

8 - Outer plate -B2-
- Quantity ⇒ from Page J 00-6
- Determining thickness of outer plate ⇒ Adjusting 2nd and 4th gear brake -B2-, Page J 37-103

9 - Inner plate -B2-
- Quantity ⇒ from Page J 00-6
- Before installing place new inner plates in ATF for 15 min.

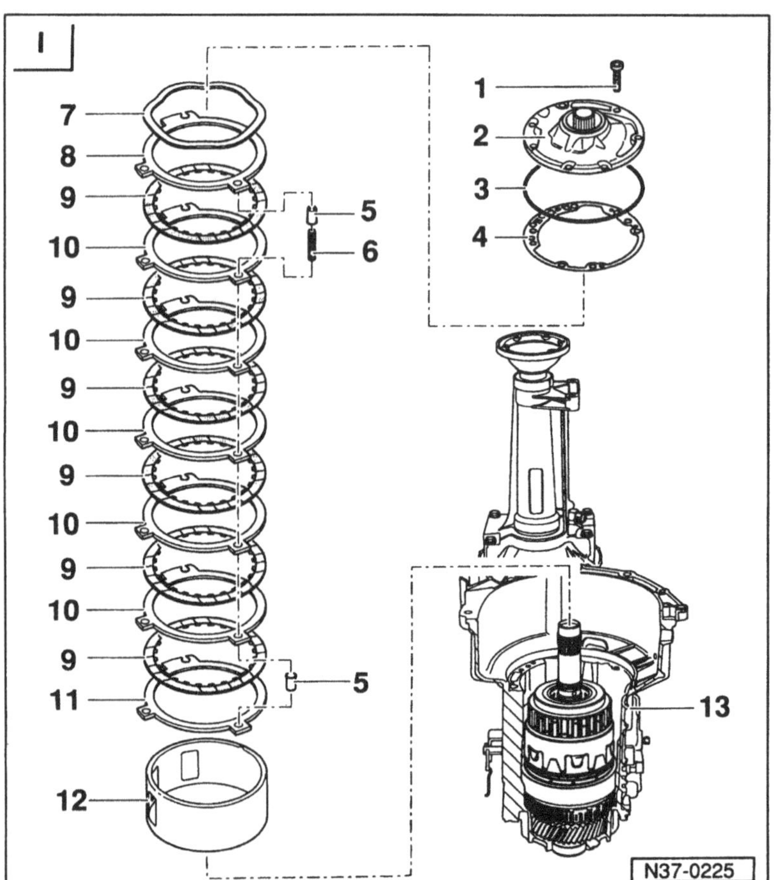

10 - Outer plate -B2-
- Always install 2 mm thick outer plates
- Quantity ⇒ from Page J 00-6

11 - Outer plate -B2-
- Install 3 mm thick outer plate to supporting tube

12 - Supporting tube -B2-
- For -B2- plate package
- Length of -B2- with

4 inner plates	72.3 mm
5 inner plates	68.6 mm
6 inner plates	64.9 mm

- Quantity of -B2- inner plates ⇒ from Page J 00-6
- Install so that groove engages in wedge of free wheel

13 - Transmission housing with clutches installed

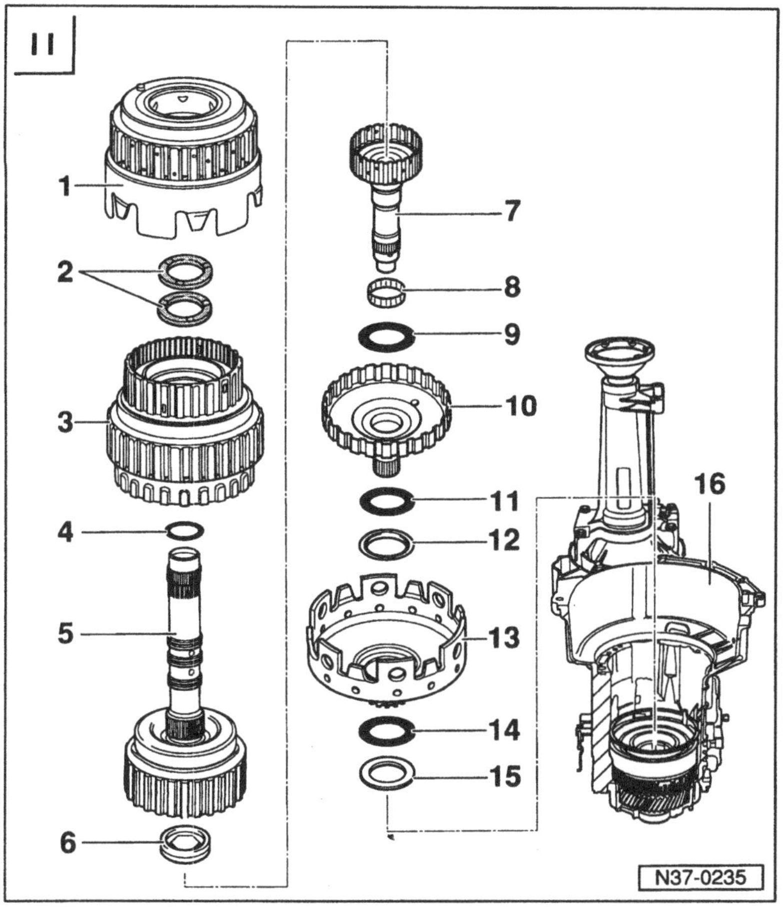

II – Reverse gear clutch -K2- to small sun gear, removing and installing

⇒ Overview, Page J 37-52

Note:

Disassembling planetary gearbox ⇒ Page J 37-64,

Assembling ⇒ Page J 37-71.

1 - Reverse gear clutch -K2-
- Disassembling and assembling ⇒ Page J 38-12

2 - Shim
- Determining thickness ⇒ Adjusting clutch play between -K1- and -K2-, Page J 37-96
- 1 or 2 shims can be installed

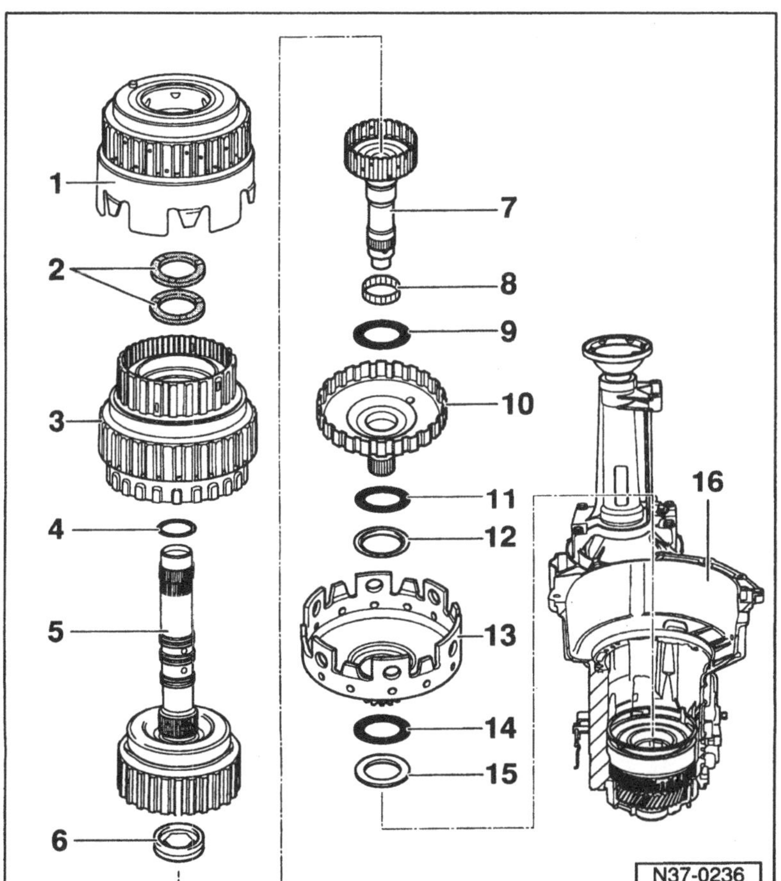

3 - 1st to 3rd gear clutch -K1-
- Disassembling and assembling ⇒ Page J 38-7

4 - Seal
- Always replace

5 - 4th gear clutch -K3- with turbine shaft
- Disassembling and assembling ⇒ Page J 38-15

6 - Axial needle bearing with washer
- Axial needle bearing faces towards small drive shaft

7 - Small drive shaft

8 - Needle bearing

9 - Axial needle bearing

10 - Large drive shaft

11 - Axial needle bearing

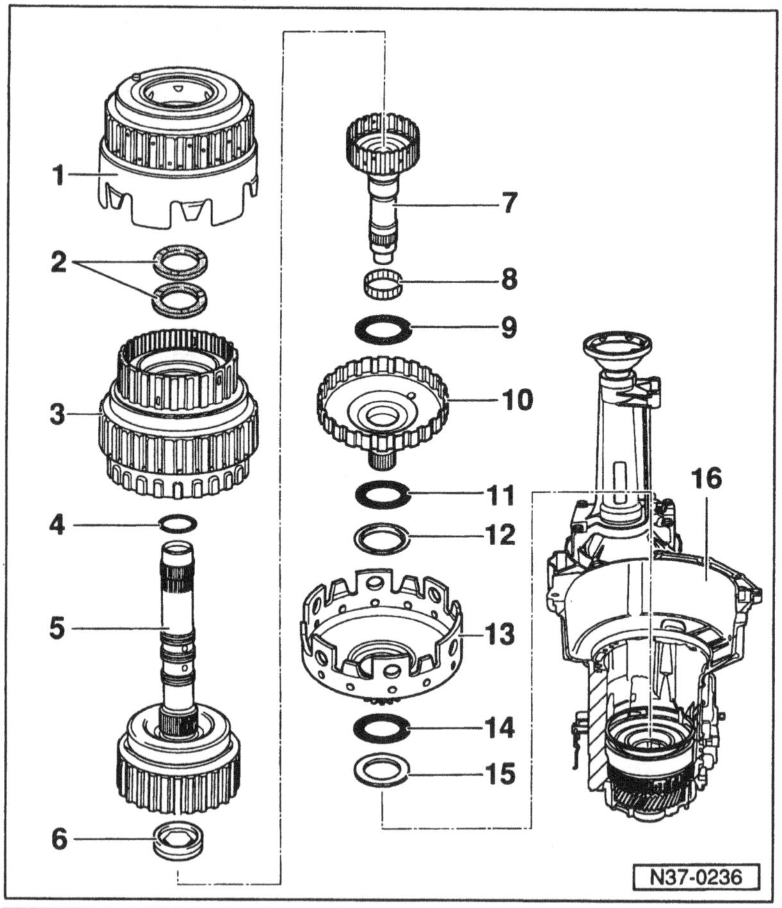

12 - **Axial needle bearing washer**
- With shoulder

13 - **Large sun gear**

14 - **Axial needle bearing**

15 - **Axial needle bearing washer**

16 - **Transmission housing with free wheel and circlips installed**

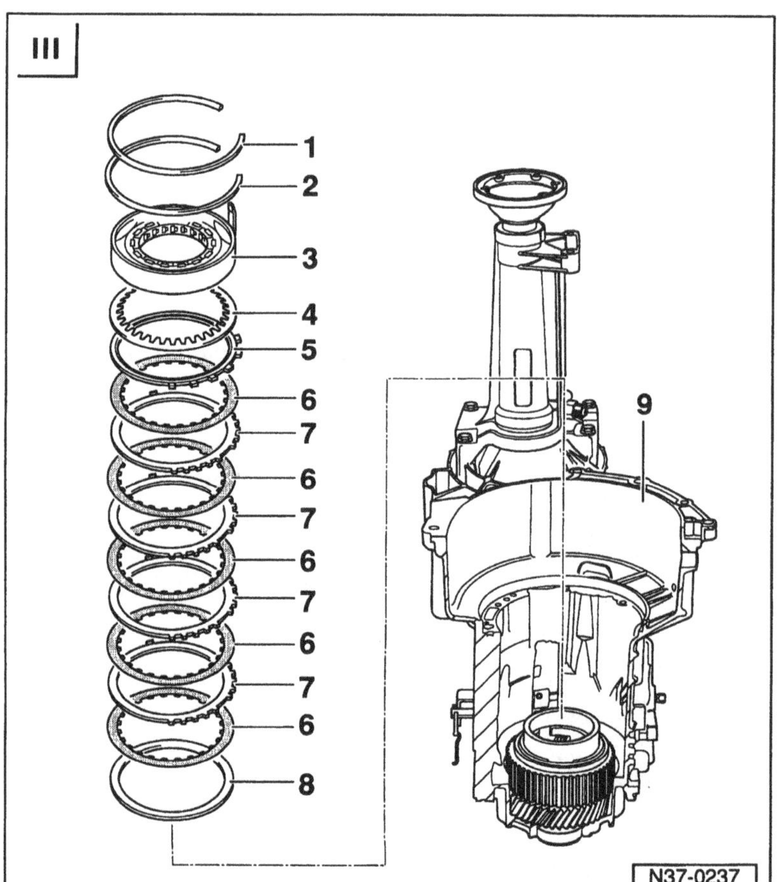

III – Free wheel and reverse gear brake -B1-, removing and installing

⇒ Overview, Page J 37-52

Note:

Disassembling planetary gearbox ⇒ Page J 37-64,

Assembling ⇒ Page J 37-71.

1 - **Circlip**
- For supporting tube -B2-

2 - **Circlip**
- For free wheel

3 - **Free wheel with -B1- piston**
- Before removing free wheel, remove valve body and sealing plugs ⇒ Page J 38-20
- Disassembling and assembling ⇒ Page J 38-4

4 - **Dished spring**
- Install with convex facing free wheel

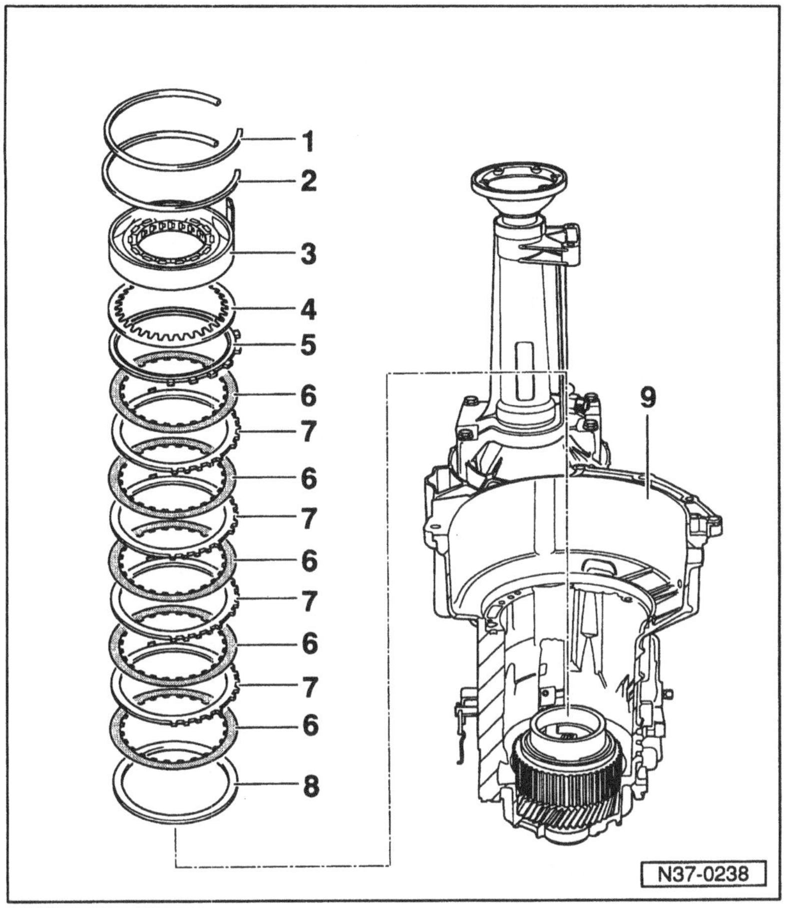

5 - **Pressure plate -B1-**
- Install with flat side facing plates
- 7.5 mm thick and a 2 mm thick outer plate
- 13.5 mm thick and a 1.5 mm thick outer plate

6 - **Inner plate, -B1-**
- Quantity of inner plates -B1- from Page J 00-6
- Before installing place plates in ATF for 15 minutes

7 - **Outer plate, -B1-**
- Quantity of inner plates -B1- from Page J 00-6

8 - **Shim**
- Determining thickness ⇒ Page J 37-89, adjusting reverse gear brake -B1-

9 - **Transmission housing with planet carrier installed**

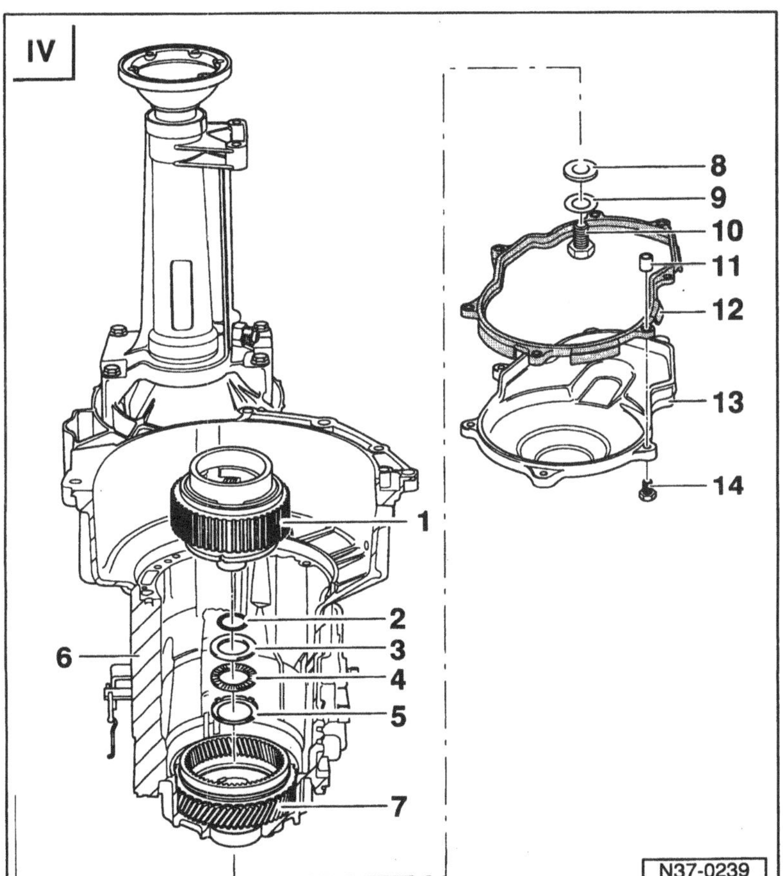

IV – Planet carrier and transmission housing with input gear and cover

⇒ Overview, Page J 37-52

Note:

Disassembling planetary gearbox ⇒ Page J 37-64,

Assembling ⇒ Page J 37-71.

1 - **Planet carrier**
- Adjusting ⇒ Page J 37-81

2 - **O-ring**
- Always replace
- Install in planet carrier

3 - **Axial needle bearing washer**

4 - **Axial needle bearing**

5 - **Axial needle bearing washer**
- Install smooth side in input gear

6 - **Transmission housing with input gear installed**

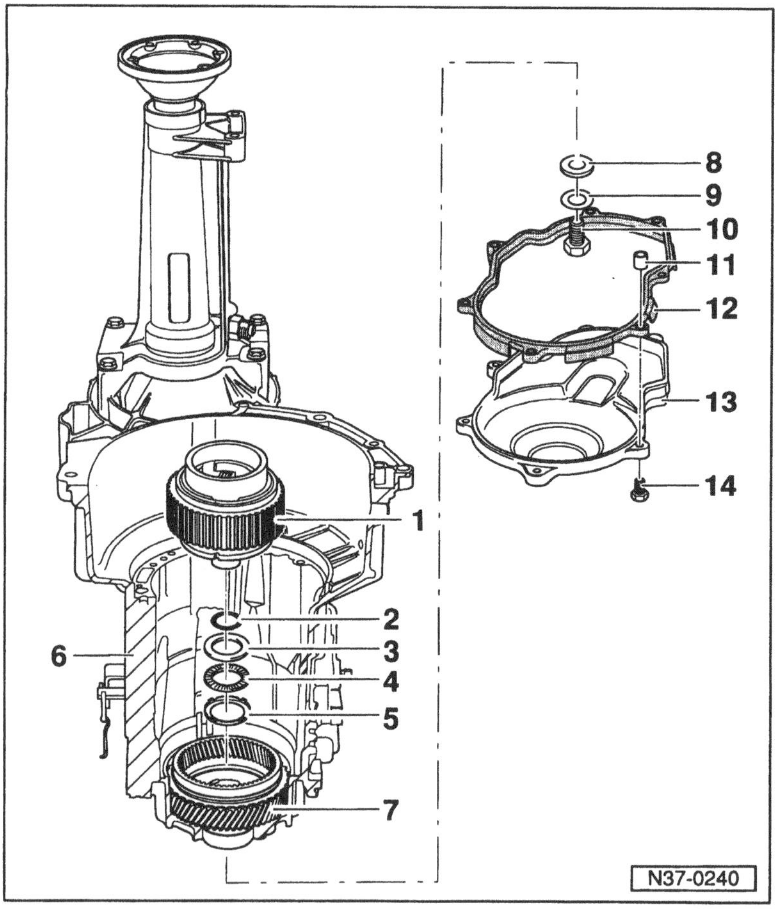

7 - **Input gear**
- Do not remove to disassemble planetary gearbox
- Removing and installing input gear or axial needle bearing for planet carrier ⇒ Page J 39-22

8 - **Planet carrier shim**
- Determining thickness, ⇒ Adjusting planet carrier, Page J 37-81

9 - **Washer**

10 - **Bolt**
- Tightening torque: 30 Nm (22 ft lb)
- For small input shaft

11 - **Spacer bushing**
- Clip qty. 7 into gasket

12 - **Gasket**
- Always replace

13 - **Cover**

14 - **Bolt**
- Tightening torque: 30 Nm (22 ft lb)

Planetary gearbox, disassembling and assembling

Disassembling

- Remove torque converter.

- Seal ATF cooler connections.

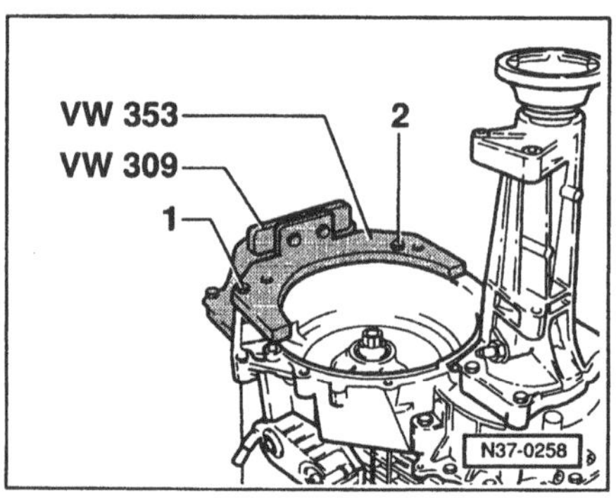

◄ – Secure transmission to assembly support with bolts -1- and -2-.

- Drain ATF.

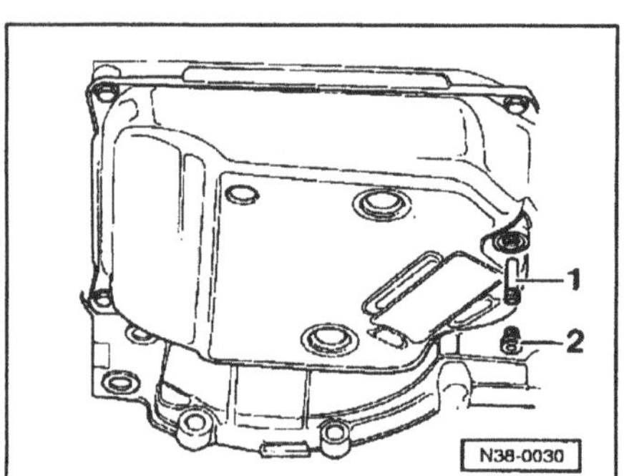

◄ – Remove ATF overflow pipe -1- and plug -2-.

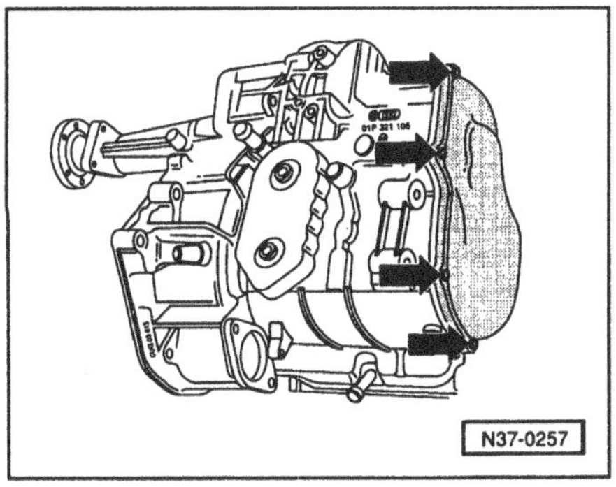

- Remove transmission housing cover with gasket -arrows-.

- Remove oil pan.
- Remove ATF screen.

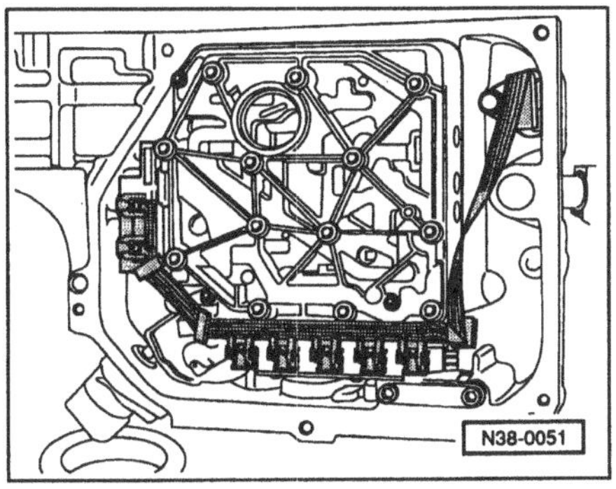

- Remove valve body with conductor strip. ⇒ Page J 38-20

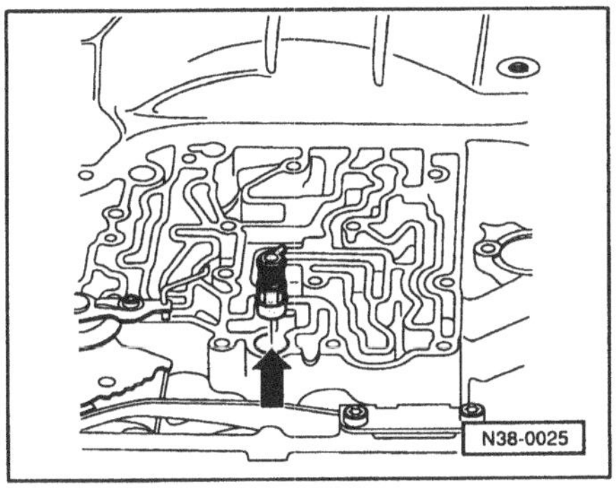

- Remove sealing plugs for -B1-.

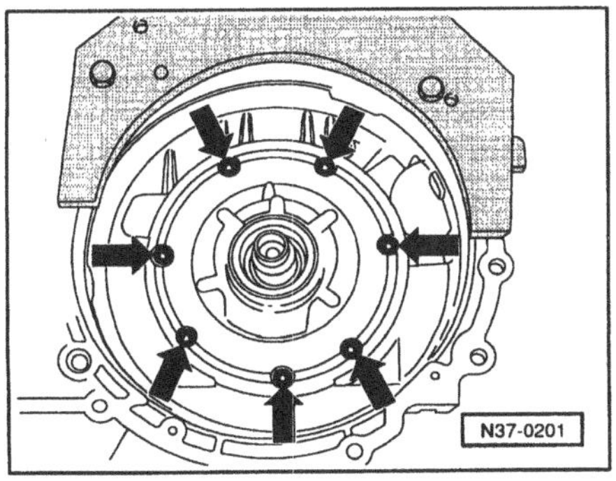

- Remove ATF pump bolts (arrows).

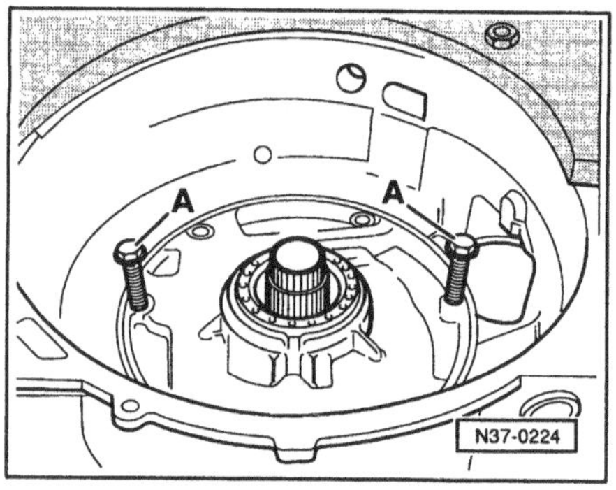

To remove ATF pump:

◄ – Thread bolts -A- (M8) into tapped holes of ATF pump.

– Press ATF pump off transmission by turning bolts -A- evenly.

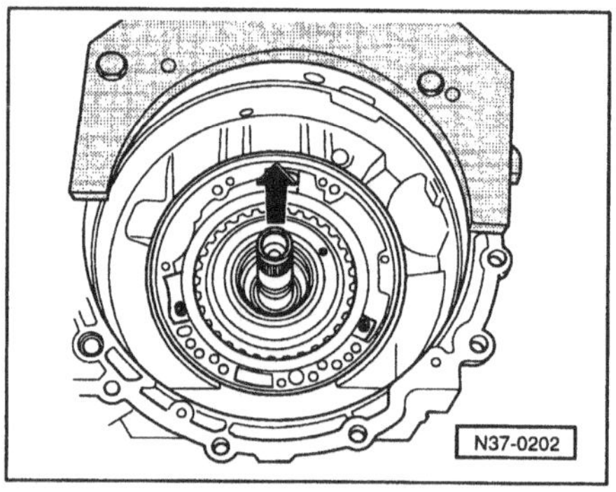

◄ – Remove all clutches with supporting tube, -B2- plates, springs and spring caps together.

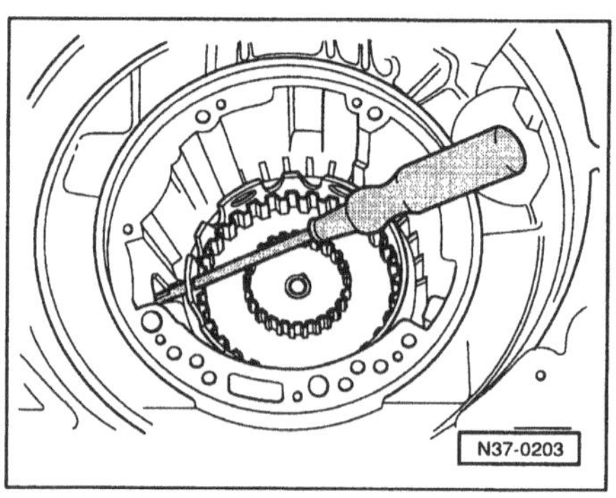

◄ – Insert screwdriver through hole of large sun gear to loosen and tighten bolt for small drive shaft.

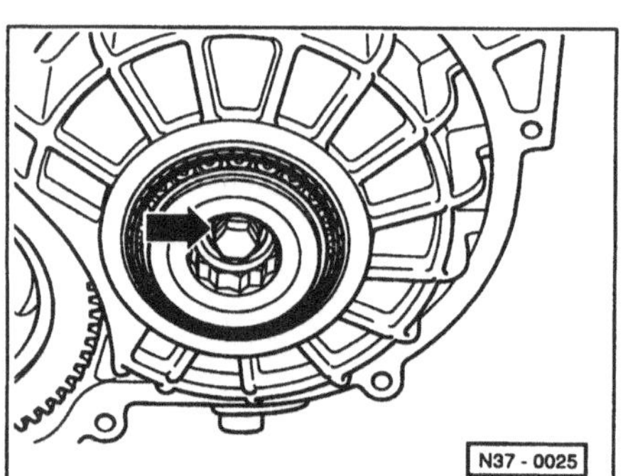

◄ – Loosen small drive shaft bolt -arrow-.

– Remove small drive shaft bolt, washer and shim.

● Planet carrier axial needle bearing remains in transmission/ input gear.

– Pull out small drive shaft.

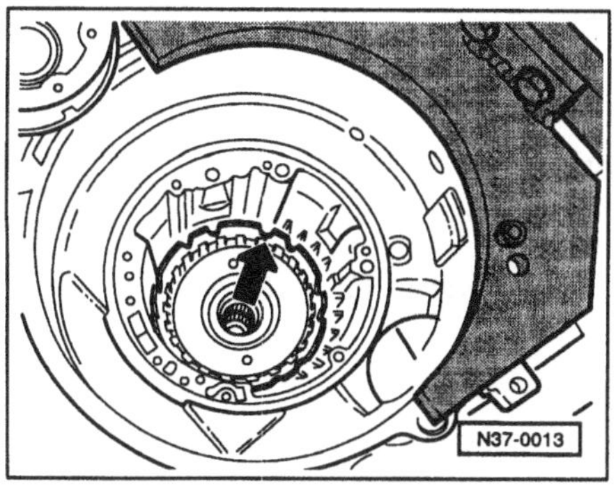

– Pull out large drive shaft -arrow-.

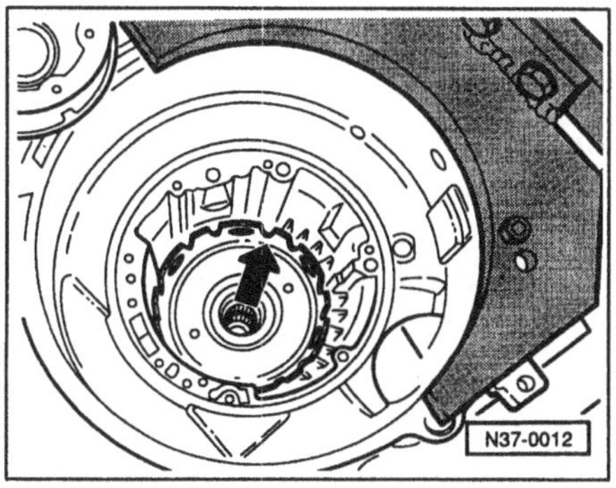

– Pull out large sun gear -arrow-.

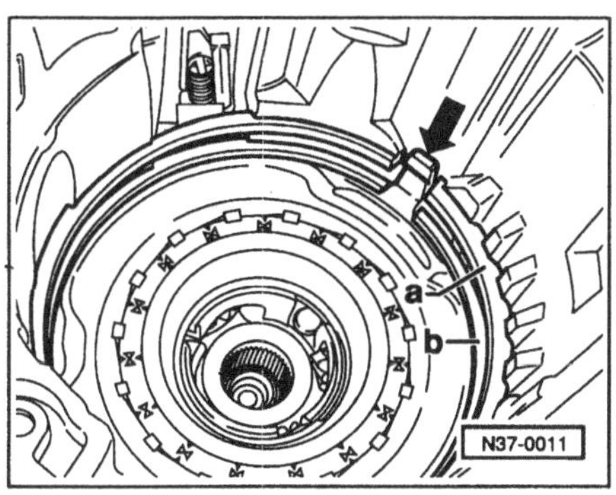

– Remove Transmission Vehicle Speed Sensor -G38- ⇒ Page J 01-19, Fig. 10

– Remove supporting tube circlip -a- and free wheel circlip -b-.

– Take free wheel out of transmission housing by pulling on free wheel retaining wedge -arrow- with pliers.

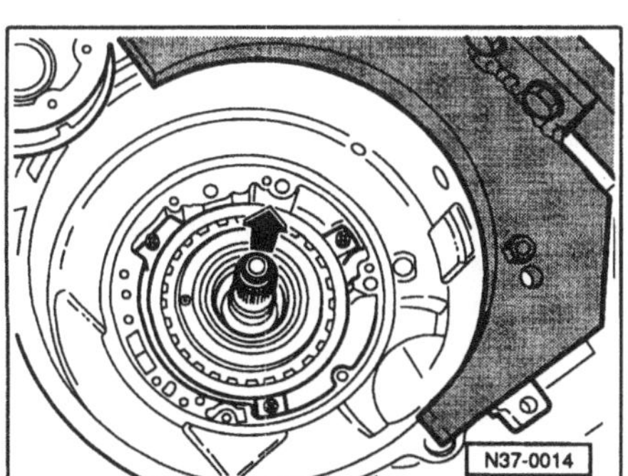

– Pull out planet carrier with dished spring.

– Remove reverse gear brake -B1- plates.

Note:

- *It is not necessary to remove the input gear to disassemble the planetary gearbox.*
- *Planetary gearbox, assembling ⇒ Page J 37-71.*

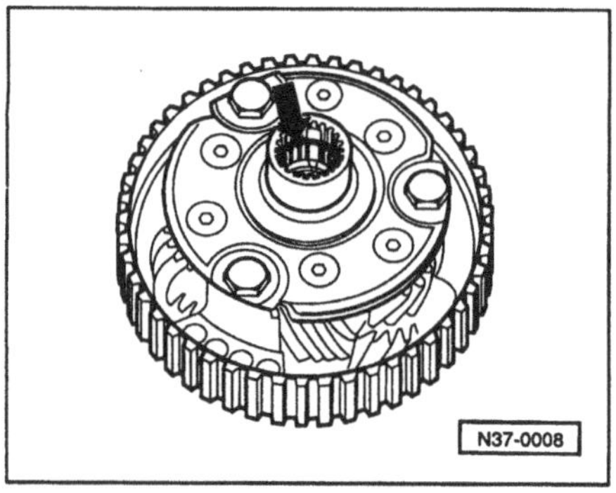

Assembling

- Install O-ring into planet carrier.

Note:

After replacing planet carrier, adjust ⇒ Page J 37-81.

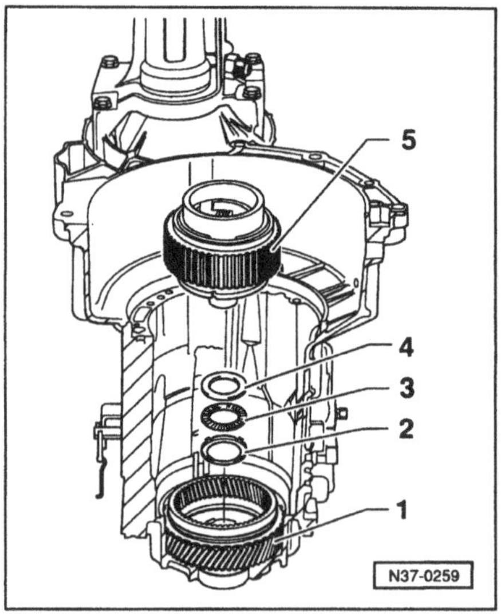

- Install axial needle bearing with washers in input gear.

1 – Input gear, installed in transmission housing ⇒ Page J 37-52.

2 – Axial needle bearing washer, install smooth side into input wiring, routing gear.

3 – Axial needle bearing

4 – Axial needle bearing washer

5 – Planet carrier, install with seal fitted

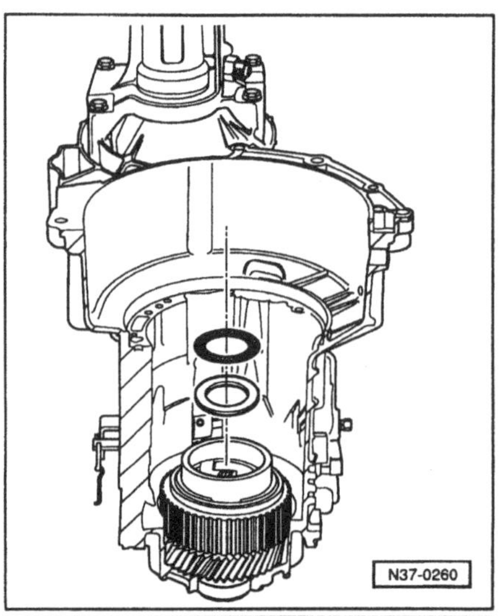

- Place washer and axial needle bearing on small sun gear in planet carrier.
- Centralize washer and axial needle bearing in small sun gear.

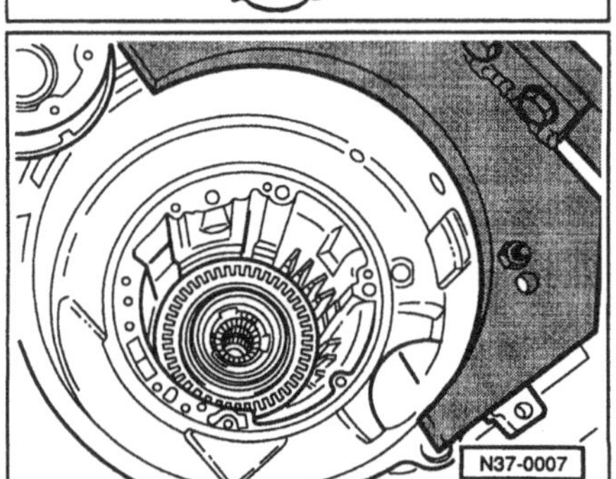

- Install inner and outer -B1- plates.
- Install pressure plate with flat side facing plates.

Thickness of pressure plate varies according to number of plates ⇒ Page J 37-61.

- Install dished washer with convex side facing toward free wheel.

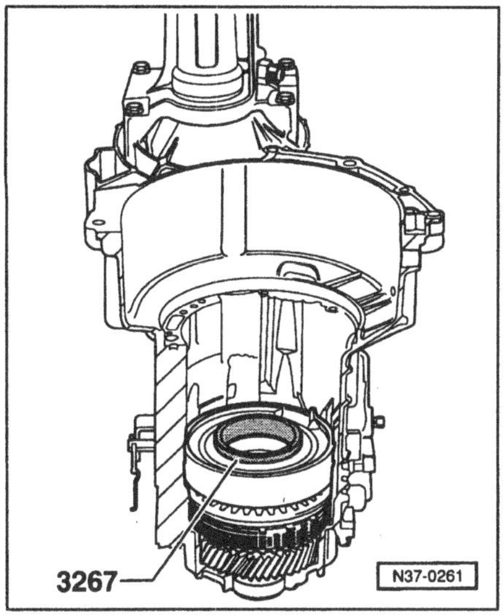

Note:

When replacing

- *transmission housing*
- *Free wheel*
- *Piston for reverse gear brake -B1-*
- *Plates*

adjust -B1-, ⇒ Page J 37-89.

◄ – Preload free wheel rollers with 3267 and install free wheel.

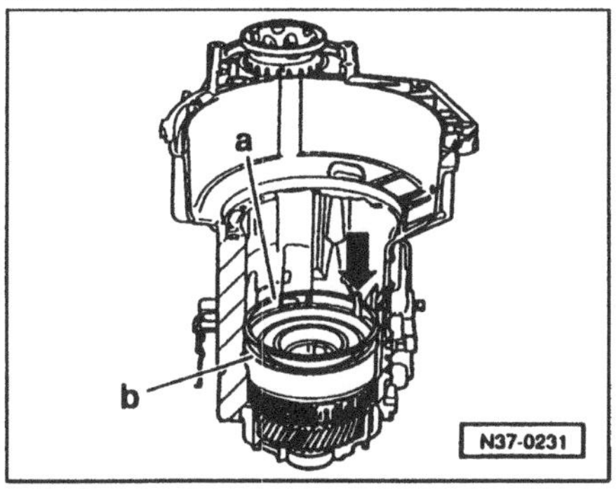

◄ – Install circlip for free wheel -b- and circlip for supporting tube -a-.

– Install circlip gaps over free wheel retaining wedge -arrow-.

– Install Transmission Vehicle Speed Sensor -G38- ⇒ Page J 01-19, Fig. 10

– Perform check measurement of -B1 ⇒ Page J 37-95

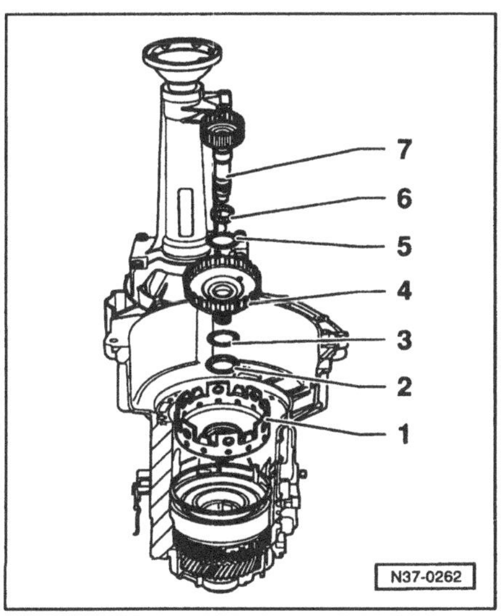

◄ – Install components into transmission housing beginning with large sun wheel down to small input shaft.

1 – Large sun wheel

2 – Needle bearing washer (collar faces large sun wheel)

3 – Axial needle bearing

4 – Large input shaft

5 – Axial needle bearing

6 – Needle bearing

7 – Small input shaft

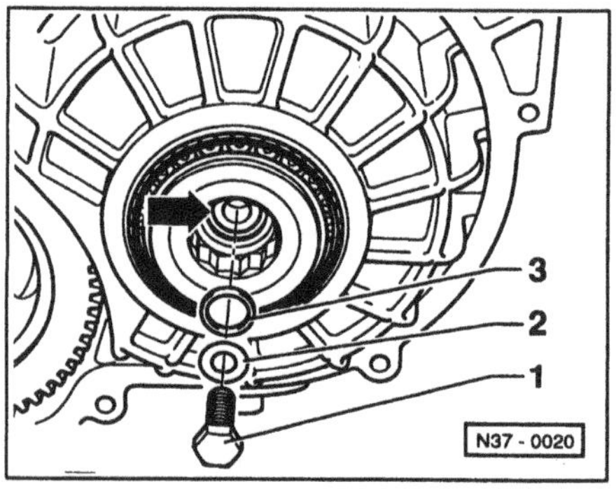

◄ – Install small drive shaft bolt -1- with washer -2- and shim -3-.

- Tightening torque: 30 Nm (22 ft lb)

Notes:

- *Insert shim -3- onto small drive shaft shoulder -arrow-.*
- *Determining shim thickness ⇒ adjusting planet carrier, Page J 37-81.*

– Perform planet carrier check measurement ⇒ Page J 37-88.

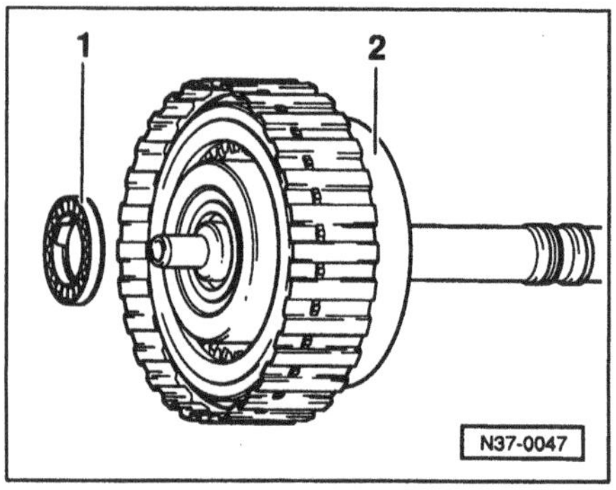

◄ – Install axial needle bearing with washer on 4th gear clutch -K3-.

1 – Axial needle bearing with washer

2 – 4th gear clutch -K3-

– Moisten axial needle bearing washer with ATF so that bearing, when installed, sticks to -K3-.

– Install bearing washer on -K3-.

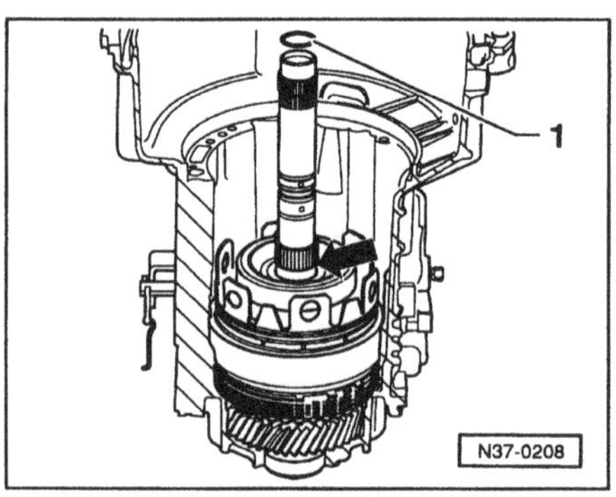

◄ – Install 4th gear clutch -K3-.

– Install seal -1- in groove -arrow-.

Note:

Make sure the piston rings seat correctly.

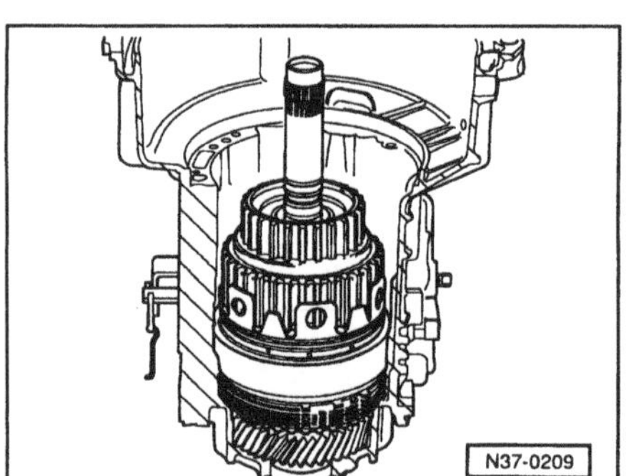

◄ – Install 1st to 3rd gear clutch -K1-.

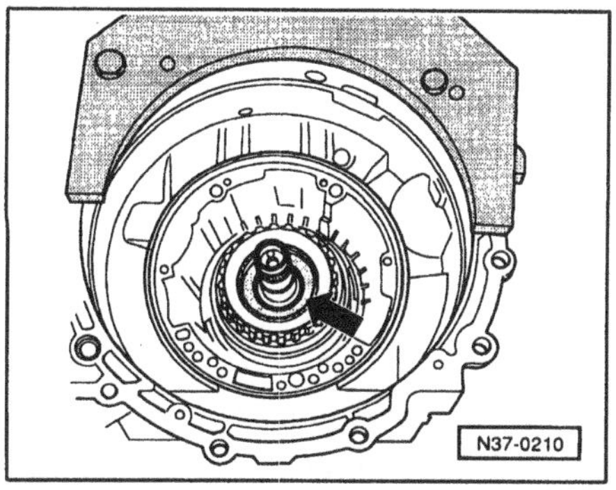

N37-0210

◄ – Install shims -arrow- in -K1-.

Notes:

- *Remeasure shim when replacing -K1-, -K2- or ATF pump ⇒ adjusting clutch play between -K1- and -K2-, Page J 37-96.*
- *1 or 2 shims can be installed.*

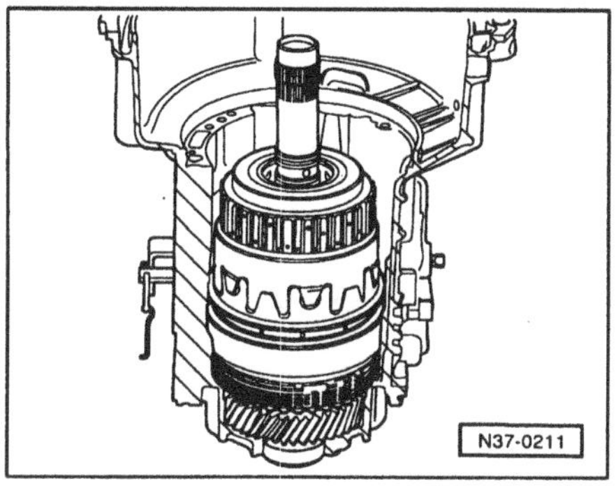

N37-0211

– Install reverse gear clutch -K2-.

◄ – Install supporting tube for -B2- plate set -arrow- so that supporting tube groove engages in free wheel wedge.

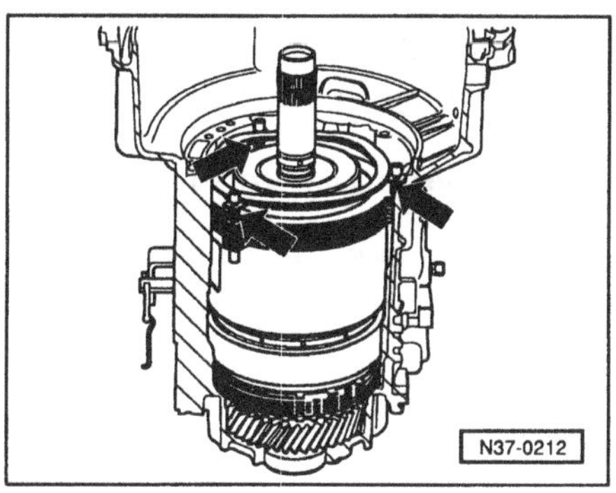

N37-0212

◄ Install -B2- plates as follows:

– First install 3 mm thick outer plate

– Install three spring caps in outer plate

– Install compression springs -arrows

– Install all but last outer plate

– Before installing last (measured) outer plate, install three spring caps onto compression springs.

– Replace corrugated washer

Note:

When replacing

- *Supporting tube*
- *ATF pump*
- *Plates*

then adjust 2nd and 4th gear brakes -B2- ⇒ Page J 37-103.

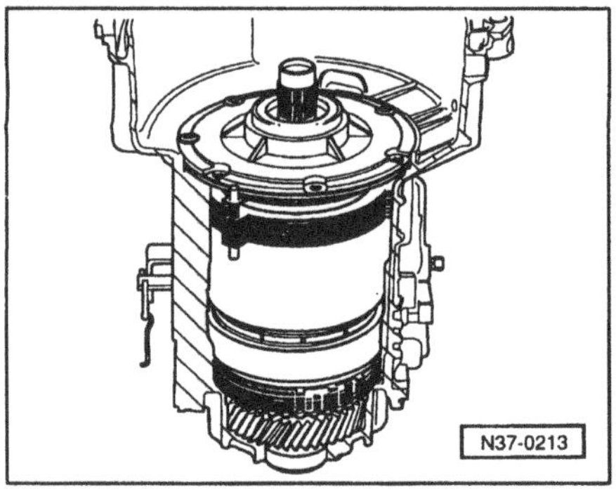

- Replace ATF pump seal.
- Install O-ring on ATF pump..
- Install ATF pump.
- Tighten bolts evenly in a crosswise pattern.

Note:

- *Make sure that the O-ring is not damaged.*
- *Tightening torque: 8 Nm (71 in. lb) plus an additional 90° further (1/4 turn). Additional 1/4 turn can be done in stages.*

- Check measurement of clutch play ⇒ Page J 37-102.

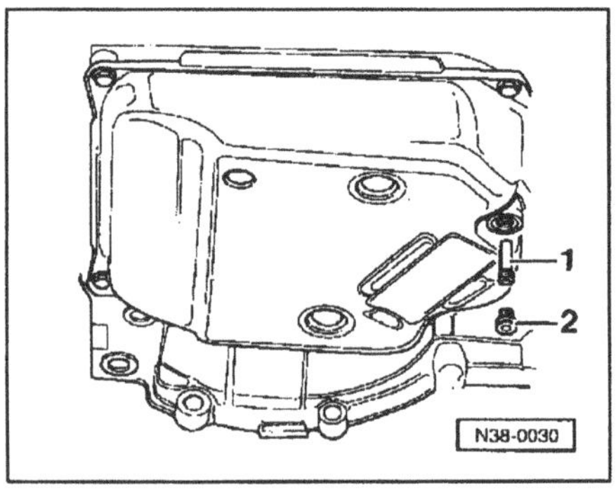

- Install sealing plugs, valve body and oil pan ⇒ Page J 38-20.
- Install cover with gasket and spacer bushings.
- Install ATF overflow pipe -1- and level plug -2-.

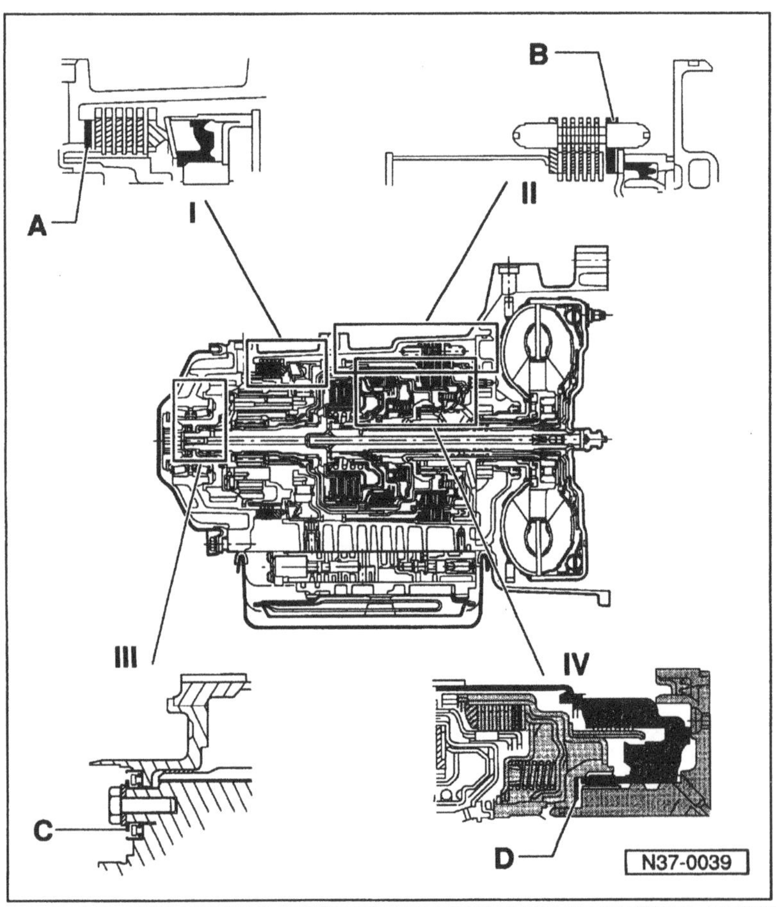

Overview of all planetary gearbox adjustments

I = Reverse gear brake -B1-
- A – Shim
- Determining thickness ⇒ adjusting reverse gear brake -B1-, Page J 37-89

II = 2nd and 4th gear brake -B2-
- B – Outer plate
- Determining thickness ⇒ adjusting 2nd and 4th gear brake -B2-, Page J 37-103

III = Planet carrier
- C – Shim
- Determining thickness ⇒ adjusting planet carrier, Page J 37-81

IV = Clutch play
- D – Shim
- Determining thickness ⇒ adjusting clutch play between -K1- and -K2-, Page J 37-96

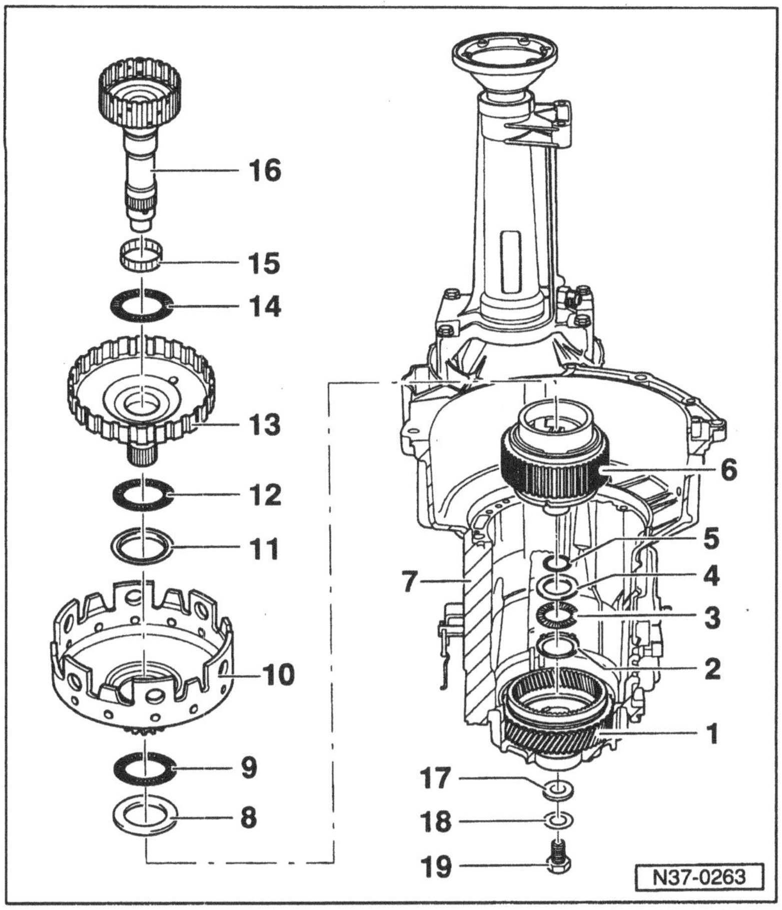

Planet carrier, adjusting

Overview of components to adjust planet carrier

When adjusting planet carrier install components in transmission housing without shim (item 17).

1 - Input gear
- ♦ Do not remove to adjust planet carrier
- ♦ Removing and installing input gear or axial needle bearing for planet carrier ⇒ Page J 39-22

2 - Axial needle bearing washer
- ♦ Install smooth side in input gear

3 - Axial needle bearing

4 - Axial needle bearing washer

5 - O-ring
- ♦ Install into planet carrier ⇒ Planet carrier, assembling, Page J 37-71

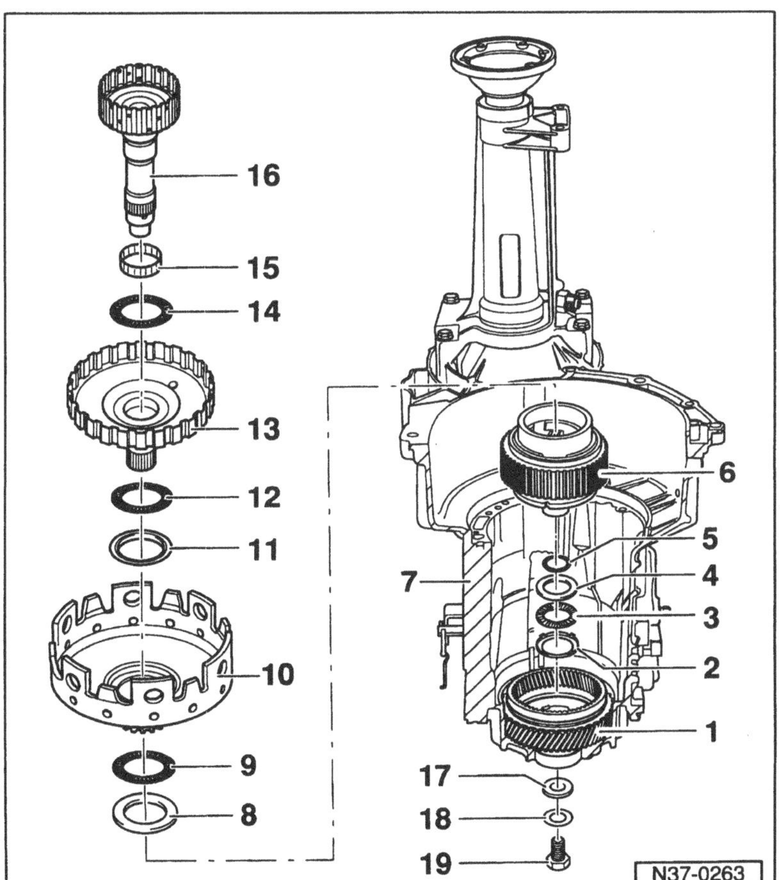

6 - Planet carrier

7 - Transmission housing
- ♦ With input gear and axial needle bearing
- ♦ Axial needle bearing remains in input gear
- ♦ If axial needle bearing is damaged ⇒ Input gear, removing and installing Page J 39-22

8 - Washer

9 - Axial needle bearing

10 - Large sun gear

11 - Washer
- ♦ Install into large sun gear ⇒ Planetary gearbox, assembling, Page J 37-71

12 - Axial needle bearing

13 - Large drive shaft

14 - Axial needle bearing

15 - Needle bearing

16 - Small drive shaft

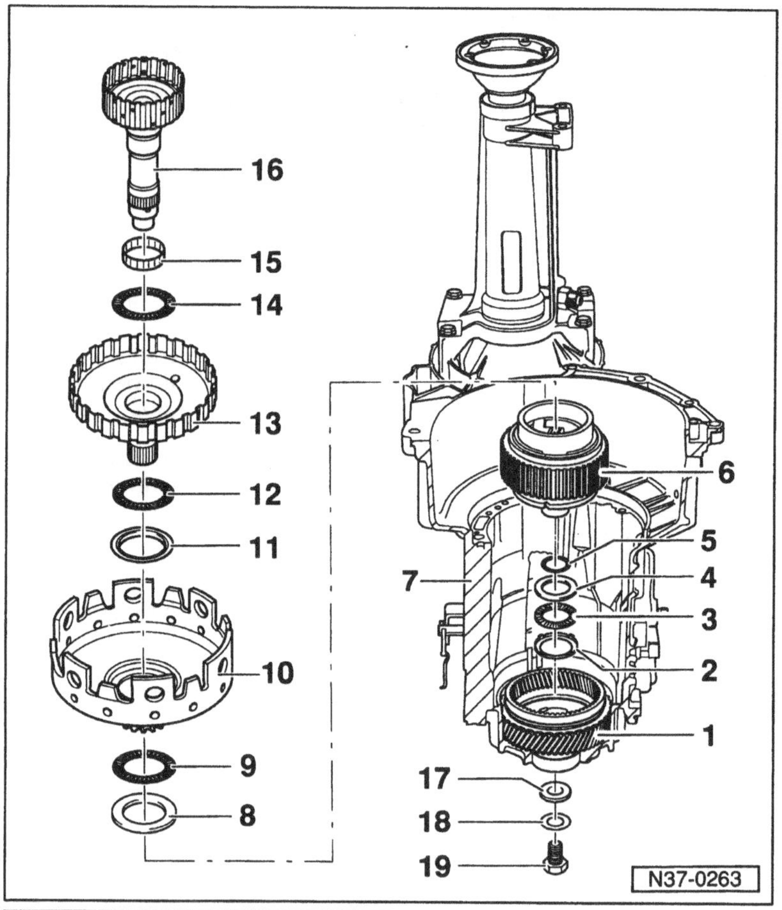

17 - **Shim**
- Do not install in planet carrier when adjusting

18 - **Washer**

19 - **Small drive shaft bolt**
- Tightening torque: 30 Nm (22 ft lb)

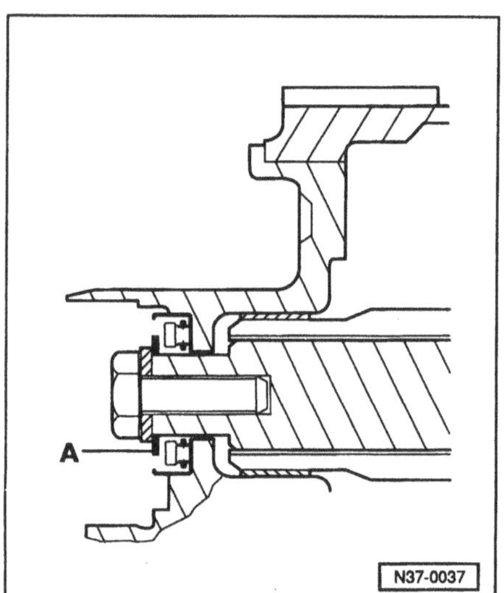

◄ **Determining shim -A**

– Install all components to adjust planet carrier (items 2 to 16) into transmission housing ⇒ Page J 37-81.

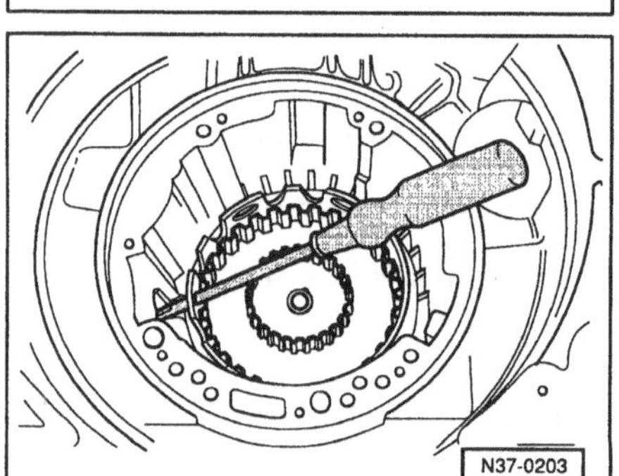

◄ – Insert screwdriver through hole of large sun gear to loosen and tighten bolt for small drive shaft.

- Install small input shaft bolt -1- with washer -2- but without shim.

● Tightening torque: 30 Nm (22 ft lb)

- Set up dial indicator so that top of gauge is positioned on bolt head center with 1 mm preload.
- Zero dial gauge.
- Move small input shaft up and take reading.

Example: 2 mm (0.079 in.)

- Specify thickness of shim according to table and determine Part Number from parts catalog.

Table of shims:

Indicator reading mm (in.)	Shims mm (in.)
1.26 (0.0496) ... 1.35 (0.0531)	1.0 (0.0393)
1.36 (0.0535) ... 1.45 (0.0571)	1.1 (0.0433)
1.46 (0.575) ... 1.55 (0.0610)	1.2 (0.0472)
1.56 (0.0614) ... 1.65 (0.0650)	1.3 (0.0512)
1.66 (0.0653) ... 1.75 (0.0689)	1.4 (0.0551)
1.76 (0.0693) ... 1.85 (0.0728)	1.5 (0.0591)
1.86 (0.0732) ... 1.95 (0.0768)	1.6 (0.0630)
1.96 (0.0772) ... 2.05 (0.0807)	1.7 (0.0669)
2.06 (0.0811) ... 2.15 (0.0846)	1.8 (0.0709)
2.16 (0.0850) ... 2.25 (0.0886)	1.9 (0.0748)
2.26 (0.0890) ... 2.35 (0.0925)	2.0 (0.0787)
2.36 (0.0929) ... 2.45 (0.0965)	2.1 (0.0827)
2.46 (0.0969) ... 2.55 (0.1004)	2.2 (0.0866)
2.56 (0.1008 ... 2.65 (0.1043)	2.3 (0.0906)
2.66 (0.1047) ... 2.75 (0.1083)	2.4 (0.0945)
2.76 (0.1087) ... 2.85 (0.1122)	2.5 (0.0984)
2.86 (0.1126) ... 2.95 (0.1161)	2.6 (0.1024)
2.96 (0.1165) ... 3.05 (0.1201)	2.7 (0.1063)
3.06 (0.1205) ... 3.15 (0.1240)	2.8 (0.1102)
3.16 (0.1244) ... 3.25 (0.1280)	2.9 (0.1142)

Example:

Measurement 2.00 mm (0.079 in.)

Install 1.7 mm (0.067 in.) thick shim.

– Remove small input shaft bolt.

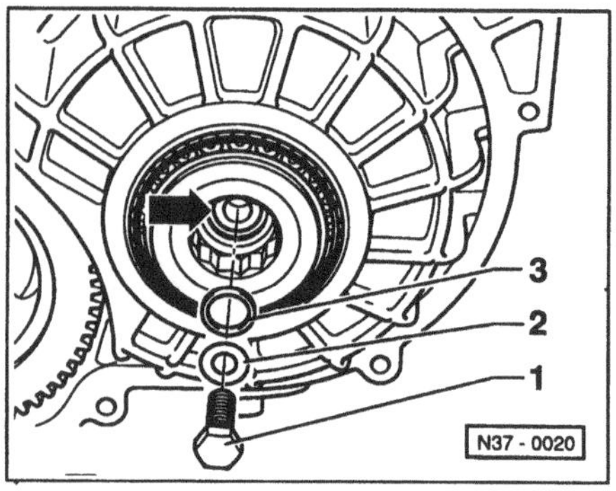

◄ – Install correct shim -3- on small input shaft -arrow-.

– Tighten small input shaft bolt -1- with washer -2-.

● Tightening torque: 30 Nm (22 ft lb)

– Perform planet carrier check measurement.

Planet carrier check measurement

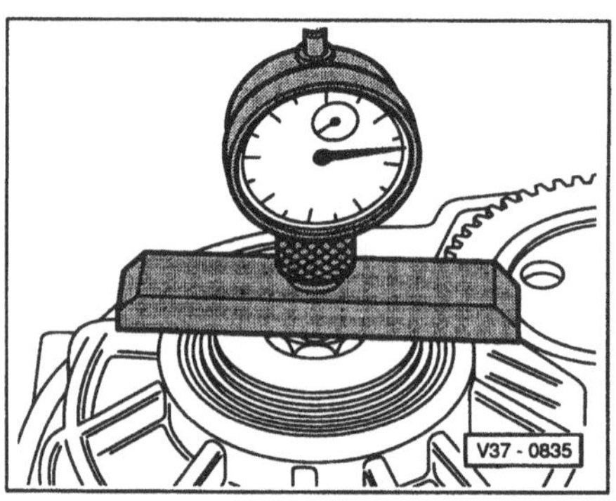

◄ – Install dial indicator in VW 382/7 measuring bar and place tip on bolt of small input shaft.

– Move small input shaft up and down and read play on dial indicator.

● Play:

min. = 0.23 mm (0.0091 in.)

max. = 0.37 mm (0.0146 in.)

Note:

If the planet carrier has been adjusted with reverse gear brake -B1- and free wheel removed, then before installing planet carrier, install reverse gear brake -B1-,
⇒ Planetary gearbox, assembling, Page J 37-71.

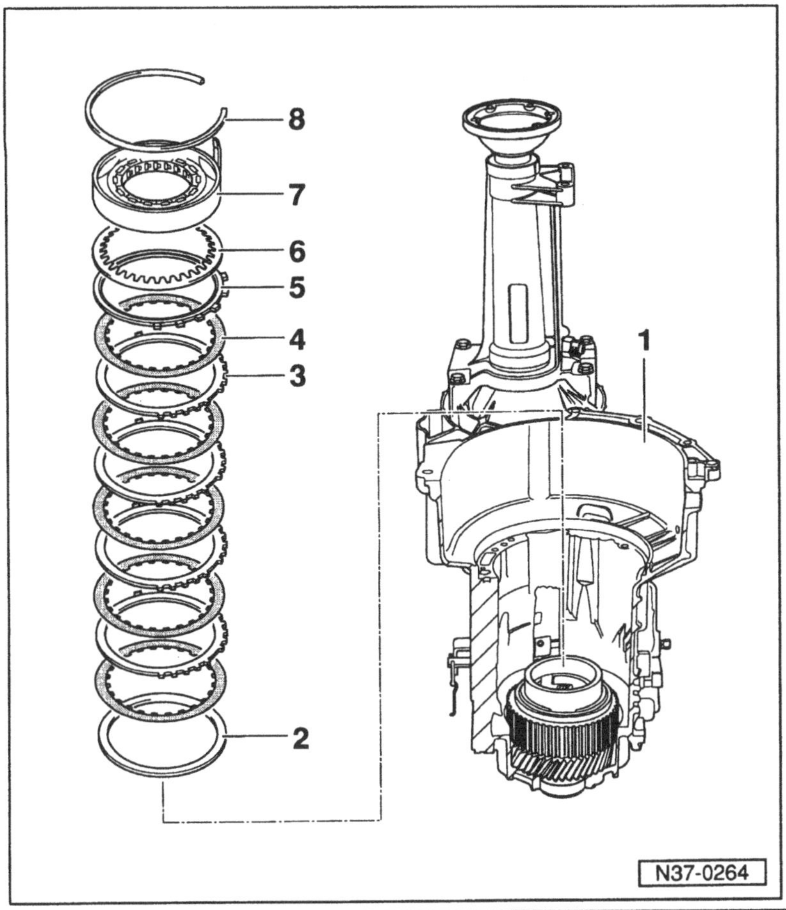

Reverse gear brake -B1-, adjusting

Overview of components to adjust reverse gear brake -B1-

1 - **Transmission housing**
- ♦ With input gear and planet carrier installed

2 - **Shim**

3 - **Outer plate -B1-**
- ♦ Quantity ⇒ from Page J 00-6

4 - **Inner plate -B1-**
- ♦ Quantity ⇒ from Page J 00-6

5 - **Pressure plate**
- ♦ Install with flat side facing plates
- ♦ Different thicknesses depending upon the number of inner plates installed

Transmission with 4 inner plates
- ♦ 13.5 mm (0.5315 in.)

Transmission with 5 inner plates
- ♦ 10.5 mm (0.413 in.)

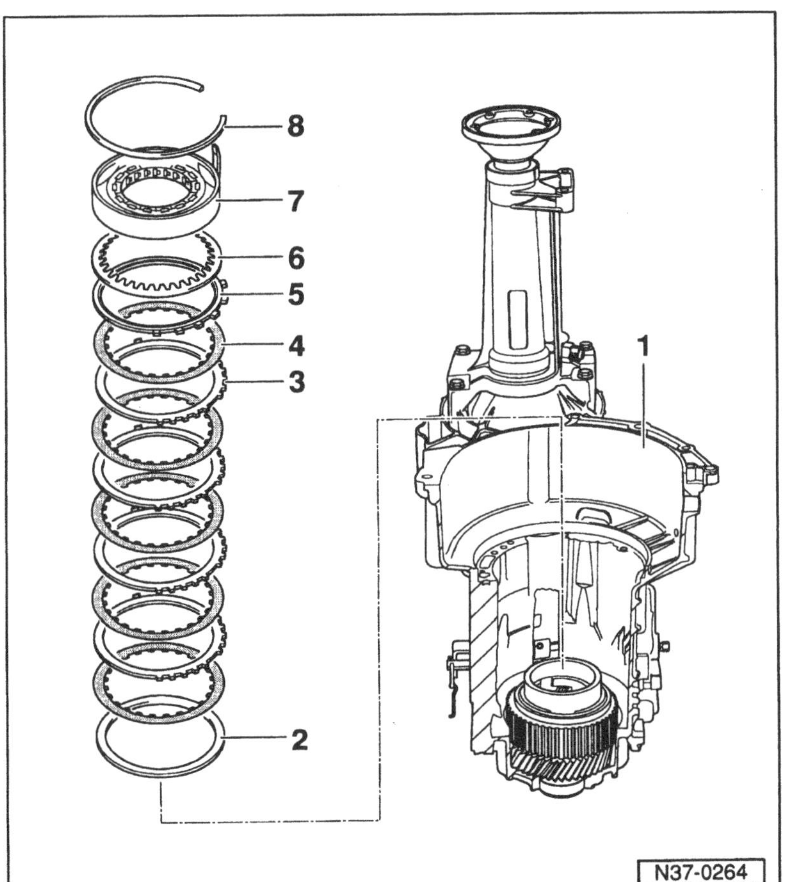

6 - **Dished washer**
- ♦ Convex side faces toward free wheel

7 - **Free wheel with -B1- piston**

Note:

Before removing or installing free wheel remove valve body and take out sealing plugs ⇒ Page J 38-20; also remove Transmission Vehicle Speed Sensor -G38- Page J 01-19, Fig. 1 0.

8 - **Circlip**

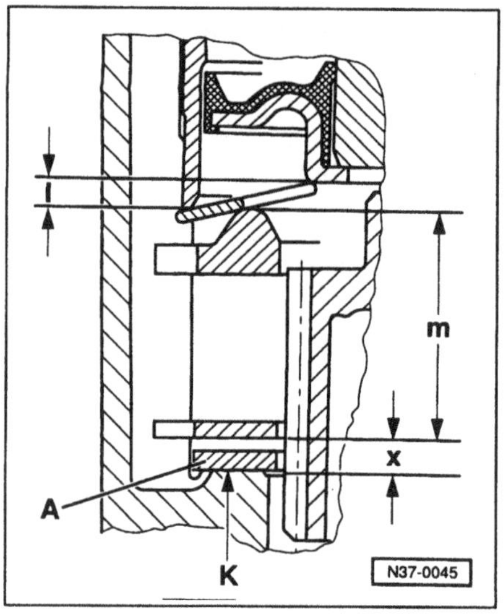

◄ **Determining shim -A- thickness**

Shim thickness is determined by gap "x" and selected from table, ⇒ Page J 37-94.

Gap x = k + (l ÷ 2) – m

- A = Shim
- x=Gap
- l = Position of piston in free wheel
- m = Height of plate set including pressure plate
- K = Constant = 30.5 mm (1.2008 in.)

  "K" is determined by installation height in transmission housing and is not adjustable.

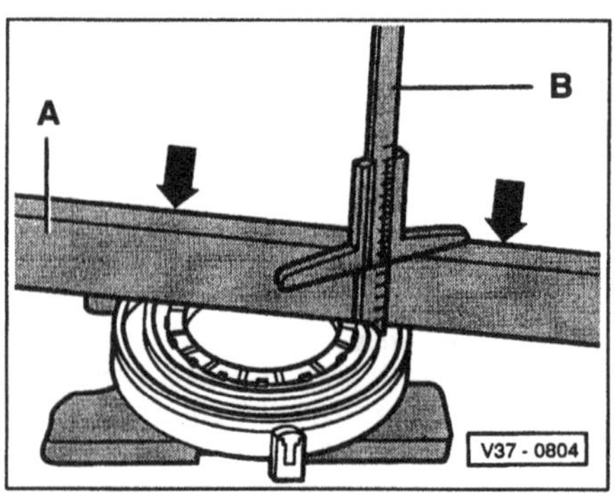

Calculating "l":

◄

- Press piston in direction of arrow up to stop.
- Place straightedge -A- onto outer race of free wheel.
- Measure to inner edge of piston with depth gauge -B-.

Example:

Reading	=	51.8 mm (2.0393 in.)
– Straightedge	=	–48.2 mm (1.8976 in.)
Calculated size "l"	=	3.6 mm (0.1417 in.)

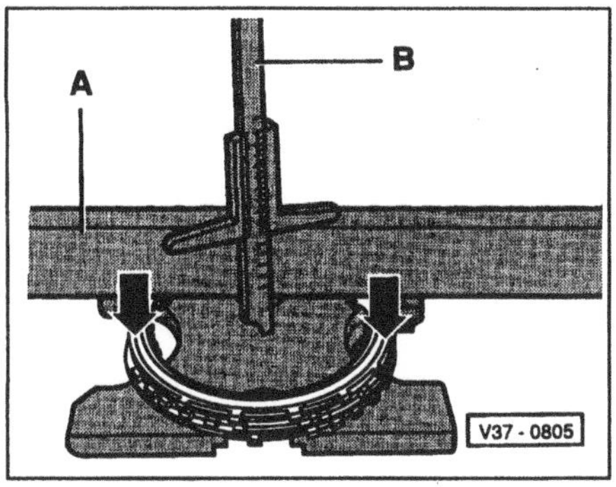

Calculating "m":

- Place straightedge -A- on thrust plate.
- Compress set of plates including pressure plate in *direction* of arrow and measure thickness of plate set with depth gauge -B-.

Example:

Reading	=	77.2 mm (3.0393 in.)
– Straightedge	=	–48.2 mm (1.8976 in.)
Calculated size "m"	=	29.0 mm (1.1417 in.)

Gap x = k + (l ÷ 2) – m

= 30.5 + (3.6 ÷ 2) – 29.0 = 3.3 mm

= (1.2008 + [0.1417 ÷ 2] – 1.1417 = 0.1299 in.)

- Determine size of shims according to table:

Table of shims:

Gap "x" mm (in.)	Shims mm (in.)
2.36 (0.0929) ... 2.45 (0.0965)	1.0 (0.0393)
2.46 (0.0969) ... 2.55 (0.1004)	1.1 (0.0433)
2.56 (0.1008) ... 2.65 (0.1043)	1.2 (0.0472)
2.66 (0.1047) ... 2.75 (0.1083)	1.3 (0.0512)
2.76 (0.1087) ... 2.85 (0.1122)	1.4 (0.0551)
2.86 (0.1126) ... 2.95 (0.1161)	1.5 (0.0591)
2.96 (0.1165) ... 3.05 (0.1201)	1.6 (0.0630)
3.06 (0.1205) ... 3.15 (0.1240)	1.7 (0.0669)
3.16 (0.1244) ... 3.25 (0.1280)	1.8 (0.0709)
3.26 (0.1283) ... 3.35 (0.1319)	1.9 (0.0748)
3.36 (0.1323) ... 3.45 (0.1358)	1.0 + 1.0 (0.0393 +0.0393)
3.46 (0.1362) ... 3.55 (0.1398)	1.0 + 1.1 (0.0393 + 0.0433)
3.56 (0.1402) ... 3.65 (0.1437)	1.1 + 1.1 (0.0433 + 0.0433)
3.66 (0.1441) ... 3.75 (0.1476)	1.1 + 1.2 (0.0433 + 0.0472)
3.76 (0.1480) ... 3.85 (0.1516)	1.2 + 1.2 (0.0472 + 0.0472)
3.86 (0.1520) ... 3.95 (0.1555)	1.2 + 1.3 (0.0472 + 0.0512)
3.96 (0.1559) ... 4.05 (0.1594)	1.3 + 1.3 (0.0512 + 0.0512)
4.06 (0.1598) ... 4.15 (0.1634)	1.3 + 1.4 (0.0512 + 0.0551)
4.16 (0.1638) ... 4.25 (0.1673)	1.4 + 1.4 (0.0551 +0.0551)

- Determine thickness of shims according to table and obtain Part No. from parts catalog.

- After determining shim for -B1- perform check measurement.

Reverse gear brake -B1- check measurement

- Install parts up to free wheel and secure with circlip ⇒ Planetary gearbox, assembling, Page J 37-71.

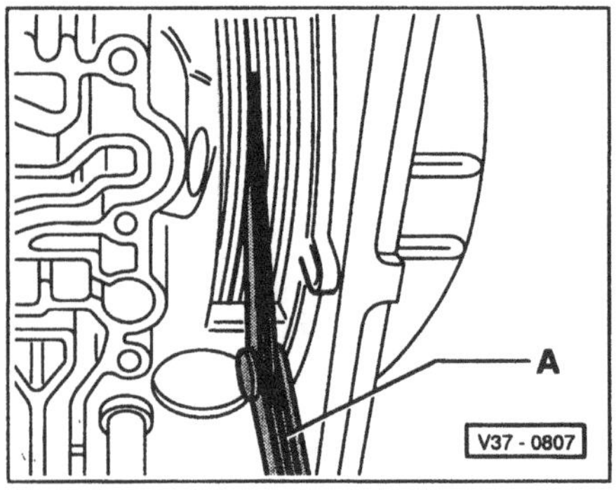

◄ - Measure play between plates with feeler gauge -A-.

- Play:

min. = 1.20 mm (0.0472 in.)

max. = 1.80 mm (0.0708 in.)

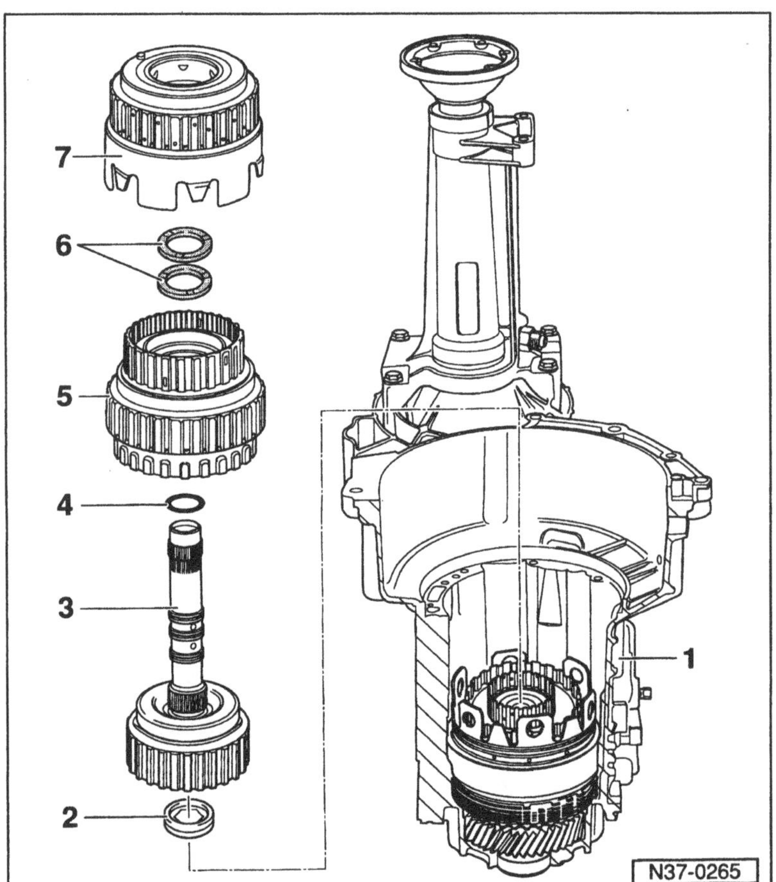

Clutch play between -K1- and -K2-, adjusting

To adjust clutch play install components without shim (item 6) ⇒ Planetary gearbox, assembling, Page J 37-71.

1 - **Transmission housing**
- ♦ Planetary gearbox installed up to small input shaft ⇒ Planetary gearbox, assembling, Page J 37-71

2 - **Axial needle bearing with washer**
- ♦ Axial needle bearing faces toward small input shaft
- ♦ Washer faces toward -K3-

3 - **4th gear clutch -K3- with pump shaft**

4 - **O-ring**

5 - **1st to 3rd gear clutch -K1- with turbine shaft**

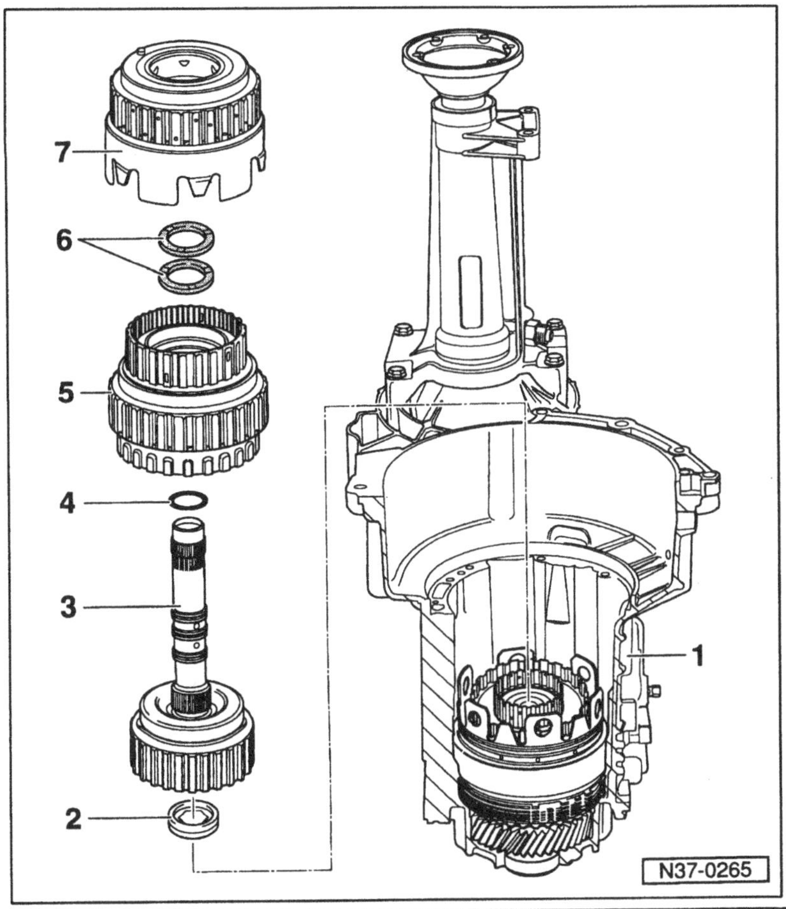

6 - Shim

- Do not install when adjusting clutch play
- 1 or 2 shims can be installed

7 - Reverse gear clutch -K2-

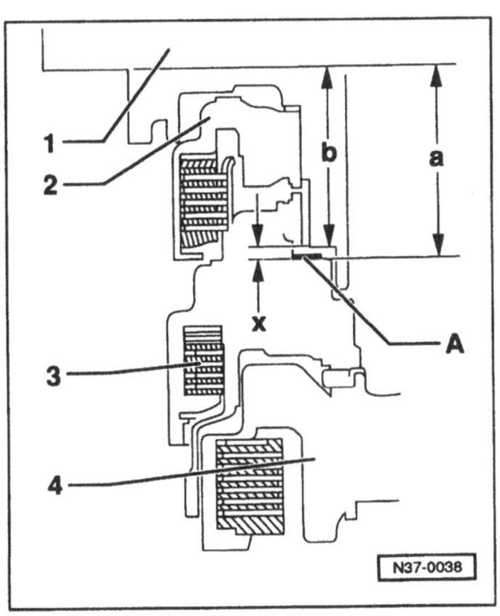

◄ **Determining shim -A**

A – Shim

1 – ATF pump

2 – Reverse gear clutch -K2-

3 – 1st to 3rd gear clutch -K1-

4 – 4th gear clutch -K3-

– Determine gap "x" and calculate shim thickness.

Gap x = a – b

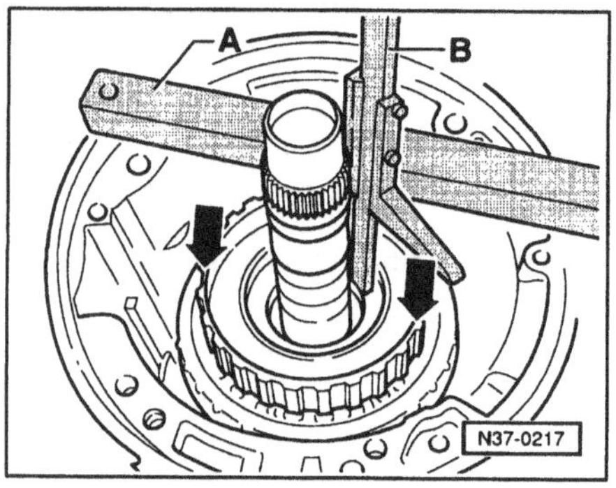

Calculating "a":

◄ – Place straightedge -A- onto transmission housing.

– Press -K1- down in direction of arrow and measure distance into -K1- with depth gauge -B-.

Example:

Reading 1 = 88.5 mm (3.4842 in.)

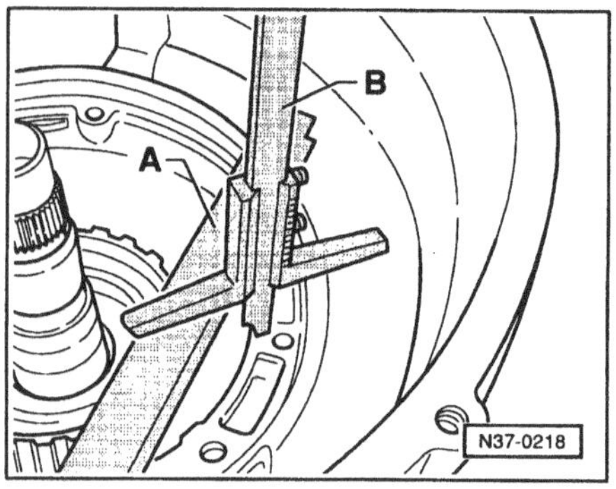

◄ – Measure with depth gauge -B- to pump flange on transmission housing.

Reading 2 = 34.3 mm (1.3504 in.)

Reading 1	=	88.5 mm (3.4842 in.)
– Reading 2	=	–34.3 mm (1.3504 in.)
Calculated size "a"	=	54.2 mm (2.1338 in.)

= Size measured from pump flange/transmission housing into -K1-

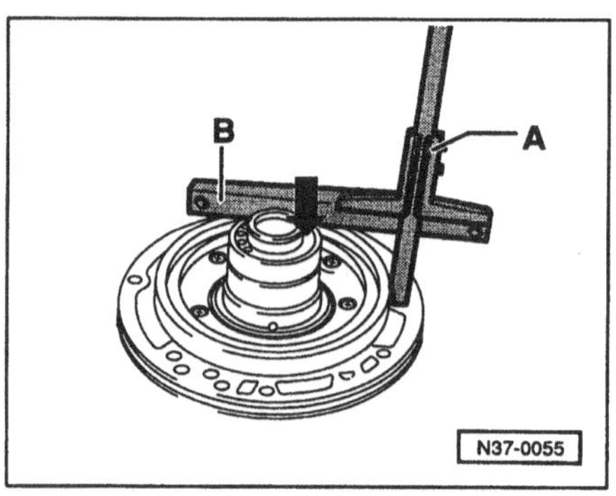

Calculating "b":

◄ – Place straightedge -B- onto stator support -arrow- and measure with depth gauge -A- to gasket of pump flange.

Example:

Reading	=	70.5 mm (2.7756 in.)
– Straightedge	=	–19.5 mm (0.7677 in.)
Calculated size "b"	=	51.0 mm (2.0079 in.)

Gap x = a–b

= 54.2 mm – 51.0 mm = 3.2 mm

= (2.1339 in. – 2.0079 in. = 0.01260 in.)

– Determine size of shim according to table:

Table of shims:

Gap "x" mm (in.)	Shims mm (in.)
... 2.54 (0.1000)	1.4 (0.0551)
2.55 (0.1004) ... 3.09 (0.1217)	1.0 + 1.0 (0.0393 + 0.0393)
3.10 (0.1378) ... 3.49 (0.1374)	1.2 + 1.2 (0.0472 + 0.0472)
3.50 (0.1378) ... 3.89 (0.1531)	1.4 + 1.4 (0.0551 + 0.0551)
3.90 (0.1535) ... 4.29 (0.1689)	1.6 + 1.6 (0.0630 + 0.0630)
4.30 (0.1693) ... 4.69 (0.1846)	1.8 + 1.8 (0.0709 + 0.0709)
4.70 (0.1850) ... 5.04 (0.1984)	1.2 + 1.2 + 1.6 (0.0472 +0.0472 + 0.0630)
5.05 (0.1988) ... 5.25 (0.2067)	1.2 + 1.2 + 1.8 (0.0472 +0.0472 +0.0709)

– Determine thickness of shims according to table and obtain Part No. from parts catalog.

– Perform clutch play check measurement after installing ATF pump.

Check measurement (clutch play)

The clutch play cannot be measured until the ATF pump has been installed.

– Attach dial indicator holder to transmission housing and place dial indicator on turbine shaft with a preload of 1.0 mm (0.0393 in.).

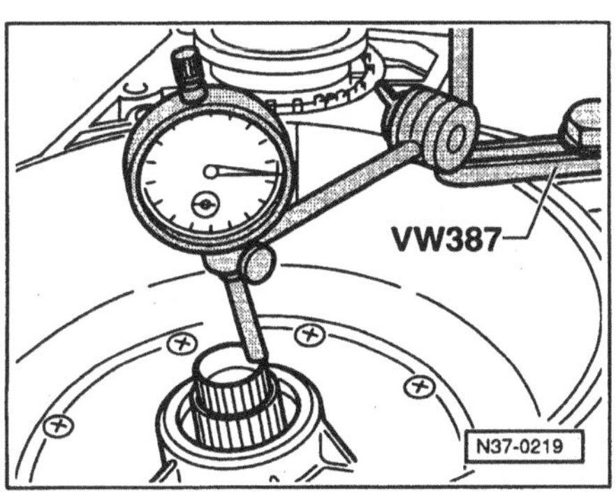

◄ – Move turbine shaft up and down and read play on dial indicator.

• Play:

min. = 0.5 mm (0.0197 in.)

max. = 1.2 mm (0.0472 in.)

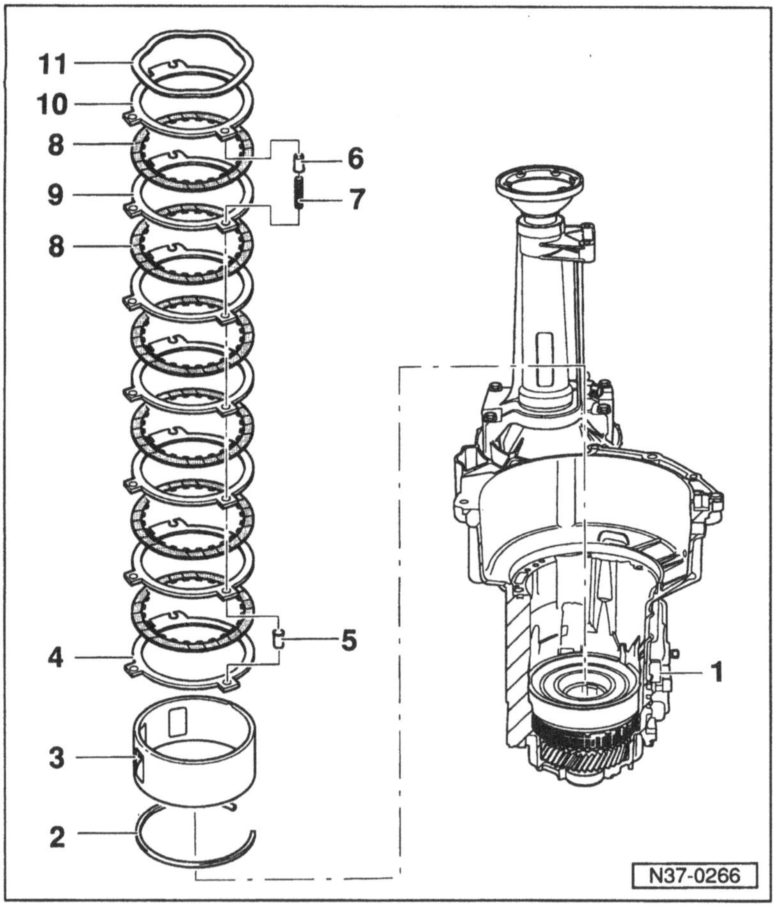

2nd and 4th gear brake -B2-, adjusting

Install components to adjust 2nd and 4th gear brake -B2- without corrugated washer (item 11), without last outer plate (item 10) and without spring caps (item 6)
⇒ Planetary gearbox, assembling, Page J 37-71.

1 - Transmission housing

2 - Circlip
- For supporting tube

3 - Supporting tube -B2-
- Length of -B2- with

5 inner plates	68.6 mm (2.70 in.)
6 inner plates	64.9 mm (2.55 in.)

- Quantity of inner plates -B2- ⇒ from Page J 00-6
- Install so that notch locates on free wheel wedge

4 - Outer plate -B2-
- 3 mm (0.118 in.) thick

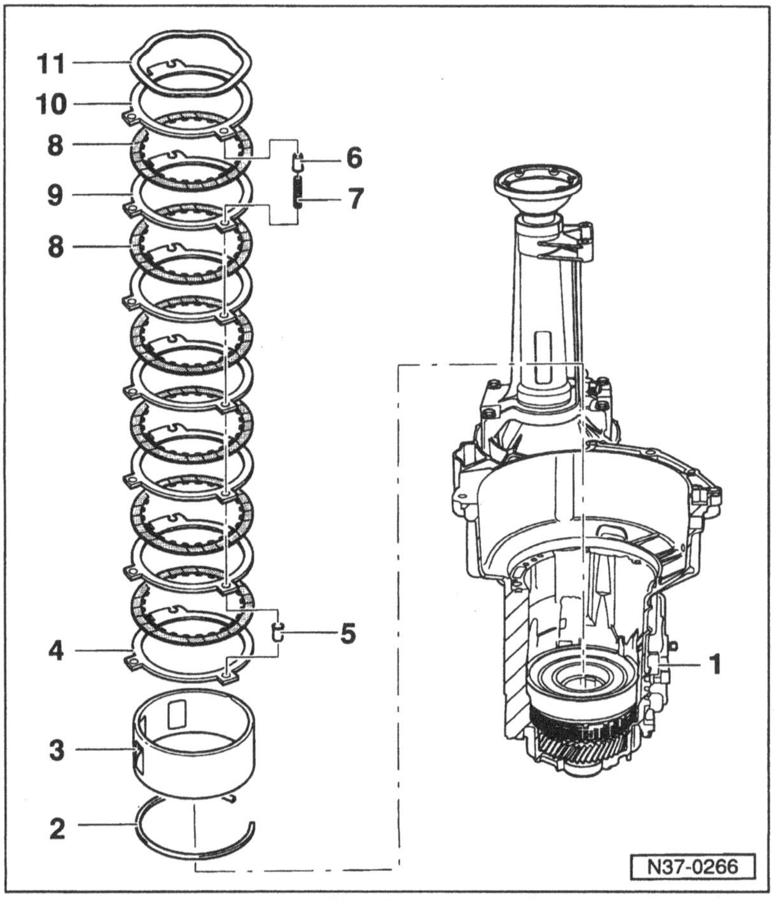

5 - Spring cap
- Install after installing first outer plate

6 - Spring cap
- Do not install when adjusting -B2-
- Install before installing last outer plate

7 - Spring

8 - Inner plate -B2-
- Quantity ⇒ from Page J 00-6

9 - Outer plate -B2-
- Always install 2 mm (0.0797 in.) thick outer plates
- Quantity ⇒ from Page J 00-6

10 - Outer plate -B2-
- Do not install when adjusting -B2-
- 2 outer plates can be installed.

11 - Corrugated washer
- Do not install when adjusting -B2-

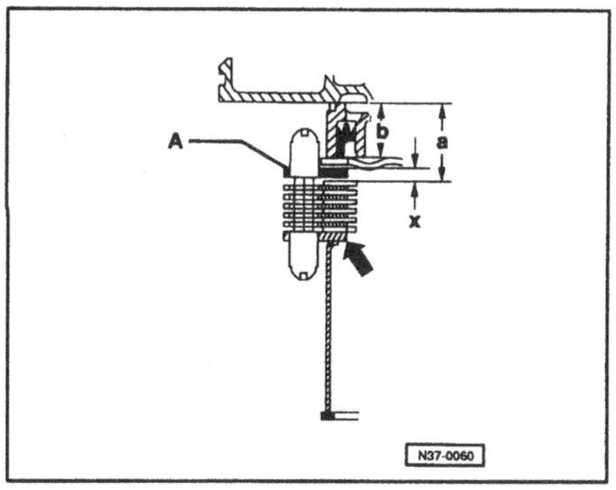

◄ **Calculating thickness of last outer plate -A-**

The thickness of the outer plate is determined by gap "x" and is selected from table, ⇒ Page J 37-108.

Gap x = a – b – 3.2 mm (0.1260 in)

A – Outer plate

Note:

First outer plate -arrow- is always 3 mm (0.0118 in.) thick.

Calculating "a":

(Clutches installed ⇒ Planetary gearbox, assembling, Page J 37-71)

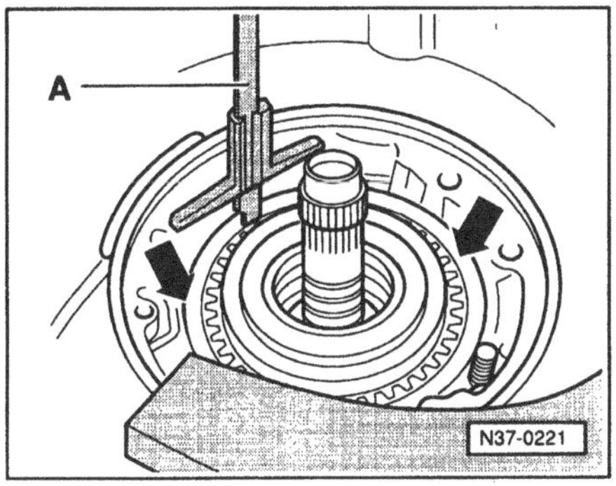

◄ – Measure from pump flange/transmission housing to last inner plate with depth gauge -A-.

Example:

Reading for "a"	=	30.2 mm (1.1890 in.)

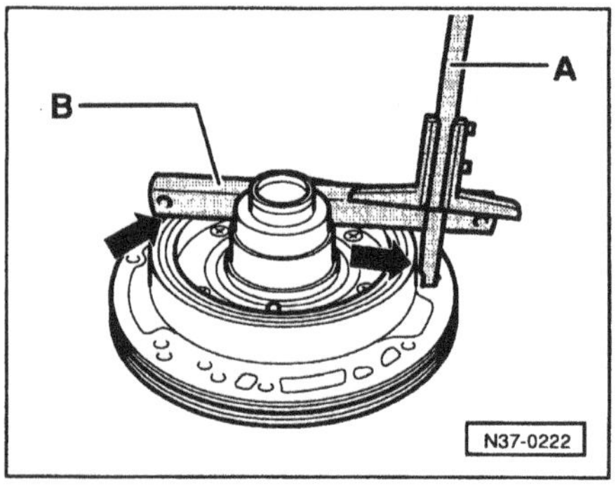

Calculating "b":

◄ – Place straightedge -B- onto stator support -arrow- and measure with depth gauge -A- to gasket of pump flange.

Example:

Reading	=	40.1 mm (1.5787 in.)
– Straightedge	=	–19.5 mm (0.7677 in.)
Calculated size "b"	=	20.6 mm (0.8110 in.)

Gap x = a – b – 3.2 mm (0.1260 in.)

= 30.2 – 20.6 – 3.2 = 6.4 mm

= (1.1890 – 0.8110 – 0.1260 = 0.2520 in .)

– Determine thickness of outer plate according to table:

Table of plates:

Gap "x" mm (in.)	Plate mm (in.)
4.25 (0.1673) ... 4.49 (0.1768)	2.75 (0.1083)
4.50 (0.1772) ... 4.74 (0.1866)	3.00 (0.1180)
4.75 (0.1870) ... 4.99 (0.1965)	3.25 (0.1280)
5.00 (0.1969) ... 5.24 (0.2063)	3.50 (0.1378)
5.25 (0.2067) ... 5.49 (0.2161)	3.75 (0.1476)
5.50 (0.2165) ... 5.74 (0.2260)	2.00 + 2.00 (0.0787 + 0.0787)
5.75 (0.2264) ... 5.99 (0.2358)	2.00 + 2.25 (0.0787 + 0.0886)
6.00 (0.2362) ... 6.24 (0.2457)	2.25 + 2.25 (0.0886 + 0.0886)
6.25 (0.2461) ... 6.49 (0.2555)	2.25 + 2.50 (0.0886 + 0.0984)
6.50 (0.2559) ... 6.74 (0.2654)	2.50 + 2.50 (0.0984 + 0.0984)
6.75 (0.2657) ... 7.00 (0.2756)	2.50 + 2.75 (0.0984 + 0.1082)

– Determine thickness of outer plate according to table and obtain Part No. from parts catalog.

Note:
Two outer plates can be installed.

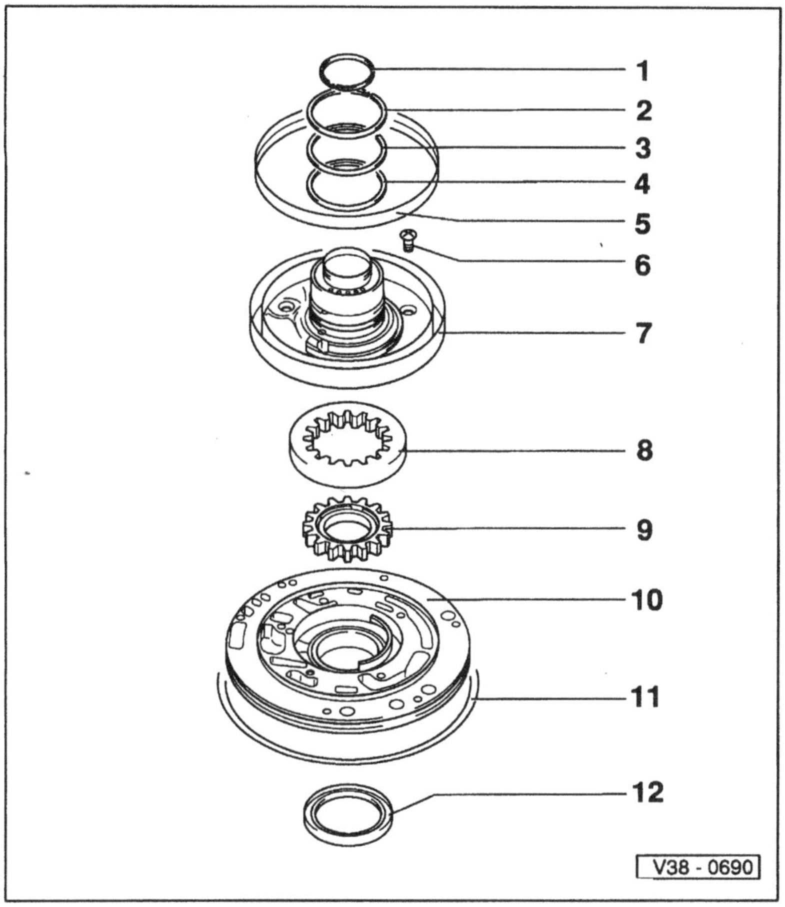

ATF pump, disassembling and assembling

1 - Piston ring
- Check for correct location
- Installing ⇒ Fig. 2

2 - Piston ring
- Check for correct location
- Installing ⇒ Fig. 2

3 - Piston ring
- Check for correct location
- Installing ⇒ Fig. 2

4 - Thrust washer

5 - Piston
- Sealing lips are vulcanized to the piston
- Moisten sealing lips with ATF before inserting
- Turn piston slightly when installing

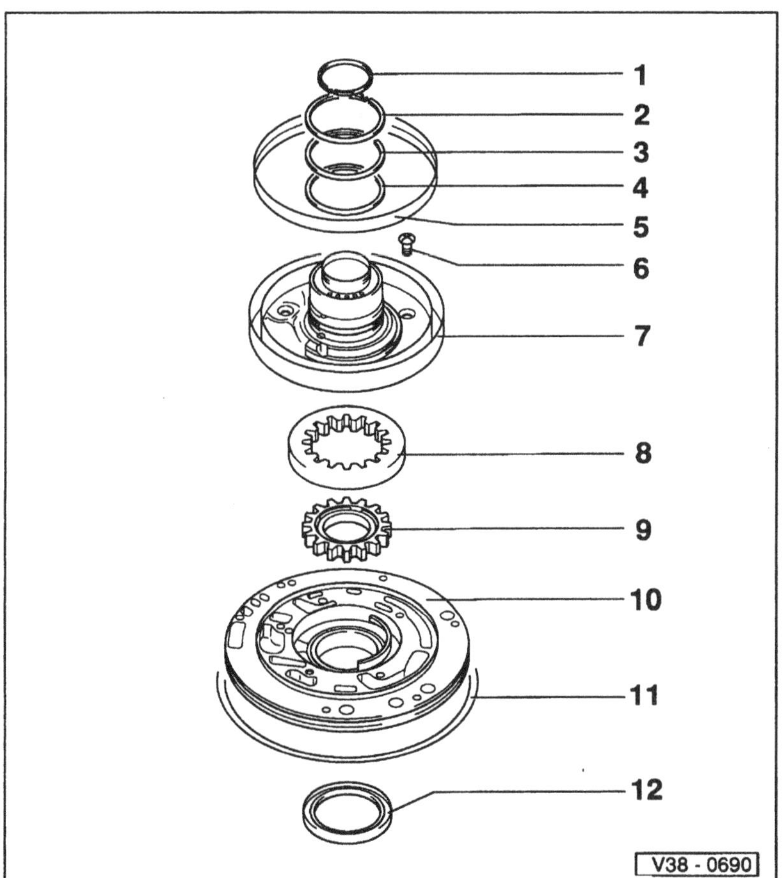

6 - Bolt
- 10 Nm (7 ft lb) plus 45° (1/8 turn) further

7 - Stator support

8 - Outer gear
- Manufacturer's marking faces stator support
- If outer gear is incorrectly installed, the ATF pump may not operate freely in the installed position

9 - Inner gear
- Larger groove faces stator support

10 - Housing for ATF pump

11 - O-ring
- Always replace

12 - Torque converter oil seal
- Removing and installing ⇒ Page J 32-2

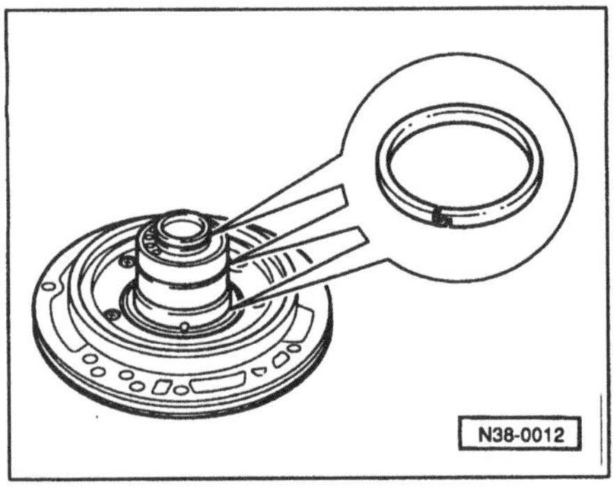

◄ **Fig. 1 Piston rings, checking for correct location**

Ensure that ends of piston rings are hooked together.

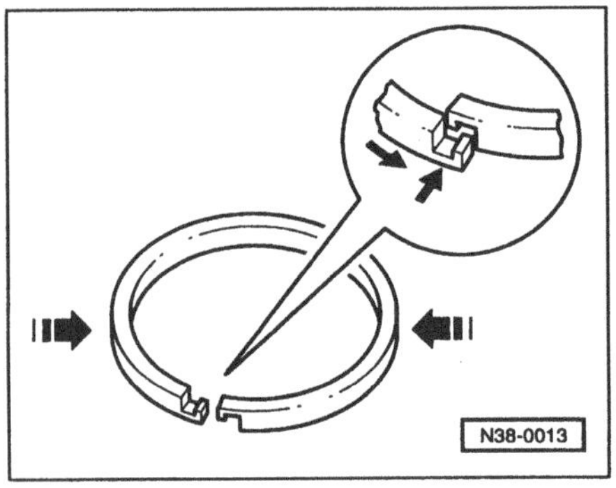

◄ **Fig. 2 Piston rings, installing together**

- Place piston ring in groove.
- Hook ends together by pressing sides of ring in and guiding ends over one another.

> ***CAUTION!***
> ***Do not twist piston ring out of groove on one side.***

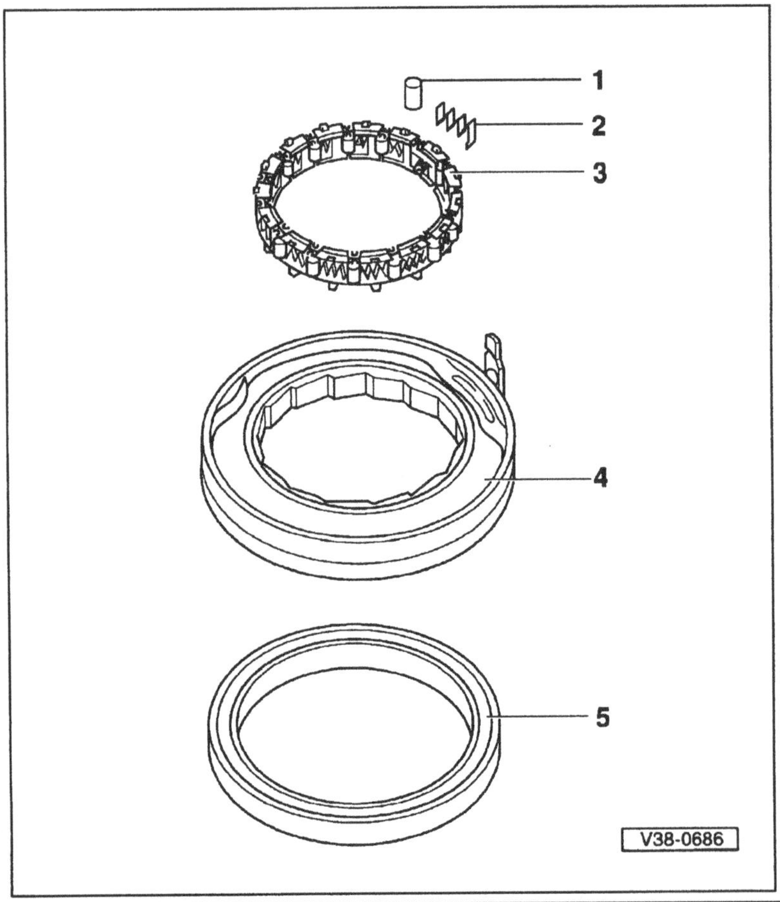

Free wheel with -B1- piston, disassembling and assembling

1 - Rollers
- ♦ Installing ⇒ Fig. 1

2 - Springs
- ♦ Installing ⇒ Fig. 1

3 - Cage
- ♦ Installing ⇒ Fig. 2
- ♦ Securing ⇒ Fig. 3

4 - Outer ring

5 - Piston
- ♦ Sealing lips are vulcanized to the piston
- ♦ Moisten sealing lips with ATF before installing
- ♦ Turn piston slightly when installing
- ♦ Installed position ⇒ Fig. 4

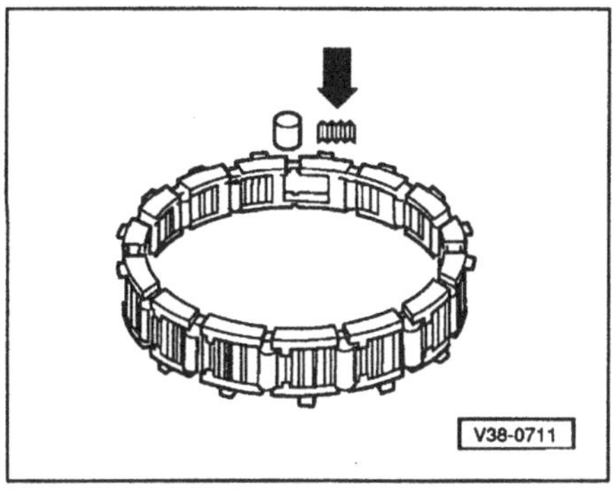

◄ **Fig. 1 Rollers and springs, installing**

Insert springs so that spring -arrow- engages in cage.

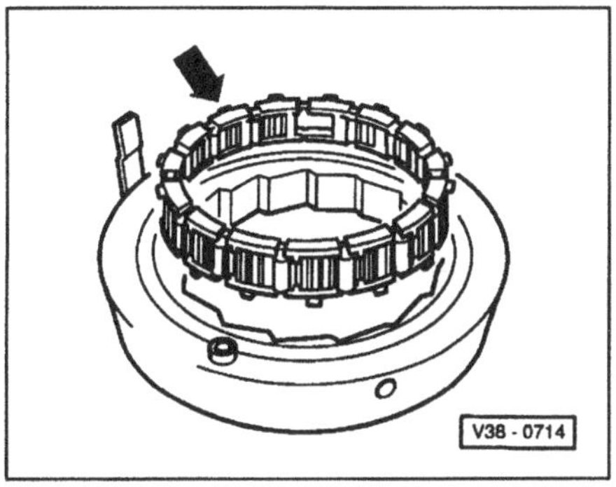

◄ **Fig. 2 Cage with springs and rollers, installing**

Install with large lugs -arrow- facing up.

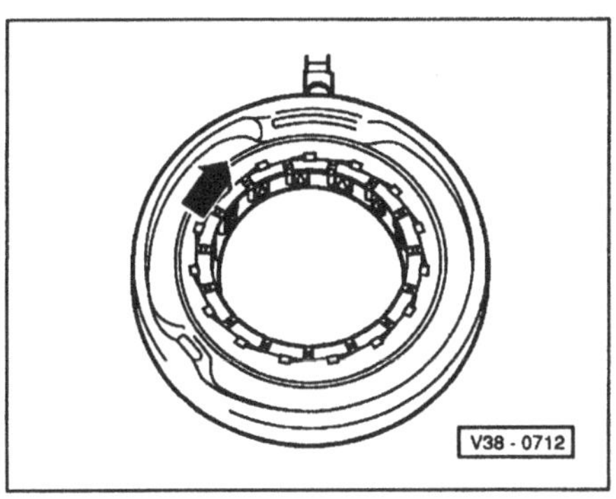

◄ **Fig. 3 Cage, securing**

– Secure cage by turning in direction of arrow as far as stop.

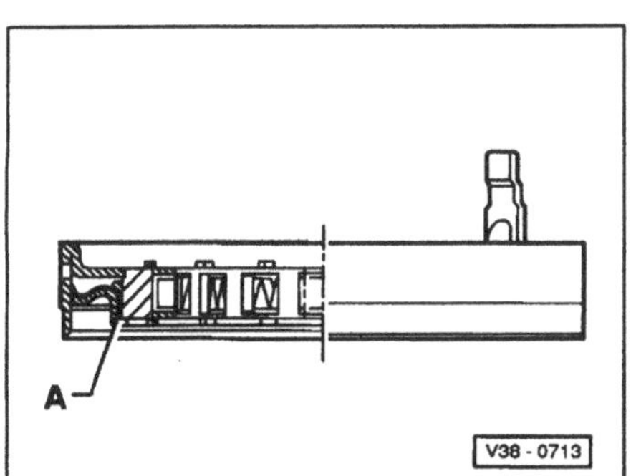

◄ **Fig. 4 Piston, installing**

Insert piston -A- as shown.

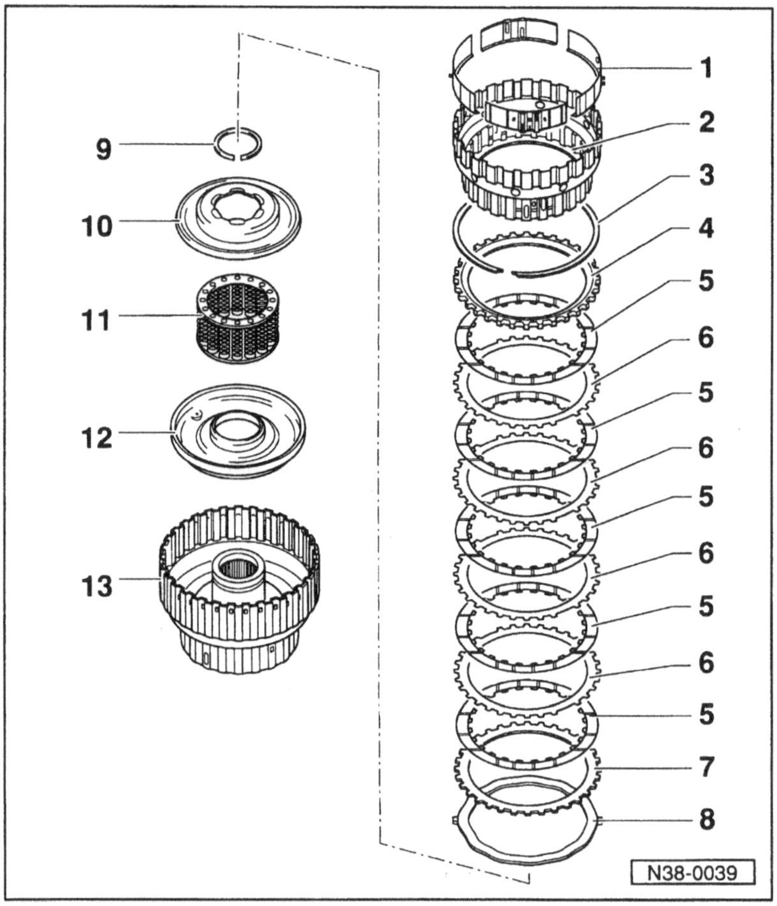

1st to 3rd gear clutch -K1-, disassembling and assembling

1 - Supporting ring
- ♦ 4-piece, unclip from inner plate carrier
- ♦ Installing ⇒ Fig. 2

2 - Inner plate carrier
- ♦ Installing ⇒ Fig. 4
- ♦ Assembling ⇒ Fig. 2

3 - Circlip
- ♦ Different thicknesses
- ♦ Mark after removing and insert in same position

4 - Thrust plate
- ♦ Smooth side faces inner plate
- ♦ Install together with inner plate carrier ⇒ Fig. 2

5 - Inner plate
- ♦ Quantity ⇒ Page J 00-6
- ♦ Installing ⇒ Fig. 2

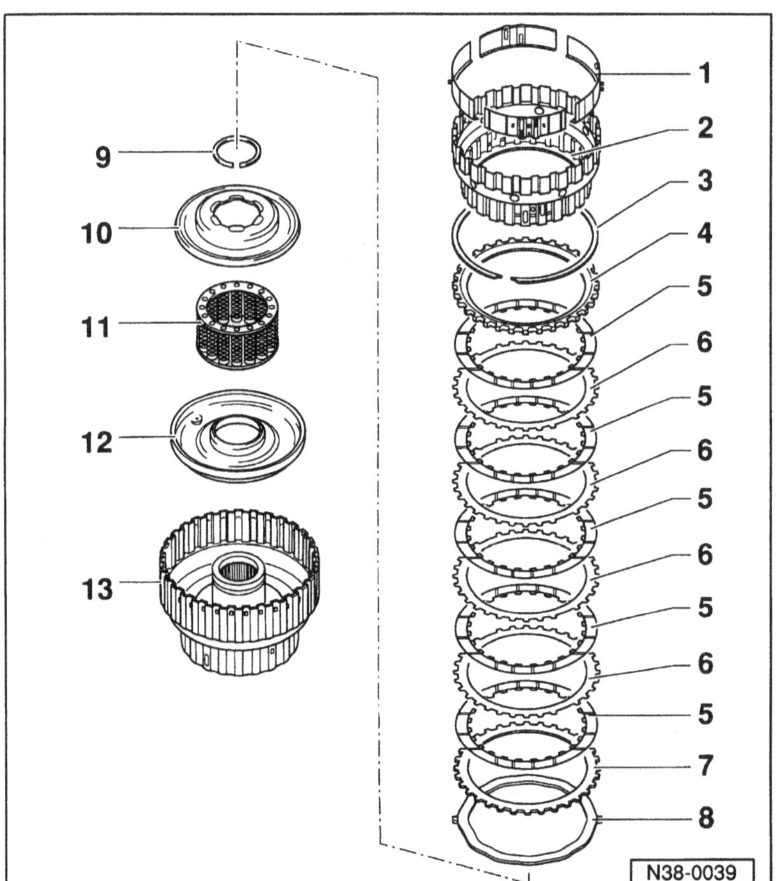

6 - Outer plate
- ♦ Quantity ⇒ Page J 00-6
- ♦ Installing ⇒ Fig. 2
- ♦ Always 1.5 mm thick

7 - Outer plate
- ♦ Always 2 mm thick
- ♦ Installing ⇒ Fig. 2

8 - Corrugated washer

9 - Circlip
- ♦ Removing and installing ⇒ Fig. 1

10 - Piston cap
- ♦ Sealing lips are vulcanized to cap
- ♦ Moisten sealing lips with ATF before installing
- ♦ Turn cap slightly when installing

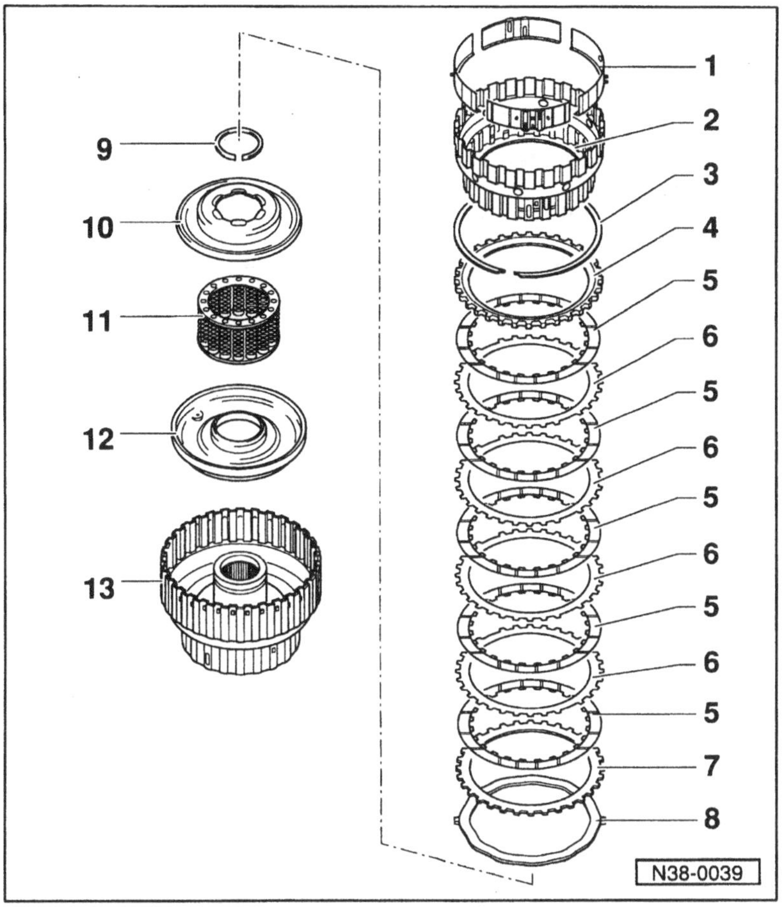

11 - Intermediate spring ring

- With spring supporting shell and compression springs

12 - Piston

- Sealing lips are vulcanized to piston
- Moisten sealing lips with ATF before installing
- Turn piston slightly when installing

13 - Clutch housing

- Before installing the inner plate carrier corrugated washer (item 8), insert inner and outer plates in the clutch ⇒ Fig. 3

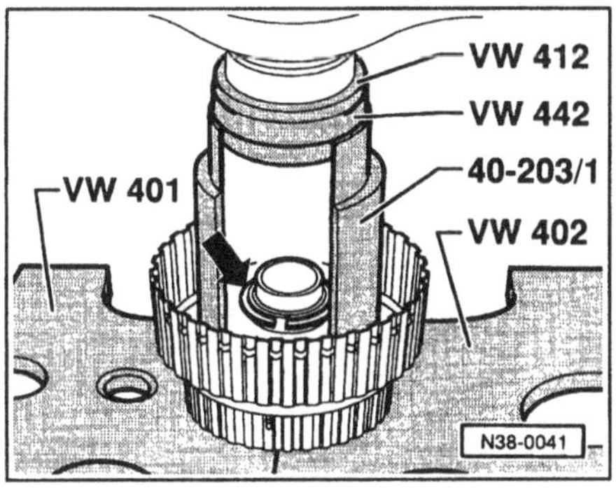

Fig. 1 Circlip, removing and installing

– Press piston cap with intermediate spring ring down far enough for circlip -arrow- to be inserted in groove.

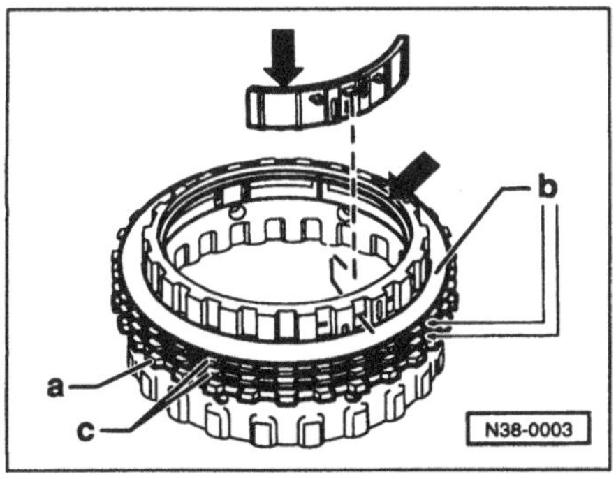

Fig. 2 Thrust plate and plates, installing onto inner plate carrier

– Install thrust plate -a- with smooth side facing inner plate and stepped side towards inner plate carrier.

– Install 3 inner plates -b- and 2 outer plates -c-.

– Clip in supporting ring with step -arrow-.

Notes:

- *Locate lugs correctly*
- *Installing corrugated washer and further plates in clutch housing ⇒ Fig. 3.*

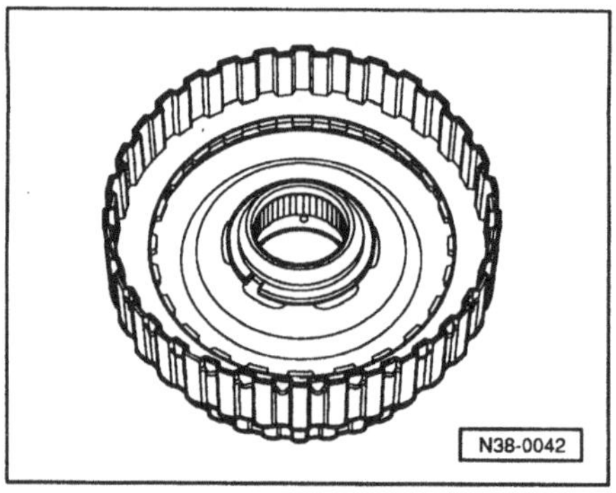

Fig. 3 Corrugated washer and inner and outer plates, installing in clutch housing

- Install corrugated washer.
- Install 2 mm thick outer plate.
- Install remaining inner and outer plates.

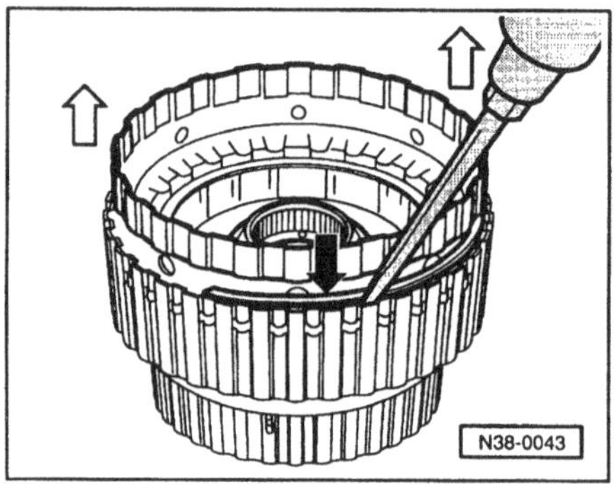

Fig. 4 Inner plate carrier, inserting and installing circlip -arrow-

- Lift inner plate carrier slightly to install circlip.

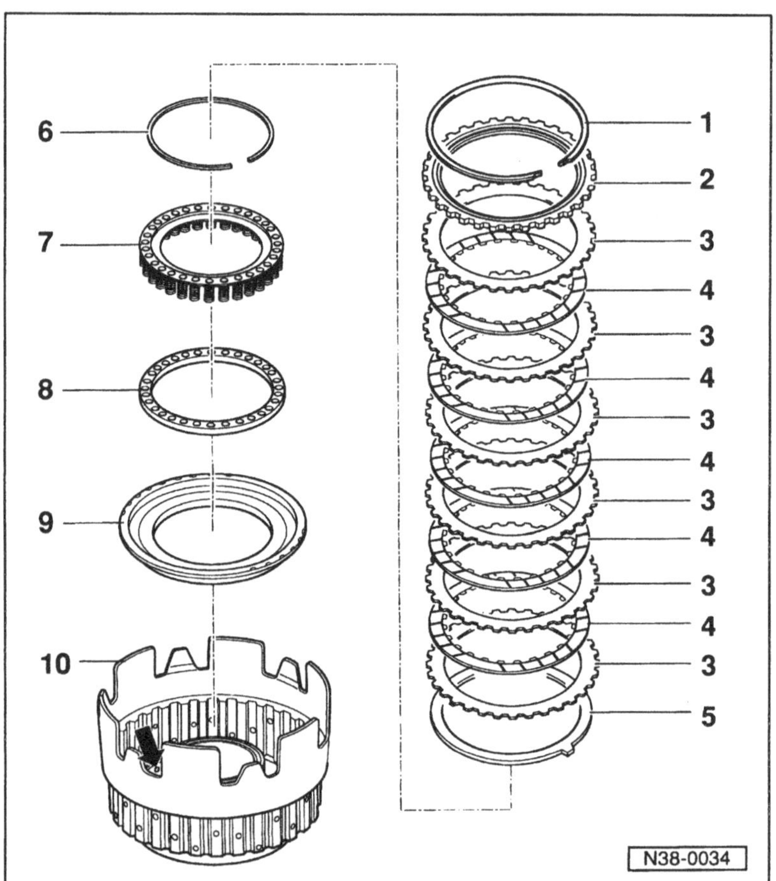

Reverse gear clutch -K2-, disassembling and assembling

Note:

When performing repairs to the clutch, ensure that the ball valve -arrow- is not damaged.

1 - Circlip
- ♦ Mark after removing and install in same position

2 - Thrust plate
- ♦ Smooth side facing inner plates
- ♦ Thickness with clutches of

4 inner plates	6.5 mm
5 inner plates	4.9 mm

3 - Outer plate
- ♦ Quantity ⇒ Page J 00-6

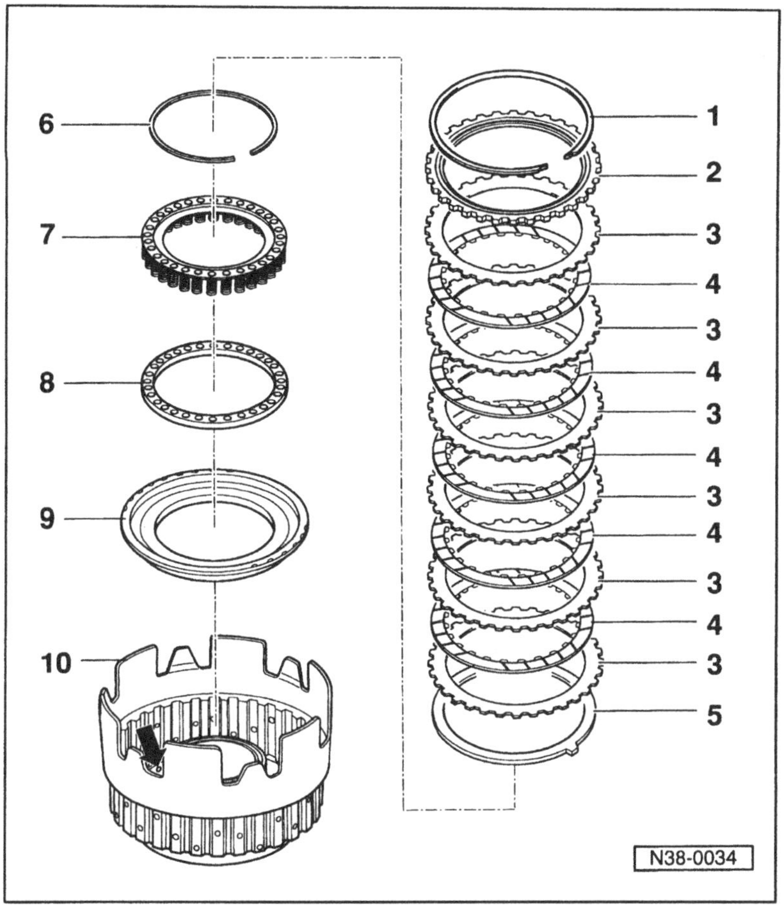

4 - Inner plate
- Quantity ⇒ Page J 00-6

5 - Corrugated washer

6 - Circlip
- Removing and installing ⇒ Fig. 1
- Installation position ⇒ Fig. 2

7 - Spring supporting plate
- With springs
- Installation position ⇒ Fig. 2

8 - Spring supporting ring

9 - Piston
- Sealing lips are vulcanized to piston
- Moisten sealing lips with ATF before installing
- Turn piston slightly when installing

10 - Clutch housing

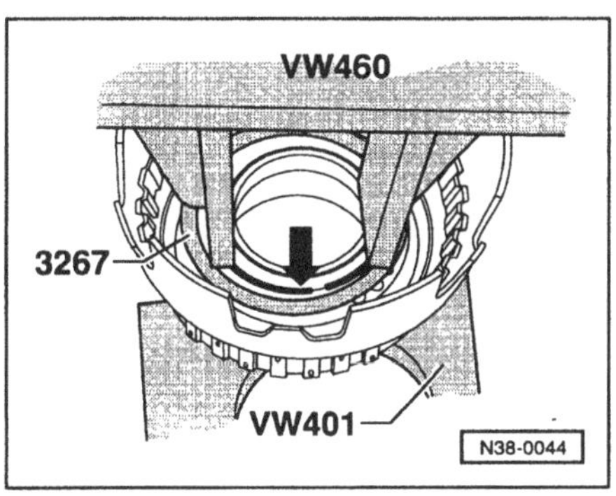

◄ **Fig. 1 Circlip –arrow–, removing and installing**

– Press spring supporting plate down with 3267 far enough to be able to remove and install circlip.

> ***CAUTION!***
>
> ***Do not damage clutch housing ball valve.***

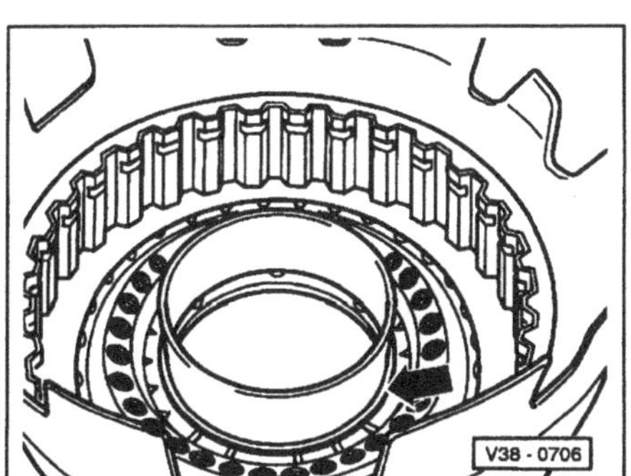

◄ **Fig. 2 Spring supporting plate, installation position /checking circlip**

Circlip -arrow- must be seated in groove of clutch housing.

Spring supporting plate must be centered relative to hub.

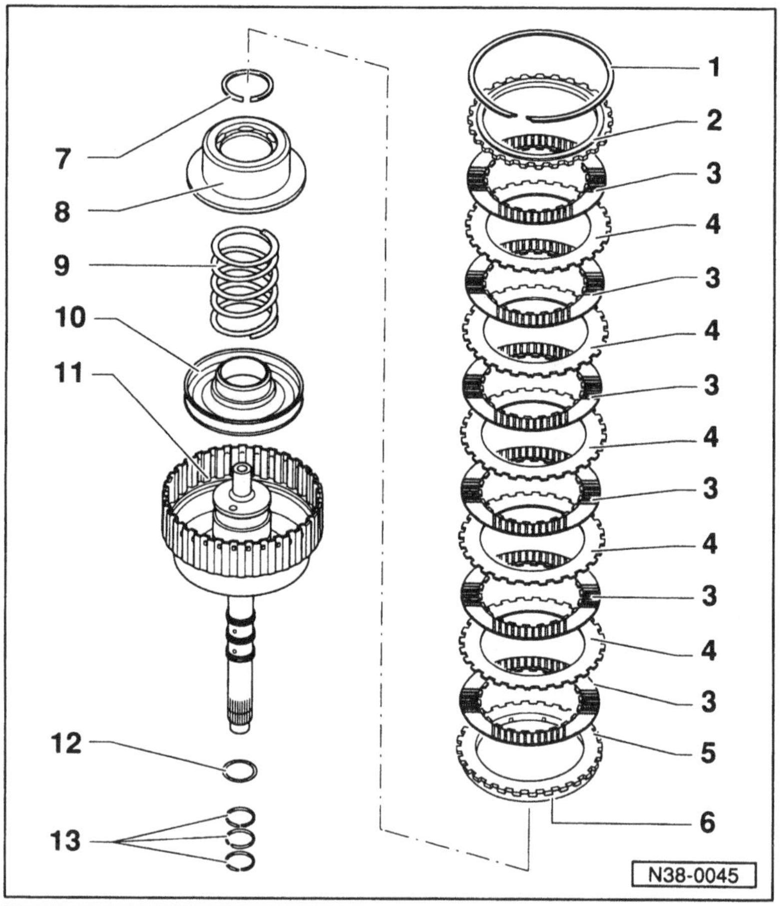

4th gear clutch -K3- with turbine shaft, disassembling and assembling

1 - Circlip
- Different thicknesses
- Mark after removing and insert at same point

2 - Thrust plate
- Install with stepped side facing circlip

3 - Inner plate
- Quantity ⇒ Page J 00-6

4 - Outer plate
- Quantity ⇒ Page J 00-6

5 - Pressure plate
- Install with peened corrugated washer facing piston

6 - Corrugated washer
- Peened to pressure plate (item 5)

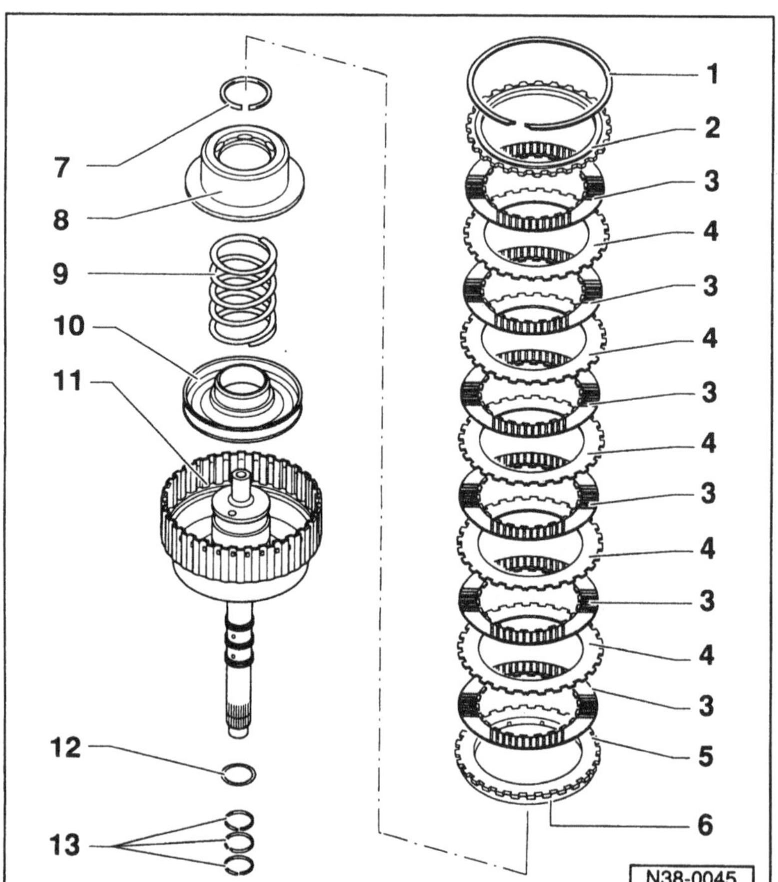

7 - Securing ring
- Removing and installing ⇒ Fig. 1

8 - Piston cap
- Sealing lips are vulcanized to piston cap
- Moisten with ATF before installing
- Turn piston cap slightly when installing
- When removing and installing press down and install securing ring (item 7)

9 - Spring
- Seat between piston and piston cap

10 - Piston
- Sealing lips are vulcanized to piston
- Moisten with ATF before installing
- Turn piston slightly when installing

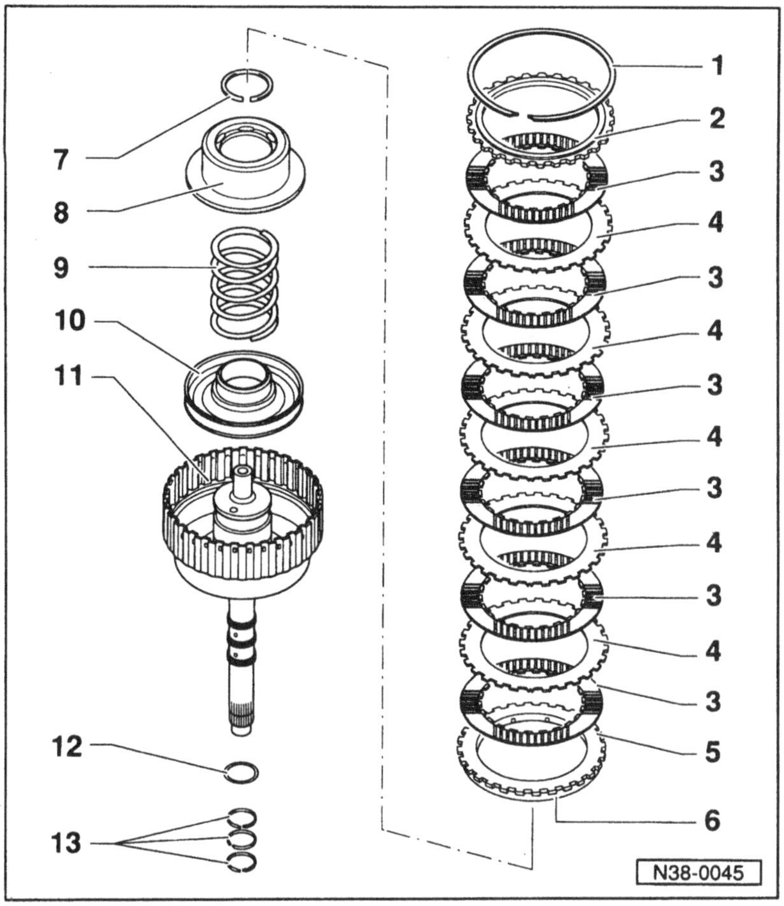

11 - Clutch housing with turbine shaft
- Height varies according to number of inner and outer plates

12 - O-ring
- Always replace
- Installing ⇒ Fig. 4

13 - Piston ring
- Check that it is correctly located ⇒ Fig. 2
- Installing ⇒ Fig. 3

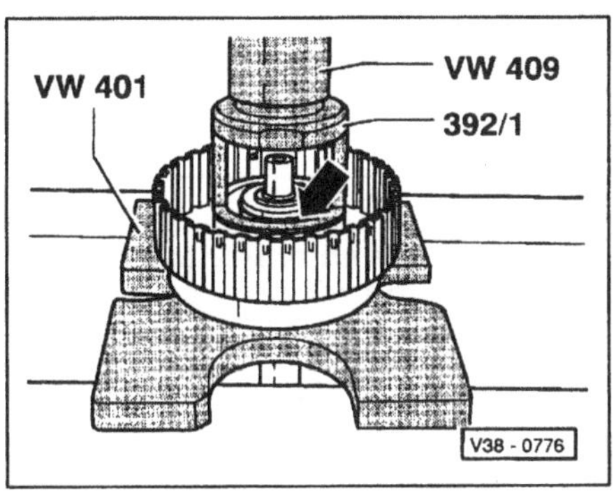

◄ **Fig. 1 Circlip -arrow-, removing and installing**

– Press down carefully on piston cap until circlip can be installed.

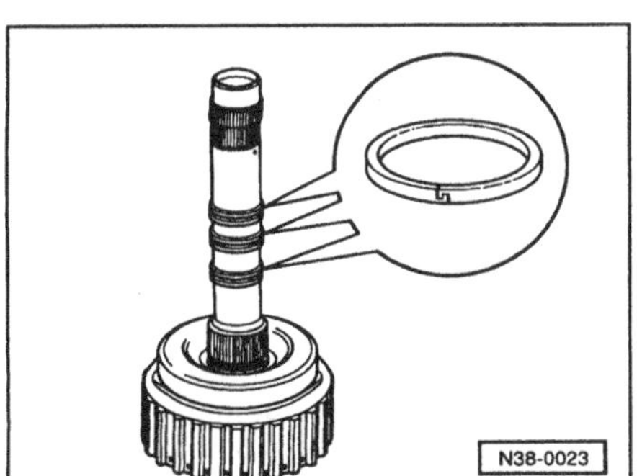

◄ **Fig. 2 Seating of piston rings, checking**

Ensure that ends of piston rings are hooked together.

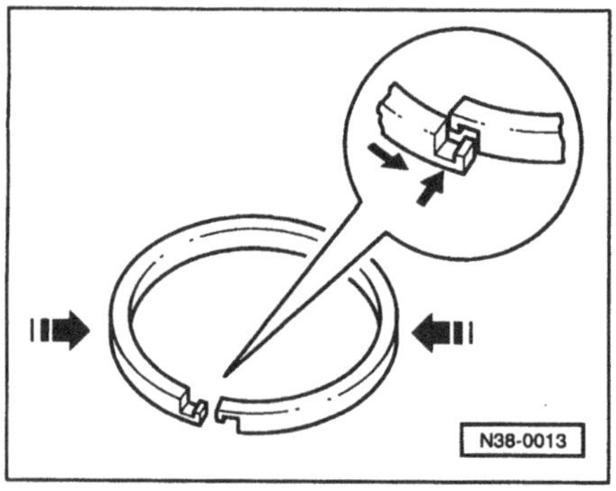

Fig. 3 Piston ring, installing and hooking ends together

- Place piston ring in groove.
- Press sides of ring in and guide ends over one another.
- Do not twist piston ring out of groove on one side.

> ***Caution!***
> ***Do not twist piston ring out of groove on one side.***

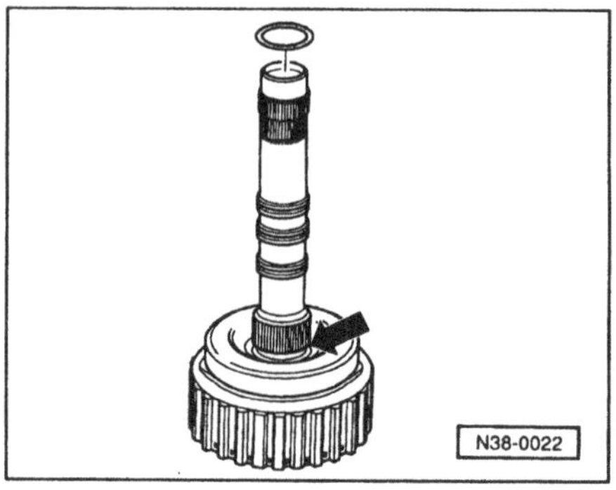

Fig. 4 O-ring, installing

- Install O-ring in recess in turbine shaft -arrow-.

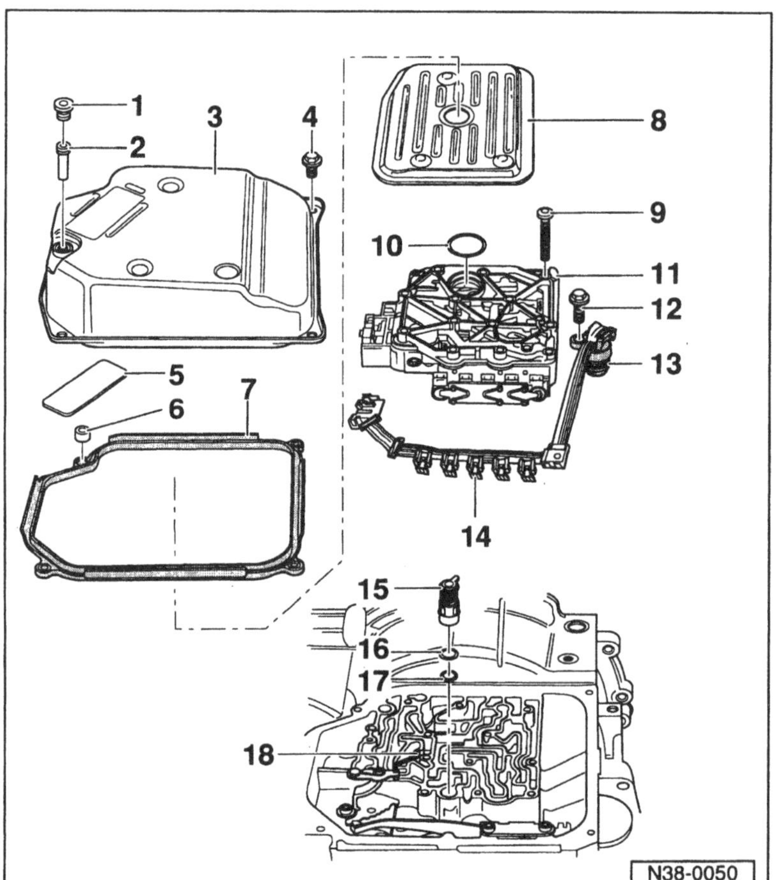

Valve body, removing and installing

> ***CAUTION!***
> ***Do not run the engine or tow the vehicle with the oil pan removed or when there is no ATF in the transmission.***

Notes:

- *The valve body or the conductor strip can also be removed when the transmission is installed.*
- *The conductor strip can be removed individually.*
- *After re-installing the oil pan, fill with ATF ⇒ Page J 37-42.*
- *Always replace a valve body which is clogged or otherwise faulty.*

1 - Plug
- Always replace seal ⇒ Fig. 2
- Checking ATF level and topping-up ⇒ Page J 37-42

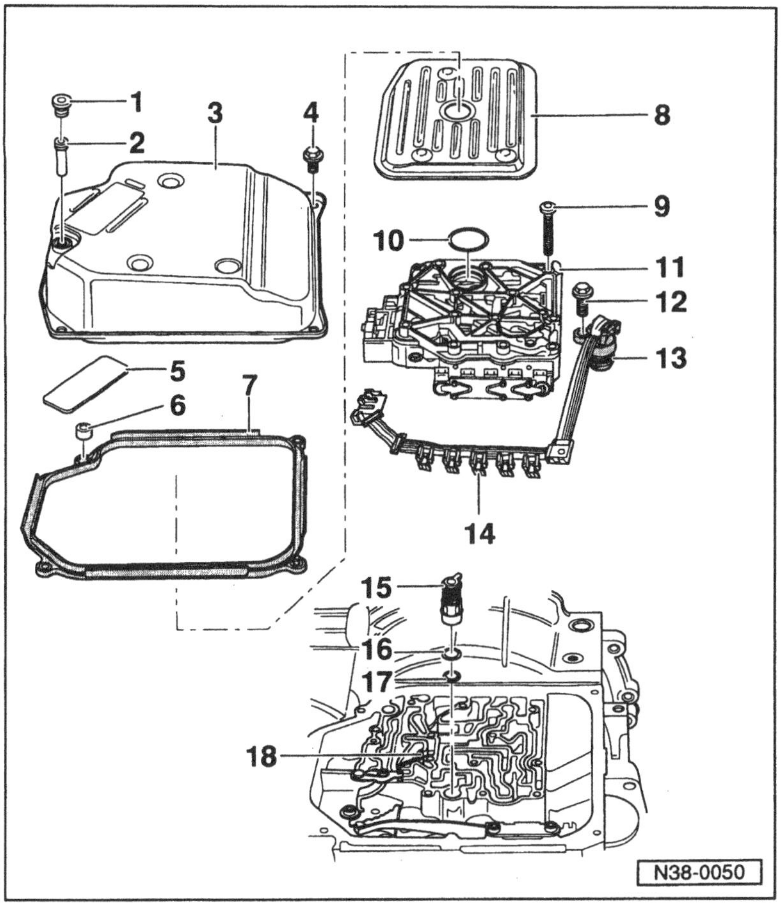

2 - Overflow pipe
- Remove to drain ATF ⇒ Fig. 1
- Thread in onto stop
- Checking ATF level and topping-up ⇒ Page J 37-42

3 - Oil pan

4 - Bolt, 12 Nm (9 ft lb)

5 - Magnet
- Install in recess in oil pan

6 - Spacer bushing
- Press into gasket

7 - Gasket
- Always replace
- Press spacer bushings into gasket

8 - ATF screen
- Installing ⇒ Fig. 10

9 - Bolt, 5 Nm (44 in. lb)

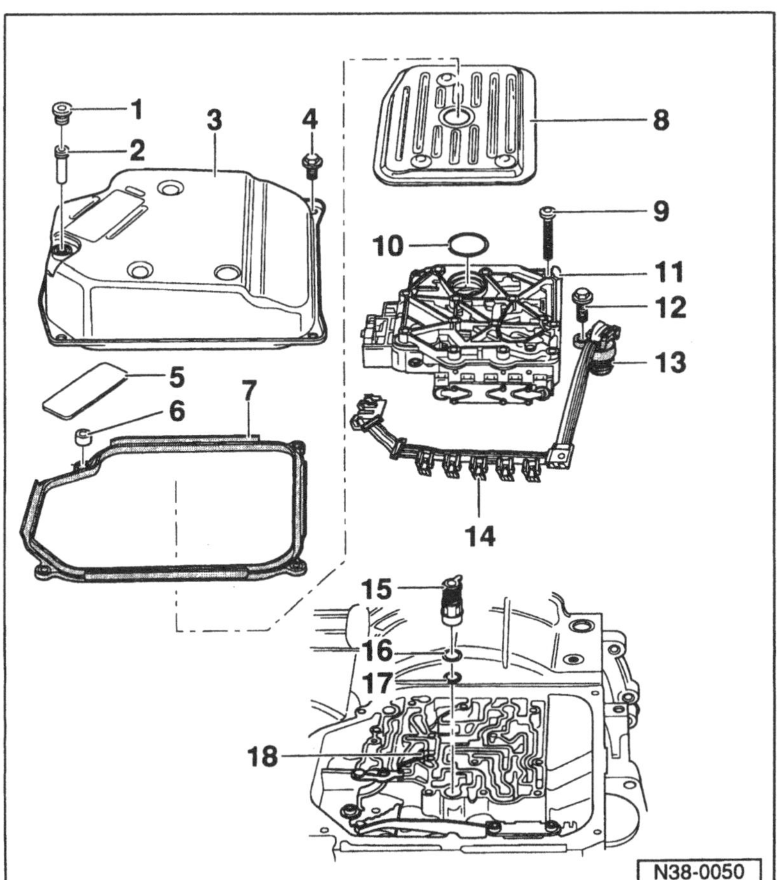

10 - Gasket
- Always replace
- Install in valve body

11 - Valve body
- Removing ⇒ Fig. 4
- Installing ⇒ Fig. 9
- Marking ⇒ Fig. 6
- Installing seal -item 10-

12 - Bolt, 10 Nm (7 ft lb)

13 - O-ring
- For cable transition piece
- Always replace

14 - Conductor strip with connector for solenoid valves
- With cable transition piece and connector
- Can be removed and installed with transmission installed
- Remove solenoid valves with tool 3373 or equivalent ⇒ Fig. 3

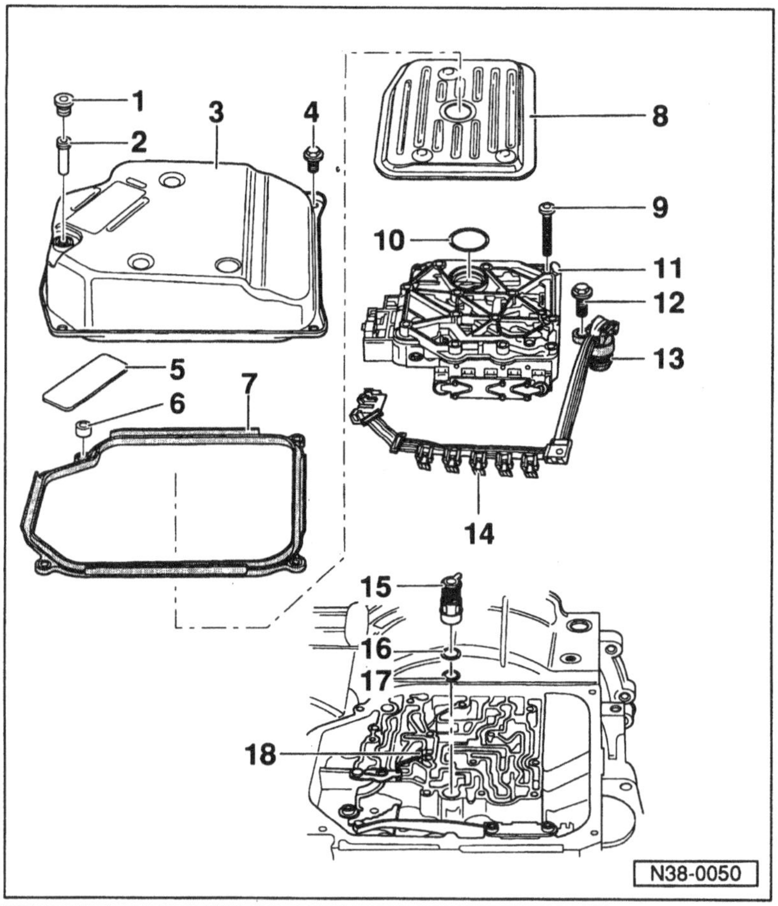

15 - Sealing plug

- Take out before removing and installing free wheel
- Insert so that the lug is located in the groove on housing, ⇒ Fig. 5
- Place O-rings onto sealing plug

16 - O-ring

- Always replace

17 - O-ring

- Always replace

18 - Operating rod for manual selector valve

- Unhooking ⇒ Fig. 4
- Install with stepped side facing manual selector valve ⇒ Fig. 7
- Adjusting ⇒ Fig. 8

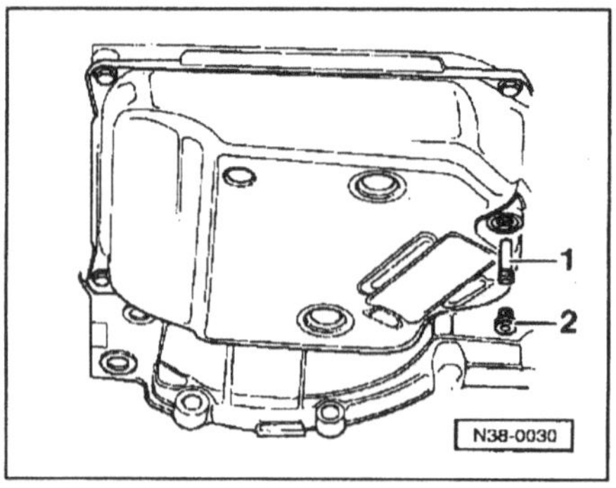

◄ **Fig. 1 ATF, draining**

- Place container under transmission.
- Remove plug -2-.
- Remove overflow pipe -1- and allow ATF to drain.
- Thread overflow pipe in onto stop.

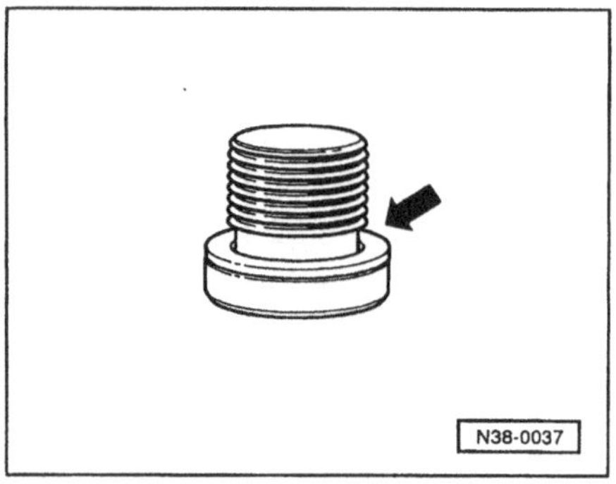

◄ **Fig. 2 Plug seal, replacing**

- Cut open seal -arrow- with side cutting pliers (always replace seal).
- After repairs install new seal to plug and screw in handtight when filling with ATE
- Pour in 3 liters of ATF
- Check ATF level and top-up ⇒ Page J 37-42.

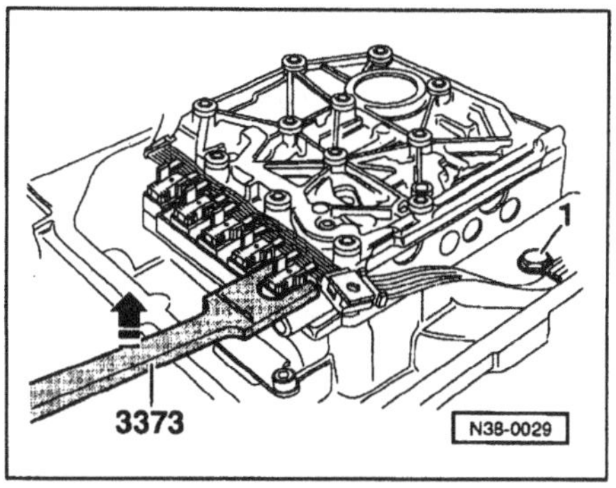

◄ **Fig. 3 Conductor strip, removing**

– Place 3373 onto stop under the solenoid valve plug.

– Remove with 3373 in direction of arrow.

– Remove bolt -1-.

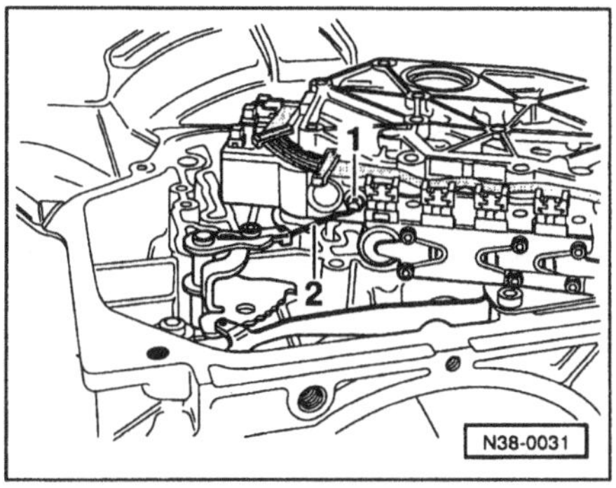

◄ **Fig. 4 Valve body, removing / unhooking operating rod**

When removing the valve body the manual selector valve -1- remains in the valve body.

– Pull manual selector valve out until it is just possible to unhook operating rod -2-.

– Secure manual selector valve so that it cannot fall out.

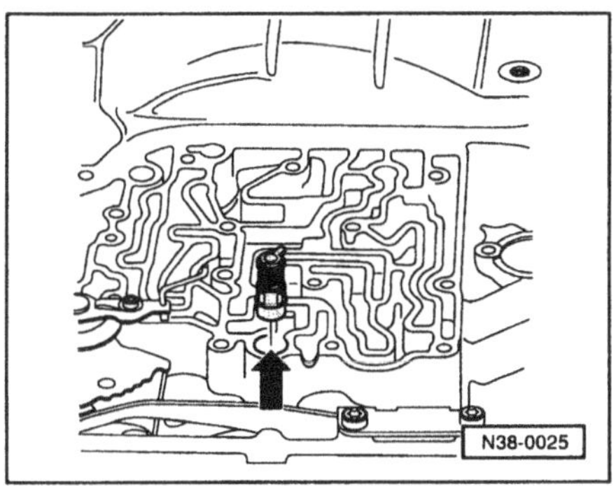

◄ **Fig. 5 Sealing plug, installing**

Pull sealing plug out of housing before removing and installing free wheel otherwise the sealing plug and the O-ring will be damaged.

– Install O-rings to sealing plug.

– Install sealing plug in hole in transmission housing -arrow-.

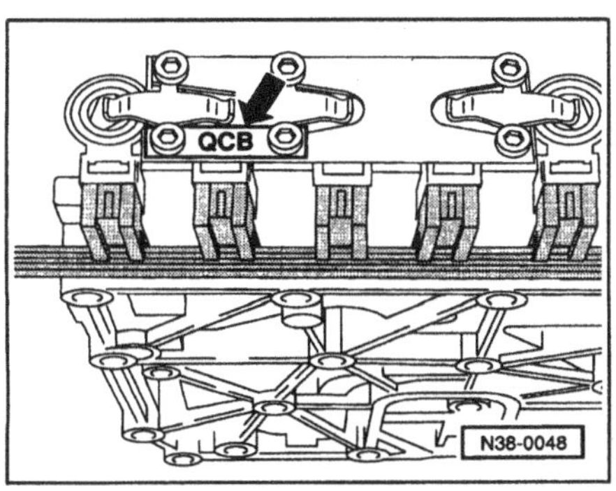

◄ **Fig. 6 Identification of valve body**

Code letters are stamped on a metal tab. Metal tab must remain on the valve body.

Valve body / transmission applications ⇒ from Page J 00-6

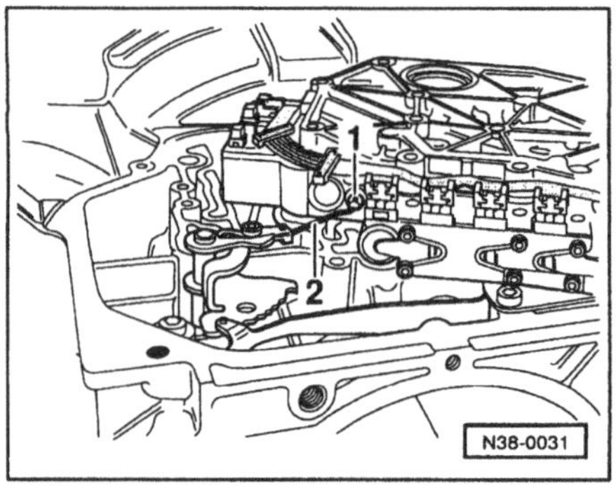

Fig. 7 Operating rod, unhooking from manual selector valve

- Turn manual selector valve -1- so that shoulder is to operating rod.
- Guide operating rod -2- with manual selector valve into valve body.

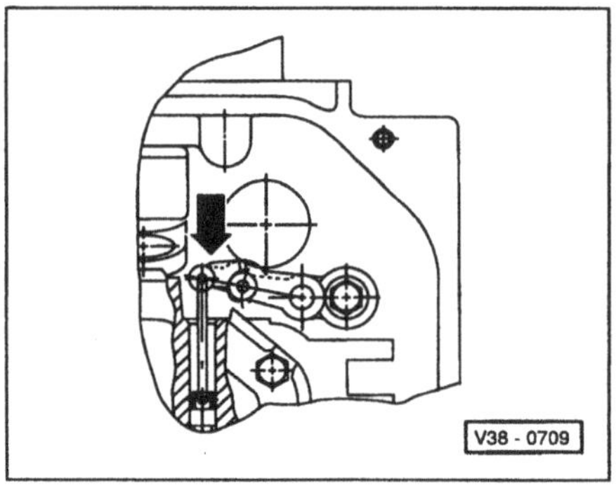

Fig. 8 Operating rod for manual selector valve, adjusting

- Move selector shaft to position "P".
- Push operating rod with manual selector valve into valve body up to stop (direction of arrow) and tighten bolt to 4 Nm (35 in. lb).

Note:

- *Manual selector valve must always be resting against the stop. Counter-hold in direction of arrow when tightening bolt.*
- *Always replace bolt and manual selector valve retaining clip.*

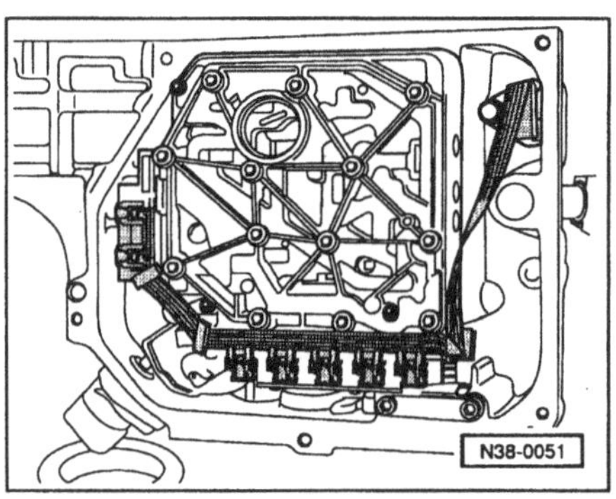

Fig. 9 Valve body, installing

- Route conductor strip as illustrated.

CAUTION!
Do not kink or twist conductor strip.

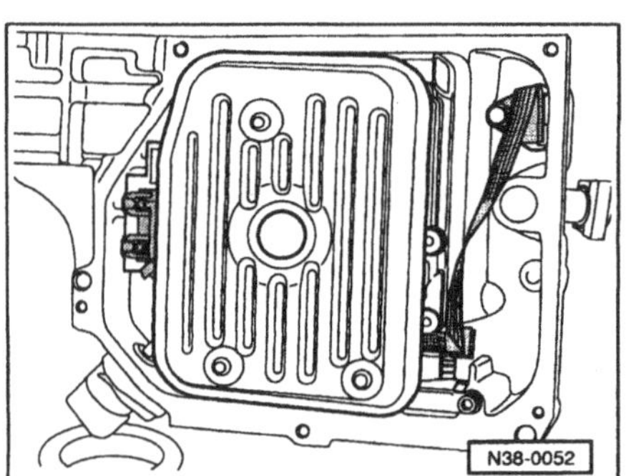

Fig. 10 ATF screen, installing

- Before installing the ATF screen, insert seal in valve body.
- Firmly press ATF screen onto valve body.

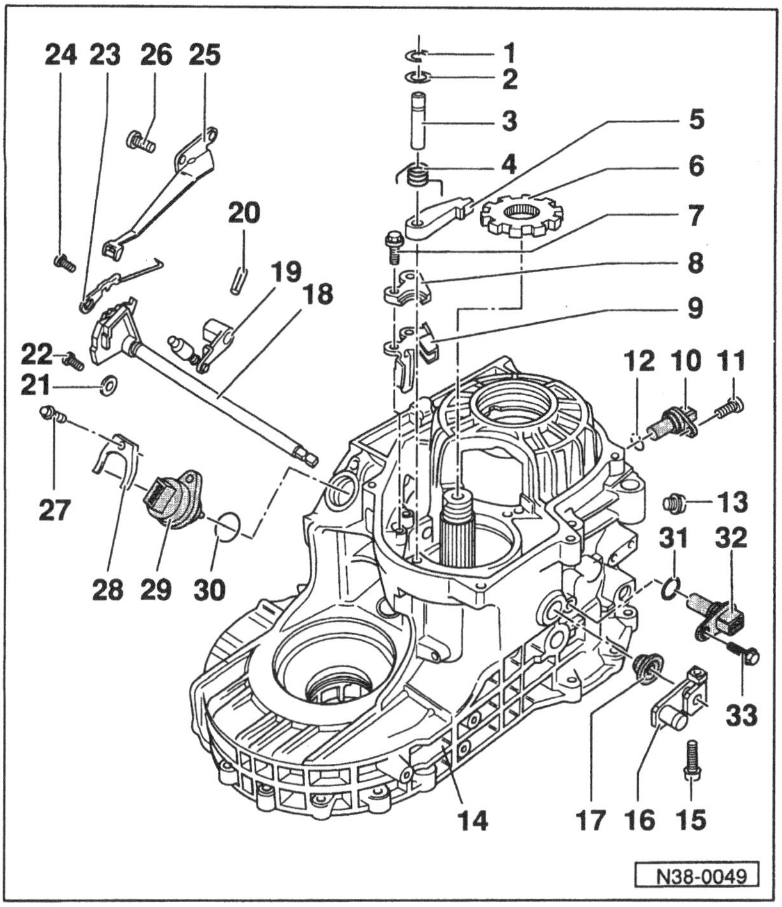

Parking lock, disassembling and assembling

Note:

- *Multi-Function Transmission (TR) Switch -F125- and Vehicle Speed Sensor -G68- can be tested in the vehicle by On Board Diagnostic (OBD)*
- *The output gear must be removed to disassemble and assemble the parking lock*
  ⇒ Removing and installing drive pinion, Page J 39-27.

1 - Circlip
- Always replace

2 - Washer

3 - Detent lever shaft
- Is secured by circlip and bearing cap of drive pinion
- According to version with groove for circlip

4 - Return spring
- Installing ⇒ Fig. 4

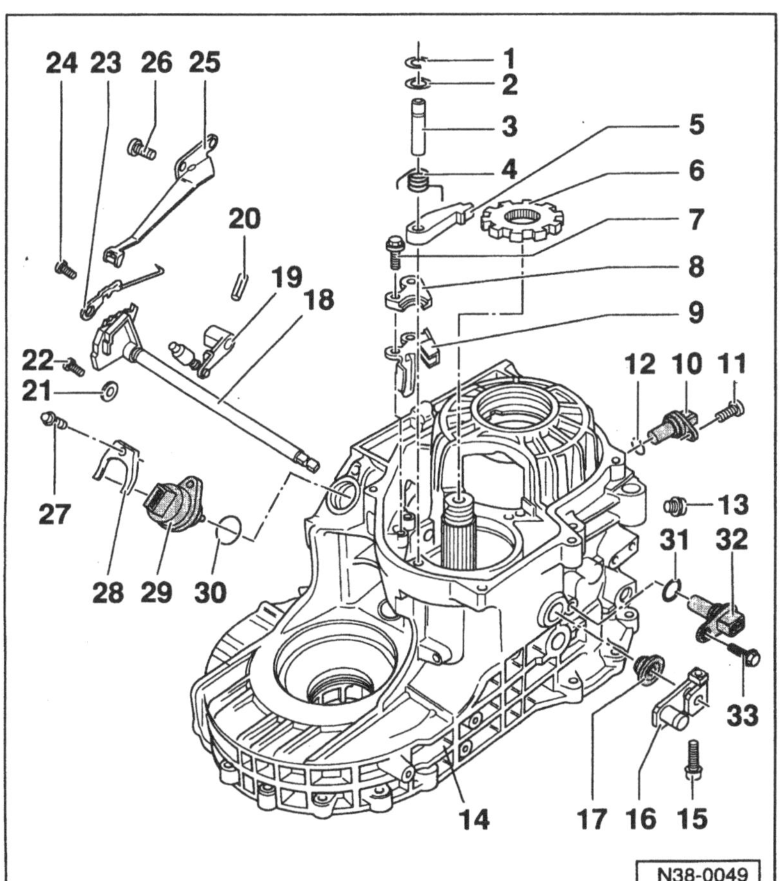

5 - Detent lever
- Installing with return spring ⇒ Fig. 4

6 - Parking lock gear
- Rounded side faces teeth of drive pinion

7 - Bolt, 14 Nm (10 ft lb)

8 - Supporting plate
- Installing ⇒ Fig. 3

9 - Guide plate
- Install before supporting plate ⇒ Fig. 3

10 - Vehicle Speed Sensor -G68-
- Can be tested in the vehicle by On Board Diagnostic (OBD) ⇒ Page J 01-25

11 - Bolt, 10 Nm (7 ft lb)

12 - O-ring
- Always replace

13 - Bolt, 15 Nm (11 ft lb)
- With sealing ring
- Always replace

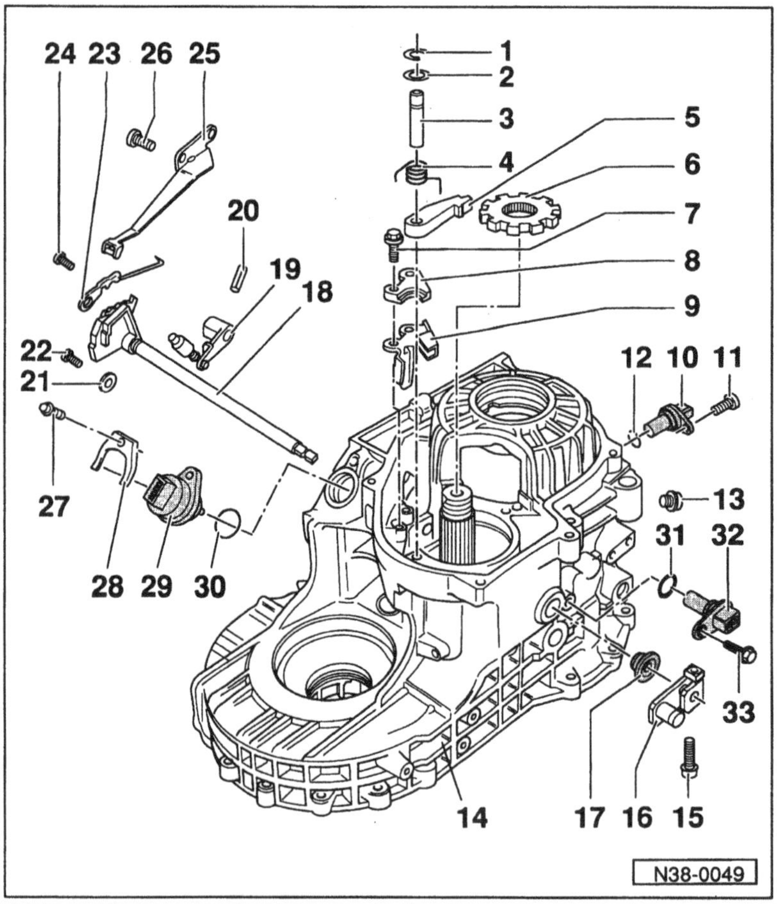

14 - Transmission housing

15 - Bolt,10 Nm (7 ft lb)

16 - Lever
- ♦ For selector shaft

17 - O-ring
- ♦ Pry out with screwdriver
- ♦ Drive in flush with VW 423

18 - Selector shaft with selector segment and detent segment
- ♦ Install together with engaging lever
- ♦ Removing and installing ⇒ Fig. 2

19 - Engaging lever
- ♦ Install together with selector shaft and detent segment

20 - Spring pin
- ♦ Drive in after installing engaging lever ⇒ Fig. 1
- ♦ Drive out with drift

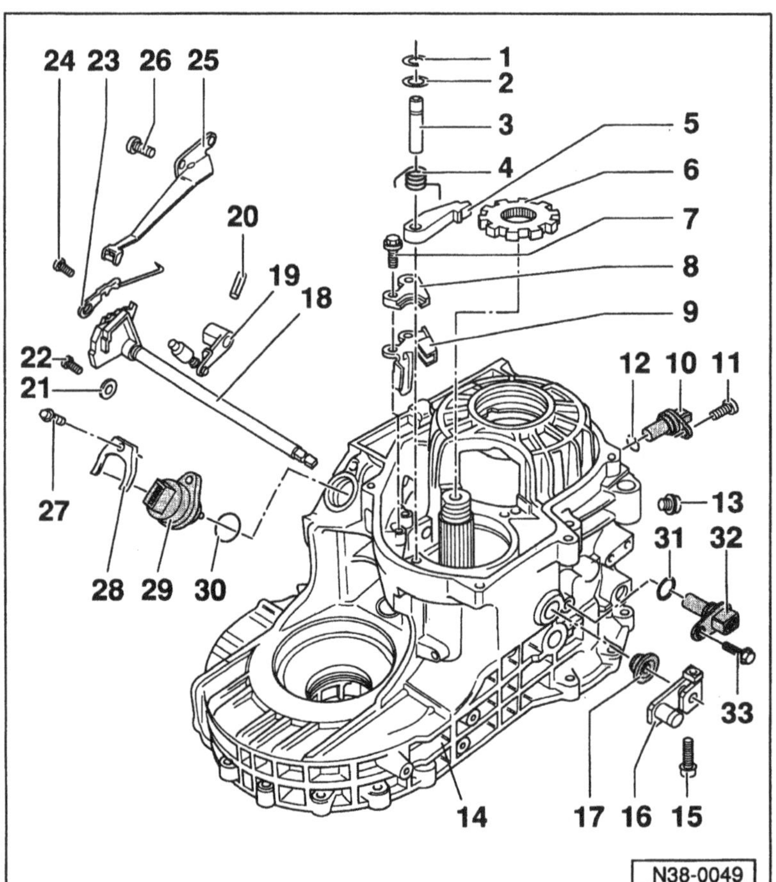

21 - Locking washer
- ♦ Removing and installing ⇒ Fig. 2

22 - Bolt,10 Nm (7 ft lb)

23 - Manual selector valve control
- ♦ Adjusting ⇒ Page J 38-20, Removing and installing valve body

24 - Bolt, 10 Nm (7 ft lb)

25 - Spring for selector segment

26 - Bolt, 10 Nm (7 ft lb)

27 - Bolt, 10 Nm (7 ft lb)

28 - Bracket
- ♦ For Muiti-Function TR Switch -F125-

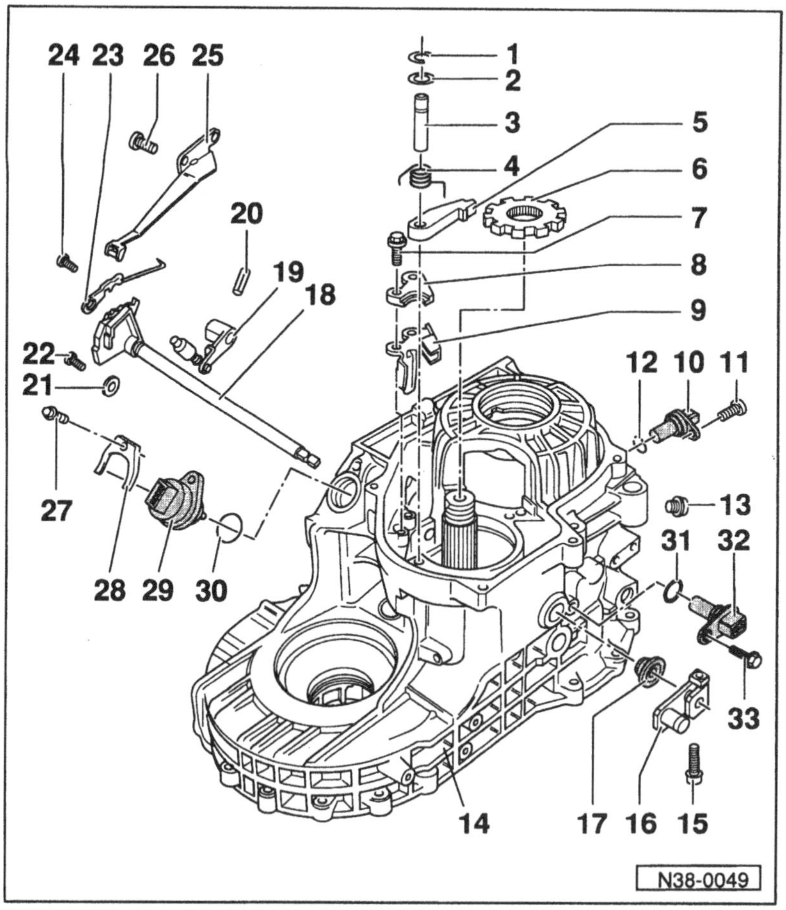

29 - Multi-Function Transmission (TR) Switch -F125-

- ♦ To remove:
- – Remove insulation tray
- – Pull connector off and place oil tray under transmission
- – When installing ensure connector is correctly sealed
- – After installing top up ATF ⇒ Page J 37-44
- ♦ Install so that 1 pin each is resting on a slope of detent segment
- ♦ Can be tested in the vehicle by On Board Diagnostic (OBD) ⇒ Page J 01-25

30 - O-ring

- ♦ Always replace

31 - O-ring

- ♦ Always replace

32 - Transmission Vehicle Speed Sensor -G38-

- ♦ Can be tested in the vehicle by On Board Diagnostic (OBD) ⇒ Page J 01-25

33 - Bolt, 10 Nm (7 ft lb)

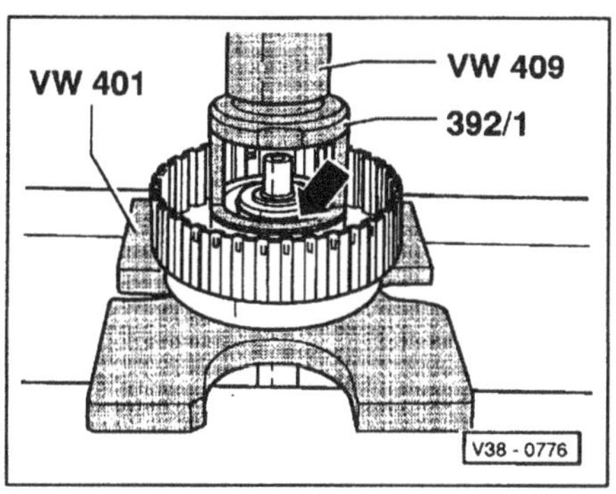

◄ **Fig. 1 Spring pin -arrow-, driving out and in**

Driving out

– Drive engaging lever spring pin out only far enough to remove selector shaft.

> ***CAUTION!***
>
> ***Engaging lever spring pin may drop into planetary gearbox.***

Driving in

– Drive in spring pin, flush, after installing selector shaft and engaging lever.

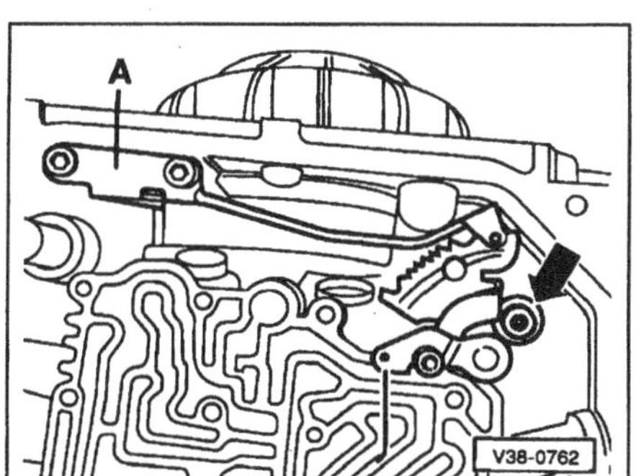

◄ **Fig. 2 Selector shaft with detent and selector segments, removing and installing**

– Remove bolt -arrow- and take out locking washer.

Notes:

- ♦ *The locking washer is placed into the groove of the selector shaft and secures the shaft from dropping out.*
- ♦ *The selector shaft can be removed or installed after taking out the spring -A- for the selector segment.*

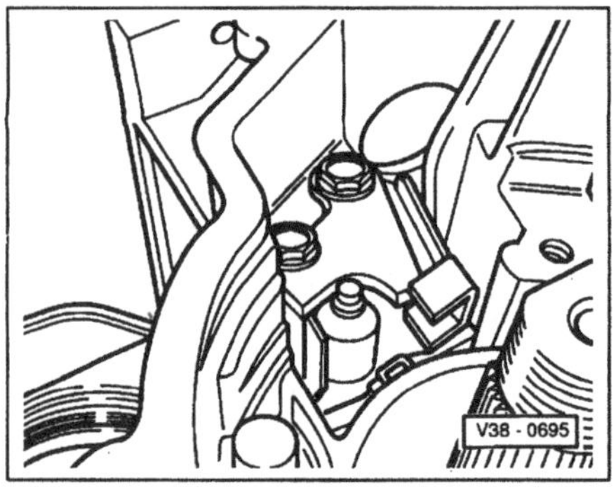

Fig. 3 **Guide plate and supporting plate, installing**

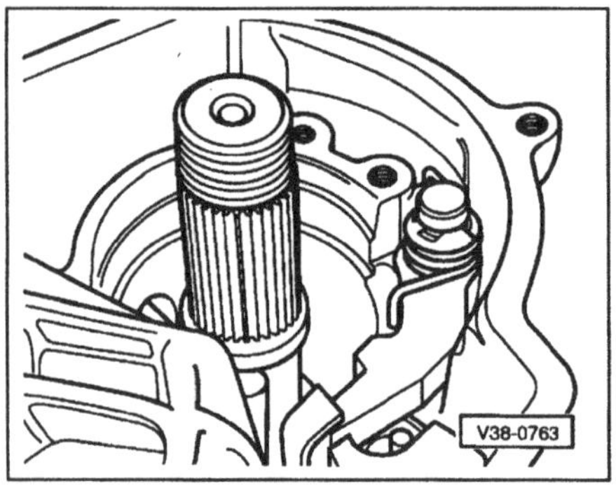

◄ **Fig. 4** **Pawl with return spring, installing**

Final drive oil level, checking

The oil level is checked with the transmission installed.

Checking:

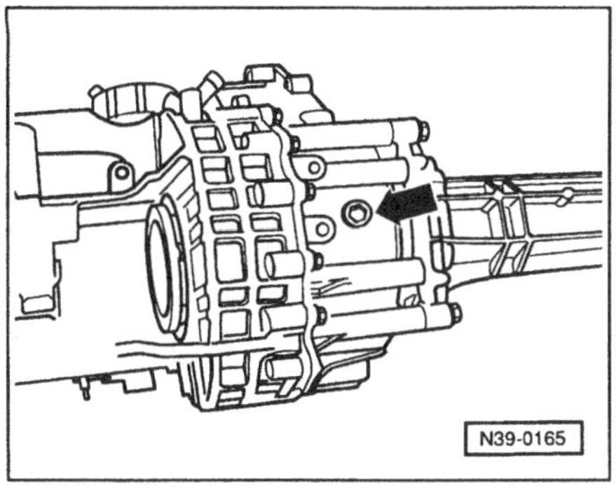

◄ – Remove plug -arrow- to check gear oil level and top gear oil up to lower edge of hole if necessary.

Specification ⇒ Page J 00-14

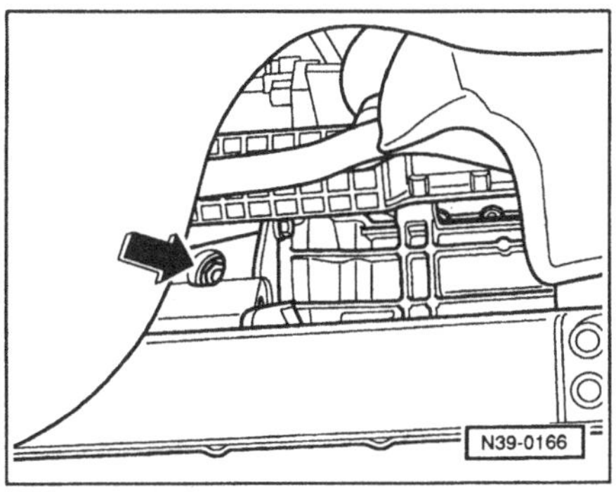

◄ – Remove and install plug -arrow- through hole in subframe.

- Tightening torque: 20 Nm (15 ft lb)

Oil seal for flange shaft or drive flange, removing and installing

Note:

When replacing right side oil seal on vehicles produced from 01.96, ⇒ page J 39-8.

(Transmission installed)

– Install support device 3250 before lifting vehicle.
⇒ Front Suspension; Repair group 40

– Remove sound deadener pan.

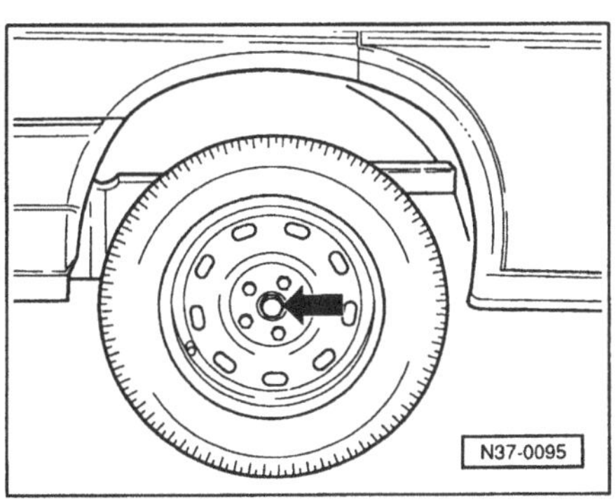

◄ – Remove hex bolt -arrow- for drive shaft/wheel hub.

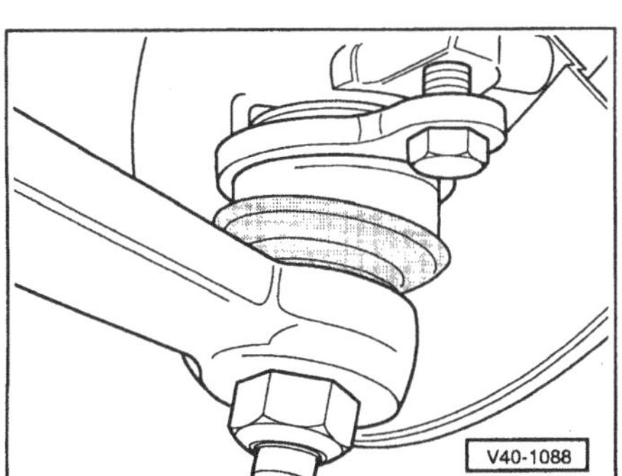

◄ – Disconnect wheel bearing housing/lower ball joint bolted connection.

– Remove bolt for left-hand shock absorber/stabilizer bar coupling rod.

– Press drive shaft out of wheel bearing housing.

– Compress shock absorber.

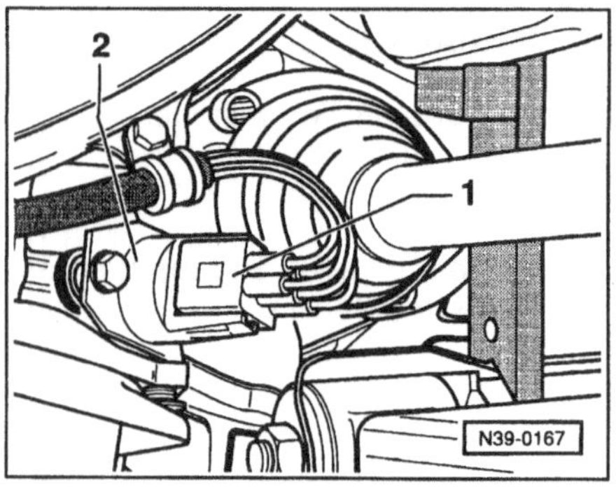

- Unbolt drive shaft from flange shaft/drive flange.

When replacing left-hand oil seal:

(Right-hand oil seal, replacing ⇒ Page J 39-6)

◄ – Disconnect connector -1- from multi-function switch -2-.

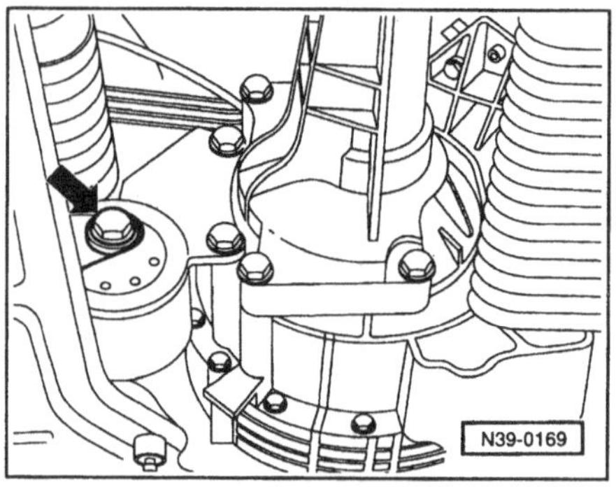

◄ – Remove bolt -arrow-.

- Remove left-hand drive shaft by pressing engine/transmission slightly forward.

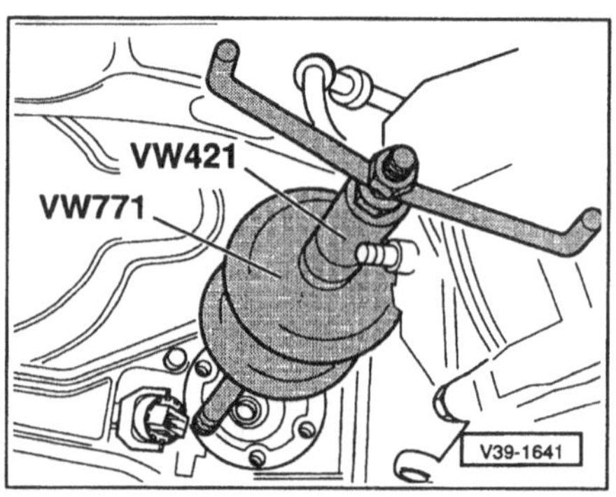

◄ – Remove flange shaft using VW 771 and VW 421.

Note:
Always replace flange shaft circlip.

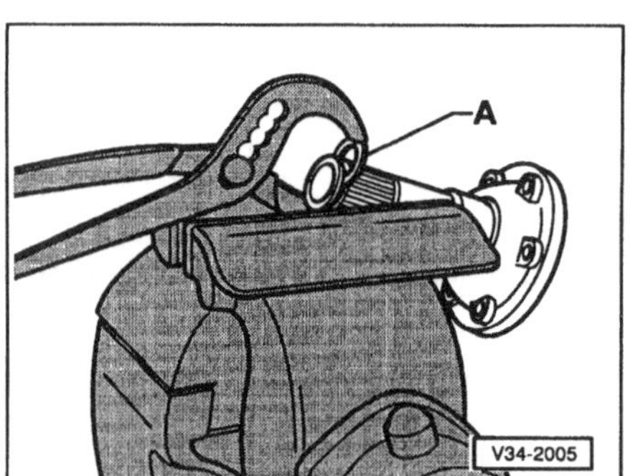

◄ – Clamp flange shaft in vise using protective vise clamps.

- Press circlip out of flange shaft groove using new circlip -A-.

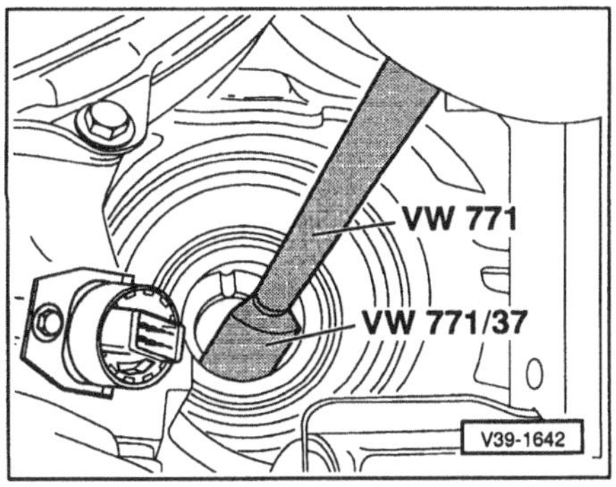

- Pull flange shaft oil seal out using VW 771 and VW 771/37.

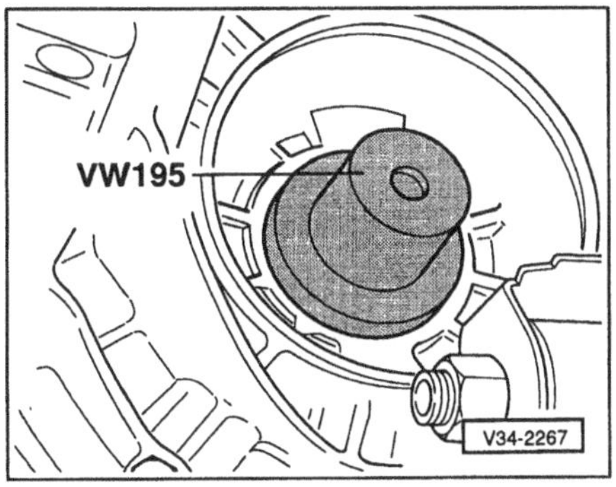

- Drive in oil seal up to stop with VW 195.
- Fill space between seal and dust lip with multi-purpose grease.
- Install flange shaft.
- Install drive shaft.
- With transmission installed, check final drive oil level ⇒ Page J 39-1.

When replacing right-hand oil seal:

Note:

When replacing right side oil seal on vehicles produced from 01.96, *⇒ page J 39-8.*

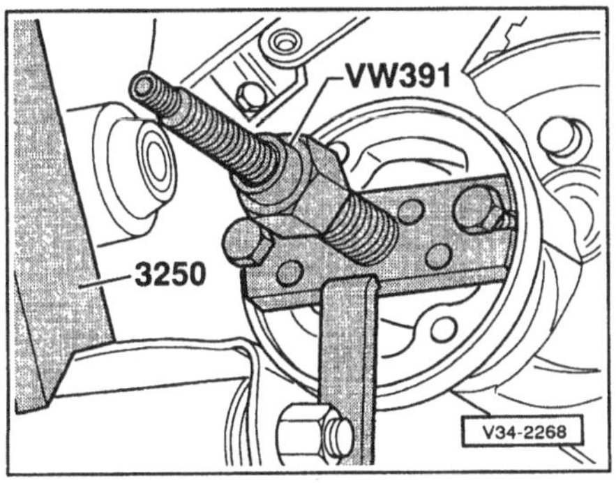

- Remove drive shaft.
- Pierce sealing cap in drive flange with a screwdriver and pry out.
- Remove drive flange circlip.
- Remove drive flange with VW 391.
- Remove drive flange oil seal with VW 681.

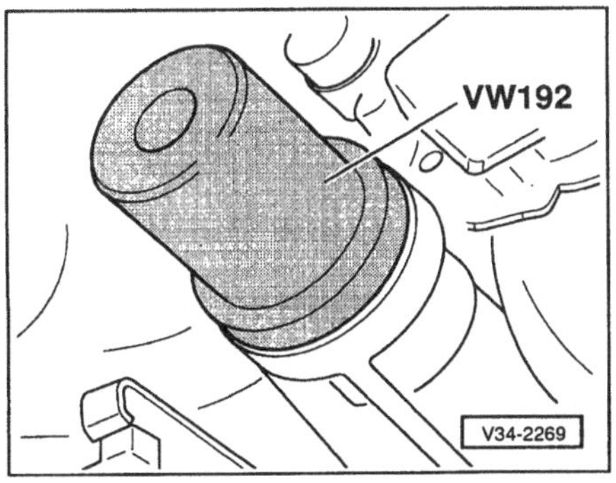

- Drive in oil seal up to stop with VW 192.
- Fill space between seal and dust lip with multi-purpose grease.

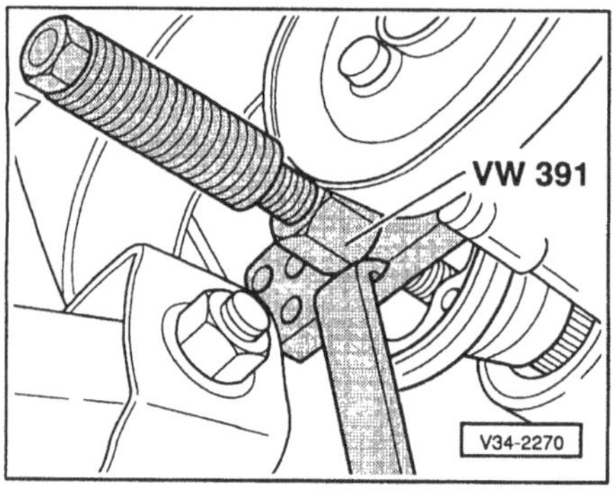

◄ – Pull off drive flange with VW 391.

– Install new circlip and new sealing cap.

– Install drive shaft.

– With transmission installed, check final drive oil level ⇒ Page J 39-1.

Tightening torque:

Wheel bearing housing to lower ball joint (always replace bolt)	55 Nm (41 Nm) +90° (1/4 turn)
Shock absorber to axle link	160 Nm (118 ft lb)
Hex bolt to drive shaft (always replace bolt)	150 Nm (111 ft lb) + 90° (1/4 turn)
Bonded rubber mounting to subframe	200 Nm (148 ft lb)
Wheel bolts to wheel hub	160 Nm (118 ft lb)
Drive shaft to drive flange	70 Nm (52 ft lb)

Oil seal for right side flange shaft or drive flange, removing and installing (vehicles produced from 01.96)

– Remove right drive shaft far enough so that drive flange is accessible and can be removed.
⇒ Suspension, Wheels, Brakes, Steering Repair Group 40

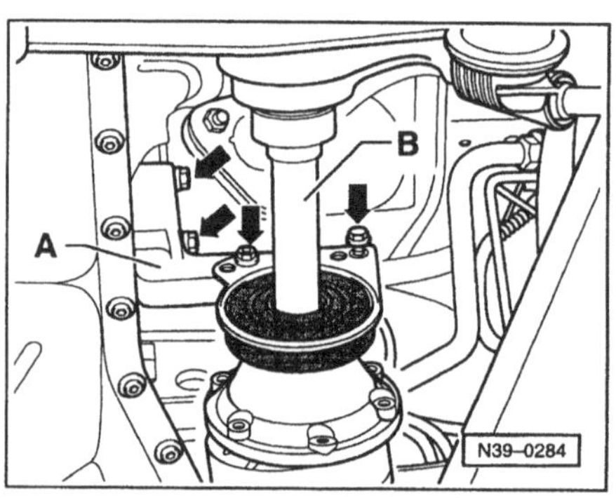

◄ – Remove bracket -A- of right drive flange bearing -arrows-.

– Move steering wheel fully to right and remove drive flange -B-.

– Remove O-ring from output shaft.

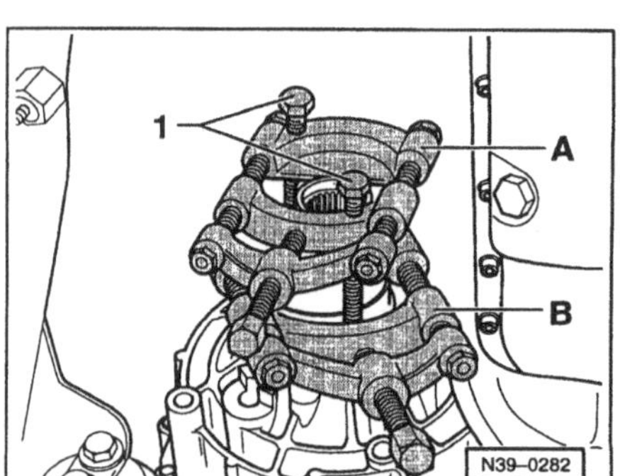

◄ – Remove output shaft.

A = Separating device, e.g. Kukko 15–0

B = Separating device, e.g. Kukko 17–1

1 = M10 x 80 bolts (qty. 2)

– Turn bolts -1- uniformly against separating device -B-.

– Pull seal out with VW 681.

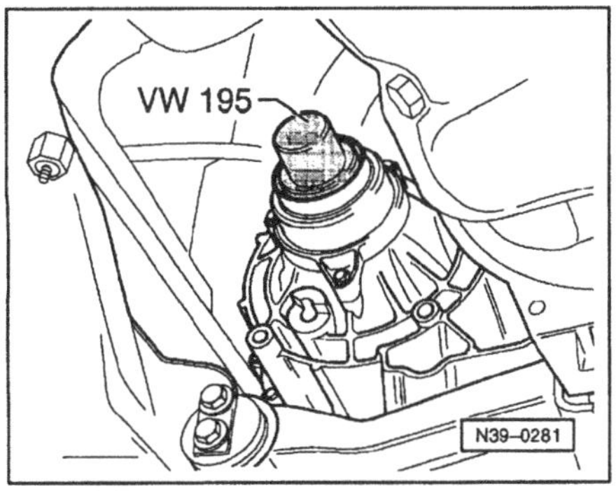

- Drive seal in up to stop with VW 195.
- Fill space between seal lips with multi-purpose grease.

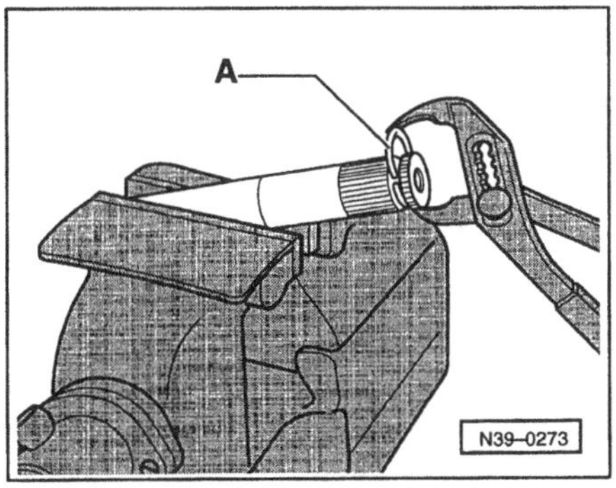

- Clamp output shaft in vise with protective jaw covers.
- Use new circlip -A- to push circlip out of shaft groove.
- Use plastic hammer to drive in shaft with new circlip and O-ring.
- Install drive flange, bracket and drive shaft.
- Check fluid in final drive with transmission installed in vehicle, ⇒ page J 39-1.
- Install sound deadener pan.

Tightening torque:

Drive shaft to drive flange	80 Nm (59 ft lb)
Bracket to engine	50 Nm (37 ft lb)
Bracket to flange shaft bearing	27 Nm (20 ft lb)

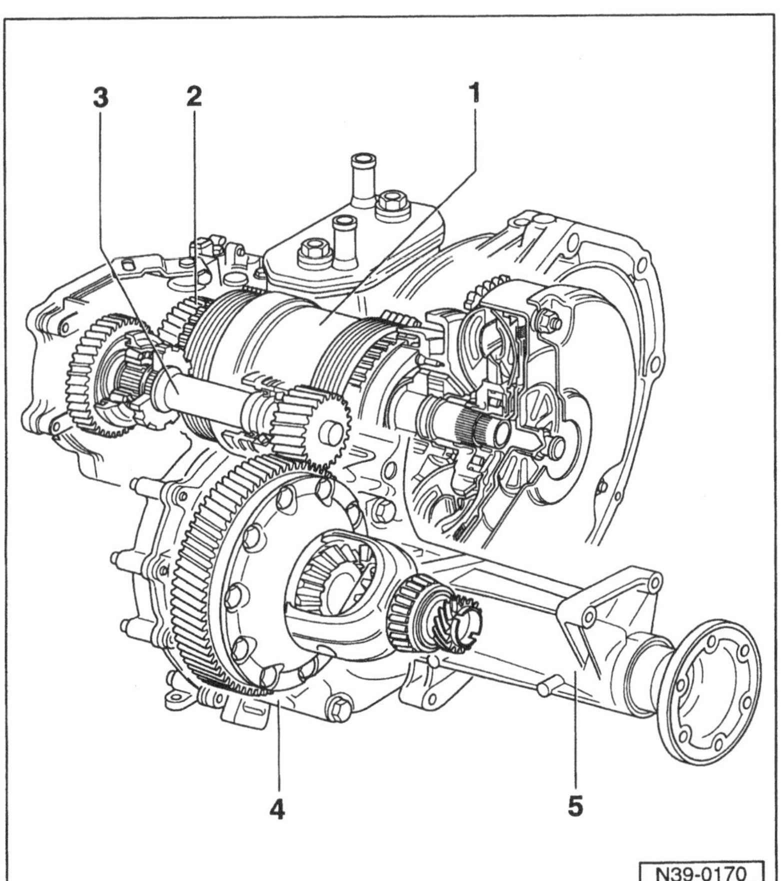

Final drive, disassembling and assembling

1 - Planetary gear
- Disassembling and assembling ⇒ Page J 37-52

2 - Input gear
- Removing and installing ⇒ J 39-22
- Adjusting ⇒ Page J 39-56
- Adjusting final drive ⇒ PageJ 39-47

Only repair the input gear after the component parts of the planetary gear have been removed.

3 - Drive pinion
- Removing ⇒ J 39-27
- Adjusting ⇒ Page J 39-51
- (overview) ⇒ Page J 39-47

4 - Differential
- ⇒ J 39-17
- Disassembling and assembling ⇒ Page J 39-37
- Adjusting ⇒ Page J 39-60
- Adjusting final drive (overview) ⇒ Page J 39-47

5 - Bearing housing
- Removing and installing ⇒ J 39-11

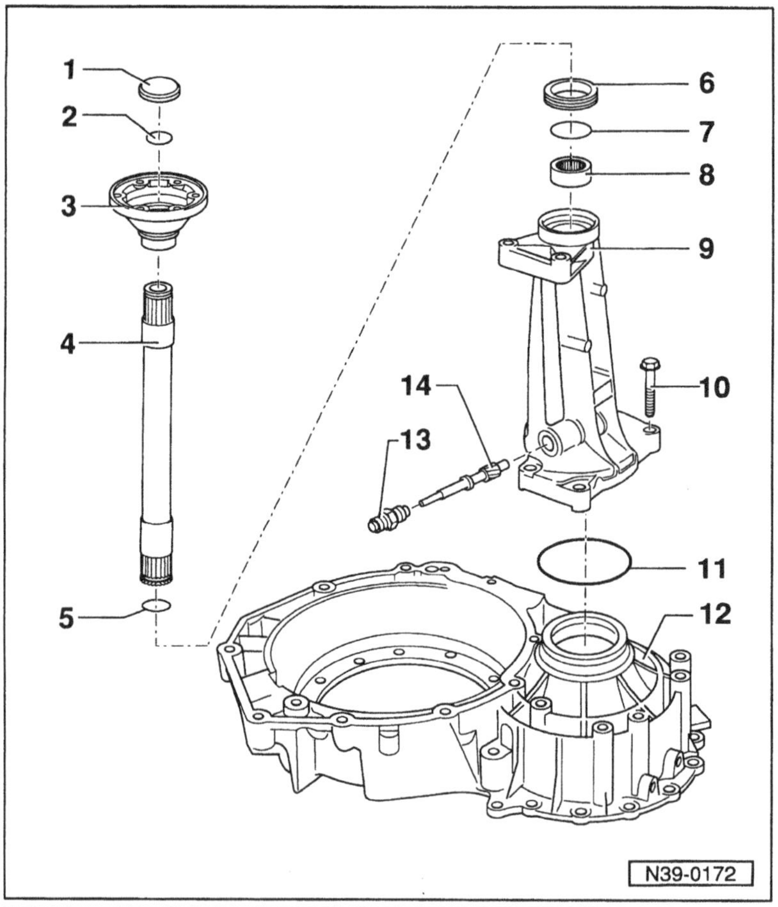

Bearing housing, removing and installing

1 - **Sealing cap**
- ♦ Always replace
- ♦ Pry out with screwdriver

2 - **Circlip**

3 - **Drive flange**
- ♦ Pressing off ⇒ Fig. 2
- ♦ Pressing on ⇒ Fig. 3

4 - **Output shaft**
- ♦ Drive out with ⇒ Fig. 1

5 - **Circlip**
- ♦ Always replace
- ♦ Removing and installing ⇒ Fig. 4

6 - **Oil seal for drive flange**
- ♦ Before installing, pack space between sealing lips with multi-purpose grease
- ♦ Can be replaced with transmission installed ⇒ Page J 39-2
- ♦ Remove with VW 681
- ♦ Driving in ⇒ Fig. 7

7 - **Circlip**
- ♦ For needle roller bearing
- ♦ Always replace

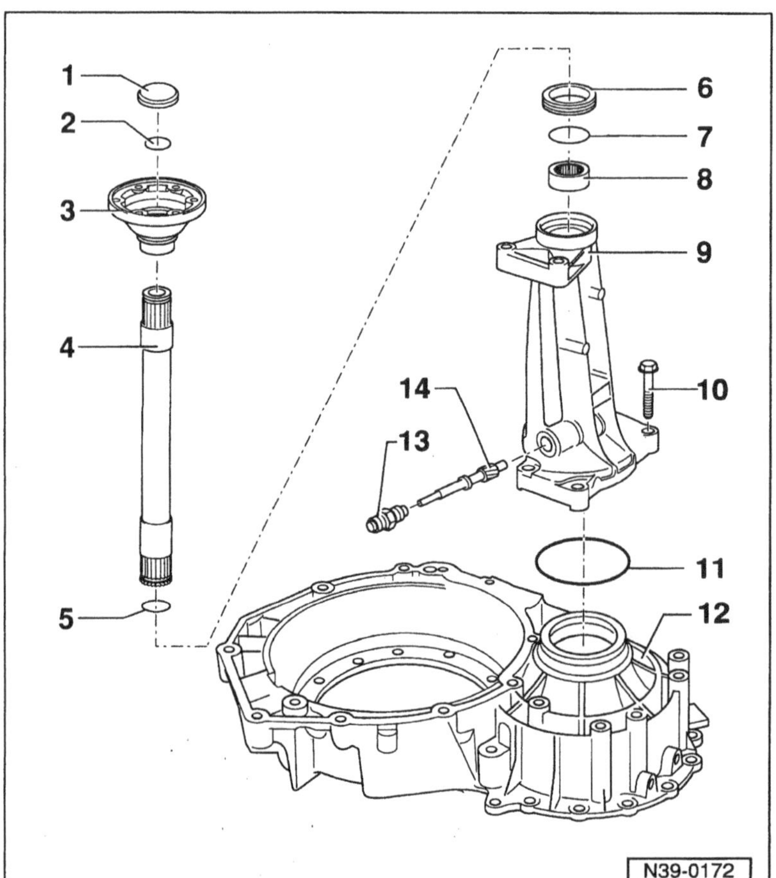

8 - **Needle roller bearing**
- ♦ Pulling out ⇒ Fig. 5
- ♦ Pressing in ⇒ Fig. 6

9 - **Bearing housing**

10 - **Bolt**
- ♦ Tighten to 40 Nm (30 ft lb) and then turn 90° (1/4 turn) further

11 - **O-ring**
- ♦ Always replace
- ♦ Insert with grease

12 - **Torque converter housing**
- ♦ Removing ⇒ Page J 39-17 Removing and installing differential

13 - **Guide bushing**
- ♦ For speedometer drive

14 - **Speedometer drive**
- ♦ Remove before removing bearing housing

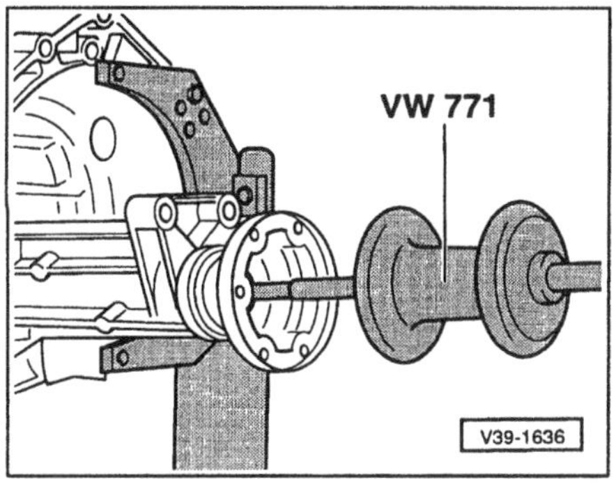

◄ **Fig. 1 Output shaft with drive flange, driving out**

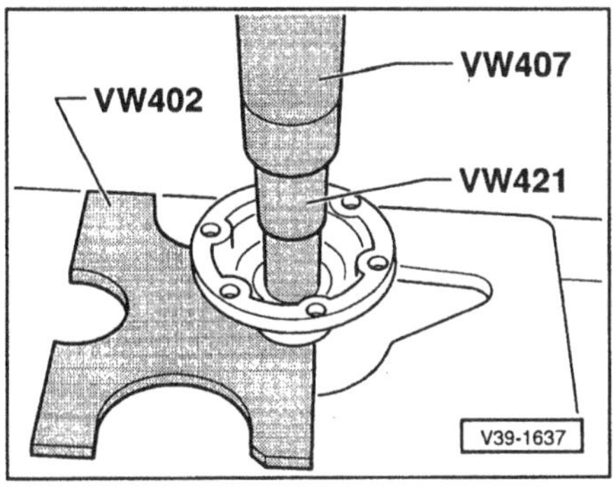

◄ **Fig. 2 Drive flange, pressing off**

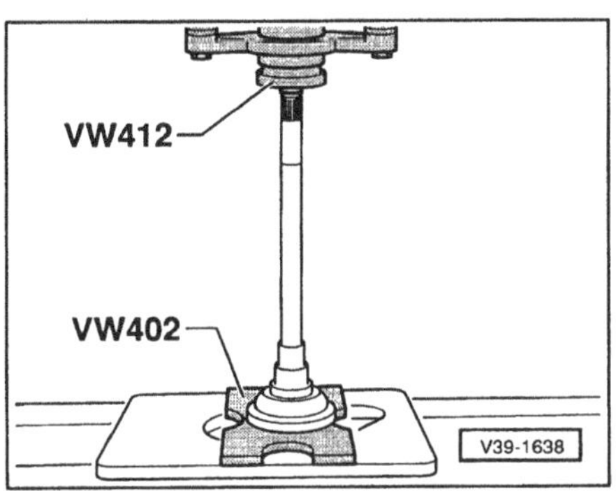

◄ **Fig. 3 Drive flange, pressing on**

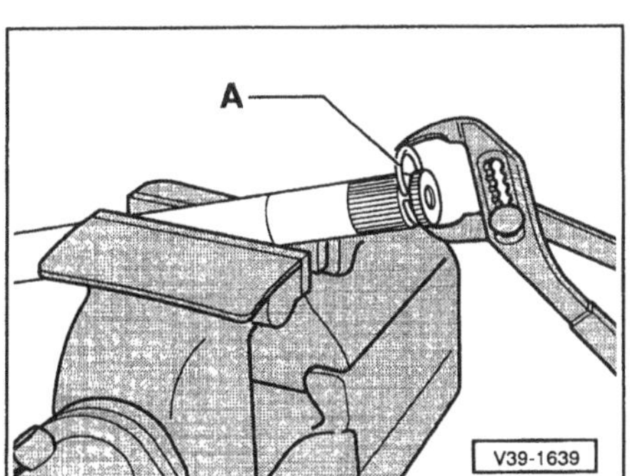

◄ **Fig. 4 Circlip, removing and installing**

- Clamp output shaft in vise using protective vise clamps.
- Press circlip out of output shaft groove using new circlip -A-

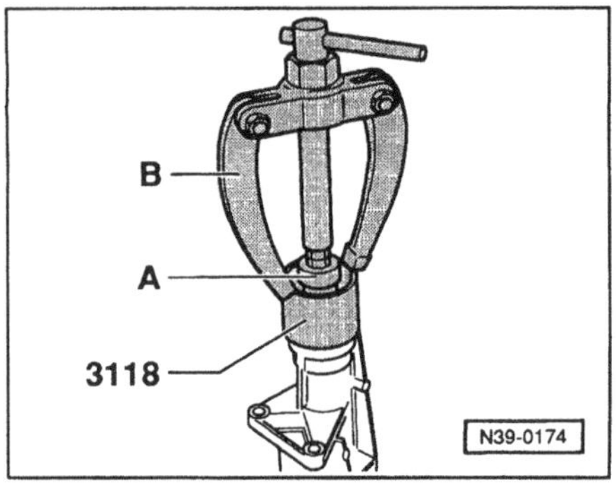

◄ **Fig. 5 Needle roller bearing, pulling out**

A – Internal puller, e.g. Kukko 21/5

B – Counter support, e.g. Kukko 22–2

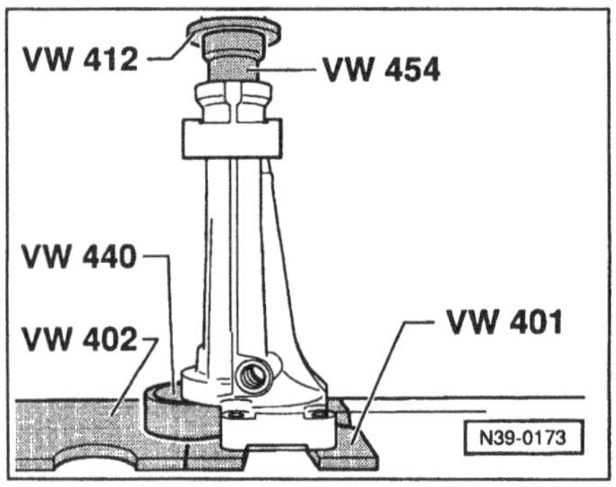

◄ **Fig. 6 Needle roller bearing, pressing in**

– Press needle roller bearing in onto stop.

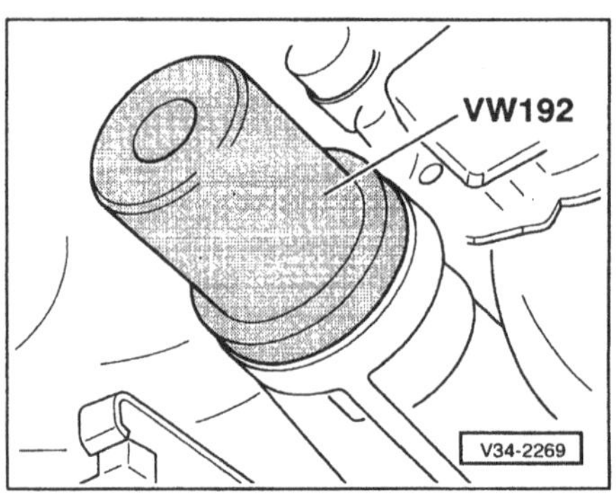

◄ **Fig. 7 Seal, driving in onto stop using VW 192**

– Fill space between seal and dust lip with multi-purpose grease.

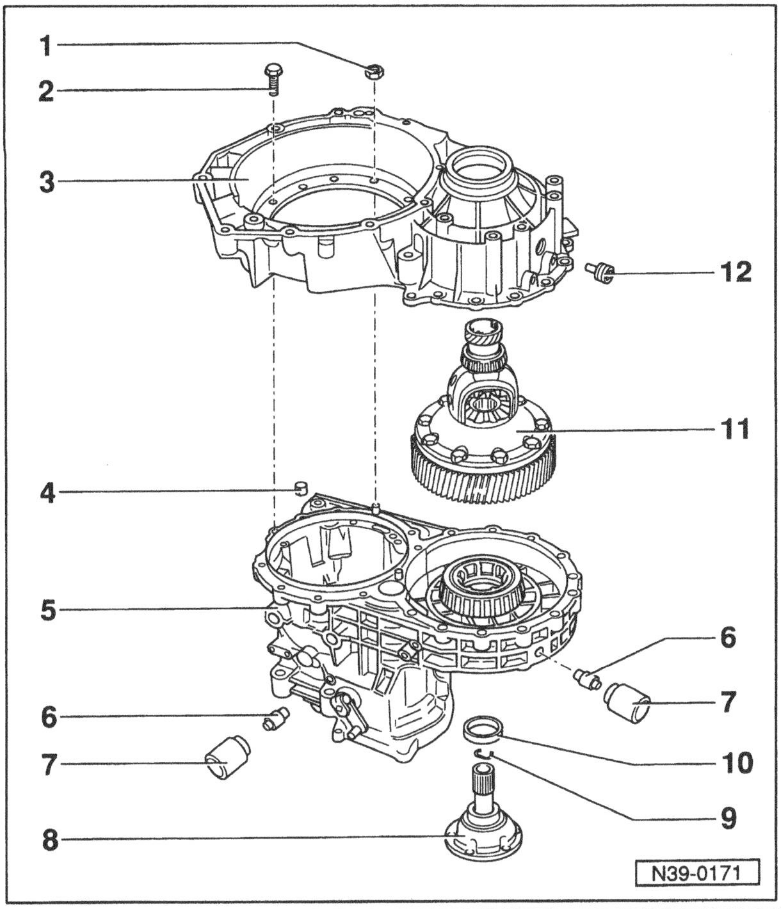

Differential, removing and installing

Note:

- *The drive pinion does not need to be removed for removing and installing the differential.*

1 - Nut, 25 Nm (18 ft lb)
- Qty. 2

2 - Screw, 25 Nm (18 ft lb)

3 - Torque converter
- Is allocated to transmission housing according to engine version
- As a replacement part, it is allocated according to transmission code letters

⇒ Replacement parts catalog
- On vehicles with diesel engine the Engine Speed (RPM) Sensor -G28- is installed in the top of the torque converter housing ⇒ Fig. 3

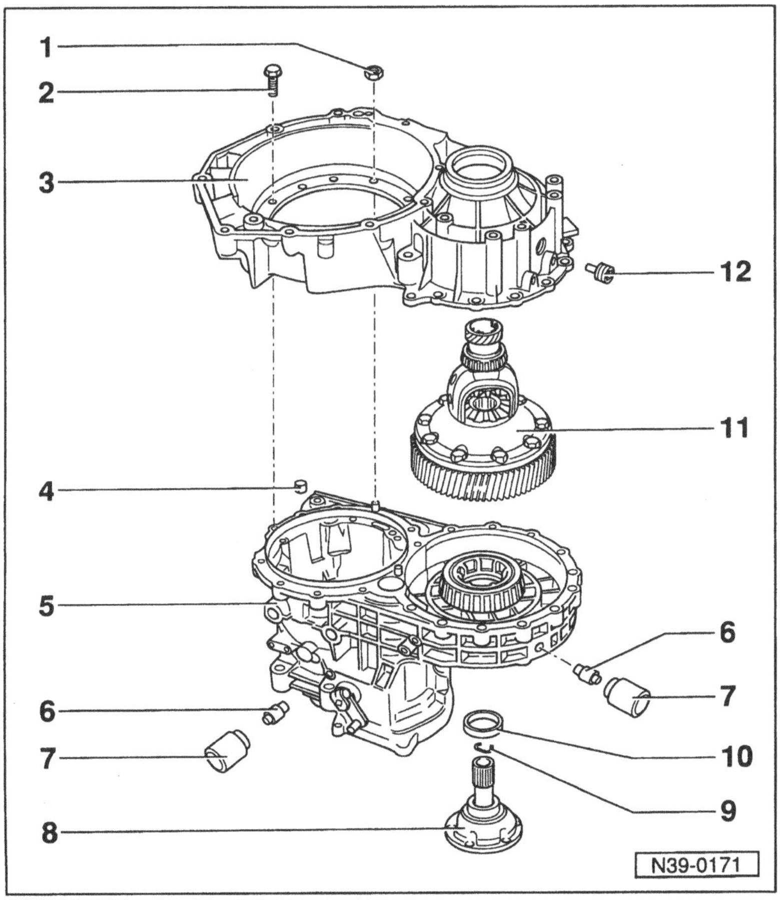

4 - Sleeve
- To centralize the transmission and torque converter housing

5 - Transmission housing
- Before assembling transmission and torque converter housings together, coat sealing surfaces with AMV 188 200

6 - Breather tube
- Remove using pliers
- Drive onto stop using drift VW 460/2

7 - Breather cap

8 - Flange shaft
- Remove using VW 771
- Drive in using plastic hammer

9 - Circlip
- Always replace
- Removing and installing ⇒ Fig. 1

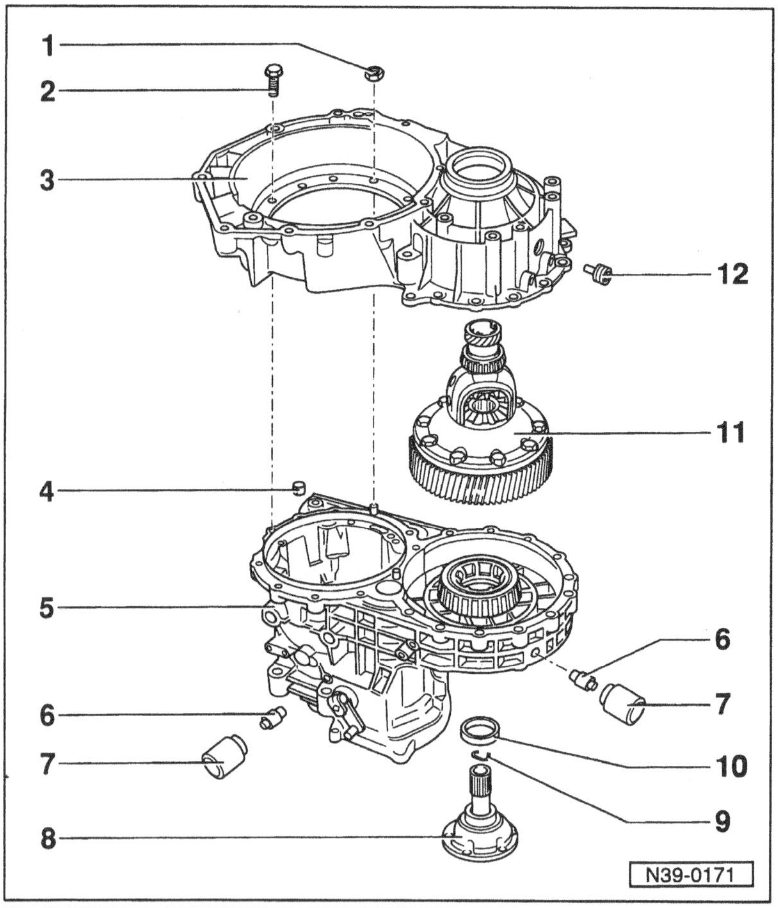

10 - Oil seal for flange shaft

- Before installing, pack space between sealing lips with multi-purpose grease
- Can be replaced with transmission installed ⇒ Page J 39-2
- Remove with VW 681
- Driving in ⇒ Fig. 2

11 - Plug, 20 Nm (15 ft lb)

- Checking gear oil with transmission installed ⇒ Page J 39-1

12 - Differential

- Disassembling and assembling ⇒ Page J 39-37

13 - Plug, 20 Nm (15 ft lb)

- Checking gear oil with transmission installed ⇒ Page J 39-1

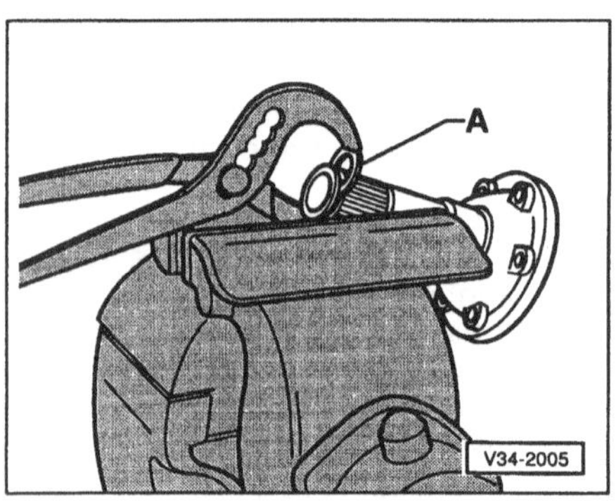

◄ **Fig. 1 Removing and installing circlip**

- Clamp flange shaft in vice using protective vice clamps.
- Press circlip out of flange shaft groove using new circlip -A-.

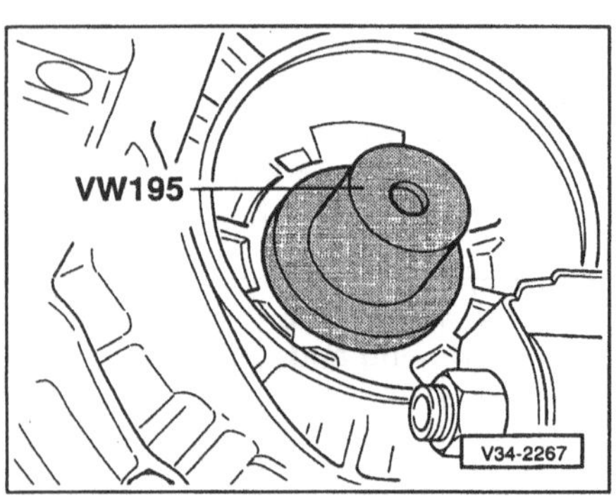

◄ **Fig. 2 Drive seal in onto stop using VW 195**

- Fill space between seal and dust lip with multi-purpose grease.

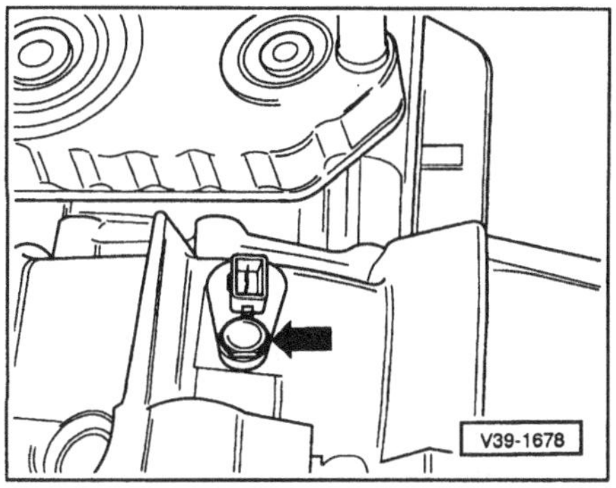

◄ **Fig. 3 Engine speed sensor**

(only for diesel engines)

– Remove bolt -arrow- and remove Engine Speed Sensor -G28- from torque converter housing.

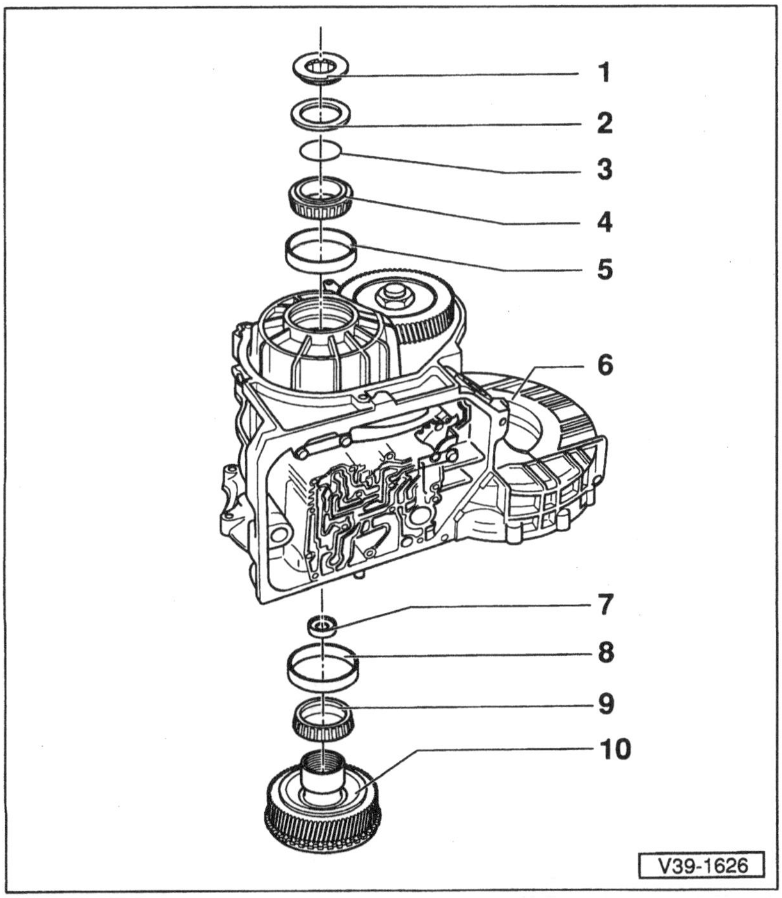

Input gear, removing and installing

Note:

Drive pinion and differential do not need to be removed.

1 - Socket-head fastener, 250 Nm (184 ft lb)
- ♦ Engage parking lock to remove and install
- ♦ Use 22 mm socket
- ♦ Before installing socket-head fastener install axial needle bearing (item 7)

2 - Dished washer
- ♦ Curved side faces socket-head fastener

3 - Shim
- ♦ Determining thickness ⇒ Adjusting input gear, Page J 39-56

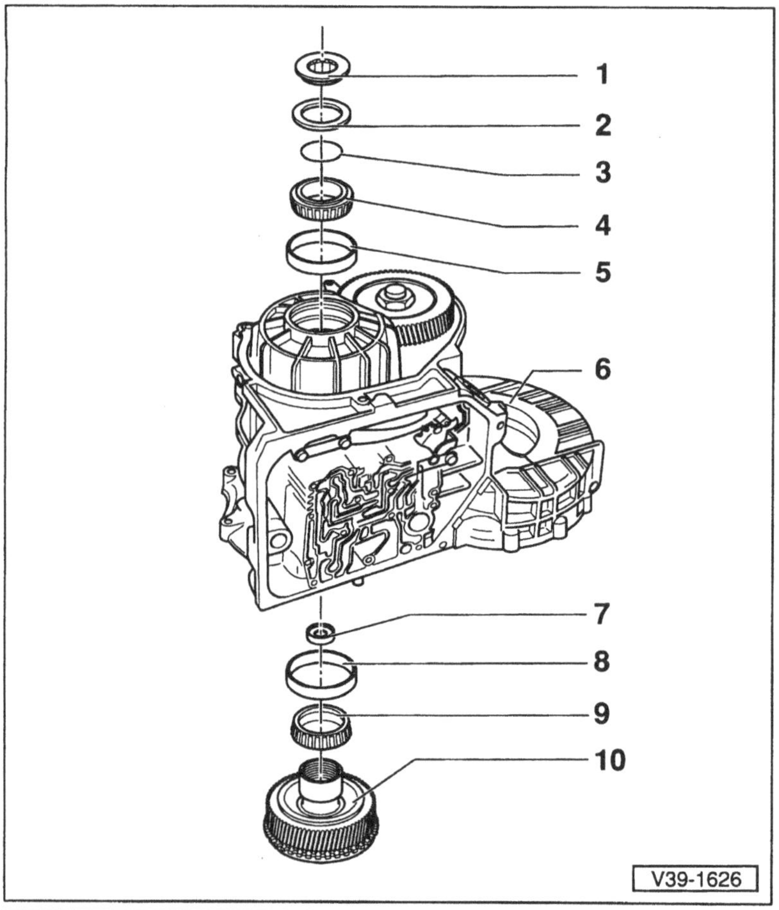

4 - Tapered roller bearing inner race
- After measuring thickness of shim, install on input gear with locking fluid AMV 185 101 A1

5 - Tapered roller bearing outer race
- Drive out with drift
- Drive in with 30–205
- Install with locking fluid AMV 185 101 A1

6 - Transmission housing

7 - Axial needle bearing
- Install with flat side facing input gear
- Insert into input gear before installing socket-head fastener (item 1)

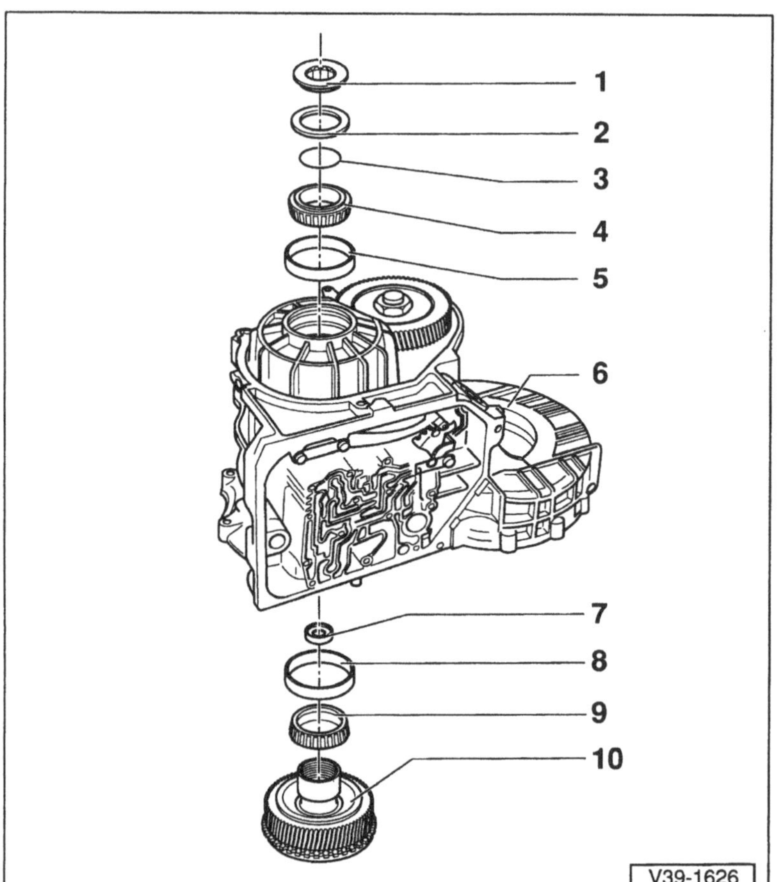

8 - Tapered roller bearing outer race
- Drive out with drift
- Drive in with 30–205 and appropriately long drift
- Install with locking fluid AMV 185 101 A1

9 - Tapered roller bearing inner race
- Pulling off ⇒ Fig. 1
- Pressing on ⇒ Fig. 2
- Install with locking fluid AMV 185 101 A1

10 - Input gear
- With impulse wheel for Vehicle Speed Sensor -G68-
- Number of teeth ⇒ Page J 00-6
- Adjusting ⇒ Page J 39-56
- If signs of damage exist, always replace input gear and output gear together
- Removing ⇒ Fig. 4

Note:

If input gear is replaced, adjust planet carrier ⇒ Page J 37-81.

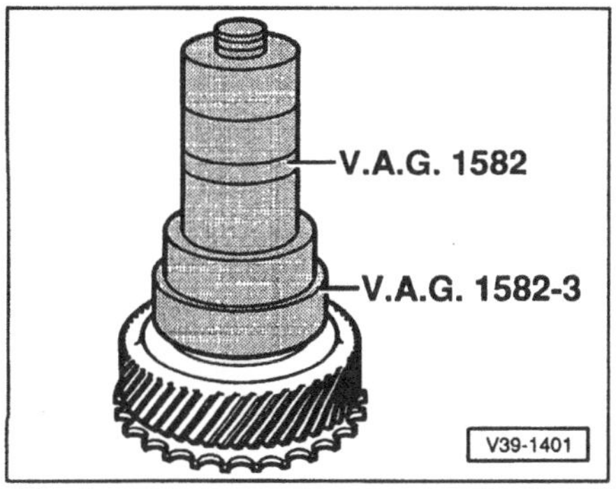

Fig. 1 Tapered roller bearing inner race, pulling off

– Place press piece on input gear.

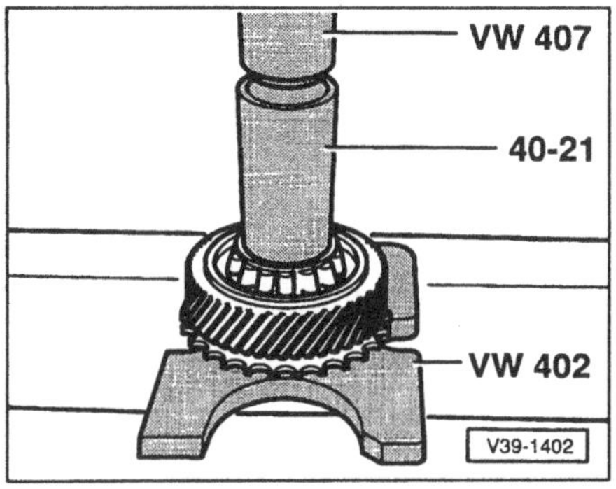

Fig. 2 Tapered roller bearing inner race, pressing on

Note:

A press piece must be placed below the hub of the input gear to avoid damaging it.

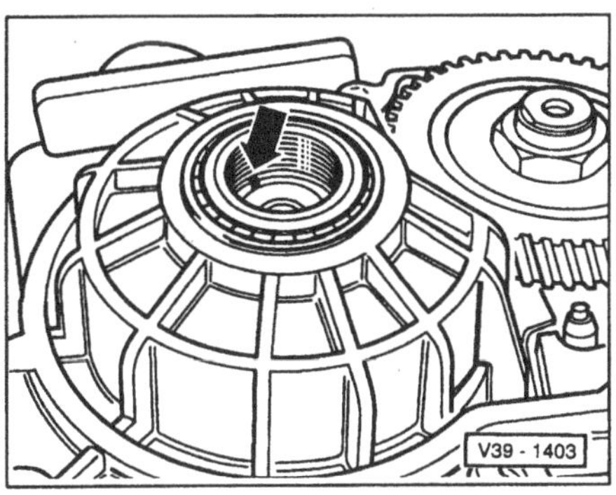

Fig. 3 Tapered roller bearing inner race, installing

– Install tapered roller bearing inner race with AMV 185101 A1.

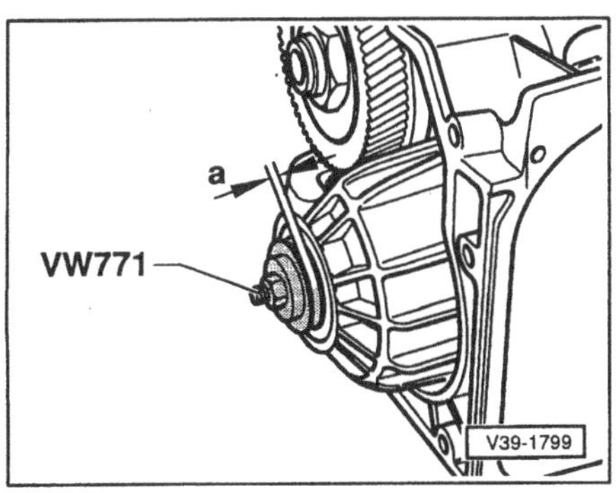

Fig. 4 Input gear, removing

– Insert socket-head fastener without dished washer, shim and axial needle bearing far enough for gap (size -a- = 3 mm) to exist between socket-head fastener and tapered roller bearing inner race.

– Drive out input gear with VW 771 up to stop of tapered roller bearing inner race.

– Remove VW 771 and take out input gear.

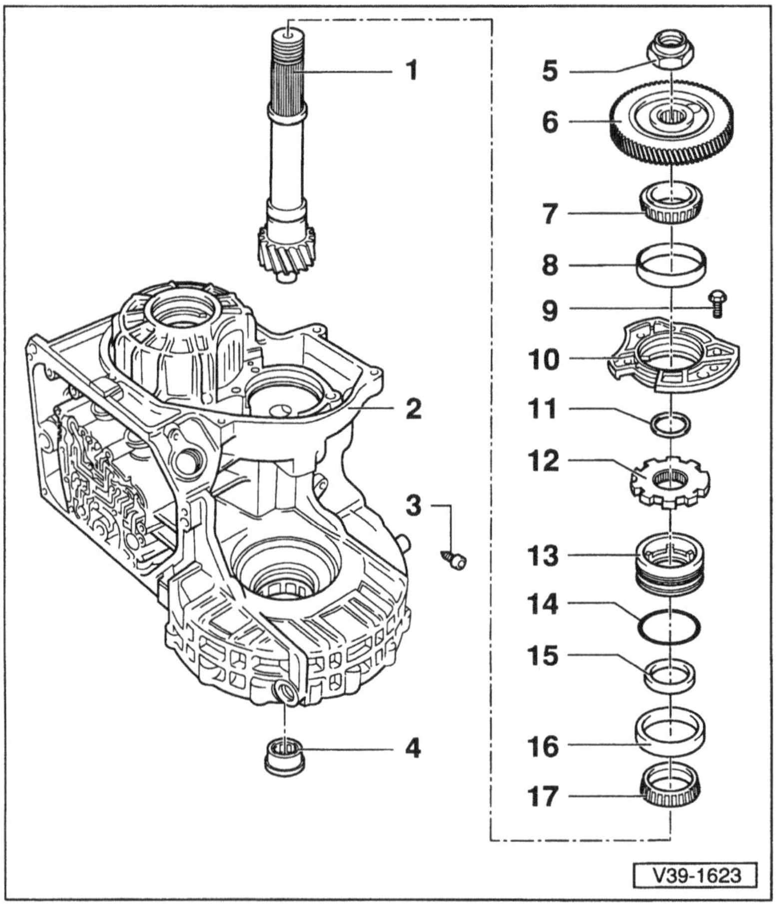

Drive pinion, removing and installing

To remove the drive pinion, the selector shaft must be removed after taking out the parking lock ⇒ Page 38–29.

1 - Drive pinion
- Number of teeth ⇒ from Page J 00-6
- Adjusting ⇒ Page J 39-51
- Removing ⇒ Fig. 8

2 - Transmission housing

3 - Screw, 12 Nm (9 ft lb)
- Remove screw before removing or installing bearing supporting ring -item 13-.
- Install with sealing agent AKD 456 000 01

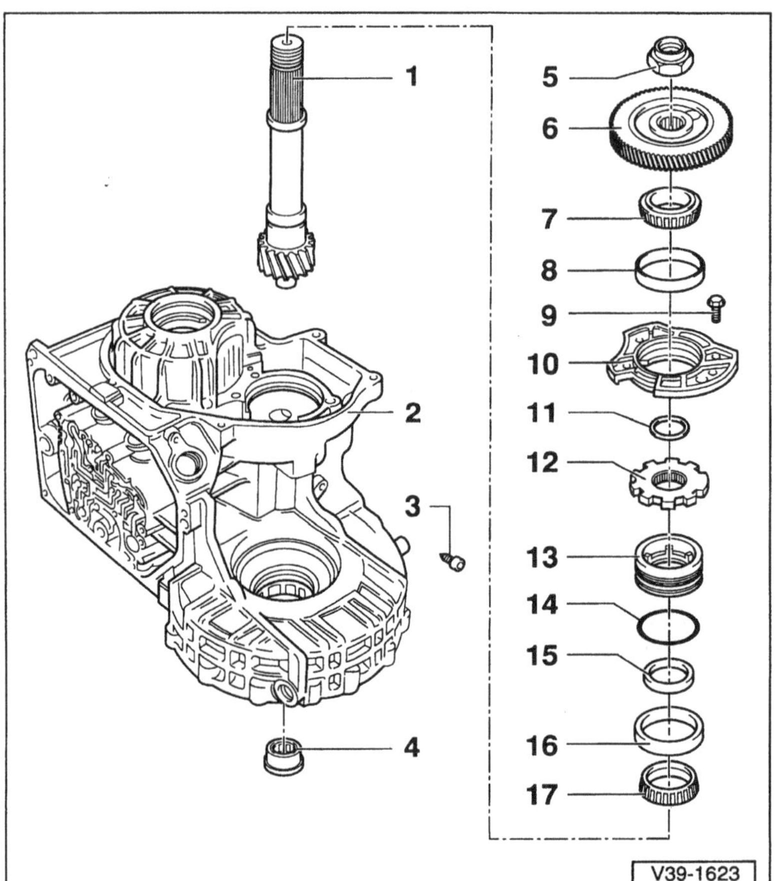

4 - Needle roller bearing
- Pressing out ⇒ Fig. 10
- Pressing in ⇒ Fig. 11

5 - Hex nut, 250 Nm (184 ft lb)
- Engage parking lock to remove or install
- Secure with drift

6 - Output gear
- Number of teeth ⇒ from Page J 00-6
- Pulling off ⇒ Fig. 1
- If signs of damage exist, always replace input gear and output gear together

7 - Tapered roller bearing inner race
- Pulling off ⇒ Fig. 2
- Pressing on ⇒ Fig. 3

8 - Tapered roller bearing outer race
- Drive out with drift
- Pressing in ⇒ Fig. 4

9 - Bolt, 25 Nm (18 ft lb)

10 - Bearing cap

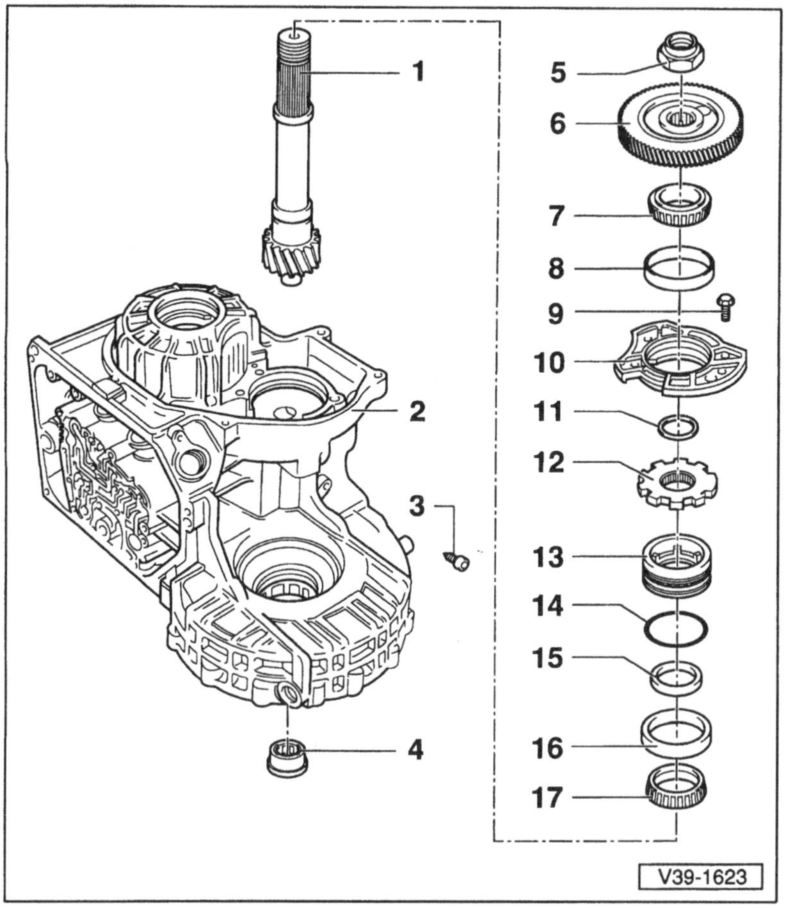

11 - Shim

- Determining thickness ⇒ Page J 39-51, Adjusting drive pinion

12 - Parking lock gear

- Rounded side faces teeth of drive pinion

13 - Bearing supporting ring, 200 Nm (148 ft lb)

- Remove screw -item 3- before removing or installing bearing supporting ring.
- Removing or installing ⇒ Fig. 5
- Bearing supporting ring cannot be unfastened until selector shaft has been removed ⇒ Page J 38-29, Disassembling and assembling parking lock
- Install bearing supporting ring carefully. Sealing lip and spring must not be damaged.
- Supporting ring with seal ⇒ Fig. 6

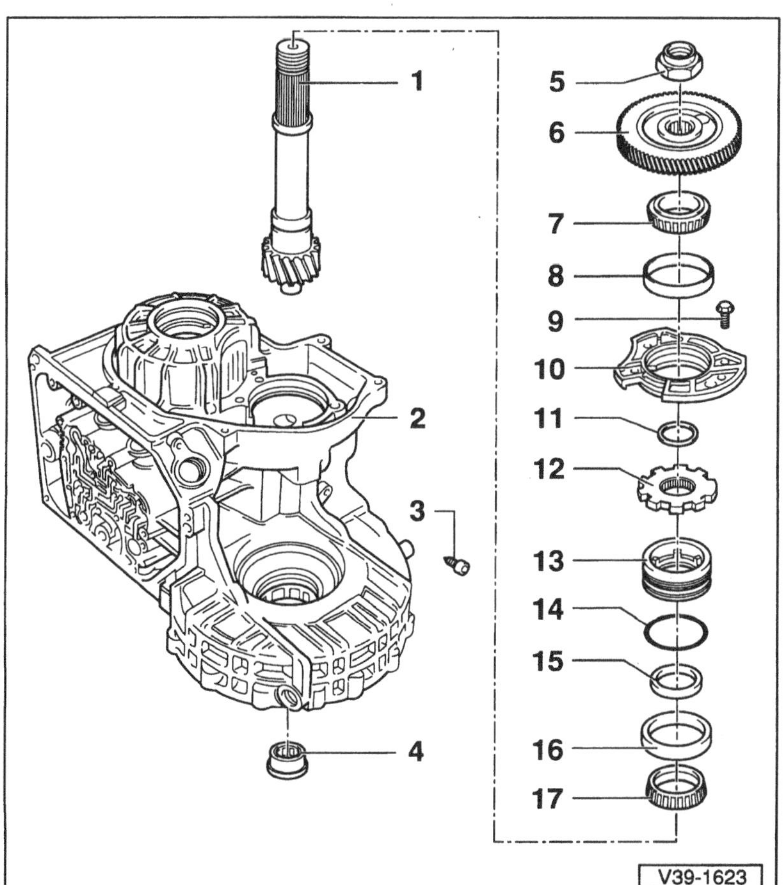

14 - O-ring

- Always replace
- Place onto bearing supporting ring -item 13-

15 - Oil seal for drive pinion

- Remove with screwdriver
- Installation position ⇒ Fig. 6
- Driving in ⇒ Fig. 7

16 - Tapered roller bearing outer race

- Driving in ⇒ Fig. 9

17 - Tapered roller bearing inner race

- Drive off with drift
- Pressing on ⇒ Fig. 12

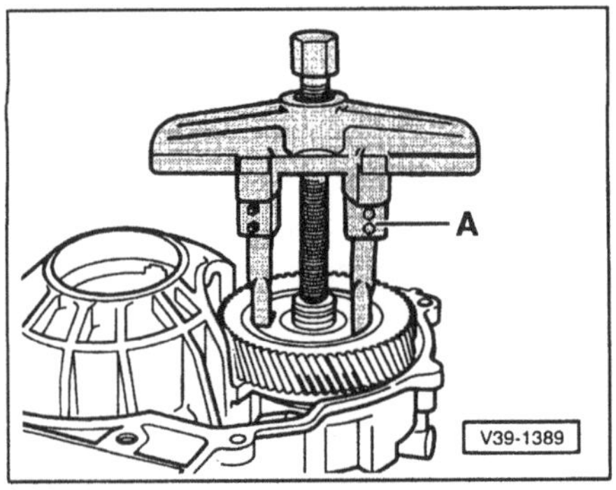

◄ **Fig. 1 Output gear, pulling off**

A – Kukko 20–10 and Matra V 172 pulling hooks.

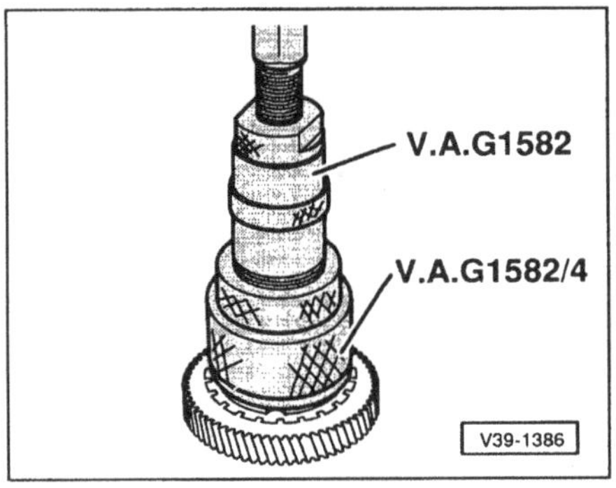

◄ **Fig. 2 Tapered roller bearing inner race, pulling off**

– Place press piece onto collar of output gear.

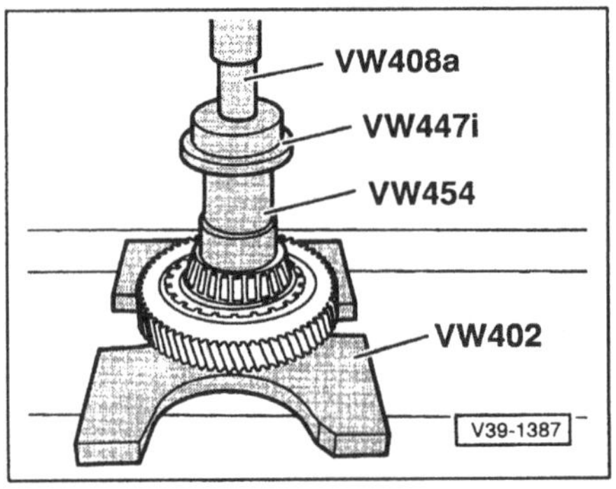

◄ **Fig. 3 Tapered roller bearing inner race, pressing on**

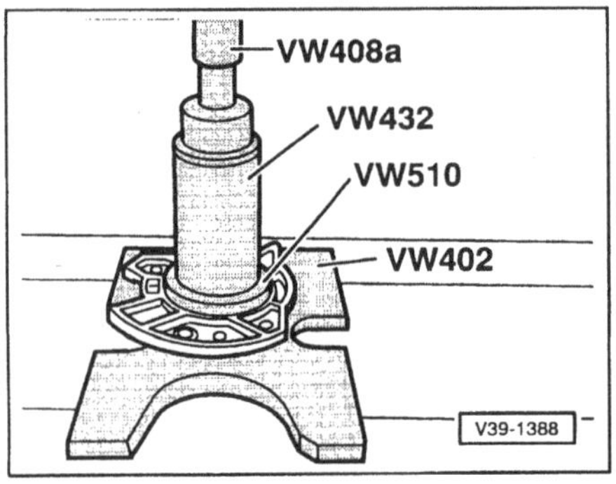

◄ **Fig. 4 Tapered roller bearing outer race, pressing in**

– Press tapered roller bearing outer race in onto stop in bearing cap.

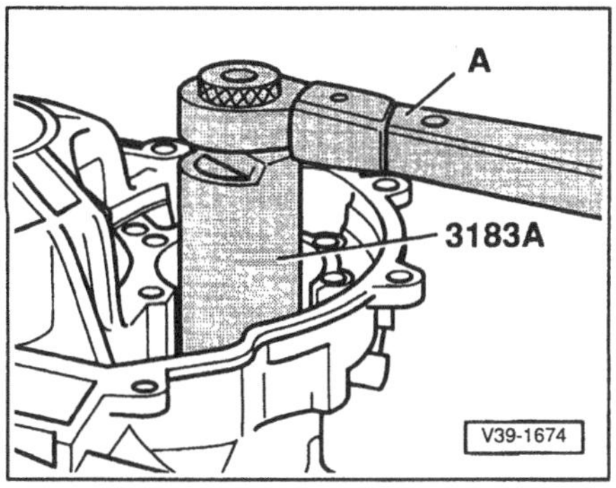

Fig. 5 Bearing supporting ring, removing and installing

A – Torque wrench

– Tighten bearing supporting ring to 200 Nm (148 ft lb) and secure with locking screw. Insert screw with AKD 456 000 01.

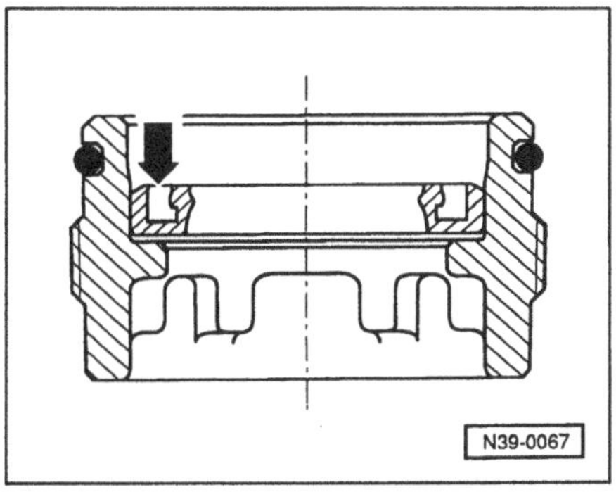

Fig. 6 Bearing supporting ring, installation position of seal

Open side -arrow- faces towards tapered roller bearing outer race

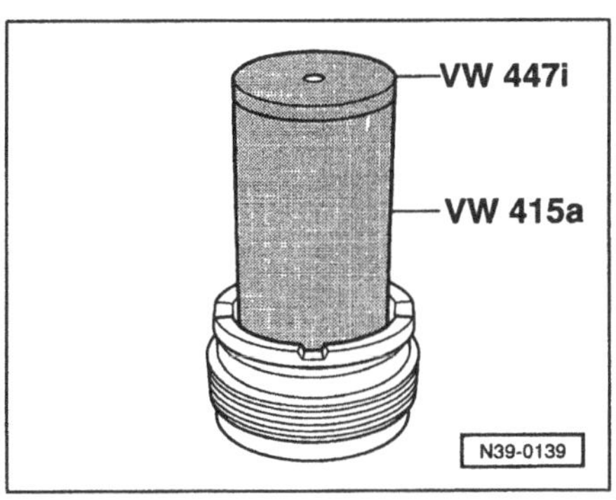

Fig. 7 Drive pinion oil seal, driving in

– Install oil seal so that sealing lip faces special tool.

> ***CAUTION!***
> ***Install oil seal carefully. Sealing lip and spring must not be damaged.***

– Drive oil seal in onto stop.

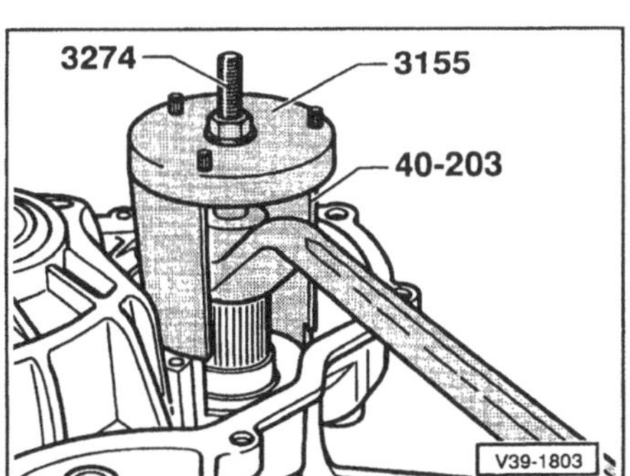

Fig. 8 Drive pinion, removing

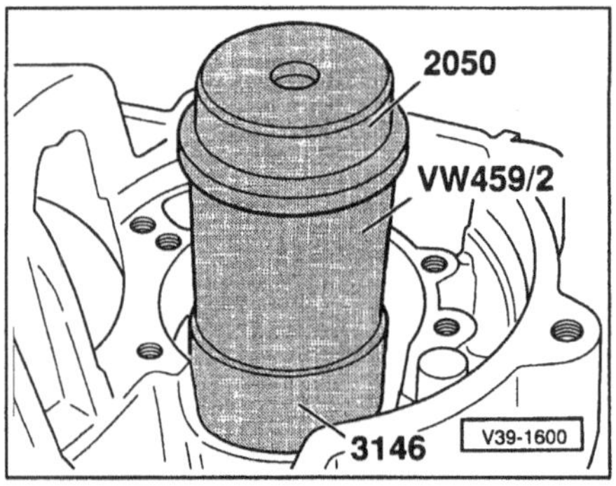

◄ **Fig. 9 Tapered roller bearing outer race, driving in**

- Install drive pinion with tapered roller bearing inner race, then drive in outer race.

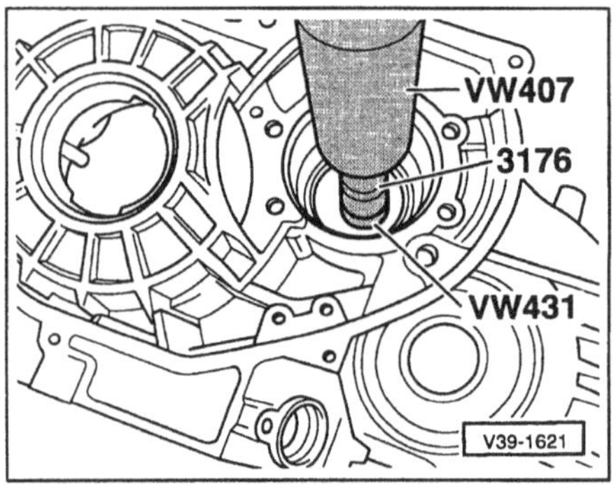

◄ **Fig. 10 Needle roller bearing, pressing out**

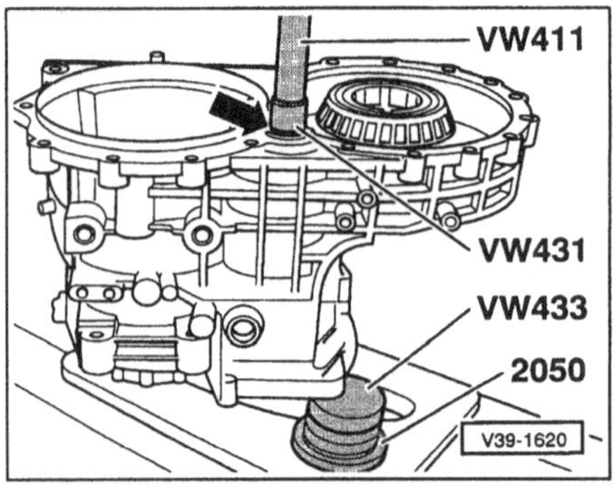

◄ **Fig. 11 Needle roller bearing, pressing in**

- Position reduced side -arrow- to outside and press needle roller bearing onto stop.

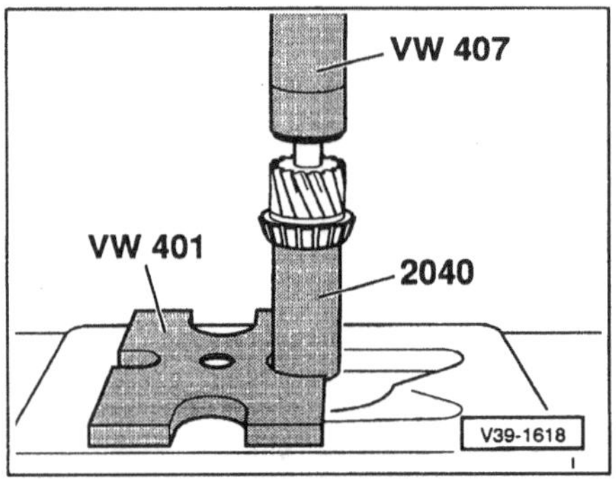

◄ **Fig. 12 Tapered roller bearing inner race, pressing on**

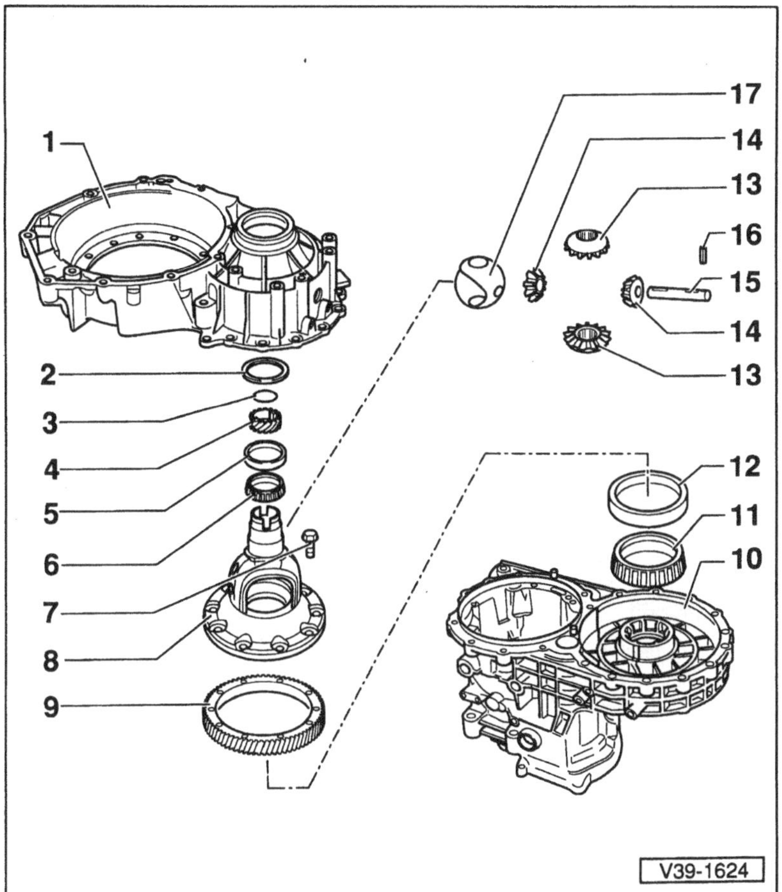

Differential, disassembling and assembling

Notes:

- *Heat tapered roller bearing inner race to 100°C (212°F) before pressing on.*

1 - Torque converter

- Is allocated to transmission according to engine version.
- Removing and installing ⇒ Page J 39-17
- As a replacement part differential is allocated according to transmission code letters

⇒ Replacement parts catalog

- On vehicles with diesel engine the Engine Speed (RPM) Sensor -G28- is installed in the top of the torque converter housing ⇒ Page J 39-21, Fig. 3

2 - Shim

- Determining thickness ⇒ Page J 39-59, Adjusting differential

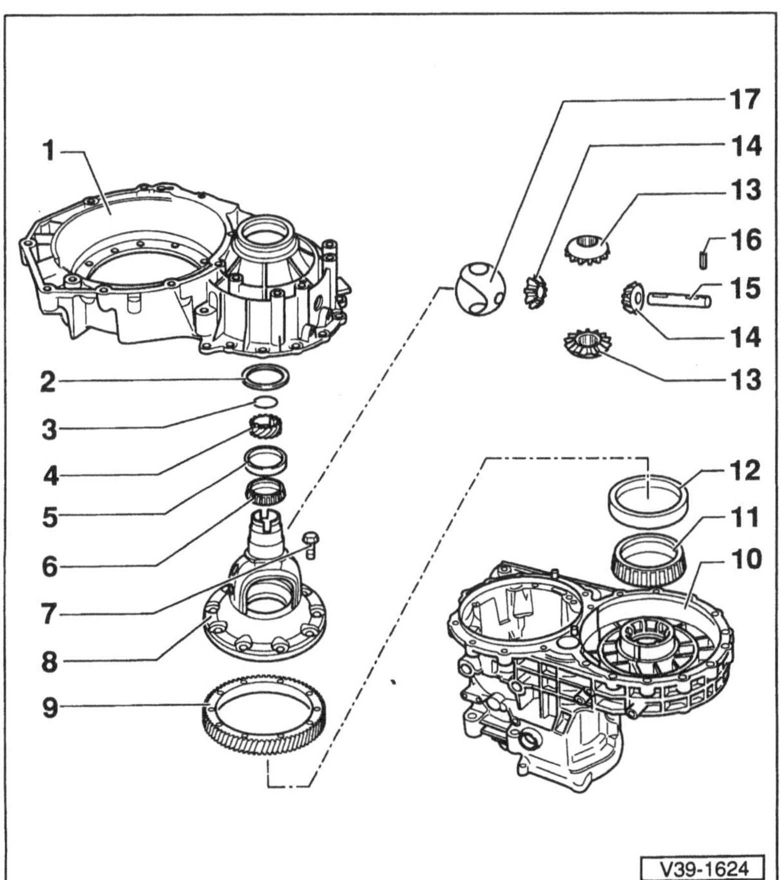

3 - Circlip

- For securing speedometer drive gear

4 - Speedometer drive gear

- Seat drive gear lugs into differential housing grooves, straight face outwards

5 - Tapered roller bearing outer race

- Remove and install only when differential housing heated
- Pressing out ⇒ Fig. 1
- Pressing in ⇒ Fig. 2

6 - Tapered roller bearing inner race

- Pulling off ⇒ Fig. 3
- Pressing on ⇒ Fig. 4

7 - Bolt

- Tighten all bolts to 80 Nm (59 ft lb), then afterwards turn 90° further
- Always replace

8 - Differential housing

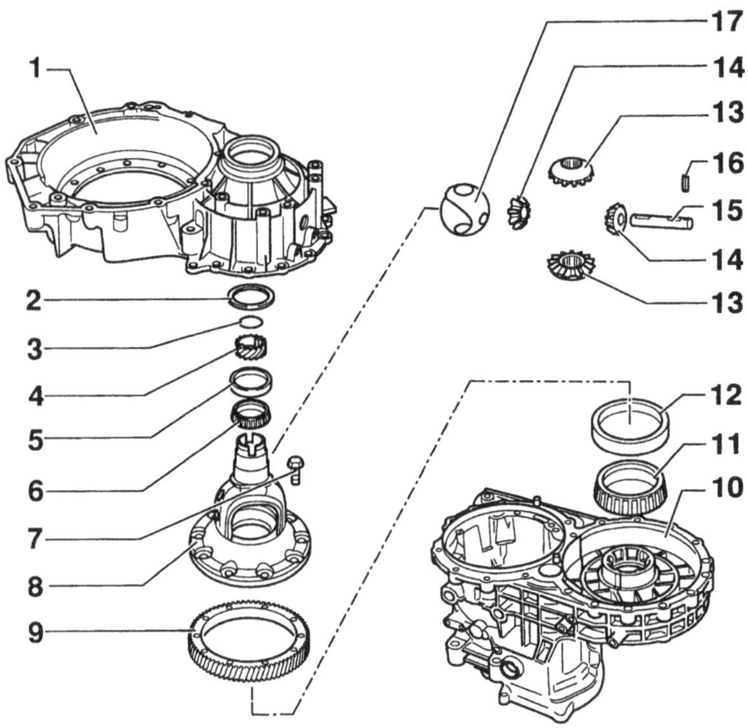

V39-1624

9 - Final drive gear wheel
- Number of teeth ⇒ Page J 00-6
- Install heated ⇒ Fig. 5

10 - Transmission housing
- Removing and installing ⇒ Removing and installing differential, Page J 39-17
- As a replacement part housing is allocated according to transmission code letters

⇒ Replacement parts catalog
- Before assembling transmission and torque converter housings together, coat sealing surfaces with AMV 188 200

11 - Tapered roller bearing inner race
- Pulling off ⇒ Fig. 6
- Pressing into transmission housing ⇒ Fig. 7

12 - Tapered roller bearing outer race
- Pulling off ⇒ Fig. 8
- Pressing into differential housing ⇒ Fig. 9

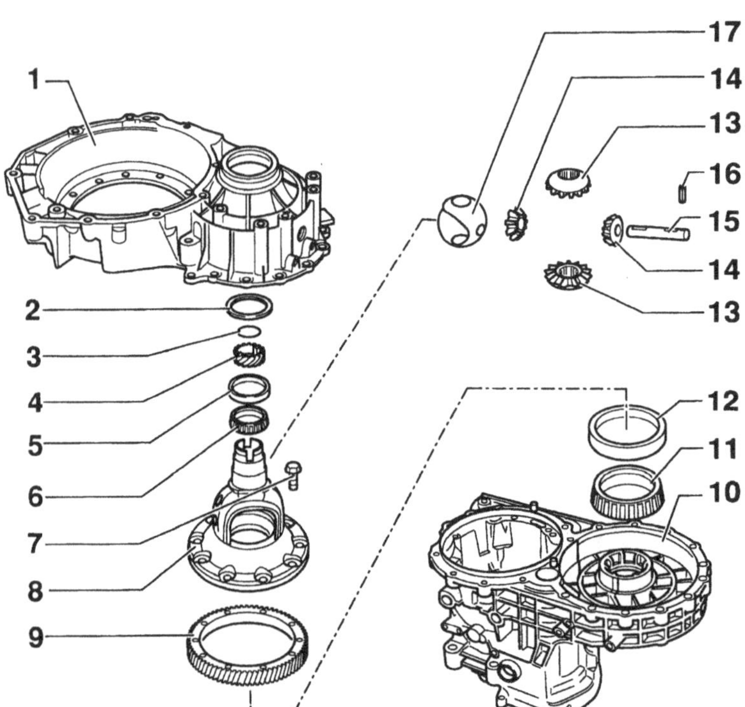

V39-1624

13 - Large bevel gears
- Installing ⇒ Fig. 10

14 - Small bevel gears
- Installing ⇒ Fig. 10

15 - Shaft for bevel gears
- Drive out with a drift
- When driving in, do not damage one-piece thrust washer

16 - Spring pin
- For securing shaft for bevel gears
- Removing and installing spring pin with circumferential groove ⇒ Fig. 11

17 - One-piece thrust washer
- Place into differential housing before installing bevel gears

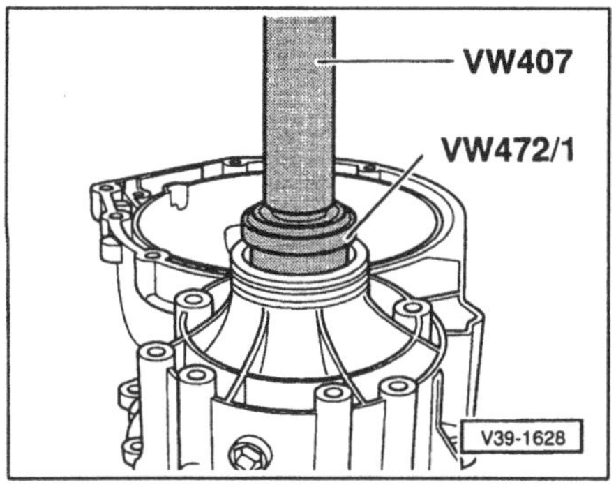

◄ **Fig. 1 Tapered roller bearing outer race, pressing out**

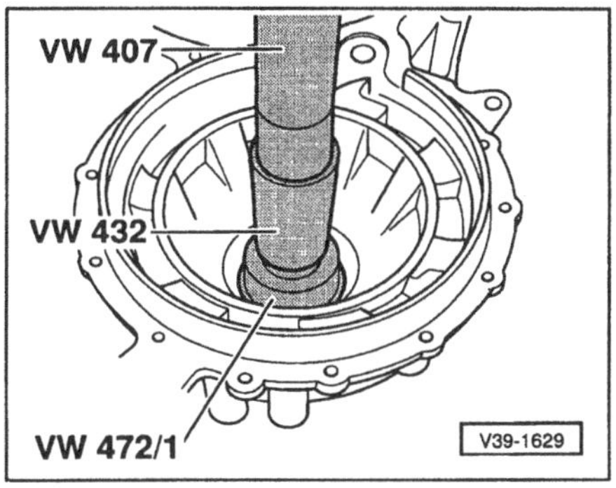

Note:

Only press taper roller bearing outer race in when torque converter housing is heated.

◄ **Fig. 2 Tapered roller bearing outer race, pressing into torque converter housing**

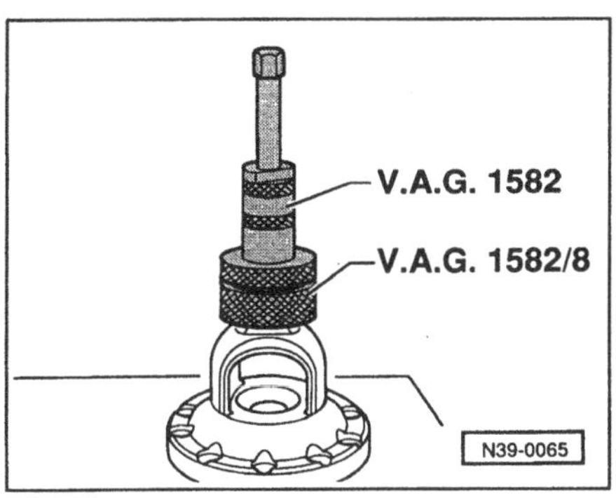

◄ **Fig. 3 Tapered roller bearing inner race, pressing off**

– Place press piece 40–105 on differential housing

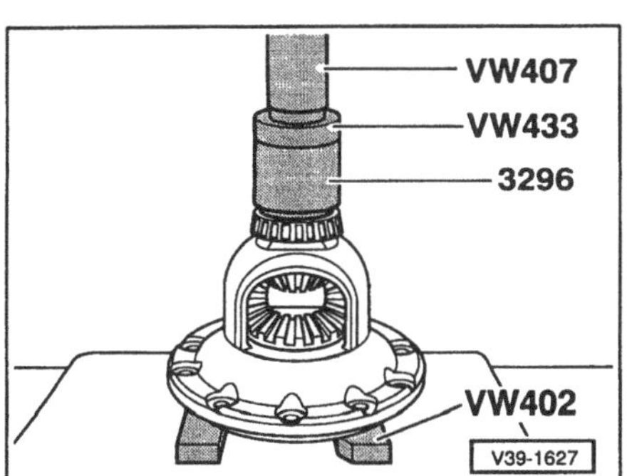

◄ **Fig. 4 Tapered roller bearing inner race, pressing on**

– Press tapered roller bearing inner race onto stop on differential housing

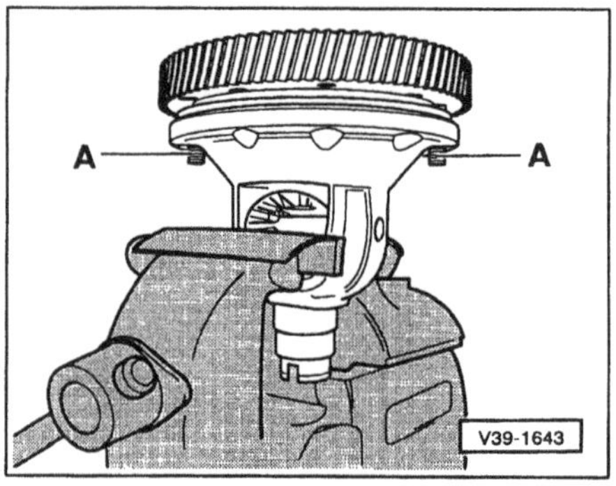

◄ **Fig. 5 Heated final drive gear wheel, installing**

- Heat final drive gear wheel to approx. 100°C (212°F).
- Install final drive gear wheel over centralizing pins -A- (local manufacturer) on differential housing.

Note:

Recess in gear wheel faces towards differential housing.

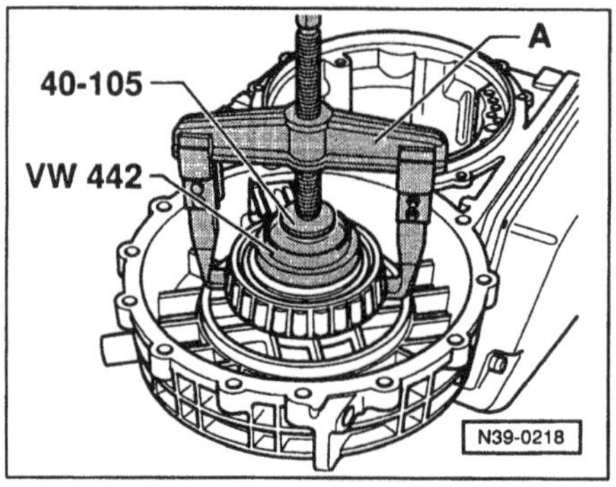

◄ **Fig. 6 Tapered roller bearing inner race, removing**

A – Kukko 20–10 and Matra V 172 pulling hooks.

- Insert puller hooks into cut-outs in transmission housing.

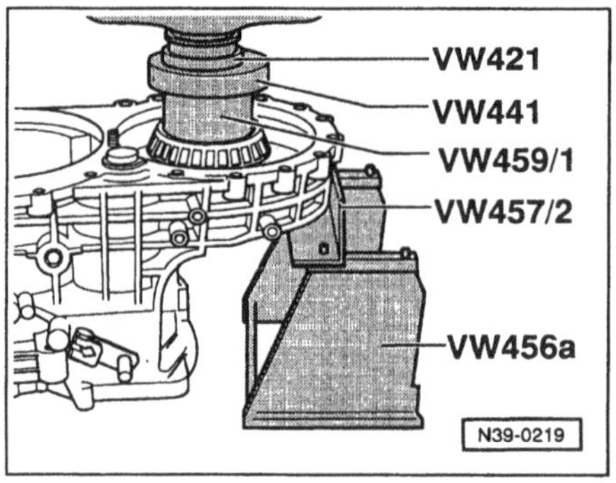

◄ **Fig. 7 Tapered roller bearing inner race, pressing into transmission housing**

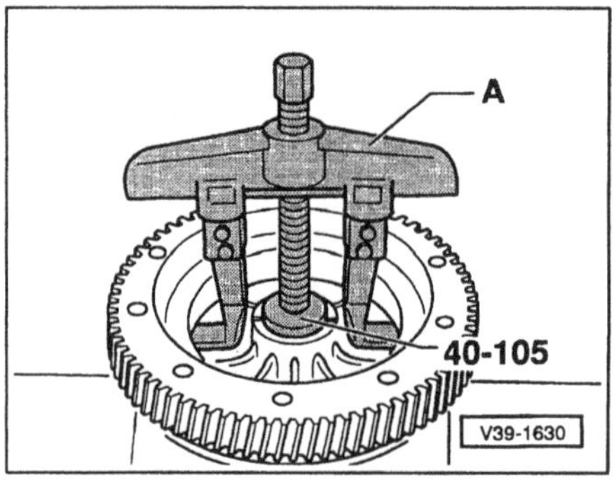

◄ **Fig. 8 Tapered roller bearing outer race, removing**

A – Kukko 20–10

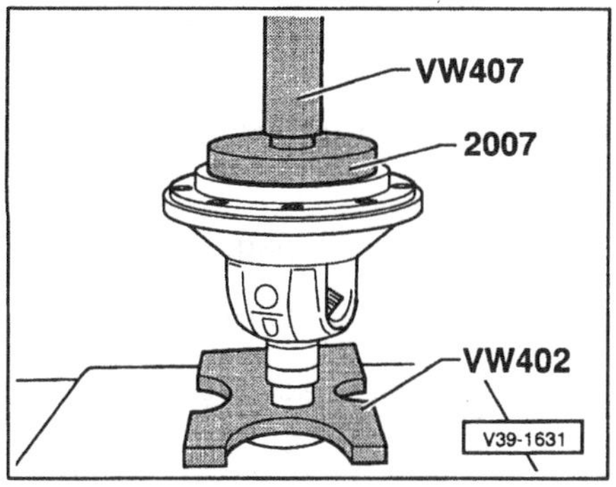

◄ **Fig. 9 Tapered roller bearing outer ring, pressing into differential housing and driving fully onto stop with drift**

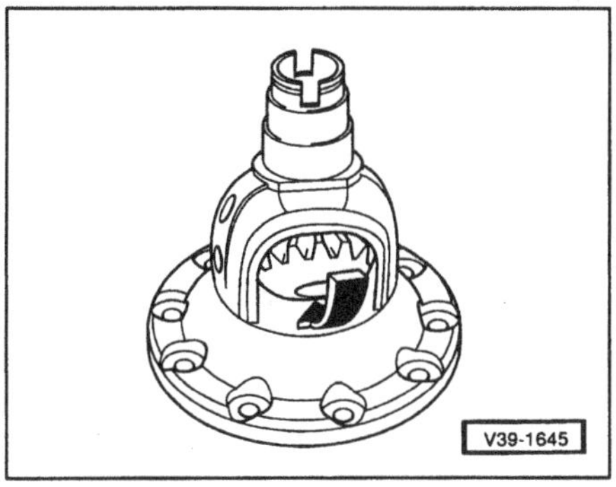

◄ **Fig. 10 Differential bevel gears, installing**

- Install one-piece thrust washer with gear oil.
- Install small bevel gears, drive in shaft and secure with spring pin.
- Install large bevel gears offset 180° and swivel in in direction of arrow.

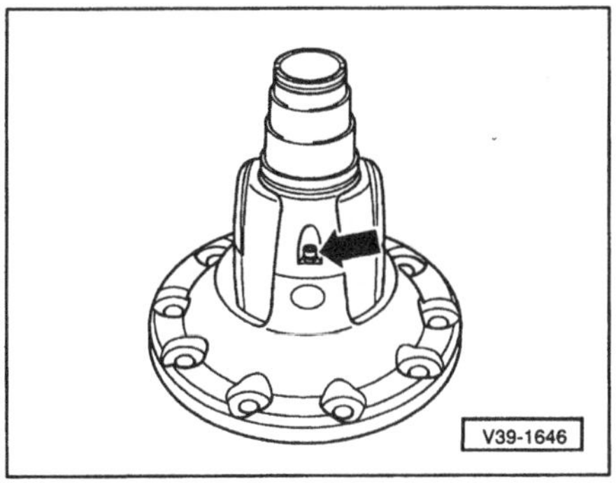

◄ **Fig. 11 Spring pin with circumferential groove, removing and installing**

Removing

- Drive out spring pin using a chisel.

Installing

- Drive in spring pin with groove facing outward.
- Drive in spring pin up to stop.

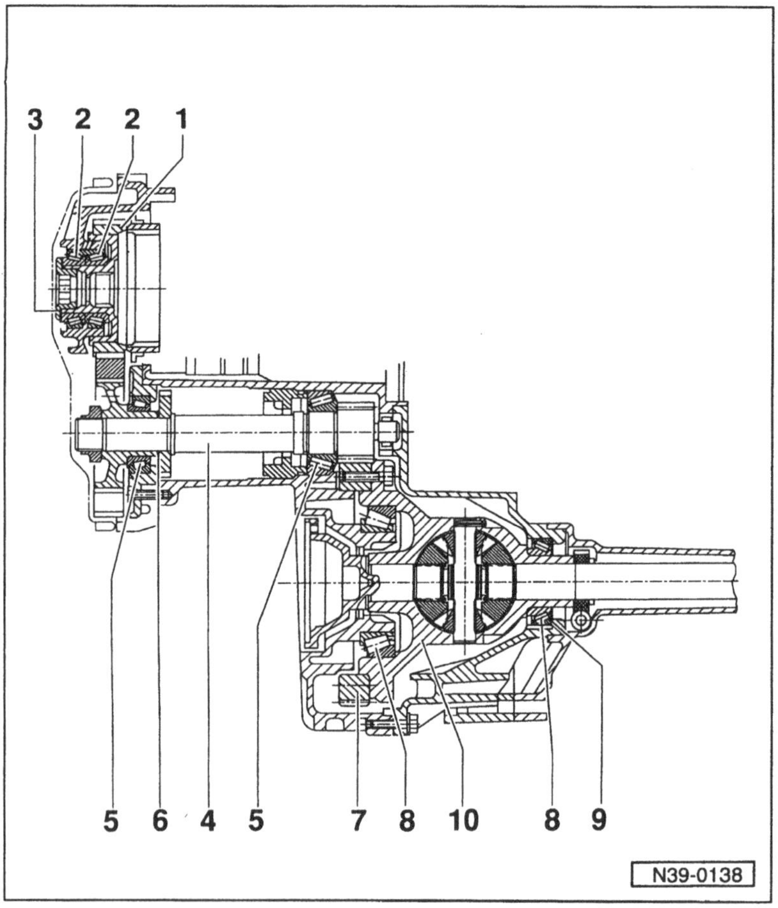

Final drive, adjusting

Overview of adjustments

1 - Input gear
- Adjusting ⇒ page J 39-57

2 - Tapered roller bearings for input gear
- Always replace together
- Measure friction torque via drive pinion. Friction torque of new tapered roller bearings must be 180 – 220 Ncm (16–20 in. lb) greater than friction torque of pinion shaft

3 - Shim for input gear
- Determine thickness on basis of table ⇒ page J 39-54
- Determine part number from parts catalog microfiche

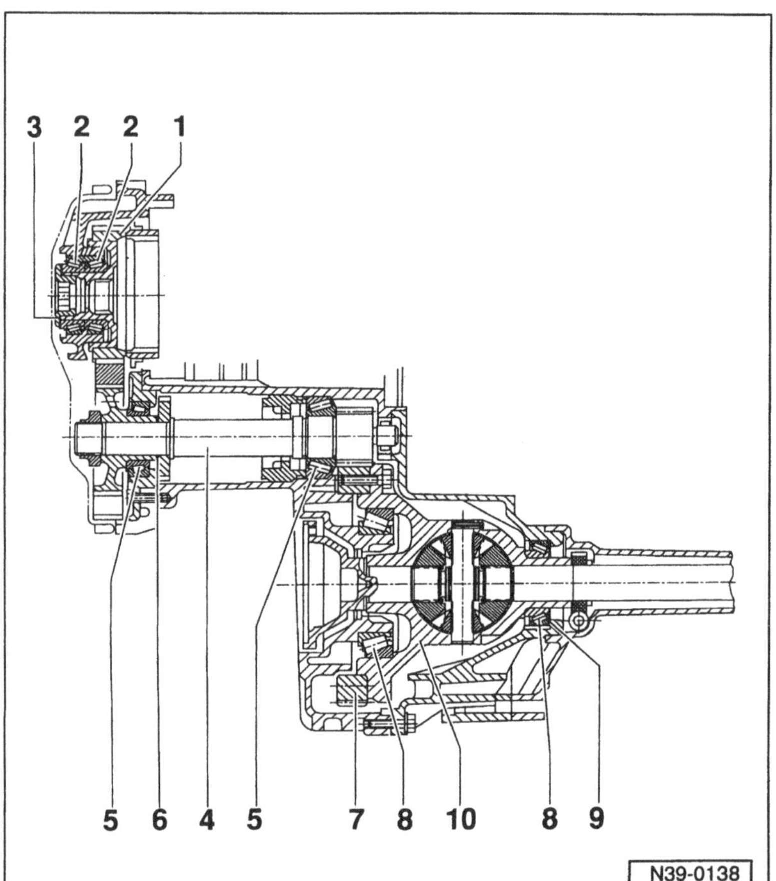

4 - Drive pinion
- Adjusting ⇒ page J 39-51
- If input gear, differential and drive pinion tapered roller bearings are replaced, then the overall friction torque must be 220 – 420 Ncm (19 – 37 in. lb)

5 - Tapered roller bearings for drive pinion
- Always replace together
- Re–determine shim when replacing tapered roller bearings
- Friction torque of new tapered roller bearings must be 80 – 120 Ncm (7 – 11 in. lb)

6 - Shim for drive pinion
- Determine thickness on basis of table ⇒ page J 39-54
- Determine part number from parts catalog microfiche

7 - Final drive gear wheel

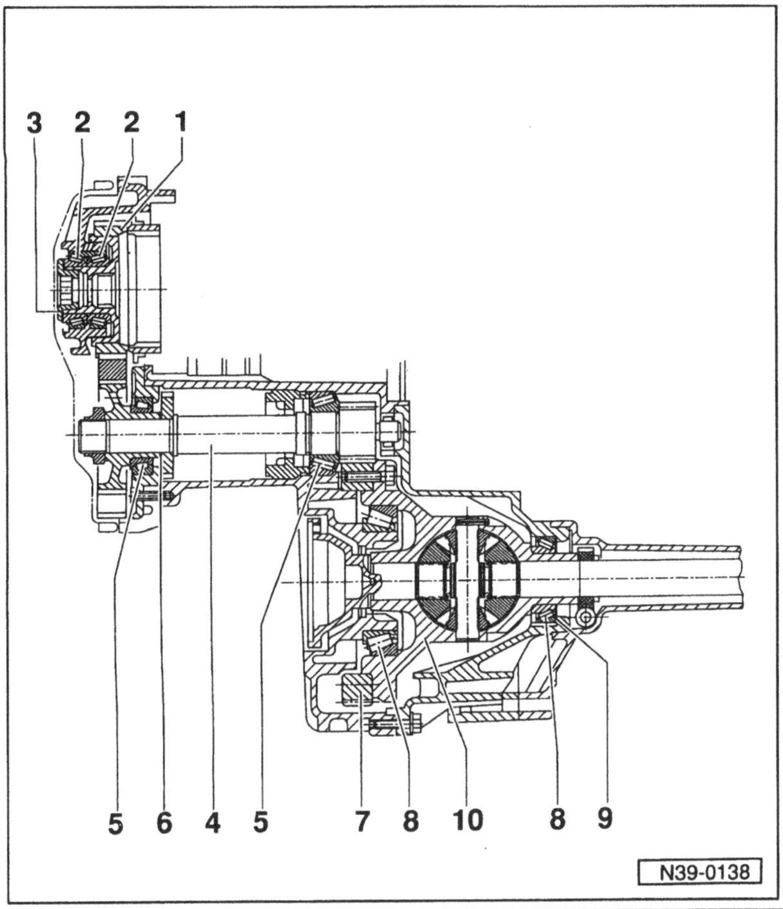

8 - Tapered roller bearings for differential
- Always replace together
- Measure friction torque through the drive *pinion*. Friction torque of new tapered roller bearings must be 50 – 90 Ncm (4 – 8 in. lb) greater than drive pinion friction torque.

9 - Shim for differential
- Determine thickness from table ⇒ page J 39-64

10 - Differential
- Adjusting ⇒ page J 39-60

Adjustments to be completed after replacing individual components

When working on the transmission,if individual components have been replaced that directly affect the final drive adjustment, only the following components in this table require adjustment:

Components being replaced	To adjust		
	Differential ⇒ **page J 39-60**	**Adjusting input gear**[1)] ⇒ **page J 39-57**	**Adjusting drive pinion with output gear**[1)] ⇒ **page J 39-51**
Transmission housing	X	X	X
Torque converter housing	X		
Input gear[1)]		X	X
Input gear tapered roller bearing		X	
Output gear		X	X
Drive pinion			X
Drive pinion tapered roller bearing			X
Differential housing with gear wheel	X		
Tapered roller bearing for differential, bearing body or shim	X		
Supporting ring, park lock gear or bearing cover			X

1) Always replace input gear and output gear together

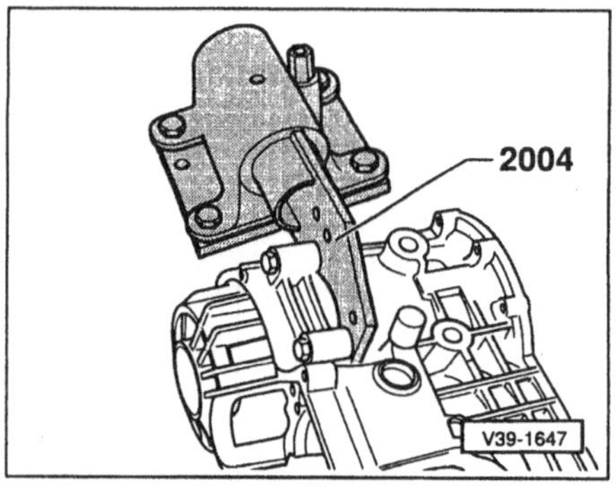

Drive pinion, adjusting

– Secure transmission housing to 2004 transmission holder .

– Place washers between transmission housing and transmission holder 2004 and secure transmission holder with 8 mm bolts.

Conditions:

- Supporting ring for drive pinion bearing is installed and secured.
- Parking lock mechanism and parking lock gear are installed.

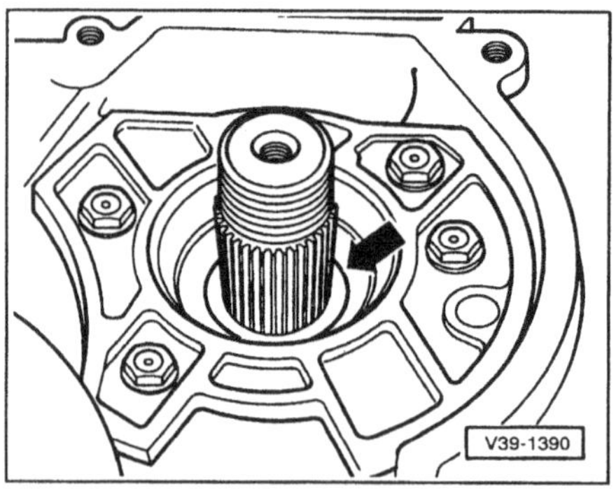

Adjusting:

– Place two shims (arrow), each 1.5 mm (0.591 in.) thick –arrow– onto drive pinion.

– Place output gear with tapered roller bearing onto drive pinion and tighten hex nut (41 mm AF) to 250 Nm (184 in. lb); engage parking lock when performing this step..

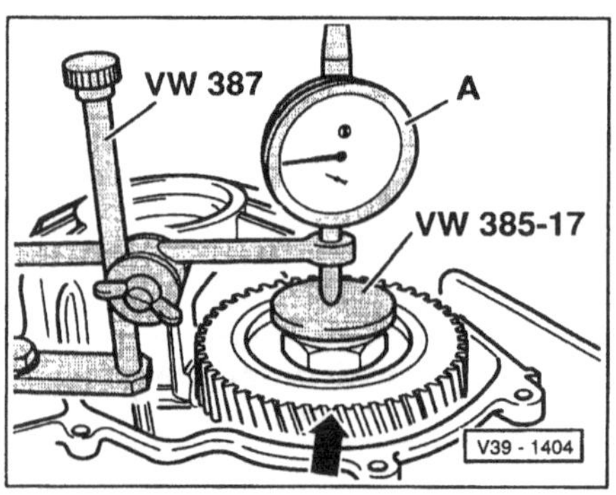

– Insert dial indicator -A- and zero gauge with a preload of 1 mm (0.0393in.) .

– Move drive pinion up and down. Read off and record axial play at dial indicator (example 0.93 mm (0.0366 in.)).

Note:

Do not turn the drive pinion while performing measurement because the bearings will settle and give a false measurement.

Determining shims

The specified bearing preload is achieved when the reading, 0.93 mm (0.0366 in.), the preload on the bearings, 0.12 mm (0.0047 in.), and the amount for compression 0.10 mm (0.0039 in.) are deducted from the 2 x 1.5 mm (0.0590 in.) thick shims.

Example:

Shims installed 2 x 1.5 mm (0.0590 in.)	3.00 mm (0.1181 in.)
– Reading	– 0.93 mm (0.0366 in.)
– Preload	– 0.12 mm (0.0047 in.)
– Amount for compression	– 0.10 mm (0.0039 in.)
= Thickness of shims required	1.85 mm (0.0728 in.)

– Remove output gear. Determine size of shim from table and install shim onto drive pinion. Table of shims ⇒ page J 39-54.

– Re-install output gear and tighten hex nut to 250 Nm (184 ft lb) and secure with drift.

Table of shims for drive pinion and input gear

Gauged thickness – mm (in.)	Shim to be installed mm (in.)	Gauged thickness mm (in.)	Shim to be installed mm (in.)
0.975 (0.03838) – 0.999 (0.03933)	1.000 (0.03937)	1.550 (0.06102) – 1.574 (0.06196)	1.575 (0.06200)
1.000 (0.03937) – 1.024 (0.04031)	1.025 (0.04035)	1.575 (0.06200) – 1.599 (0.06295)	1.600 (0.06299)
1.025 (0.04035) – 1.049 (0.04129)	1.050 (0.04133)	1.600 (0.06299) – 1.624 (0.06393)	1.625 (0.06397)
1.050 (0.04133) – 1.074 (0.04228)	1.075 (0.04232)	1.625 (0.06397) – 1.649 (0.06492)	1.650 (0.06496)
1.075 (0.04232) – 1.099 (0.04326)	1.100 (0.04330)	1.650 (0.06496) – 1.674 (0.06590)	1.675 (0.06594)
1.100 (0.04330) – 1.124 (0.04425)	1.125 (0.04429)	1.675 (0.06594) – 1.699 (0.06688)	1.700 (0.06692)
1.125 (0.04429) – 1.149 (0.04523)	1.150 (0.04527)	1.700 (0.06692) – 1.724 (0.06787)	1.725 (0.06791)
1.150 (0.04527) – 1.174 (0.04622)	1.175 (0.04625)	1.725 (0.06791) – 1.749 (0.06885)	1.750 (0.06889)
1.175 (0.04625) – 1.199 (0.04720)	1.200 (0.04724)	1.750 (0.06889) – 1.774 (0.06984)	1.775 (0.06988)
1.200 (0.04724) – 1.224 (0.04818)	1.225 (0.04822)	1.775 (0.06988) – 1.799 (0.07082)	1.800 (0.07086)
1.225 (0.04822) – 1.249 (0.04917)	1.250 (0.04921)	1.800 (0.07086) – 1.824 (0.07181)	1.825 (0.07185)
1.250 (0.04921) – 1.274 (0.05015)	1.275 (0.05019)	1.825 (0.07185) – 1.849 (0.07279)	1.850 (0.07283)
1.275 (0.05019) – 1.299 (0.05114)	1.300 (0.05118)	1.850 (0.07283) – 1.874 (0.07377)	1.875 (0.07381)
1.300 (0.05118) – 1.324 (0.05212)	1.325 (0.05216)	1.875 (0.07381) – 1.899 (0.07476)	1.900 (0.07480)
1.325 (0.05216) – 1.349 (0.05311)	1.350 (0.05314)	1.900 (0.07480) – 1.924 (0.07574)	1.925 (0.07578)
1.350 (0.05314) – 1.374 (0.05409)	1.375 (0.05413)	1.925 (0.07578) – 1.949 (0.07673)	1.950 (0.07677)
1.375 (0.05413) – 1.399 (0.05507)	1.400 (0.05511)	1.950 (0.07677)– 1.974 (0.07771)	1.975 (0.07775)
1.400 (0.05511) – 1.424 (0.05606)	1.425 (0.05610)	1.975 (0.07775) – 1.999 (0.07870)	2.000 (0.07874)
1.425 (0.05610) – 1.449 (0.05704)	1.450 (0.05708)	2.000 (0.07874) – 2.024 (0.07968)	2.025 (0.07972)
1.450 (0.05708) – 1.474 (0.05803)	1.475 (0.05807)	2.025 (0.07972) – 2.049 (0.08066)	2.050 (0.08070)
1.475 (0.05807) – 1.499 (0.05901)	1.500 (0.05905)	2.050 (0.08070) – 2.074 (0.08165)	2.075 (0.08169)
1.500 (0.05905) – 1.524 (0.05999)	1.525 (0.06003)	2.075 (0.08169) – 2.099 (0.08263)	2.100 (0.08267)
1.525 (0.06003)– 1.549 (0.06098)	1.550 (0.06102)	2.100 (0.08267)– 2.124 (0.08362)	2.125 (0.08366)

Note:

- *Different tolerances enable the required shim thickness to be measured precisely.*
- *Part numbers for shims ⇒ parts catalog microfiche.*

Table of shims for drive pinion and input gear

Gauged thickness – mm	Shim to be installed mm (in.)
2.125 (0.08366) – 2.149 (0.08460)	2.150 (0.08464)
2.150 (0.08464) – 1.274 (0.08559)	2.175 (0.08562)
2.175 (0.08562) – 2.199 (0.08657)	2.200 (0.08661)
2.200 (0.08661) – 2.224 (0.08755)	2.225 (0.08759)
2.225 (0.08759) – 2.249 (0.08854)	2.250 (0.08858)
2.250 (0.08858) – 2.274 (0.08952)	2.275 (0.08956)
2.275 (0.08956) – 2.299 (0.09051)	2.300 (0.09055)
2.230 (0.09055) – 2.324 (0.09149)	2.325 (0.09153)
2.325 (0.09153) – 2.349 (0.09248)	2.350 (0.09251)
2.350 (0.09251) – 2.374 (0.09346)	2.375 (0.09350)
2.375 (0.09350) – 2.399 (0.09444)	2.400 (0.09448)
2.400 (0.09448) – 2.424 (0.09543)	2.425 (0.09547)
2.425 (0.09547) – 2.449 (0.09641)	2.450 (0.09645)
2.450 (0.09645) – 2.474 (0.09740)	2.475 (0.09744)
2.475 (0.09744) – 2.499 (0.09838)	2.500 (0.09842)
2.500 (0.09842) – 2.524 (0.09936)	2.525 (0.09940)
2.525 (0.09940) – 2.549 (0.10035)	2.550 (0.10039)
2.550 (0.10039) – 2.574 (0.10133)	2.575 (0.10137)
2.575 (0.10137) – 2.599 (0.10232)	2.600 (0.10236)
2.600 (0.10236) – 2.624 (0.10330)	2.625 (0.10334)
2.625 (0.10334) – 2.649 (0.10429)	2.650 (0.10433)
2.650 (0.10433) – 2.674 (0.10527)	2.675 (0.10531)
2.675 (0.10531) – 2.700 (0.10629)	2.700 (0.10629)

Note:

- *Different tolerances enable the required shim thickness to be measured precisely.*
- *Part numbers for shims ⇒ parts catalog microfiche.*

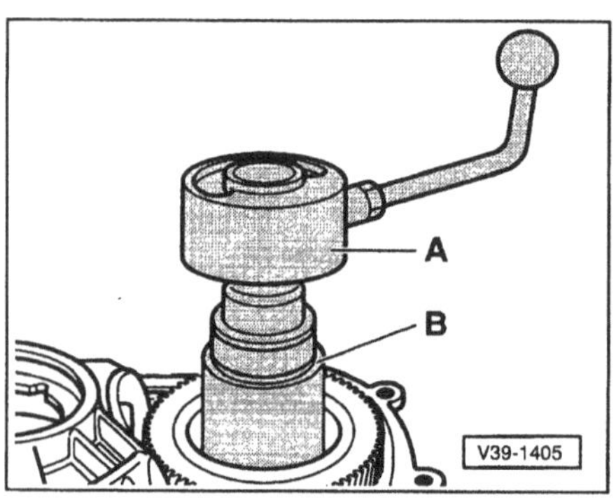

◄ **Checking friction torque**

A - Torque wrench, commercially available, 0–600 Ncm (0–53 in. lb)

B - Wrench socket (41 mm AF)

– Oil bearing with gear oil.

- Friction torque of new tapered roller bearings must be 80–120 Ncm (7–11 in. lb).

Input gear, adjusting

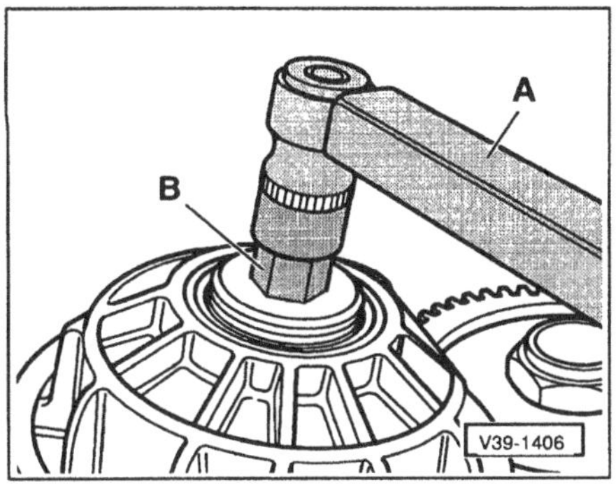

◄ – Tighten socket-head fastener to 100 Nm (74 ft lb) without dished washer and without shim.

A - Torque wrench

B - 22 mm socket

The parking lock must be engaged.

– Remove socket-head fastener.

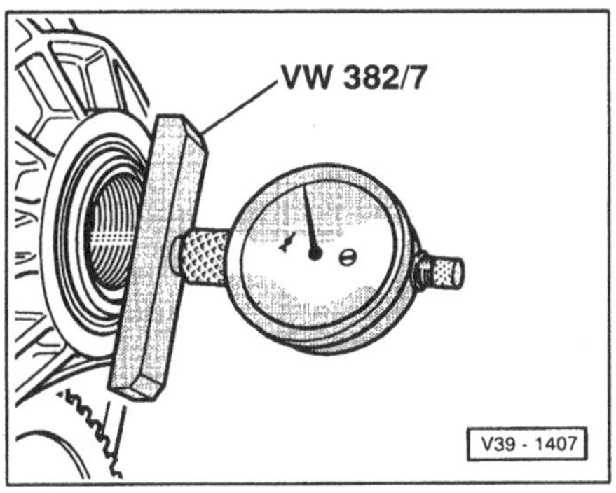

◄ – Measure distance between input gear and tapered roller bearing inner race.

– Insert dial indicator and zero gauge with a 3 mm (0.1181 in.) preload. Record reading Example 1.00 mm (0.0393 in.).

The thickness of the dished washer, 1.50 mm (0.0590 in.), must be added to the reading: 1.00 mm (0.393 in.) + 1.50 mm (0.0590 in.) = 2.50 mm (0.0984 in.).

Determining shim thickness

The specified bearing preload is achieved when the bearing preload of 0.18 mm (0.0070 in.) is subtracted from the measured value – (dished washer + reading) of 2.50 mm (0.9842 in.).

Example:

Dished washer + reading	2.50 mm (0.0984 in.)
– Bearing preload	– 0.18 mm (.0.0070 in.)
= Thickness of shim required	2.32 mm (0.0913 in.)

– Determine size of shim from table and install onto drive shaft. Table of shims ⇒ page J 39-54.

– Insert axial needle bearing and tighten socket-head fastener including dished washer to 250 Nm (184 ft lb).

Note:

Insert axial needle bearing for input shaft before tightening socket-head fastener. Installation position ⇒ Removing and installing input gear, page J 39-22.

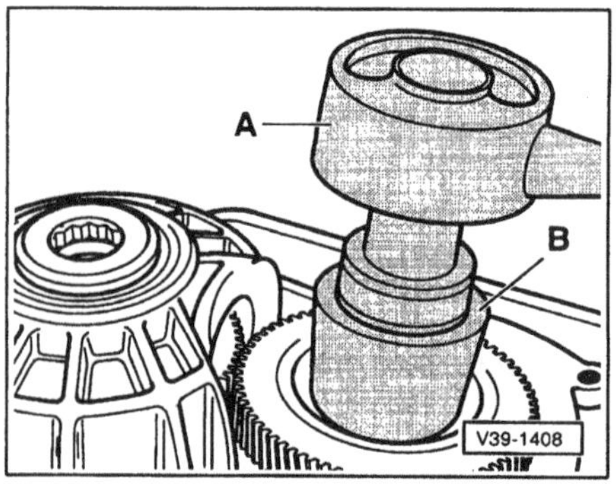

Checking friction torque

A - Torque wrench, commercially available 0–600 Ncm (0–53 in. lb)

B - Wrench Socket (41 mm AF)

The friction torque of new tapered roller bearings must be 180–220 Ncm (16–20 in. lb) greater than friction torque at the drive pinion.

Example:

	Friction torque
Drive pinion (from measurement)	100 Ncm[1)] (9 in. lb)
Input gear (average)	200 Ncm (18 in. lb)
Friction torque of input gear + drive pinion	= 300 Ncm (27 in. lb)

– Moisten bearings beforehand with ATF.

1) Insert the friction torque measured for the drive pinion if used bearings are reinstalled.

Differential, adjusting

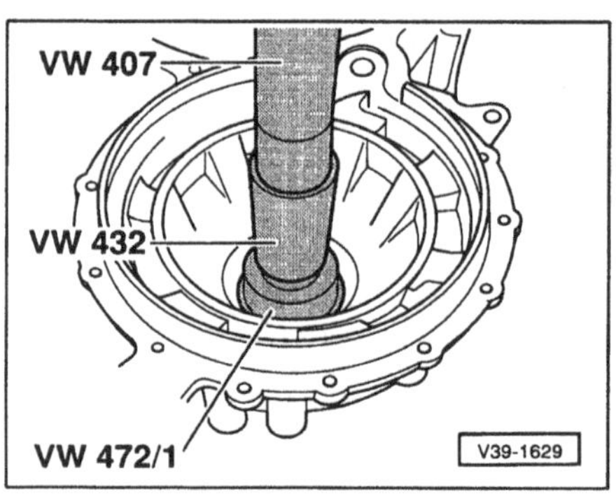

– Press tapered roller bearing outer race, without shim, into torque converter housing.

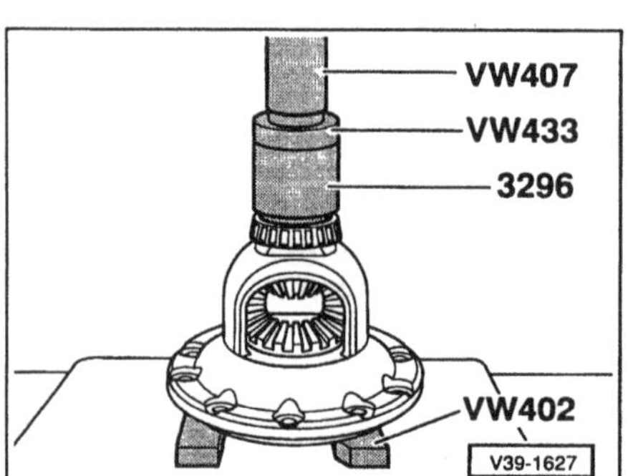

– Press tapered roller bearing inner race onto differential housing.

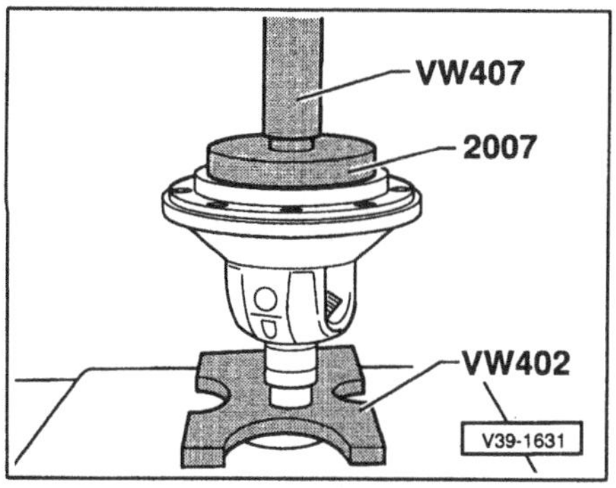

- Press tapered roller bearing outer race fully onto stop in differential housing.

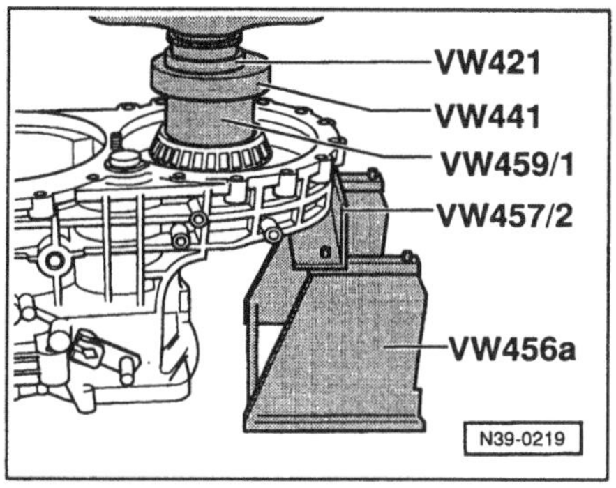

- Press tapered roller bearing inner race into differential housing.

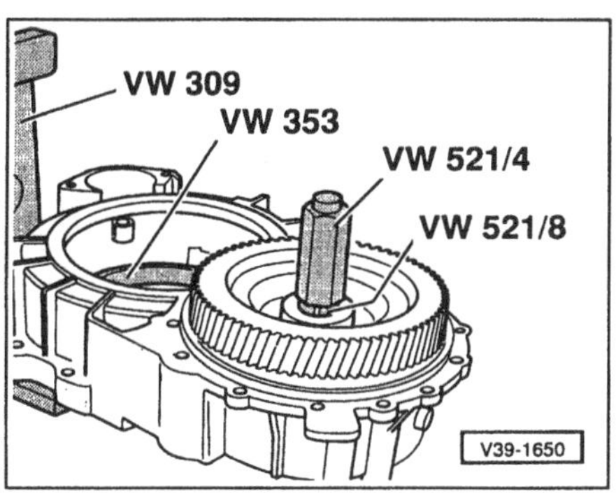

- Bolt torque converter housing onto VW 353 transmission support and install differential.
- Install special tool into differential housing.
- Install transmission housing on differential and install every second bolt. Tighten to 25 Nm (18 ft lb).

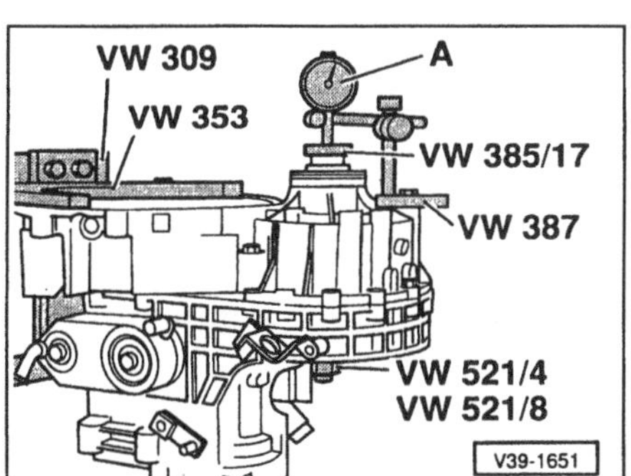

- Assemble special tool to measure play.
- Zero dial indicator with 1 mm (0.0393 in.) preload.
- Lift and lower differential.
- Read and record axial play from dial indicator.

Note:

Do not turn differential while performing measurement because the bearings will settle and give a false measurement.

Determining shim thickness

The specified bearing preload is achieved if the bearing preload of 0.12 mm (0.0047 in.) and the settling amount of 0.10 mm (0.0039 in.) is added to the figure obtained from the measurement.

Example:

Measured value	0.60 mm (0.0236 in.)
+ preload	0.12 mm (0.0047 in.)
+ settling allowance	0.10 mm (0.0039 in.)
= Thickness of shims to be installed	0.82 mm (0.0322 in.)

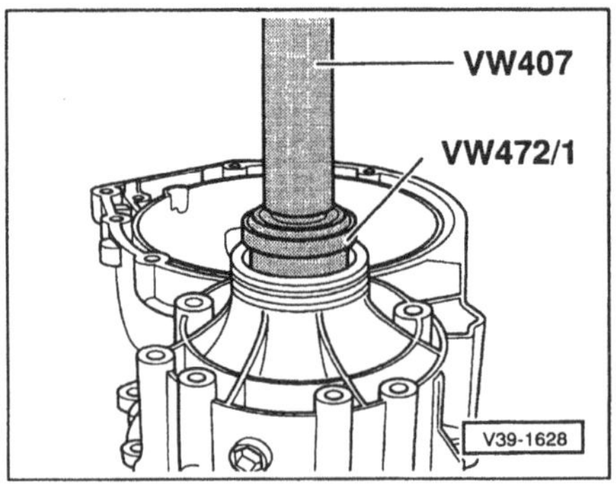

◄ – Unbolt transmission housing and remove differential from torque converter housing.

– Press tapered roller bearing, outer race, out of torque converter housing.

Table of shims for differential

Thickness mm (in.)	Part number	Thickness mm (in.)	Part number
0.65 (0.0255)	02B 409 210	1.05 (0.0413)	02B 409 210 H
0.70 (0.0275)	02B 409 210 A	1.10 (0.0255)	02B 409 210 J
0.75 (0.0295)	02B 409 210 B	1.15 (0.0255)	02B 409 210 K
0.80 (0.0314)	02B 409 210 C	1.20 (0.0255)	02B 409 210 L
0.85 (0.0334)	02B 409 210 D	1.25 (0.0255)	02B 409 210 M
0.90 (0.0354)	02B 409 210 E	1.30 (0.0255)	02B 409 210 N
0.95 (0.0374)	02B 409 210 F	1.35 (0.0255)	02B 409 210 P
1.00 (0.0393)	02B 409 210 G	1.40 (0.0255)	02B 409 210 Q

If the thickness measured is greatest than the sizes in the table then 2 shims equal to the thickness of the shim to be installed may be used.

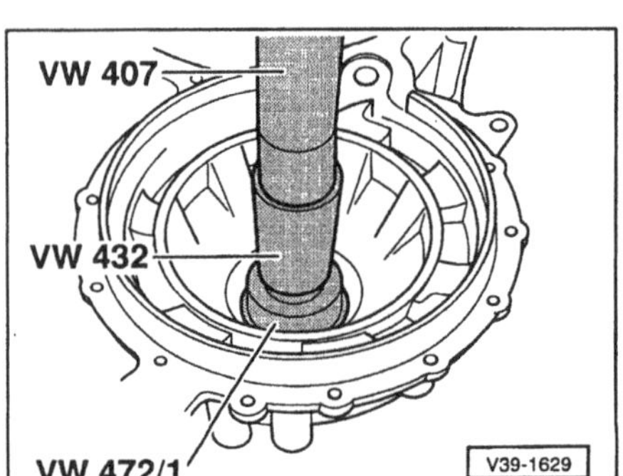

◄ – Press tapered roller bearing outer ring, with shim, into torque converter housing.

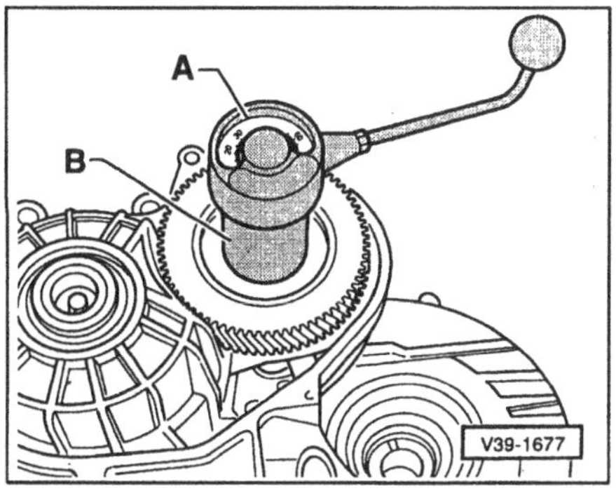

Checking friction torque

A - Torque wrench, commercially available 0 – 600 Ncm (0–53 in. lb)

B - Wrench socket (41 mm)

The friction torque of new tapered roller bearings must be 50 – 90 Ncm (4 – 8 in. lb) greater than at the drive pinion.

Example:

	Friction torque
Drive pinion (measured value)	300 Ncm[1)] (27 in. lb)
Differential (average)	70 Ncm (6 in. lb)
Friction torque of differential + drive pinion	= 370 Ncm (33 in. lb)

– Oil bearing beforehand with gear oil.

1) Install the friction torque measured for the drive pinion in the case of used bearings.

Manual Transmission 02B

Section K

> ***WARNING!***
> ***Before proceeding with any work, please read the warnings and cautions immediately following this index. Also read any warning and caution that accompanies a procedure in the manual.***

Please read these WARNINGS and CAUTIONS before proceeding with maintenance and repair work.

WARNING—

- Some repairs may be beyond your capability. If you lack the skills, tools and equipment, or a suitable workplace for any procedure described in this manual, we suggest you leave such repairs to an authorized Volkswagen dealer service department, or other qualified shop.

- Volkswagen is constantly improving its cars. Sometimes these changes, both in parts and specifications, are made applicable to earlier models. Therefore, before starting any major jobs or repairs to components on which passenger safety may depend, consult your authorized Volkswagen dealer about the Technical Bulletins that may have been issued since the editorial closing of this manual.

- Do not re-use any fasteners that are worn or deformed in normal use. Many fasteners are designed to be used only once and become unreliable and may fail when used a second time. This includes, but is not limited to, nuts, bolts, washers, self–locking nuts or bolts, circlips and cotter pins. Always replace these fasteners with new parts.

- Never work under a lifted car unless it is solidly supported on stands designed for the purpose. Do not support a car on cinder blocks, hollow tiles or other props that may crumble under continuous load. Never work under a vehicle that is supported solely by a jack. Never work under the car while the engine is running.

- If you are going to work under a car on the ground, make sure that the ground is level. Block the wheels to keep the car from rolling. Disconnect the battery negative terminal (ground strap) to prevent others from starting the vehicle while you are under it.

- Never run the engine unless the work area is well ventilated. Carbon monoxide (CO) kills.

- Do not attempt to work on your vehicle if you do not feel well. You increase the danger of injury to yourself and others if you are tired, upset or have taken medicine or any other substance that may impair you or keep you from being fully alert.

- Disconnect the battery negative (–) terminal (ground strap) whenever you work on the fuel system or the electrical system. Do not smoke or work near heaters or other fire hazards. Keep an approved fire extinguisher handy.

- Batteries produce explosive hydrogen gas. Keep sparks, lighted matches and open flame away from the top of the battery. If hydrogen gas escaping from the cap vents is ignited, it will ignite gas trapped in the cells and cause the battery to explode.

- Connect and disconnect battery cables, jumper cables or a battery charger only with the ignition switched off, to prevent sparks. Do not quick–charge the battery (for boost starting) for longer than one minute, and do not allow charging voltage to exceed 16.5 volts. Wait at least one minute before boosting the battery a second time.

- Illuminate your work area adequately but safely. Use a portable safety light for working inside or under the car. Make sure the bulb is enclosed by a wire cage. The hot filament of an accidentally broken bulb can ignite spilled fuel or oil.

- Tie long hair behind your head. Do not wear a necktie, a scarf, loose clothing, or a necklace when you work near machine tools or running engines. If your hair, clothing, or jewelry were to get caught in the machinery, severe injury could result.

- Finger rings, bracelets and other jewelry should be removed so that they cannot cause electrical shorts, get caught in running machinery, or be crushed by heavy parts.

- Catch draining fuel, oil, or brake fluid in suitable containers. Do not use food or beverage containers that might mislead someone into drinking from them. Store flammable fluids away from fire hazards. Wipe up spills at once, but do not store the oily rags, which can ignite and burn spontaneously.

- Always observe good workshop practices. Wear goggles when you operate machine tools or work with battery acid. Gloves or other protective clothing should be worn whenever the job requires working with harmful substances.

- The Air-Conditioning (A/C) system is filled with a chemical refrigerant, which is hazardous. The A/C system should be serviced only by trained technicians using approved refrigerant recovery/recycling equipment, trained in related safety precautions, and familiar with regulations governing the discharging and disposal of automotive chemical refrigerants.

- Do not expose any part of the A/C system to high temperatures such as an open flame. Excessive heat will increase system pressure and may cause the system to burst.

- Some aerosol tire inflators are highly flammable. Be extremely cautious when repairing a tire that may have been inflated using an aerosol tire inflator. Keep sparks, open flame or other sources of ignition away from the tire repair area. Inflate and deflate the tire at least four times before breaking the bead from the rim. Completely remove the tire from the rim before attempting any repair.

- Friction materials such as brake or clutch discs may contain asbestos fibers. Do not create dust by grinding, sanding, or by cleaning with compressed air. Avoid breathing asbestos fibers and asbestos dust. Breathing asbestos can cause serious diseases such as asbestosis or cancer, and may result in death.

- Greases, lubricants and other automotive chemicals contain toxic substances, many of which are absorbed directly through the skin. Read manufacturer's instructions and warnings carefully. Use hand and eye protection. Avoid direct skin contact.

- An airbag is operated by an explosive device. Handled improperly or without adequate safeguards, it can be accidently activated and cause serious personal injury. To prevent personal injury or airbag system failure, **only trained Volkswagen dealer service technicians** should test, disassemble or service the airbag system.

WARNINGS and CAUTIONS continued on next page ▶

Please read these WARNINGS and CAUTIONS before proceeding with maintenance and repair work.

WARNING (continued) —

- ALWAYS disconnect the airbag power supply before using a computer memory saver. Failure to disconnect the airbag power supply may result in accidental activation of the airbag.
- The airbag unit should not be removed, nor should any work be performed on the airbag system, without first reviewing the warnings and cautions that accompany the airbag system repair information in this manual.
- Never apply stickers or any other coverings on the steering wheel. Do not let chemical cleaners, oil or grease come into contact with the vinyl covering of the airbag unit.
- Never open or otherwise attempt to repair airbag system parts. Always use new parts. Never leave airbag parts or the partially disassembled airbag system unattended.
- Never use a test light or volt meter to conduct electrical tests on the airbag system. The airbag system must only be tested by trained Volkswagen Service technicians using special airbag tester equipment. The airbag system must only be tested while it is installed in the car.
- Do not expose an airbag unit to temperatures above 194°F (90°C) even for a brief period, while handling it during the repair process. Keep the unit clear of heat sources such as hot plates, soldering irons, heat lamps, welding equipment, or the like.
- Always replace airbag system components that have been mechanically damaged (bubbles, cracks, etc.).
- Always set down a removed airbag unit so that the horn pad is facing upward.
- When driving or riding in an airbag-equipped vehicle, never hold test equipment in your hands or lap while the vehicle is in motion. Objects between you and the airbag can increase the risk of injury in an accident.
- The storage, transportation and disposal of airbag units are subject to the laws for flammable solids.

CAUTION—

- If you lack the skills, tools and equipment, or a suitable workshop for any procedure described in this manual, we suggest you leave such repairs to an authorized Volkswagen dealer or other qualified shop. We especially urge you to consult an authorized Volkswagen dealer before beginning repairs on any car that may still be covered wholly or in part by any of the extensive warranties issued by Volkswagen of America, Inc.
- Volkswagen is constantly improving its cars and sometimes these changes, both in parts and specifications, are made applicable to earlier models. Therefore, part numbers listed in this manual are for reference only. Always check with your authorized Volkswagen dealer parts department for the latest information.
- Before starting a job, make certain that you have all the necessary tools and parts on hand. Read all the instructions thoroughly, do not attempt shortcuts. Use tools that are appropriate to the work and use only replacement parts meeting Volkswagen specifications. Makeshift tools, parts and procedures will not make good repairs.
- Use pneumatic and electric tools only to loosen threaded parts and fasteners. Never use these tools to tighten fasteners, especially on light alloy parts. Always use a torque wrench to tighten fasteners to the tightening torque specification listed.
- Be mindful of the environment and ecology. Before you drain the crankcase, find out the proper way to dispose of the oil. Do not pour oil onto the ground, down a drain, or into a stream, pond, or lake. Consult local ordinances that govern the disposal of wastes.
- When connecting cables or ground straps, all connections must be cleaned to bare metal. Be sure mounting bolts are tight.
- Do not replace fuse links with fuse links of other gauge or with standard wire. Always replace the entire length of the fuse link.
- Before doing any electrical welding on vehicles equipped with Anti-lock Brakes (ABS), disconnect the battery negative terminal (ground strap) and the ABS Control Module connector.
- On cars equipped with ABS, the ABS control unit cannot withstand temperatures from a paint–drying booth or a heat lamp in excess of 203°F (95°C) and should not be subjected to temperatures in excess of 185°F (85°C) for more than two hours.
- On cars equipped with anti-theft radios, make sure you know the correct radio activation code before disconnecting the battery or removing the radio. If the wrong code is entered when power is restored, the radio may lock up and be rendered inoperable, even if the correct code is then entered.
- When boost-charging the battery, first remove the fuses for the Engine Control Module (ECM), the Transmission Control Module (TCM), the ABS Control Module, and the On-Board computer. In cases where one or more of these components is not separately fused, disconnect the Control Module connector(s).
- Do not quick-charge the battery (for boost starting) for longer than one minute, and do not exceed 16.5 volts at the battery with the boosting cables attached. Wait at least one minute before boosting the battery a second time.

◀ **WARNINGS and CAUTIONS continued from previous page**

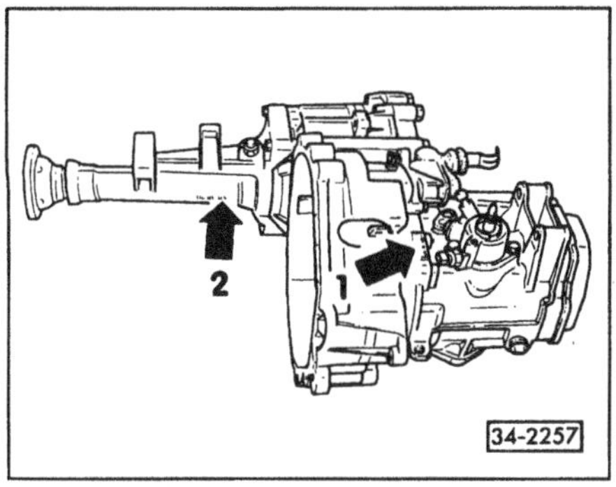

Transmission identification

The manual transmission 02B is installed in the EuroVan 1992 ➤ as a 5-speed transmission in conjunction with the 5-Cylinder engine. Application ⇒ Page K 00-3.

Location on transmission

◄ **Code letters and date of manufacture -arrow 1-**
Manual transmission 02B -arrow 2-

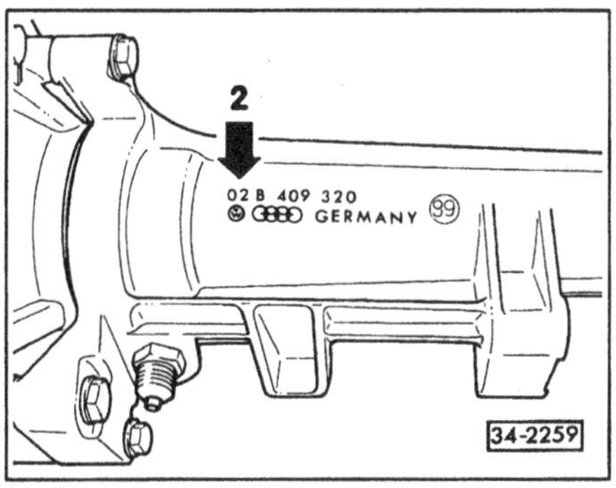

◄ **Manual transmission 02B -arrow 2-**

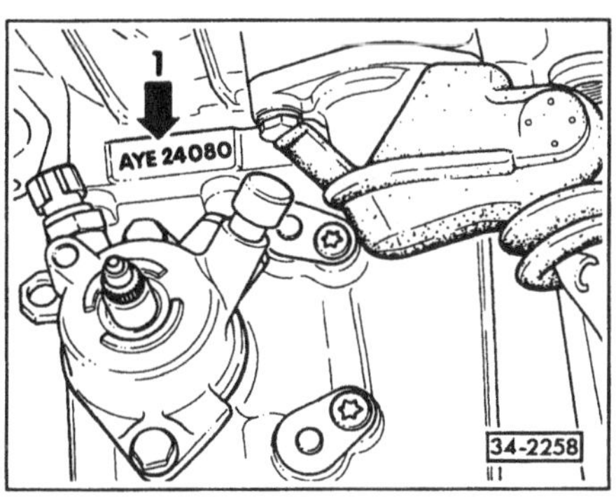

◄ **Code letters and date of manufacture of transmission -arrow 1-**

Example:	**AYD**	**24**	**08**	**3**
	I	I	I	I
	I	I	I	I
	Code letters	Day	Month	Year (1993) of manufacture

Additional data depends on manufacture.

Note:

The code letters of the transmission are also given on the vehicle data plates.

Code letters, transmission application, ratios, capacities

Manual transmission		5 speed 02B		
Code letters		**AYD**	**AYF**	**CCW**
Manufactured	from to	06.90 08.91	06.90 08.91	08.91 08.92
Application	Model	**EuroVan 1992**	**EuroVan 1992**	**EuroVan 1992**
	Engine	2.4 L - 57 kW (76 hp) Diesel 2.5 L - 81 kW (109 hp)	2.4 L - 57 kW (76 hp) Diesel 2.5 L - 81 kW (109 hp)	2.4 L - 57 kW (76 hp) Diesel 2.5 L - 81 kW (109 hp)
Ratio: $Z_2 : Z_1$	Final drive	83 : 18 = 4.611	83 : 18 = 4.611	83 : 18 = 4.611
	1st speed gear	34 : 9 = 3.778	34 : 9 = 3.778	34 : 9 = 3.778
	2nd speed gear	36 : 17 = 2.118	36 : 17 = 2.118	36 : 17 = 2.118
	3rd speed gear	40 : 28 = 1.429	39 : 29 = 1.345	35 : 24 = 1.458
	4th speed gear	35 : 34 = 1.029	34 : 35 = 0.971	35 : 34 = 1.029
	5th speed gear	36 : 43 = 0.837	34 : 45 = 0.756	36 : 43 = 0.837
	Reverse gear	19 : 9 x 36 : 20 = 3.800[1)]	19 : 9 x 36 : 20 = 3.800[1)]	18 : 9 x 36 : 20 = 3.600
	Speedometer drive	12 : 20 = 0.600	12 : 20 = 0.600	12 : 20 = 0.600

Code letters	**AYD**	**AYF**	**CCW**
Capacity	2.5 liters (2.7 qt)[2)]		
Specification	Gear oil G50 SAE 75W90 (synthetic oil)[4)]		
Clutch control	3)		
Clutch plate diameter mm (in.)	228 mm (8.98)	228 mm (8.98)	228 mm (8.98)
Drive shaft flange diameter mm (in.)	108 mm (4.25)	108 mm (4.25)	108 mm (4.25)
iOverall ratio in top gear	3.895	3.485	3.859

1) On all transmissions with a reverse gear with 19 teeth, a reverse gear with18 teeth can also be installed "i" - Reverse gear is then 3.600

2) Oil level below filler hole ⇒ Page K 34-59

3) Clutch control:	**Hydraulic**	**Mechanical**
Engine	2.5 L 2.4 L Diesel 1.9 L Diesel	2.5 L (except Bus GL and Camper) 2.4 L Diesel (only drop-side, van and Camper
	all vehicles with additional water heater all vehicles for USA/Canada	

4) Gear oil SAE 75W90 (synthetic oil) is available as a replacement part: Container 1.0 L (1.06 qt.) - Part No. - G052 145 A2

Manual transmission		5 speed 02B		
Code letters		**CCY**	**CHP**[1]	**CHS**[1]
Manufactured	from to	08.91 08.92	08.92 08.93	09.92 08.93
Application	Model	**EuroVan 1993**	**EuroVan 1993**	**EuroVan 1993**
	Engine	2.4 L - 57 kW (76 hp) Diesel	2.4 L - 57 kW (76 hp) Diesel 2.5 L - 81 kW (109 hp)	2.4 L - 57 kW (76 hp) Diesel 2.5 L - 81 kW (109 hp)
Ratio: $Z_2 : Z_1$	Final drive	83 : 18 = 4.611	83 : 18 = 4.611	83 : 18 = 4.611
	1st speed gear	34 : 9 = 3.778	34 : 9 = 3.778	34 : 9 = 3.778
	2nd speed gear	36 : 17 = 2.118	36 : 17 = 2.118	36 : 17 = 2.118
	3rd speed gear	34 : 25 = 1.360	35 : 24 = 1.458	34 : 25 = 1.360
	4th speed gear	34 : 35 = 0.971	35 : 34 = 1.029	34 : 35 = 0.971
	5th speed gear	34 : 45 = 0.756	36 : 43 = 0.837	34 : 45 = 0.756
	Reverse gear	18 : 9 x 36 : 20 = 3.600	18 : 9 x 36 : 20 = 3.600	18 : 9 x 36 : 20 = 3.600
	Speedometer drive	12 : 20 = 0.600	12 : 20 = 0.600	12 : 20 = 0.600

Code letters	**CCY**	**CHP**	**CHS**
Capacity	2.5 liters (2.7 qt)[2]		
Specification	Gear oil G50 SAE 75W90 (synthetic oil)[4]		
Clutch control	[3]		
Clutch plate diameter mm (in.)	228 mm (8.98)	228 mm (8.98)	228 mm (8.98)
Drive shaft flange diameter mm (in.)	108 mm (4.25)	108 mm (4.25)	108 mm (4.25)
Overall ratio in top gear	3.485	3.859	3.485

1) Rear transmission mounting modified ⇒ Page K 34-52

2) Oil level below filler hole ⇒ Page K 34-59

3) Clutch control:	**Hydraulic**	**Mechanical**
Engine	2.5 L 2.4 L Diesel	2.5 L (except Bus GL and Camper) 2.4 L Diesel (only drop-side, van and Camper)
	all vehicles with additional water heater all vehicles for USA/Canada	

4) Gear oil SAE 75W90 (synthetic oil) is available as a replacement part:
Container 1.0 L (1.06 qt.) - Part No. - G052 145 A2

Manual transmission		5 speed 02B		
Code letters		**CRL[1]**	**CRN[1]**	**CPV[1]**
Manufactured	from to	09.93 01.94	09.93	01.94
Application	Model	**EuroVan 1993**	**EuroVan 1993**	**EuroVan 1993**
	Engine	2.4 L - 57 kW (76 hp) Diesel 2.5 L - 81 kW (109 hp)	2.4 L - 57 kW (76 hp) Diesel 2.5 L - 81 kW (109 hp)	2.4 L - 57 kW (76 hp) Diesel 2.5 L - 81 kW (109 hp)
Ratio: $Z_2 : Z_1$	Final drive	83 : 18 = 4.611	83 : 18 = 4.611	73 : 16 = 4.563
	1st speed gear	34 : 9 = 3.778	34 : 9 = 3.778	34 : 9 = 3.778
	2nd speed gear	36 : 17 = 2.118	36 : 17 = 2.118	33 : 16 = 2.063
	3rd speed gear	35 : 34 = 1.029	34 : 35 = 0.971	31 : 23 = 1.348
	4th speed gear			29 : 30 = 0.967
	5th speed gear	36 : 43 = 0.837	34 : 45 = 0.756	30 : 39 = 0.769
	Reverse gear	18 : 9 x 36 : 20 = 3.600	18 : 9 x 36 : 20 = 3.600	18 : 9 x 36 : 20 = 3.600
	Speedometer drive	12 : 20 = 0.600	12 : 20 = 0.600	12 : 20 = 0.600

Code letters	**CRL**	**CRN**	**CPV**
Capacity	2.5 liters (2.7 qt)[2]		
Specification	Gear oil G50 SAE 75W90 (synthetic oil)[4]		
Clutch control	[3]		
Clutch plate diameter mm (in.)	228 mm (8.98)	228 mm (8.98)	228 mm (8.98)
Drive shaft flange diameter mm (in.)	108 mm (4.25)	108 mm (4.25)	108 mm (4.25)
^{i}**Overall ratio in top gear**	3.859	3.485	3.509

[1] Rear transmission mounting modified ⇒ Page K 34-52

[2] Oil level below filler hole ⇒ Page K 34-59

[3] Clutch control:	**Hydraulic**	**Mechanical**
Engine	2.5 L 2.4 L Diesel 1.9 L Diesel	2.5 L (except Bus GL and Camper 2.4 L Diesel (only drop-side, van and Camper)
	all vehicles with additional water heater all vehicles for USA/Canada	

[4] Gear oil SAE 75W90 (synthetic oil) is available as a replacement part:
Container 1.0 L (1.06 qt.) - Part No. - G052 145 A2

Manual transmission		5 speed 02B		
Code letters		**CPW**[1)]	**DCU**[1)]	**DCW**[1)]
Manufactured	from to	01.94	01.96	01.96
Application	Model	**EuroVan 1992**	**EuroVan 1992**	**EuroVan 1992**
	Engine	2.4 L - 57 kW (76 hp) Diesel 2.5 L - 81 kW (109 hp)	1.9 L - 50kW (67 hp) Turbo Diesel 2.4 L - 57 kW (76 hp) Diesel 2.5 L -81 kW (109 hp)	1.9 L - 50kW (67hp) Turbo Diesel 2.4 L - 57 kW (76 hp) Diesel 2.5 L -81 kW (109 hp)
Ratio: $Z_2 : Z_1$	Final drive	73 : 16 = 4.563	73 : 16 = 4.563	73 : 16 = 4.563
	1st speed gear	34 : 9 = 3.778	34 : 9 = 3.778	34 : 9 = 3.778
	2nd speed gear	33 : 16 = 2.063	33 : 16 = 2.063	33 : 16 = 2.063
	3rd speed gear	32 : 22 = 1.455	32 : 22 = 1.455	31 : 23 = 1.348
	4th speed gear	30 : 29 = 1.034	30 : 29 = 1.034	29 : 30 = 0.967
	5th speed gear	31 : 37 = 0.838	31 : 37 = 0.838	30 : 39 = 0.769
	Reverse gear	18 : 9 x 36 : 20 = 3.600	18 : 9 x 36 : 20 = 3.600	18 : 9 x 36 : 20 = 3.600
	Speedometer drive	12 : 20 = 0.600	12 : 20 = 0.600	12 : 20 = 0.600

Code letters	CPW	DCU[1)]	DCW[1)]
Capacity	2.5 liters [2)]		
Specification	Gear oil G50 SAE 75W90 (synthetic oil)[6)]		
Clutch control	[3)]		
Clutch plate diameter mm (in.)	228 mm (8.98)[4)5)]	228 mm (8.98)[4)5)]	
Drive shaft flange diameter mm (in.)	108 mm (4.25)	108 mm (4.25)	
ⁱOverall ratio in top gear	3.485	3.824	

[1)] Rear transmission mounting modified ⇒ Page K 34-52

[2)] Oil level below filler hole ⇒ Page K 34-59

[3)] Clutch control:	**Hydraulic**	**Mechanical**
Engine	2.5 L (GL and Camper only) 2.4 L Diesel	2.5 L (except Bus GL and Camper 2.4 L Diesel (only drop-side, van and Camper
	all vehicles for USA/Canada all vehicles with additional water heater	

[4)] 1.9 L Diesel – 45 kW (60 hp), 1.9 L Turbo Diesel – 50 kW (67 hp), 215 mm

[5)] 2.4 L Diesel – 57 kW (67 hp) from engine number: AAB 369 935 with inertia flywheel, clutch plate 219 mm

[6)] Gear oil SAE 75W90 (synthetic oil) is available as a replacement part:
Container 1.0 L (1.06 qt.) - Part No. - G052 145 A2

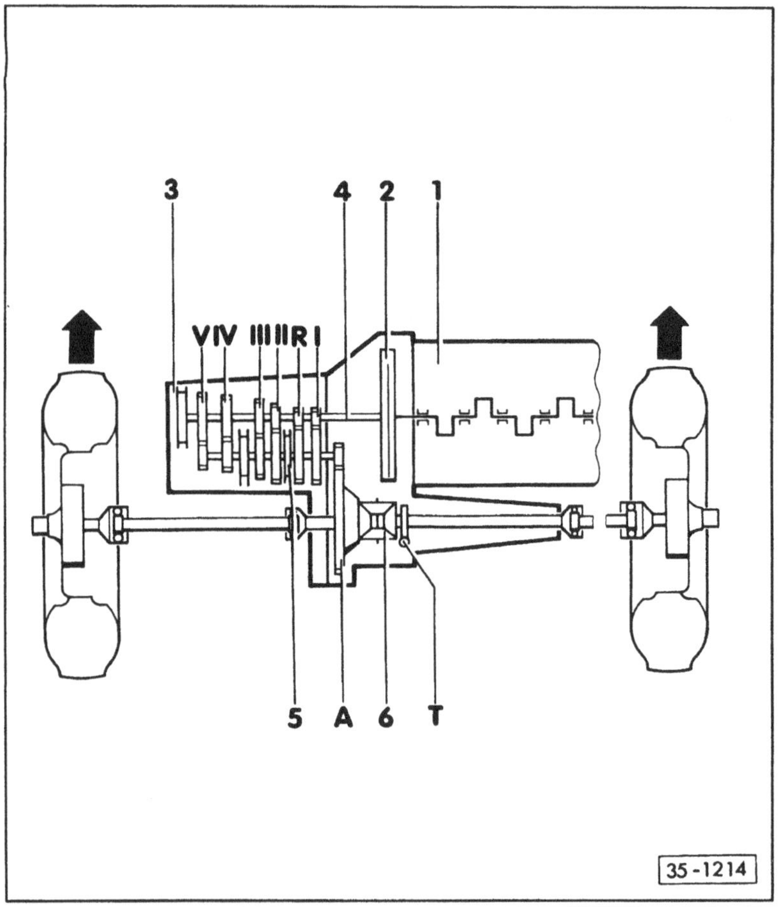

Transmission components

Designation

1 - Engine

2 - Clutch

3 - Transmission

4 - Input shaft

5 - Output shaft/pinion shaft

6 - Differential

Arrows point in direction of travel

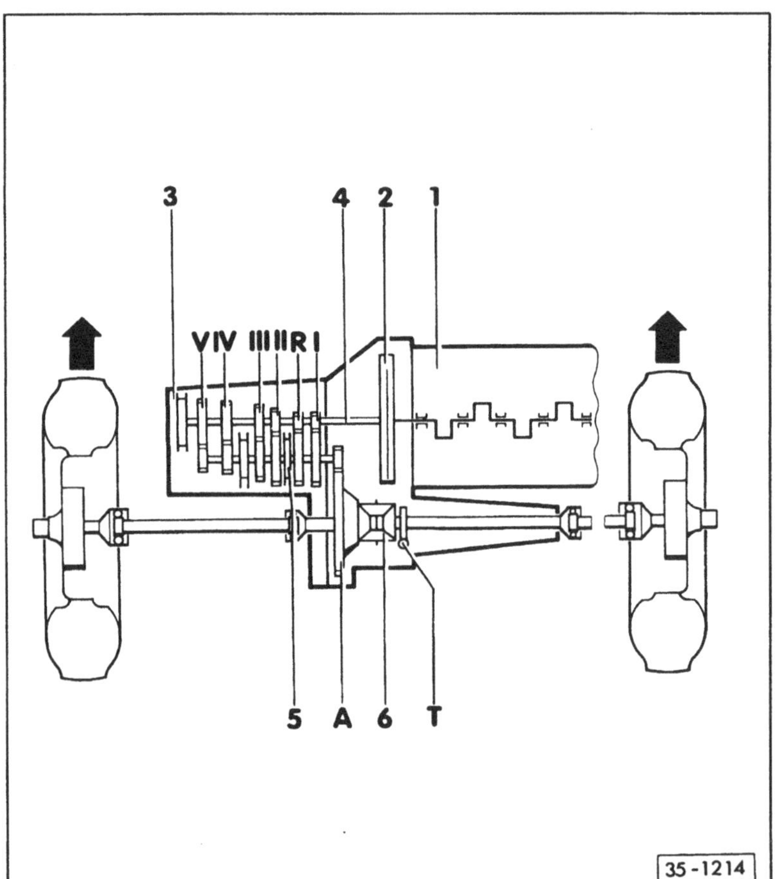

Gear ratios

I	-	1st speed gear
II	-	2nd speed gear
III	-	3rd speed gear
IV	-	4th speed gear
V	-	5th speed gear
R	-	Reverse gear
A	-	Final drive
T	-	Speedometer drive

Arrows point in direction of travel

Computing overall ratio (i)

Example:

	5th speed gear	Final drive
Drive gear	$ZG_1 = 45$	$ZA_1 = 17$
Driven gear	$ZG_2 = 34$	$ZA_2 = 84$

$i = Z_2 : Z_1$ [1)]

i_G = Gear ratio = $ZG_2 : ZG_1 = 34 : 45 = 0.756$

i_A = Axle ratio = $ZA_2 : ZA_1 = 84 : 17 = 4.941$

Overall ratio = total ratio
= $i_G \times i_A = 0.756 \times 4.941 = 3.735$

[1)] Z_1 = No. of teeth drive gear, Z_2 = No. of teeth driven gear

Repair instructions

Proper tools and the maximum possible care and cleanliness are essential for satisfactory and successful transmission repairs. Basic safety precautions apply when doing vehicle repairs.

Standard procedures used throughout the Repair Manual are summarized here.

Special tools

A summary of the special tools used in this Repair Manual can be found in the Special Tools/Catalog microfiche.

Transmission

- When installing transmission, make sure that the dowel sleeves for engine/transmission alignment are correctly located.
- When changing a transmission, transmission oil is to be checked and filled as necessary.
- Capacities and specifications ⇒ Page K 00-3.

Notes:

- *If the oil level is topped off with the transmission installed, check oil level as follows:*
- *Fill with oil, as described on page K 34-59 and replace filler plug. Start engine, engage a gear, and allow transmission to turn for approximately 2 minutes. Turn off engine. Check transmission oil ⇒ Page K 34-59 and if necessary top-off transmission oil again.*

Gaskets, oil seals

- Thoroughly clean joint surfaces and apply AMV 188 200 03 sealant.
- Apply sealant uniformly but not too thick.
- Replace O-rings.
- Radial shaft oil seals.

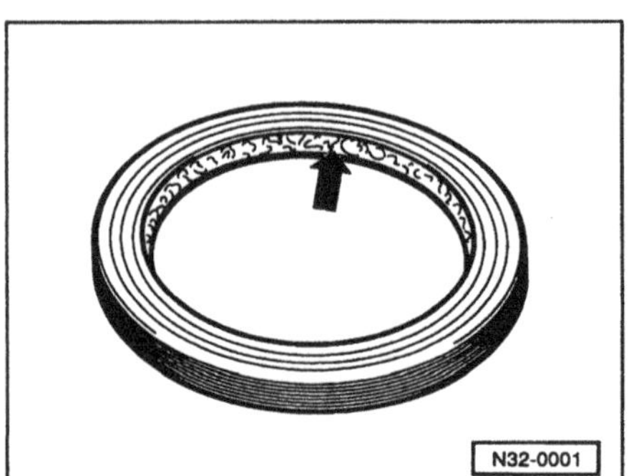

Before installing:

◄ – Lightly oil outer edge, pack space between sealing lips -arrow- with grease before installing.

After installation:

– Check transmission oil and, if necessary, top-off, specification ⇒ Page K 00-3.

Locking devices

- Always replace circlips.
- Do not overstretch circlips.
- Circlips must be properly seated in groove.

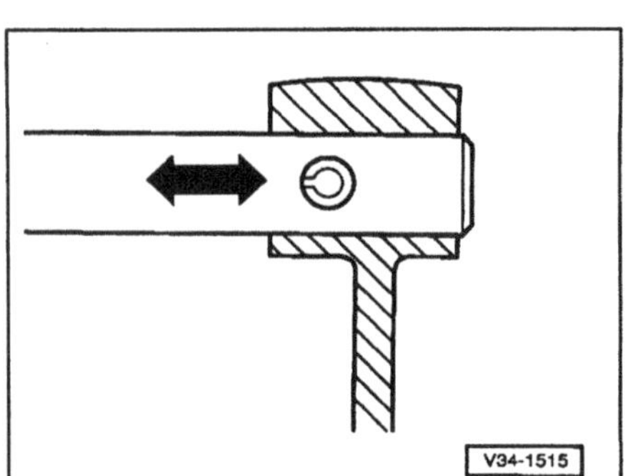

◄ ♦ Always replace spring pins
Installation position: Slot longitudinal to line of force.

Nuts and bolts

- Nuts and bolts for attaching covers and housings must be loosened and tightened in a diagonal sequence.
- Especially sensitive parts, such as clutch pressure plates, must not be distorted. Loosen and tighten nuts and bolts in stages in a diagonal sequence.
- Tightening torques as specified are for dry nuts and bolts.
- Always replace self-locking nuts.

Bearings

- Install needle bearings with lettered side (thicker metal) facing drift.
- Lubricate all transmission bearings with transmission oil before installing. Lubricate with special care when measuring friction torque.
- Tapered roller bearings installed on one shaft must be replaced as a set. Use bearings from same manufacturer, if possible.
- Heat inner races of tapered roller bearings to approximately 100°C (212°F) before installing.
- Do not interchange inner and outer races of bearings of same size; bearings are paired.

Shims

- Measure shims at several points with a micrometer. Different tolerance in the same size shim make it possible to find the exact shim thickness required.
- Check for signs of burrs and damage. Only install shims that are in perfect condition.

Synchronizer rings

- Do not interchange. When reusing always install on same gear.
- Check for signs of wear and replace if necessary.
- Coat with transmission oil before installing.

Gears

- Clean and heat gears to approximately 100°C (212°F) before installing.
- Temperature can be checked with VAG 1558 Temperature Tester or equivalent.

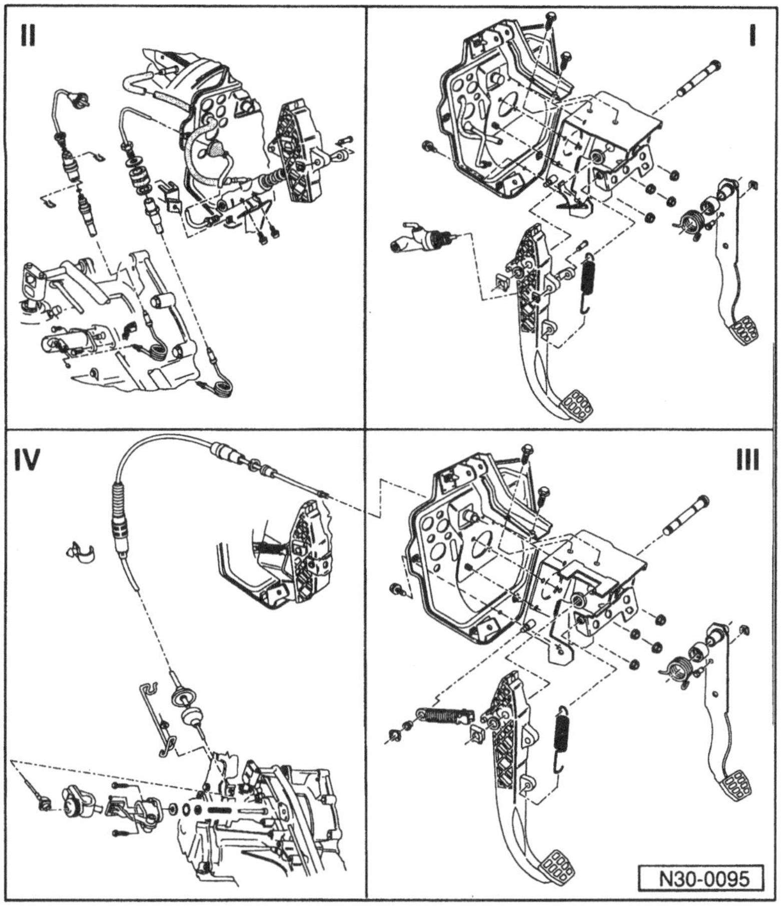

Clutch control, servicing

I - **Pedal cluster - hydraulic clutch control, assembly overview** ⇒ **Page K 30-3**

II - **Hydraulics, assembly overview** ⇒ **Page K 30-8**

III - **Pedal cluster - mechanical clutch control, assembly overview** ⇒ **Page K 30-19**

IV - **Mechanics, assembly overview** ⇒ **Page K 30-24**

Notes:

- *Disconnect Battery Ground (GND) strap.*
- *For vehicles with coded radio, obtain radio code.*
- *Lubricate all bearings and contact surfaces with MoS_2 grease.*

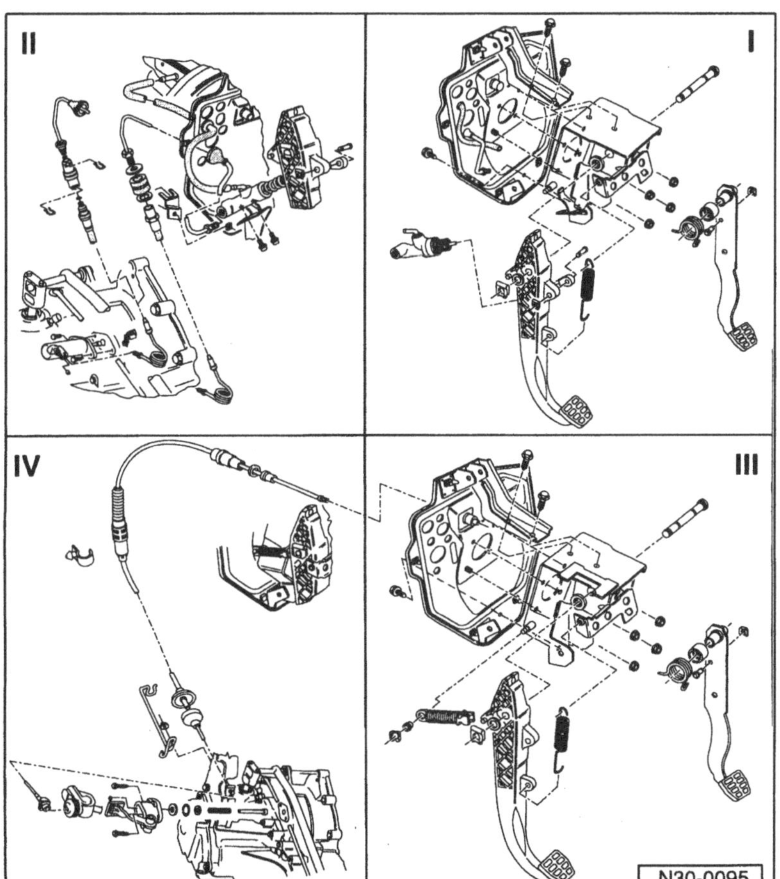

- *Before starting repairs to the pedal cluster, remove the following components:*

Driver's side storage compartment

⇒ General body repairs; Repair group 70; Trim/insulation, removing and installing

Steering column trim

⇒ Suspension, Wheels, Brakes, Steering; Repair group 48; Steering column and column tube, removing and installing

Relay plate (releasing from mounting)

⇒ Electrical system; Repair group 97; wiring

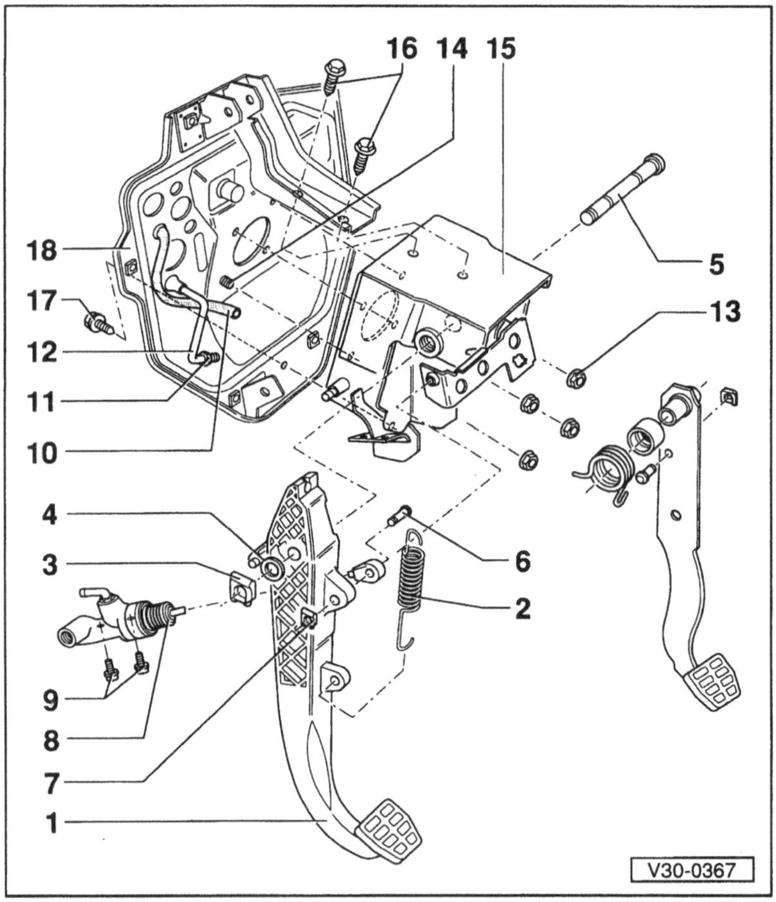

Pedal cluster / hydraulic clutch control, assembly overview

1 - **Clutch pedal**
- Removing and installing ⇒ Page K 30-7
-

2 - **Spring**

3 - **Locking clip**

4 - **Washer**

5 - **Pivot pin for clutch and brake pedal**
- Drive out with drift

6 - **Pivot pin for master cylinder operating rod**

7 - **Locking clip**

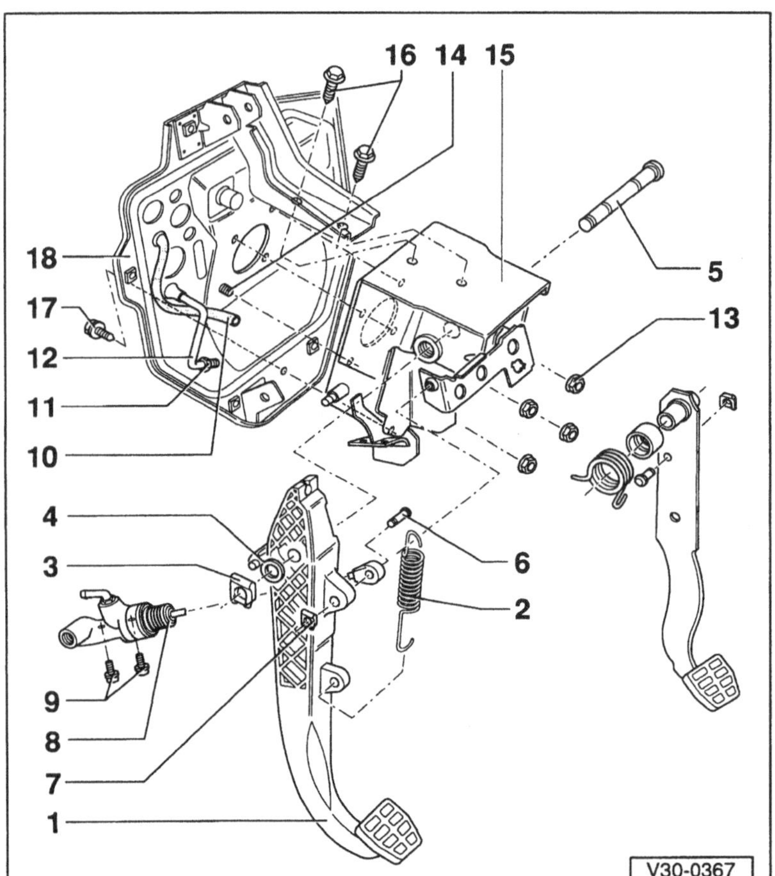

8 - **Master cylinder**

9 - **Bolt**
- Tightening torque: 20 Nm (15 ft lb)
- For master cylinder to mounting bracket

10 - **Supply hose**

11 - **Clutch hudraulic line union nut**
- Tightening torque: 20 Nm (15 ft lb)

12 - **Pressure line**

13 - **Hex nut**
- Tightening torque: 25 Nm (18 ft lb)
- Always replace
- For mounting bracket and brake servo to mounting plate

14 - **Brake servo stud**

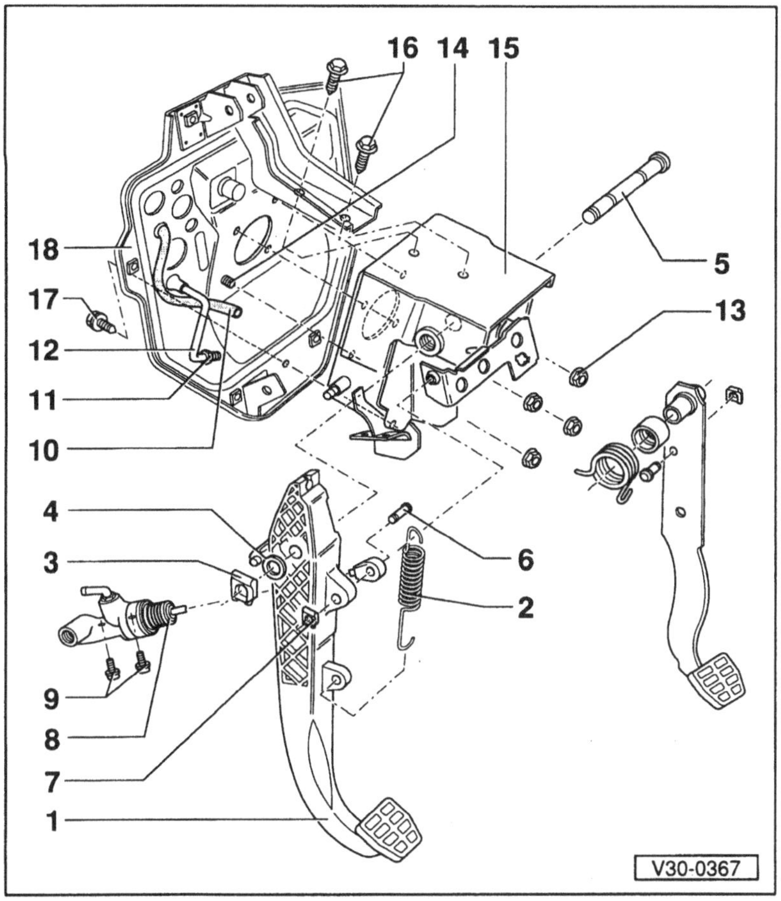

15 - Mounting bracket
- Replacing ⇒ Fig. 1
- Before removing, unbolt steering column and lay to side

⇒ Suspension, Wheels, Brakes, Steering; Repair group 48; Steering column and column tube, removing and installing

16 - Hex bolt
- Tightening torque: 25 Nm (18 ft lb)

17 - Hex bolt
- Tightening torque: 25 Nm (18 ft lb)

18 - Mounting plate
- Need not be disassembled to remove pedal cluster
- Seal to dash panel with AKD 497 010 04 R 10

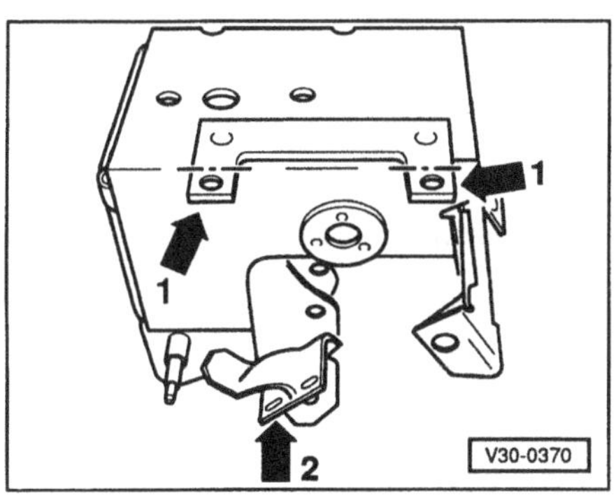

◄ **Fig. 1 Mounting bracket, removing and installing**

As a replacement part, the mounting bracket is only available with both bearing plates -arrow 1- (for the clutch pedal on vehicles with mechanical clutch control) and mounting bracket -arrow 2- (for the master cylinder on vehicles with hydraulic clutch control).

Note:

- *The bearing plates -arrow 1- are to be reduced along the line as marked when installing in vehicles with hydraulic clutch control, otherwise insufficient pedal travel will affect function of the clutch.*
- *When installing the mounting bracket in vehicles with mechanical clutch control, the mounting bracket for master cylinder can remain in place.*

Clutch pedal - Hydraulic clutch control, removing and installing

Removing

- Separate clutch pedal from master cylinder ⇒ Page K 30-13.
- Unhook clutch pedal spring.
- Remove clutch pedal securing clip from pivot pin.
- Drive pivot pin out to the right until clutch pedal can be removed.
- Remove clutch pedal.

Installing

Installation is carried out in reverse sequence.

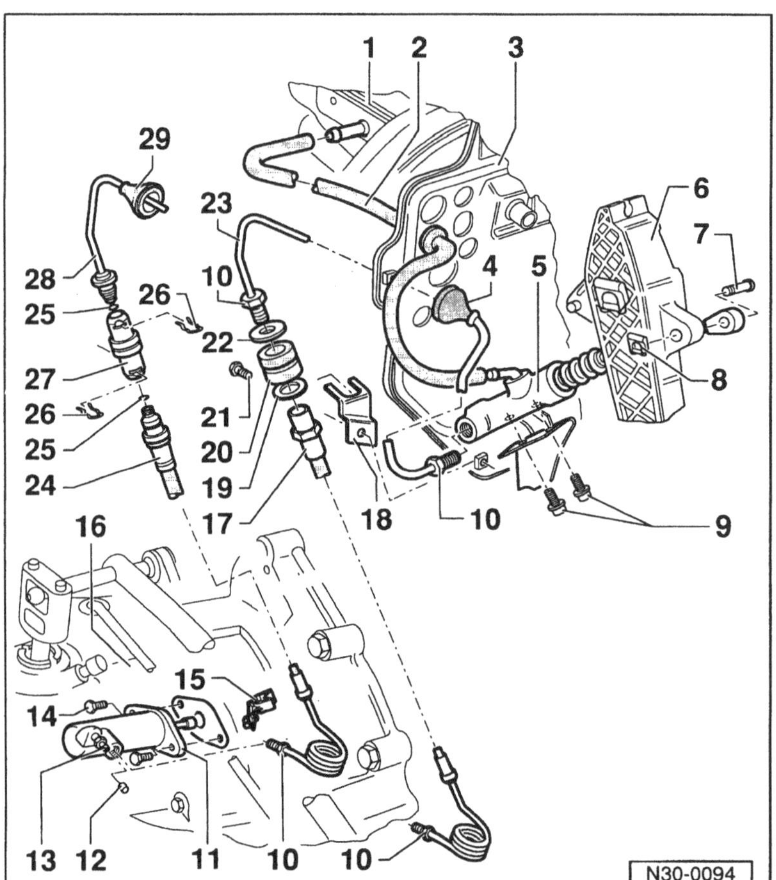

Hydraulic system, assembly overview

1 - **Brake fluid reservoir**

2 - **Supply hose**

3 - **Mounting plate**

4 - **Grommet**
- For pressure line -23-
- Modification: From June 1991, new installation position ⇒ -29-

5 - **Master cylinder**

6 - **Clutch pedal**
- Removing and installing ⇒ Page K 30-7

7 - **Pivot pin for master cylinder operating rod**

8 - **Locking clip**

9 - **Bolt**
- Tightening torque: 20 Nm (15 ft lb)
- For master cylinder to mounting bracket

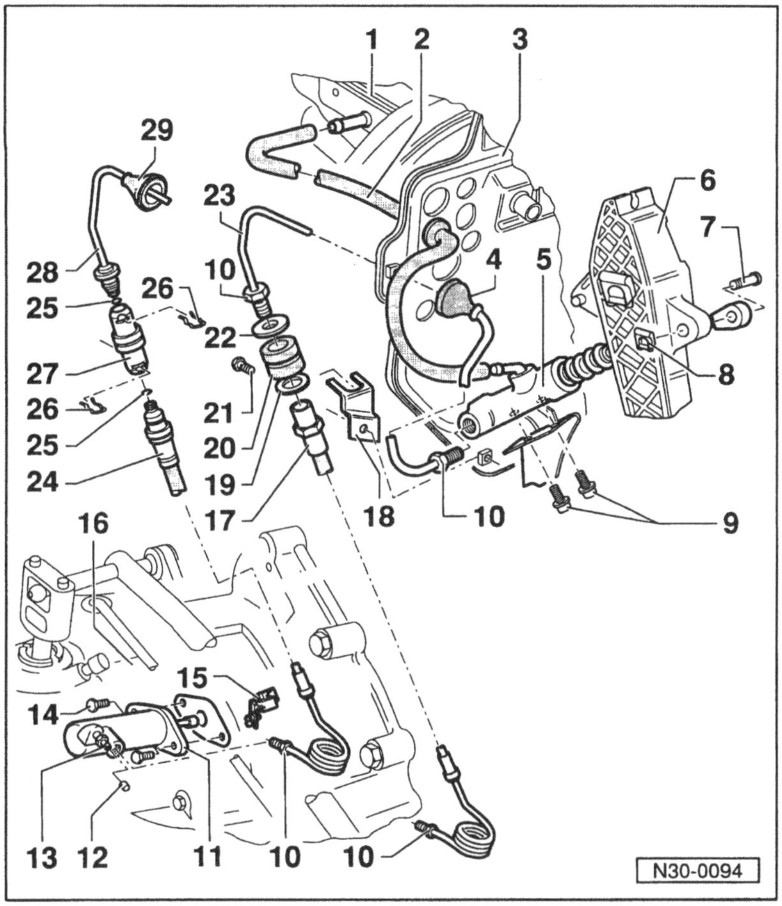

10 - Clutch hydraulic line union nut
- Tightening torque: 20 Nm (15 ft lb)

11 - Slave cylinder
- Removing and installing ⇒ Page K 30-13
- Do not operate clutch pedal after removing the slave cylinder

12 - Dust cap

13 - Bleed valve/master cylinder
- Bleeding clutch system ⇒ Page K 30-18

14 - Hex bolt
- Tightening torque: 25 Nm (18 ft lb)

15 - Hose bracket
- Secure to transmission mounting bracket

16 - Transmission

17 - Pressure hose
- Modification: Discontinued from May 1992

18 - Hose bracket
- Secured to mounting plate

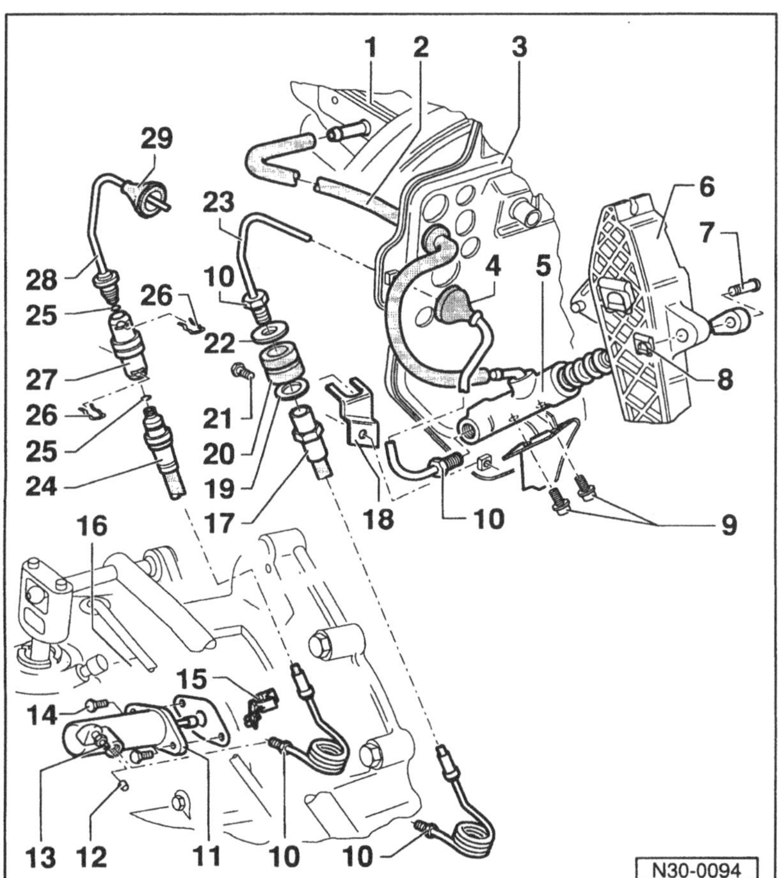

19 - Washer
- Larger inner diameter
- Convex towards pressure hose
- Modification: Discontinued from May 1992

20 - Bushing
- Modification: Discontinued from May 1992

21 - Hex bolt
- Tightening torque: 25 Nm (18 ft lb)

22 - Washer
- Smaller inner diameter
- Convex towards pressure pipe -23-
- Modification: Discontinued from May 1992

23 - Pressure line
- Modification: Discontinued from May 1992

24 - Pressure hose
- Introduced from May 1992

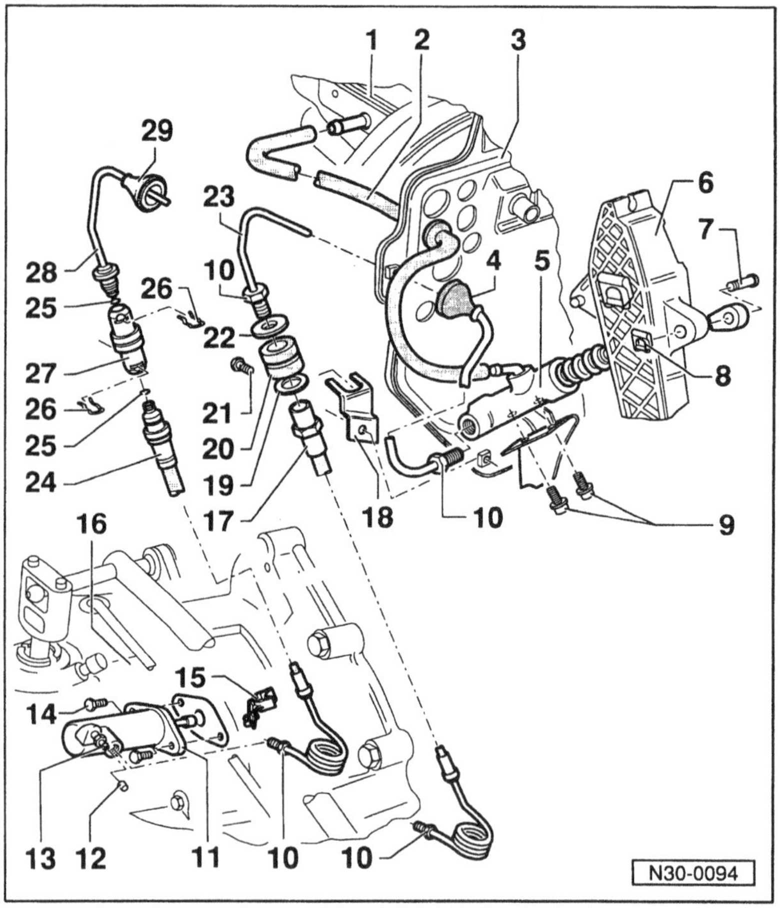

25 - O-ring

- ♦ Always replace
- ♦ Install onto line/hose connection
- ♦ Install with brake fluid

26 - Clip

- ♦ Removing ⇒ Fig. 1
- ♦ Installing: press in by hand

27 - Connecting piece

- ♦ Introduced from May 1992

28 - Pressure line

- ♦ Introduced from May 1992

29 - Grommet

- ♦ Introduced from May 1992

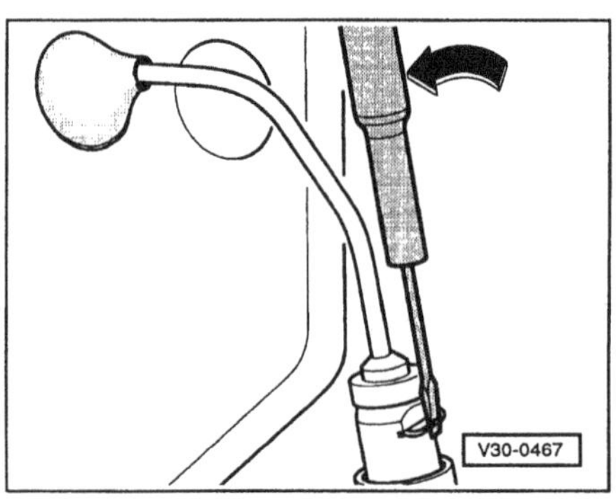

◄ **Fig. 1 Clip, removing**

– Guide screwdriver between clip and connecting piece and turn connection piece so that flattened section in area of clip is pointing forward.

– Carefully pry clip out and remove.

Slave cylinder, removing and installing

Removing

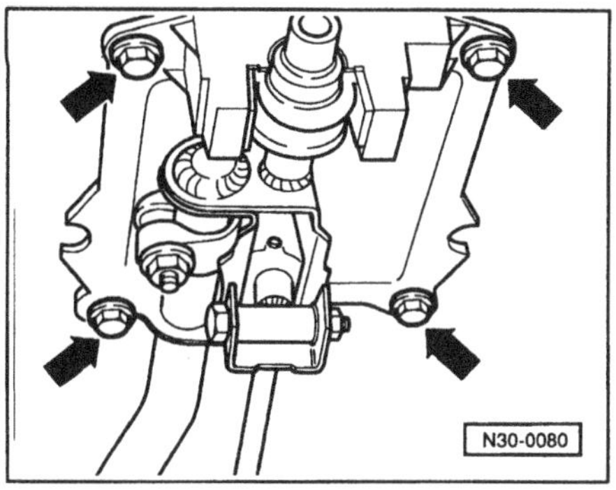

◄ – Loosen hex bolts -arrows- securing shift housing.

The shift housing must be able to move. If necessary, remove boot frame and gear lever boot.

– Remove noise insulation tray below engine/transmission, if installed.

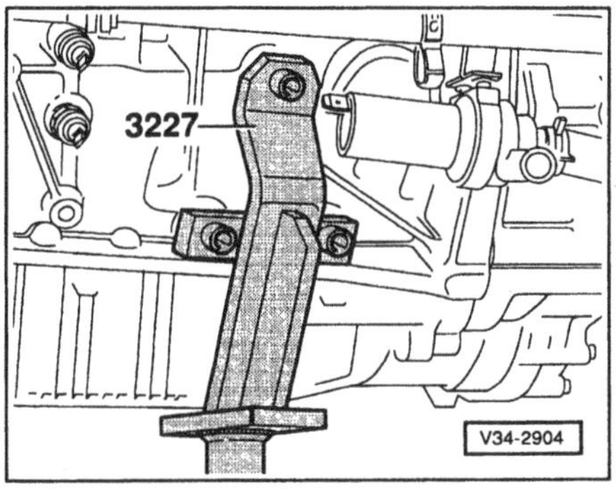

– Attach engine bracket to engine block.

◄ 3227 engine bracket for vehicles with 5-cylinder engine.

3230 engine bracket for vehicles with 4-cylinder engine.

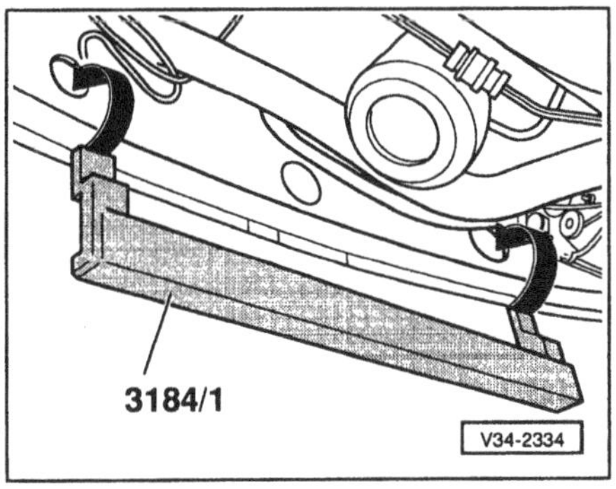

◄ – Attach 3184/1 auxiliary support for 3184 removal device into cross panel -arrow-. If necessary, remove wiring harness and lines from cross panel.

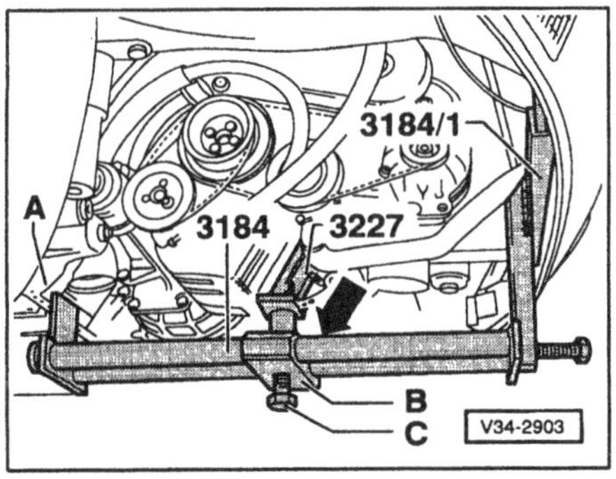

◄ – Install 3184 removal device into 3184/1 auxiliary support and into subframe -A-. Screw spindle -C- into hole on engine mounting.

– Align 3184 removal device so that slide -B- is resting against the right-hand tube -arrow-.

– Secure 3184 removal device to subframe -A- and to 3184/1 auxiliary support by tightening the angle plates.

– Release weight from subframe using spindle -C- to raise engine/transmission.

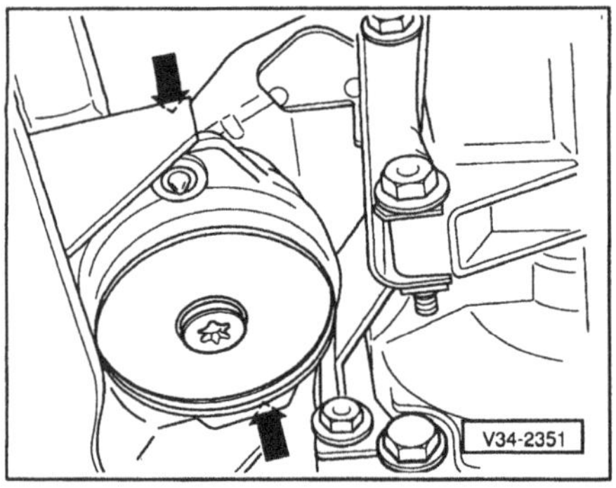

Transmission to build date 09 08 2

◄ – Remove rear bonded rubber mounting bolts -arrows- on assembly carrier.

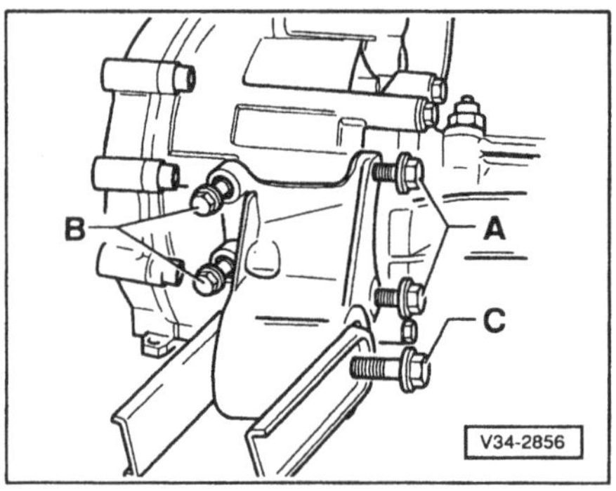

Transmission from build date 10 08 2

◄ – Remove bolt -C- from rear of assembly carrier.

Note:

Bolts -A- and -B- are not removed.

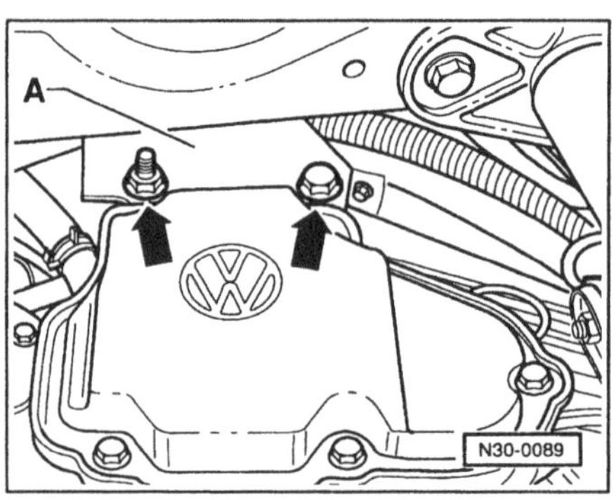

◄ – Unbolt -arrows- mounting bracket -A- from transmission.

– Lower transmission using support device spindle.

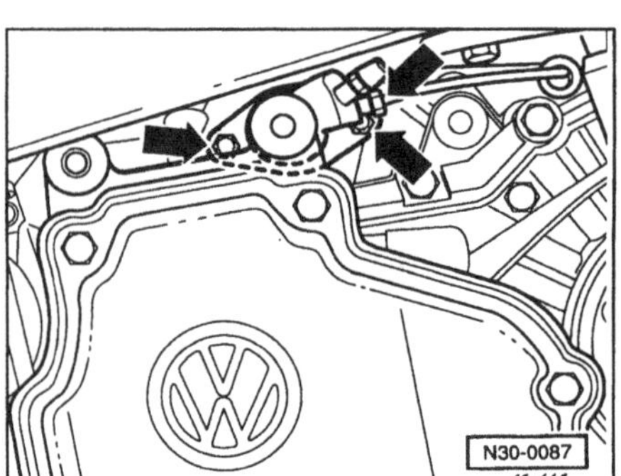

◄ – Remove line/hose and bolts for securing slave cylinder -arrows-.

– Remove slave cylinder.

Installing

Install in reverse sequence and note the following:

– Bleed clutch system after installing slave cylinder ⇒ Page K 30-18.

Tightening torques

Left transmission mounting to transmission	65 Nm (48 ft lb)
Bonded rubber mounting to assembly carrier	45 Nm (33 ft lb)
Rear assembly mounting to assembly carrier	200 Nm (148 ft lb)
Shift housing to floor plate	9 Nm (7 ft lb)
Slave cylinder to transmission	25 Nm (18 ft lb)
Line/hose to slave cylinder	20 Nm (15 ft lb)

Clutch hydraulic system, bleeding

Bleed clutch system using the US 1116 B brake filling and bleeding appliance.

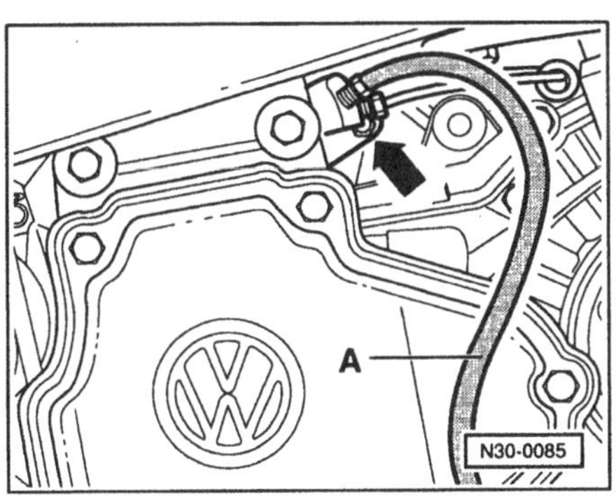

– Remove noise insulation tray below engine/transmission, if installed.

– Connect bleeder hose -A- to slave cylinder -arrow- and open bleeder valve.

– Push clutch pedal several times after completion of bleeding process.

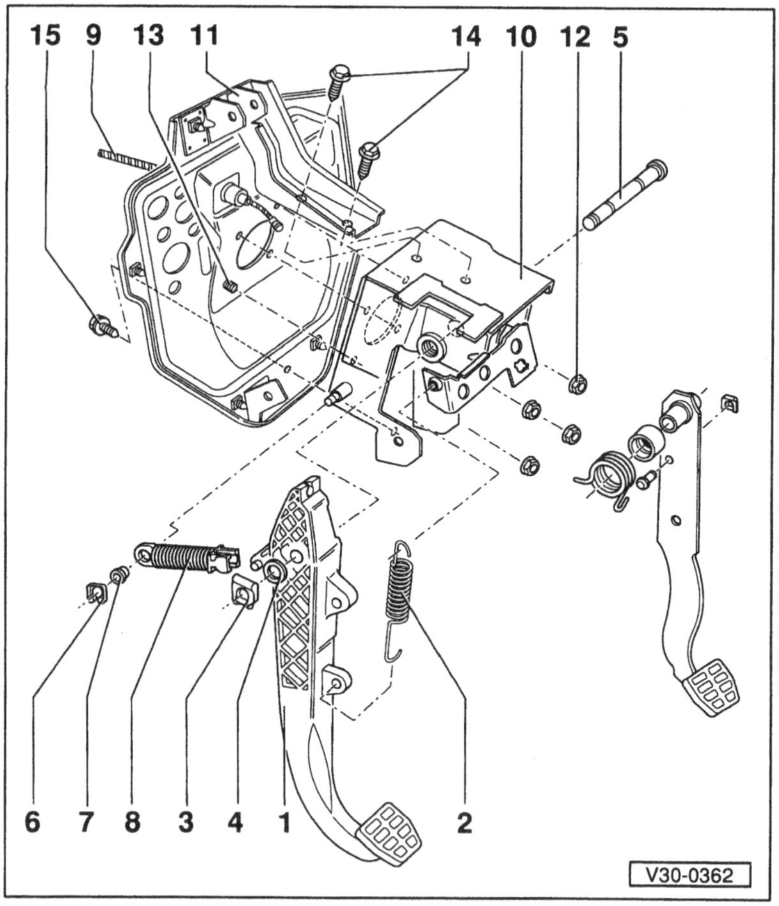

Pedal cluster / mechanical clutch control, assembly overview

1 - Clutch pedal
- Removing and installing ⇒ Page K 30-23

2 - Spring
- If installed, remove before removing clutch pedal

3 - Locking clip

4 - Washer

5 - Pivot pin for brake and clutch pedal
- Drive out with drift

6 - Locking clip

7 - Bushing

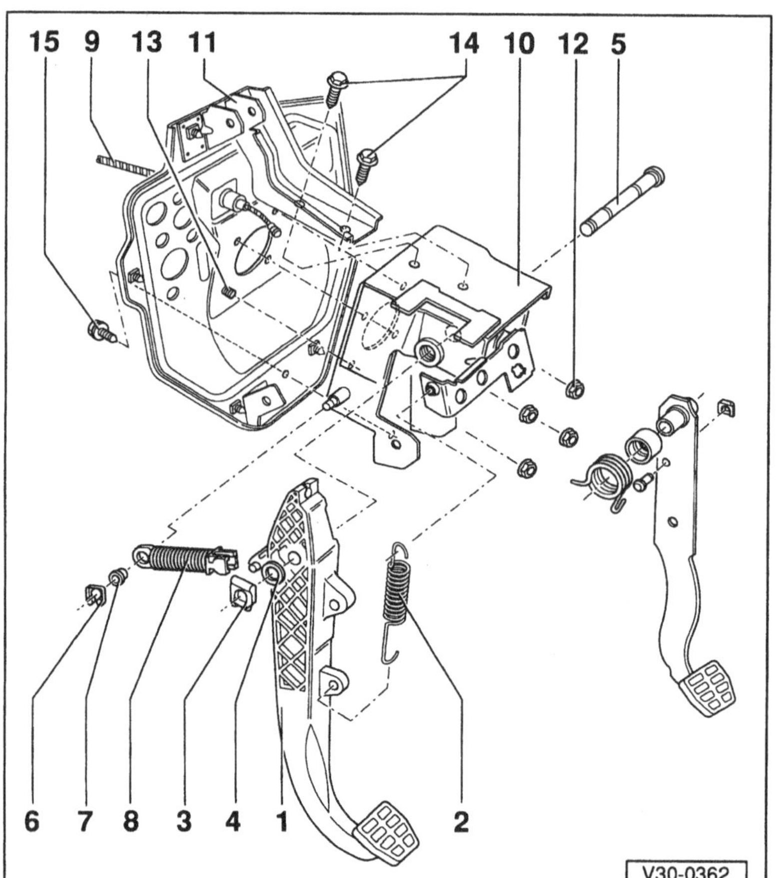

8 - Over-center spring
- Removing and installing ⇒ Page K 30-22

9 - Clutch cable
- Removing and installing ⇒ Page K 30-28

10 - Mounting bracket
- Before removing unbolt steering column and lay to side

⇒ Suspension, Wheels, Brakes, Steering; Repair group 48; Removing and installing steering column and column tube

11 - Mounting plate
- Need not be removed to remove pedal cluster
- Seal to dash panel with AKD 497 010 04 R 10

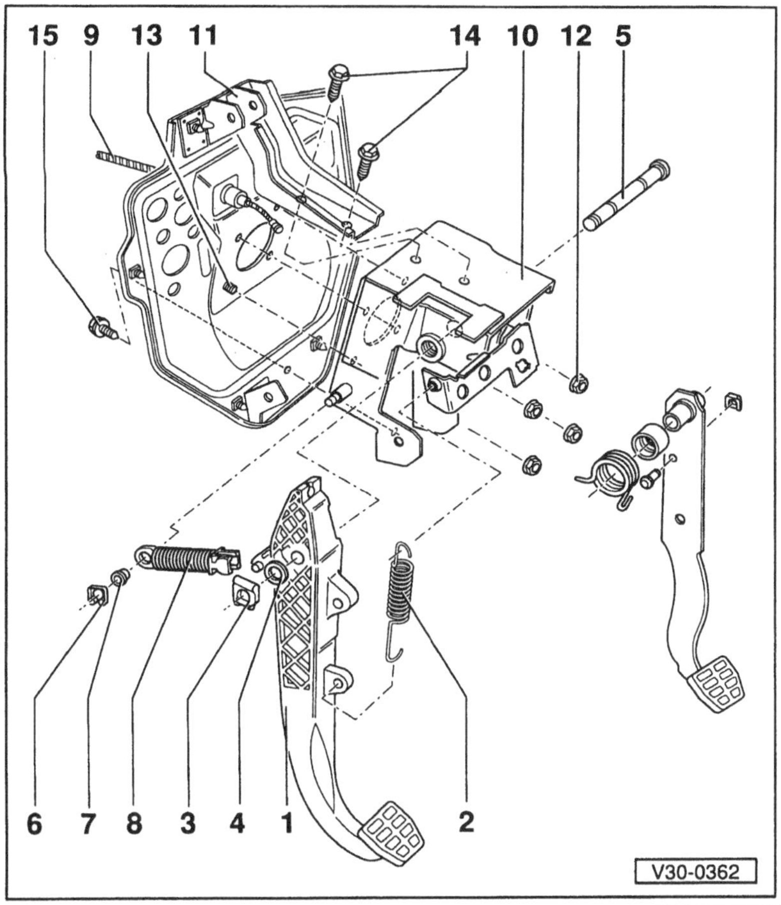

12 - Hex nut
- Tightening torque: 25 Nm (18 ft lb)
- Always replace
- For mounting bracket and brake servo to mounting plate

13 - Brake servo stud

14 - Hex bolt
- Tightening torque: 25 Nm (18 ft lb)

15 - Hex bolt
- Tightening torque: 25 Nm (18 ft lb)

Over-center spring, removing and installing

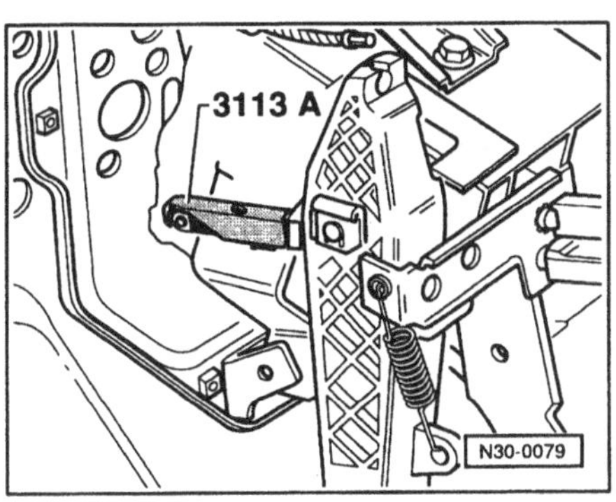

Removing

– Remove carpet in area of over-center spring.

– Pry securing clip off with a screwdriver and remove washer.

◄ – Tension over-center spring with clutch pedal and secure 3113A retaining clamp from above.

> ***CAUTION!***
>
> ***Do not press retaining clamp on from below: there is a danger that the over-center spring may pop out of the retaining clamp during installation.***

– Move clutch pedal onto stop.

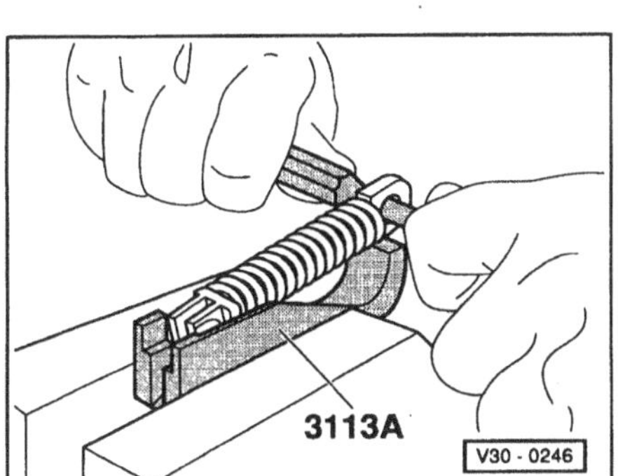

– Remove over-center spring together with 3113A retaining clamp.

◄ – Remove over-center spring from the retaining clamp and install.

Installing

Install in reverse sequence.

Clutch pedal / Mechanical clutch control, removing and installing

Removing

– Remove over-center spring ⇒ Page K 30-22.

– Unhook clutch cable ⇒ Page K 30-28.

– Unhook clutch pedal spring, if equipped ⇒ Page K 30-19.

– Remove clutch pedal locking clip from pivot pin.

– Drive pivot pin out to right until clutch pedal can be removed.

– Remove clutch pedal from pivot pin.

Installing

– Install in the reverse sequence.

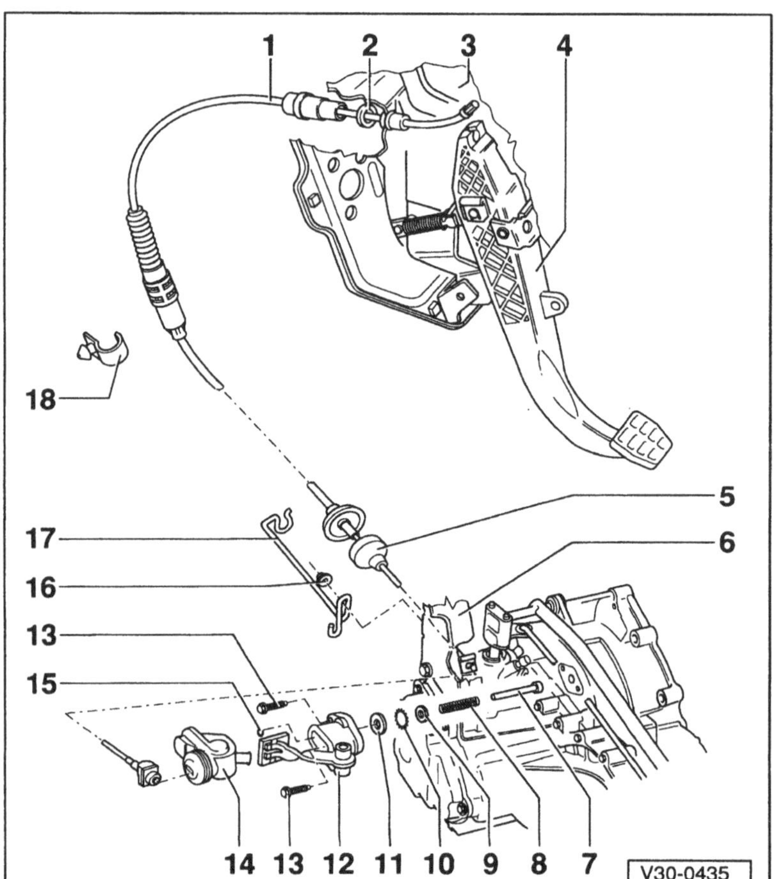

Clutch cable assembly, components

1 - **Clutch cable with adjustment mechanism**
   - ♦ Removing and installing ⇒ Page K 30-28

2 - **O-ring**

3 - **Mounting plate with mounting bracket**

4 - **Clutch pedal**
   - ♦ Removing and installing ⇒ Page K 30-23

5 - **Rubber buffer**

6 - **Support with transmission mounting bracket**

7 - **Plunger**
   - ♦ Install after installing lockwasher -10- and flat washer -11-

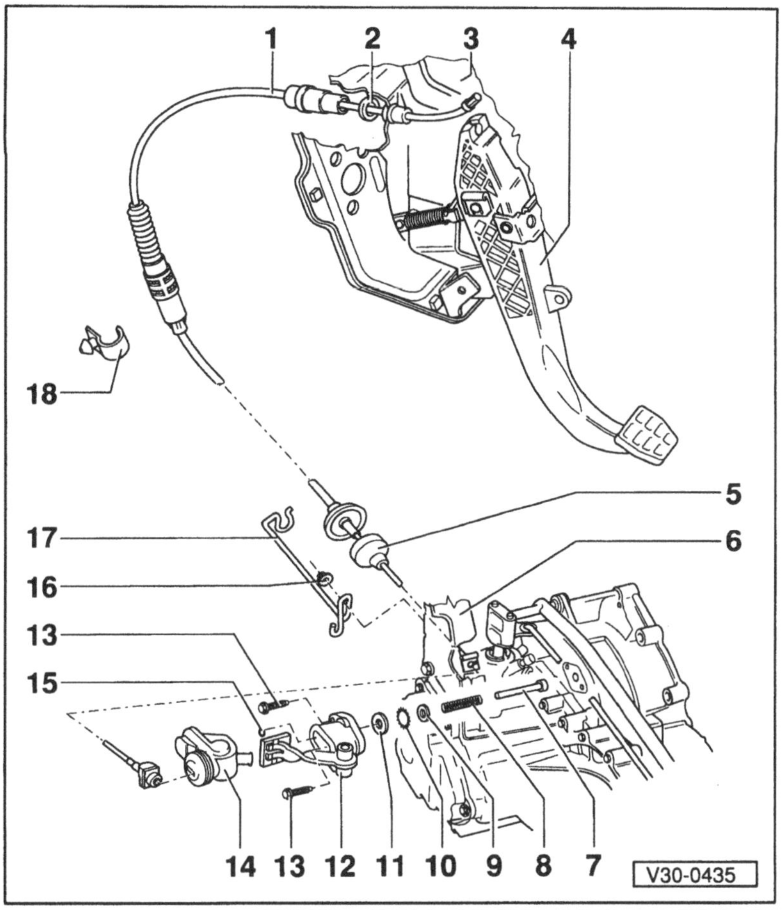

8 - Spring

9 - Washer

10 - Lockwasher
- ♦ Always replace
- ♦ Drive off over washer -11- with 30-24 A drift
- ♦ Remove plunger -7- first
- ♦ Installing ⇒ Fig. 1

11 - Washer
- ♦ Always replace
- ♦ Must be able to move freely

12 - Operating lever and housing
- ♦ Disconnect clutch cable before removing ⇒ Page K 30-23
- ♦ Modification:
  With balance weight from August 93

13 - Bolt
- ♦ Tightening torque: 25 Nm (18 ft lb)

14 - Boot

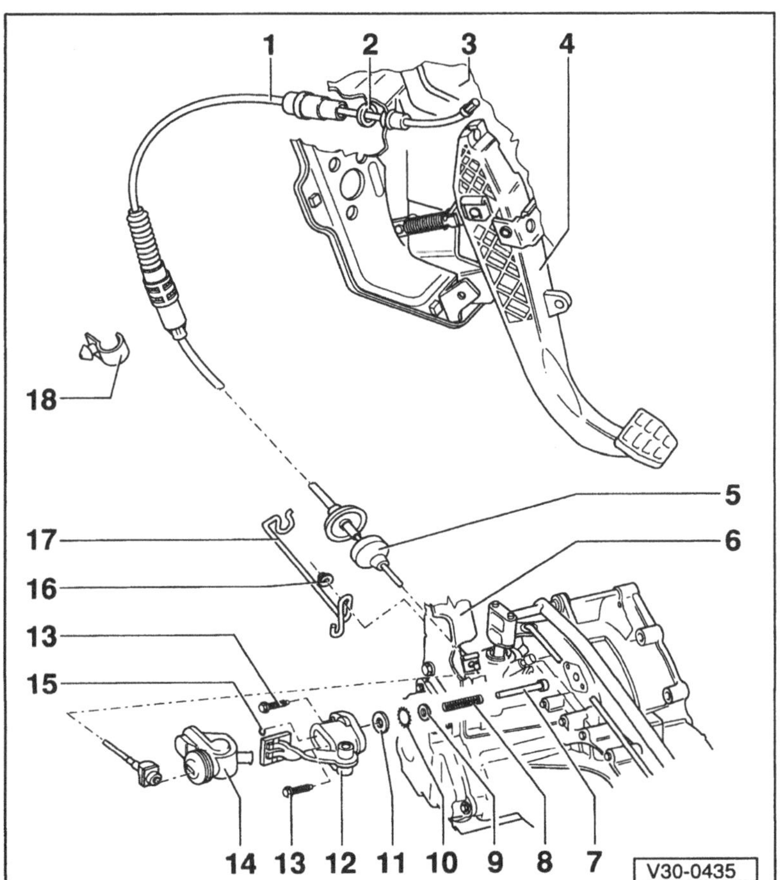

15 - Circlip
- ♦ Install approximately 8 mm (5/16 in.) from end of plunger

16 - Hex nut
- ♦ For retainer to front transmission mounting bracket

17 - Retainer

18 - Cable clip

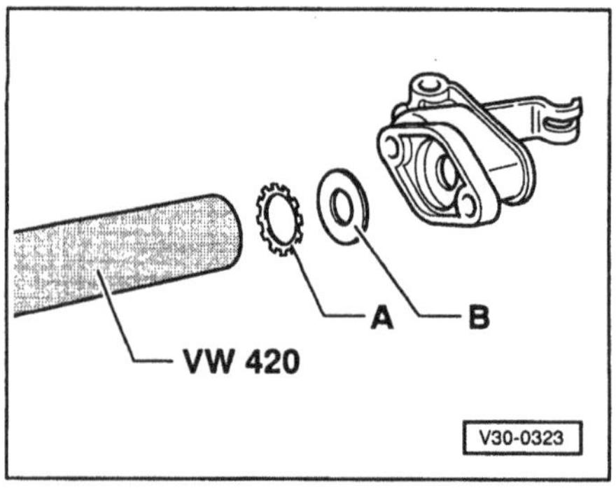

◄ **Fig. 1 Lockwasher -A-, installing**

– Drive in carefully

Installation position:

Curved side to housing

Washer -B- must remain free to move

Clutch cable, removing and installing

When removing and installing the clutch cable, the adjuster mechanism must be tensioned with the retaining strap. The strap is used for tensioning the adjusting mechanism when installing new cables. It is included with new cables.

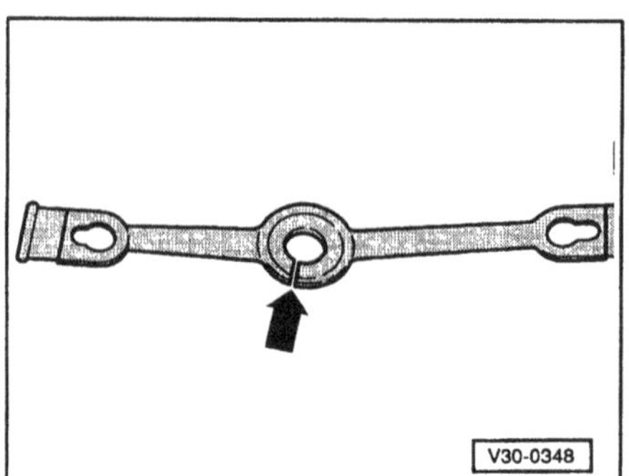

◄ – Remove retaining strap. If necessary carefully cut eye open -arrow-.

> ***CAUTION!***
>
> ***Do not kink the clutch cable or correct functioning of the adjuster mechanism cannot be assured.***

Removing

– In cases of complaint, check function of clutch cable before removal, ⇒ Page K 30-31. Additionally, check shift mechanism adjustment ⇒ Page K 34-23.

– Fully push clutch pedal onto stop at least 5 times.

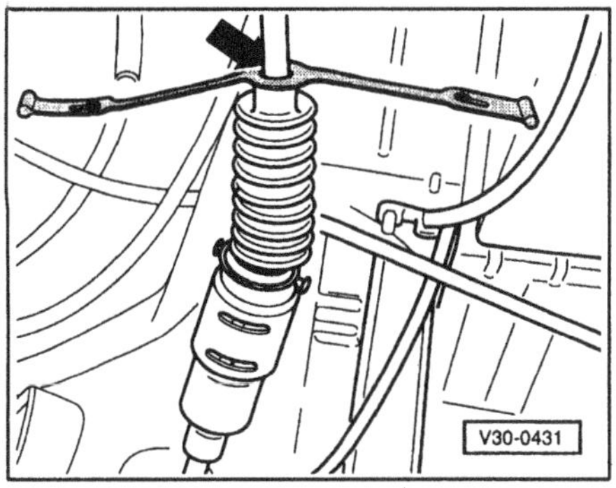

◄ – Hook retaining strap eye on above protective sleeve.

Installation position:

Slit -arrow- towards dash panel (obstructed here by clutch cable).

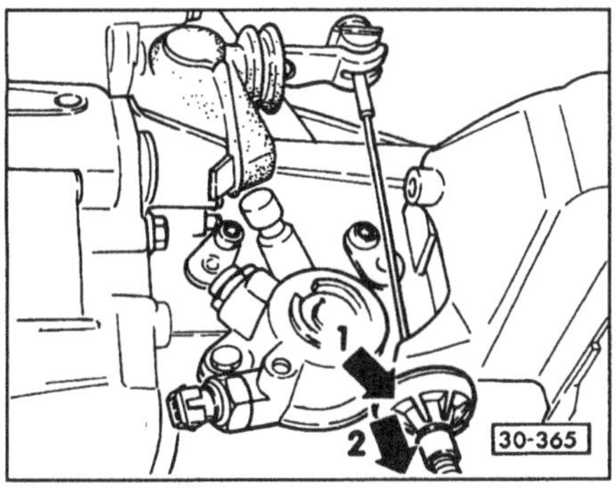

◄ – Install clutch cable, using rubber buffer plastic holder -arrow 1-, in direction of travel onto stop -arrow 2-.

This will compress adjustment mechanism.

If the adjustment mechanism cannot be compressed, a malfunction has occurred. Removal of clutch cable is necessary, resulting in destruction of the clutch cable.

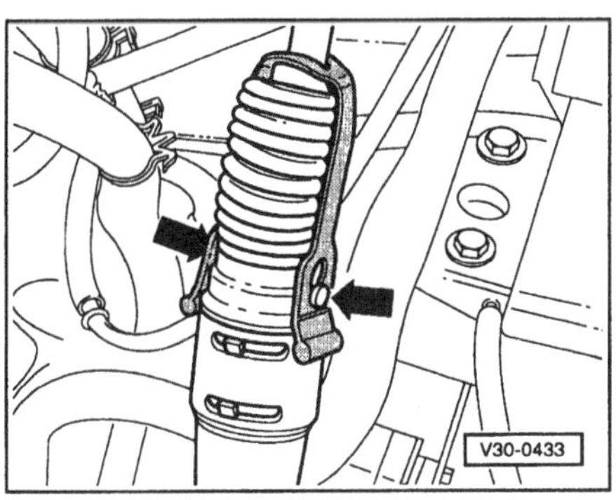

◄ – Hook retaining strap tabs onto adjustment mechanism pins -arrows- (2nd technician required).

– Unhook clutch cable at transmission operating lever.

– Unhook eye at clutch pedal.

– Remove rubber buffer from support and remove cable.

– Unhook clutch cable from retainer and cable clip.

Installing

– Lubricate clutch cable attachment point at clutch pedal with MoS_2 grease.

– Hook clutch cable onto operating lever.

– Install rubber buffer into support.

– Hook clutch cable into clutch pedal and install into retainer and cable clip.

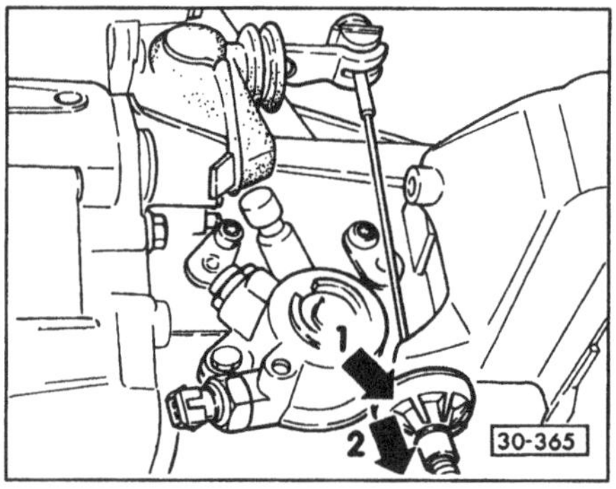

– After installing clutch pedal remove retaining strap in a manner so that it can be used again ⇒ Page K 30-28.

Perform a functional check after installing clutch cable:

Functional check

♦ Clutch cable adjustment mechanism must be undamaged.

– Push clutch pedal fully onto stop at least 5 times.

◄ – Remove clutch cable sleeve, at plastic holder, -arrow 1- approximately 10 mm (3/8 in.) in direction of travel -arrow 2-. It must be free to move easily.

Notes:

♦ *If this functional check is to be repeated several times, push clutch pedal at least 5 times before each additional check because adjustment mechanism is locked after each check and can not be compressed.*

♦ *Do not open automatic adjuster mechanism housing. It contains a number of balls which cannot be inserted by hand.*

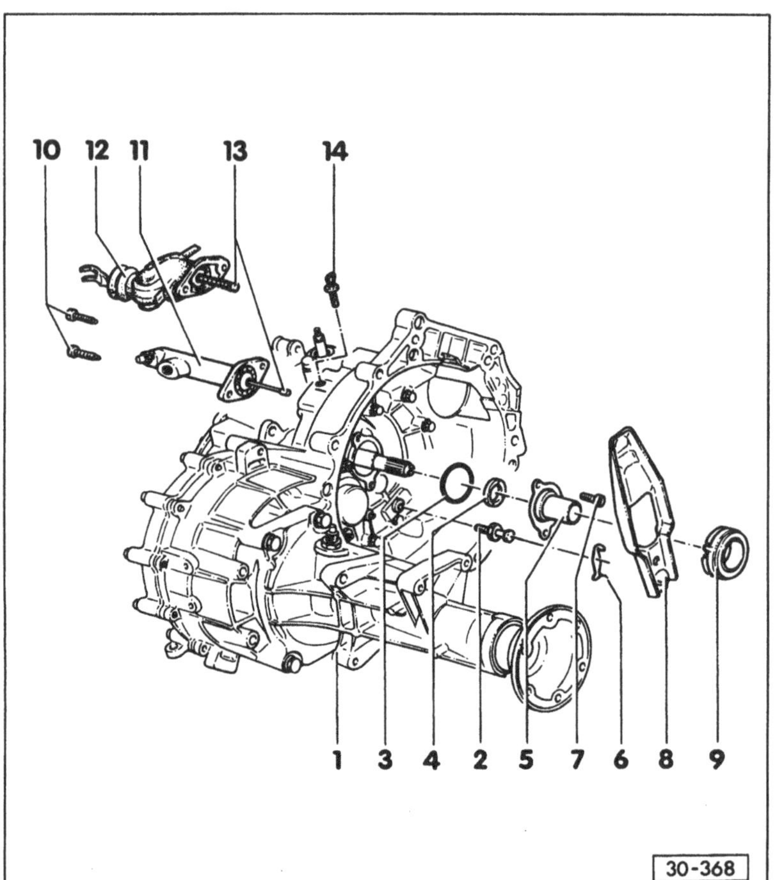

Clutch release mechanism, servicing

1 - Transmission

2 - Ball stud
- ♦ Tightening torque: 25 Nm (18 ft lb)
- ♦ Lubricate with MoS_2 grease

3 - O-ring
- ♦ Always replace

4 - Input shaft oil seal
- ♦ Replacing ⇒ Page K 34-99

5 - Guide sleeve
- ♦ Lubricate in area of release bearing with MoS_2 grease

6 - Retaining spring
- ♦ Secure to clutch release lever

7 - Bolt
- ♦ Tightening torque: 20 Nm (15 ft lb)

8 - Clutch release lever

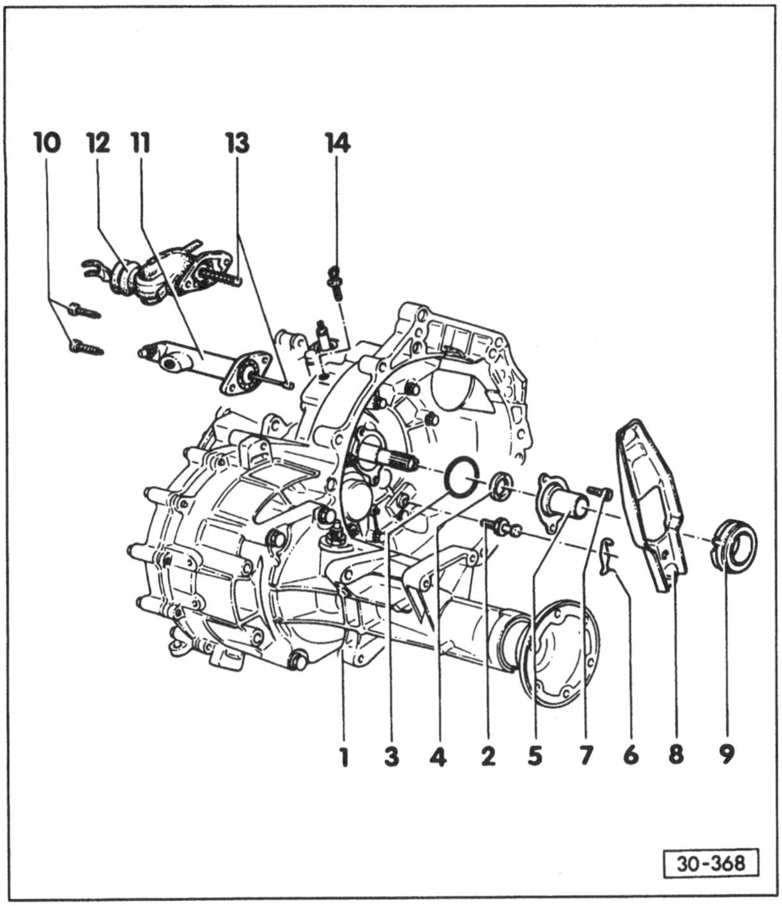

9 - Release bearing
- Do not wash bearing, wipe only
- Replace noisy bearings
- Lubricate surfaces which contact release lever with MoS_2 grease

10 - Bolt
- Tightening torque: 25 Nm (18 ft lb)

11 - Slave cylinder
- Removing and installing ⇒ Page K 30-13

12 - Mechanical clutch mechanism
- Assembly overview ⇒ Page K 30-19

13 - Plunger
- Grease end of plunger with MoS_2 grease

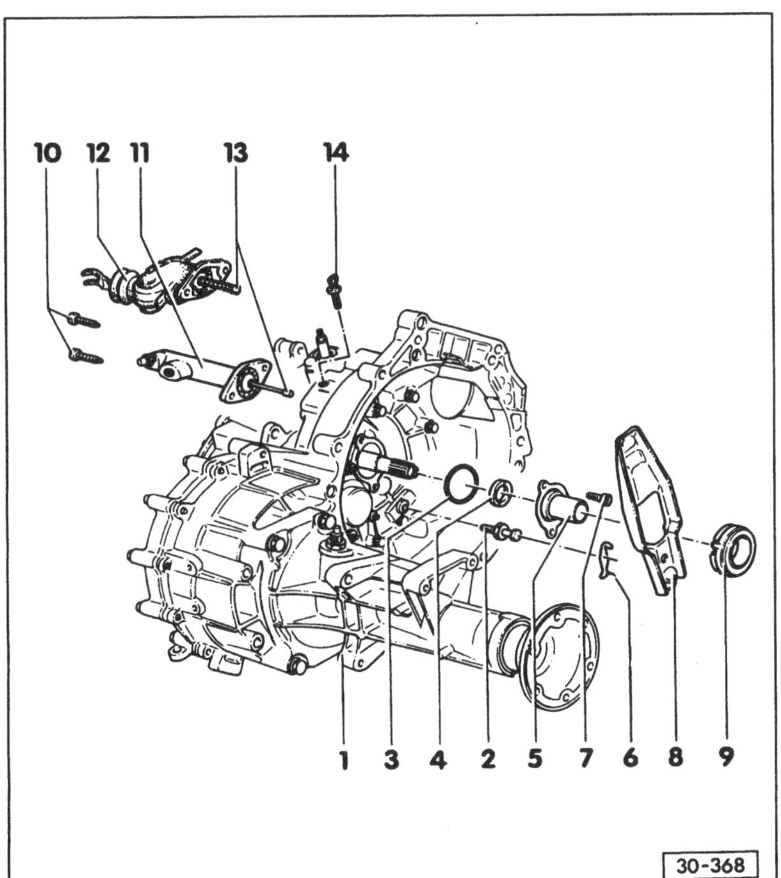

14 - Assembly bolt
- Secures clutch release lever when installing transmission, ⇒ Page K 34-46
- Remove after transmission has been installed

Note:

If the assembly bolt -14- is not available, an M8 x 35 bolt can be used.

Clutch, servicing

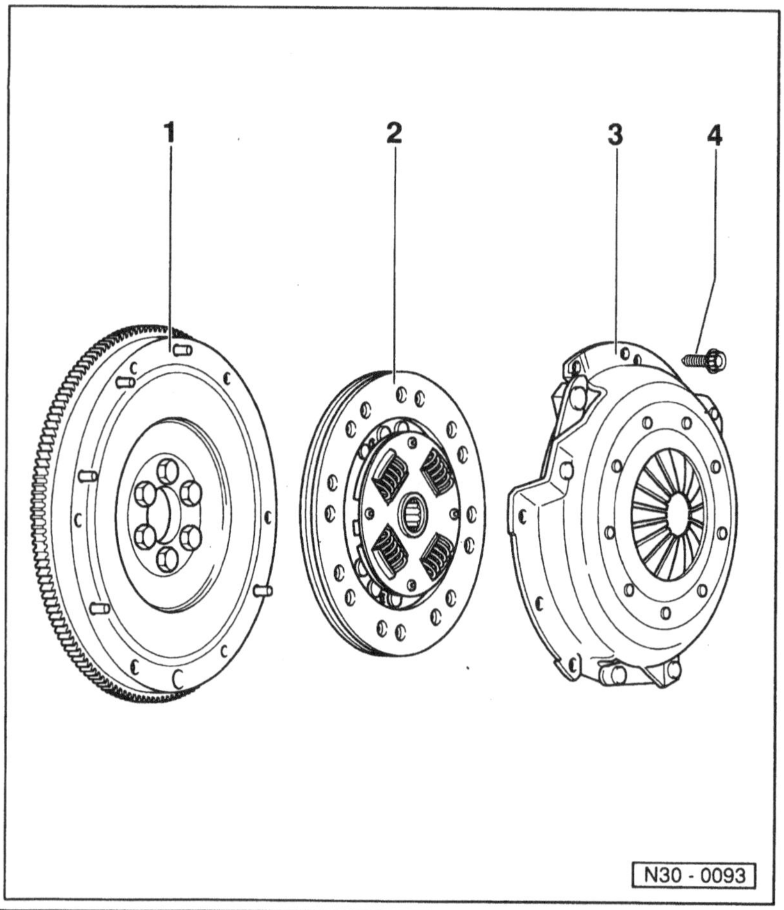

(Transmission installed)

Notes:

- *Replace clutch and pressure plate if rivets are damaged or loose.*
- *Make sure correct clutch and pressure plate are selected from parts catalog microfiche.*

1 - Flywheel

- Make sure centering pins are a tight fit
- Contact surface for clutch lining must be free of grooves, oil and grease
- Removing and installing

⇒ Repair group 13

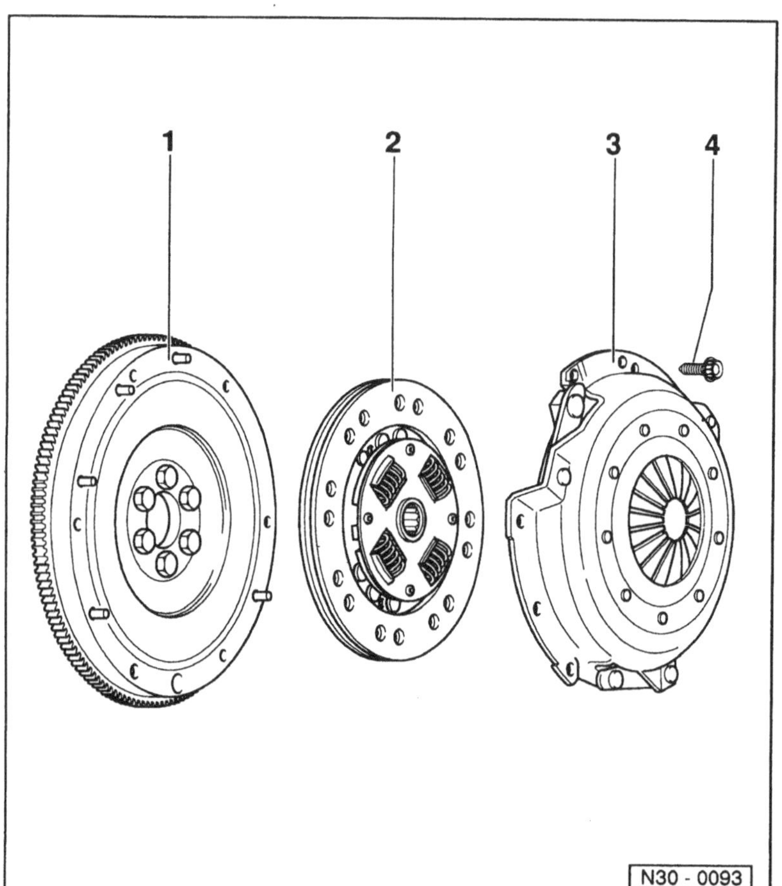

2 - Clutch plate

- Diameter ⇒ from page K 00-3
- Note installation position, spring cage faces to pressure plate
- Centering ⇒ Fig. 1 to Fig. 3
- Lightly grease splines

Note:

Clean input shaft and hub splines on used clutch plates. Remove corrosion and apply only a very thin coating of grease, Part No: G 000 100, to the input shaft splines. Move clutch plate back and forth on input shaft until hub moves freely on shaft. Excess grease must be removed.

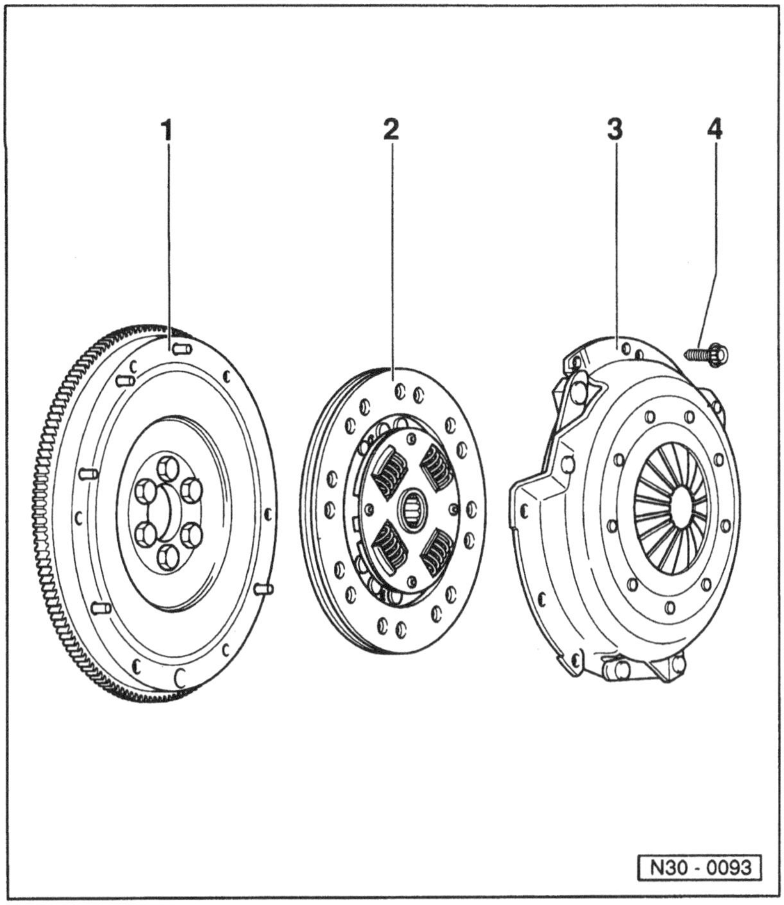

3 - Pressure plate

- Removing and installing ⇒ Fig. 1
- Check ends of diaphragm spring ⇒ Fig. 4

Note:

Pressure plates are protected against corrosion and greased. Only the contact surface may be cleaned, otherwise the service life of the clutch will be considerably reduced.

4 - Twelve-point bolt

- Tightening torque: 20 Nm (15 ft lb)
- Loosen and tighten gradually and diagonally

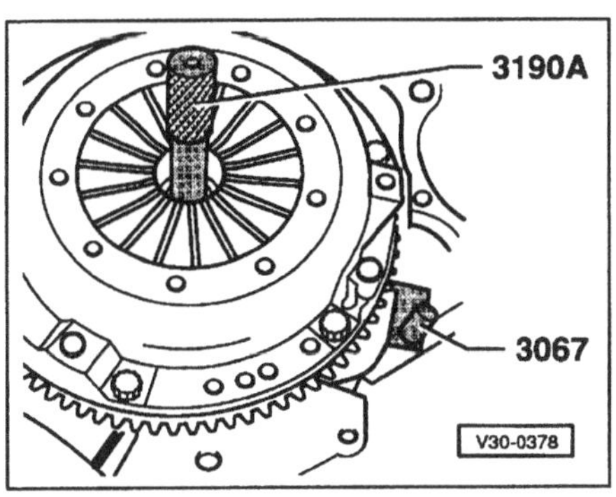

Fig. 1 Clutch plate, centering, removing and installing pressure plate

- Loosen and tighten bolts gradually and diagonally.
- Reverse position of 3067 retainer when removing.

Note:

Pressure plate contact surface and clutch plate lining must make full contact with the flywheel. Tighten securing bolts diagonally and evenly so as to prevent damage to pressure plate centering hole and the flywheel centering pins.

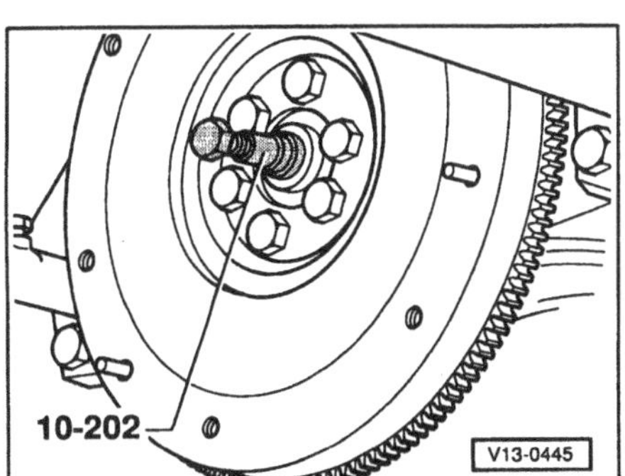

Fig. 2 Needle roller bearing, removing

- The first new Transporters with a 5-cylinder engine were equipped with a needle roller bearing instead of a pilot bushing.
- To enable clutch plate to be centered, needle roller bearing must be removed and a pilot bushing installed before new clutch plate is installed ⇒ Fig. 3.

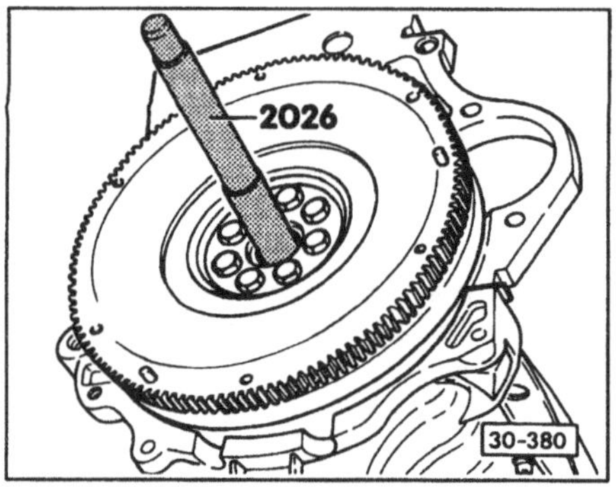

◄ **Fig. 3 Pilot bushing, installing**

Drive Part No. 074 105 321 pilot bushing on to stop.

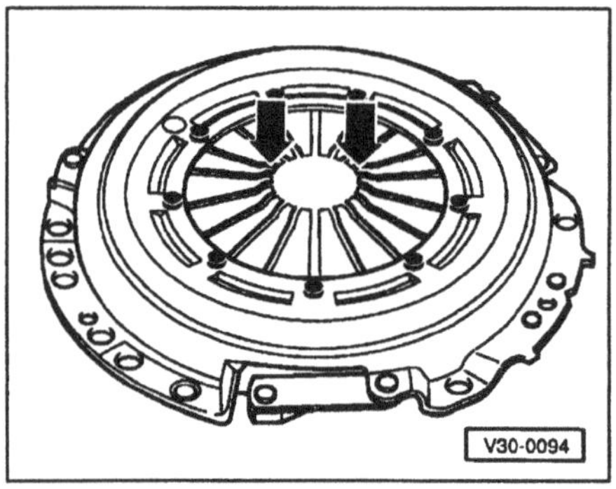

◄ **Fig. 4 Diaphragm spring ends -arrows-, checking**

- Wear of up to half the thickness of diaphragm spring is permitted.

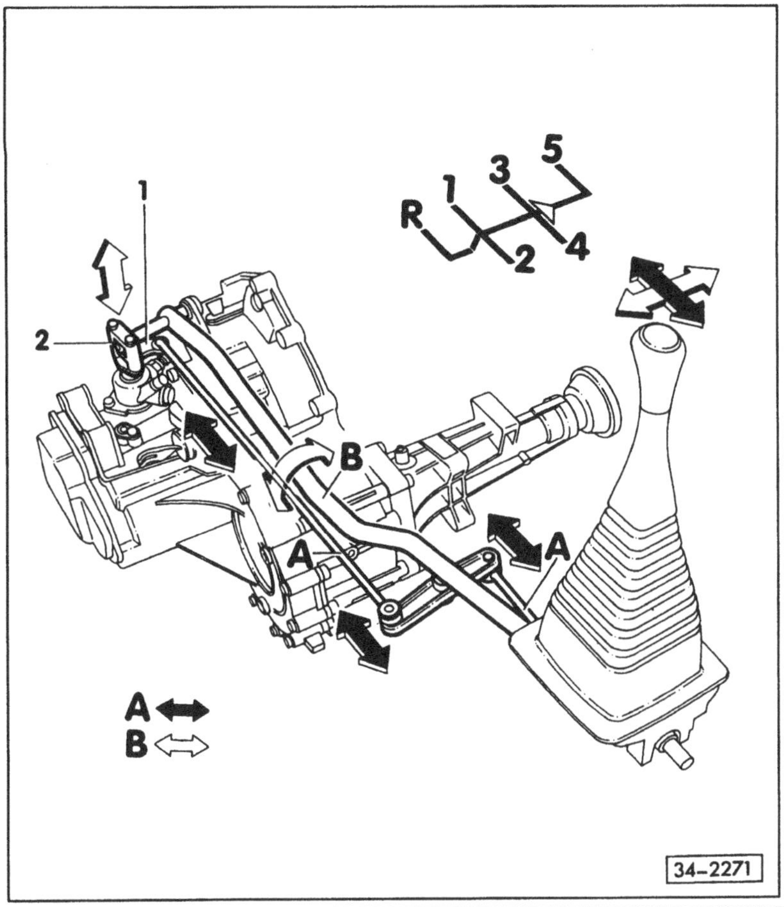

Gearshift linkage, servicing

Installation position of gearshift linkage

A - Gear shift rod

B - Gate shift rod

Double-headed arrow -A- Gear shift movement

Double-headed arrow -B- Gate shift movement

1 - Gear shift lever for shift movement

2 - Swivel head for gate movement

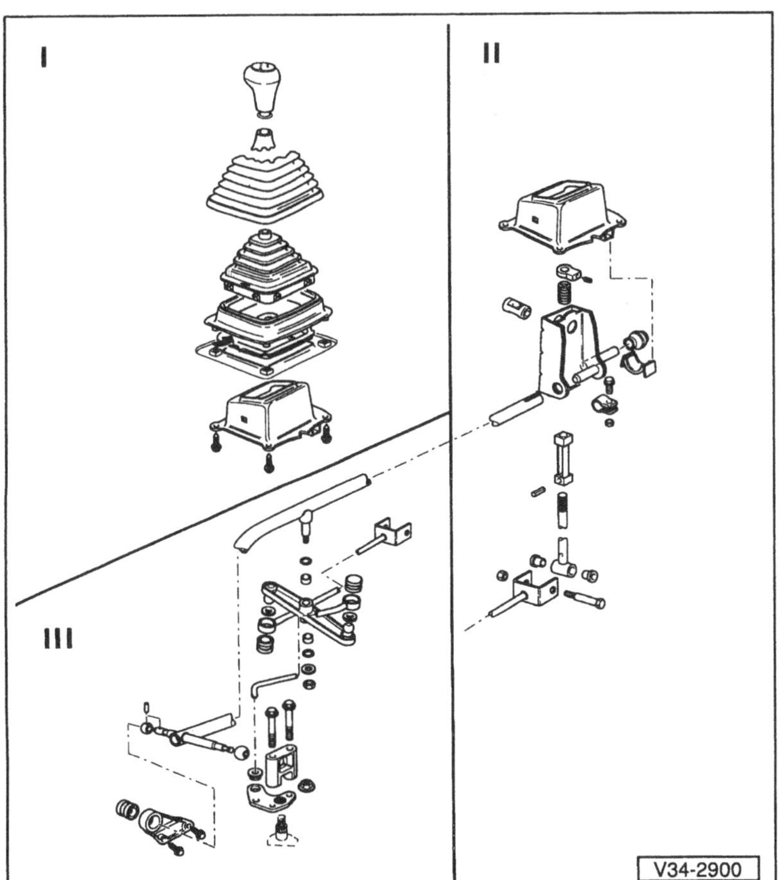

I - Gearshift housing, removing and installing **⇒ Page K 34-3**

II - Gearshift, removing and installing **⇒ Page K 34-6**

III - Shift and gate rods, removing and installing **⇒ Page K 34-9**

Gearshift linkage, removing and installing ⇒ Page K 34-19

Notes:

- *Before working on the shift change mechanism, disconnect Ground (GND) strap from the battery in the engine compartment.*
- *Note radio code on vehicles with coded radio.*
- *Lubricate all mountings and contact surfaces with MoS_2 grease.*
- *Adjusting shift cables* *⇒ Page K 34-23*

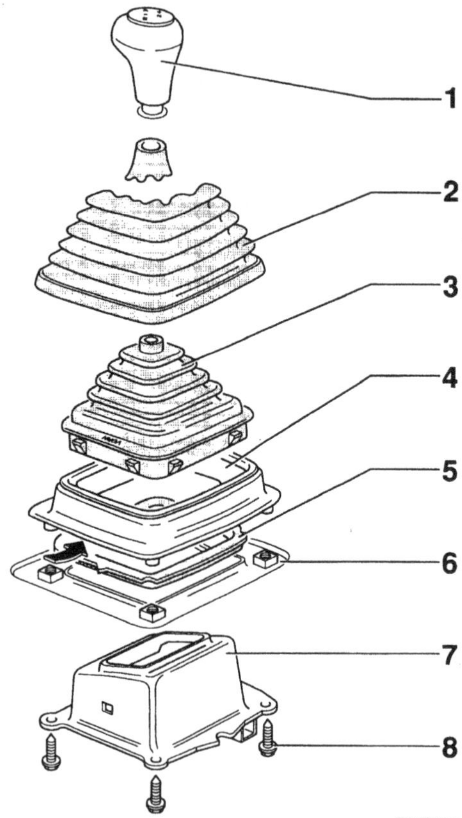

V34-2901

Gearshift housing, removing and installing

1 - Gearshift knob
- Thread on and off together with boot -2-
- When threading on hold boot in position

2 - Boot
- To remove, unclip from frame -4-
- Remove and install together with gearshift knob -1-
- After installing, clip into frame-4-
- Installation position
  ⇒ Page K 34-22, Fig. V34-2980

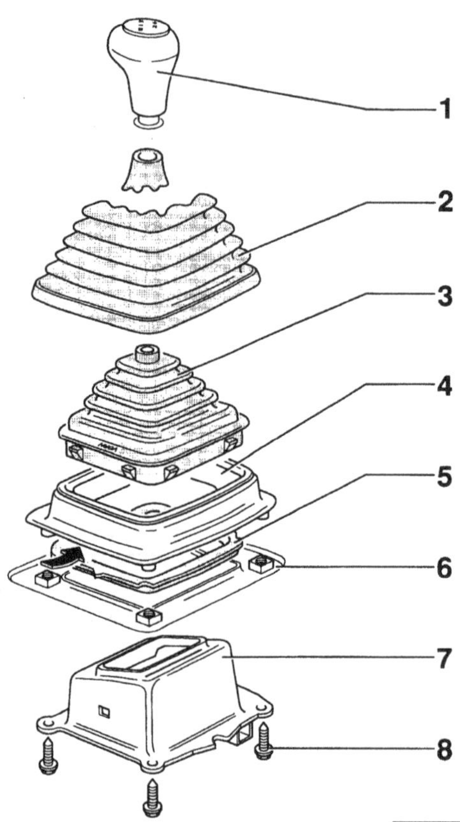

V34-2901

3 - Boot
- Remove and install together with frame -4-
  ⇒ Page K 34-19 and K 34-22
- Clip and unclip into frame -4-
  ⇒ Page K 34-22, Fig. V34- 2978
- Installation position: Inscription "VORN" (front) faces in direction of travel

4 - Frame
- Remove and install together with boot -3-
  ⇒ Page K 34-19 and K 34-22
- Installation position: Inscription "VORN" (front) faces in direction of travel
  ⇒ Page K 34-22, Fig. V34-2978

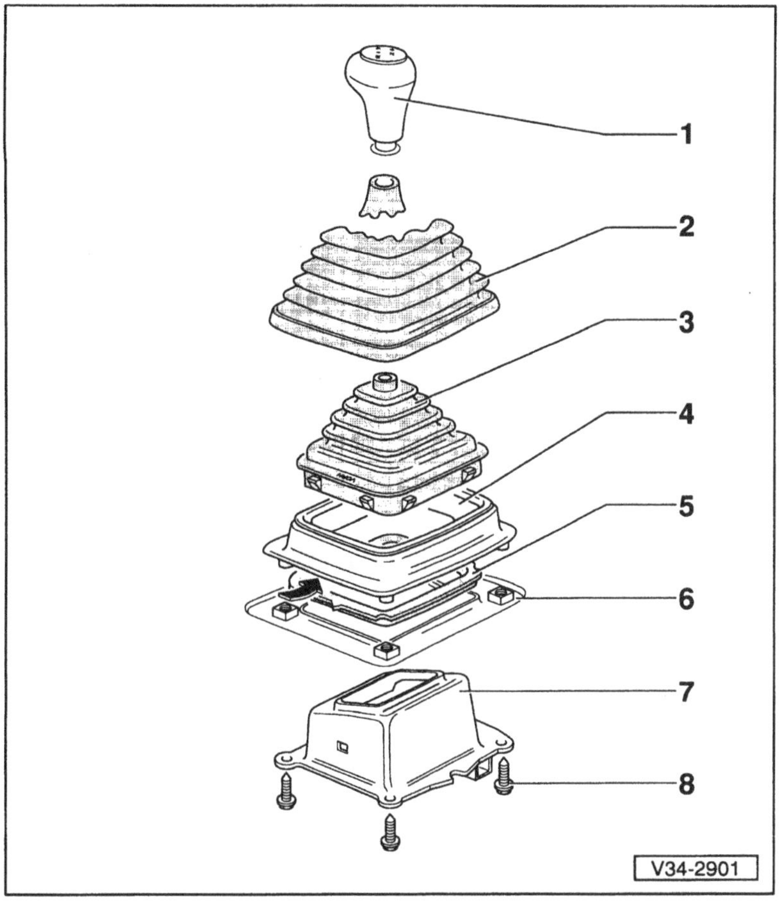

5 - **Gasket**
- When installing, place on edge of floor plate
- Installation position: Arrow type markings face in direction of travel, lip towards floor plate

6 - **Floor plate**

7 - **Gearshift housing**

8 - **Hex bolt**
- Tightening torque: 9 Nm (7 ft lb)

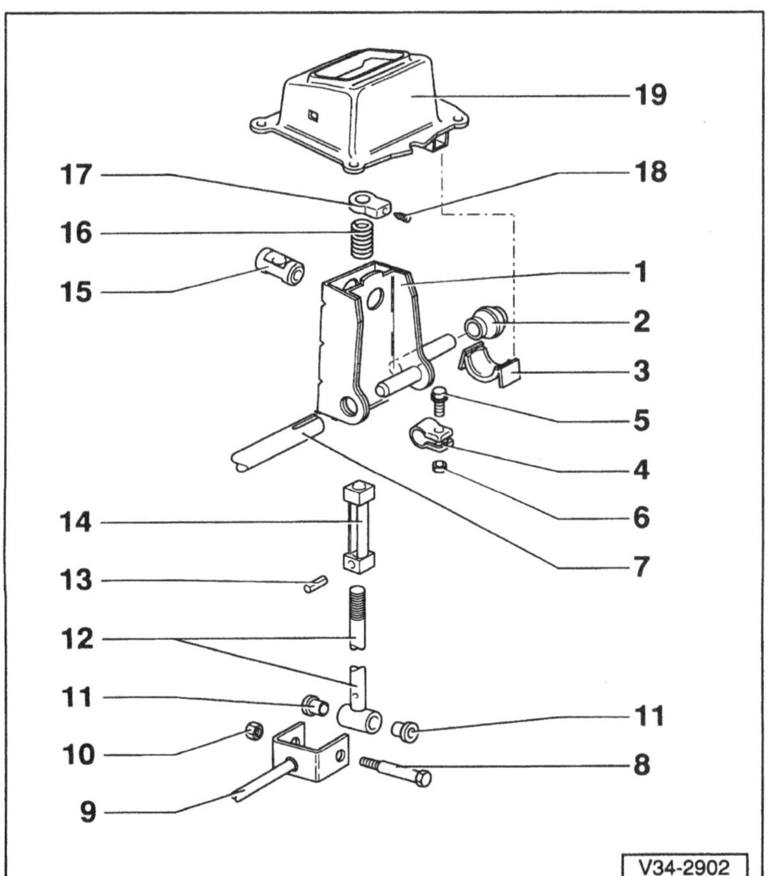

Gearshift, removing and installing

1 - **Gearshift housing**
- Remove with gearshift
- First remove following components: rear shift rod -9-, gate rod -7-, bushing shell -3-

2 - **Bushing**

3 - **Bushing shell**
- Removing
  ⇒ Page K 34-20,Fig. V34-2289
- When installing, insert with sealant AMV 188 200

4 - **Clamp**
- Loosen to adjust gearshift linkage and to remove shift and gate rods

5 - **Hex bolt**

6 - **Hex nut**
- Tightening torque: 20 Nm (15 ft lb)

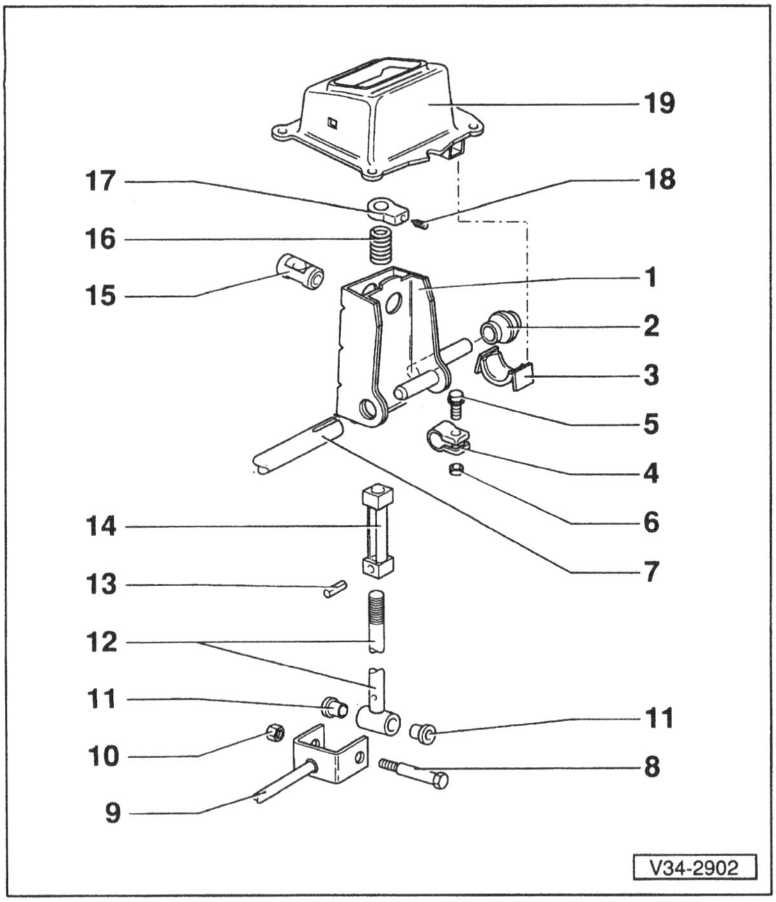

7 - Gate rod

- Removing and installing to transmission ⇒ Page K 34-19 and K 34-21

8 - Bolt

9 - Rear shift rod

- Disassembling and assembling to relay lever ⇒ Page K 34-20 and K 34-21

10 - Hex nut

- Tightening torque: 5 Nm (44 in. lb)

11 - Bushing

12 - Gearshift

- Top section angled rearward.

13 - Spring pin

- Drive out with drift

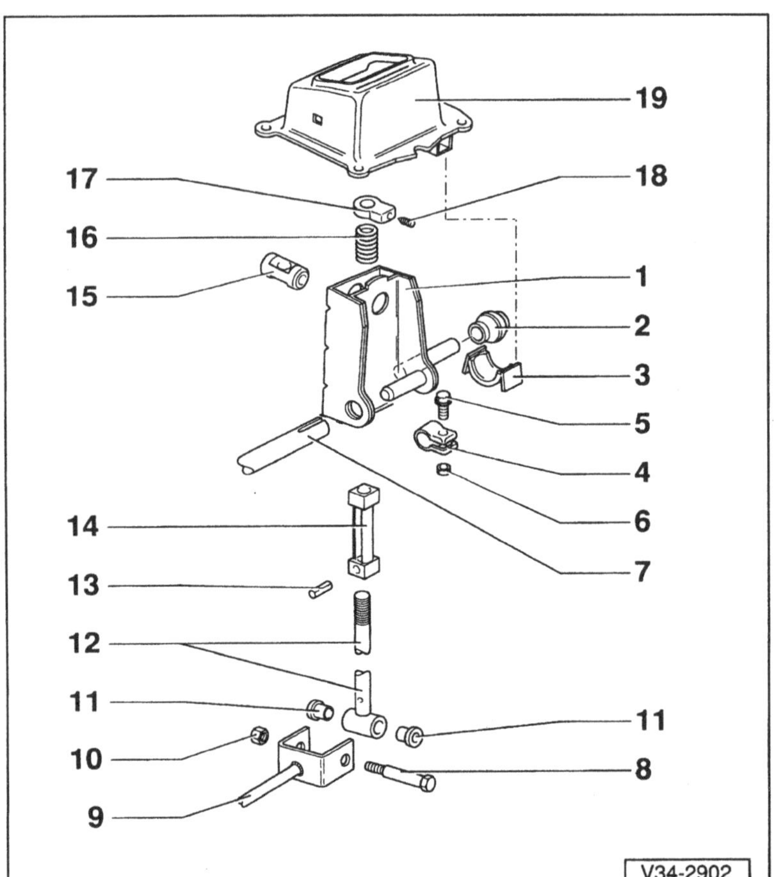

14 - Guide sleeve

15 - Bearing pin

16 - Spring

17 - Stop

- Installation position ⇒ Page K 34-23

18 - Screw

- Tightening torque: 4 Nm (35 in. lb)

19 - Gearshift housing

- Disassembling and assembling to floor plate ⇒ Page K 34-20

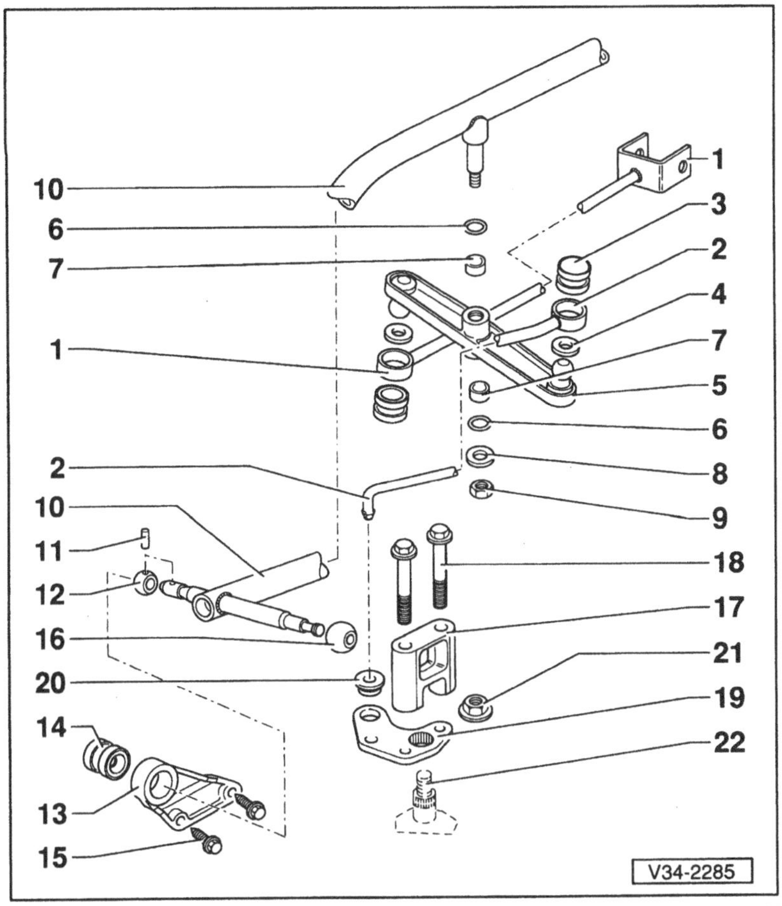

Shift and gate rods, removing and installing

1 - Rear shift rod
- Removing from gearshift ⇒ Page K 34-20
- Prying off relay lever -5-
- Installing relay lever ⇒ Page K 34-21

2 - Front shift rod
- Prying off relay lever -5- and from gear shift lever -19-
- Installing relay lever ⇒ Page K 34-21

Note:

The shift rods can be removed and installed with the gearshift linkage installed.

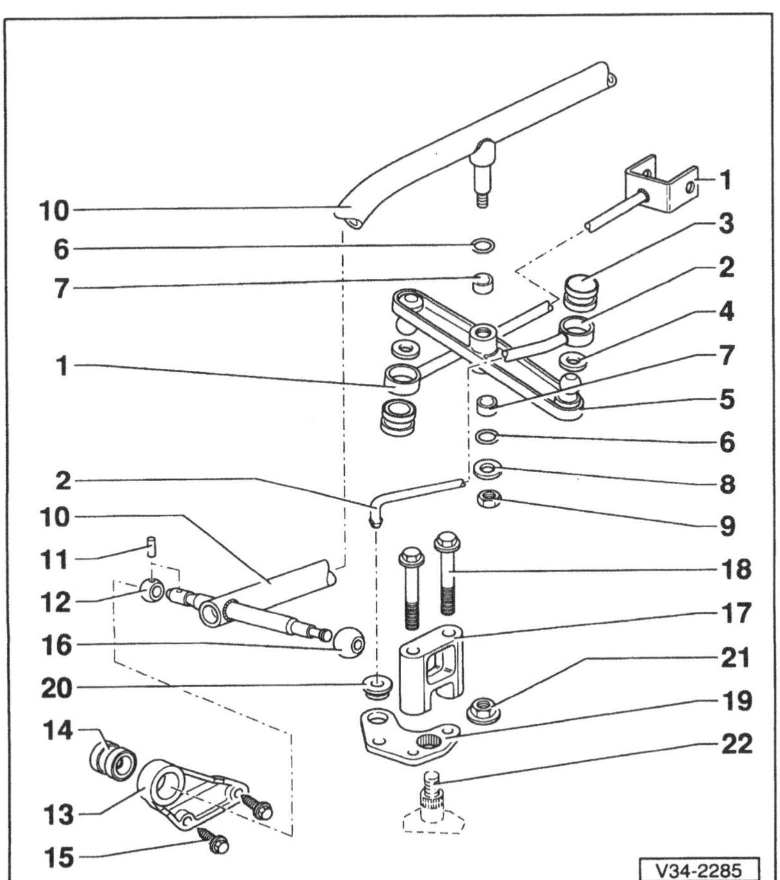

3 - Bushing
- Removing ⇒ Fig. 1
- Installing ⇒ Fig. 2

4 - Sealing washer

5 - Relay lever
- Installation position ⇒ Fig. 12

6 - O-ring

7 - Bushing
- Removing ⇒ Fig. 3
- Installing ⇒ Fig. 4

8 - Washer

9 - Hex nut
- Tightening torque: 25 Nm (18 ft lb)
- Always replace

10 - Gate rod
- Remove from transmission together with mounting bracket -13-
- Removing from gearshift bracket ⇒ Page K 34-20

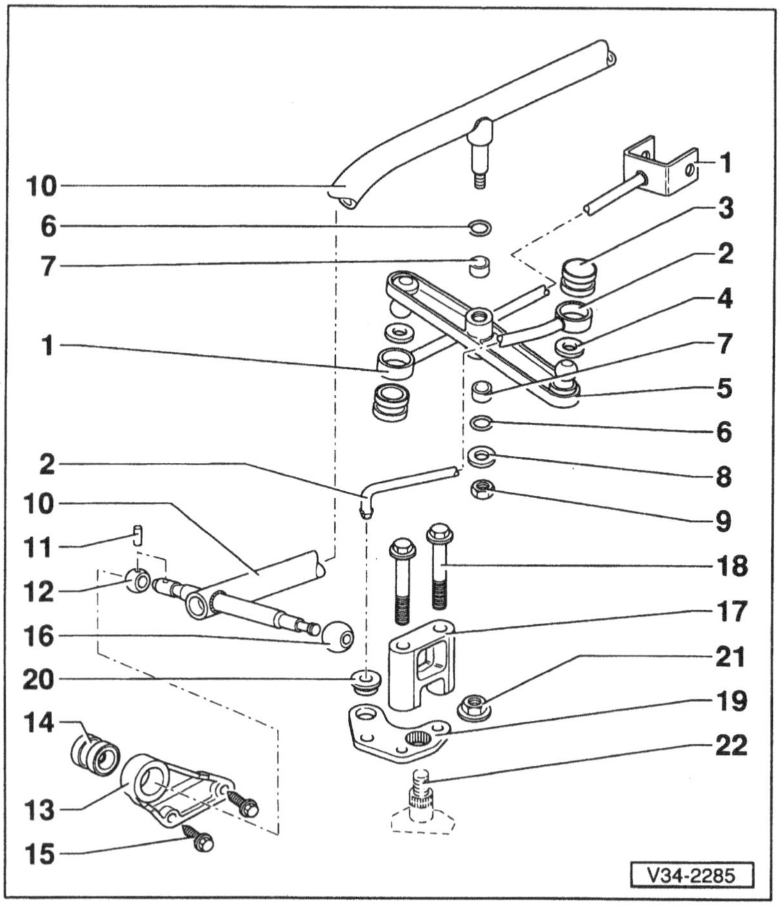

11 - Spring pin
- Drive out with drift

12 - Ball bearing

13 - Mounting bracket
- Removing from gate rod ⇒ Fig. 5
- Installing ⇒ Fig. 6

14 - Bushing
- Removing ⇒ Fig. 7
- Installing ⇒ Fig. 8

15 - Hex bolt
- Tightening torque: 20 Nm (15 ft lb)
- For mounting bracket to transmission

16 - Ball bearing
- Mounts gate rod in swivel head
- Removing ⇒ Fig. 9
- Installing ⇒ Fig. 10

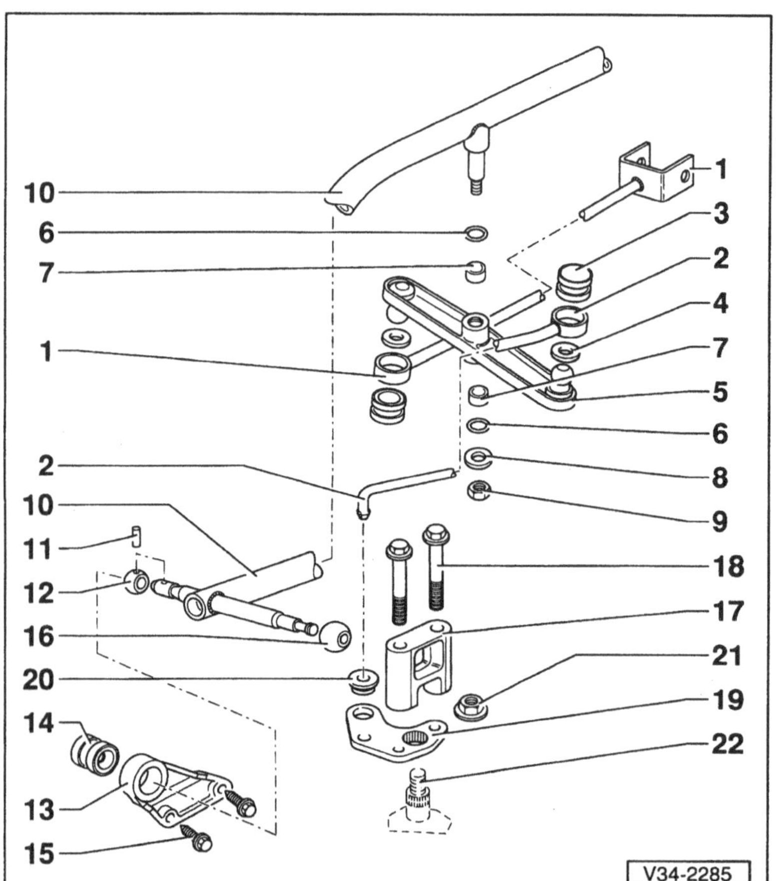

17 - Swivel head

18 - Hex nut
- Tightening torque: 9 Nm (7 ft lb)
- Always replace

19 - Gear shift lever
- Install so that the master spline aligns with shift shaft

20 - Bushing
- For front shift rod
- Remove with 10-206 drift
- Installing ⇒ Fig. 11

21 - Hex nut
- Tightening torque: 25 Nm (18 ft lb)
- Always replace

22 - Shift shaft

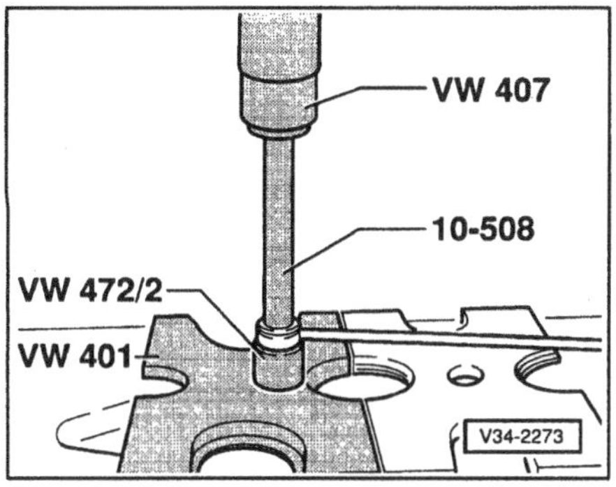

◄ **Fig. 1** **Shift rod bushing, removing**

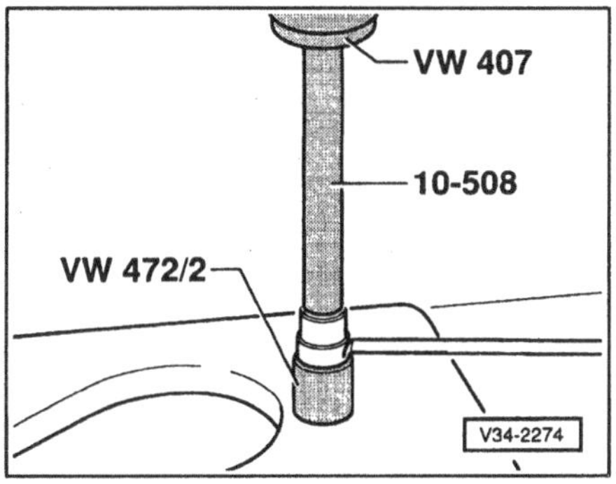

◄ **Fig. 2** **Shift rod bushing, installing**

– Apply a light coat of MoS_2 grease to bushing and carefully install. Do not tilt shift rod.

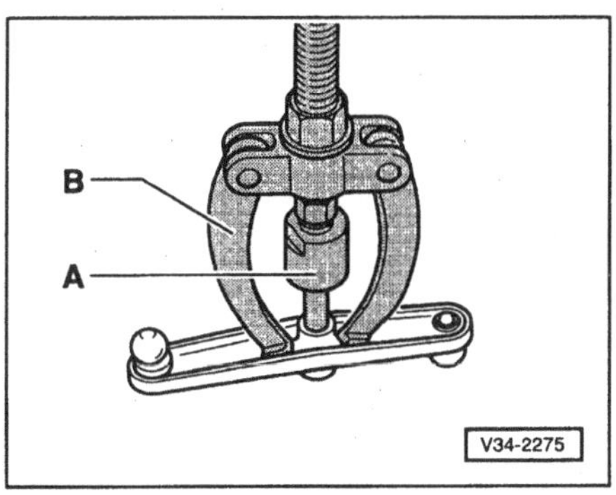

◄ **Fig. 3** **Relay lever bushings, removing**

A – 21/1 Kukko puller - 12 ... 14.5 mm, or equivalent

B – 22/1 Kukko support, or equivalent

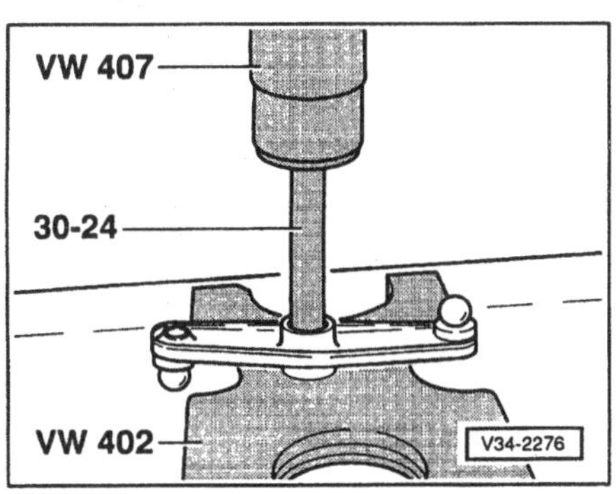

◄ **Fig. 4** **Relay lever bushings, installing**

– Install carefully into tool stop

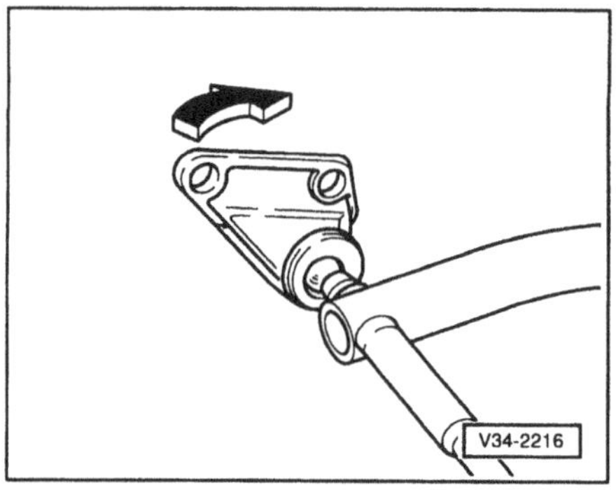

◄ **Fig. 5 Gate rod mounting bracket, removing**

- Remove in direction of arrow

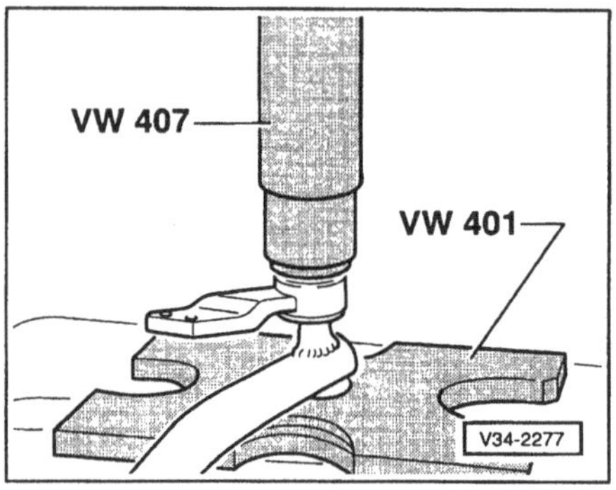

◄ **Fig. 6 Gate rod mounting bracket, installing**

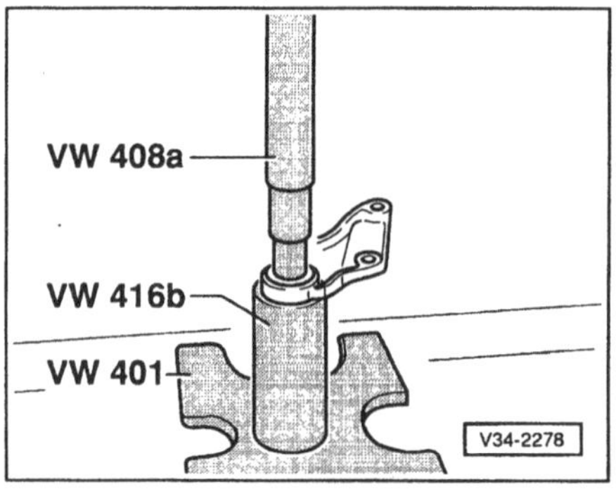

◄ **Fig. 7 Mounting bracket bushing, removing**

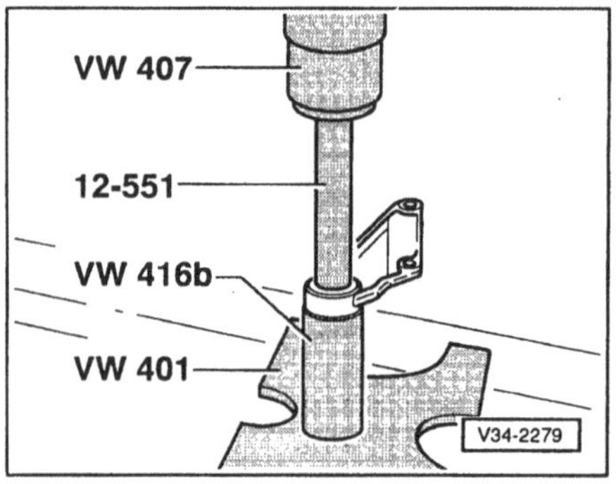

◄ **Fig. 8 Mounting bracket bushing, installing**

- Apply a light coat of MoS_2 grease to bushing and carefully install.

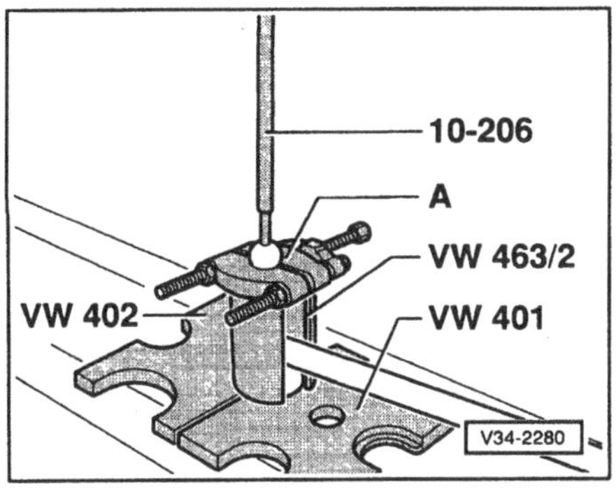

◄ **Fig. 9 Ball, removing from gate rod**

A – 17/0 Kukko separating tool - 5 ... 60 mm, or equivalent

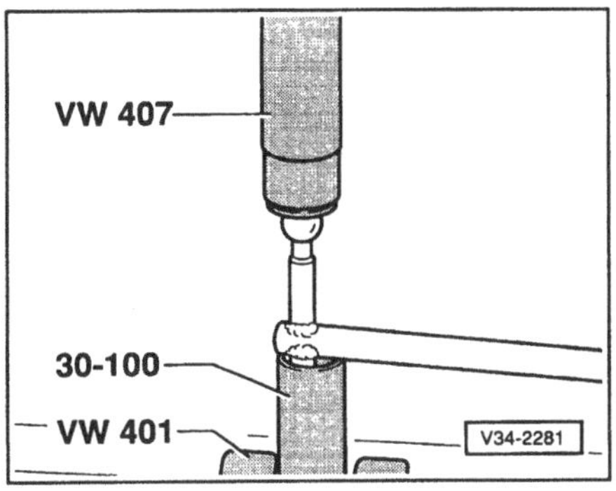

◄ **Fig. 10 Ball, installing on gate rod**

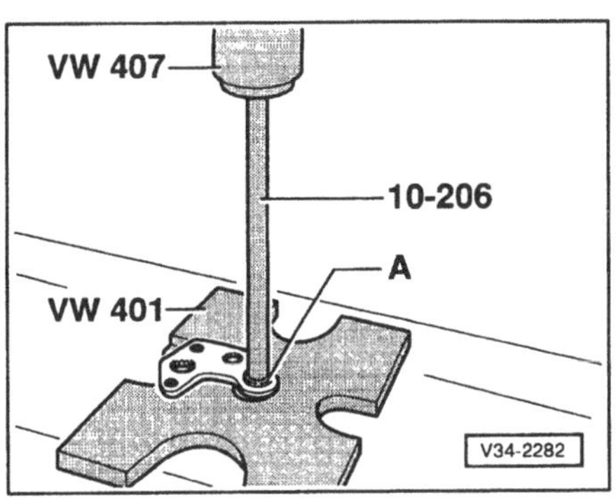

◄ **Fig. 11 Transmission shift lever bushing, pressing in**

A - Washer 10 mm (0.3937 in.)

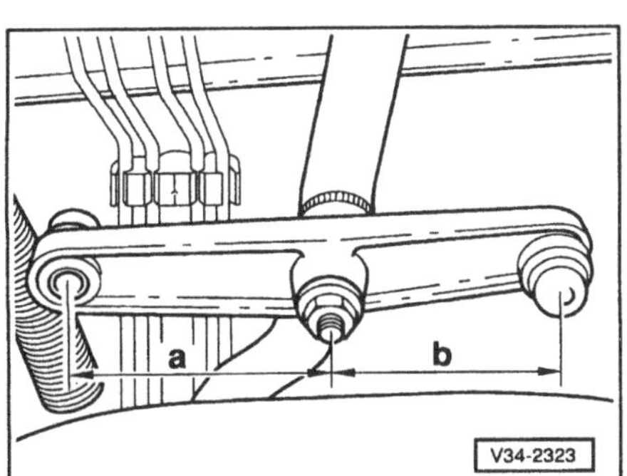

◄ **Fig. 12 Relay lever, installation position**

Dimension "a" = 80 mm (3.150 in.) - points to left-hand side of vehicle

Dimension "b" = 67 mm (2.638 in.) - points to right-hand side of vehicle

Gearshift linkage, removing and installing

Removing

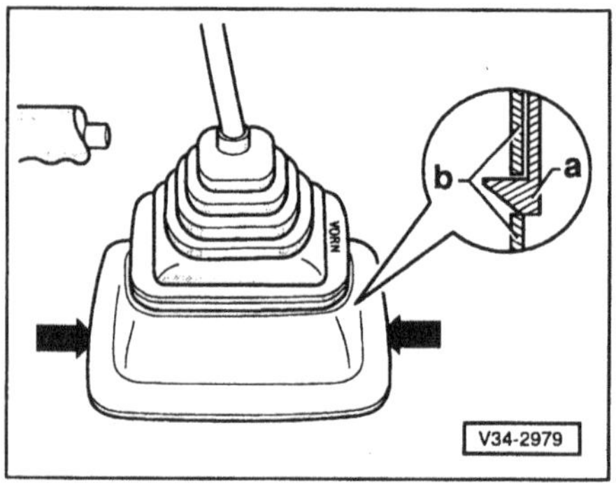

- Unclip boot from frame, unscrew and remove from gearshift along with gearshift knob ⇒ Page K 34-3, -1- and -2-.
- Release frame -a- from gearshift housing -b- by prying upwards in front and rear areas -arrows-.

Note:

Remove and install boot and frame together.

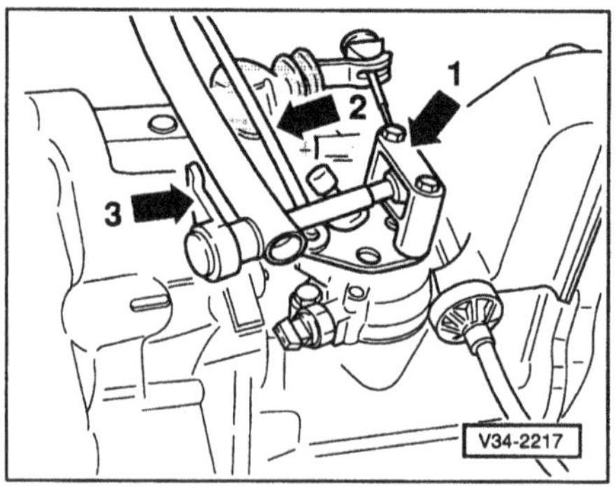

- Remove seal from floor plate.
- Unbolt swivel head -arrow 1-.
- Pry front shift rod off transmission shift lever -arrow 2-.
- Unbolt mounting bracket with gate rod -arrow 3-.

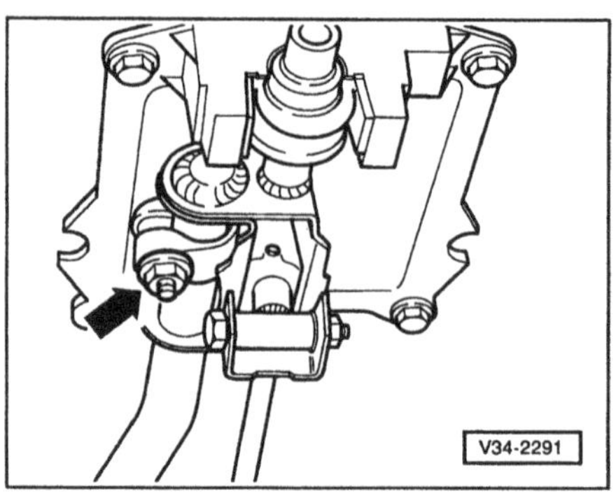

- Pry rear shift rod off relay lever.
- Remove rear shift rod from shift lever.
- After loosening the clamp -arrow- pull gate rod forward, off gearshift bracket.

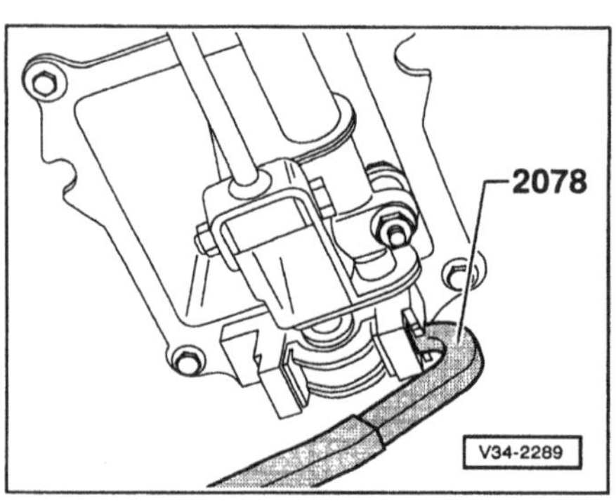

- Carefully pry bushing shell out. If necessary, remove AMV 188 200 sealant.
- Remove gearshift bracket with gearshift downwards.
- Guide gate rod with mounting bracket, relay lever, and front shift rod out to the rear.
- Remove gearshift housing from floor plate.

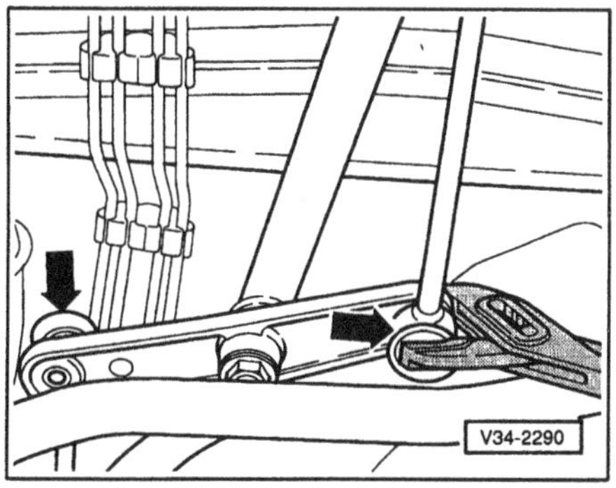

Installing

Install in reverse order.

– Install gearshift bracket with gearshift lever from below in gearshift housing.

– Install bushing shell for mounting the gearshift bracket to gearshift housing with AMV 188 200 sealant.

– Guide gate rod into gearshift bracket and slide onto guide pin. First install clamp (⇒ Page K 34-20, Fig. V34-2291, -arrow-).

Note:

Make sure of free movement at gate rod/gearshift bracket joint, so that the gearshift linkage can be adjusted as described on Page K 34-23.

◄ – Install rear shift rod (right-hand arrow) on relay lever.

– Adjust gearshift linkage ⇒ Page K 34-23.

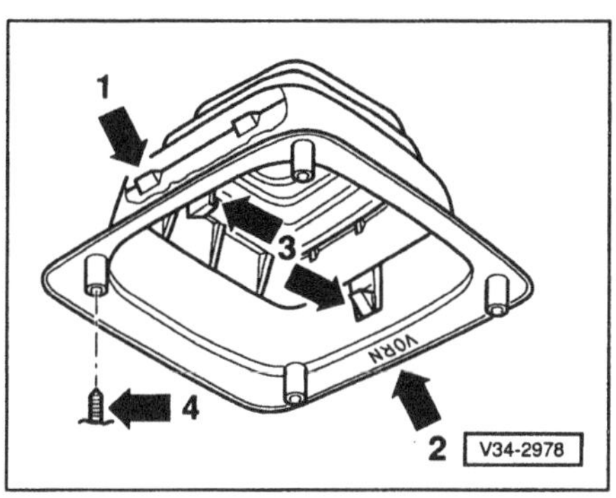

◄ – Engage lugs -arrow 1- located around circumference of boot into frame.

– Place seal around edge of floor plate.

– Place frame mounting tubes on gearshift housing securing screws -arrow 4- and then press, first front and then rear, fully home.

Inscription “VORN” (front) -arrow 2- points in direction of travel.

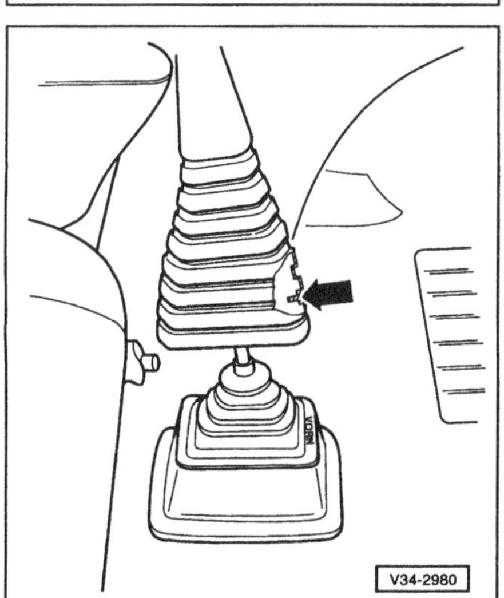

– Engage frame locking devices -arrow 3- in gearshift housing cutout ⇒ Page K 34-19, Fig. V34-2979.

– Install gearshift knob with boot, holding boot in position.

– Clip boot into frame.

◄ One lug -arrow- on inside of boot is located in direction of travel.

Gearshift linkage, adjusting

Notes:

The following points are essential for correct gearshift linkage adjustment:

- *Perfect operation of gearshift linkage, operating, and relay elements.*
- *Free movement of gearshift linkage.*

Also transmission, clutch, and clutch controls must be in perfect condition.

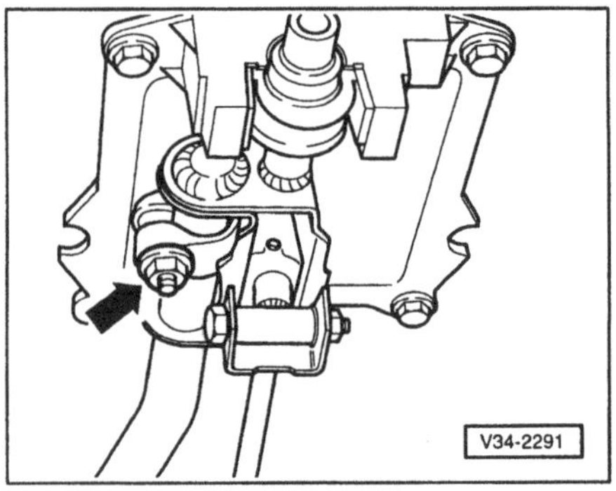

- Transmission in neutral position.

– Remove gearshift knob, boot, frame, and boot ⇒ Page K 34-19.

◄ – Loosen nut -arrow-.

- Gate rod/gearshift bracket connection must move freely.

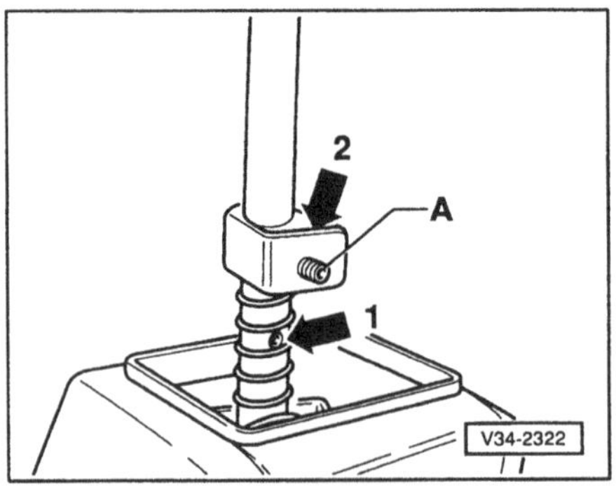

Note:

Check whether contact surfaces on gearshift lever and gearshift housing stops are aligned parallel to one another. Adjust, if necessary.

◄ – Align stop so that grub screw -A- locates in gearshift depression -arrow 1-. Chamfered side of stop -arrow 2- points towards gearshift knob.

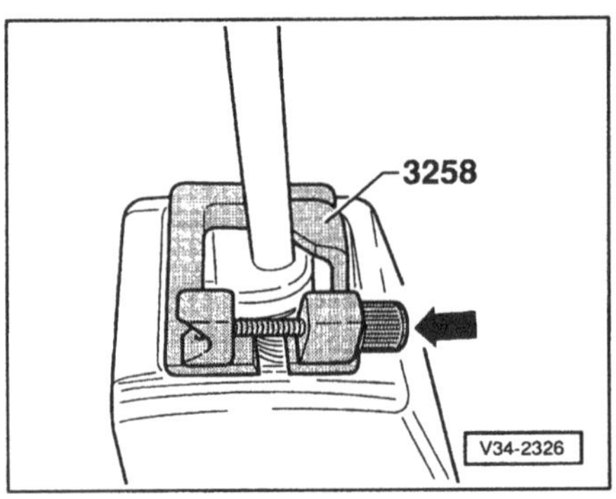

◄ – Insert 3258 gearshift lever gauge until it sits on stop and lock by tightening knurled wheel -arrow-.

– Align gate rod free of stress (transmission in neutral) and tighten nut to 20 Nm (15 ft lb).
(⇒ Page K 34-23, Fig. V34-2291, -arrow-).

– Pull gearshift lever towards seat, loosen knurled nut and remove 3258 gearshift lever gauge.

Functional check

- With transmission in neutral, gearshift lever must rest in 3rd/4th gear plane. Lower area of gearshift must be vertical.

– Operate clutch.

– Select all gears several times. Pay particular attention to the operation of the reverse gear lock.

Should a gear fail to engage smoothly after repeated selection, continue as follows:

– Engage 1st gear.

– Remove play in gearshift linkage by carefully moving gearshift to left.

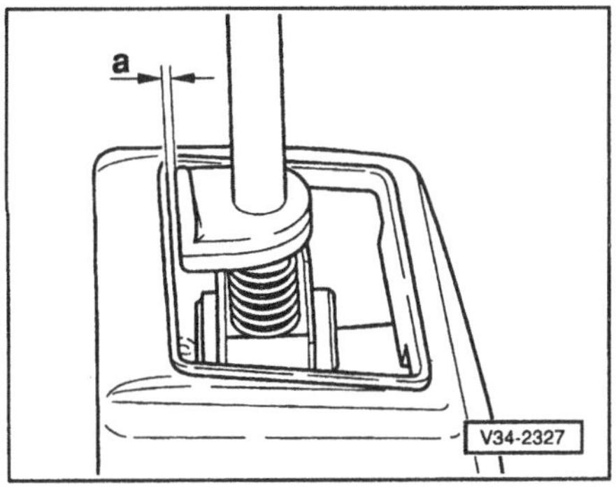

- The distance between gearshift stop and gearshift housing stop must be approximately 2 mm (0.0797 in.) (dimension "a").

– If not, repeat adjustment sequence. Reinstate dimension "a" with 1st gear engaged, if necessary.

– Install gearshift knob and covering ⇒ Page K 34-22

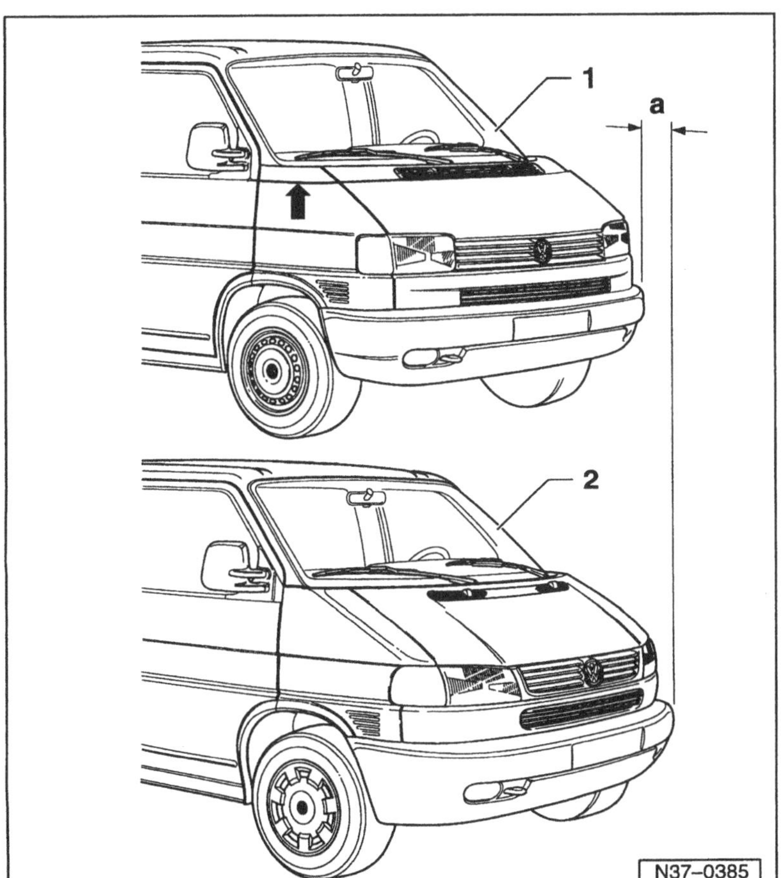

Differences between Transporter with long and short front ends

1 - Transporter with short front end
- ♦ Character line -arrow- on upper fender.

2 - Transporter with long front end
- ♦ 90 mm (dimension -a-) longer than short front end vehicle.
- ♦ No character line on upper fender

Transmission, removing and installing

Removing

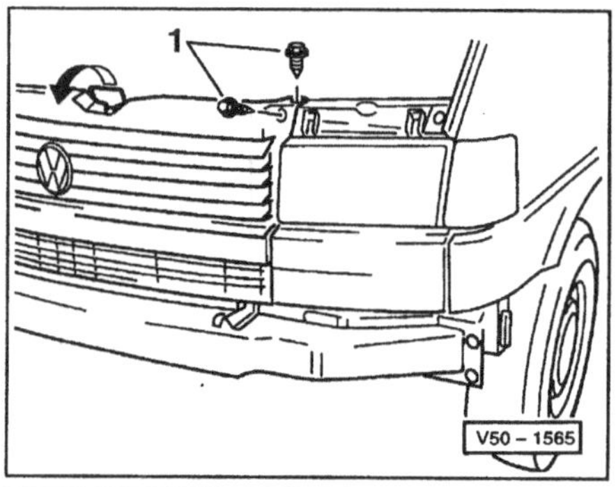

- With ignition turned off, disconnect Battery Ground (GND) strap (in engine compartment and, if installed, under driver's side).
⇒ Electrical system; Repair Group 27

Note:

For vehicles with coded radio obtain radio code.

Vehicles up to 12.95 only

◄ – Remove bolts -1- and tilt radiator forward in direction of arrow.

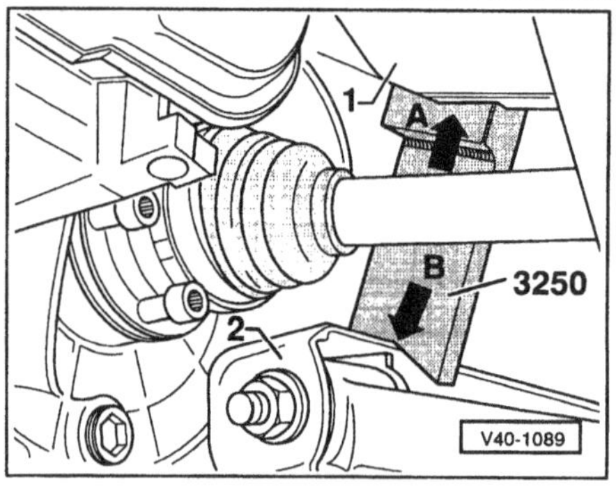

◄ – With vehicle in an unloaded condition, attach 3250 support frames to both sides before vehicle is raised to enable the left-hand drive shaft (figure) and the left and right-hand shock absorber/stabilizer bar coupling link to be removed.

Installation position:

The 3250 support frame bearing surface -arrow A- must be located below the upper control arm -1-.

The V-shape opening -arrow B- bears on the subframe -2-.

Vehicles from 01.96

Support frame 3250 can no longer be used.

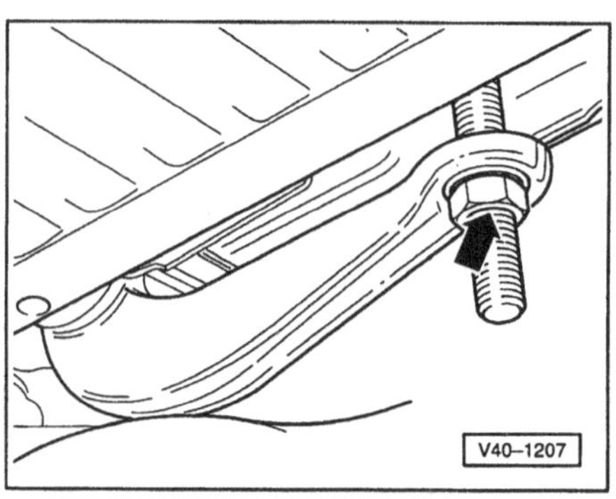

◄ – Mark position of torsion bar setting on thread -arrow- or measure thread projection from nut to end of threaded rod and note.

– Relieve tension on torsion bar with special tool 3257.
⇒ Suspension, Wheels, Brakes, Steering; Repair Group 40

With long front end:

Radiator and hood lock carrier no longer fold forward.

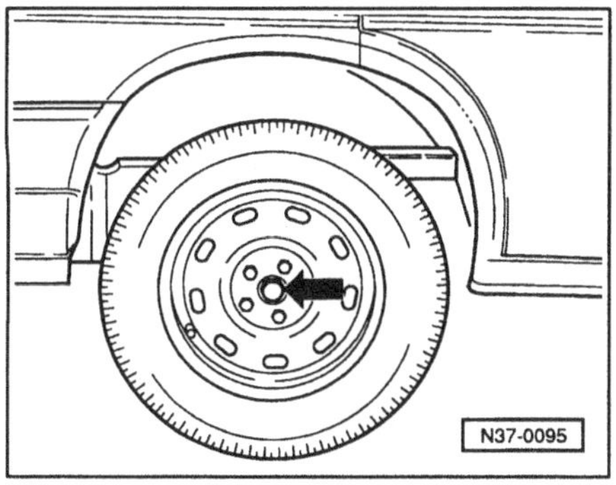

Continuation for all vehicles

- With vehicle resting on its wheels loosen left-hand wheel hub/ drive shaft bolt and left and right-hand wheel lug bolts.
- Raise vehicle.
- Remove noise insulation tray below engine/transmission, if installed.

Remove left-hand drive shaft as follows:

- Remove bolt for left-hand shock absorber/stabilizer bar coupling rod/stabilizer.

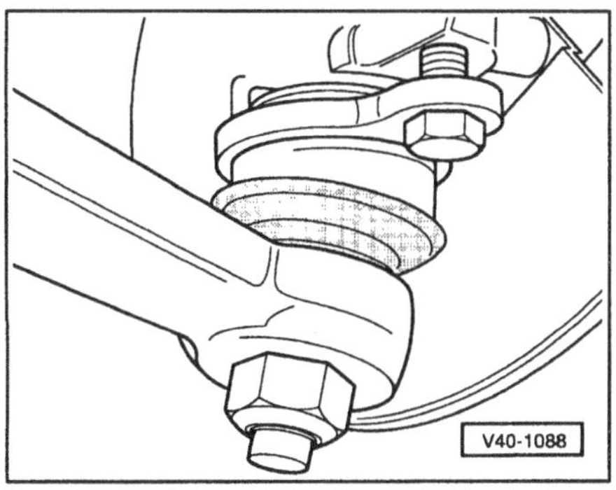

- Fully compress left-hand shock absorber.
- Disconnect drive shaft at flange.
- Remove wheel.
- Separate left-hand lower wheel bearing housing ball joint bolted connection. Support ball joint with VAG 1383 A transmission jack to prevent it from falling.
- Remove drive shaft.

- Disconnect right-hand drive shaft at flange.
- Remove right-hand wheel.
- Remove bolt for right-hand shock absorber/stabilizer bar coupling rod/stabilizer.
- Fully compress right-hand shock absorber.
- Remove Ground (GND) strap from transmission and mounting bracket.
- Remove all electrical connections from transmission and starter motor.

⇒ Electrical system; Repair Group 27

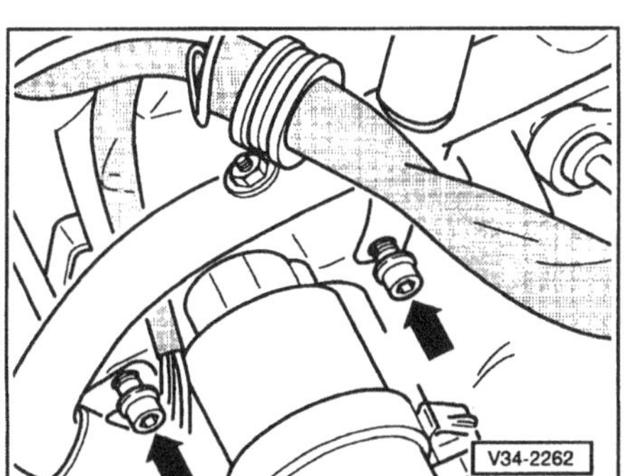

- Remove speedometer drive shaft cable or sensor connector and speedometer sensor.
- If attached, separate electrical wiring from coolant hoses.
- Separate coolant pipes at starter motor –arrows– (from expansion tank to water pump).
- Remove wiring harness from coolant pipework.
- Remove starter.

⇒ Electrical system; Repair Group 27

Remove gearshift linkage from transmission as follows:

- Remove swivel head -arrow 1-.
- Pry front gear shift rod off transmission shift lever -arrow 2-.
- Unbolt mounting bracket with gate rod -arrow 3-.
- Secure shift and gate rods with wire.

Vehicles with mechanical clutch control

- Disconnect clutch cable at transmission and at transmission mounting bracket ⇒ Page K 30-28

Vehicles with gasoline engine and air conditioner

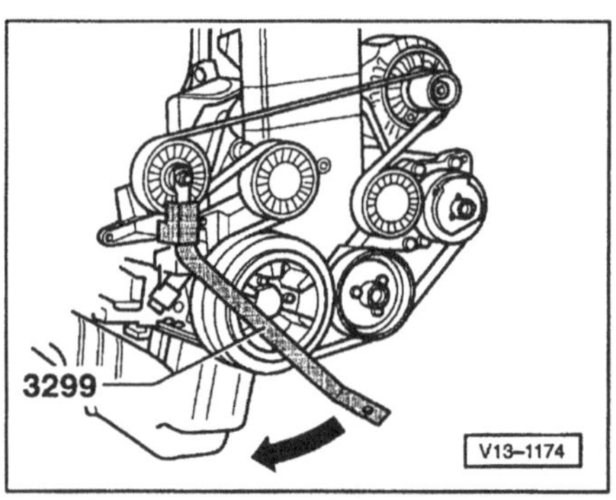

- Release tension in ribbed V-belt and take belt off generator and compressor.

⇒ 5-Cylinder injection engine Mechanics; Repair Group 13

- Remove Generator.

⇒ Electrical system; Repair Group 27

Vehicles with air conditioning system

- Remove compressor from mounting bracket and secure to body so that hoses are stress-free.

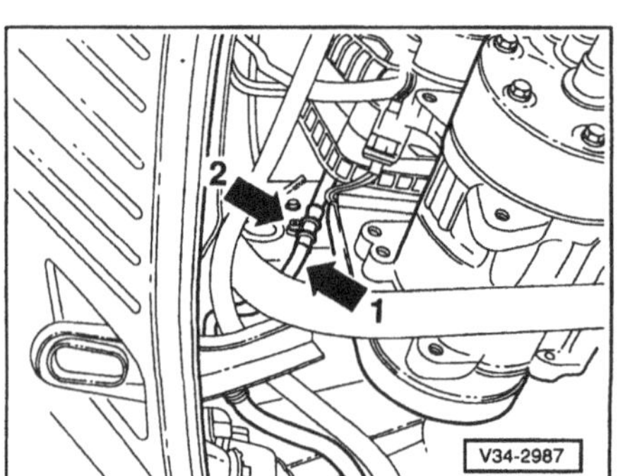

CAUTION!

- ***The air conditioning refrigerant circuit must not be opened.***
- ***Refrigerant hoses are prone to kinking.***

⇒ Heating, air conditioning system; Repair Group 87

- Disconnect refrigerant hoses -arrow 1- from front right-hand longitudinal member between air conditioning system sight glass/ pressure switch and fluid container by removing bolt for clamp -arrow 2-.

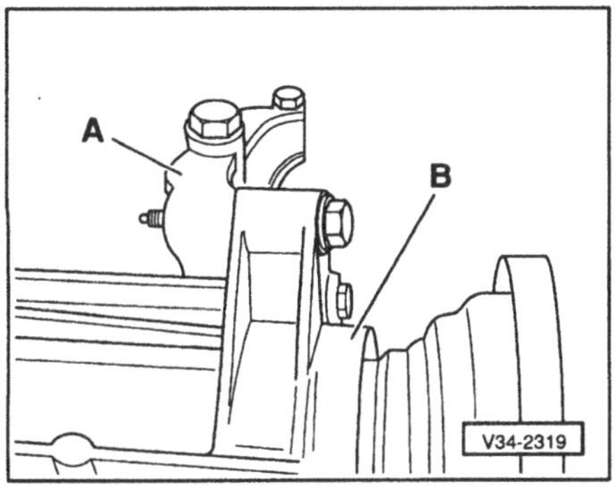

- Remove transmission carrier -A- from engine and bearing housing -B-.

Vehicles with gasoline engine

- Remove oxygen sensor from front exhaust pipe, if installed.
- Remove front exhaust stop plate.
⇒ 5-Cylinder injection engine, Mechanics; Repair Group26
- Separate front exhaust pipe at Threeway Catalytic Converter.
⇒ 5-Cylinder injection engine, Mechanics; Repair Group26

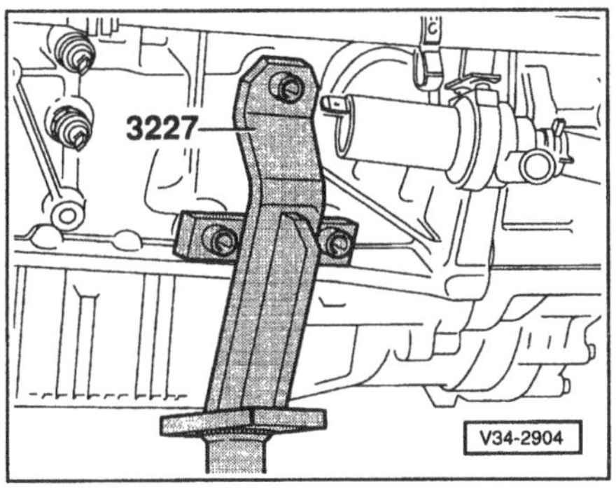

Vehicles with Diesel engine

- Separate front exhaust pipe at muffler
⇒ 5-Cylinder Diesel engine, Mechanics; Repair Group26
- Bolt engine bracket to engine block.

3227 engine bracket for vehicles with 5-cylinder engine

3230 engine bracket for vehicles with 4-cylinder engine

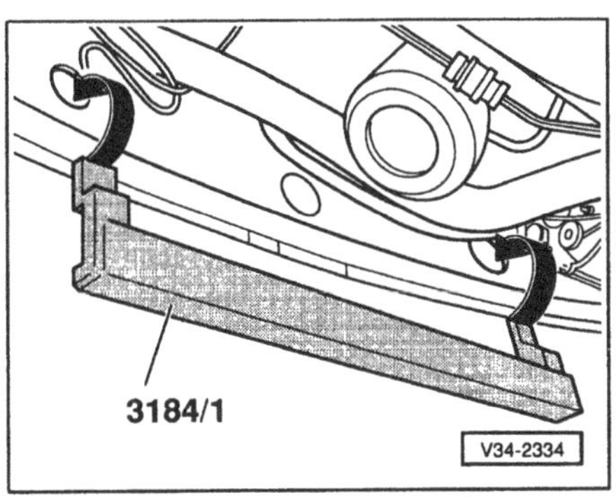

Vehicles up to 12.95

- Attach 3184/1 auxiliary support for 3184 removal device into cross panel -arrow-. If necessary, remove wiring harness and pipes from cross panel.

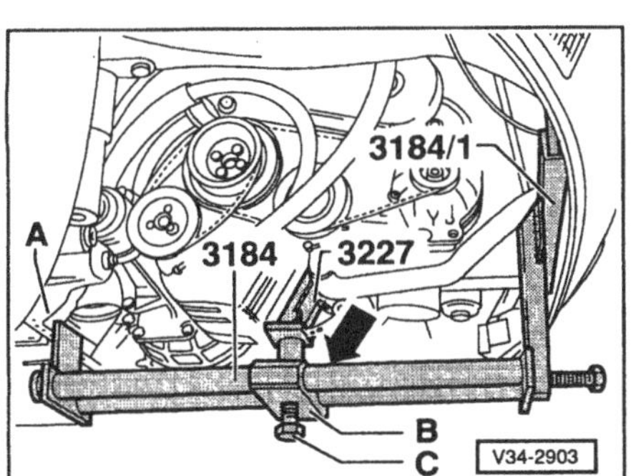

- Attach 3184 removal device with 3184/1 auxiliary support to subframe -A-. Screw spindle -C- into hole on engine mounting.
- Align 3184/1 auxiliary support so that slide -B- is resting against the right-hand tube -arrow-.
- Secure 3184 removal device to subframe -A- and to 3184/1 auxiliary support by tightening the angle plates.
- Release weight from subframe using spindle -C- to raise engine/transmission.

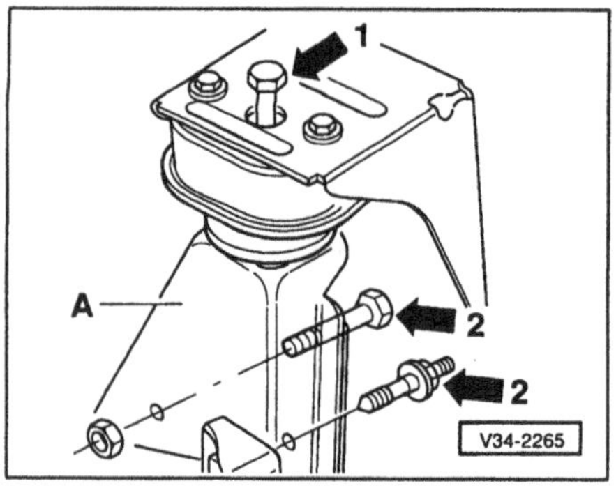

◄ – Unbolt transmission bearing bracket -A- from bonded rubber mounting at body -arrow 1-.

– Lower engine using removal device spindle.

– Unbolt bearing block -A- from transmission -arrow 2-.

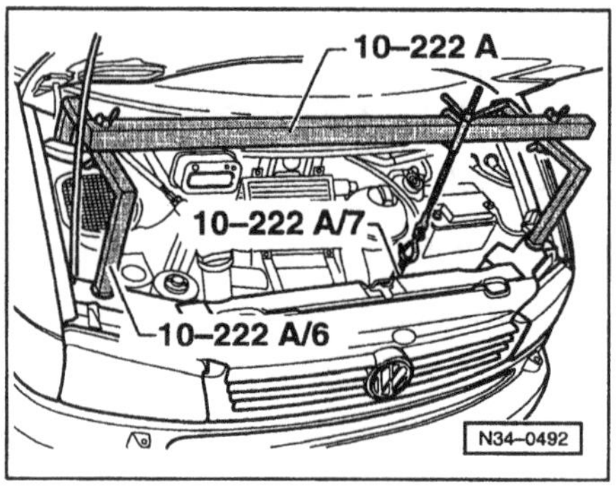

Vehicles from 01.96 with long front end

◄ – Support engine with 10-222 A and supports 10-222 A/6 and additional hooks 10-222 A/7 if necessary.

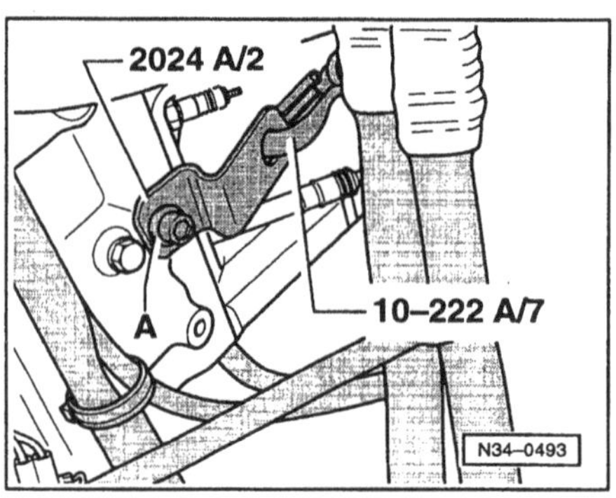

Vehicles from 01.96 with long front end and 2.5 L gas engine

◄ – Pull connector off spark plug for no. 2 cylinder and bolt eye 2024 A/2 onto engine block

A = 2 mm thick washer

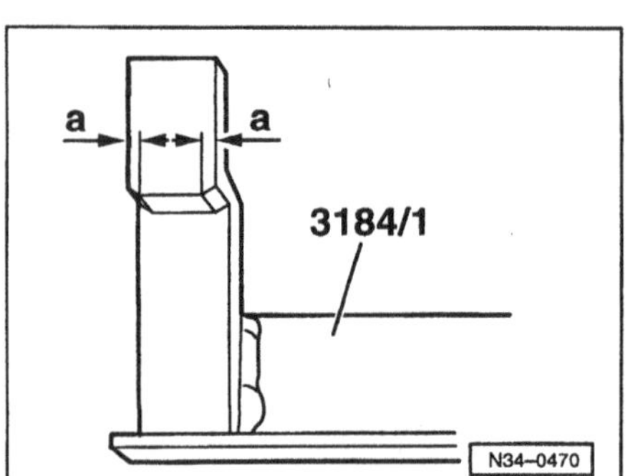

Vehicles from 01.96 with short front end

◄ – Create two 45° chamfers on each arm so that auxilliary support 3184/1 for removal device 3184 can continue to be hooked into cross panel.

Dimension a = 5 mm

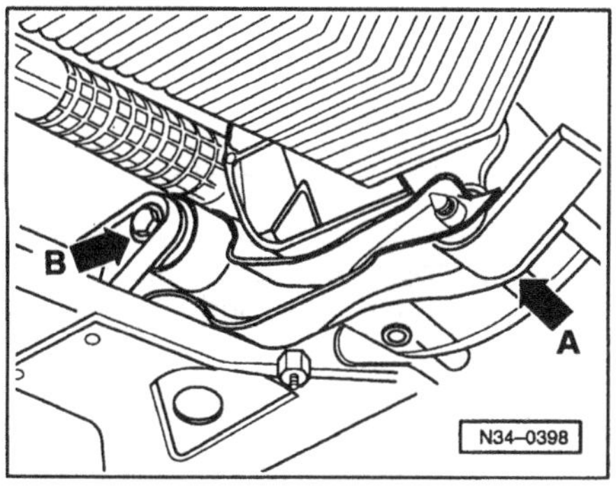

Vehicles from 01.96 with short front end

Modified rear transmission mounting:

◄ The rear transmission mounting is now a pendulum support. Also ⇒ page K 34-94.

Tightening torques

Bolt -A-	80 Nm (59 ft lb) + 90° (1/4 turn)
Bolt -B-	200 Nm (148 ft lb)

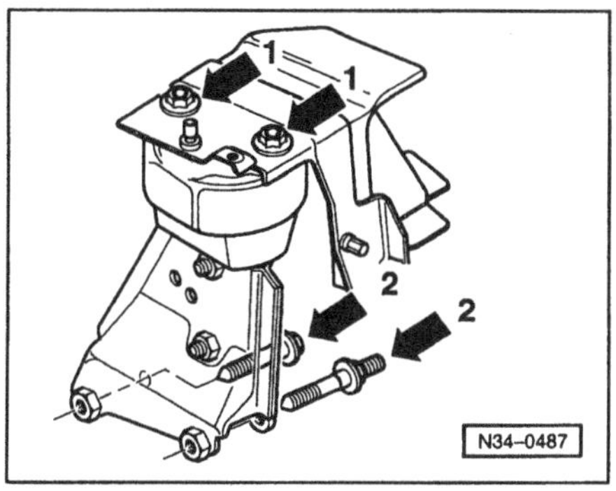

Modified left transmission mounting:

– Remove complete mounting by removing nuts -1- and bolts-2-.

Tightening torques

Nuts -1-	55 Nm (41 ft lb)
Bolts -2-	50 Nm (37 ft lb) + 90° (1/4 turn)

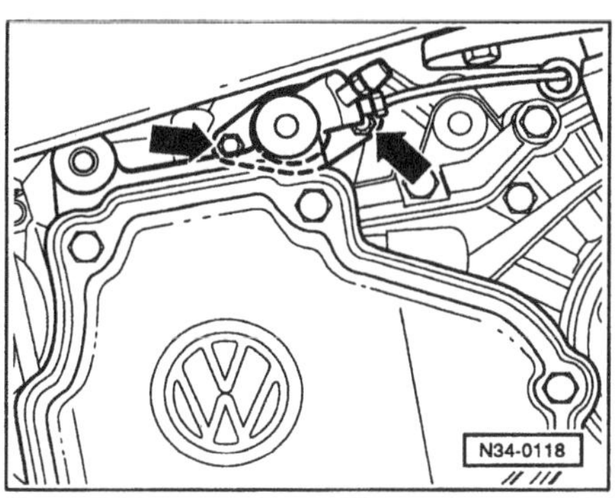

◄ **Vehicles with hydraulic clutch control**

– Remove slave cylinder -arrows- place to side and secure with wire, do not disconnect pipes.

Note:

Do not push clutch pedal.

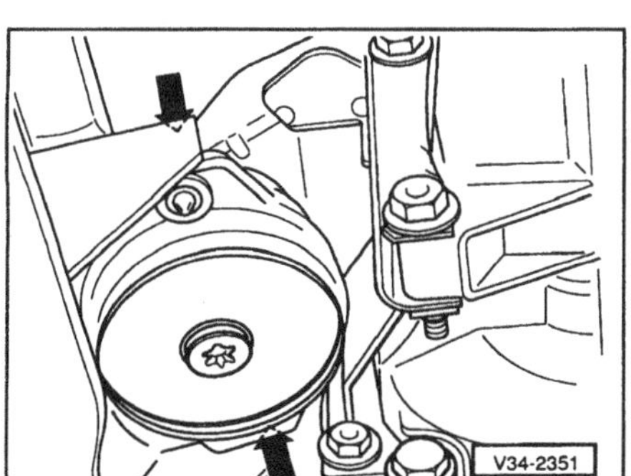

Vehicles up to 07.92

◄ – Remove mounting with bonded rubber bushing by unbolting at rear from subframe -arrow- and at front from transmission.

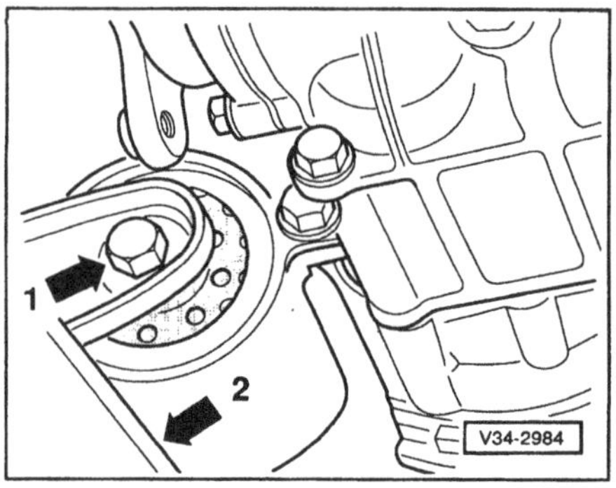

Vehicles from 08.92

◄ – Remove bolt -arrow 1- from rear assembly mounting.

– Carefully swing engine/transmission forward and lower until the V-belt pulley lightly contacts the longitudinal member (gasoline engine) while guiding assembly mounting around forward area -arrow 2- of subframe.

Notes:

- *When moving engine/transmission assembly observe air conditioning system refrigerant hoses if installed.*
- *Make sure right-hand drive shaft has sufficient clearance.*
- *Take care not to damage coolant system hoses and cable pipes on cross member when swinging engine/transmission assembly forward.*

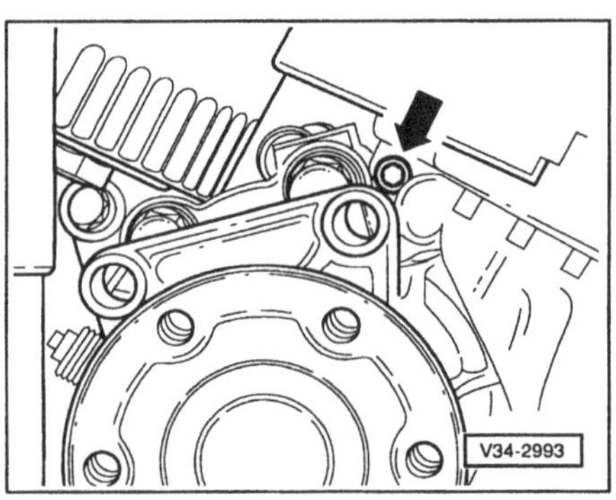

Vehicles with 5-Cylinder engine

◄ – Remove engine/transmission hex socket head bolt -arrow-.

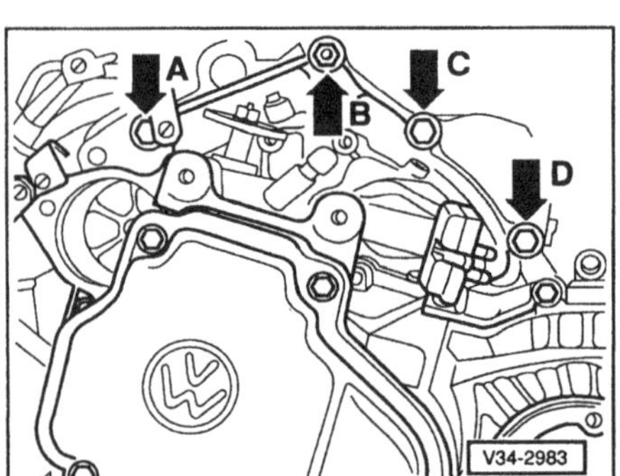

Continuation for all vehicles

◄ – Remove upper engine/transmission bolts -arrows A-, B-, C-, and D-.

Note:

Remove cover plate when removing bolts -C- and -D-.

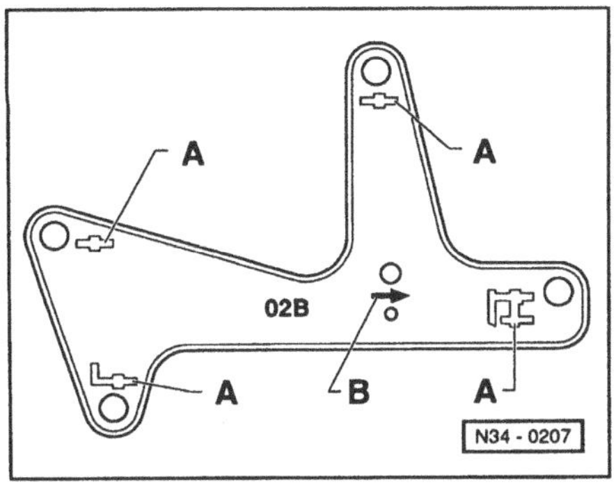

- Assemble VAG 1383A jack with 3282 transmission support and 3282/1 adjustment plate and support elements as follows:
- Place 3282/1 adjustment plate on 3282 transmission support. (Adjustment plate only installs in one position).
- Align arms of transmission support according to holes in adjustment plate.
- Attach support elements -A-, as illustrated on adjustment plate.

Note:

Support element in area of bearing housing/clutch housing on rib at rear right of clutch housing.

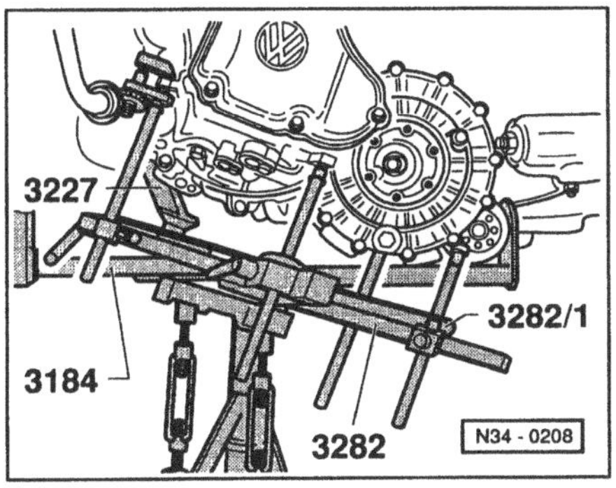

- Place VAG 1383A jack with 3282 transmission support under vehicle, arrow symbol -B- (⇒ Fig. N34-0207) on 3282/1 adjustment plate points to front of vehicle.
- Align 3282/1 adjustment plate parallel to transmission and lock safety supports on transmission.
- Remove lower engine/transmission bolts. Locating the bolts ⇒ Page K 34-50 and K 34-51.

- Remove transmission from engine and remove cover plate from between engine/transmission.
- Carefully lower transmission and swing out.
- To avoid contact and provide enough clearance to subframe and 3184 removal device, correct transmission position using transmission jack spindles.

CAUTION!

If the vehicle is to be moved with the 3184 removal device installed, the following must be observed:

- ***To lift or lower the engine/transmission, screw the spindle in as far as the stop.***
- ***Exercise extreme care in the vicinity of the engine/transmission.***

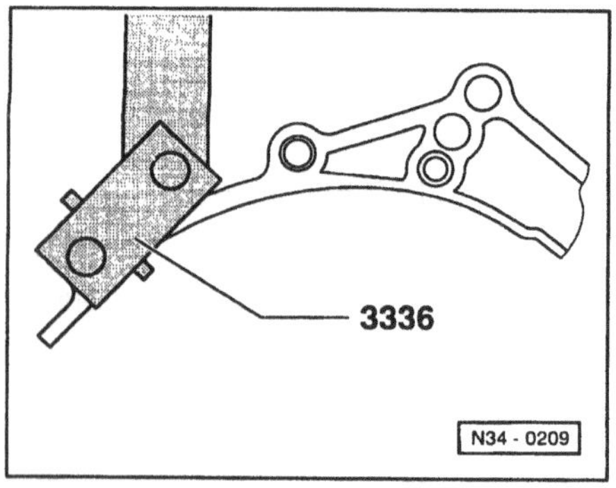

Transporting the transmission

3336 special tool may be used to transport transmission and also when setting up 3282 transmission support.

◄ – Secure 3336 special tool to clutch housing.

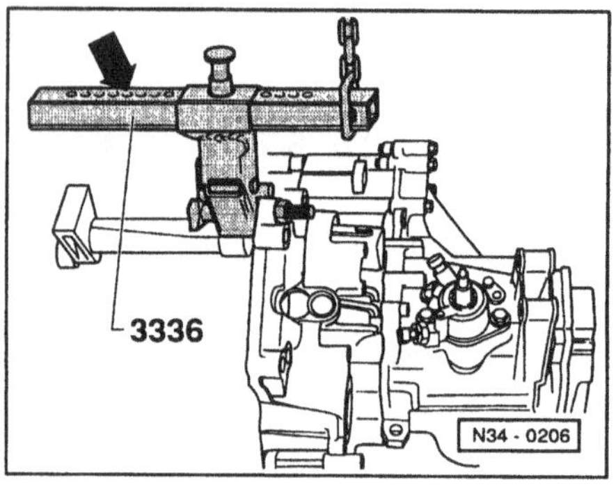

◄ – Set support arm on sliding piece with locking pin -arrow-.

– Position pins so that eight holes are visible in chain direction.

– Remove transmission with workshop crane and 3336 special tool.

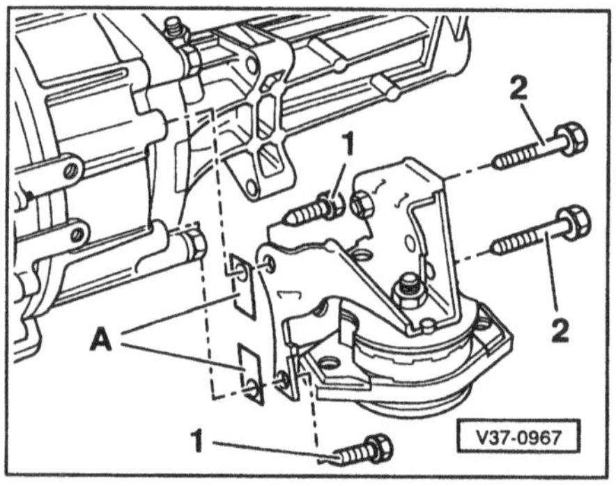

Installing

Installation is performed in reverse order.

Transmissions to build date 09 08 2

◄ *If transmission, bearing housing, or mount have been replaced, the air gap between transmission housing and mount is to be measured with a feeler gauge with the transmission removed and, if necessary, shims -A- added. To measure, remove hex bolts -1- and tighten hex bolts -2- to specified torque.*

The following shims are available for air gaps above 0.5 mm:

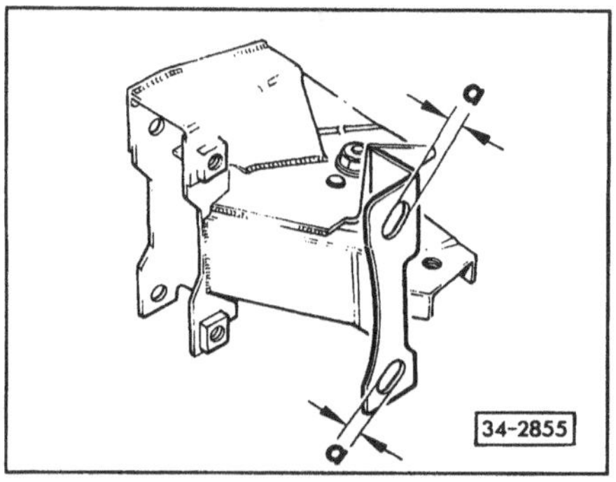

Thickness	Part No.
0.5 mm (0.0197 in.)	701 399 219
1.0 mm (0.0394 in.)	701 399 219 A

The following modifications are to be made to transmissions with M12 threaded holes in the clutch housing:

◄ – Enlarge holes in console for hex bolt -1- (Fig. V37-0967) to 13 mm (dimension -a-).

– Install bolts M12 x 30 Part No. N 010 127 8 and washers 13.0 x 24 x 2.5 Part No. N 011 531.9.

Tightening torque: 100 Nm (74 ft lb).

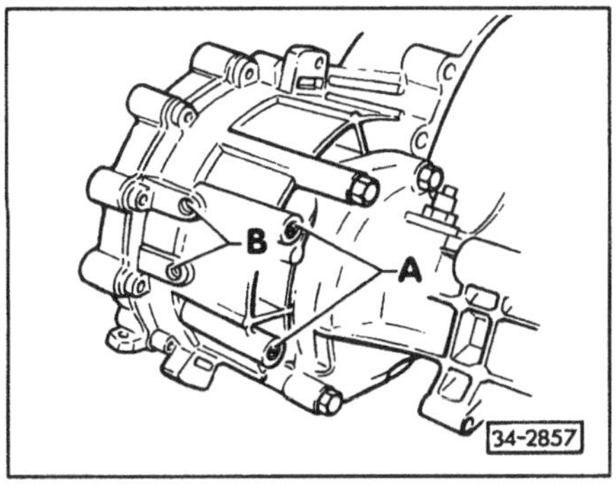

Transmissions from build date 10 08 2

From August 1992, the rear transmission mounting console and bonded rubber bushing have been incorporated into one unit.

The securing points to transmission and subframe have been changed.

Holes -A-

Transmission build date	mm (in.)
from 10 08 92	12.0 (0.447)
to 09 08 92	10.0 (0.394)

Holes -B-, new securing point to clutch housing.

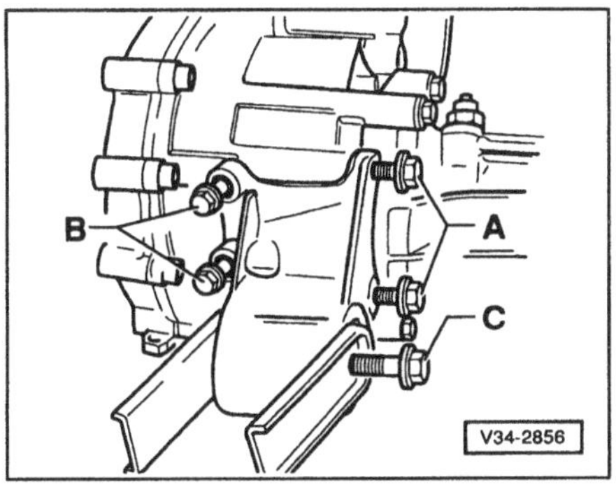

Bolt	A	B	C
Size	M12 x 35	M10 x 40	M16 x 116

Notes:

- *The new transmission mount cannot be installed in previous vehicles.*
- *Select transmission mount according to parts catalog microfiche.*

- When replacing transmission, vehicle speed sensor and gearshift lever must be transferred to new assembly.
- When replacing transmission, select mount according to parts catalog microfiche.
- Clean input shaft splines and lightly grease with G000 100 grease.

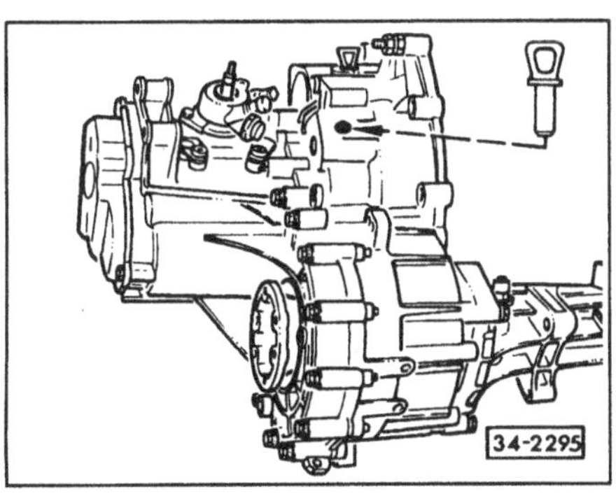

- Before installing transmission, push clutch release lever towards transmission housing and secure with assembly bolt or M 8 x 35 bolt. Remove bolt after installation of transmission. Seal hole with a plug.
- Check whether dowel sleeves for aligning engine/transmission are in cylinder block, install if necessary.
- When replacing transmission, make sure intermediate plate is correctly positioned.

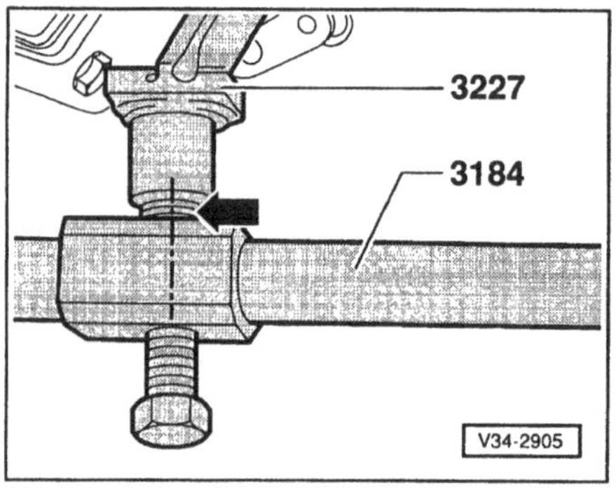

- Before installing engine/transmission assembly, attach mounting housing onto transmission carrier with bolts into bearing housing holes.

- Alternatively swing in and raise engine/transmission assembly. The right-hand drive shaft is to be guided so that it bears on the transmission drive flange.

Install engine/transmission mounts, stress-free, as follows:

- Install bolts for rear transmission mount.

- Before installing bolts to secure left-hand mounting bracket to body, make sure that engine mounting pin and removal device spindle -arrow- are, as much as possible, in line.

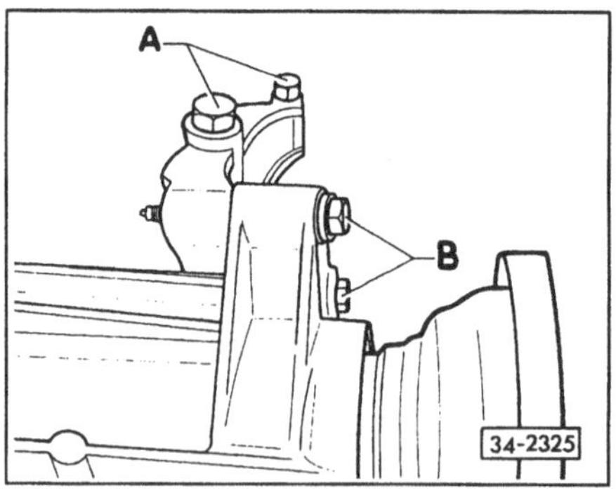

Bolts for transmission carrier to engine and mounting housing are tightened as follows:

- Install bolts -A- and -B-.

- Tighten bolts -A-.

- Tighten bolts -B-.

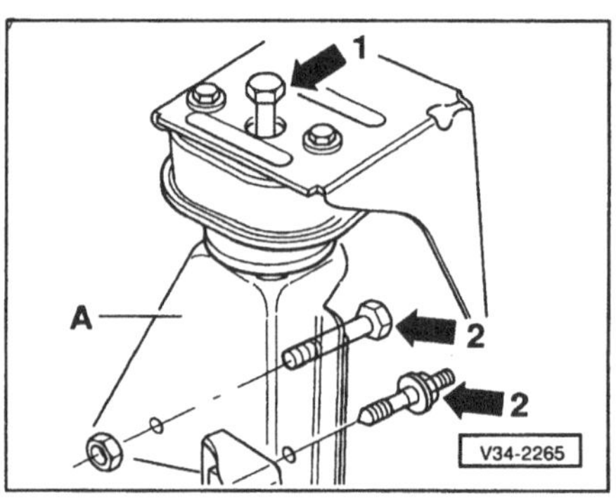

- Before lifting engine/transmission, tighten bearing block -A- bolts -arrows 2-.

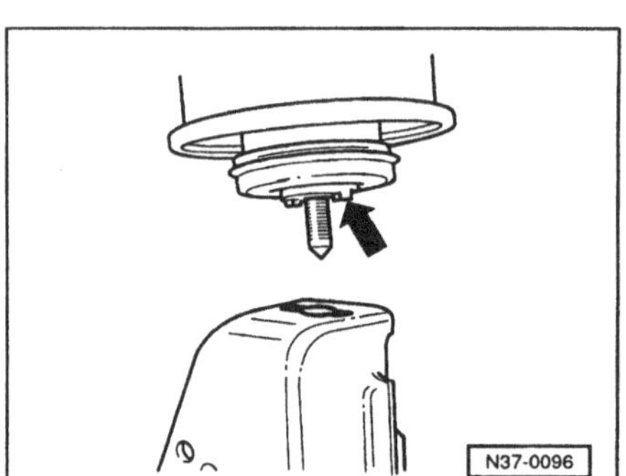

- Make sure the bonded rubber mounting lugs -arrow- are seated in bearing bracket.

- After lifting engine/transmission against body mounted bracket, tighten bolt (-arrow 1-, Fig. V34-2265).

- Tighten rear transmission mount to specified torque.

CAUTION!

Do not remove 3184 removal device until bolts securing left-hand mounting bracket have been securely tightened.

- Attach front exhaust pipe to Three Way Catalytic Converter or front muffler.

⇒ Repair Group 26; Exhaust system, removing and installing.

- Always replace seal between flange shaft and drive shaft, but first remove protective foil before bolting drive shafts to flanges.

⇒ Suspension, Wheels, Brakes, Steering; Repair Group 40; Drive shaft, servicing

- Install starter motor/wiring.

⇒ Repair Group 27; Starter motor, removing and installing.

- Fully push clutch pedal at least 5 times onto stop on vehicles with mechanical clutch control. Check function of adjustment mechanism ⇒ Page K 30-31.

- Checking transmission oil ⇒ Page K 34-59.

- Adjusting gearshift linkage ⇒ Page K 34-23.

Vehicles with air conditioner:

- Install air conditioner compressor

⇒ Heating, air conditioning system, Repair Group 87

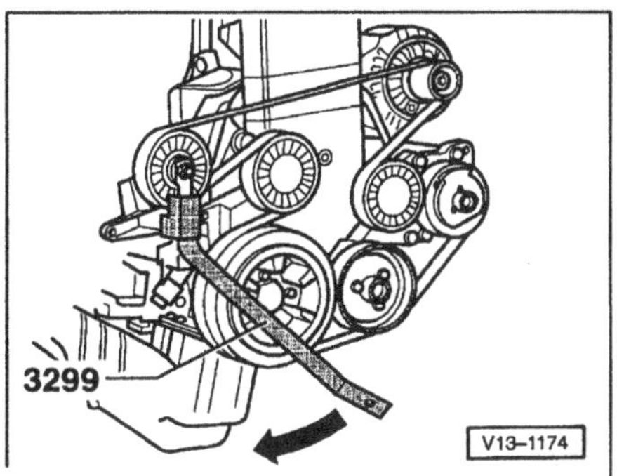

◄ – Install ribbed V-belts making sure they are correctly seated in belt pulleys.

Tightening torques

Transmission to engine

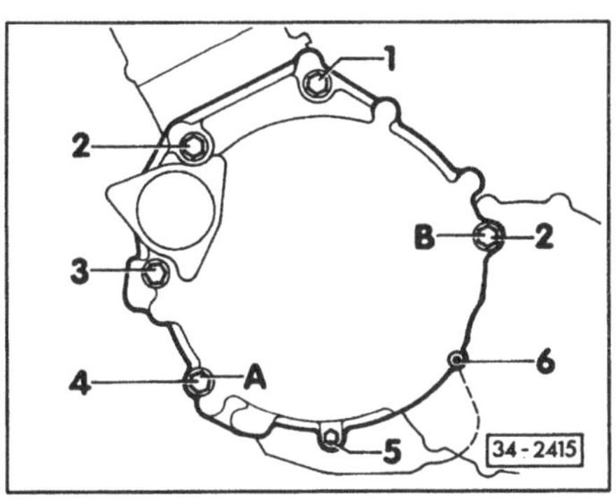

◄ **Vehicles with 4-Cylinder engine**

Pos.	Bolt	Qty.	Nm (ft lb)
1	M 12 x 55	1	80 (59)
2	M 12 x 60	2	80 (59)
3	M 12 x 100	1	80 (59)
4	M 12 x 130	1	80 (59)
5[1)]	M 8 x 35	1	25 (18)
6[1)]	M 8 x 12	1	25 (18)

1) Flywheel shield securing bolts

Position A and B = Dowel sleeves

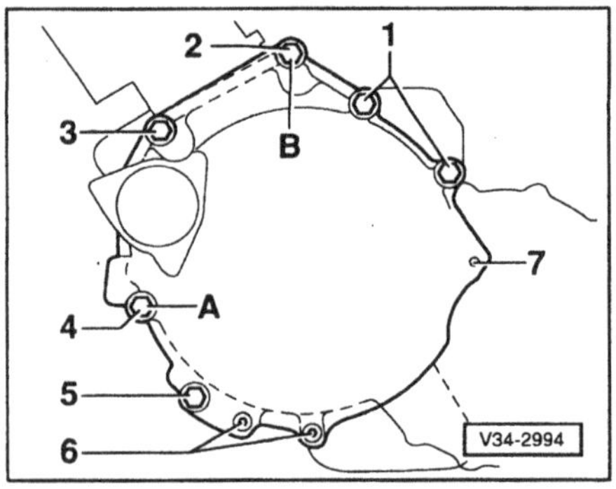

Vehicles with 5-Cylinder engine

Pos.	Bolt	Qty.	Nm (ft lb)
1[1)]	M 12 x 70	2	80 (59)
2	M 12 x 70	1	80 (59)
3	M 12 x 60	1	80 (59)
4	M 12 x 100	1	80 (59)
5	M 12 x 130	1	80 (59)
6	M 8 x 45	1	25 (18)
7[2)]	M 8 x 45	1	25 (18)

[1)] Additionally for flywheel cover plate

[2)] Tighten from engine side

Position A and B = Dowel sleeves

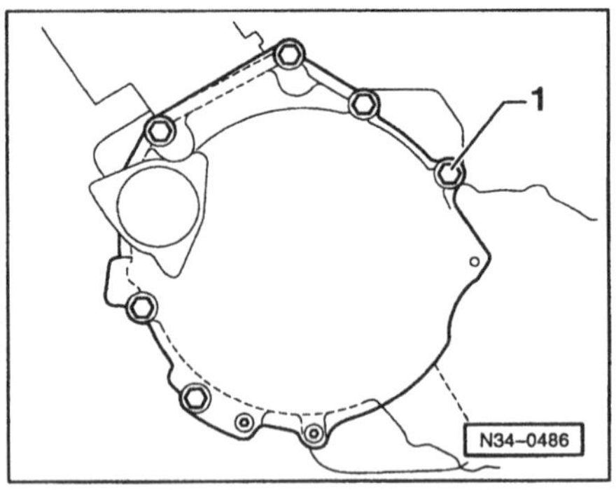

Vehicles with 5-Cylinder engine from 01.96 (from transmission 02 01 6)

The transmission to engine bolted flange has been reinforced in the area of bolt -1-. The bolt in this location is now longer (M12 x 130) than previously used.

Tightening torque

80 Nm (59 ft lb)

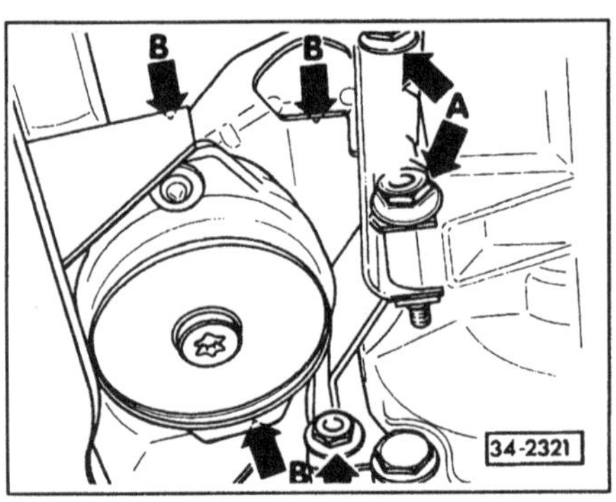

Rear transmission mount

Transmission to 09 08 2

Bolts -A-	65 Nm (48 ft lb)
Bolts -B-	45 Nm (33 ft lb)

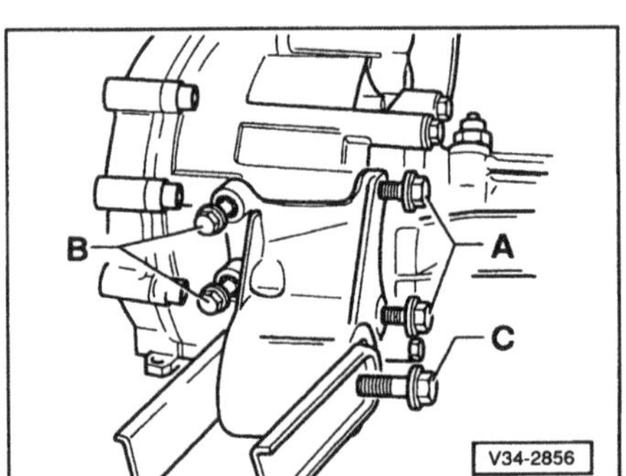

Transmission from 10 08 2

Bolts -A-	100 Nm (74 ft lb)
Bolts -B-	40 Nm (30 ft lb)
Bolts -C-	200 Nm (148 ft lb)

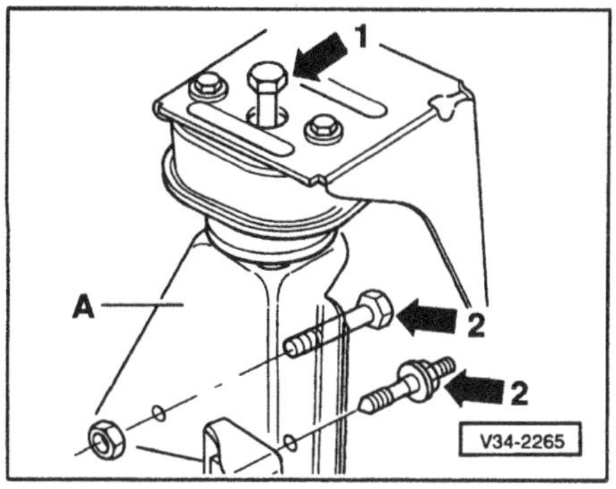

◄ Front left transmission mounting

Bolt -1-	65 Nm (48 ft lb)
Bolts -2-	65 Nm (48 ft lb)

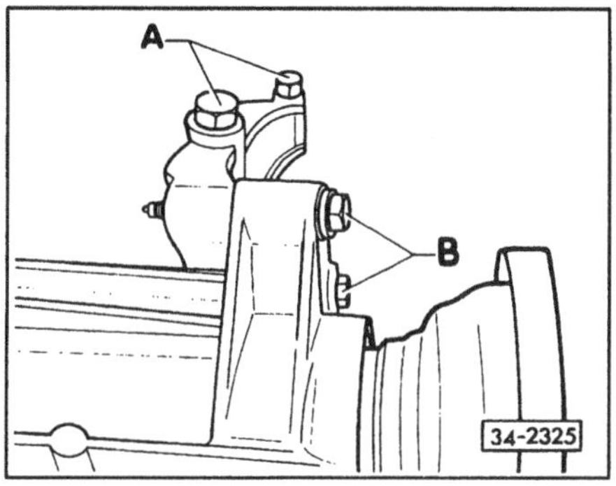

◄ Transmission carrier

Bolt -A-	60 Nm (44 ft lb)
Bolts -B-	25 Nm (18 ft lb)
Clutch controls to transmission	25 Nm (18 ft lb)
Gearshift linkage mounting bracket to transmission	20 Nm (15 ft lb)
Gearshift linkage swivel head to transmission	9 Nm (7 ft lb)
Speedometer sensor to drive	10 Nm (7 ft lb)

Bearing housing with right-hand output shaft/drive flange, removing and installing

(transmission installed)

Removing

Remove right-hand drive shaft as follows:

Note:

The figure shows the left-hand side of the vehicle.

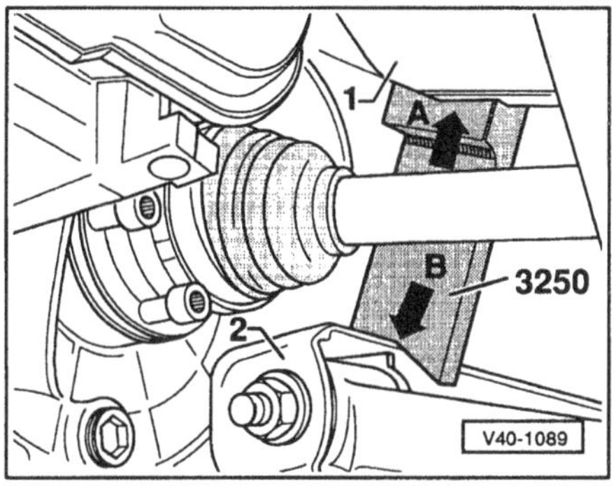

◄ – To enable left-hand drive shaft –figure– and left and right-hand shock absorber/stabilizer bar coupling link to be removed, install 3250 support frame, with vehicle in an unloaded condition, to both sides before vehicle is raised.

Installation position:

The 3250 support frame bearing surface -arrow A- must be positioned below the upper control arm -1-.

The V-shape opening -arrow B- is positioned on subframe -2-.

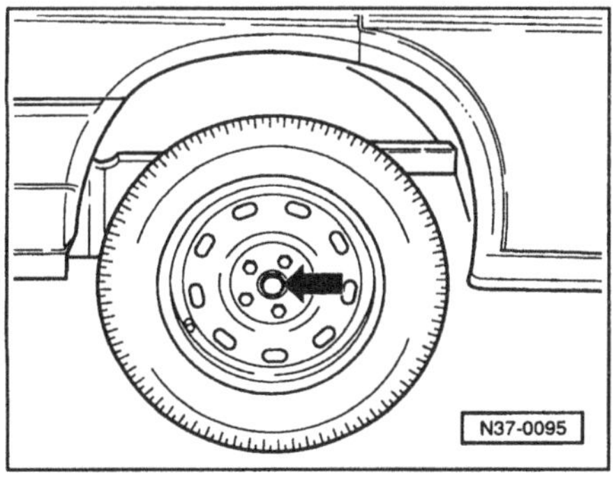

◄ – With vehicle resting on its wheels, loosen right-hand drive shaft/wheel hub bolt and wheel lug bolts.

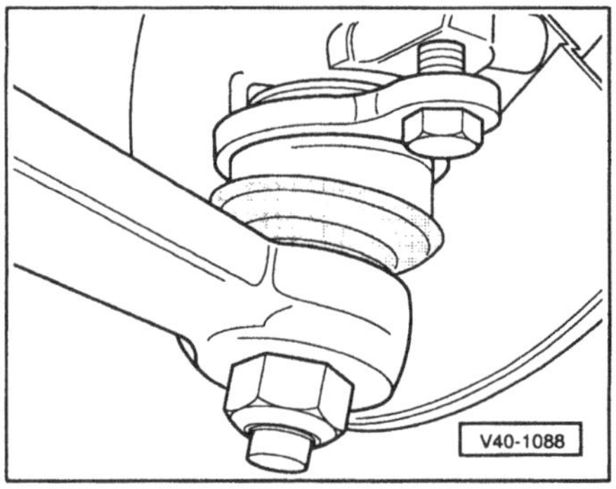

- Raise vehicle.
- Remove noise insulation tray below engine/transmission, if installed.
- Remove bolt for right-hand shock absorber/stabilizer bar coupling rod/stabilizer.
- Fully compress shock absorber.
- Disconnect drive shaft from flange.
- Remove wheel.
- ◄ Using VAG 1383 A transmission jack for support, separate right-hand lower wheel bearing housing ball joint bolted connection.

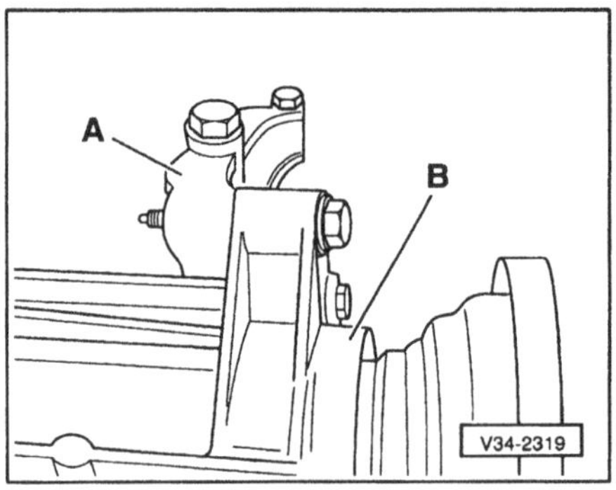

- Remove drive shaft.
- ◄ Unbolt transmission carrier -A- from engine block and mounting housing -B-.

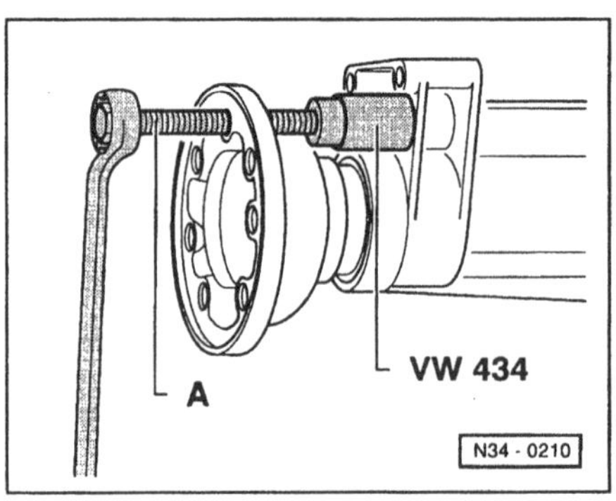

- Place drain pan under flange.
- ◄ Remove drive flange/output shaft from locating device in differential with VW 434 arbor and hex bolt -A- (M10 x 75).

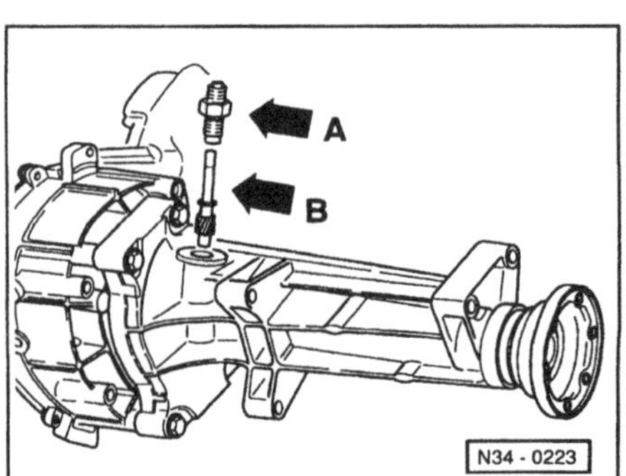

- Remove speedometer drive shaft.
- ◄ Remove speedometer drive:

  Unbolt guide sleeve -arrow A- and remove speedometer drive -arrow B-.
- Unbolt and remove bearing housing from transmission and remove together with output shaft/flange shaft. Push engine/transmission unit slightly forward, if required.

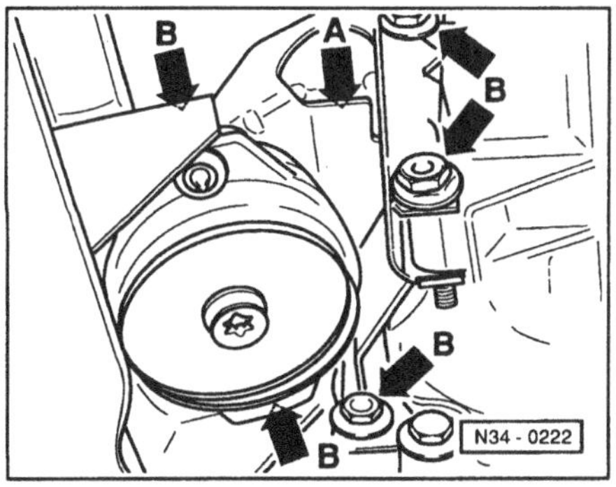

Transmission to build date 09 08 2

◄ ***Note:***

Removal of transmissions prior to 09 08 2 is possible only if there is a hole in the mount in the area of arrow -A-.

– Remove rear transmission mount arrows -A- and -B-.

Installing

Install in reverse order. Always replace O-ring to seal bearing housing with clutch housing.

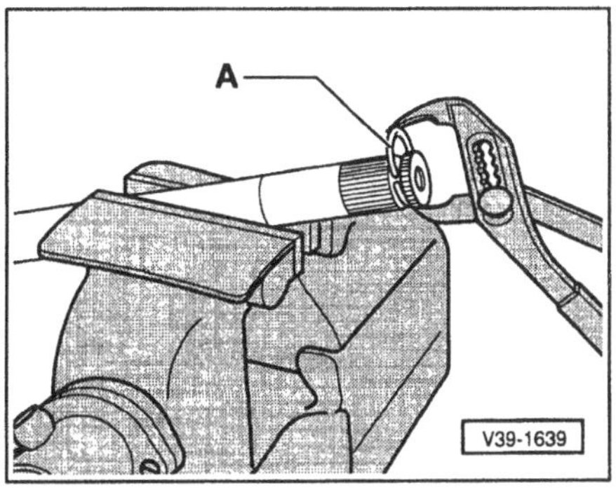

Note:

◄ *Drive the drive flange/output shaft in with a non–marring hammer. First: replace securing ring -A-.*

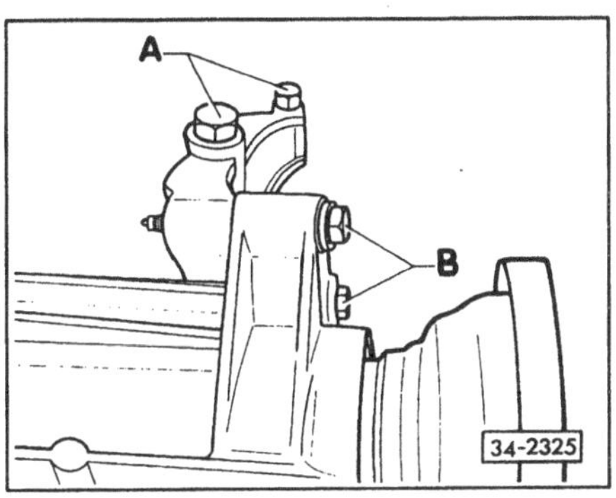

Bolts for transmission carrier to engine and bearing housing are tightened as follows:

– Install bolts -A- and -B-.

– Tighten bolts -A-.

– Tighten bolts -B-.

Transmission oil, checking ⇒ Page K 34-59

Tightening torques

Bearing housing to transmission	40 Nm (30 ft lb) plus an additional 1/4 turn (90°)
Transmission carrier to engine	60 Nm (44 ft lb)
Transmission carrier to bearing housing	25 Nm (18 ft lb)
Guide sleeve to bearing housing	30 Nm (22 ft lb)

Rear transmission mounting to subframe ⇒ Page K 34-52

Transmission oil, checking

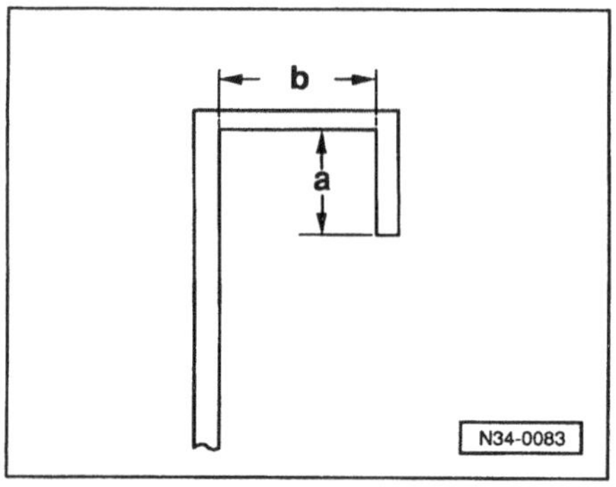

Transmission oil specification ⇒ Page K 00-3

– It is necessary to manufacturer a tool to check transmission oil level:

Dimension “a” 15 mm (0.5905 in.)

Dimension “b” 22 mm (0.8661 in.)

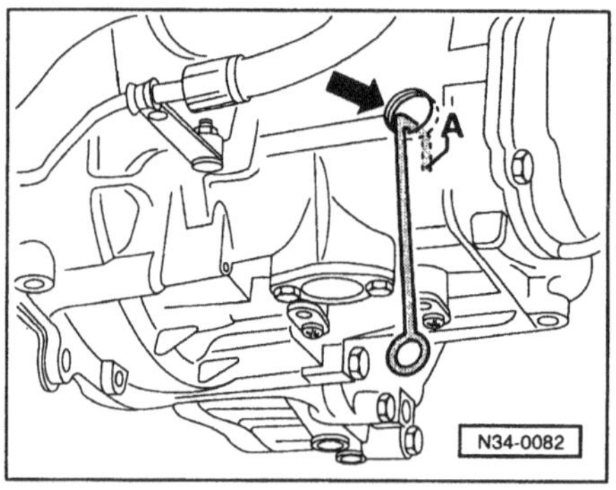

– Remove plug -arrow- to check oil.

– Carefully insert tool -arrow-.

Arm -A- must lie against inner transmission wall.

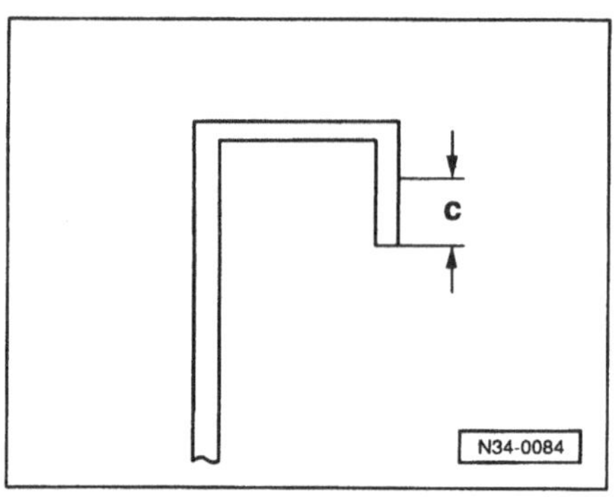

Oil level is correct when transmission oil covers lower area of arm (dimension “c”).

- Dimension “c” 10 mm (0.3937 in.)

– Install plug.

When filling with new transmission oil note following:

– Remove plug.

– Top up transmission oil.

- Capacity: 2.5 liter (2.7 qt)

– Install plug.

– Start engine, engage transmission and allow transmission to turn for approximately 2 minutes.

– Stop engine and remove plug.

– Insert tool for checking.

– If necessary top up with transmission oil until transmission oil covers lower area of arm (dimension “c”).

- Dimension “c” 10 mm (0.3937 in.)

– Install plug.

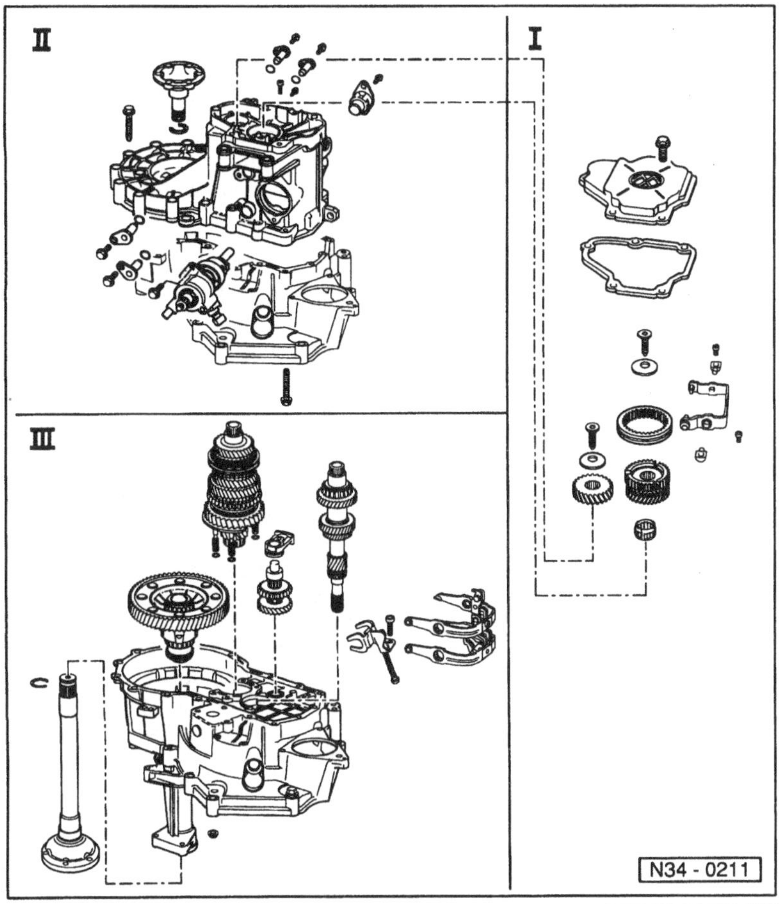

Transmission, disassembling and assembling

Disassembling and assembling sequence
⇒ Page K 34-72

I – Cover for transmission housing and 5th gear, removing and installing
⇒ Page K 34-62

II – Transmission housing and shift mechanism, disassembling and assembling ⇒ Page K 34-66

III – Input shaft, output shaft, differential, and shift forks, disassembling and assembling
⇒ Page K 34-69

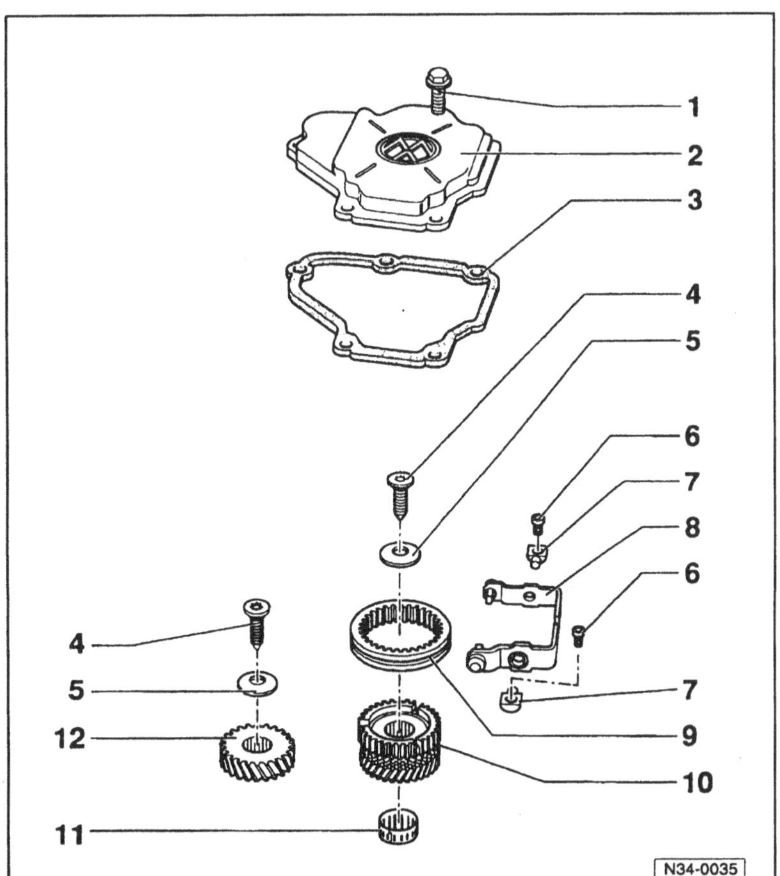

Cover for transmission housing and 5th gear, removing and installing

1 - Hex bolt
- Tightening torque: 10 Nm (7 ft lb)
- For transmission housing cover

2 - Transmission housing cover

Note:

If transmission housing cover is in place with the transmission installed, check oil level as follows:

– *Fill with oil as described on Page K 34-59 and install plug. Start engine, engage gear and allow transmission to turn for approximately 2 minutes. Stop engine and check transmission oil level ⇒ Page K 34-59, top off with transmission oil, if necessary.*

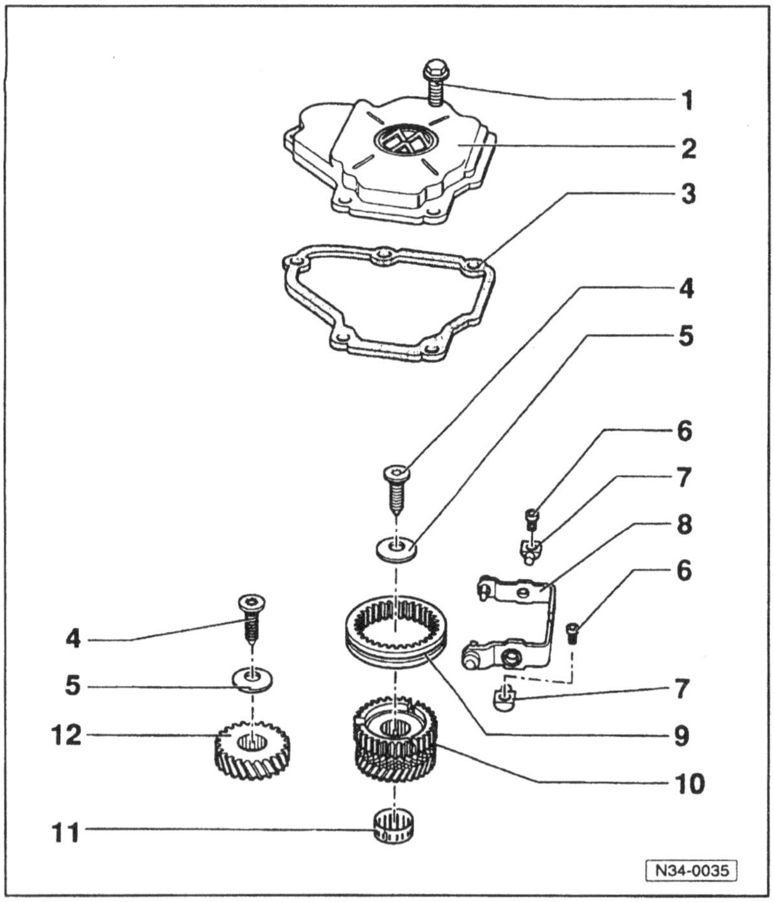

3 - **Gasket**

4 - **Torx socket head bolt M10**
- Tightening torque: 80 Nm (59 ft lb)
- Bolt head holds diaphragm spring in place with shape of bolt head

5 - **Diaphragm spring**
- Installing ⇒ Page K 34-87

6 - **Bolt**
- Tightening torque: 25 Nm (18 ft lb)
- For pivot pin to transmission housing

7 - **Pivot pin**

8 - **5th gear shift fork**
- Disassembling and assembling ⇒ Page K 34-117
- Adjusting ⇒ Page K 34-88
- Modification:
  From transmission build date 31 10 1, modified in conjunction with modified 5th gear synchronization, dimensions modified ⇒ Page K 34-123
- Adjusting ⇒ Page K 34-89

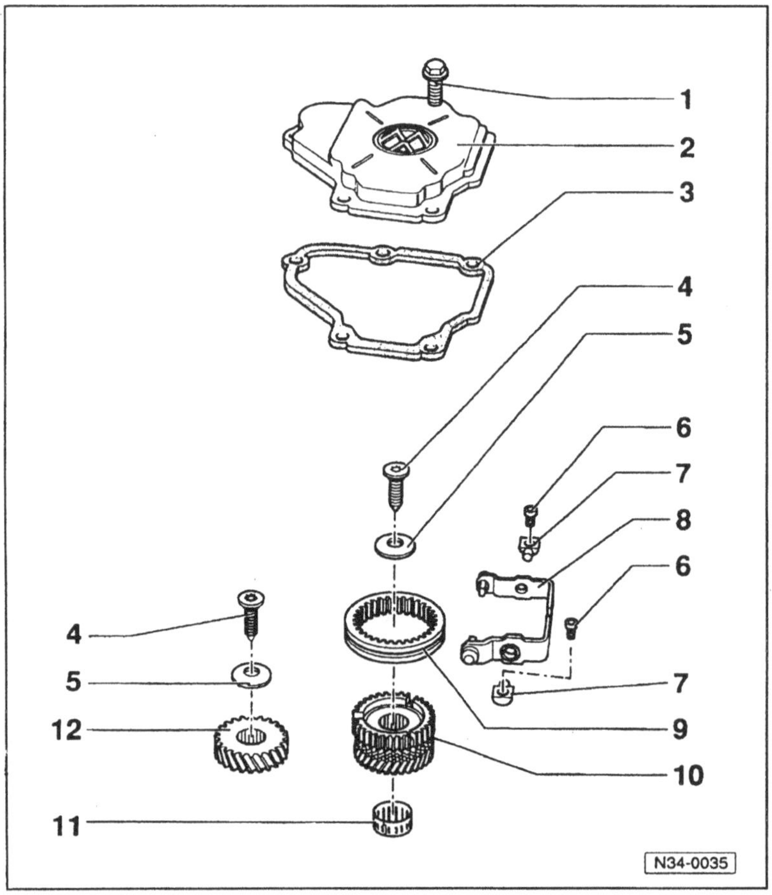

9 - **5th gear locking collar**
- Installation position ⇒ Page K 34-85
- Adjusting ⇒ Page K 34-88
- Modification:
  From transmission build date 31 10 1 synchronization with additional locking pieces ⇒ Page K 35-12
- Adjusting ⇒ Page K 34-89

10 - **Synchro-hub with 5th gear and synchronizing ring**
- Disassembling and assembling ⇒ Page K 35-12
- Modification:
  From transmission build date 31 10 1 synchronization with additional locking pieces ⇒ Page K 35-12

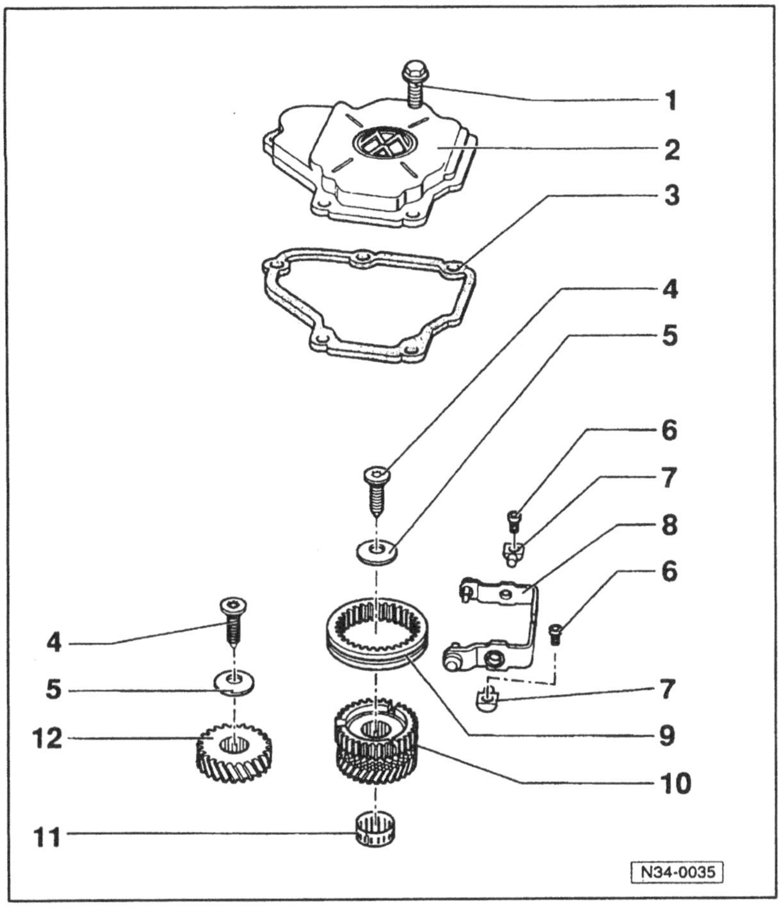

Note:

If 5th gear synchronizing unit with additional locking pieces is to be installed in a transmission built before 29 10 1, the modified 5th gear shift fork must be retrofitted ⇒ Page K 34-123.

11 - Needle roller bearing

12 - 5th gear
- ♦ Installation position ⇒ Page K 34-83

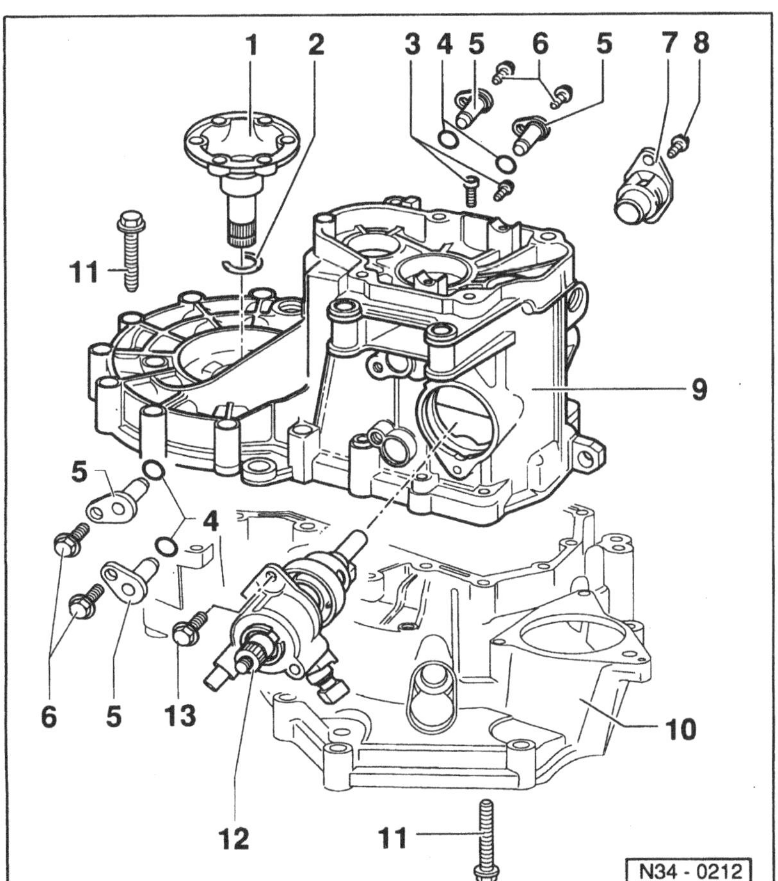

Transmission housing and shift mechanism, removing and installing

1 - Flange shaft

2 - Circlip
- ♦ Always replace

3 - Torx socket head bolt
- ♦ Tightening torque: 30 Nm (22 ft lb)
- ♦ For reverse shaft support
- ♦ Self-locking
- ♦ Always replace

4 - O-ring
- ♦ Always replace

5 - Pivot pin

6 - Bolt
- ♦ Tightening torque: 25 Nm (18 ft lb)

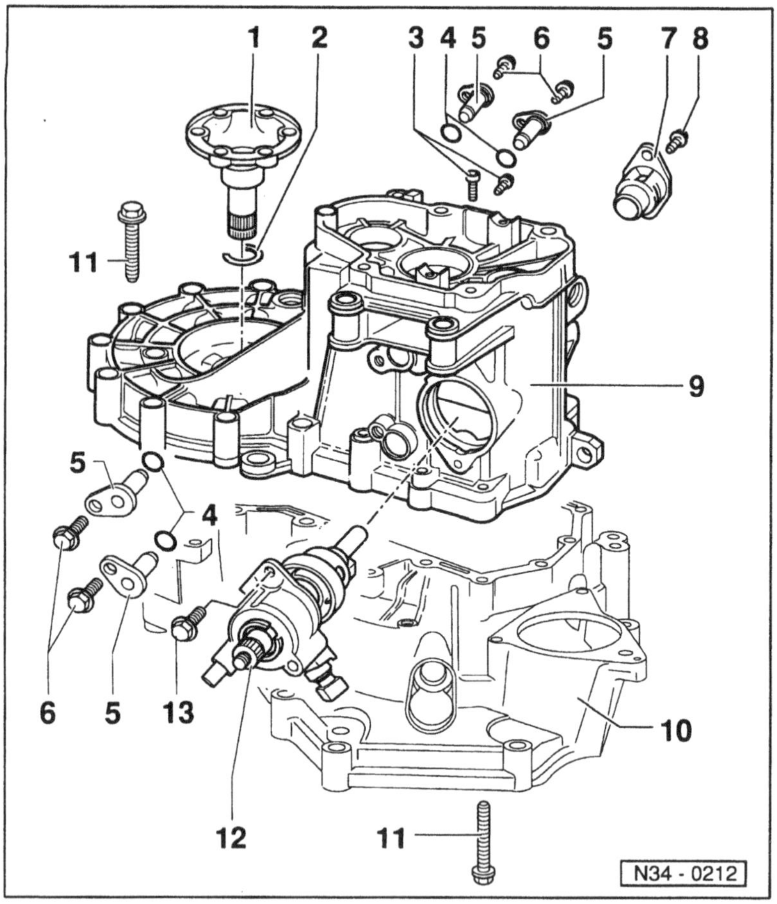

7 - Cover plate
- With bearing sleeve
- Removing bearing sleeve ⇒ Page K 34-115
- Installing bearing sleeve ⇒ Page K 34-116
- Modification:
  From transmission build date 31 10 1 modified in conjunction with bearing surface in cover plate ⇒ Page K 34-116

8 - Bolt
- Tightening torque: 25 Nm (18 ft lb)

9 - Transmission housing
- Servicing ⇒ Page K 34-90
- Modification:
  From transmission build date 05 06 2 modified transmission housing ⇒ Page K 34-92

10 - Clutch housing
- Servicing ⇒ Page K 34-94

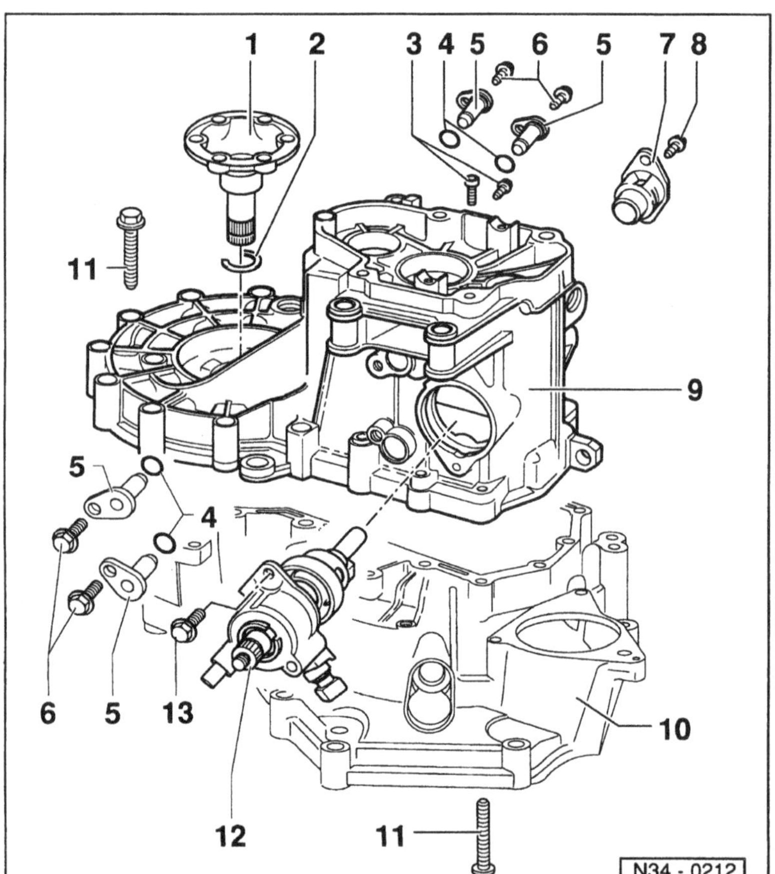

11 - Hex bolt
- Tightening torque: 25 Nm (18 ft lb) plus an additional 1/4 turn (90°)
- Modification:
  From transmission build date 05 06 2 in conjunction with modified shortened transmission housing
  new: 60 mm (2.3622 in.) long
  previous: 65 mm (2.5591 in.) long

Note:

New type bolts must not be installed in earlier transmissions.

12 - Shift mechanism
- Disassembling and assembling ⇒ from Page K 34-109

13 - Hex bolt
- Tightening torque: 25 Nm (18 ft lb)

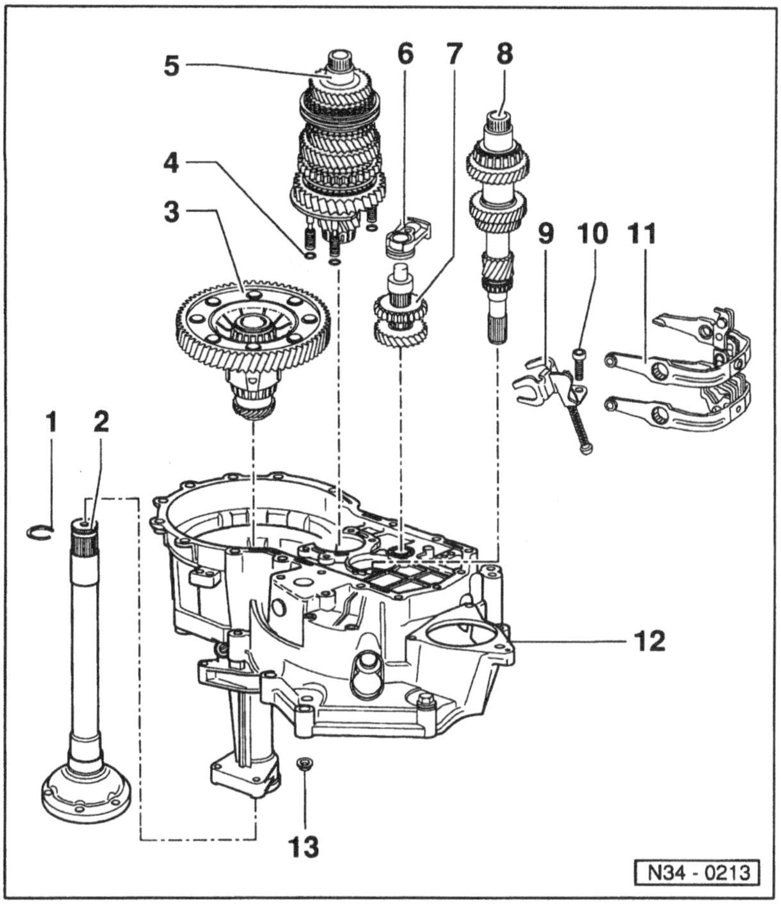

Input shaft, output shaft (pinion shaft), differential, and shift forks, removing and installing

1 - Circlip
- Always replace
- Disassembling and assembling ⇒ Page K 34-107

2 - Output shaft with right-hand shaft flange
- Removing ⇒ Page K 34-107
- Drive in with non–marring hammer
- Separating and assembling output shaft and shaft flange ⇒ Page K 34-108

3 - Differential
- Disassembling and assembling ⇒ Page K 39-6

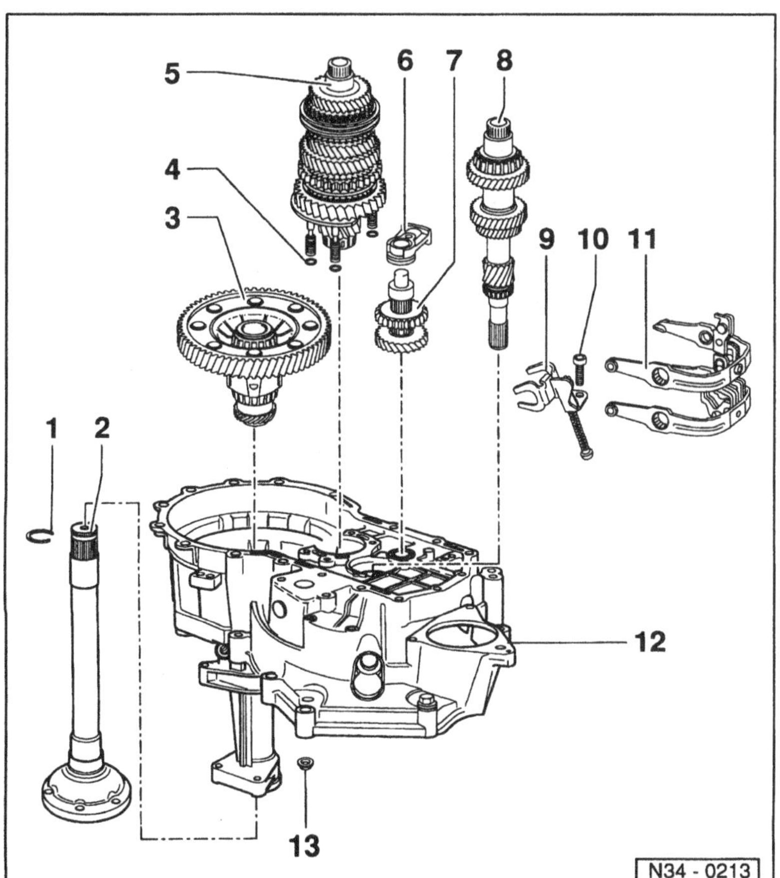

4 - Sealing ring
- Quantity 4
- Always replace

5 - Output shaft
- Disassembling and assembling ⇒ Page K 35-20

6 - Reverse shaft support
- Disassembling and assembling ⇒ Page K 35-51

7 - Reverse shaft
- Disassembling and assembling ⇒ Page K 35-49

8 - Input shaft
- Disassembling and assembling ⇒ Page K 35-1

9 - Reverse gear shift fork
- Disassembling and assembling ⇒ Page K 34-120
- Installation position ⇒ Page K 34-80

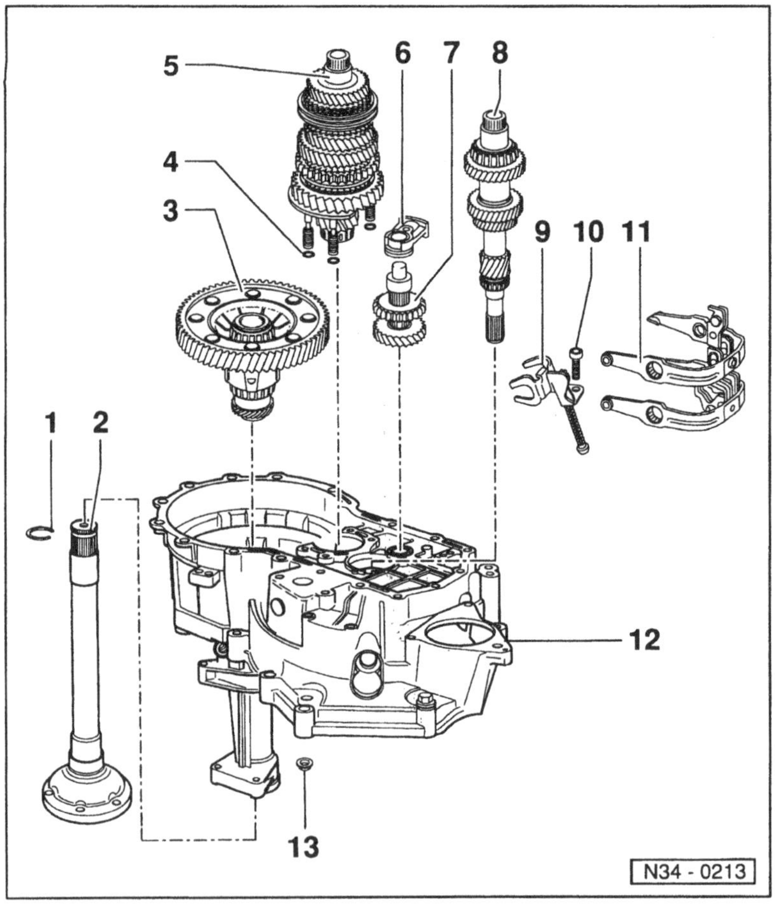

10 - Torx socket-head bolt
- Tightening torque: 25 Nm (18 ft lb)

11 - Shift mechanism
- (Shift forks)
- Disassembling and assembling ⇒ Page K 34-117

12 - Clutch housing
- Servicing ⇒ Page K 34-94

13 - Hex nut
- Tightening torque: 25 Nm (18 ft lb) plus an additional 1/4 turn (90°)
- 4 nuts for bearing support

Disassembling and assembling sequence

Disassembling and assembling transmission housing cover, transmission housing, shift mechanism, input shaft, output shaft (pinion shaft), differential, and shift forks

Disassembling

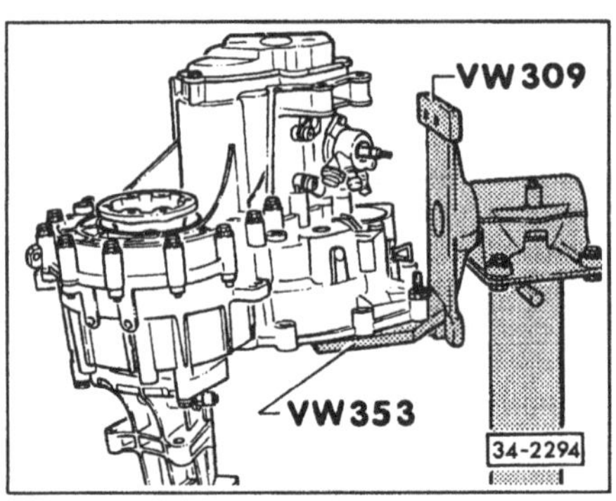

– Drain transmission oil.

◄ – Mount transmission on repair stand.

– Remove clutch release lever and release bearing ⇒ Page K 30-32.

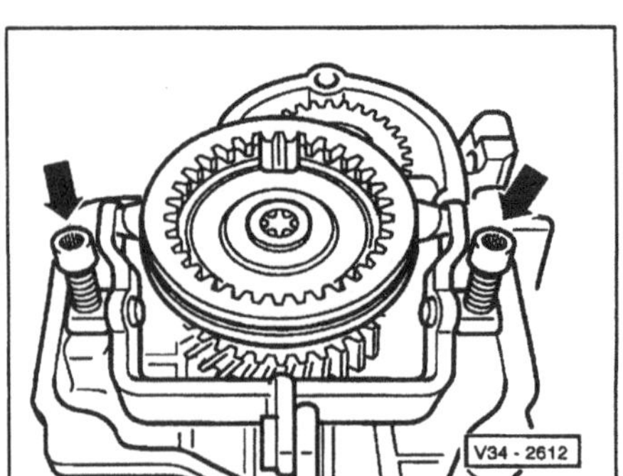

– Remove guide sleeve ⇒ Page K 30-32.

– Remove transmission housing cover.

◄ – Remove 5th gear shift fork -arrow-.

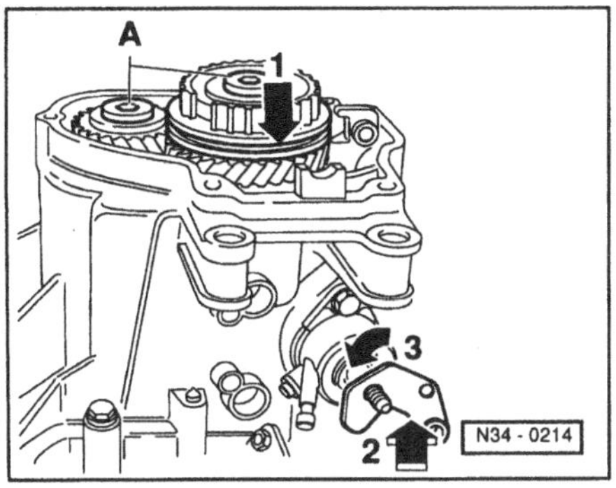

◄ – Remove bolts -A- for synchro-hub and 5th gear. Do this by selecting 1st gear –arrows 2- and 3– and 5th gear -arrow 1-.

Note:

After engaging both gears, the input shaft and output shaft are locked. Synchro-hub and gear wheel cannot turn. Both bolts can now be loosened or tightened.

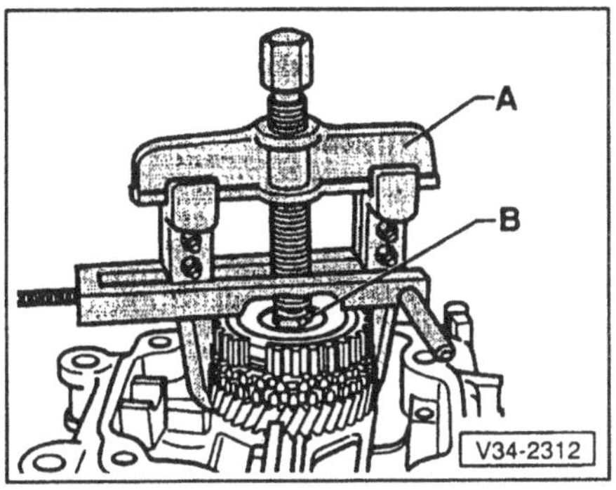

◄ – Remove synchro-hub with 5th gear and remove needle roller bearing.

A – 20/10 Kukko two arm puller with V/170 Matra puller hooks w/tensioner.

B – Hex bolt M10 x 20

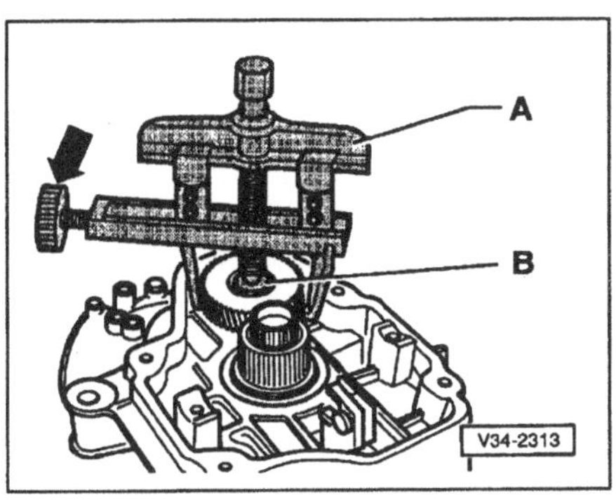

◄ – Remove 5th gear

A – 20/10 Kukko two arm puller with V/170 Matra puller hooks w/tensioner.

B – Hex bolt M10 x 20

Removal sequence:

– Assemble two arm puller and align by turning bolt -arrow-.

– Heat gear with heat gun and remove with two arm puller.

Note:

When removing gear, make sure that hooks do not bend outward; if necessary, tighten screw -arrow-. Check 5th gear for damage after removal.

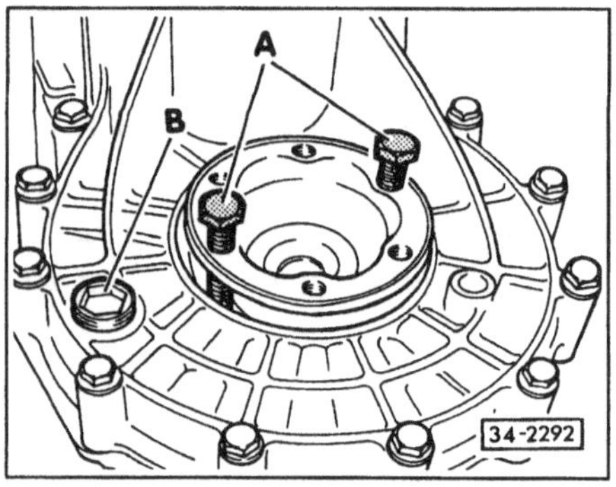

◄ – Remove left-hand drive flange.

Install M10 x 75 bolts -A- and turn alternately.

Note:

Locate bolts -A- so that they align with oil drain plug -B-. They then bear on two reinforcements of the housing.

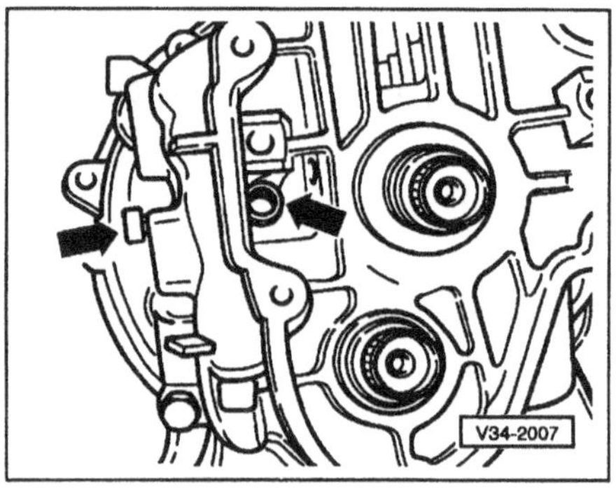

◄ – Remove both reverse shaft support bracket bolts -arrows-.

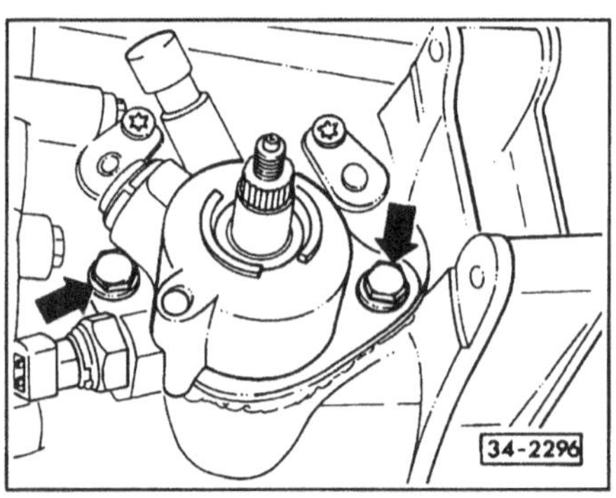

◄ – Place shift shaft in neutral position.

– Remove bolts -arrows- and remove shift shaft from transmission housing.

– Remove shift shaft with shift cover.

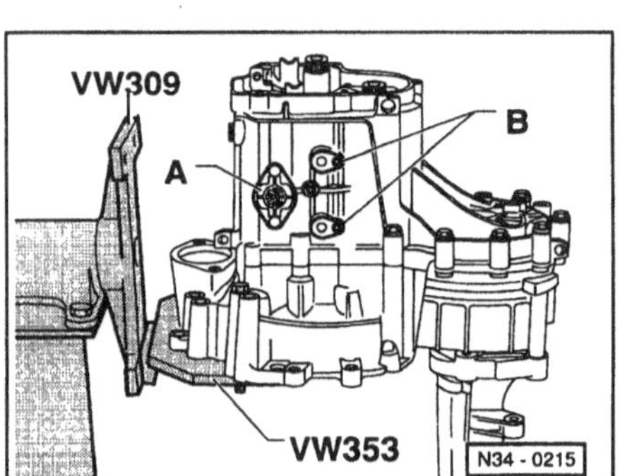

◄ – Remove cover plate -A- and pivot pins -B- from underside of transmission.

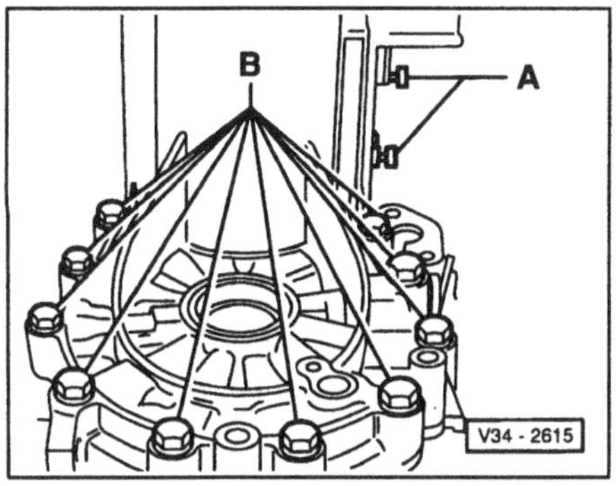

- Remove pivot pins -A- on top side of transmission and transmission housing to clutch housing bolts -B- in area of differential.
- Removing output shaft with shaft flange ⇒ Page K 34-107.

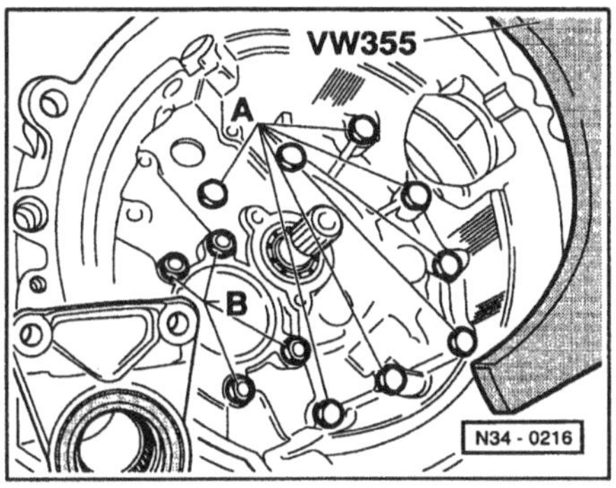

- Remove bolts -A-, securing transmission housing on clutch housing side.

Note:

Do not remove nuts -B- for output shaft bearing support.

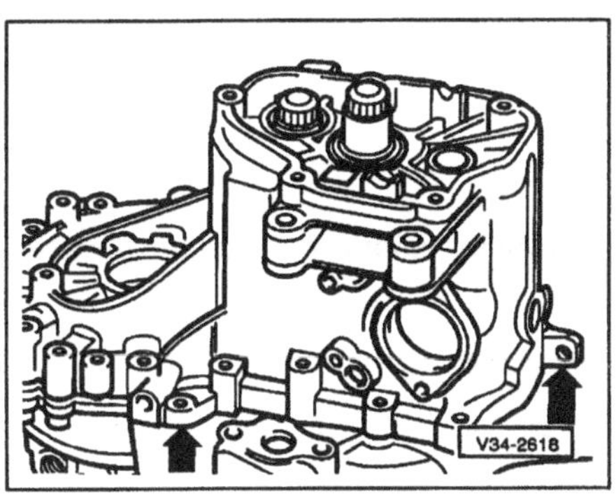

- Remove transmission housing. If necessary, carefully pry up all around, alternating between sides on protruding housing flange -arrows-. Take care not to damage sealing surfaces.

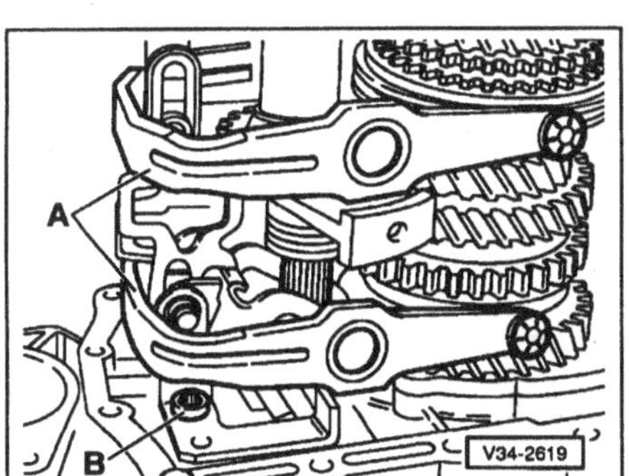

- Remove shift forks -A- together with shift plates.
- Remove bolt for reverse gear shift mechanism -B-.

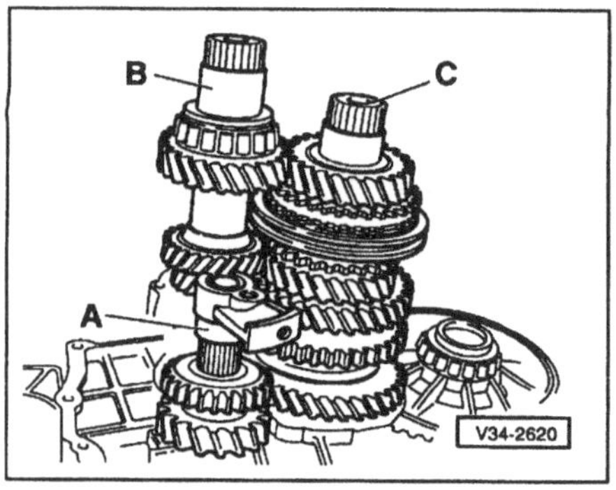

– Remove output shaft bearing support nuts (⇒ Page K 34-77, Fig. N34-0216, -B-).

◄ – Remove reverse gear -A-, input shaft -B- and output shaft -C-, in that order, from clutch housing.

– Remove differential.

Assembling

– Install differential.

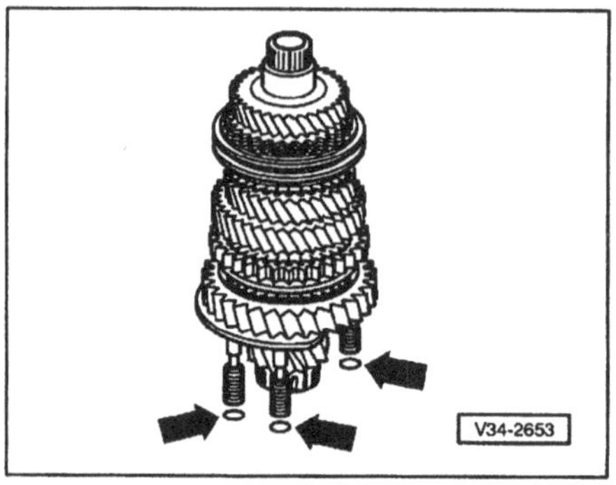

◄ – Always replace O-rings -arrows- for output shaft bearing support.

Note:

The figure only shows 3 of the 4 O-rings.

– Install input shaft, output shaft, and reverse shaft together.

– Tighten nuts for output shaft bearing support.

– Clean locking fluid from threads in reverse shaft support using a tap (⇒ Page K 34-81, Fig. V34-2293, -arrow-).

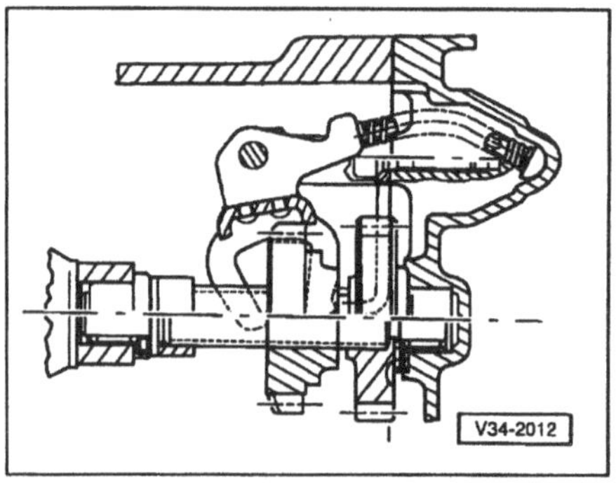

– Install reverse shaft support.

– Install reverse gear shift mechanism.

◄ Installation position reverse gear

– Install shift forks together with shift plates.

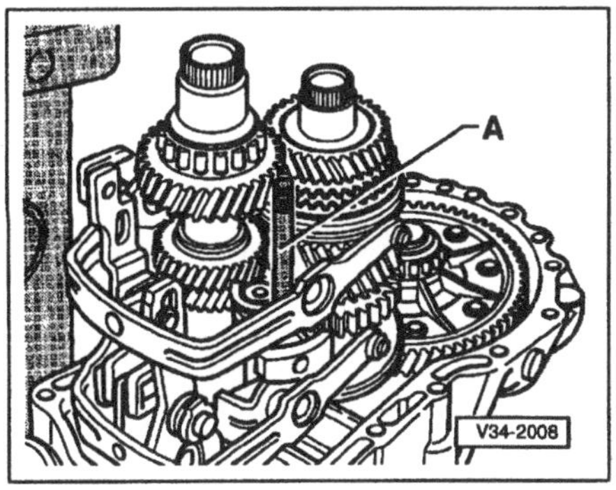

◄ – After installing transmission housing, screw M 8 x 100 mm studs -A- into reverse shaft support to align shaft.

– Align shift plates.

Note:

The shift fork segments must be positioned in grooves on locking collars.

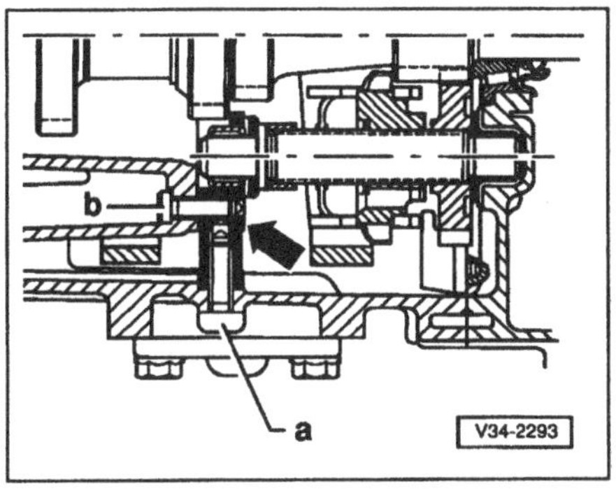

– Install transmission housing.

Install reverse shaft support bolts -arrow- as follows:

◄ – Insert bolt -a-, remove stud (⇒ Fig. V34-2008), install bolt -b- and hand tighten.

Torquing sequence:

1 – Bolt -a- 30 Nm (22 ft lb)
2 – Bolt -b- 30 Nm (22 ft lb)

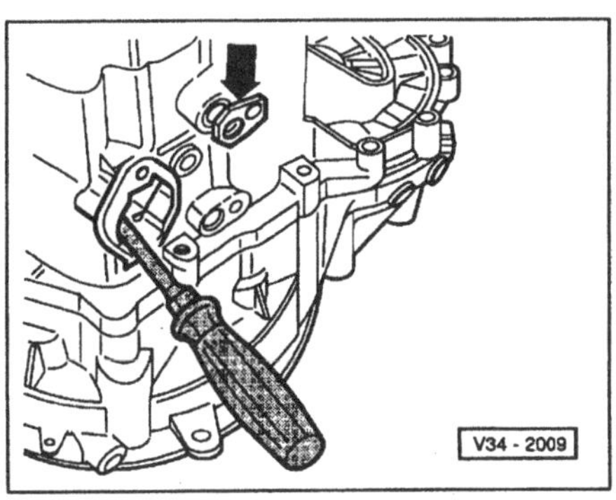

◄ – To install locating pin -arrow- for shift forks, align shift mechanism with a screwdriver so the relevant pivot pin can be installed.

– Install shift shaft cover.

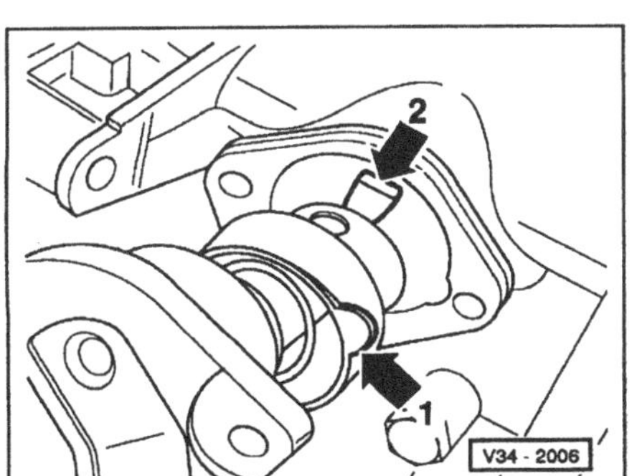

◄ Install shift shaft as follows:

– Place shift plates in neutral.

– Locate lug -arrow 1- in recess in transmission housing. Position shift shaft so that shift finger -arrow 2- can be inserted into shift plates.

– Install shift shaft cover.

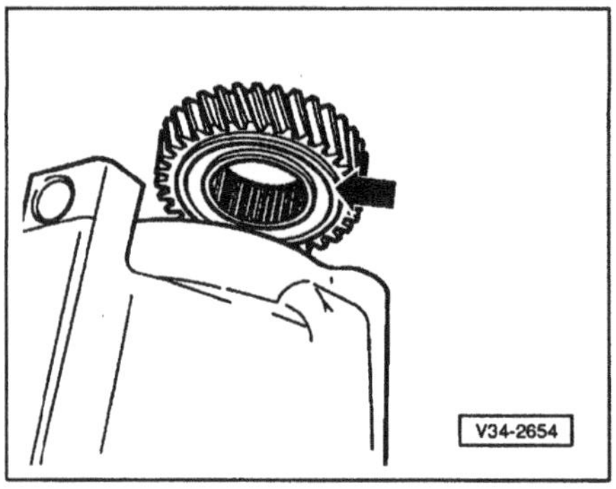

5th gear, installation position

The circumferential groove -arrow- faces transmission housing.

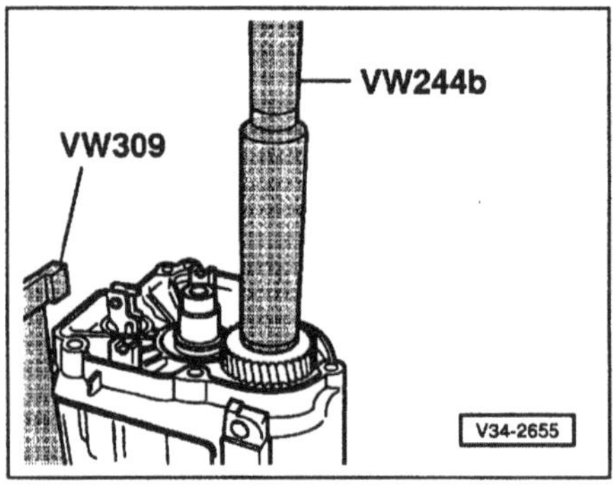

– Drive on 5th gear with VW 244 B driving sleeve.

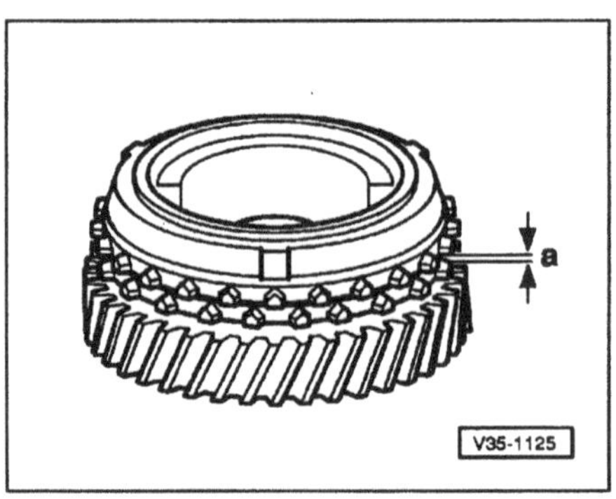

5th gear synchro-ring, inspecting

– Before installing 5th gear and synchro-ring, push syncro-ring onto cone of gear wheel and measure gap -A- with a feeler gauge.

Gap -A-	Installation mm (in.)	Wear limit mm (in.)
5th gear	1.1 ... 1.7 (0.0433 ... 0.0669)	0.5 (0.020)

– Install 5th gear with needle roller bearing.

– Place 5th gear synchro-ring on gear.

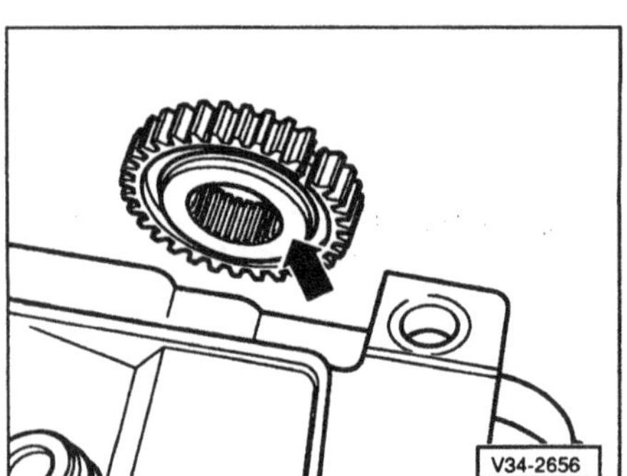

Installation position of 5th gear synchro-hub to transmission build date 30 10 1:

The high shoulder -arrow- faces transmission housing.

Notes:

- *From transmission build date 31 10 1, assemble synchro hub/5th gear operating sleeve before installing ⇒ Page K 35-13.*
- *Installation position of synchro hub/5th gear operating sleeve from transmission build date 31 10 1 ⇒ Page K 34-86.*

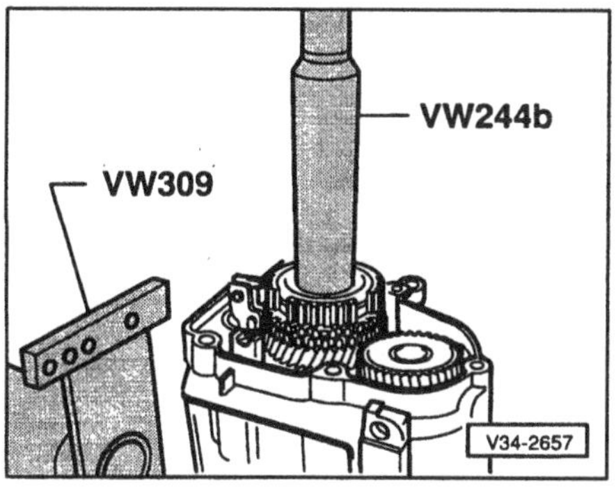

◄ – Drive on 5th gear synchro–hub with VW 244 B driving sleeve.

To transmission build date 30 10 1.

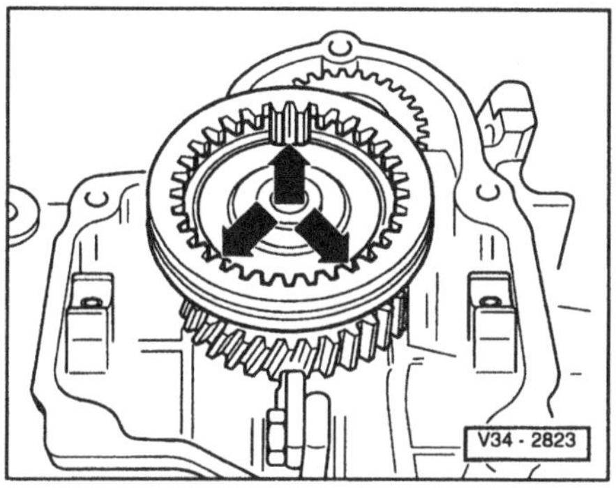

◄ **Installation position 5th gear operating sleeve to transmission build date 30 10 1:**

Install with chamfered teeth towards 5th gear. Slide locking collar over synchro-hub so the teeth indicated -arrows- face slots in synchro-hub.

– Install 5th gear operating sleeve.

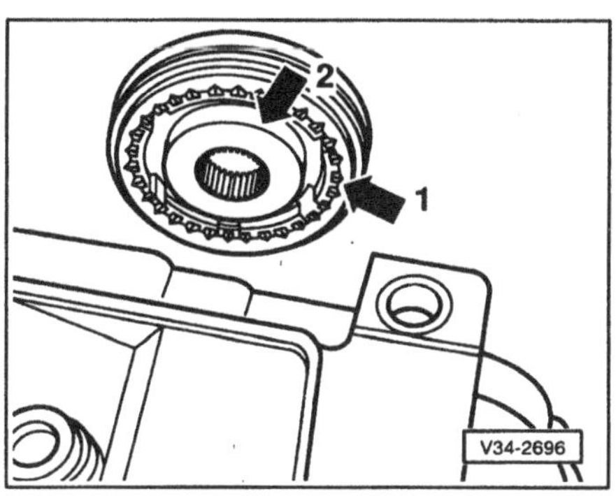

◄ **Installation position synchro-hub/5th gear operating sleeve from transmission build date 31 10 1:**

Chamfered teeth of operating sleeve -arrow 1- and high shoulder of synchro-hub -arrow 2- face transmission housing.

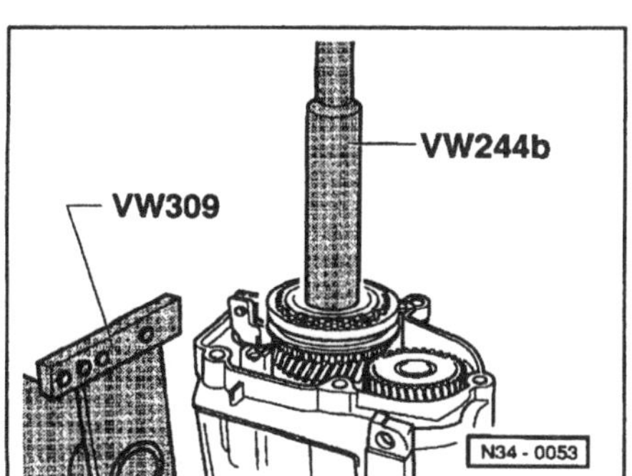

◄ – Drive on 5th gear synchro–hub with VW 244 B driving sleeve.

From transmission build date 31 10 1.

Note:

When driving on, make sure synchro-ring is free to move.

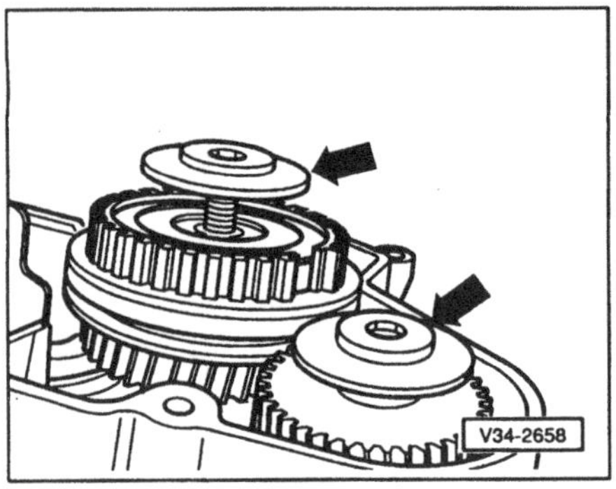

– Install bolts for synchro-hub and 5th gear.

◄ **Installation position of diaphragm spring for gear wheel and 5th gear synchro-hub bolts:**

Outer diameter (concave side -arrows-) faces 5th gear.

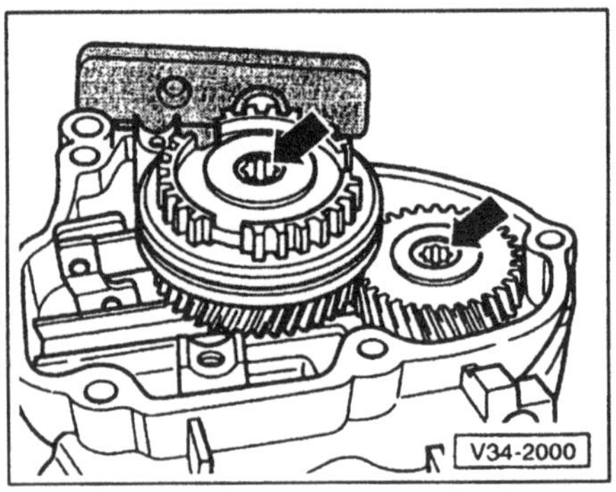

◄ – Engage two gears before tightening 5th gear synchro-hub and gear wheel securing bolts -arrows-
(⇒ Page K 34-73, Fig. N34-0214)

– Install 5th gear shift fork.

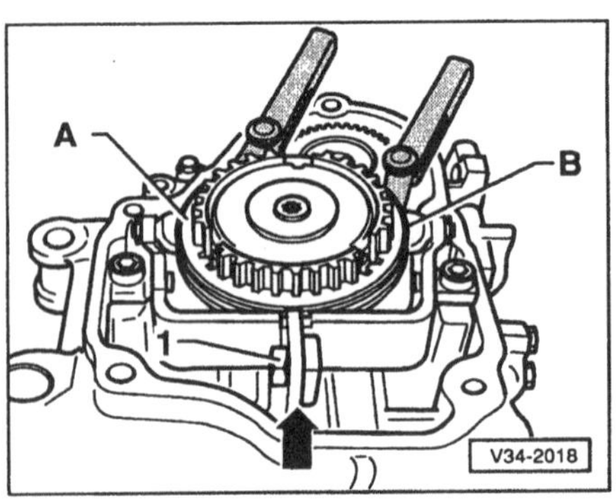

◄ **Adjusting 5th gear up to transmission build date 30 10 1:**

Sequence:

– Engage 5th gear. Loosen bolt -1-. Position 1.3 mm (0.0512 in.) feeler gauge between locking collar and gear at positions -A- and -B-. Install shift fork into 5th gear and, using a screwdriver, push shift fork jaws in direction of arrow and torque bolt -1- to 25 Nm (18 ft lb).

– Disengage 5th gear. Operating sleeve must now be in neutral position. Spring and synchro-ring must be free to move.

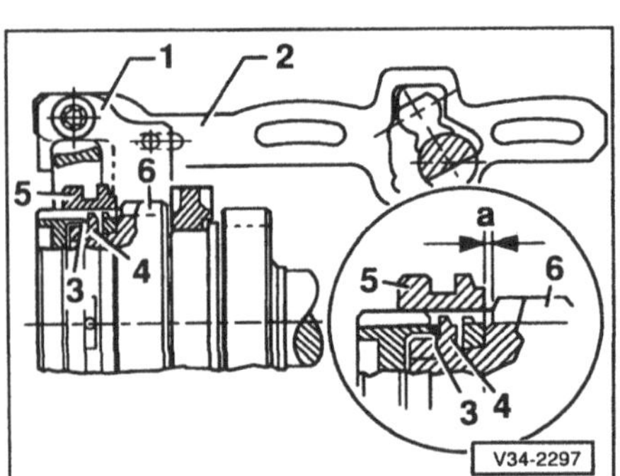

◄ **5th gear locking collar installation position:**

1 - Shift fork jaws

2 - Shift plate

3 - Spring

4 - Synchro-ring

5 - Operating sleeve

6 - Gear

Dimension -a- = 1.3 mm (0.0512 in.) with 5th gear engaged.

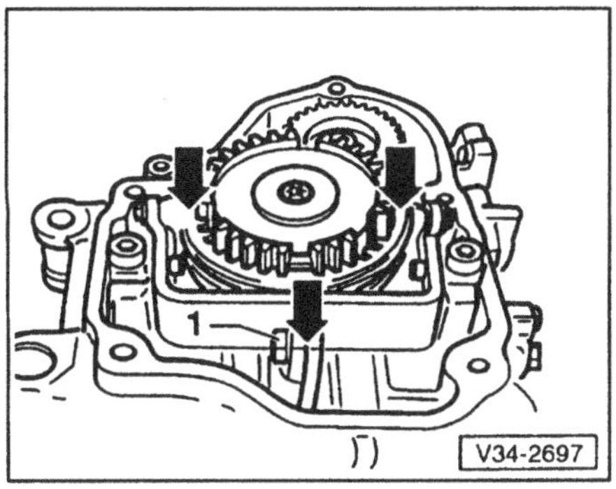

Adjusting 5th gear from transmission build date 31 10 1:

– Engage 5th gear. Loosen bolt -1-. Press operating sleeve and shift fork jaws in direction of arrow, then torque bolt -1- to 25 Nm (18 ft lb).

- Check measurement :
  It must not be possible to slide a 0.2 mm (0.0079 in.) feeler gauge between operating sleeve and gear wheel. If necessary, repeat adjustment.

– Disengage 5th gear. Operating sleeve must now be in neutral position. Synchro-ring must move freely.

– Install transmission housing cover.

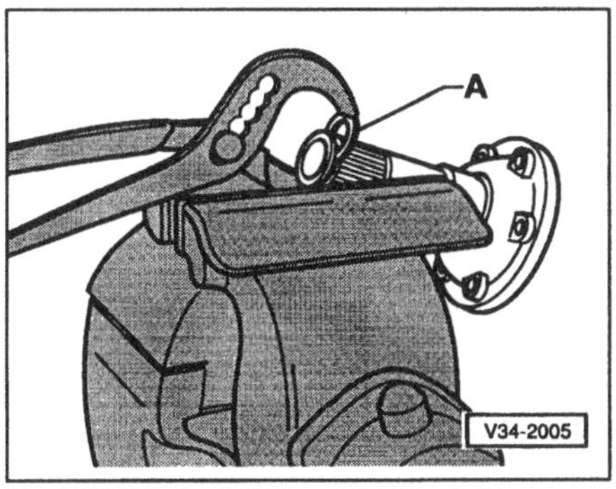

Flange shaft or output shaft/drive flange circlips, replacing

– Using vice clamps, clamp flange shaft or output shaft/shaft flange in vice. Remove old circlip from groove with new circlip -A-.

– Drive in left-hand flange shaft and/or right-hand output shaft/drive flange using non–marring hammer.

– Install clutch release lever and release bearing ⇒ Page K 30-32.

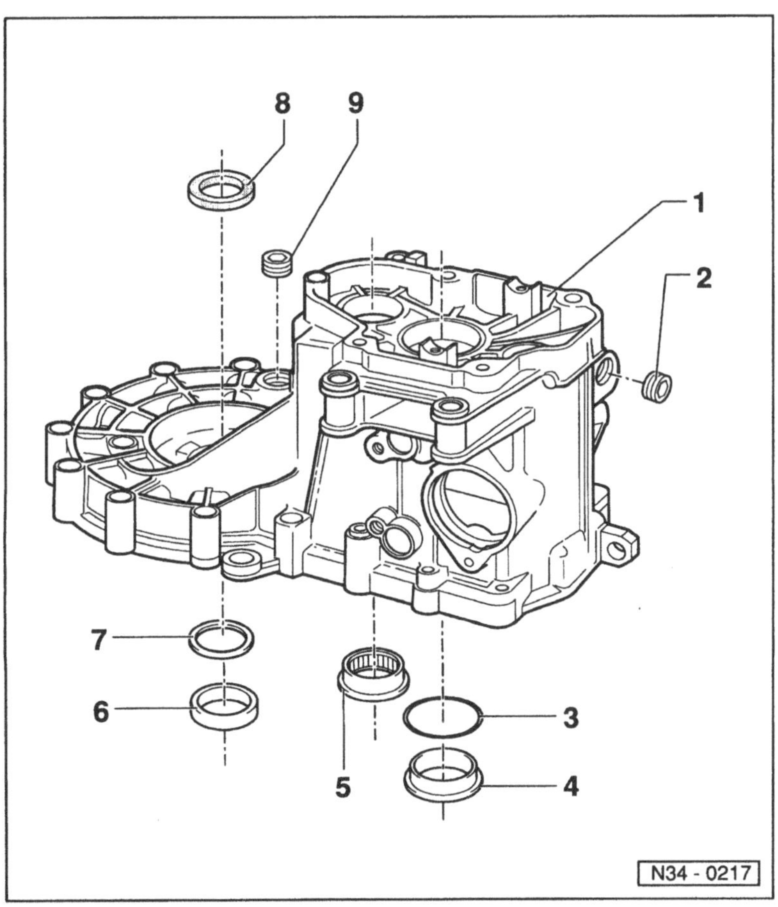

Transmission housing, servicing

1 - Transmission housing
- If replaced: Adjust input shaft and differential ⇒ Page K 39-20
- Modification:
  From transmission build date 05 06 2, bolt holes for securing transmission housing to clutch housing are shallow ⇒ Fig. 1

2 - Oil filler plug
- Tightening torque: 25 Nm (18 ft lb)
- Without magnet

3 - Shim
- For input shaft
- Adjustment overview ⇒ Page K 39-20

4 - Outer race/tapered roller bearing
- For input shaft
- Removing and installing ⇒ from Page K 35-10
- If replaced: adjust input shaft ⇒ Page K 35-14

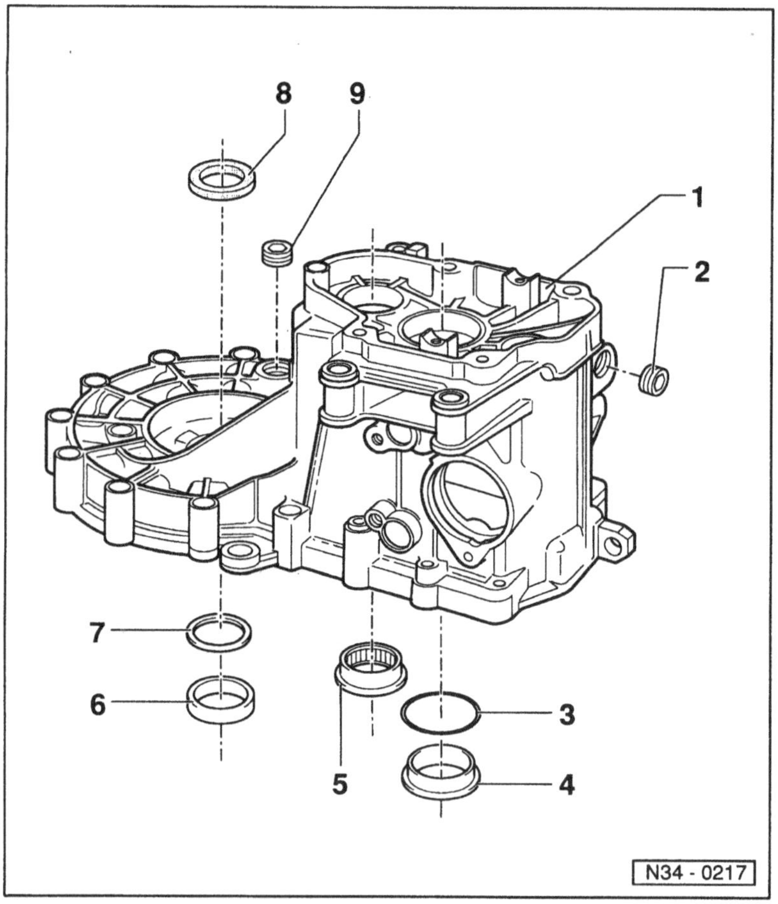

5 - Needle roller bearing

- For output shaft
- Removing ⇒ Fig. 2
- Installing and securing ⇒ Fig. 3

6 - Outer race/tapered roller bearing

- For differential
- Removing and replacing ⇒ Page K 39-9
- If replaced: Adjust differential ⇒ Page K 39-21

7 - Shim

- For output shaft
- Adjustment overview ⇒ Page K 39-20

8 - Oil seal

- Replacing ⇒ Page K 39-1

9 - Oil drain plug

- Tightening torque: 25 Nm (18 ft lb)
- Without magnet

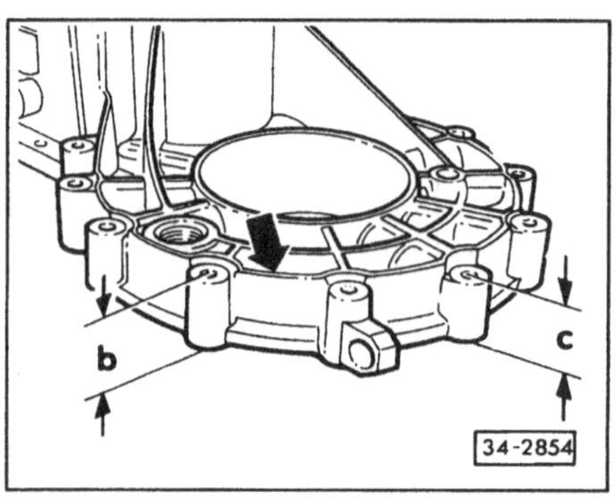

◄ **Fig. 1 Holes for securing transmission housing to clutch housing reduced in height from transmission build date 05 06 2**

(e.g., in area of differential):

Dimension -b-	
new:	37.0 mm (1.4567 in.)
previous:	42.0 mm (1.6535 in.)
Dimension -c-	
new:	33.0 mm (1.2992 in.)
previous:	42.0 mm (1.6535 in.)

Holes are now below the web -arrow-.

Notes:

- *Earlier and later transmission securing bolts are not interchangeable.*
- *Always use shorter securing bolts when installing a modified transmission housing.*

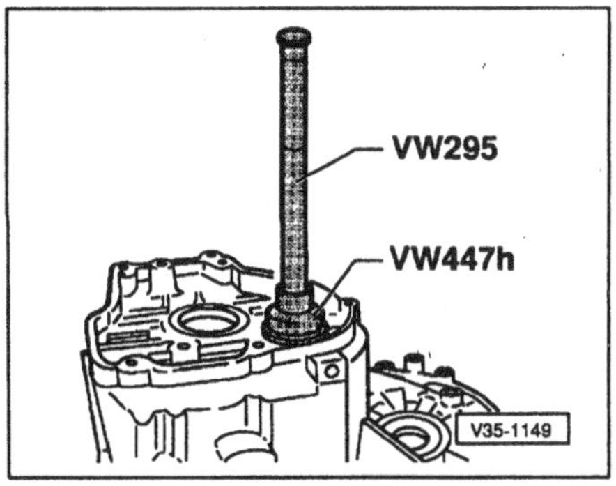

Fig. 2 **Needle roller bearing, removing using VW 295 needle bearing drift and VW 447 H thrust pad**

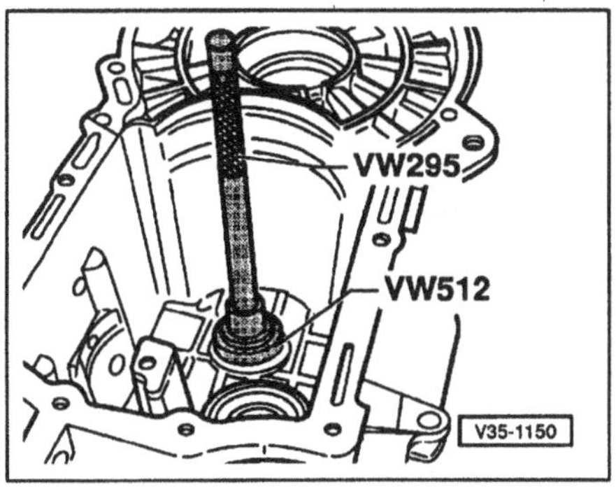

Fig. 3 **Needle roller bearing, installing onto stop using VW 295 needle bearing drift and VW 512 thrust pad**

- Secure needle roller bearing in transmission housing at three points 120° apart using a punch.

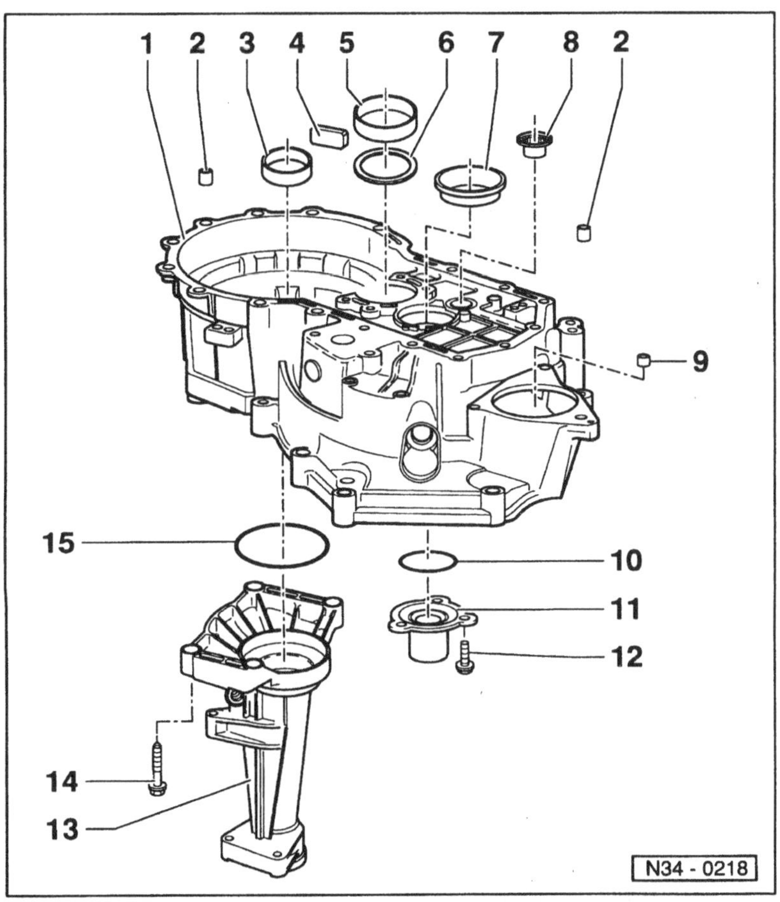

Clutch housing, servicing

1 - Clutch housing
- If replaced: Adjustment overview ⇒ Page K 39-20
- Modification:
  From transmission build date 10 08 2, modified bolt-on points in clutch housing for new rear transmission mounting ⇒ Fig. 5
- Modification:
  Vehicles from 01.96 have a bonded rubber mounting added for new transmission mount ⇒ Fig. 6 and Fig. 7

2 - Bushing
- Quantity 2

3 - Outer race/tapered roller bearing
- For differential
- Removing and replacing ⇒ Page K 39-9
- If replaced: Adjust differential ⇒ Page K 39-21

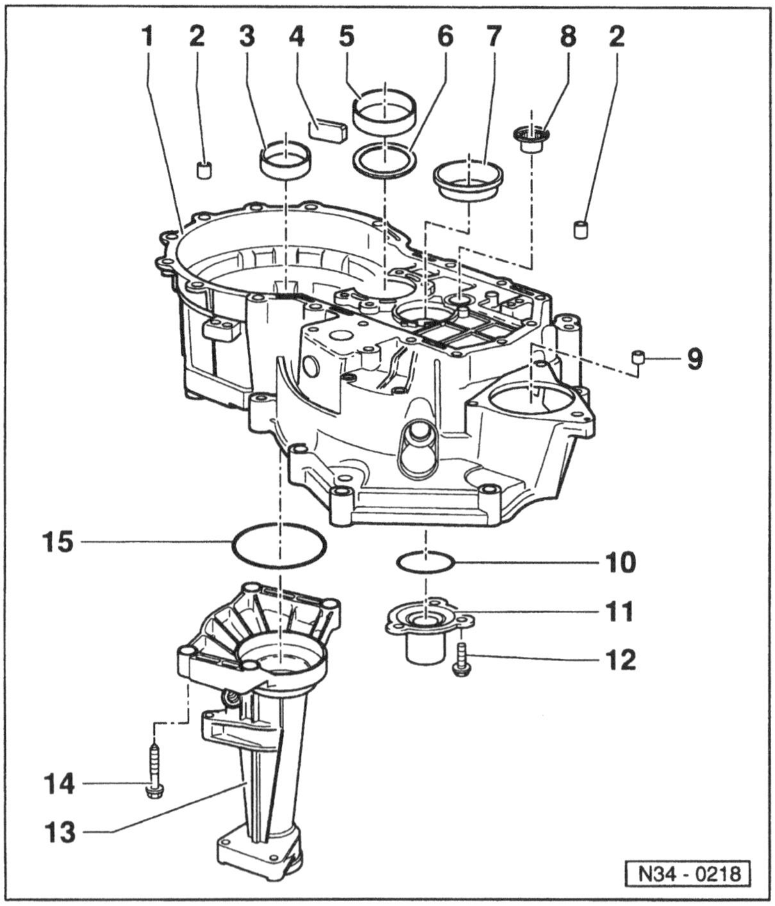

4 - Magnet
- Retained by housing joint surface

5 - Outer race/tapered roller bearing
- For output shaft
- Removing and installing ⇒ Page K 35-30
- If replaced: Adjust output shaft ⇒ Page K 35-45

6 - Shim
- For output shaft
- Adjustment overview ⇒ Page K 39-20

7 - Outer race/tapered roller bearing
- For input shaft
- Removing and installing ⇒ Page K 35-6
- If replaced: Adjust input shaft ⇒ Page K 35-14

8 - Needle bearing
- Removing and installing ⇒ Page K 35-52

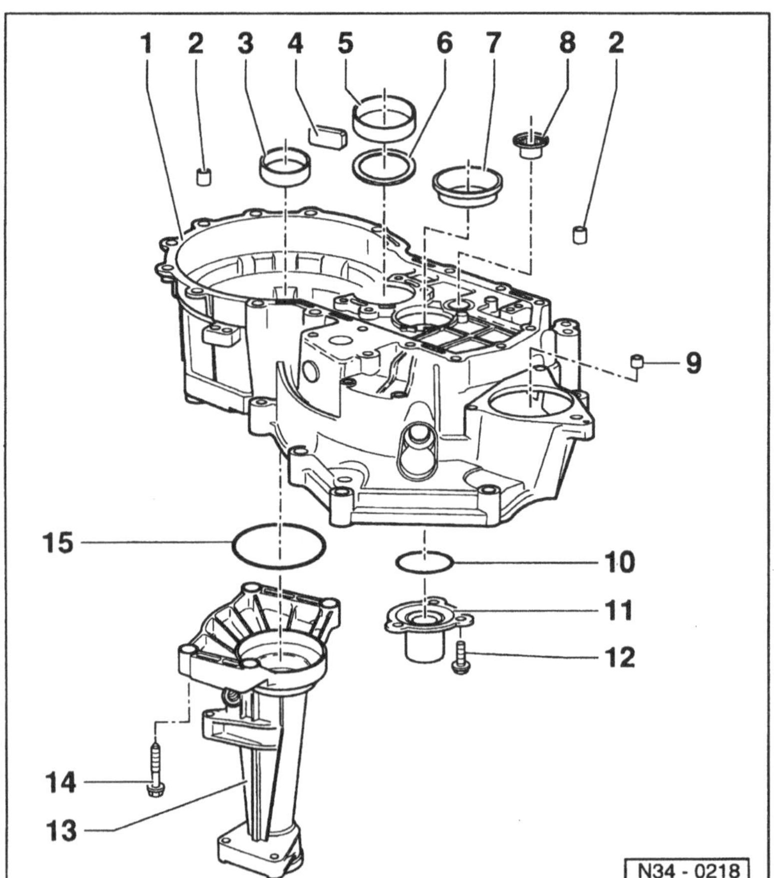

9 - Starter bushing
- Removing ⇒ Fig. 1
- Installing ⇒ Fig. 2
- Can be installed without removing transmission

10 - O-ring
- Always replace
- Place on guide sleeve and apply multi-purpose grease

11 - Guide sleeve
- With oil seal
- Removing oil seal ⇒ Fig. 3
- Installing oil seal ⇒ Fig. 4
- Remove guide sleeve to change the oil seal

12 - Socket head bolt
- Tightening torque: 20 Nm (15 ft lb)
- Self-locking
- Always replace

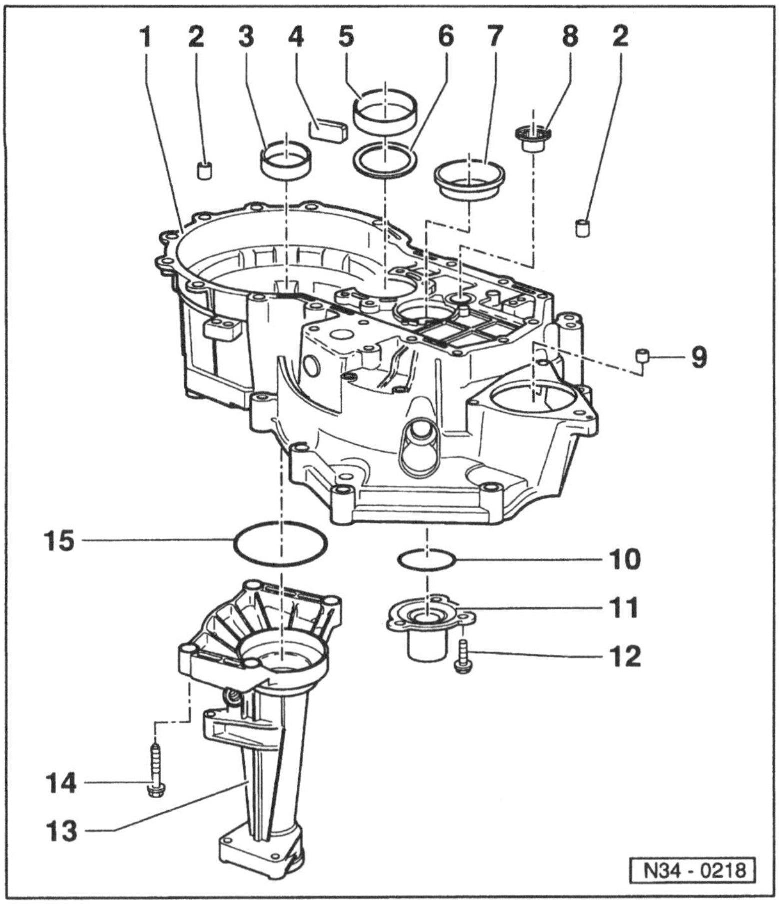

13 - Bearing housing

- Before removing, remove speedometer drive and output shaft with right-hand drive flange
- Removing and installing with transmission installed ⇒ Page K 34-54
- Disassembling and assembling ⇒ Page K 34-102

14 - Hex bolt

- Tightening torque: 40 Nm (30 ft lb) plus an additional 1/4 turn (90°)

15 - O-ring

- Always replace
- Apply multi–purpose grease and insert into groove in clutch housing

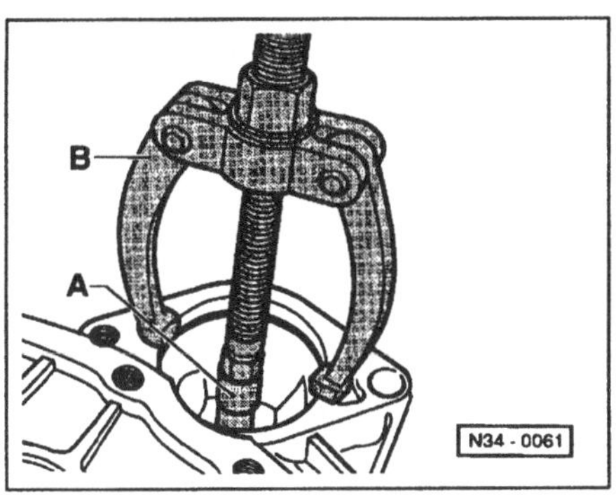

◄ **Fig. 1 Starter bushing, removing**

A – 21/1 Kukko 12 ... 14.5 mm, or equivalent

B – 22/1 Kukko support, or equivalent

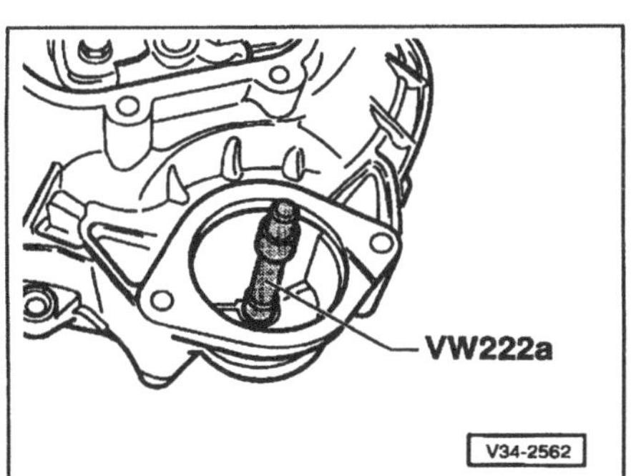

◄ **Fig. 2 Starter bushing, driving in with VW 222 A pilot drift.**

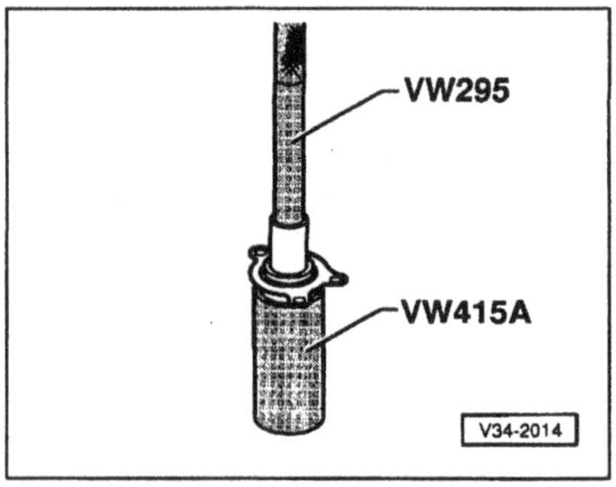

Fig. 3 Oil seal, removing from guide sleeve.

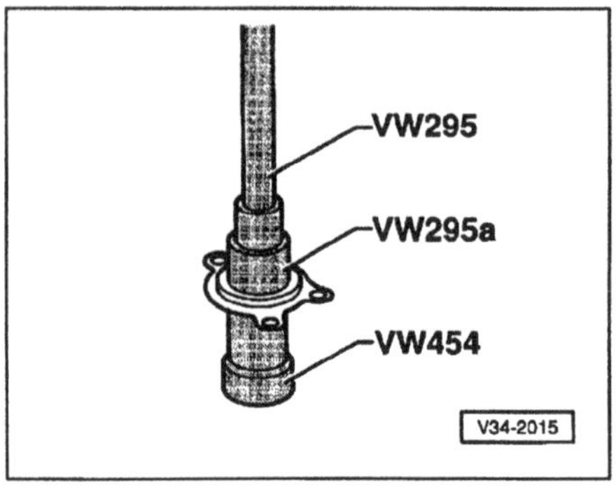

Fig. 4 Oil seal, driving into guide sleeve onto stop.

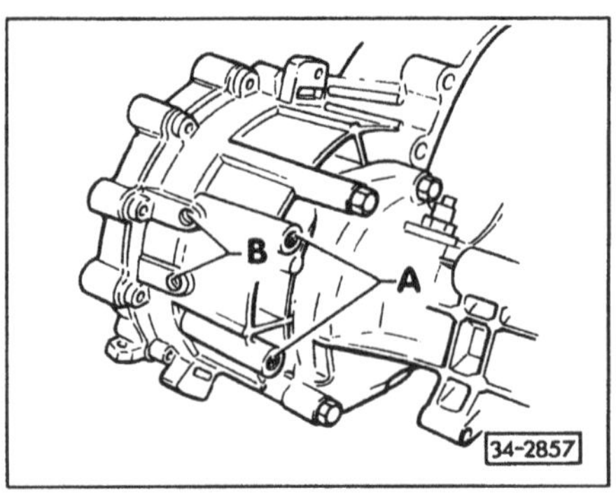

Fig. 5 Modified attachment points on clutch housing for new rear transmission mounting

Holes -A-, larger diameter

Transmission build date	mm (in.)
from 10 08 2	12.0 (0.472)
to 09 08 2	10.0 (0.394)

Holes -B- for new transmission mounting
⇒ Page K 34-45.

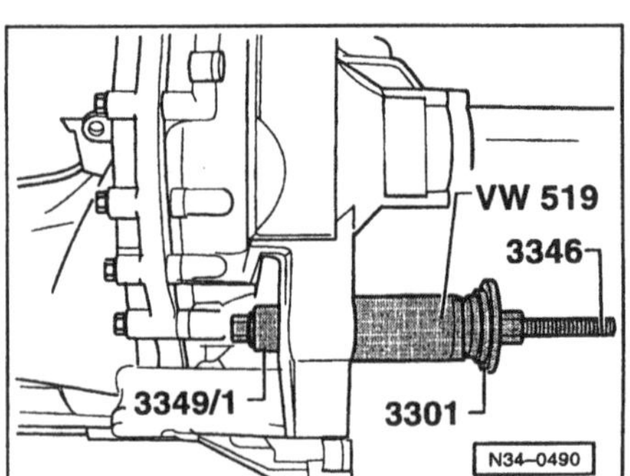

Fig. 6 Bonded rubber mounting, removing

A bonded rubber mounting has been added to the clutch housing with the introduction of the pendulum support (vehicles from 01.96), ⇒ page K 34-37

Note:

It is important to grease the spindle with MoS_2 grease before using.

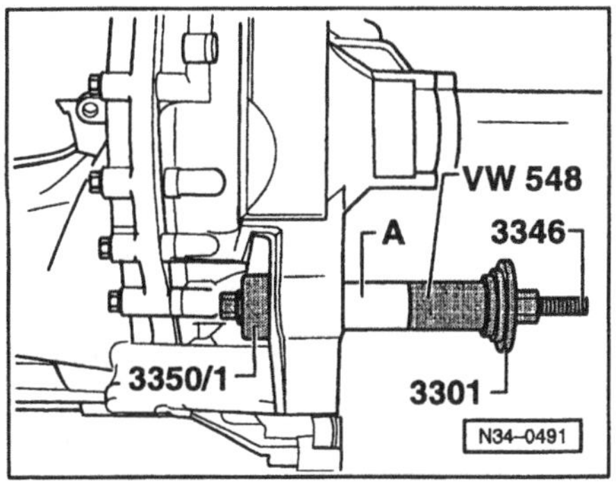

◄ **Fig. 7 Bonded rubber mounting, pulling in**

A = bonded rubber mounting

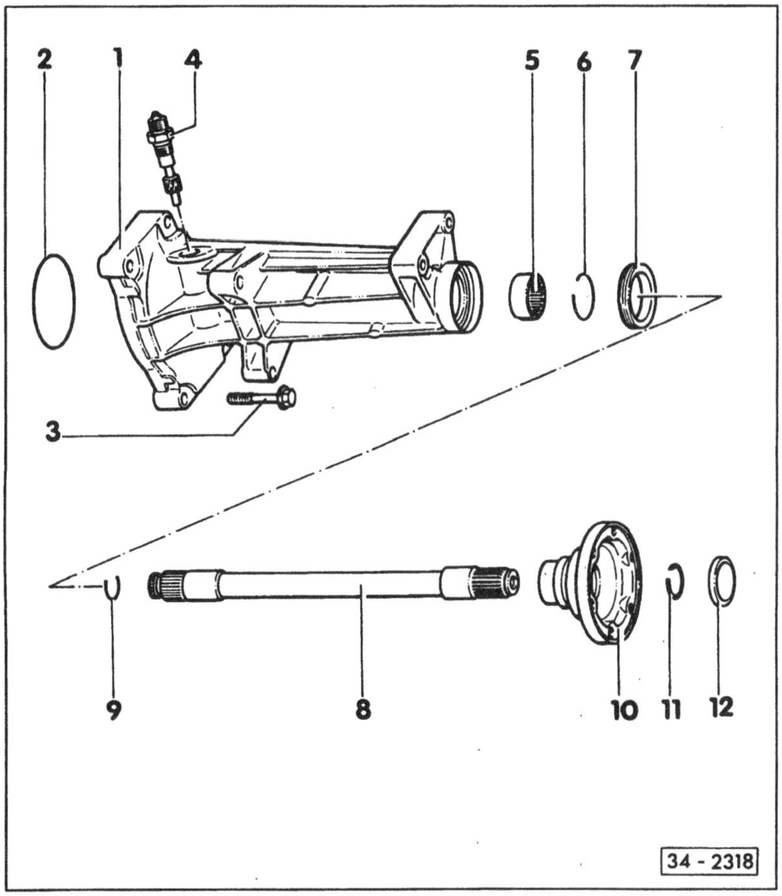

Bearing housing, disassembling and assembling

1 - Bearing housing
- ♦ Removing and installing ⇒ Page K 34-54
- ♦ Modification: Modified from transmission build date 01 09 3 ⇒ Fig. 1

2 - O-ring
- ♦ Always replace
- ♦ Install with multi–purpose grease

3 - Bolt
- ♦ Tightening torque: 40 Nm (30 ft lb) plus an additional 1/4 turn (90°)

4 - Speedometer drive
- ♦ Remove before removing bearing housing

5 - Needle roller bearing
- ♦ Removing ⇒ Fig. 3
- ♦ Installing ⇒ Fig. 4

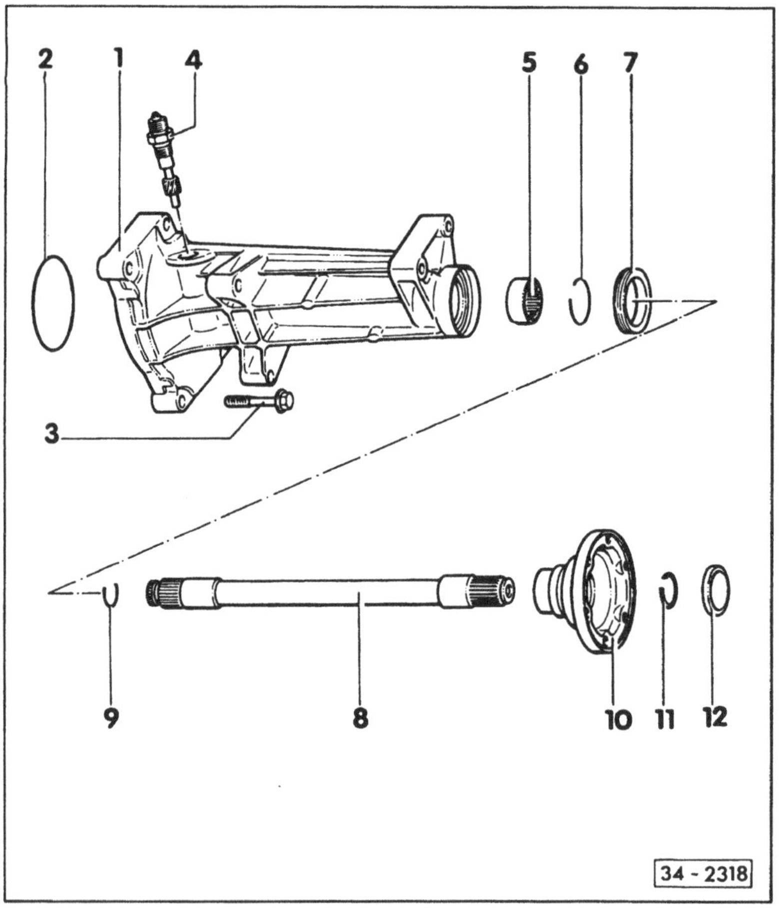

6 - Circlip
- For needle roller bearing
- Always replace

7 - Oil seal for drive flange
- Before installing, pack space between sealing lips with multi-purpose grease
- Can be replaced with transmission installed ⇒ Page K 39-1
- Remove with VW 681
- Driving in ⇒ Fig. 2

8 - Output shaft
- Remove with drive flange ⇒ Fig. 5

9 - Circlip
- Always replace
- Removing and installing ⇒ Fig. 6

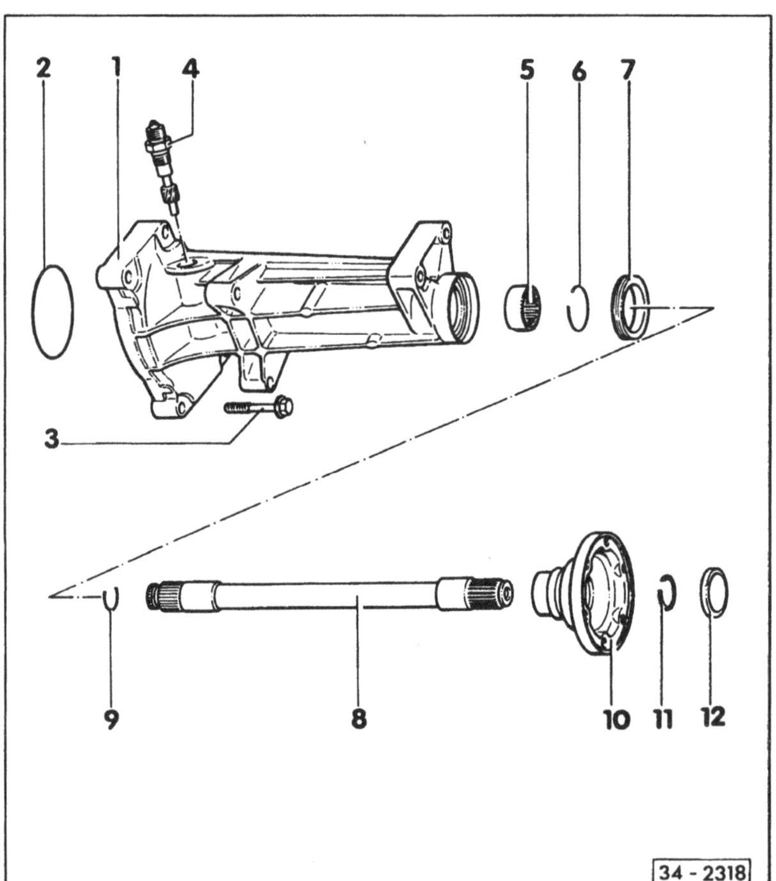

10 - Drive flange
- Removing ⇒ Fig. 7
- Installing ⇒ Fig. 8
- Remove and install with output shaft installed ⇒ from Page K 39-3
- Heat to approximately 90°C (194°F) before installing

11 - Circlip
- Always replace

12 - Sealing cap
- Always replace
- Pry out with screwdriver

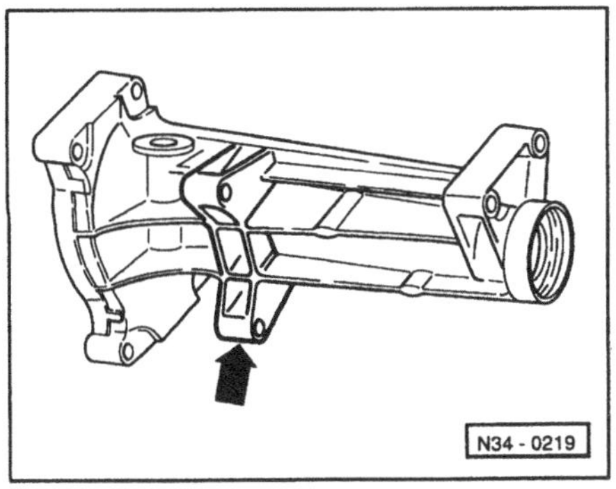

◄ **Fig. 1 Modified bearing housing from transmission build date 01 09 3**

Mounting -arrow- has been deleted. Bearing housing cannot be installed to earlier transmissions.

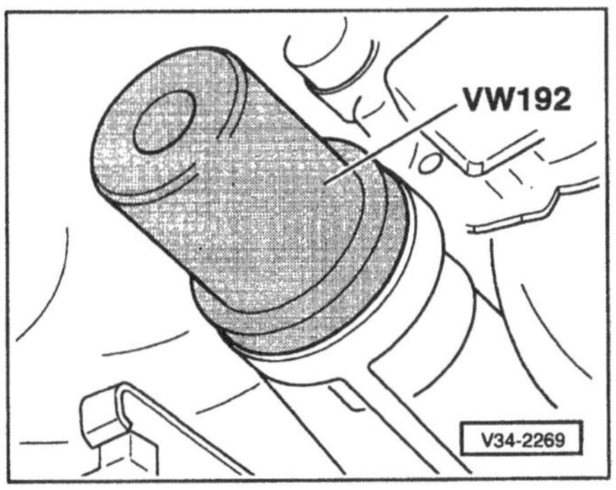

◄ **Fig. 2 Seal, driving in onto stop using VW 192 arbor**

– Fill space between seal and dust lip with multi-purpose grease.

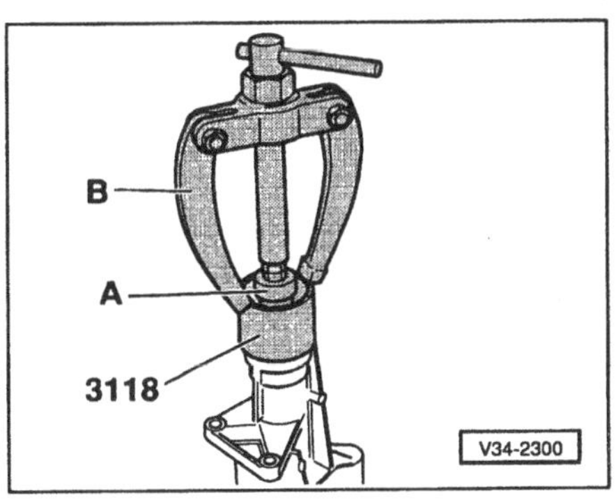

◄ **Fig. 3 Needle roller bearing, removing**

A – 21/5 Kukko extractor 30 ... 37 mm, or equivalent

B – 22/2 Kukko support, or equivalent

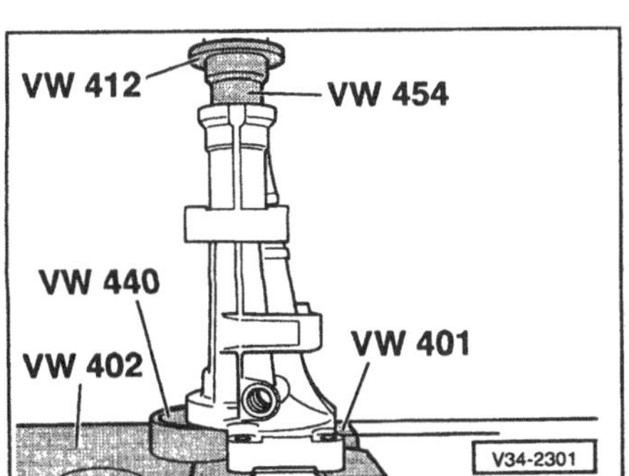

◄ **Fig. 4 Needle roller bearing, installing into bearing housing**

– Install needle roller bearing in onto stop.

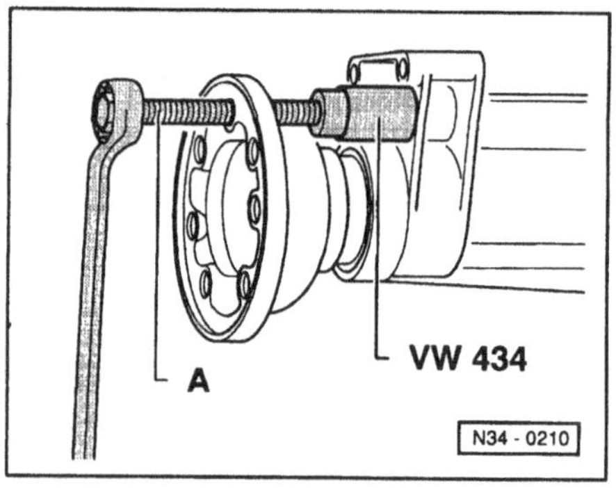

Fig. 5 Output shaft with drive flange, removing

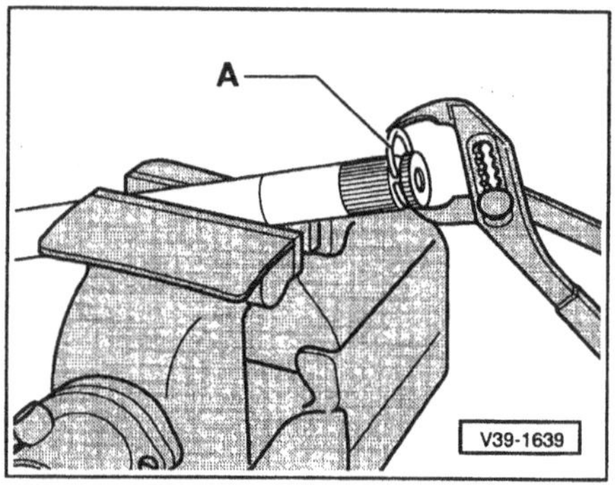

Fig. 6 Circlip, removing and installing

- Clamp output shaft in vice using protective vice clamps.
- Install circlip out of output shaft groove using new circlip -A-.

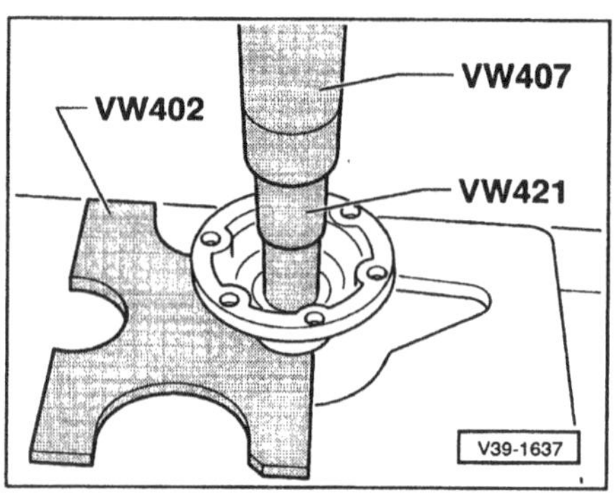

Fig. 7 Drive flange, removing from output shaft

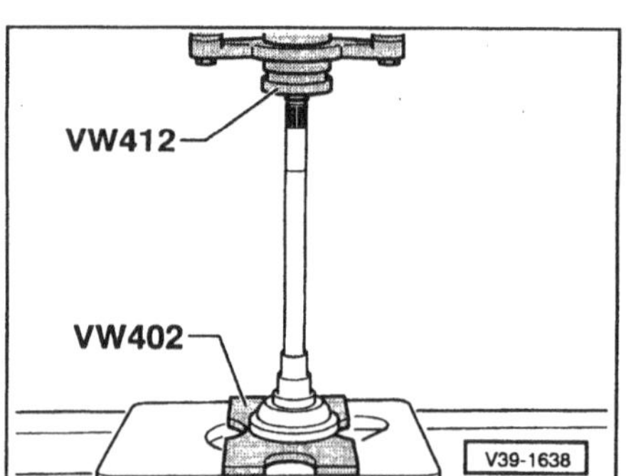

Fig. 8 Drive flange, installing onto output shaft

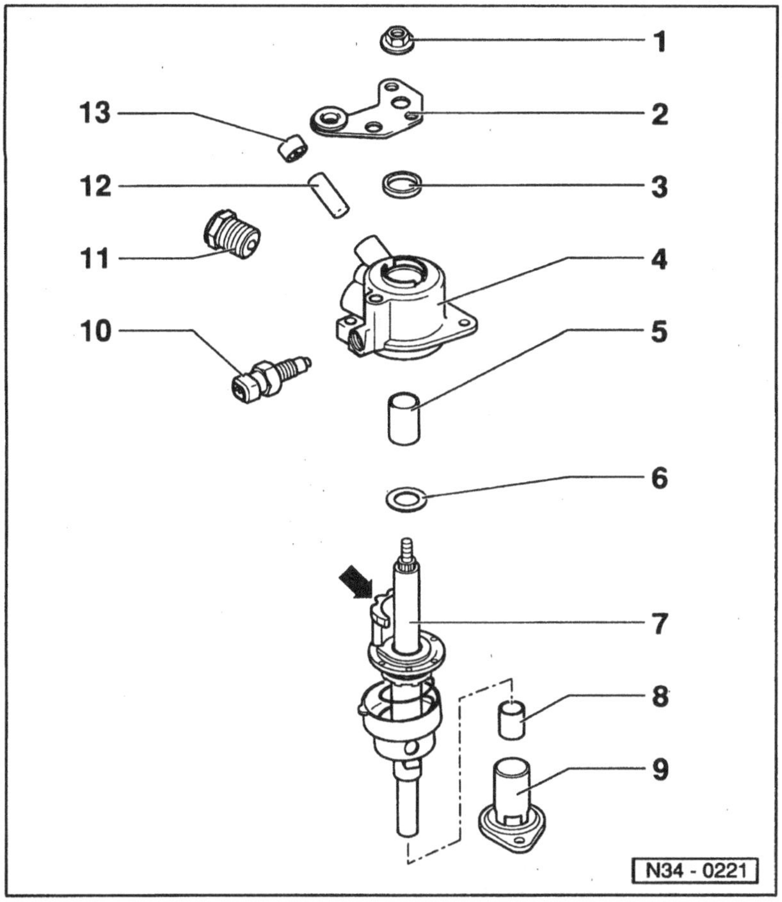

Shift mechanism, transmission side, disassembling and assembling

1 - Hex nut
- ♦ Tightening torque: 25 Nm (18 ft lb)
- ♦ Self-locking
- ♦ Always replace

2 - Gear shift lever
- ♦ Install so that master spline aligns with shift shaft
- ♦ Replacing bushings for shift rods ⇒ Page K 34-13

3 - Oil seal
- ♦ Pry out with screwdriver
- ♦ Installing ⇒ Fig. 1

4 - Shift mechanism cover

5 - Bearing sleeve
- ♦ Removing ⇒ Fig. 2
- ♦ Installing ⇒ Fig. 3

6 - Stop ring

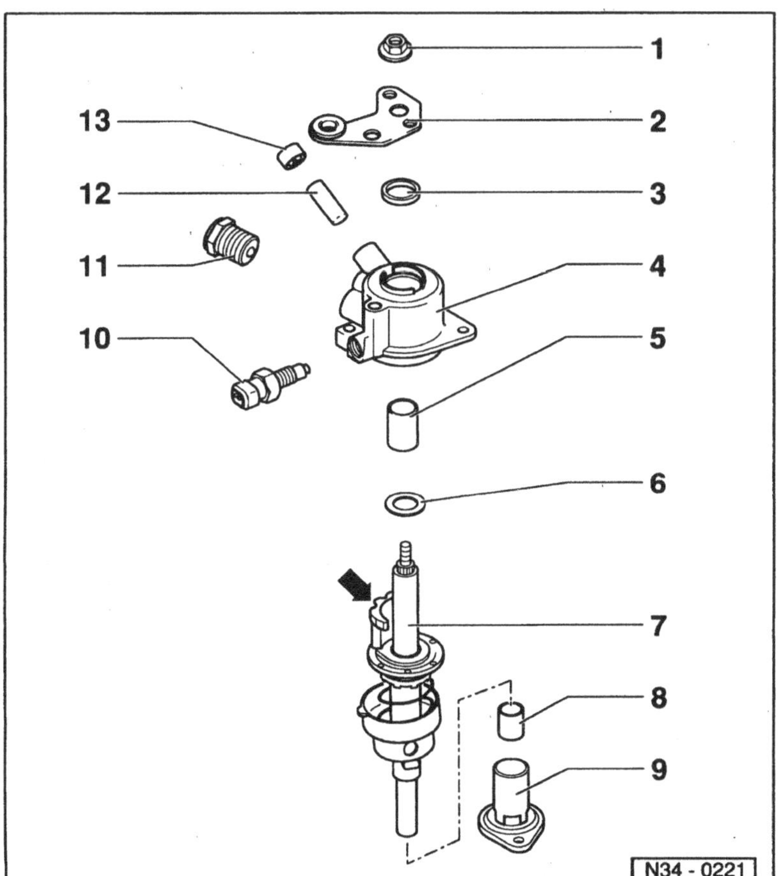

7 - Shift shaft
- ♦ Modification:
  From transmission build date 31 10 1, in conjunction with modified end cover plate bearing surface -9- ⇒ Fig. 4.
- ♦ Modification:
  From transmission build date 04 12 1, shift shaft with spring-loaded ball, installed ⇒ Fig. 5, to brake input shaft by modifying 5th gear shift plate chamfer (⇒ Page K 34-124) to ensure correct reverse gear engagement.
- ♦ Modification:
  From transmission build date 11 01 3 locking element -arrow- replaced by circumferent locking element. Modified shift shaft can be installed in transmissions from build date 31 10 1.

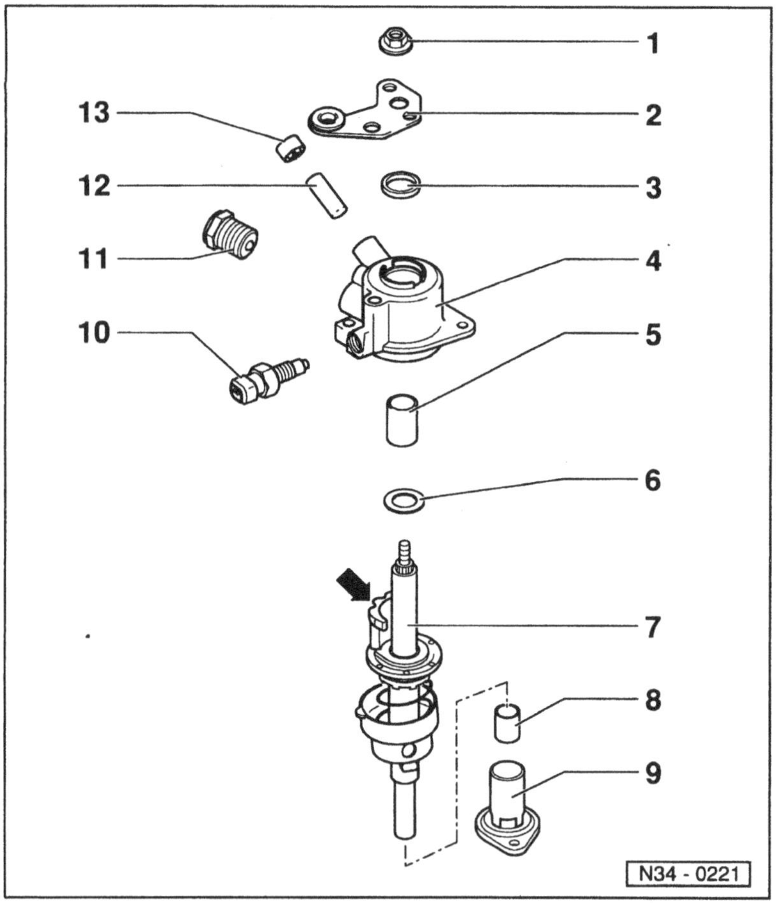

8 - Bearing sleeve
- Removing ⇒ Fig. 6
- Installing ⇒ Fig. 7
- Modification:
  From transmission build date 31 10 1, deleted because of shift shaft bearing surface in end cover plate ⇒ Fig. 8

9 - End cover plate
- Modification:
  From transmission build date 31 10 1, modified in conjunction with bearing surface ⇒ Fig. 8

10 - Reversing light switch
- Tightening torque: 20 Nm (15 ft lb)
- Apply a light coat of MoS_2 grease to contact area of switch and shift shaft locking device

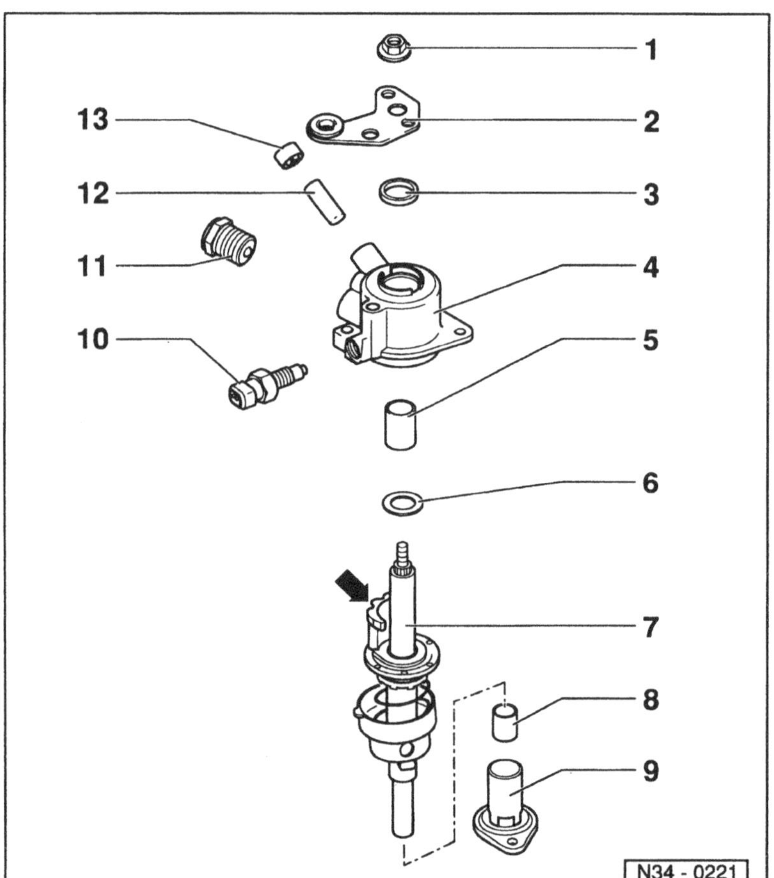

11 - Locking bolt
- Tightening torque: 40 Nm (30 ft lb)

12 - Sleeve
- Drive onto stop

13 - Cap
- For transmission breather

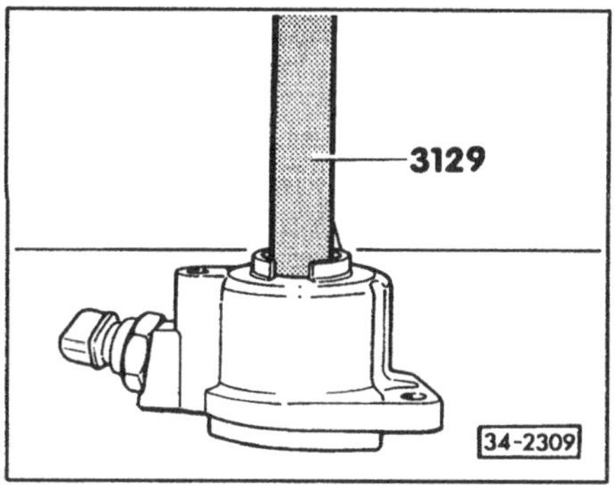

◄ **Fig. 1 Seal, installing into shift cover onto stop with 3129 seal driver**

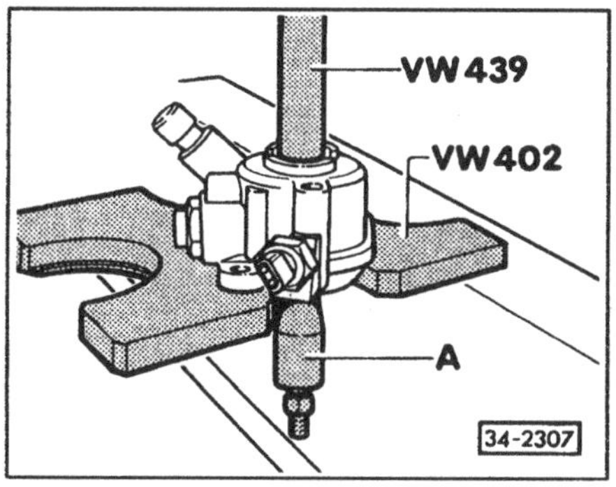

◄ **Fig. 2 Bearing sleeve, removing from shift cover**

– Before pulling out bearing sleeve, destroy plastic cage and remove balls.

A – 21/3 Kukko extractor 18.5 ... 23.5 mm, or equivalent

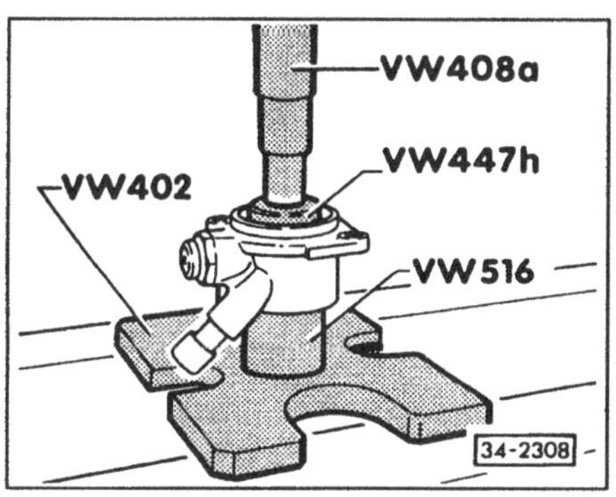

◄ **Fig. 3 Bearing sleeve, installing onto stop**

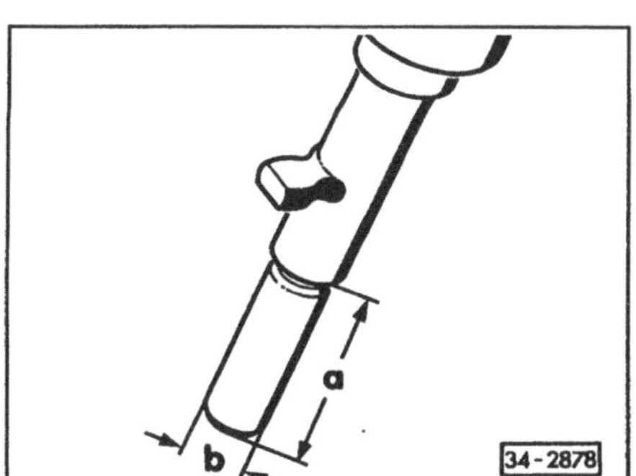

◄ **Fig. 4 Modified shift shaft**

In conjunction with bearing surface, lower area shortened and diameter larger ⟨⇒ Fig. 8).

Transmission build date	Dimension a mm (in.)	Dimension b mm (in.)
from 31 10 1	41.0 (1.614)	16.0 (0.630)
up to 30 10 1	57.0 (2.244)	15.0 (0.591)

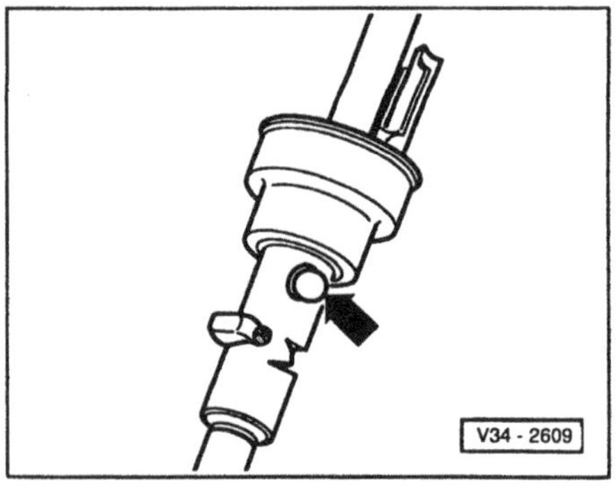

Fig. 5 Shift shaft with spring-loaded ball -arrow-

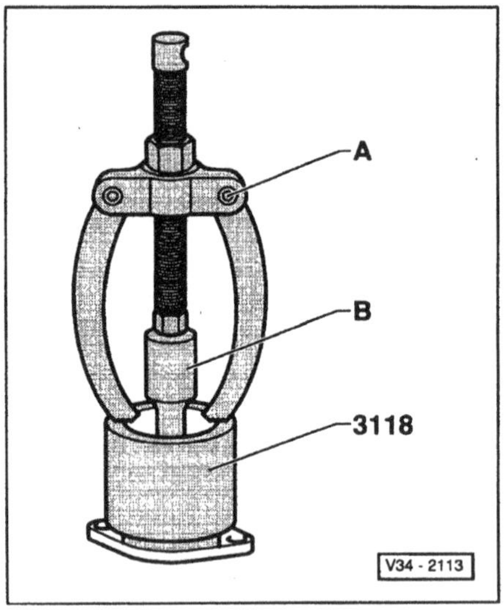

Fig. 6 Bearing sleeve, removing from end cover

– Before removing bearing sleeve, destroy plastic cage and remove balls.

A – 22/1 Kukko support, or equivalent

B – 21/3 Kukko extractor 18.5 ... 23.5 mm, or equivalent

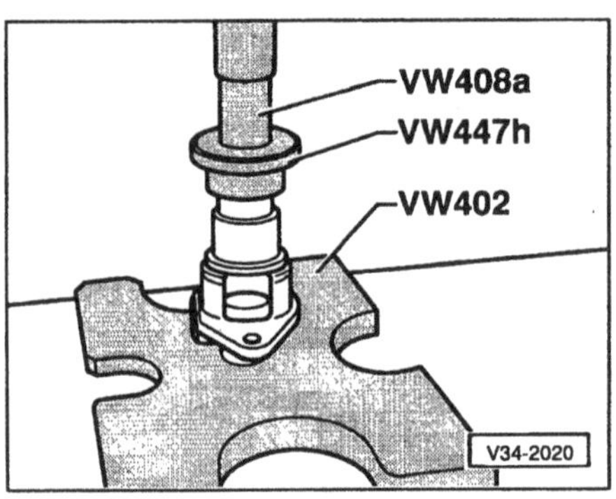

Fig. 7 Bearing sleeve, installing into end cover

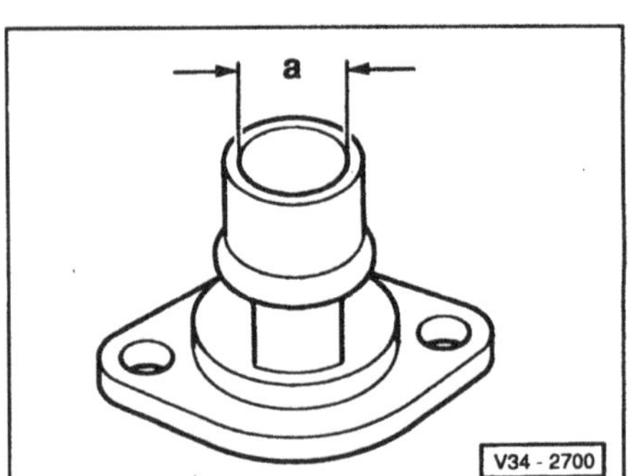

Fig. 8 End cover with bearing surface for shift shaft

Transmission build date	Dimension a mm (in.)
from 31 10 1	16.0 (0.630)
up to 30 10 1	21.0 (0.827)

Note:

- *The bearing sleeve for mounting bearing shaft is deleted.*
- *Only install together with modified shift shaft with a bearing surface ⇒ Fig. 4.*

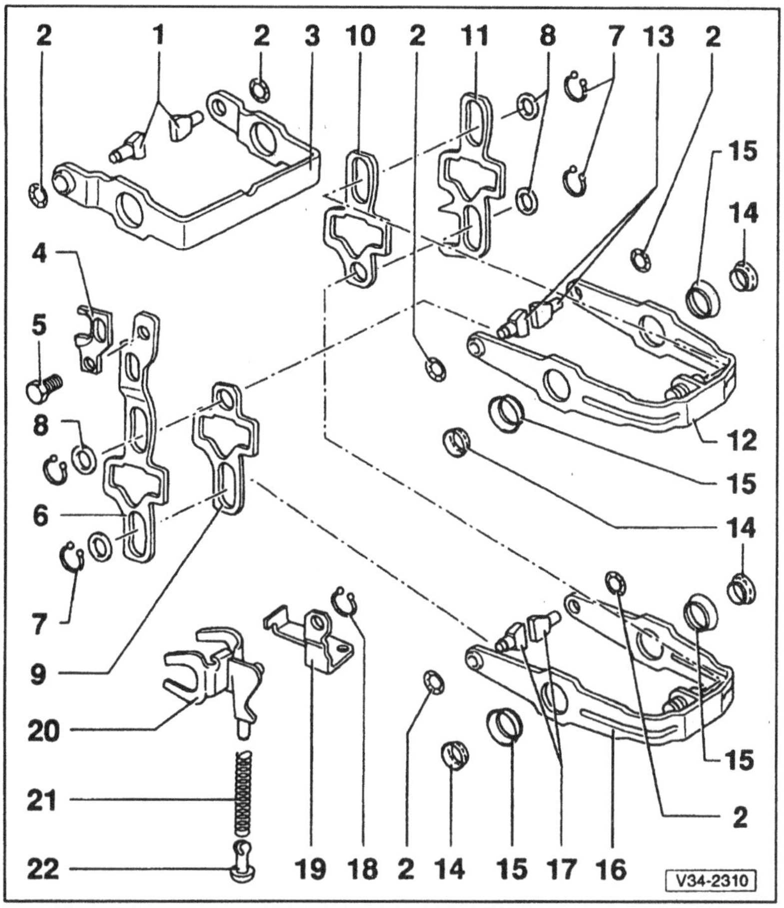

Shift forks, disassembling and assembling

1 - Shift fork segment
- Identification ⇒ Fig. 1
- After installing circlip, shift fork segment must rotate freely

2 - Circlip
- Pry off with screwdriver
- Installing ⇒ Fig. 2
- Always replace

3 - 5th gear shift fork
- Adjusting ⇒ Page K 34-88
- Modification:
  From transmission build date 31 10 1, in conjunction with modified 5th gear synchronization ⇒ Page K 35-12, new dimensions ⇒ Fig. 6

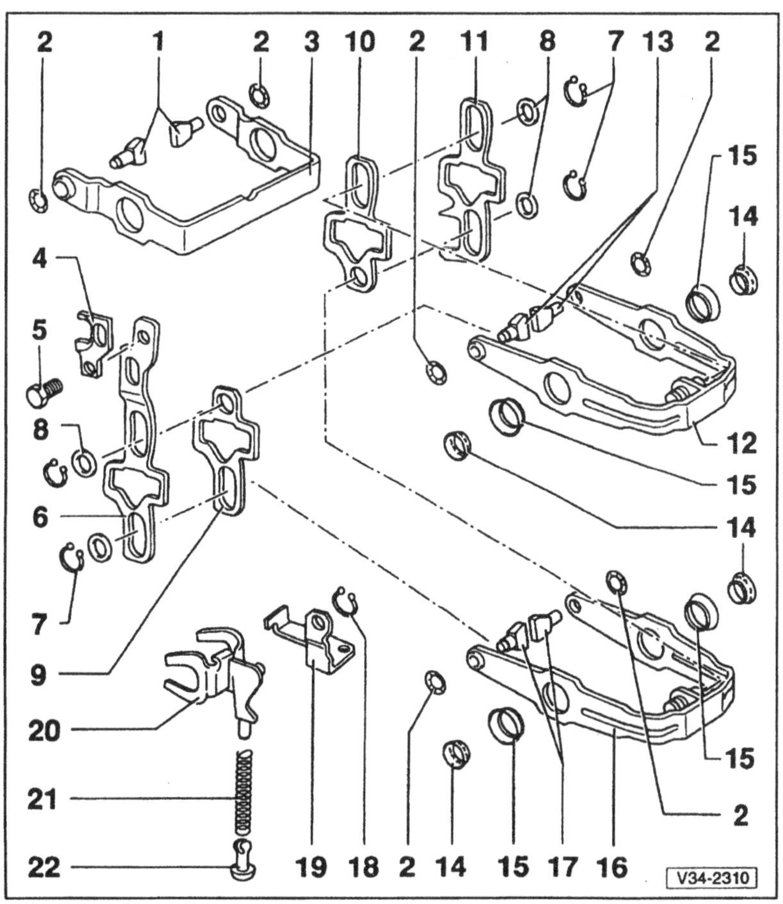

4 - 5th gear shift jaw

5 - Bolt
- Tightening torque: 25 Nm (18 ft lb)

6 - 5th gear shift plate
- Modification:
  From transmission build date 17 05 1, with chamfer which with modified shift shaft (–7–, Page K 34-110) enables reverse gear to be selected immediately ⇒ Fig. 8.

7 - Circlip

8 - Washer

9 - 3rd/4th gear shift plate

10 - 1st/2nd gear shift plate

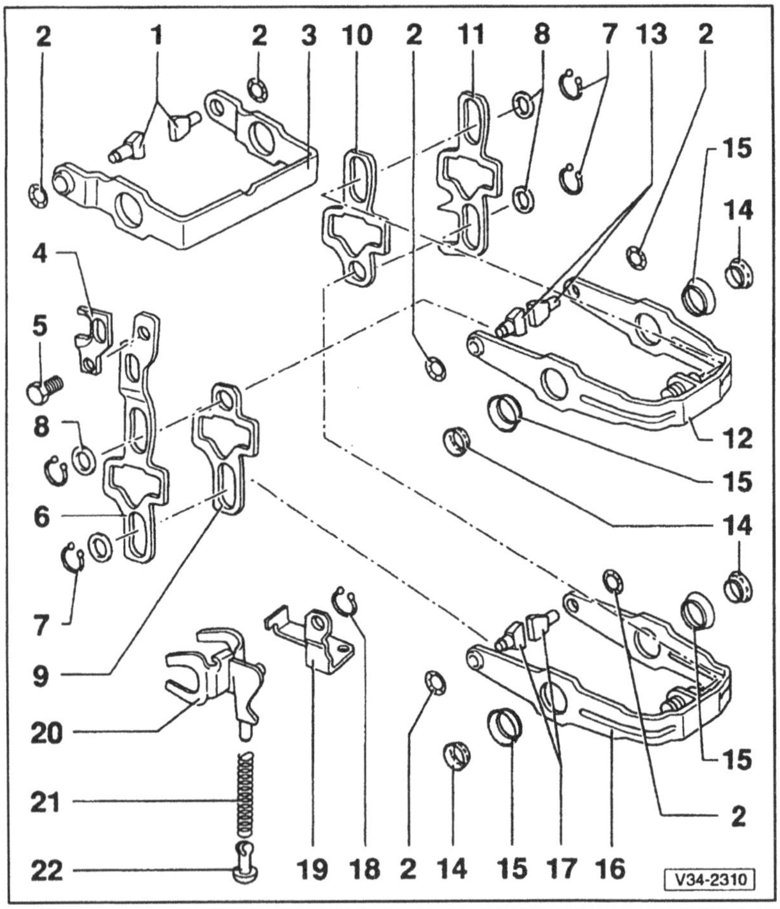

11 - Reverse gear shift plate

12 - 3rd/4th gear shift fork
- With shift fork segments ⇒ Fig. 5
- Modification:
  Bearings modified from transmission build date 06 02 2 ⇒ Fig. 7

13 - Shift segment
- Identification ⇒ Fig. 1
- After installing circlip, segment must rotate freely

14 - Bearing

15 - Bearing inner race
- Removing ⇒ Fig. 3
- Installing ⇒ Fig. 4
- Modification:
  Larger diameter from transmission build date 06 02 2 ⇒ Fig. 7

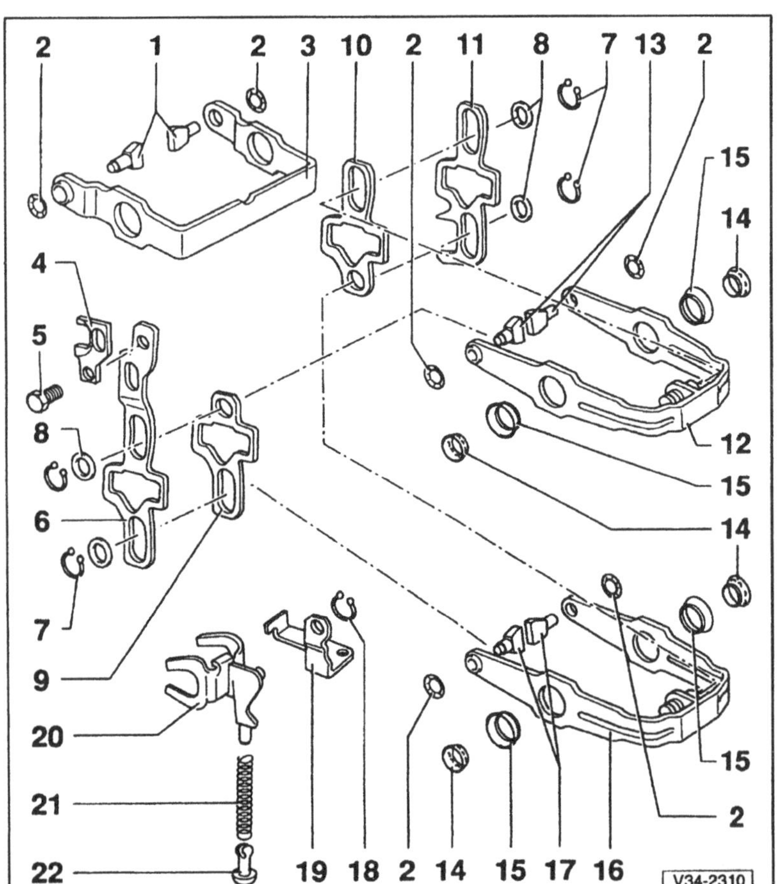

16 - 1st/2nd gear shift fork
- With shift fork segments ⇒ Fig. 5
- Modification:
  Bearings modified from transmission build date06 02 2 ⇒ Fig. 7

17 - Shift fork segment
- Identification ⇒ Fig. 1
- After installing circlip, segment must rotate freely

18 - Circlip

19 - Reverse gear shift fork support

20 - Reverse gear shift fork

21 - Spring

22 - Slide pin

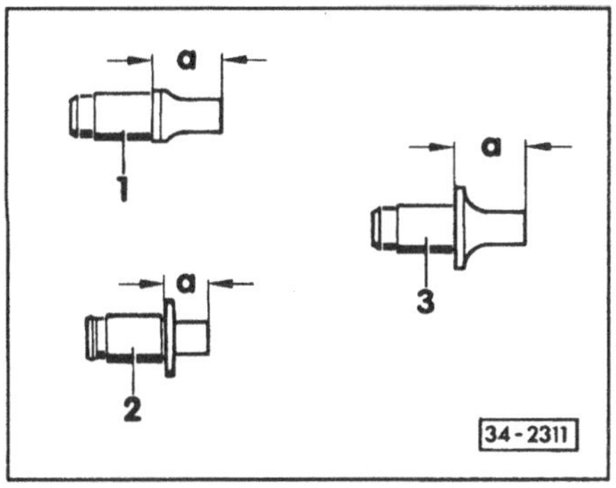

Fig. 1 Shift fork segments, identifying

Dimension -a-

1 – 1st and 2nd gear shift fork segments = 12.1 mm (0.476 in.)

2 – 3rd and 4th gear shift fork segments = 7.7 mm (0.303 in.)

3 – 5th gear shift fork segments = 12.1 mm (0.476 in.)

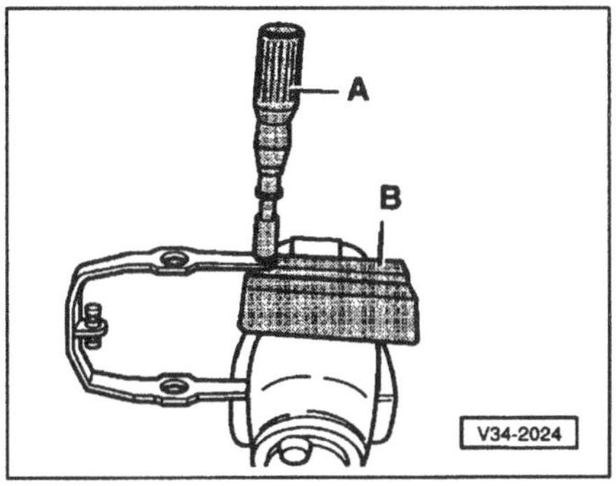

Fig. 2 Circlip, installing

– Install circlip into shift segment groove with socket.

Note:
After installing circlip, shift segment must rotate freely.

A – Socket, A/F 10 mm (0.394 in.)

B – Vice clamps

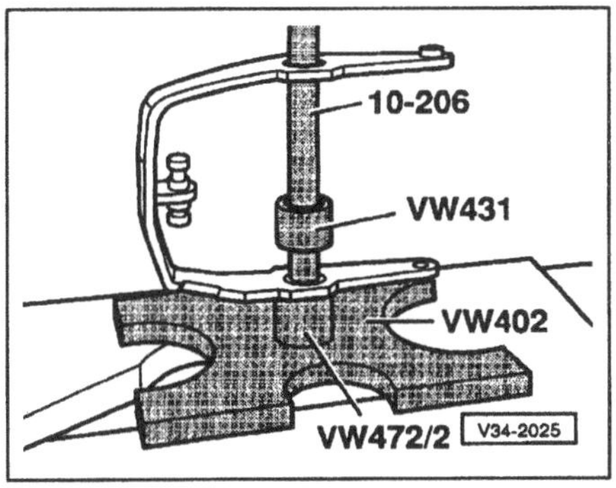

Fig. 3 Bearing inner race, installing

Note:
When removing and installing inner races do not bend shift forks.

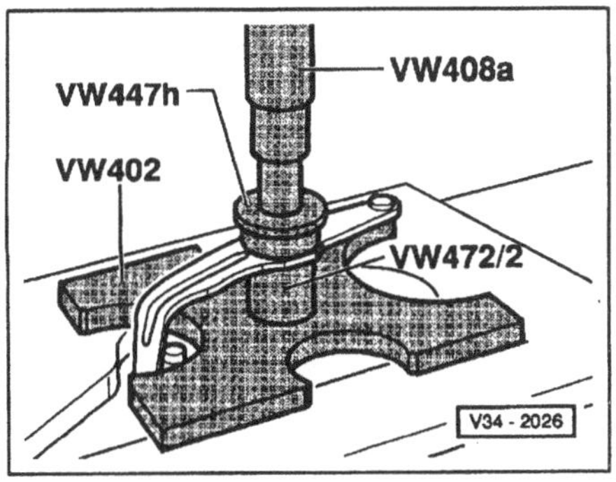

Fig. 4 Inner race, installing onto stop

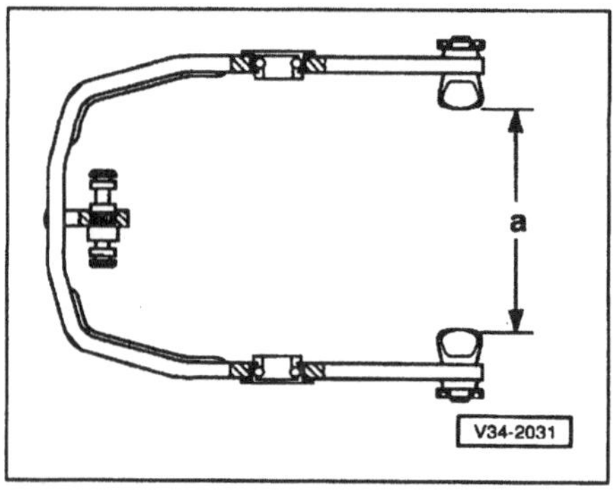

Fig. 5 Shift fork with shift segments installed

Shift fork	Dim. -a- mm (in.)
1st/2nd gear shift fork	87.2 ... 87.9 (3.433 ... 3.461)
3rd/4th gear shift forks	96.0 ... 96.7 (3.780 ... 3.807)

Shift segment assignment ⇒ Fig. 1.

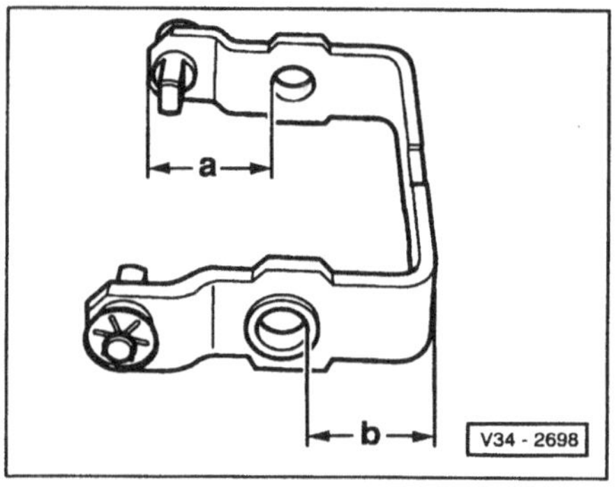

Fig. 6 5th gear shift fork with new dimensions

Transmission build date	Dim. -a- mm (in.)	Dim. -b- mm (in.)
from 31 10 1	33.7 (1.326)	24.3 (0.957)
up to 30 10 1	35.0 (1.378)	23.3 (0.917)

Note:

The modified shift fork can also be installed in previous transmissions.

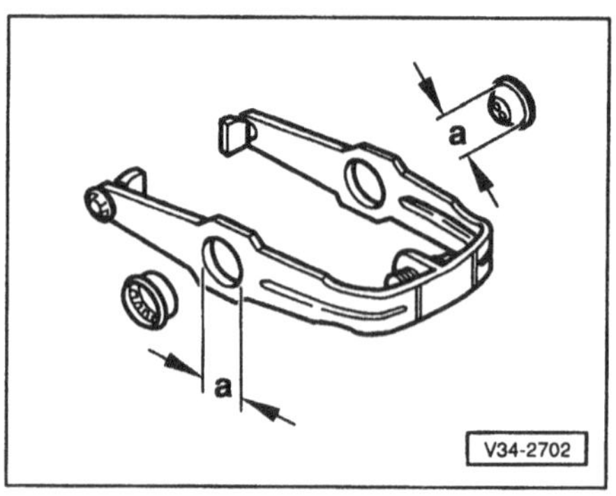

Fig. 7 Modification of 1st - 4th gear shift fork and bearings

Transmission build date	Dimension -a- mm (in.)
from 06 02 2	21.0 (0.827)
up to 05 02 2	20.0 (0.787)

Note:

Shift forks and bearings can be installed together in previous transmissions.

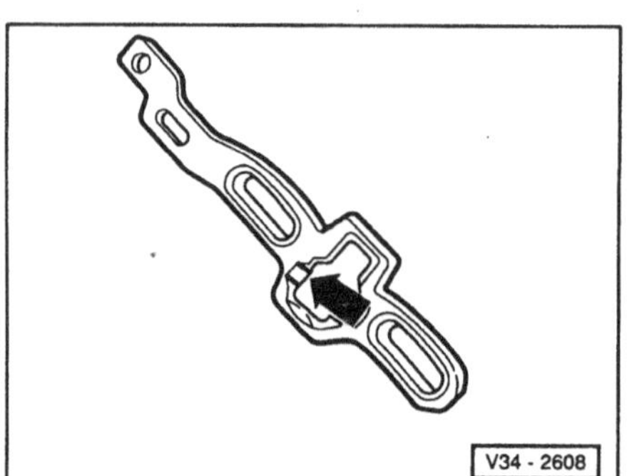

Fig. 8 Modified 5th gear shift plate

From transmission build date 17 05 1, shift shaft opening now has a chamfer -arrow-.

Notes:

- *The modified shift plate can also be installed in previous transmissions.*
- *Only the shift plate with chamfer will be supplied as a replacement part.*

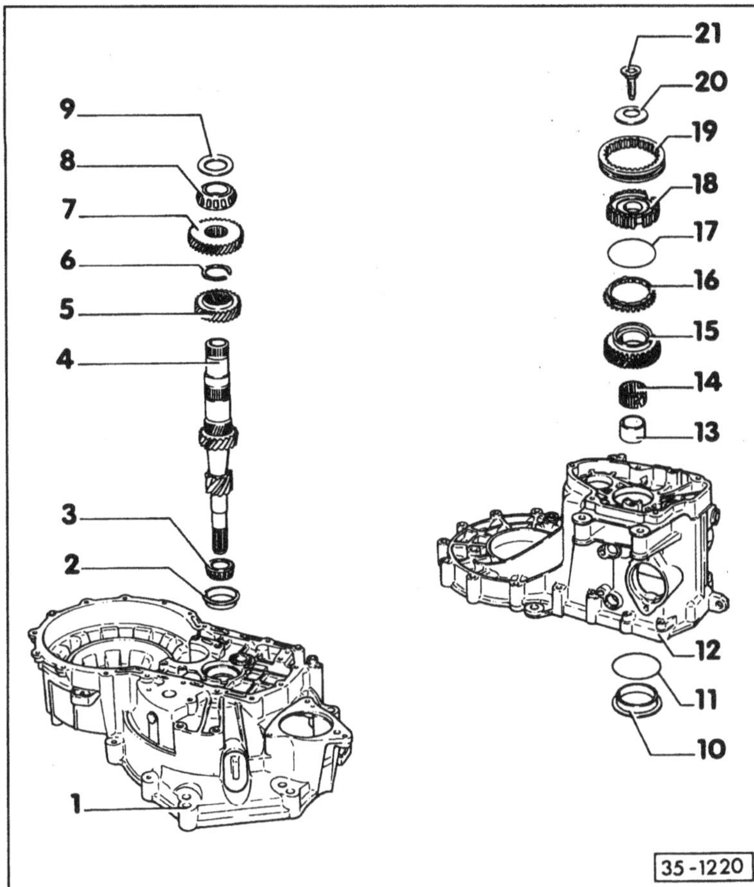

Input shaft, disassembling and assembling

Notes:

- *When installing new gears, consult technical data ⇒ Page K 00-3.*
- *If position of tapered roller bearings is affected when exchanging parts, input shaft must be re–adjusted. See adjustment overview ⇒ Page K 39-20.*

1 - Clutch housing

2 - Tapered roller bearing outer race
- Removing ⇒ Fig. 1
- Installing ⇒ Fig. 2

3 - Tapered roller bearing inner race
- Drive off input shaft with a drift
- Installing ⇒ Fig. 3

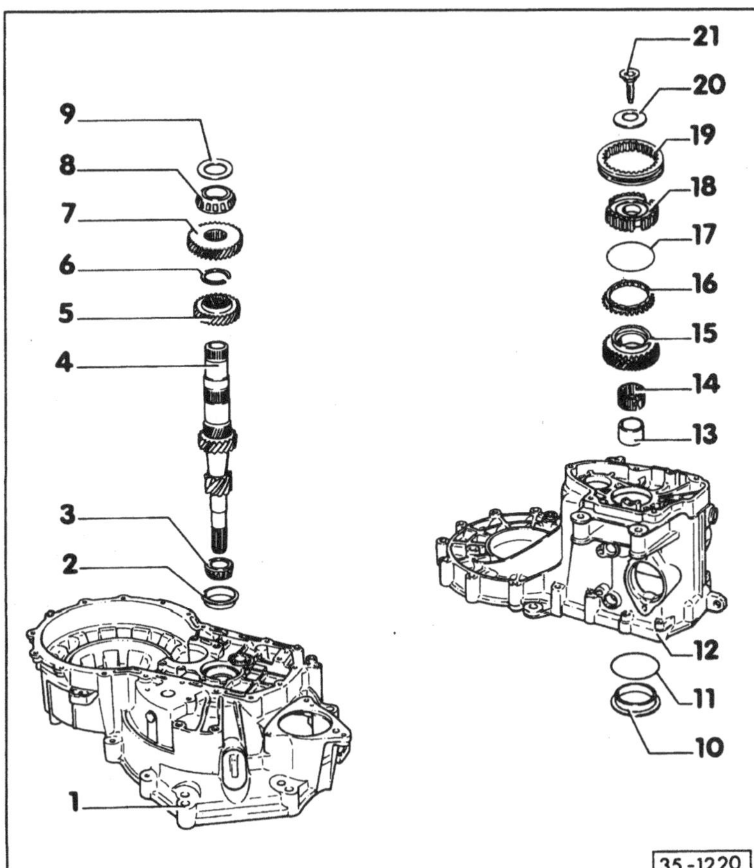

4 - Input shaft
- Adjusting ⇒ Page K 35-14

5 - 3rd gear
- Shoulder faces 4th gear
- Removing ⇒ Fig. 5
- Installing ⇒ Fig. 6

6 - Circlip
- Always replace

7 - 4th gear
- Removing with tapered roller bearing inner race and sleeve ⇒ Fig. 4
- Installing ⇒ Fig. 7
- Shoulder faces 3rd gear

8 - Tapered roller bearing inner race
- Removing with 4th gear and sleeve ⇒ Fig. 4
- Installing ⇒ Fig. 8

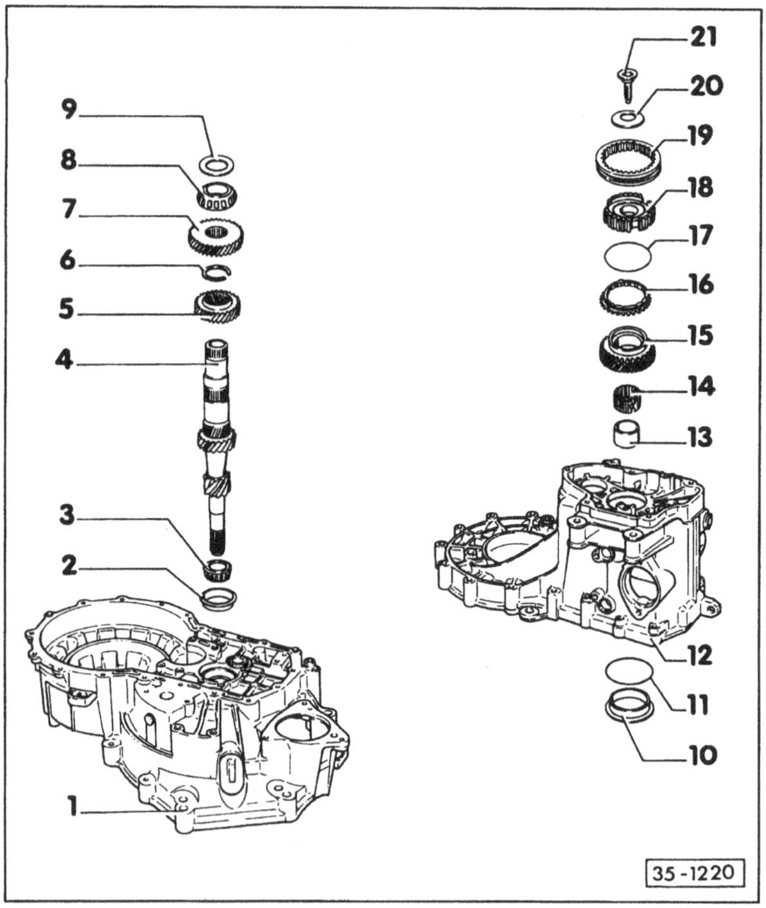

9 - **Thrust washer**

10 - **Tapered roller bearing outer race**
- Removing ⇒ Fig. 10
- Installing ⇒ Fig. 11

11 - **Shim**
- Determining thickness ⇒ input shaft adjusting, Page K 35-14

12 - **Transmission housing**

13 - **Sleeve**
- For needle roller bearing
- Remove with 4th gear and tapered roller bearing inner race ⇒ Fig. 4
- Installing ⇒ Fig. 9
- Place thrust washer -9- on before installing

14 - **Needle roller bearing**

15 - **5th gear**
- Remove with 5th gear synchro-hub ⇒ from Page K 34-72

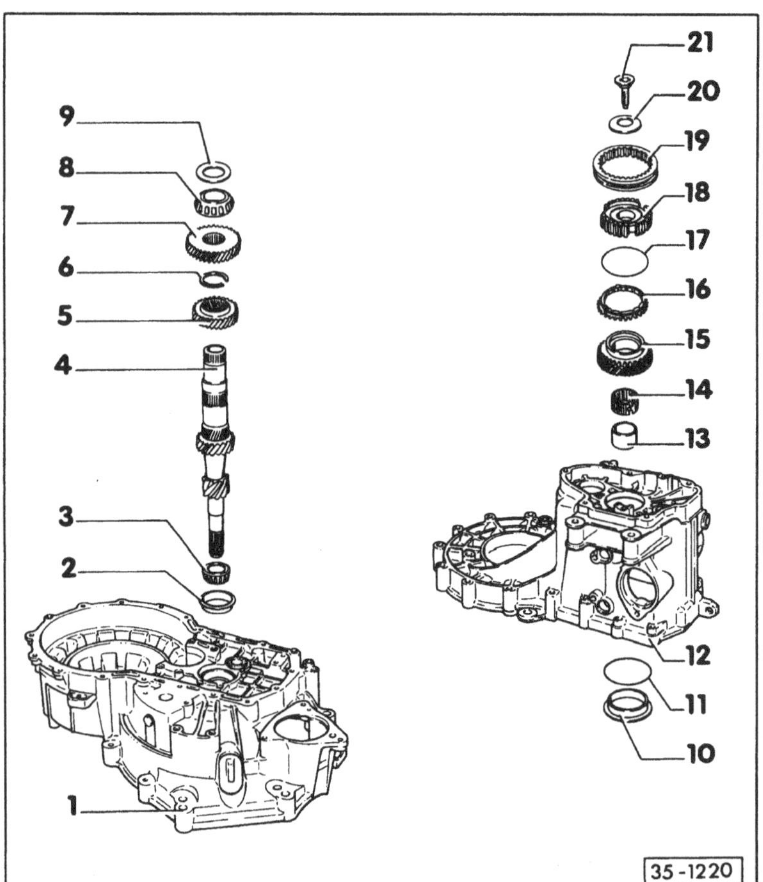

16 - **5th gear synchro-ring**
- Locking keys are on synchro-ring
- Inspect for wear ⇒ Page K 34-84
- Modification:
  From transmission build date 31 10 1 ⇒ Fig. 12

17 - **Spring**
- Place on synchro-ring
- Replace if worn
- Modification:
  Discontinued from transmission build date 31 10 1

18 - **5th gear synchro-hub**
- Removing and installing ⇒ from Page K 34-72
- Modification:
  From transmission build date 31 10 1, additional locking keys ⇒ Fig. 13

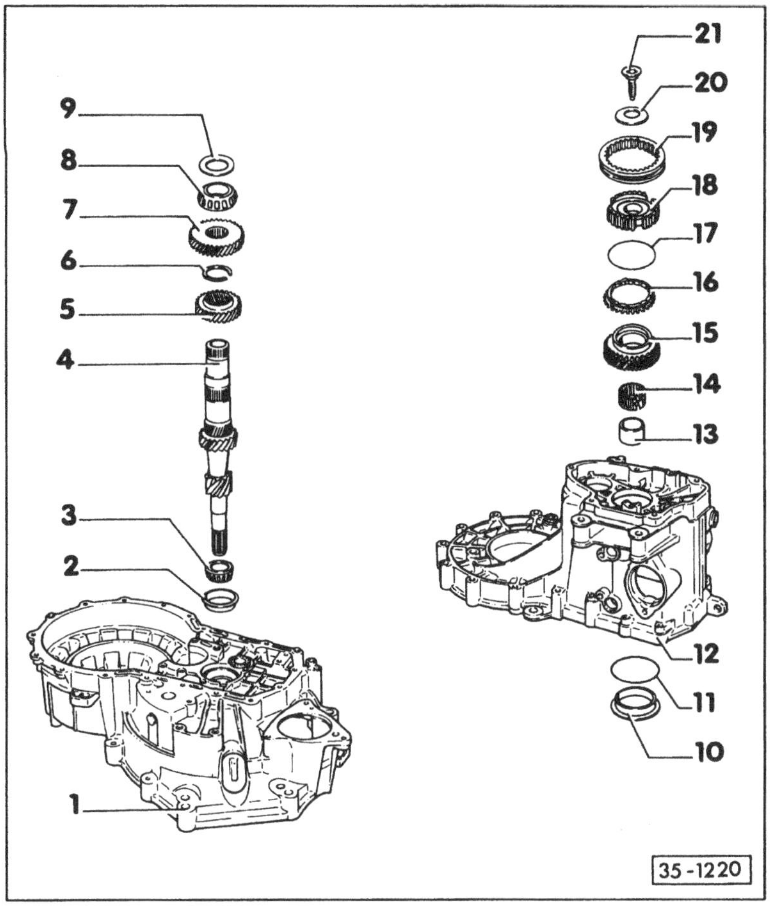

19 - 5th gear locking collar

- Removing and installing
⇒ from Page K 34-72
- Modification:
From transmission build date 31 10 1, with additional locking keys
⇒ Fig. 13

Notes:

- *Disassembling and assembling 5th gear synchronization with additional locking keys, from transmission build date 31 10 1*
⇒ Fig. 13 and Fig. 14.
- *Installing ⇒ Page K 34-86*

20 - Diaphragm spring

- Installation position
⇒ Page K 34-87

21 - Torx bolt M 10

- Tightening torque: 80 Nm (59 ft lb)
- Fitted piece on bolt head locates diaphragm spring

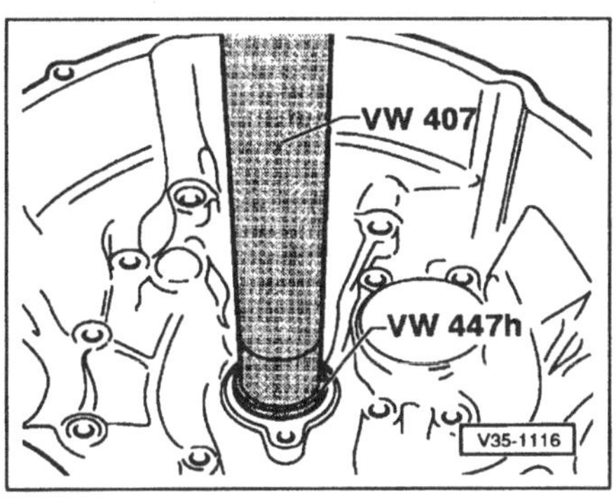

◄ **Fig. 1 Tapered roller bearing outer race, removing**

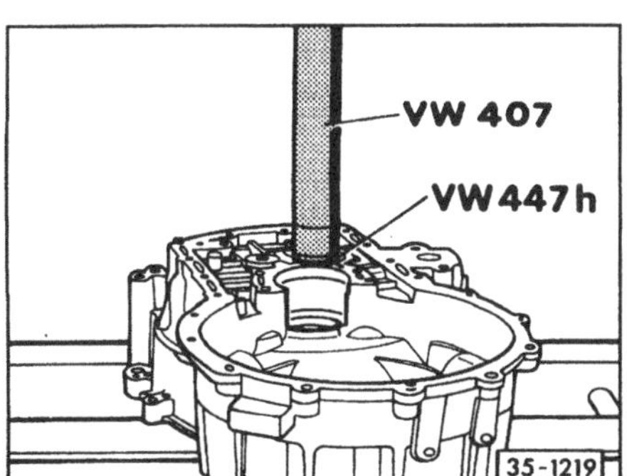

◄ **Fig. 2 Tapered roller bearing outer race, installing**

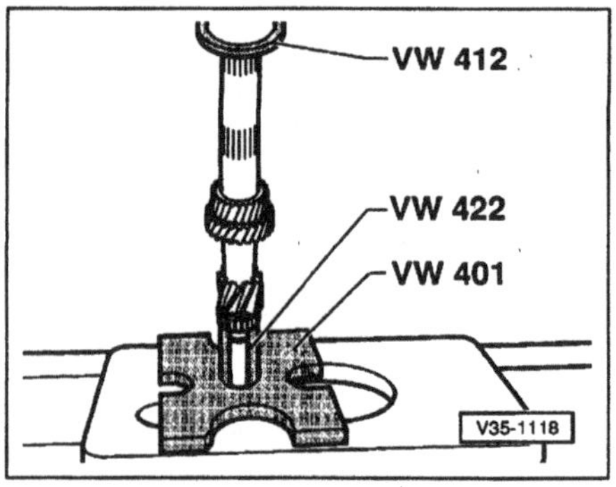

Fig. 3 Tapered roller bearing inner race, installing

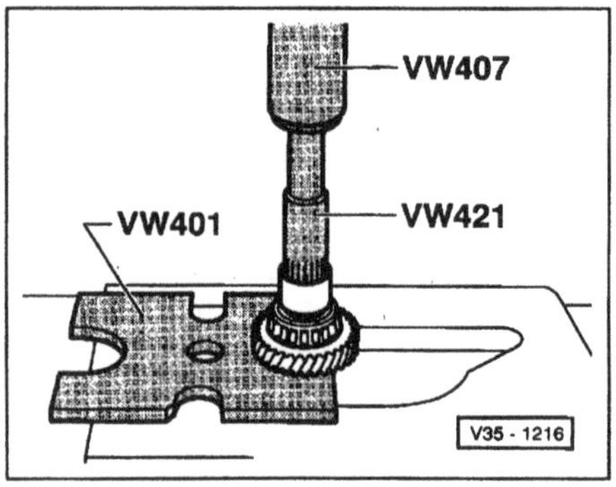

Fig. 4 4th gear, removing with tapered roller bearing inner race and sleeve

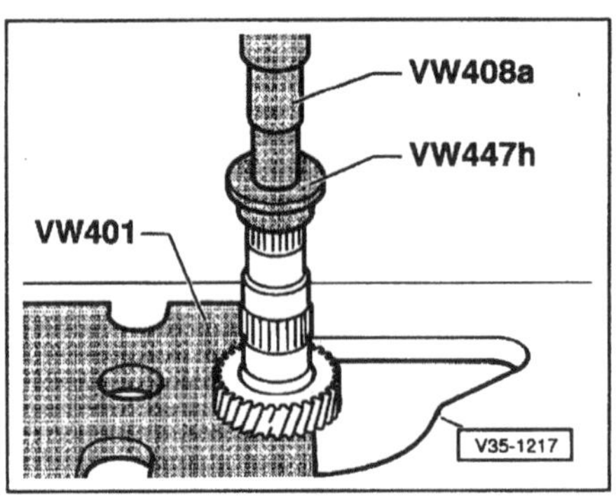

Fig. 5 3rd gear, removing

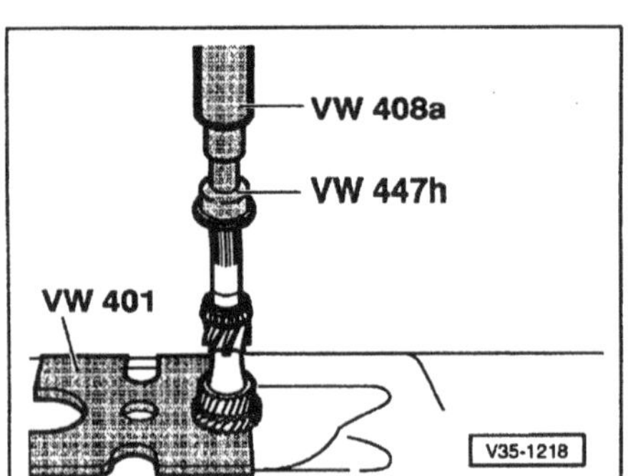

Fig. 6 3rd gear, installing

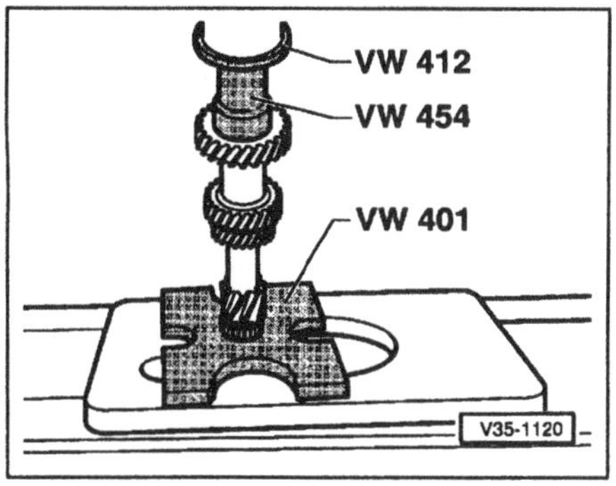

◄ **Fig. 7** **4th gear, installing**

Shoulder faces 3rd gear.

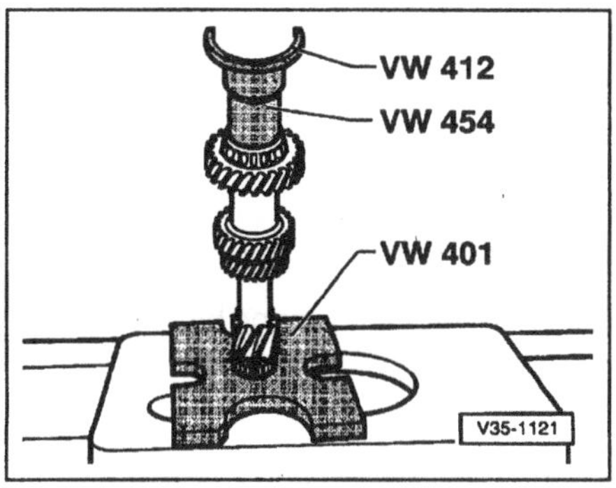

◄ **Fig. 8** **Tapered roller bearing inner race, installing**

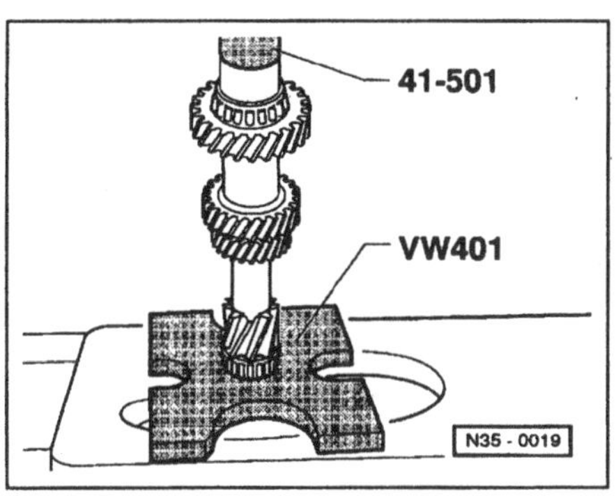

◄ **Fig. 9** **Needle roller bearing sleeve, installing**

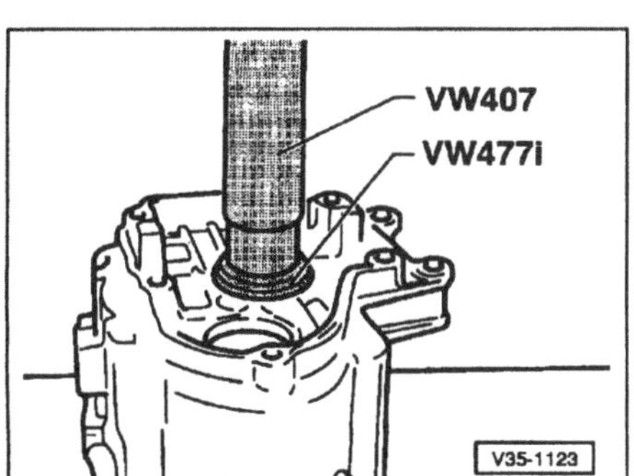

◄ **Fig. 10** **Tapered roller bearing outer race, removing**

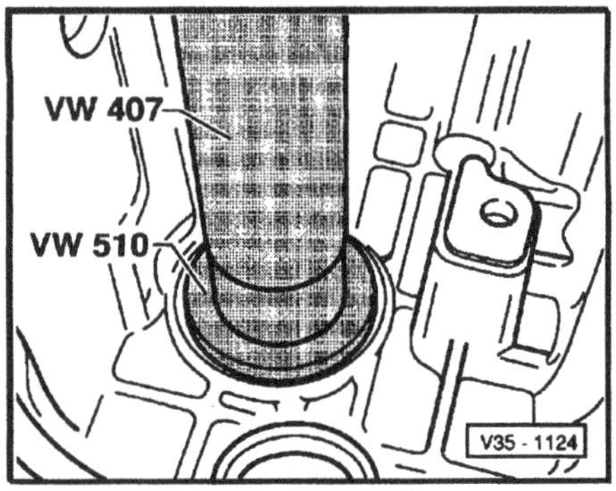

◄ **Fig. 11 Tapered roller bearing outer race, installing**

– Install with shim after adjusting input shaft.

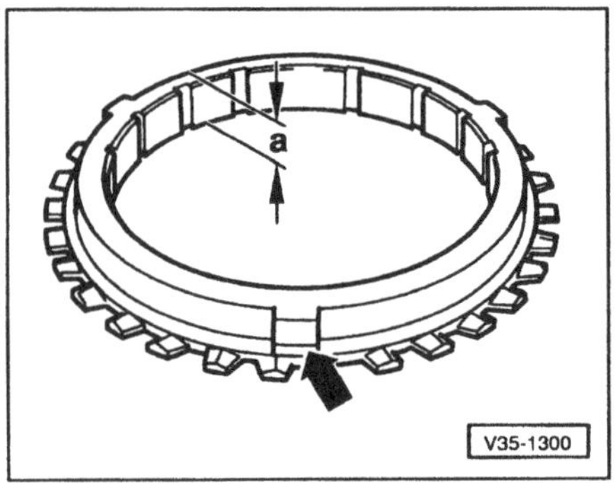

◄ **Fig. 12 5th gear synchro-ring modified**

Transmission build date	Dimension -a- mm (in.)
from 31 10 1	8.6 (0.340)
up to 30 10 1	9.0 (0.354)

Slots in cast locking keys -arrows- for previous spring have been deleted.

Note:

Do not interchange modified and previous synchro-rings.

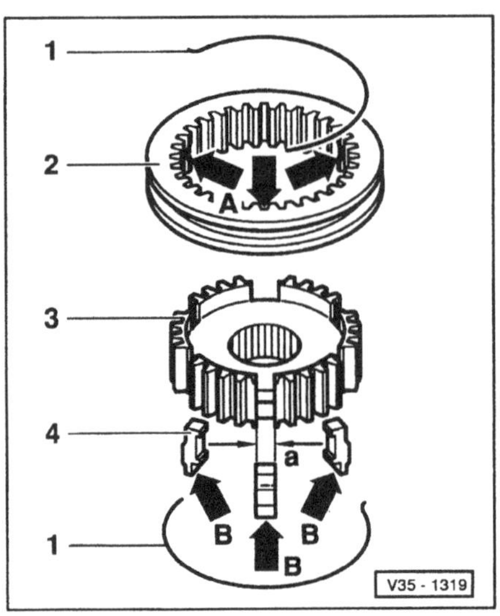

◄ **Fig. 13 5th gear synchronization with additional locking keys**

From transmission build date 31 10 1

1 - Springs
Replace previous spring on synchro-ring

2 - Operating sleeve
With 3 slots -arrow A- for additional locking keys -4-

3 - Synchro-hub
With 3 smaller additional slots for additional locking keys -4-
Dimension -a- 8.6 mm (0.339 in.)

4 - Locking keys
Installation position: Thin extension -arrow B- faces chamfered teeth of operating sleeve -2-

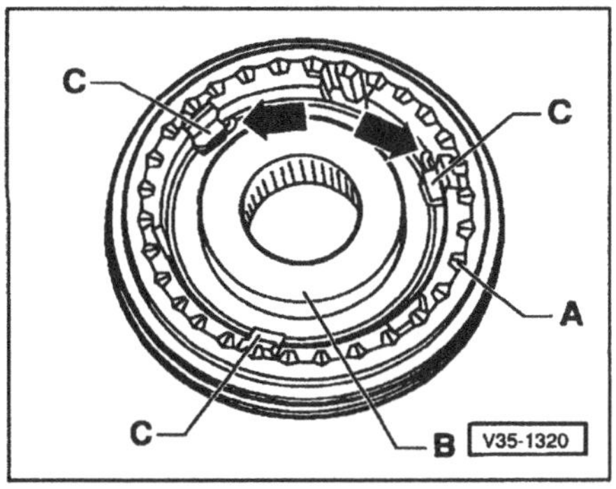

Fig. 14 5th gear operating sleeve/synchro-hub, assembling

From transmission build date 31 10 1

– Push operating sleeve over synchro-hub.

Chamfered teeth -A- and shoulder -B- of synchro-hub face in same direction. Slots for locking keys in operating sleeve and synchro-hub must be aligned (Fig. 13).

– Insert locking keys (Installation position: Fig. 13).

– Insert springs offset by 120° underneath lug -C-. Spring must act on all three locking keys, angled ends directly behind two of locking keys -arrows-.

Input shaft, adjusting

(Determining thickness of shim for input shaft)

It is necessary to readjust the input shaft when:

♦ Transmission housing

♦ Clutch housing

♦ Input shaft

♦ 4th gear

or the

♦ Tapered roller bearing

are replaced.

Adjustment overview ⇒ Page K 39-20

– Press tapered roller bearing outer race, without shim, into transmission housing onto stop, using VW 510 thrust pad.

– Install input shaft in clutch housing and assemble transmission housing. Tighten hex head bolts to 25 Nm (18 ft lb) plus an additional 1/4 turn (90°).

– Insert VW 387 dial gauge holder, VW 385–17 magnetic plate, and dial indicator in clutch housing (large tapered roller bearing in transmission housing).

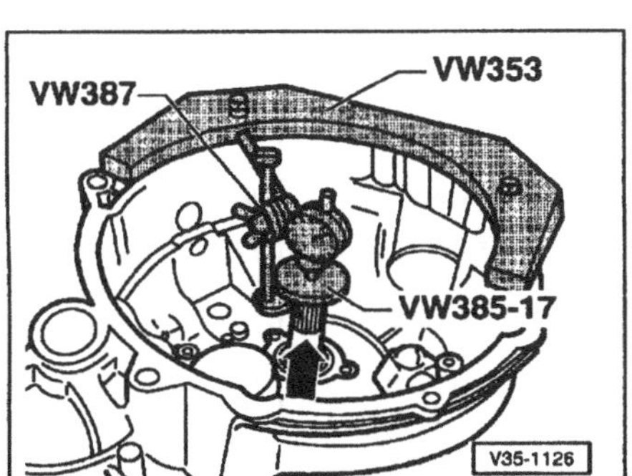

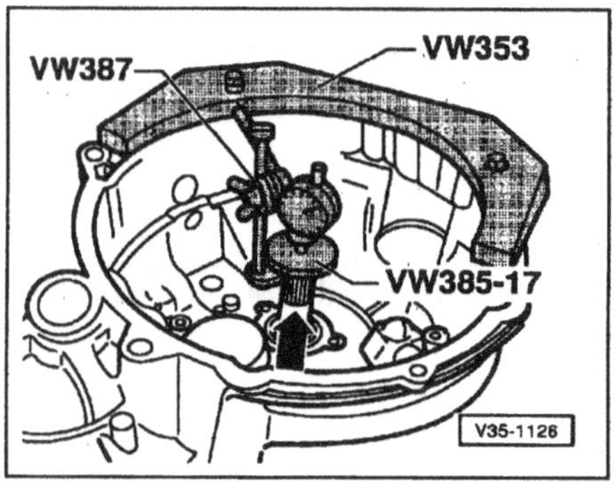

- Before taking any measurements, rotate input shaft to allow bearing to settle. Zero dial indicator with 1 mm (0.03937 in.) preload.

Note:

This preparation is to be repeated for each subsequent measurement, otherwise gauge will not return to starting position.

- Push input shaft in direction of dial indicator -arrow-.
- Read off and note clearance indicated on dial indicator (example, 1.21 mm [0.0476 in.]). Record axial play at dial indicator and note.

Note:

Dial indicator does not return to original position.

- Determine thickness of shim from table ⇒ from Page K 35-16 to Page K 35-18 (example, 1.175 mm [0.0462 in.]).
- Remove input shaft and remove tapered roller bearing outer race from transmission housing using VW 447 I thrust pad.
- Install tapered roller bearing outer race together with shim (example, 1.175 mm [0.462 in.]) into transmission housing.
- Install transmission housing and tighten hex head bolts to 25 Nm (18 ft lb) plus an additional 1/4 turn (90°).

Table of adjustment shims

Measured bearing clearance mm (in.)	Shim	
	Thickness mm (in.)	Part No.
0.671 ... 0.699 (0.02641 ... 0.02751)	0.650 (0.02559)	02A 311 140
0.700 ... 0.724 (0.02755 ... 0.02850)	0.675 (0.02657)	02A 311 140 A
0.725 ... 0.749 (0.02854 ... 0.02948)	0.700 (0.02755)	02A 311 140 B
0.750 ... 0.744 (0.02066 ... 0.02929)	0.725 (0.02854)	02A 311 140 C
0.775 ... 0.799 (0.03051 ... 0.03145)	0.750 (0.02066)	02A 311 140 D
0.800 ... 0.824 (0.03149 ... 0.03244)	0.775 (0.03051)	02A 311 140 E
0.825 ... 0.849 (0.03248 ... 0.03342)	0.800 (0.03149)	02A 311 140 F
0.850 ....0.874 (0.03346 ... 0.03340)	0.825 (0.03248)	02A 311 140 G
0.875 ... 0.899 (0.03444 ... 0.03539)	0.850 (0.03346)	02A 311 140 H
0.900 ... 0.924 (0.03543 ... 0.03637)	0.875 (0.03444)	02A 311 140 J
0.925 ... 0.949 (0.03641 ... 0.03736)	0.900 (0.03543)	02A 311 140 K
0.950 ... 0.974 (0.03740 ... 0.03834)	0.925 (0.03641)	02A 311 140 L
0.975 ... 0.999 (0.03838 ... 0.03933)	0.950 (0.03740)	02A 311 140 M
1.000 ... 1.024 (0.03937 ... 0.04031)	0.975 (0.03838)	02A 311 140 N
1.025 ... 1.049 (0.04035 ... 0.04129)	1.000 (0.03937)	02A 311 140 P
1.050 ... 1.074 (0.04133 ... 0.04228)	1.025 (0.04035)	02A 311 140 Q
1.075 ... 1.099 (0.04232 ... 0.04326)	1.050 (0.04133)	02A 311 140 R
1.100 ... 1.124 (0.04330 ... 0.04425)	1.075 (0.04232)	02A 311 140 S
1.125 ... 1.149 (0.04429 ... 0.04523)	1.100 (0.04330)	02A 311 140 T
1.150 ... 1.174 (0.04527 ... 0.04622)	1.125 (0.04429)	02A 311 140 AA
1.175 ... 1.199 (0.04625 ... 0.04720)	1.150 (0.04527)	02A 311 140 AB

Measured bearing clearance mm (in.)	Shim	
	Thickness mm (in.)	Part No.
1.200 ... 1.224 (0.04724 ... 0.04818)	1.175 (0.04625)	02A 311 140 AC
1.225 ... 1.249 (0.04822 ... 0.04917)	1.200 (0.04724)	02A 311 140 AD
1.250 ... 1.274 (0.04921 ... 0.05015)	1.225 (0.04822)	02A 311 140 AE
1.275 ... 1.229 (0.05019 ... 0.04838)	1.250 (0.04921)	02A 311 140 AF
1.300 ... 1.324 (0.05118 ... 0.05212)	1.275 (0.05019)	02A 311 140 AG
1.325 ... 1.349 (0.05216 ... 0.05311)	1.300 (0.05118)	02A 311 140 02A 311 140
1.350 ... 1.374 (0.05314 ... 0.05409)	1.325 (0.05216)	02A 311 140 02A 311 140 A
1.375 ... 1.399 (0.05413 ... 0.05507)	1.350 (0.05314)	02A 311 140 A 02A 311 140 A
1.400 ... 1.424 (0.05511 ... 0.05606)	1.375 (0.05413)	02A 311 140 A 02A 311 140 B
1.425 ... 1.449 (0.05610 ... 0.05704)	1.400 (0.05511)	02A 311 140 B 02A 311 140 B
1.450 ... 1.474 (0.05708 ... 0.05803)	1.425 (0.05610)	02A 311 140 B 02A 311 140 C
1.475 ... 1.499 (0.05807 ... 0.05901)	1.450 (0.05708)	02A 311 140 C 02A 311 140 C
1.500 ... 1.524 (0.05905 ... 0.05999)	1.475 (0.05807)	02A 311 140 C 02A 311 140 D

Measured bearing clearance mm (in.)	Shim	
	Thickness mm (in.)	Part No.
1.525 ... 1.549 (0.06003 ... 0.06098)	1.500 (0.05905)	02A 311 140 D 02A 311 140 D
1.550 ... 1.574 (0.06102 ... 0.06196)	1.525 (0.06003)	02A 311 140 D 02A 311 140 E
1.575 ... 1.599 (0.06200 ... 0.06295)	1.550 (0.06102)	02A 311 140 E 02A 311 140 E
1.600 ... 1.624 (0.06299 ... 0.06393)	1.575 (0.06200)	02A 311 140 E 02A 311 140 F
1.625 ... 1.649 (0.06397 ... 0.06492)	1.600 (0.06299)	02A 311 140 F 02A 311 140 F
1.650 ... 1.674 (0.06496 ... 0.06590)	1.625 (0.06397)	02A 311 140 F 02A 311 140 G
1.675 ... 1.699 (0.06594 ... 0.06688)	1.650 (0.06496)	02A 311 140 G 02A 311 140 G
1.700 ... 1.724 (0.06692 ... 0.06787)	1.675 (0.06594)	02A 311 140 G 02A 311 140 H
1.725 ... 1.749 (0.06791 ... 0.06885)	1.700 (0.06692)	02A 311 140 H 02A 311 140 H
1.750 ... 1.774 (0.06889 ... 0.06984)	1.725 (0.06791)	02A 311 140 H 02A 311 140 J
1.775 ... 1.791 (0.06988 ... 0.07051)	1.750 (0.06889)	02A 311 140 J 02A 311 140 J

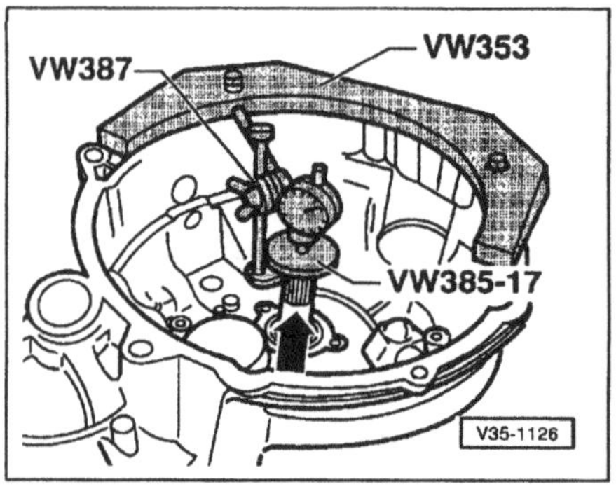

Measurement check

◄ – Insert VW 387 dial gauge holder, VW 385-17 magnetic plate, and dial indicator.

– Rotate input shaft so that tapered roller bearing settles.

– Push input shaft in direction of arrow.

- Bearing play should be minimum 0.01mm (0.0004 in.), maximum 0.09 mm (0.0035 in.).

Note:

If bearing clearance cannot be measured but input shaft play is perceptible, and input shaft turns freely, the adjustment is acceptable.

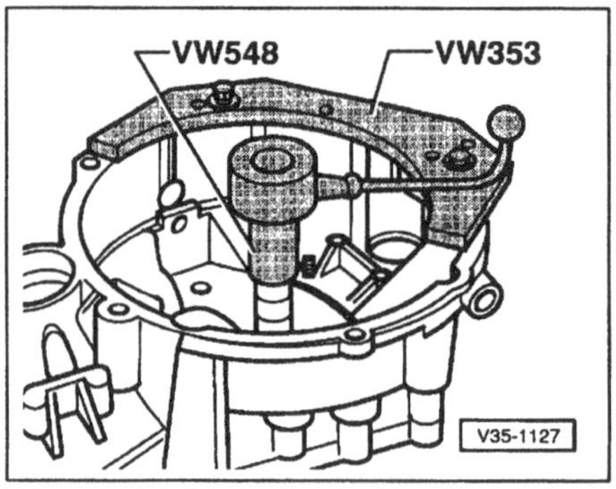

◄ Frictional torque = maximum 20 Ncm (2 in. lb).

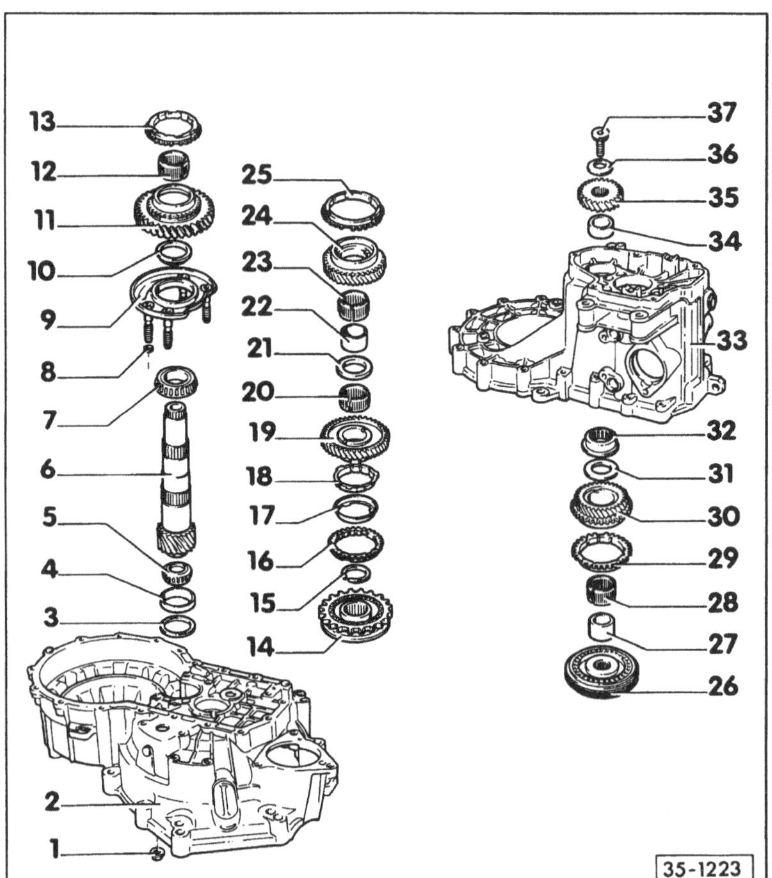

Output shaft (drive pinion), disassembling and assembling

Notes:

- *Output shaft is disassembled as follows:*
  - *Position 17/2 Kukko separating tool 22 ... 115 mm, or equivalent, under 2nd gear -19- and push off as described in Fig. 5.*
  - *Remove circlip -15-.*
  - *Position 20/10 Kukko two-arm puller, or equivalent, under bearing support -9- and remove operating sleeve with 1st and 2nd gear synchro-hub as described in Fig. 6.*
- *When installing new gears or a new output shaft consult technical data ⇒ Page K 00-3.*
- *Output shaft must be re-adjusted if position of tapered roller bearings is affected when interchanging parts. See adjustment overview ⇒ Page K 39-20*

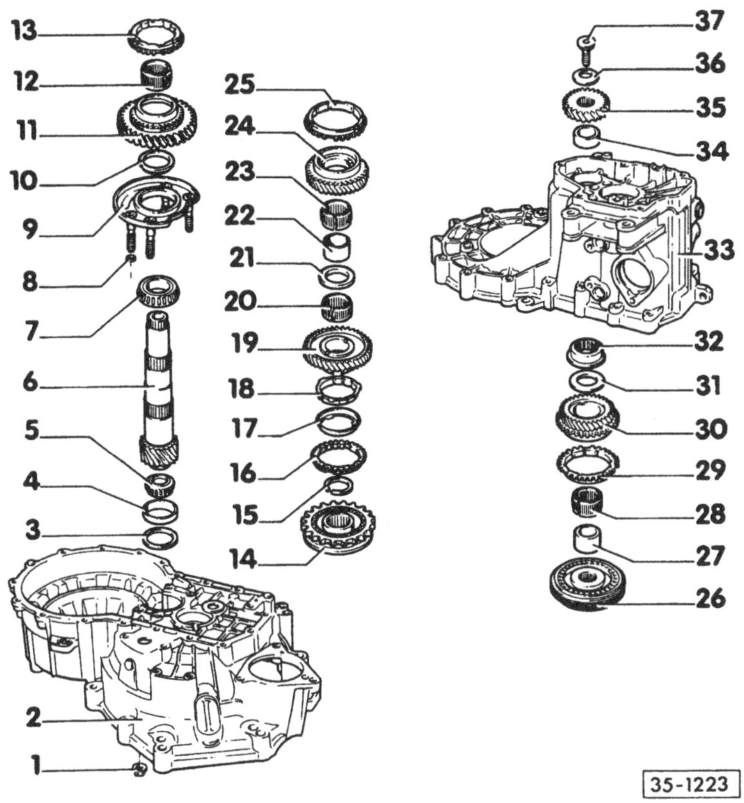

- *Removing and installing output shaft ⇒ from Page K 34-72.*
- *Both tapered roller bearings must be replaced together as a set.*

1 - Hex nut
- Tightening torque: 80 Nm (59 ft lb) plus an additional 1/4 turn (90°)
- 4 nuts for bearing support

2 - Clutch housing

3 - Shim
- Determining thickness ⇒ Page K 35-45 Output shaft, adjusting

4 - Small tapered roller bearing outer race
- Removing ⇒ Fig. 1
- Installing ⇒ Fig. 2

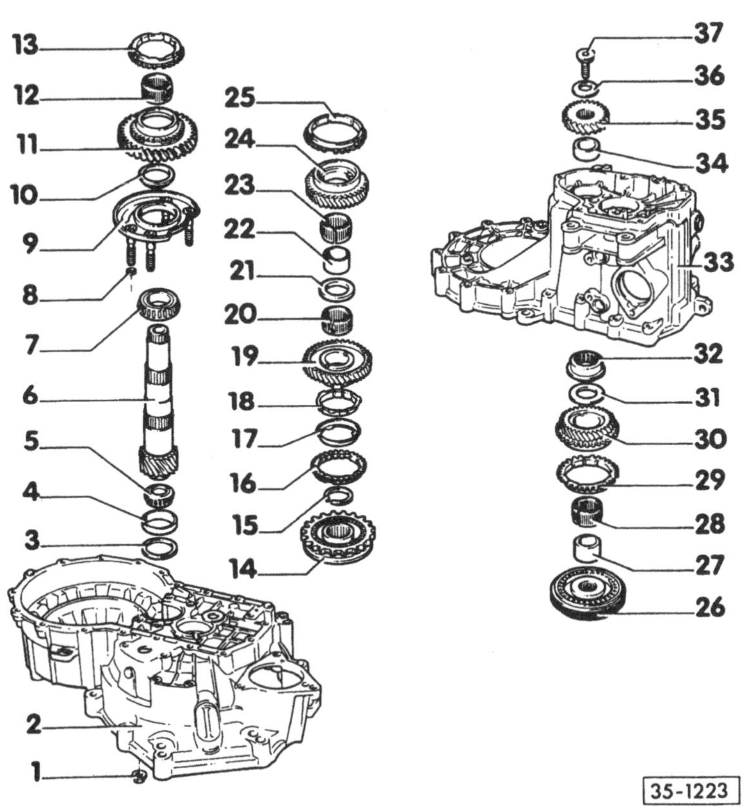

5 - Small tapered roller bearing, inner race
- Removing ⇒ Fig. 3
- Installing ⇒ Fig. 4

6 - Output shaft
- Is paired with final drive gear, always replace as a set
- Adjusting ⇒ Page K 35-45

7 - Large tapered roller bearing, inner race
- Removing ⇒ Fig. 7
- Installing ⇒ Fig. 8

8 - O-ring
- Place O-rings (Quantity 4) on bearing carrier bolts
- Always replace

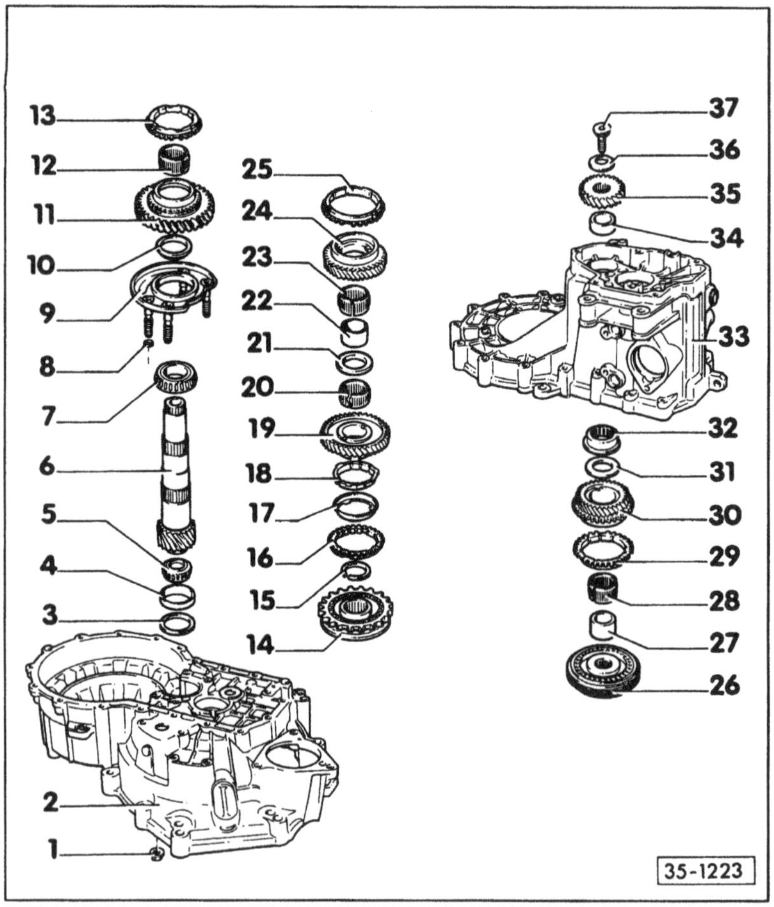

9 - Bearing carrier
- With large tapered roller bearing outer race and bolts
- Only change outer race together with large tapered roller bearing and bearing carrier

10 - Thrust washer
- Shoulder on thrust washer faces tapered roller bearing

11 - 1st gear

12 - Needle roller bearing
- 1st gear

13 - 1st gear synchro-ring
- Inspecting for wear ⇒ Fig. 9
- Identification ⇒ Fig. 11 and Fig. 12

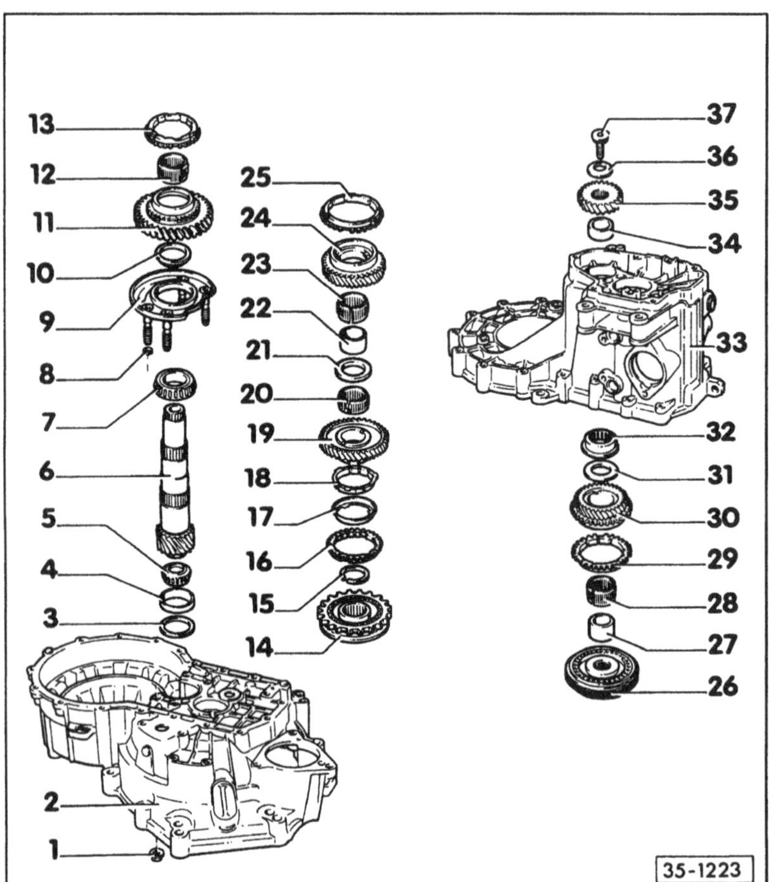

14 - 1st and 2nd gear locking collar with synchro-hub
- Remove over bearing carrier after removing circlip -15- ⇒ Fig. 6
- Modification: From transmission build date 31 08 2, larger slots for synchro hub ⇒ Fig. 13
- Disassembling ⇒ Fig. 14
- Assembling locking collar/synchro hub ⇒ Fig. 14 and Fig. 15
- Installation position ⇒ Fig. 16
- Installing ⇒ Fig. 17

15 - Circlip

16 - 2nd gear synchro-ring
- Identification ⇒ Fig. 11
- Inspecting for wear ⇒ Fig. 18
- Assemble so that slots locate in locking collar -14- locking keys

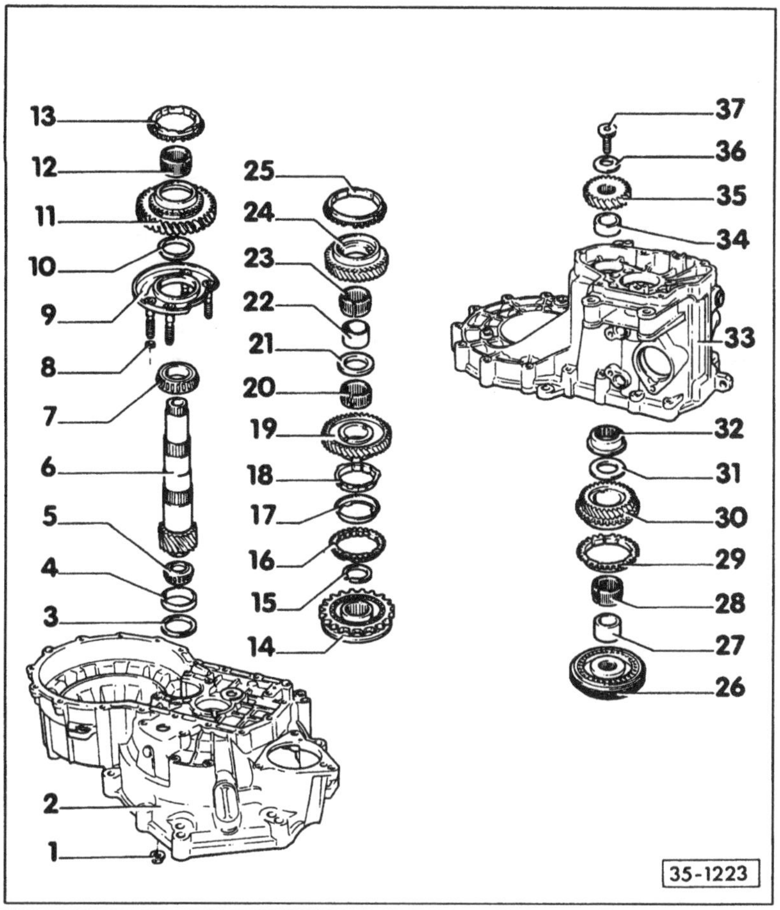

17 - Outer ring
- Install in synchro-ring -16-, installation position ⇒ Fig. 19
- Replace if scored

18 - Synchro-ring
- (Inner ring)
- Inspecting for wear ⇒ Fig. 20
- Inspect lugs for scoring
- Installation position ⇒ Fig. 21

19 - 2nd gear
- Installation position ⇒ Fig. 22

20 - Needle roller bearing
- 2nd gear

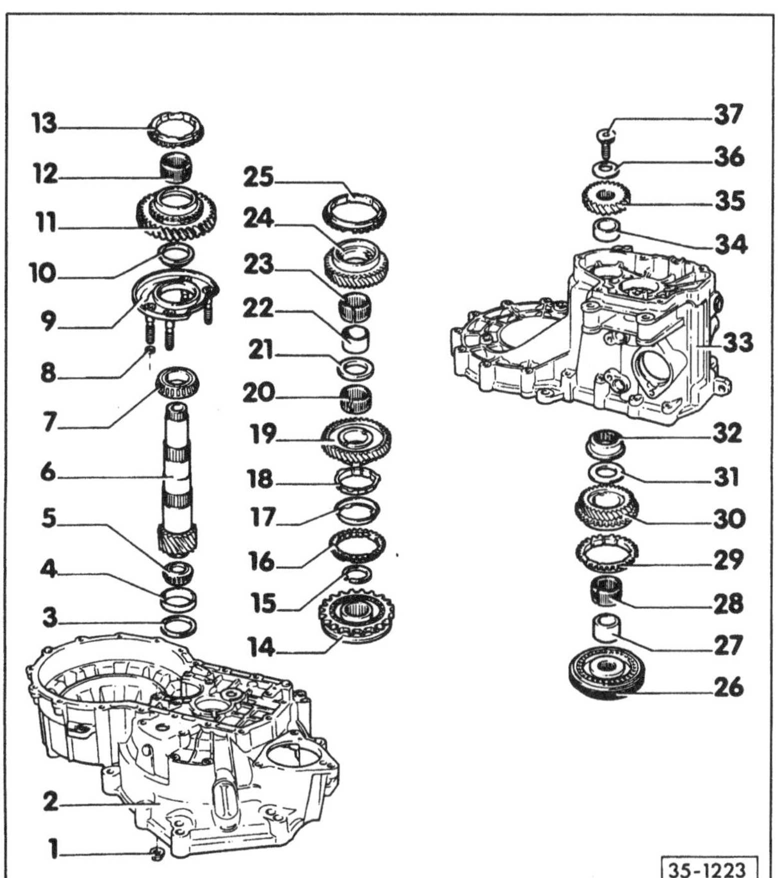

21 - Thrust washer

22 - 3rd gear needle roller bearing sleeve
- Removing with 2nd gear ⇒ Fig. 5
- Installing ⇒ Fig. 23

23 - Needle roller bearing
- 3rd gear

24 - 3rd gear

25 - 3rd gear synchro-ring
- Inspecting for wear ⇒ Fig. 9

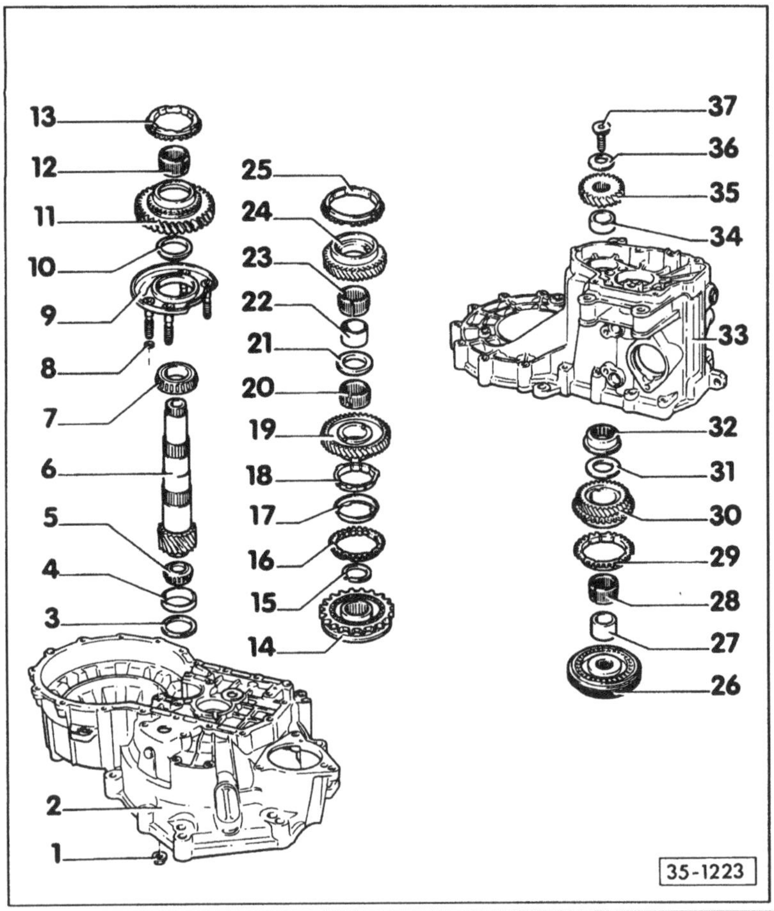

26 - Locking collar with 3rd and 4th gear synchro-hub

- Remove with 2nd gear -19- and 3rd gear -24- ⇒ Fig. 5
- Disassembling ⇒ Fig. 24
- Installation position locking collar/synchro-hub ⇒ Fig. 26
- Assembling locking collar/synchro-hub ⇒ Fig. 24 and Fig. 25
- Installing ⇒ Fig. 27

27 - Needle roller bearing sleeve

- Remove with locking collar and 3rd and 4th gear synchro-hub -26- ⇒ Fig. 5
- Installing ⇒ Fig. 28

28 - Needle roller bearing

- 4th gear

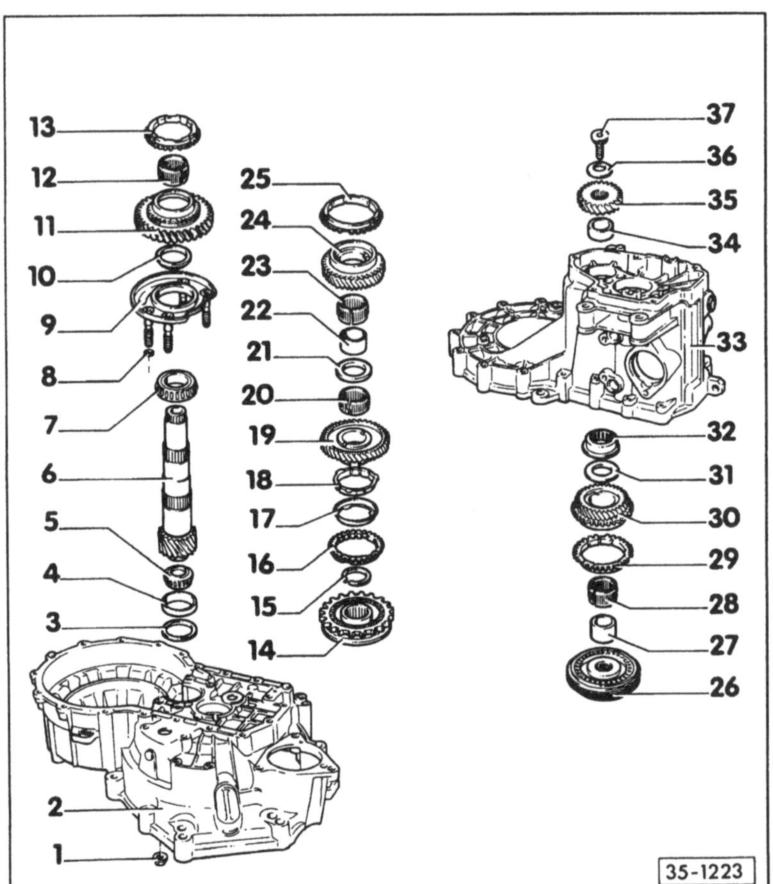

29 - 4th gear synchro-ring

- Inspecting for wear ⇒ Fig. 9

30 - 4th gear

31 - Thrust washer

32 - Output shaft needle roller bearing

- Removing and installing ⇒ Page K 34-93

33 - Transmission housing

34 - Output shaft needle roller bearing sleeve

- Removing ⇒ Fig. 5
- Installing ⇒ Fig. 29

35 - 5th gear

- Removing and installing ⇒ from Page K 34-74

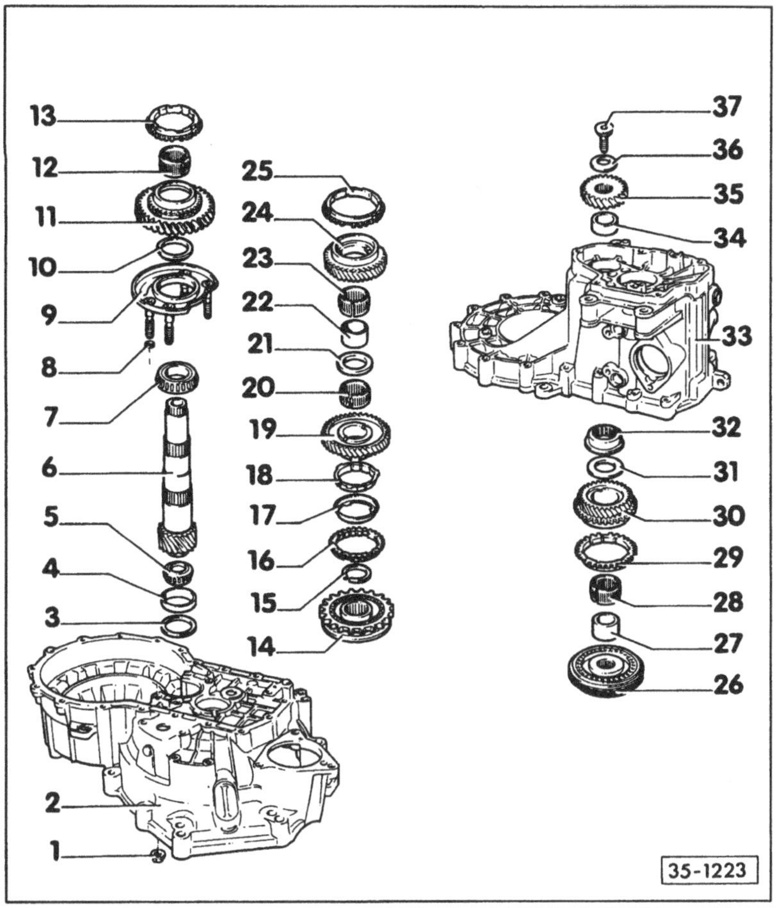

36 - Diaphragm spring

- Installation position ⇒ Page K 34-87

37 - Torx bolt M10

- Tightening torque: 80 Nm (59 ft lb)
- Fitted portion on bolt head holds diaphragm spring in position

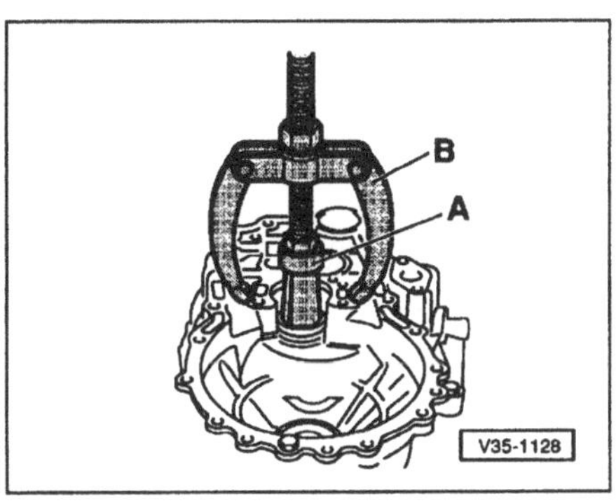

◄ **Fig. 1 Small tapered roller bearing outer race, removing**

A - 21/6 Kukko extractor 37 ... 46 mm, or equivalent

B - 22/2 Kukko support, or equivalent

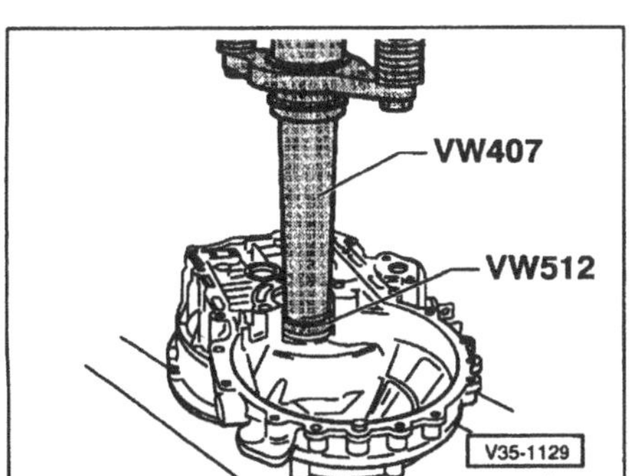

◄ **Fig. 2 Tapered roller bearing outer race, installing**

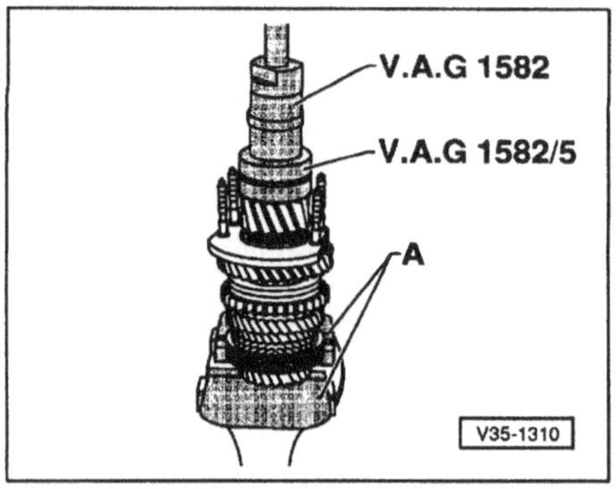

Fig. 3 Small tapered roller bearing inner race, removing

A - Vice clamps

– Install extractor and clamp behind bearing rollers, then turn bearing and retighten extractor.

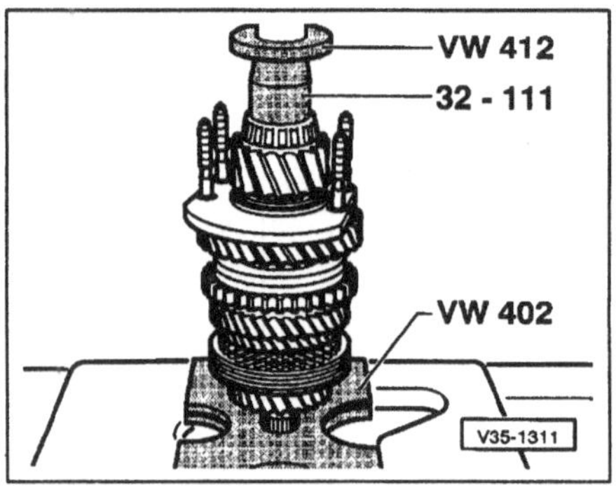

Fig. 4 Small tapered roller bearing inner race, installing

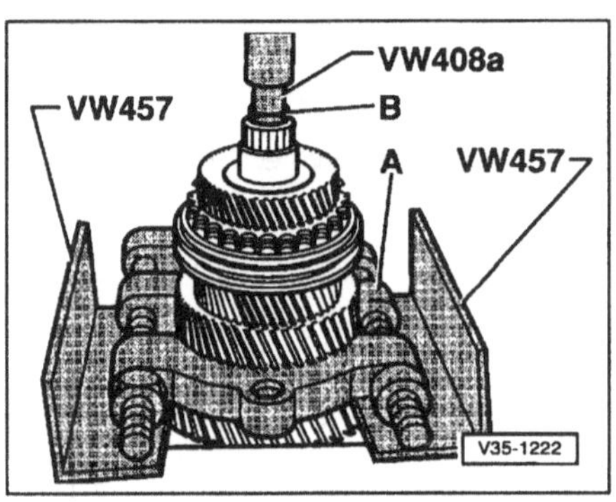

Fig. 5 3rd and 4th gear synchro-hub/locking collar, 2nd, 3rd, and 4th gears with transmission housing needle roller bearing sleeve, removing

A – 17/2 Kukko separating tool 22 ... 115 mm, or equivalent

B – M10 x 20 hex bolt (17 mm [0.6692 in.] A/F)

Note:

Support separating tool so that 1st and 2nd gear locking collar is not removed at the same time.

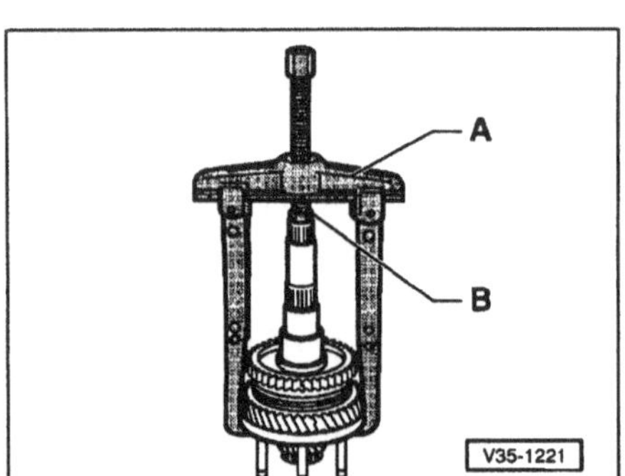

Fig. 6 Locking collar with synchro-hub over bearing carrier, removing

A – 20/10 Kukko 2-arm puller with US 1079 puller hooks 250 mm (9.8425 in.) long

B – M10 x 20 hex bolt (17 mm [0.6692 in.] A/F)

Note:

Remove circlip before removing individual components.

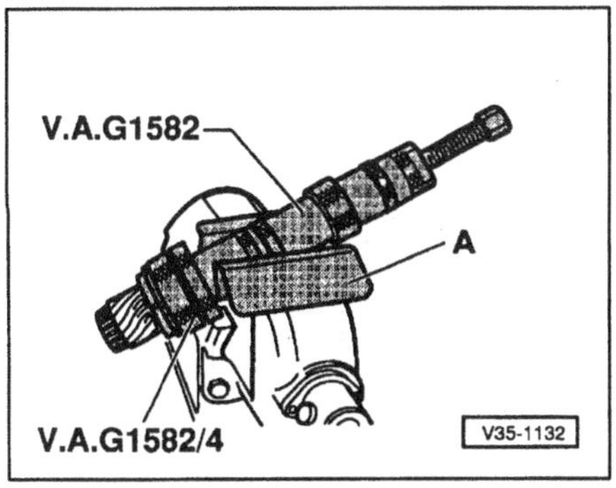

Fig. 7 Large tapered roller bearing inner race, removing

A - Vice clamps

– Insert M 10 x 20 hex bolt in output shaft bore before installing puller.

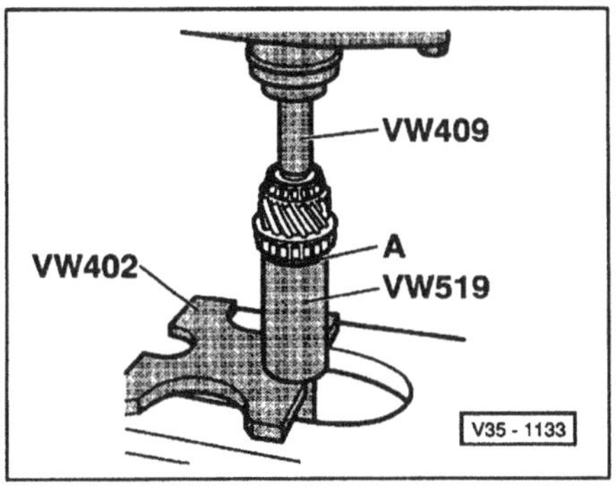

Fig. 8 Large tapered roller bearing inner race, installing

A - Thrust washer

– Install thrust washer before inserting inner race.

● Shoulder faces inner race.

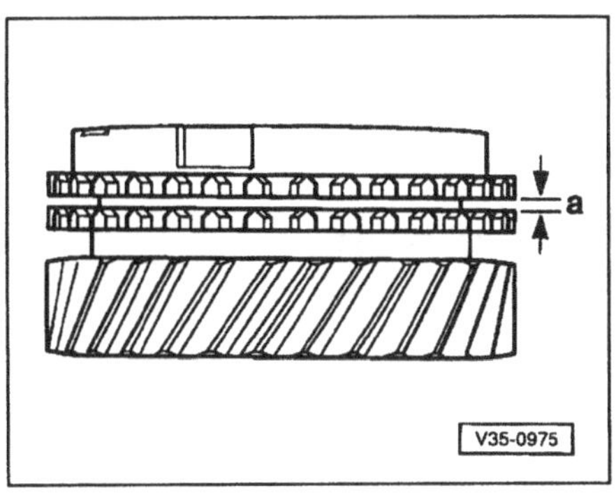

Fig. 9 Synchro-ring, inspecting for wear

– Install synchro-ring onto gear cone and measure gap “a” with a feeler gauge.

Gap “a”	Installation dimension mm (in.)	Wear limit mm (in.)
1st gear	1.0 ... 1.7 (0.0393 ... 0.0669)	
3rd gear	1.0 ... 1.7 (0.0393 ... 0.0669)	0.5 (0.0196)
4th gear	1.0 ... 1.7 (0.0393 ... 0.0669)	

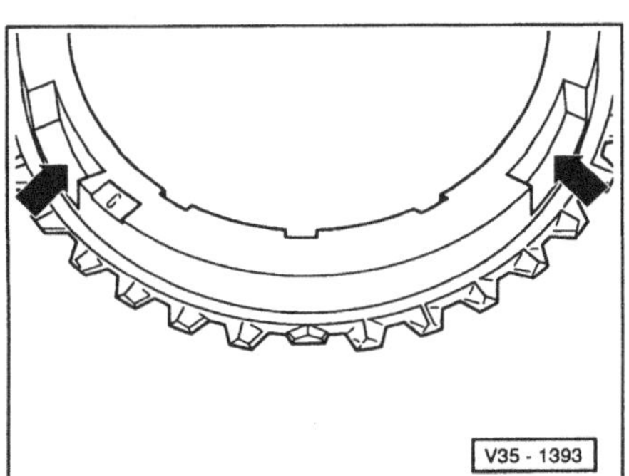

Fig. 10 1st and 2nd gear synchro-rings with smaller slots -arrows- for locking key

From transmission build date 31 08 2

Note:

In earlier transmissions, install operating sleeve (Fig. 13) and new synchro-ring together.

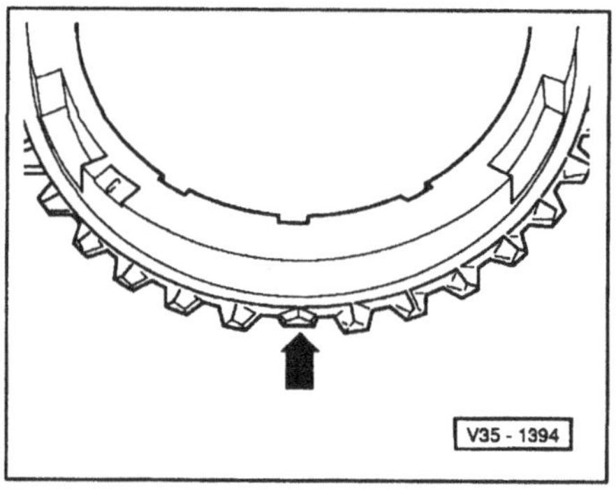

Fig. 11 1st and 2nd gear synchro-rings, identifying

1st gear Synchro-ring

Transmission build date	Identification
from 31 08 2	3 x $^1/_2$-height teeth -arrow-
to 30 08 2	3 x 1 tooth missing

These synchro-rings must only be installed to 1st gear.

2nd gear Synchro-ring

Transmission build date	Identification
from 31 08 2	3 x $^1/_2$ height teeth -arrow- or 2 x $^1/_2$ height teeth
to 30 08 2	no teeth missing or 2 x 1 tooth missing

Only synchro-ring with "2 x $^1/_2$ height teeth" is available as a spare part.

Note:

In earlier transmissions, install operating sleeve (Fig. 15) and new synchro-ring together.

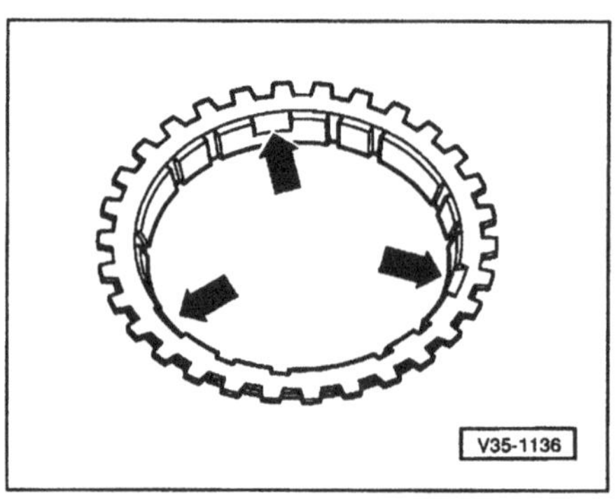

Fig. 12 1st and 2nd gear synchro-rings, identifying

- 1st gear synchro-ring - no slots
- 2nd gear synchro-ring - three slots -arrows-

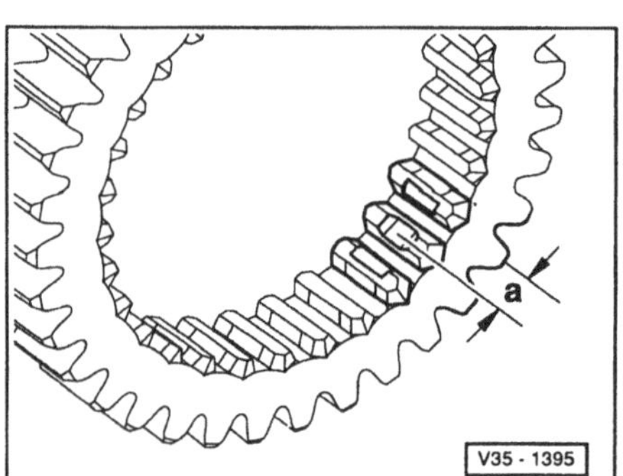

Fig. 13 1st and 2nd gear operating sleeve with larger slot

Transmission build date	Dimension -a- mm (in.)
from 31 08 2	16.1 (0.6338)
to 30 08 2	16.6 (0.6535)

Note:

In earlier transmissions, install operating sleeve (Fig. 12 and Fig. 13) and new synchro-ring together.

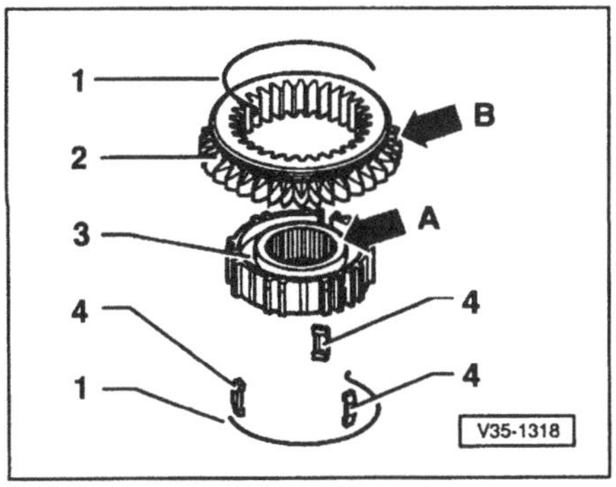

Fig. 14 1st and 2nd gear operating sleeve and synchro-hub disassembling and assembling

1 - Spring

2 - Operating sleeve

3 - Synchro-hub

4 - Locking key

– Slide operating sleeve over synchro-hub.

Wide shoulder of synchro-hub -arrow A- and outer splines of locking collar -arrow B- face opposite directions after assembly. Slots for locking keys on synchro-hub and operating sleeve must align.

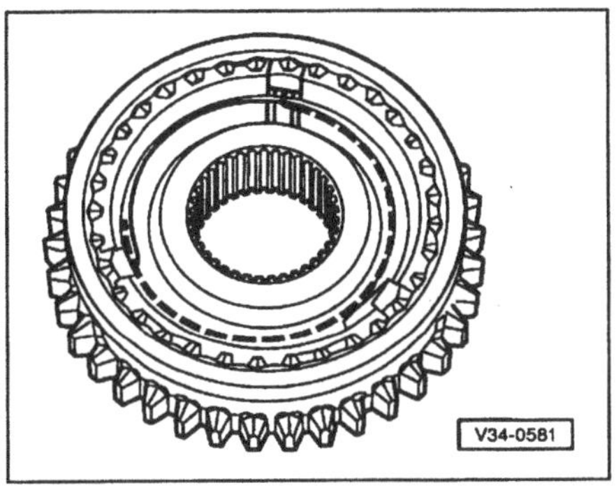

Fig. 15 1st and 2nd gear operating sleeve and synchro-hub, assembling

- Operating sleeve is pushed over synchro-hub.

– Install locking keys and springs 120° offset. Angled end of springs must hook into hollow locking key.

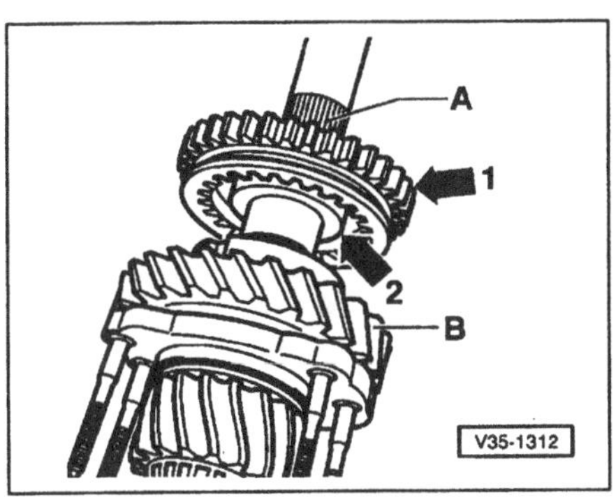

Fig. 16 1st and 2nd gear operating sleeve/synchro-hub, Installation position

Operating sleeve splines -arrow 1- face splines for 3rd/4th gear synchro-hub -A-.

The wide shoulder of the synchro-hub -arrow 2- then faces 1st gear -B-.

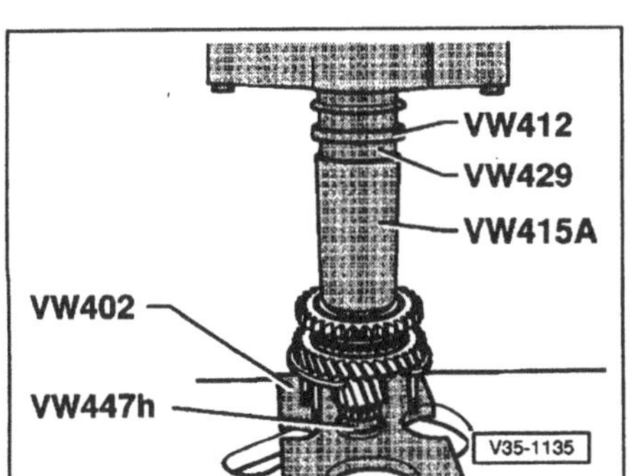

Fig. 17 1st and 2nd gear operating sleeve/synchro-hub, installing

Note:

Before installing, place 1st gear synchro-ring on 1st gear.

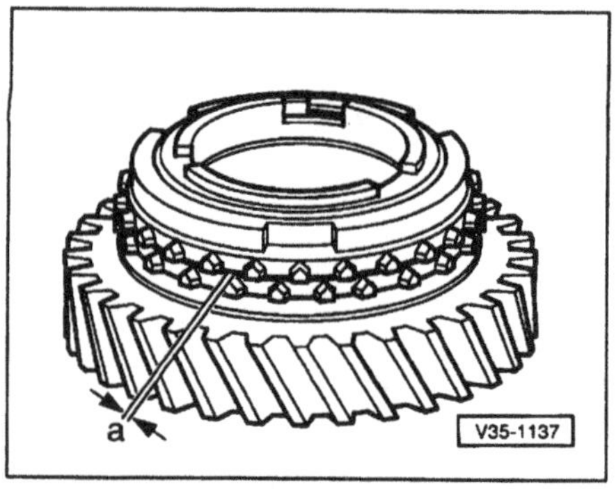

Fig. 18 Inspecting 2nd gear synchro-ring for wear

– Press synchro-ring, outer- and inner-ring onto 2nd gear cone and measure gap "a" with feeler gauge.

Gap "a"	Installation dimension mm (in.)	Wear limit mm (in.)
2nd gear	1.2 ... 1.8 (0.0472 ... 0.0708)	0.5 (0.0196)

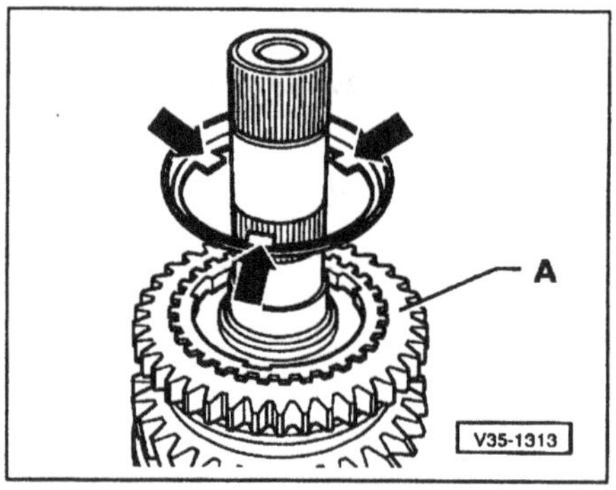

Fig. 19 Outer-ring, installation position

Lugs -arrows- face 1st gear -A-

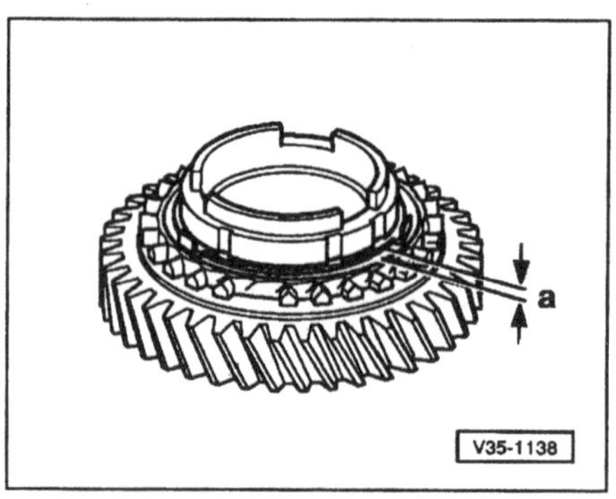

Fig. 20 Inner ring, inspecting for wear

– Press inner ring onto 2nd gear cone and measure gap "a" with a feeler gauge.

Gap "a"	Installation dimension mm (in.)	Wear limit mm (in.)
2nd gear	0.75 ... 1.25 (0.0295 ... 0.0492)	0.3 (0.0118)

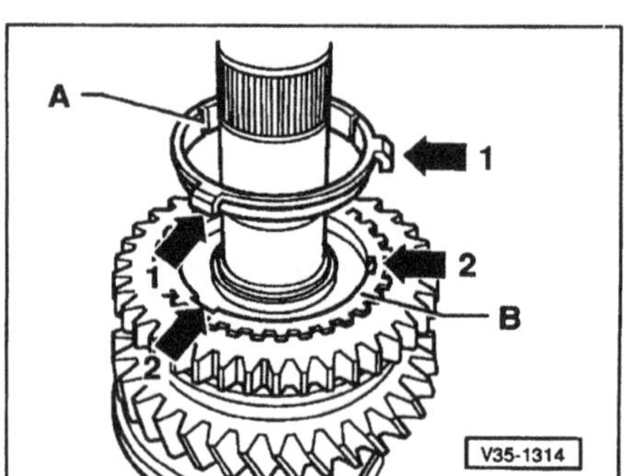

Fig. 21 Synchro-ring (inner ring) -A-, installation position

Lugs -arrow 1- install in slots -arrow 2- in synchro-ring -B-.

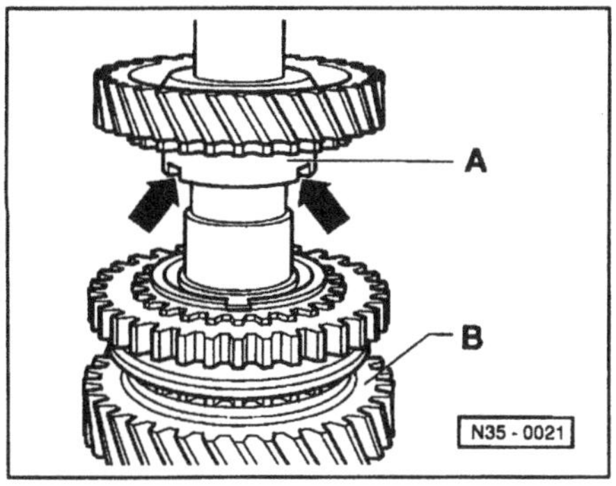

◄ **Fig. 22 2nd gear, installation position**

Higher shoulder -A- faces 1st gear -B-. Slots in shoulder -arrows- position in lugs of outer-ring (-arrows- ⇒ Fig. 19)

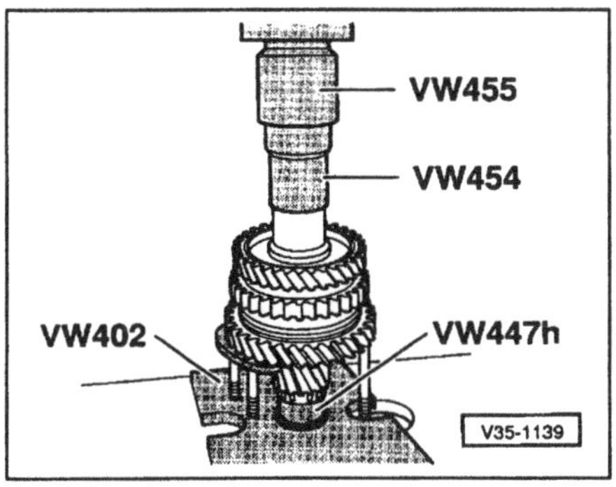

◄ **Fig. 23 Needle roller bearing sleeve for 3rd gear, installing**

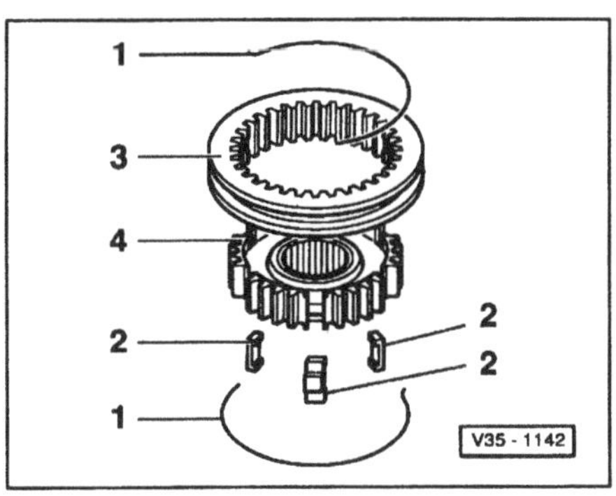

◄ **Fig. 24 3rd and 4th gear operating sleeve and synchro-hub, disassembling and assembling**

1 - Spring

2 - Locking key

3 - Operating sleeve

4 - Synchro-hub

– Slide operating sleeve over synchro-hub.

- Slots for locking keys in synchro-hub and operating sleeve must align.

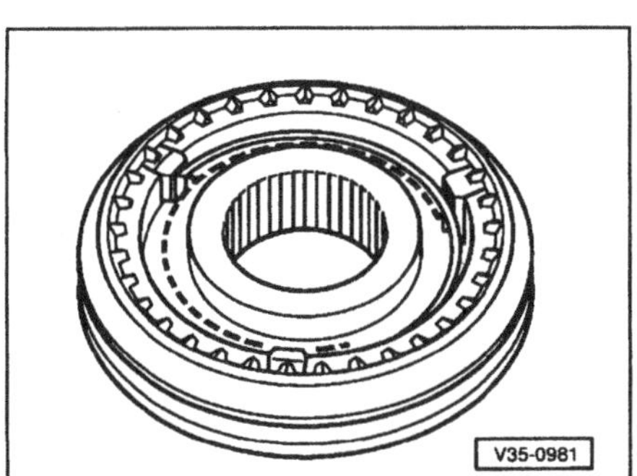

◄ **Fig. 25 3rd and 4th gear operating sleeve/synchro-hub, assembling**

Operating sleeve has been pushed over the synchro-hub.

– Install locking keys and springs 120° offset. Angled end of spring must hook into hollow locking key.

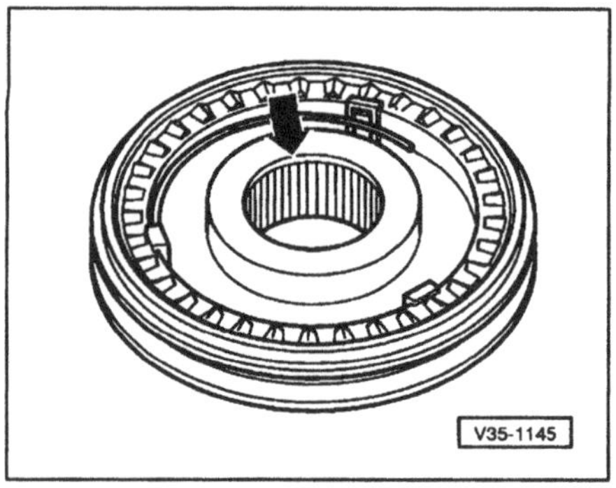

◄ **Fig. 26 3rd and 4th gear operating sleeve/synchro-hub, installation position**

Chamfer -arrow- faces 4th gear.

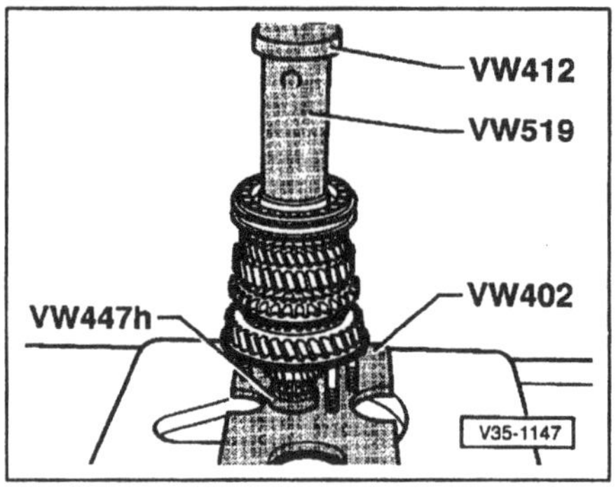

◄ **Fig. 27 3rd and 4th gear synchro-hub with operating sleeve, installing**

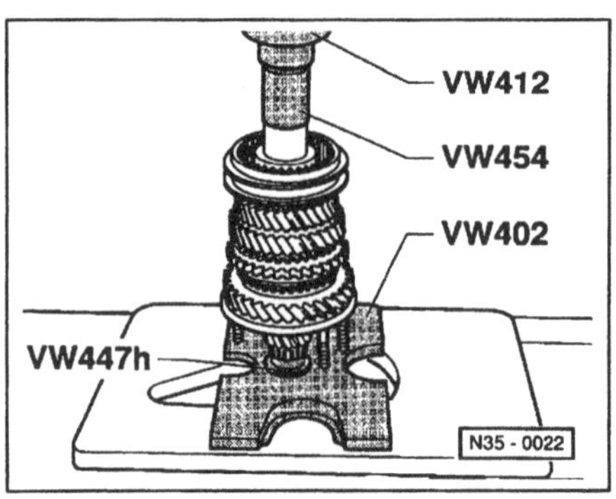

◄ **Fig. 28 4th gear needle roller bearing sleeve, installing**

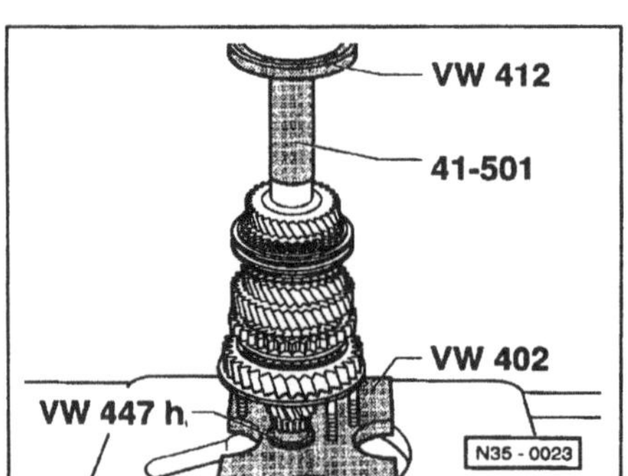

◄ **Fig. 29 Output shaft needle roller bearing sleeve, installing**

Output shaft, adjusting

(Determining shim thickness for output shaft)

It is necessary to readjust output shaft when:

- ♦ Final drive
- ♦ Clutch housing

or

- ♦ Tapered roller bearing

have been changed.

Adjustment overview ⇒ Page K 39-20

– Install small tapered roller bearing outer race together with a 0.65 mm (0.0256 in.) thick shim fully into clutch housing ⇒ Page K 35-30.

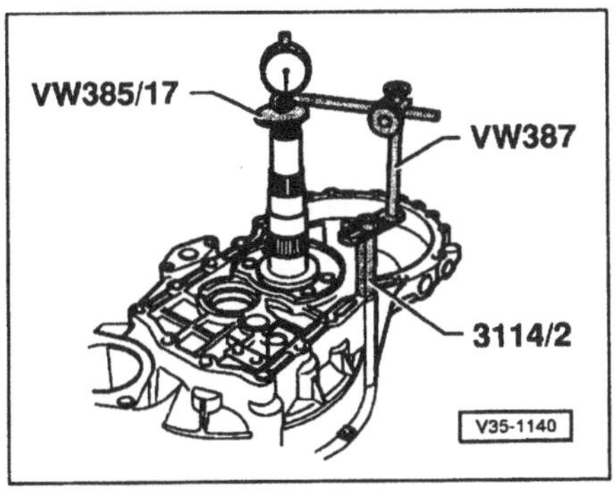

– Install output shaft and tighten bearing carrier to 25 Nm (18 ft lb) plus an additional 1/4 turn (90°).

– Mount dial indicator (3 mm (0.1181 in.) measuring range) and zero with 1 mm (0.0393 in.) preload.

– Move output shaft up and down and note dial indicator reading (example, 0.30 mm [0.0118 in.])

Note:

Do not rotate the output shaft when measuring or the bearing will settle and a false reading will be given.

Determining thickness of shim

The specified bearing preload is obtained by adding a constant figure to the reading obtained (0.30 mm [0.0118 in.]) and the thickness of the shim installed (0.65 mm [0.0256 in.]).

Example:

Shim installed	0.65 mm (0.0256 in.)
+ Reading	0.30 mm (0.0118 in.)
+ Preload (constant)	0.20 mm (0.0079 in.)
Thickness of shim	1.15 mm (0.0453 in.)

– Remove output shaft and remove small tapered roller bearing outer race ⇒ Page K 35-30

The following shims are available:

Size mm (in.)	Part No.
0.65 (0.0256)	020 311 391 P
0.70 (0.0276)	020 311 391 Q
0.75 (0.0295)	020 311 391
0.80 (0.0315)	020 311 391 A
0.85 (0.0335)	020 311 391 B
0.90 (0.0354)	020 311 391 C
0.95 (0.0374)	020 311 391 D
1.00 (0.0394)	020 311 391 E
1.05 (0.0413)	020 311 391 F
1.10 (0.0433)	020 311 391 G
1.15 (0.0453)	020 311 391 H
1.20 (0.0472)	020 311 391 J
1.25 (0.0492)	020 311 391 K
1.30 (0.0512)	020 311 391 L
1.35 (0.0351)	020 311 391 M
1.40 (0.0551)	020 311 391 N

Varying tolerances make it possible by measuring the shims to obtain the exact shim thickness required.

- Install small tapered roller bearing outer race together with correct shim (example, 1.15 mm [0.0453 in.] ⇒ Page K 35-30) and install output shaft.

- Tighten bearing carrier nuts in clutch housing to 25 Nm (18 ft lb) plus an additional 1/4 turn (90°).

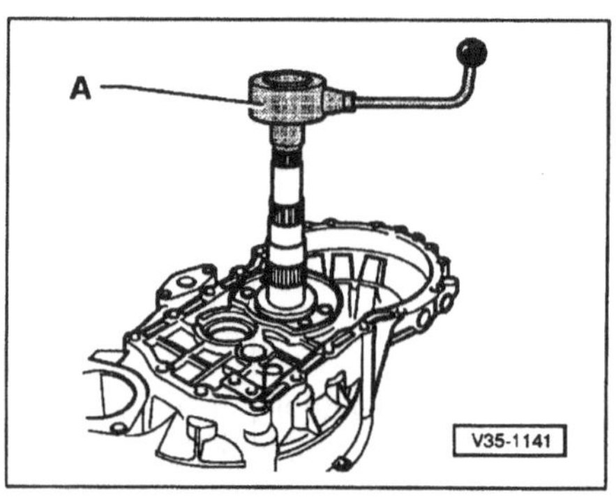

Inspecting bearing preload

◄ A – Torque wrench 0 ... 600 Ncm (0 ... 53 in. lb)

For new tapered roller bearings the bearing preload must be 130 ... 180 Ncm (11.5 ... 16 in. lb).

For used tapered roller bearings the bearing preload must be at least 30 Ncm (3 in. lb).

Note:

Lightly coat bearings with transmission oil before checking bearing preload.

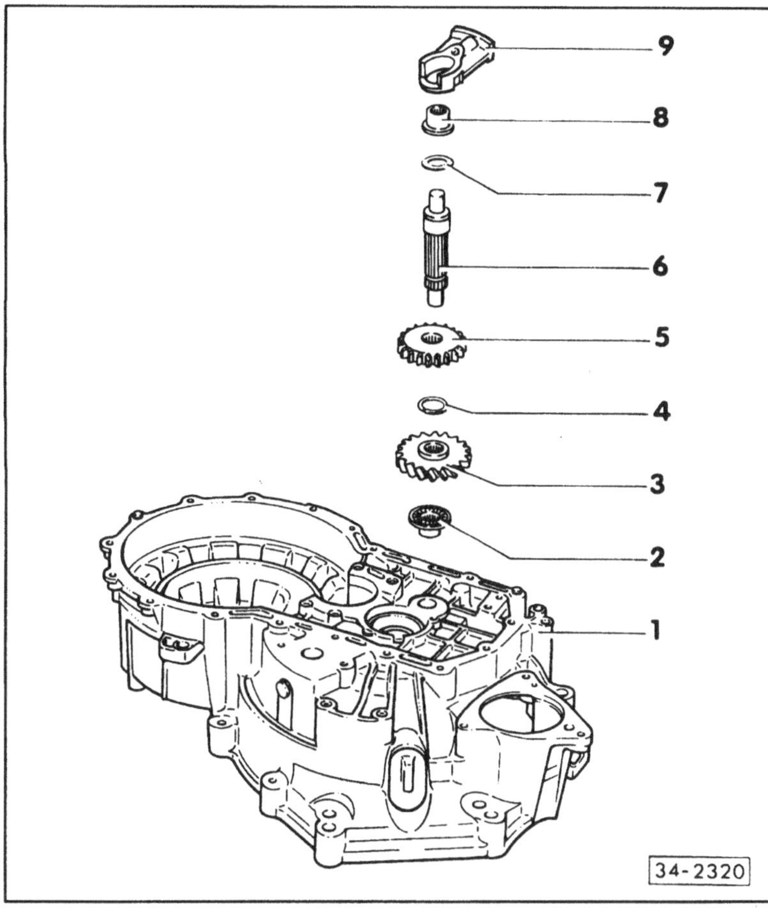

Reverse shaft, disassembling and assembling

Note:

After disassembling, always replace needle roller bearing in clutch housing and reverse shaft support.

1 - Clutch housing

2 - Needle roller bearing
- Removing ⇒ Fig. 1
- Installing ⇒ Fig. 2
- Modification: for modified reverse gear from transmission build date 18 09 0 ⇒ Fig. 3
- Installing ⇒ Fig. 4

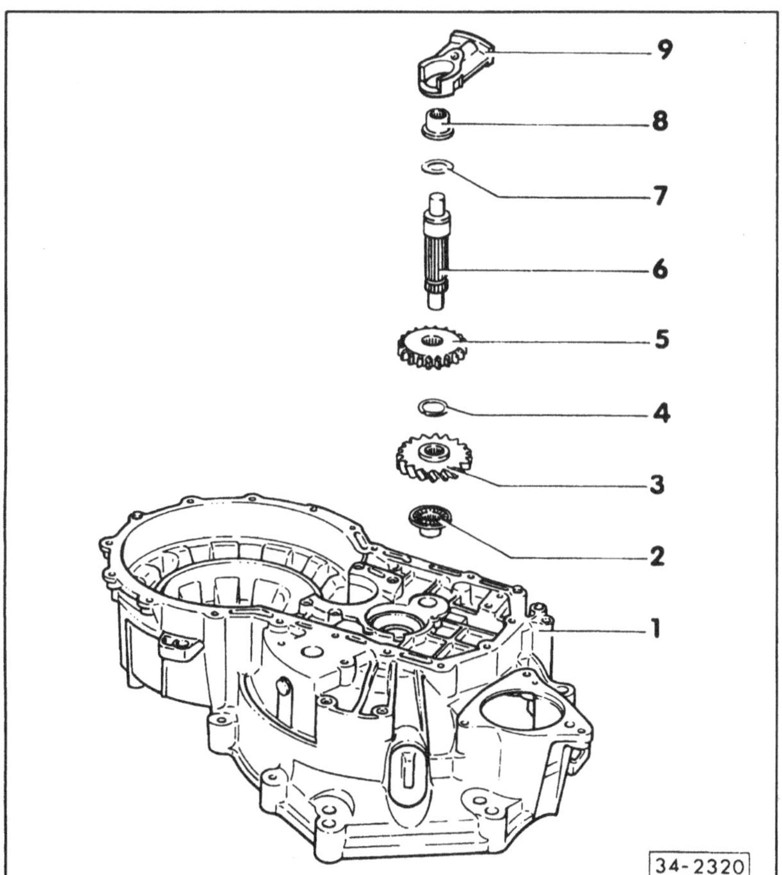

3 - Reverse gear
- Modification: Modified in area of needle roller bearing from transmission build date 18 09 0 ⇒ Fig. 3

Note:

When replacing previous reverse gear for new type, always install new needle roller bearing.

4 - Circlip
- Always replace after removing

5 - Reverse gear
- Before replacing remove circlip
- Shoulder faces reverse gear

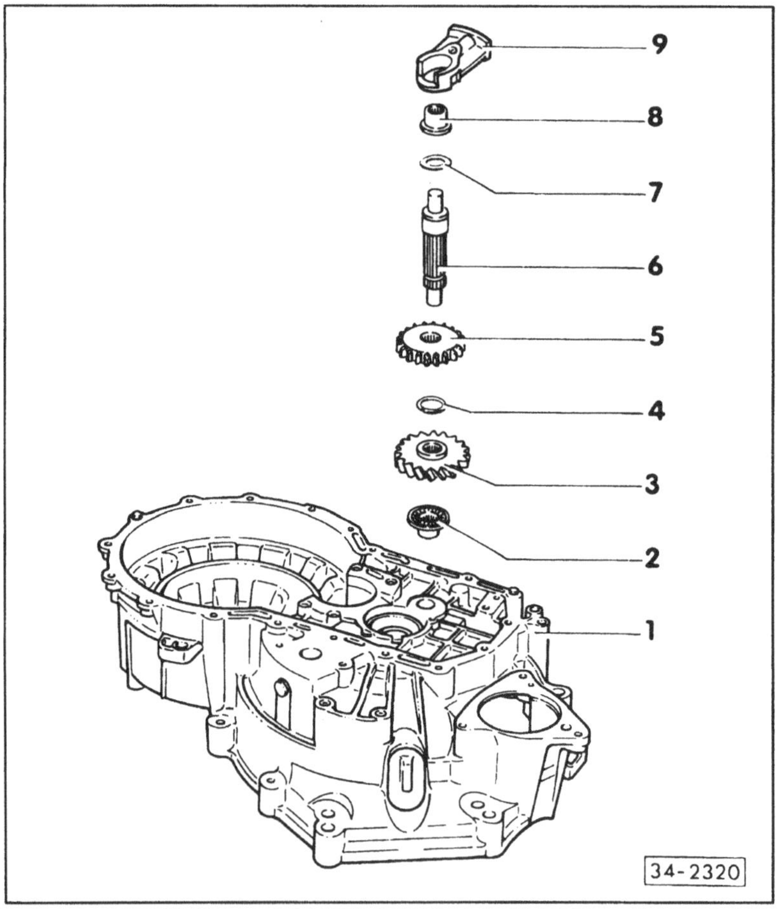

6 - Reverse gear shaft
- Removing ⇒ from Page K 34-75
- Modification:
  From transmission build date 15 08 0, modified in area of needle roller bearing -8- ⇒ Fig. 5

7 - Thrust washer
- Modification:
  Deleted from transmission build date 15 08 0 ⇒ Fig. 5

8 - Needle roller bearing
- Removing ⇒ Fig. 6
- Installing ⇒ Fig. 7

9 - Reverse gear shaft support
- Removing and installing ⇒ from Page K 34-75

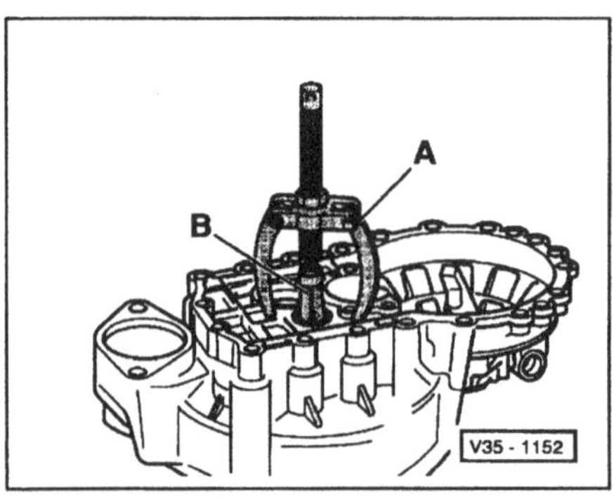

◄ **Fig. 1 Needle roller bearing, removing from clutch housing**

A – 22/1 Kukko support, or equivalent

B – 21/2 Kukko extractor 14.5 ... 18.5 mm, or equivalent

Note:

The needle bearing is destroyed during removal and must be replaced.

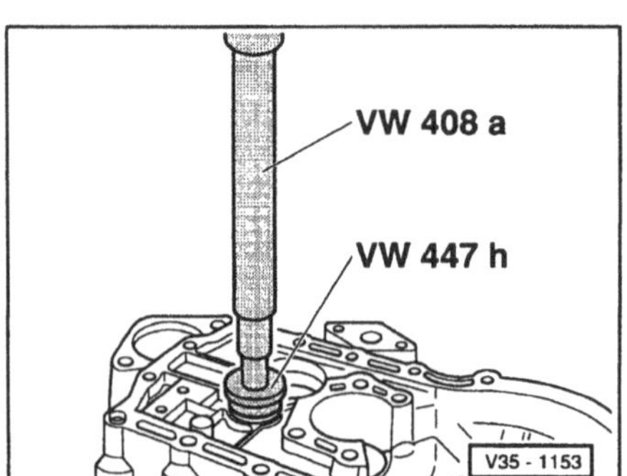

◄ **Fig. 2 Needle roller bearing, installing into clutch housing**

Up to transmission build date 17 09 0

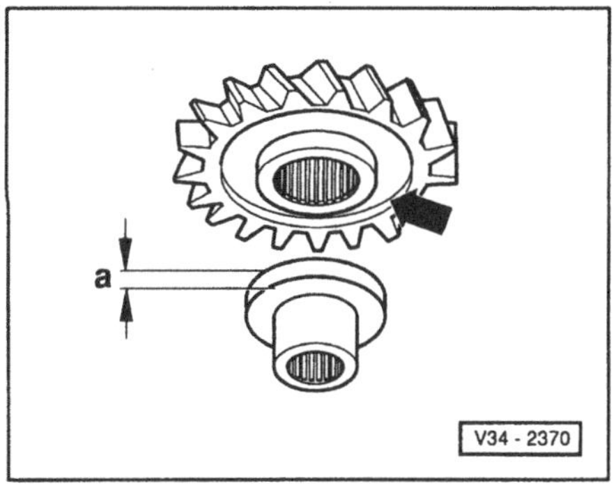

◄ **Fig. 3 Modified needle roller bearing and reverse gear**

From transmission build date 18 09 0.
The groove -arrow- on the clutch housing side of reverse gear has been deleted.
Also, the depth of the needle roller bearing sleeve collar has been reduced to prevent it making contact with the gear pinion.

Transmission build date	Dimension -a- mm (in.)
from 18 09 0	3.85 (0.1516)
up to 17 09 0	5.00 (0.1969)

Note:

Modified gear can be installed in earlier transmissions. When replacing an earlier reverse gear for a new type gear always install with a new type needle roller bearing.

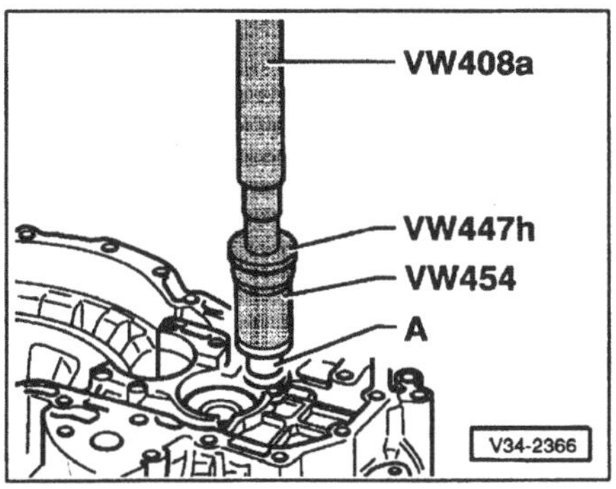

◄ **Fig. 4 Needle roller bearing -A-, installing into clutch housing**

From transmission build date 18 09 0

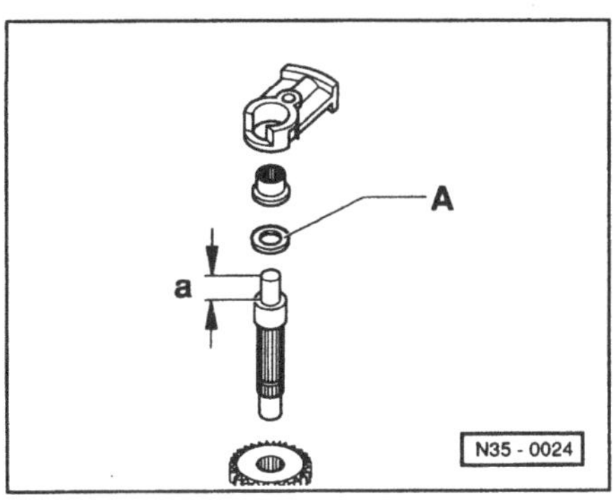

◄ **Fig. 5 Modified reverse gear shaft**

Reverse gear shaft, needle roller bearing spigot has been reduced (dimension -a-)

Transmission build date	Dimension -a- mm (in.)
from 15 08 0	20.0 (0.7874)
up to 14 08 0	21.0 (0.8661)

Washer -A- is then deleted

Note:

Modified reverse gear shaft can also be installed in previous transmissions. In this case the washer -A- must not be installed.

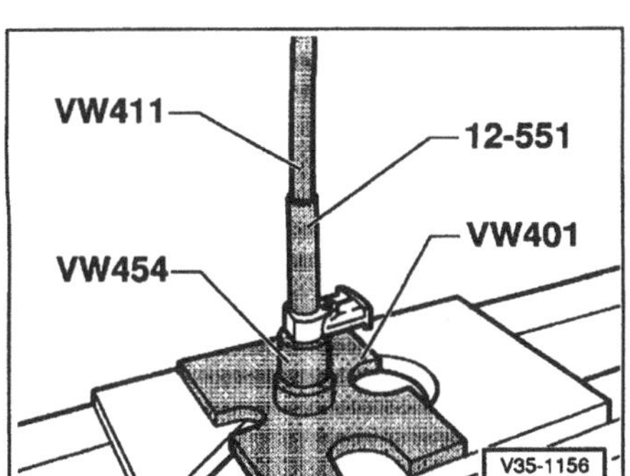

◄ **Fig. 6 Needle roller bearing, removing from reverse shaft support**

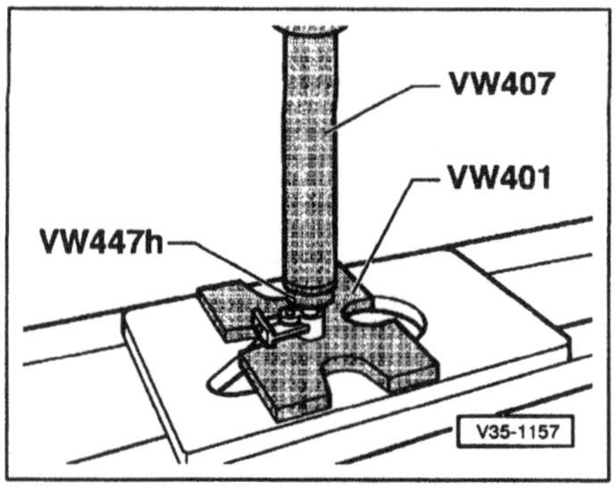

◄ Fig. 7 **Needle roller bearing, installing into reverse shaft support**

Oil seal for flange shaft or drive flange, removing and installing

(Transmission installed)

To replace left or right hand flange oil seals, corresponding drive shaft must be removed:

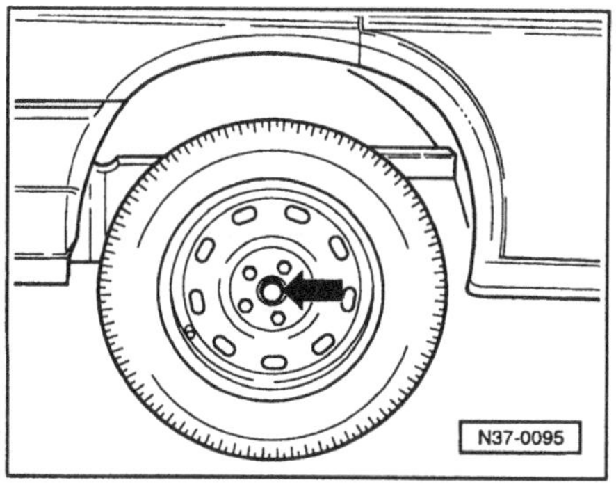

– Install 3250 support frame before lifting vehicle.
⇒ Suspension, Wheels, Brakes, Steering; Repair group 40; Servicing front suspension

◄ – Remove hex bolt -arrow- for drive shaft/wheel hub and loosen wheel securing bolts

– Remove noise insulation tray.

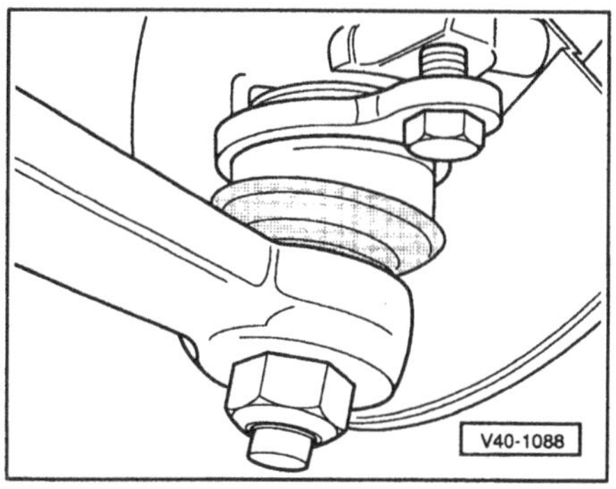

– Disconnect drive shaft from drive flange or flanged shaft.

– Remove wheel.

◄ – Disconnect wheel bearing housing/lower ball joint bolted connection.

– Remove bolt for shock absorber/stabilizer bar coupling link.

– Remove drive shaft from wheel bearing housing.

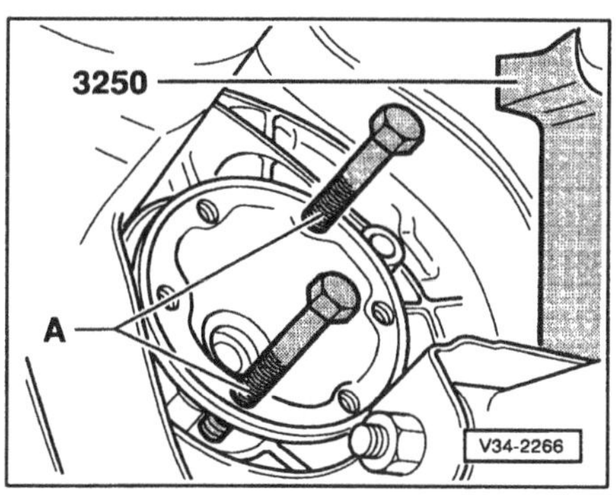

– Fully compress shock absorber.

When replacing left-hand oil seal:

– Removing drive shaft ⇒ Page K 39-1.

◄ – Remove flange shaft by turning 2 bolts -A- in alternately.

Note:

Align bolts vertically over oil drain plug. Transmission is equipped with 2 reinforcements, and bolts bear on reinforcements when being turned in.

– Always replace flange shaft circlip ⇒ Page K 34-89.

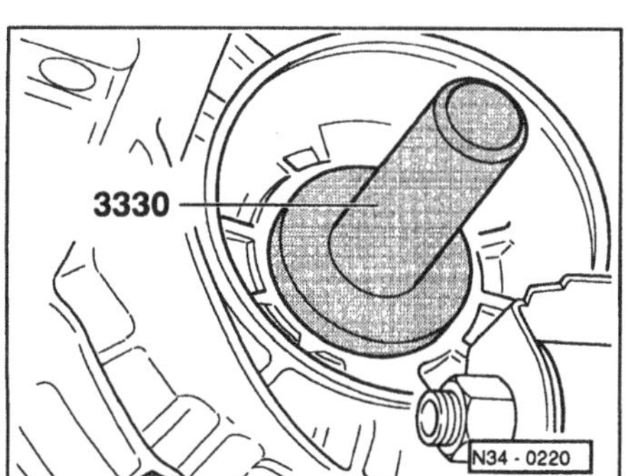

– Pry seal out with a screw driver.

◄ – Drive new seal into stop with 3330 press piece.

– Pack space between seal and dust lip with multi-purpose grease.

– Install flange shaft.

– Install drive shaft.

– Inspect transmission oil ⇒ Page K 34-59

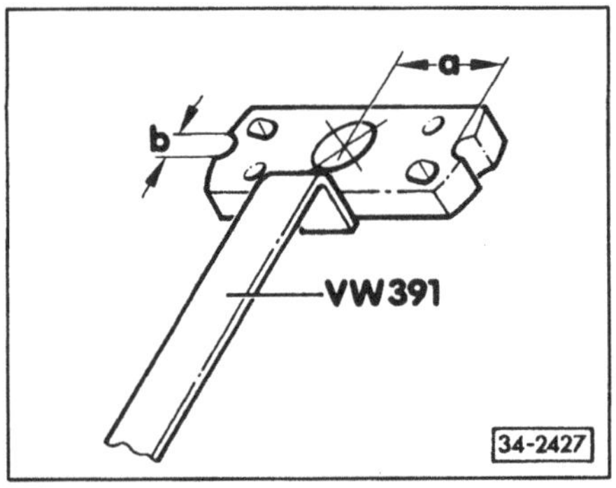

Replacing right-hand drive flange oil seal:

– Remove drive shaft ⇒ Page K 39-1.

– Pierce sealing cap in drive flange with a screwdriver and pry out.

– Remove drive flange circlip.

◄ Use VW 391 puller to remove drive flange.

– Before positioning VW 391 puller, cut notches in base plate sides.

Dimension -a- = 42 mm (1.1635 in.)

Dimension -b- = 10.3 mm (0.4055 in.)

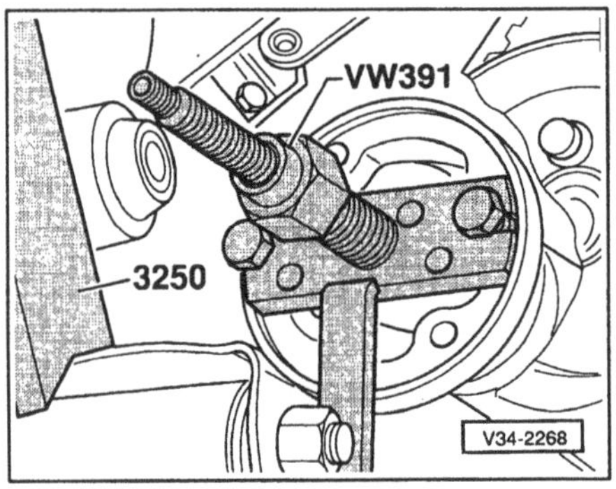

◄ – Remove drive flange with VW 391 puller.

– Carefully remove drive flange oil seal with VW 681 extractor lever.

Note:

Bearing housing must be replaced when seal seat is damaged.

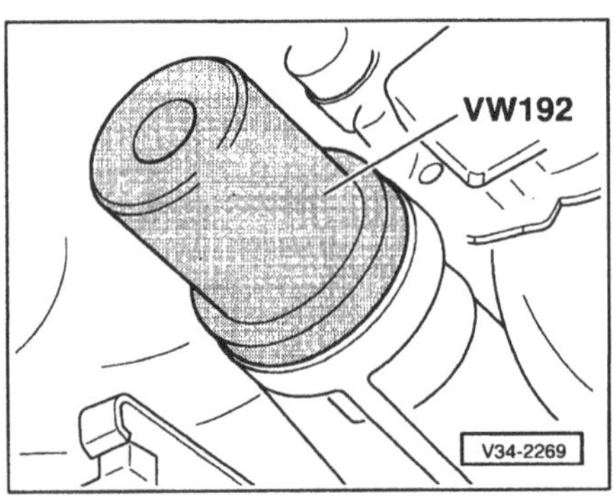

◄ – Drive oil seal onto stop with VW 192 arbor.

– Pack space between seal and dust lip with multi-purpose grease.

Installing right-hand drive flange.

– Heat drive flange in area of splines to 90°C (194°F). Temperature can be measured with VAG 1558 temperature tester.

> ***CAUTION!***
> ***Temperatures higher than 90°C (194°F) will damage the seal.***

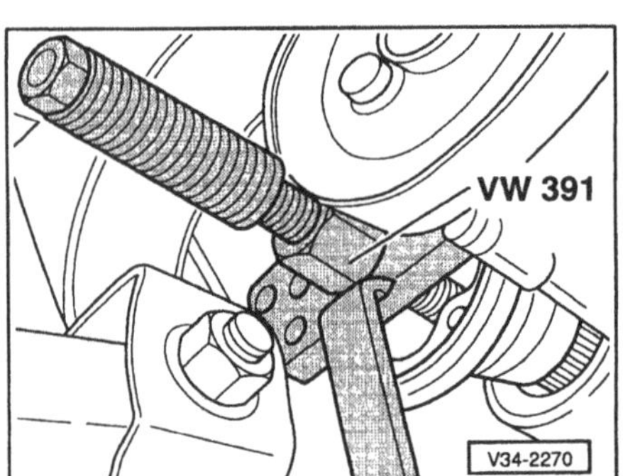

◄ – Install drive flange.

Note:

Before installing or removing, coat VW 391 puller spindle contact surfaces on base plate and output shaft with multi–purpose grease to prevent damage.

– Install new circlip and new end cap.

– Install drive shaft.

– Check transmission oil level ⇒ Page K 34-59

Tightening torques

Drive shaft to drive flange	45 Nm (33 ft lb)
Shock absorber to stabilizer	170 Nm (125 ft lb)
Wheel bearing housing to lower shaft joint	55 Nm (41 ft lb) plus an additional 1/4 turn (90°)
Drive shaft to wheel hub	200 Nm (148 ft lb)
Wheel bolts to wheel hub	160 Nm (118 ft lb)

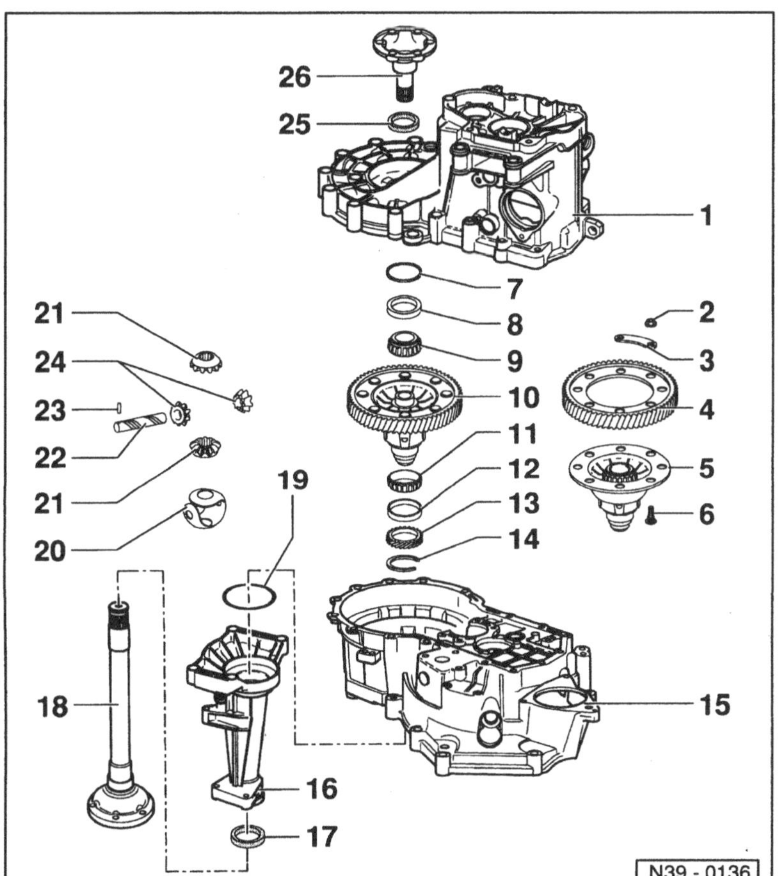

Differential, disassembling and assembling

Notes:

- *Heat tapered roller bearing inner race to 100°C (212°F) before installing.*
- *Tapered roller bearings are a matched set. Always replace together.*
- *Adjust differential if tapered roller bearing, differential housing, transmission housing, and/or clutch housing have been replaced ⇒ Page K 39-21.*

1 - Transmission housing

2 - Nut

- Tightening torque: 60 Nm (44 ft lb) plus an additional 1/4 turn (90°)

3 - Locking plate

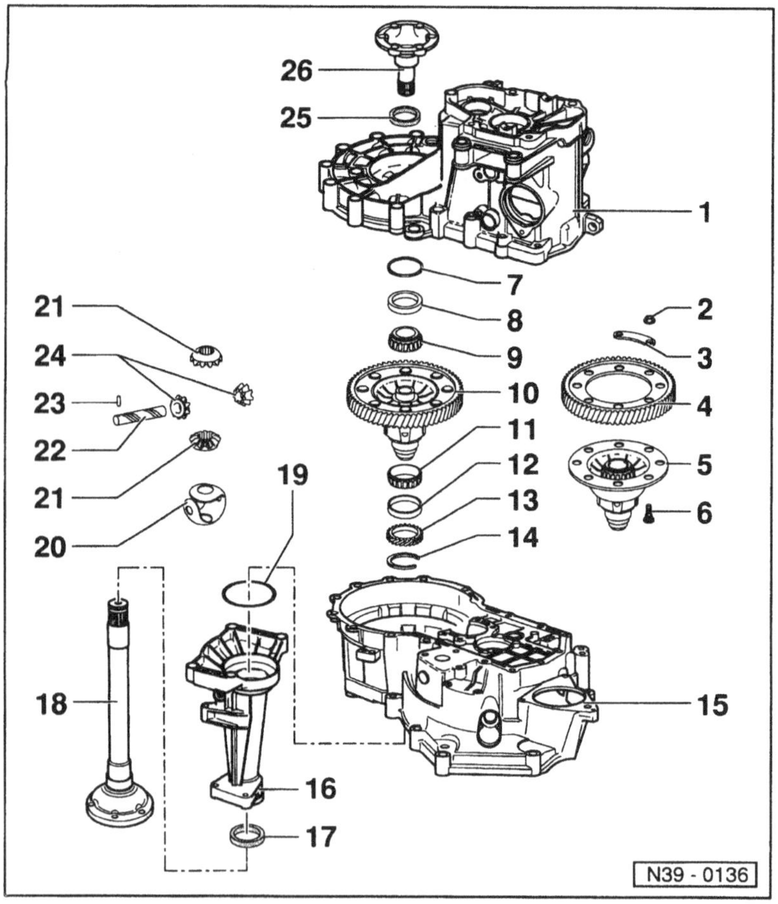

4 - Final drive gear
- Riveted in production
- Bolted ⇒ Fig. 12
- Heat to 100°C (212°F) before installing
- Removing ⇒ Fig. 10
- Install to differential housing ⇒ Fig. 11
- Is matched to output shaft, always replace as a set

5 - Differential housing
- Bolt to final drive gear wheel ⇒ Fig. 12

6 - Bolt

7 - Shim
- For differential
- Determining thickness ⇒ Page K 39-22

8 - Tapered roller bearing outer race
- Removing ⇒ Fig. 1
- Installing ⇒ Fig. 2

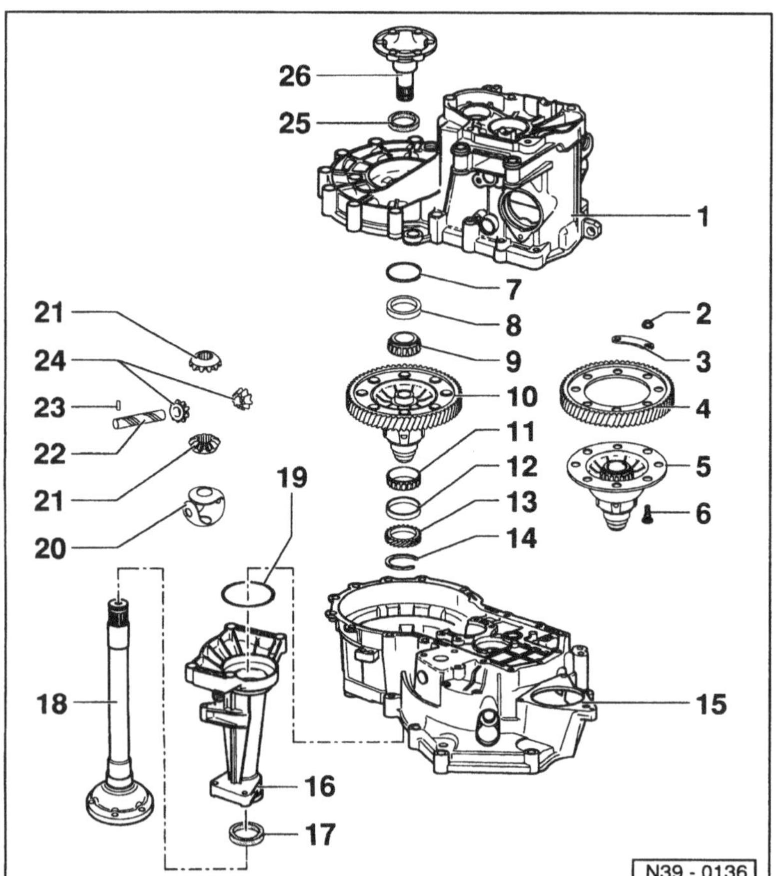

9 - Tapered roller bearing inner race
- Removing ⇒ Fig. 3
- Installing ⇒ Fig. 4

10 - Differential housing
- Before removing, remove output shaft -18- with drive flange out of locking device in differential
- With riveted final drive gear
- When replacing final drive gear, drill out rivet heads ⇒ Fig. 9
- Attaching final drive gear with bolts ⇒ Fig. 9

11 - Tapered roller bearing inner race
- Before removing, remove circlip -14- and speedometer drive gear -13-
- Removing ⇒ Fig. 5
- Installing ⇒ Fig. 4

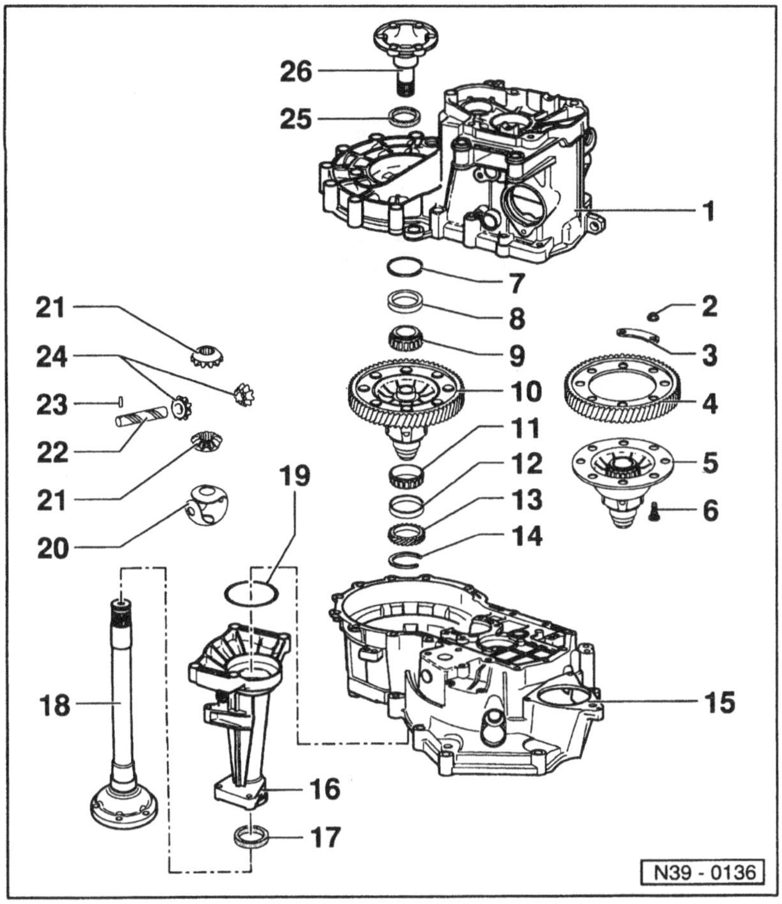

12 - Tapered roller bearing outer race
- Before removing, remove bearing housing -16-
- Removing ⇒ Fig. 7
- Installing ⇒ Fig. 8

13 - Speedometer drive gear
- After installing inner race, install onto stop of differential housing

14 - Circlip
- For speedometer drive gear
- Always replace

15 - Clutch housing

16 - Bearing housing
- For output shaft/right-hand drive flange
- Removing and installing ⇒ Page K 34-54

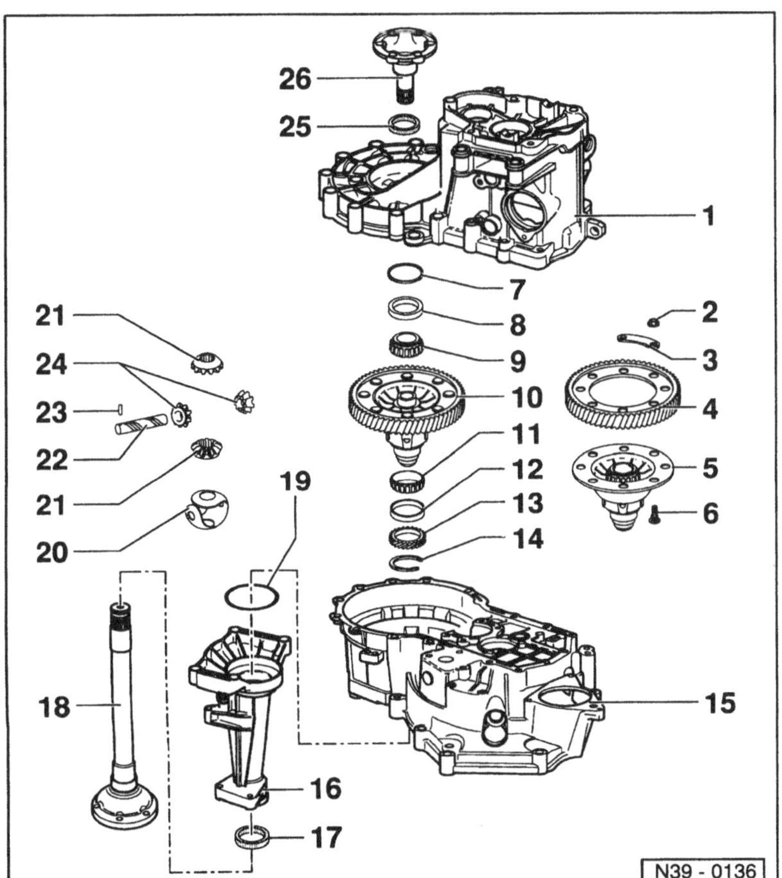

17 - Oil seal
- Replacing ⇒ Page K 39-1

18 - Output shaft with right-side drive flange
- Pull out of differential locking device together ⇒ Page K 34-107
- Separating and assembling output shaft and drive flange ⇒ Page K 34-108

19 - O-ring
- Always replace
- Insert into clutch housing groove with multi-purpose grease

20 - One-piece thrust washer
- Coat with transmission oil when installing

21 - Differential gears
- Installing ⇒ Fig. 16

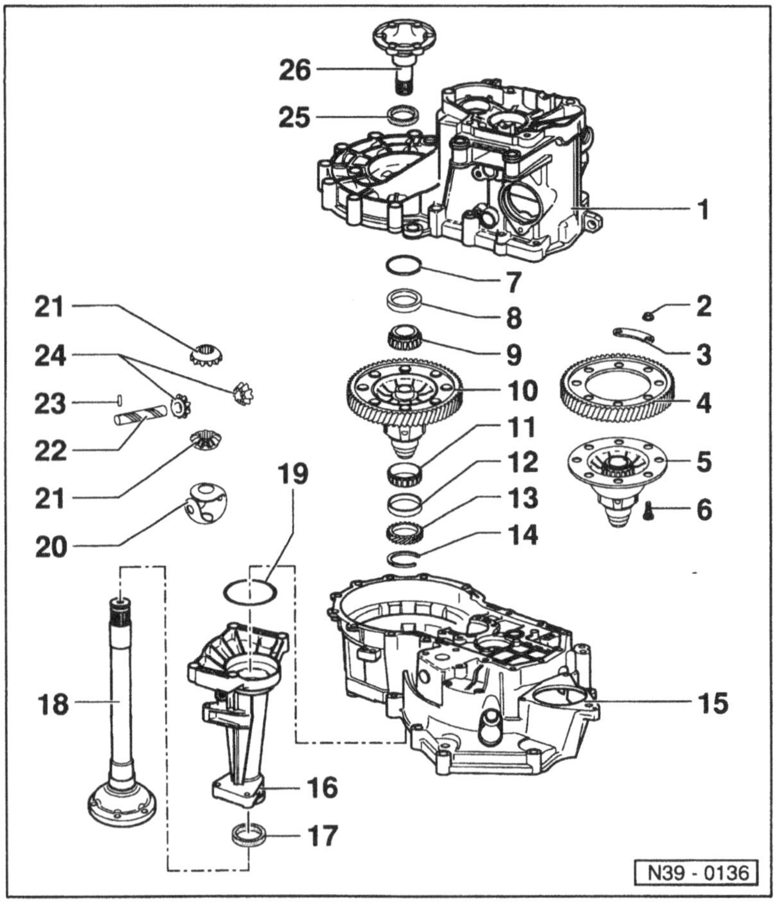

22 - Pinion shaft
- Drive out with drift

23 - Spring pin
- For securing pinion shaft
- Removing and installing ⇒ Fig. 13
- Modification: Longer from transmission build date 08 11 0 ⇒ Fig. 14
- Removing and installing ⇒ Fig. 15

24 - Differential pinion
- Installing ⇒ Fig. 16

25 - Oil seal
- Replacing ⇒ Page K 39-1

26 - Drive flange
- Removing and installing ⇒ from Page K 39-1

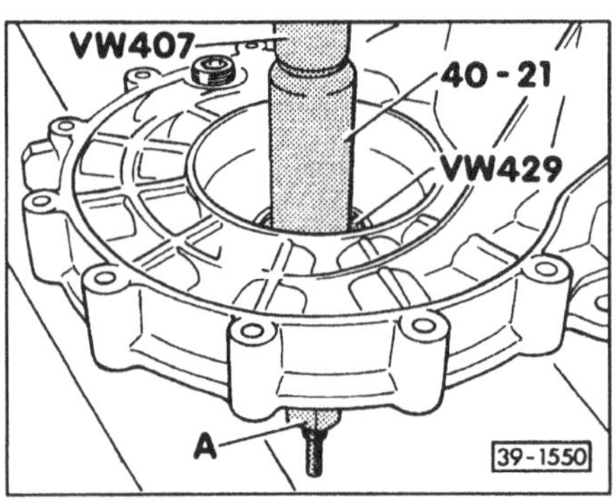

◄ **Fig. 1 Tapered roller bearing outer race, removing from transmission housing**

A - 21/7 Kukko extractor 46...58 mm, or equivalent

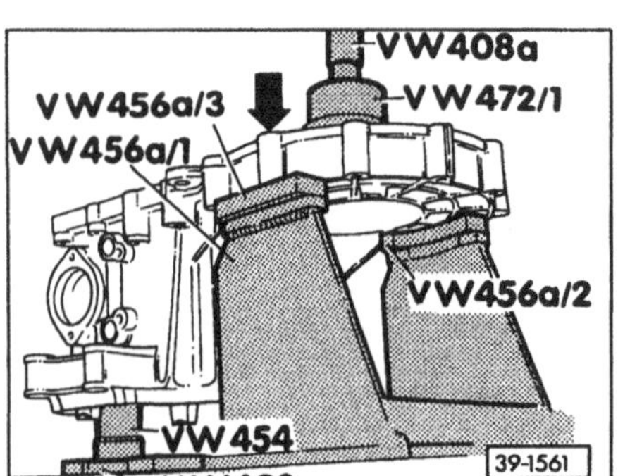

◄ **Fig. 2 Tapered roller bearing outer race, installing into transmission housing**

- Insert VW 456a/3 mounting with a web in transmission housing bolt hole -arrow-. Both webs of VW 456a/2 mounting will then grip in slots on both sides of oil drain hole.
- Install VW 472/1 thrust piece with cone inserted in outer race.

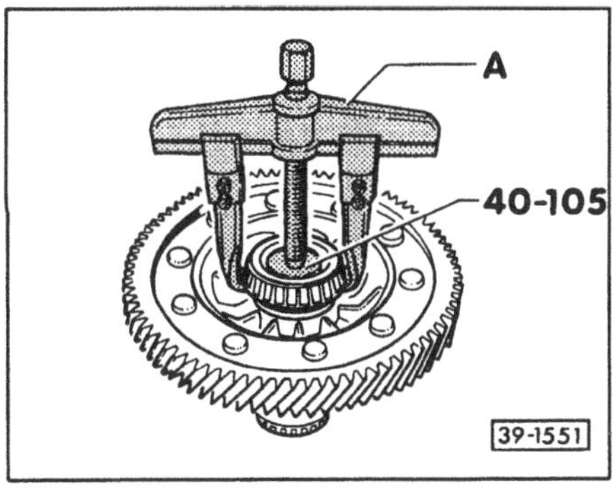

◄ **Fig. 3 Tapered roller bearing inner race, removing**

A - 20/10 Kukko two arm puller, or equivalent

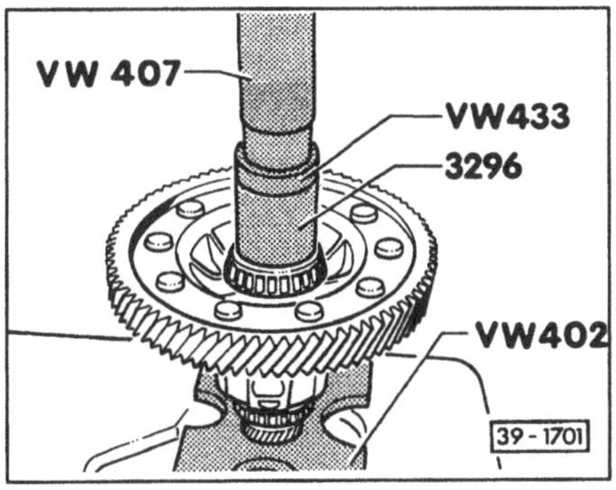

◄ **Fig. 4 Tapered roller bearing inner race, installing**

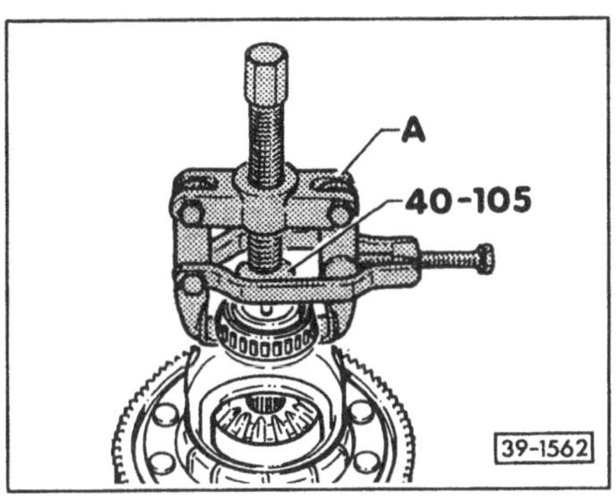

◄ **Fig. 5 Tapered roller bearing inner race, removing**

A - 204/2 Kukko two arm puller, or equivalent

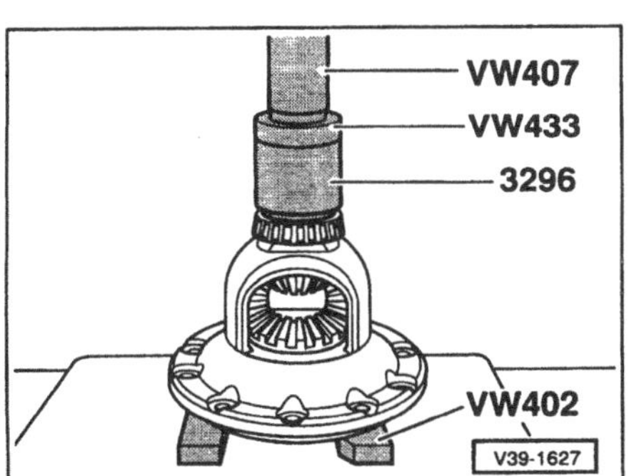

◄ **Fig. 6 Tapered roller bearing inner race, installing**

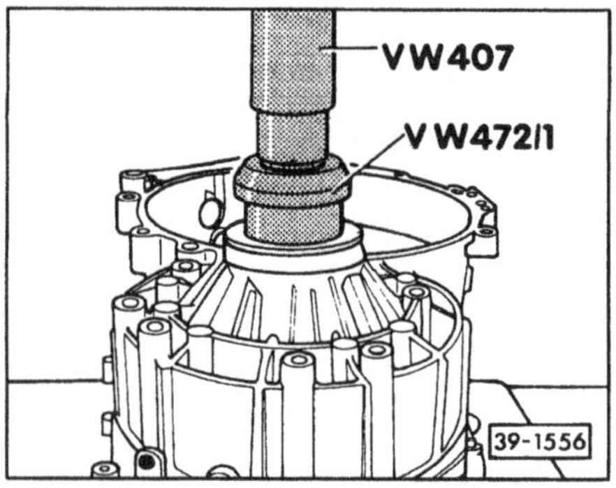

◄ **Fig. 7 Tapered roller bearing outer race, removing from clutch housing**

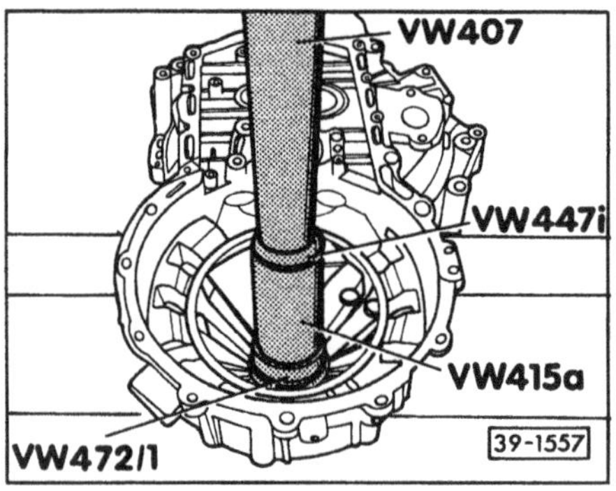

◄ **Fig. 8 Tapered roller bearing outer race, installing into clutch housing**

Shims are not installed on clutch housing end.

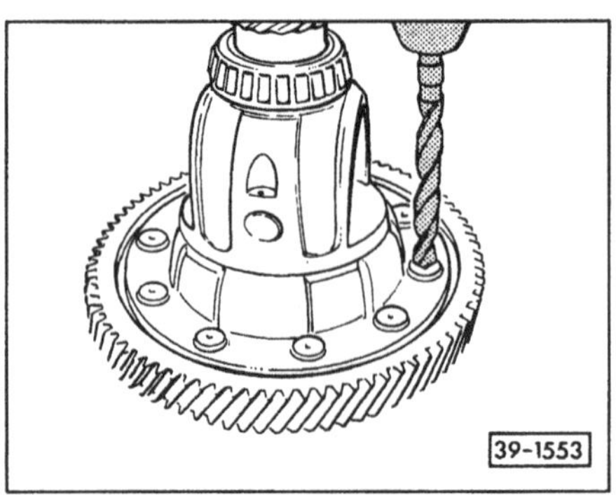

◄ **Fig. 9 Rivet heads, drilling out**

– Drill out rivet heads with 12 mm (0.472 in.) drill from countersunk side and drive rivets out with drift, or, if necessary, use 10-508 assembly mandrel to remove.

Note:

Clean differential before and after drilling to protect tapered roller bearing from drilling burrs.

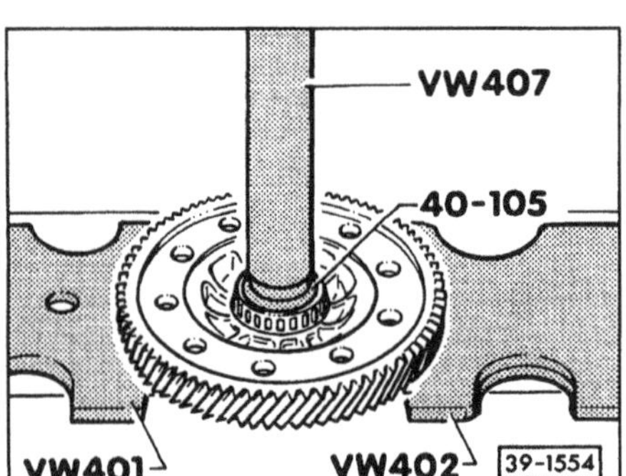

◄ **Fig. 10 Final drive gear, removing**

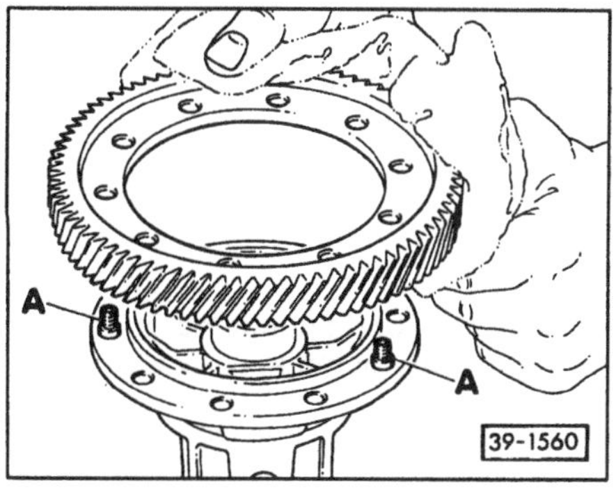

Fig. 11 Heating final drive gear wheel to approximately 100°C (212°F) and installing

Use bolts -A- from repair kit to guide final drive gear into position during installation.

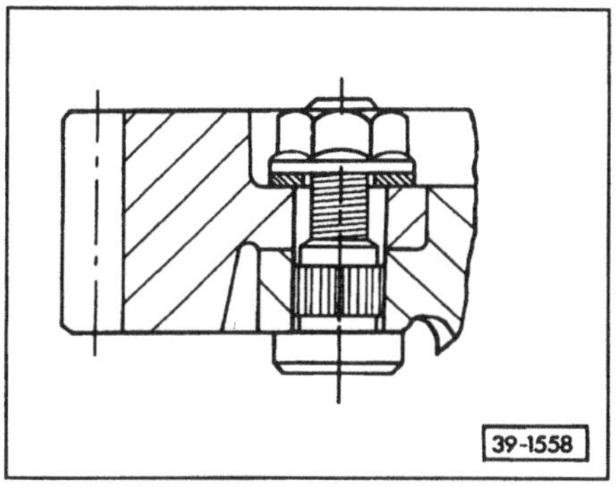

Fig. 12 Bolting final drive gear and differential housing together

Use special bolts with locking plates and nuts ⇒ parts catalog.

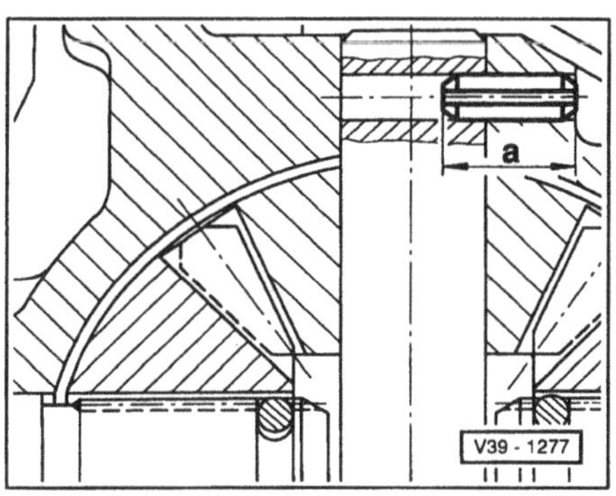

Fig. 13 Spring pin for pinion shaft, removing and installing

Up to transmission build date 07 11 0

Removing:

- Drive into pinion shaft.
- Drive-out of pinion shaft after shaft has been removed.

Installing:

- Drive-in flush with differential housing.

Length of spring pin -a-14 mm (0.552 in.)

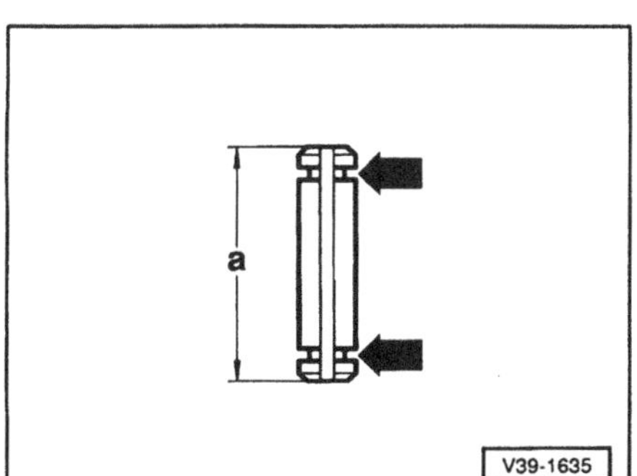

Fig. 14 Longer spring pin for pinion shaft

Transmission build date	Dimension -a- mm (in.)
from 08 11 0	28.5 (1.1220)
up to 07 11 0	14.0 (0.5512)

- The new spring pin has two circumferential grooves -arrows-.

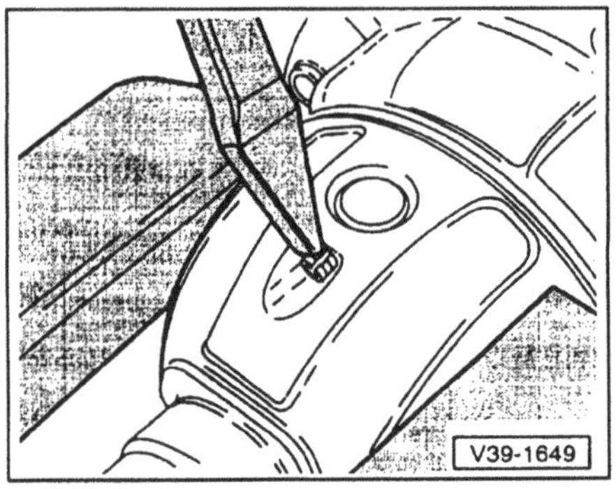

Fig. 15 Spring pin for pinion shaft, removing and installing

From transmission build date 08 11 0

Removing:

- Cover tapered roller bearing inner race and speedometer drive gear to avoid possible damage and entry of metal particles.
- Drive-out spring pin with chisel, inserting chisel into circumferential groove.

Installing:

- Drive fully into differential housing.

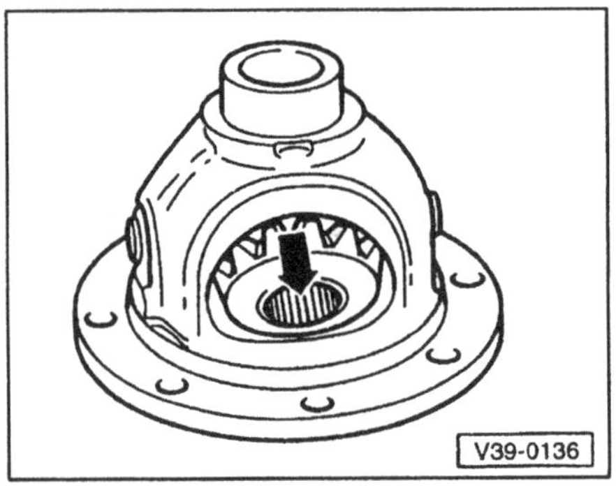

Fig. 16 Differential bevel gears, installing

- Coat one-piece thrust washer with transmission oil and install.
- Install differential gears, drive in shaft and secure with spring pin
- Drive in new spring pin ⇒ Fig. 13 or Fig. 15
- Install differential pinions with 180° offset and pivot in direction of -arrow-.

Adjustment overview

Note:

If repairs have been carried out to transmission, it is only necessary to adjust input shaft, output shaft, or differential if components have been replaced which have a direct effect on adjustment of transmission. Refer to the following table to avoid unnecessary adjustment:

		Required adjustment:		
		Input shaft ⇒ Page K 35-14	Output shaft ⇒ Page K 35-45	Differential ⇒ Page K 39-21
Parts replaced:	Transmission housing	x		x
	Clutch housing	x	x	x
	Input shaft	x		
	Output shaft (final drive)		x	
	Differential housing			x
	Input shaft tapered roller bearing	x		
	Output shaft tapered roller bearing		x	
	Differential tapered roller bearing			x
	4th speed gear	x		

Differential, adjusting

The differential must be adjusted when:

- ♦ Transmission housing
- ♦ Clutch housing
- ♦ Differential housing

or the

- ♦ Differential tapered roller bearing

have been replaced.

Adjustment overview ⇒ Page K 39-20.

– Installing tapered roller bearing outer race without shim into clutch housing using 472/1 thrust piece ⇒ Page K 39-15.

Note:

Inner and outer tapered roller bearing races are a matched pair. Do not interchange.

– Installing tapered roller bearing outer race without shim into transmission housing using 472/1 thrust piece ⇒ Page K 39-12.

– Install differential into clutch housing

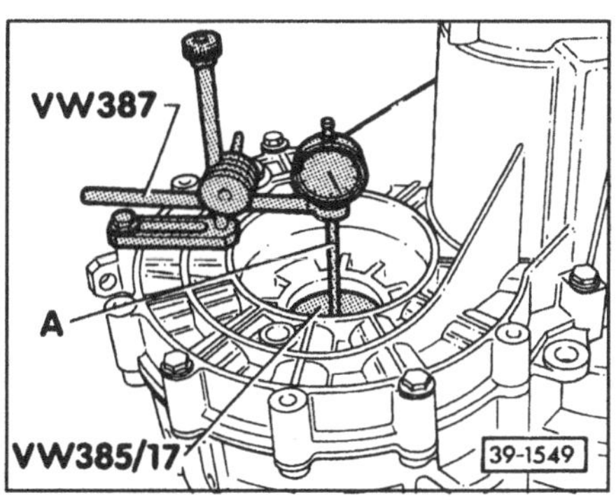

– Install transmission housing and torque 5 bolts to 25 Nm (18 ft lb) plus an additional 1/4 turn (90°).

◄ – Mount dial indicator and Zero with 1 mm (0.0393 in.) preload.

– Move differential up and down and note dial indicator reading (Example, 0.70 mm [0.0276 in.]).

A - 85 mm (3.3456 in.) dial indicator extension

Note:

Do not rotate differential when measuring or bearing will settle and a false reading will be given.

Determining shim thickness

The specified bearing preload is obtained by adding a constant figure (0.40 mm [0.0157 in.]) to the reading obtained.

Example:

Reading	0.70 mm (0.0276 in.)
+ Preload (constant figure)	0.40 mm (0.0157 in.)
Thickness of shim =	1.10 mm (0.0433 in.)

– Remove transmission housing and remove tapered roller bearing outer race from transmission housing
⇒ Page K 39-12

– Insert shims of correct thickness, thickest shim first.

The following shims are available:

Size mm (in.)	Part No.
0.65 (0.0256)	02A 409 210
0.70 (0.0276)	02A 409 210 A
0.75 (0.0295)	02A 409 210 B
0.80 (0.0315)	02A 409 210 C
0.85 (0.0335)	02A 409 210 D
0.90 (0.0354)	02A 409 210 E
0.95 (0.0374)	02A 409 210 F
1.05 (0.0413)	02A 409 210 G
1.05 (0.0413)	02A 409 210 H
1.10 (0.0433)	02A 409 210 J
1.15 (0.0453)	02A 409 210 K
1.20 (0.0472)	02A 409 210 L
1.25 (0.0492)	02A 409 210 M

If size of shim required is larger than those listed in table, two shims totalling the correct figure can be installed.

Varying tolerances make it possible by measuring shims to obtain exact shim thickness required.

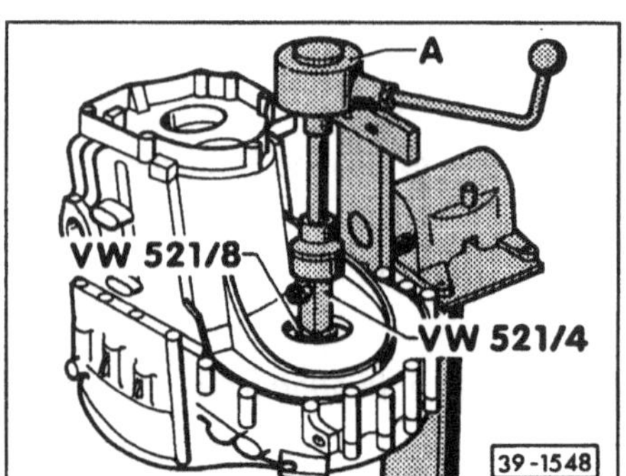

– Install outer race again and tighten transmission housing to torque specifications.

◄ **Checking bearing preload**

A – Torque wrench 0 ... 600 Ncm (0 ... 53 in. lb)

– Coat tapered roller bearing with gear oil.

Bearing preload:

♦ New tapered roller bearings 120 ... 320 Ncm (11 ... 28 in. lb)

♦ Used tapered roller bearings at least 30 Ncm (3 in. lb)

Manual Transmission 02G

Section L

WARNING!
Before proceeding with any work, please read the warnings and cautions immediately following this index. Also read any warning and caution that accompanies a procedure in the manual.

Please read these WARNINGS and CAUTIONS before proceeding with maintenance and repair work.

WARNING—

- Some repairs may be beyond your capability. If you lack the skills, tools and equipment, or a suitable workplace for any procedure described in this manual, we suggest you leave such repairs to an authorized Volkswagen dealer service department, or other qualified shop.

- Volkswagen is constantly improving its cars. Sometimes these changes, both in parts and specifications, are made applicable to earlier models. Therefore, before starting any major jobs or repairs to components on which passenger safety may depend, consult your authorized Volkswagen dealer about the Technical Bulletins that may have been issued since the editorial closing of this manual.

- Do not re-use any fasteners that are worn or deformed in normal use. Many fasteners are designed to be used only once and become unreliable and may fail when used a second time. This includes, but is not limited to, nuts, bolts, washers, self–locking nuts or bolts, circlips and cotter pins. Always replace these fasteners with new parts.

- Never work under a lifted car unless it is solidly supported on stands designed for the purpose. Do not support a car on cinder blocks, hollow tiles or other props that may crumble under continuous load. Never work under a vehicle that is supported solely by a jack. Never work under the car while the engine is running.

- If you are going to work under a car on the ground, make sure that the ground is level. Block the wheels to keep the car from rolling. Disconnect the battery negative terminal (ground strap) to prevent others from starting the vehicle while you are under it.

- Never run the engine unless the work area is well ventilated. Carbon monoxide (CO) kills.

- Do not attempt to work on your vehicle if you do not feel well. You increase the danger of injury to yourself and others if you are tired, upset or have taken medicine or any other substance that may impair you or keep you from being fully alert.

- Disconnect the battery negative (–) terminal (ground strap) whenever you work on the fuel system or the electrical system. Do not smoke or work near heaters or other fire hazards. Keep an approved fire extinguisher handy.

- Batteries produce explosive hydrogen gas. Keep sparks, lighted matches and open flame away from the top of the battery. If hydrogen gas escaping from the cap vents is ignited, it will ignite gas trapped in the cells and cause the battery to explode.

- Connect and disconnect battery cables, jumper cables or a battery charger only with the ignition switched off, to prevent sparks. Do not quick–charge the battery (for boost starting) for longer than one minute, and do not allow charging voltage to exceed 16.5 volts. Wait at least one minute before boosting the battery a second time.

- Illuminate your work area adequately but safely. Use a portable safety light for working inside or under the car. Make sure the bulb is enclosed by a wire cage. The hot filament of an accidentally broken bulb can ignite spilled fuel or oil.

- Tie long hair behind your head. Do not wear a necktie, a scarf, loose clothing, or a necklace when you work near machine tools or running engines. If your hair, clothing, or jewelry were to get caught in the machinery, severe injury could result.

- Finger rings, bracelets and other jewelry should be removed so that they cannot cause electrical shorts, get caught in running machinery, or be crushed by heavy parts.

- Catch draining fuel, oil, or brake fluid in suitable containers. Do not use food or beverage containers that might mislead someone into drinking from them. Store flammable fluids away from fire hazards. Wipe up spills at once, but do not store the oily rags, which can ignite and burn spontaneously.

- Always observe good workshop practices. Wear goggles when you operate machine tools or work with battery acid. Gloves or other protective clothing should be worn whenever the job requires working with harmful substances.

- The Air-Conditioning (A/C) system is filled with a chemical refrigerant, which is hazardous. The A/C system should be serviced only by trained technicians using approved refrigerant recovery/recycling equipment, trained in related safety precautions, and familiar with regulations governing the discharging and disposal of automotive chemical refrigerants.

- Do not expose any part of the A/C system to high temperatures such as an open flame. Excessive heat will increase system pressure and may cause the system to burst.

- Some aerosol tire inflators are highly flammable. Be extremely cautious when repairing a tire that may have been inflated using an aerosol tire inflator. Keep sparks, open flame or other sources of ignition away from the tire repair area. Inflate and deflate the tire at least four times before breaking the bead from the rim. Completely remove the tire from the rim before attempting any repair.

- Friction materials such as brake or clutch discs may contain asbestos fibers. Do not create dust by grinding, sanding, or by cleaning with compressed air. Avoid breathing asbestos fibers and asbestos dust. Breathing asbestos can cause serious diseases such as asbestosis or cancer, and may result in death.

- Greases, lubricants and other automotive chemicals contain toxic substances, many of which are absorbed directly through the skin. Read manufacturer's instructions and warnings carefully. Use hand and eye protection. Avoid direct skin contact.

- An airbag is operated by an explosive device. Handled improperly or without adequate safeguards, it can be accidently activated and cause serious personal injury. To prevent personal injury or airbag system failure, **only trained Volkswagen dealer service technicians** should test, disassemble or service the airbag system.

WARNINGS and CAUTIONS continued on next page ▶

Please read these WARNINGS and CAUTIONS before proceeding with maintenance and repair work.

WARNING (continued) —

- ALWAYS disconnect the airbag power supply before using a computer memory saver. Failure to disconnect the airbag power supply may result in accidental activation of the airbag.
- The airbag unit should not be removed, nor should any work be performed on the airbag system, without first reviewing the warnings and cautions that accompany the airbag system repair information in this manual.
- Never apply stickers or any other coverings on the steering wheel. Do not let chemical cleaners, oil or grease come into contact with the vinyl covering of the airbag unit.
- Never open or otherwise attempt to repair airbag system parts. Always use new parts. Never leave airbag parts or the partially disassembled airbag system unattended.
- Never use a test light or volt meter to conduct electrical tests on the airbag system. The airbag system must only be tested by trained Volkswagen Service technicians using special airbag tester equipment. The airbag system must only be tested while it is installed in the car.
- Do not expose an airbag unit to temperatures above 194°F (90°C) even for a brief period, while handling it during the repair process. Keep the unit clear of heat sources such as hot plates, soldering irons, heat lamps, welding equipment, or the like.
- Always replace airbag system components that have been mechanically damaged (bubbles, cracks, etc.).
- Always set down a removed airbag unit so that the horn pad is facing upward.
- When driving or riding in an airbag-equipped vehicle, never hold test equipment in your hands or lap while the vehicle is in motion. Objects between you and the airbag can increase the risk of injury in an accident.
- The storage, transportation and disposal of airbag units are subject to the laws for flammable solids.

CAUTION—

- If you lack the skills, tools and equipment, or a suitable workshop for any procedure described in this manual, we suggest you leave such repairs to an authorized Volkswagen dealer or other qualified shop. We especially urge you to consult an authorized Volkswagen dealer before beginning repairs on any car that may still be covered wholly or in part by any of the extensive warranties issued by Volkswagen of America, Inc.
- Volkswagen is constantly improving its cars and sometimes these changes, both in parts and specifications, are made applicable to earlier models. Therefore, part numbers listed in this manual are for reference only. Always check with your authorized Volkswagen dealer parts department for the latest information.
- Before starting a job, make certain that you have all the necessary tools and parts on hand. Read all the instructions thoroughly, do not attempt shortcuts. Use tools that are appropriate to the work and use only replacement parts meeting Volkswagen specifications. Makeshift tools, parts and procedures will not make good repairs.
- Use pneumatic and electric tools only to loosen threaded parts and fasteners. Never use these tools to tighten fasteners, especially on light alloy parts. Always use a torque wrench to tighten fasteners to the tightening torque specification listed.
- Be mindful of the environment and ecology. Before you drain the crankcase, find out the proper way to dispose of the oil. Do not pour oil onto the ground, down a drain, or into a stream, pond, or lake. Consult local ordinances that govern the disposal of wastes.
- When connecting cables or ground straps, all connections must be cleaned to bare metal. Be sure mounting bolts are tight.
- Do not replace fuse links with fuse links of other gauge or with standard wire. Always replace the entire length of the fuse link.
- Before doing any electrical welding on vehicles equipped with Anti-lock Brakes (ABS), disconnect the battery negative terminal (ground strap) and the ABS Control Module connector.
- On cars equipped with ABS, the ABS control unit cannot withstand temperatures from a paint–drying booth or a heat lamp in excess of 203°F (95°C) and should not be subjected to temperatures in excess of 185°F (85°C) for more than two hours.
- On cars equipped with anti-theft radios, make sure you know the correct radio activation code before disconnecting the battery or removing the radio. If the wrong code is entered when power is restored, the radio may lock up and be rendered inoperable, even if the correct code is then entered.
- When boost-charging the battery, first remove the fuses for the Engine Control Module (ECM), the Transmission Control Module (TCM), the ABS Control Module, and the On-Board computer. In cases where one or more of these components is not separately fused, disconnect the Control Module connector(s).
- Do not quick-charge the battery (for boost starting) for longer than one minute, and do not exceed 16.5 volts at the battery with the boosting cables attached. Wait at least one minute before boosting the battery a second time.

◀ **WARNINGS and CAUTIONS** continued from previous page

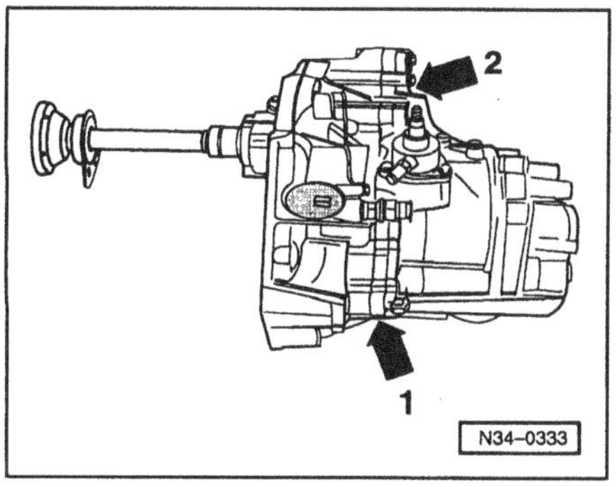

Transmission identification

Manual transmission 02G is installed as a 5-speed transmission with 5-cylinder engine.

Applications
⇒ Page L 00-3.

◄ **Location on transmission**

Code letters and date of manufacture –arrow 1– Manual transmission 02G –arrow 2–

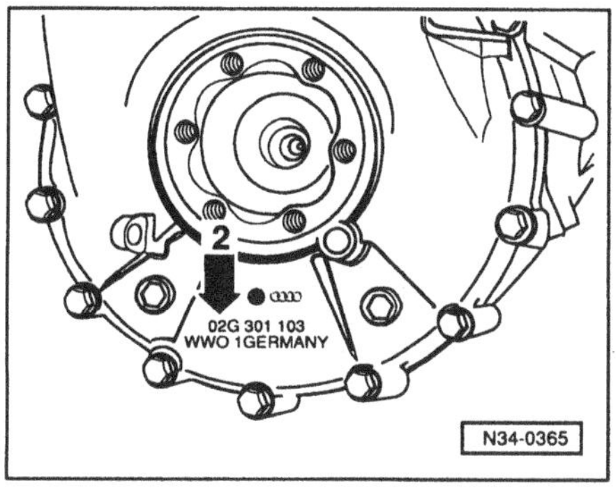

◄ **Manual transmission 02G –arrow–**

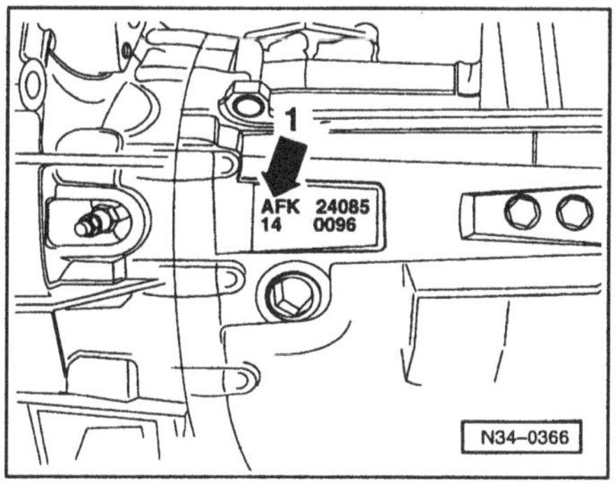

◄ **Code letters and date of of transmission manufacture –arrow 1–**

Example:	AFK	24	08	5
	I	I	I	I
	I	I	I	I
	Code letters	Day	Month	Year (1995) of manufacture

Additional data depends on manufacture.

Note:

Transmission code letters are also shown on vehicle identification plates.

Code letters, transmission allocation, ratios, capacities

Manual transmission		5-speed 02G		
Engine code		**AFK**		
Manufactured	from to	08.95		
Application	Model	EuroVan 1997		
	Engine	2.5 L (100 hp) Turbo Diesel		
Ratio: $Z_2 : Z_1$	Final drive	82 : 21 = 3.905		
	1st gear	43 : 11 = 3.909		
	2nd gear	36 : 17 = 2.118		
	3rd gear	35 : 26 = 1.346		
	4th gear	35 : 36 = 0.972		
	5th gear	35 : 48 = 0.729		
	6th gear	–		
	Reverse gear	37 : 23 x 23 : 10 = 3.700		
	Speedometer drive	13 : 22 = 0.591		

Engine code	**AFK**		
Capacity			
Specification	Gear oil G50 SAE 75W90 (synthetic oil)		
Clutch control	hydraulic		
Clutch plate	219 mm		
Drive shaft flange diameter	108 mm		
Overall ratio in top gear	2.847		

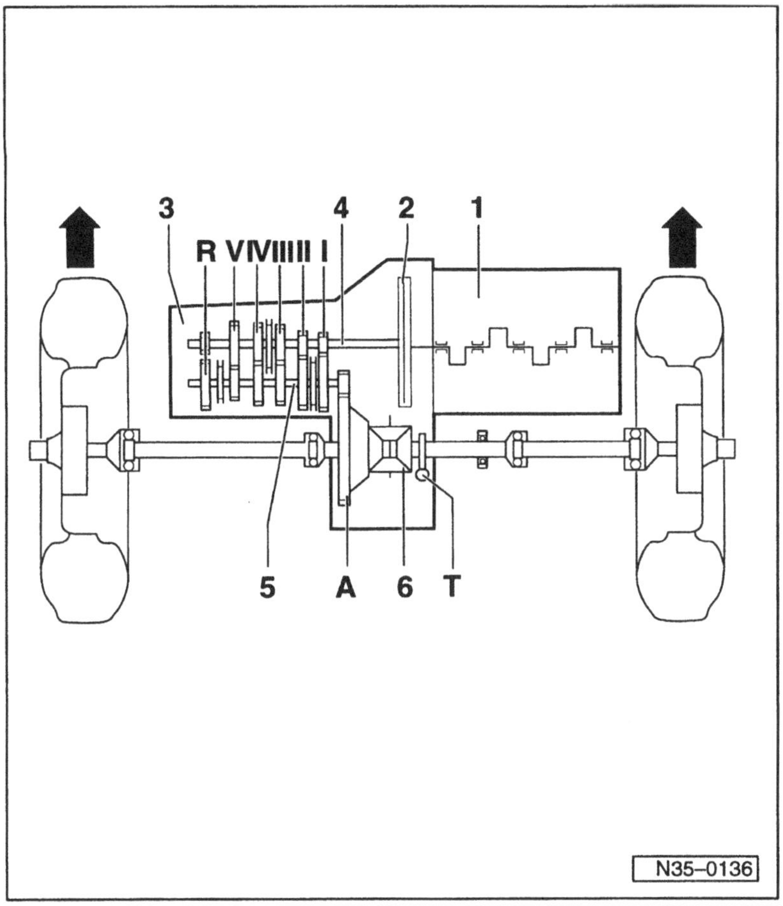

Transmission layout

Designation

1 - Engine

2 - Clutch

3 - Transmission

4 - Input shaft

5 - Output shaft/pinion shaft

6 - Differential

Arrows point in direction of travel

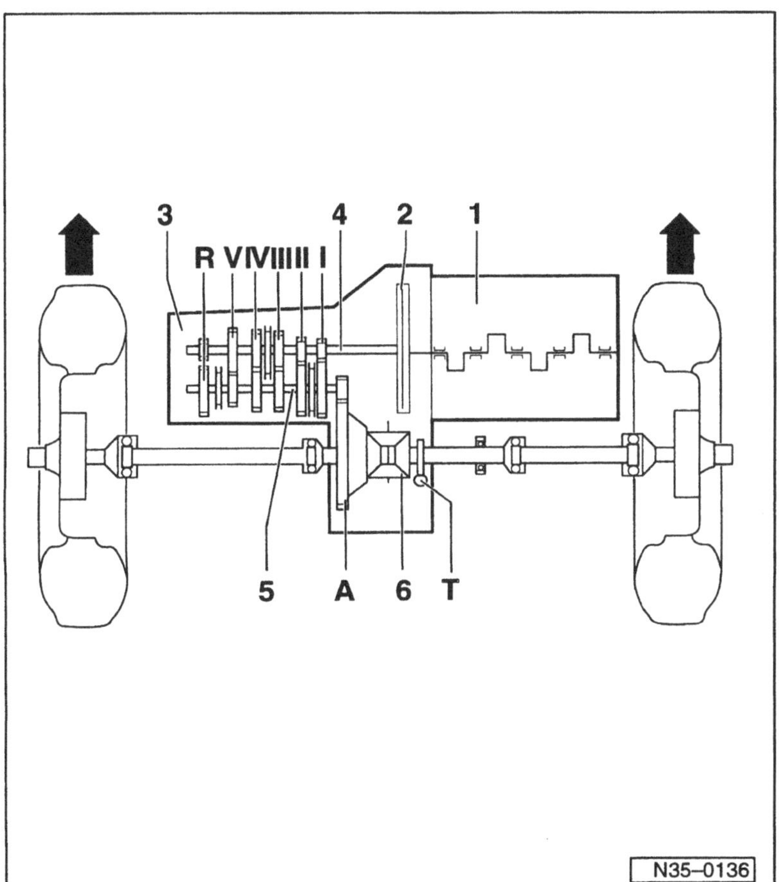

Gears

I – 1st gear

II – 2nd gear

III – 3rd gear

IV – 4th gear

V – 5th gear

R – Reverse gear

A – Final drive

T – Speedometer drive

Arrows point in direction of travel

Calculating overall ratio (i)

Example:

	5th gear	Final drive
Drive gear	$ZG_1 = 48$	$ZA_1 = 21$
Driven gear	$ZG_2 = 35$	$ZA_2 = 82$

$i = Z_2 : Z_1$ [1)]

i_G = Gear ratio = $ZG_2 : ZG_1 = 35 : 48 = 0.729$

i_A = Axle ratio = $ZA_2 : ZA_1 = 82 : 21 = 3.905$

$i_{overall\ ratio}$ = Total ratio

$= i_G \times i_A = 0.729 \times 3.905 = 2.847$

[1)] Z_1 = No. of teeth drive gear, Z_2 = No. of teeth driven gear

General repair instructions

To ensure perfect and successful transmission repairs, the greatest care and cleanliness as well as use of good and proper tools is essential. Obviously, basic rules for safety are also applicable during repair work.

A number of generally valid instructions applicable for various repair procedures – which were previously repeated a number of times at numerous places in the Repair Manual – are summarized here.

Transmission

- When installing ensure that dowel sleeves between engine and transmission are correctly located.
- When changing a transmission fill with gear oil to lower edge of filler hole.
- Capacities and specifications ⇒ Page L 00-3.

Gaskets, seals

- Thoroughly clean joint surfaces and apply sealant AMV 188 200 03.
- Apply sealant uniformly but not too thick.
- Always replace "O-rings".
- Radial shaft oil seals –

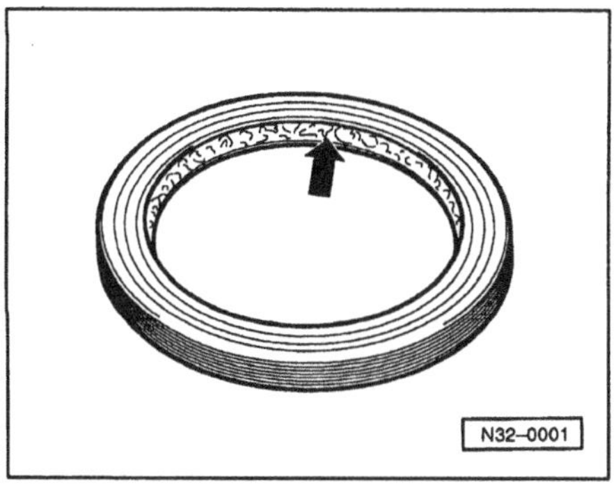

Before installing:

◄ – Lightly oil outer edge, fill space between sealing lips -arrow- with grease.

After installation:

- Check transmission oil and top-up if necessary
- Specification
⇒ Page L 00-3.

Locking devices

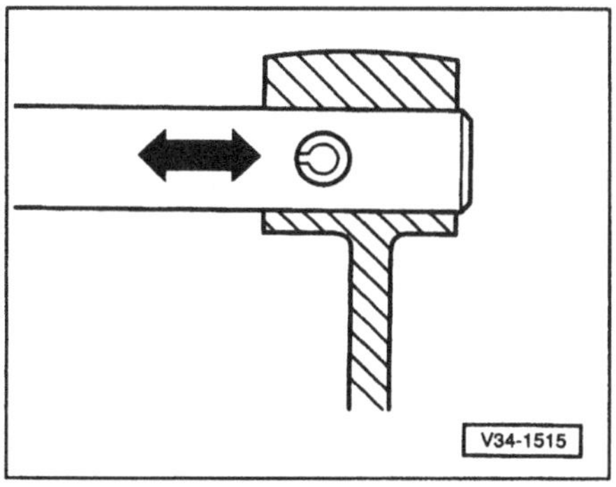

- Always replace circlips.
- Do not overstretch circlips.
- Must locate circlips properly in groove.

◄ – Replace spring pins Installation position: Slot longitudinal to line of force.

Bolts and nuts

- Tighten and loosen bolts and nuts for securing covers and housings in a diagonal sequence.
- Especially delicate parts, such as clutch pressure plates, must not be distorted. Loosen and tighten nuts and bolts in stages in a diagonal sequence.
- Tightening torques as specified are for unoiled bolts and nuts.
- Always replace self-locking bolts and nuts.

Bearings

- Install needle bearings with lettered side (thicker metal) toward installation tool.
- Lubricate all transmission bearings with gear oil before installing. Lubricate with special care when measuring turning torque.
- Tapered roller bearings installed on one shaft must be replaced as a set. Use bearings from same manufacturer if possible!
- Heat inner tapered roller bearing races to approx. 100°C (212°F) before pressing on.
- Do not interchange same size outer and inner bearing races; bearings are paired.

Shims

- Measure shims at several points with a micrometer. Tolerance variations make it possible to find the exact shim thickness required.
- Check for burrs and damage.
- Only install perfect shims.

Synchronizer rings

- Do not interchange. When reusing always match to same gear.
- Check for wear and replace if necessary.
- Coat with gear oil before installing.

Gears

- Before installing clean and heat on a hot plate to approx. 100 °C (212°F).
- Temperature can be checked with temperature tester VAG 1558 or equivalent.

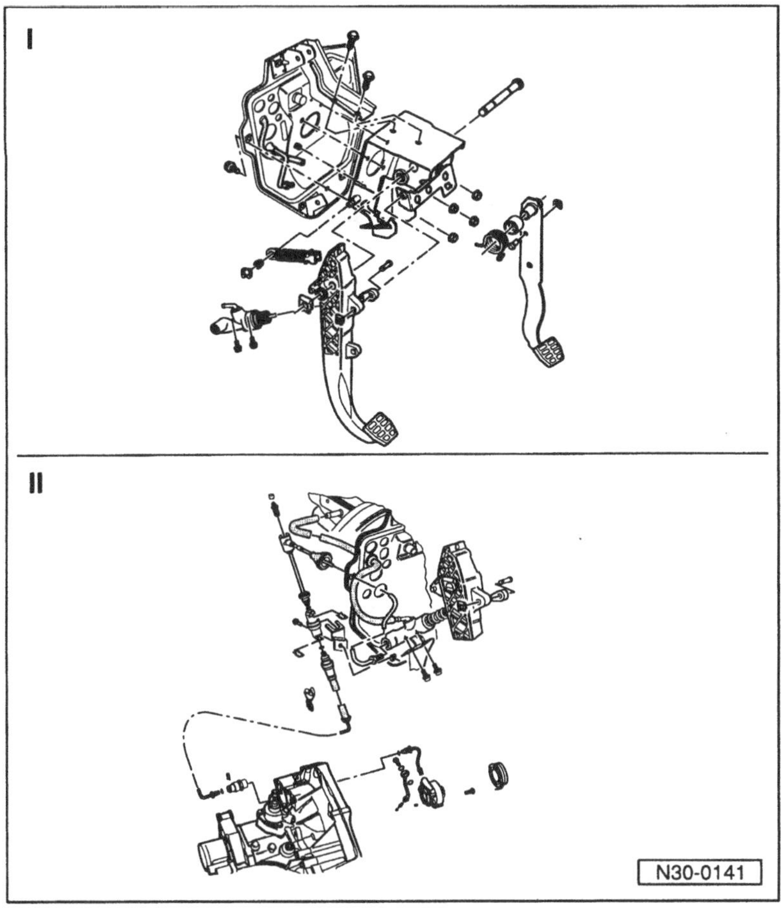

Clutch control, servicing

I - Assembly overview - pedal cluster
⇒ page L 30-2

II - Assembly overview - hydraulics and clutch release
⇒ Page L 30-6

Notes:

– *Disconnect battery ground strap.*

– *Obtain radio code for vehicles with coded radio.*

– *Lubricate all bearings and contact surfaces with MoS_2 grease.*

♦ *Before working on the pedal cluster remove the following components:*

Driver's side storage compartment
⇒ Body Interior; Repair group 70;

Steering column trim
⇒ Suspension, Wheels, Brakes, Steering; Repair group 48

Relay plate (releasing from mounting)
⇒ Electrical system; Repair group 97; wiring

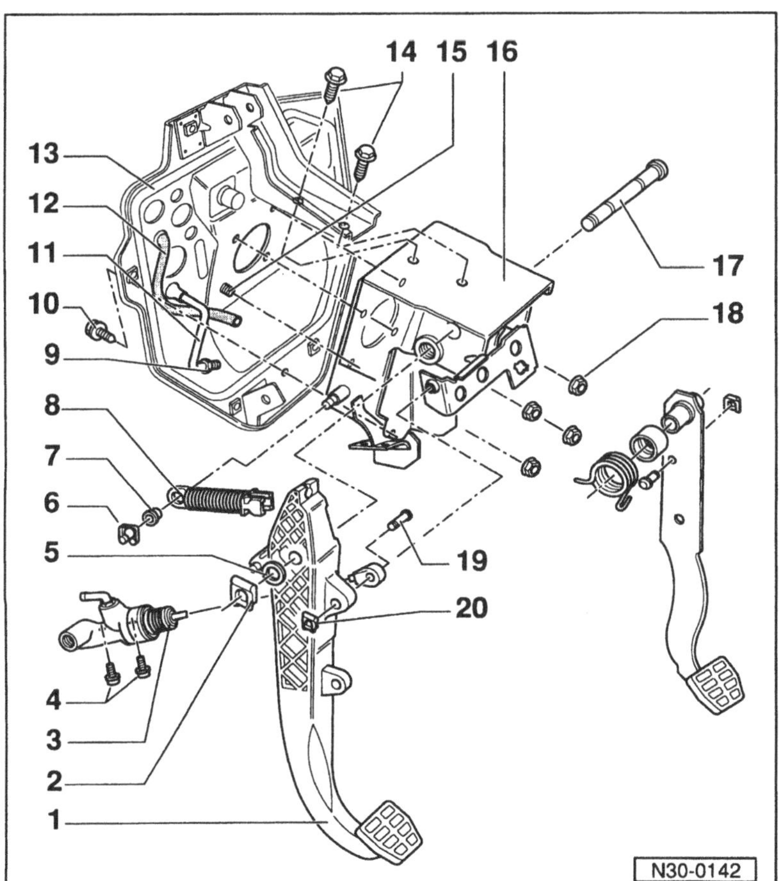

Pedal cluster - assembly overview

1 - Clutch pedal
♦ Removing/installing
⇒ Page L 30-5

2 - Locking clip

3 - Master cylinder

4 - Combi-bolt, 20 Nm (15 ft lb)
♦ For master cylinder to mounting bracket

5 - Washer

6 - Locking clip

7 - Bushing

8 - Over-center spring
♦ Removing/Installing
⇒ Page L 30-11

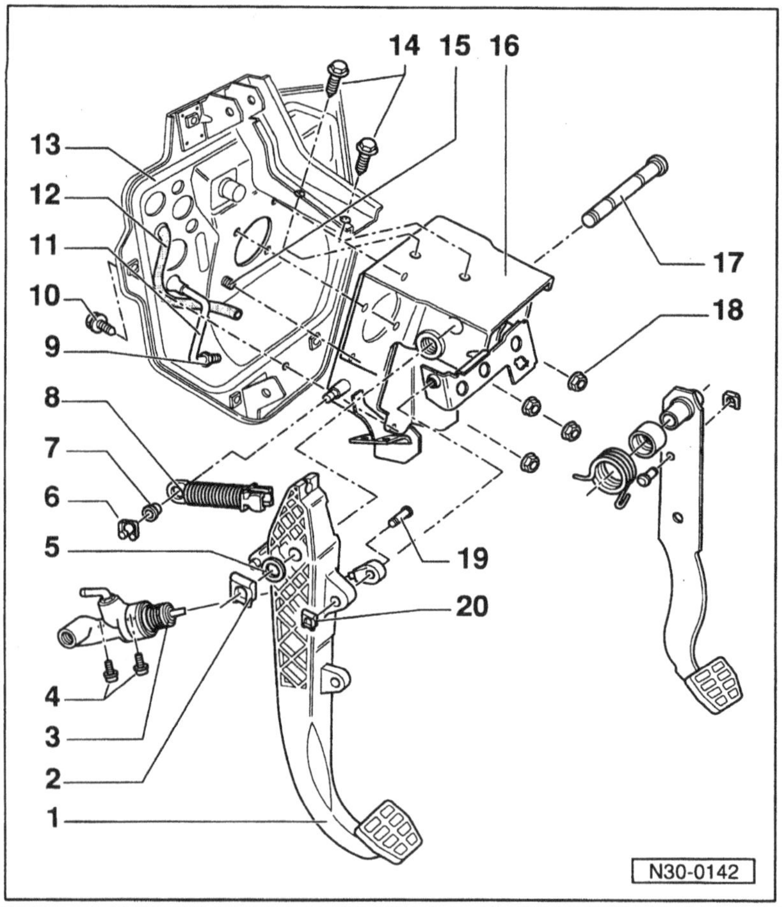

9 - Line union nut, 20 Nm (15 ft lb)

10 - Hex bolt, 25 Nm (18 ft lb)

11 - Pressure line

12 - Supply hose

13 - Mounting plate
- ♦ Need not be disassembled to remove pedal cluster
- ♦ Seal to bulkhead with AKD 497 010 04 R 10

14 - Hex bolt, 25 Nm (18 ft lb)

15 - Brake servo stud

16 - Mounting bracket
- – Before removing unbolt steering column and lay to side

⇒ Suspension, Wheels, Brakes, Steering; Repair group 48

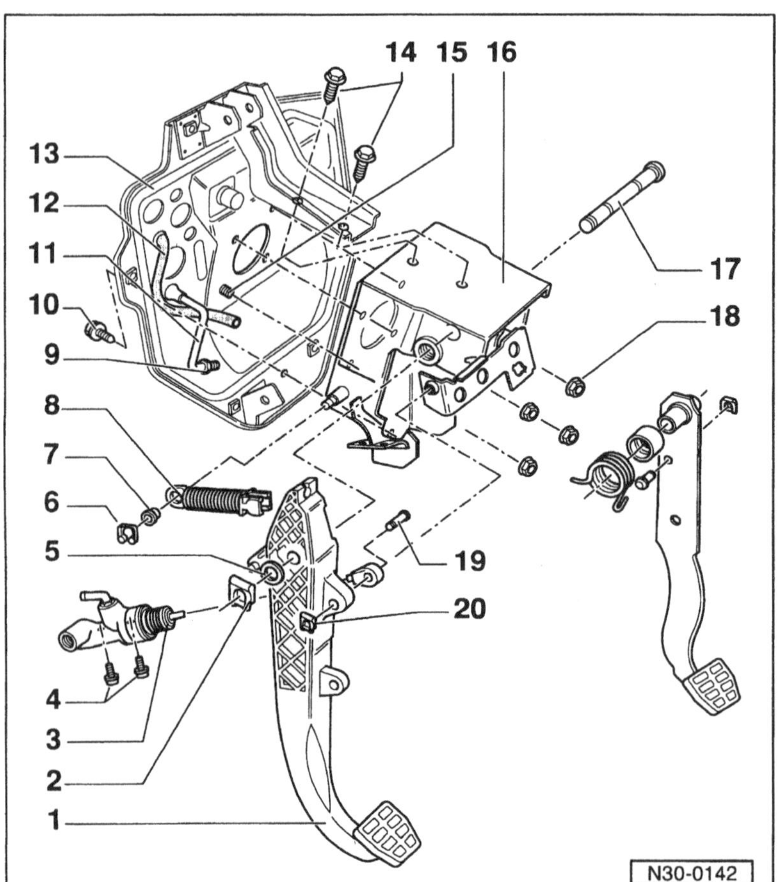

17 - Pivot pin
- ♦ For clutch and brake pedal
- – Knock out with drift

18 - Self-locking hex nut, 25 Nm (18 ft lb)
- – Always replace
- ♦ For mounting bracket and brake servo to mounting plate

19 - Pivot pin
- ♦ For master cylinder operating rod

20 - Locking clip

Clutch pedal, removing and installing

Removing

– Separate clutch pedal from master cylinder
⇒ Page L 30-2.

– Remove over-center spring
⇒ Page L 30-11.

– Remove clutch pedal locking clip from pivot pin.

– Drive pivot pin out to right until clutch pedal can be pulled off.

– Remove clutch pedal.

Installing

Install in reverse sequence.

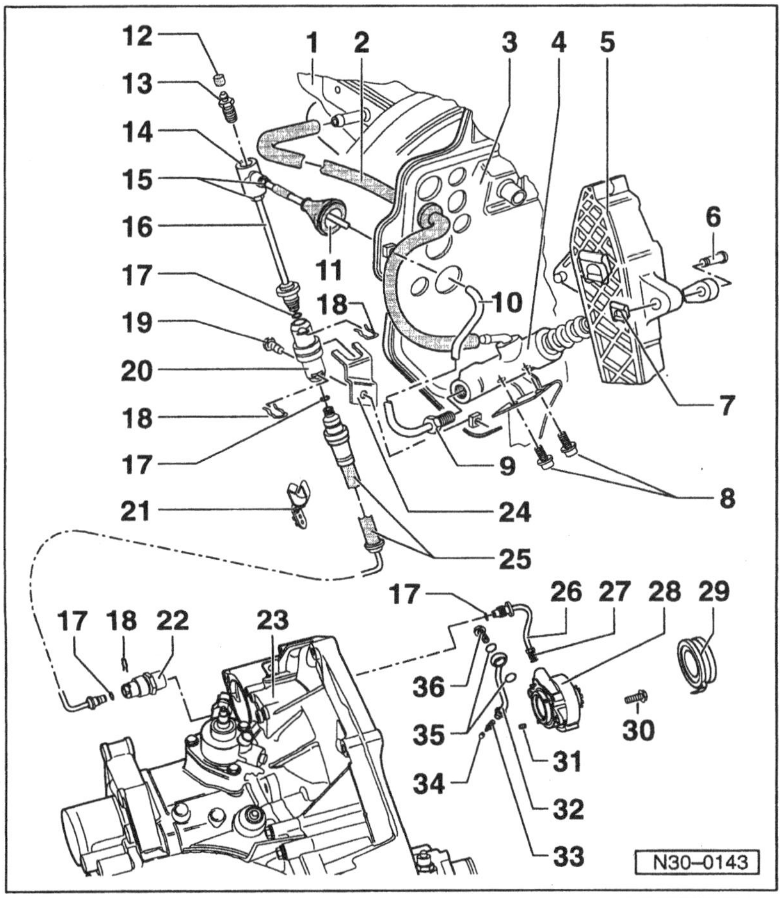

Hydraulic clutch release - assembly, overview

1 - **Brake fluid reservoir**

2 - **Supply hose**

3 - **Mounting plate**

4 - **Master cylinder**

5 - **Clutch pedal**
- ♦ Removing/installing ⇒ Page L 30-5

6 - **Pivot pin**
- ♦ For master cylinder operating rod

7 - **Locking clip**

8 - **Securing bolt, 20 Nm (15 ft lb)**
- ♦ For master cylinder to mounting bracket

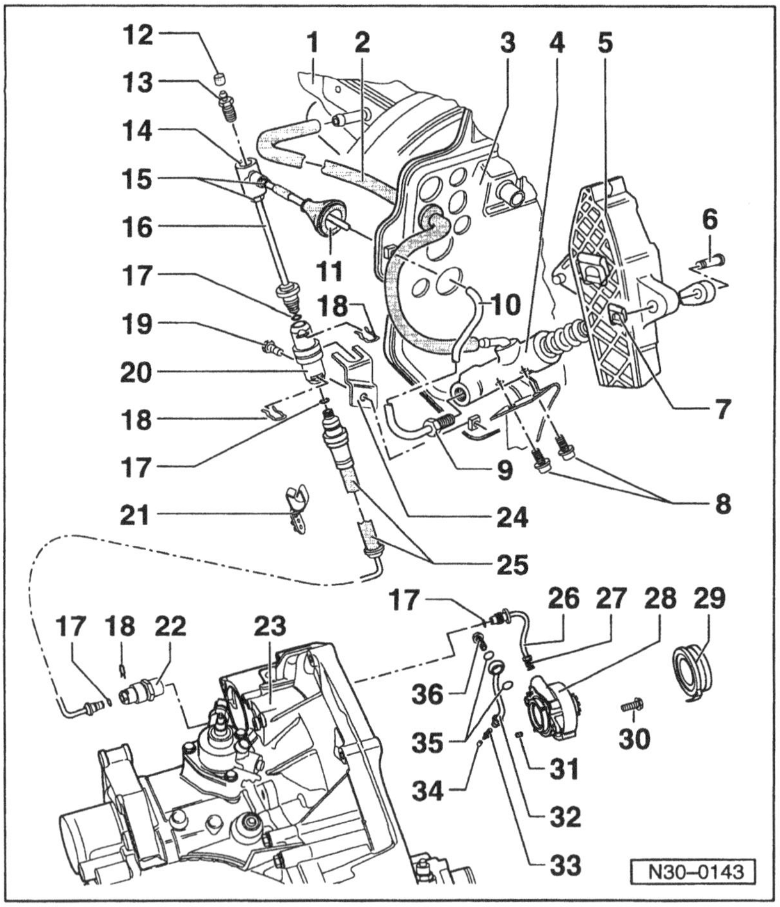

9 - **Line union nut, 20 Nm (15 ft lb)**

10 - **Pressure line**

11 - **Grommet**

12 - **Dust cap**

13 - **Bleed valve/master cylinder**
- ♦ Bleeding clutch system

⇒ Page L 30-12

14 - **Intermediate piece**

15 - **Line union nut, 20 Nm (15 ft lb)**

16 - **Pressure line**

17 - **O-ring**
- Always replace
- Pull onto line/hose connection
- Coat with brake fluid before installing

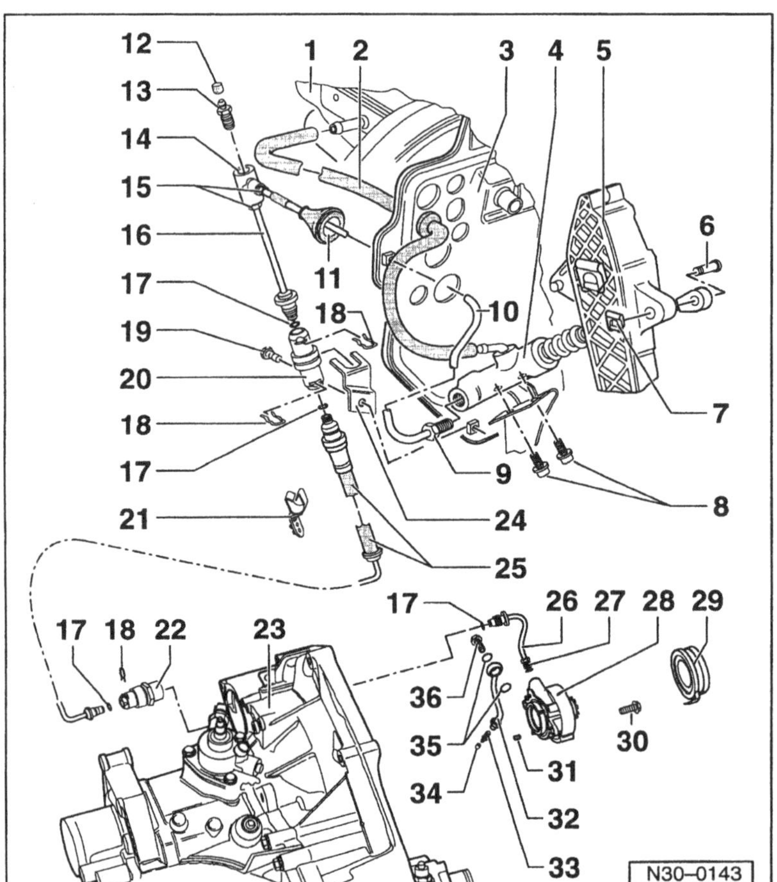

18 - **Clip**
- ♦ Removing

⇒ Fig. 1

- Installing: press in by hand

19 - **Hex bolt, 25 Nm (18 ft lb)**

20 - **Connecting piece**

21 - **Hose bracket**
- Secure to left transmission support

22 - **Valve unit, 5 Nm (44 in. lb)**

23 - **Transmission**

24 - **Hose bracket**
- Secured to mount plate

25 - **Hose assembly**

26 - **Pressure line**

27 - **Line union nut, 20 Nm (15 ft lb)**

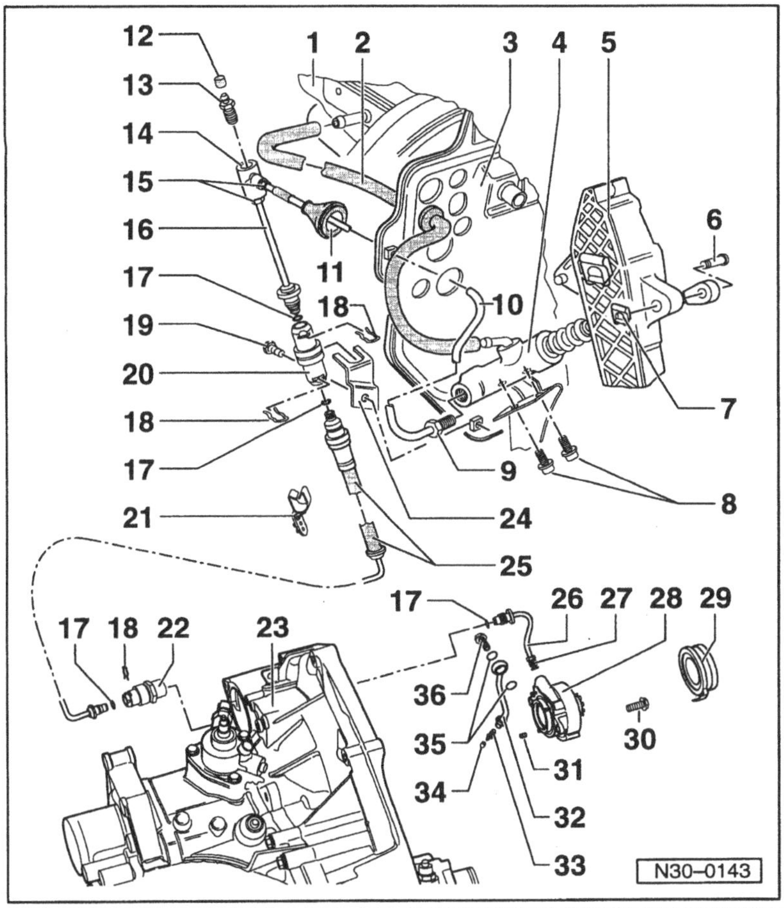

28 - Slave cylinder

29 - Release bearing
- Pry out with a screwdriver

30 - Hex bolt, 12 Nm (9 ft lb)
- ♦ Quantity 2

31 - Rubber cap
- ♦ Quantity 3
- Install on slave cylinder flange

32 - LIne

33 - Bleed valve/master cylinder
- ♦ Bleeding clutch system

⇒ Page L 30-12

34 - Dust cap

35 - O-ring
- Always replace

36 - Union bolt

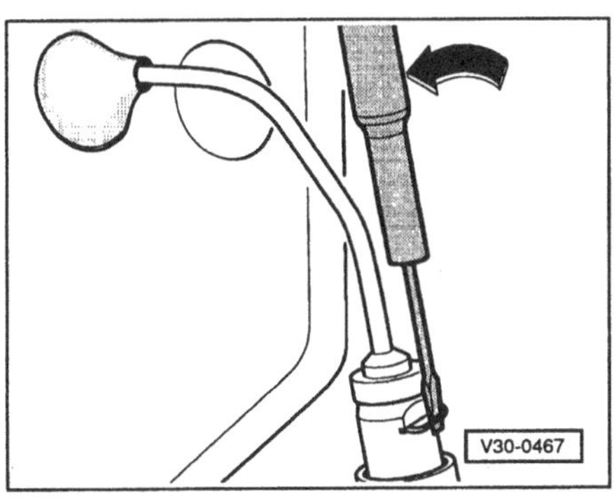

◄ **Fig. 1 Clip, removing**

- Guide screwdriver between clip and connecting piece, if necessary turn connecting piece first so flattened section of connecting piece in area of clip is pointing forward.
- Carefully pry clip out and remove.

Over-center spring, removing/installing

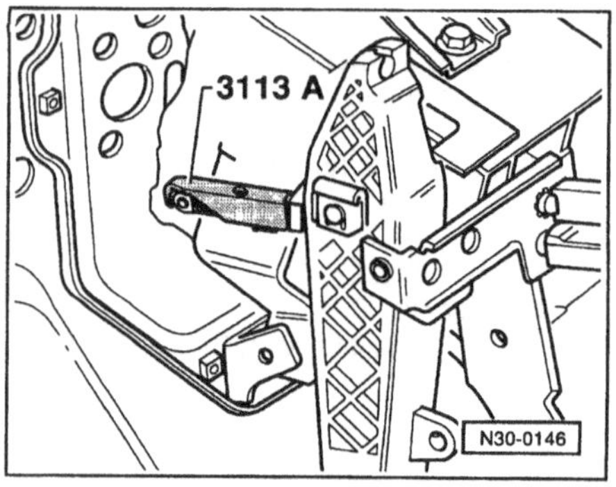

Removing

- Remove carpet in area of over-center spring.
- Pry securing clip off with a screwdriver and remove washer.
- Tension over-center spring with clutch pedal and install retaining clamp 3113A from above.

> ***CAUTION!***
>
> ***Do not press retaining clamp on from below; there is a danger over-center spring may jump out of retaining clamp during installation.***

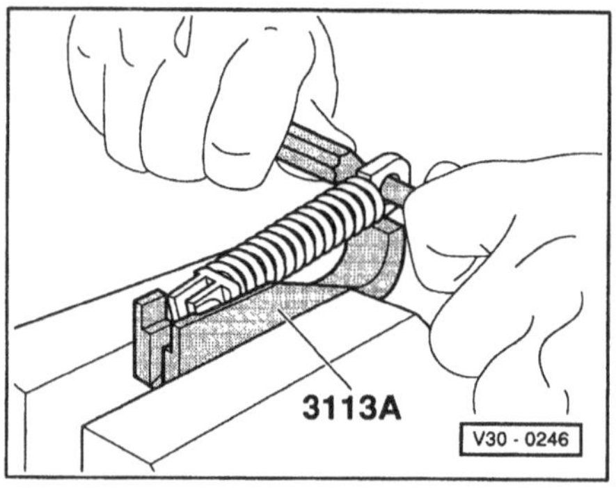

- Move clutch pedal onto stop.
- Remove over-center spring together with retaining clamp 3113A.
- Remove over-center spring from retaining clamp and install.

Installing

Install in reverse sequence.

Clutch system, bleeding

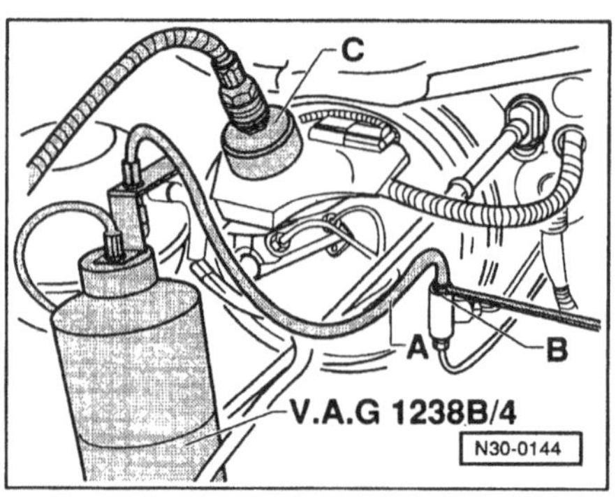

Bleed clutch system using brake filling/bleeding tool VAG1238 B and bleed hose VAG 1238 B/3 or use Fluid Evacuator EZ1 or equivalent.

- Remove insulation tray below engine/transmission.
- Connect pressure hose -C- to brake fluid reservoir.
- Step 1: Connect bleed hose -A- to bleed valve -B- in engine compartment and open bleed valve.
- Switch on bleeder unit.

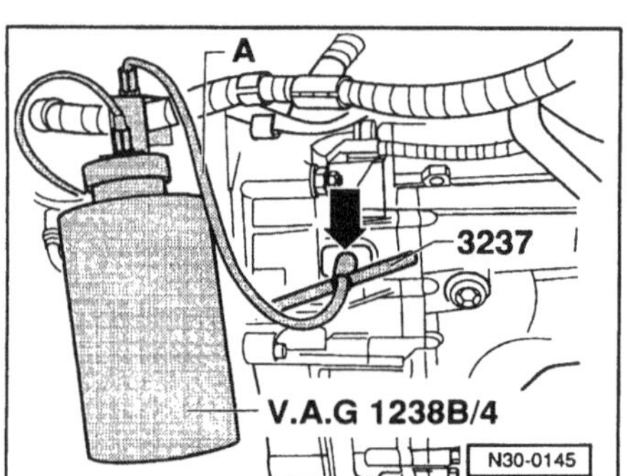

- Step 2: Connect bleed hose -A- to bleed valve -arrow- below transmission and open bleed valve.
- Switch on bleeder unit.
- Step 3: Repeat bleed sequence at bleed valve -B- in engine compartment,
⇒ illustration N30-0144.
- Depress clutch pedal several times after completion of bleeding process.

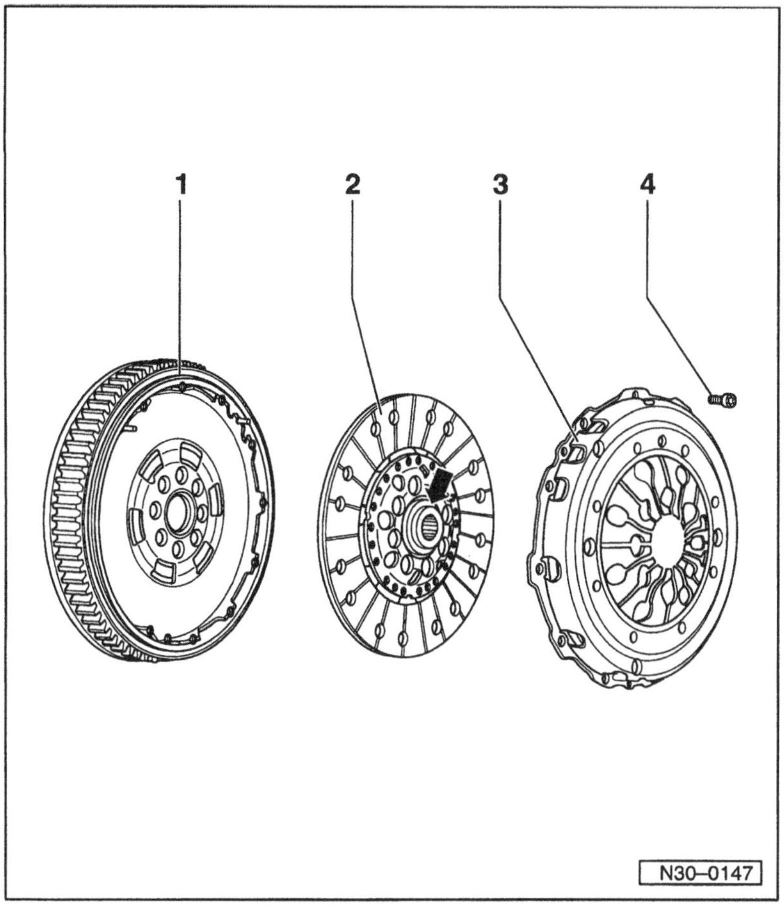

Clutch, servicing

Transmission installed

Notes:

– *Replace clutch plates and pressure plates having damaged or loose rivets*

– *Ensure correct clutch plate and pressure plate is selected from parts list*

1 - Flywheel

– Ensure centering pins are a tight fit
– Contact surface for clutch lining must be free of grooves, oil and grease
♦ Removing and installing
⇒ Repair Manual, General Engine, Repair Group13

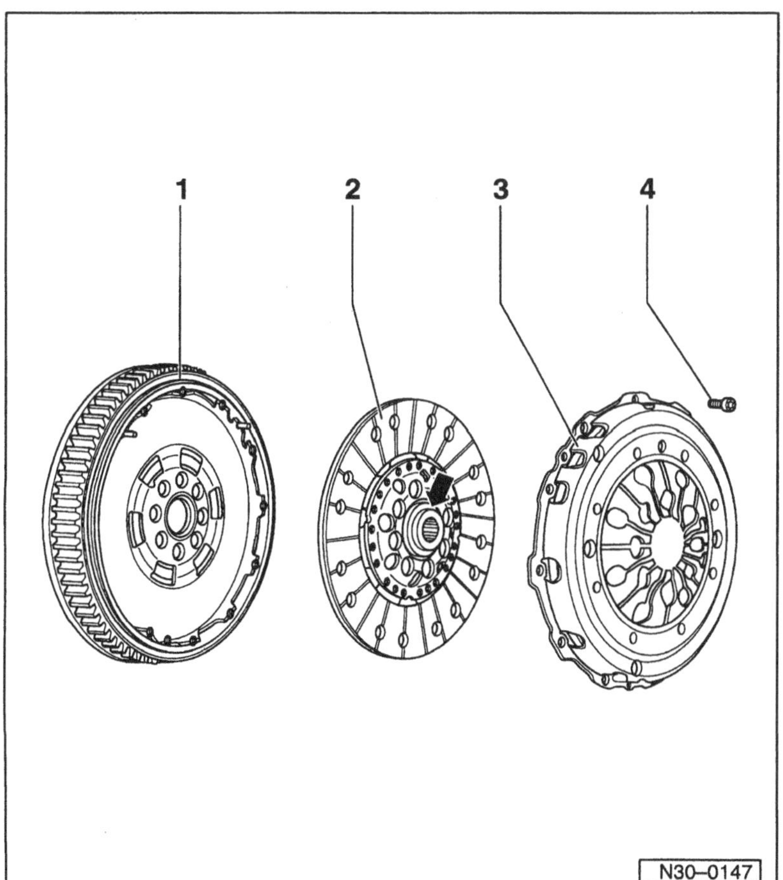

2 - Clutch plate

♦ Diameter
⇒ from page L 00-3
♦ Installation position: Shorter hub end (arrow) points toward pressure plate
♦ Centering
⇒ Fig. 1
– Lightly grease splines

Note:

Clean input shaft splines and, on used clutch plates, hub splines. Remove corrosion and apply only a very thin coating of grease Part No: G 000 100 to the input shaft splines. Then move the clutch plate back and forth on the input shaft until the hub moves freely on the shaft. Excess grease must be removed.

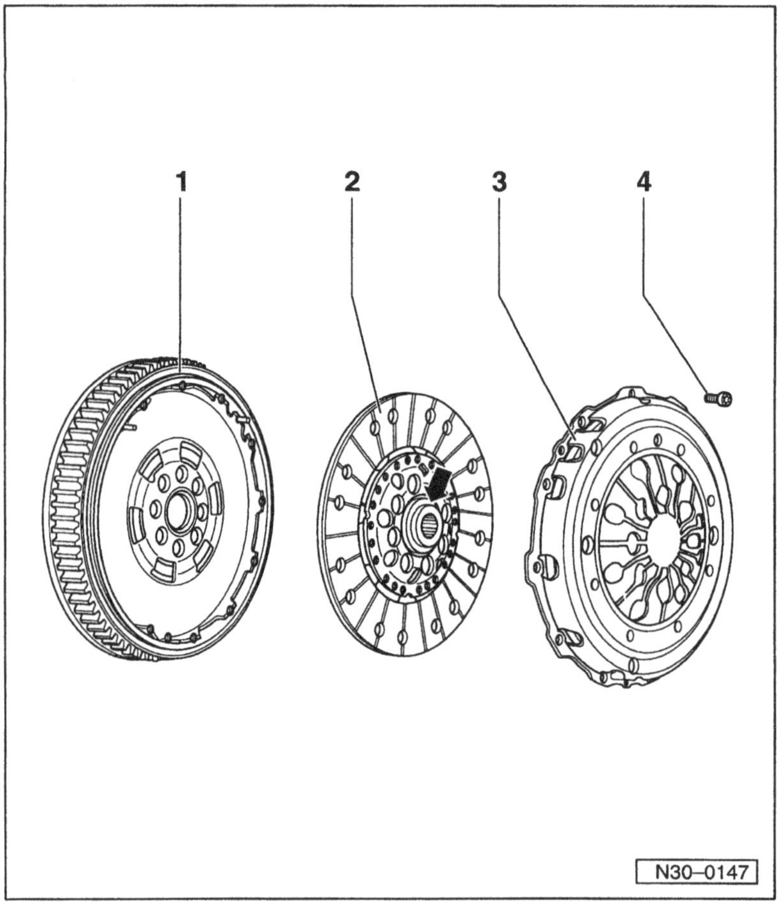

3 - Pressure plate

- ♦ Removing/installing

⇒ Fig. 1

- – Check ends of diaphragm spring

⇒ Fig. 2

Note:

Pressure plates are protected against corrosion and greased. Only the contact surface may be cleaned, otherwise service life of the clutch will be considerably reduced.

4 - Bolt, 13 Nm (10 ft lb)

- – Loosen and tighten gradually and diagonally

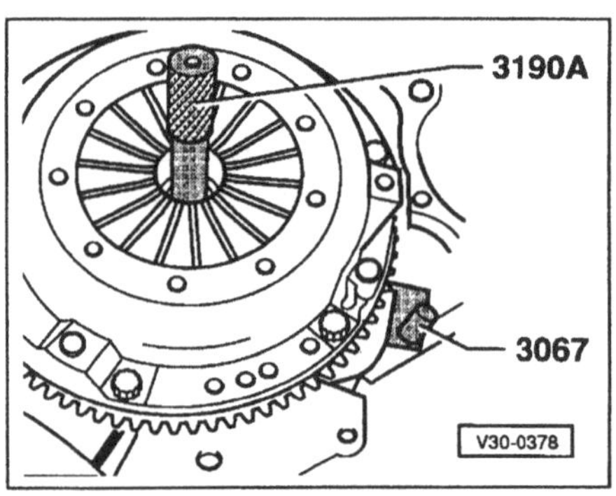

◄ **Fig. 1 Clutch plate, centering; pressure plate, removing and installing**

– Loosen and tighten bolts gradually and diagonally.

– Reverse position of retainer 3067 when removing.

Note:

The pressure plate contact surface and the clutch plate lining must make full contact with the flywheel. Tighten securing bolts diagonally and evenly so as to prevent damage to the pressure plate centering hole and the flywheel centering pins.

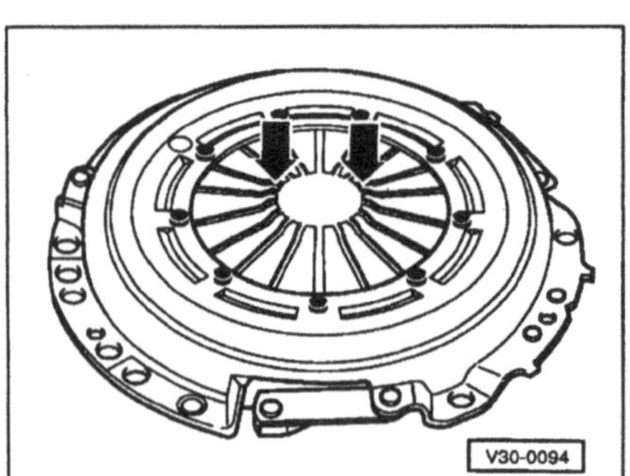

◄ **Fig. 2 Diaphragm spring, checking ends –arrows–**

- Wear up to half the thickness of the diaphragm spring is permitted.

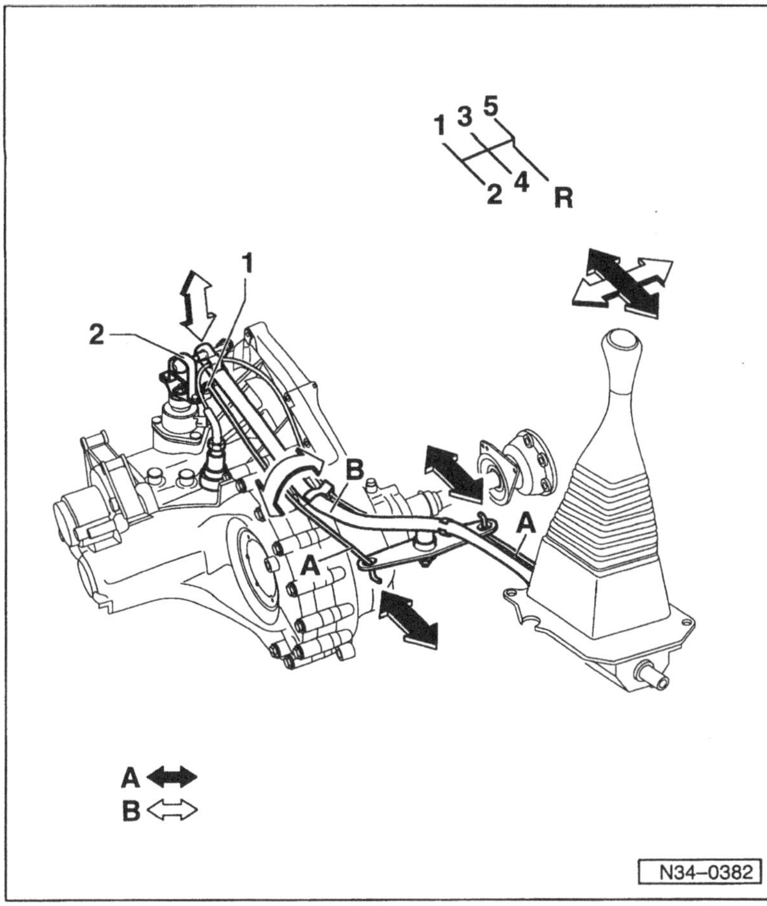

Gear selector, servicing

Installation position of gear selector

A - Gear selector rod

B - Gate selector rod

Double-headed arrow -A- Gear selection movement

Double-headed arrow -B- Gate selection movement

1 - Gear selector lever

2 - Swivel head with selector rod

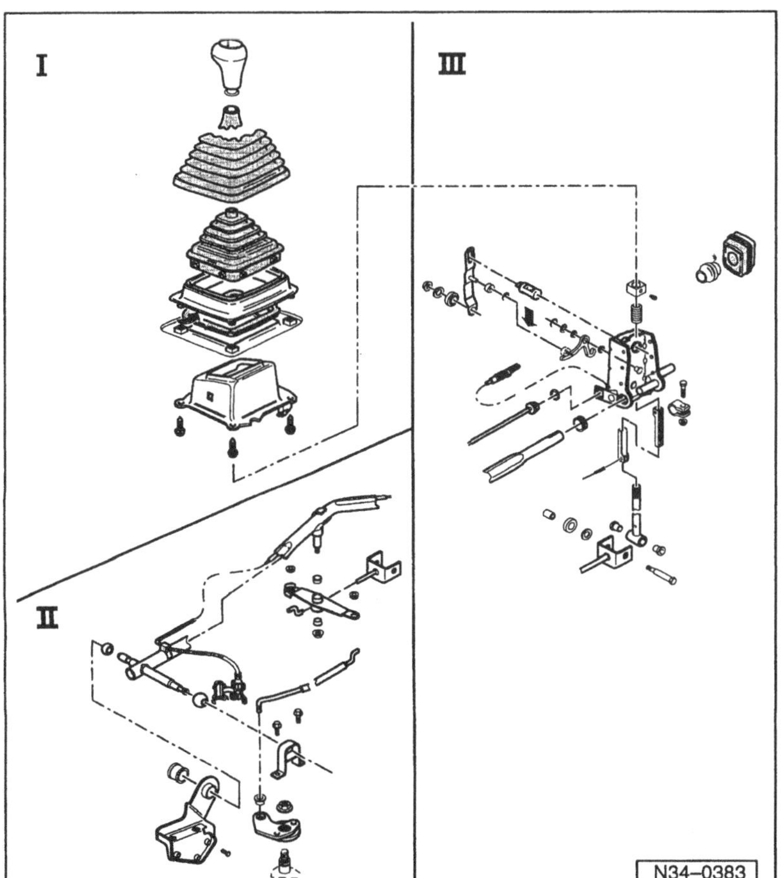

I - Removing/installing gear shift housing
⇒ **Page L 34-3**

II - Removing/installing selector rods and gate rod
⇒ **Page L 34-6**

III - Removing/installing gear lever
⇒ **Page L 34-15**

Removing/installing selector mechanism
⇒ Page L 34-19

Notes:

- *Before working on the selector mechanism, disconnect the battery ground strap in the engine compartment.*
- *Note radio code on vehicles with coded radio.*
- *Lubricate all mountings and contact surfaces with MoS_2 grease.*

♦ *Adjusting selector cables*
⇒ *Page L 34-23*

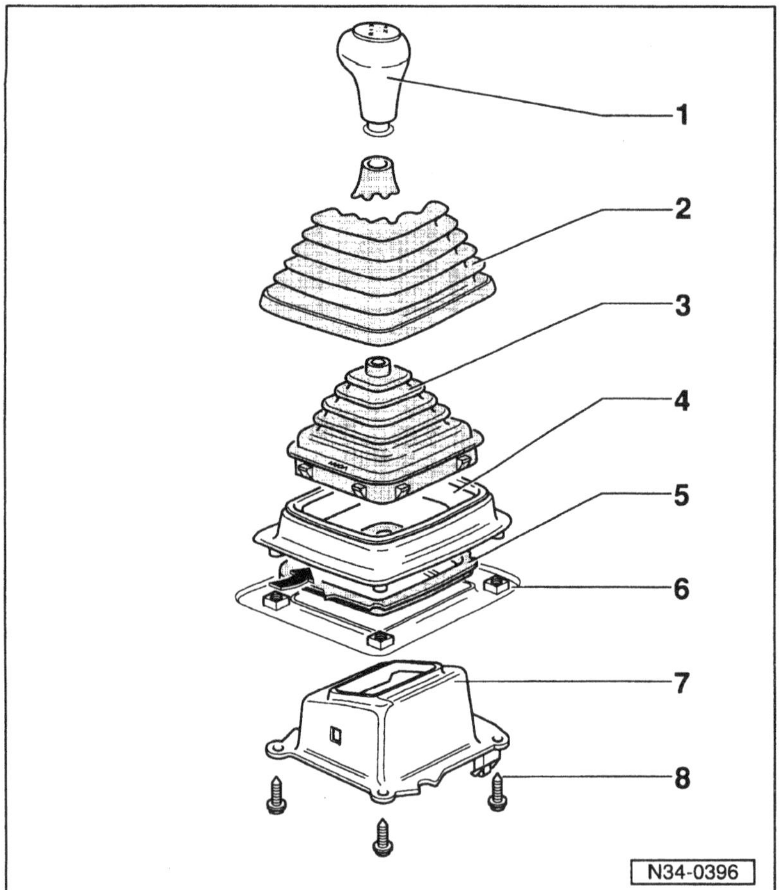

Gear shift housing, removing/installing

1 - Gear shift knob
- Thread on and off together with boot (item 2)
- When threading on hold boot in position

2 - Boot
- To remove, unclip from frame (item 4)
- Remove and install together with gear shift knob (item 1)
- After installing clip into frame (item 4)
- ♦ Installation position

⇒ Page L 34-22, illustration V34-2980

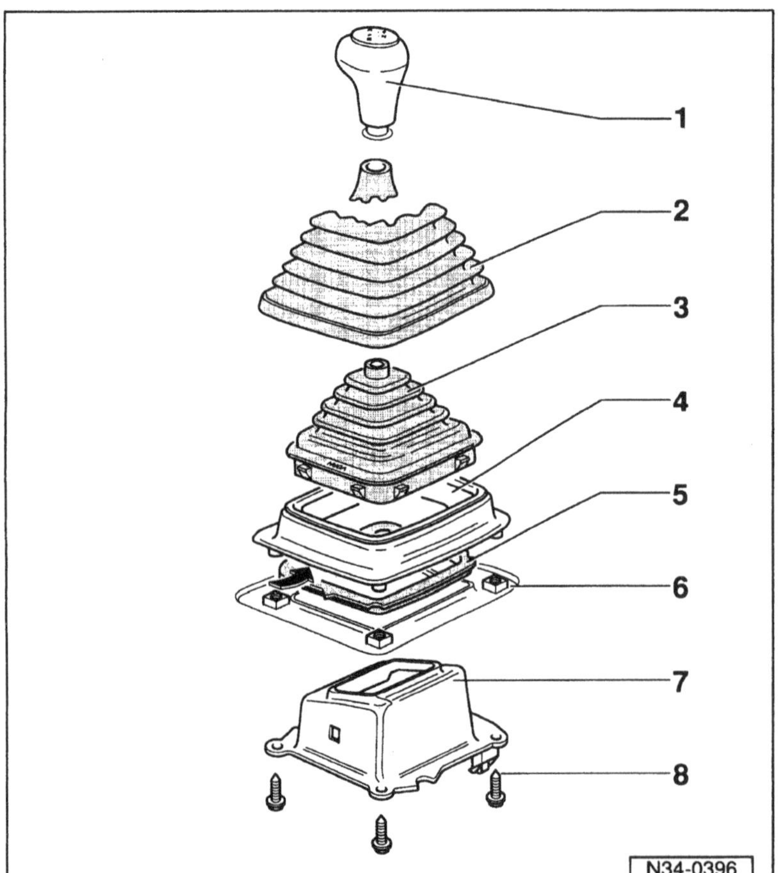

3 - Boot w/frame
- Remove/install together with frame (item 4)

⇒ Page L 34-19 and L 34-22

- Clip and unclip into frame (item 4)

⇒ Page L 34-22, Fig. 34-2978

- ♦ Installation position: Inscription "VORN" (front) points in direction of travel

4 - Frame
- Remove and install together with boot (item 3)

⇒ Page L 34-19 and L 34-22

- ♦ Installation position: Inscription "VORN" (front) -arrow- points in direction of travel

⇒ Page L 34-22, illustration V34-2978

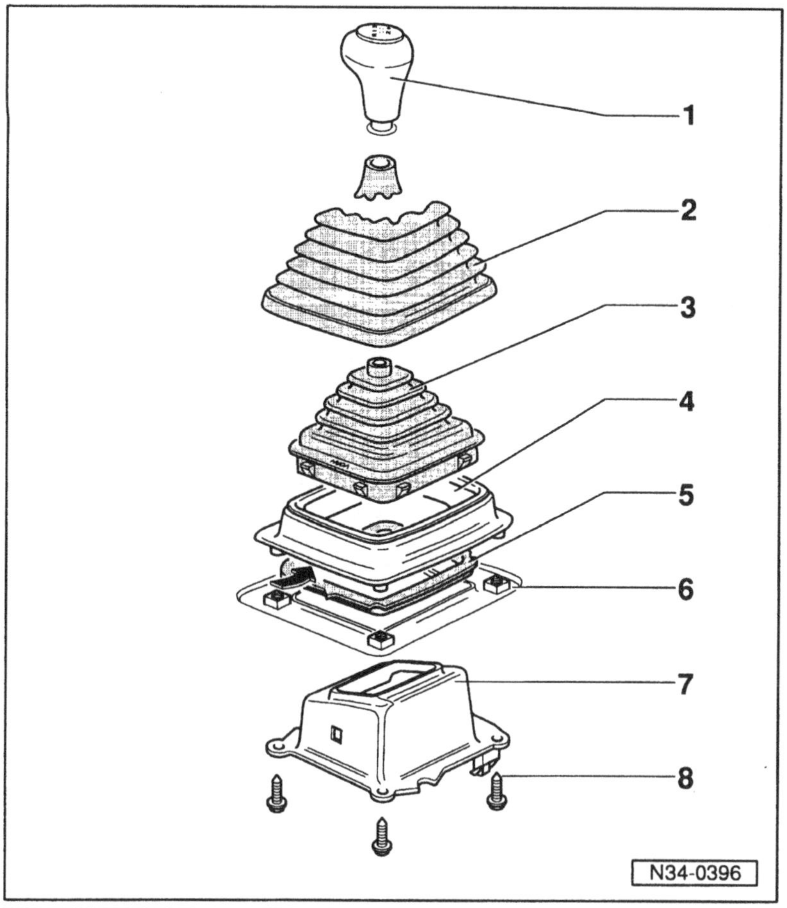

5 - Gasket
- ♦ When installing, place on edge of floor plate
- ♦ Installation position: Arrow type markings point in direction of travel, lip toward floor plate

6 - Floor plate

7 - Gear shift housing

8 - Hex bolt, 5 Nm (44 in. lb)

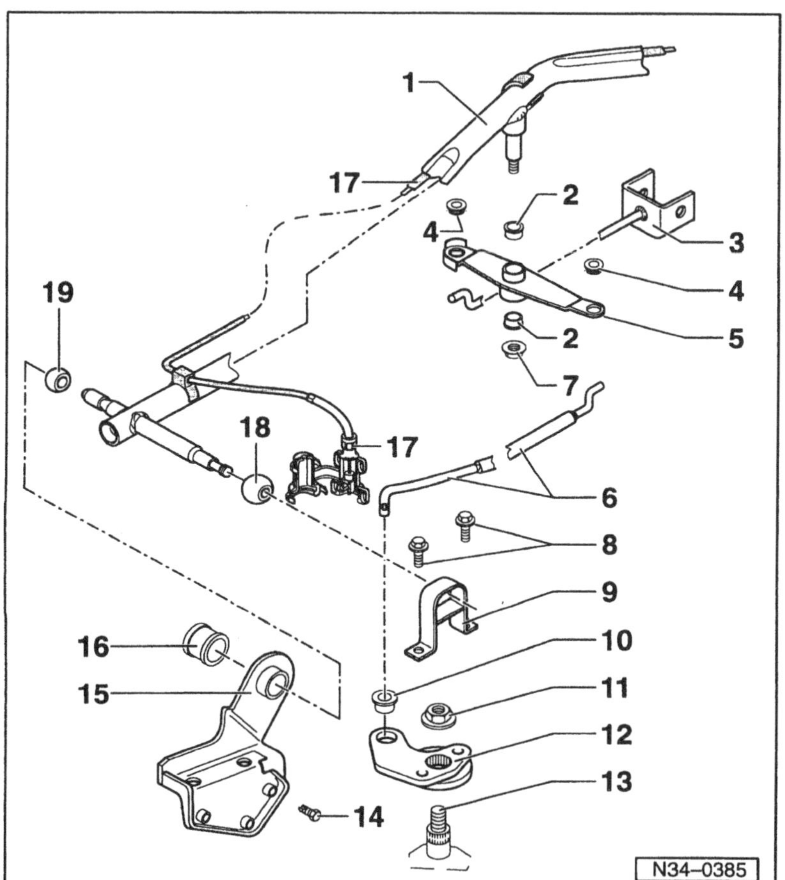

Selector rods and gate rod, removing/installing

1 - Gate rod
- Remove from transmission together with mounting bracket (item 15)
- Removing from gearshift bracket

⇒ Page L 34-20

2 - Bushing

3 - Rear selector rod
- Remove from gear shift and relay lever

⇒ Page L 34-20

4 - Bushing
- Pry out with screwdriver
- ♦ Installing

⇒ Fig. 1

5 - Relay lever
- ♦ Installation position

⇒ Fig. 9

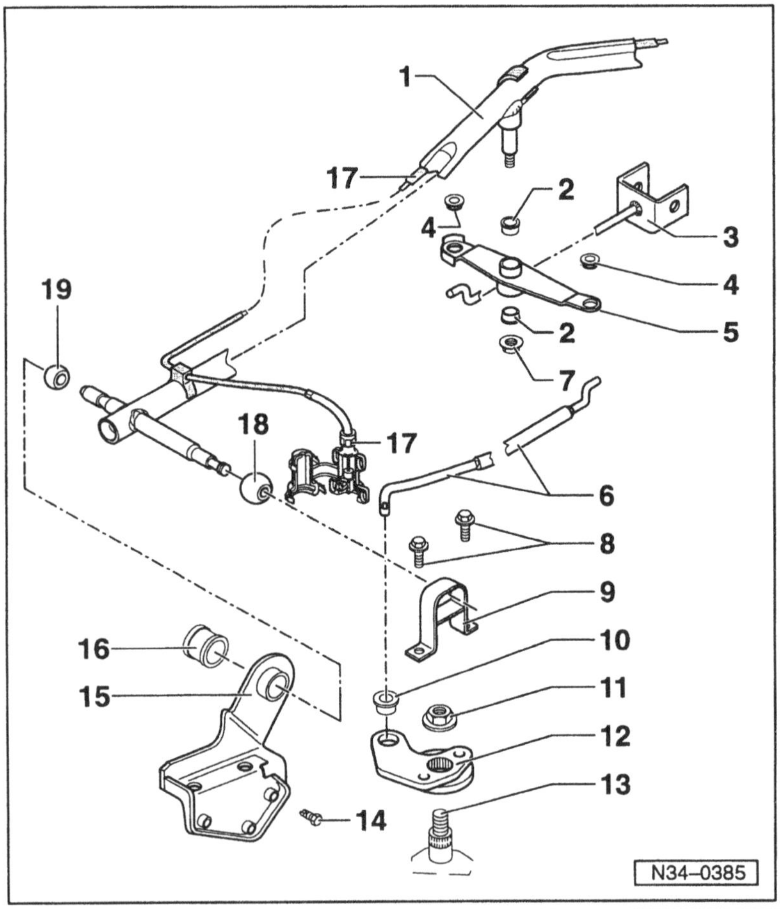

6 - Front selector rod
- Pry off and press into gear selector lever (item 12)

7 - Self-locking hex nut, 25 Nm (18 ft lb)

8 - Self-locking hex bolt, 9 Nm (7 ft lb)

9 - Swivel head

10 - Bushing
- ♦ For front selector rod
- Press into gear selector lever

⇒ Fig. 8

11 - Self-locking hex nut, 25 Nm (18 ft lb)

12 - Gear selector lever
- Install so that master spline aligns with selector shaft

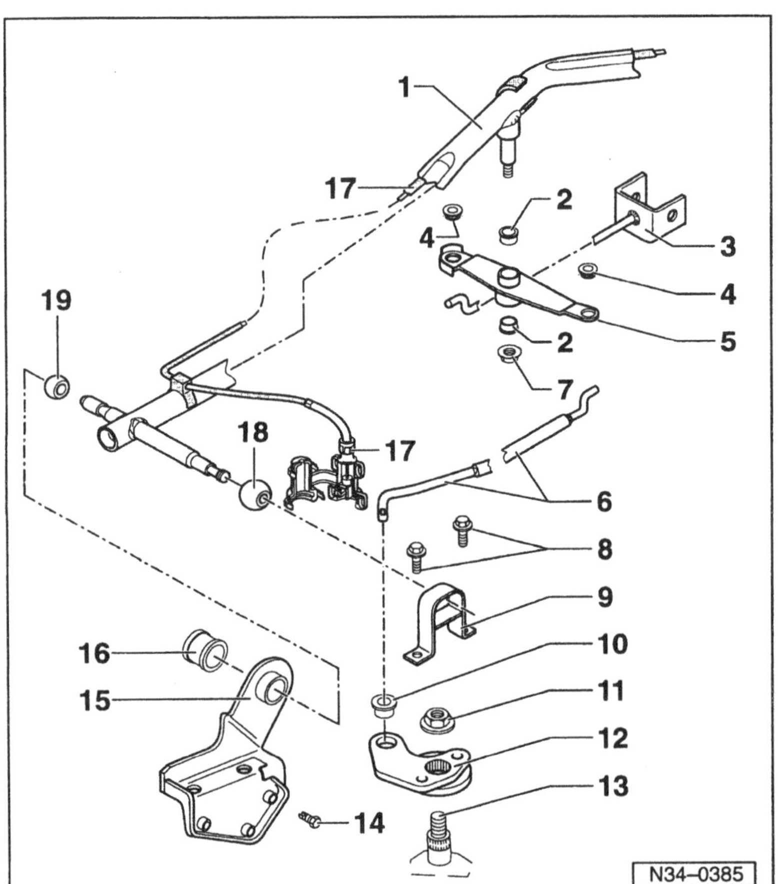

13 - Selector shaft

14 - Hex bolt, 20 Nm (15 ft lb)
- ♦ For mounting bracket to transmission

15 - Mounting bracket
- ♦ Removing from gate rod

⇒ Fig. 2
- ♦ Pressing on

⇒ Fig. 3

16 - Bushing
- ♦ Removing

⇒ Fig. 4
- ♦ Installing

⇒ Fig. 5

17 - Reverse gear cable
- ♦ Transmission side: engage in reverse gear locking unit
- ♦ Secure to gate rod (item1)
- ♦ Engage onto gear shift housing and hook onto lever

⇒ Page L 34-15

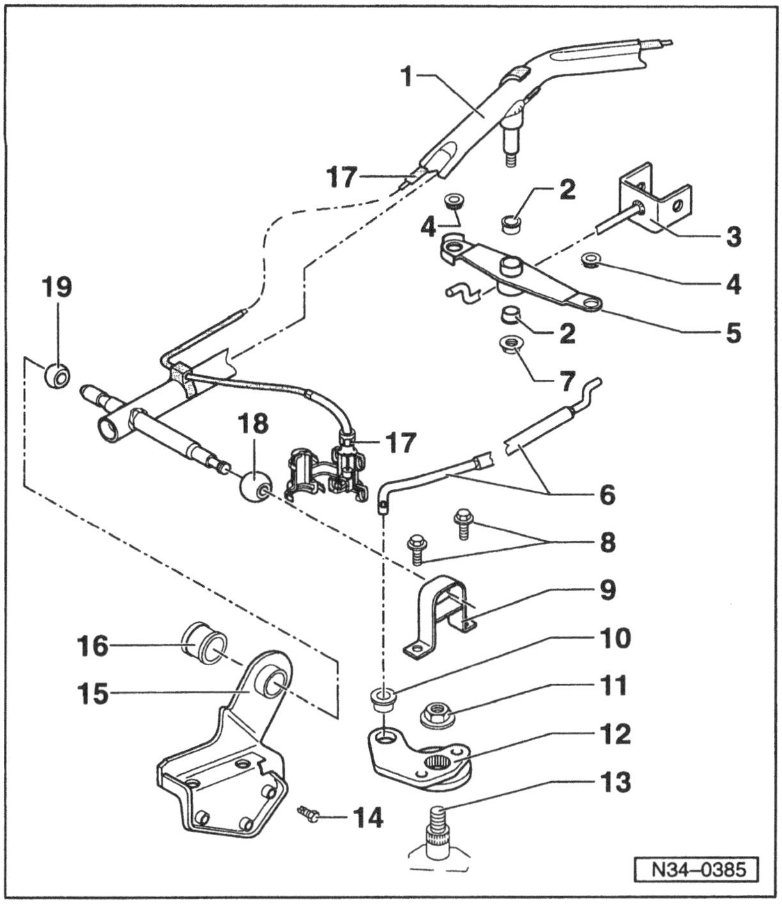

18 - Bearing ball

- Mounts gate rod in swivel head
- Pressing off

⇒ Fig. 6

- Pressing on

⇒ Fig. 7

19 - Bearing ball

- Mounts gate rod in mounting bracket
- Pressing off

⇒ Fig. 6

- Pressing on

⇒ Fig. 7

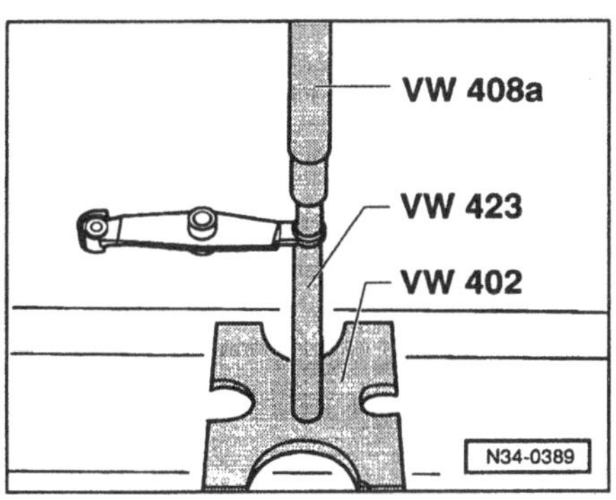

◄ **Fig. 1 Mount bushings, pressing in**

Installation position:

– Press in from smaller shoulder side of center mount.

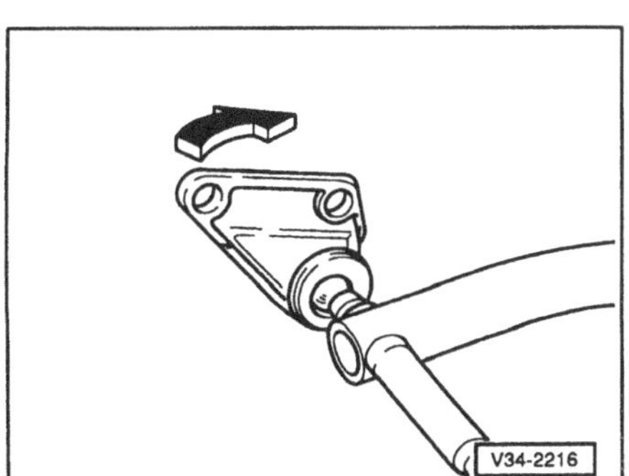

◄ **Fig. 2 Gate rod mount bracket, removing**

– Remove in direction of arrow.

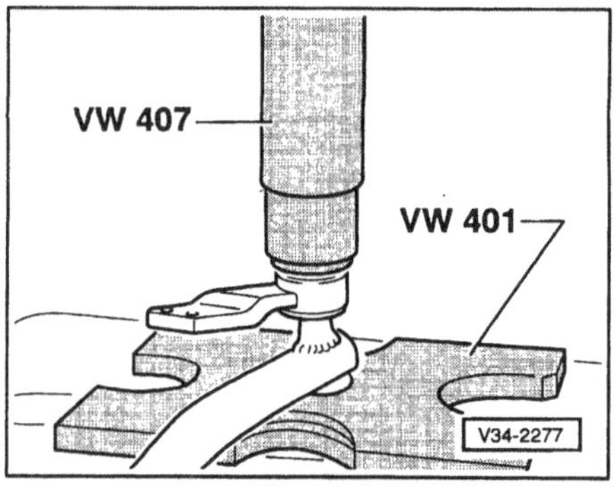

Fig. 3 Gate rod mounting bracket, pressing on

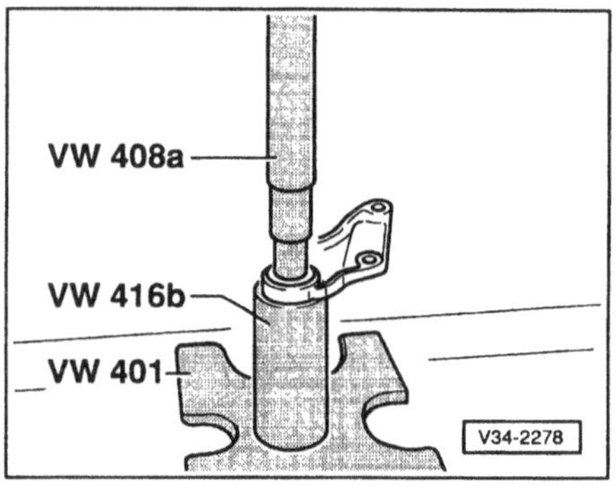

Fig. 4 Bushing, pressing out of mounting bracket

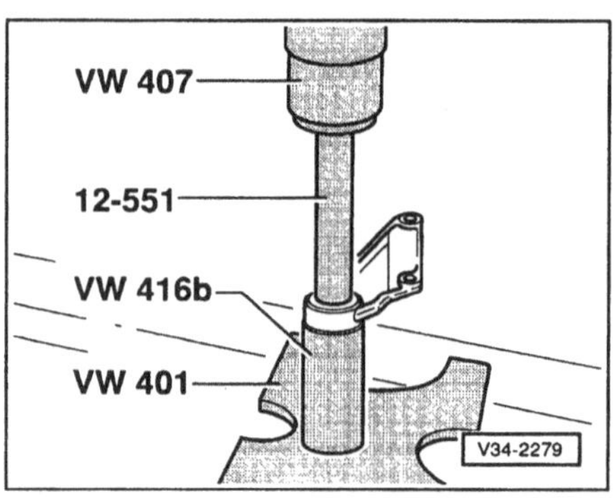

Fig. 5 Bushing, pressing into mounting bracket

– Coat bushing lightly with MoS_2 grease and carefully press in.

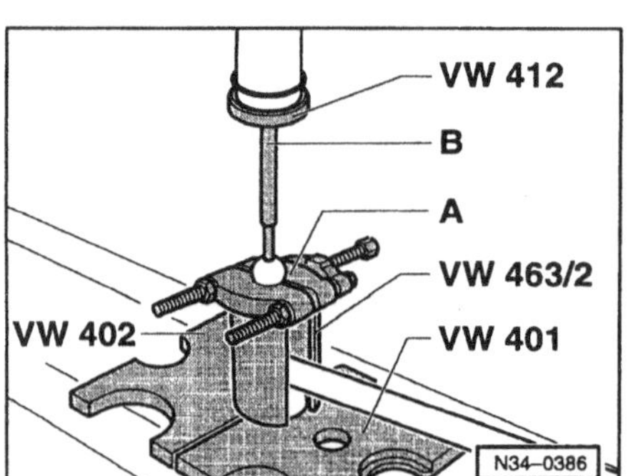

Fig. 6 Ball, pressing off gate rod

A – Separating device 5 ... 60 mm, e.g., Kukko 17/0

B – Driver, 6 mm diameter

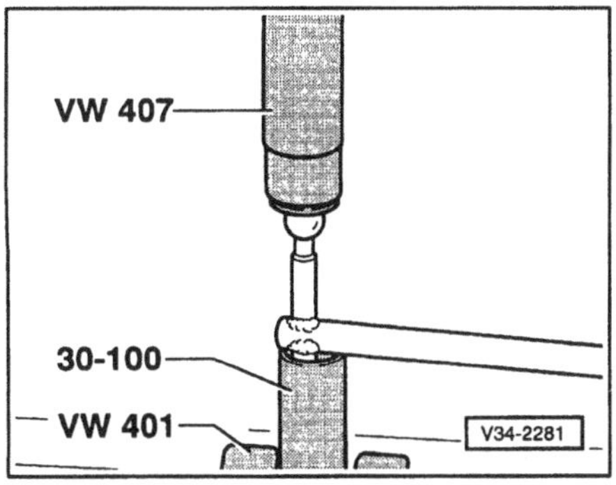

◄ **Fig. 7 Ball, pressing on gate rod**

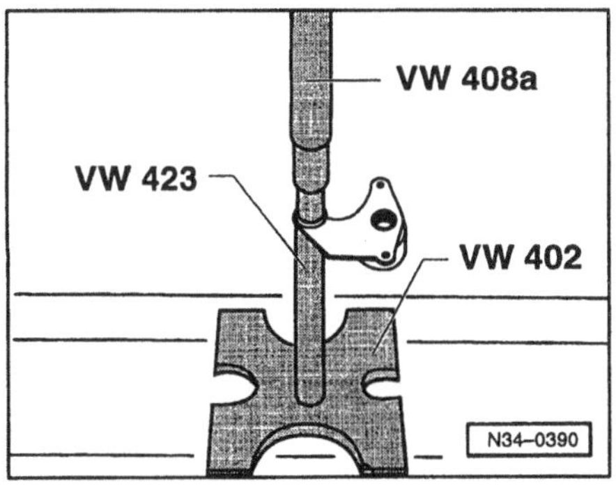

◄ **Fig. 8 Bushing, pressing into transmission selector lever**

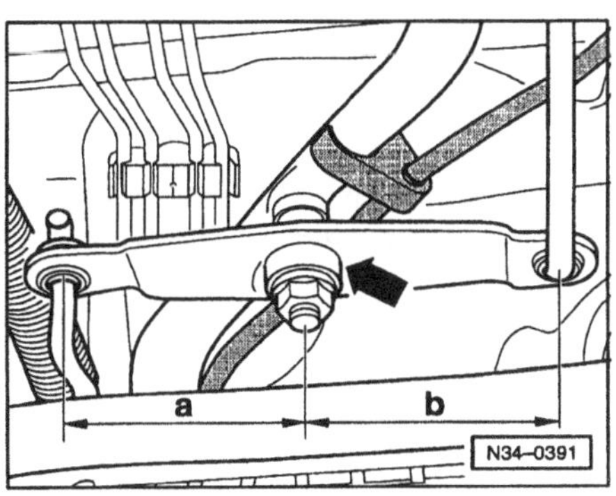

◄ **Fig. 9 Relay lever, installation position**

Dimension "a" = 78 mm - points to left side of vehicle

Dimension "b" = 80 mm - points to right side of vehicle

Thicker shoulder -arrow- points downward.

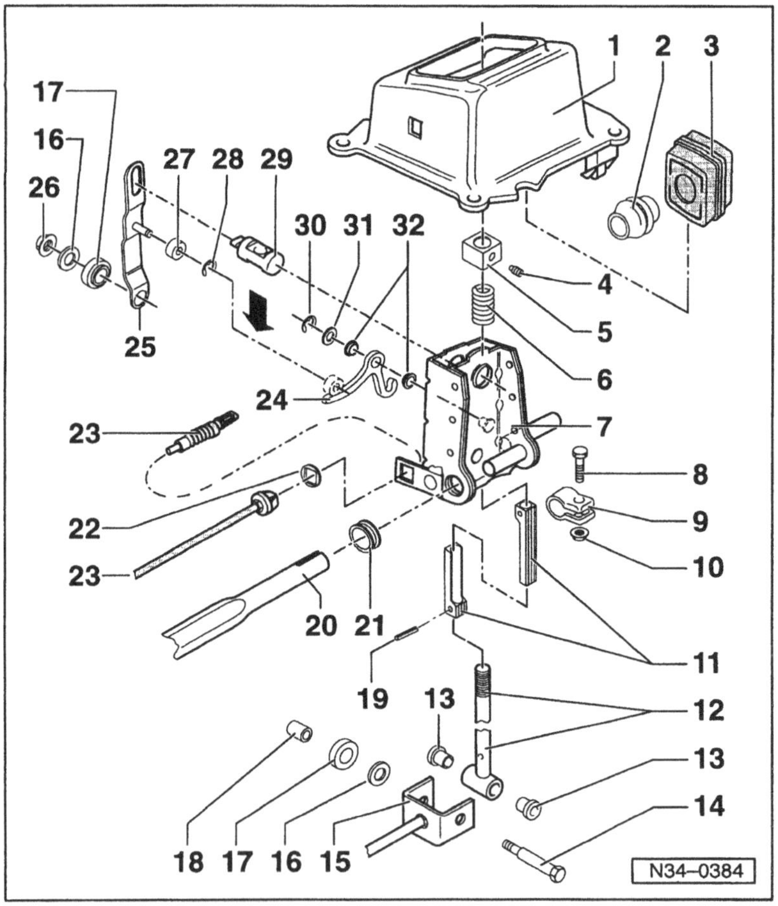

Gear shift mechanism, removing/installing

1 - Gear shift housing
- ♦ Disassembling/assembling to floor

⇒ Page L 34-20

2 - Bushing
- – Pry out with a screwdriver before removing rubber mount (item 3)

3 - Rubber mounting
- – Pry out with a screwdriver after removing bushing

4 - Grub screw, 4 Nm (35 in. lb)

5 - Stop
- ♦ Installation position

⇒ Page L 34-23

6 - Spring

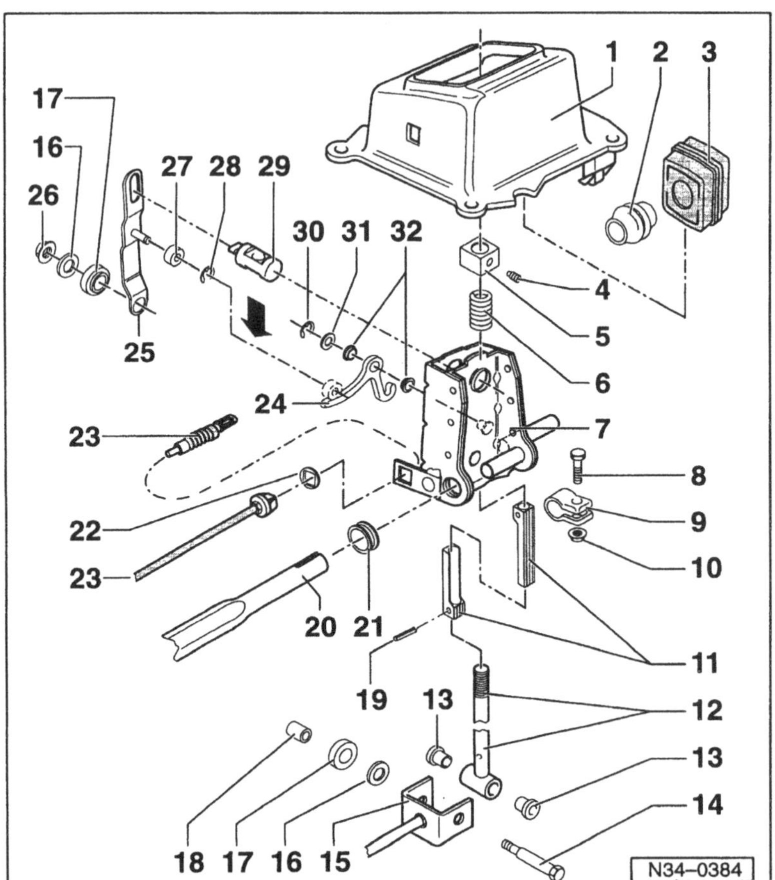

7 - Gear shift bracket
- ♦ Remove with gear shift
- – First remove following components: rear selector rod (item 15), gate rod (item 20)

8 - Hex bolt

9 - Clamp
- – Loosen to adjust gear selector mechanism and to remove selector and gate rods

10 - Self-locking hex nut, 20 Nm (15 ft lb)

11 - Guide sleeve
- ♦ Two part

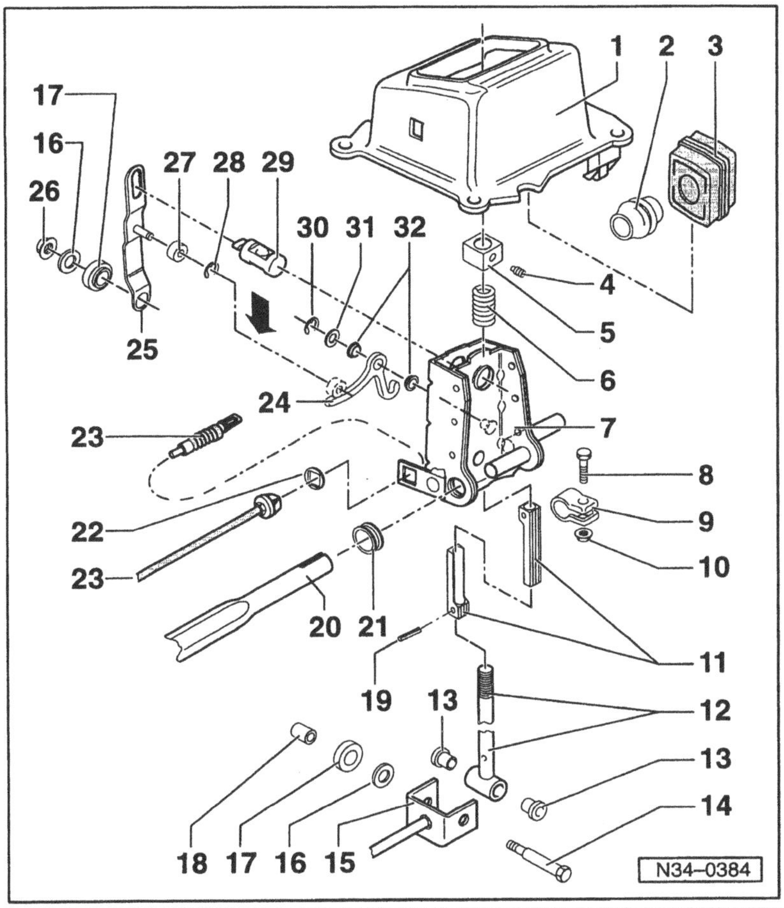

12 - Gear shift
- ♦ Top section angled rearward

13 - Bushing

14 - Special bolt

15 - Rear selector rod

16 - Washer

17 - Bushing

18 - Bushing

19 - Spring pin
- – Knock out with drift

20 - Gate rod
- ♦ Removing/installing to transmission

⇒ Page L 34-19 and L 34-21

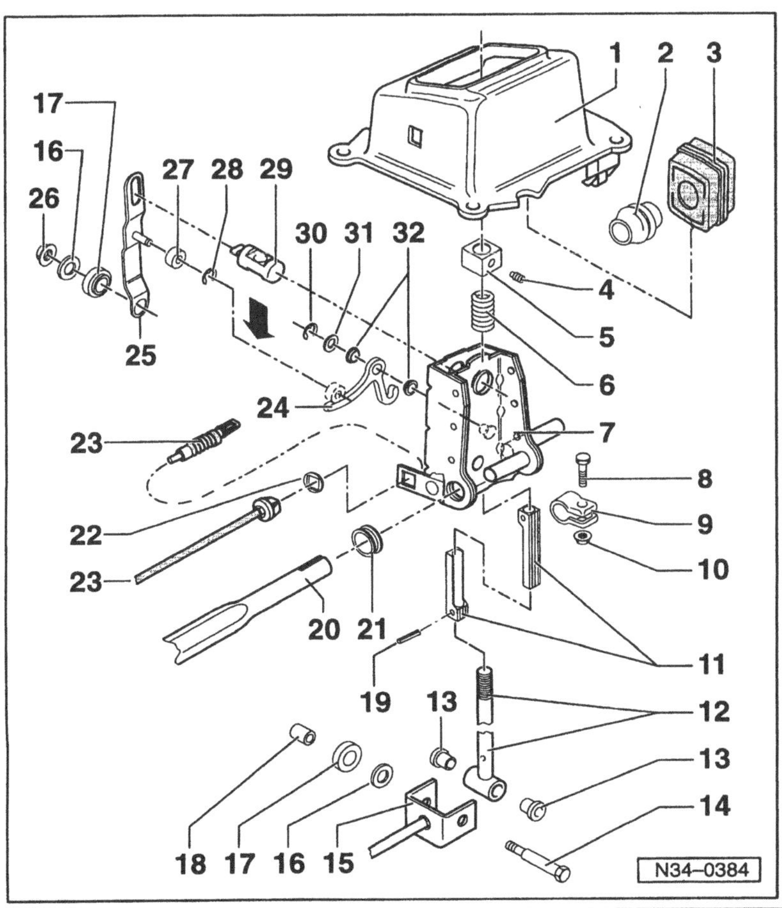

21 - Bushing

22 - Rubber washer

23 - Reverse gear cable
- – Hook into lever (item 24)

24 - Lever
- ♦ Changes downward motion -arrow- of gear shift to cable movement

25 - Pull rod
- ♦ Mounted on bushing (item 18)

26 - Self-locking hex nut

27 - Roller

28 - Circlip

29 - Mounting pin

30 - Circlip

31 - Washer

32 - Bushing

Gear selector mechanism, removing/installing

Removing

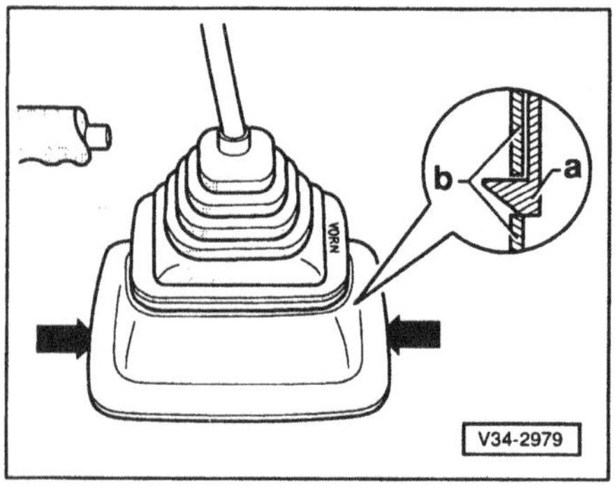

- Unclip boot from frame, unscrew and remove from gearshift together with gear knob
  ⇒ Page L 34-3, items 1 and 2.
- ◄ Release frame -a- from gearshift housing -b- by prying upward in front and rear areas -arrows-.

Note:

Only remove and install boot and frame together.

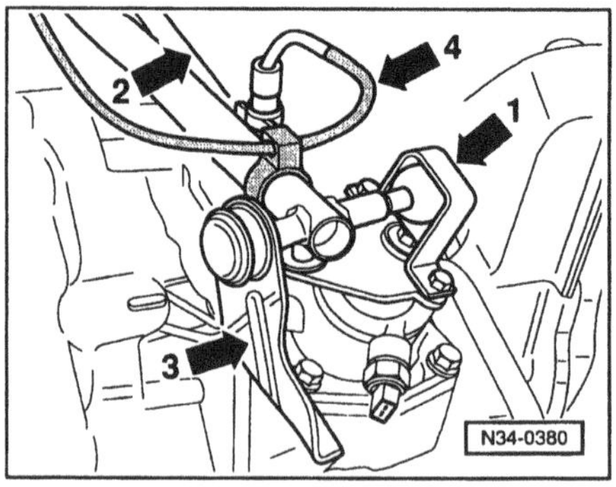

- Remove seal from floor plate.
- ◄ Unbolt swivel head -arrow 1-.
- Pry front selector rod off transmission selector lever -arrow 2-.
- Unbolt mounting bracket with gate rod -arrow 3-.
- Unclip reverse gear lock cable -arrow 4- from reverse gear locking unit.

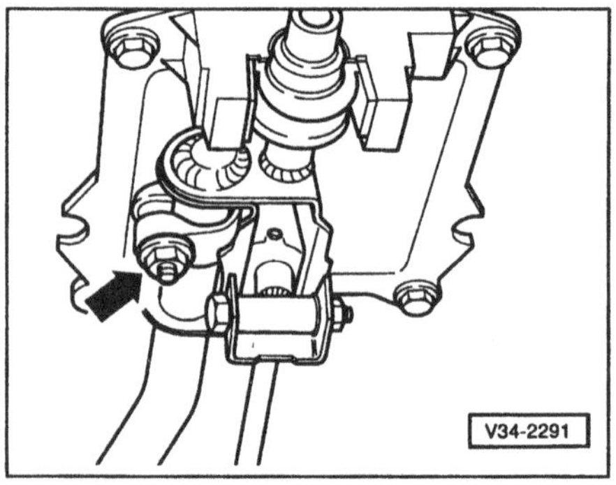

- Remove rear selector rod from relay lever.
- Remove rear selector rod from selector lever.
- ◄ After loosening the clamp -arrow- pull gate rod forward, off gearshift bracket.
- Remove gearshift housing from floor plate.
- Press gear shift bracket together with gear shift forward, so as to guide around fuel tank and then remove downward.
- Guide gate rod with mounting bracket, relay lever and front selector rod out rearward.

Installing

Install in reverse order.

- Guide gear shift housing with gear shift bracket and gear shifter into position from below and secure.

- Guide gate rod into gearshift bracket and slide onto guide pin. First install clamp
⇒ Page L 34-20, illustration V34-2291, arrow.

Note:

Ensure freedom of movement at gate rod/gearshift bracket joint, so that the selector mechanism can be adjusted
⇒ Page L 34-23.

- Hook rear selector rod into relay lever, right
⇒ illustration N34-0391, Page L 34-14.

- Adjust selector mechanism
⇒ Page L 34-23.

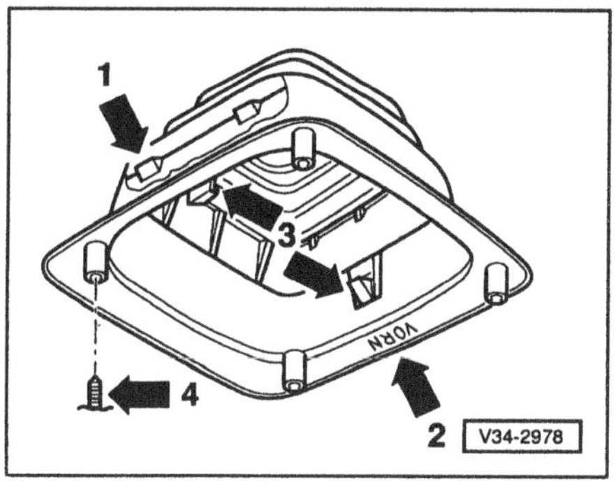

◄ – Engage lugs -arrow 1- around circumference of boot into frame.

- Place seal around edge of floor plate.

- Place frame mounting tubes on gearshift housing securing screws -arrow 4- and then press, first front and then rear, fully home.

Inscription "VORN" (front) -arrow 2- points in direction of travel

- Engage frame locking devices -arrow 3- by pressing into gearshift housing cut-outs
⇒ Page L 34-19, illustration V34-2979.

- Thread on gearshift knob together with boot, holding boot in position.

- Clip boot into frame.

◄ One lug -arrow- on inside of boot is applied on front (direction of travel).

Gear selector mechanism, adjusting

Notes:

The following points are essential to ensure correct selector adjustment:

- ♦ *Perfect operation of selector mechanism, operating and relay elements.*
- ♦ *Light selector mechanism operation.*
- ♦ *Transmission, clutch and clutch operating mechanisms must be in perfect condition.*

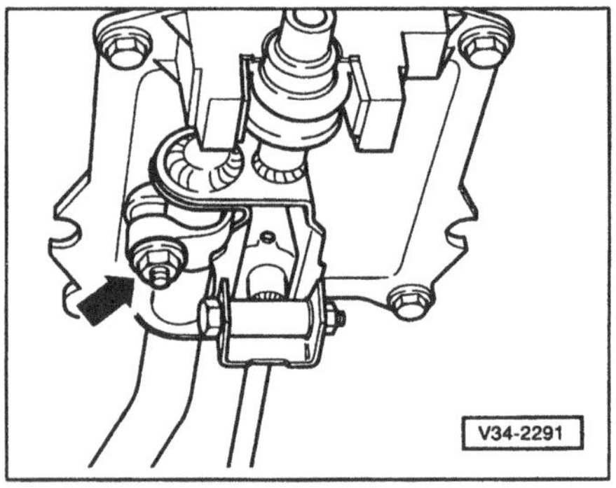

– Place transmission in neutral position.

– Remove gear shift knob, boot, and boot frame
⇒ Page L 34-19.

◄ – Loosen nut -arrow-.

● The gate rod/gearshift bracket connection must be free to move.

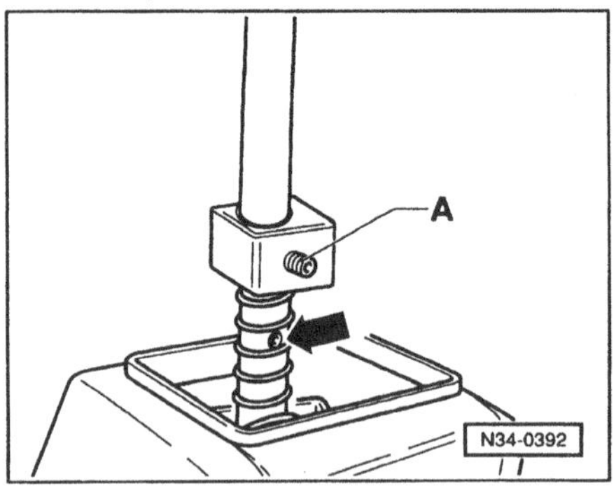

Note:

Check whether contact surfaces on selector lever and gearshift housing stops are aligned parallel to one another. Adjust if necessary.

◄ – Align stop so that set screw -A- locates in gearshift depression -arrow-.

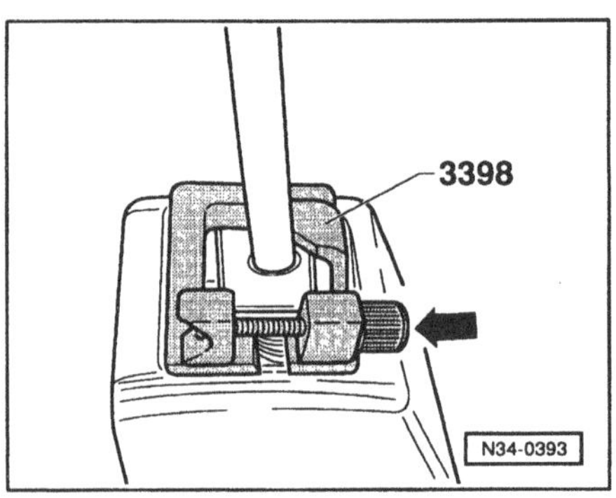

◄ – Install gear shift jig until it sits on stop and lock by tightening knurled wheel -arrow-.

– Align gate rod free of stress (transmission in neutral) and tighten nut to 20 Nm (15 ft lb).
⇒ Page L 34-23, illustration V34-2291, –arrow–.

– Pull gear lever towards seat, loosen knurled nut and remove jig.

Functional check

● The gear shift lever must rest in the 3rd/4th gear plane with the transmission in neutral. The lower area of the gearshift must be vertical.

– Operate clutch.

– Select all gears several times. Pay particular attention to operation of reverse gear catch.

Should a gear fail to engage smoothly after repeated selection then continue as follows:

– Engage 1st gear.

– Remove freeplay in selector mechanism by carefully moving gearshift to left.

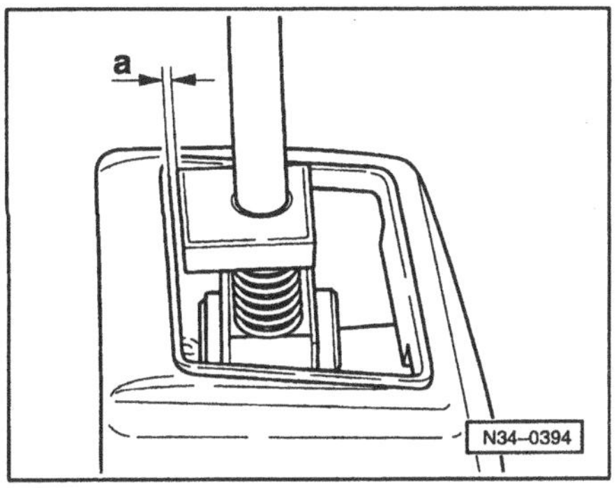

◄ • The distance between gearshift stop and gearshift housing must be approx. 2 mm (dimension "a").

– Repeat adjustment sequence if necessary. Reinstate dimension "a" with 1st gear engaged, if necessary.

– Install gear shift knob and covering
⇒ Page L 34-22.

Transmission, removing/installing

Removing

– Switch ignition off and disconnect battery strap (in engine compartment and, if equipped, under driver's side).
⇒ Electrical system; Repair group 27

Note:

Obtain radio code for vehicles with coded radio .

– Remove radiator grille with trim frame.
⇒ General body repairs; Repair group 66

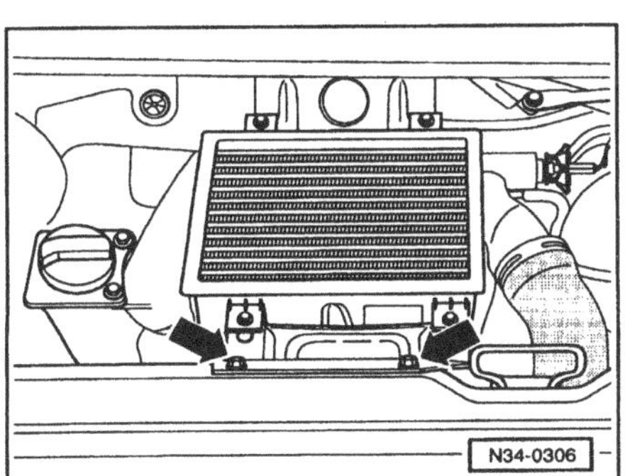

◄ – Remove charge air cooler from lock carrier -arrows- if equipped.

– Fold radiator and lock carrier out forward.
⇒ General body repairs; Repair group 66

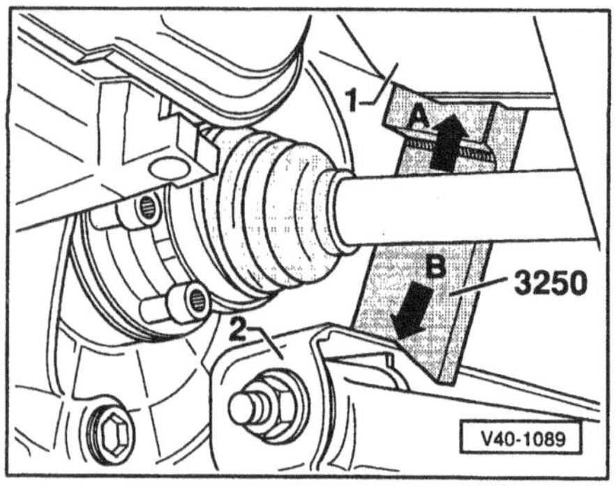

◄ – Install support frames 3250, with vehicle in unloaded condition, to both sides before vehicle is raised.

Note:

Support frames 3250 allow removal of left drive shaft (illustration) and left and right shock absorber/stabilizer bar coupling link.

Installation position:

Support frame bearing surface -arrow A- must be located below upper control arm -1-.

V-shaped opening -arrow B- bears on assembly carrier -2-.

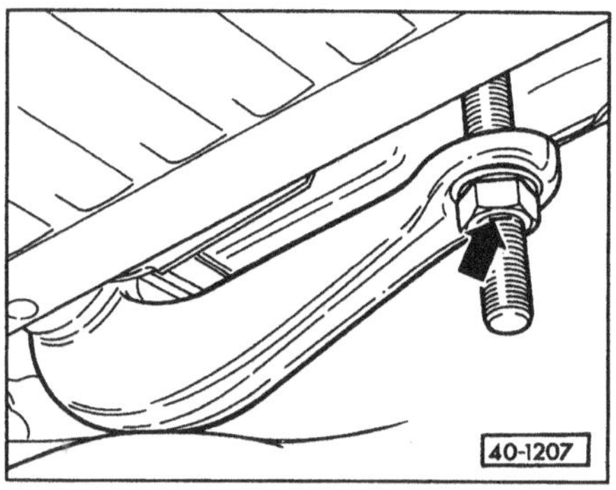

Note:

◄ *Relieve torsion bar tension with special tool 3257 if necessary. First measure thread length (arrow).*
⇒ Suspension, Wheels, Brakes, Steering; Repair group 40

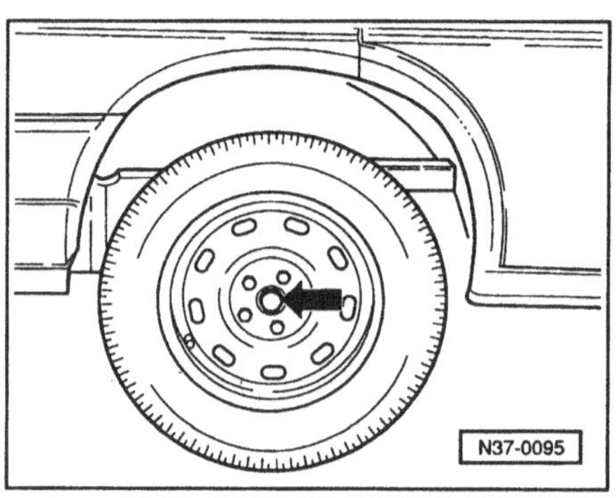

◄ – With vehicle weight on wheels loosen left and right drive shaft/wheel hub bolt and left and right wheel bolts.

– Raise vehicle.

– Remove noise insulation tray below engine/transmission..

Remove left and right drive shafts as follows:

– Remove bolt for left shock absorber/stabilizer bar coupling rod.

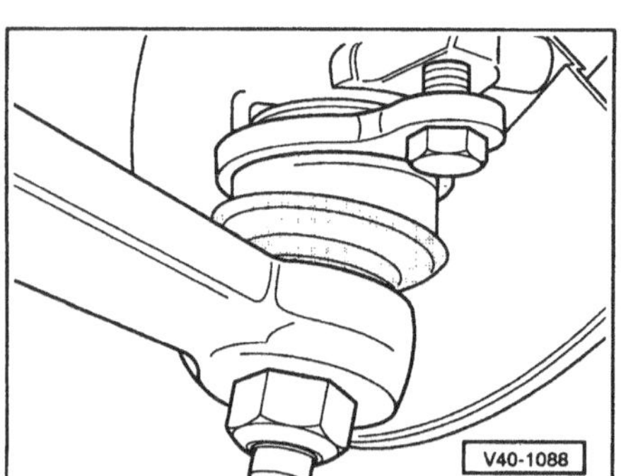

– Compress left shock absorber completely.

– Disconnect drive shaft at flange.

– Remove wheel.

◄ – Separate lower wheel bearing housing ball joint bolted connection. Support ball joint when separating

– Remove drive shaft.

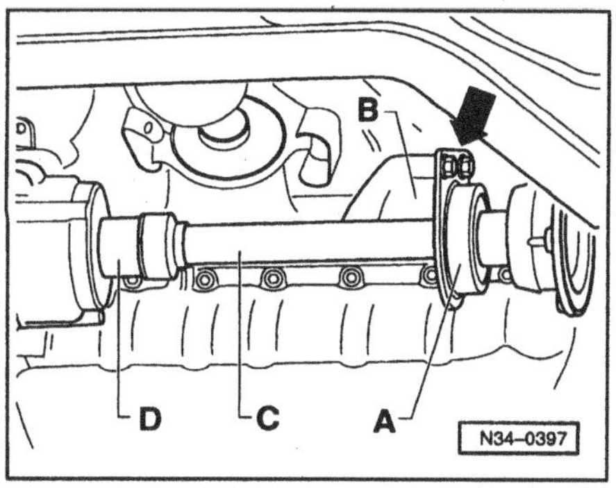

◄ – Unbolt flange shaft mounting -A- from bracket -B- -arrow-.

– Pull right-hand flange shaft -C- out from stub shaft -D-.

– Remove battery ground strap from transmission and mount bracket.

– Remove all electrical connections from transmission and starter motor.
⇒ Electrical system; Repair group Gr. 27

– Unscrew speedometer drive shaft cable or remove sensor connector and speedometer sensor.

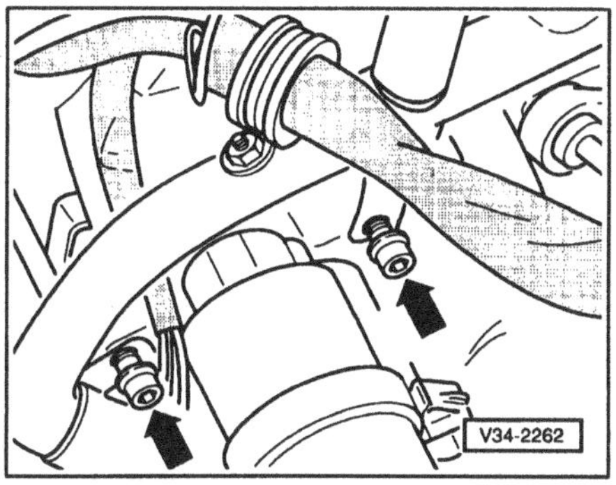

– If so attached, unfasten electrical wiring from coolant hoses.

◄ – Unfasten coolant lines at starter motor -arrows- (from expansion tank to water pump).

– Remove wiring harness from lines.

– Remove starter.
⇒ Electrical system; Repair group 27

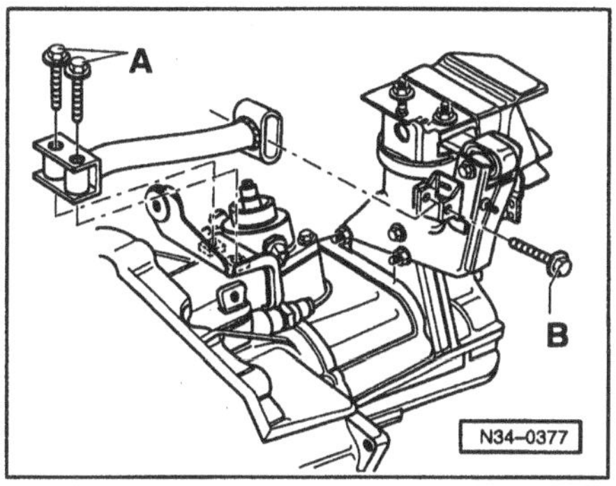

– Remove front exhaust stop plate.
⇒ 5-Cyl. Diesel engine, Mechanics; Repair group 26

– Separate front exhaust pipe at catalyst.
⇒ 5-Cyl. Diesel engine, Mechanics; Repair group 26

◄ – Remove transverse support after removing bolts on selector mounting bracket -A- and bolt on left-hand transmission support -B-.

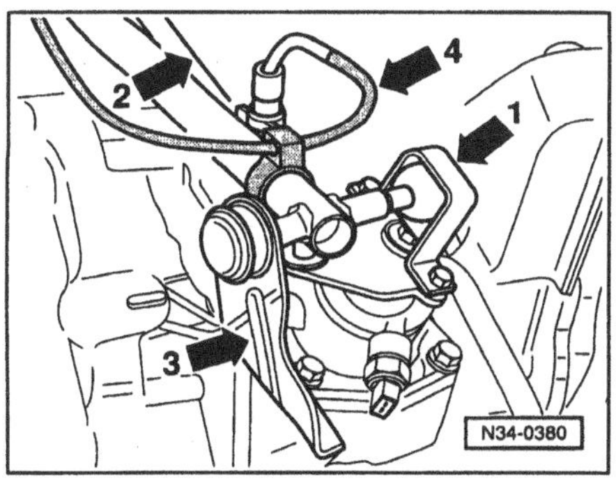

◄ Remove selector mechanism from transmission as follows:

– Remove back-up light switch.

– Remove swivel head -arrow 1-.

– Pry front selector rod off transmission selector lever -arrow 2-.

– Unbolt mounting bracket with gate rod -arrow 3-.

– Unclip reverse gear lock cable -arrow 4- from transmission.

– Tie-up selector and gate rods.

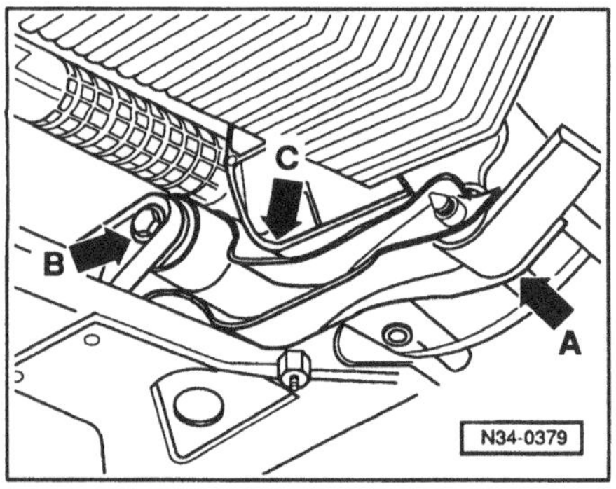

- Remove clutch mechanism line from transmission, release from transmission support retainer and tie-up.
- Remove auxiliary heater line, if necessary.
- ◄ Remove pendulum support after removing bolts arrows -A- and -B- by pushing up lightly on rear of transmission restriction -arrow C-.

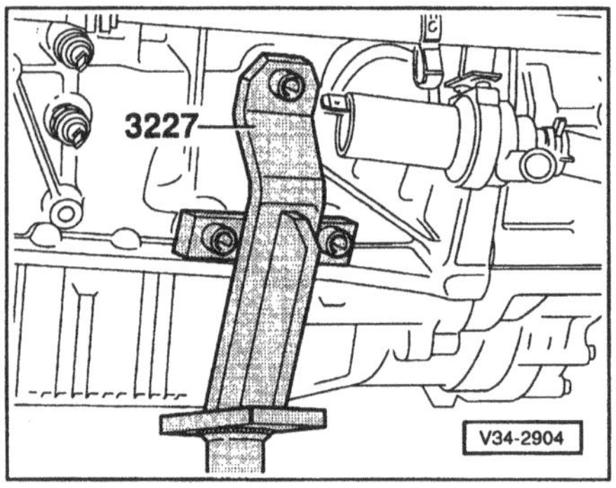

- Remove retaining plate for electrical connectors from crankcase.
- ◄ Bolt engine support to block.

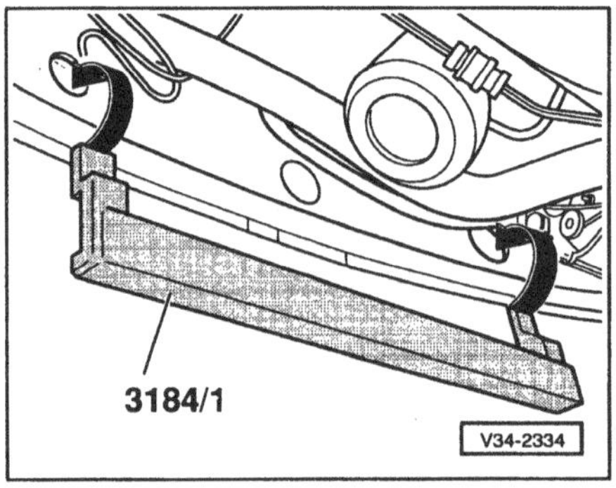

- ◄ Attach auxiliary support 3184/1 for removal device 3184 into cross panel -arrows-.

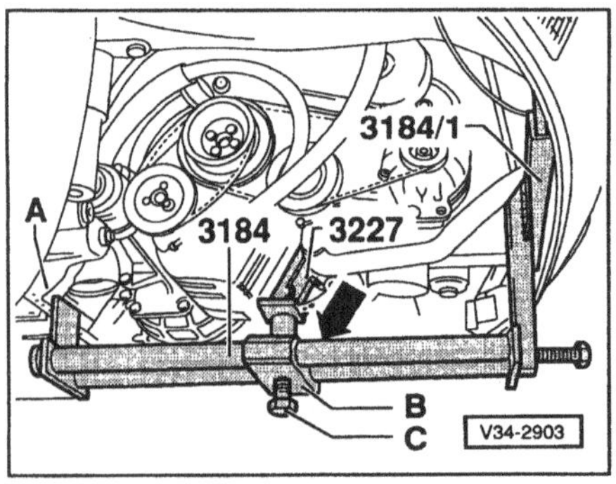

- ◄ Install removal device 3184 into auxiliary support 3184/1 and into subframe -A-. Screw spindle -C- into hole on engine mounting.
- Align removal device so that slide -B- is resting against right-hand tube -arrow-.
- Secure removal device to subframe -A- and to auxiliary support 3184/1 by tightening angle plates.
- Relieve subframe loading by raising engine/transmission by means of spindle -C-.

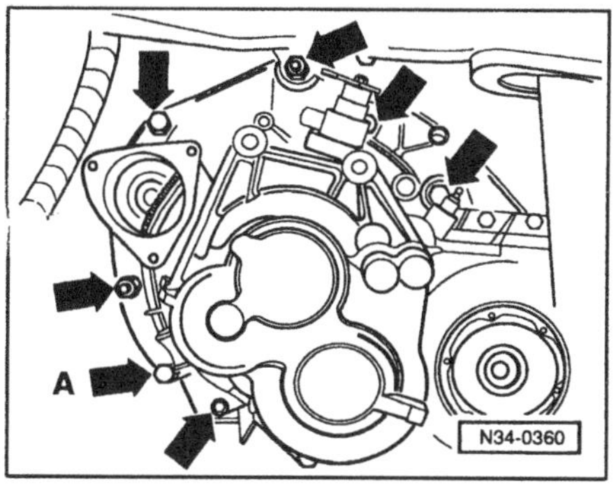

◄ – Remove engine/transmission connecting bolts -arrows- except for bolt -arrow a-. Also remove flywheel small cover plate.

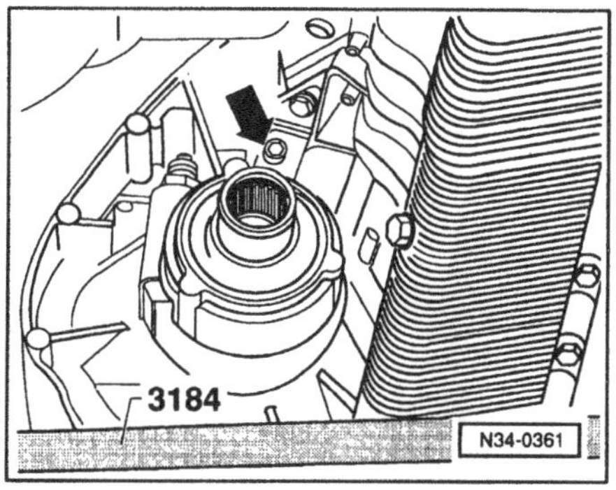

◄ – Remove engine/transmission hex socket-head bolt -arrow-.

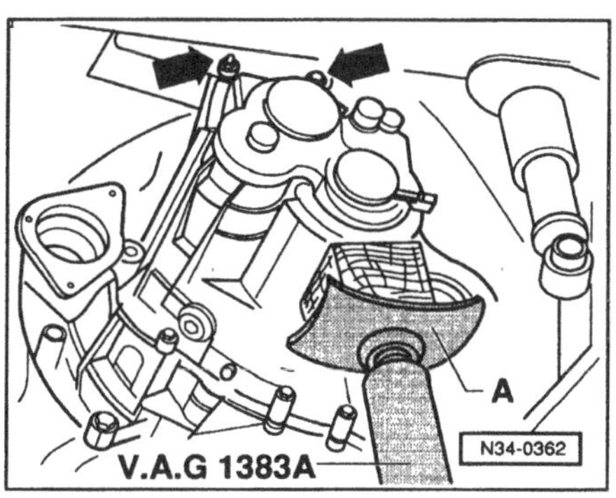

◄ – Support transmission, left, with transmission jack VAG 1383 A.

A – Trolley jack insert

– Unbolt left-hand transmission support -arrows- on transmission.

– Lower engine using removal device spindle.

– Carefully rock engine/transmission alternately forward and then lower until V-belt pulley almost contacts long member.

Note:

Do not damage coolant system hoses and lines on cross member when rocking engine/transmission assembly forward.

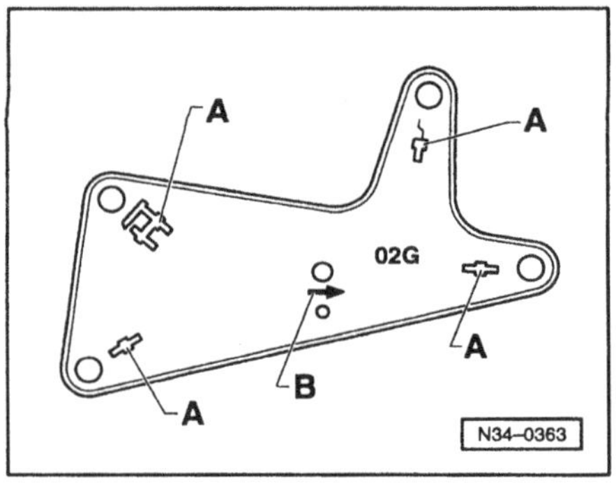

- Assemble transmission jack VAG 1383A with transmission support 3282, adjustment plate 3282/23 (for "02G" transmission) and support elements as follows:
- Place adjustment plate 3282/23 on transmission support 3282. (Adjustment plate only fits in one position).
- Align arms of transmission support according to holes in adjustment plate.
- ◄ Bolt on support elements -A-, as illustrated on adjustment plate.

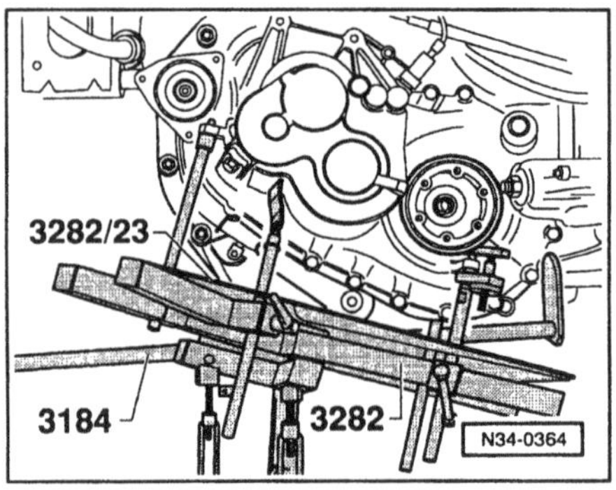

- ◄ Place transmission jack, with transmission support, under vehicle, arrow symbol -B- ⇒ Illustration N34-0363) on adjustment plate points to front of vehicle.
- Align adjustment plate parallel to transmission and lock safety supports on transmission.
- Remove lower engine/transmission securing bolts ⇒ Page L 34-33 arrow -A- on illustration N34-0360).

- Press transmission off engine and remove cover plate from between engine/transmission.
- Carefully lower transmission and swing out.
- Constantly correct transmission position via transmission jack spindles. Ensure there is sufficient clearance to assembly carrier and removal device 3184.

Note:

If the vehicle is to be moved with the removal device fitted then the following must be observed:

- ♦ *Screw spindle in onto stop, to lift or lower the engine/transmission.*
- ♦ *Exercise extreme care in the vicinity of a hoist.*

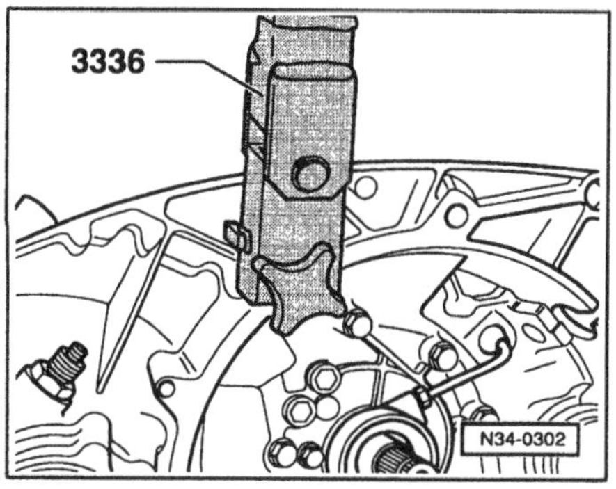

Transmission, transporting

Special tool 3336 may be used to transport the transmission and also when setting up transmission mount 3282.

◄ – Bolt transmission lifting device 3336 to clutch housing.

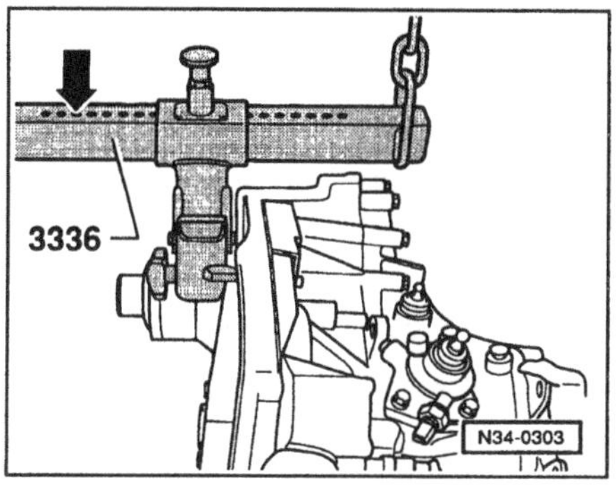

◄ – Set support arm on sliding piece with locking pin (arrow).

Number of holes visible = 7

– Remove transmission with workshop crane and transmission lifting device 3336.

Installing

Install in reverse order.

Note:

– *Determine transmission mounting parts via parts catalog.*

– *Check whether dowel sleeves for aligning engine/transmission are installed in cylinder block, install if necessary.*

– *When replacing transmission ensure that intermediate plate is correctly positioned.*

– *When replacing transmission, transfer gear selector lever to new assembly.*

– *Clean input shaft splines and lightly grease with G000 100 grease.*

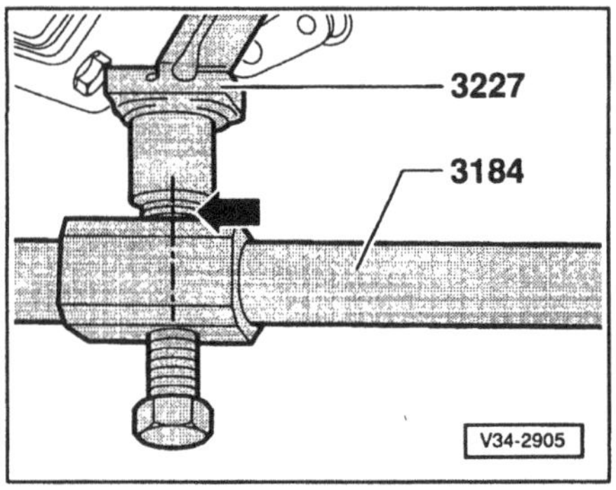

– Alternately swing in and raise engine/transmission assembly.

Install engine/transmission mountings stress-free, as follows:

◄ – Before installing bolts to secure left-hand mounting bracket to body ensure that engine mounting pin and removal device spindle -arrow- are in line.

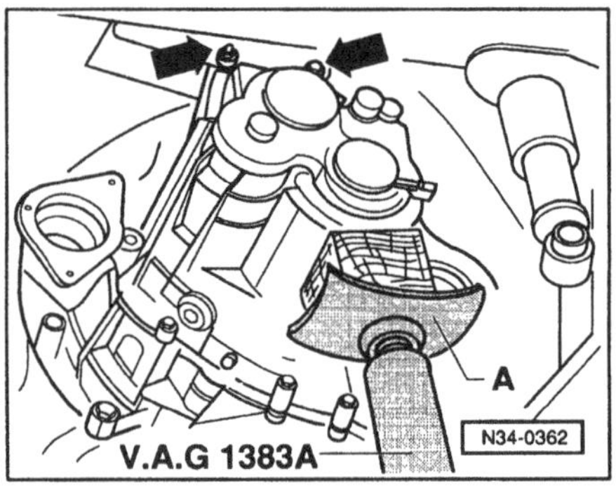

◄ – Use transmission support VAG 1383 A to support transmission when guiding transmission mounting bracket.

A - Trolley jack insert

– Tighten rear transmission mounting to torque setting.

– Tighten pendulum support to torque.

> ***CAUTION!***
>
> ***Do not remove removal device 3184 until the bolts securing the left-hand mounting bracket have been securely tightened.***

– Install front exhaust pipe to catalyst
⇒ Repair group 26; Removing and installing parts of exhaust system.

– Always replace seal between flange shaft and drive shaft, but first remove protective foil before bolting drive shafts to flanges.
⇒ Running gear; Repair group 40; Drive shaft, servicing

– Install starter/wiring.
⇒ Repair group 27; Starter removing/installing .

– Check transmission oil
⇒ Page L 00-3

– Adjust gear change mechanism
⇒ Page L 34-23.

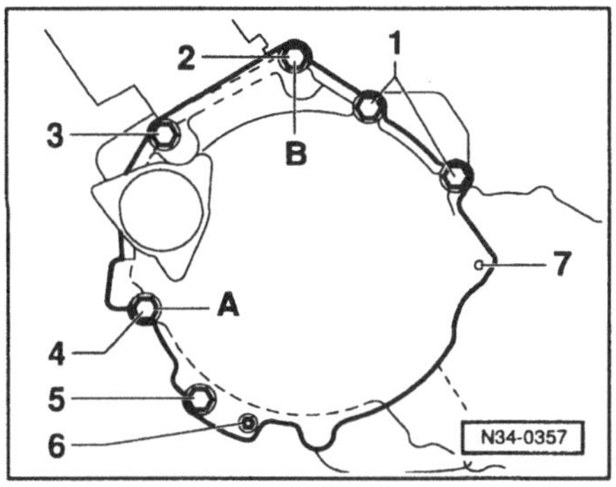

Tightening torques

Transmission to Engine

Pos.	Bolt	Qty.	Nm (ft lb)
1[1]	M 12 x 70	2	80 (59)
2	M 12 x 70	1	80 (59)
3	M 12 x 60	1	80 (59)
4	M 12 x 100	1	80 (59)
5	M 12 x 130	1	80 (59)
6	M 8 x 45	1	25 (18)
7[2]	M 8 x 45	1	25 (18)

[1] Additionally for flywheel cover plate

[2] Tighten from engine side

Positions A & B = Dowel sleeves

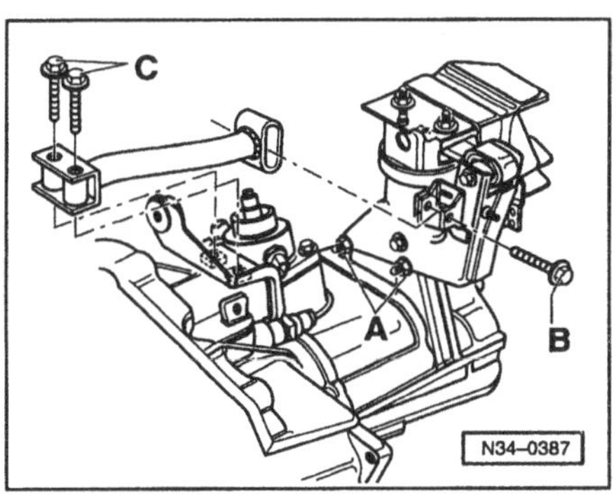

Front left transmission mounting

Bolts -A-	50 Nm (37 ft lb) + 90°
Bolt -B-	30 Nm (22 ft lb)
Bolts -C-	30 Nm (22 ft lb)

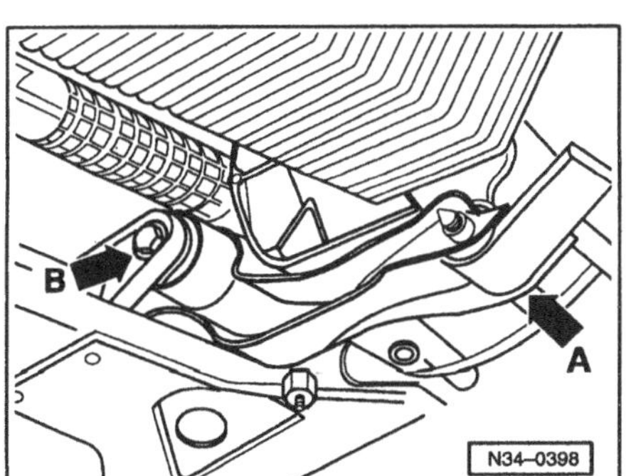

Rear transmission mounting

Bolt -A-	80 Nm (59 ft lb) + 90°
Bolt -B-	200 Nm (148 ft lb)

Flange shaft mounting to bracket	25 Nm (18 ft lb)
Clutch mechanism valve unit to transmission	5 Nm (44 in. lb)
Selector mechanism mounting bracket to transmission	20 Nm (15 ft lb)
Selector mechanism swivel head to transmission	9 Nm (7 ft lb)
Back-up light switch to switch unit	20 Nm (15 ft lb)

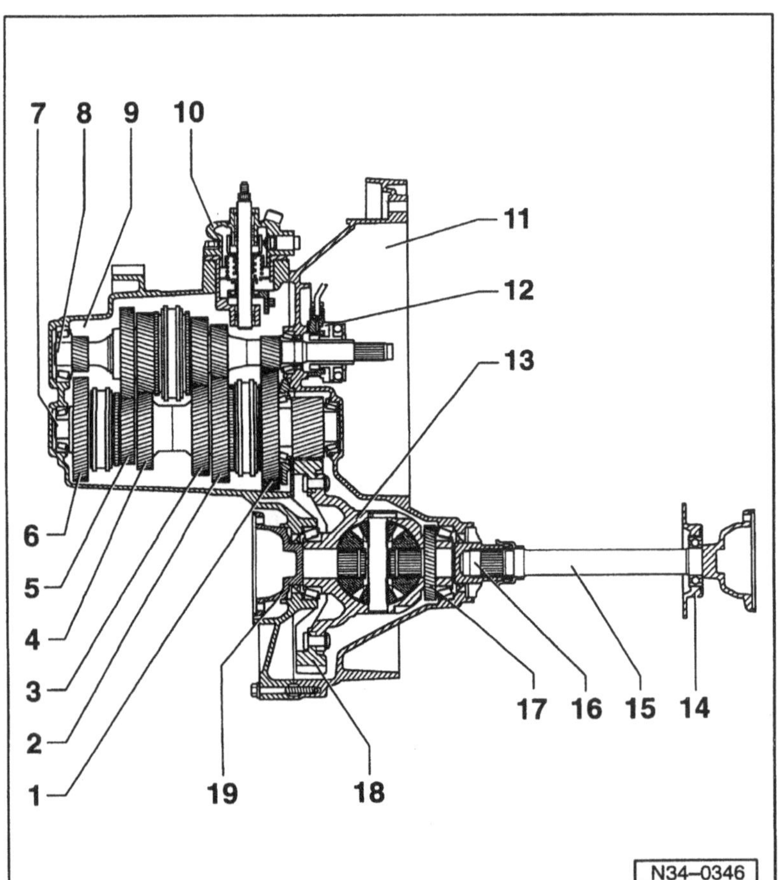

Transmission, disassembling/assembling

Sequence for dismantling/assembling
⇒ Page L 34-52

Assembly overview

1 - 1st gear

2 - 2nd gear

3 - 3rd gear

4 - 4th gear

5 - 5th gear

6 - Reverse gear

- Dismantling/assembling reverse gear wheel

⇒ Page L 35-49

- Removing/installing

⇒ Page L 34-52

7 - Output shaft

- Dismantling/assembling

⇒ Page L 35-18

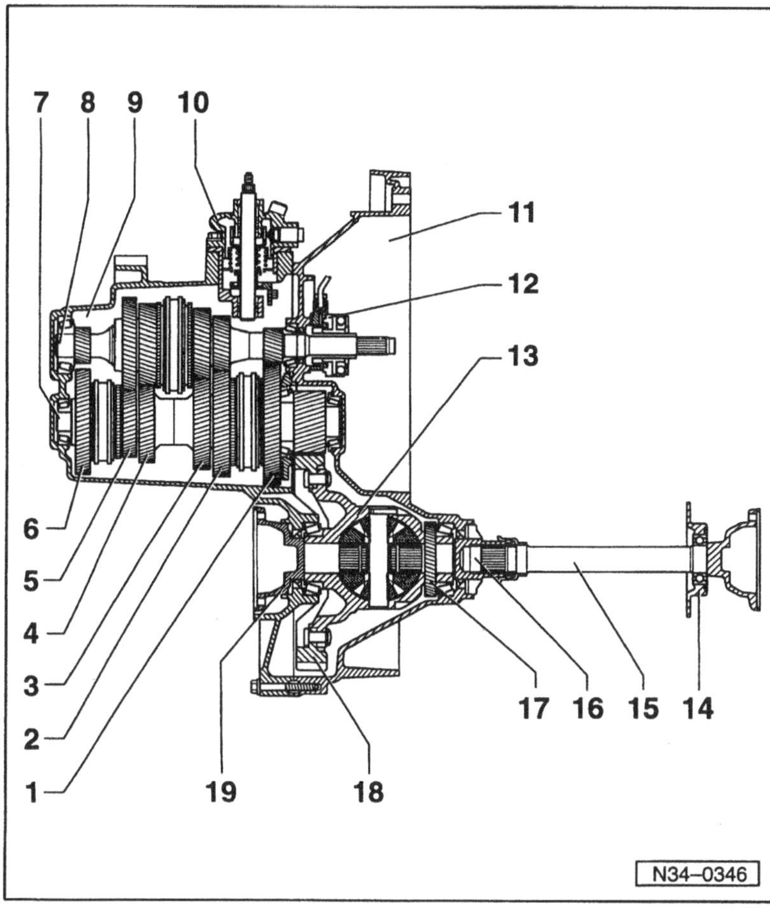

8 - Input shaft
- ♦ Dismantling/assembling

⇒ Page L 34-49

9 - Transmission housing
- ♦ Disassembling from clutch housing

⇒ Page L 34-52
- ♦ Servicing

⇒ Page L 34-59

10 - Selector mechanism
- ♦ Disassembling/assembling

⇒ from Page L 34-70

11 - Clutch housing
- ♦ Servicing

⇒ Page L 34-64

12 - Slave cylinder with release bearing
- ♦ Removing/installing

⇒ Page L 34-52

13 - Differential
- ♦ Disassembling/assembling

⇒ Page L 39-7

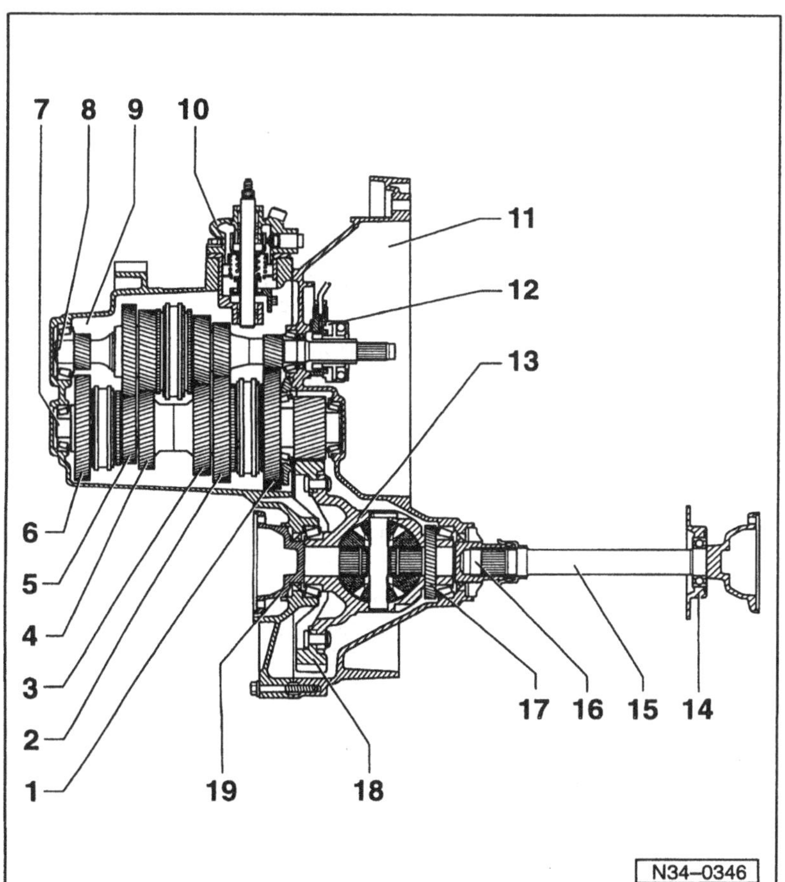
7
8
9
10
11
12
13
6
5
4
3
2
1
19
18
17
16
15
14
N34–0346

14 - Flange shaft bearing
- ♦ Removing/installing

⇒ Page L 34-29

15 - Right-hand flange shaft
- ♦ Removing/installing

⇒ Page L 34-29

16 - Stub shaft
- ♦ Removing/installing

⇒ Page L 39-1

17 - Speedometer drive gear
- ♦ Removing/installing

⇒ Page L 34-52

18 - Final drive ring gear
- ♦ Removing/installing

⇒ Page L 39-7

19 - Left-hand flange shaft
- ♦ Removing/installing

⇒ Page L 34-52

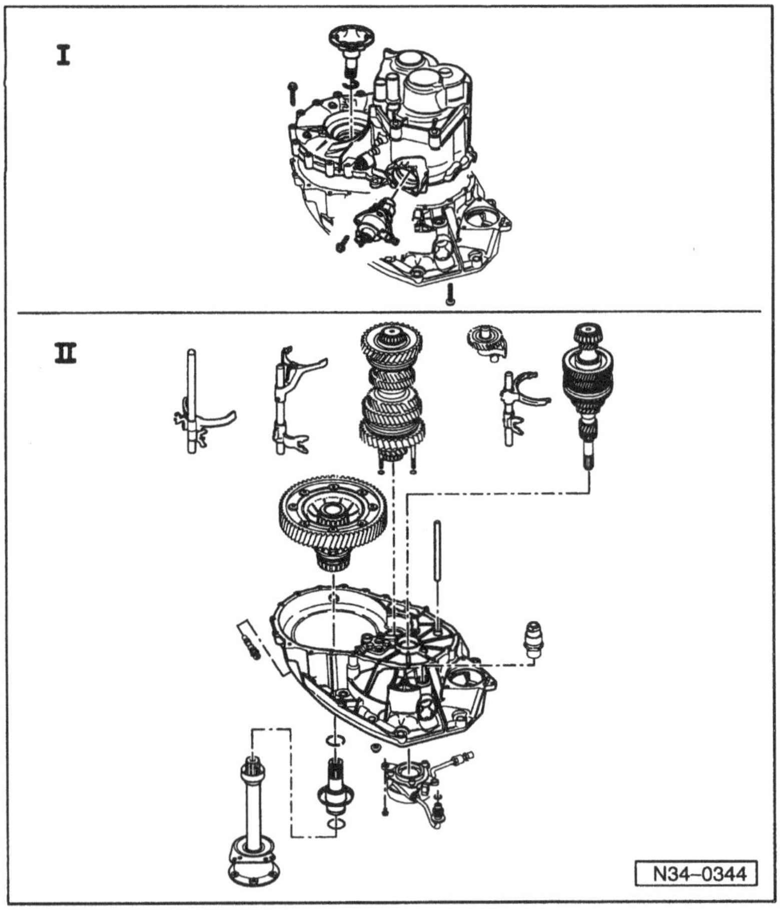

I – Dismantling and assembling transmission housing and selector mechanism
⇒ page L 34-48

II – Dismantling and assembling input shaft, output shaft, differential and selector rods
⇒ page L 34-49

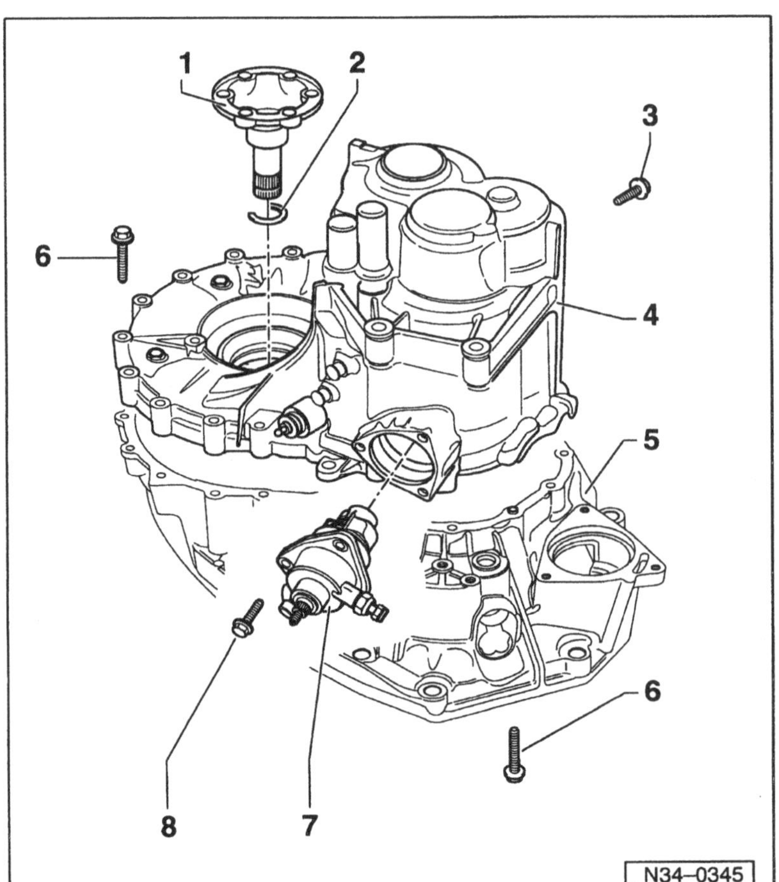

Transmission housing and selector mechanism, removing/installing

1 - Flange shaft
- ♦ Removing

⇒ Page L 34-52
- ♦ Drive in with synthetic–head hammer

2 - Circlip
- – Always replace

⇒ page L 34-58

3 - Bolt, 25 Nm (18 ft lb)

4 - Transmission housing
- ♦ Servicing

⇒ Page L 34-59

5 - Clutch housing
- ♦ Servicing

⇒ Page L 34-64

6 - Hex bolt, 25 Nm (18 ft lb) plus 90° (1/4 turn) further

7 - Selector mechanism
- ♦ Dismantling/assembling

⇒ from page L 34-70

8 - Hex bolt, 25 Nm (18 ft lb)

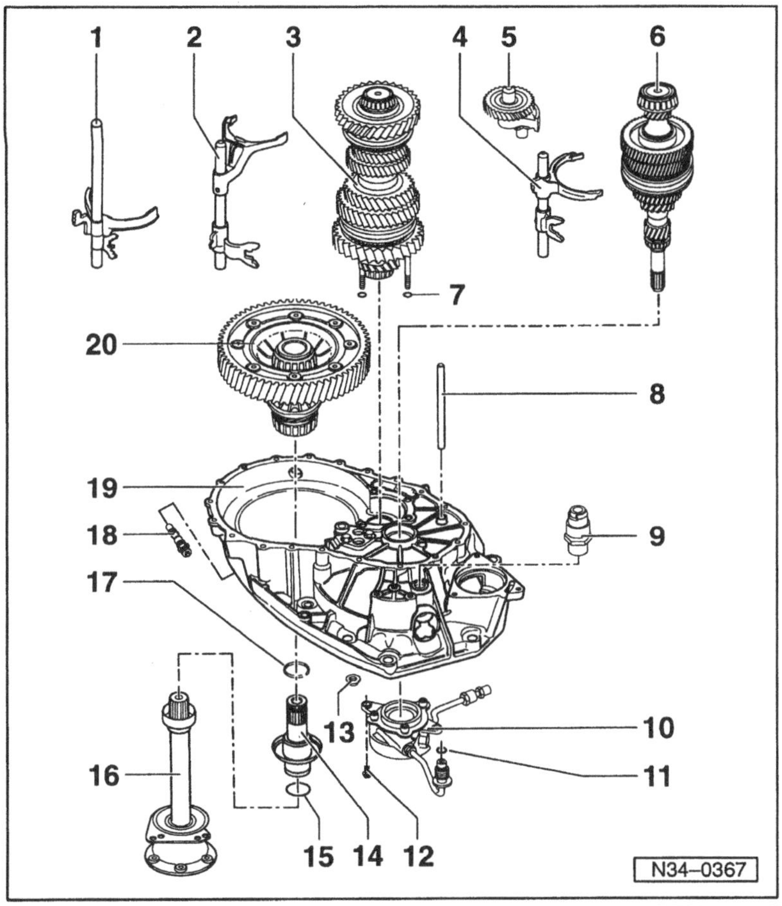

Input shaft, output shaft (pinion shaft), differential and selector rods, disassemble/assembling

1 - 1st and 2nd gear selector rod with selector fork

2 - 5th and reverse gear selector rod with selector fork

3 - Output shaft
- ♦ Disassembling/assembling

⇒ Page L 35-18

4 - 3rd and 4th gear selector rod with selector fork

5 - Reverse gear
- ♦ Removing/installing

⇒ Page L 35-49

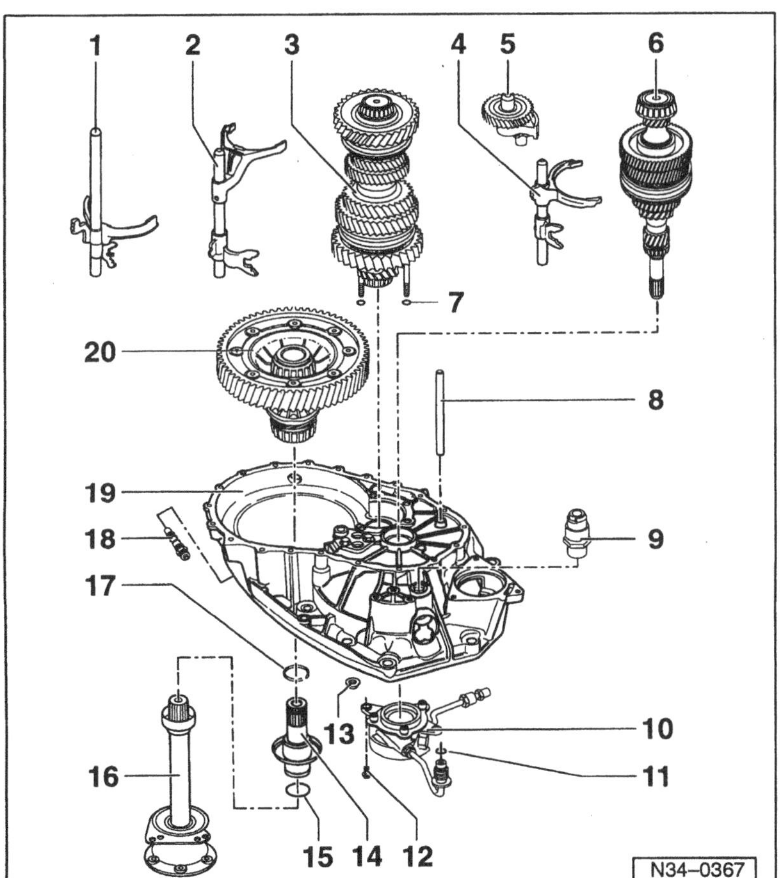

6 - Input shaft
- ♦ Dismantling/assembling

⇒ Page L 35-1

7 - Sealing ring
- ♦ Qty. 4
- – Always replace

8 - Reverse gear guide tube

9 - Valve unit, 5 Nm (44 in. lb)
- ♦ Bolt to slave cylinder pipe

10 - Slave cylinder with release bearing

11 - O-ring
- – Always replace

12 - Hex bolt, 12 Nm (9 ft lb)
- ♦ Qty. 2

13 - Hex nut, 25 Nm (18 ft lb) plus 90° (1/4 turn) further
- ♦ 4 bearing support nuts

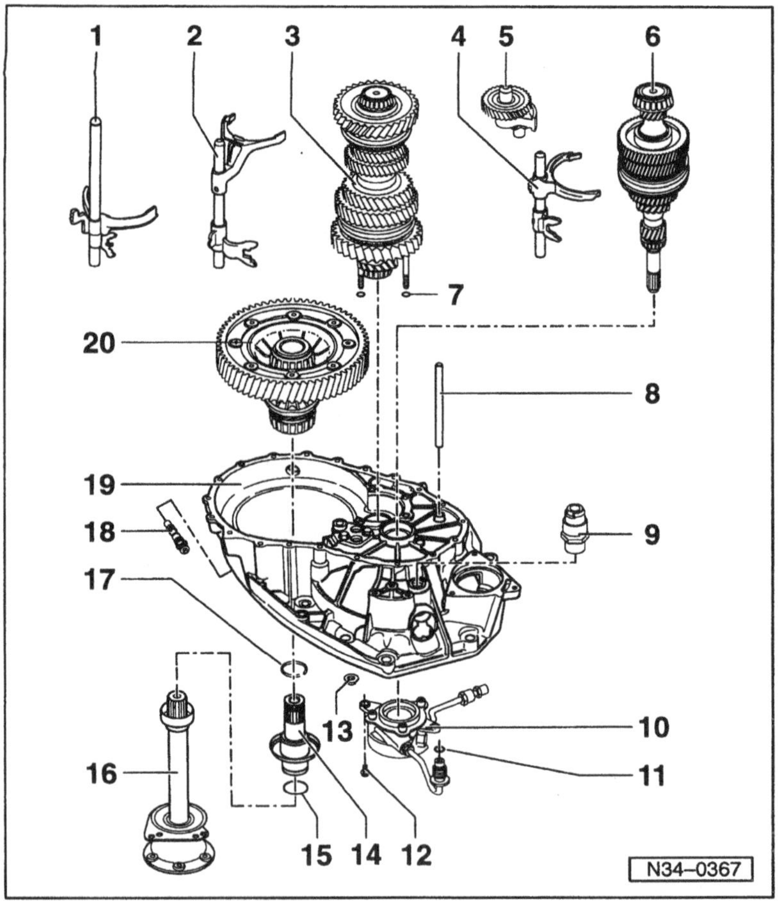

14 - Stub shaft
- ♦ Removing

⇒ Page L 34-52
- ♦ Drive in with synthetic-head hammer

15 - O-ring
- – Always replace

16 - Flange shaft

17 - Circlip
- – Always replace

⇒ Page L 34-58

18 - Speedometer drive

19 - Clutch housing
- ♦ Servicing

⇒ Page L 34-64

20 - Differential
- ♦ Disassembling/assembling

⇒ Page L 39-7

Disassembling/assembling sequence

Disassembling/assembling transmission housing, selector mechanism, input shaft, output shaft (pinion shaft), differential and selector rods

Disassembling

– Drain transmission oil.

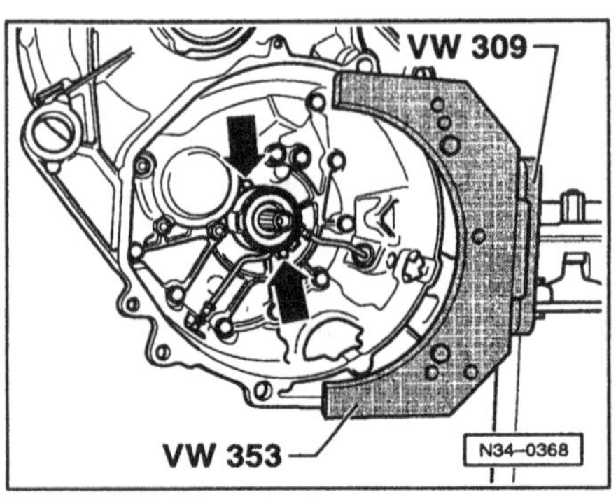

◄ – Mount transmission on repair stand.

– Remove slave cylinder with release bearing –arrows–.

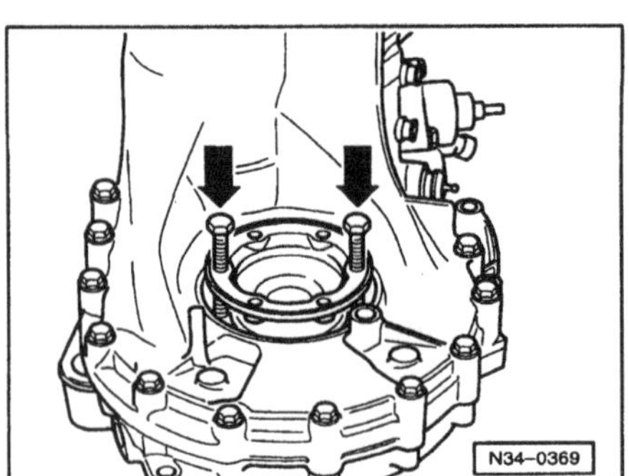

◄ – Remove left flange shaft

Screw in M10 x 75 bolts –arrows– alternately.

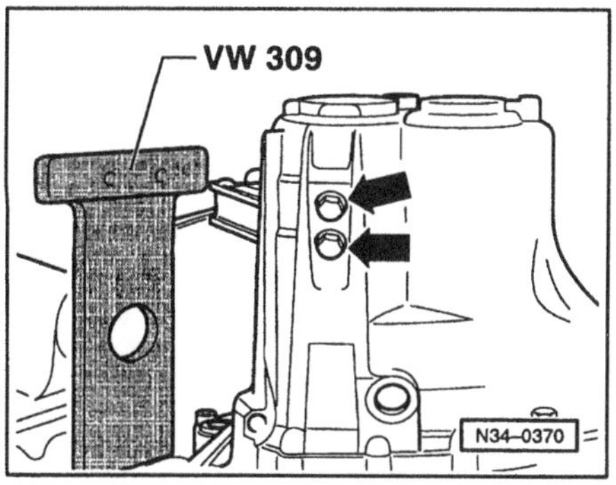

◄ – Remove both bolts –arrows– for reverse gear wheel mount bracket.

◄ – Remove selector mechanism. To do so place selector shaft in neutral position. Remove bolts –arrows– and pull selector shaft out from transmission housing.

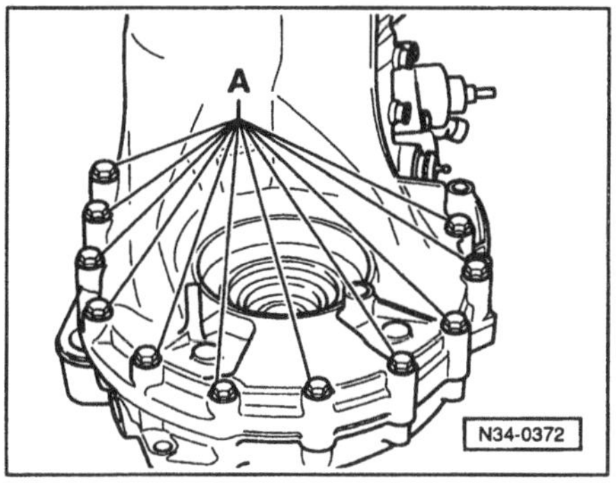

◄ – Remove transmission housing to clutch housing bolts -A- in area of differential.

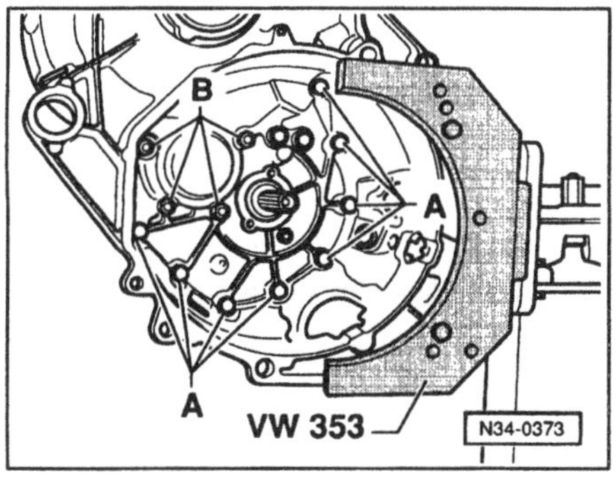

◄ – Remove bolts -A-, from inside clutch housing (secures clutch housing to transmission housing).

Note:

Do not remove nuts -B- for output shaft bearing support.

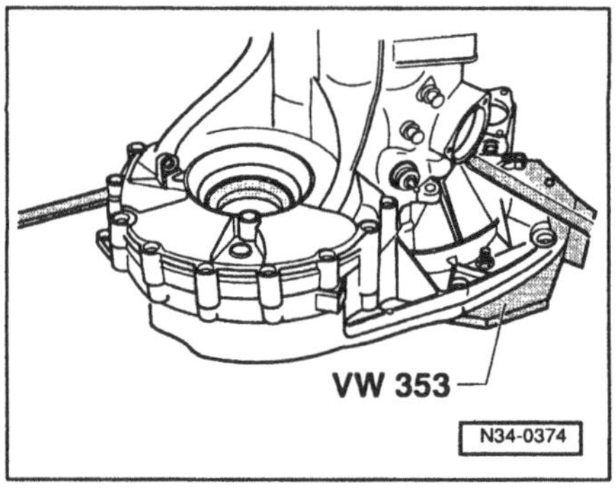

◄ – Remove transmission housing, carefully lift up all around and alternating between sides on protruding flange –arrows–, if necessary (do not damage sealing surfaces).

– When removing, pull pull rod out from locking unit.

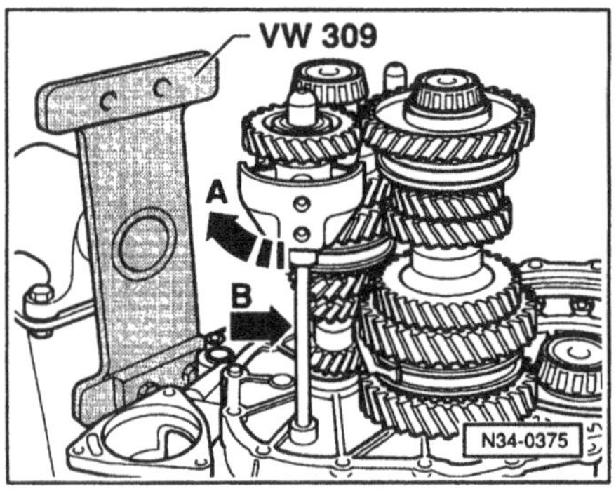

◄ – Swing reverse gear in direction of arrow -A- and remove.

– Remove reverse gear guide tube, arrow B-.

– Remove output shaft bearing support nuts
⇒ Page L 34-54 Item -B-, Illustration N34-0373.

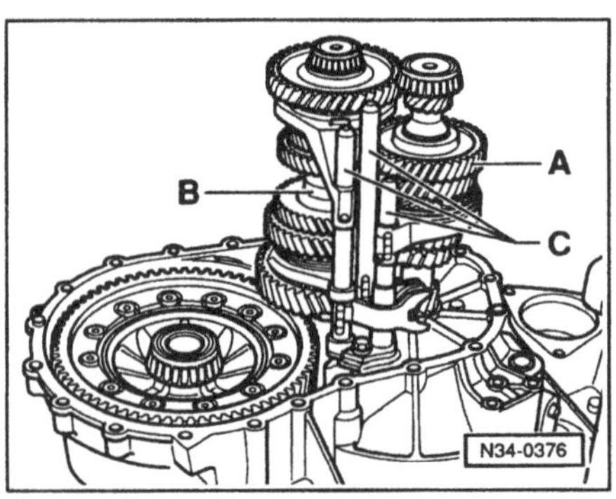

◄ – Remove input shaft -A-, output shaft -B- and selector rods -C- together from clutch housing.

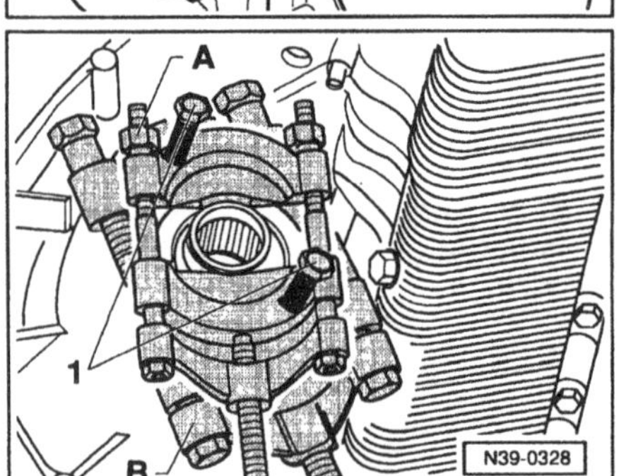

◄ – Remove stub shaft, right (shown with transmission installed).

A - Separating device Kukko 17/0

B - Separating device Kukko 17/1

1 - Bolt M10 x 80 (Qty. 2)

– Turn bolts -1- evenly against separating device -B-.

– Remove speedometer drive.

– Remove differential.

Assembling

- Install differential.
- Install speedometer drive.

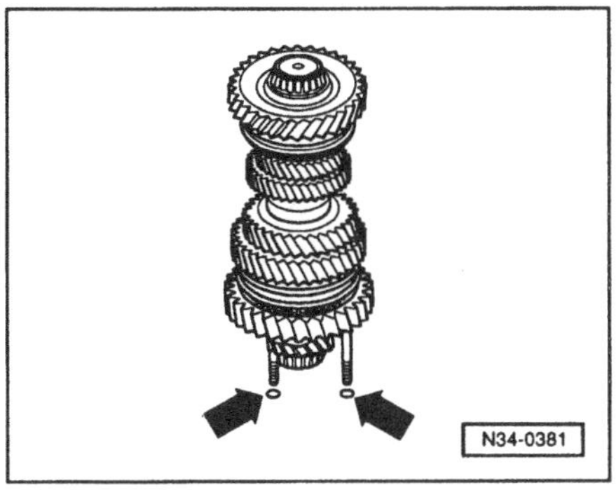

◄ – Always replace sealing rings –arrows– for output shaft bearing support.

Note:

Illustration shows 2 of 4 seal rings.

- Install input shaft, output shaft and selector rods together.

Note:

Align with holes in clutch housing when inserting output shaft bearing support. Do not damage input shaft seal.

- Tighten output shaft bearing support nuts.
- Install reverse gear guide tube.
- Install reverse gear.

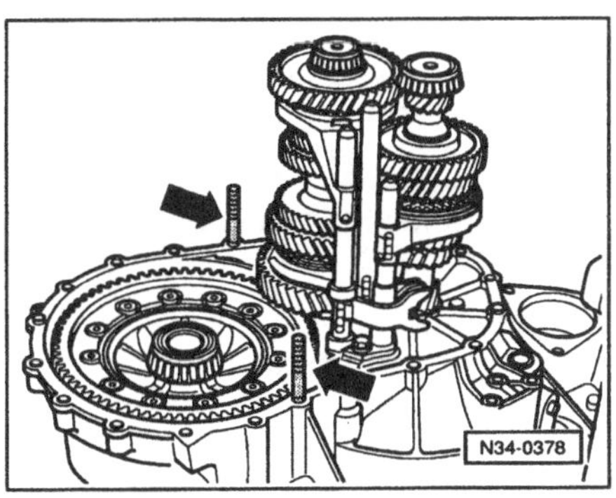

◄ – Install M8 x 80 mm studs in clutch housing -arrows-, so that when installing transmission housing, stop sleeves are aligned to selector rods.

- Install transmission housing.
- Place selector plates in neutral.
- Install selector mechanism.

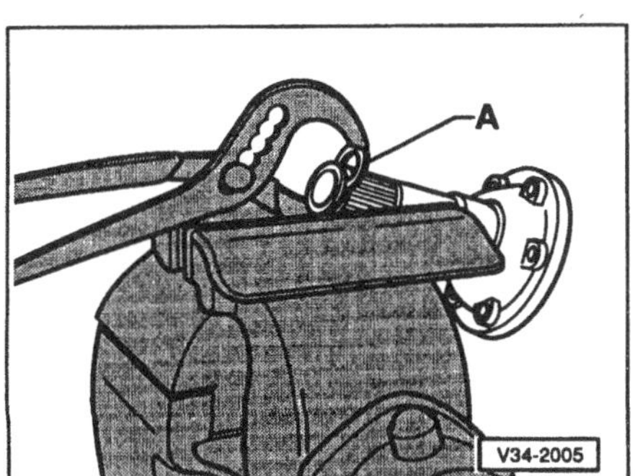

◄ – Replace securing rings for flange shaft and/or stub shaft.

- Secure flange shaft or stub shaft in a vice. Press old circlip out from groove with new circlip -A-.
- Drive in left flange shaft as well as right stub shaft with a plastic–head hammer.
- Install slave cylinder with release bearing.

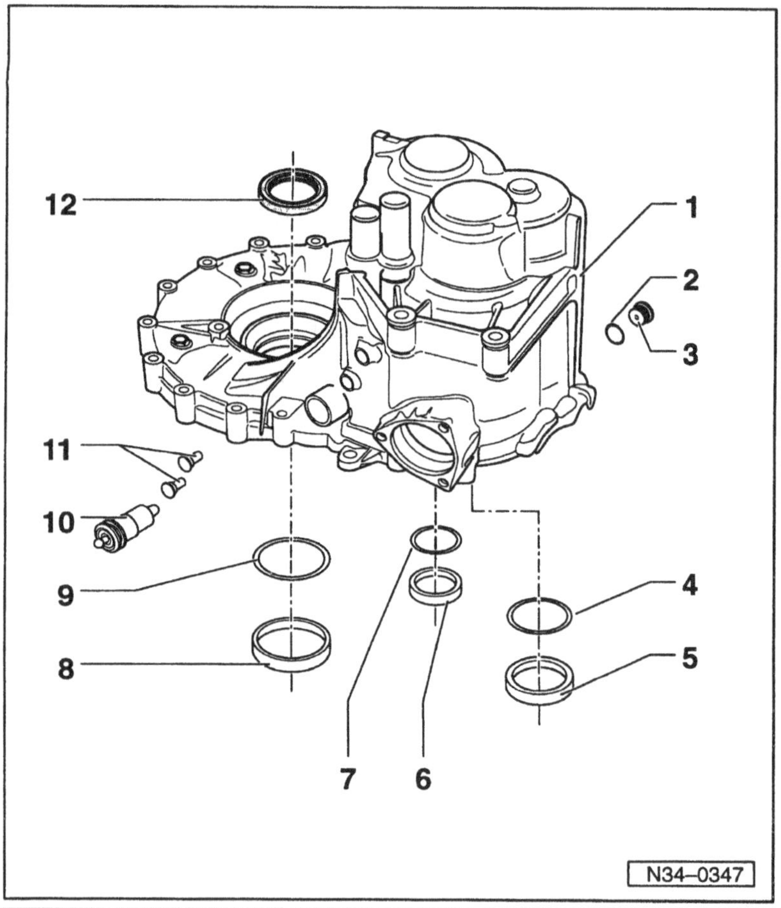

Transmission housing, repairing

1 - Transmission housing
- If replaced: Adjust input/output shaft and differential

⇒ Page L 39-24

2 - Sealing ring
- Always replace

3 - Oil filler plug, 25 Nm (18 ft lb)

4 - Shim
- Output shaft
- Adjustment overview

⇒ Page 39–36

5 - Outer race/tapered roller bearing
- For input shaft
- Removing/installing

⇒ from Page L 35-11
- If replaced: Adjust input shaft

⇒ Page L 35-13

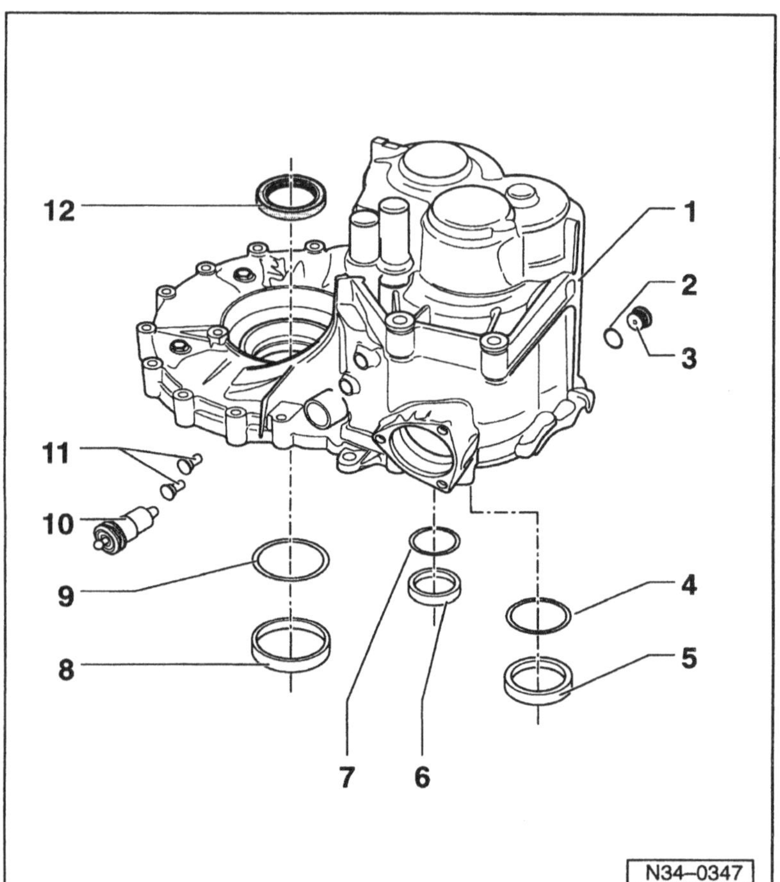

6 - Outer race/tapered roller bearing
- For output shaft
- Removing/installing

⇒ from Page L 35-40
- If replaced: Adjust output shaft

⇒ Page L 35-42

7 - Shim S2
- For output shaft
- Adjustment overview

⇒ Page L 39-24

8 - Outer race/tapered roller bearing
- For differential
- Removing/installing

⇒ from Page L 39-7
- If replaced, Adjust differential

⇒ Page L 39-25

9 - Shim
- For differential
- Adjustment overview

⇒ Page L 39-24

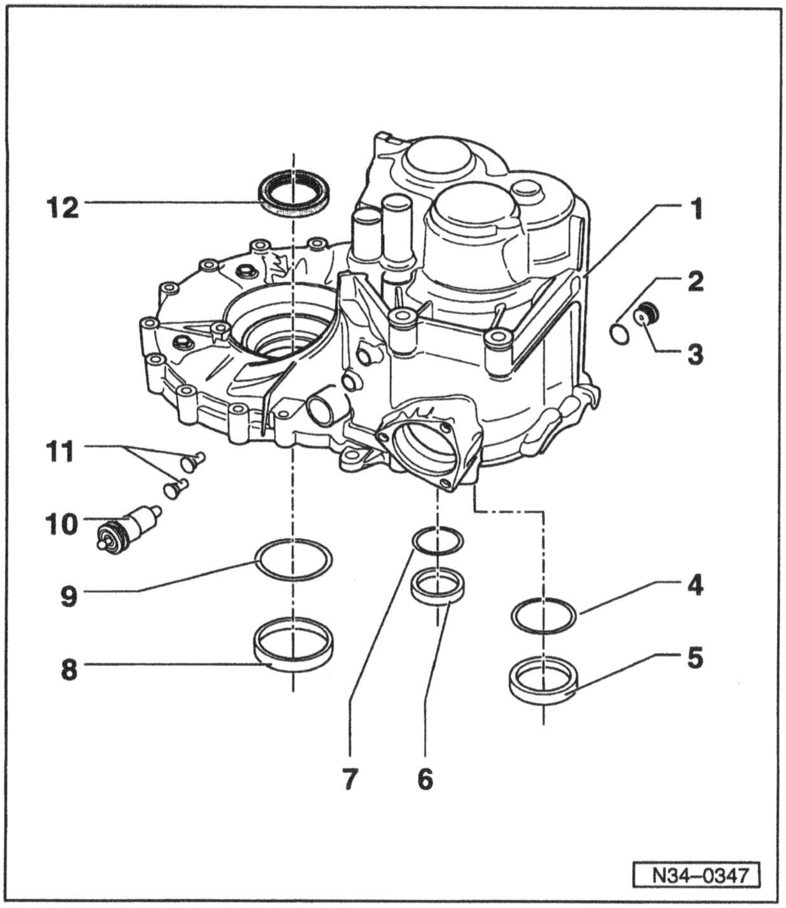

10 - Locking unit
- For reverse gear
- Removing

⇒ Fig. 1
- Driving in

⇒ Fig. 2

11 - Stop sleeves
- Removing

⇒ Fig. 3
- Drive in with plastic–head hammer

12 - Sealing ring
- Replacing

⇒ Page L 39-1

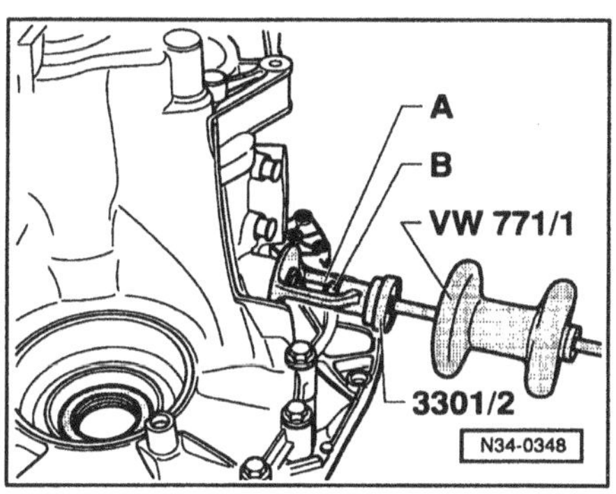

◄ **Fig. 1 Reverse gear locking unit, removing**

A - Ball joint separator (commercial type) e.g. Kukko No. 128/2

B - Hex nut M10

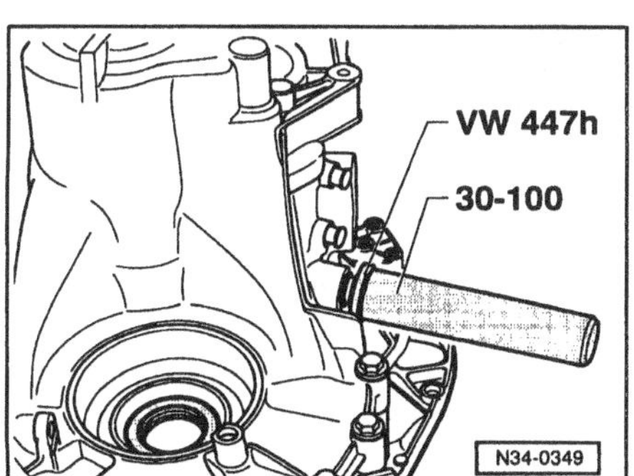

◄ **Fig. 2 Reverse gear locking unit, driving in onto stop**

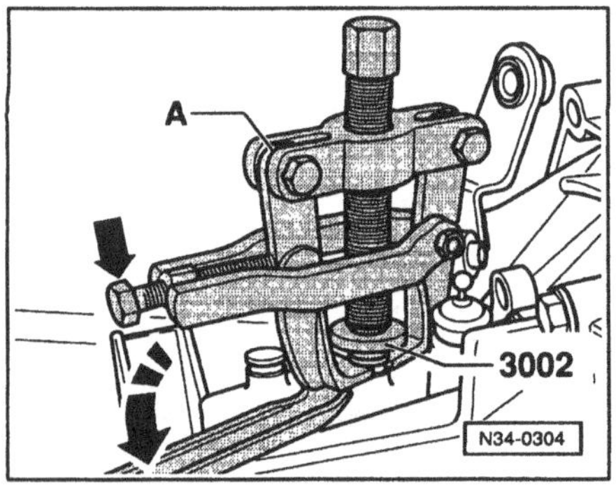

Fig. 3 Stop sleeves, removing

A - Puller Kukko 204-1

- Tension puller when assembling –arrow– then using a lever under puller, remove stop sleeve.

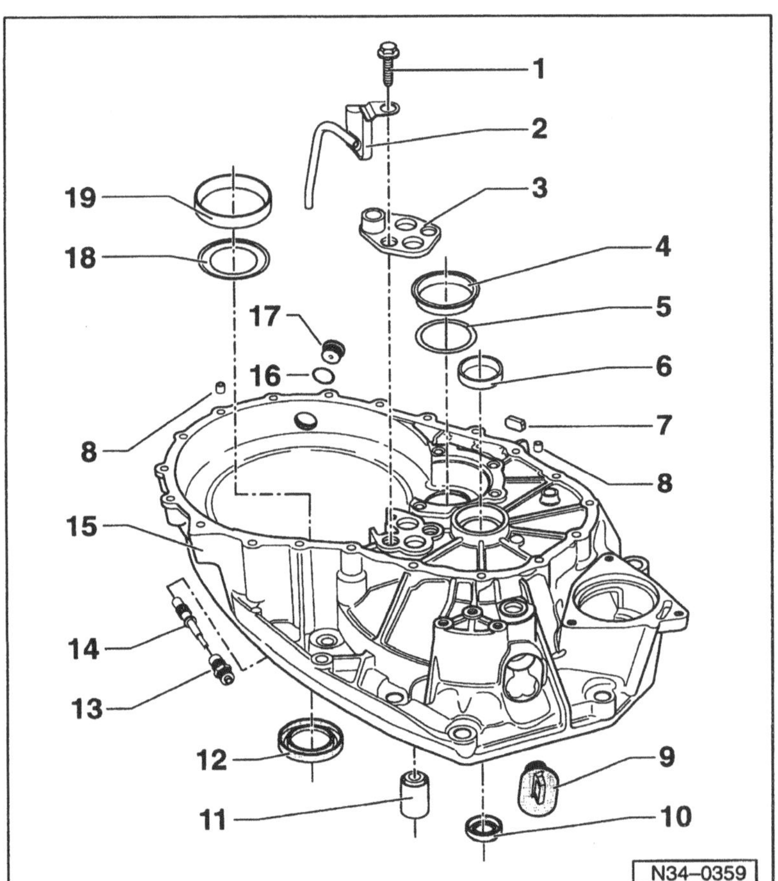

Clutch housing, servicing

1 - Hex bolt, 25 Nm (18 ft lb)

2 - Oil pan
- ♦ Installing

⇒ Fig. 1

3 - Selector rod guide
- ♦ Installing

⇒ Fig. 1

4 - Outer race/tapered roller bearing
- ♦ For output shaft
- ♦ Removing/installing

⇒ Page L 35-26
- ♦ If replaced: Adjust output shaft

⇒ Page L 35-42

5 - Shim S1
- ♦ For output shaft
- ♦ Adjustment overview

⇒ Page 39–52

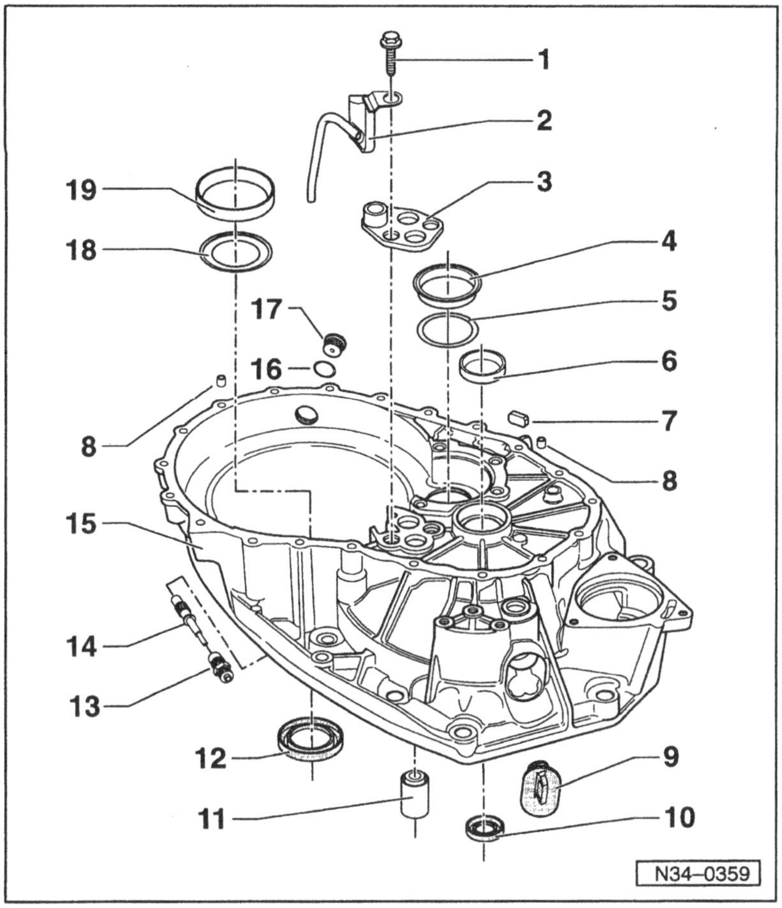

6 - Outer race/tapered roller bearing
- For input shaft
- Removing/installing

⇒ Page L 35-5
- If replaced: Adjust input shaft

⇒ Page L 35-13

7 - Magnet
- Held in place by housing joint surface

8 - Dowel sleeve
- (Quantity 2)

9 - Protective cap

10 - Input shaft oil seal
- Removing

⇒ Fig. 2
- Installing

⇒ Fig. 3

11 - Bonded rubber mounting
- Pulling out

⇒ Fig. 4
- Pulling in

⇒ Fig. 5

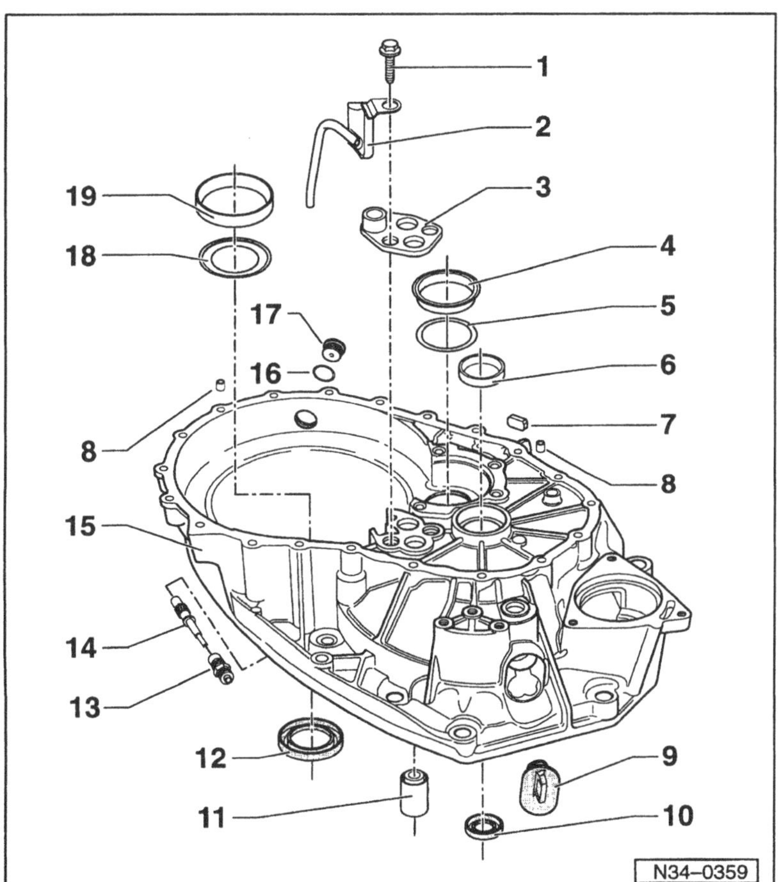

12 - Oil seal
- Replacing

⇒ Page L 39-1

13 - Guide bushing

14 - Drive shaft with pinion
- For speedometer

15 - Clutch housing
- If replaced:

⇒ Adjustment overview, Page L 39-24

16 - Sealing ring
- Always replace

17 - Oil drain plug, 25 Nm (18 ft lb)

18 - Oil deflector ring
- Installation position: Step on internal diameter points to seal

19 - Outer race/tapered roller bearing
- For differential
- Removing/installing

⇒ from Page L 39-7
- If replaced: Adjust differential

⇒ Page L 39-25

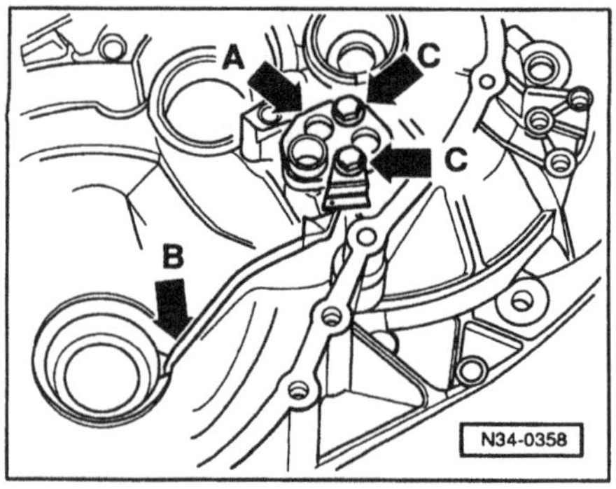

◄ **Fig. 1 Oil pan and selector rod guide, installing**

- Insert selector rod guide -A- with locating shoulder in clutch housing hole.
- Insert oil pan -B- into housing cut-out next to differential tapered roller bearing outer race.
- Secure oil pan and selector rod guide with a hex bolt -C-.

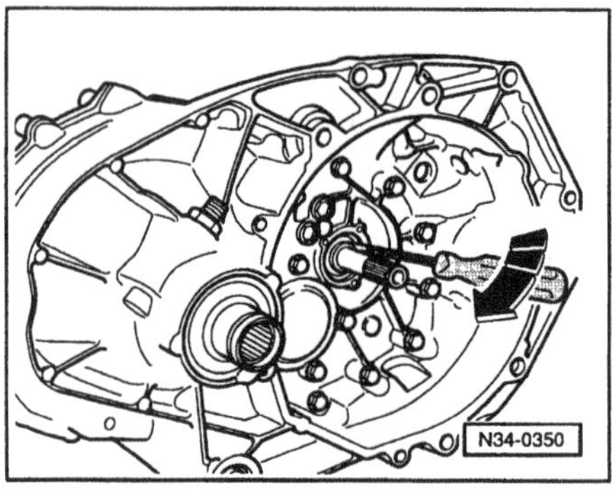

◄ **Fig. 2 Output shaft oil seal, removing**

- Lift oil seal out carefully with a screwdriver.

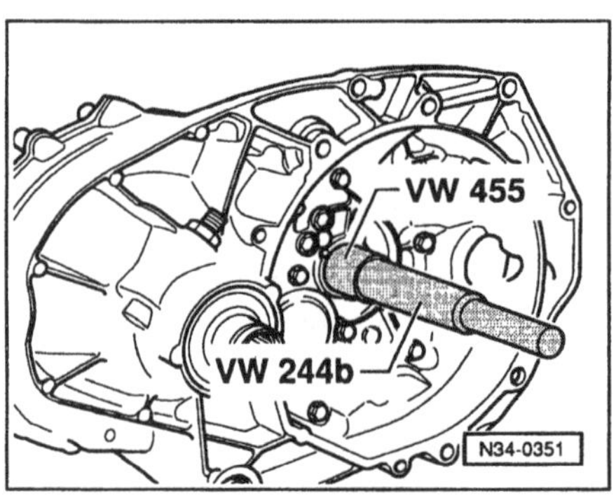

◄ **Fig. 3 Input shaft oil seal, driving in**

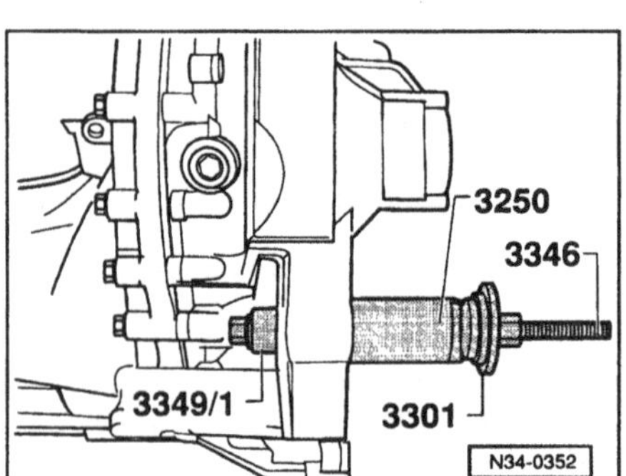

◄ **Fig. 4 Bonded rubber mount, pulling out**

Note:

Must grease spindle with Molybdenum grease before using.

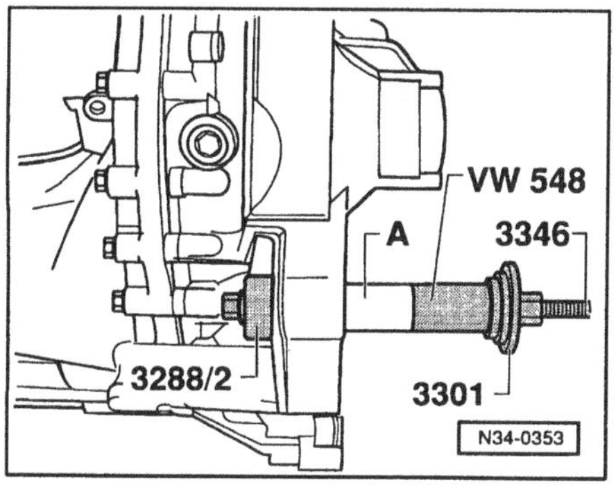

◄ **Fig. 5 Bonded rubber mount, pulling in**
A - Bonded rubber mount

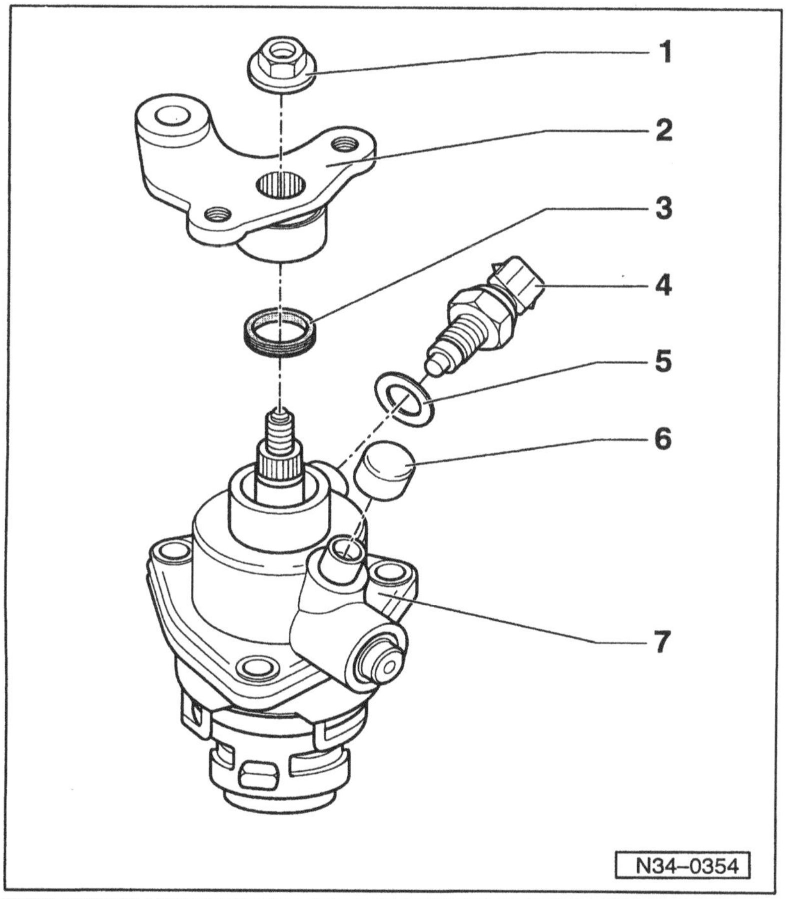

Transmission selector, disassembling/assembling

1 - Hexagon nut, 20 Nm (15 ft lb)

2 - Gear selector lever
- Install so master spline aligns with selector shaft
- Can be replaced with selector mechanism installed

3 - Oil seal
- Removing

⇒ Fig. 1
- Installing

⇒ Fig. 2

4 - Reverse light switch, 20 Nm (15 ft lb)

5 - Sealing ring
- Always replace

6 - Cap
- Transmission breather

7 - Selector unit

◄ **Fig. 1 Removing selector shaft seal**

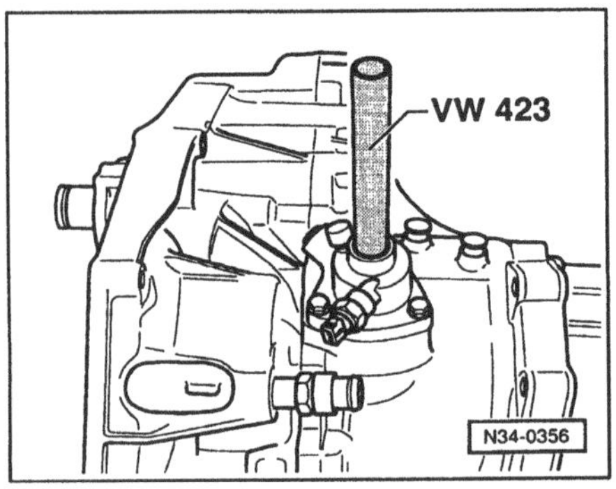

◄ **Fig. 2 Installing oil seal onto stop**

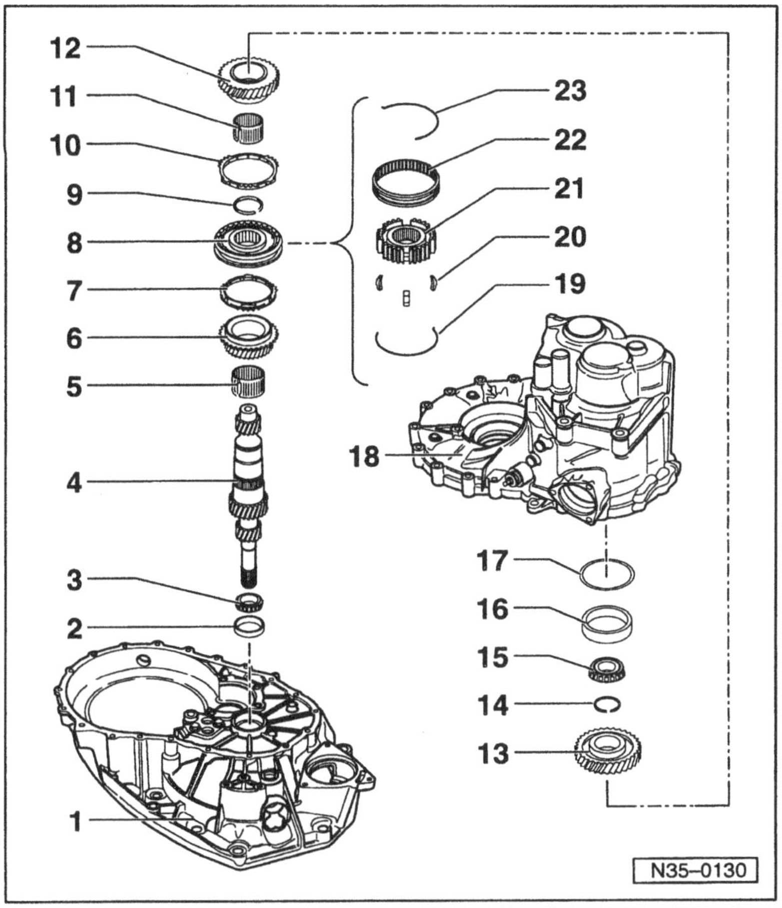

Input shaft, disassembling/ assembling

Note:

- *When installing new gears, consult technical data*
⇒ Page L 00-3.
- *If the position of tapered roller bearings is affected when exchanging parts, input shaft must be re-adjusted. See adjustment overview*
⇒ *Page 39–36*

1 - Clutch housing

2 - Tapered roller bearing outer race
- Pulling out
⇒ Fig. 1
- Pressing in
⇒ Fig. 2

3 - Tapered roller bearing inner race
- Pressing off
⇒ Fig. 6
- Pressing on
⇒ Fig. 7

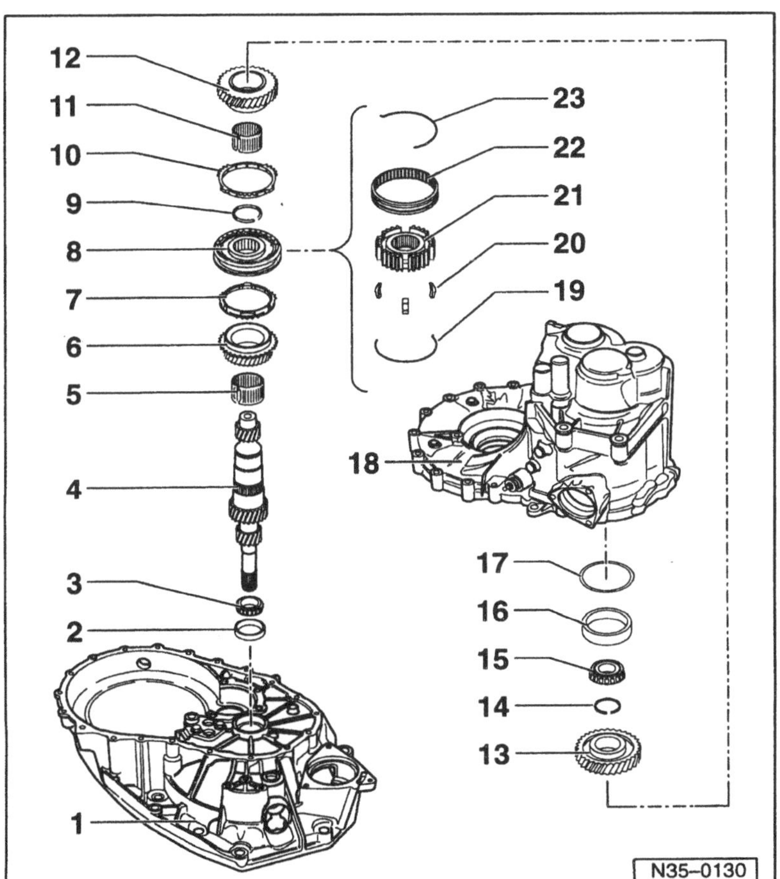

4 - Input shaft
- Adjusting
⇒ Page L 35-13

5 - Needle roller bearing
- For 3rd gear

6 - 3rd gear

7 - 3rd gear synchro-ring
- Checking for wear
⇒ Fig. 8

8 - Operating sleeve with 3rd and 4th gear synchro-hub
– Press off with 3rd gear
⇒ Fig. 5
- Disassembling
⇒ Fig. 9
- Assembling operating sleeve/ synchro-hub
⇒ Fig. 9 and Fig. 10
- Pressing on
⇒ Fig. 11

9 - Circlip

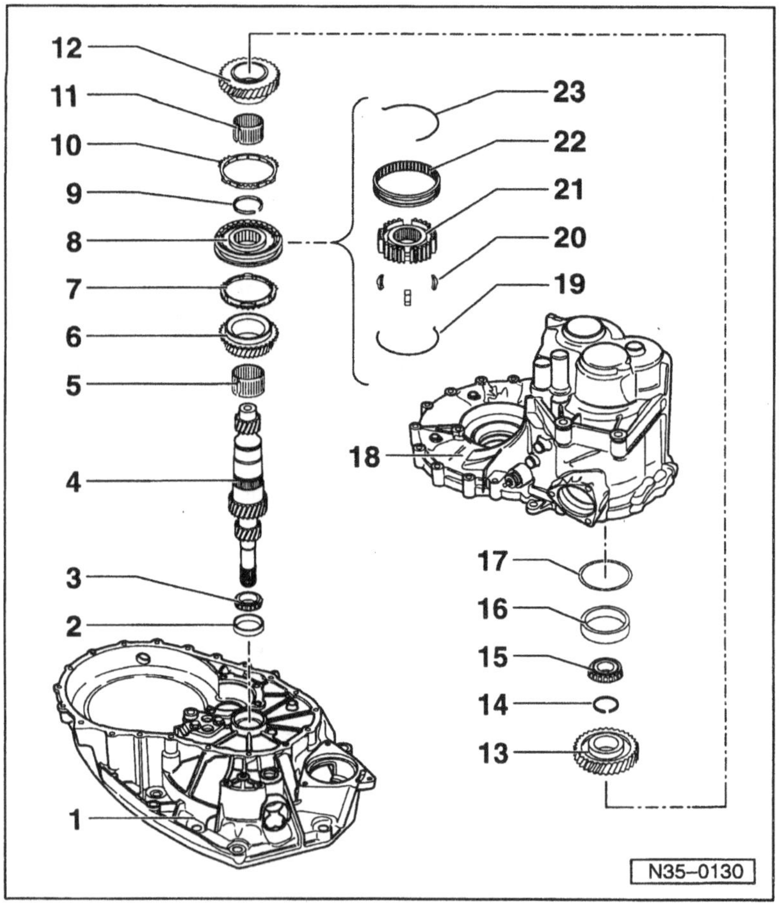

10 - 4th gear synchro-ring
- Checking for wear

⇒ Fig. 8

11 - Needle roller bearing
- For 4th gear

12 - 4th gear

13 - 5th gear
- Pressing off

⇒ Fig. 4
- Pressing on

⇒ Fig. 12

14 - Circlip

15 - Tapered roller bearing inner race
- Pulling off

⇒ Fig. 3
- Pressing on

⇒ Fig. 13

16 - Outer race/tapered roller bearing
- Pulling out

⇒ Fig. 14
- Pressing in

⇒ Fig. 15

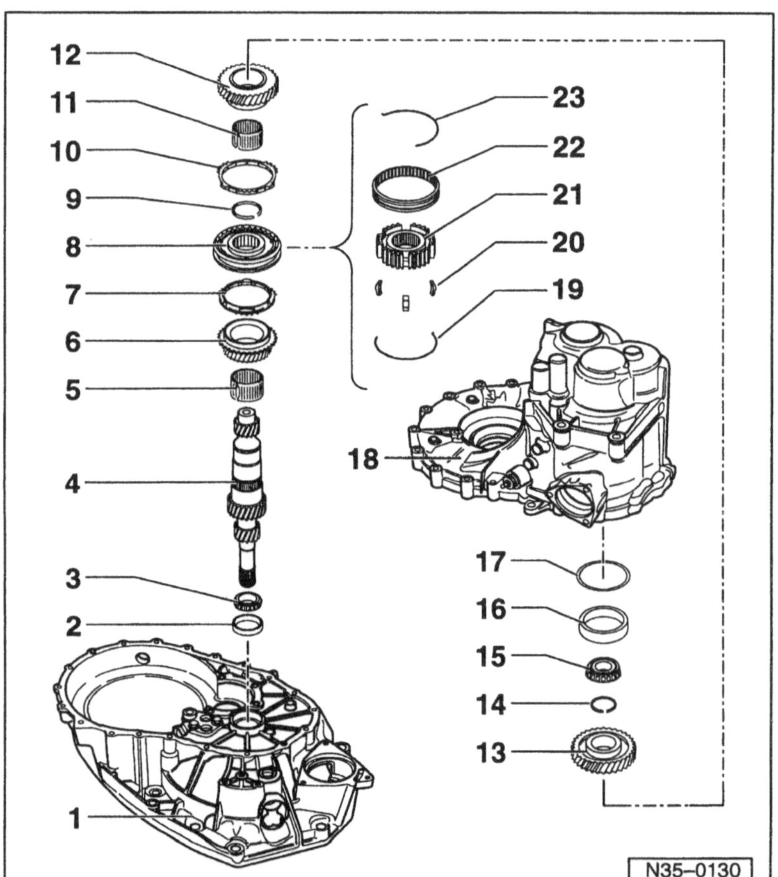

17 - Shim
- Determining thickness

⇒ Page L 35-13

18 - Transmission housing

19 - Spring
- Installation position

⇒ Fig. 10

20 - Lock pieces (Qty. 3)
- Installation position

⇒ Fig. 9

21 - Synchro-hub

22 - Operating sleeve

23 - Spring
- Installation position

⇒ Fig. 10

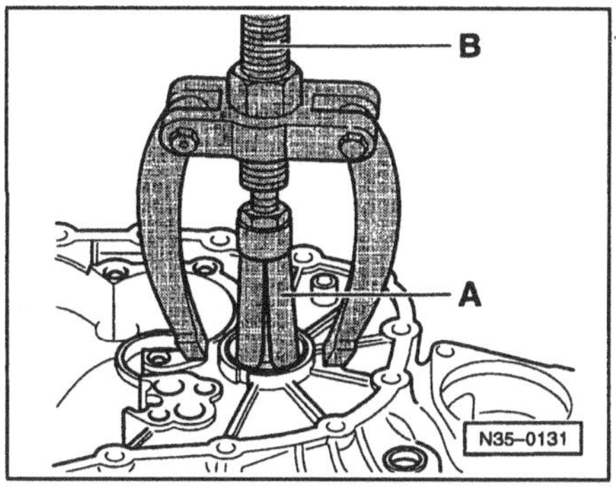

◄ **Fig. 1 Tapered roller bearing outer race, pulling out**

A – Internal extractor, e.g. Kukko 21/6, 37 ... 46 mm

B – Counter support, e.g. Kukko 22/2

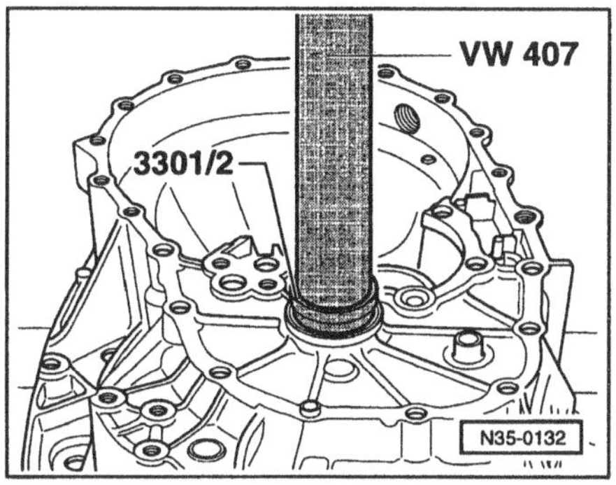

◄ **Fig. 2 Tapered roller bearing outer race, pressing in**

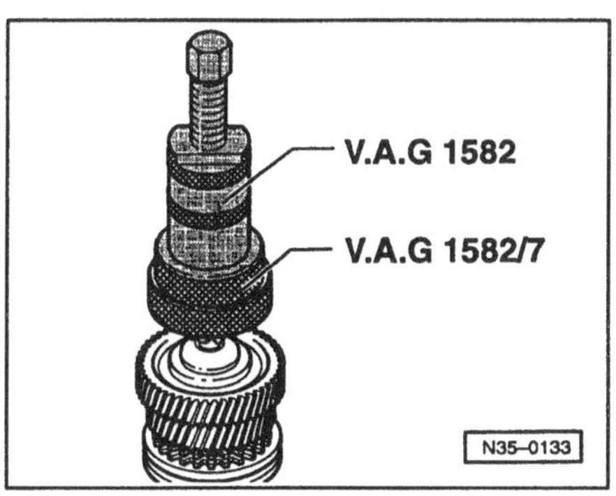

◄ **Fig. 3 Tapered roller bearing inner race, pulling off**

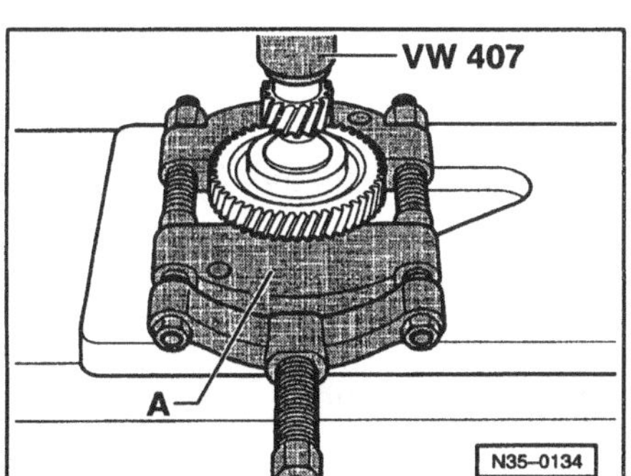

◄ **Fig. 4 5th gear, pressing off**

A – Separating device 22 ... 115 mm, e.g. Kukko 17/2

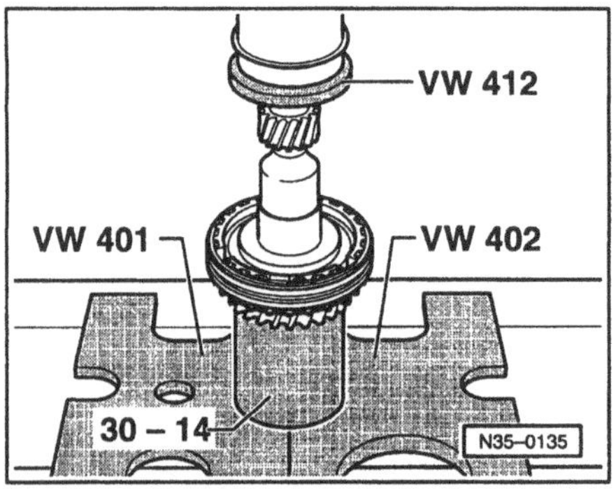

Fig. 5 **3rd and 4th gear synchro hub and operating sleeve with 3rd gear sliding wheel, pressing off**

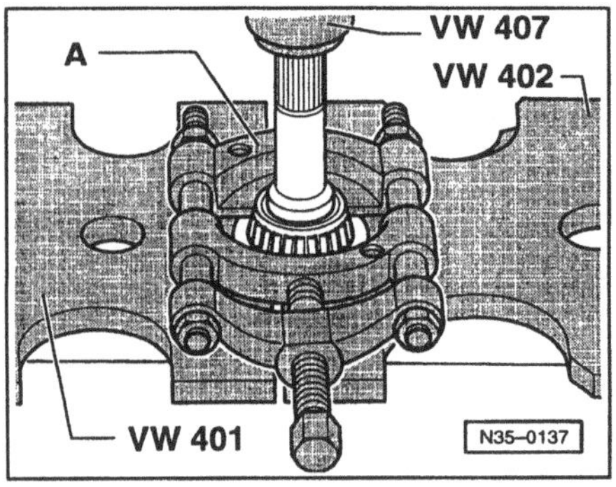

Fig. 6 **Tapered roller bearing inner race, pressing off**

A - Separating device 12 ... 75 mm, e.g., Kukko 17/1

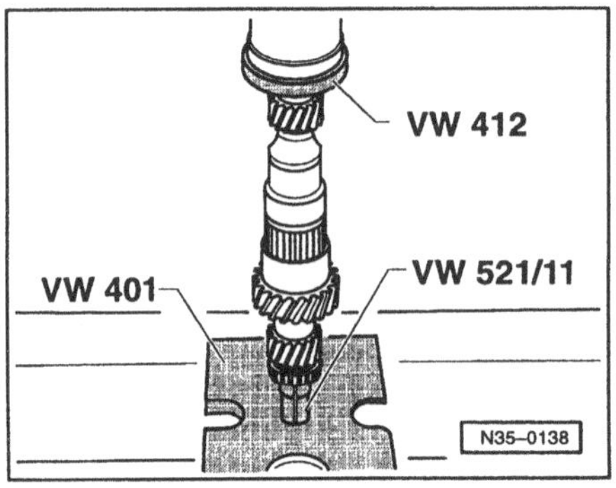

Fig. 7 **Tapered roller bearing inner race, pressing on**

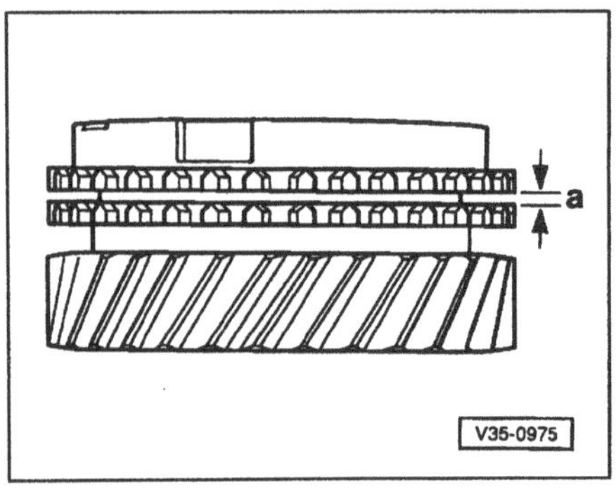

Fig. 8 **Synchro-ring, checking for wear**

– Press synchro-ring onto gear and measure gap "a" with feeler gauge.

Gap "a"	New dimension	Wear limit
3rd gear 4th gear	1.0 ... 1.7 mm 1.0 ... 1.7 mm	0.5 mm

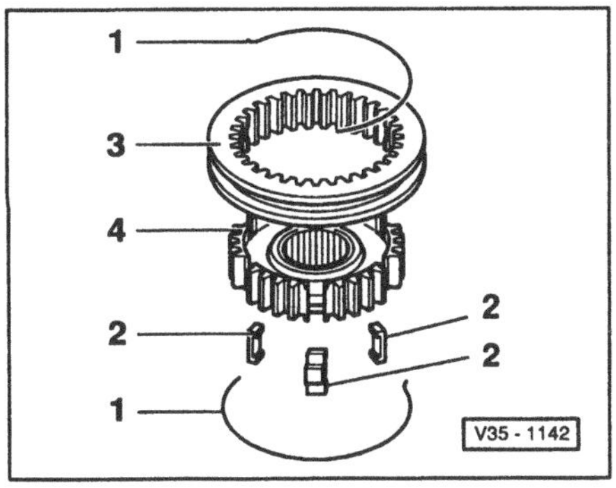

Fig. 9 3rd and 4th gear operating sleeve and synchro-hub, disassembling/assembling

1 - Spring

2 - Locking piece

3 - Operating sleeve

4 - Synchro-hub

– Slide operating sleeve over synchro-hub.

- Recesses for locking pieces in synchro-hub and operating sleeve must align.

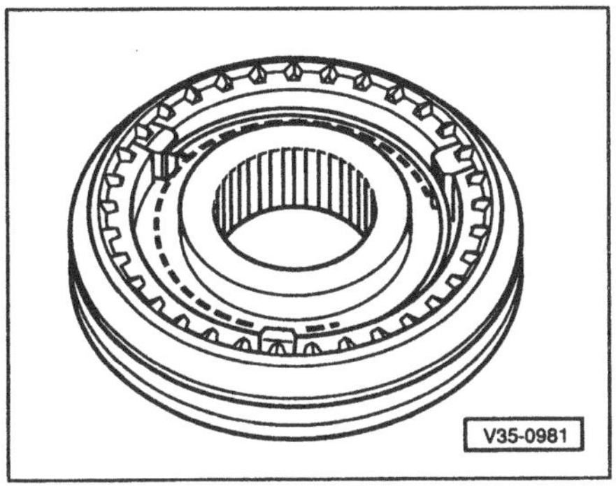

Fig. 10 3rd and 4th gear operating sleeve/synchro-hub, assembling

operating sleeve has been pushed over synchro-hub.

– Insert locking pieces and install springs 120° off-set.

- Angled end of spring must locate in hollow locking piece.

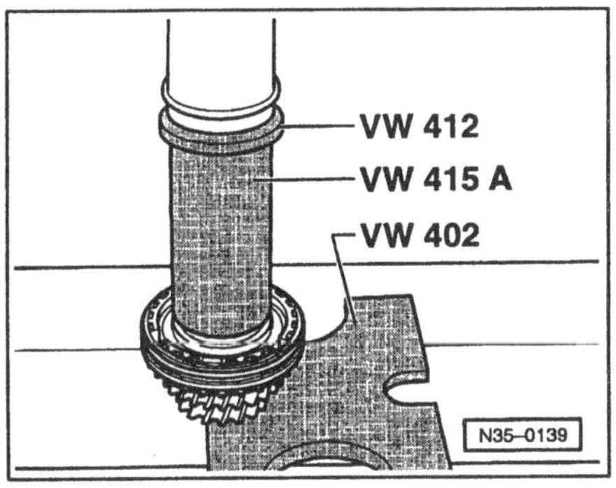

Fig. 11 3rd and 4th gear synchro-hub with operating sleeve, pressing on

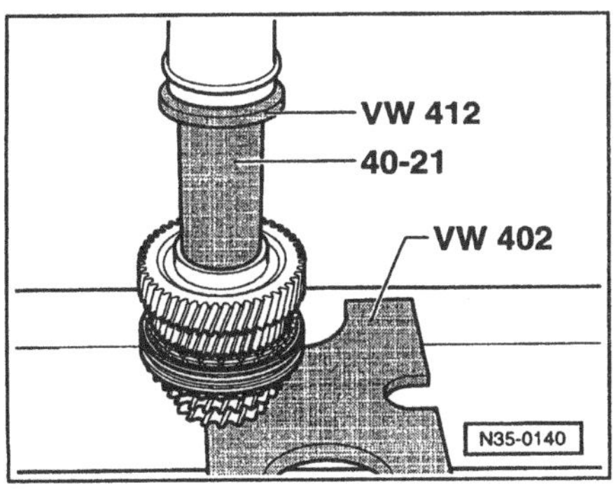

Fig. 12 5th gear, pressing on

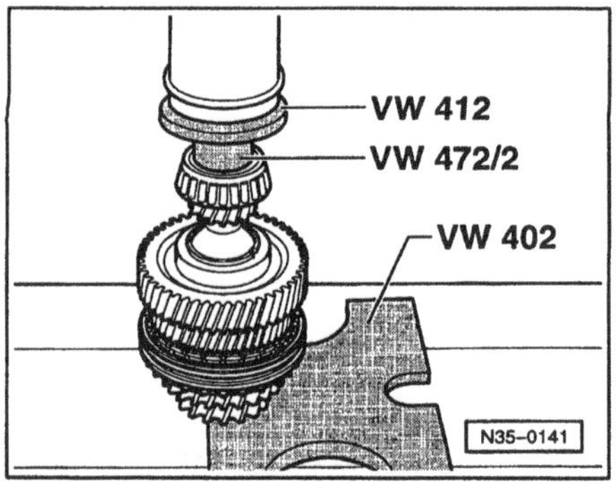

◄ **Fig. 13 Tapered roller bearing inner race, pressing on**

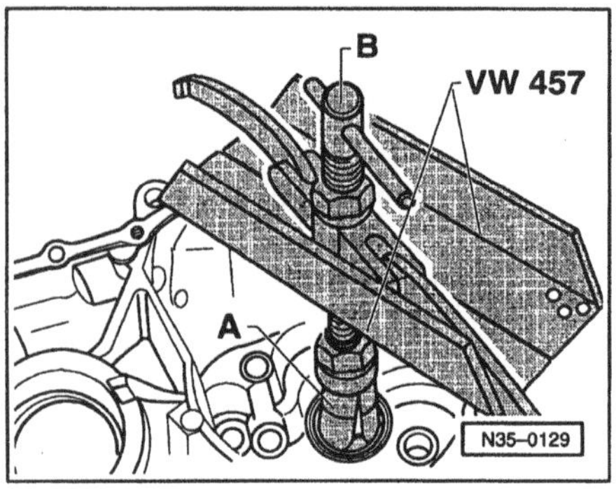

◄ **Fig. 14 Tapered roller bearing outer race, pulling out of transmission housing**

A – Internal puller 46 ... 58 mm, e.g., Kukko 21/7

B – Counter support, e.g. Kukko 22/2

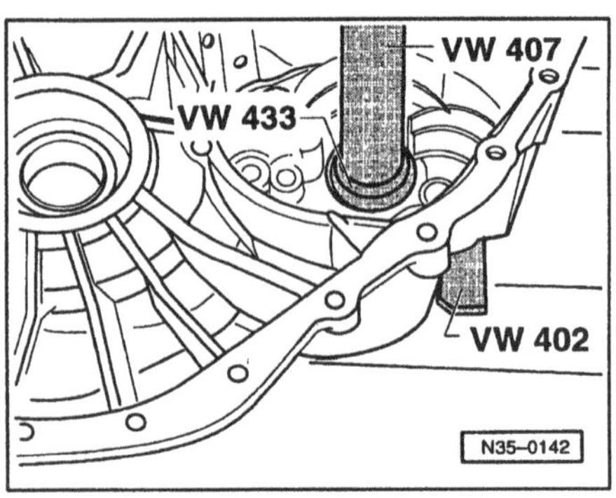

◄ **Fig. 15 Tapered roller bearing outer race, pressing into transmission housing**

- Position shim under outer race.
- Support transmission housing with press piece 3062 directly below bearing support.

Input shaft, adjusting

(Determining thickness of shim for input shaft)

It is only necessary to readjust the input shaft when transmission housing, clutch housing, Input shaft, or tapered roller bearing have been replaced

Adjustment overview
⇒ Page L 39-24

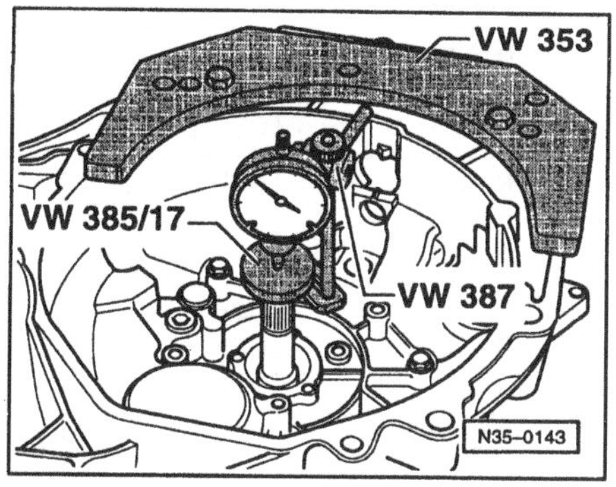

- Press tapered roller bearing outer race without shim into transmission housing onto stop, using VW 433.
- Install input shaft in clutch housing and assemble transmission housing. Tighten hex head bolts to 25 Nm (18 ft lb) and then turn 90° further.
- ◄ Install measuring fixture and dial indicator in clutch housing (large tapered roller bearing in transmission housing).

- Before measuring, rotate input shaft to allow bearing to settle. Set dial indicator to "0" with 1 mm (0.0393 in.) preload.

Note:

Repeat this preparation for each subsequent measurement, or gauge will not return to starting position.

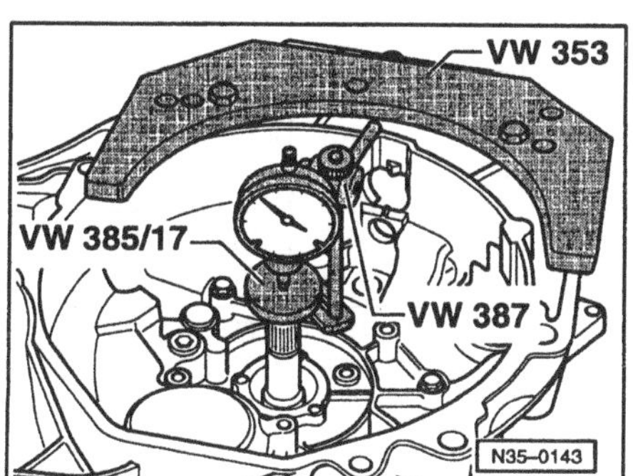

- ◄ Press input shaft in direction of dial indicator.
- Read off and note play indicated on dial indicator (example, 0.975 mm [0.038 in.]).

Note:

Dial indicator *does not return to original position.*

- Determine thickness of shim from table
⇒ Page L 35-15 (example, 0.875 mm [0.034 in.]).
- Remove input shaft and pull tapered roller bearing outer race from transmission housing.
- Press tapered roller bearing outer race together with 0.875 mm (0.034 in.) shim into transmission housing using VW 433.
- Assemble transmission housing and tighten hex bolts to 25 Nm (18 ft lb) then turn an additional 90° (1/4 turn).

Table of shims

Bearing play Measured value (mm)	Shim Thickness (mm)	Part No.
0.695 ... 0.769	0.650	02F 311 391
0.770 ... 0.794	0.675	02F 311 391 A
0.795 ... 0.819	0.700	02F 311 391 B
0.820 ... 0.844	0.725	02F 311 391 C
0.845 ... 0.869	0.750	02F 311 391 D
0.870 ... 0.894	0.775	02F 311 391 E
0.895 ... 0.919	0.800	02F 311 391 F
0.920 ... 0.944	0.825	02F 311 391 G
0.945 ... 0.969	0.850	02F 311 391 H
0.970 ... 0.994	0.875	02F 311 391 J
0.995 ... 1.019	0.900	02F 311 391 K
1.020 ... 1.044	0.925	02F 311 391 L
1.045 ... 1.069	0.950	02F 311 391 M
1.070 ... 1.094	0.975	02F 311 391 N
1.095 ... 1.119	1.000	02F 311 391 P
1.120 ... 1.144	1.025	02F 311 391 Q
1.145 ... 1.169	1.050	02F 311 391 R
1.170 ... 1.194	1.075	02F 311 391 S
1.195 ... 1.219	1.100	02F 311 391 T
1.220 ... 1.244	1.125	02F 311 391 AA
1.245 ... 1.269	1.150	02F 311 391 AB

Bearing play Measured value (mm)	Shim Thickness (mm)	Part No.
1.270 ... 1.294	1.175	02F 311 391 AC
1.295 ... 1.319	1.200	02F 311 391 AD
1.320 ... 1.344	1.225	02F 311 391 AE
1.345 ... 1.369	1.250	02F 311 391 AF
1.370 ... 1.394	1.275	02F 311 391 AG
1.395 ... 1.419	0.650 + 0.650 = 1.300	02F 311 391 + 391
1.420 ... 1.444	0.650 + 0.675 = 1.325	02F 311 391 + A
1.445 ... 1.469	0.675 + 0.675 = 1.350	02F 311 391 A + A
1.470 ... 1.494	0.675 + 0.700 = 1.375	02F 311 391 A + B
1.495 ... 1.519	0.700 + 0.700 = 1.400	02F 311 391 B + B
1.520 ... 1.544	0.700 + 0.725 = 1.425	02F 311 391 B + C
1.545 ... 1.569	0.725 + 0.725 = 1.450	02F 311 391 C + C
1.570 ... 1.590	0.725 + 0.750 = 1.475	02F 311 391 C + D

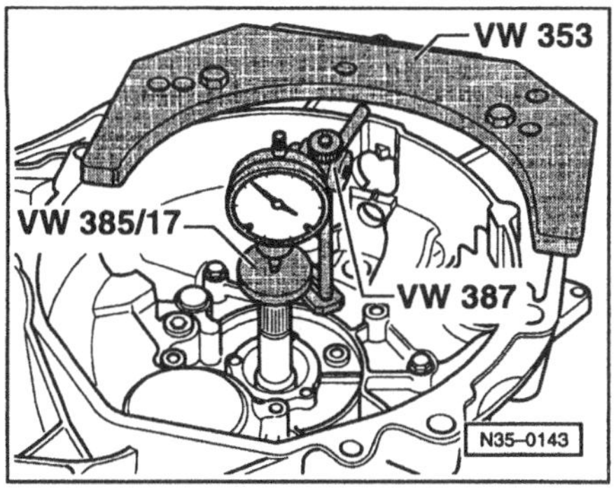

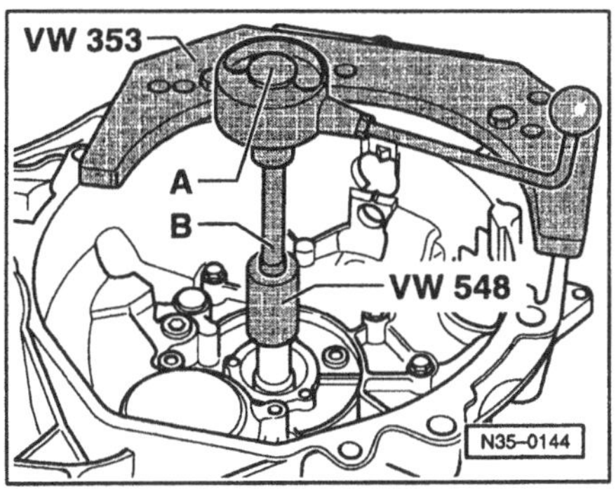

Checking measurement

- Install measuring fixture and dial indicator.
- Rotate input shaft so that the taper roller bearing settles.
- Press input shaft in direction of arrow.

● Minimum bearing play = 0.04 mm (0.00157 in.).

● Maximum bearing play = 0.12 mm (0.00472 in.).

Note:

If bearing clearance cannot be measured, but input shaft play is perceptible and input shaft turns freely, adjustment is acceptable.

- Frictional moment (turning torque at bearing preload) maximum 20 Ncm (1.8 in. lb).

A – Commercial torque gauge 0 ... 600 Ncm (0...53 in. lb)

B – Extension for 1/2 " square drive socket

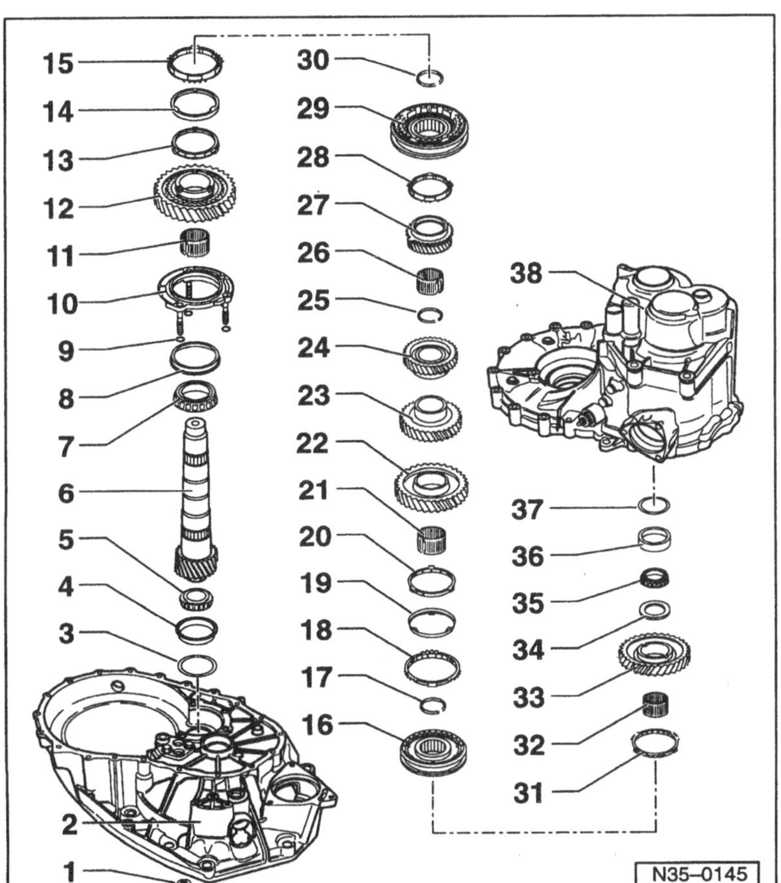

Output shaft, dis–assembling/assembling

Notes:

♦ *When installing new gears or a new output shaft, consult technical data*
⇒ Page L 00-3.

♦ *If the output shaft or tapered roller bearings have been replaced, then adjust output shaft*
⇒ Page L 35-42.

1 - Hex nut, 25 Nm (18 ft lb) plus 90° (1/4 turn) further
♦ 4 nuts for bearing support

2 - Clutch housing

3 - Shim S1
♦ Determining thickness
⇒ Page L 35-42

4 - Small tapered roller bearing outer race
♦ Pulling out
⇒ Fig. 1
♦ Pressing in
⇒ Fig. 2

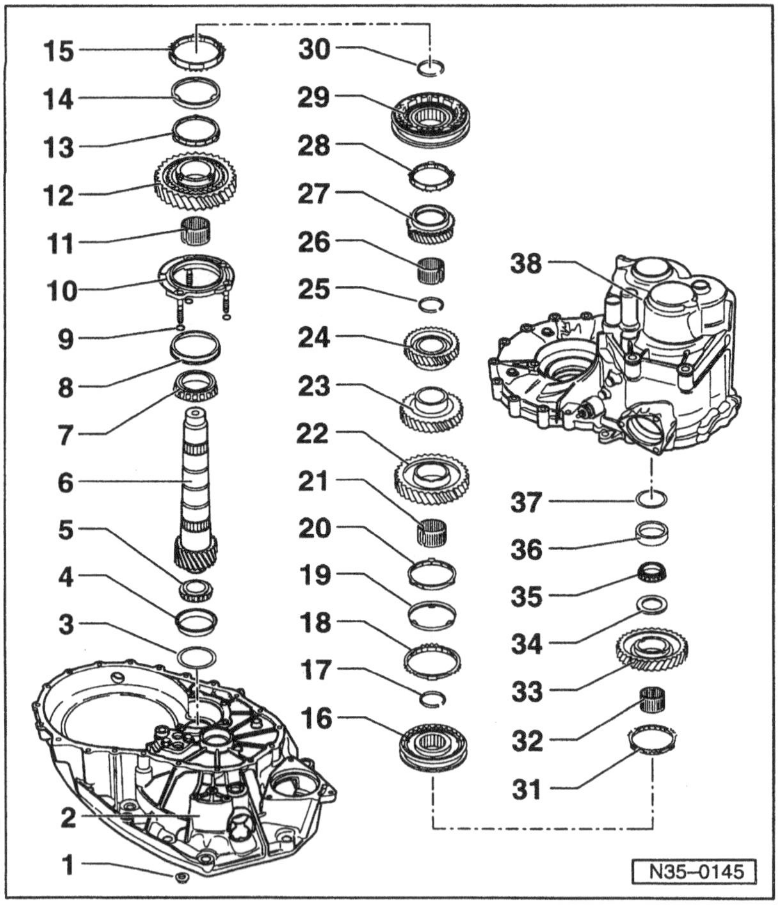

5 - Small tapered roller bearing inner race
- Pulling off

⇒ Fig. 3
- Pressing on

⇒ Fig. 4

6 - Output shaft
- Is paired with final drive gear, always replace together as a set
- Adjusting

⇒ Page L 35-42

7 - Large tapered roller bearing inner race
- Pressing off

⇒ Fig. 10
- Pressing on

⇒ Fig. 13

8 - Large tapered roller bearing outer race
- Pressing out

⇒ Fig. 11
- Pressing in

⇒ Fig. 12

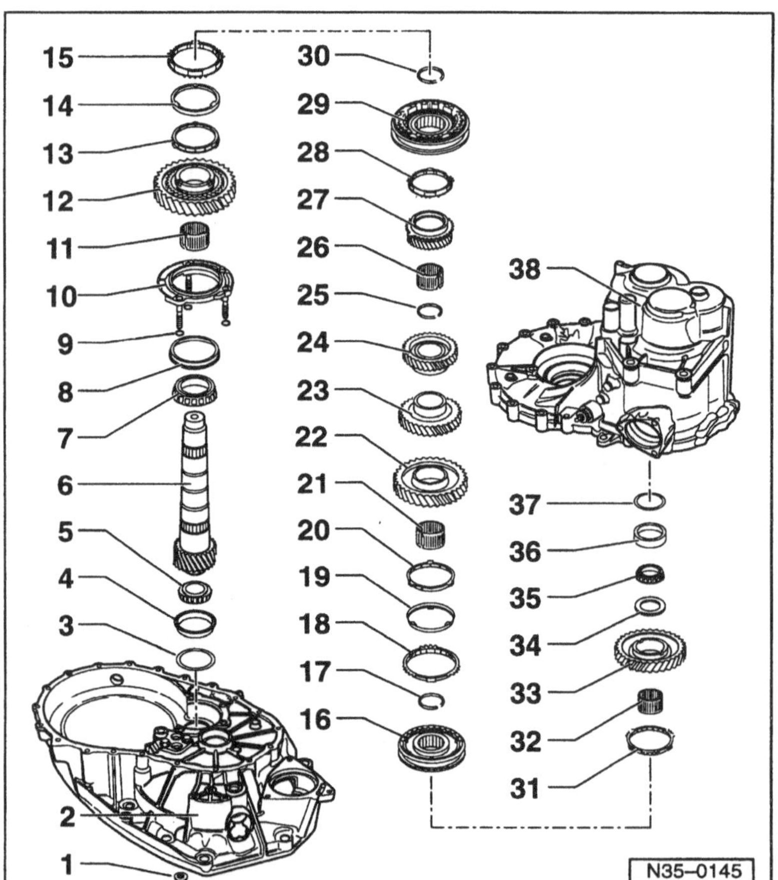

9 - Sealing ring
- Place sealing rings (Quantity 4) on bearing support bolts

10 - Bearing support

11 - Needle roller bearing
- For 1st gear

12 - 1st gear
- Installation position

⇒ Fig. 16

13 - 1st gear synchronizer (synchro-ring)
- Checking for wear

⇒ Fig. 14
- Check lugs for scoring
- Installation position

⇒ Fig. 16

14 - 1st gear outer ring
- Insert in synchro-ring (item13)
- Installation position

⇒ Fig. 16
- Replace if scored

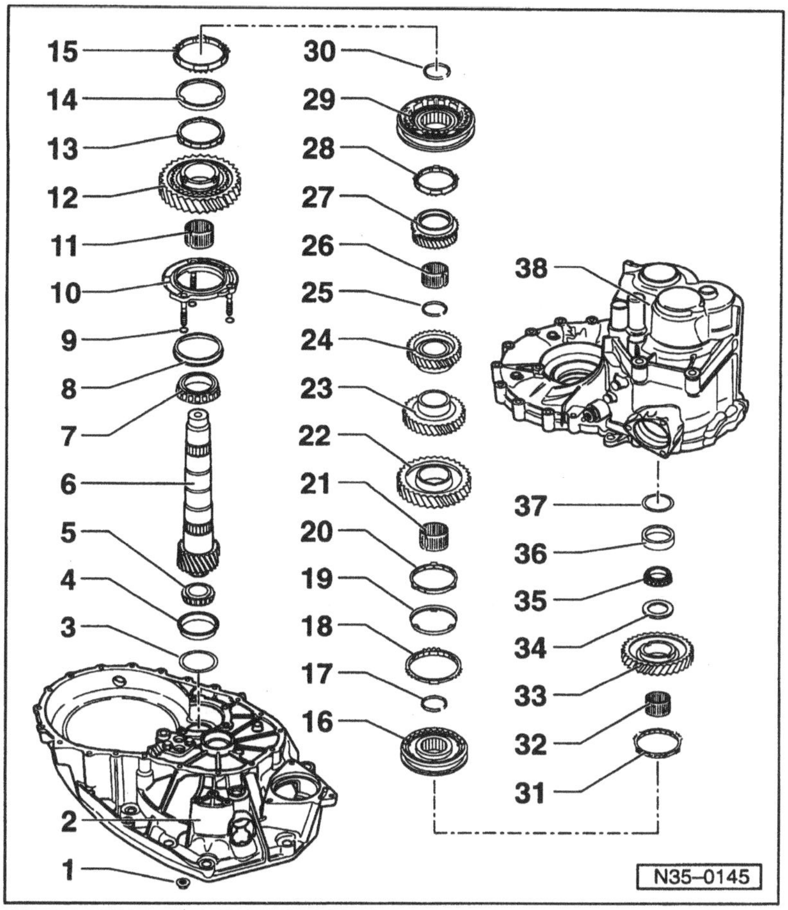

15 - 1st gear (1GR) synchro-ring
- ♦ Checking for wear

⇒ Fig. 15
- ♦ Installation position

⇒ Fig. 16

16 - 1st/2nd gear operating sleeve with synchro-hub
- – After removing securing clip (item 17) pull off over bearing support

⇒ Fig. 9
- ♦ Disassembling

⇒ Fig. 17
- ♦ Assembling operating sleeve/synchro hub

⇒ Fig. 17 and Fig. 18
- ♦ Installation position

⇒ Fig. 19
- ♦ Pressing on

⇒ Fig. 20

17 - Securing clip

18 - 2nd gear synchro-ring
- ♦ Checking for wear

⇒ Fig. 15
- – Assemble so recesses locate in operating sleeve (item16)

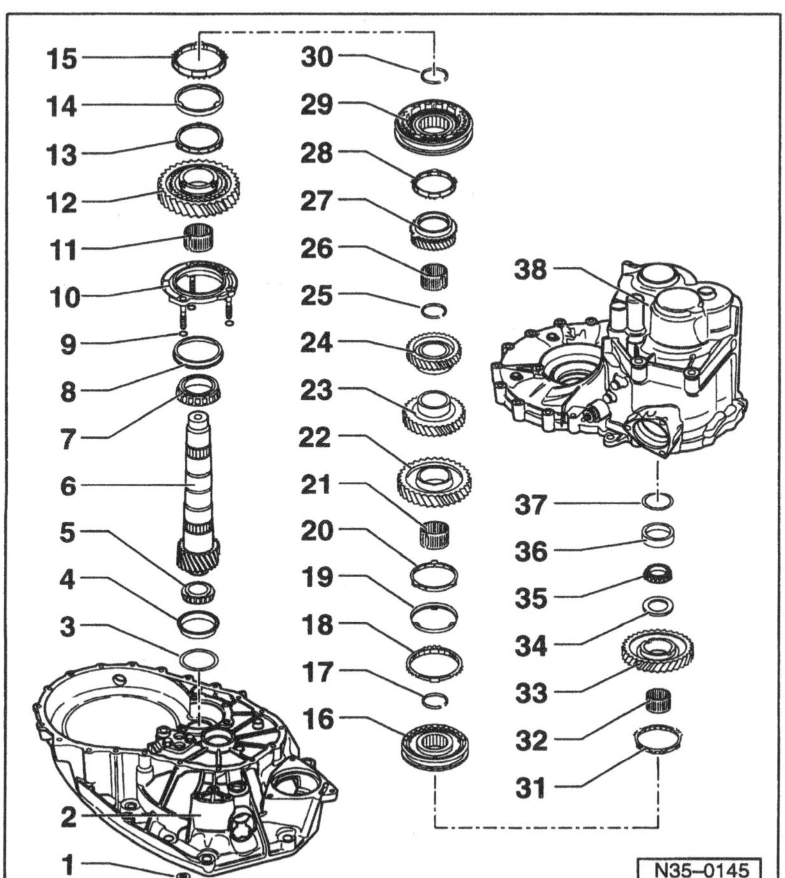

19 - Outer ring
- ♦ Insert in synchro-ring (item18), installation position

⇒ Fig. 21
- – Replace if scored

20 - Synchro-ring
- ♦ (Inner ring)
- ♦ Checking for wear

⇒ Fig. 14
- – Check lugs for scoring
- ♦ Installation position

⇒ Fig. 22

21 - 2nd gear
- ♦ Installation position

⇒ Fig. 23

22 - Needle roller bearing
- ♦ For 2nd gear

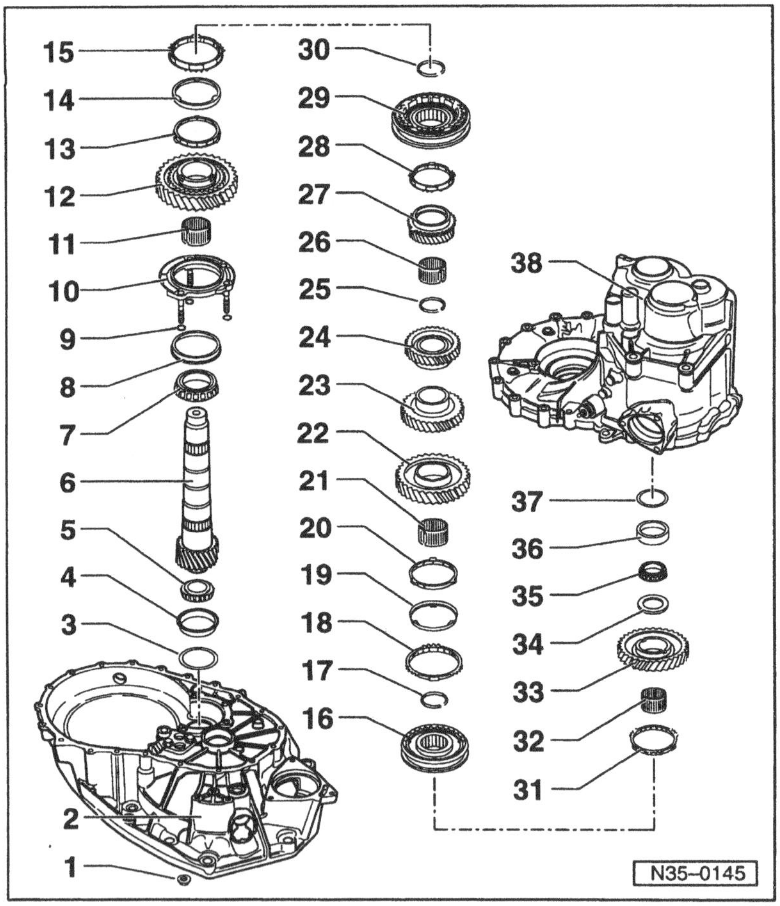

23 - 3rd gear
- Press off with 2nd gear

⇒ Fig. 8
- ♦ Pressing on

⇒ Fig. 24
- ♦ Shoulder points to 4th gear

24 - 4th gear
- ♦ Pressing off

⇒ Fig. 7
- ♦ Pressing on

⇒ Fig. 25
- ♦ Shoulder points to 3rd gear

25 - Circlip

26 - Needle roller bearing
- ♦ For 5th gear

27 - 5th gear
- Press off together with operating sleeve/synchro hub

⇒ Fig. 6

28 - 5th gear synchro-ring
- Check for wear

⇒ Fig. 26

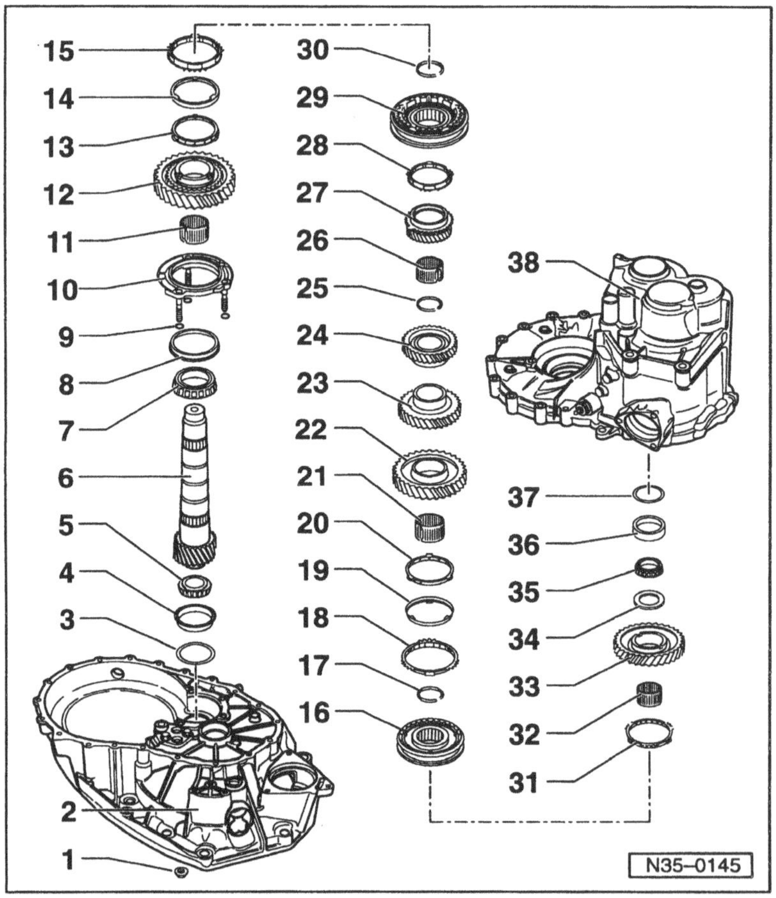

29 - 5th and reverse gear operating sleeve with synchro-hub
- Press off with 5th gear (item 27)

⇒ Fig. 6
- ♦ Disassembling

⇒ Fig. 17
- ♦ Installation position operating sleeve/synchro-hub

⇒ Fig. 27
- ♦ Assembling operating sleeve/ synchro hub

⇒ Fig. 17 and Fig. 18
- ♦ Pressing on

⇒ Fig. 28

30 - Circlip

31 - Reverse gear synchronizer (synchro-ring)
- ♦ Check for wear

⇒ Fig. 26

32 - Needle roller bearing
- ♦ For reverse gear

33 - Reverse sliding gear

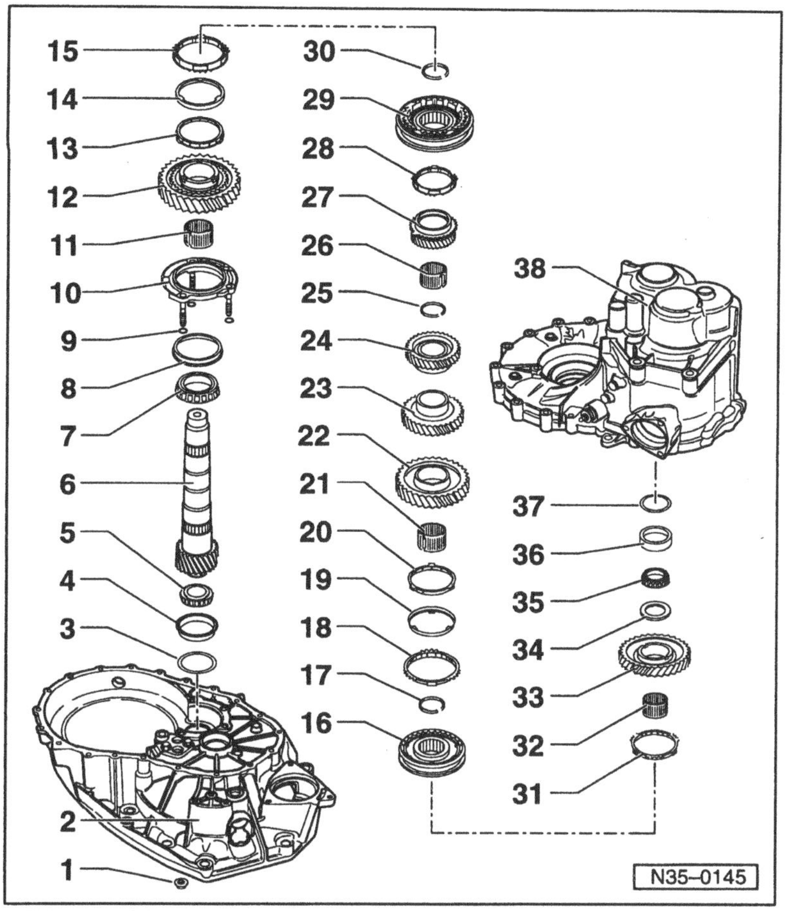

34 - Thrust washer

35 - Tapered roller bearing inner race
- Press off with reverse gear

⇒ Fig. 5
- ♦ Pressing on

⇒ Fig. 29

36 - Outer race/tapered roller bearing
- ♦ Pulling out

⇒ Fig. 30
- ♦ Pressing in

⇒ Fig. 31

37 - Shim S2
- ♦ Determining thickness

⇒ Page L 35-45

38 - Transmission housing

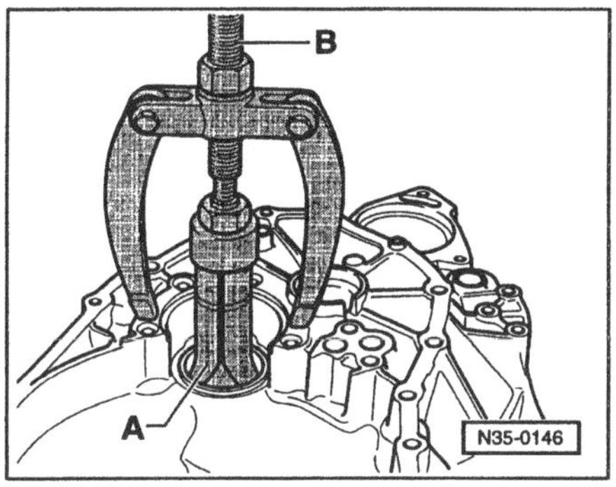

◄ **Fig. 1 Small tapered roller bearing outer race, removing**

A – Internal puller 46 ... 58 mm, e.g. Kukko 21/7

B – Counter support, e.g. Kukko 22/2

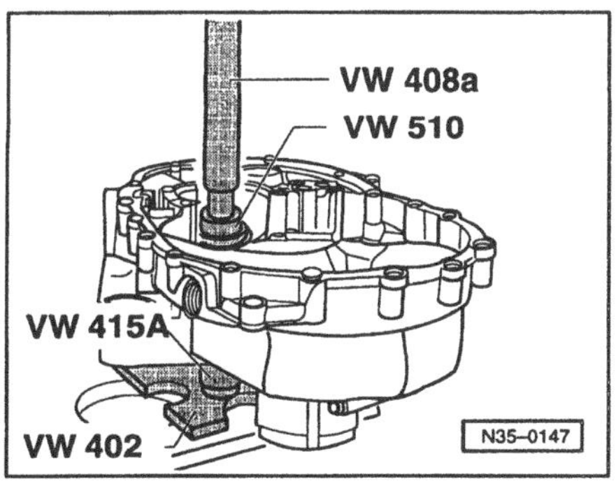

◄ **Fig. 2 Tapered roller bearing outer race, installing (pressed on)**

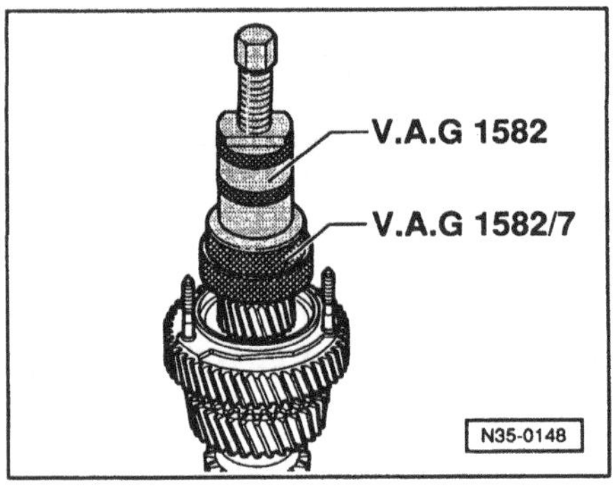

◄ **Fig. 3 Small tapered roller bearing inner race, pressing off**

– Install puller well behind bearing rollers, then turn bearing and puller.

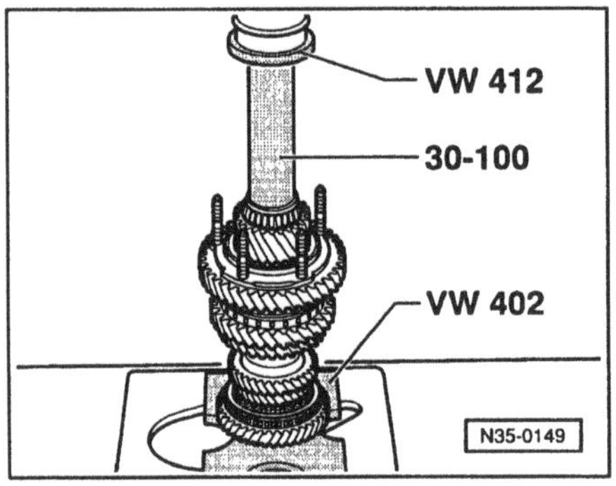

◄ **Fig. 4 Small tapered roller bearing inner race, pressing on**

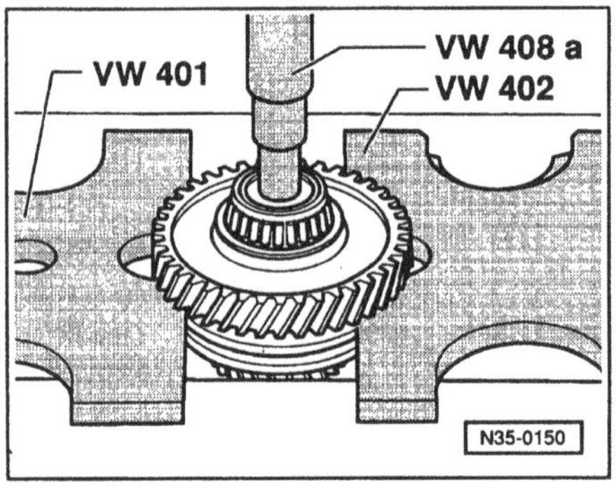

◄ **Fig. 5 Tapered roller bearing inner race, pressing off**

– Press off tapered roller inner race via reverse gear.

Note:
Before pressing off, slide operating sleeve over to 5th gear.

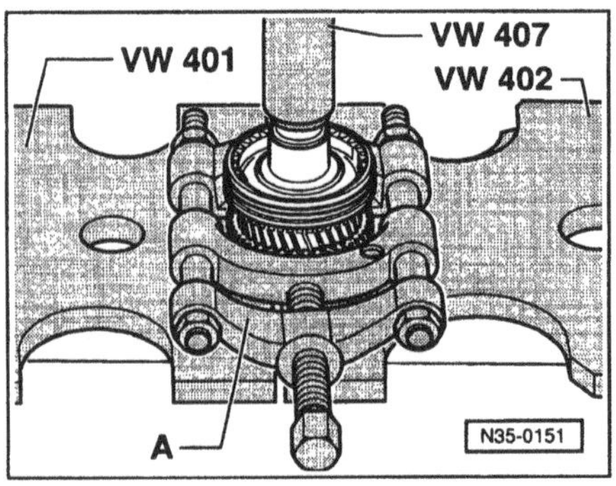

◄ **Fig. 6 5th gear and reverse gear operating sleeve/synchro hub, pressing off**

A – Separating device 22 ... 115 mm, e.g. Kukko 17/2

After removing securing ring, press 2nd gear and operating sleeve/synchro-hub off together.

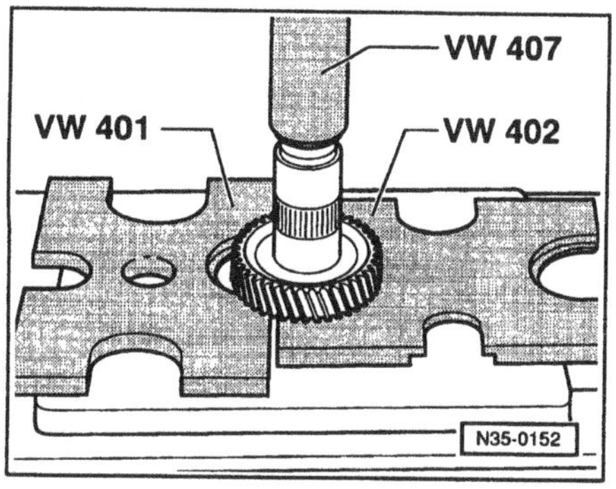

◄ **Fig. 7 4th gear, pressing off**

- First remove securing clip.

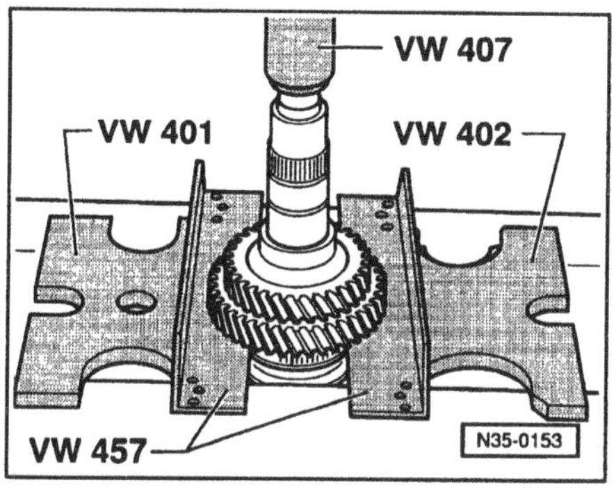

◄ **Fig. 8 3rd gear, pressing off**

- Press off 3rd gear via 2nd gear.

Note:

Before pressing off, slide operating sleeve over to 1st gear.

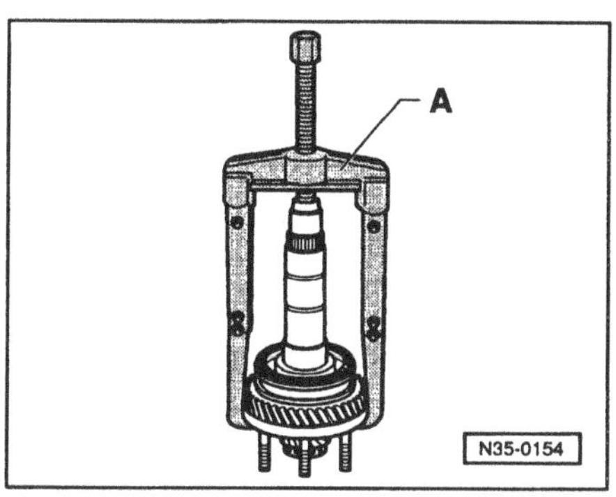

◄ **Fig. 9 Operating sleeve with synchro-hub, pulling off via bearing support**

A – Two-arm puller, e.g., Kukko 20/10 with 250 mm (9.8 in.) reach hooks

Note:

Remove circlip before pulling off individual components.

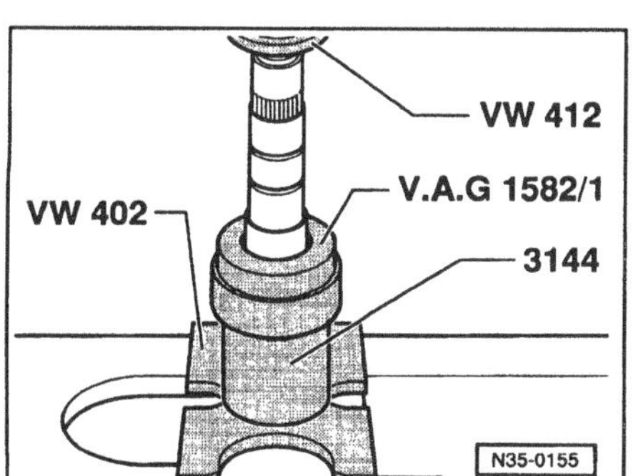

◄ **Fig. 10 Large tapered roller bearing inner race, pressing off**

- Before starting, tension gripping device V.A.G. 1582/1 onto inner race.

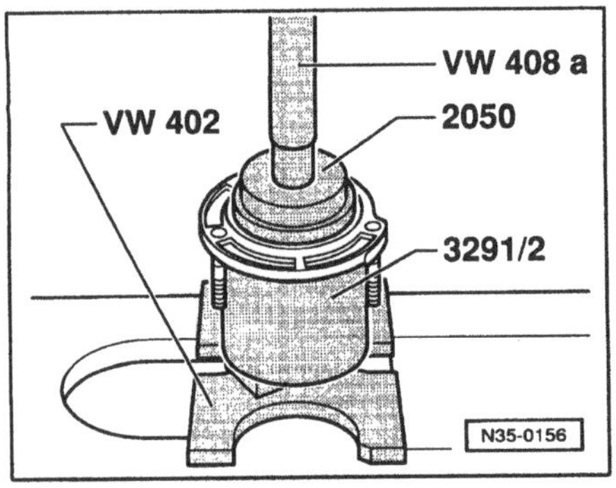

Fig. 11 Large tapered roller bearing outer race, pressing out from bearing support

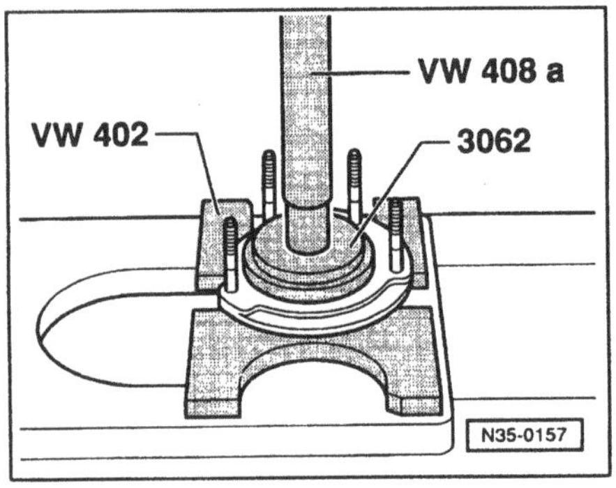

Fig. 12 Large tapered roller bearing outer race, pressing into bearing support

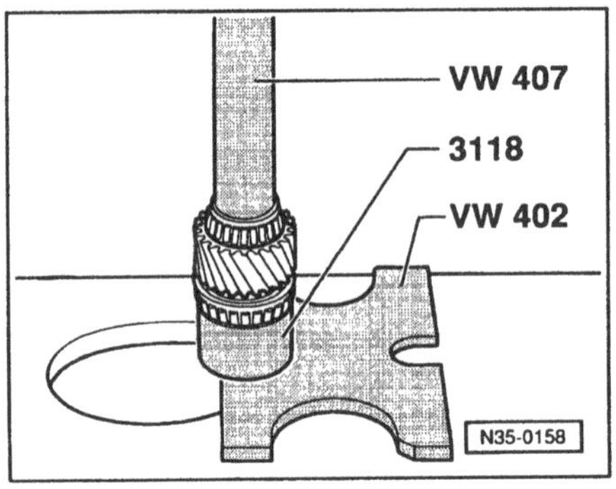

Fig. 13 Large tapered roller bearing inner race, pressing on

Note:

Press piece 3118 step points towards taper roller bearing inner race.

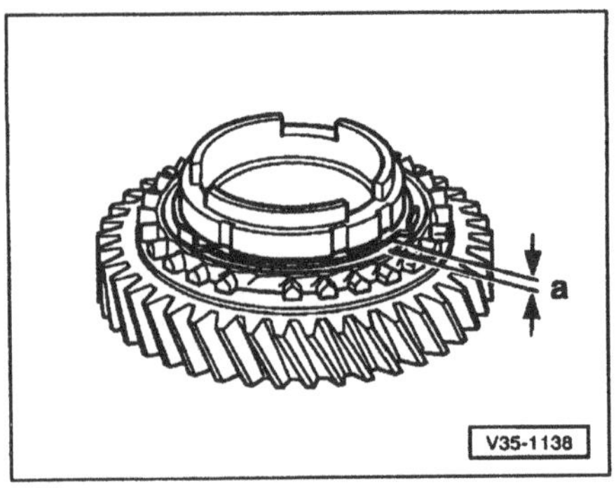

Fig. 14 1st and 2nd gear inner-ring, checking for wear

- Press inner ring onto gear and measure gap "a" with feeler gauge.

Gap "a"	New dimension	Wear limit
1st and 2nd gear	0.75 ... 1.25 mm	0.3 mm

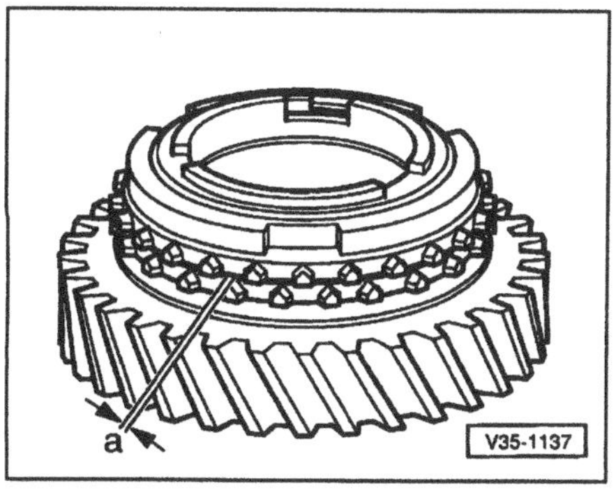

◄ **Fig. 15 1st and 2nd gear synchro-ring, checking for wear**

– Press synchro-ring, outer-ring and inner-ring onto gear cone and measure gap "a" with feeler gauge.

Gap "a"	New dimension	Wear limit
1st and 2nd gear	1.2 ... 1.8 mm	0.5 mm

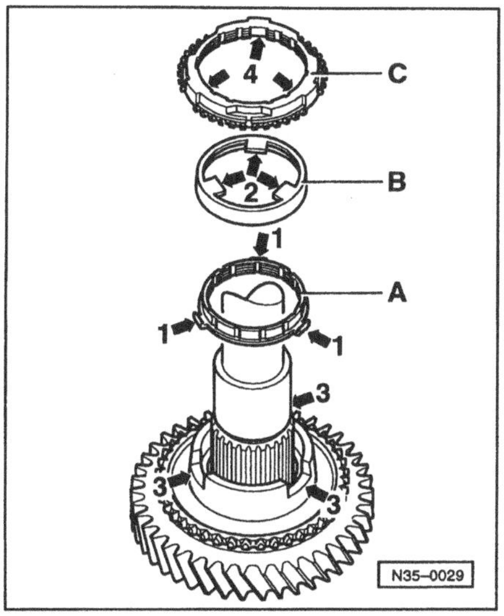

◄ **Fig. 16 Outer ring, inner ring and 1st gear synchro ring, installation position**

– Place inner ring -A- on 1st gear.

● Angled lugs –arrow– point toward outer ring -B-.

– Position outer ring -B-.

● Lugs –arrow 2– engage in sliding wheel recesses –arrow 3–.

– Install synchro-ring -C-.

● Grooves –arrow 4– engage in inner ring -A- lugs –arrow 1–.

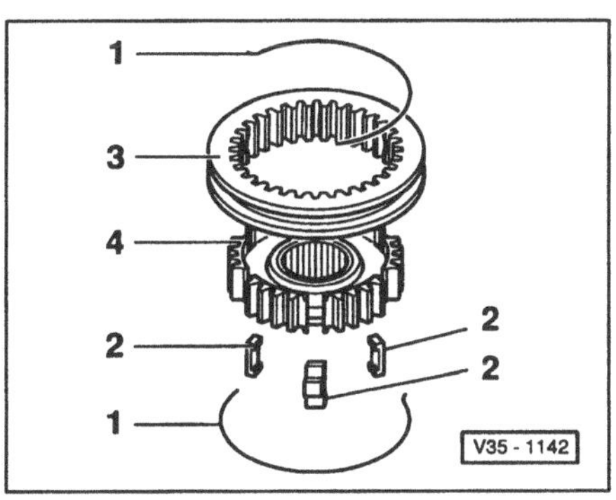

◄ **Fig. 17 Operating sleeve and synchro hub, disassembling/ assembling**

1 - Spring
2 - Operating sleeve
3 - Synchro-hub
4 - Locking piece

– Slide locking collar over synchro-hub.

● Recesses for locking pieces in synchro-hub and operating sleeve must align.

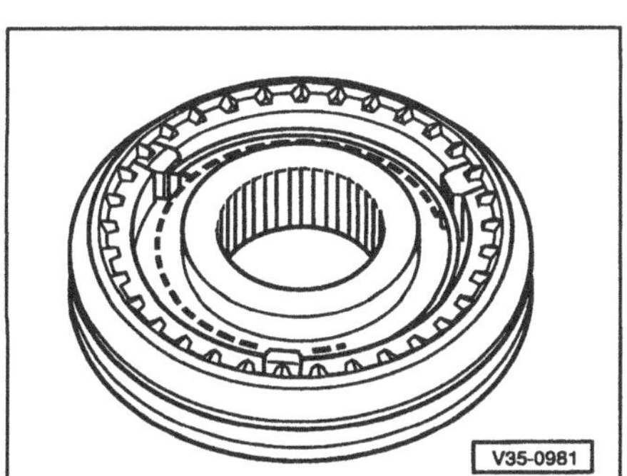

◄ **Fig. 18 Operating sleeve and synchro hub, assembling**

Operating sleeve has been pushed over synchro-hub.

– Insert locking pieces and install springs 120° off-set.

● Angled end of spring must locate in hollow locking piece.

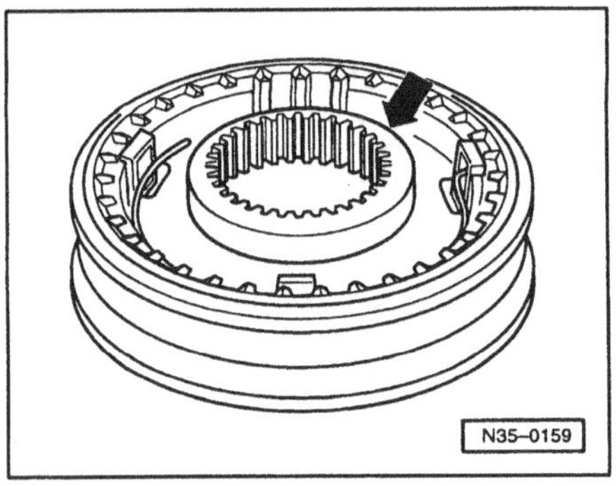

◄ **Fig. 19 1st and 2nd gear operating sleeve/synchro-hub, installation position**

Wider shoulder of synchro-hub –arrow– faces toward 2nd gear.

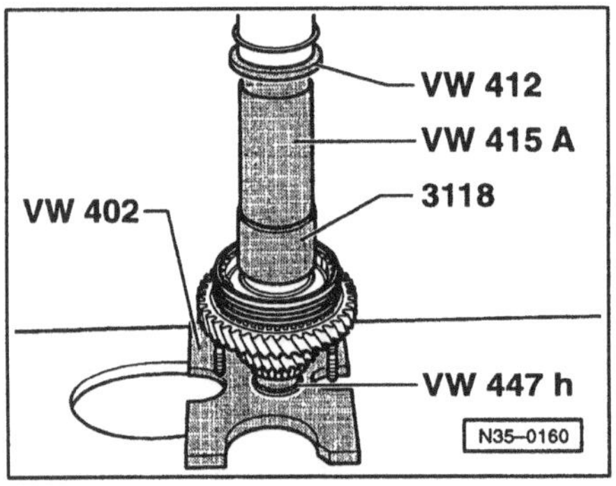

◄ **Fig. 20 1st and 2nd gear operating sleeve/synchro-hub, pressing on**

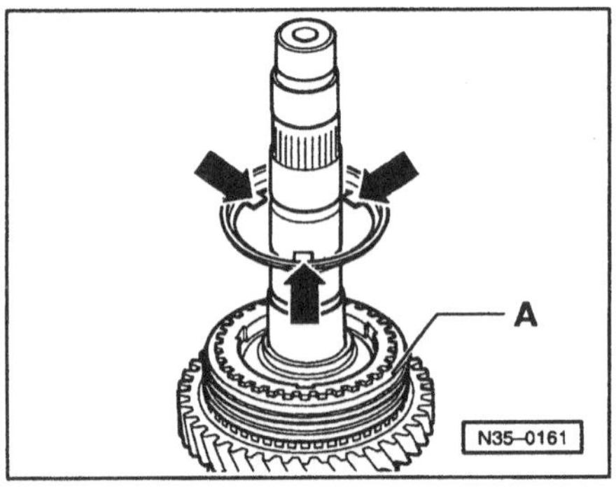

◄ **Fig. 21 2nd gear outer-ring, installation position**

Lugs –arrows– face 1st gear -A-

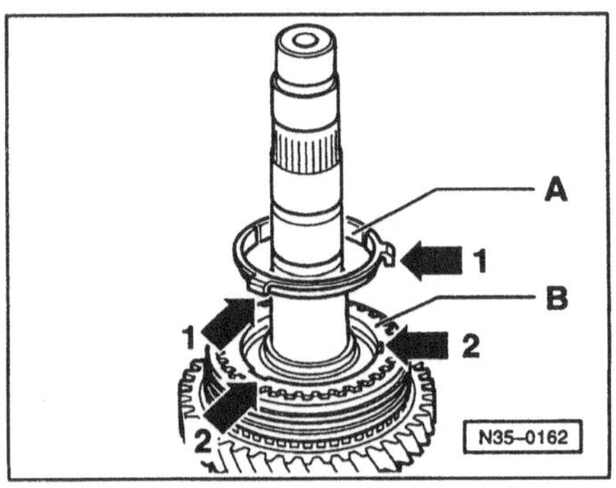

◄ **Fig. 22 Synchro ring -A- (inner ring for 2nd gear), installation position**

Lugs –arrow 1– locate in recesses –arrow 2– in synchro-ring -B-.

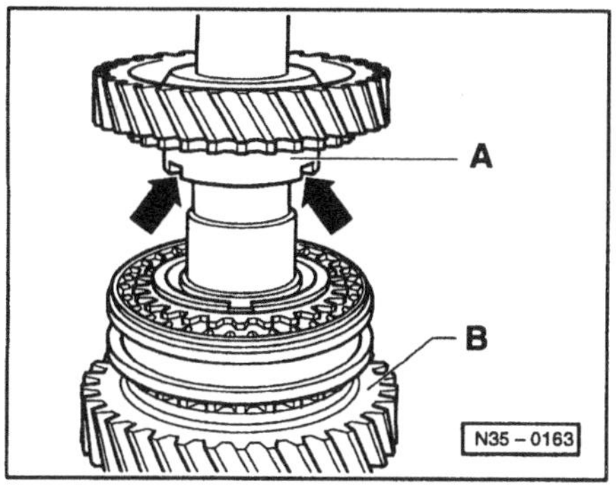

◄ **Fig. 23 2nd gear, installation position**

Higher shoulder -A- faces 1st gear -B-. Recesses in shoulder –arrows– locate in lugs of outer ring –arrow– ⇒ Fig. 21

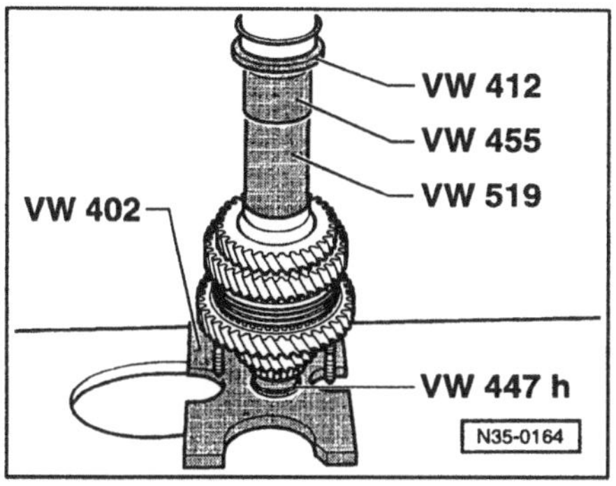

◄ **Fig. 24 3rd gear, pressing on**

Shoulder points towards 4th gear.

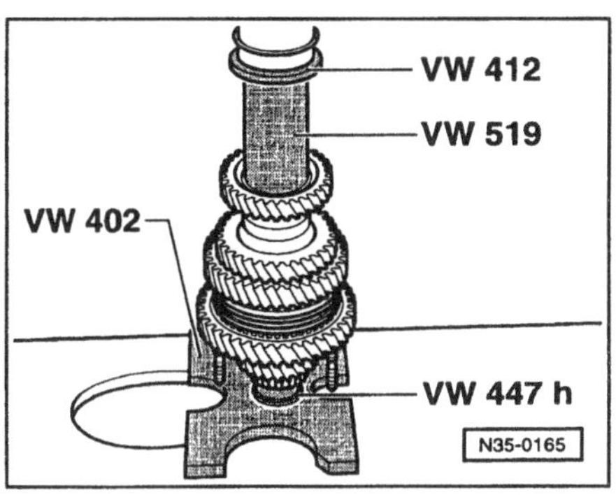

◄ **Fig. 25 4th gear, pressing on**

Shoulder points to 3rd gear.

– Position securing ring.

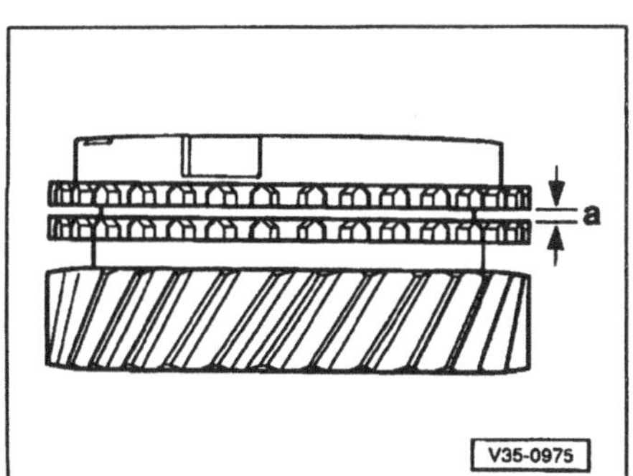

◄ **Fig. 26 Synchro ring, checking for wear**

– Press synchro-ring onto gear and measure gap "a" with feeler gauge.

Gap "a"	New dimension	Wear limit
5th gear and reverse gear	1.0 ... 1.7 mm	0.5 mm

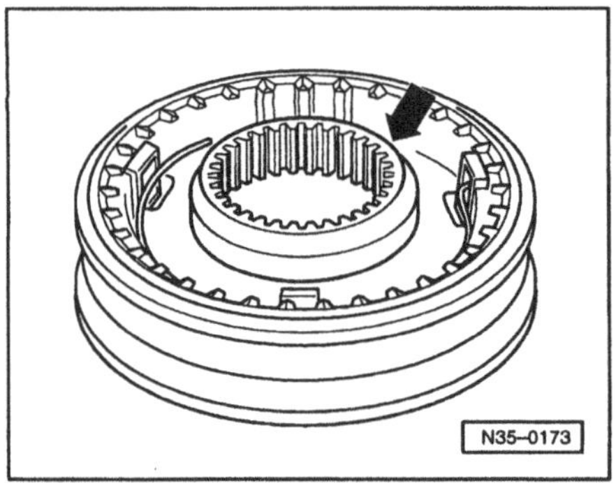

◄ **Fig. 27 5th gear and reverse gear operating sleeve/synchro hub, installation position**

Wider synchro hub shoulder with chamfer –arrow– faces toward 5th gear.

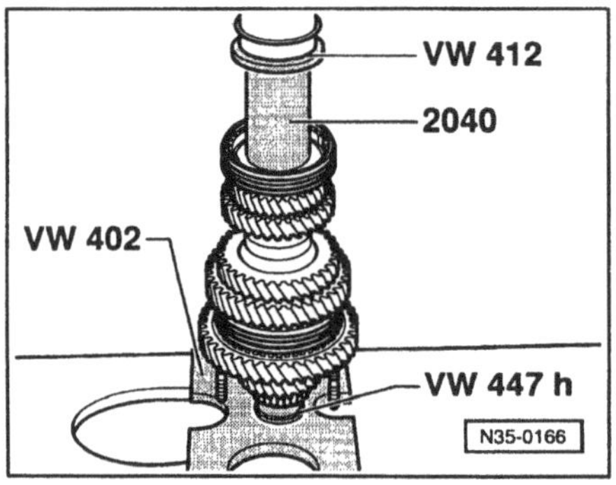

◄ **Fig. 28 5th gear and reverse gear operating sleeve with synchro hub, pressing on**

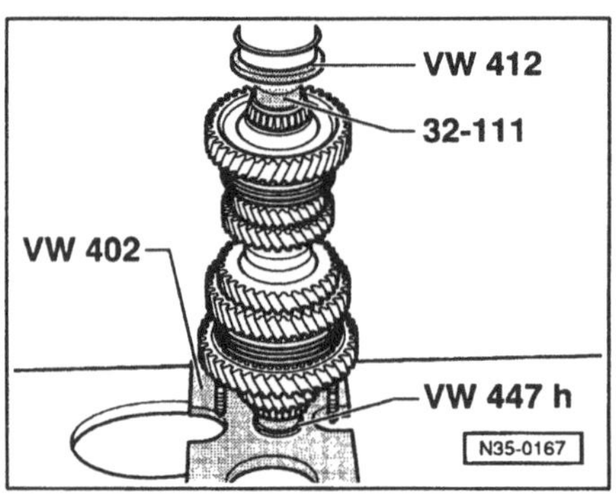

◄ **Fig. 29 Tapered roller bearing inner race, pressing on**

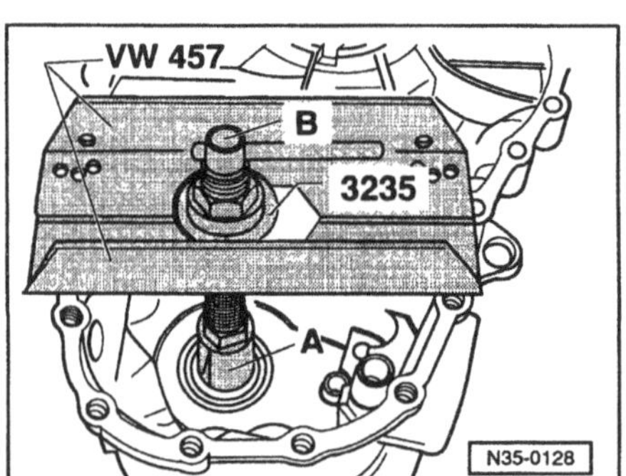

◄ **Fig. 30 Tapered roller bearing outer race, pulling out from transmission housing**

A – Internal extractor 37 ... 46 mm, e.g., Kukko 21/6

B – Counter support, e.g., Kukko 22/2

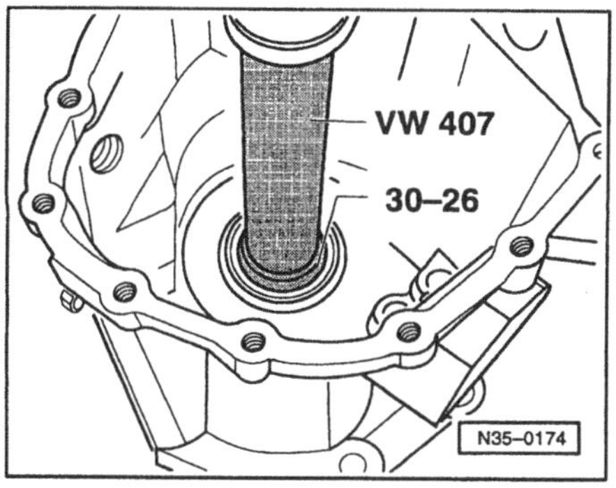

◄ **Fig. 31 Tapered roller bearing outer race, pressing into transmission housing**

– Position shim under outer race.

– Support transmission housing with press piece 3062 directly below bearing support.

Output shaft, adjusting

Determining output shaft shims S1 and S2

Output shaft must be adjusted when, final drive, clutch housing, transmission housing, bearing support, or tapered roller bearing have been replaced.

Adjustment overview

⇒ Page L 39-24

Shim S1, determining

– Press small tapered roller bearing outer race, together with a 0.65–mm–thick shim, fully into clutch housing

– ⇒ Page L 35-26.

– Insert output shaft and tighten bearing support to 25 Nm (18 ft lb) then turn 90° further.

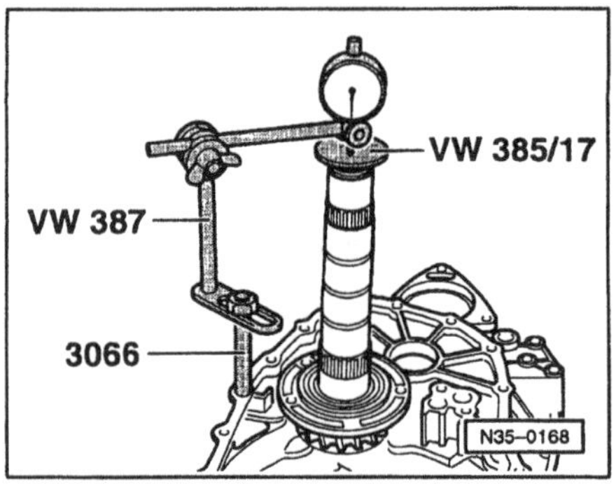

- Mount D.T.I. (3 mm measuring range) and "zero" with 1 mm (0.0393 in.) preload.
- Move output shaft up and down and record dial gauge reading (example, 0.30 mm)

Note:

Do not turn the output shaft while measuring, otherwise the bearing will settle and a false reading will be given.

Determining thickness of shim

The specified bearing preload is obtained by adding a constant figure to the reading obtained (0.30 mm) and the thickness of the shim installed (0.65 mm).

Example:

Shim installed	0.65 mm
+ Reading	0.30 mm
+ Preload (constant)	0.20 mm
S1 shim thickness	1.15 mm

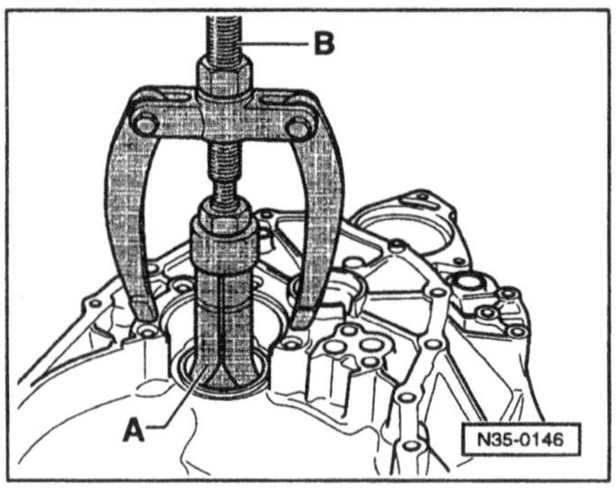

- Remove output shaft and pull out small tapered roller bearing outer race.

The following shims are available:

Size (mm)	Part No.
0.65	02A 409 210
0.70	02A 409 210 A
0.75	02A 409 210 B
0.80	02A 409 210 C
0.85	02A 409 210 D
0.90	02A 409 210 E
0.95	02A 409 210 F
1.00	02A 409 210 G
1.05	02A 409 210 H
1.10	02A 409 210 J
1.15	02A 409 210 K
1.20	02A 409 210 L
1.25	02A 409 210 M
1.30	02A 409 210 N
1.35	02A 409 210 P
1.40	02A 409 210 Q

Varying tolerances make it possible by measuring the shims to obtain the exact shim thickness required.

- Press in small tapered roller bearing outer race together with the determined shim (example, 1.15 mm) ⇒ Page L 35-26.

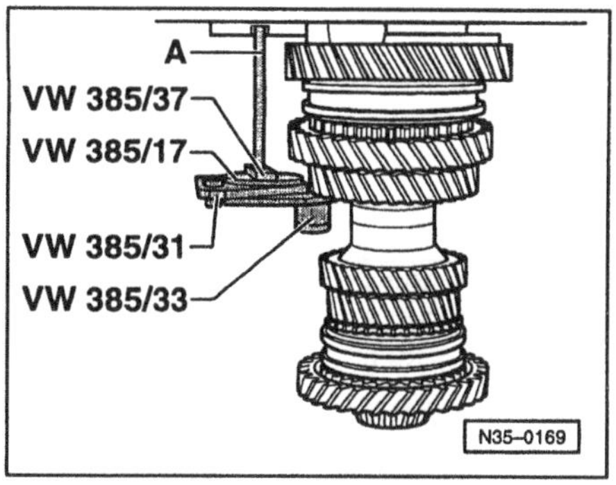

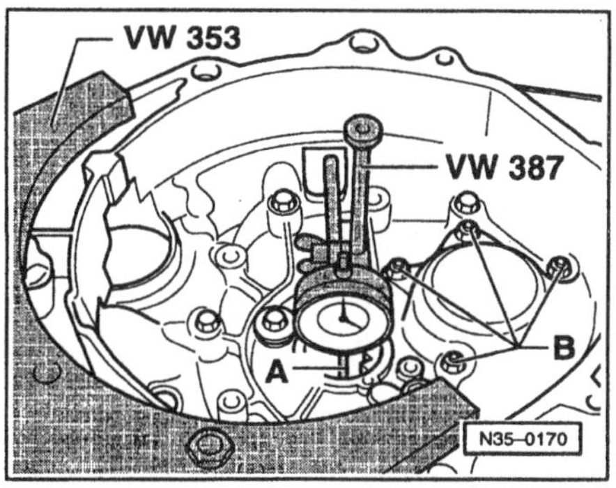

Shim S2, determining

– Insert output shaft without bearing support O-ring and tighten bearing support nut to 25 Nm (18 ft lb) then turn 90° further.

◄ – Attach measuring fixture to 3rd gear.

A – Dial indicator extension from VW 2066

– Press tapered roller bearing outer race with a 0.65 mm shim into transmission housing onto stop, using 30-26.

– Position transmisssion housing and tighten bolts to torque setting.

◄ – Mount dial indicator and "zero" with 1 mm preload.

A – Dial indicator extension from VW 2066

Notes:

- ♦ *Dial indicator extension must be against output shaft measuring assembly as illustrated in illustration N35-0169.*
- ♦ *Do not turn output shaft when measuring.*

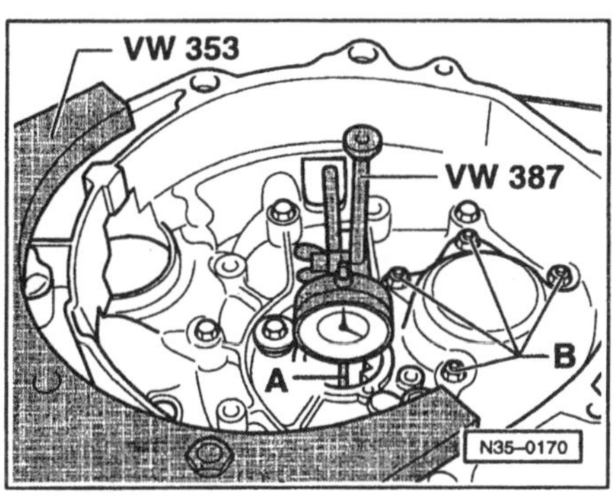

◄ – Loosen bearing support hex nuts -B- using cross-over sequence until nuts release output shaft.

– Read dial indicator and note reading (example, 0.30 mm)

Determining thickness of shim

Specified bearing preload is obtained by adding to reading obtained (0.30 mm) and the thickness of the shim installed (0.65 mm) a preload constant of (0.10 mm).

Example:

Shim installed	0.65 mm
+ Reading	0.30 mm
+ Preload (constant)	0.10 mm
S2 shim thickness	1.05 mm

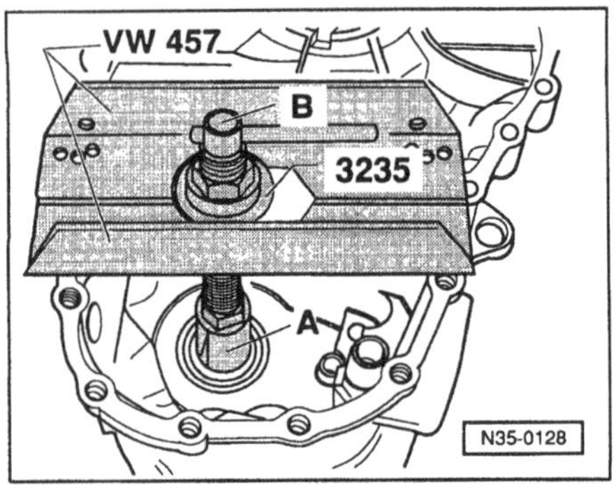

- Remove transmission housing
- Insert shims of correct thickness, thickest shim first.

Following shims are available:

Size (mm)	Part No.
0.65	020 311 391 P
0.70	020 311 391 Q
0.75	020 311 391
0.80	020 311 391 A
0.85	020 311 391 B
0.90	020 311 391 C
0.95	020 311 391 D
1.00	020 311 391 E
1.05	020 311 391 F
1.10	020 311 391 G
1.15	020 311 391 H
1.20	020 311 391 J
1.25	020 311 391 K
1.30	020 311 391 L
1.35	020 311 391 M
1.40	020 311 391 N
0.70+0.75 = 1.45	020 311 391 Q/391
0.75+0.75 = 1.50	020 311 391/391
0.75+0.80 = 1.55	020 311 391/391 A

- Press outer race in again.
- Remove output shaft and engage 4th gear with operating sleeve.
- Insert input shaft together with output shaft. Tighten bearing support nut in clutch housing to 25 Nm (18 ft lb) then turn 90° further.
- Install transmission housing and tighten hex bolts to 25 Nm (18 ft lb) and then turn 90° further.

Checking bearing preload

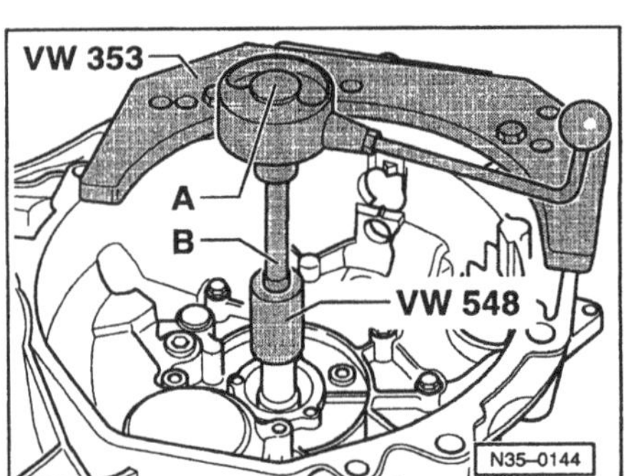

A – Torque gauge 0 ... 600 Ncm (0...53 in. lb)

B – Extension for 1/2 " square drive socket

For new tapered roller bearings, friction moment must be 130 ... 180 Ncm (11.5...16 in. lb).

For used tapered roller bearings, friction moment must be at least 30 Ncm (3 in. lb).

Note:

Lightly coat bearings with transmission oil before checking bearing preload.

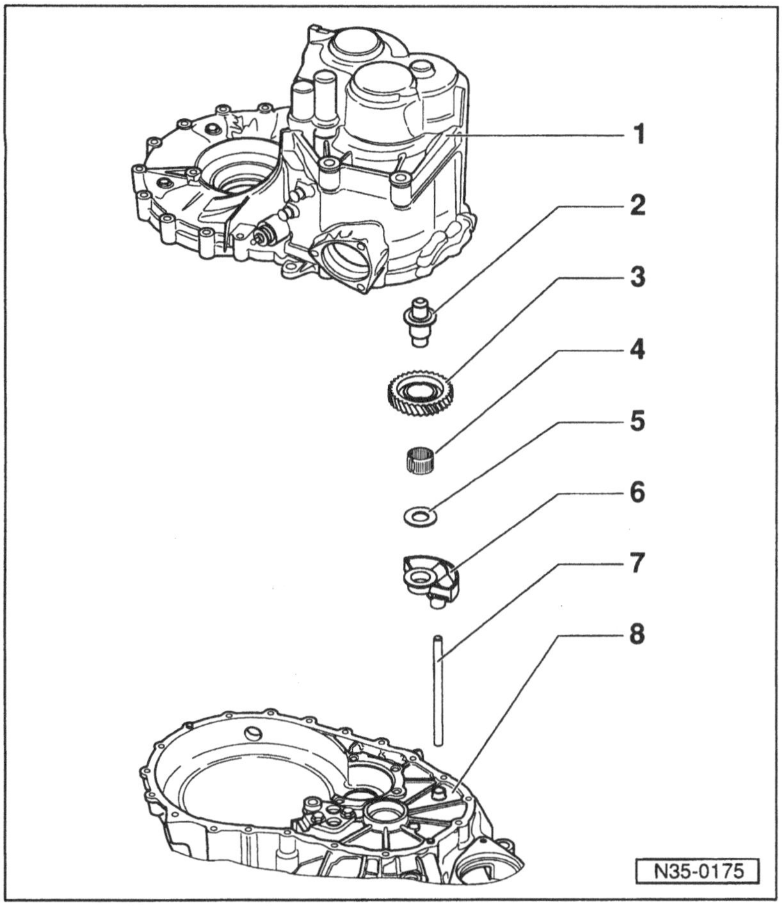

Reverse gear, removing/ installing

1 - Transmission housing

2 - Shaft

– Press out from mount (item 6) with reverse gear

⇒ Fig. 1

♦ Pressing in

⇒ Fig. 2

3 - Reverse gear

4 - Needle roller bearing

5 - Thrust washer

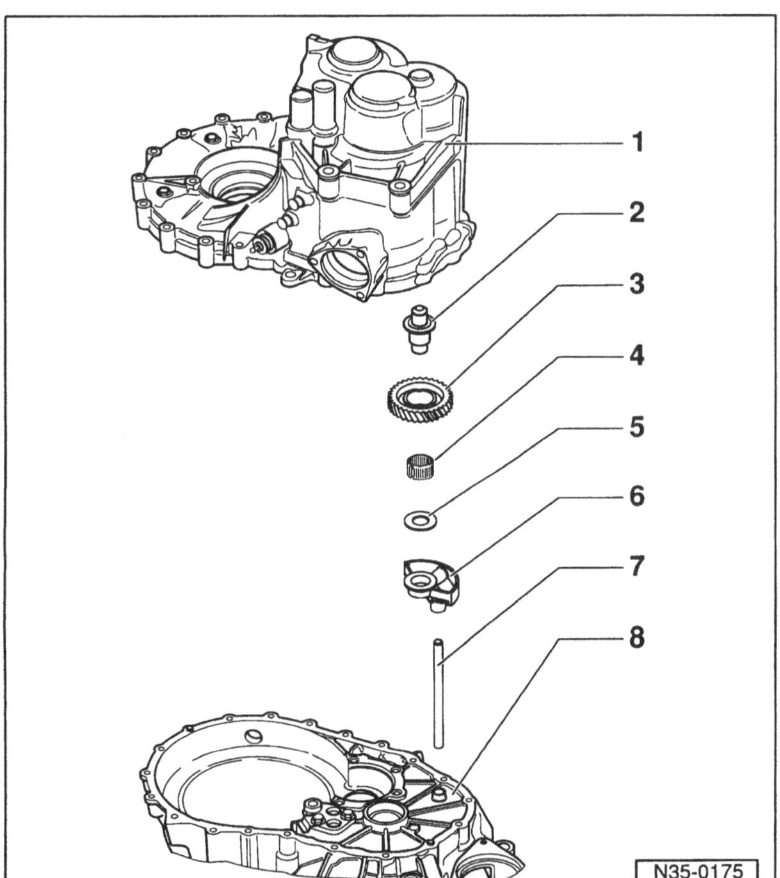

6 - Reverse gear mount

♦ Removing/installing

⇒ Page L 34-52

7 - Reverse gear guide tube

♦ Removing/installing

⇒ Page L 34-52

8 - Clutch housing

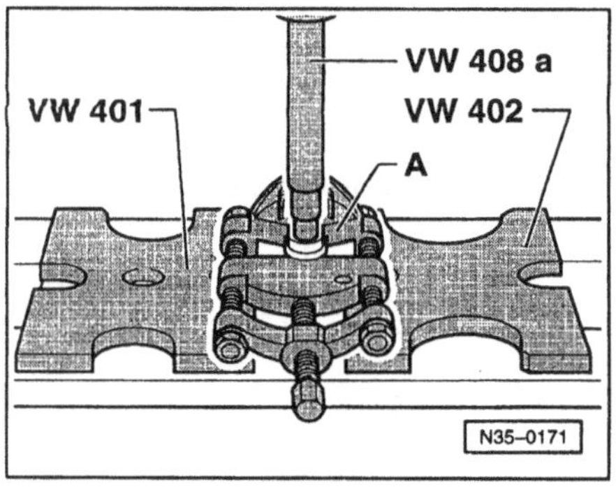

◄ **Fig. 1 Reverse gear, pressing out from mount**

A - Separating device 12 ... 75 mm, e.g., Kukko 17/1

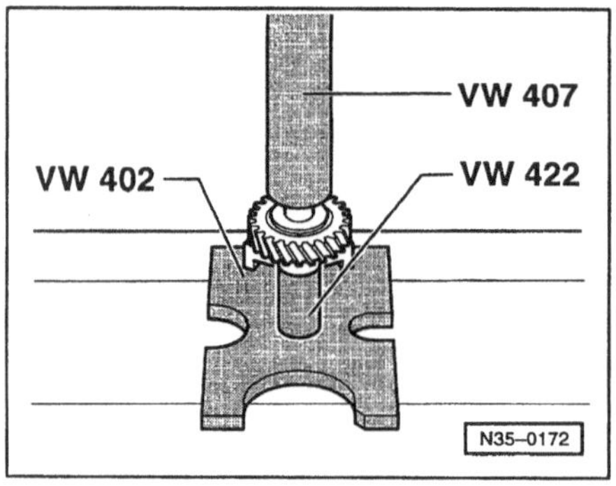

◄ **Fig. 2 Reverse gear shaft, pressing into mount**

Stub shaft/flange shaft oil seal, removing/installing

Transmission installed

To replace left flange shaft or right stub shaft seal, remove relevant drive shaft as follows:

– Position support device 3250 before lifting vehicle.
⇒ Suspension, Wheels, Brakes, Steering; Repair group 40, Servicing front suspension

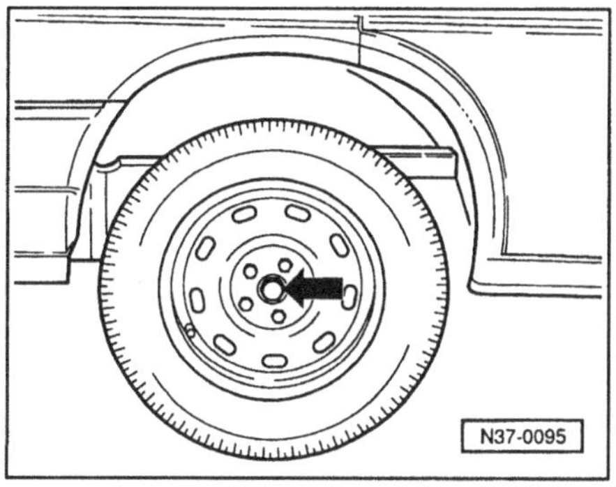

◄ – Remove hexagon bolt –arrow– for drive shaft/wheel hub and loosen wheel bolts

– Remove noise insulation tray.

– Disconnect drive shaft from differential flange.

– Remove wheel.

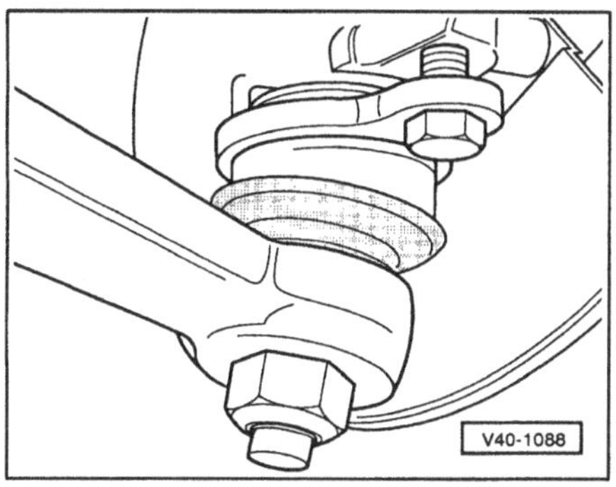

◄ – Disconnect lower ball joint from steering knuckle.

– Remove bolt for shock absorber/stabilizer bar coupling link.

– Remove drive shaft from steering knuckle.

Left oil seal, replacing

– Remove drive shaft
⇒ Page L 39-1.

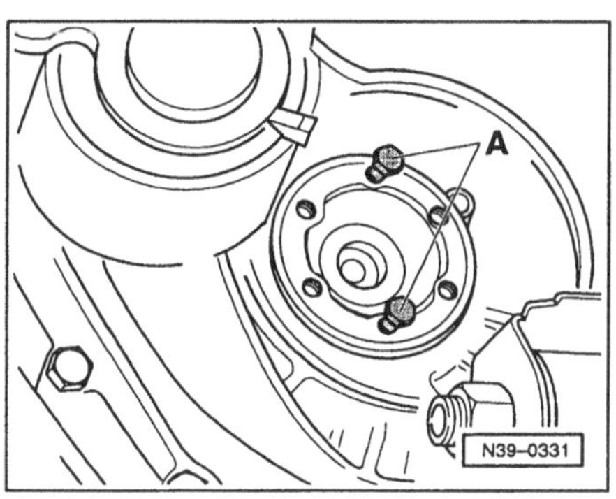

◄ – Remove flange shaft. To do this, screw 2 bolts -A- in, alternately.

– Always replace flange shaft circlip
⇒ Page L 34-58.

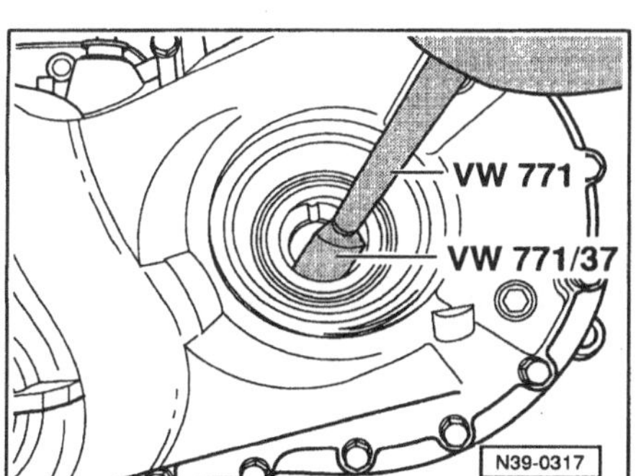

◄ – Remove (pull) flange shaft oil seal using VW 771 and VW 771/37.

Note:

Do not damage oil deflector ring when pulling out.

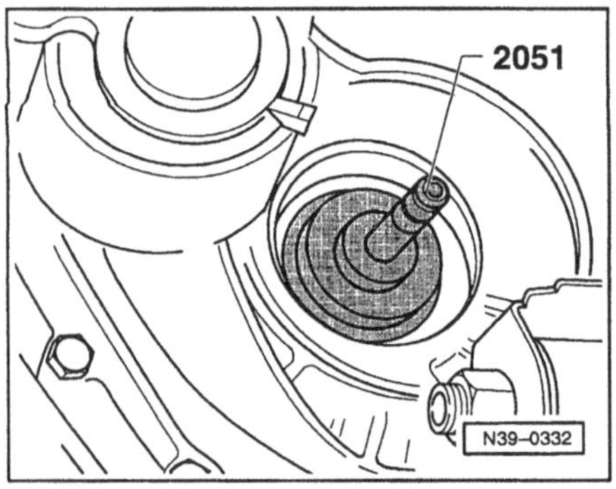

◄ – Drive new seal in to stop.

– Fill space between seal and dust lip with multi-purpose grease.

– Drive in flange shaft.

– Install drive shaft.

– Check transmission oil
⇒ Page L 00-3

– Install noise insulation tray.

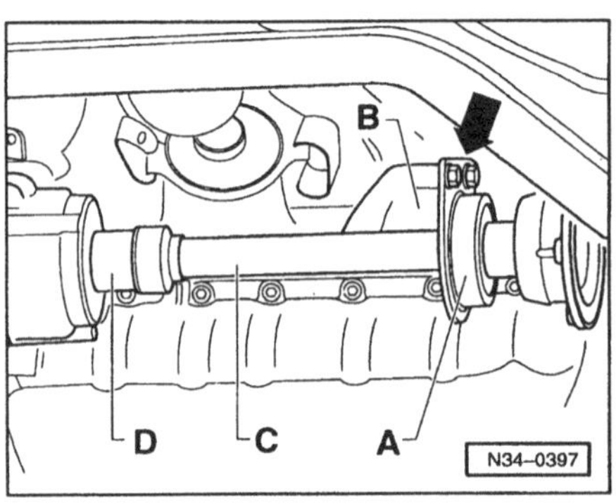

Right oil seal, replacing

– Remove drive shaft
⇒ Page L 39-1.

◄ – Unbolt flange shaft mount -A- from bracket -B-.

– Pull right flange shaft -C- from stub shaft -D-.

– Remove O ring from stub shaft.

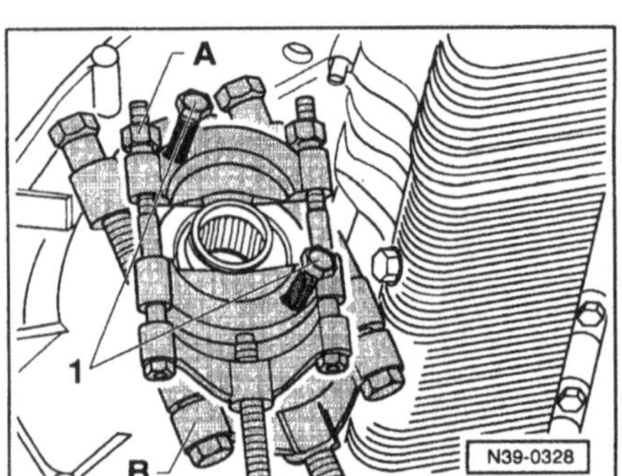

◄ – Axle shaft, removing

A - Separating tool Kukko 17/0

B - Separating tool Kukko 17/1

1 - Bolt M10 x 80 (Qty. 2)

– Turn bolts -1- evenly against separating device -B-.

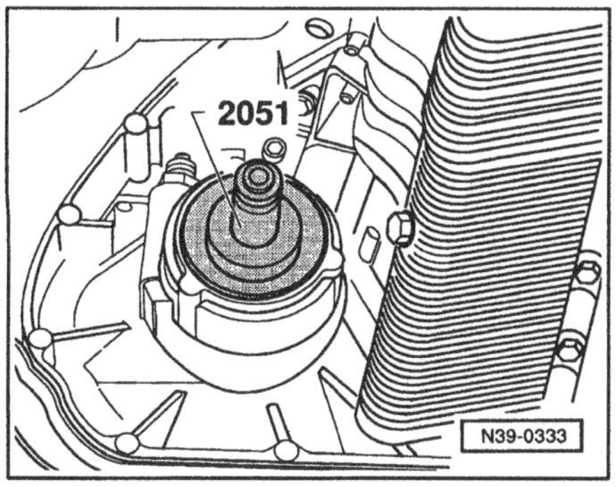

◄

- Remove seal
- Drive new seal in to stop.
- Fill space between seal and dust lip with multi-purpose grease.

Note:

Always replace stub shaft circlip
⇒ Page L 34-58.

- Drive stub shaft in complete with new circlip and O ring with plastic–head hammer.

Note:

Grease stub shaft internal splines with drive shaft grease G 000 603.

- Install right flange shaft.
- Install drive shaft.
- Check transmission oil
⇒ Page L 00-3
- Install noise insulation tray.

Tightening torque:

Drive shaft to drive flange	80 Nm (59 ft lb)
Wheel bearing housing to lower ball joint (steering knuckle)	55 Nm (41 ft lb)
Hex bolt on drive shaft	150 Nm (111 ft lb) + 90°
Flange shaft mount to bracket	25 Nm (18 ft lb)

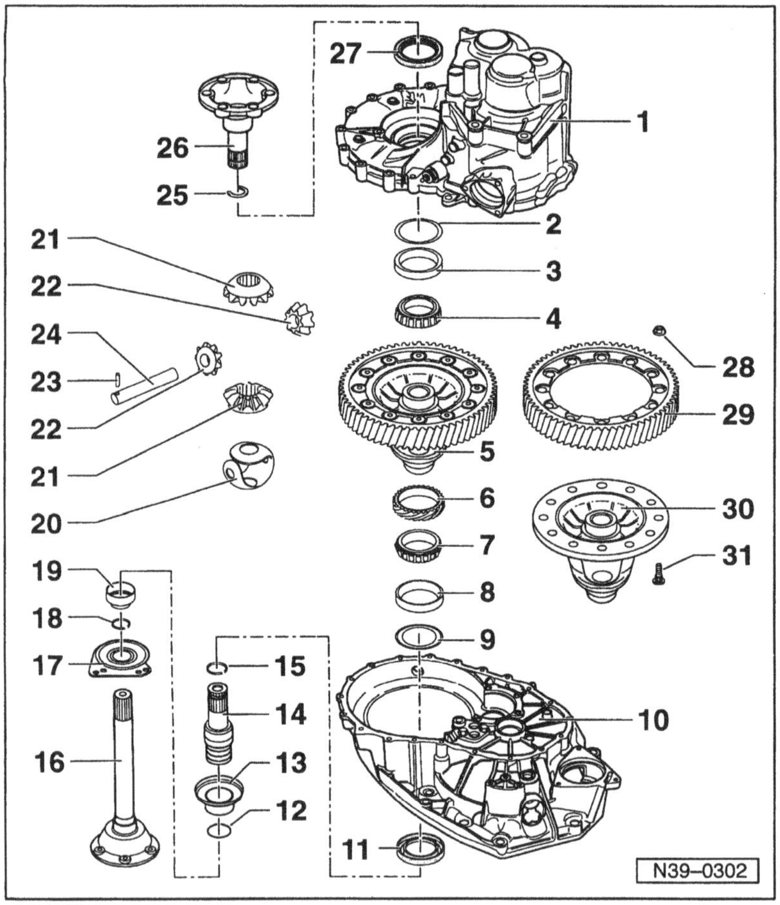

Differential disassembling/ assembling

Notes:

- *Heat tapered roller bearing inner race to 100°C (212°F) before installing.*
- *Always replace both tapered roller bearings together as a set.*
- *If differential housing, transmission housing, or clutch housing tapered roller bearings are replaced, adjust differential*

⇒ Page L 39-25.

1 - Transmission housing

2 - Shim
- For differential
- Determining thickness

⇒ Page L 39-25

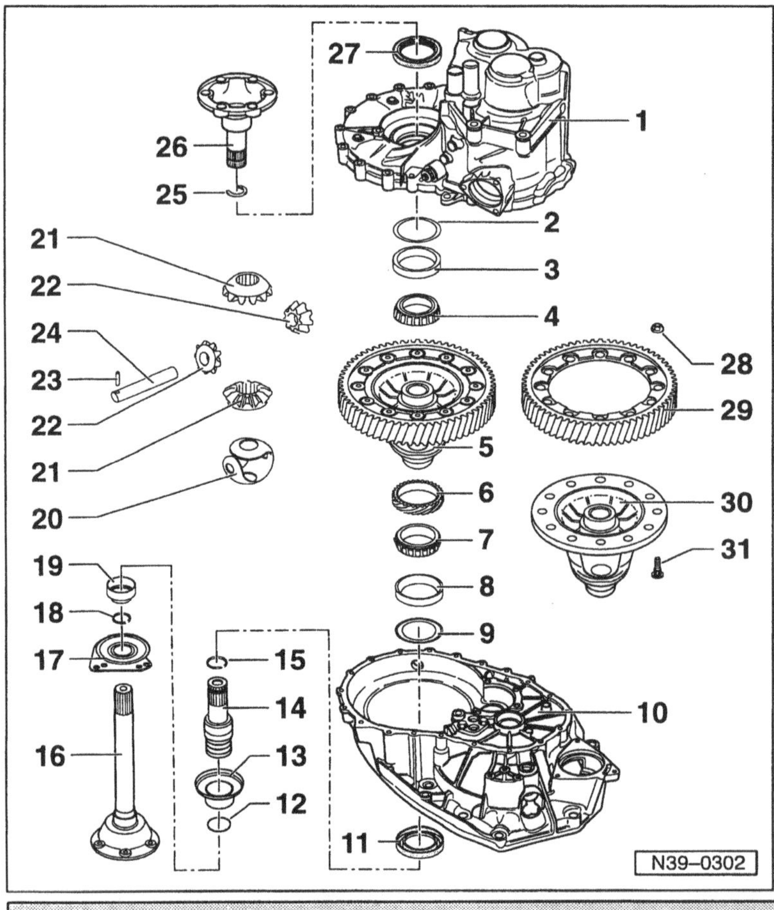

3 - Tapered roller bearing outer race
- Pulling out

⇒ Fig. 5
- Pressing in

⇒ Fig. 6

4 - Tapered roller bearing inner race
- Pulling off

⇒ Fig. 3
- Pressing on

⇒ Fig. 4

5 - Differential housing
- Rivetted final drive gear
  - Drill out rivet heads when replacing final drive gear

⇒ Fig. 7
  - Attach final drive gear with bolts

⇒ Fig. 10

6 - Speedometer drive gear
  - Install on differential housing before pressing on inner race

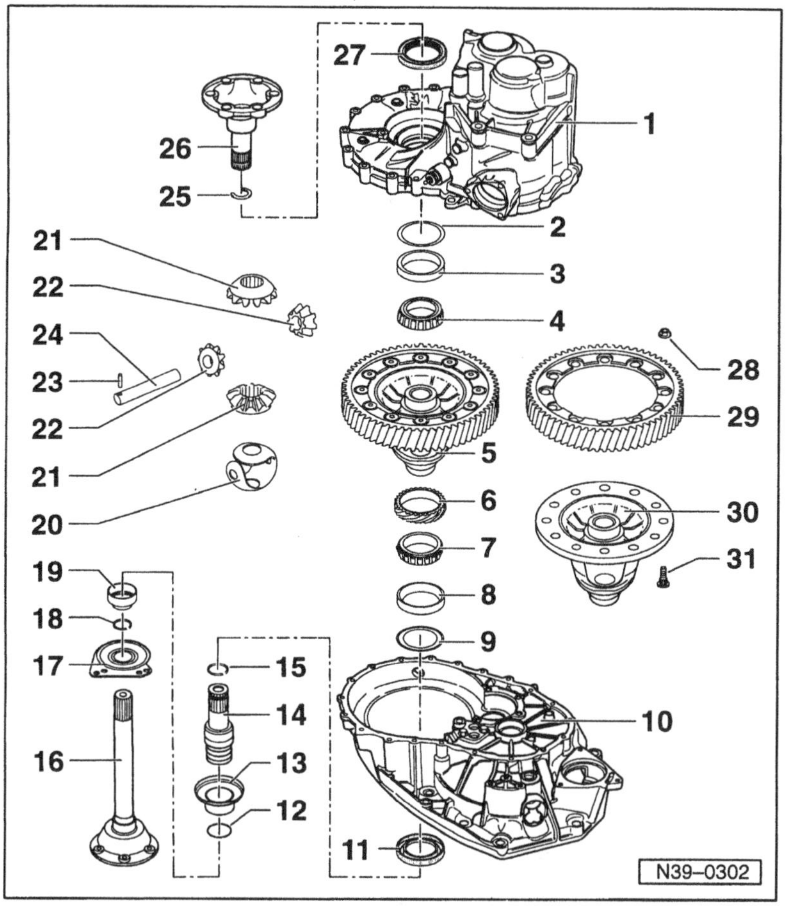

7 - Tapered roller bearing inner race

- ♦ Pulling off

⇒ Fig. 3

- ♦ Pressing on

⇒ Fig. 4

8 - Tapered roller bearing outer race

- ♦ Pressing out

⇒ Fig. 1

- ♦ Pressing in

⇒ Fig. 2

9 - Oil deflector ring

- ♦ Press out with tapered roller bearing outer race (item 8)

⇒ Fig. 1

10 - Clutch housing

11 - Oil seal

- ♦ replacing

⇒ Page L 39-1

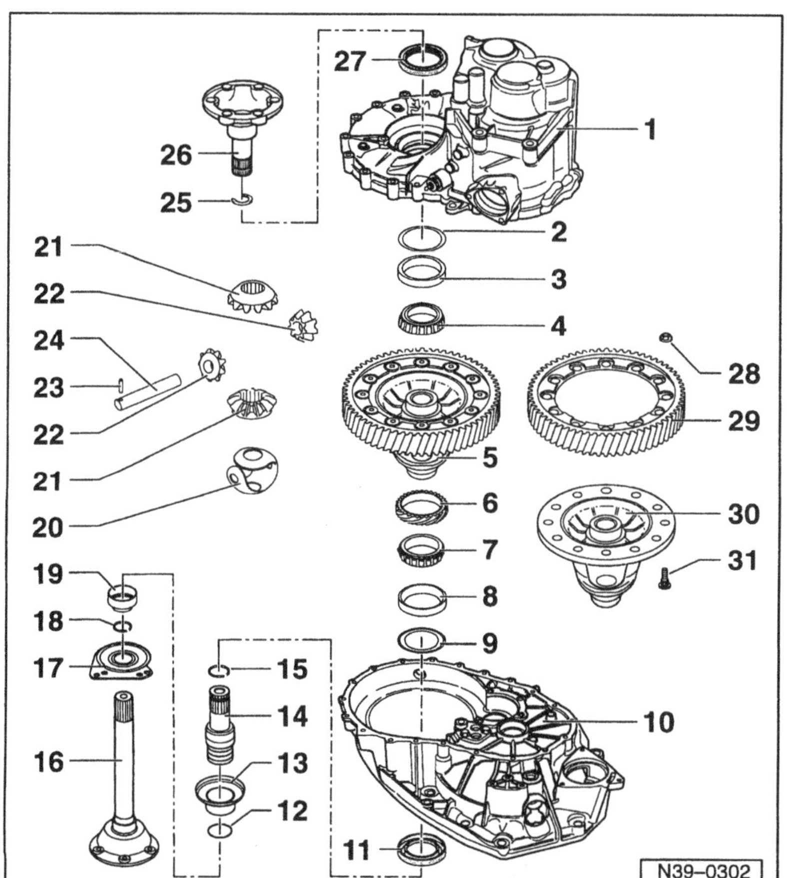

12 - O-ring

- – Always replace
- – Insert into stub shaft groove (item 14)

13 - Dust plate

- ♦ For stub shaft
- ♦ Pressing off

⇒ Fig. 17

- ♦ Pressing on

⇒ Fig. 18

14 - Stub shaft

- ♦ Removing/installing

⇒ Page L 39-1

- – Grease internal splines with drive shaft grease G 000 603

15 - Circlip

- – Always replace

16 - Right-hand flange shaft

- ♦ Removing/installing

⇒ Page L 39-1

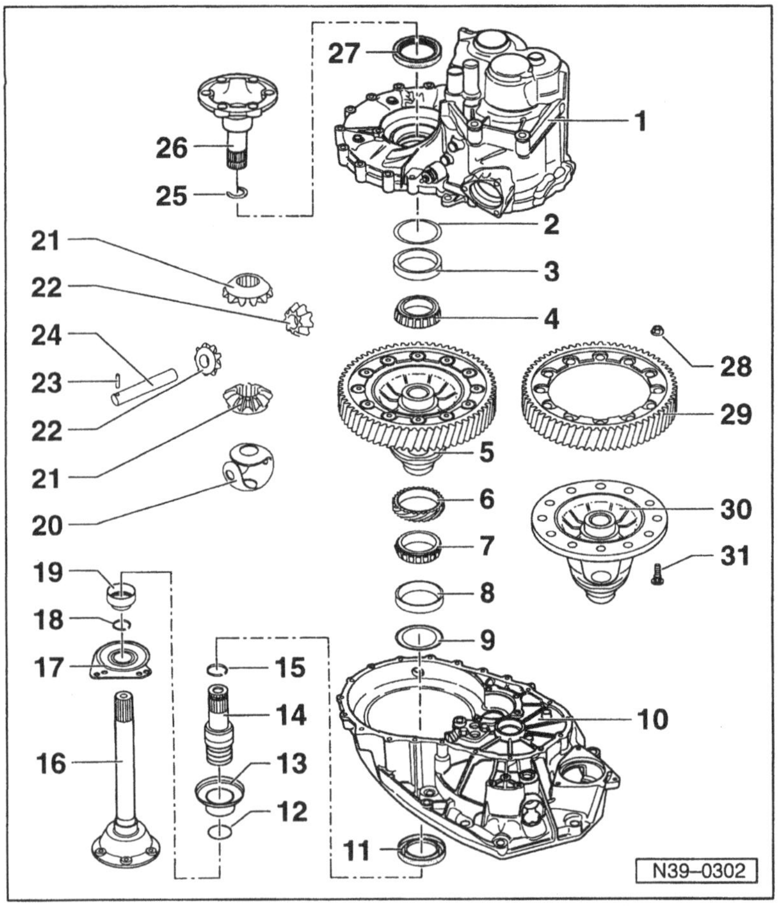

17 - Flange shaft bearing
- Pressing off

⇒ Fig. 15
- Pressing on

⇒ Fig. 16

18 - Circlip

19 - Dust plate
- For flange shaft
- Pressing off

⇒ Fig. 13
- Pressing on

⇒ Fig. 14

20 - One-piece thrust washer
- Coat with transmission oil when installing

21 - Differential gear
- Installing

⇒ Fig. 12

22 - Differential pinion
- Installing

⇒ Fig. 12

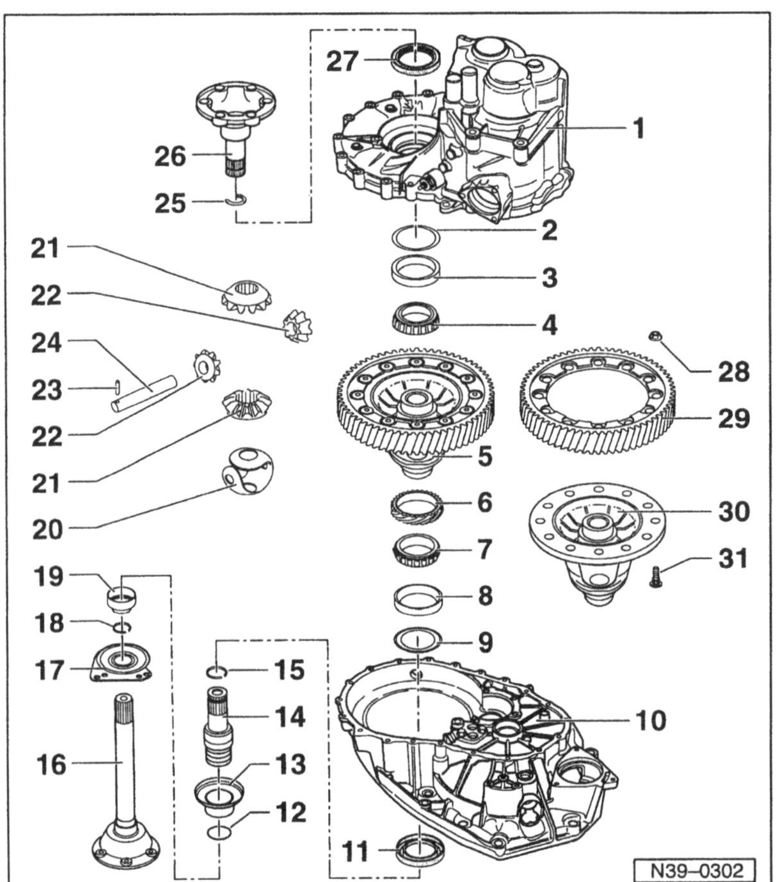

23 - Spring pin
- For securing planet pinion axis shaft
- Removing/installing

⇒ Fig. 11

24 - Pinion shaft
- Remove with drift

25 - Circlip
- Always replace

26 - Flange shaft
- Removing/installing

⇒ Page L 39-1

27 - Oil seal
- Replacing

⇒ Page L 39-1

28 - Hexagon nut, 60 Nm (44 ft lb) and turn 90° further

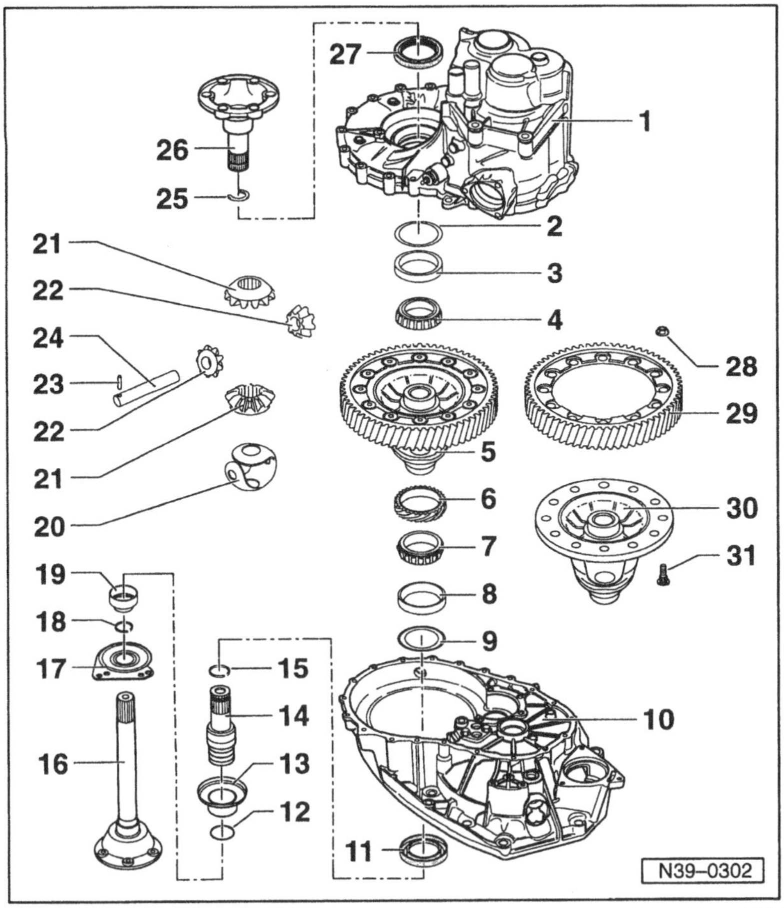

29 - Final drive gear
- ♦ Riveted in production
- ♦ Bolted

⇒ Fig. 10
- – Heat to 100°C (212°F) before installing
- ♦ Pressing off

⇒ Fig. 8
- – Install on differential housing

⇒ Fig. 9
- ♦ Is paired with output shaft, only replace as a set

30 - Differential housing
- – Bolt to final drive gear

⇒ Fig. 10

31 - Bolt

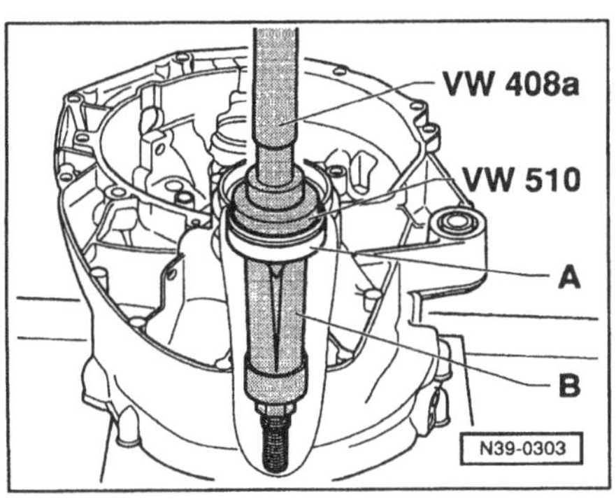

◄ **Fig. 1 Tapered roller bearing outer race -A- , pressing out from clutch housing**

B - Internal puller 46...58 mm, e.g. Kukko 21/7

Note:

After pressing out, check oil deflector ring for damage and replace, if necessary.

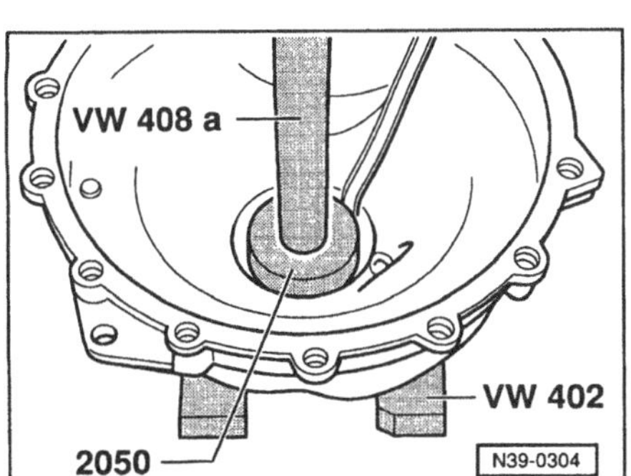

◄ **Fig. 2 Tapered roller bearing outer race, pressing into clutch housing**

– First insert oil deflector ring.

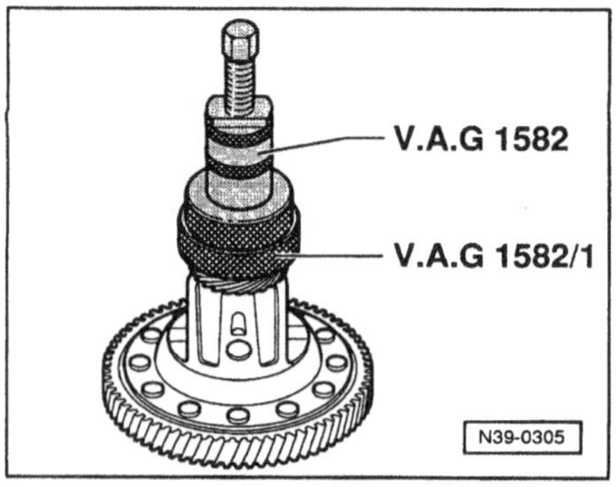

◄ **Fig. 3 Tapered roller bearing inner race, pulling off**

- Before installing extractor position press piece 40-105 on differential housing.

Note:

Pulling off both taper roller bearing inner races from differential housing is identical.

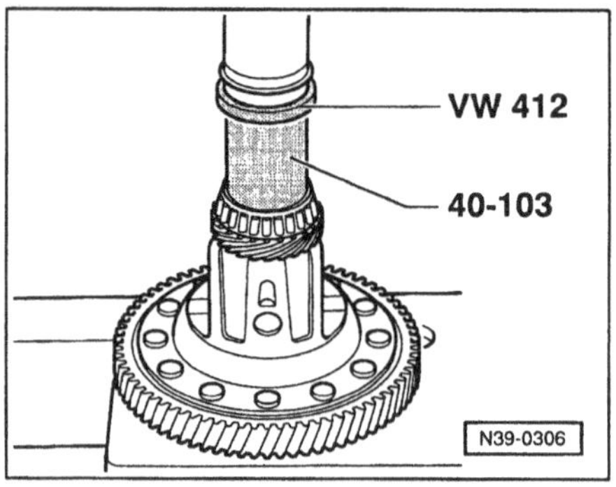

◄ **Fig. 4 Tapered roller bearing inner race, pressing on**

Note:

tapered roller bearing inner races for transmission housing and clutch housing are pressed on with the same press tools.

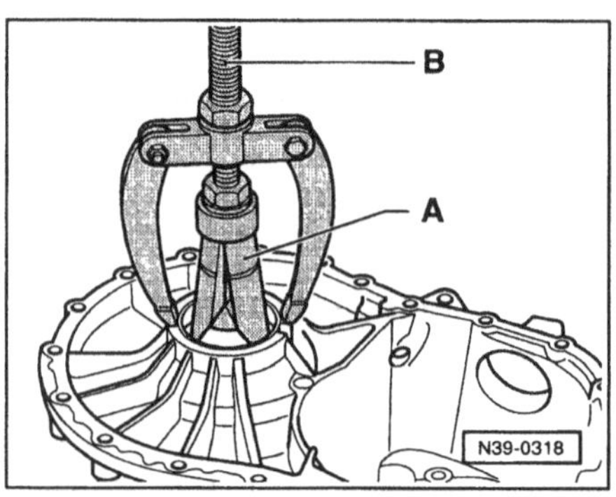

◄ **Fig. 5 Tapered roller bearing outer race, pulling from transmission housing**

A - Internal puller 56...70 mm, e.g. Kukko 21/8

B – Counter support, e.g. Kukko 22/2

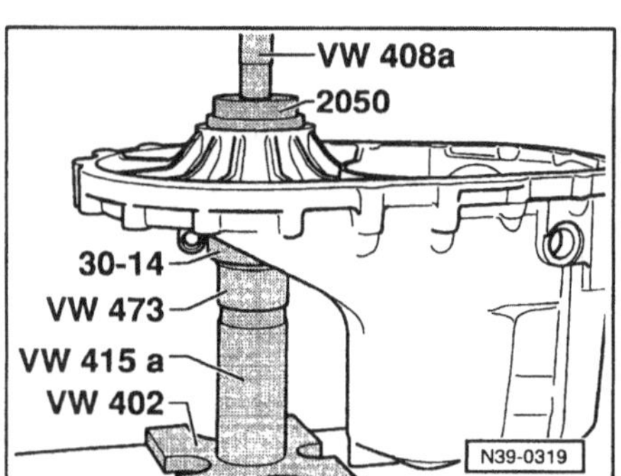

◄ **Fig. 6 Tapered roller bearing outer race, pressing into transmission housing**

- Position shim under outer race.
- Support transmission housing with sleeve 30-14 directly below bearing support.

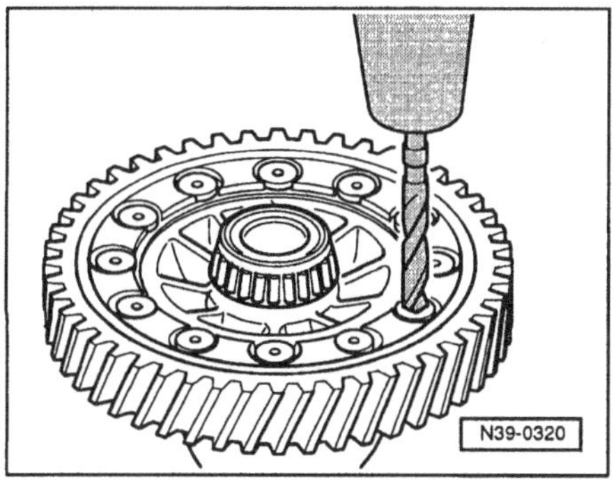

◄ **Fig. 7 Rivet heads, drilling out**

- Drill out rivet heads with 12 mm (0.47 in.) drill from countersunk side and knock-out with drift.

Note:

Clean differential before and after drilling, protect taper roller bearing from drillings.

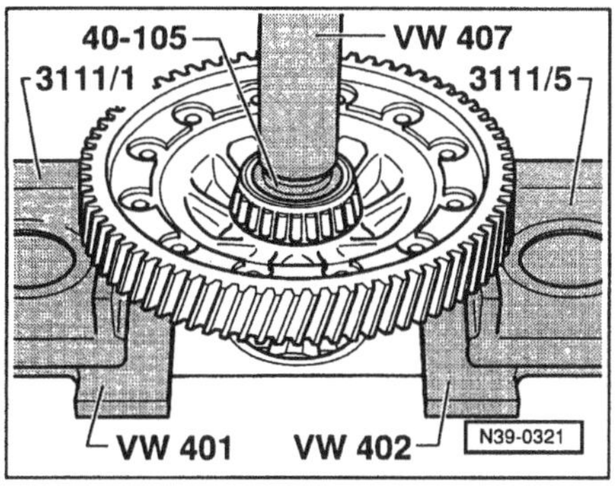

◄ **Fig. 8 Final drive gear, pressing off**

- Place thrust piece 40-105 on differential housing.

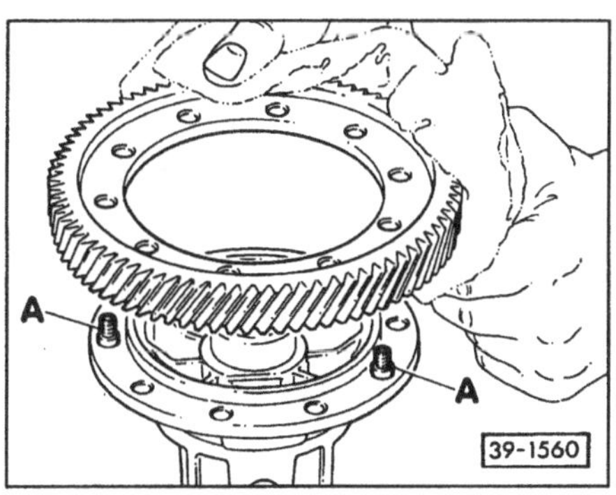

◄ **Fig. 9 Final drive gear , heating/installing (to approximately 100°C [212°F])**

- Use bolts -A- from repair kit to guide final drive gear into position during installation.

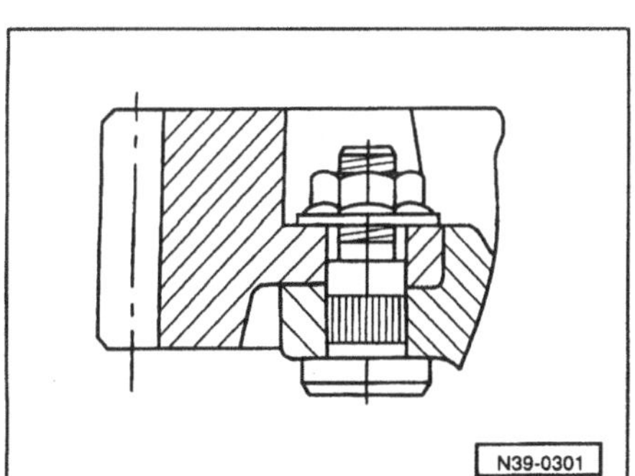

◄ **Fig. 10 Final drive gear and differential housing, fastening together**

Use special bolts with nuts according to parts microfiche.

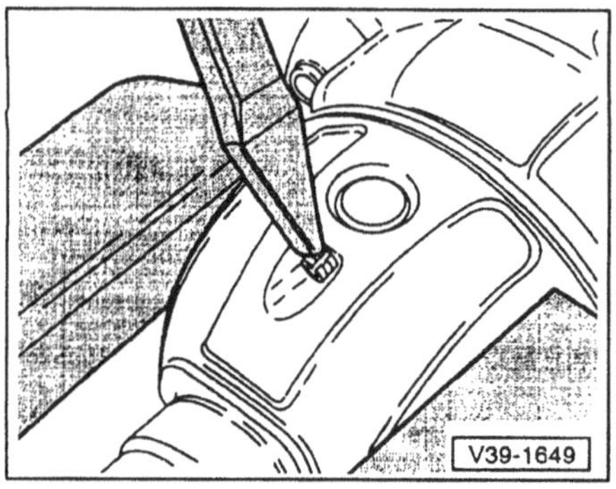

Fig. 11 Planet gear axis shaft spring pin, removing/installing

Removing

- Cover tapered roller bearing inner race and speedometer drive gear to avoid possible damage and contamination with metal particles.
- Knock-out spring pin with chisel, inserting chisel into circumferential groove.

Installing

- Drive fully into differential housing.

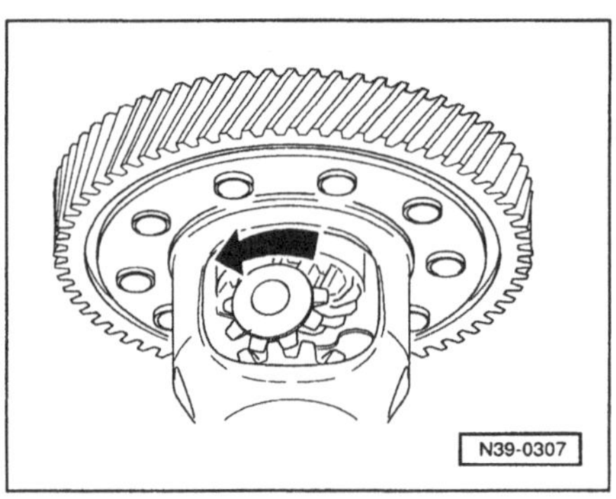

Fig. 12 Differential gears and pinions, installing

- Lubricate one piece thrust washer with transmission oil and install.
- Install both differential gears and secure (e.g., with flange shaft).
- Insert differential gears turned through 180° and pivot into position.
- Drive pinion shaft into final position and secure with spring pin.

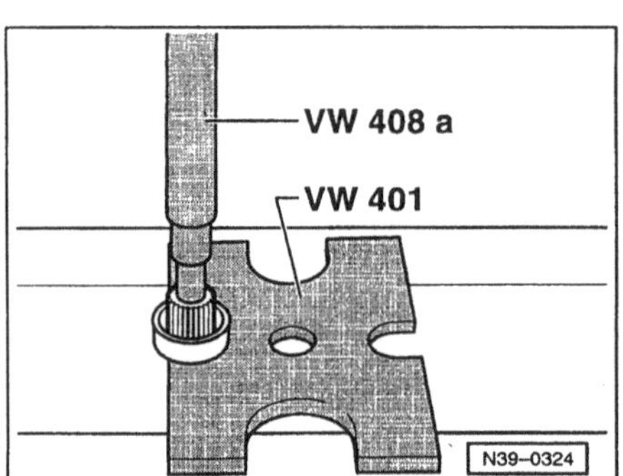

Fig. 13 Flange shaft dust plate, pressing off

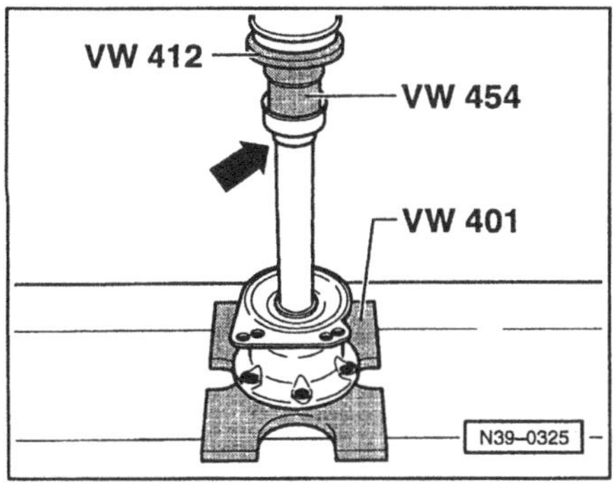

◄ **Fig. 14 Flange shaft dust plate, pressing on**

– Press new dust plate on until flush with shoulder -arrow-.

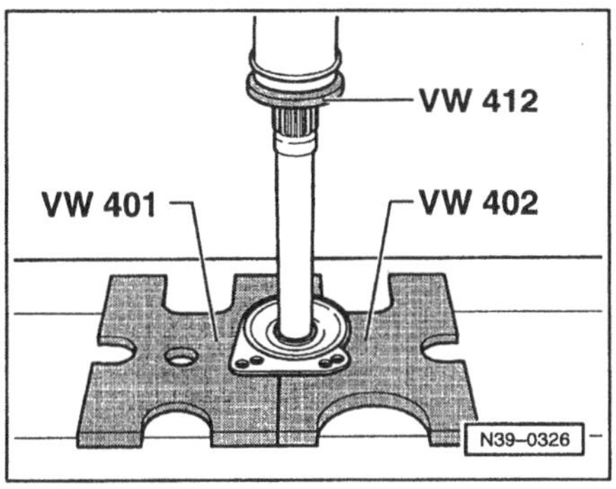

◄ **Fig. 15 Flange shaft mounting, pressing off**

– Remove circlip on bearing first.

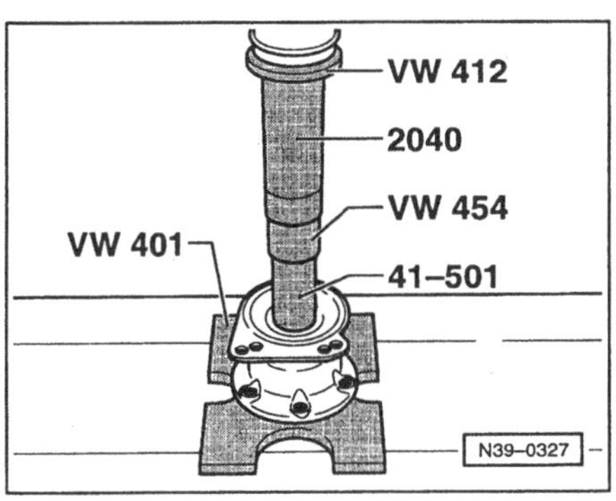

◄ **Fig. 16 Flange shaft bearing, pressing on**

– Install circlip.

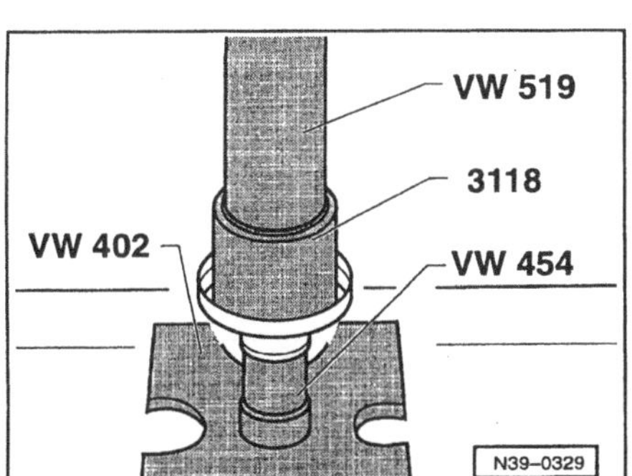

◄ **Fig. 17 Stub shaft dust plate, pressing off**

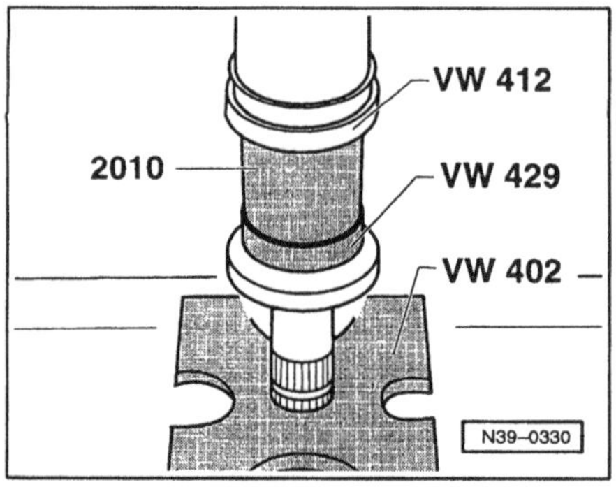

◄ **Fig. 18 Stub shaft dust plate, pressing on**

Adjustment overview

Note:

If transmission repairs have been performed it is only necessary to adjust the input shaft, output shaft or differential if components have been replaced which have a direct effect on the transmission. See table below for adjustment information.

		to be adjusted:		
		Input shaft ⇒ Page L 35-13	Output shaft (S1 and S2) ⇒ Page L 35-42	Differential ⇒ Page L 39-25
Parts renewed:	Transmission housing	x	x	x
	Clutch housing	x	x	x
	Input shaft	x		
	Output shaft (final drive)		x	
	Differential housing			x
	Input shaft tapered roller bearing	x		
	Output shaft tapered roller bearing		x	
	Differential tapered roller bearing			x
	Bearing support		x	

Adjusting differential

Differential must be adjusted, when transmission housing, clutch housing, differential housing, or differential tapered roller bearing have been replaced.

Adjustment overview
⇒ Page L 39-24.

- Press tapered roller bearing outer race with oil deflector ring into clutch housing using thrust piece 2050
⇒ Page L 39-9.

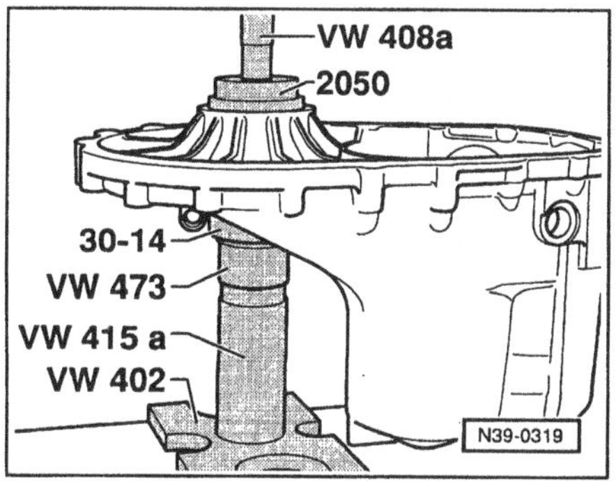

Note:

Inner and outer tapered roller bearing races are a pair. Do not interchange.

◄ – Press tapered roller bearing outer race, without shim, in transmission housing using thrust piece 2050
⇒ Page L 39-11.

- Position differential in clutch housing

- Position transmission housing and torque 5 bolts to specification.

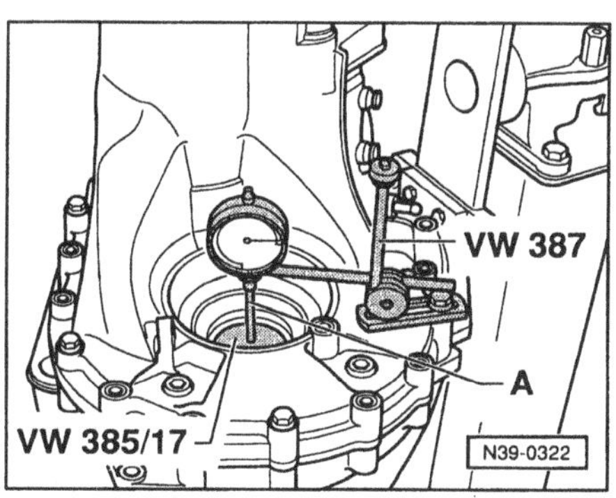

◄ – Mount dial indicator and zero 1 mm (0.0393 in.) preload.

- Move differential up and down and note play indicated on dial indicator(Example, 0.70 mm)

A - 60 mm gauge extension

Note:

Do not turn differential when measuring otherwise a false measurement will result.

Shim thickness

Specified bearing preload is obtained by adding a constant figure (0.40 mm) (0.0157 in.) to reading.

Example:

Reading	0.70 mm
+ Preload (constant figure)	0.40 mm
Thickness of shim =	1.10 mm

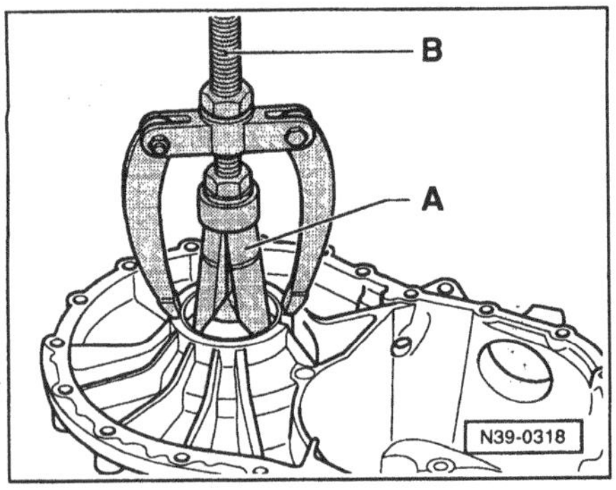

- Remove transmission housing and remove tapered roller bearing outer race from transmission housing.
- Insert correct shims, thickest shim first.

Following shims are available:

Size (mm)	Part No.
0.65	02G 409 210
0.70	02G 409 210 A
0.75	02G 409 210 B
0.80	02G 409 210 C
0.85	02G 409 210 D
0.90	02G 409 210 E
0.95	02G 409 210 F
1.00	02G 409 210 G
1.05	02G 409 210 H
1.10	02G 409 210 J
1.15	02G 409 210 K
1.20	02G 409 210 L
1.25	02G 409 210 M
1.30	02G 409 210 N
1.35	02G 409 210 P
1.40	02G 409 210 Q
0.70+0.75 = 1.45	02G 409 210 A + B
0.75+0.75 = 1.50	02G 409 210 B + B
0.75+0.80 = 1.55	02G 409 210 B + C

- Press outer race in again and tighten transmission housing to torque specification.

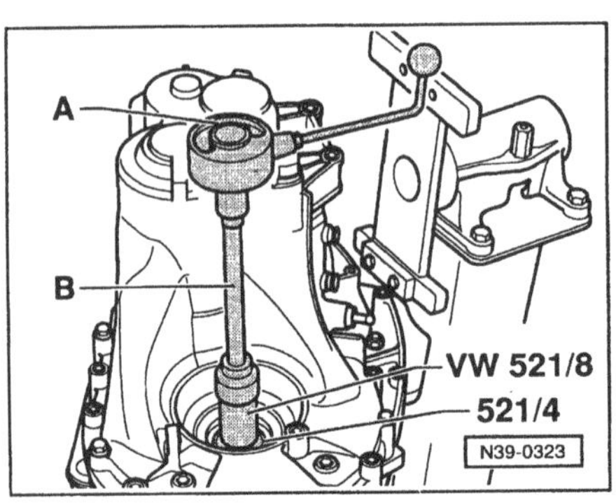

Bearing preload, checking

A – Commercial torque gauge 0 ... 600 Ncm (0...53 in. lb)

B – Extension for 1/2–inch square drive socket

- Lubricate tapered roller bearing with gear oil.

Bearing preload:

- New tapered roller bearings 120 ... 320 Ncm (11...28 in. lb)
- Used tapered roller bearings at least 30 Ncm (3 in. lb).

Suspension, Wheels, Brakes, Steering (1992 ➤ 1996)

Section **M**

> ***WARNING!***
> ***Before proceeding with any work, please read the warnings and cautions immediately following this index. Also read any warning and caution that accompanies a procedure in the manual.***

47 Brakes—Hydraulic components

48 Steering

Please read these WARNINGS and CAUTIONS before proceeding with maintenance and repair work.

WARNING—

- Some repairs may be beyond your capability. If you lack the skills, tools and equipment, or a suitable workplace for any procedure described in this manual, we suggest you leave such repairs to an authorized Volkswagen dealer service department, or other qualified shop.

- Volkswagen is constantly improving its cars. Sometimes these changes, both in parts and specifications, are made applicable to earlier models. Therefore, before starting any major jobs or repairs to components on which passenger safety may depend, consult your authorized Volkswagen dealer about the Technical Bulletins that may have been issued since the editorial closing of this manual.

- Do not re-use any fasteners that are worn or deformed in normal use. Many fasteners are designed to be used only once and become unreliable and may fail when used a second time. This includes, but is not limited to, nuts, bolts, washers, self–locking nuts or bolts, circlips and cotter pins. Always replace these fasteners with new parts.

- Never work under a lifted car unless it is solidly supported on stands designed for the purpose. Do not support a car on cinder blocks, hollow tiles or other props that may crumble under continuous load. Never work under a vehicle that is supported solely by a jack. Never work under the car while the engine is running.

- If you are going to work under a car on the ground, make sure that the ground is level. Block the wheels to keep the car from rolling. Disconnect the battery negative terminal (ground strap) to prevent others from starting the vehicle while you are under it.

- Never run the engine unless the work area is well ventilated. Carbon monoxide (CO) kills.

- Do not attempt to work on your vehicle if you do not feel well. You increase the danger of injury to yourself and others if you are tired, upset or have taken medicine or any other substance that may impair you or keep you from being fully alert.

- Disconnect the battery negative (–) terminal (ground strap) whenever you work on the fuel system or the electrical system. Do not smoke or work near heaters or other fire hazards. Keep an approved fire extinguisher handy.

- Batteries produce explosive hydrogen gas. Keep sparks, lighted matches and open flame away from the top of the battery. If hydrogen gas escaping from the cap vents is ignited, it will ignite gas trapped in the cells and cause the battery to explode.

- Connect and disconnect battery cables, jumper cables or a battery charger only with the ignition switched off, to prevent sparks. Do not quick–charge the battery (for boost starting) for longer than one minute, and do not allow charging voltage to exceed 16.5 volts. Wait at least one minute before boosting the battery a second time.

- Illuminate your work area adequately but safely. Use a portable safety light for working inside or under the car. Make sure the bulb is enclosed by a wire cage. The hot filament of an accidentally broken bulb can ignite spilled fuel or oil.

- Tie long hair behind your head. Do not wear a necktie, a scarf, loose clothing, or a necklace when you work near machine tools or running engines. If your hair, clothing, or jewelry were to get caught in the machinery, severe injury could result.

- Finger rings, bracelets and other jewelry should be removed so that they cannot cause electrical shorts, get caught in running machinery, or be crushed by heavy parts.

- Catch draining fuel, oil, or brake fluid in suitable containers. Do not use food or beverage containers that might mislead someone into drinking from them. Store flammable fluids away from fire hazards. Wipe up spills at once, but do not store the oily rags, which can ignite and burn spontaneously.

- Always observe good workshop practices. Wear goggles when you operate machine tools or work with battery acid. Gloves or other protective clothing should be worn whenever the job requires working with harmful substances.

- The Air-Conditioning (A/C) system is filled with a chemical refrigerant, which is hazardous. The A/C system should be serviced only by trained technicians using approved refrigerant recovery/recycling equipment, trained in related safety precautions, and familiar with regulations governing the discharging and disposal of automotive chemical refrigerants.

- Do not expose any part of the A/C system to high temperatures such as an open flame. Excessive heat will increase system pressure and may cause the system to burst.

- Some aerosol tire inflators are highly flammable. Be extremely cautious when repairing a tire that may have been inflated using an aerosol tire inflator. Keep sparks, open flame or other sources of ignition away from the tire repair area. Inflate and deflate the tire at least four times before breaking the bead from the rim. Completely remove the tire from the rim before attempting any repair.

- Friction materials such as brake or clutch discs may contain asbestos fibers. Do not create dust by grinding, sanding, or by cleaning with compressed air. Avoid breathing asbestos fibers and asbestos dust. Breathing asbestos can cause serious diseases such as asbestosis or cancer, and may result in death.

- Greases, lubricants and other automotive chemicals contain toxic substances, many of which are absorbed directly through the skin. Read manufacturer's instructions and warnings carefully. Use hand and eye protection. Avoid direct skin contact.

- An airbag is operated by an explosive device. Handled improperly or without adequate safeguards, it can be accidently activated and cause serious personal injury. To prevent personal injury or airbag system failure, **only trained Volkswagen dealer service technicians** should test, disassemble or service the airbag system.

WARNINGS and CAUTIONS continued on next page ▶

Please read these WARNINGS and CAUTIONS before proceeding with maintenance and repair work.

WARNING (continued) —

- ALWAYS disconnect the airbag power supply before using a computer memory saver. Failure to disconnect the airbag power supply may result in accidental activation of the airbag.
- The airbag unit should not be removed, nor should any work be performed on the airbag system, without first reviewing the warnings and cautions that accompany the airbag system repair information in this manual.
- Never apply stickers or any other coverings on the steering wheel. Do not let chemical cleaners, oil or grease come into contact with the vinyl covering of the airbag unit.
- Never open or otherwise attempt to repair airbag system parts. Always use new parts. Never leave airbag parts or the partially disassembled airbag system unattended.
- Never use a test light or volt meter to conduct electrical tests on the airbag system. The airbag system must only be tested by trained Volkswagen Service technicians using special airbag tester equipment. The airbag system must only be tested while it is installed in the car.
- Do not expose an airbag unit to temperatures above 194°F (90°C) even for a brief period, while handling it during the repair process. Keep the unit clear of heat sources such as hot plates, soldering irons, heat lamps, welding equipment, or the like.
- Always replace airbag system components that have been mechanically damaged (bubbles, cracks, etc.).
- Always set down a removed airbag unit so that the horn pad is facing upward.
- When driving or riding in an airbag-equipped vehicle, never hold test equipment in your hands or lap while the vehicle is in motion. Objects between you and the airbag can increase the risk of injury in an accident.
- The storage, transportation and disposal of airbag units are subject to the laws for flammable solids.

CAUTION—

- If you lack the skills, tools and equipment, or a suitable workshop for any procedure described in this manual, we suggest you leave such repairs to an authorized Volkswagen dealer or other qualified shop. We especially urge you to consult an authorized Volkswagen dealer before beginning repairs on any car that may still be covered wholly or in part by any of the extensive warranties issued by Volkswagen of America, Inc.
- Volkswagen is constantly improving its cars and sometimes these changes, both in parts and specifications, are made applicable to earlier models. Therefore, part numbers listed in this manual are for reference only. Always check with your authorized Volkswagen dealer parts department for the latest information.
- Before starting a job, make certain that you have all the necessary tools and parts on hand. Read all the instructions thoroughly, do not attempt shortcuts. Use tools that are appropriate to the work and use only replacement parts meeting Volkswagen specifications. Makeshift tools, parts and procedures will not make good repairs.
- Use pneumatic and electric tools only to loosen threaded parts and fasteners. Never use these tools to tighten fasteners, especially on light alloy parts. Always use a torque wrench to tighten fasteners to the tightening torque specification listed.
- Be mindful of the environment and ecology. Before you drain the crankcase, find out the proper way to dispose of the oil. Do not pour oil onto the ground, down a drain, or into a stream, pond, or lake. Consult local ordinances that govern the disposal of wastes.
- When connecting cables or ground straps, all connections must be cleaned to bare metal. Be sure mounting bolts are tight.
- Do not replace fuse links with fuse links of other gauge or with standard wire. Always replace the entire length of the fuse link.
- Before doing any electrical welding on vehicles equipped with Anti-lock Brakes (ABS), disconnect the battery negative terminal (ground strap) and the ABS Control Module connector.
- On cars equipped with ABS, the ABS control unit cannot withstand temperatures from a paint–drying booth or a heat lamp in excess of 203°F (95°C) and should not be subjected to temperatures in excess of 185°F (85°C) for more than two hours.
- On cars equipped with anti-theft radios, make sure you know the correct radio activation code before disconnecting the battery or removing the radio. If the wrong code is entered when power is restored, the radio may lock up and be rendered inoperable, even if the correct code is then entered.
- When boost-charging the battery, first remove the fuses for the Engine Control Module (ECM), the Transmission Control Module (TCM), the ABS Control Module, and the On-Board computer. In cases where one or more of these components is not separately fused, disconnect the Control Module connector(s).
- Do not quick-charge the battery (for boost starting) for longer than one minute, and do not exceed 16.5 volts at the battery with the boosting cables attached. Wait at least one minute before boosting the battery a second time.

◀ **WARNINGS and CAUTIONS** continued from previous page

Technical data

Suspension

Wheelbase	mm	2920	3320
Track, empty	mm	1607 front, wheel offset 35 1572 rear, wheel offset 35	1607 front, wheel offset 35 1572 rear, wheel offset 35
		1589 front, wheel offset 44 1554 rear, wheel offset 44	1589 front, wheel offset 44 1554 rear, wheel offset 44
		1575 front, wheel offset 51 1540 rear, wheel offset 51	1575 front, wheel offset 51 1540 rear, wheel offset 51
Turning circle	m	11.7	12.9
Wheel lock angle	inner	43°	43°
Ground clearance at permissible gross weight – normal suspension – with lowered suspension	 mm mm	 180 160	 180 160

Steering

Steering gear		Power steering gear
Steering roll radius (negative)	mm	– 5.0
Turns of steering wheel from lock to lock	approx.	3.8
Overall steering ratio		17.5
Wheel alignment specifications ⇒ Page M 44-1		

Brakes

Version		Dim.	EuroVan with 15" Wheels	
Optional equipment No. (PR-No.)[1]			1LE (floating frame caliper)	1LP (floating caliper)
Brake master cylinder	dia.	mm	22.2/23.81[3]	22.2/23.81[3]
Brake booster unit	dia.	inch	9	9
Front brakes				
Brake caliper piston	dia.	mm	54	54
Brake disc	dia.	mm	280	282
Brake disc thickness		mm	24	18
Brake pad thickness		mm	13	13.5
Rear brakes				
Brake drum		mm	268	268
Wheel cylinder	dia.	mm	19.05[2]/20.64	20.64
Brake lining thickness		mm	6.3	6.3

[1] Optional equipment number; Explanation ⇒ Page M 00-3
[2] Only up to vehicle No. 70m 033 887
[3] Only with Tandem Brake Booster

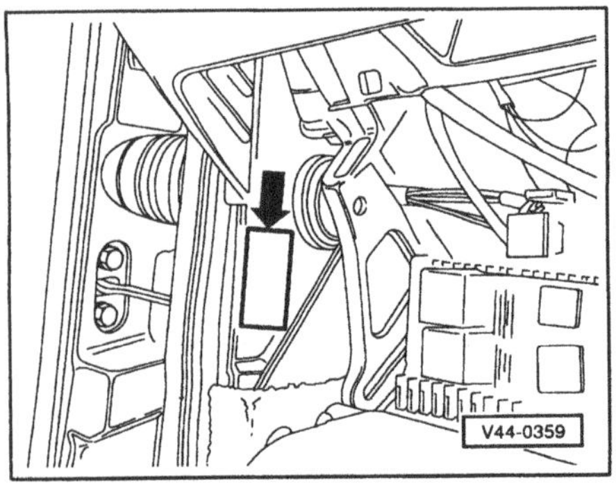

Identification of brake caliper type via optional equipment number (PR No.)

Different PR numbers are printed on the vehicle data plate for each brake caliper version.

◄ **Vehicle data plate location (arrow)**

Left-hand drive vehicles: Left side A-pillar next to central electric unit.

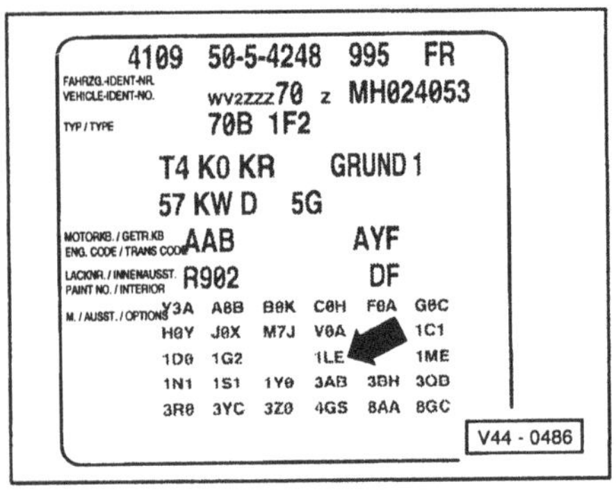

◄ **Location of optional equipment number (PR No.) on vehicle data plate (arrow)**

- 1LP = Floating caliper
- 1LE = Floating frame caliper

Note:

Depending on equipment level, the brake caliper code is printed in different locations on the vehicle data plate.

Wheels, tires

(1992 ➤ 1995)

Tire size[1]	Wheel	Offset
195/70 R15 97 S reinforced	6 J x 15	44 mm
205/65 R15 98 S reinforced	6 J x 15	44 mm
205/65 R15 98 S reinforced	7 J x 15	35 mm
215/65 R15 100 T reinforced	7 J x 15	35 mm

(1996 ➤)

Tire size[1]	Wheel	Offset
195/70 R15 97 S reinforced	6 J x 15	55 mm
195/70 R15 C 104/102 R	6 J x 15	55 mm
205/65 R15 99 S reinforced	6 J x 15	55 mm
205/65 R15 99 S reinforced	7 J x 15	46 mm
205/65 R15 C 100 T	7 J x 15	46 mm
205/65 R15 C 100 T	6 J x 15	55 mm

[1] Tire pressures ⇒ sticker on vehicle

> ***Caution!***
> ***Wheels from vehicles up to 12.95 cannot be used on vehicles produced after 12.95.***

On Board Diagnostic (OBD) of anti-lock brake system (ABS)

Function

The ABS Control Module -J104- is located behind the trim in the lower left-hand A-pillar. It is equipped with a Diagnostic Trouble Code (DTC) Memory. The Data Link Connectors (DLC) are located behind the relay plate trim.

On Board Diagnostic refers to the electrical/electronic components of the ABS; i.e., only malfunctions which produce electrical signals are recognized.

Up to 3 malfunctions can be stored if they occur in the monitored sensors or components. The type of malfunction is indicated. The ABS control in the control module is then switched off at the same time as the ABS Warning Light in the instrument cluster lights up. The conventional brake system remains operational.

The contents of the DTC Memory must always be checked before commencing troubleshooting as, in some cases, the ABS Warning Light -K47- goes out after the ignition is switched off although the DTC Memory retains its contents.

A malfunction which occurs only occasionally (e.g., loose contact), is likewise stored. If the malfunction no longer exists after the ignition is again switched on (sporadic malfunction), then the ABS Warning Light -K47- goes out after approx. 2 seconds.

If a malfunction is not registered again after the ignition is switched on 50 times, the relevant DTC Memory is automatically erased.

Interrogating the DTC Memory ⇒ Page M 01-15 is performed after initiating On Board Diagnosis (OBD) via the Scan Tool VAG 1551. The vehicle must not be moving and must be in the "rapid data transfer" mode.

On Board Diagnostic (OBD) is cancelled if the vehicle is moved at a speed of more than 2.5 km/h (1.56 mph).

The ABS system is not operational during OBD.

The DTC information displayed, is used in conjunction with a DTC table ⇒ Page M 01-17 which has information on the possible causes of malfunctions.

Erase the DTC Memory after correcting malfunctions, ⇒ Page M 01-30.

Technical data of On Board Diagnostic (OBD)

Diagnostic Trouble Code (DTC) Memory

- Contents of permanent memory are retained even after the supply voltage is switched off.

Data output

- Rapid data transfer (operating mode 1)
- Blink code output (do not use)
- When the ABS is switched off the ABS Warning Light in the instrument panel lights up.

Functions

- Check Control Module version
- Check DTC Memory
- Erase DTC Memory
- Read measured value block
- Read individual measured value

On Board Diagnostic (OBD), safety and precautionary measures

General troubleshooting

- *The ABS is a vehicle safety system; the appropriate knowledge is necessary to work on the system.*
- *Do not drive with the connector disconnected from the ABS Control Module -J104-.*
- *In order to check complaints and to be able to carry out troubleshooting, the DTC Memory must be checked before commencing work on the ABS system.*
- *Do not separate connectors when the ignition is switched on.*
- *Do not loosen screws or bolts on the ABS Hydraulic Unit with the exception of the relay cover.*
- *Observe the appropriate instructions regarding the handling of brake fluid* *⇒ Page M 47-9.*
- *ABS malfunctions are indicated by the the ABS Warning Light lighting up; certain malfunctions will only be recognized when exceeding a speed of 12 km/h (7.5 mph) (road test).*

On Board Diagnostic (OBD) test conditions

- The tires on all wheels must be of the same (approved) size.
- Conventional brake system including Brake Light Switch and Brake Lights OK.
- Hydraulic lines and connections not leaking (visual check of ABS Hydraulic Unit, brake cylinders).
- Wheel bearings and wheel bearing play OK.
- Ground (GND) connection for ABS Return Flow Pump -V39- at ABS Hydraulic Unit OK.
- ABS Return Flow Pump Relay -J105- and ABS Solenoid Valve Relay -J106- are correctly connected.
- Connector on ABS Control Module -J104-, close to bottom left A-pillar, correctly connected (retainer is engaged).
- Control Module -J104- Ground (GND) connection OK.
- All fuses according to wiring diagram OK.
- Supply voltage OK (at least 10.5 V).

Required technical information

♦ *ABS wiring diagram*
⇒ Electrical wiring diagrams.

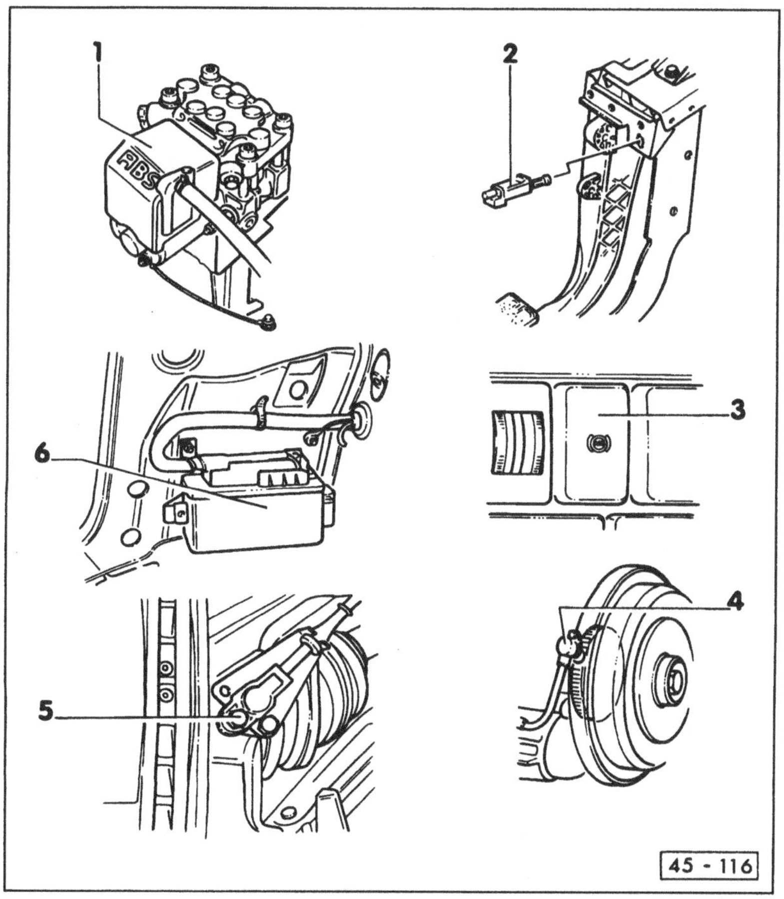

ABS electrical components

All components marked with 1) are monitored by On Board Diagnostic (OBD)

1 - ABS Hydraulic Unit -N55-
- ♦ Use adapter set VAG 1594 for testing the electrical connections to the ABS Hydraulic Unit
- ♦ ABS Hydraulic Unit multi-pin connector
⇒ Page M 01-43
- ♦ The ABS Hydraulic Unit contains the following components:ABS Return Flow Pump Relay -J105-
ABS Solenoid Valve Relay -J106-1)
ABS Return Flow Pump -V39-1)
Inlet/Outlet Valves -N137-1) to -N140-1)

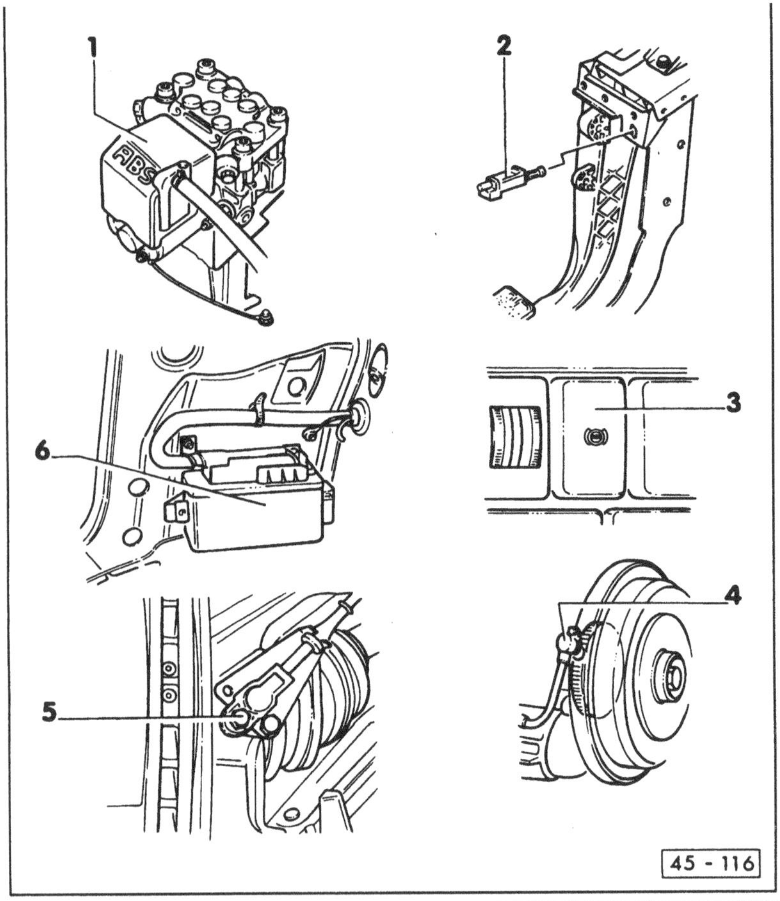

Notes:

- *Both relays are located below a protective cover; Torx bolt, T15 drive.*
- *ABS Return Flow Pump and solenoid valves cannot be replaced individually. Replace hydraulic unit if a malfunction exists.*

2 - Brake Light Switch -F- [1)]
- The Brake Light Switch plunger must have switched the brake lights off when the brake pedal is in the rest position

3 - ABS Warning Light -K47-
- In center of instrument panel
- Lights up for approx. 2 sec. after ignition is switched on

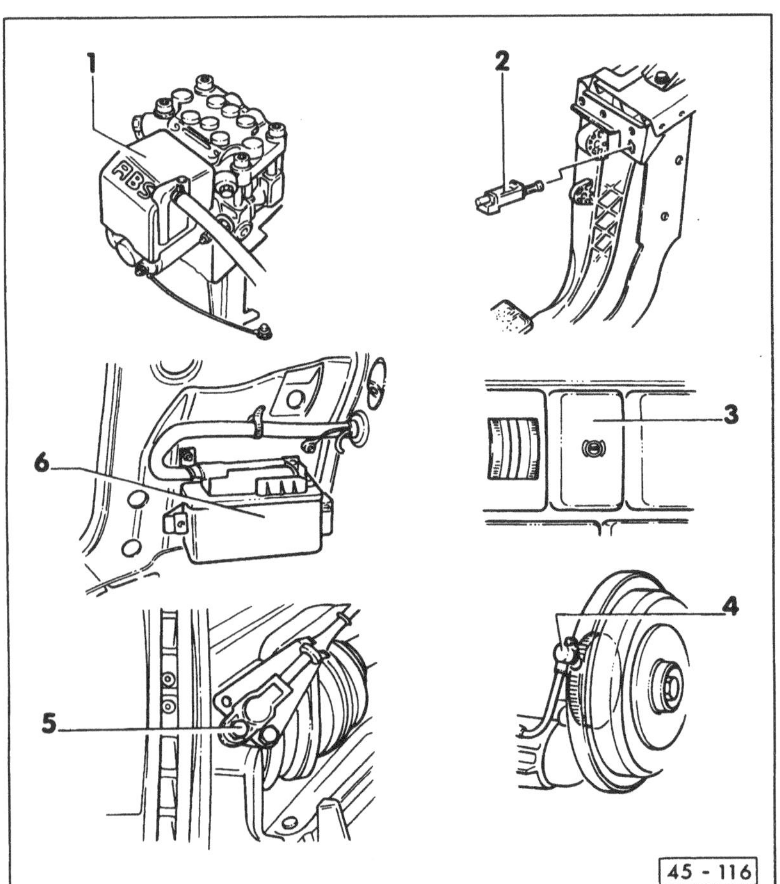

4 - Right Rear/Left Rear ABS Wheel Speed Sensor -G44-/-G46- [1)]
- Before inserting the sensor, clean the inner surface of the sensor mounting and coat with lubricating paste G 000 650
- Observe assembly sequence: Insert sensor first and then clip cable into the appropriate holder, starting from the sensor.

5 - Right Front/Left Rear ABS Wheel Speed Sensor -G45-/-G47- [1)]
- Installation instructions ⇒ Item 4

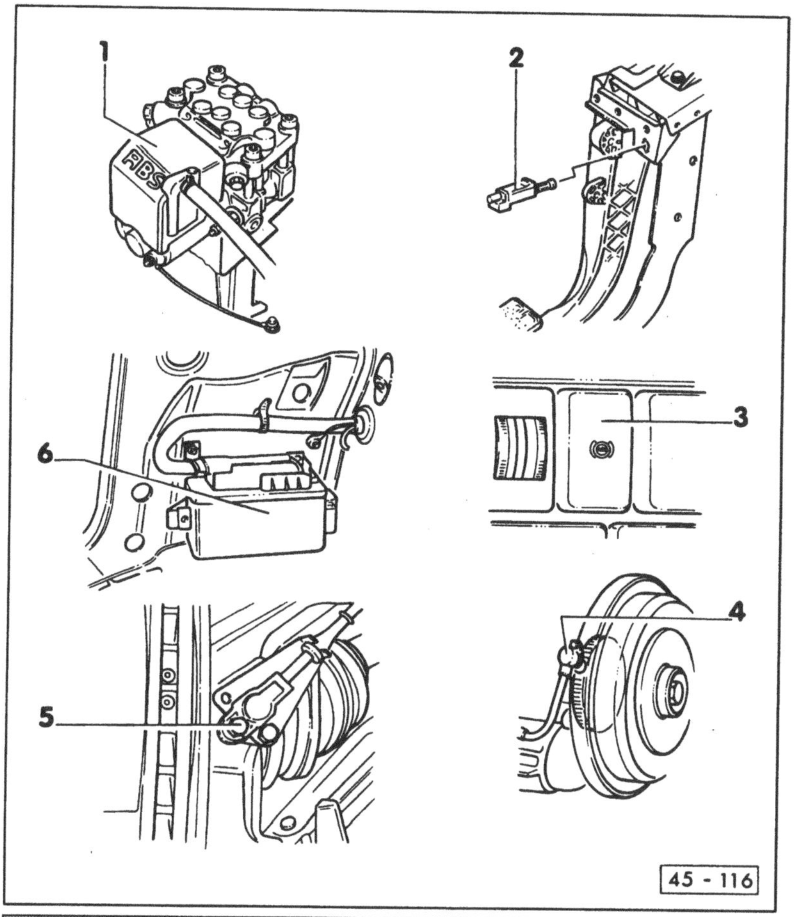

6 - **ABS Control Module -J104-** [1)]
- Location: close to left-hand A-pillar lower section
- Do not disconnect connector before checking DTC Memory, switch ignition off first.
- Use test box VAG 1598 with adapter VAG 1598/3 to check multi-pin connector to ABS Control Module ⇒ Page M 01-43

VAG 1551 Scan Tool, connecting and selecting functions

- Test conditions ⇒ Page M 01-5.

Notes:

- *The ABS function is switched off in the ABS Control Module during the On Board Diagnostic (OBD).*
- *The Diagnostic Trouble Code (DTC) Memory can be erased after successful interrogation and repair.*

– Fold down cover in front of relay plate.

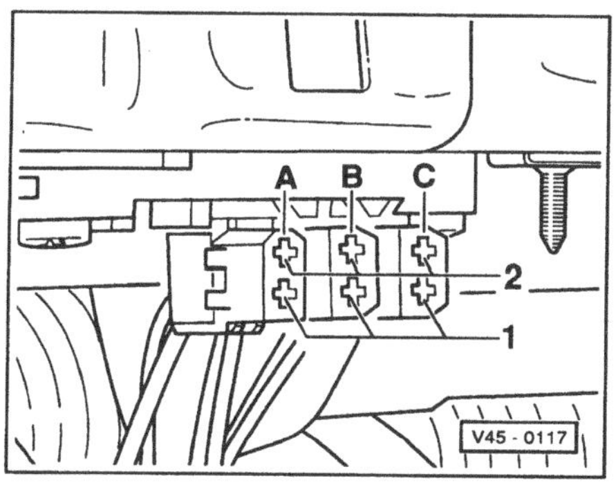

◄

– Switch ignition off and connect VAG 1551 Scan Tool using harness VAG 1551/1 to the Data Link Connectors (DLC) in the instrument panel opening as follows:

  – First connect black VAG 1551/1 connector to black DLC -A- (voltage supply).

VAG On Board Diagnostic HELP 1-Rapid data transfer [1] 2-Blink code output [1]

◄ Indicated on display

[1] Is displayed alternately

Note:

If the display remains blank, do not connect white or blue connector! Check the DLC (black), Electrical check ⇒ Page M 01-38.

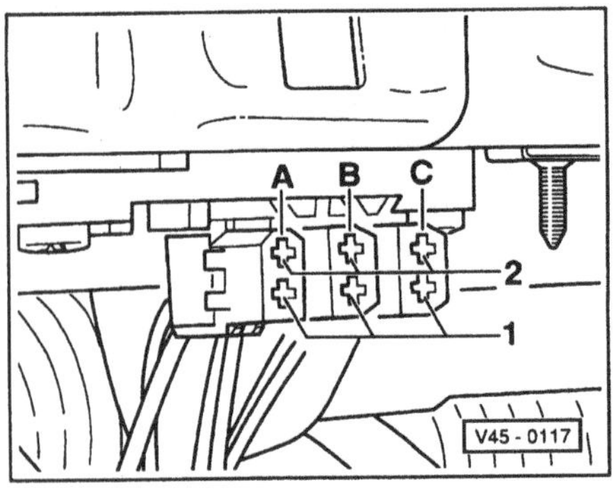

◄ – Connect VAG 1551 white connector to white DLC -B- or -C- in vehicle. The blue connector is not required.

Notes:

- *Depending on the program, additional operating information can be printed out by pressing the HELP key of VAG 1551.*
- *The → key is used for moving forward within the program*
- *The PRINT key is used for switching on the printer (warning lamp in key comes on).*

– Switch on ignition.

– Press key -1- for "Rapid data transfer" operating mode.

Rapid data transfer HELP Enter address word XX

◄ Indicated on display

– Press keys -0- and -3-;
03 inputs the address word of the system (brake electronics) to be tested.

Rapid data transfer Q 03 Brake electronics

◄ Indicated on display

– Confirm input with Q key.

701907379 B Brake electronics →

◄ and then the following appears in the display:

The ABS Control Module identification number is displayed.

Allocation of ABS Control Modules
⇒ Parts catalog

If the ABS Control Module identification number does not appear:
⇒ Summary of functions.

– Press → button.

Rapid data transfer HELP Select function XX

◄ Indicated on display

Note:

A list of the possible functions is printed out after pressing the HELP key:

List of functions

	Page
01 - Check Control Module version	M 01-12
02 - Check DTC Memory	M 01-15
05 - Erase DTC Memory	M 01-30
06 - End output	M 01-30
08 - Read measured value block	M 01-33
09 - Read individual measured value	M 01-36

Currently only the listed functions can be used for OBD.

After pressing the → key, the VAG 1551 program returns to the start position.

Rapid data transfer HELP Select function XX	

◄ Indicated on display
(Function selection, e.g., 02 – check DTC Memory)

Notes:

Rapid data transfer HELP Control Module does not answer!	

◄ ♦ *If the adjacent fault information appears in the display, the possible causes of the fault can be printed out by pressing the HELP key.*

Rapid data transfer HELP K-wire not switching to positive!	

◄ (Ignition must be switched on!)

Rapid data transfer → No signal from Control Module!	

◄ ♦ *Faults have occurred at the start or during the program (external interference?)*

Rapid data transfer → Fault in communication build-up	

– Check diagnostic wires and also voltage supply and Ground connection near the ABS Control Module -J104, Electrical test ⇒ Page M 01-38.

– After correcting the possible causes of the malfunction, again enter the address word 03 for Brake electronics and confirm with Q key.

Rapid data transfer Tester sends address word 03	

◄ Indicated on display after entering the address word 03

701907379 B Brake electronics →	

◄ And then the following appears on the display:

The ABS Control Module identification number is displayed.

– Press → button.

Rapid data transfer HELP Select function XX	

◄ Indicated on display

Diagnostic Trouble Code (DTC) Memory, checking

Rapid data transfer HELP Select function XX	◄ Indicated on display (Function selection)

– Press keys -0- and -2-;
  02 enters the “Check DTC Memory” function.

Rapid data transfer Q 02 – Check DTC Memory	◄ Indicated on display

– Confirm entry with Q key.

X DTCs recognized!	◄ The number of DTCs stored or “No DTCs recognized” will be displayed.
No DTCs recognized! →	

– Press → key.

The stored DTCs are displayed and printed out one after the other.

– Go to the DTC table on Page M 01-17 with the DTC information print-out.

As with “No DTCs recognized,” the program returns to the start position after pressing the → key:

Rapid data transfer HELP Select function XX	◄ Indicated on display (Function selection)

– End output (Function 06) ⇒ Page M 01-30.

– Switch off ignition and disconnect Data Link Connectors.

Notes:

- *If a DTC was recognized:*
  1. Correct malfunction
  2. Check DTC Memory (function 02)
  3. Erase DTC Memory (function 05)
- *If a malfunction continues to exist, which has not been recognized by the On Board Diagnostic (OBD), then perform troubleshooting with ABS tester VAG 1710*

Diagnostic Trouble Code (DTC) table

Notes:

- *All the possible malfunctions which can be recognized by the ABS Control Module -J104- and printed-out on the Scan Tool VAG 1551, are listed on the following pages according to the 5 digit DTC code.*
- *DTC code and blink code are only printed out in "rapid data transfer" mode.*

Example:

DTC code	Blink code
5 digit	4 digit
00277	1221

- *Sporadic malfunctions are identified in the display with "/ SP". During the testing of sporadic malfunctions move the wiring running to the component (loose contact).*
- *Items marked with an * in the DTC table provide additional information regarding localizing the malfunction.*
- *Reference is made in the column "Correction" to individual test steps from "ABS, checking function with VAG 1710," beginning on page M 01-44.*
- *Additional aids for troubleshooting/repairs are:*

a) Functions:

-08- Reading measured value block
-09- Reading individual measured value

(Notes for these functions are given in the DTC table, column "Correction")

b) Electrical troubleshooting ⇒ Page M 01-38

- *Before replacing components indicated as being faulty, check all the appropriate connectors, wiring and Ground connections using the wiring diagram.*
- *After repairs, always test the ABS system with the ABS tester VAG 1710. Then, once again check and erase the DTC Memory with Scan Tool VAG 1551. Also, road test the vehicle (exceeding a speed of 30 km/h (19 mph)) for a brief period. If during the road test the ABS Warning Light -K47- does not light up, then the system is OK.*

VAG 1551 print-out	Possible cause	Correction
00000 4444 No DTC recognized	If "No DTC recognized" appears after repairing, On Board Diagnostic (OBD) is ended. If, despite "No DTC recognized" appearing the display, the ABS system does not operate properly, then proceed as follows: 1. Road test vehicle and exceed 12 km/h (7.5 mph). 2. Check DTC Memory again if there is still no DTC stored. 3. Troubleshoot with VAG 1710, ⇒ ABS, checking function with VAG 1710 on page M 01-44.	
00277 1221 Left Front ABS Inlet/Outlet Valve -N137-	– Open circuit or short to positive/Ground between ABS Hydraulic Unit -N55-, contact 1 and ABS Control Module -J104-, contact 2. – -N137- faulty.	– Locate open or short circuit according to wiring diagram and correct. – Check ABS Hydraulic Unit, functional check with VAG 1710 (test steps 1 and 5)

VAG 1551 print-out	Possible cause	Correction
00283 1233 Left Front ABS Wheel Speed Sensor -G47- * Open circuit / short to positive * no DTC recognized![1]	– Open circuit or short to positive between -G47- and ABS Control Module -J104-, contacts 4 and 5 – -G47- faulty – Short to Ground between -G47- and ABS Control Module -J104-, contacts 4 and 5 – Rotor dirty or damaged – Play in wheel bearing – -G47- not installed correctly – -G47 -faulty	– Locate open or short circuit according to wiring diagram and correct. – Replace -G47-. – Locate short circuit according to wiring diagram and correct. – Check rotor, clean or replace. – Service front suspension ⇒ Page M 40-1. – Check installation position – Check -G47-; (1.0 – 1.25 kΩ) functional check with VAG 1710 (test step 6). (Read measured value block, function 08.)

[1] It may not be possible to localize malfunction, or it may be erroneously displayed if a wheel is rotated at more than 6 km/h (3.75 mph) with ignition on.

VAG 1551 print-out	Possible cause	Correction
00284 1234 Right Front ABS Inlet/Outlet Valve -N138-	– Open circuit or short to positive/Ground between ABS Hydraulic Unit -N55-, contact 3 and ABS Control Module -J104-, contact 35 – -N138- faulty	– Locate open or short circuit according to wiring diagram and correct. – Check ABS Hydraulic Unit, functional check with VAG 1710 (test steps 1 and 5).
00285 1241 Right Front ABS Wheel Speed Sensor -G45- * Open circuit / short to positive Continued on next page	– Open circuit or short to positive between -G45- and -J104-, contacts 11 and 21	– Locate open or short circuit according to wiring diagram and correct.

VAG 1551 print-out	Possible cause	Correction
00285 1241 * no DTC recognized![1]	– -G45- faulty – Short to Ground between -G45- and ABS Control Module -J104-, contacts 11 and 21 – Rotor dirty or damaged – Play in wheel bearing – -G45- not installed correctly – -G45- faulty	– Replace -G45- – Locate short circuit according to wiring diagram and correct. – Check rotor, clean or replace. – Service front suspension ⇒ Page M 40-1. – Check installation position. – Check -G45-; (1.0 – 1.25 kΩ) functional check with VAG 1710 (test step 6).
00286 1242 Left Rear ABS Inlet/Outlet Valve -N139-	– Open circuit or short to positive/Ground between ABS Hydraulic Unit -N55-, contact 5 and ABS Control Module -J104-, contact 18 – -N139- faulty	– Locate open or short circuit according to wiring diagram and correct. – Check ABS Hydraulic Unit, functional check with VAG 1710 (test steps 1 and 5). (Read measured value block, function 08.)

[1] It may not be possible to localize malfunction or may be erroneously displayed if a wheel is rotated at more than 6 km/h (3.75 mph) with ignition on.

VAG 1551 print-out	Possible cause	Correction
00287 1243 Right Rear ABS Wheel Speed Sensor -G44- * Open circuit / short to positive	– Open circuit or short to positive between -G44- and ABS Control Module -J104-, contacts 24 and 26	– Locate open or short circuit according to wiring diagram and correct.
	– -G44- faulty	– Replace -G44-.
* no DTC recognized![1]	– Short to Ground between -G44- and -J104-, contacts 24 and 26	– Locate short circuit according to wiring diagram and correct.
	– Rotor dirty or damaged	– Check rotor, clean or replace.
	– Play in wheel bearing	– Service rear suspension ⇒ Page M 42-1.
	– -G44- not installed correctly	– Check installation position
	– -G44- faulty	– Check -G44-; (1.0 – 1.25 kΩ) functional check with VAG 1710 (test step 6). (Read measured value block, function 08.)

[1] It may not be possible to localize malfunction or may be erroneously displayed if a wheel is rotated at more than 6 km/h (3.75 mph) with ignition on.

VAG 1551 print-out	Possible cause	Correction
00289 1244 Right Rear ABS Inlet/Outlet Valve -N140-	– Open circuit or short to positive/Ground between ABS Hydraulic Unit -N55-, contact 7 and ABS Control Module -J104-, contact 19	– Locate open or short circuit according to wiring diagram and correct.
	– -N140- faulty	– Check ABS Hydraulic Unit, functional check with VAG 1710 (test steps 1 and 5).
00290 1311 Left Rear ABS Wheel Speed Sensor -G46- * Open circuit / short to positive Continued on next page	– Open circuit or short to positive between -G46- and ABS Control Module -J104-, contacts 7 and 9	– Locate open or short circuit according to wiring diagram and correct.

VAG 1551 print-out	Possible cause	Correction
00290 1311 * no DTC recognized![1]	– -G46- faulty – Short to Ground between -G46- and ABS Control Module -J104-, contacts 7 and 9 – Rotor dirty or damaged – Play in wheel bearing – -G46- not installed correctly – -G46- faulty	– Replace -G46- – Locate short circuit according to wiring diagram and correct. – Check rotor, clean or replace. – Servicing rear suspension ⇒ Page M 42-1. – Check installation position – Check -G46-; (1.0 – 1.25 kΩ) functional check with VAG 1710 (test step 6). (Read measured value block, function 08.)

[1] It may not be possible to localize malfunction or may be erroneously displayed if a wheel is rotated at more than 6 km/h (3.75 mph) with ignition on.

VAG 1551 print-out	Possible cause	Correction
00301 1334 ABS Return Flow Pump -V39-	– Open circuit or contact resistance in Ground or positive connection to ABS Return Flow Pump -V39-: ABS Hydraulic Unit -N55-, contact 10 – Open circuit or short to positive/Ground in wiring between ABS Return Flow Pump Relay -J105- and ABS Control Module -J104-, contacts 14, 17 and 28 – ABS Return Flow Pump Relay -J105-, ABS Return Flow Pump -V39- or ABS Hydraulic Unit -N55- faulty	– Locate open circuit or contact resistance according to wiring diagram and correct. – Locate open or short circuit according to wiring diagram and correct. – Check relay (-J105-); (Coil resistance: 45 – 55 Ω). – Check ABS Return Flow Pump -V39- and ABS Hydraulic Unit -N55-, functional check with VAG 1710, (test step 3). (Read measured value block, function 08.)

VAG 1551 print-out	Possible cause	Correction
00302 1341 ABS Solenoid Valve Relay -J106-	– Open circuit or contact resistance in B+ connection to ABS Solenoid Valve Relay -J106-; ABS Hydraulic Unit -N55-, contact 6 – Open circuit or short to positive/Ground in wiring between ABS Solenoid Valve Relay -J106- and ABS Control Module -J104-, contacts 17, 27 and 32 – -J106- or ABS Hydraulic Unit -N55- faulty	– Locate open circuit or contact resistance according to wiring diagram and correct. – Locate open or short circuit according to wiring diagram and correct. – Check relay (-J106-); (Coil resistance: 50 – 65 Ω) functional check with VAG 1710, (test step 1). (Read measured value block, function 08.)

VAG 1551 print-out	Possible cause	Correction
00526 2131 Brake Light Switch -F-	– Open circuit between Brake Light Switch/brake lights and ABS Control Module -J104-, contact 25 – Open circuit in brake light system	– Locate open circuit according to wiring diagram and correct. – Check brake lights, functional check with VAG 1710, (test step 2). (Read measured value block, function 08.)
00532 2234 Voltage supply (at solenoid valves)	– Insufficient supply voltage at solenoid valves: Voltage at ABS Control Module -J104-, contact 32, is only 7.5 – 9.0 Volts	– Check supply voltage and determine contact resistance between Battery and ABS Hydraulic Unit -N55-, contact 6, as well as between ABS Hydraulic Unit -N55-, contact 4, and ABS Control Module -J104-, contact 32. Functional check with VAG 1710, (test steps 1 and 5). (Read individual measured value, function09.)

VAG 1551 print-out	Possible cause	Correction
00597 Different wheel speed impulses[1]	No signal or incorrect impulses from one of the 4 speed sensors: – 1 Rotor with incorrect number of teeth installed – Rotors dirty or damaged – 1 Wheel Speed Sensor not installed correctly – Play in wheel bearing or wheel bearing faulty – 1 Wheel Speed Sensor faulty	– Check all rotors. – Check all rotors, clean or replace. – Check all Wheel Speed Sensor installation positions. – Service front suspension ⇒ Page M 40-1. – Service rear suspension ⇒ Page M 42-1. – Check Wheel Speed Sensor, functional check with VAG 1710 (test step 6). (Read measured value block, function 08.)
65535 1111 ABS Control Module -J104- faulty[2]	– Contact resistance in Ground or positive wiring to ABS Control Module -J104- – ABS Control Module -J104- faulty	– Locate contact resistance according to wiring diagram and correct. – Replace ABS Control Module -J104-.

1) Wheel information not possible or may be erroneously displayed if a wheel is rotated at more than 6 km/h (3.75 mph) with ignition on.

2) If, simultaneously, the DTC "ABS Return Flow Pump -V39-" is displayed, then correct this first.

DTC Memory, ending output

Requirements:

- DTC Memory has been checked.

1. Erasing DTC Memory

– Press → key,

Rapid data transfer	HELP
Select function XX	

◄ Indicated on display
(Clear for function selection)

– Press keys -0- and -5-; 05 enters the "Erase DTC Memory" function.

Rapid data transfer	Q
05 Erase DTC Memory	

◄ Indicated on display.

– Confirm entry with Q key.

Rapid data transfer	→
DTC Memory is erased!	

◄ Indicated on display.

– Press → key.

Rapid data transfer	HELP
Select function XX	

◄ Indicated on display
(Clear for function selection).

Attention! DTC Memory has not been checked

Notes:

- *If this appears in the display, the test sequence is faulty.*
- *Adhere exactly to test sequence:*
  First check DTC Memory, then erase.
- *The DTC Memory is not erased if:*
  - *The ignition was switched off between "Checking DTC Memory" and "Erasing DTC Memory."*
  - *The vehicle has been moved at more than 2.5 km/h (1.56 mph).*

2. Ending output

– Press keys -0- and -6-, this ends the output.

Rapid data transfer	Q
06 End output	

Indicated on display.

– Confirm entry with Q key.

Rapid data transfer	Help
Enter address word XX	

Indicated on display

– Switch off ignition.

– Disconnect Scan Tool VAG 1551.

– Switch on ignition.

ABS Warning Light -K47- must go out after about 2 seconds.

– Road test vehicle at a speed which briefly exceeds 30 km/h (19 mph).

- The ABS Warning Light -K47- must not light up again.

Measured value block, reading

Rapid data transfer HELP Select function XX	

◄ Indicated on display.
(Clear for function selection.)

– Press keys -0- and -8-; 08 enters the "Read measured value block" function.

Rapid data transfer Q
08 Read measured value block

◄ Indicated on display.

– Confirm entry with Q key.

Read measured value block HELP

Enter display group number XX

◄ Indicated on display.

Note (print out also by pressing HELP key):

To output measured values without a physical quantity enter display group number "00".

– Press key -0- twice.

– Confirm entry with Q key.

Read measured value block →
2 2 2 2 0 0

◄ Indicated on display.

Notes:

- *To convert the individual readouts in the 6 display zones* *⇒ Test table, Page M 01-35.*
- *If the printer is switched on, the current measured value block is printed out.*
- *If the measured values obtained indicate malfunctions, troubleshoot according to wiring diagram.*

– Press → key.

Rapid data transfer HELP
Select function XX

◄ Indicated on display.
(Clear for function selection.)

Test table (Read measured value block)

Display zone	Designation	• Test requirements	Displayed on VAG 1551
1	Left front wheel speed (km/h)		2 ... 255
2	Right front wheel speed (km/h)		2 ... 255
3	Left rear wheel speed (km/h)		2 ... 255
4	Right rear wheel speed (km/h)		2 ... 255
5	Brake Light Switch -F- (contact 25 to -J104-)	• -F- not operated Wiring open circuit • -F- operated	0 1 [1)] 2
6	Voltage at ABS Retrun Flow Pump motor -V39- (contact 14 to -J104-)	• No voltage (OK.) • Voltage present (faulty)	0 255

[1)] An open circuit in the wiring of a vehicle with a cruise control system cannot currently be recognized; instead, -2- (for Brake Light Switch -F- operated) is displayed.

Individual measured value (voltage at solenoid valves), reading

Rapid data transfer HELP Select function XX

◄ Indicated on display.
(Clear for function selection.)

– Press keys -0- and -9-; 09 enters the "Read individual measured value" function.

Rapid data transfer Q 09 Read individual measured value

◄ Indicated on display.

– Confirm entry with Q key.

Read measured value block Enter channel number XX

◄ Indicated on display.

Note:

An accessible measured value can only be read from channel 03 (voltage at solenoid valves).

– Press keys -0- and -3-.

– Confirm entry with Q key.

Read individual measured value → Channel 03 Measured value 245

◄ Indicated on display.

Notes:

- *Converting measured values;*
  ⇒ Test table, Page M 01-37.
- *If the printer is switched on the current measured value block is printed out.*
- *If the measured value indicates a malfunction, troubleshoot according to wiring diagram.*

– Press → key.

Rapid data transfer	HELP
Select function XX	

◄ Indicated on display.
(Clear for function selection.)

Test table (reading individual measured value)

Display zone	Designation	• Test requirements – additional work	Displayed on VAG 1551
3	Voltage at the solenoid valves (monitoring undervoltage, contact 32 at -J104-)	• Ignition switched on	0 – 255 [1]

[1] 1 Unit corresponds to 47.33 mV, ⇒ Conversion table. Voltages over 11.3 Volts (over 239 units) are no longer displayed linearly. Voltages over 9.0 Volts (190 units) are OK.

Conversion table

Unit	0	106	148	158	169	180	190	201	211	222	232	239
Voltage (V)	0	5.0	7.0	7.5	8.0	8.5	9.0	9.5	10.0	10.5	11.0	11.3

Electrical test

Notes:

- *Use the following additional equipment for the test:*
  - *Digital multimeter Fluke 83 (US1119) or equivalent.*
  - *Adapter set VAG 1594.*
  - *Test box VAG 1598 and adapter VAG 1598/3 (connect only to the multi-pin connector which has been disconnected from ABS Control Module -J104-).*
- *If the readings obtained deviate from the specifications, locate malfunction according to wiring diagram and correct.*
- *The socket designation of test box VAG 1598 is identical to the contact designations on ABS Control Module -J104-.*
- *To prevent destruction of electrical components, select relevant multimeter scale before connecting to test cables/leads.*

Test requirements:

- Battery Positive Voltage (B+) OK.
- Fuses OK.
- Ground (GND) connections OK.

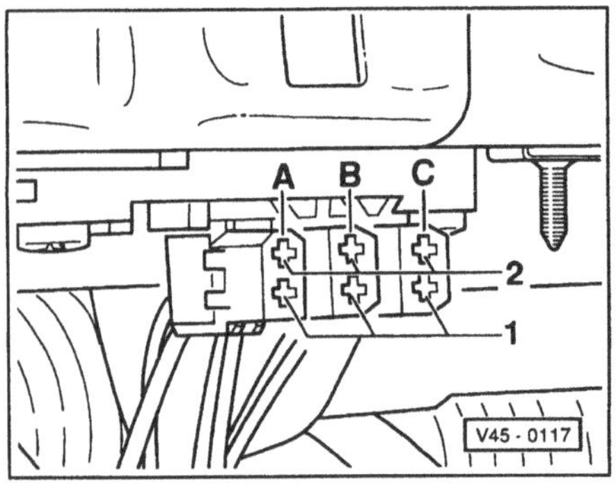

1. Checking Data LInk Connectors (DLC)

Notes:

- *The 2-pin Data Link Connectors (black/white), to connect Scan Tool VAG 1551, are located behind the relay plate.*
- *Harness color codes and information relating to further systems which are also connected to the white DLC,* ⇒ Wiring diagrams.

– Checking 2 pin DLC (black) -A- for B+:

Contact -1- (lower): Ground.

Contact -2- (upper): B+ via fuse S21 at terminal 30.

– Checking 2 DLC (white) -B- or -C- for data transfer:

Contact -1- (lower): connected via an additional connection with contact 30 on ABS Control Module -J104-.

Contact -2- (upper): connected via an additional connection with contact 31 on ABS Control Module -J104-.

Note:

If a "Fault in the communication build-up" occurs in a vehicle with several On Board Diagnostic (OBD) systems, the fault may influence the other systems.

Remedy:

– Separate all 2 pin connectors of the other vehicle systems between the DLC (white) and the Control Modules, ⇒ Wiring diagrams.

– Enter the address word "03" (brake electronics), again in the "Rapid data transfer" mode and repeat the test.

– If no further fault is found, reconnect all connectors after completing troubleshooting.

2. Checking voltage supply at ABS Control Module -J104- and ABS Warning Light -K47-

Note:

The following checks relate to wiring which cannot be monitored with On Board Diagnostic (OBD):

Switch to measuring range:
Voltage measurement (20 V =)

Test step	VAG 1598 sockets	Item tested	• Test conditions – Additional operations	Specification	Measures for deviations from specification
1	1 + 20	Terminal X voltage supply	• Ignition switched on	approx. Battery voltage	– Check wiring from contact 1 via fuse S51 to relay plate W/3 – Check wiring from contact 20 to Ground near -J104-
2	29 + 34	Terminal 15 voltage supply	• Ignition switched on	approx. Battery voltage	– Check wiring from contact 29 to relay plate W/2 – Check wiring from contact 34 to Ground near -J104-

Functional check

Test step	VAG 1598 sockets	Item tested	• Test conditions – Additional operations	Specification	Measures for deviations from specification
3	-	ABS Warning Light -K47-, function	• Ignition switched off	-K47- does not light up	
			– Switch ignition on	-K47- lights up	– Check wiring from -K47- to relay plate X/4 and X/8. – Check -K47-. – Check wiring from relay plate W/2 to ABS Hydraulic Unit -N55-, contact 11. – Check ABS Solenoid Valve Relay -J106- -rest position-. – Check Ground strap on ABS Return Flow Pump -V39-.

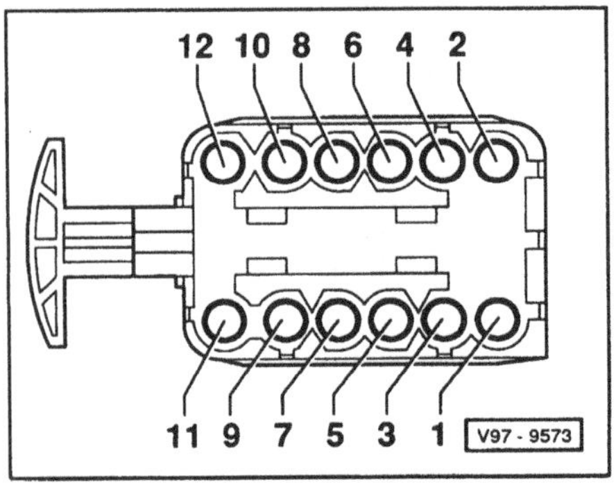

3. Multi-pin connectors

◄ ABS Control Module -J104-

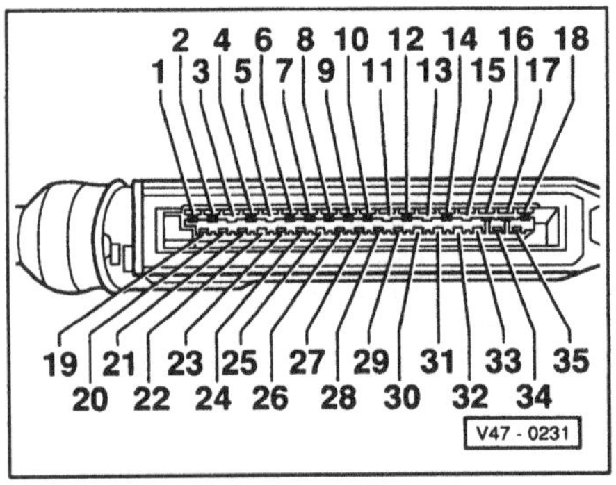

◄ ABS Hydraulic Unit -N55-

Note:

Before separating the connector, pull side–locking device out of housing.

ABS, checking function with VAG 1710

Test sequence

Note:

On vehicles with On-Board Diagnostic (OBD), check the ABS Diagnostic Trouble Code (DTC) Memory.

– *Drive vehicle onto hoist platform and raise far enough until all wheels can be turned freely.*

Note:

- *No gear engaged (automatic transmission in position "N").*
- *Parking brake not applied.*

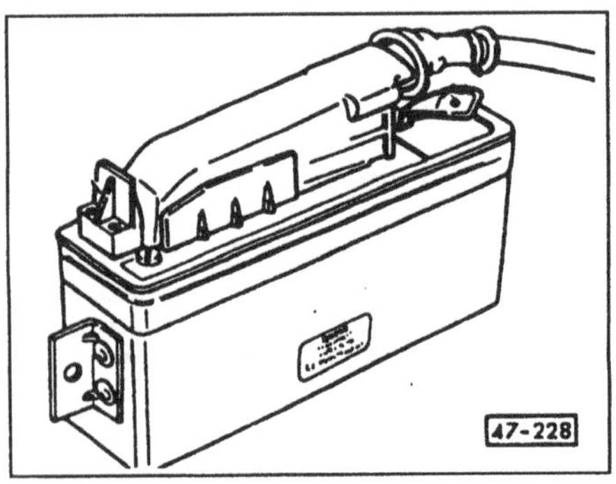

– *Switch off ignition.*

◄ – *Remove plug connection to ABS Control Module -J104- and plug into connector of cable of ABS tester VAG 1710.*

– *Set coding switch at connector of cable to correct setting for this vehicle, EuroVan=2.*

> ***CAUTION!***
>
> ***DO NOT drive vehicle with testers connected.***
>
> ***DO NOT test the brake system on a brake dynamometer.***

Note:

The ABS Warning Light -K47- lights up in test steps 1 to 4 and 6 when the ignition is switched on.

If the LED on the ABS tester (LED -1-) goes out during the ABS test, test power supply to ABS Control Module -J 104-.

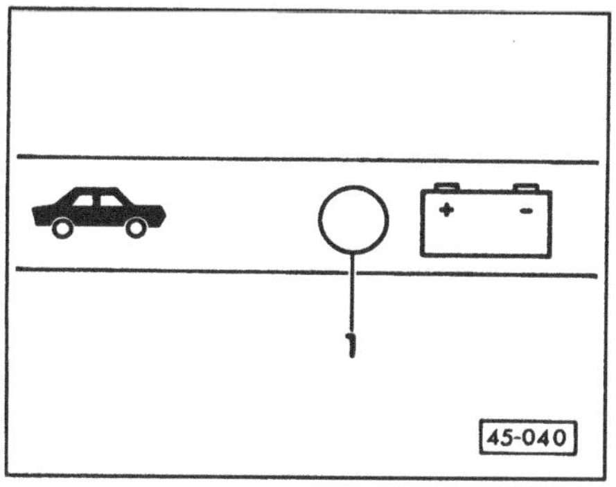

Test step 1

Subject of test:

- Power supply to ABS Control Module -J104-.

No.	• Test conditions – Additional operations	Specification	Correction
1.1	• Ignition switched on. – Program switch in position 1.	– LED -1- lights up – Warning Light -K47- lights up.	Page M 01-57, item 1.1 Page M 01-57, item 1.2
1.2	not applicable	–	–
1.3	not applicable	–	–
1.4	not applicable	–	–
1.5	not applicable	–	–

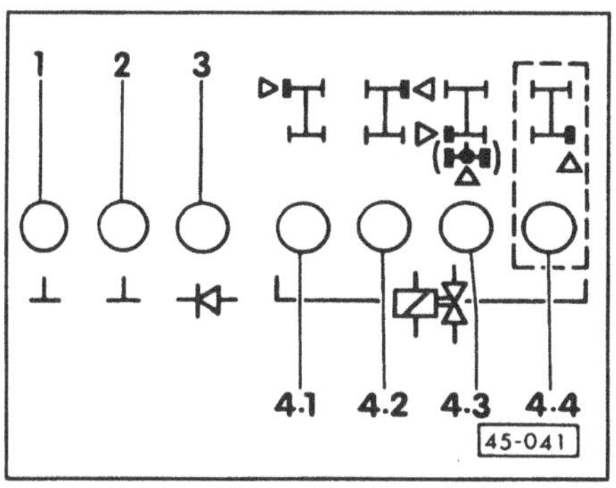

Subject of test:

- Ground connection to ABS Control Module -J104-.
- Diode for ABS Warning Light -K47-.
- Solenoid valves/inlet/outlet valves* in ABS Hydraulic Unit -N 55- (resistance).

Note:

If one of the LEDs (from -1- to -4.4-) illuminates weakly, this indicates contact resistance; test relevant plug connections and wiring on basis of wiring diagram.

No.	• Test conditions – Additional operations	Specification	Correction
1.6	• Ignition switched on. • Program switch in position 1.	– LEDs -1- to -4.4- light up with same intensity ** (FWD models).	Page M 01-57, item 1.7

* Designation for models with ABS On Board Diagnostic (OBD)

** On models with ABS, OBD LED -2- does not light up; if it does light up, check wiring to ABS Control Module -J104- (pin 10) using Wiring Diagram (pin must not be connected).

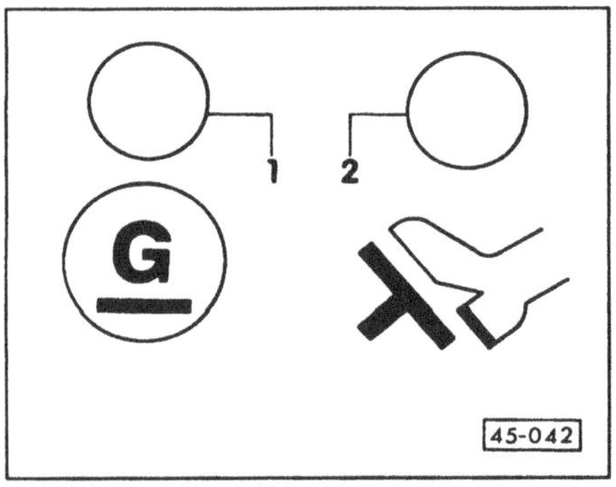

Test step 2

Subject of test:

- Terminal 61 (D+) to ABS Control Module -J104-*
- Brake Light Switch -F-.

No.	• Test conditions – Additional operations	Specification	Correction
2.1	• Ignition switched on. • Program switch in position 2.	– LED -1- lights up.* – LED -2- lights up.*	Page M 01-59, item 2.1
2.2	– Start engine (If LED -1- does not go out fully, briefly increase engine speed to approx. 2000 rpm).	– LED -1- goes out.* – LED -2- lights up.	Page M 01-60, item 2.2
2.3	– Depress brake pedal. – Switch off engine	– LED -2- goes out.	Page M 01-60, item 2.3

* Not on models with ABS On-Board Diagnostic (OBD); otherwise check wiring to ABS Control Module -J104- (pin 15), using Wiring Diagram.

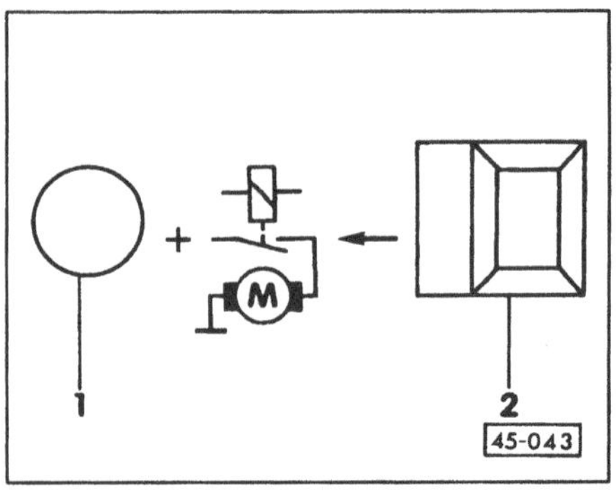

Test step 3

Subject of test:

- ABS Return Flow Pump Relay -J105-.
- ABS Return Flow Pump -V39- and related Ground connection.

No.	• Test conditions – Additional operations	Specification	Correction
3.1	– Ignition switched ON. – Program switch in position 3. – Operate button -2-.	– LED -1- lights up and ABS Return Flow Pump -V39- (in hydraulic unit) operates.	Page M 01-60, item 3.1
3.2	– Release button -2-.	– After button -2- is released, LED -1- remains on for short time and then goes out.	Page M 01-61, item 3.2

Test step 4

Note:

Test step 4 is not applicable.

Test step 5

Note:

Assistance of a second technician required.

Subject of test:

- ABS Solenoid Valve Relay -J106-.
- Operation of solenoid valves and correct connection of related electrical and hydraulic components.

Note:

The 4 wheels are selected individually using the wheel selector switch 4.

Pos. 5 = Left Front ABS Solenoid Valve -N59-
*Left Front ABS Inlet/Outlet Valve -N 137-**

Pos. 6 = Right Front ABS Solenoid Valve -N58
*Right Front ABS Inlet/Outlet Valve -N138-**

Pos. 7 = Left Rear ABS Solenoid Valve -N 57-
*Left Rear ABS Inlet/Outlet Valve -N139-**

Pos. 8 = Right Rear ABS Solenoid Valve -N56-
Right Rear ABS Inlet/Outlet Valve -N140-

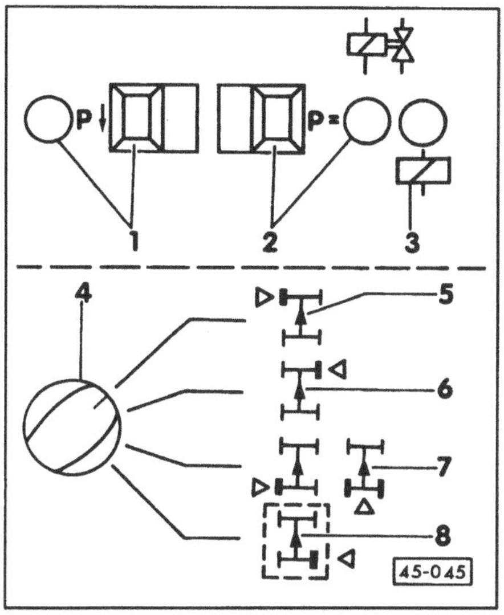

No.	• Test conditions – Additional operations	Specification	Correction
5.1	• Ignition switched on – Program switch in position 5.	– LED -3- lights up.	Page M 01-62, item 5.1
		– ABS Warning Light (K47) lights up.	
5.2	– Wheel selector switch -4- in position 5.		

* Designation on models with ABS On Board Diagnostic (OBD)

No.	• Test conditions – Additional operations	Specification	Correction
5.3	– Press button -2- and hold.	– LED 2 lights up.	Page M 01-62, item 5.3
5.4	– Button -2- is pressed. – Depress brake pedal and hold.	– The selected wheel (front left) can be rotated by hand.	Page M 01-62, item 5.4
5.5	• Brake pedal is depressed. – Release button -2-.	– The selected wheel (front left) is blocked.	Page M 01-62, item 5.5

No.	• Test conditions – Additional operations	Specification	Correction
5.6	• Brake pedal is depressed.		
	– Press button -1- and hold.	– LED -1- lights up.	Page M 01-62, item 5.3
		– The selected wheel (front left) can be rotated and the return pump runs.	Page M 01-62, item 5.6 ***Note:*** *If ABS Return Flow Pump -V39- does not run, see test step 3.*
	– Release button -1-. – Release brake pedal.		

- Wheel selector switch -4- in position 6 and perform test on **front right wheel** from test step 5.3.
- Wheel selector switch -4- in position 7 and perform test on **rear left wheel** from test step 5.3
- Wheel selector switch -4- in position 8 and perform test on **rear right wheel** from test step 5.3.

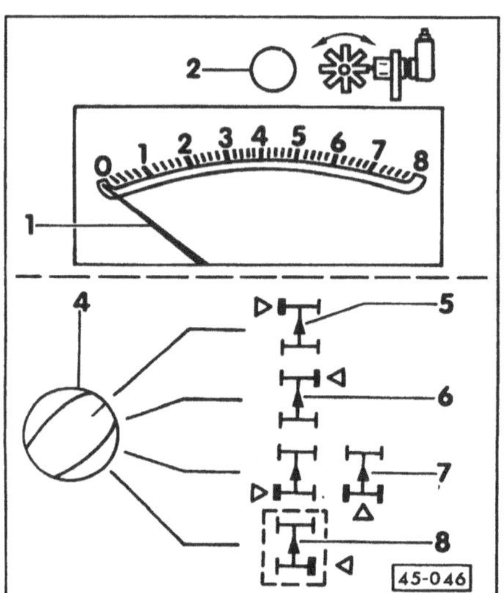

Test step 6

Subject of test:

- Operation and correct connection of wheel speed sensor.

Note:

- *Do not read pointer -1- until LED -2- shows a steady light, without flickering.*
- *The dial of instrument -1- consists of 8 units.*
- *The LED -2- goes out if wheel speed is excessive.*
- *The wheel speed sensors are selected individually with the wheel selector switch -4-.*

Pos. 5 = Left Front ABS Wheel Speed Sensor -G47-

Pos. 6 = Right Front ABS Wheel Speed Sensor -G45-

Pos. 7 = Left Rear ABS Wheel Speed Sensor -G46-

Pos. 8 = Right Rear ABS Wheel Speed Sensor -G44-

No.	• Test conditions – Additional operations	Specification	Correction
6.1	• Ignition switched on. – Program selector switch in position 6.	–	–
6.2	– Wheel selector switch -4- in position 5- (-G47-).	–	–

No.	• Test conditions – Additional operations	Specification	Correction
6.3	– Lock other wheel of driven axles to prevent it turning. – Rotate selected wheel (front left) **evenly** at approx. 1 revolution per second.	– LED -2- lights up without flickering. Pointer of instrument -1- moves.	Page M 01-63, item 6.1
		– Reading on instrument -1- more than: units (for rotor with 96 teeth). units (for rotor with 48 teeth). units (for rotor with 45 teeth).	Page M 01-63, item 6.2
		– Fluctuation range per wheel revolution less than 25% from maximum reading.	Page M 01-63, item 6.3
		– Differences between readings on one axle less than 2.0 units.	Page M 01-63, item 6.2

- Wheel selector switch -4- in position 6 and perform test of speed sensor **(G 45)** on front right wheel from test step 6.3.
- Wheel selector switch -4- in position 7 and perform test of speed sensor **(G 46)** on rear left wheel from test step 6.3.
- Wheel selector switch -4- in position 8 and perform test of speed sensor **(G 44)** on rear right wheel from test step 6.3.

- If no malfunction was found when testing the ABS system in test steps 1 to 6, see page M 01-46:
- Switch off ignition.
- Separate connector to ABS tester **VAG 1710**.
- Connect ABS Control Module -J104- and install.
- If vehicle equipped with ABS On Board Diagnostic (OBD), interrogate ABS Diagnostic Trouble Code Memory (DTC) and erase.
- Perform test drive.

Note:

- *The ABS Warning Light -K47- must go out when engine is running (only on models without ABS OBD).*
- *The ABS Warning Light -K47- must go out 2 seconds after the ignition is switched on (on models with ABS OBD).*
- *Drive vehicle at a speed of at least 30 km/h for more than 20 seconds. The ABS Warning Light -K47- must not light up during this time.*

Additional malfunctions, checking

- **ABS Warning Light -K47- lights up sporadically and goes out again by itself:**
- Test vehicle voltage (electrical loads switched on).

Note:
Supply voltage must be greater than 10.5 V.

- **ABS Warning Light -K47- lights up intermittently and goes out after restarting**
- **ABS Return Flow Pump -V39- starts at low speed (up to 10 km/h)(6 mph):**
- Test wheel speed sensor and sensor wiring in succession, with test box VAG 1598, adapter cable VAG 1598/3, and digital multimeter Fluke 83, for loose contacts according to wiring diagram.

Note:

- *Set 2 k ohms range at digital multimeter Fluke 83 (Specification: 0.8 k ohms up to 1.8 k ohms)*
- *Move wheel speed sensor wiring during test (to check for loose contacts).*

- **ABS Warning Light -K47- shows steady light:**
- Replace ABS Control Module -J104-

Note:
If vehicle has additional outlets for speed sensing (e.g., to Transmission Control Module -J217- or Anti-Slip Control Module -J231- –⇒ wiring diagram); test voltage supply for chamber 33 at 35-pin connector before replacing ABS Control Module -J104-.

Troubleshooting (after checking function with VAG 1710)

Note:

See page M 01-44 for test sequence with VAG 1710.

Test step 1

Item	• Deviation from specification – Possible causes of malfunction*
1.1	• **LED -1- does not light up** – Battery not sufficiently charged (supply voltage less than 10.5V); test. – Open circuit in ground connections to ABS Control Module -J104-, test. – Fuse on ABS Combination Relay -J156- faulty; replace. – Open circuit in voltage supply or Ground connection for -J156-, test. – Wiring connecting -J156- and -J104- has open circuit; test. – ABS Combination Relay -J156- faulty; replace.
1.2	• **ABS Warning Light -K47- does not light up** – ABS Warning Light -K47- faulty; replace. – Wiring connecting ABS Warning Light -K47-/ABS Control Module -J104- has open circuit; test.
1.3	• not applicable
1.4	• not applicable
1.5	• **not applicable**
1.6	• **not applicable**
1.7	• **LED -1- and/or LED -2- does not light up**** – Open circuit in Ground connection to ABS Control Module -J104-; test.

* Determine faulty wiring (short circuit or open circuit) according to wiring diagram and repair.
** LED -2- does not light up; not a malfunction on models with ABS On Board Diagnostic (OBD).

Item	• Deviation from specification – Possible causes of malfunction*
1.7	• **LED -3- does not light up** – Ground connection for ABS Hydraulic Unit -N55- has open circuit; test. – Ground connection for ABS Control Module -J104- has open circuit; test. – ABS Solenoid Valve Relay faulty; replace. – Wiring connecting ABS Warning Light -K47- and ABS Hydraulic Unit -N55- has open circuit; test. – Wiring connecting diode for ABS Warning Light -K47- (in ABS Hydraulic -N55- or in ABS Solenoid Valve Relay -J106- — see wiring diagram) and ABS Control Module -J104- has open circuit; test. – Diode for ABS Warning Light -K47- faulty; replace ABS Solenoid Valve Relay or ABS Hydraulic Unit -N55- ⇒ wiring diagram.
	• **LED -3- lights up, none of the LEDs -4.1- to -4.4- lights up** – Ground connections for ABS Hydraulic Unit -N55- has open circuit; test. – ABS Solenoid Valve Relay -J106- faulty; replace.
	• **LED -4.1- does not light up** – Wiring between ABS Control Module -J104- and Left Front ABS Solenoid Valve -N59-/Left Front ABS Inlet/Outlet Valve -N137-** has open circuit; test. – Resistance for Left Front ABS Solenoid Valve -N59-/Left Front ABS Inlet/Outlet Valve -N137-** not OK (specification 0.7 ohms up to 1.8 ohms); replace hydraulic unit.

* Determine faulty wiring (short circuit or open circuit) according to wiring diagram and repair.
** Designation on models with ABS On Board Diagnostic (OBD).

Item	• Deviation from specification – Possible causes of malfunction*
1.7	• **LED -4.2- does not light up** – Wiring between ABS Control Module -J104-, Right Front ABS Solenoid Valve -N58-, and Left Rear ABS Inlet/Outlet Valve -N139-** has open circuit; test. – Resistance for Right Front ABS Solenoid Valve -N58- and Right Front ABS Inlet/Outlet Valve -N138-** not OK (specification 0.7 ohms up to 1.8 ohms); replace hydraulic unit.
	• **LED -4.3- does not light up** – Wiring between ABS Control Module -J104-, Left Rear ABS Solenoid Valve -N57- and Left Rear ABS INlet/Outlet Valve -N139-** – Resistance for Left Rear ABS Solenoid Valve -N57- and Left Rear ABS Inlet/Outlet Valve -N139-** not OK (specification 0.7 ohms up to 1.8 ohms); replace hydraulic unit.
	• **LED -4.4- does not light up** – Wiring between ABS Control Module -J104-, Right Rear ABS Solenoid Valve -N56- and Right Rear ABS Inlet/Outlet Valve -N140-** has open circuit; test. – Resistance for Right Rear ABS Solenoid Valve -N56-, Right Rear ABS Inlet/Outlet Valve -N104-** not OK (specification 0.7 ohms up to 1.8 ohms); replace hydraulic unit.

* Determine faulty wiring (short circuit or open circuit) according to wiring diagram and repair.
** Designation on models with ABS On Board Diagnostic (OBD).

Test step 2

Item	• Deviation from specification – Possible causes of malfunction*
2.1	• **LED -1- does not light up** – Wiring between ABS Control Module -J104- and Generator has open circuit; test.
	• **LED -2- does not light up** – Wiring between ABS Control Module -J104- and Brake Light Switch -F- has open circuit or short circuit to positive; test.

Item	• Deviation from specification – Possible causes of malfunction*
2.2	• **LED -1- does not go out after engine starts** – Wiring between ABS Control Module -J104- and Generator has open circuit or short circuit to Ground; test.
2.3	• **LED -2- does not go out when brake pedal depressed** – Wiring between Brake Light Switch -F- and ABS Control Module -J104- has open circuit; test. – Brake Light Switch -F- faulty; replace.

* Determine faulty wiring (short circuit or open circuit) according to wiring diagram and repair.

Test step 3

Item	• Deviation from specification – Possible causes of malfunction*
3.1	• **LED -1- does not light up/ABS Return Flow Pump -V39- does not run** – Wiring between ABS Control Module -J104-, ABS Combination Relay -J156- and ABS Return Flow Pump Relay -J105- has open circuit; test. – Voltage supply for ABS Return Flow Pump Relay has open circuit; test. – ABS Return Flow Pump Relay -J105- faulty; replace.
	• **LED -1- does not light up/ABS Return Flow Pump -V39- runs** – Wiring between ABS Return Flow Pump Relay -J105- and ABS Control Module -J104- has open circuit; test.
	• **LED -1- lights up/ABS Return Pump -V39- does not run** – Ground connection for ABS Return Flow Pump -V39- has open circuit; test. – ABS Return Flow Pump -V39- faulty; replace hydraulic unit.

Item	• Deviation from specification – Possible causes of malfunction*
3.2	• **LED -1- goes out as soon as button -2- is released** – Contact resistance in Ground connection for ABS Return Flow Pump -V39-; test – ABS Return Flow Pump -V39- faulty; replace hydraulic unit.
	• **LED -1- remains on after button -2- is released/return pump -V39- continues running** – Short to Ground in wiring between ABS Control Module -J104- and ABS Return Flow Pump Relay -J105-; test. – ABS Return Flow Pump Relay -J105- faulty; replace.

* Determine faultly wiring (short circuit of open circuit) according to wiring diagram and repair.

Test step 4

Item	• Deviation from specification – Possible causes of malfunction*
4.1	• **LED -1- does not light up** – Wiring between ABS Control Module -J104- and ABS Acceleration Switch -F113- has open circuit; test. – ABS Acceleration Switch -F113- faulty; replace.
4.2	• **LED -2- lights up** – Check wiring to ABS Control Module -J104- (contact **13** must not be connected).
4.3	• **LED -1- does not light up** – Open circuit in power supply for ABS Control Module -J104- (contact **16**).
4.4	• **not applicable**

* Determine faulty wiring (short circuit or open circuit) according to wiring diagram and repair.

Test step 5

Item	• Deviation from specification – Possible causes of malfunction*
5.1	• **LED -3- does not light up** – Open circuit in voltage supply for ABS Solenoid Valve Relay -J106-; test. – Open circuit in wiring between ABS Control Module -J104- and ABS Solenoid Valve Relay -J106-; test. – ABS Solenoid Valve Relay -J106- faulty; replace.
5.2	• **ABS Warning Light -K47- lights up** – Wiring between ABS Warning Light -K47-, ABS Control Module -J104- and ABS Hydraulic Unit -N55- has short circuit to Ground; test.
5.3	• **LED -2- (when button -2- pressed) or LED -1- (when button -1- pressed) does not light up or goes out when performing test step 5** – Test Battery voltage and state of charge (see Wiring diagrams, Troubleshooting and Component Locations). ***Note:*** Start engine to continue test.
5.4	• **The selected wheel blocks when button -2- is pressed** – Brake lines at hydraulic unit or at brake pressure regulator mixed up; check — see Group 47. – Individual leads in wiring between ABS Control Module -J104- and the solenoid valves/inlet/outlet valves** of the ABS Hydraulic Unit -N55- mixed up; check. – Solenoid valve/inlet/outlet valve** faulty; replace hydraulic unit.
5.5	• **The selected wheel does not block after button -2- is released** – Solenoid valve/inlet/outlet valve** faulty; replace hydraulic unit.
5.6	• **The selected wheel blocks when button -1- is pressed** – Solenoid valve/inlet/outlet valve** faulty; replace hydraulic unit.

* Determine faulty wiring (short circuit or open circuit) according to wiring diagram and repair.
** Designation on models with ABS On Board Diagnostic (OBD).

Test step 6

Item	• **Deviation from specification** – **Possible causes of malfunction***
6.1	• **LED -2- does not light up/no reading on instrument** – Coding switch on connector of cable **VAG 1710;** check — – Wheel rotated too slowly or too fast (approximately 1 revolution per second). – Wiring between wheel speed sensor and ABS Control Module -J 104- has open circuit or leads mixed up; test. – Wheel speed sensor not correctly installed; check. – Wheel speed sensor faulty (internal resistance specification: 0.8 k ohms up to 1.8 k ohms); replace.
	• **LED -2- flickers/pointer of instrument moves** – Wheel rotated too slowly (approximately 1 revolution per second). – Wheel speed sensor not correctly installed; check. – Rotor fouled or loose; check and clean if necessary or replace.
6.2	• **Reading on instrument (when LED -2- illuminated) smaller than** **1.0 units (for rotor with 96 teeth)** **1.6 units (for rotor with 48 teeth)** **2.4 units (for rotor with 45 teeth)** • **Deviation between speed sensors of one axle greater than 2.0 units** – Rotor fouled; clean. – Wheel speed sensor not correctly installed; check. – Wheel speed sensor faulty; replace.
6.3	• **Fluctuation range greater than 25% from maximum reading during single rotation of wheel** – Rotor fouled; clean. – Play of wheel bearings excessive; adjust. – Rotor loose; check.

* Determine faulty wiring (short circuit or open circuit) according to wiring diagram and repair.

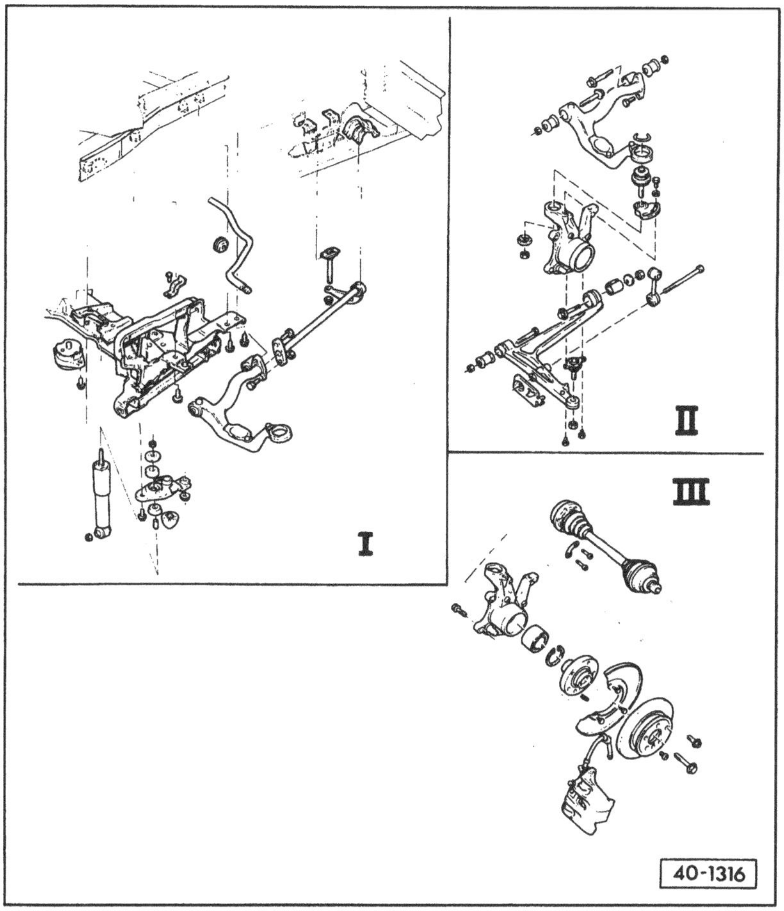

Front suspension assembly

I - Removing and installing subframe, stabilizer bar, torsion bar and shock absorber,
⇒ Page M 40-4.

II - Removing and installing upper and lower control arms,
⇒ Page M 40-20.

III - Removing and installing wheel mountings and axle shafts,
⇒ Page M 40-38.

Notes:

- *The upper control arm is pushed down by the force of the torsion bar. During the following operations always install the support 3250, or relieve pressure on the torsion bar:*
- *Removing and installing shock absorber.*
- *Removing and installing wheel bearing/wheel bearing housing.*
- *Removing and installing axle shaft*
- *Removing and installing lower ball joint/control arm.*

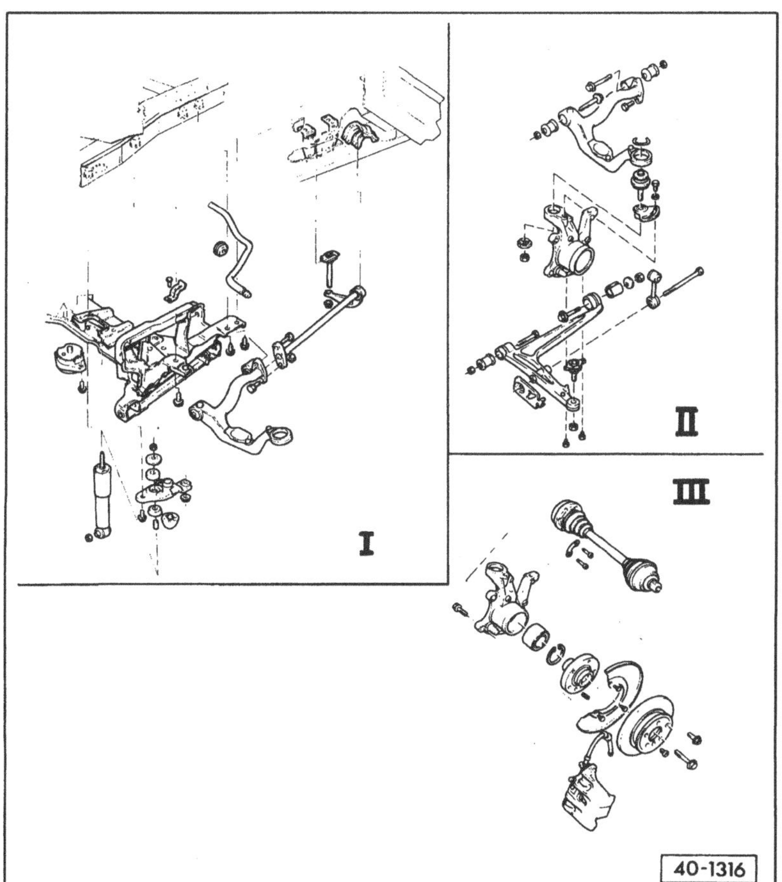

- *Removing and installing upper control arm.*
- *Removing and installing eccentric bushing on upper control arm.*
- *Removing and installing stabilizer bar.*
- *Installing support 3250,* *⇒ Page M 40-3.*

Support 3250, installing to support upper control arm

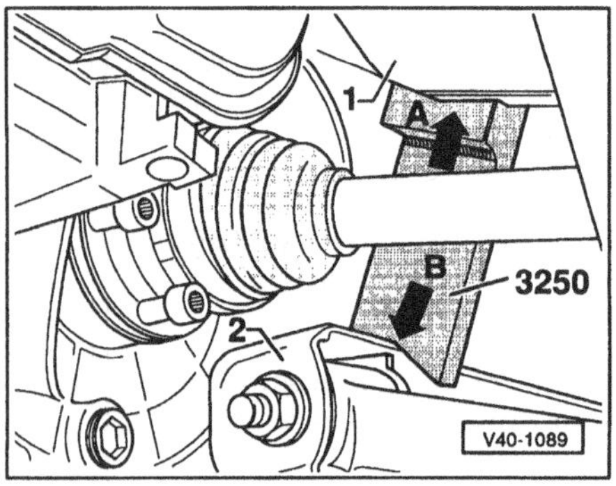

Install support 3250 before lifting the vehicle.

If the support 3250 cannot be installed, even after loading the vehicle, torsion bar tension must be relieved ⇒ Page M 40-12. Removing and installing torsion bar/adjusting torsion bar.

◄ – Install support 3250 so that surface (arrow A) is positioned under upper control arm -1- and V-shaped cut-out (arrow B) is resting on subframe -2-.

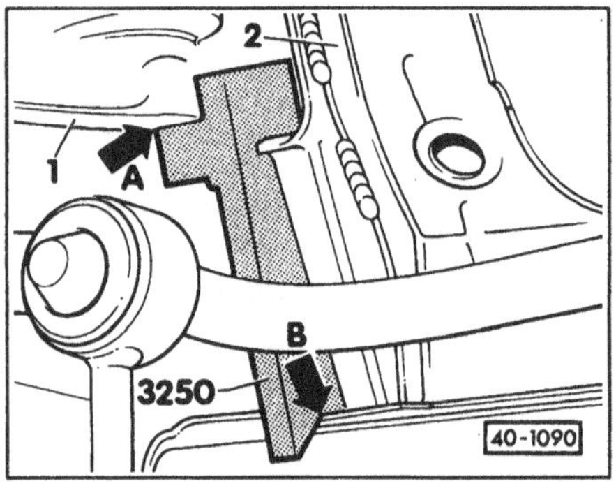

Note:

If the vehicle is fully loaded, raise it on a hoist just far enough to install the support.

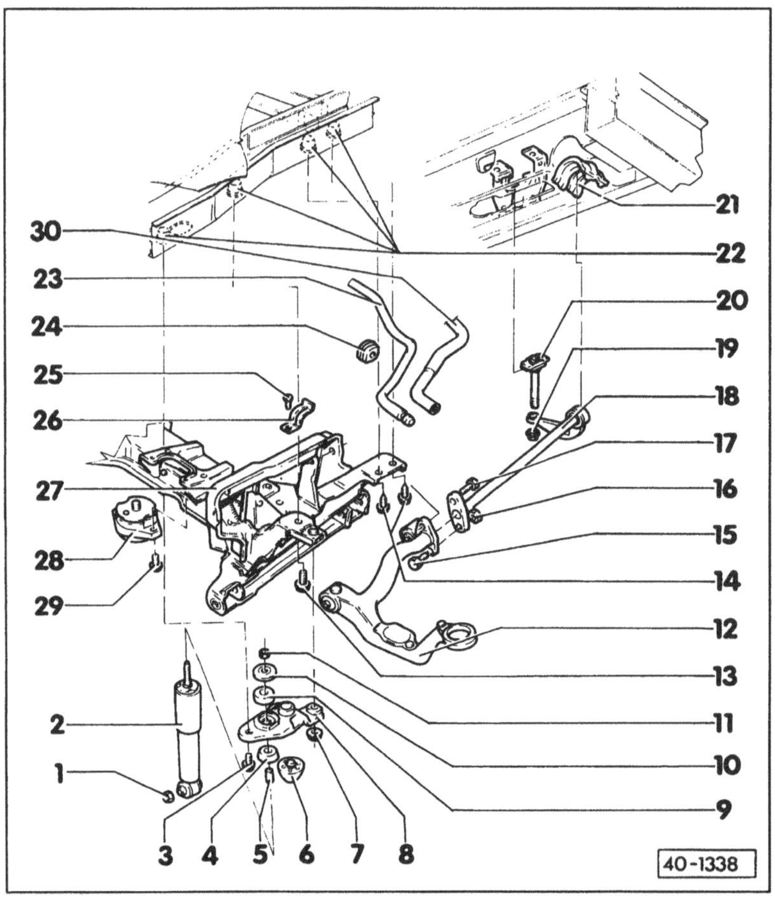

Subframe assembly, stabilizer bar, torsion bar and control arm, removing and installing

Notes:

- *If the vehicle has to be moved after removing an axle shaft, first install an outer CV joint in place of the axle shaft or the wheel bearing will be damaged.*
- *Specifications for wheel alignment ⇒ Page M 44-6.*
- *Do not weld or straighten load bearing or wheel locating components of the front suspension.*
- *Always replace self-locking nuts.*
- *Always replace corroded fasteners.*

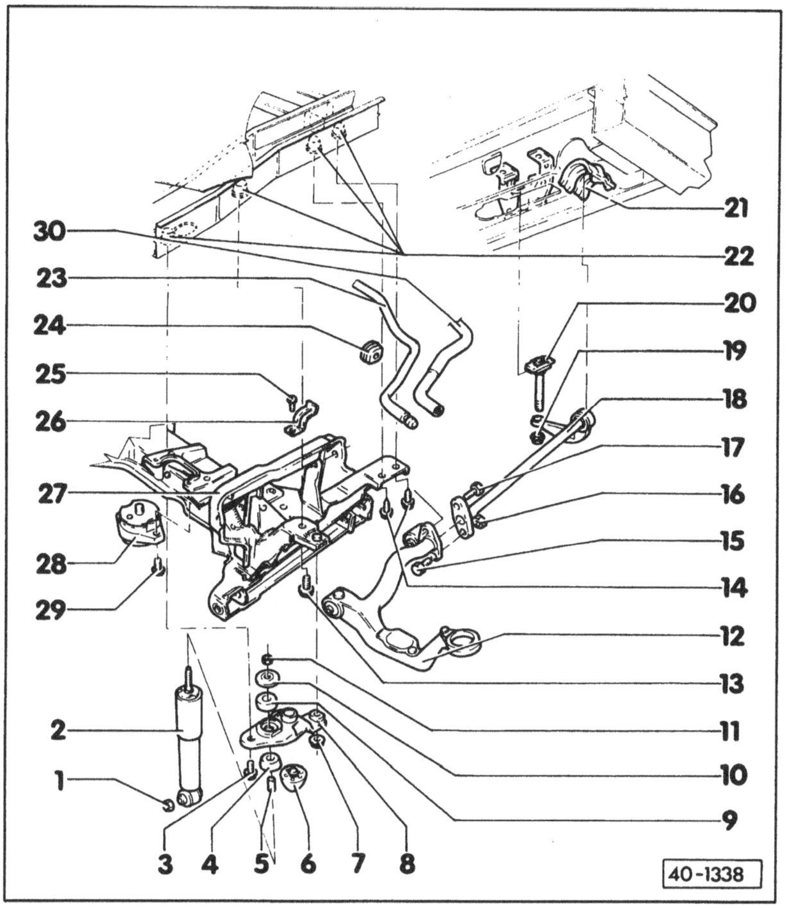

1 - Hex nut, 160 Nm (118 ft lb)

2 - Shock absorber
- Removing and installing ⇒ Page M 40-11.
- Identification ⇒ Page M 40-11.

3 - Bolt, 100 Nm (74 ft lb)

4 - Lower bearing bushing

5 - Spacer tube

6 - Bump stop

7 - Hex nut, 100 Nm (74 ft lb)

8 - Shock absorber mounting

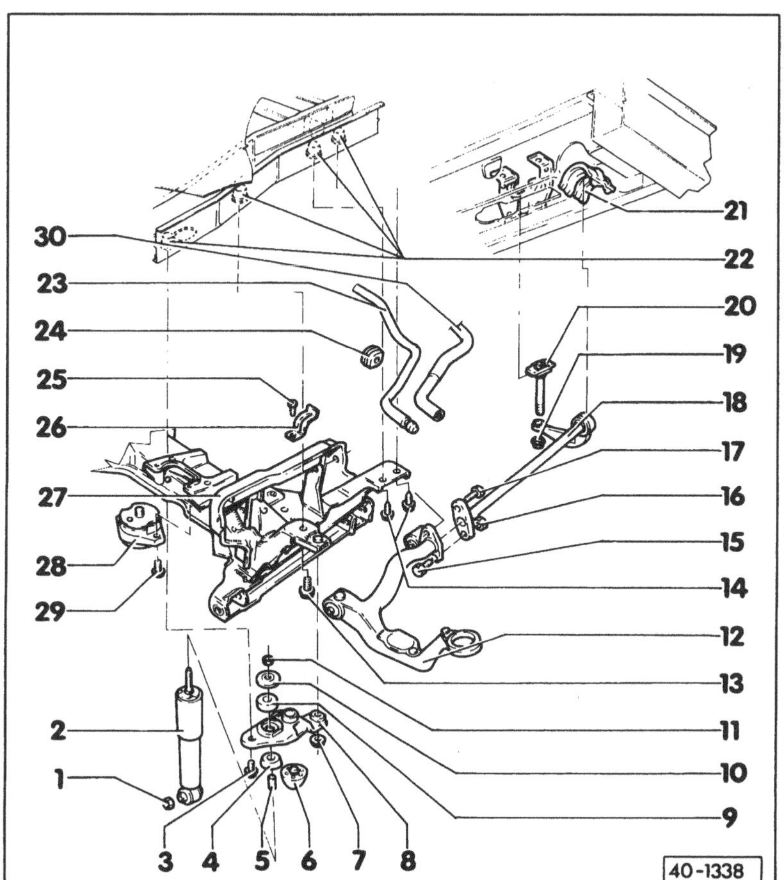

9 - Upper bearing bushing

10 - Washer
- Collar faces downward.

11 - Hex nut, 25 Nm (18 ft lb)

12 - Upper control arm
- Can only be removed when subframe assembly has been removed.

13 - Hex bolt, 160 Nm (118 ft lb)
- M 14 x 1.5 x 40.

14 - Hex bolt, 100 Nm (74 ft lb)
- M 12 x 1.5 x 32.

15 - Hex bolt
- M 14 x 1.5 x 42.

16 - Hex nut, 160 Nm (118 ft lb)

17 - Hex bolt, 160 Nm (118 ft lb)
- M 14 x 1.5 x 30.

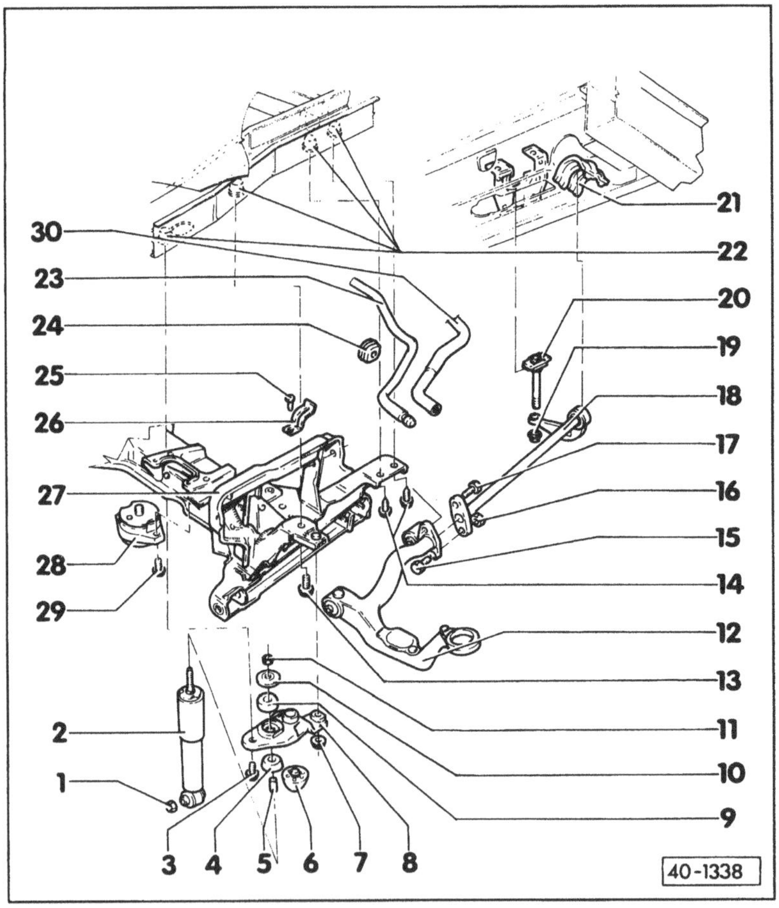

18 - Torsion bar
- Right-hand marked with "R" on end face.
- Left-hand marked with "L" on end face.
- Right and Left must not be interchanged.
- Supplied as a spare part, with tensioning lever.
- Removing ⇒ Page M 40-12.
- Check for paint damage, repair paint damage if necessary.
- Allocation and torsion bar markings ⇒ Page M 40-17.

19 - Nut for tensioning torsion bar
- Always replace.
- Mark position before removing ⇒ Page M 40-12.

20 - Torsion bar adjustment bolt

21 - Rubber mounting

22 - Welded nut
- If damaged, repair ⇒ Page M 40-19.

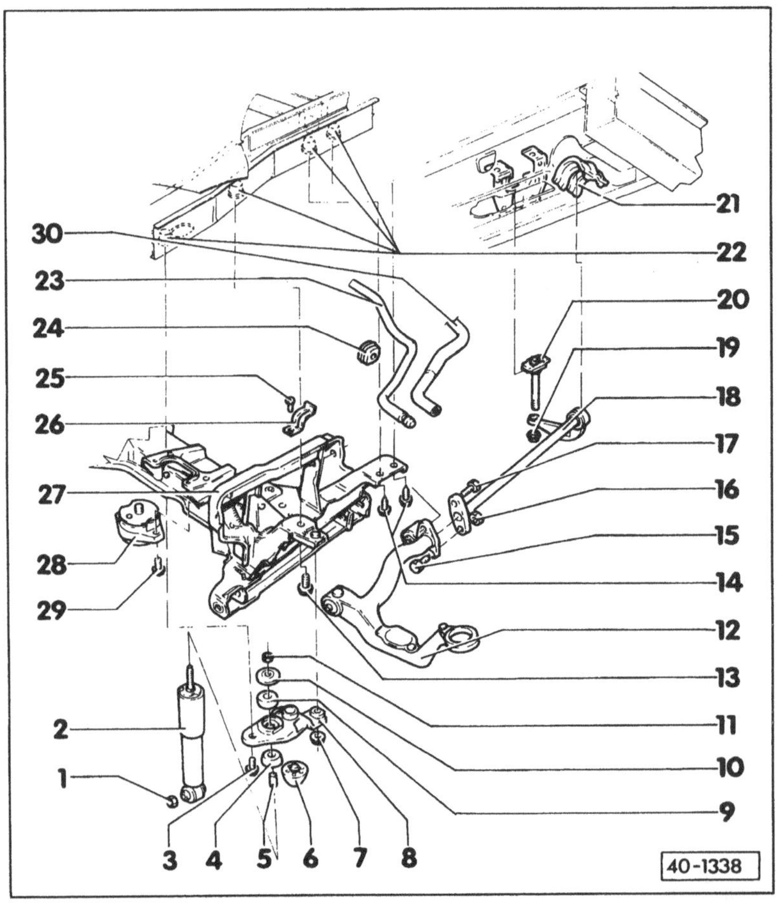

23 - Stabilizer bar
- 23 mm diameter
- Removing and installing ⇒ Page M 40-13.

24 - Stabilizer bar mounting

25 - Hex bolt, 55 Nm (41 ft lb)

26 - Clamp

27 - Subframe
- Removing complete ⇒ Fig. 1
- Up to VIN 70 M 083 341: 12 mm elongated hole for mounting the caster adjustment bolt
- From VIN 70 M 083 342: 14 mm elongated hole for mounting the caster adjustment bolt
- Only subframes with a 14 mm elongated hole are supplied as a spare part. Installing instead of subframe with 12 mm elongated hole ⇒ Page M 40-15.
- From VIN 70 PH 004 343: Parts on subframe modified ⇒ Page M 40-16.

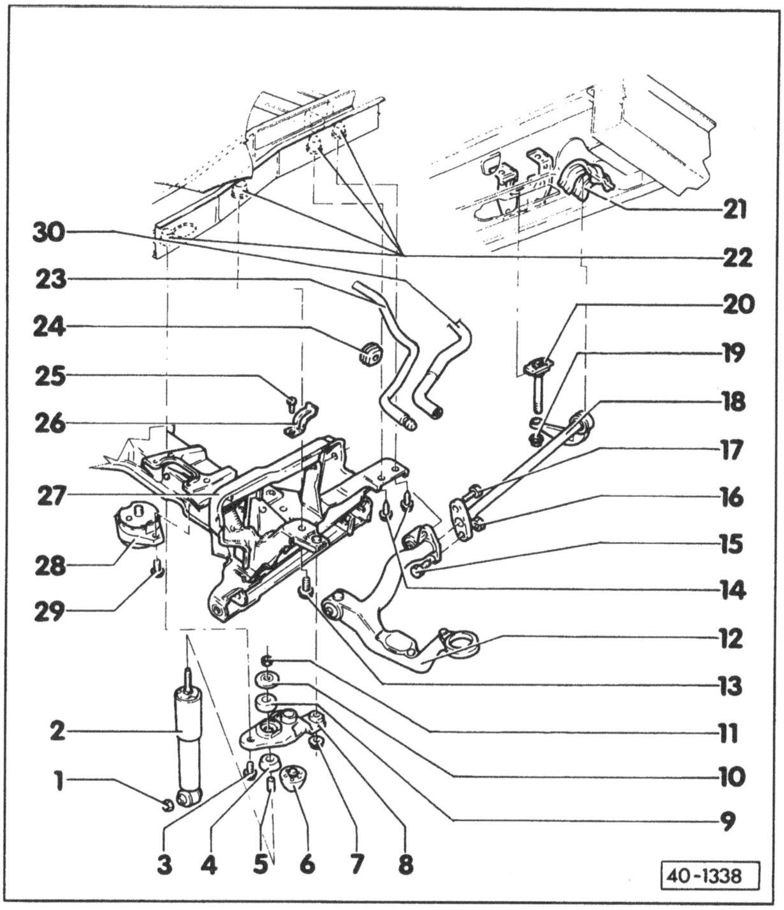

28 - **Bonded rubber mounting**

29 - **Hex bolt, 45 Nm (33 ft lb)**

30 - **Stabilizer bar**
- ♦ 27 mm diameter
- ♦ Removing and installing, ⇒ Page M 40-13.

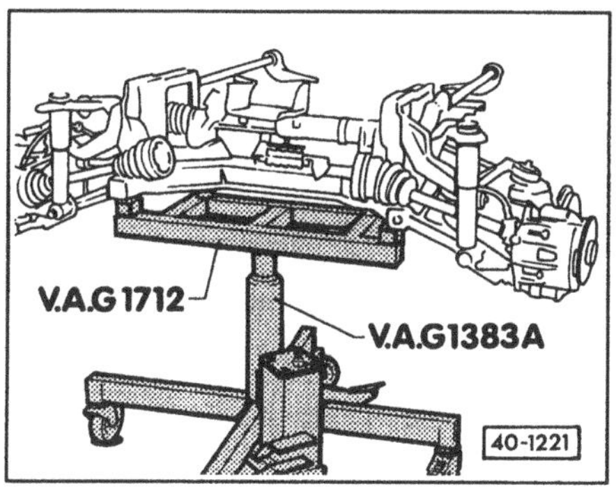

◄ **Fig. 1 Subframe, removing complete**

- Relieve torsion bar tension ⇒ Page M 40-12, removing and installing torsion bar/adjusting torsion bar.
- Remove universal joint bolt at steering gear and remove lines/hoses from steering gear if necessary.
- Disconnect brake lines.
- Disconnect axle shafts at final drive.
- Remove bolted connections at transmission bonded rubber mountings.
- Remove subframe bolts.

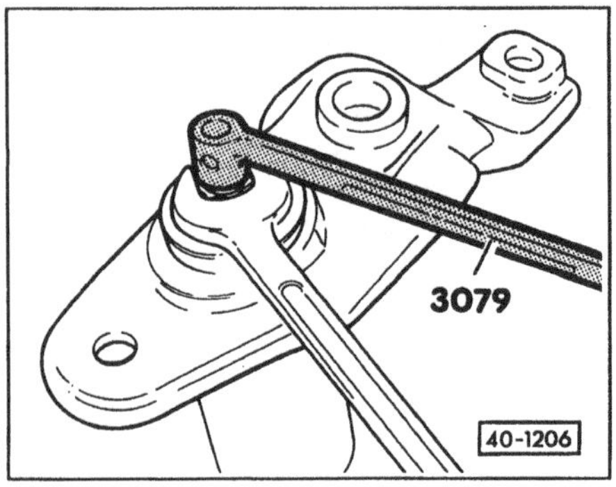

Shock absorbers, removing and installing

Removing

– Install support 3250 on left and right-hand sides ⇒ Page M 40-3.

– Remove lower bolt for shock absorber.

– Remove shock absorber mounting from body.

◄ – Remove shock absorber, as illustrated, from mounting.

Installing

Install in reverse order.

Shock absorber color coding:

- Black: standard.
- Brown: ambulance.
- Green: heavy-duty.

Torsion bar, removing and installing /adjusting

Removing

– Remove fuel tank,
⇒ Engine, Repair group 20.

– Remove exhaust system heat shield and loosen exhaust system, if necessary.

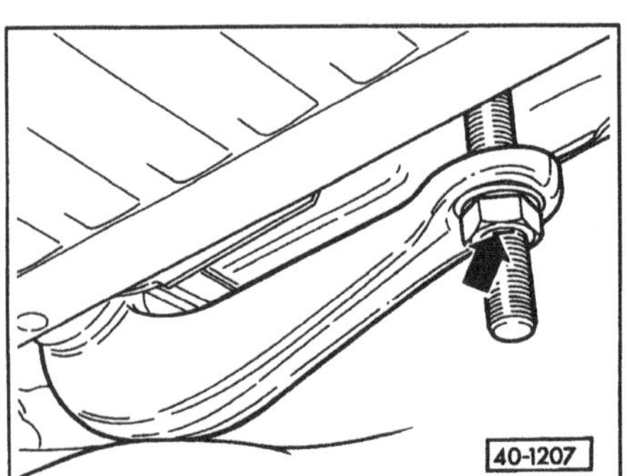

◄ – Mark position of torsion bar setting on threads (arrow) or measure thread projection from nut to end of threaded rod and note.

– Relieve tension on torsion bar with special tool 3257.

– Unbolt torsion bar from control arm and remove.

Notes:

- *The length of the threaded rod projection on the right-hand side may differ from that on the left-hand side.*
- *Measuring ride height, ⇒ Page M 44-4.*
- *Ride height adjustment figures, ⇒ from Page M 44-6.*

Installing

Install in reverse order.

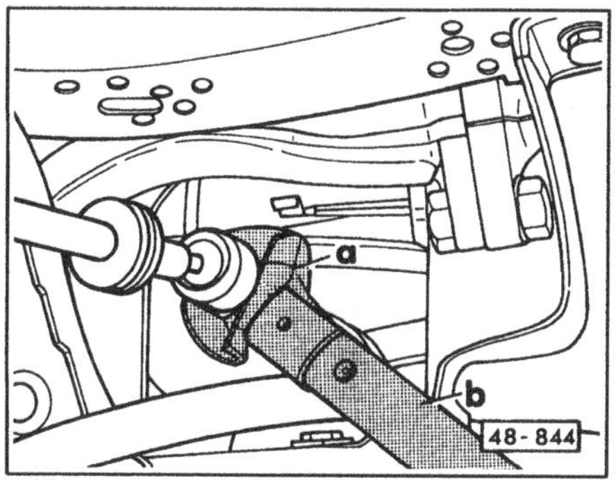

Stabilizer bar, removing and installing

- Install support 3250 to support upper control arm ⇒ Page M 40-3.
- Remove link from stabilizer bar (23 mm diameter stabilizer bar, ⇒ Page M 40-31; 27 mm diameter heavy-duty stabilizer bar, ⇒ Page M 40-32).
- Unbolt both shock absorbers from control arms and push completely upward.
- ◄ Disconnect tie rods from steering gear.

  a - VAG 1332/5
  b - torque wrench VAG 1332
- Unbolt steering gear from subframe.

Note:

On vehicles with a manual steering gear, remove the steering damper.

- Remove universal joint bolt on steering pinion and separate steering gear from steering column.
- Remove clamps for stabilizer bar from subframe.

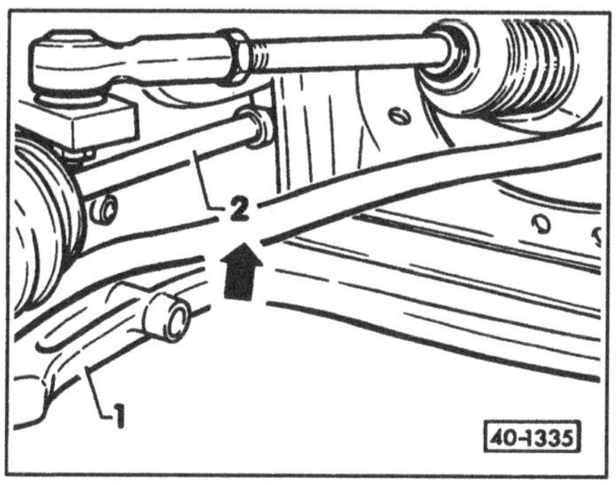

- ◄ Pull stabilizer bar (arrow) out to side between control arm -1- and axle shaft -2-. Push steering gear as far as possible to achieve necessary clearance.

Installing

Install in reverse order.

Link to stabilizer bar tightening torque:
100 Nm (74 ft lb).

Notes:

- *After installing stabilizer bar links, ensure that the link boots are not twisted.*
- *When installing stabilizer bar, ensure that the surface is not damaged. If necessary, repair paint damage and protect against corrosion with underbody sealer.*

Modified subframe

1. Subframe with 14 mm elongated hole to adjust caster.

Notes:

When installing a new subframe (with 14 mm elongated hole) and continued use of original control arms, the following parts are to be replaced:

- *Rear bearing (aluminium bushing, 14 mm inner diameter). Pressing in and out, ⇒ from Page M 40-29.*
- *Hex bolt M 14 x 1.5*
- *Hex nut M 14 x 1.5*
  Tightening torque: 160 Nm (118 ft lb).
- *Eccentric disc, 14 mm diameter.*

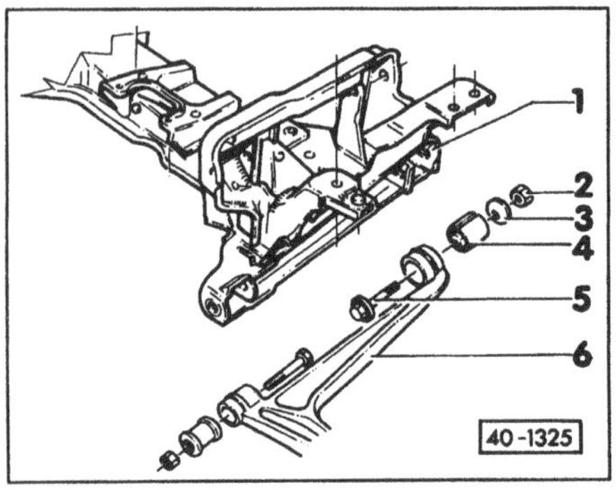

◄

1 - Subframe
2 - Hex nut M14 x 1.5
3 - Eccentric disc (14 mm diameter) to adjust caster
4 - Lower control arm rear bushing
5 - Eccentric bolt M 14 x 1.5, to adjust caster
Need not be replaced after loosening.
6 - Lower control arm

2. Subframe with modified mounting points for manual transmission and steering gear from chassis No. 70 PH 004 343

The following parts of the subframe have been modified:

- Thread (arrows) for securing steering gear – from M 8 to M 10.
- Transmission mounting brackets -A- and bolt -B-.

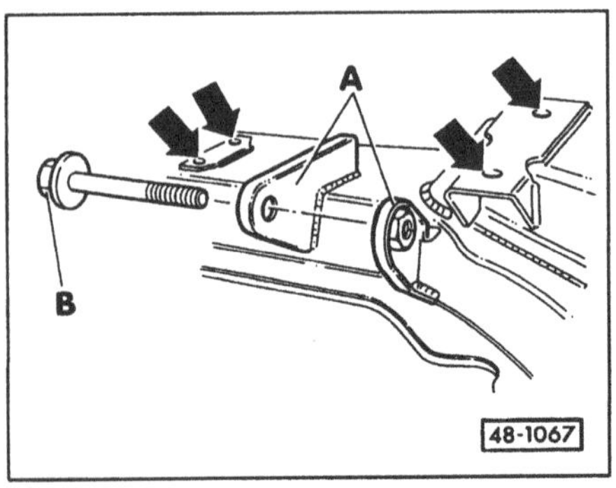

◄

A - Modified transmission mounting brackets.
B - Hex bolt, 200 Nm (148 ft lb).

Notes:

- *The new subframe cannot be installed in vehicles prior to VIN 70 PH 004 343.*
- *Subframe for vehicles up to VIN 70 PH 004 343 available as replacement part while stocks last.*
- *When stocks of this subframe have been depleted, a subframe with M 10 threads for securing the steering gear will be made available for vehicles up to chassis Number 70 PH 004 343. In this case, a new rubber bushing (10 mm diameter) in steering gear and M 10 bolts are to be used.*

Torsion bar, identifying

Model	Equipment + (net load)[1]	Torsion bar	Color[2]	dia. (mm)
Weekender	short wheelbase (1000 kg) long wheelbase (1000 kg)	701 411 103 G/104 G	green	25.3
	short wheelbase (1200 kg) long wheelbase (1200 kg)	701 411 103 H/104 H	brown	26.4

Bus, MultiVan, Weekender w/seats	short wheelbase (800 kg)	701 411 103 F/104 F	white	23.8
	short wheelbase 1000 kg long wheelbase 1000 kg	701 411 103 G/104 G	green	25.3
	short wheel base 1200 kg long wheelbase 1200 kg	701 411 103 H/104 H	brown	26.4

[1] With standard suspension/shock absorbers

[2] Color code: approx.75 mm from flange (2.95 in.)

Model	Equipment + (net load)[1]	Torsion bar	Color[2]	dia. (mm)
Flat bed, Double cab	short wheelbase 1000 kg long wheelbase 1000 kg	701 411 103 G/104 G	green	25.3
	short wheelbase 1200 kg long wheelbase 1200 kg	701 411 103 H/104 H	brown	26.4
	short wheelbase 1000 kg long wheelbase 1000 kg	701 411 103 H/104 H	brown	26.4
Campers	short wheelbase 1000 kg long wheelbase 1000 kg	701 411 103 G/104 G	green	25.3

[1] With standard suspension/shock absorbers

[2] color code: 75 mm from flange (2.95 in.)

Weld nut in long member, repairing

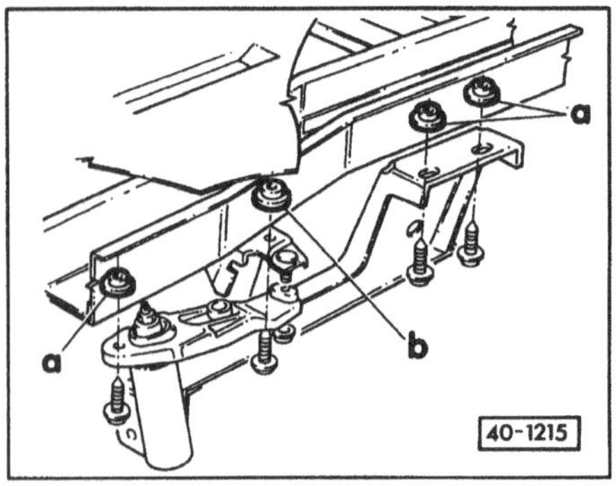

If the threads of a weld nut in the long member are damaged, they can be repaired with Heli-coil® thread inserts.

– Remove subframe, ⇒ Page M 40-10.

◄ – Drill through the damaged welded nut -a- with a 14.5 mm drill and -b- with a 16.5 mm drill.

CAUTION!
Wear protective goggles.

Notes:

- *Do not tilt drill when drilling.*
- *Only use inserts with a thread pitch of 1.5.*
- *Install Heli-coil® threaded inserts according to manufacturer's instructions.*

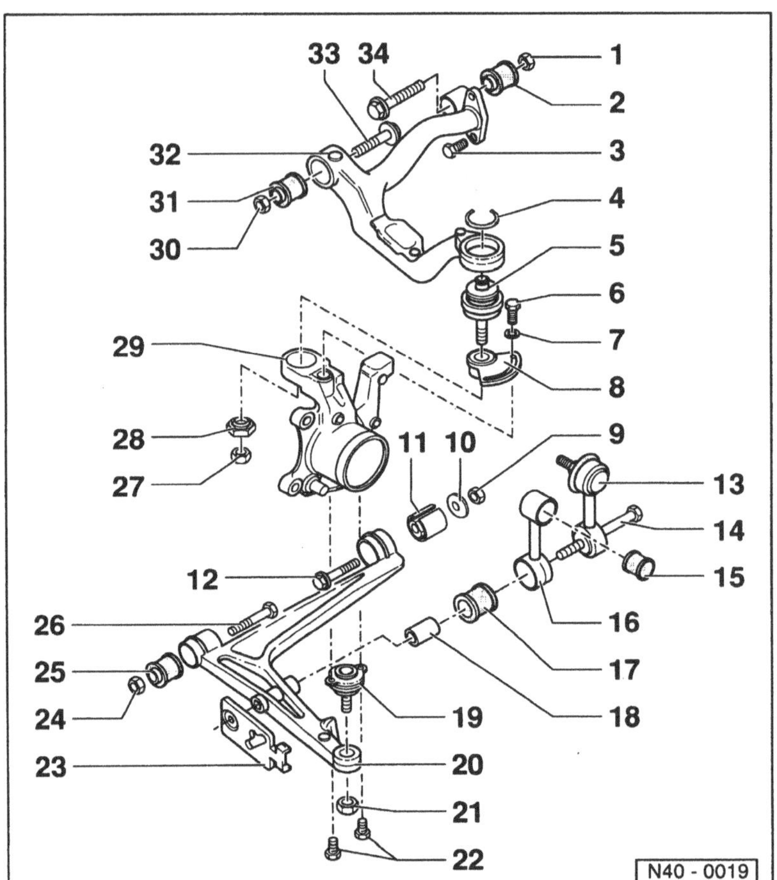

Control arms, removing and installing

1 - Hex nut, 100 Nm (74 ft lb)

2 - Upper control arm rear bushing
- Pressing out, ⇒ Fig. 3.
- Pressing in, ⇒ Fig. 4.

3 - Hex bolt
- M 14 x 1.5 x 42

4 - Circlip
- Pry out with a screwdriver.
- Always replace.

5 - Ball joint, upper
- Removing and installing, ⇒ Page M 40-35.

6 - Hex bolt, 60 Nm (44 ft lb)
- M 12 x 1.5 x 22

7 - Washer

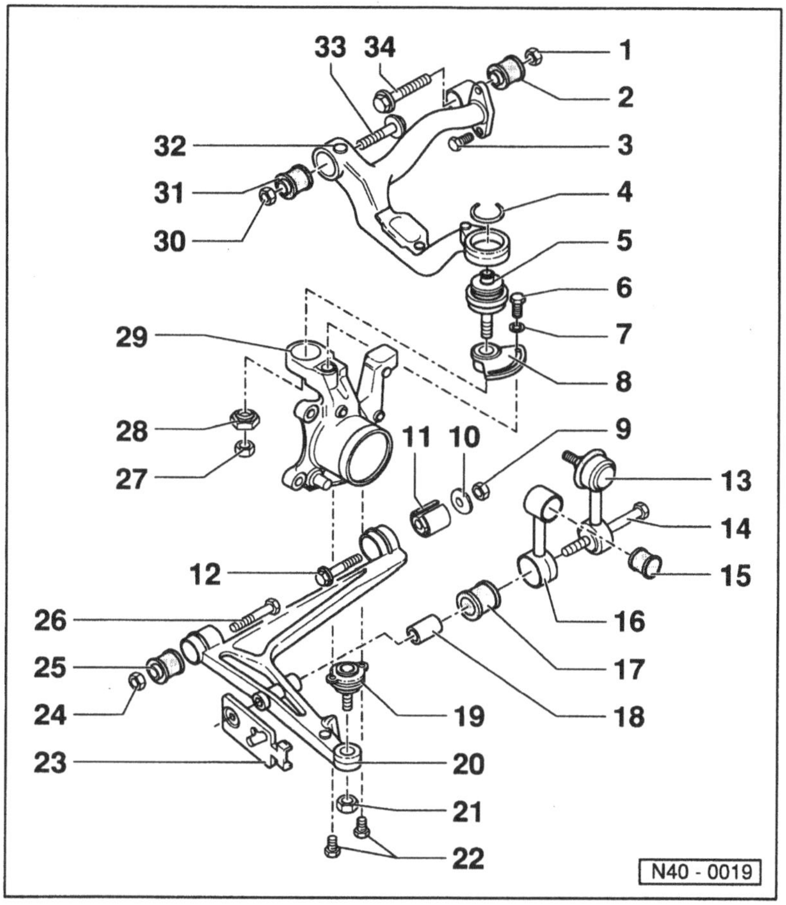

8 - Eccentric bushing
- Adjusting camber, ⇒ Page M 44-14.
- Pressing off, ⇒ Page M 40-36.
- Mark installation position before removing.

9 - Hex nut
- Up to VIN 70 M 083 341, M 12 x 1.5.
- Tightening torque: tighten to 75 Nm (55 ft lb) and turn 90° further.
- From 70 M 083 342, M 14 x 1.5. Tightening torque: 160 Nm (118 ft lb).
- Tighten with vehicle standing on its wheels.

10 - Eccentric washer
- To VIN 70 M 083 341, 12 mm diameter.
- From VIN 70 M 083 342, 14 mm diameter.

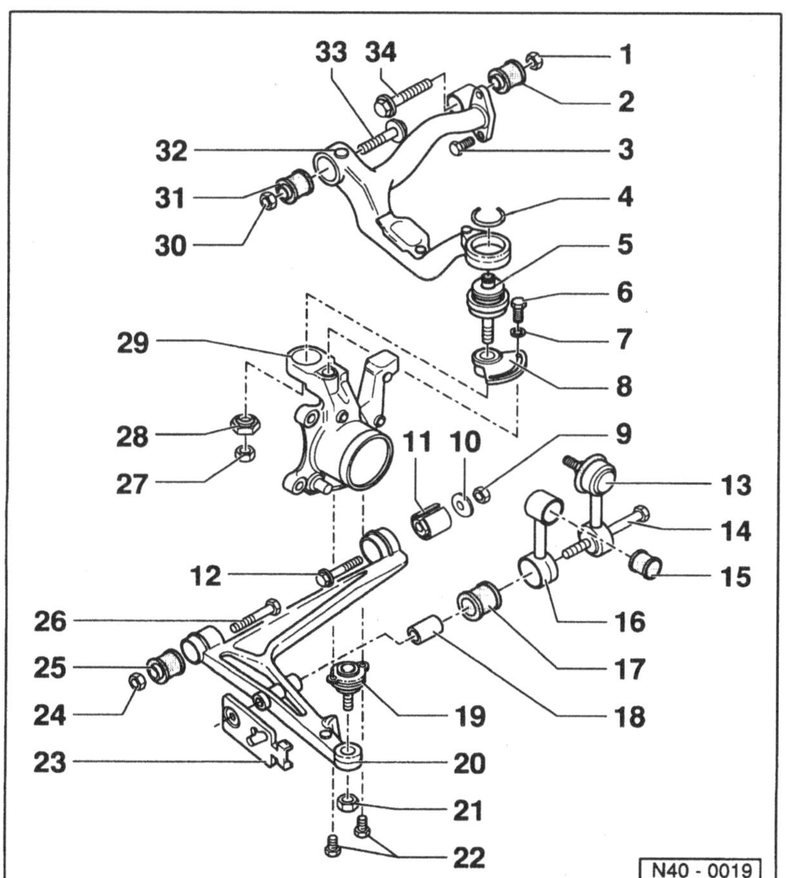

11 - Lower control arm rear bushing
- To VIN 70 M 083 341, aluminum bushing 12 mm diameter
- From VIN 70 M 083 342, aluminum bushing 14 mm diameter.
- Pressing out, ⇒ Fig. 7.
- Installation position, ⇒ Fig. 8.
- Pretensioning, ⇒ Fig. 9.
- Pressing in, ⇒ Fig. 10.

12 - Eccentric bolt
- To VIN 70 M 083 341, M 12 x 1.5 x 90.
- Replace if nut has been loosened.
- From VIN 70 M 083 342, M 14 x 1.5 x 90.
- Need not be replaced.
- Adjusting caster, ⇒ Page M 44-16.

13 - Connecting link
- For 27 mm diameter stabilizer bar
- Bolting to and from stabilizer bar, ⇒ Fig. 13.

14 - Hex bolt
- M 14 x 1.5 x 175.

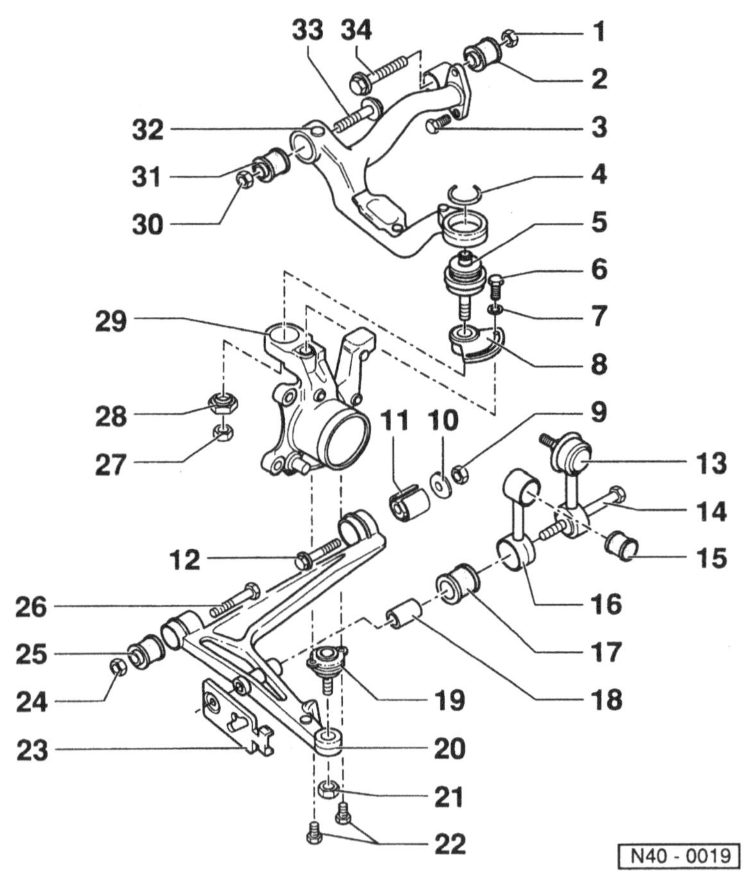

15 - Rubber mounting
- Only on stabilizer link for 23 mm dia. stabilizer bar.
- Removing and installing, ⇒ Fig. 14.

16 - Connecting link
- For 23 mm diameter stabilizer bar.
- Pressing off stabilizer bar, ⇒ Fig. 12.

17 - Rubber mounting
- Removing and installing, ⇒ Fig. 15 and Fig. 16.

18 - Bushing
- Removing and installing, ⇒ Fig. 15 and Fig. 16.

19 - Lower ball joint
- Removing and installing, ⇒ Page M 40-34.

20 - Lower control arm
- When tightening, hold bolts in installation position.

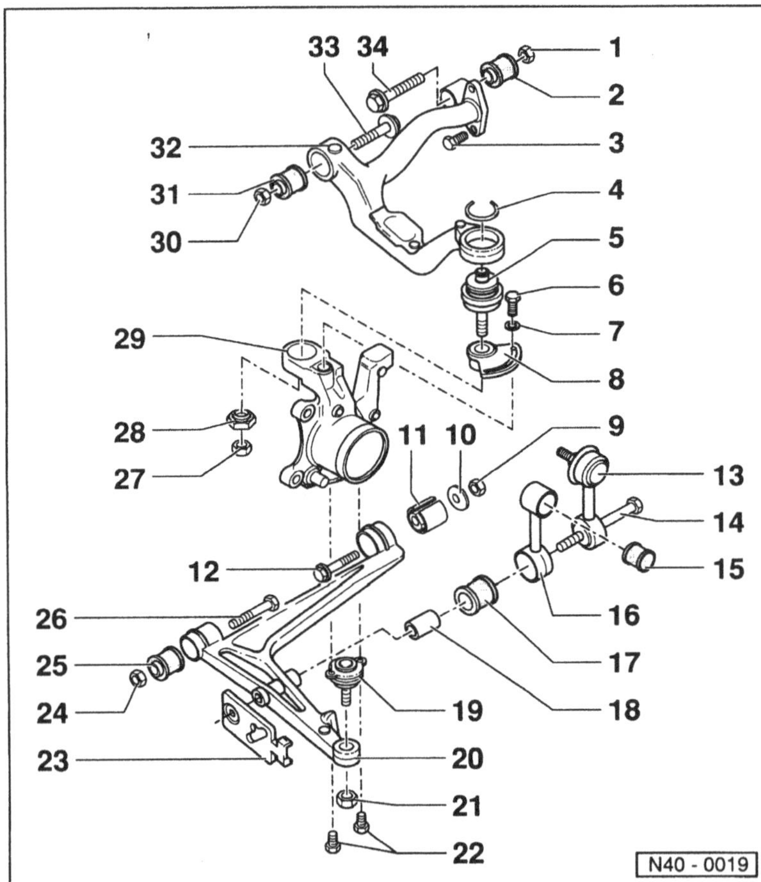

21 - Self-locking hex nut, 110 Nm (81 ft lb)

22 - Hex bolt
- Always replace.
- 55 Nm (41 ft lb) and turn 90° further.

23 - Air deflector plate
- Only on vehicles with 14" wheels

24 - Hex nut,160 Nm (118 ft lb)

25 - Lower control arm front bushing
- Pressing out, ⇒ Fig. 5.
- Pressing in, ⇒ Fig. 6.

26 - Hex bolt
- M 14 x 1.5 x 82.

27 - Self-locking hex nut, 110 Nm (118 ft lb)

28 - Eccentric washer

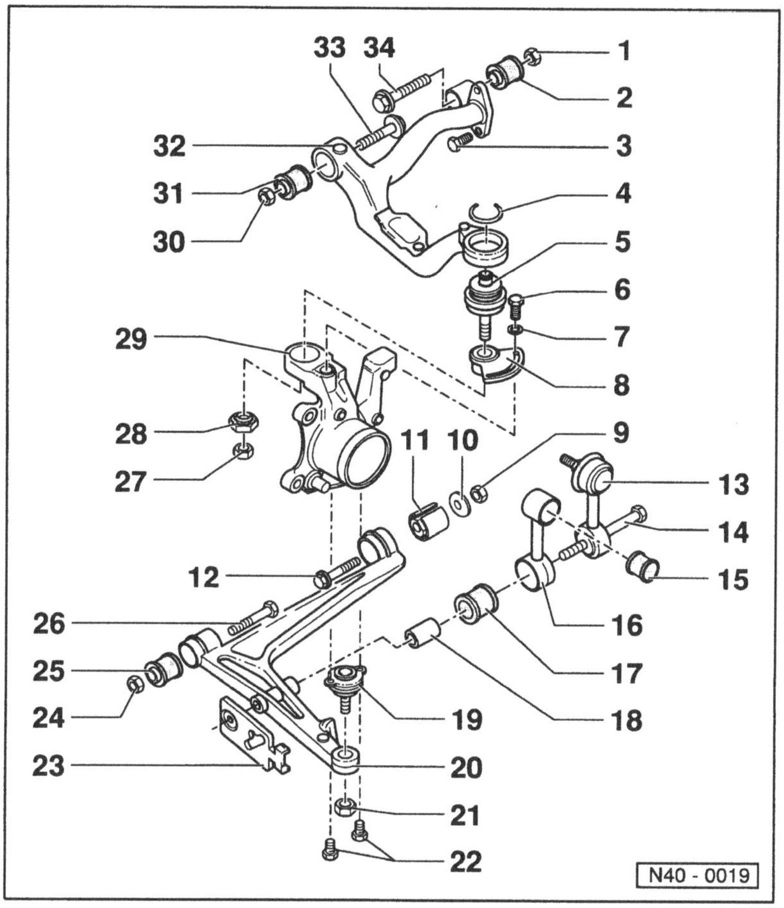

29 - Wheel bearing housing
- Removing and installing, ⇒ Page M 40-34.

30 - Hex nut, 100 Nm (74 ft lb)

31 - Upper control arm front bushing
- Pressing out, ⇒ Fig. 1.
- Pressing in, ⇒ Fig. 2.

32 - Upper control arm
- Can only be removed when subframe has been removed.
- When tightening, hold bolts in installation position.

33 - Hex bolt
- M 14 x 1.5 x 80.

34 - Hex bolt
- M 14 x 1.5 x 80.

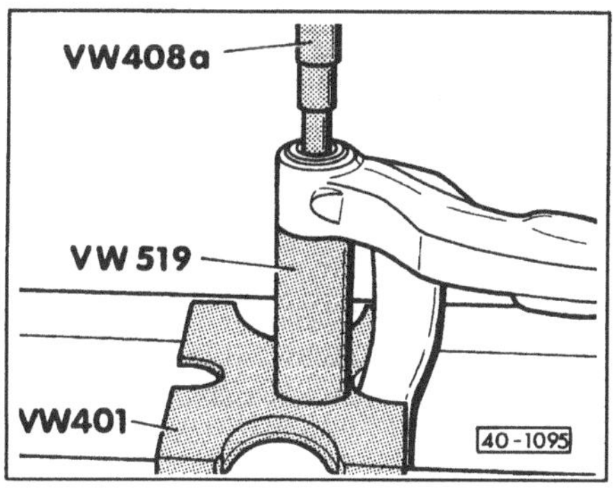

◄ **Fig. 1 Upper control front bushing, pressing out**

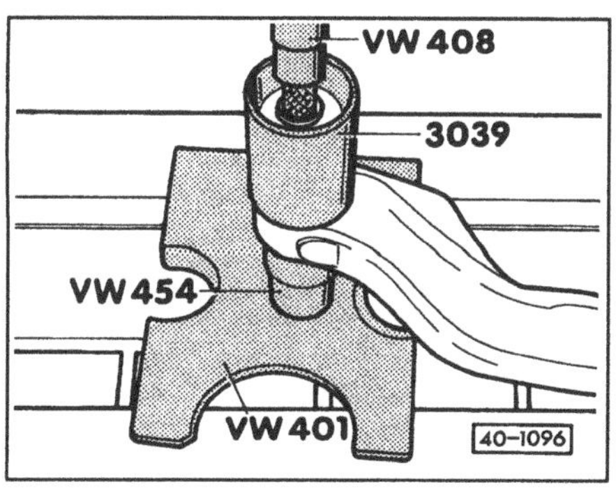

◄ **Fig. 2 Upper control arm front bushing, pressing in**

– Press in until shoulders of bushing project evenly on both sides.

Note:

Before pressing in, coat with lubricant, e.g., soft soap.

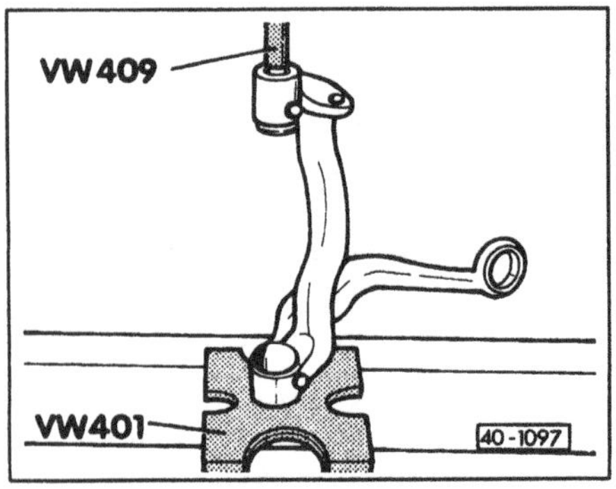

◄ **Fig. 3 Upper control arm rear bushing, pressing out**

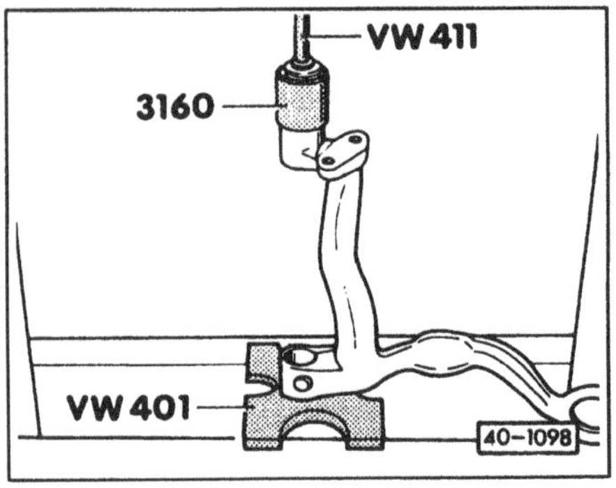

◄ **Fig. 4 Upper control arm rear bushing, pressing in**

– Press in until shoulders of bushing project evenly on both sides.

Note:

Before pressing in, coat with lubricant, e.g., soft soap.

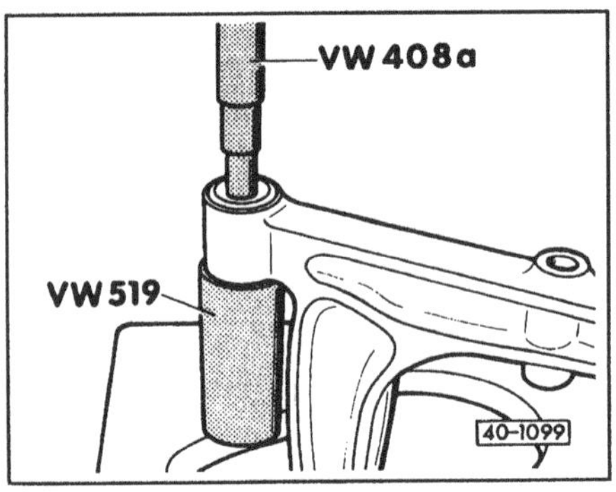

◄ **Fig. 5 Lower control arm front bushing, pressing out**

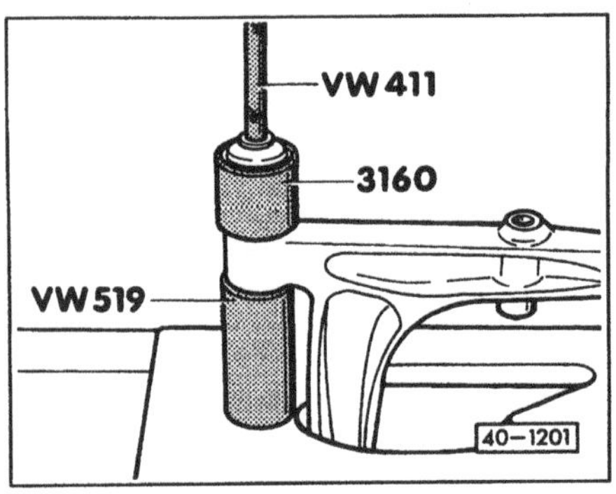

◄ **Fig. 6 Lower control arm front bushing, pressing in**

– Press in until shoulders of bushing project evenly on both sides.

Note:

Before pressing in, coat with lubricant, e.g., soft soap.

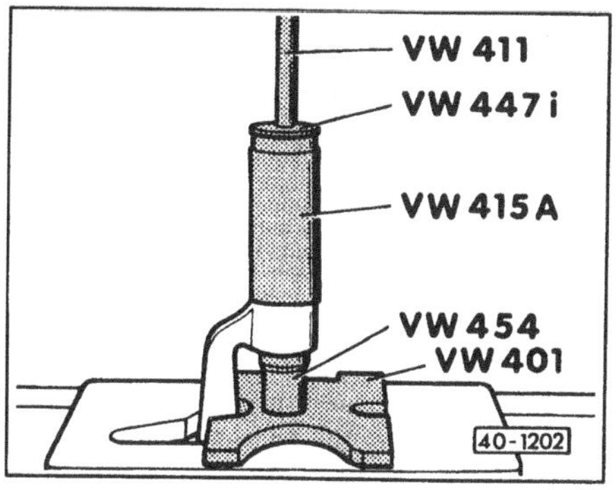

◄ **Fig. 7 Lower control arm rear bonded rubber bushing, pressing out**

– First mark installation position, e.g., by punch mark ⇒ arrow B on Fig. 8.

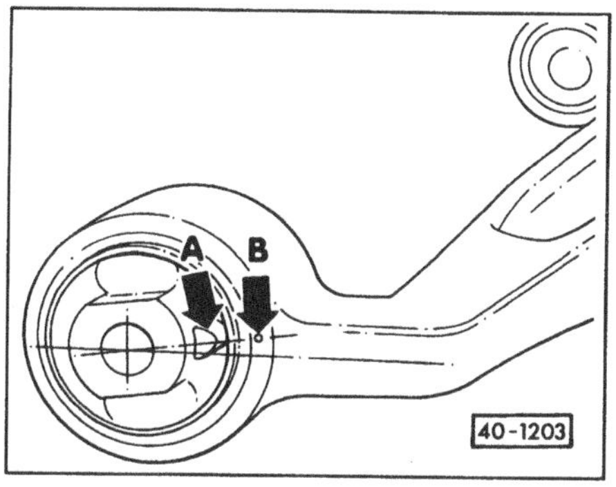

◄ **Fig. 8 Rear bonded rubber bushing, installation position**

The bonded rubber bushing is to be pressed in so that the lug is offset upwards, 11° to the left and right-hand control arm symmetrical axis.

Note:

Hold the control arm in installation position.

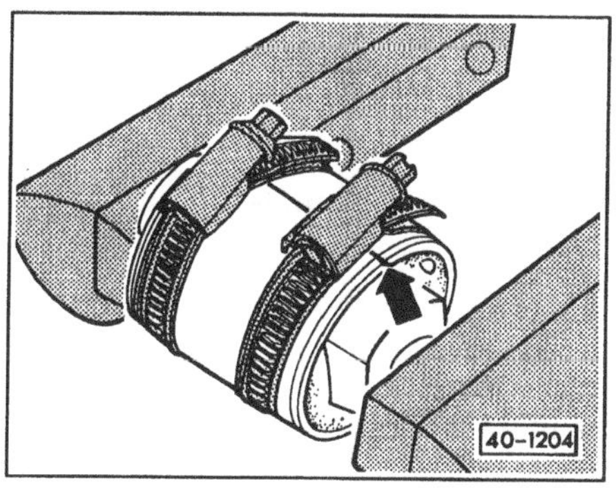

◄ **Fig. 9 Lower control arm rear bonded rubber bushing, pre-tensioning**

– First, pre–tension bonded rubber bushing with hose clamps, as shown, so that two metal shells of bushing are touching (arrow).

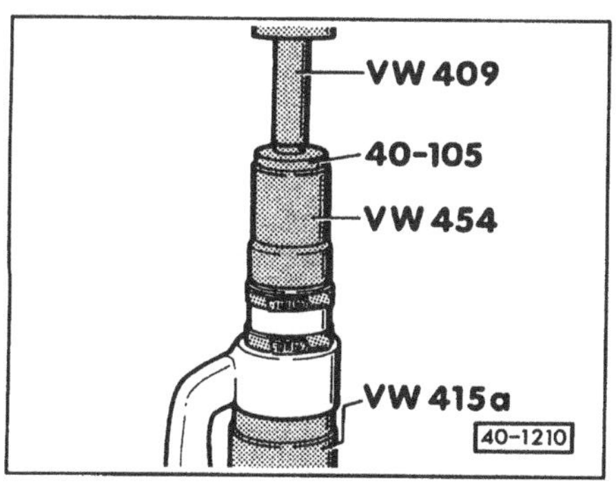

◄ **Fig. 10 Rear bonded rubber bushing, pressing into lower control arm**

– Press pre–tensioned bonded rubber bushing (⇒ Fig. 9) into installation position (⇒ Fig. 8) far enough so that a clearance of 371 ± 1 mm (14.60 ± 0.039 in.) exists between front rubber bushing inner tube and rear bonded rubber bushing inner tube ⇒ Fig. 11.

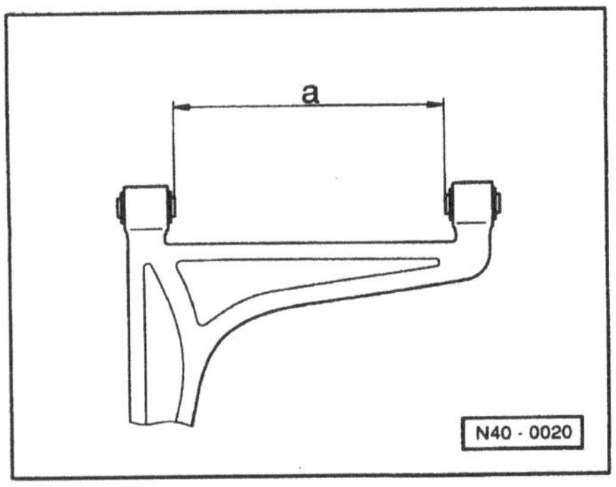

◄ **Fig. 11 Lower control arm bonded rubber bushing, installation dimension**

Dimension a = 371 ± 1 mm (14.60 ± 0.039 in.)

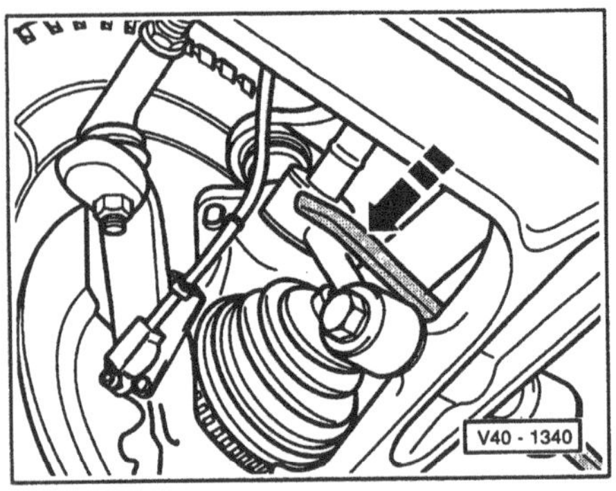

◄ **Fig. 12 Link, pressing off stabilizer bar (23 mm dia.)**

– Coat inner diameter of link rubber bushing with lubricant, e.g., soft soap, and press link off stabilizer bar.

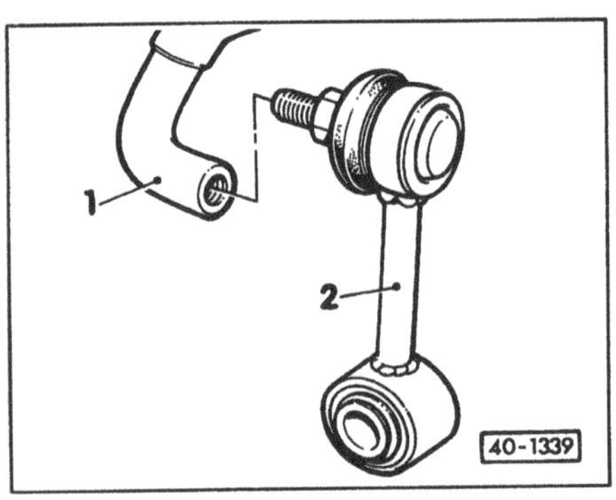

◄ **Fig. 13 Stabilizer link, removing from stabilizer bar (27 mm dia.)**

1 - Heavy-duty stabilizer bar, 27 mm diameter.
2 - Link.

Link to stabilizer bar tightening torque: 100 Nm (74 ft lb).

Note:
After installing link, ensure that link boots are not twisted.

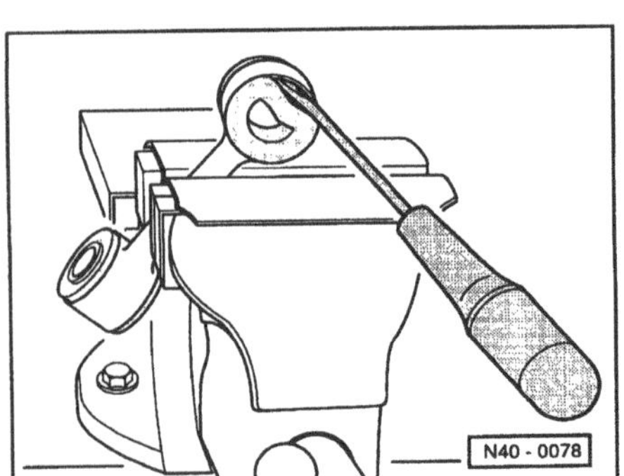

◄ **Fig. 14 Stabilizer link rubber bushing, removing and installing from stabilizer bar**

– Pry out rubber bushing with a screwdriver as shown in illustration.

– Press new rubber bushing in.

Note:
Before pressing in, coat with lubricant, e.g., soft soap.

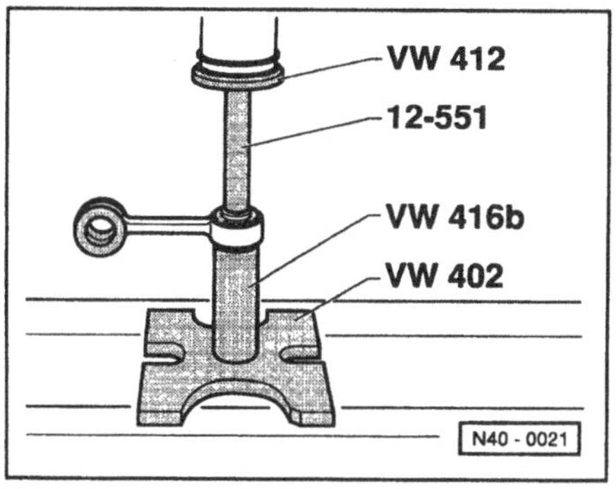

◄ **Fig. 15 Stabilizer link rubber bushing, removing from control arm end**

– Press bushing out as shown in illustration.

– Pry rubber bushing out with a screwdriver.

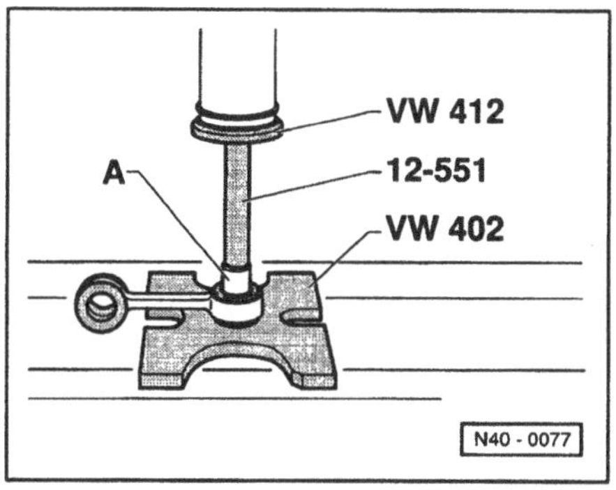

◄ **Fig. 16 Stabilizer link rubber bushing, installing in control arm end**

– Press rubber bushing into link.

– Press bushing -A- in as shown.

Note:

Before pressing in, coat parts with lubricant, e.g., soft soap.

Lower ball joint, removing and installing

Removing

– Install support 3250 to left and right-hand sides, ⇒ Page M 40-3.

– Remove bolt for shock absorber and stabilizer bar link.

– Remove bolt and separate connection between lower control arm and wheel bearing housing, ⇒ Page M 40-35.

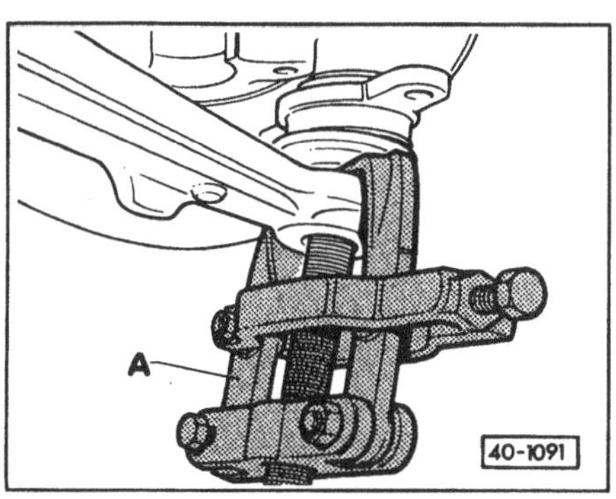

◄ – Press ball joint out.

A - Puller, commercially available, e.g., Kukko 204-2

Note:

Only use a puller with a tensioning arm.

Installing

Install in reverse order.

Upper ball joint, removing and installing

Removing

- Install support 3250 to left and right-hand sides, ⇒ Page M 40-3.
- Loosen hex bolt for axle shaft/wheel hub with vehicle standing on its wheels.

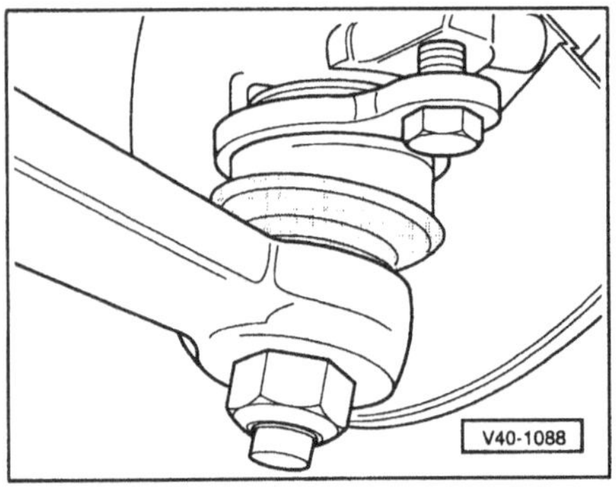

- Remove brake caliper and hang up with wire.
- ◄ Remove bolt and separate connection between wheel bearing housing and lower ball joint.
- Remove bolt for shock absorber and stabilizer bar link.
- Mark position of eccentric disc to wheel bearing housing (otherwise it will be necessary to set camber).
- Remove nut for upper ball joint.
- Remove securing bolt for eccentric disc from wheel bearing housing.
- Remove wheel bearing housing.

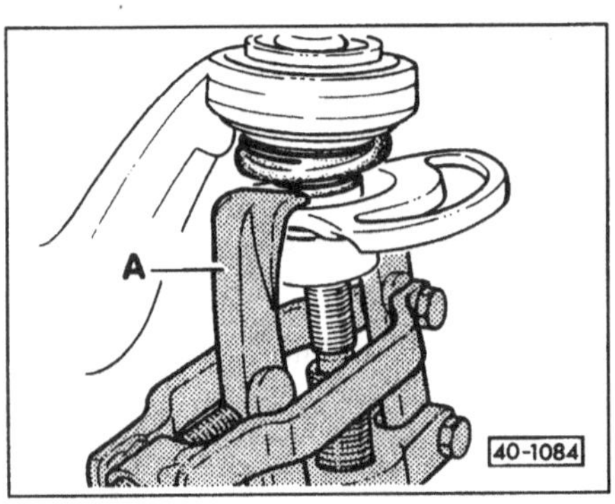

- ◄ Pull off eccentric bushing as shown in illustration.

  A - Puller, commercially available, e.g., Kukko 204-2.
- Pry out circlip with a screwdriver.

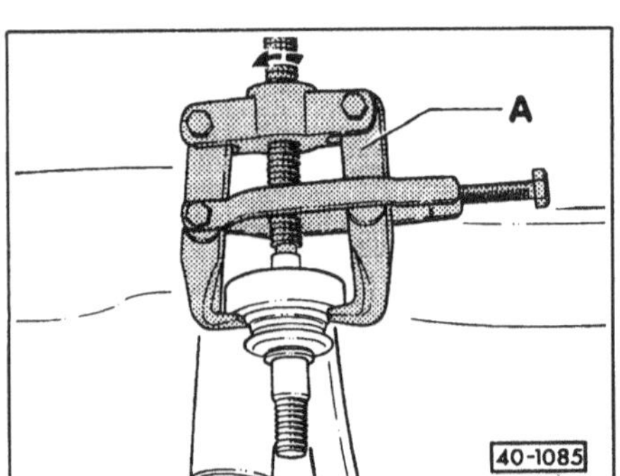

- ◄ Install puller -A- (commercially available, e.g., Kukko 204-2, with tensioning arm) as shown in illustration and press joint out.

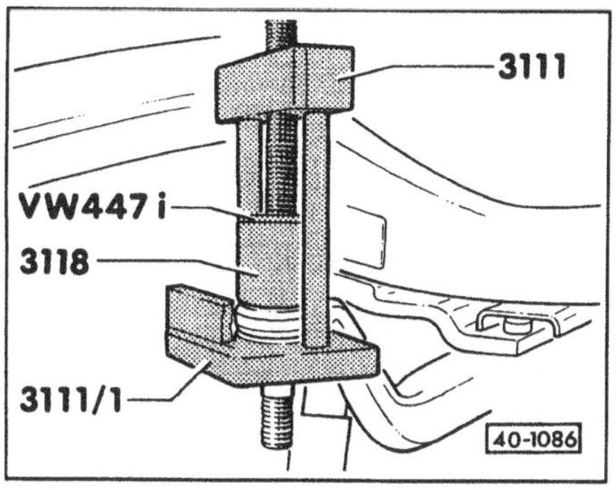

Installing

◄ – Install ball joint in control arm.

– Install special tools as shown in illustration.

– Pull ball joint in onto stop.

– Install new circlip.

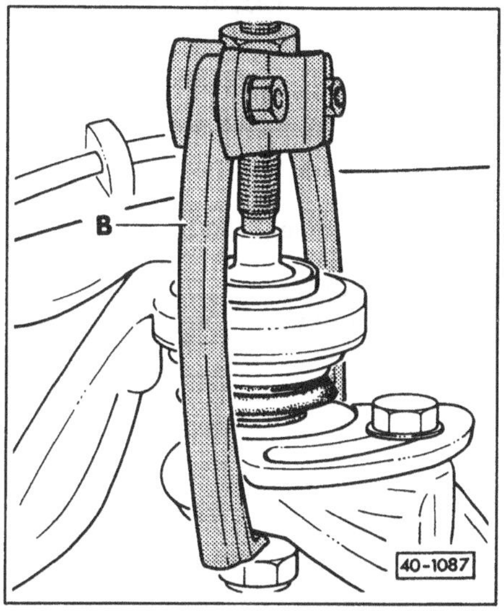

◄ – Bolt upper ball joint to wheel bearing housing.

Note:

To prevent the ball joint pin end turning when tightening the nut, ensure that the tapered surface is free of grease. Position puller -B- (commercially available e.g., Kukko 44-2) as shown in illustration, and tension only sufficiently to prevent the joint pin from turning when tightening the nut.

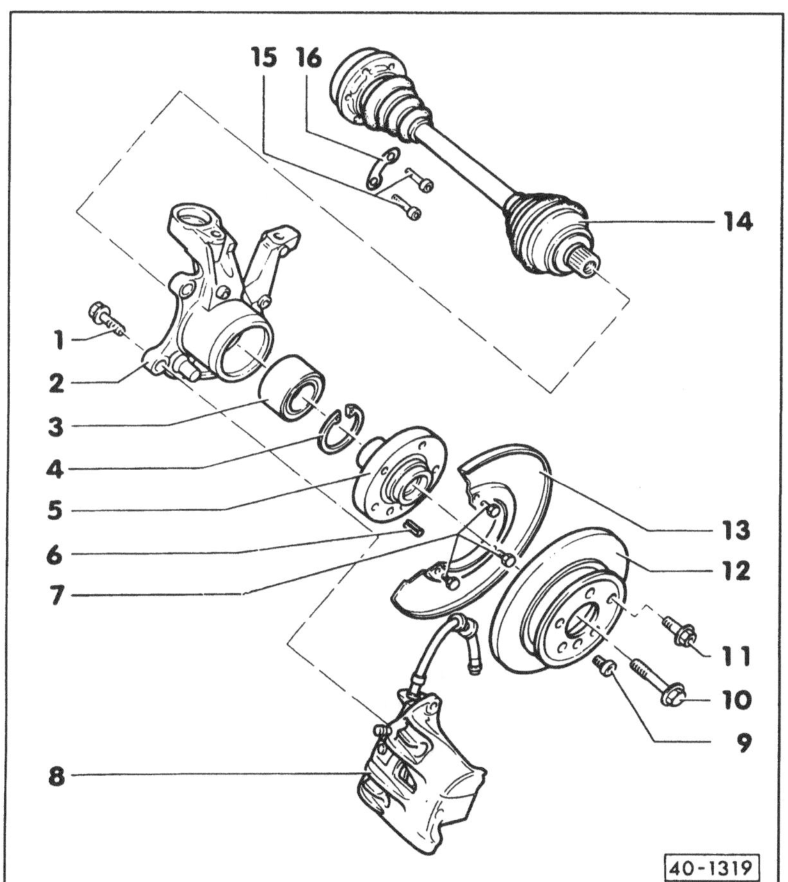

Wheel bearing and axle shaft

Notes:

- *If the vehicle has to be moved after removing an axle shaft, first install an outer CV joint in place of the axle shaft or the wheel bearing will be damaged.*
- *Do not weld or straighten load bearing suspension components or those components which locate the wheels.*
- *Always replace self-locking nuts.*
- *Always replace corroded nuts/ bolts.*

1 - Hex bolt, 270 Nm (200 ft lb)

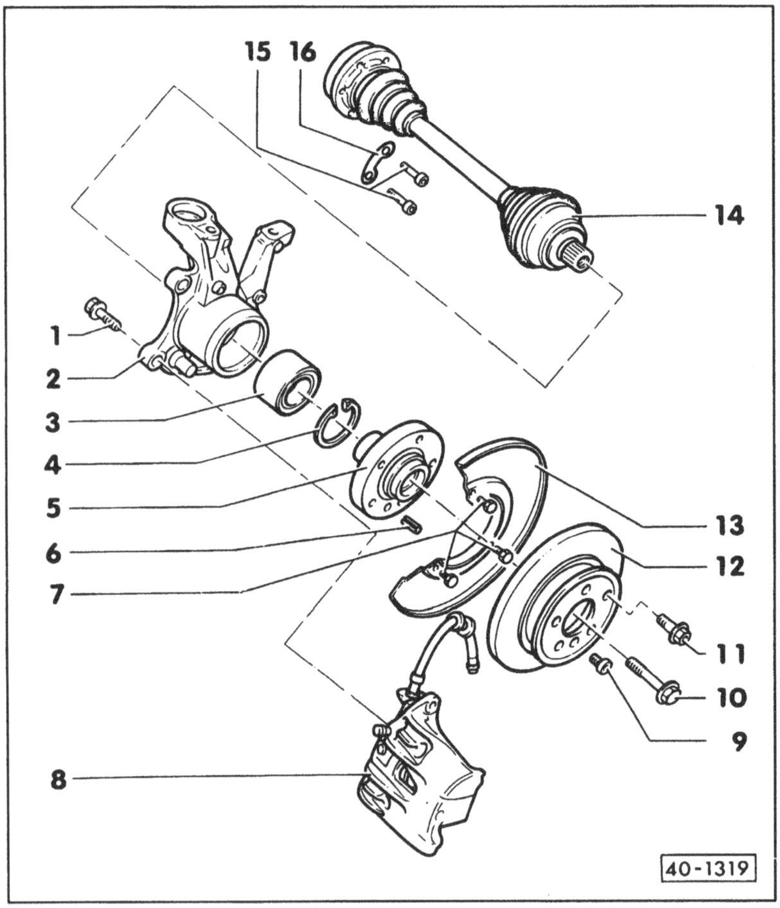

2 - Wheel bearing housing
- Removing and installing, ⇒ Page M 40-44.
- Identifying modification/ differences, ⇒ Page M 40-46.

3 - Wheel bearing
- Pressing out, ⇒ Fig. 2.
- Pressing in, ⇒ Fig. 4.

4 - Circlip
- Ensure correct seating.

5 - Hub
- Pressing out, ⇒ Fig. 1.
- Pressing in, ⇒ Fig. 5.

6 - Spring pin

7 - Hex bolt, 8 Nm (70 in. lb)

8 - Brake caliper
- Repair instructions, ⇒ Repair group 46 and 47.

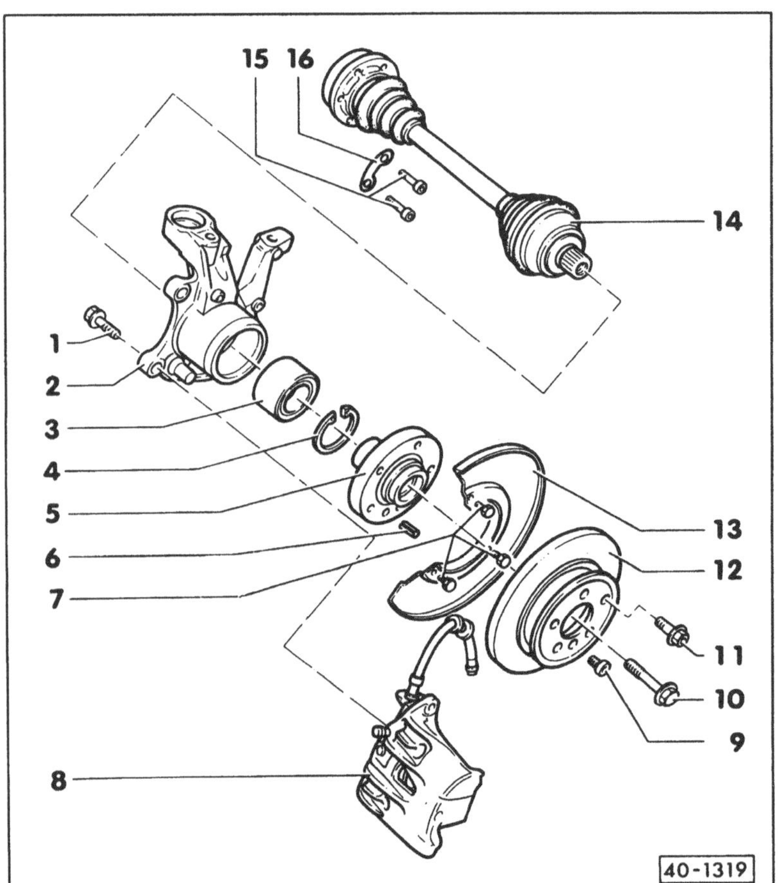

9 - Hex socket head bolt, 5 Nm (44 in. lb)
- Replace each time after removing.

10 - Hex bolt, 200 Nm (148 ft lb)
- Loosen and tighten only when vehicle is standing on its wheels (danger of accident).

11 - Wheel bolt, 160 Nm (118 ft lb)

12 - Brake disc
- Repairing, ⇒ Page 46–4 and 46–11.

13 - Splash shield

14 - Axle shaft
- Removing and installing, ⇒ Page M 40-50.

15 - Socket-head multi-point bolt, 80 Nm (59 ft lb)

16 - Backing plate

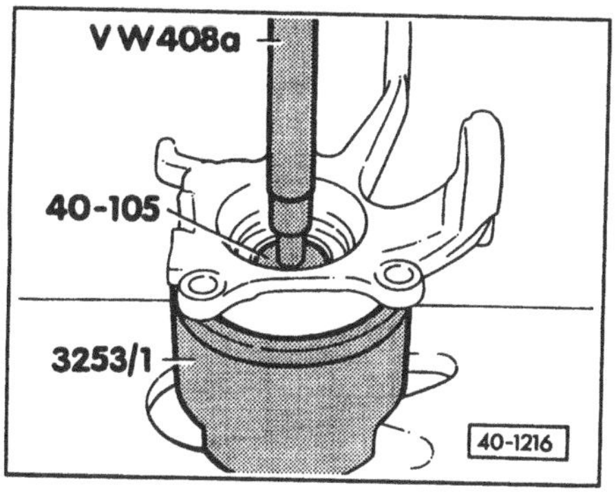

◄ **Fig. 1** **Wheel hub, pressing out**

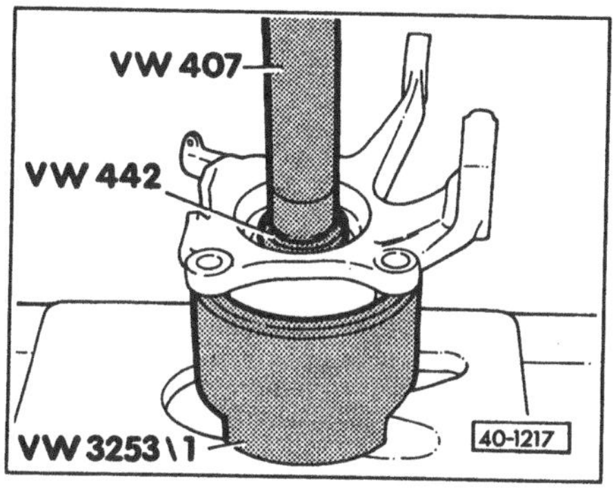

◄ **Fig. 2** **Wheel bearing, pressing out of wheel bearing housing**

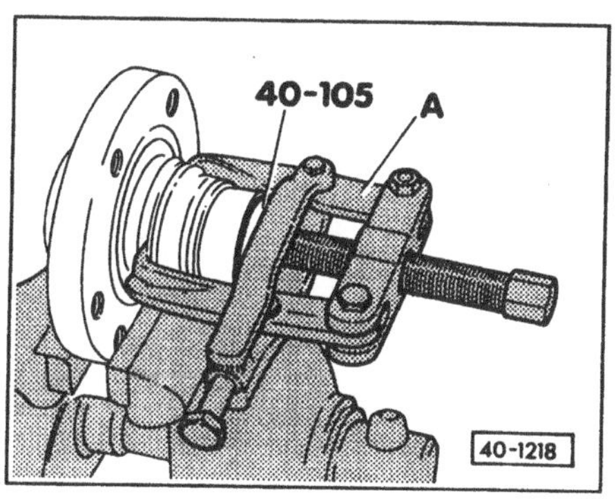

◄ **Fig. 3** **Wheel bearing inner race, pulling off of hub**

Only use puller with clamp, e.g., Kukko 204-2 (commercial type).

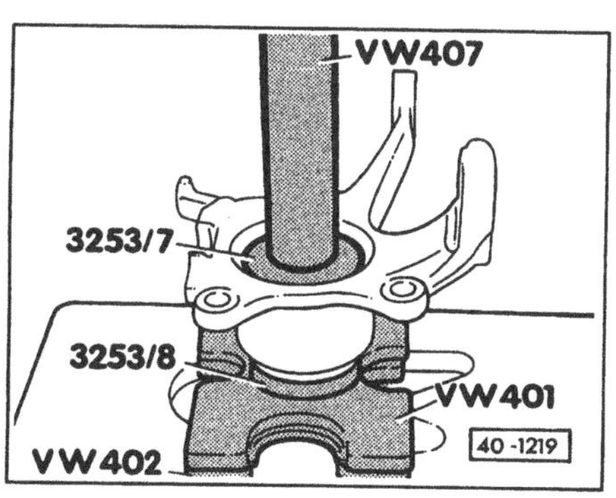

◄ **Fig. 4** **Wheel bearing, pressing into wheel bearing housing**

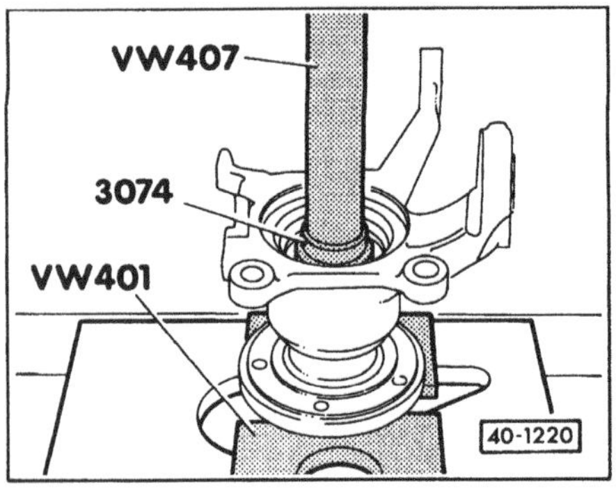

◄ **Fig. 5 Wheel hub, pressing into wheel bearing housing**

Notes:

Press in wheel hub up to stop - otherwise bearing play may result.

Wheel bearing housing, removing and installing

Removing

– Install support 3250 to left and right-hand sides, ⇒ Page M 40-3.

– Loosen hex bolt for axle shaft when vehicle is standing on its wheels.

– Remove wheel and raise vehicle.

– Remove bolt for stabilizer bar link/lower shock absorber.

– Remove brake caliper and tie to body with wire.

– Remove brake disc and cover plate.

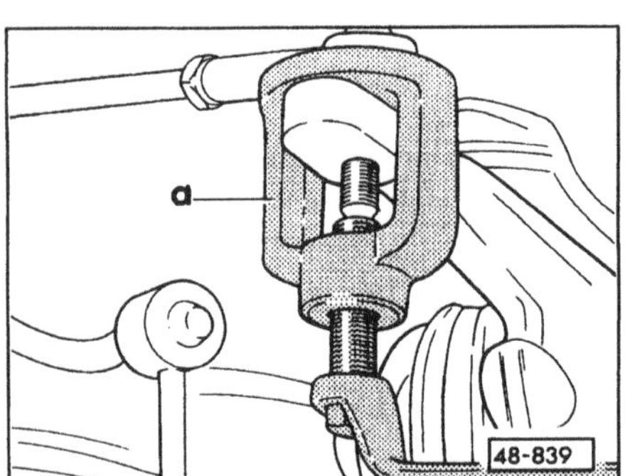

◄ – Press track rod off steering arm.

a - Ball joint splitter (commercial type), e.g., Kukko No. 128/2.

– Remove bolts for wheel bearing housing/lower ball joint, ⇒ Page M 40-35.

– Mark position of eccentric bushing to wheel bearing housing (e.g., with felt tipped pen), otherwise front axle must be realigned.

– Remove bolts for wheel bearing housing/upper ball joint.

– Remove wheel bearing housing.

Installing

Install in reverse order.

Modifications/changes to wheel bearing housing

1. Vehicles with and without power steering

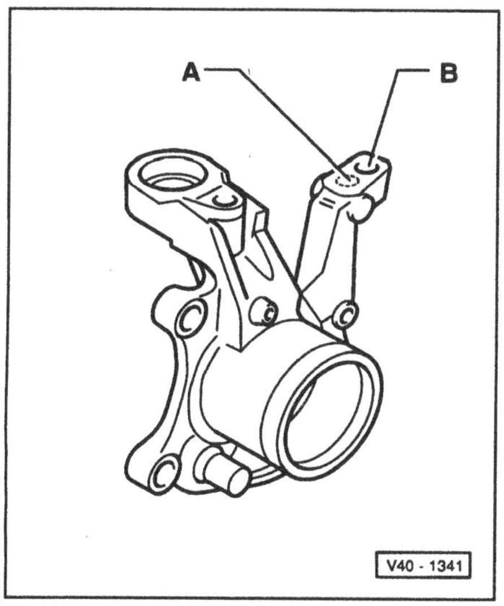

◄ On vehicles without power steering, the hole for the tie rod on the wheel bearing housing steering arm, is approximately 20 mm (0.79 in.) further outward than on vehicles with power steering.

A - Power steering tie rod mounting point

B - Non-power steering tie rod mounting point

Note:

When servicing, ensure that the correct wheel bearing housing is used. Otherwise the specifications for vehicle alignment are outside the tolerances and the steering arm stop in the steering gear will not function.

2. Modified wheel bearing housing from VIN 70 MH 096 449

The tie rod mounting point hole on the wheel bearing housing is increased to 16 mm. Only wheel bearing housings with the enlarged tie rod mounting points are supplied as spare parts. The correct tie rod end must also be used. This modification affects vehicles with and without power steering, also vehicles with and without anti-locking brake system.

Note:

A mixed installation, e.g., earlier left-hand wheel bearing housing with earlier tie rod end and right-hand new wheel bearing housing with new tie rod end is permitted.

3. Wheel bearing housing with modified tapered hole for tie rods from VIN 70 SH 002 792

> ***CAUTION!***
>
> ***The earlier wheel bearing housing must not be installed with the new tie rod end or the new wheel bearing housing installed with the earlier tie rod end.***

Before a new wheel bearing housing or a new tie rod end is installed, check dimensions -a- and -b- with a vernier caliper.

Note:

During repair, a mixed installation on vehicles prior to build date 09/94 (70 RH ...), e.g., left-hand earlier wheel bearing housing with earlier tie rod and right-hand new wheel bearing housing with new tie rod end is permitted.

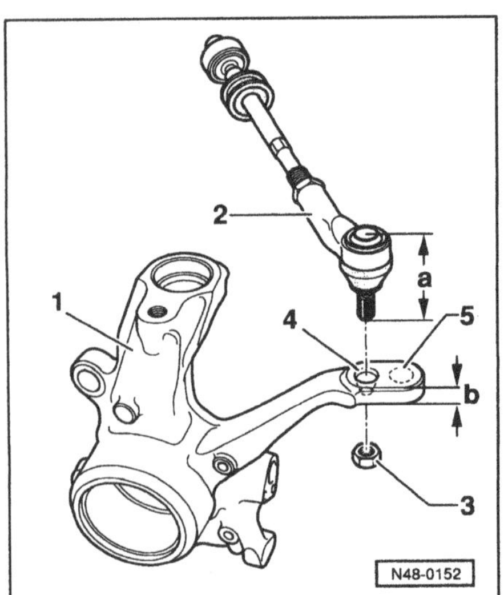

◄

1 - Wheel bearing housing:
Steering arm with modified tapered hole for tie rod end.

2 - Tie rod end:
Previous: dimension -a- 74.5 mm.
New: dimension -a- 78.5 mm, with modified taper.

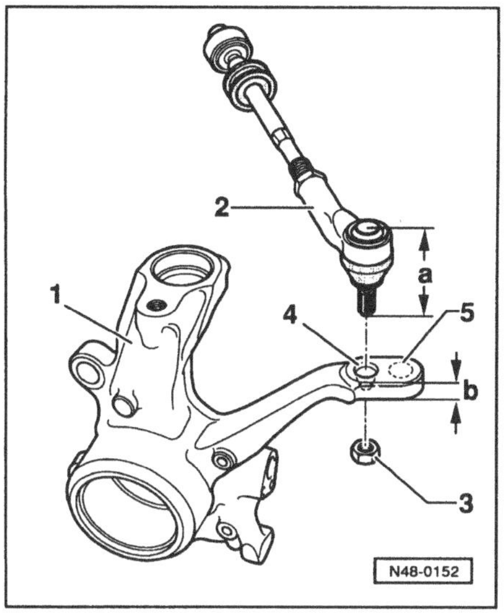

3 - Self-locking nut,
Previous: 40 Nm (30 ft lb).
New: 65 Nm (48 ft lb).

4 - Hole for tie rod end
for vehicles with power steering,
Previous: dimension -b- 18.0 mm
New: dimension -b- 22.5 mm

5 - Hole for tie rod end
for vehicles without power steering,
Previous: dimension -b- 18.0 mm.
New: dimension -b- 22.5 mm.

Axle shaft, removing and installing

Removing

Vehicles with manual transmission:

- Install support 3250 to support left and right-hand upper control arm,⇒ Page M 40-3.
- Loosen hex bolt for axle shaft/wheel hub when vehicle is standing on its wheels.
- Remove sound deadener pan.

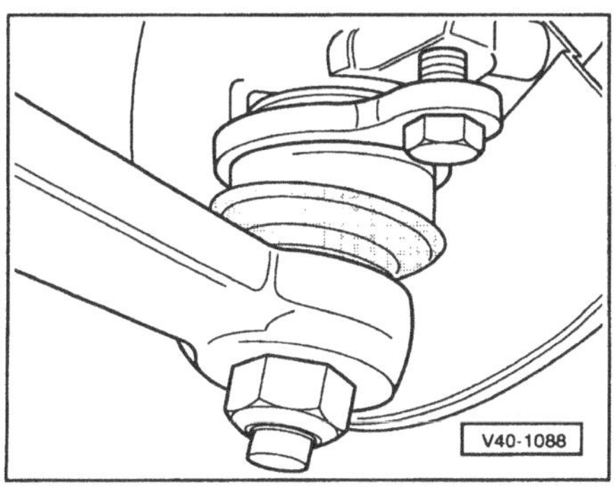

- Remove bolt for wheel bearing housing/lower ball joint.
- Remove bolt for shock absorber/stabilizer bar link.
- Separate axle shaft at transmission drive flange.
- Push shock absorber completely together.
- Remove axle shaft.

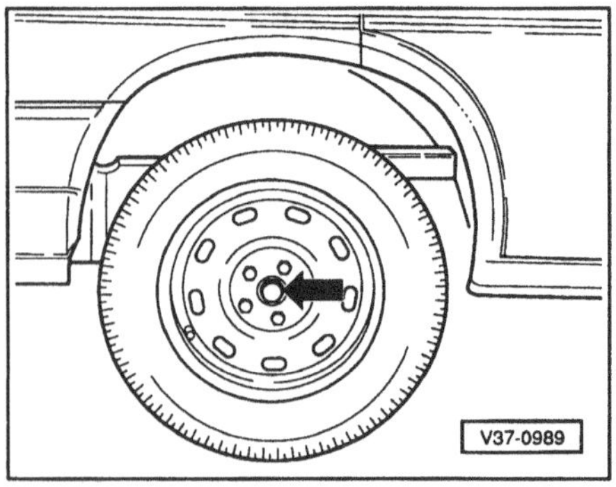

Vehicles with automatic transmission:

– Install support 3250 to support left and right-hand upper control arm, ⇒ Page M 40-3.

◄ – Loosen hex bolt for axle shaft/wheel hub when vehicle is standing on its wheels.

– Remove noise insulation tray.

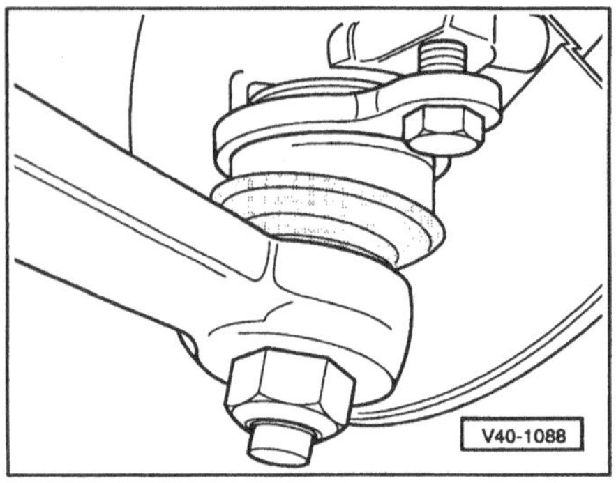

– Disconnect left-hand shock absorber from lower control arm and push completely together.

◄ – Remove bolt and separate wheel bearing housing/lower ball joint.

– Separate connector from Multi-function Switch.

– Separate axle shaft at transmission drive flange.

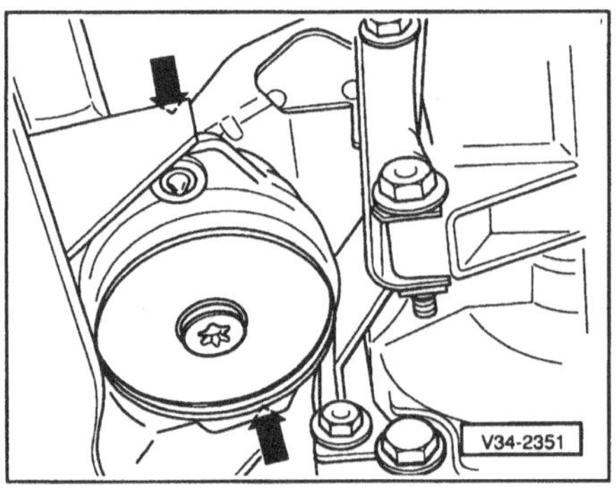

◄ – Remove bolts from rear bonded rubber mounting at subframe (arrows).

– Remove axle shaft after pushing engine/transmission slightly forward.

Installing

Install in reverse order.

Note:
EuroVans with automatic transmissions built between August 1991 and June 1992 were produced with different axle shaft bolts on the right and left sides:

Right side bolt **A**
- M10 x 48
- hardness = 12.9

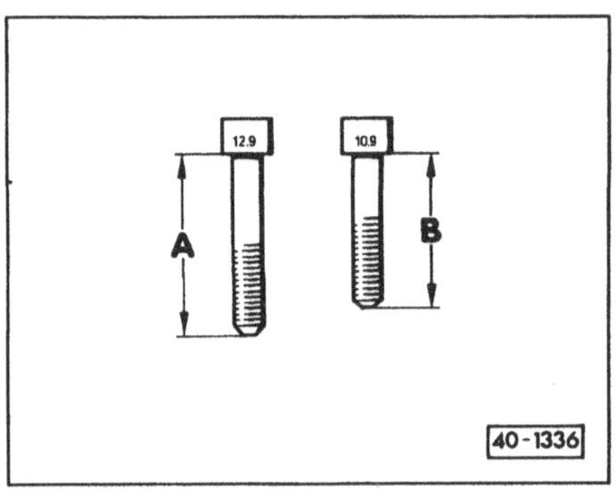

◄ Left side bolt **B**
- M10 x 45
- hardness = 10.9

> ***CAUTION!***
> ***Always use bolts of hardness 12.9 during repairs when disconnecting the axle shaft.***

– Check installed bolt **-A-** for clearance at housing.

– shorten bolt **-A-** as necessary.

Tightening torques:

Axle shaft to flange[1]	80 Nm (59 ft lb)
Wheel bearing housing to ball joint	55 Nm (41 ft lb) and turn 90° further
Wheel bolts to wheel hub	160 Nm (118 ft lb)
Shock absorber to control arm	160 Nm (118 ft lb)
Axle shaft to wheel hub	200 Nm (148 ft lb)
Bonded rubber mounting to sub-frame	45 Nm (33 ft lb)

[1] Note ⇒ Page M 40-52

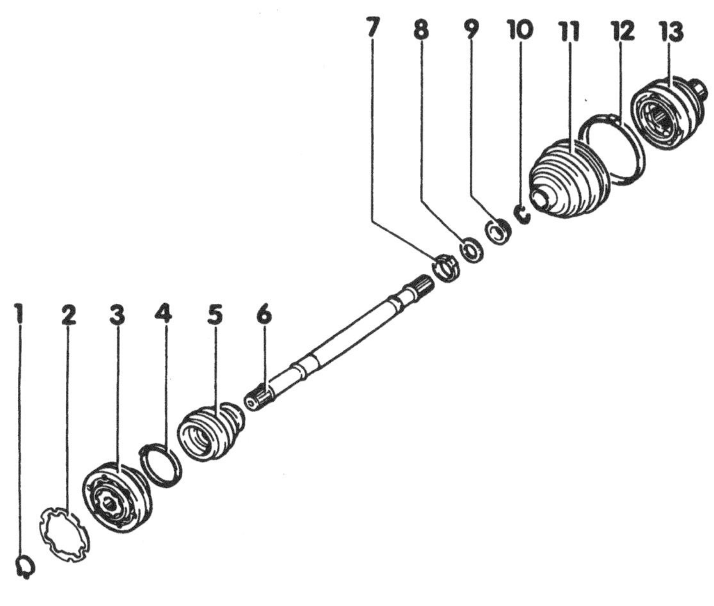

40 1208

Axle shaft, repairing

Notes:

Grease filling outer CV joints:

- *Press 120 g G-6.3 grease into joint. Re-grease joint if necessary when replacing axle shaft boot.*
- *Press approx. 80 g grease through the inner splines of the ball hub into the outer constant velocity joint.*
- *Spread approx. 40 g grease around inside of joint.*

Grease filling inner CV joints:

- *Press 120 g G-6.3 grease from each side into the joint. Re-grease joint, if necessary, when replacing axle shaft boot.*

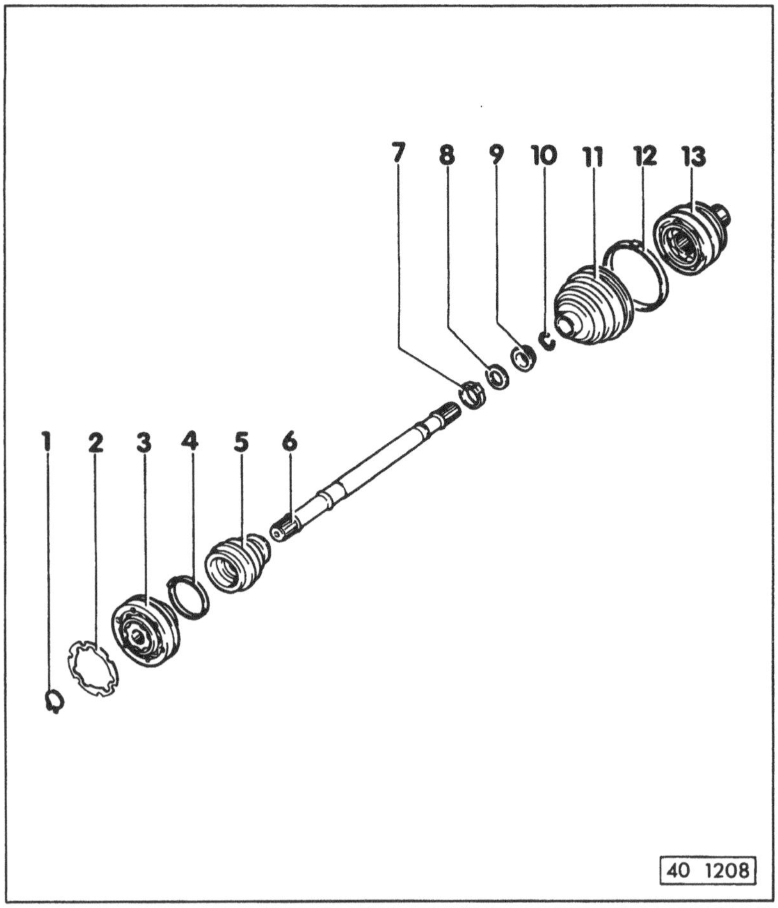

1 - Circlip
- Always replace.
- Remove and install with long-nosed pliers.

2 - Gasket
- Replacing: Pull off protective strip and stick onto joint.

3 - Inner constant velocity joint
- Only replace complete.
- Pressing off, ⇒ Fig. 5.
- Pressing on, ⇒ Fig. 6.
- Greasing,
  ⇒ Notes Page M 40-54
- Checking:
- – Clean joint.
- – Examine joint, hub, cage and balls for small indentations, (pitting) and signs of creasing.
- – Excessive play in the joint is noticeable by load–change jolts – replace joint in such cases.
- – Smooth area and signs of the balls tracking are not a reason for changing the joint.

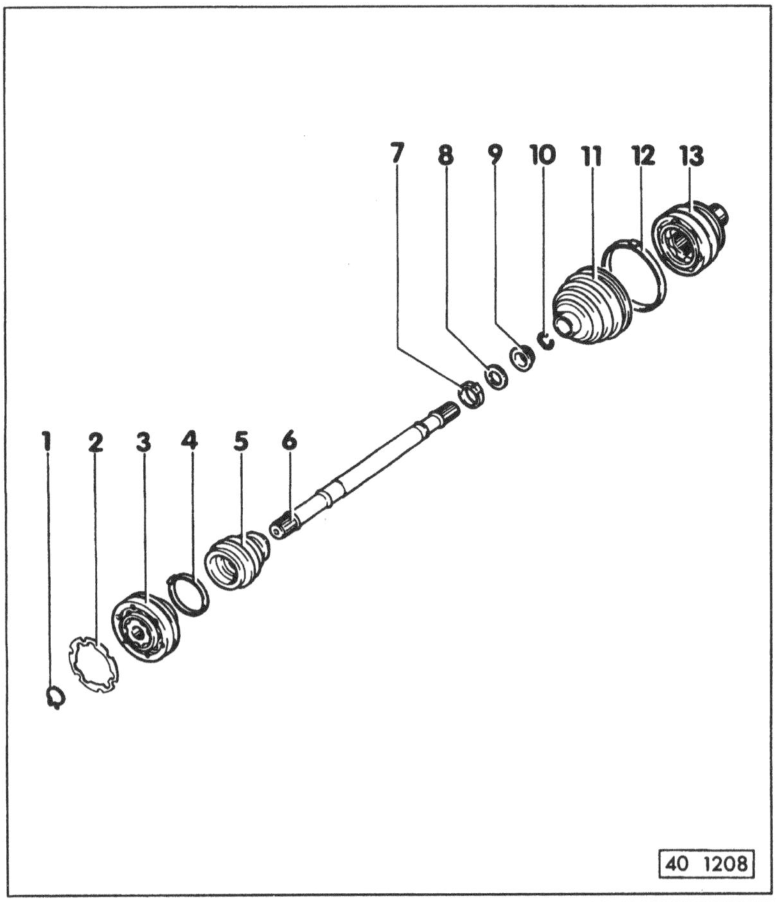

4 - Clamp
- Always replace – open with pliers.
- Tensioning, ⇒ Fig. 2.

5 - Boot for inner constant velocity joint
- Check for tears and chafing, replace if necessary.

6 - Axle shaft
- Check for paint damage.
- From VIN 70 NH 014 243, left-hand axle shaft is 10 mm shorter on vehicles with automatic transmission.
- Modification:
  From VIN 70 RH 136 291, outer constant velocity joint inner spline contour modified. Components prior to and after the above–mentioned chassis Number must not be installed assembled together.

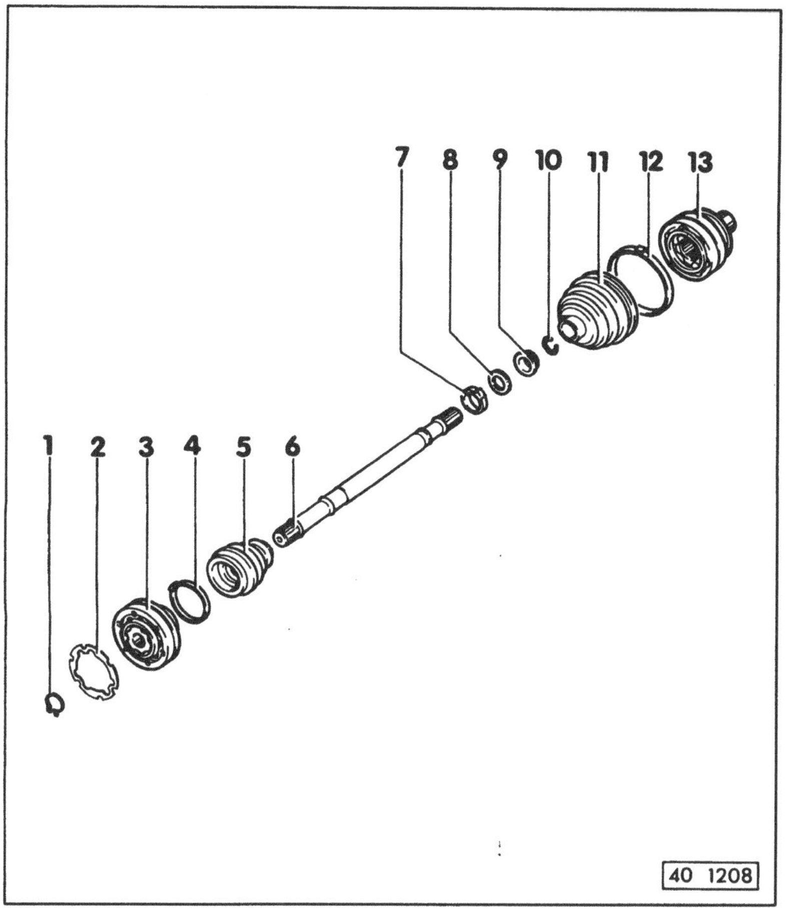

7 - Clamp
- ♦ Always replace – open with pliers.
- ♦ Tensioning, ⇒ Fig. 2.
- ♦ Modification:
  From VIN 70 RH 136 291, modified clip for protective boot is made of synthetic rubber. Tensioning, ⇒ Fig. 3.

8 - Dished washer
- ♦ Installation position: larger diameter (concave side) contacts thrust washer.

9 - Thrust washer

10 - Circlip
- ♦ Always replace.
- ♦ Insert in shaft groove.

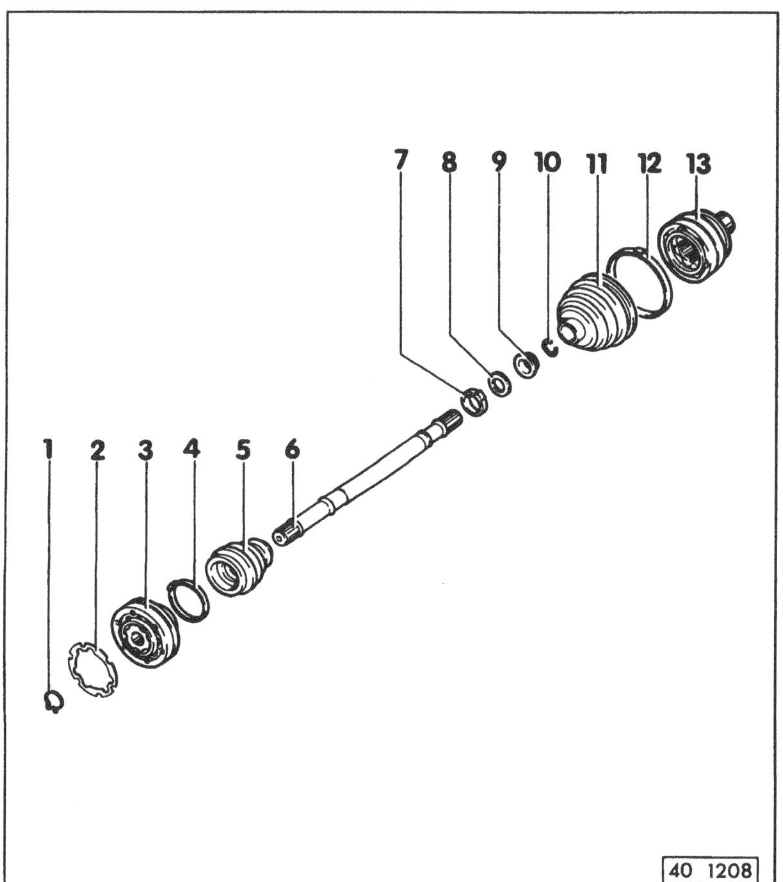

11 - Boot for outer constant velocity joint

There are two different versions:
- – Rubber version.
- – Synthetic rubber version.
- ♦ Check for tears and chafing, replace if necessary.
- ♦ To balance pressure, briefly ventilate boot before tensioning clamp, ⇒ Fig. 4.
- ♦ Modification:
  From VIN 70 RH 136 291, modified boot (synthetic rubber). Tensioning clamp for boot, ⇒ Fig. 3. Installation on axle shafts prior to above mentioned VIN is only possible when the complete axle shaft is replaced.

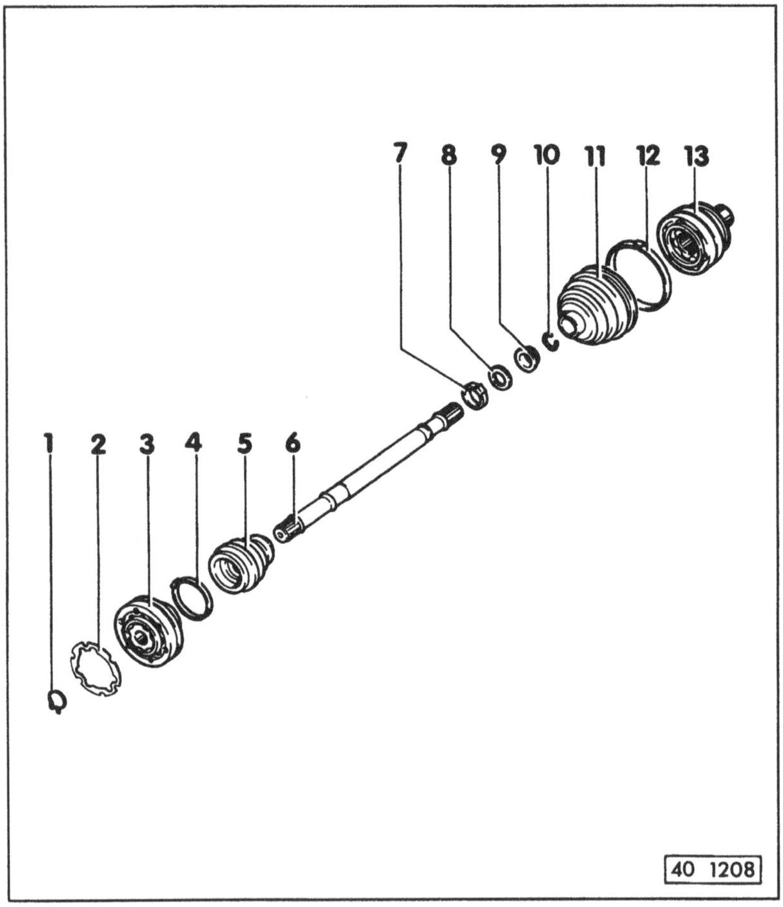

12 - Clamp

- Replace – open with pliers.
- Tensioning, ⇒ Fig. 2.
- Modification:
  From VIN 70 RH 136 291, hose clamp employed.
  Tensioning, ⇒ Fig. 3.

13 - Outer constant velocity joint

- Only replace complete.
- Removing, ⇒ Fig. 1.
- Installing: drive onto shaft with plastic hammer until compressed circlip seats.
- Greasing,
  ⇒ Notes Page M 40-54
- Checking, ⇒ Page M 40-64
- Modification:
  From VIN 70 RH 136 291, outer constant velocity joint inner spline contour modified. Components prior to and after the above mentioned VIN must not be installed assembled together.

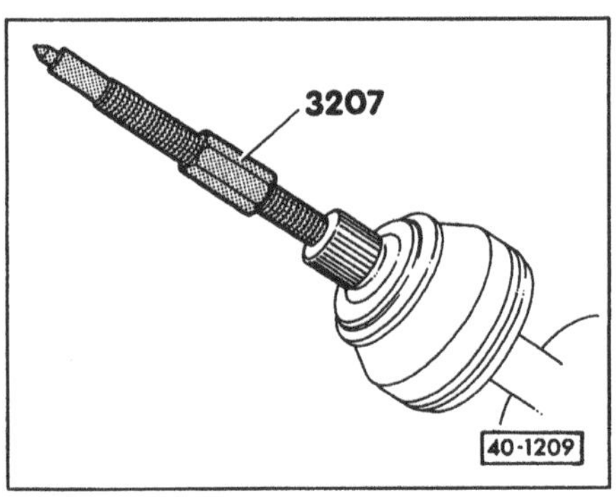

◄ **Fig. 1 Outer constant velocity joint, pressing off**

- Clamp axle shaft in vice, using vice clamps.
- Fold back boot.
- Thread special tool, using the M 16 threaded section, into the pivot pin far enough to press the constant velocity joint off the shaft.

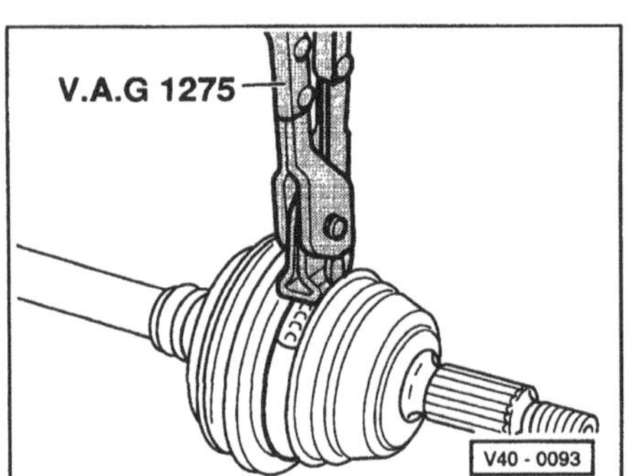

◄ **Fig. 2 Outer joint clamp on axle shafts with rubber boot, tensioning**

Use pliers VAG 1275 or equivalent.

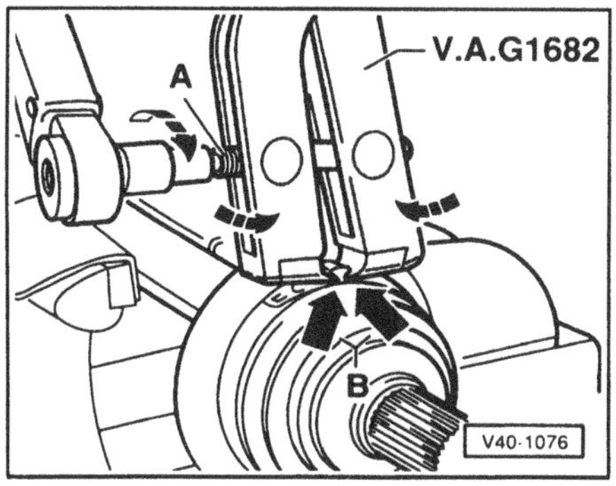

Fig. 3 Outer joint clamp on axle shafts with synthetic rubber boot, tensioning

- Position pliers VAG 1682 (Snap-On YA 3080) as shown in illustration. Ensure that jaws of pliers seat in corners (arrows B) of clamp.
- Tension clamp by turning spindle with a torque wrench (keep pliers at right angle to shaft).

Notes:

- *Due to the hardness of the boot material (compared to rubber), which makes necessary a stainless steel hose clamp, the clamp can only be properly tensioned with VAG 1682 or Snap-On YA 3080 pliers.*
- *Tightening torque: 25 Nm (18 ft lb)*
- *Use torque wrench with 5 – 50 Nm (5 – 37 ft lb) range (e.g. VAG 1331 or equivalent).*
- *Ensure that the spindle threads are not tight. If necessary, lubricate threads with MoS_2 grease*
- *If the threads are tight, e.g., dirty, the required tensioning force for the hose clamp will not be achieved despite correct torque readings.*

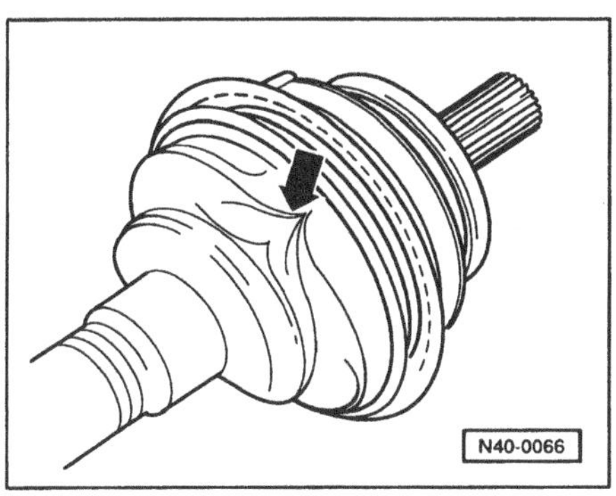

Fig. 4 CV joint boot, ventilating

The boot is often pushed in as it is installed on the joint body. This creates a vacuum inside the boot which draws in a fold (arrow) while driving. Before installing the clamp on the small diameter (inside) end of the boot, lift the boot slightly to ventilate it and equalize the pressure.

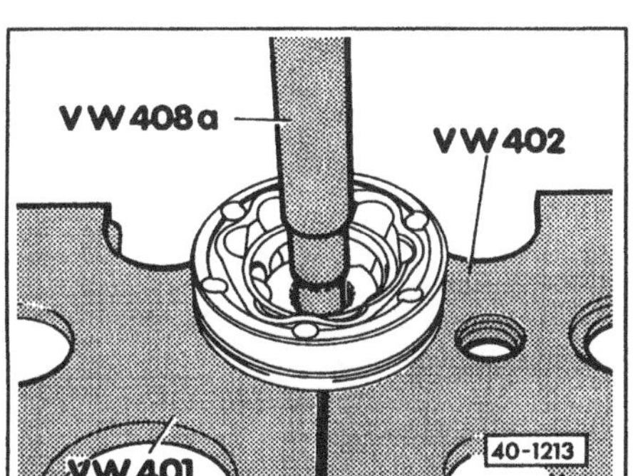

Fig. 5 Inner CV joint, pressing off

- Remove circlip with long-nosed pliers.
- Support ball hub when pressing off.

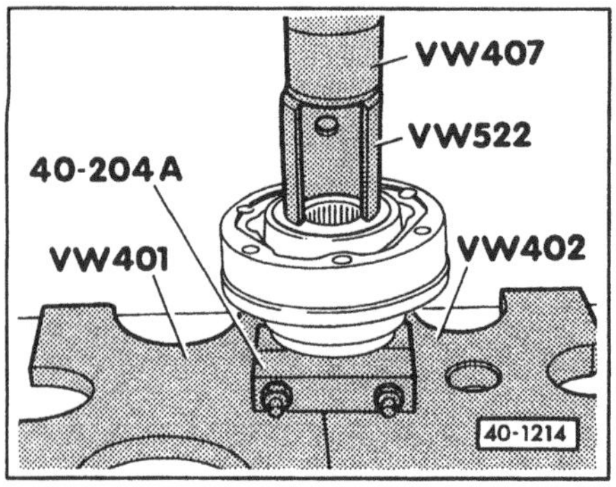

◄ **Fig. 6 Inner CV joint, pressing on**

- Press on joint up to stop.
- Install circlip.

Outer CV joint, checking

The joint should be disassembled to replace dirty grease or for checking the balls and ball tracks for wear and damage.

Removing

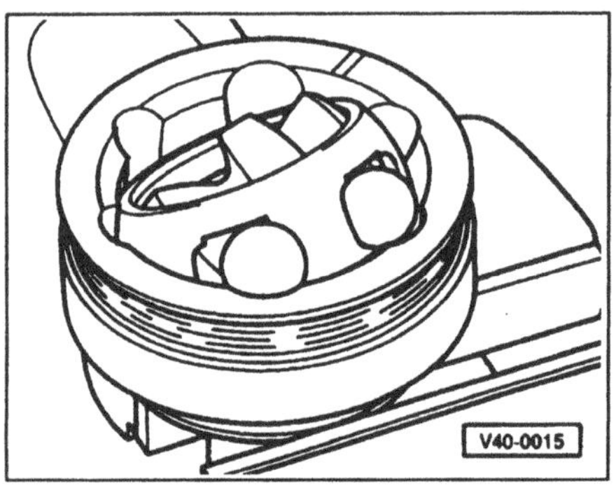

◄
- Before disassembling mark position of the ball hub in relation to the ball cage and housing with an electric scriber or oil stone.
- Swivel ball hub and ball cage.
- Remove balls one after the other.

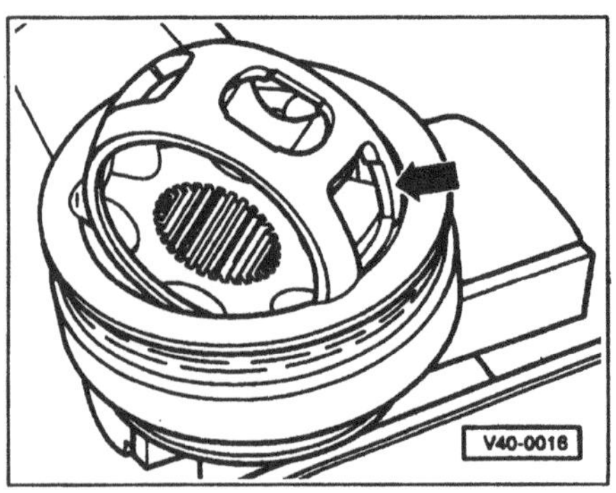

◄
- Turn cage until the two rectangular windows (arrow) are aligned with the joint housing.
- Lift out cage together with hub.

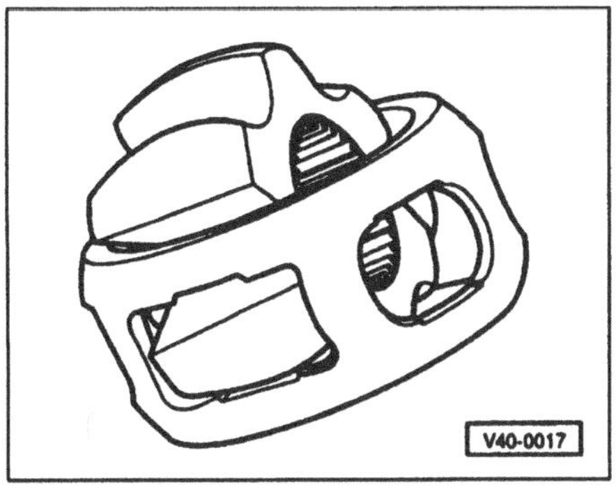

◄ – Swing segment of hub into rectangular cage window.

– Tilt hub out of cage.

Note.

- *The 6 balls for each joint belong to a tolerance group. Do not interchange.*
- *Examine axle journal, hub, cage and balls for small indentations (pitting) and signs of scoring.*
- *Excessive play in joint is noticeable by load–change jolts. Replace joint if excessive play is noticed.*
- *Do not replace joint because of a polished appearance or if ball tracks are visible.*

Installing

– Press approximately 80g of total amount of grease into joint housing.

– Install cage with hub in joint housing.

– Press in balls, one after another, from alternate sides making sure original position of hub in cage and housing is maintained.

– Install new circlip in hub.

– Press remaining grease into joint.

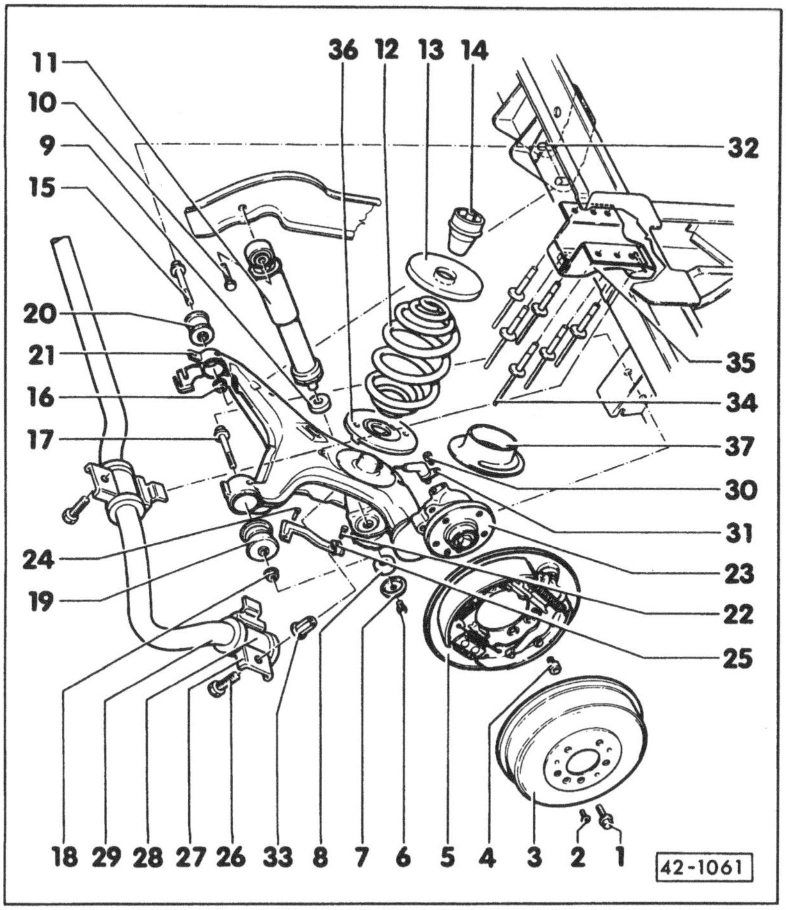

Rear suspension assembly

Notes:

- *Do not attempt to weld or straighten wheel locating components of rear suspension.*
- *Always replace self-locking nuts.*

1 - Wheel bolt 160 Nm (118 ft lb)

2 - Hex socket head bolt, 5 Nm (44 in. lb)

3 - Brake drum
- Dimensions, ⇒ Page M 46-15.

4 - Hex socket head bolt, 150 Nm (110 ft lb)

5 - Brake backing plate
- Rear wheel brake assembly, ⇒ Page M 46-15.

6 - Hex bolt, 45 Nm (33 ft lb)

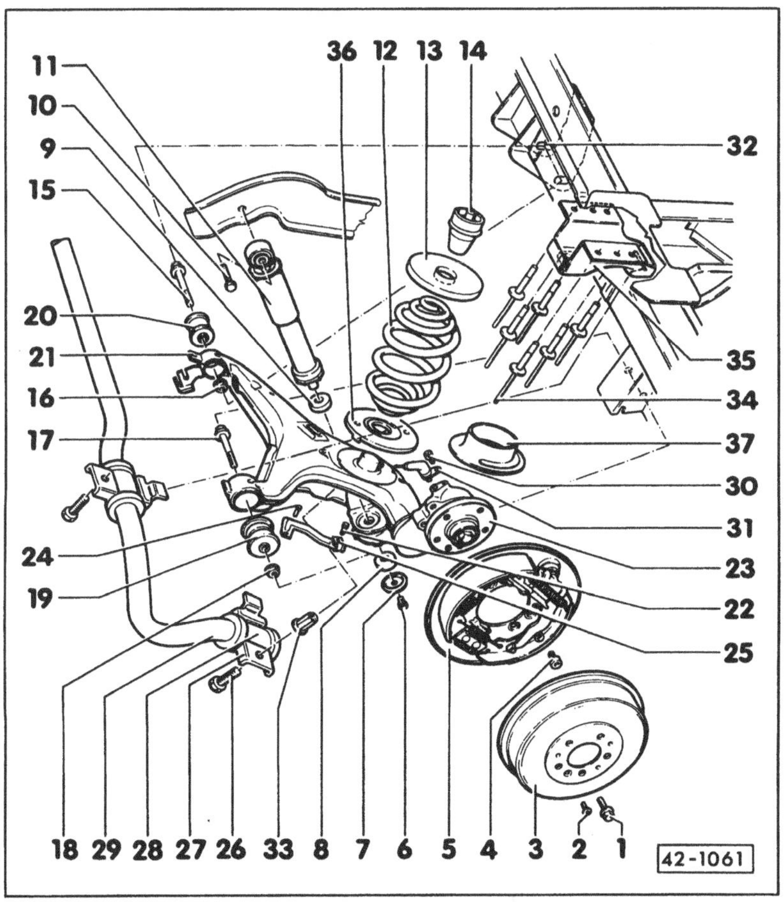

7 - Washer
- Collar points towards shock absorber.

8 - Lower shock absorber rubber bushing

9 - Upper shock absorber rubber bushing

10 - Hex bolt, 100 Nm (74 ft lb)

11 - Shock absorber
- Removing and installing, ⇒ Page M 42-16.
- Identification:
  black: standard.
  grey: standard.
  green: reinforced.

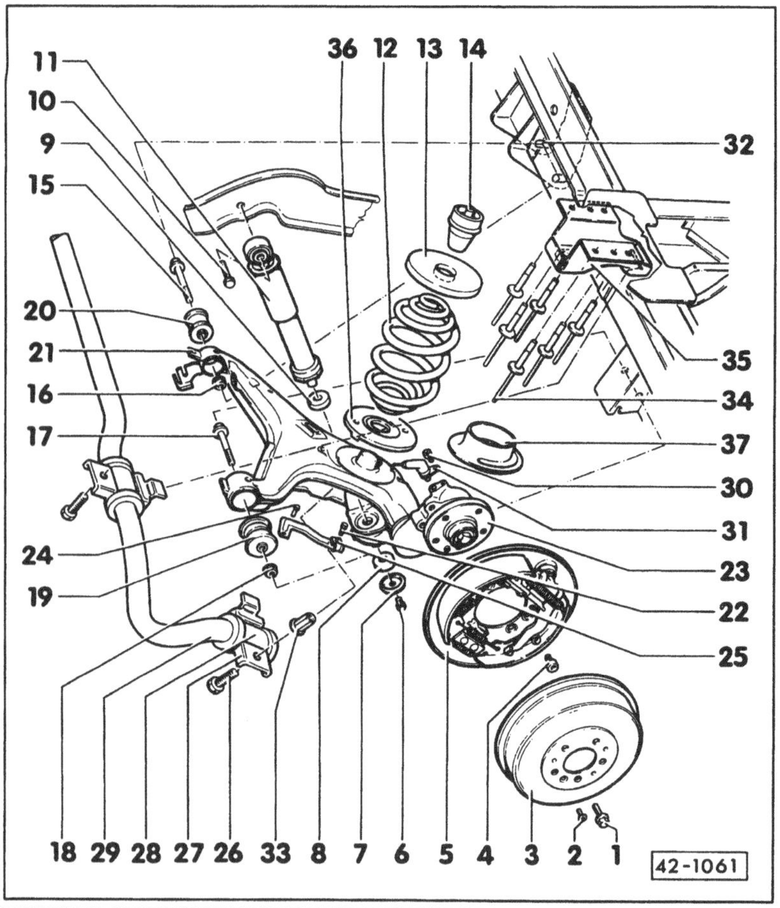

12 - Coil spring

- Removing: first remove upper bolt from shock absorber, ⇒ Page M 42-16, removing and installing shock absorber.
- Remove coil spring.
- Installing: install in reverse order
- Installation position, ⇒ Fig. 3.
- Examine for paint damage and repair paint damage, if necessary.
- Allocation, ⇒ Page 42–17.

13 - Upper washer

- Turn so that end of spring is in recess.

14 - Bump stop

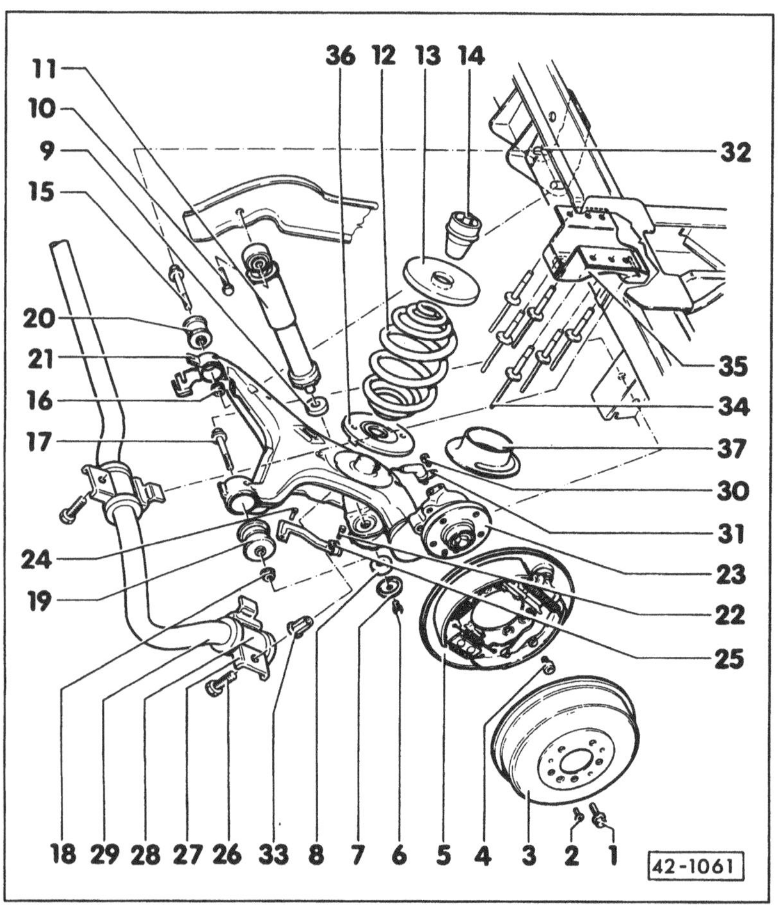

15 - Hex bolt

- M 12 x 1.5 x 108.
- Mark installation position before removal to prevent realigning rear axle.

16 - Nut, 100 Nm (74 ft lb)

17 - Hex bolt

- M 12 x 1.5 x 112.

18 - Nut, 160 Nm (118 ft lb)

19 - Outer rubber mounting

- Pulling out, ⇒ Fig. 1.
- Pulling in, ⇒ Fig. 2.

20 - Inner rubber mounting

- Pulling out, ⇒ Fig. 1.
- Pulling in, ⇒ Fig. 2.

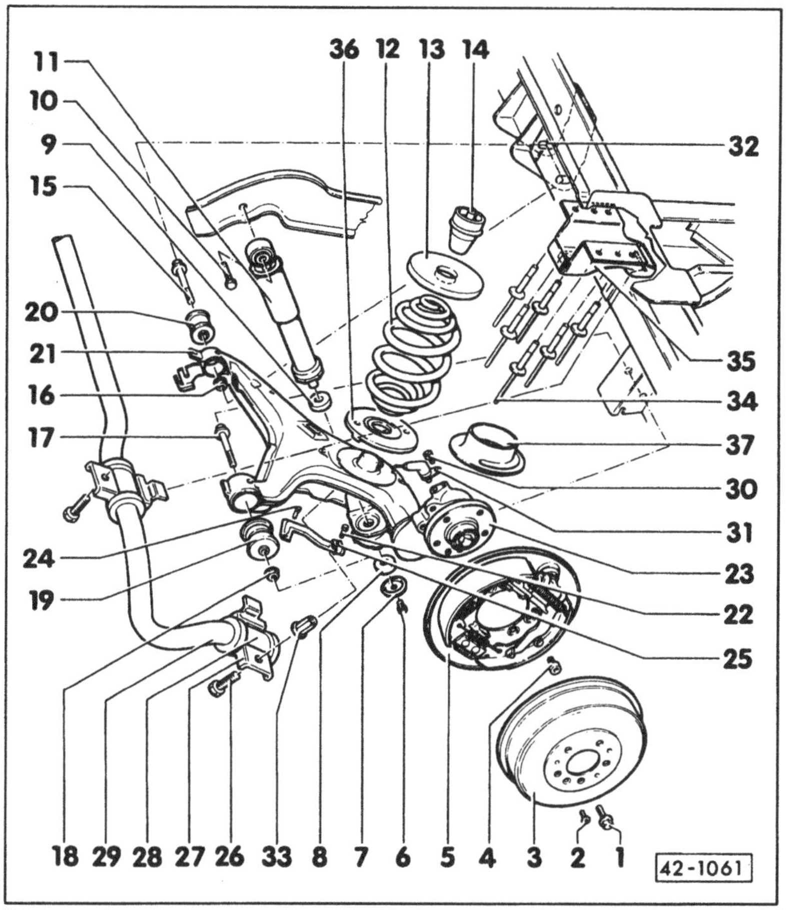

21 - Control arm
- Mark installation position on inner mounting.
- When removing, disconnect spring for brake pressure regulator at horizontal beam.
- When replacing, adjust brake pressure regulator, ⇒ Page M 47-14.
- When tightening bolts, hold control arm in installation position.

22 - Hex bolt, 7 Nm (62 in. lb)

23 - Wheel mounting
- Servicing, ⇒ Page M 42-21.

24 - Self-tapping screw, 7 Nm (62 in. lb)

25 - Parking brake cable bracket

26 - Hex bolt, 30 Nm (22 ft lb)

27 - Clamp

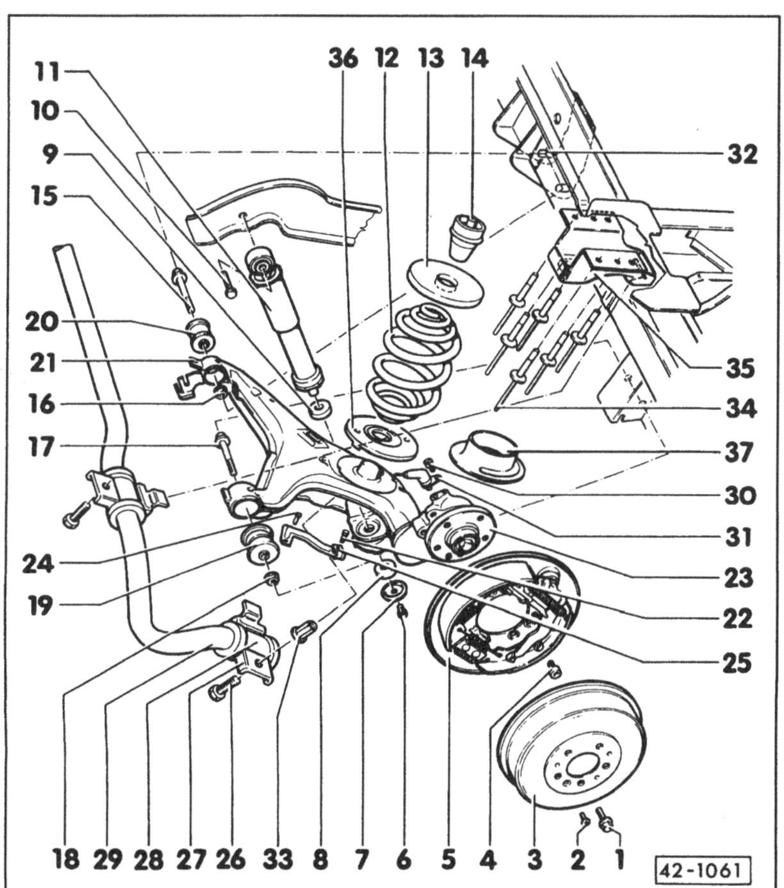

28 - Upper rubber mounting
- Coat with lubricant, e.g., soft soap, before installing.

29 - Stabilizer bar
- 26 mm diameter.
- Retrofitting, ⇒ Page M 42-10.

30 - Bolt, 10 Nm (7 ft lb)

31 - Wheel Speed Sensor
- Only on vehicles with ABS.

32 - Mounting bracket
- With horizontal elongated holes to adjust toe, Specification ⇒ Page M 44-6.

33 - Rivet nut, M8
- Only found on vehicles with stabilizer bar.
- Pulling into control arm, ⇒ Page M 42-12.

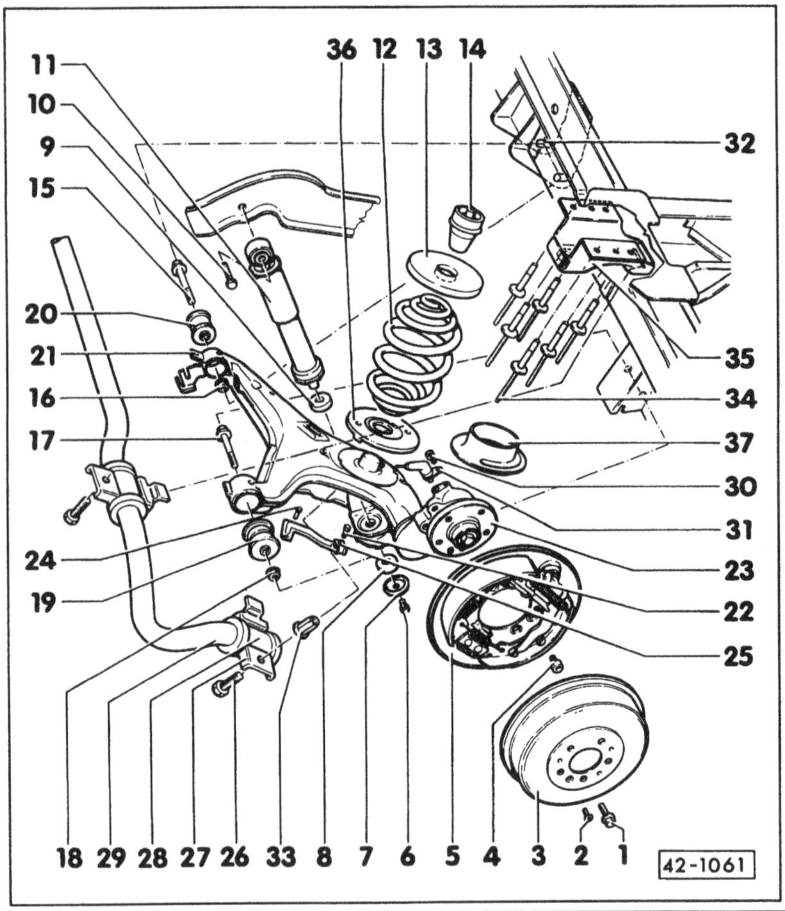

34 - Pop rivet

35 - Mounting bracket
- ♦ Installing, ⇒ Page M 42-13.

36 - Lower spring seat
- ♦ Installation position: Galvanized side faces top, rubber lug is inserted into front hole in control arm.
- ♦ Allocation, ⇒ from Page M 42-17.

37 - Cap
- ♦ Turn so that end of spring is in recess
- ♦ Introduced from VIN 70 NH 112 073
- ♦ If required, can be retrofitted in vehicles that were produced before above VIN.
- ♦ Allocation, ⇒ from Page M 42-17.

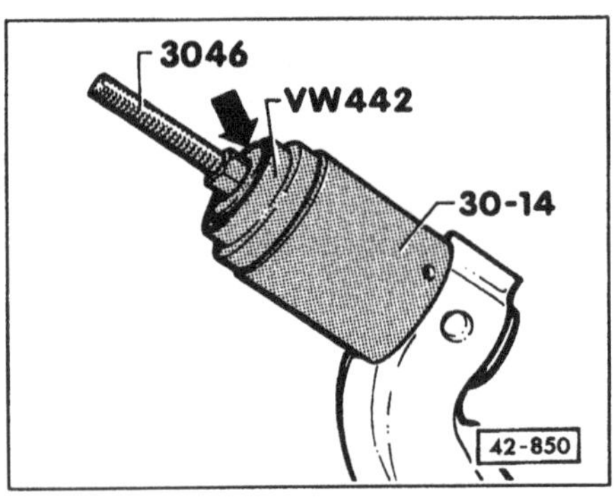

◄ **Fig. 1 Bonded rubber bushing, pulling out of control arm**

- Clamp control arm in vice.
- Coat washer with oil and put between nut and tool.
- Pull bushing out of control arm.

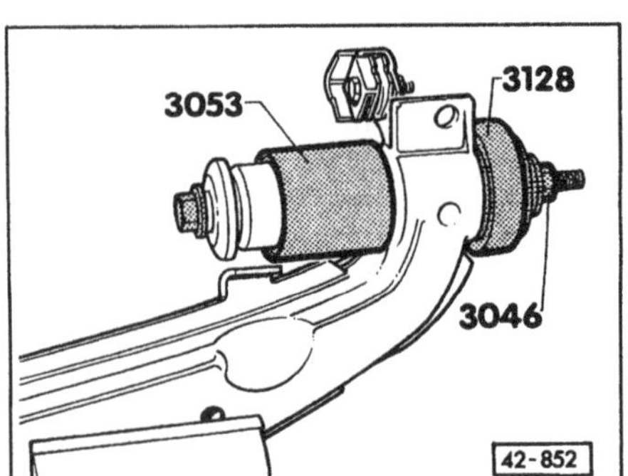

◄ **Fig. 2 Bonded rubber bushing, pulling into control arm**

- Pull in bonded rubber bushing until steel casing contacts tool.

Note:

Coat control arm mounting hole, guide sleeve and bonded rubber bushing with lubricant, e.g., soft soap, before installing.

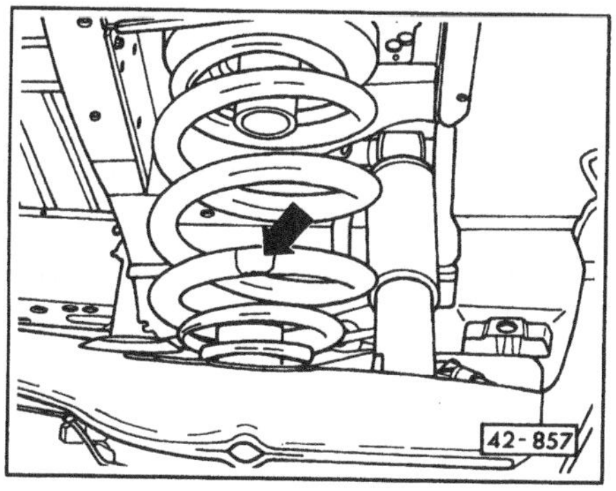

◄ **Fig. 3 Coil spring, installation position**

The colored marking (arrow) points to the rear.

Rear stabilizer bar, retrofitting

Note:

If a stabilizer bar is to be retrofitted to the rear axle then it is also necessary to have a 27 mm diameter stabilizer installed to the front axle. Retrofit 27 mm stabilizer bar to front axle if necessary ⇒ Repair Group 40.

Parts required:

- Stabilizer bar, 26 mm diameter.
- Clamp for stabilizer bar (Qty. 4).
- Rubber bushing (Qty. 4).
- Hex nut M 8 x 28 (Qty. 4).
- Rivet nut M 8 (Qty. 2).
- Mounting bracket (Qty. 2).
- Hex bolt M 6 (Qty. 8).
- Pop rivet nut M 6 (Qty. 8).

Note:

While installing the rear stabilizer bar ensure that the paintwork surface is not damaged. Repair paintwork damage, if necessary and protect against corrosion with underbody sealant.

Assembling the stabilizer bar to the control arm and mounting bracket, ⇒ from Page M 42-1.

Stabilizer bar rivet nut, installing in control arm

- Thread M 8 nut onto M 8 x 45 bolt.
- Thread M 8 x 45 mm bolt into M 8 rivet nut until bolt begins to protrude from opposite side.
- Turn M 8 nut by hand against collar of rivet nut.
- Install M 8 rivet nut into control arm.

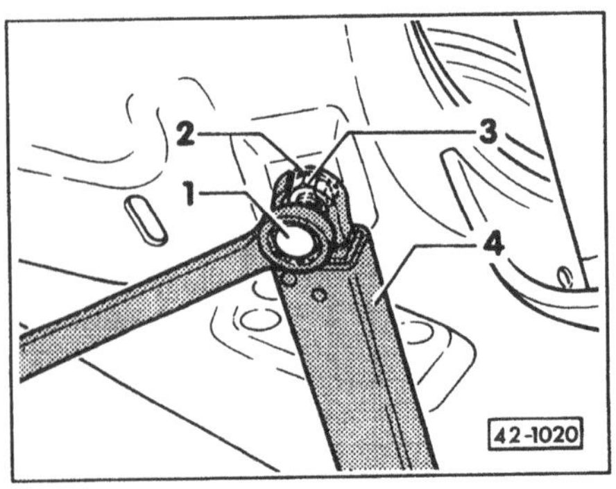

◄ - Pull pop rivet nut in under pressure by turning the M 8 nut and counter-holding at bolt head.

Tightening torque: 25 Nm (18 ft lb)

1 - Bolt M 8 x 45.

2 - Rivet nut M 8.

3 - Nut M 8.

4 - Torque wrench V.A.G 1331 or commercial equivalent.

Note:

When installing the rivet nut, ensure that the collar of the nut is resting, under pressure, against the control arm.

Stabilizer bar mounting bracket, securing with rivet nuts M 6 in crossmember

– Drill four of six holes out to 9 mm diameter .

WARNING!
Wear protective goggles.

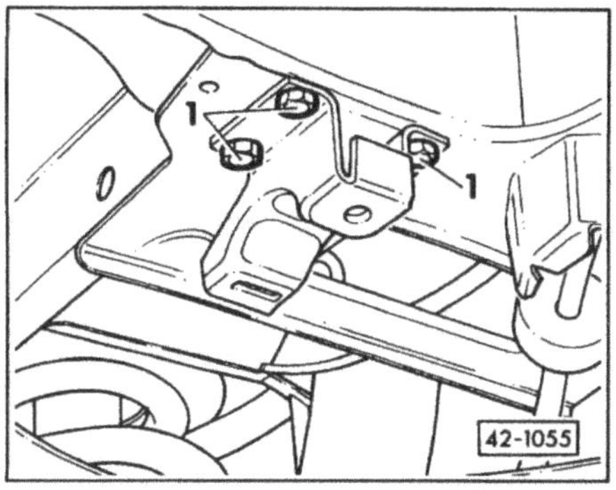

– Pull in rivet nut (using rivet nut pliers VAG 1765, or equivalent; or bolt, nut and washer, ⇒ Page M 42-14).

◄ – Secure mounting bracket with 4 bolts.

1 - Bolt, Part No. N 010 218 6.

● Tightening torque: 15 Nm (11 ft lb).

– Apply corrosion protection to crossmember as necessary.

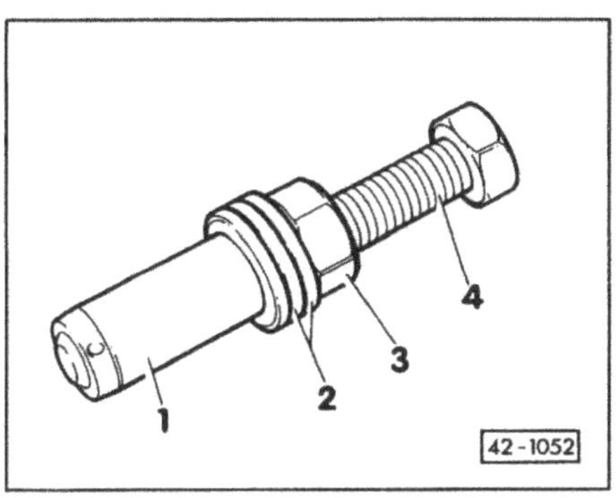

Rivet nut M 6, pulling in with bolt, nut and washer

◄ – Thread M 6 x 50 bolt with M 6 nut and two washers into rivet nut until bolt begins to protrude from opposite side.

1 - Rivet nut M 6, Part No. N 102 952 01.

2 - Washers.

3 - Nut M 6.

4 - Bolt M 6 x 50.

– Turn M 6 nut by hand against collar of pop rivet nut.

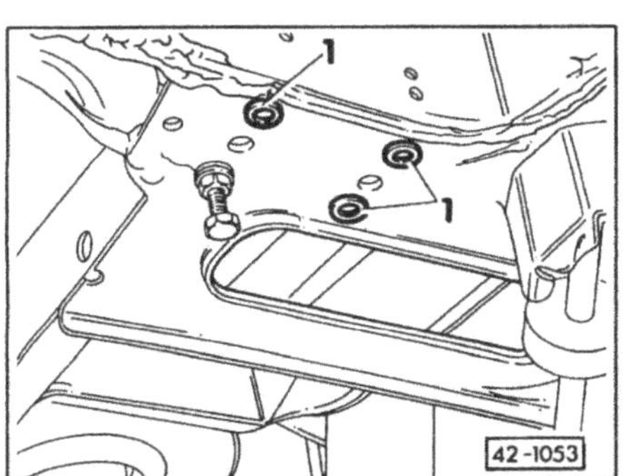

◄ – Install rivet nut M 6 into crossmember.

1 - Rivet nuts already pulled in.

– Pull rivet nuts in by turning M 6 nut and counter-holding at bolt head.

Notes:

- *Remove protective wax from the cross member before installing rivet nuts.*
- *When installing rivet nuts, ensure that the collar of the rivet nut is resting against the crossmember.*

● Tightening torque: 15 Nm (11 ft lb)

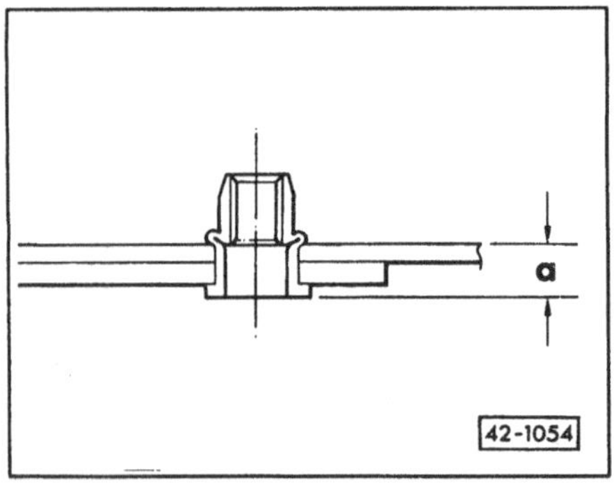

- Remove bolt and check dimension -a-.
- Dimension a: 7 – 8.5 mm (0.276 – 0.335 in.)

Shock absorber, removing and installing

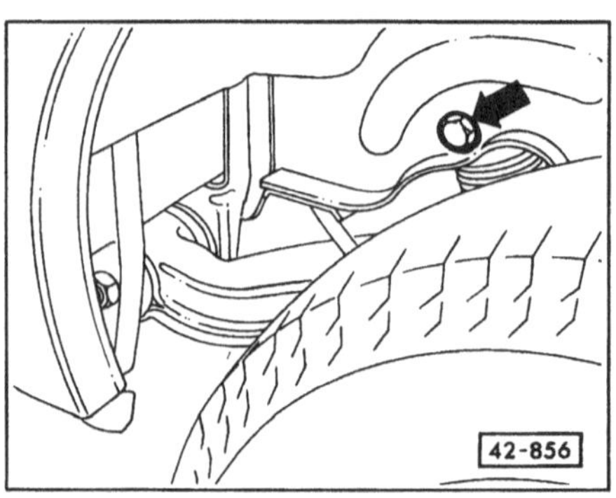

Removing

- Remove bolt (arrow) with vehicle standing on its wheels (raise vehicle, if necessary, until bolt is accessible).

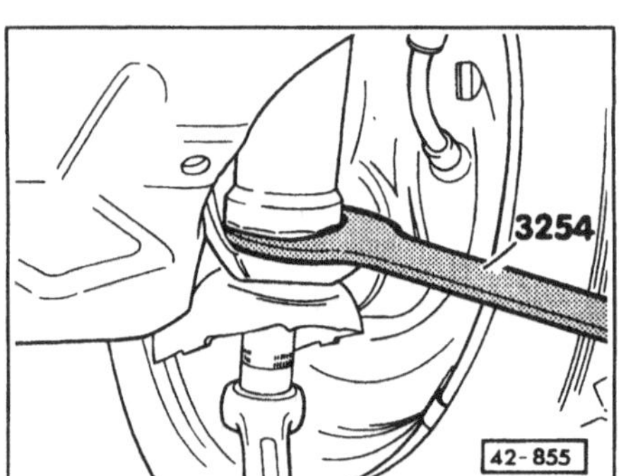

- Raise vehicle to relieve pressure on coil spring.
- Remove shock absorber at control arm by counter-holding shock absorber with special tool 3254.

Installing

Install in reverse order.

Coil springs/brake pressure regulator springs, vehicles up to VIN 70 RH 036 458

Model	Equipment + (net load)[1]	Coil spring Part No.	Color	Bump stop	Grp. num ber	Brake pressure regulator spring Part No.	Color
Bus,	Short wheelbase, lowered 15 mm	701 511 105 L[2]	gray	x	4	701 612 561 A	green
Multivan,	Short wheelbase, rear reinforced	701 511 105 A[2]	white	x	3	701 612 561 A	green
Weekender with seats	Long wheelbase, rear reinforced	701 511 105 A[2]	white	x	1	701 612 561 A	green
	Short wheelbase, front + rear reinforced	701 511 105 A[2]	white	x	1	701 612 561 A	green
	Long wheelbase, front + rear reinforced	701 511 105 A[2]	white	x	1	701 612 561 A	green

[1] The net load details are rounded figures.

[2] Vehicles with these springs are equipped with a rubber coated plate between spring and control arm, ⇒ Page M 42-7.

Model	Equipment + (net load)[1]	Coil spring Part No.	Color	Bump stop	Grp. num ber	Brake pressure regulator spring Part No.	Color
Double cab,	Short wheelbase (1200 kg)	701 511 105 B	brown	x	2	701 612 561 C	brown
Flat bed	Long wheelbase (1200 kg)	701 511 105 B	brown	x	2	701 612 561 C	brown
	Short wheelbase (1000 kg) rear reinforced	701 511 105 B	brown	x	2	701 612 561 C	brown
	Short wheelbase (1000 kg) front + rear reinforced	701 511 105 B	brown	x	2	701 612 561 C	brown
	Long wheelbase (1000 kg) rear reinforced	701 511 105 B	brown	x	2	701 612 561 C	brown
	Long wheelbase (1000 kg) front + rear reinforced	701 511 105 B	brown	x	2	701 612 561 C	brown
Camper	Short wheelbase (1000 kg)	701 511 105 K[2]	yellow	x	3	701 612 561 A	green
	Long wheelbase (1000 kg)	701 511 105 K[2]	yellow	x	1	701 612 561 A	green

[1] The net load details are rounded figures.

[2] Vehicles with these springs are equipped with a cap between spring and control arm, ⇒ Page M 42-7.

Coil springs/brake pressure regulator springs, vehicles from VIN 70 RH 036 459

Model	Equipment + (net load)[1)2)]	Coil spring Part No.	Color	Bump stop	Grp. num ber	Brake pressure regulator spring Part No.	Color
Bus,	Short wheelbase, lowered 15 mm	701 511 105 L[2)]	grey	x	4	701 612 561 A	green
Multivan,	Short wheelbase, rear reinforced	701 511 105 A[2)]	white	x	3	701 612 561 A	green
Weekender with seats	Long wheelbase, rear reinforced	701 511 105 A[2)]	white	x	1	701 612 561 A	green
	Short wheelbase, front + rear reinforced	701 511 105 A[2)]	white	x	1	701 612 561 A	green
	Long wheelbase, front + rear reinforced	701 511 105 A[2)]	white	x	1	701 612 561 A	green

[1)] The net load details are rounded figures.

[2)] Vehicles with these springs are equipped with a damper buffer (Part No. 701 511 137) between spring and control arm.

[3)] Vehicles with these springs are equipped with a cap between spring and control arm, ⇒ Page M 42-7.

Model	Equipment + (net load)[1)]	Coil spring Part No.	Color	Bump stop	Grp. num ber	Brake pressure regulator spring Part No.	Color
Double cab,	Short wheelbase (1200 kg)	701 511 105 B	brown	x	2	701 612 561 C	brown
Flat bed,	Short wheelbase (2000 kg)	701 511 105 B	brown	x	2	701 612 561 C	brown
Rialta	Long wheelbase (1200 kg)	701 511 105 B	brown	x	2	701 612 561 C	brown
	Long wheelbase (2000 kg)	701 511 105 B	brown	x	2	701 612 561 C	brown
	Short wheelbase (1000 kg) rear reinforced	701 511 105 B	brown	x	2	701 612 561 C	brown
	Short wheelbase (1000 kg) front + rear reinforced	701 511 105 B	brown	x	2	701 612 561 C	brown
	Long wheelbase (1000 kg) rear reinforced	701 511 105 B	brown	x	2	701 612 561 C	brown
	Long wheelbase (1000 kg) front + rear reinforced	701 511 105 B	brown	x	2	701 612 561 C	brown
Camper	Short wheelbase (1000 kg)	701 511 105 K[2)]	yellow	x	3	701 612 561 A	green
	Long wheelbase (1000 kg)	701 511 105 K[2)]	yellow	x	1	701 612 561 A	green

[1)] The net load details are rounded figures.

[2)] Vehicles with these springs are equipped with a rubber coated plate between spring and control arm, ⇒ Page M 42-7.

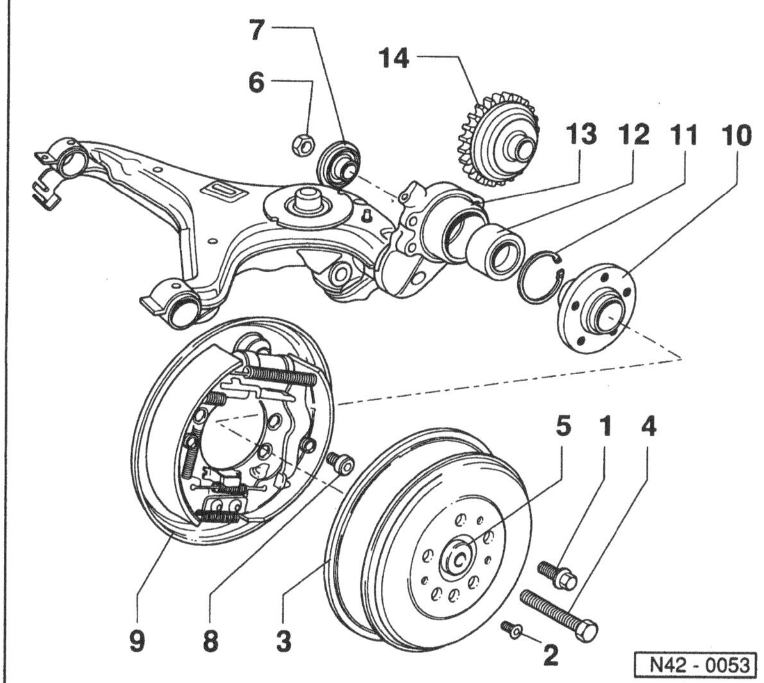

Control arm, assembly

1 - **Wheel bolt, 160 Nm (118 ft lb)**

2 - **Hex socket head bolt, 5 Nm (44 in. lb)**

3 - **Brake drum**
- ♦ Dimensions, ⇒ Page M 46-15.

4 - **Hex bolt, 200 Nm (148 ft lb)**
- ♦ Always replace.

5 - **Washer**

6 - **Self-locking hex nut, 200 Nm (148 ft lb)**
- ♦ Always replace.

7 - **Clamping plate**

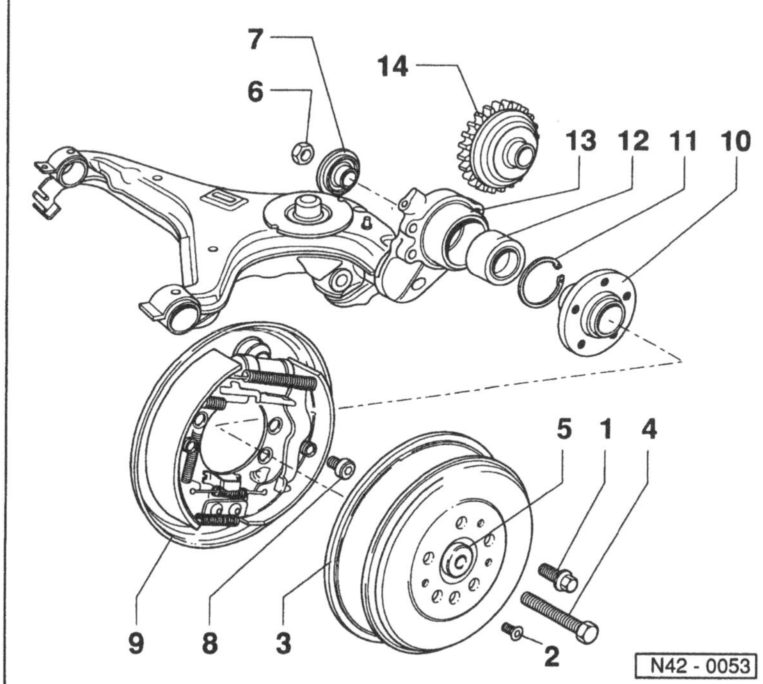

8 - **Hex socket head bolt, 150 Nm (111 ft lb)**

9 - **Backing plate**
- ♦ Servicing rear wheel brake, ⇒ Page M 46-15.

10 - **Hub**
- ♦ Pulling out, ⇒ Fig. 1.
- ♦ Pulling in, ⇒ Fig. 5.
- ♦ Pulling off inner race, ⇒ Fig. 3.

11 - **Circlip**
- ♦ Ensure correct seating.

12 - **Wheel bearing**
- ♦ Pressing out, ⇒ Fig. 2.
- ♦ Always replace, bearing destroyed during pressing out.
- ♦ Pulling in, ⇒ Fig. 4.

13 - **Control arm**

14 - **Rotor**
- ♦ Only on vehicles with ABS.

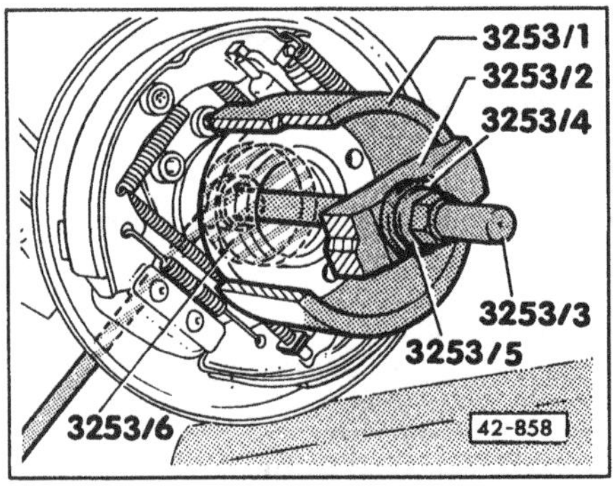

◄ **Fig. 1 Wheel hub, pulling out of control arm**

Note:

Place transmission jack VAG 1383/A or equivalent underneath.

> ***WARNING!***
> ***Parts may fall during disassembly.***

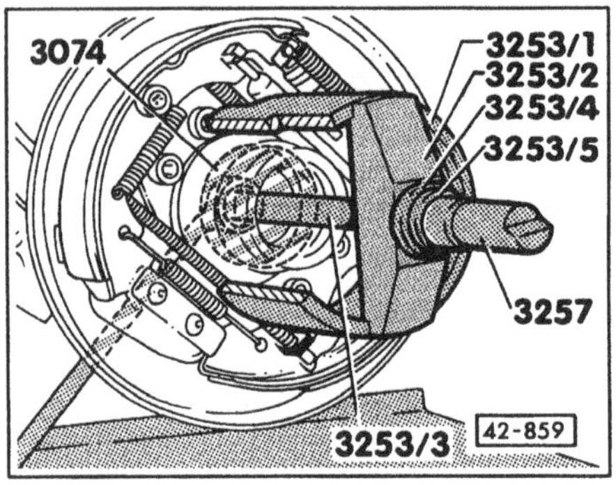

◄ **Fig. 2 Wheel bearing, pulling out of control arm**

Notes:

- *Place transmission jack VAG 1383/A or equivalent underneath.*

> ***WARNING!***
> ***Parts may fall during disassembly.***

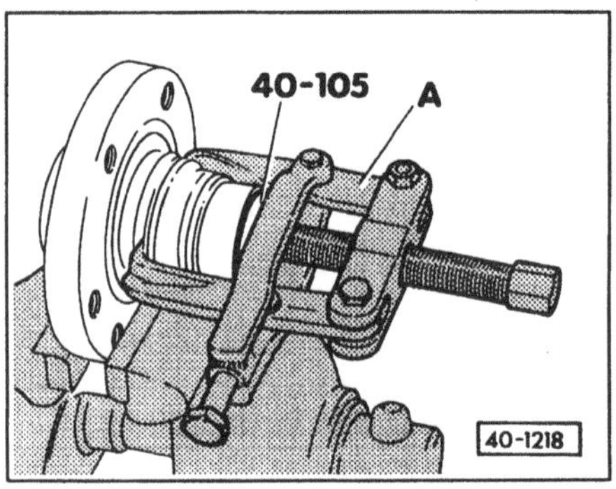

◄ **Fig. 3 Wheel bearing inner race, pulling off hub**

Only use puller with leg clamp e.g., -A- Kukko 204-2 (commercial type).

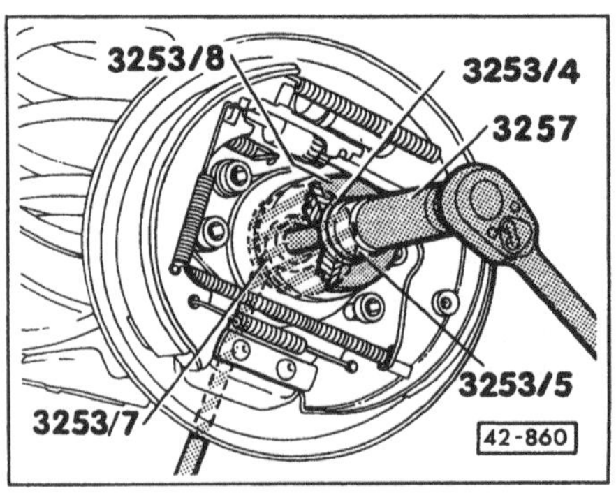

◄ **Fig. 4 Wheel bearing, pulling into control arm**

- Pull in wheel bearing up to stop.
- Insert circlip with long-nosed pliers.

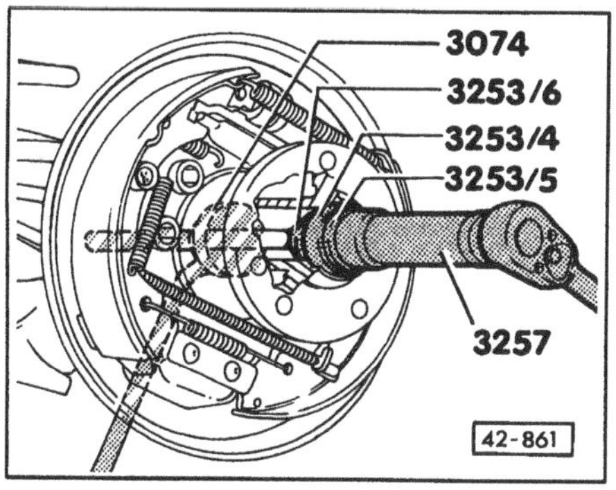

◄ **Fig. 5 Wheel hub, pulling in**

- Pull wheel hub in onto stop.

Vehicle alignment

Test prerequisites:

- Correct calibration of measuring equipment.
- Correct tire pressure (see label on vehicle).
- Vehicle accurately aligned, suspension bounced and rocked several times.
- Suspension, steering and steering linkage checked for excessive play and damage.
- Vehicle height (ride height) to specification.

Notes:

- *The specifications for vehicle alignment are divided into four groups. The group to which a vehicle is assigned is indicated on the vehicle data plate on the A-pillar, ⇒ Page M 44-3.*
- *At start of production, a number of vehicles were delivered without an indication of the alignment group number. These vehicles are in group 1.*
- *Vehicle height (ride height) and vehicle longitudinal inclination affect test values. For this reason, measure vehicle height and longitudinal inclination, and adjust vehicle height if necessary before performing check/adjustments. Measuring/adjusting vehicle height, ⇒ Page M 44-4 and measuring vehicle inclination, ⇒ Page M 44-5.*

- *The actual loaded condition at time of measuring is relative when determining the specifications (empty, ½ load, full load).*
- *If the vehicle alignment stand is approached at an acute angle, then move the vehicle backward and forward several times to relieve possible stress in the bonded rubber mountings.*
- *When adjusting, aim as closely as possible to the respective specified value.*
- *Only perform vehicle alignment after the vehicle has been driven 1000 to 2000 km (600 to 1200 miles), to ensure that the suspension has settled completely.*

WARNING!
Do not re-use any fasteners that are worn or deformed in normal use. Many fasteners are designed to be used only once and become unreliable and may fail when used a second time. This includes, but is not limited to, nuts, bolts, washers, self-locking nuts or bolts, circlips, cotter pins. Always follow recommendations given in this publication. For replacement always use new parts.

Installation position of vehicle data plate on A-pillar next to central electric

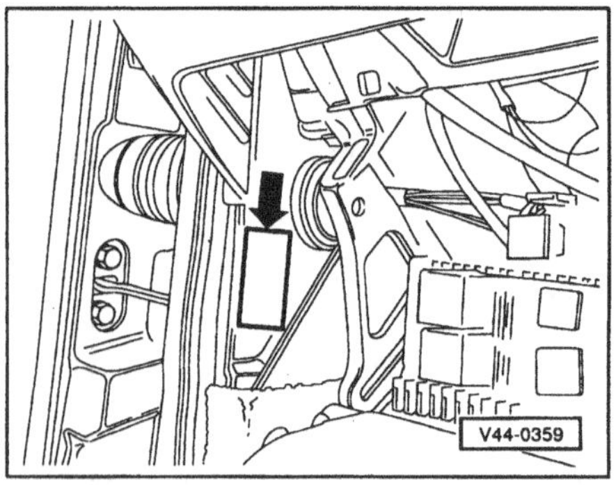

◄ On LHD vehicles - on left-hand side.

Alignment group number on vehicle data plate, locating

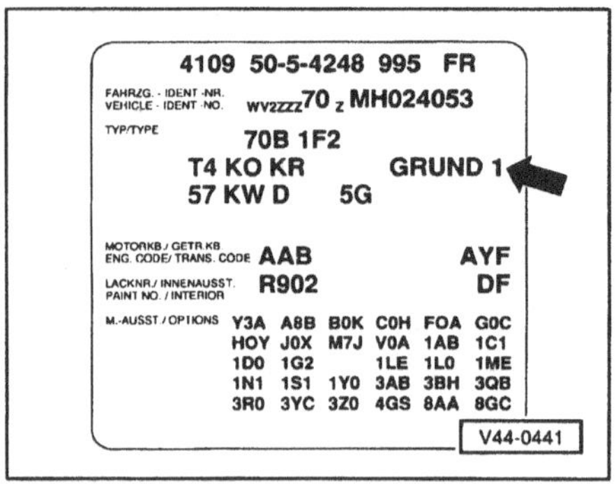

An alignment group number from 1 to 4 is indicated on the vehicle data plate.

◄ The digit (arrow), in this example 1, indicates the alignment group to which a vehicle is assigned.

Note:

The word "Grund" on the vehicle data plate is given as an example. It is possible that a different word, for example "Kom." may be printed at this point.

Vehicle height (ride height), measuring/adjusting at front axle

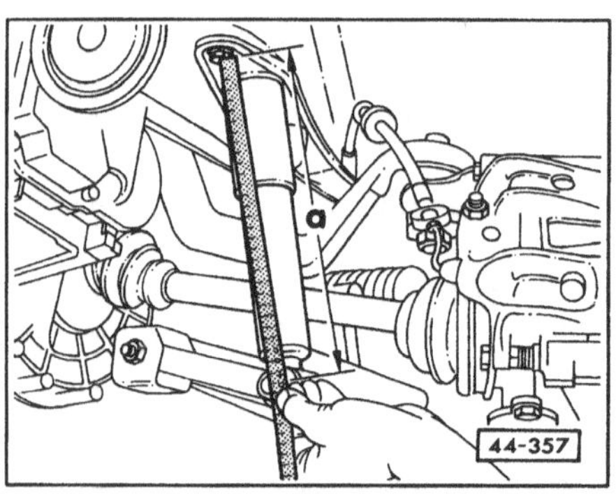

Vehicle must be standing on its wheels.

◄ The dimension -a- is measured from the upper shock absorber mounting bolt head to the center of the lower shock absorber bolt. If necessary, adjust to the prescribed height (⇒ alignment specification charts) by turning the nut on the torsion bar tensioning lever.

Adjusting torsion bar, ⇒ Page M 40-12.

Note:

For ease of illustration the vehicle is shown raised and without a wheel.

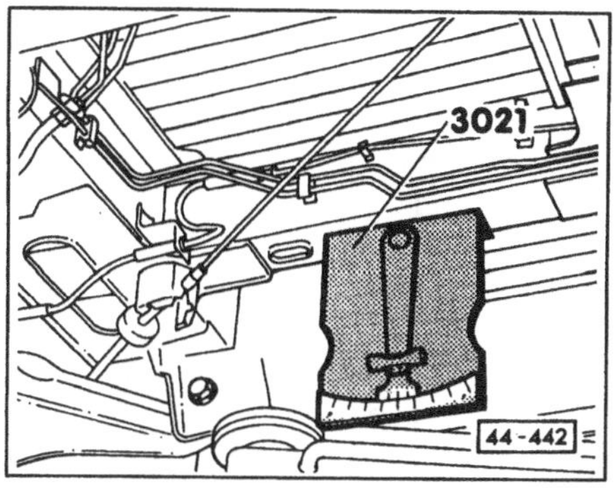

Vehicle longitudinal inclination, measuring

◄ – Measure vehicle longitudinal inclination at longitudinal member.

Note:

Ensure that the longitudinal member surface is clean and smooth in area where the measurement is to be taken.

If the vehicle is inclined to the front or rear, add to or subtract from the caster value, measured according to the direction in which the vehicle inclines.

Example:

Angle measured	1° 10′
Correction value for vehicle inclination (example: inclined to the front)	+ 1°
Caster (example: for vehicle inclination of 1° 10 to the front)	2° 10′

Notes:

- *Vehicle inclined to the front: add correction value.*
- *Vehicle inclined to the rear: subtract correction value.*

Vehicle alignment specifications

Vehicle group 1

Front axle (double control arm)			
Load state	empty [3]	1/2 load[3]	full load[3]
Vehicle height[1] front (± 2 mm)	280	267	255
Toe per wheel (not pressed)	+10′ ± 10′	0′ ± 10′	–10′ ± 10′
Total toe[1] (not pressed)	+20′ ± 20′	0′ ± 20′	–20′ ± 20′
Camber (in straight-ahead position) Maximum permissible difference between sides	15′ ± 20′ max. 30′	5′ ± 20′ max. 30′	–20′ ± 20′ max. 30′
Toe angle difference at 20° lock to left and right	–45′ ± 20′		
Caster (vehicle on level)[2] Maximum permissible difference between sides	+1° 40′ ± 30′ max. 1°	+1° 50′ ± 30′ max. 1°	+2° ± 30′ max. 1°

[1] If vehicle height (ride height) has to be adjusted, the front axle alignment should be checked in full. After adjusting the vehicle height, push vehicle forwards and backwards over approx. 2 m, then re-check height. Checking front height, ⇒ Page M 44-4.

[2] Measuring vehicle longitudinal inclination, ⇒ Page M 44-5.

[3] Load state at time of vehicle alignment measurements.

Rear axle (trailing arm)			
Load state	empty	1/2 load	full load
Toe per wheel	+10′ ± 10′	+20′ ± 10′	+30′ ± 10′
Total toe (at specified camber) Max. permissible deviation from direction of travel	+20′ ± 20′ max. 30′	+40′ ± 20′ max. 30′	+1° ± 20′ max. 30′
Camber Maximum permissible difference between sides	–30′ ± 30′ max. 30′	–1° 10′ ± 30′ max. 30′	1° 35′ ± 30′ max. 30′

Vehicle group 2

Front axle (double wishbone axle)			
Load state	empty[3]	1/2 load[3]	full load[3]
Vehicle height[1] front (± 2 mm)	280	267	255
Track per wheel (not pressed)	+10′ ± 10′	0′ ± 10′	–10′ ± 10′
Total track[1] (not pressed)	+20′ ± 20′	0′ ± 20′	–20′ ± 20′
Camber (in straight-ahead position) Maximum permissible difference between sides	+15′ ± 20′ max. 30′	–5′ ± 20′ max. 30′	–20′ ± 20′ max. 30′
Track angle difference at 20° lock to left and right		–45′ ± 20′	
Caster (vehicle inclination zero)[2] Maximum permissible difference between sides	+1° 40′ ± 30′ max. 1°	+1° 50′ ± 30′ max. 1°	+2° ± 30′ max. 1°

[1] If vehicle height (ride height) has to be adjusted, the front axle alignment should be checked in full. After adjusting the vehicle height, push vehicle forwards and backwards over approx. 2 m, then re-check height. Checking front height, ⇒ Page M 44-4.

[2] Measuring vehicle longitudinal inclination, ⇒ Page M 44-5.

[3] Load state at time of vehicle alignment measurements.

Rear axle (Trailing arm)			
Track per wheel	+3′ ± 10′	+12′ ± 10′	+20′ ± 10′
Total track (at specified camber) Maximum permissible deviation from direction of travel	+6′ ± 20′ max. 30′	+24′ ± 20′ max. 30′	+40′ ± 20′ max. 30′
Camber Maximum permissible difference between sides	0° ± 30′ max. 30′	–40′ ± 30′ max. 30′	–1° 10′ ± 30′ max. 30′

Vehicle group 3

Front axle (double wishbone axle)			
Load state	empty[3)]	1/2 load[3)]	full load[3)]
Vehicle height[1)] front (± 2 mm)	273	263	255
Track per wheel (not pressed)	+5′ ± 10′	0′ ± 10′	–10′ ± 10′
Total track[1)] (not pressed)	+10′ ± 20′	0′ ± 20′	–20′ ± 20′
Camber (in straight-ahead position) Maximum permissible difference between sides	0° ± 20′ max. 30′	–10′ ± 20′ max. 30′	–20′ ± 20′ max. 30′
Track angle difference at 20° lock to left and right	–45′ ± 20′		
Caster (vehicle inclination zero)[1)] Maximum permissible difference between sides	+1° 40′ ± 30′ max. 1°	+1° 50′ ± 30′ max. 1°	+2° ± 30′ max. 1°

1) If vehicle height (ride height) has to be adjusted, the front axle alignment should be checked in full. After adjusting the vehicle height, push vehicle forwards and backwards over approx. 2 m, then re-check height. Checking front height, ⇒ Page M 44-4.

2) Measuring vehicle longitudinal inclination, ⇒ Page M 44-5.

3) Load state at time of vehicle alignment measurements.

Rear axle (Trailing arm)			
Track per wheel	+10′ ± 10′	+20′ ± 10′	+30′ ± 10′
Total track (at specified camber) Maximum permissible deviation from direction of travel	+20′ ± 20′ max. 30′	+40′ ± 20′ max. 30′	+1° ± 20′ max. 30′
Camber Maximum permissible difference between sides	−30′ ± 30′ max. 30′	−1° 10′ ± 30′ max. 30′	−1° 35′ ± 30′ max. 30′

Vehicle group 4

Front axle (double wishbone axle)			
Load state	empty[3]	1/2 load[3]	full load[3]
Vehicle height[1] front (± 2 mm)	265	260	255
Track per wheel (not pressed)	0′ ± 10′	−5′ ± 10′	−10′ ± 10′
Total track[1] (not pressed)	0′ ± 20′	−10′ ± 20′	−20′ ± 20′
Camber (in straight-ahead position) Maximum permissible difference between sides	−10′ ± 20′ max. 30′	−15′ ± 20′ max. 30′	−20′ ± 20′ max. 30′
Track angle difference at 20° lock to left and right		−45′ ± 20′	
Caster (vehicle inclination zero)[2] Maximum permissible difference between sides	+1° 50′ ± 30′ max. 1°	+1° 55′ ± 30′ max. 1°	+2° ± 30′ max. 1°

[1] If vehicle height (ride height) has to be adjusted, the front axle alignment should be checked in full. After adjusting the vehicle height, push vehicle forwards and backwards over approx. 2 m, then re-check height. Checking front height, ⇒ Page M 44-4.

[2] Measuring vehicle longitudinal inclination, ⇒ Page M 44-5.

[3] Load state at time of vehicle alignment measurements.

Rear axle (Trailing arm)			
Track per wheel	+12′ ± 10′	+22′ ± 10′	+27′ ± 10′
Total track (at specified camber) Maximum permissible deviation from direction of travel	+24′ ± 20′ max. 30′	+44′ ± 20′ max. 30′	+54′ ± 20′ max. 30′
Camber Maximum permissible difference between sides	–40′ ± 30′ max. 30′	–1° 10′ ± 30′ max. 30′	–1° 30′ ± 30′ max. 30′

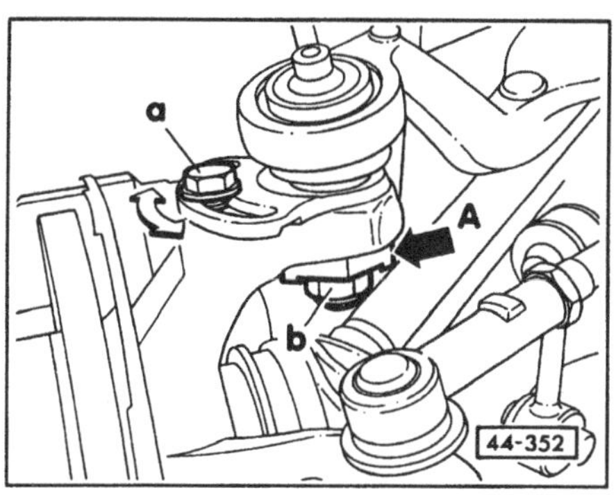

Camber, adjusting

– If necessary remove right-hand horn protective plate.

◄ – Loosen bolt -a-.

Note:

- *To loosen and tighten with wheel installed, turn steering as necessary.*
- *For ease of illustration, the bolted connection is shown without the wheel installed.*

– Loosen nut -b-.

– Turn eccentric bushing (arrow A) with special tool 3252 until correct specification is achieved.

– Tighten bolt -a- and nut -b- again.

Tightening torques:

Bolt -a-	60 Nm (44 ft lb)
Nut -b-	110 Nm (81 ft lb)

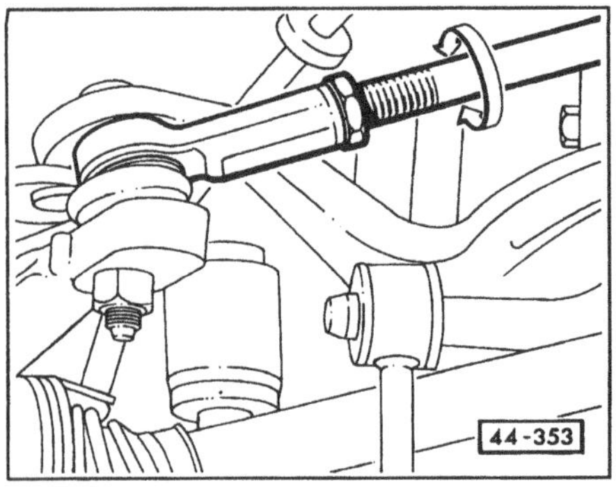

Front toe, adjusting

- Loosen lock nut (first turn steering gear to center position), ⇒ below.
- Turn both tie rods until specification is achieved.

Note:

The steering must remain in the center position.

- After turning tie rod, ensure that boots are not twisted. Twisted boots wear out quickly.

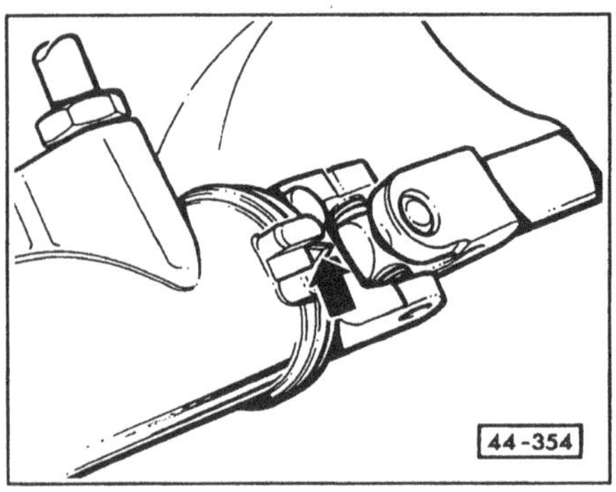

Steering gear, moving to central position

- Move steering gear to central position by turning steering wheel from lock to lock and counting the number of turns.
- Turn steering wheel back half the number of turns.
- Precisely align steering spindle with aid of lug (arrow) on rubber washer and notch in housing. Lug and notch must be aligned.

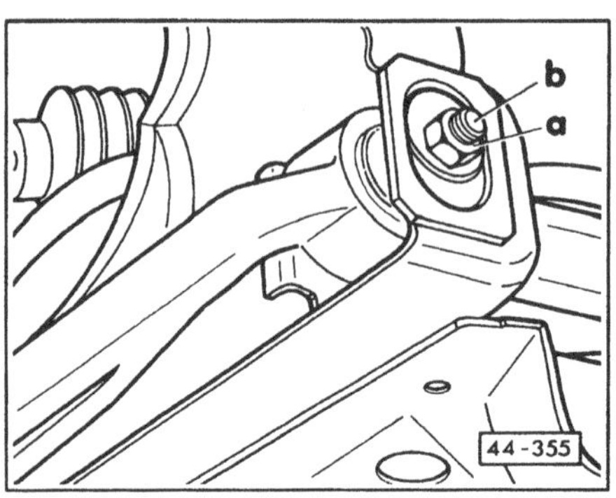

Caster, checking and adjusting

Checking

The caster should be checked in the following cases:

- ♦ Steering pulls to one side.
- ♦ Front wheel shimmy.

Adjusting

- Loosen nut -a-.
- Turn eccentric bolt -b- until correct specification is achieved.

Note:

The eccentric bolt (only M 12 x 1.5 bolt) should be replaced if the nut for adjusting the caster has been loosened.

- Tighten nut.

Tightening torques:

Nut M 12: 75 Nm (55 ft lb) and turn 90° (1/4 turn) further.

Nut M 14: 160 Nm (118 ft lb).

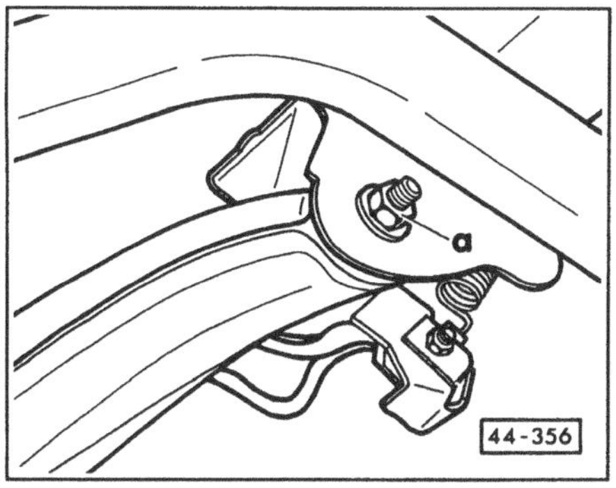

Rear axle toe, adjusting

- Loosen nut -a- on control arm.
- Move control arm forward or backward until specification is achieved (vehicle standing on its wheels).

Note:

If it is necessary to correct the toe setting at the rear wheels, check the brake pressure regulator setting or adjust the brake pressure regulator if necessary,
⇒ Page M 47-14.

Calculating the direction of travel:

- If both toe figures have the same proceeding sign (+/+ or –/–), subtract the smaller figure from the larger figure and divide by 2.

Left rear wheel track		Right rear wheel track
+ 15′		+ 5′
	15′ - 5′ = 10′ 10′ : 2 = 5′	

Deviation from direction of travel = 5′

- Add figures with different prefix (+/–) and divide by 2.

Left rear wheel track		Right rear wheel track
+ 15′		– 5′
	15′ + 5′ = 20′ 20′ : 2 = 10′	

Deviation from direction of travel = 10′

The individual result is the actual alignment deviation from the vehicle longitudinal axis.

Anti-lock brake system

General repair notes

The anti-lock brake system (ABS) is basically maintenance-free, however, note the following points when working on vehicles with ABS:

- *Before beginning repair work on the ABS system, determine the cause of the malfunction using On Board Diagnostic, ⇒ Repair group 01.*
- *Before beginning any welding work with an electric welding unit, disconnect the ABS Control Module connector.*
- *Before working on the ABS system, disconnect the Battery Ground strap. Obtain the radio code on vehicles with a coded radio.*
- *After completing work on the ABS system, reconnect and tighten Battery Ground strap.*
- *When working with brake fluid, observe the valid, relevant safety precautions and notes, ⇒ Page M 47-9.*
- *Remove as much brake fluid as possible from the reservoir with an extraction bottle before beginning work that necessitates opening the hydraulic system.*

- *Seal brake lines/hoses and connections with suitable plugs (repair kit, Part No. 1 H0 698 311) when performing work which necessitates opening the hydraulic system.*
- *After work which necessitates opening the brake hydraulic system, bleed the brake system with brake–filling–and–bleeding tool US 1116, or equivalent.*
- *Do not drive vehicle with the ABS Control Module disconnected.*
- *It is sufficient to perform a simple functional check after completing any work on the brake system which did not directly affect parts of the ABS system. The Malfunction Indicator Lamp (MIL) in the instrument cluster must not come on when a speed of 6 km/h (4 mph) is exceeded if the ABS system is functioning correctly.*
- *If work is performed on the hydraulic unit, the ABS Control Module, the Wheel Speed Sensors, the wiring harness, or if components are replaced; e.g., brake lines/hoses, brake pressure regulator, in the course of accident repairs, then the entire ABS system must be tested with ABS tester VAG 1710.*
- *During the final road test, ensure that a controlled brake test is performed at least once (pulsations must be felt at the brake pedal).*

♦ *Absolute cleanliness is required when working on the ABS brake system. Do not use any products which contain mineral oils, greases etc.*

♦ *Thoroughly clean all connections and the adjacent areas before loosening. Do not use aggressive cleaning agents such as brake cleaner, thinners, etc.*

♦ *Place removed parts on a clean surface and cover.*

♦ *Do not use fluffy cloths.*

♦ *Carefully cover or seal open components if repairs cannot be carried out immediately.*

♦ *Only remove replacement parts from packaging immediately prior to installation.*

♦ *Only use genuine packaged parts.*

♦ *When the system is open do not use compressed air in the immediate area and do not move the vehicle.*

♦ *Ensure that no brake fluid enters wiring connectors.*

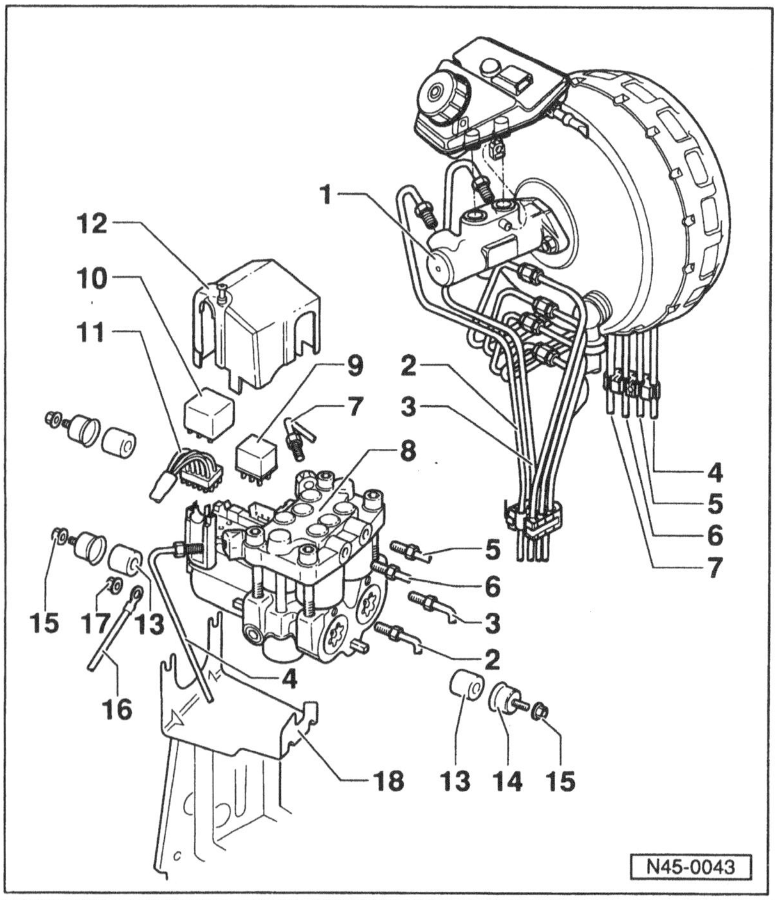

ABS hydraulic unit, removing and installing

Notes:

♦ *The hydraulic unit is installed in front left-hand side of engine compartment.*

♦ *When removing and installing brake lines/hoses, ensure that lines/ hoses are marked and installed per markings on hydraulic unit.*

♦ *Do not interchange brake lines/ hoses when reconnecting.*

1 - **Tandem brake master cylinder**
- ♦ With brake booster.

2 - **Brake line**
- ♦ Secondary piston circuit - hydraulic unit.

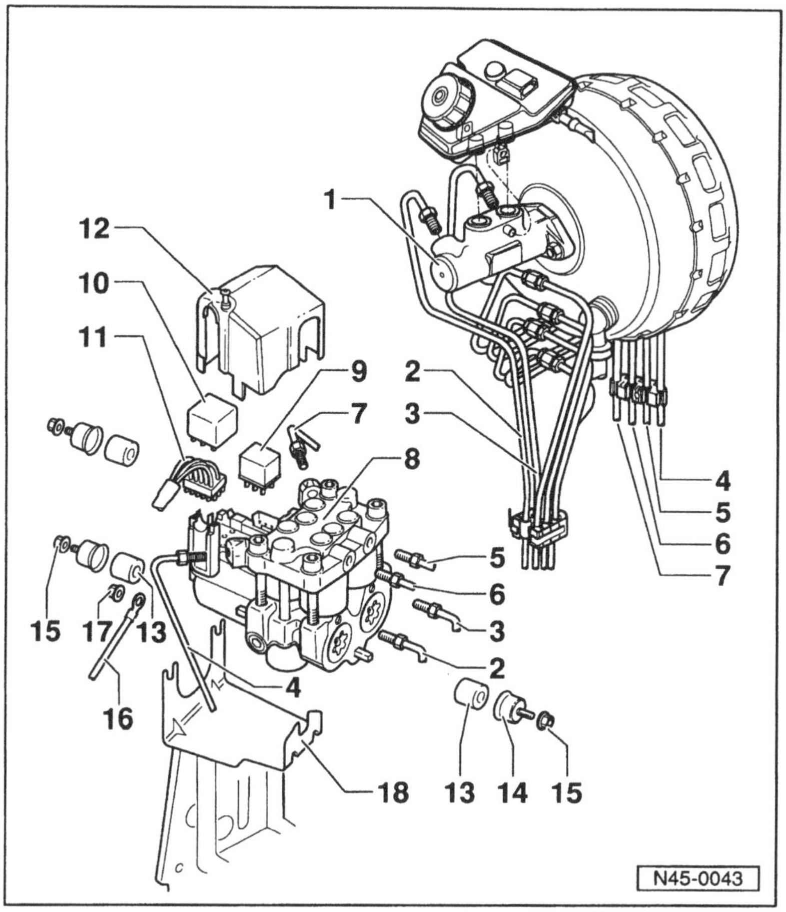

3 - Brake line
- Primary piston circuit - hydraulic unit.

4 - Brake line
- Hydraulic unit - front left brake caliper.

5 - Brake line
- Hydraluic unit - rear left wheel cylinder.

6 - Brake line
- Hydraulic unit - rear right wheel cylinder.

7 - Brake line
- Hydraulic unit - front right brake caliper.

8 - Hydraulic unit -N 55-
- No servicing possible, apart from replacing the two relays.
- Can only be replaced as a unit.

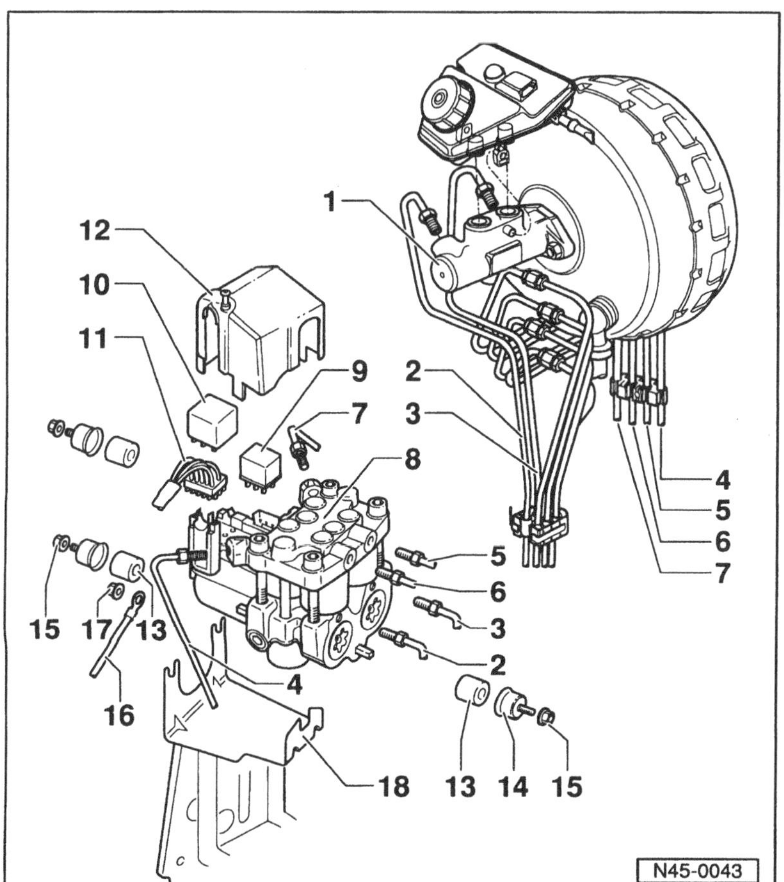

9 - ABS solenoid valve relay -J106-

10 - ABS Retrun Flow Pump Relay -J105-

11 - Wiring harness
- Release connector in order to pull off.

12 - Cover

13 - Buffer

14 - Mount

15 - Hex nut

16 - Ground cable
- Ensure good contact.

17 - Hex nut

18 - Bracket

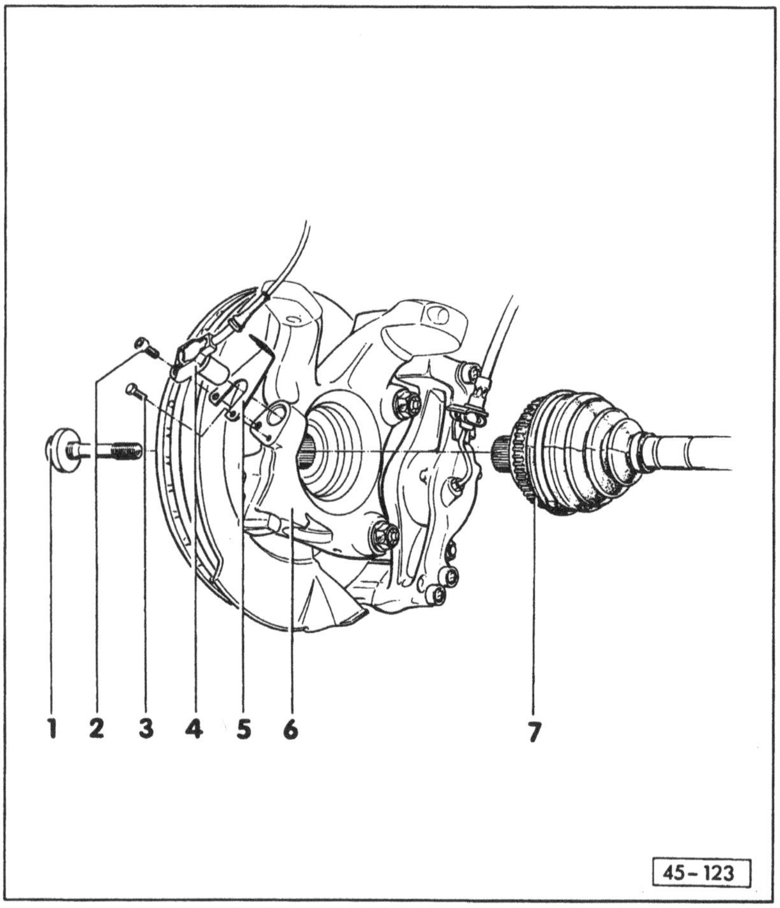

Front axle ABS components, removing and installing

1 - **Hex bolt, 200 Nm (148 ft lb)**
   - ♦ Only loosen and tighten with the vehicle standing on its wheels.

2 - **Hex socket-head bolt, 10 Nm (7 ft lb)**

3 - **Hex bolt**

4 - **Right/Left Front Wheel Speed Sensors -G45-/-G47-**
   - ♦ Before inserting sensor, clean opening inner surface and coat with lubricating paste, G 000 650.
   - ♦ Observe installation sequence: first, install speed sensor and then sensor wiring, beginning at sensor and ensuring that rubber grommets are secured in their respective mountings.
   - ♦ Routing cable, ⇒ Fig. 1.

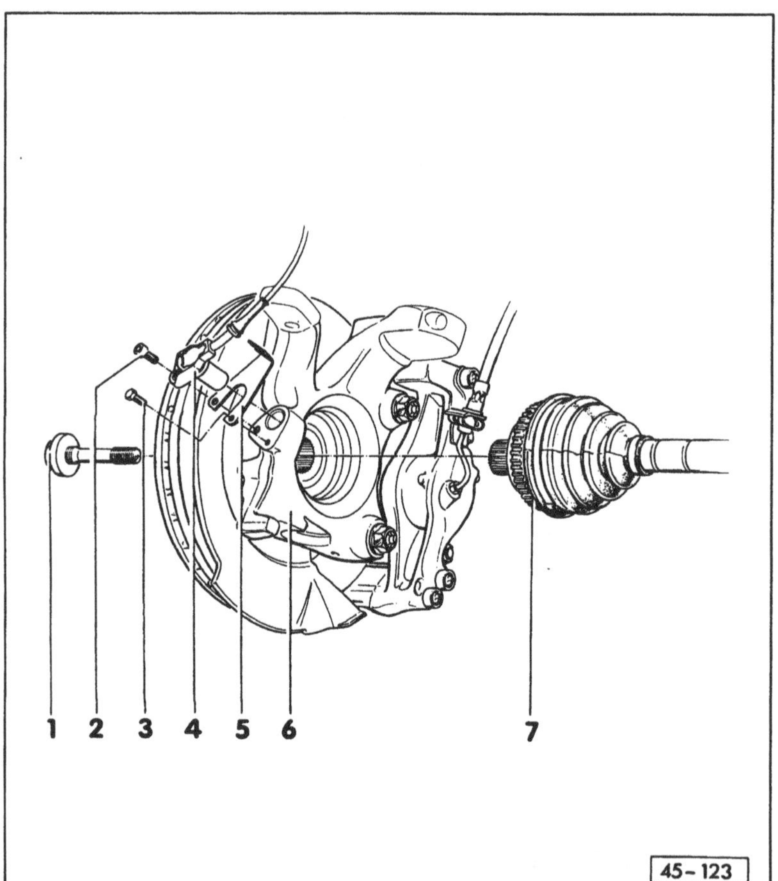

5 - **Bracket**

6 - **Wheel bearing housing**

7 - **Outer constant velocity joint**
   - ♦ Removing, ⇒ Page M 40-60.

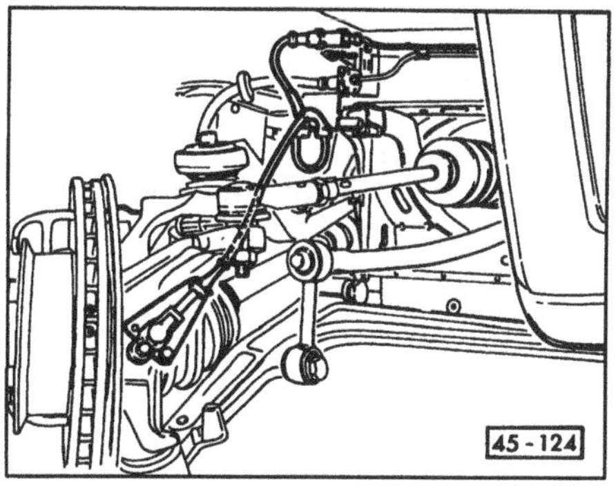

◄ **Fig. 1** **Wheel Speed Sensor harness, routing on front axle**

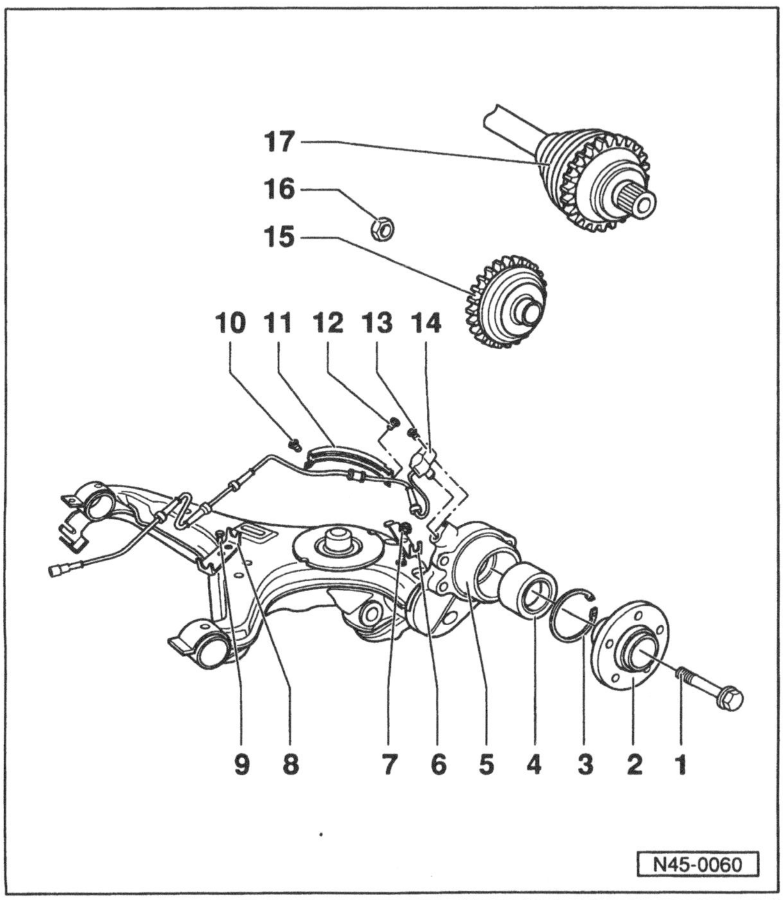

Rear axle ABS components, removing and installing

1 - Hex bolt,
200 Nm (148 ft lb)
- Always replace.

2 - Hub
- Pulling out and in, ⇒ Page M 42-21.

3 - Circlip
- Ensure correct seating.

4 - Wheel bearing
- Pulling out and in, ⇒ Page M 42-21.

5 - Control arm

6 - Bracket
- When installing, hook first into control arm.

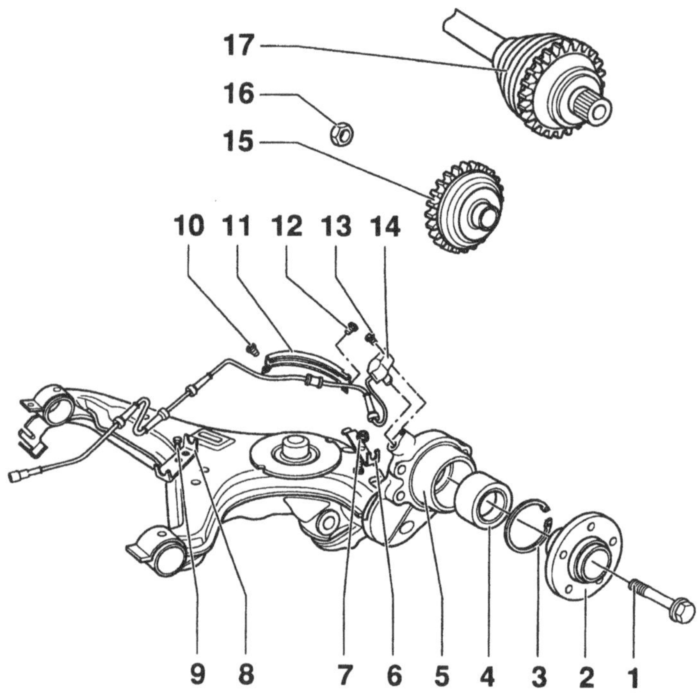

N45-0060

7 - Hex nut

8 - Bracket
- When installing, hook first into control arm.

9 - Bolt

10 - Bolt

11 - Guide tube

12 - Bolt

13 - Hex socket-head bolt, 10 Nm (7 ft lb)

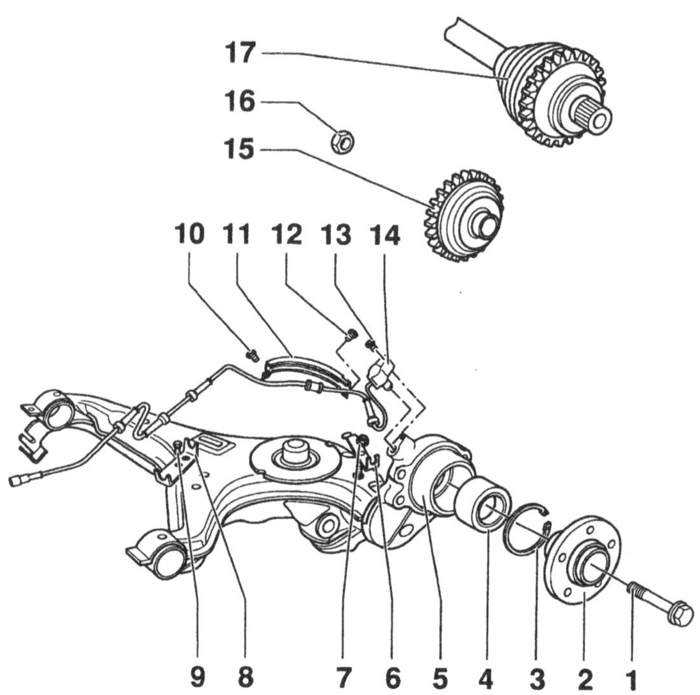

N45-0060

14 - Wheel Speed Sensor
- Before inserting sensor, clean opening inner surface and coat with lubricating paste, G 000 650.
- Observe installation sequence: First, install speed sensor and then sensor wiring, beginning at sensor and ensuring that rubber grommets are secured in their respective mountings.

15 - Rotor for Wheel Speed Sensor
- Only vehicles with front wheel drive.

16 - Self-locking hex nut, 200 Nm (148 ft lb)
- Always replace.
- Only vehicles with front wheel drive.

17 - Outer constant velocity joint
- Removing, ⇒ Page M 40-60.

ABS Control Module, removing and installing

Location: At bottom left, close to A-pillar.

Removing

– Remove cover in front of relay plate.

– Move carpeting in vicinity of left-hand A-pillar to one side.

– Pull connector off control module, upward, and release connector by pressing spring.

Note:
Switch off ignition before disconnecting and connecting connectors to control module.

– Remove control module securing clips.

– Take control module out of bracket.

Installing

Install in reverse order.

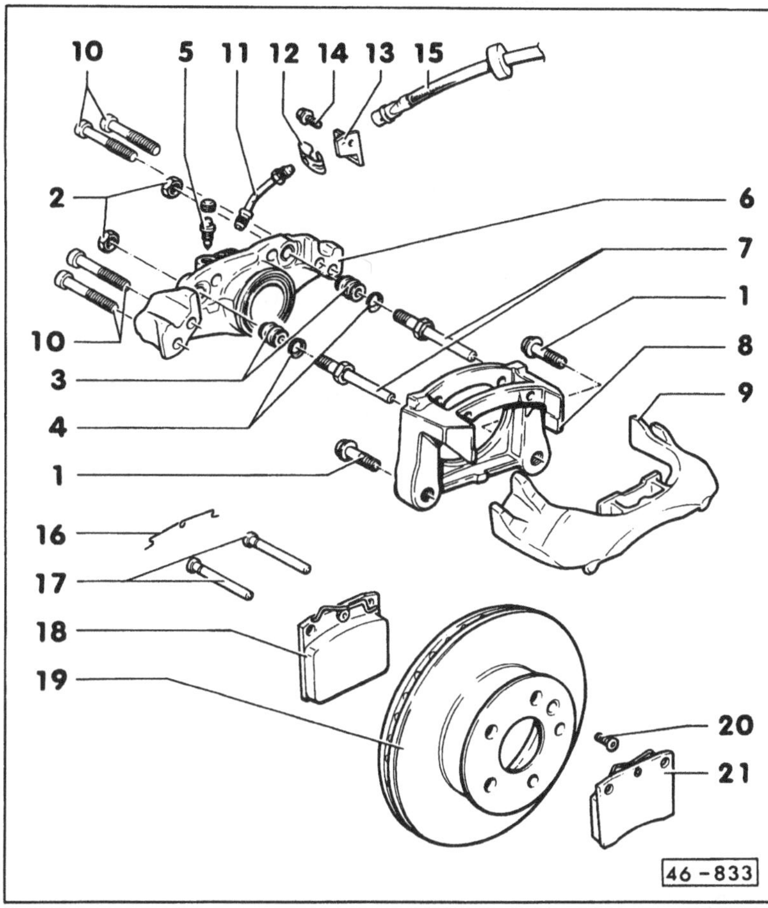

Front brake, assembly (Floating frame caliper)

From chassis No. 70 PH 102 523, two different types of floating caliper brakes were used. Distinguishing features, ⇒ Page M 46-14.

Notes:

- *After replacing brake pads, depress brake pedal firmly several times with vehicle stationary so that the brake pads are properly seated in their normal operating position.*

CAUTION!
Always remove some brake fluid from the reservoir before installing new brake pads. When the caliper piston is pushed back, fluid is forced out of the caliper into the reservoir. After pads are installed, refill the reservoir only to the MAX mark.

WARNING!
Brake fluid is poisonous and must not be siphoned by mouth.

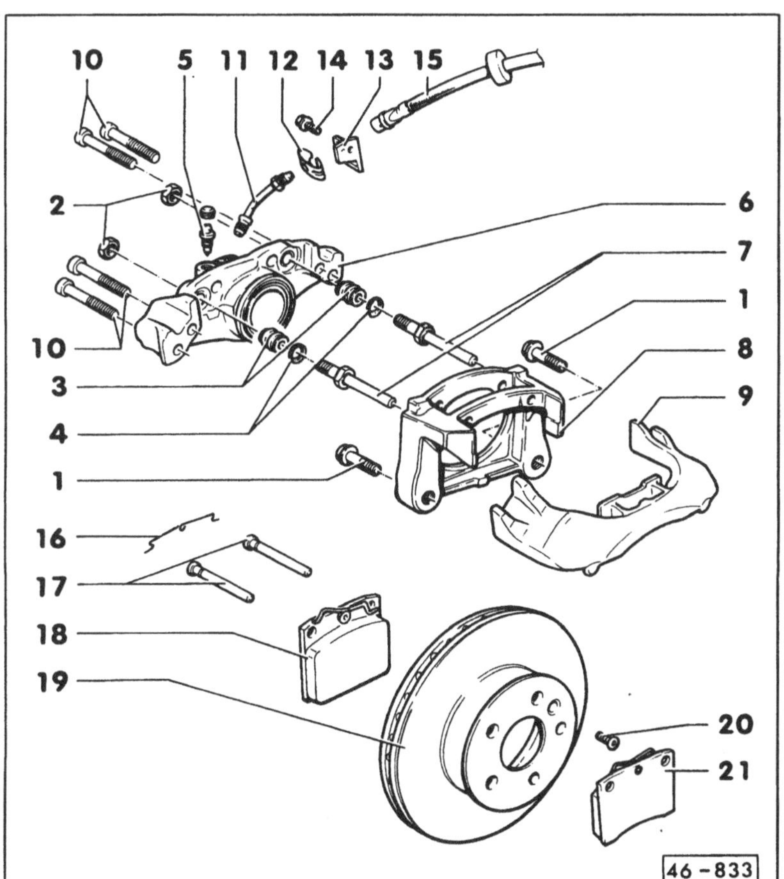

1 - **Hex bolt, 270 Nm (200 ft lb)**

2 - **Hex nut, 70 Nm (52 ft lb)**

3 - **Protective cap**
- Check for damage.

4 - **Metal ring**

5 - **Bleeder screw**

6 - **Brake caliper housing**
- Servicing, ⇒ Page M 47-5.
- Bolt together with bridge, ⇒ Fig. 1.

7 - **Guide pins**

8 - **Brake pad support**

9 - **Bridge**
- Bolt together with brake caliper housing, ⇒ Fig. 1.

10 - **Multi-point socket-head bolt, 90 Nm (66 ft lb)**

11 - **Brake line**

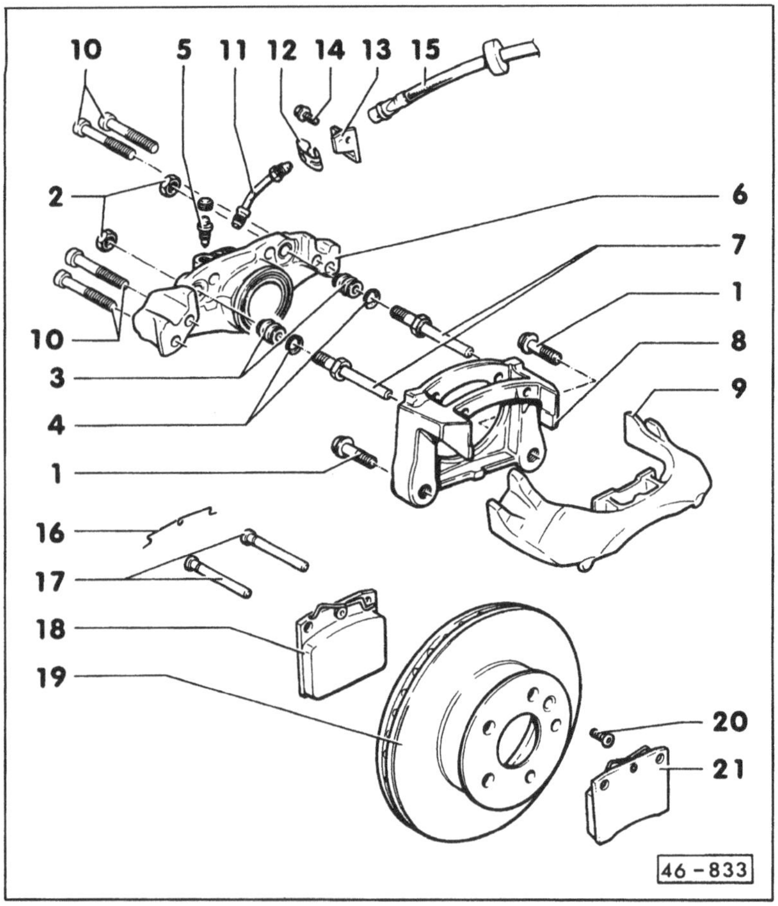

12 - Bracket

13 - Bracket for brake hose

14 - Hex bolt, 10 Nm (7 ft lb)

15 - Brake hose

16 - Securing spring

17 - Pad locating pins

18 - Inner brake pad
- ♦ When worn, always replace both sides.
- ♦ Removing and installing, ⇒ Page M 46-7.
- ♦ Pad thickness: 13 mm.
- ♦ Checking thickness,

⇒ Maintenance
- ♦ Wear limit: 2 mm (excluding backing plate).

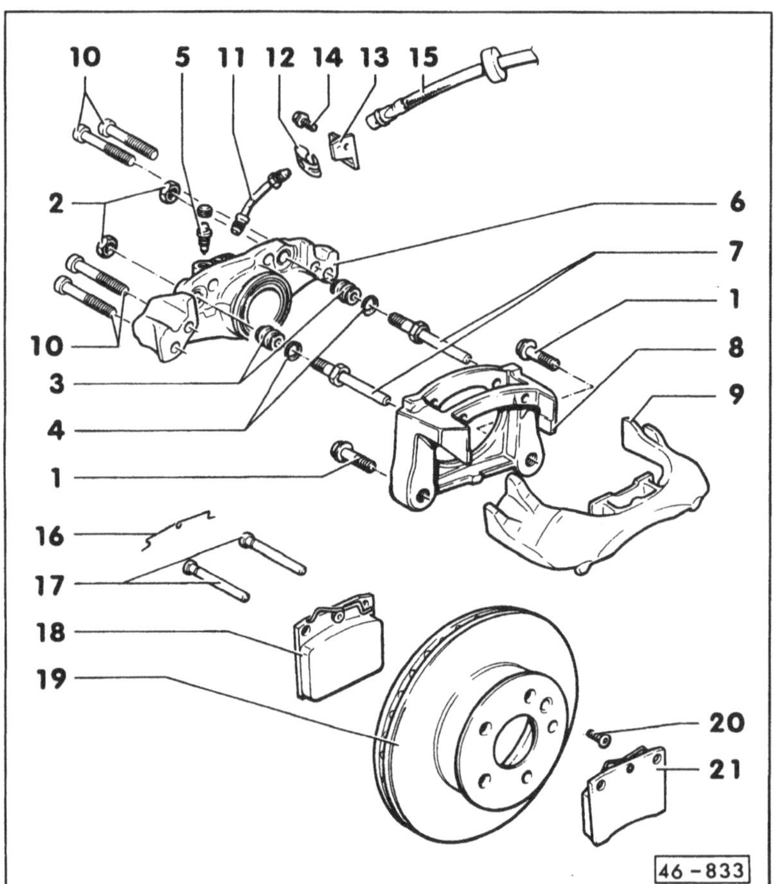

19 - Brake disc
- ♦ When worn, always replace both sides.
- ♦ To remove, first remove lower brake caliper bolt and loosen upper bolt.
- ♦ Always machine evenly on both sides, from thickness when new.
- ♦ Ensure adequate margin for wear.
- ♦ Run-out: 0.08 mm. Difference of thickness: 0.02 mm.
- ♦ Ventilated brake disc diameter: 280 mm.
- ♦ Brake disc thickness: 24 mm.
- ♦ Wear limit: 20 mm.

20 - Hex socket-head bolt, 5 Nm (44 in. lb)

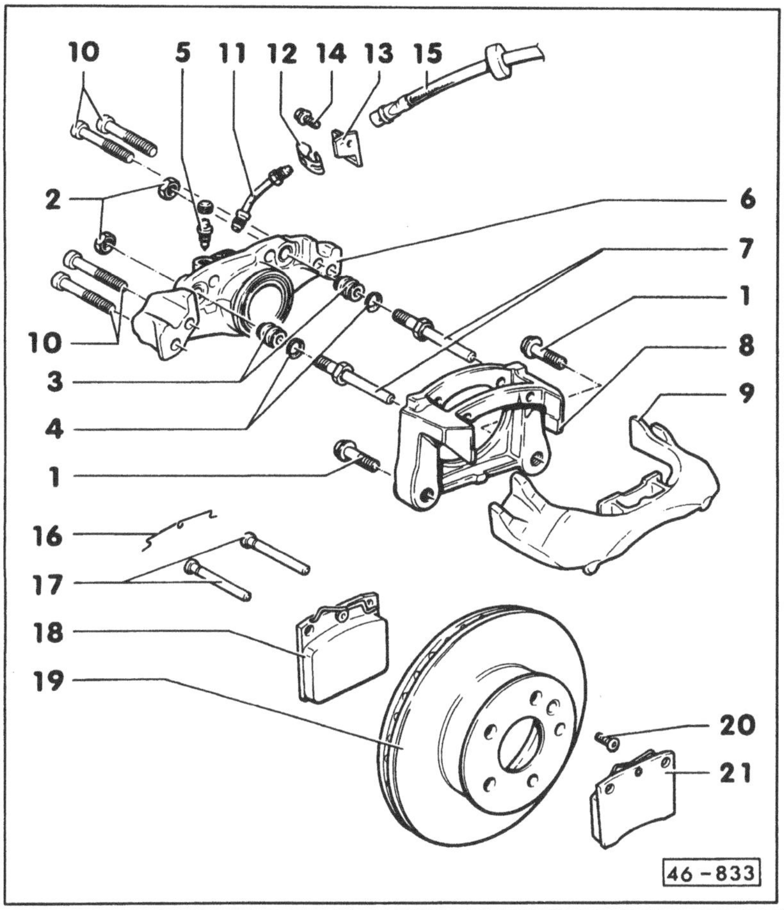

21 - Outer brake pad

- With adhesive foil on backing plate.
- When worn, always replace both sides.
- Removing and installing, ⇒ Page M 46-7.
- Pad thickness: 13 mm.
- Checking thickness, ⇒ Maintenance
- Wear limit: 2 mm (excluding backing plate).

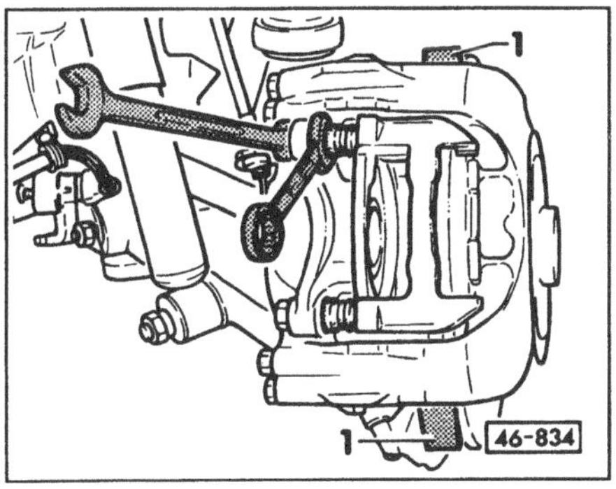

◄ **Fig. 1 Bridge and brake caliper housing, bolting together**

– Install bridge with multi-point, socket-head bolts and tighten.

– Install plate -1- (50 x 24 x 2 mm, locally manufactured) between face of brake disc and lower side of bridge. This ensures sufficient clearance between bridge and brake disc.

– Press brake caliper housing/bridge against the plate and tighten guide pin hex nuts.

– Remove plate and ensure brake disc has sufficient clearance.

Notes:

- *First tighten leading–side guide pins.*
- *Counter-hold at guide pins with an open end wrench, grind to fit, if necessary.*

Tightening torques

Multi-point socket-head bolt	90 Nm (66 ft lb)
Nuts for guide pins	70 Nm (52 ft lb)

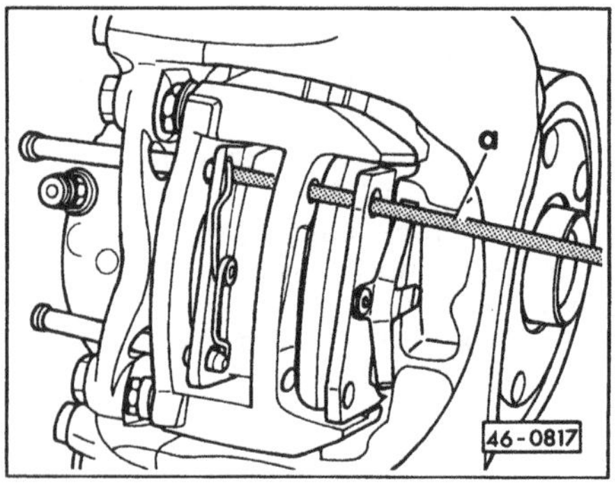

Brake pads, removing and installing

Removing

- Remove securing spring.
- Drive out pad retaining pins with drift -a-.
- Remove inner brake pad.
- Push frame outwards by approximately one pad width.
- With a screwdriver, loosen and remove outer brake pad.

> ***CAUTION!***
> ***Always remove some brake fluid from the reservoir before installing new brake pads. When the caliper piston is pushed back, fluid is forced out of the caliper and into the reservoir. After pads are installed, refill reservoir only to the MAX mark.***

> ***WARNING!***
> ***Brake fluid is poisonous and must not be siphoned by mouth.***

- Push caliper piston back.

Installing

- Clean brake caliper housing, particularly bonding surface for brake pad. It must be free of adhesive and grease residues.
- Install inner brake pad.
- Insert pad–retaining pins sufficiently to hold the inner brake pad.
- Pull protective foil off outer brake pad.
- Carefully insert outer brake pad.
- Push pad–retaining pins through fully and install securing spring.

Notes:

- *When inserting the outer brake pad, ensure that it does not stick to the brake caliper housing before it is in the correct installation position.*
- *Do not damage adhesive surface.*
- *After installing new brake pads, depress brake pedal firmly two or three times and hold pedal fully depressed for 5 – 10 seconds so that the outer pad is firmly bonded to the brake caliper housing. Firm bonding of the outer brake pad assures noise-free braking.*

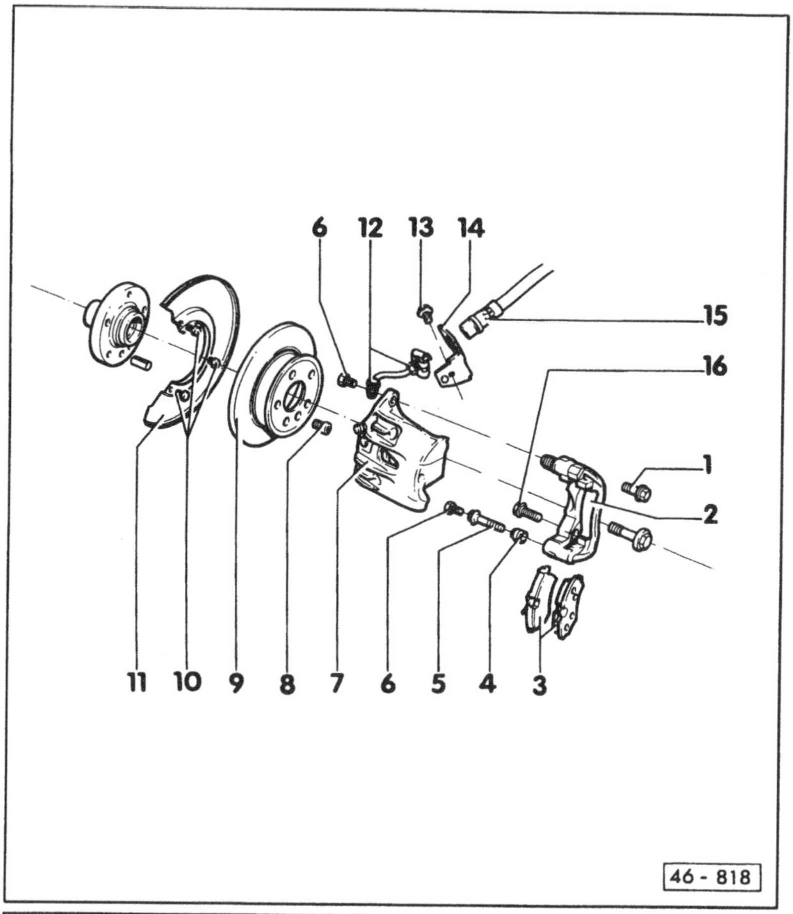

Front brake, assembly (Mark II caliper)

From chassis No. 70 PH 102 523, two different types of floating caliper brakes were used. Distinguishing features, ⇒ Page M 46-14.

Notes:

- *After replacing brake pads, depress brake pedal firmly several times with vehicle stationary, so that the brake pads are properly seated in their normal operating position.*

> ***CAUTION!***
> ***Always remove some brake fluid from the reservoir before installing new brake pads. When the caliper piston is pushed back, fluid is forced out of the caliper into the reservoir. After pads are installed, refill the reservoir only to the MAX mark.***

> ***WARNING!***
> ***Brake fluid is poisonous and must not be siphoned by mouth.***

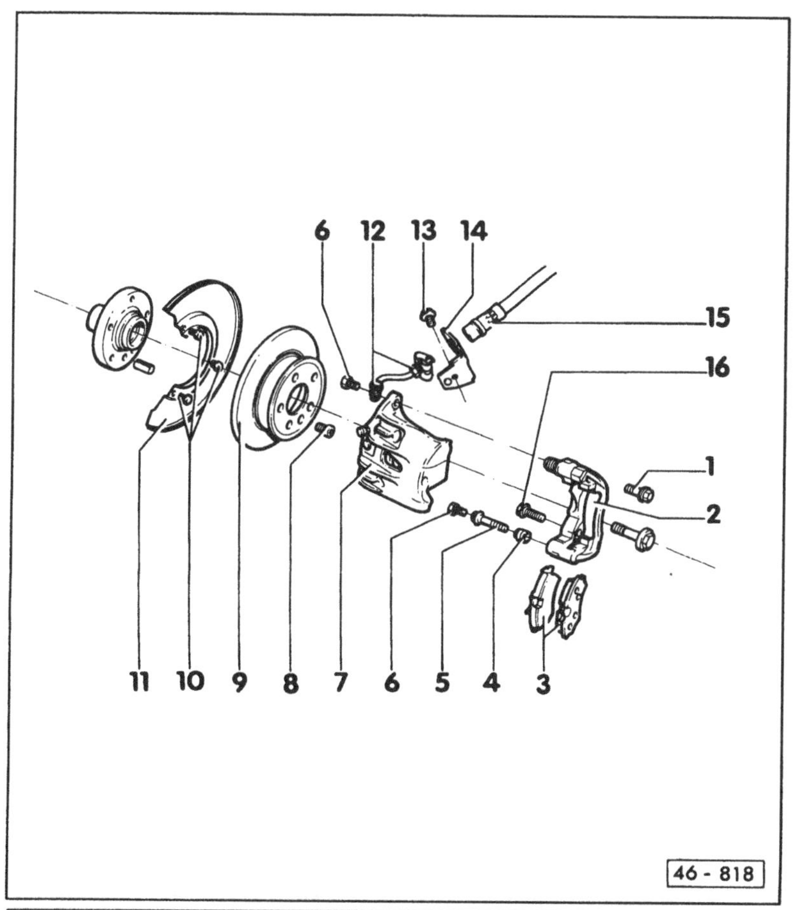

1 - Wheel bolt, 160 Nm (118 ft lb)

2 - Brake pad carrier

3 - Brake pads
- Always replace both sides.
- Pad thickness: 13.5 mm.
- Checking thickness, ⇒ Maintenance manual.
- Wear limit: 2 mm (excluding backing plate).
- Removing and installing, ⇒ Page M 46-12.

4 - Protective cap

5 - Guide pin

6 - Self-locking bolt, 35 Nm (26 ft lb)
- Always replace.

7 - Brake caliper
- Assembly, ⇒ Page M 47-1.
- Distinguishing features, ⇒ Page M 46-14.

8 - Hex socket-head bolt, 10 Nm (7 ft lb)

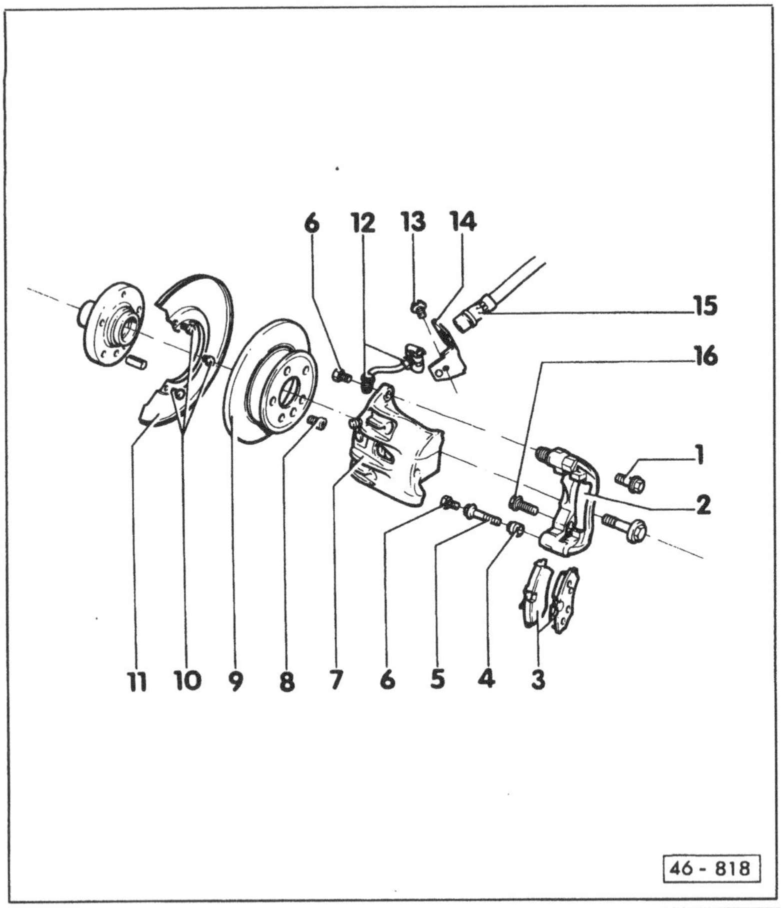

9 - Brake disc
- When worn always replace both sides.
- To remove, first remove brake caliper.
- Machine evenly, working on both sides from thickness when new.
- Ensure adequate margin for wear.
- Run-out: 0.08 mm. Difference in thickness: 0.02mm.
- Non-ventilated brake disc diameter: 282 mm.
- Brake disc thickness: 18 mm.
- Wear limit: 15 mm.

10 - Hex bolt, 10 Nm (7 ft lb)

11 - Splash shield

12 - Brake line

13 - Hex socket-head bolt, 10 Nm (7 ft lb)

14 - Bracket

15 - Brake hose

16 - Hex bolt, 270 Nm (200 ft lb)

Front brake pads, removing and installing

Removing

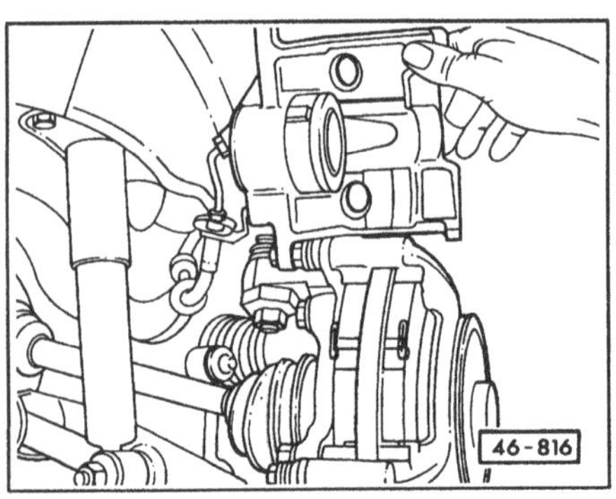

- Remove lower brake caliper housing securing bolt. Counter-hold with wrench on guide pin.
- Swing brake caliper housing upwards.
- Remove brake pads.

Installing

> ***CAUTION!***
> ***Always remove some brake fluid from the reservoir before installing new brake pads. When the caliper piston is pushed back, fluid is forced out of the caliper and into the reservoir. After pads are installed, refill reservoir only to the MAX mark.***

> ***WARNING!***
> ***Brake fluid is poisonous and must not be siphoned by mouth.***

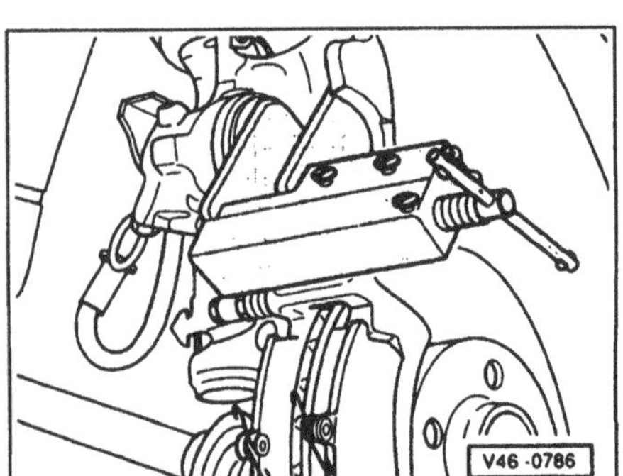

- Press piston back.
- Insert brake pads.

– Swing brake caliper housing down and tighten securing bolts to prescribed tightening torque.

Notes:

- *Always install the two self-locking hex bolts contained in the repair kit.*
- *After replacing the brake pads, depress the brake pedal firmly, several times, with the vehicle stationary so that the brake pads are properly seated in their normal operating position.*

Front brake caliper distinguishing features

Note:

Position of optional equipment number, *⇒ Page M 00-3.*

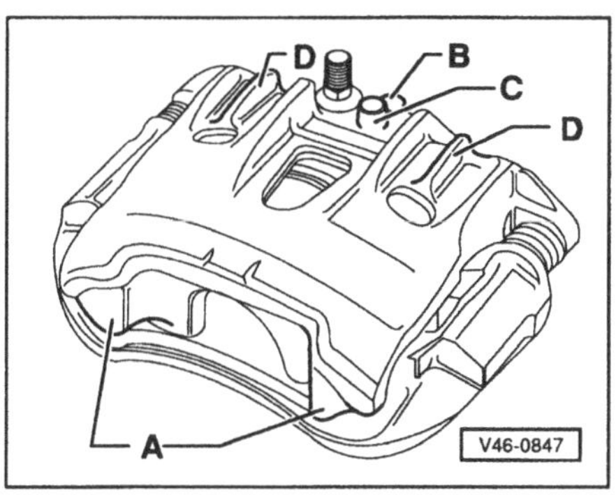

◄ **Mark II brake caliper (optional equipment number: 1LP)**

Distinguishing features:

- Symmetric balance mass -A-.
- VW symbol -B- behind projection -C-.
- Reinforcement ribs -D- extent to end.
- Further technical data, ⇒ Page M 00-2.

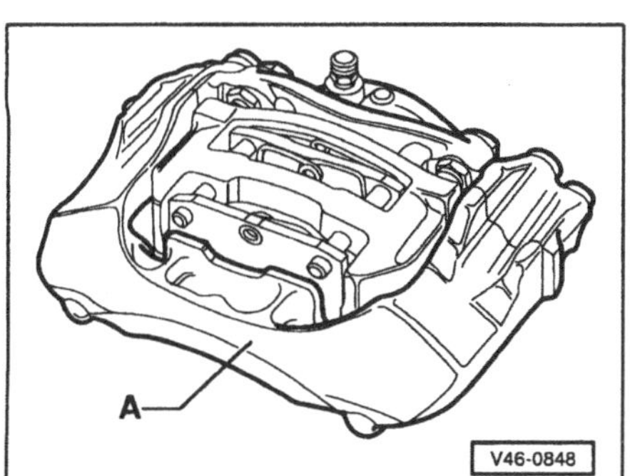

◄ **Framed brake caliper (optional equipment number: 1LE)**

Distinguishing features:

- Frame -A- is closed.
- Brake pads visible from above.
- Further technical data, ⇒ Page M 00-2.

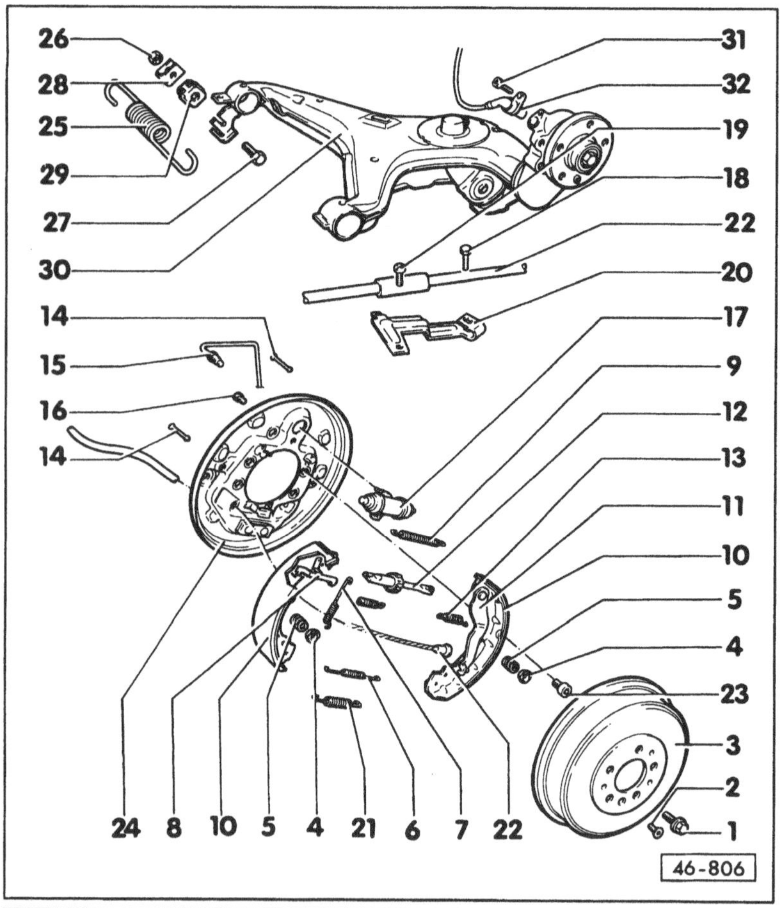

Rear wheel brake, assembly

Note:

Lightly grease push rod, brake shoe webs in the supports, brake lever at brake shoe and brake shoe contact surfaces on backing plate with lubricating paste, Part No. G 000 650.

> ***CAUTION!***
>
> - ***Always replace brake linings on both sides.***
> - ***Only use brake linings of the same quality.***

1 - Wheel bolt, 160 Nm (118 ft lb)

2 - Hex socket-head bolt, 5 Nm (44 in. lb)

3 - Brake drum
- Brake drum diameter: 268 mm.
- Wear limit: 269.5 mm.
- Removing, ⇒ Page M 46-21.

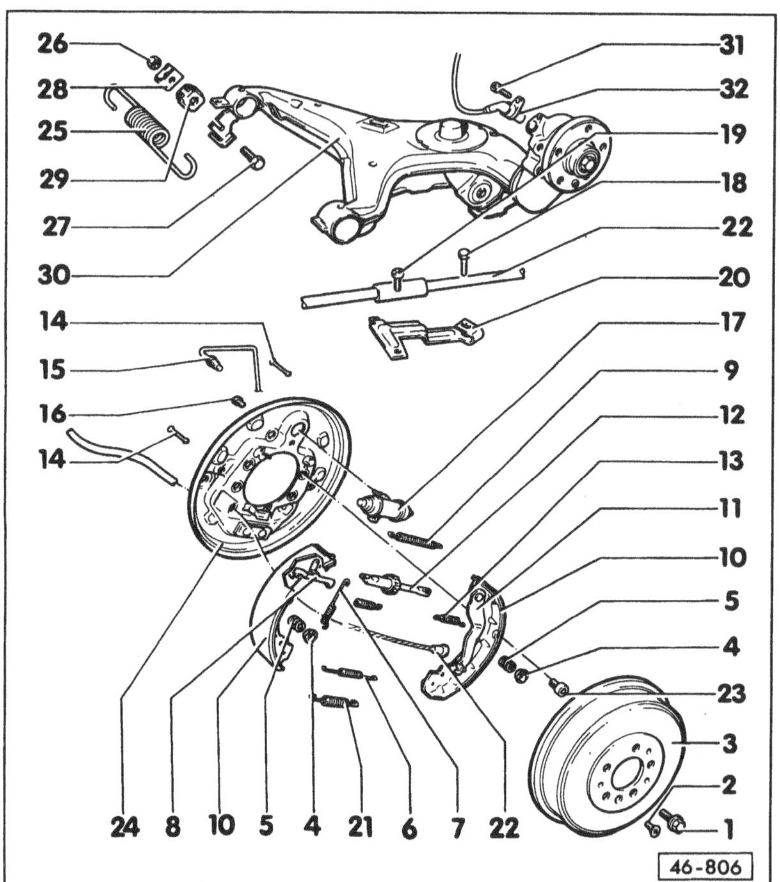

4 - Spring retainer
- To remove, push against spring and turn 90°.

5 - Spring

6 - Lower return spring

7 - Adjusting spring
- Always replace when changing linings.

8 - Adjusting lever

9 - Upper return spring

10 - Brake shoe
- Bonded brake lining.
- Secondary shoe with lever for parking brake cable.
- Lining thickness: 6.3 mm.
- Check through inspection hole in brake carrier, ⇒ Fig. 1.
- Removing and installing, ⇒ Page M 46-21.

11 - Parking brake lever

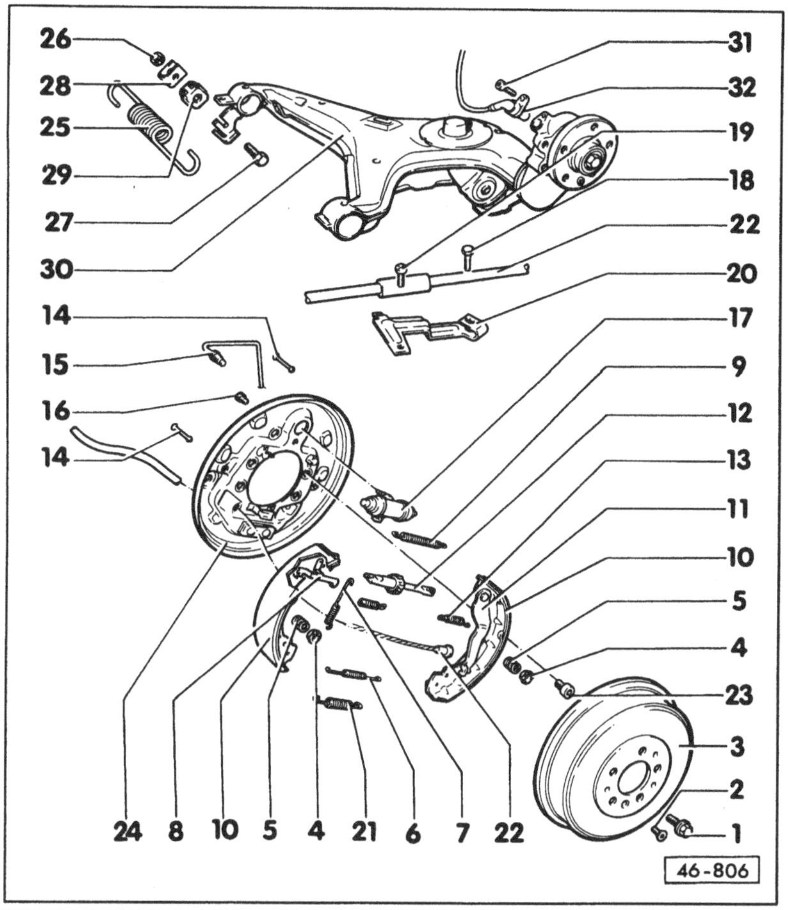

12 - Push rod with sleeve and adjustment pinion
- ♦ Left-hand rear wheel - right-hand thread.
- ♦ Right-hand rear wheel - left-hand thread.
- ♦ The chamfer on the sleeve and the longer push rod fork face towards backing plate.

13 - Return spring

14 - Pin

15 - Brake line

16 - Hex bolt, 19 Nm (14 ft lb)

17 - Wheel cylinder

18 - Hex bolt, 7 Nm (62 in.lb)

19 - Self-tapping screw, 7 Nm (62 in. lb)

20 - Parking brake cable bracket

21 - Lower return spring

22 - Parking brake cable
- ♦ Adjusting parking brake, ⇒ Page M 46-24.

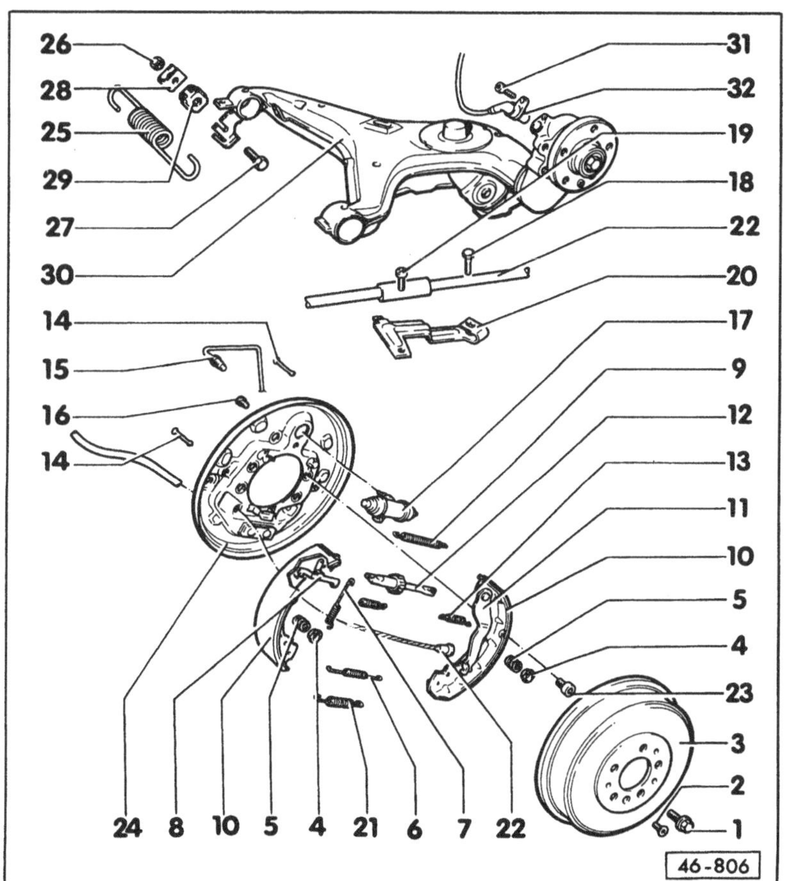

23 - Hex socket-head bolt, 160 Nm (118 ft lb)

24 - Backing plate

25 - Brake pressure regulator spring
- ♦ Adjusting brake pressure regulator, ⇒ Page M 47-14.

26 - Hex nut, 20 Nm (15 ft lb)

27 - Hex bolt

28 - Retainer
- ♦ Mark installation position before removing.

29 - Bracket

30 - Control arm
- ♦ Removing and installing, ⇒ Page M 42-1.

31 - Hex socket-head bolt, 10 Nm (7 ft lb)

32 - Wheel Speed Sensor
- ♦ Only on vehicles with ABS.
- ♦ Installing, ⇒ Page M 45-12.

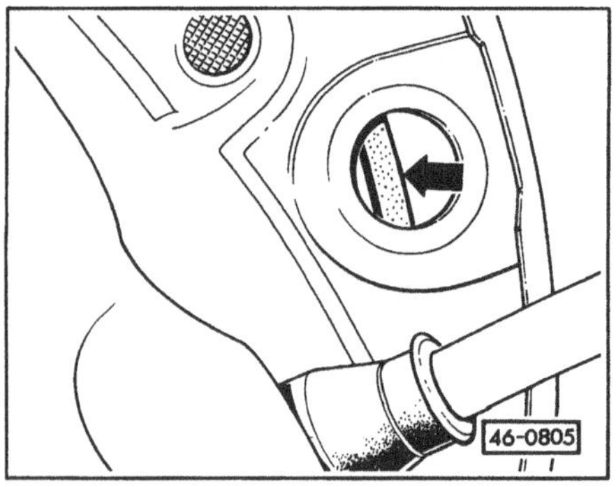

◄ **Fig. 1 Brake lining thickness, checking**

Wear limit: 1 mm (0.3937 in.).

Note:

The remaining lining is not visible when drum is worn.

Rear brakes, adjusting basic setting

Brake drum removed (removing brake drum, ⇒ Page M 46-21).

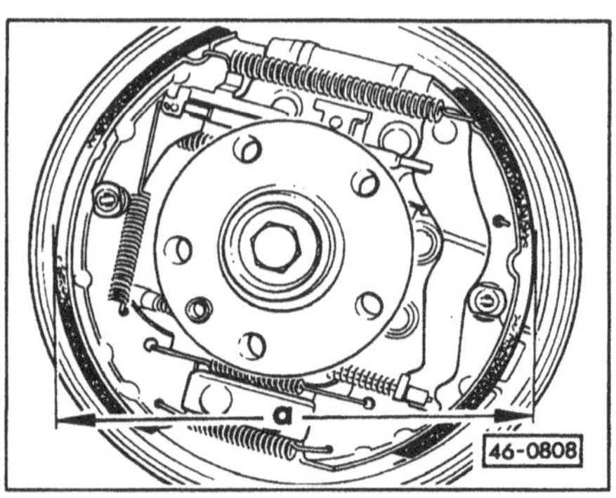

- Loosen adjusting nut for parking brake cable (parking brake lever in released position).
- Centralize brake shoes relative to backing plate.
- ◄ Adjust brake shoe to diameter -a- by turning the adjustment ratchet.

  a = Inner diameter of brake drum minus 1.0 mm.

Example:

Inner diameter of brake drum

	268.0 mm
	– 1.0 mm
a =	267.0 mm

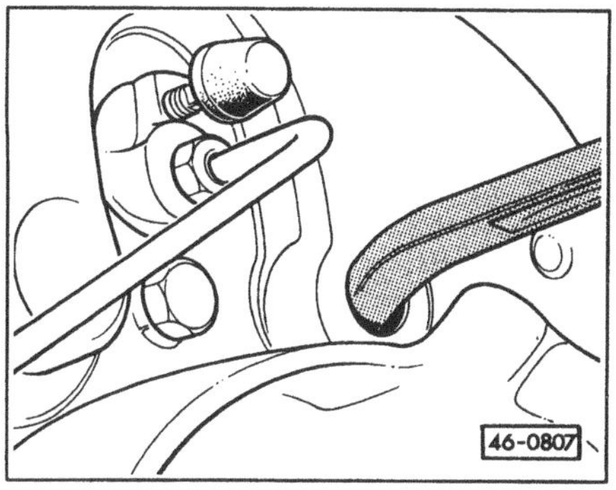

Brake drum, removing

- Remove wheel.
- Remove hex socket-head bolt.
- ◄ Push back adjusting pinion, using a lever through hole in backing plate.
- Pull off brake drum.

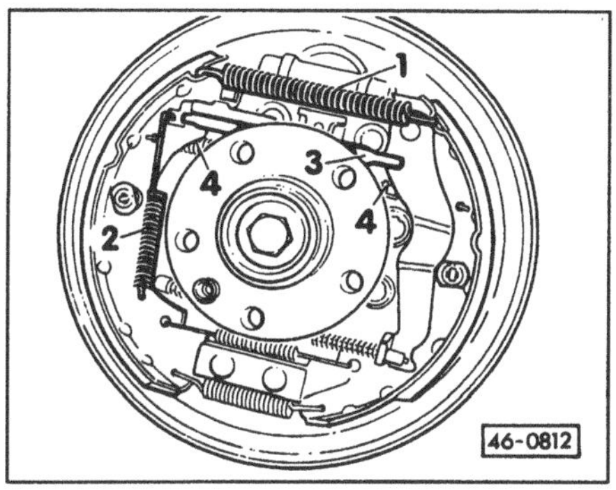

Brake shoes, removing and installing

Removing

- ◄ Remove return spring -1-.
- Remove spring for adjusting lever -2-.
- Turn back adjusting pinion -3- fully.
- Unhook return spring -4-.

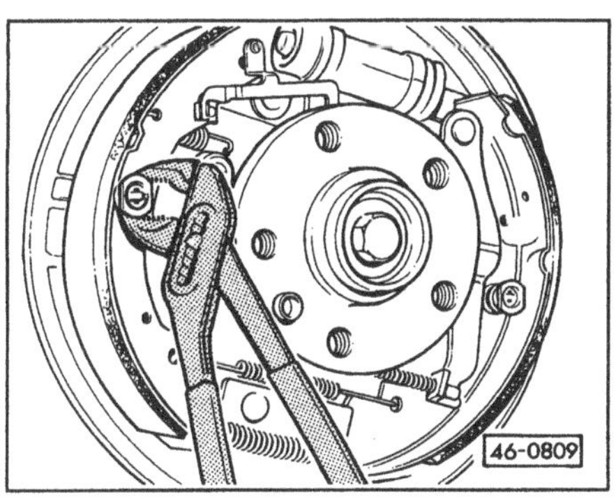

- ◄ Remove spring

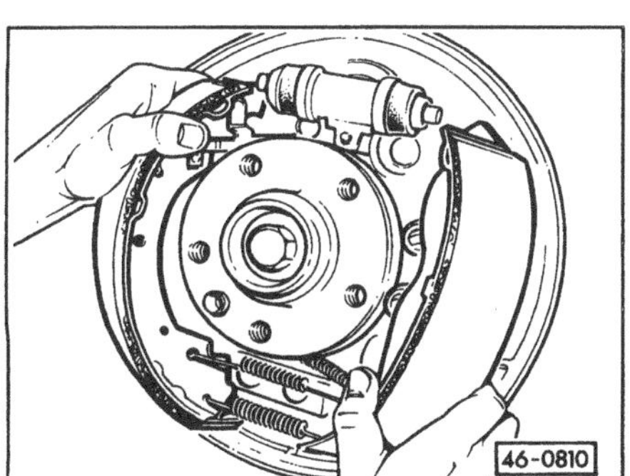

- ◄ Remove brake shoes.
- Unhook lower return spring.

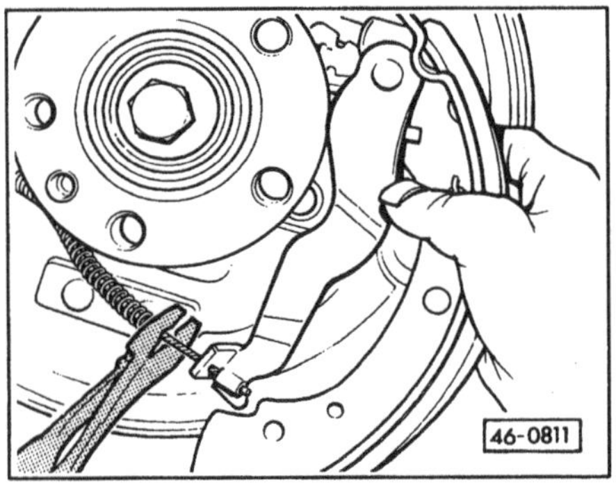

- Unhook parking brake cable from brake lever.

Installing

Install in reverse order.

Parking brake, adjusting

Note:

Due to the automatic rear wheel brake adjustment there is no requirement to adjust the parking brake. Adjustment is only necessary after replacing parking brake cables, backing plate or brake linings.

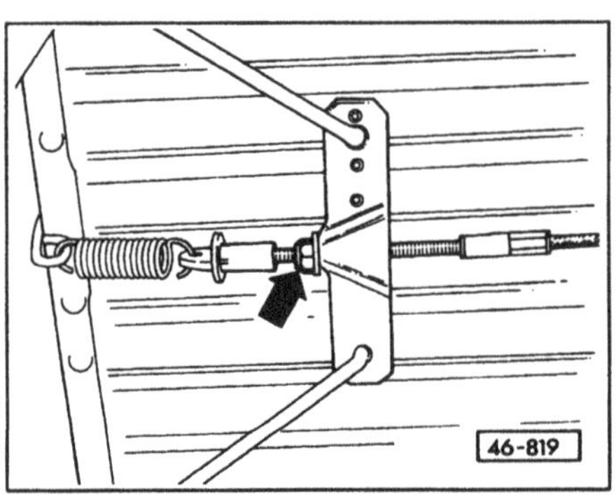

- Release handbrake.
- Remove brake drum (turn brake back first, ⇒ Page M 46-21).
- Tighten adjustment nut (arrow) until parking brake lever is lifted 2 mm off brake shoe, ⇒ Fig. 46-0820.

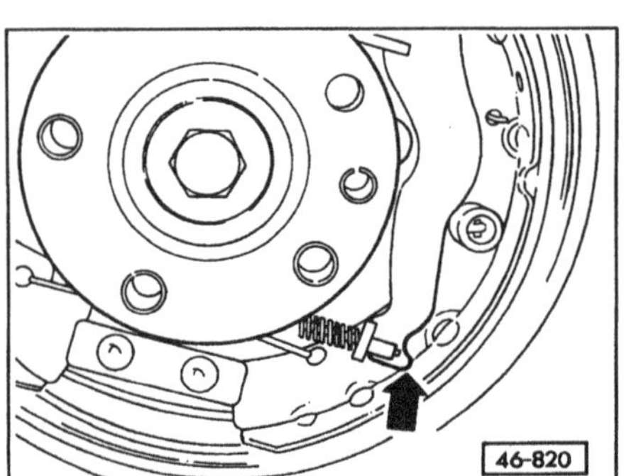

- Install brake drum.
- Check whether both wheels rotate freely.

Note:

Adjusting the rear wheel brakes will result in slightly longer travel of the parking brake lever and brake pedal. This will reduce when driving by operating the brake pedal.

Spring sleeve for parking brake lever, installed position

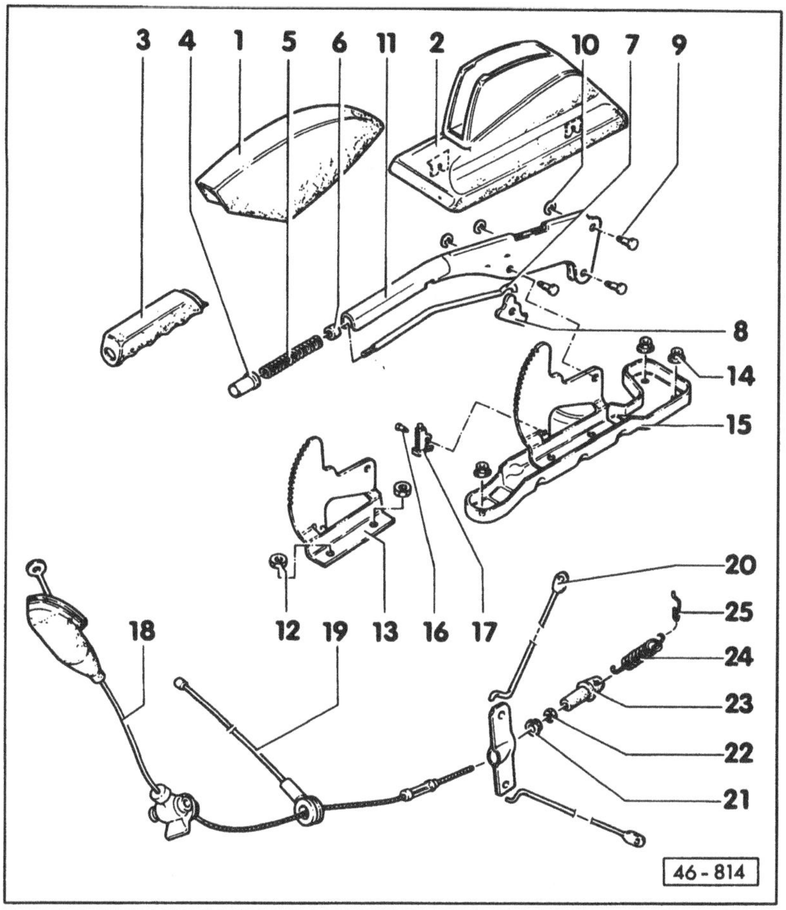

Parking brake, assembly

1 - Trim

2 - Parking brake lever trim

3 - Handle
- ♦ Press retaining tabs under handle down and pull handle forward.

4 - Push button

5 - Spring
- ♦ Installation position, ⇒ Fig. 1.

6 - Spring sleeve
- ♦ Installation position, ⇒ Fig. 1.

7 - Pawl rod

8 - Pawl

9 - Pin

10 - Securing clip

11 - Parking brake lever

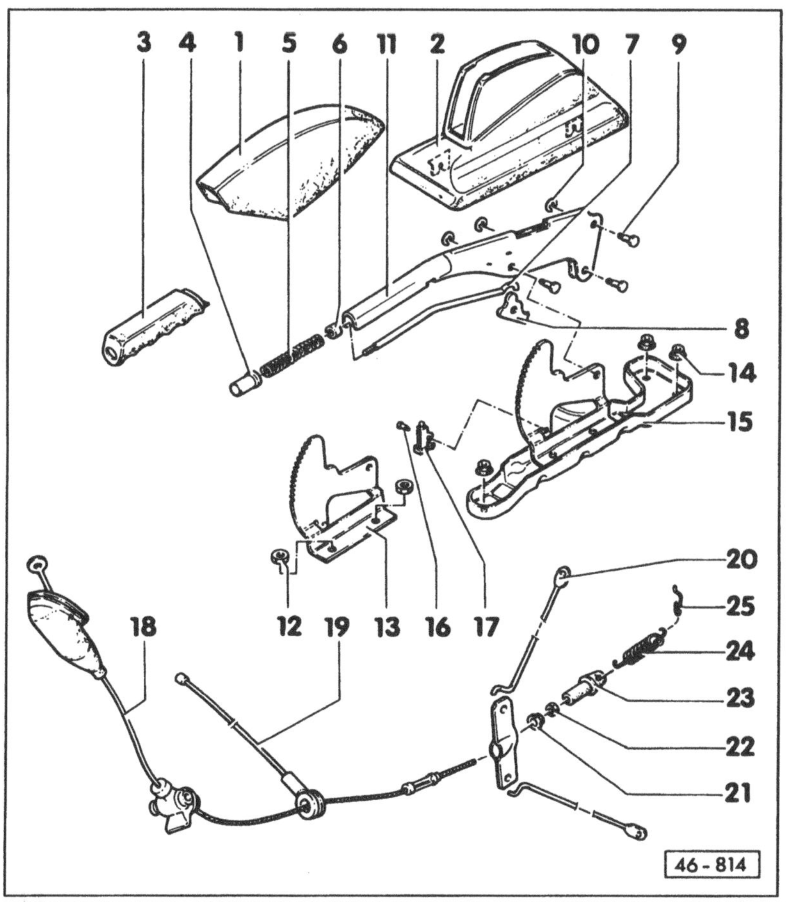

12 - Self-locking hex nut, 25 Nm (18 ft lb)
- ♦ Always replace.

13 - Toothed segment
- ♦ For RHD vehicles only.

14 - Self-locking hex nut, 25 Nm (18 ft lb)
- ♦ Always replace.

15 - Parking brake carrier

16 - Expanding rivet

17 - Contact switch

18 - Parking brake cable

19 - Tensioning cable
- ♦ Only installed to dropside flatbed vehicles.

20 - Pull rod

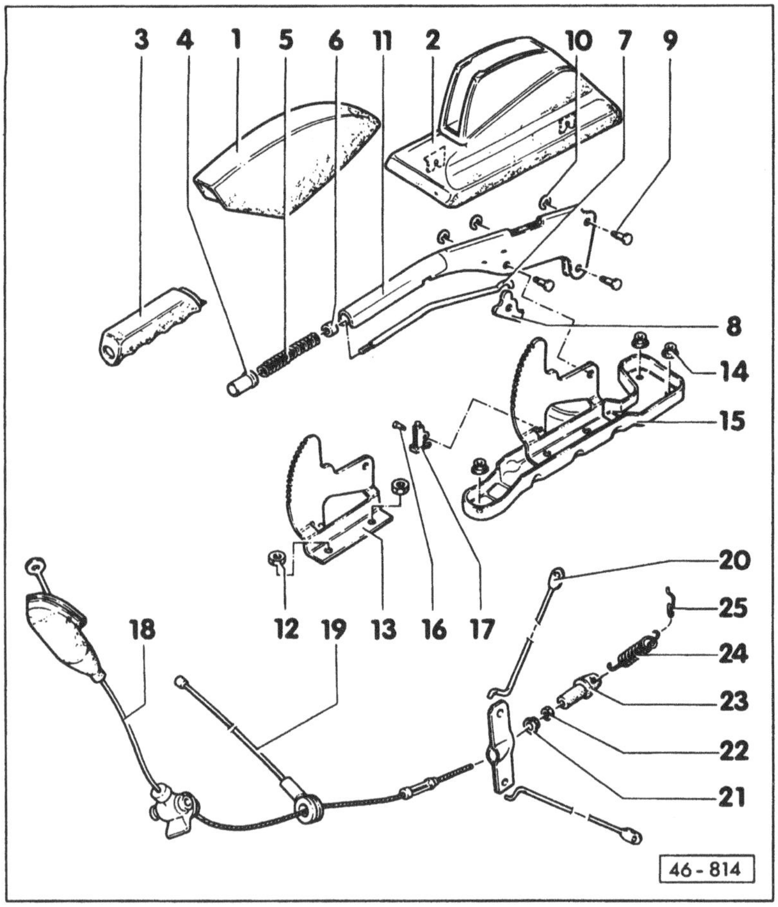

21 - Adjusting nut

- Adjusting parking brake, ⇒ Page M 46-24.

22 - Hex nut

- Lock on adjusting nut.
- Deleted with the introduction of a self-locking adjusting nut.

23 - Spring holder

24 - Return spring

25 - Bracket

26 - Parking brake cable

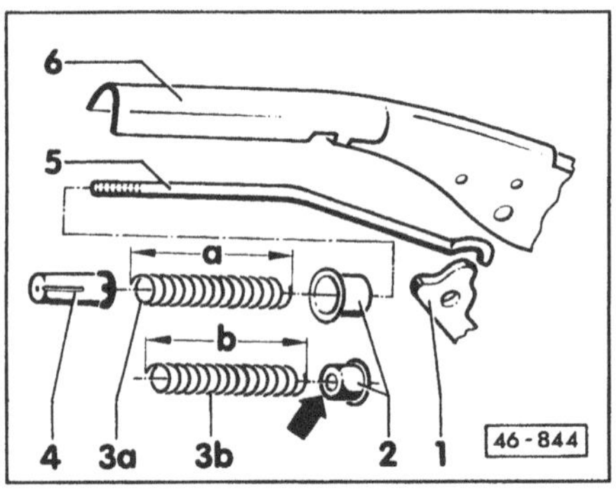

◄ **Fig. 1 Spring sleeve, installation position**

> ***CAUTION!***
>
> ***The installation position of the spring sleeve is dependent upon the length of the spring.***
>
> ***Determine length of spring before installing.***

1 - Pawl

2 - Spring sleeve

3a - Spring
Dimension -a- approx. 114 mm
(to chassisNo. 70 PM 100 936).
Dimension -a- approx. 126 mm
(from chassisNo. 70 PM 100 937).
When installing, slide spring sleeve -2- onto spring -3a-.

3b - Spring
Dimension b = approx. 104 mm.
When installing, the spring sleeve -2- small internal diameter (arrow) points towards spring -3b-.

4 - Push button
Thread on up to stop.

5 - Pawl rod

6 - Parking brake lever

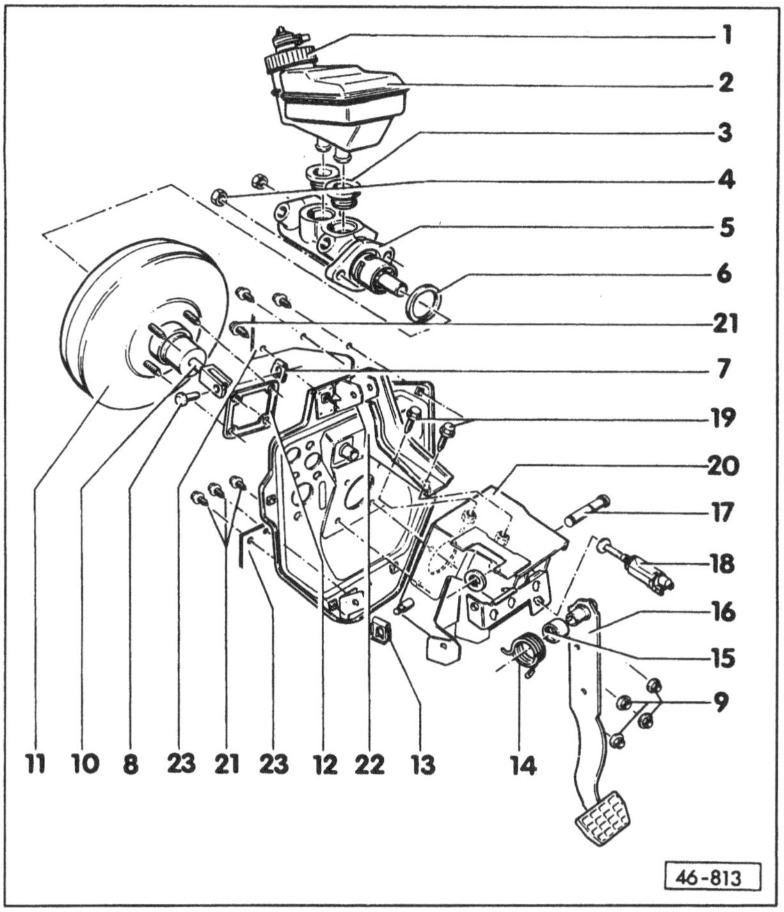

Brake pedal/brake booster/brake master cylinder, removing and installing

Tandem brake booster ➤ 5/95

> ***CAUTION!***
>
> ***Brake pedal travel must not be restricted by floor mats.***

Note:

Grease all mounts with white grease before installing (P/N AOS 126 000 06).

1 - Cap
- Contact pin in cap on vehicles with a brake fluid level warning device.
- Functional check: With parking brake released and ignition on, warning light must illuminate when contact pin is depressed.

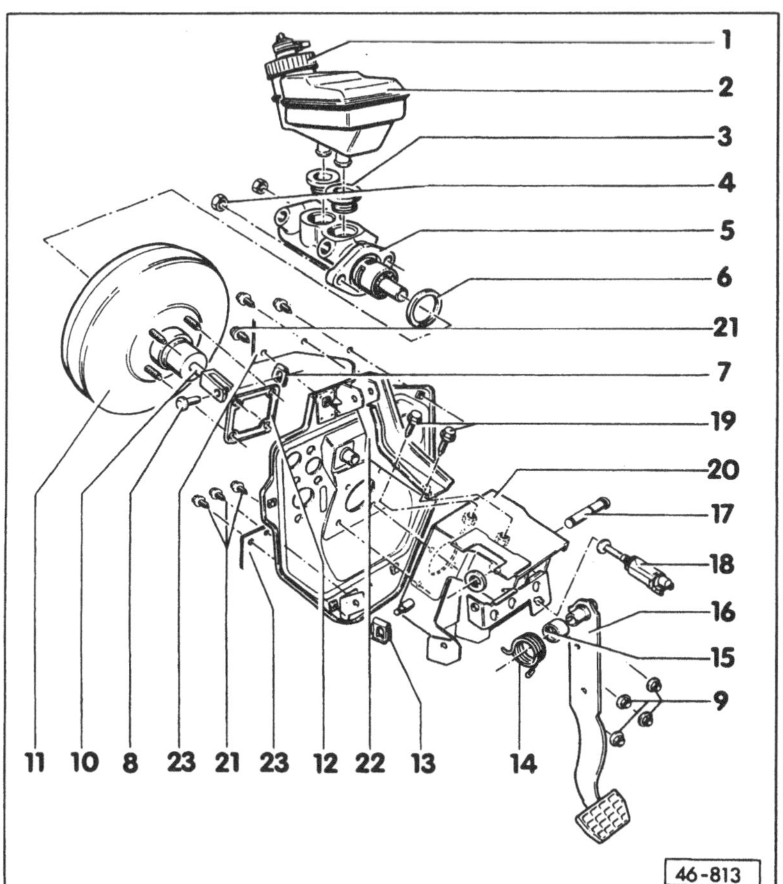

2 - Brake fluid reservoir
- With strainer

3 - Sealing plug
- Moisten with brake fluid and press into reservoir.

4 - Hex nut, 20 Nm (15 ft lb)

5 - Brake master cylinder, 22.2 mm diameter
- Cannot be repaired, replace complete.

6 - Sealing ring
- Always replace.

7 - Securing clip
- Ensure clip is seated.

8 - Pin

9 - Self-locking hex nut, 25 Nm (18 ft lb)
- Always replace.

10 - Brake booster push rod clevis

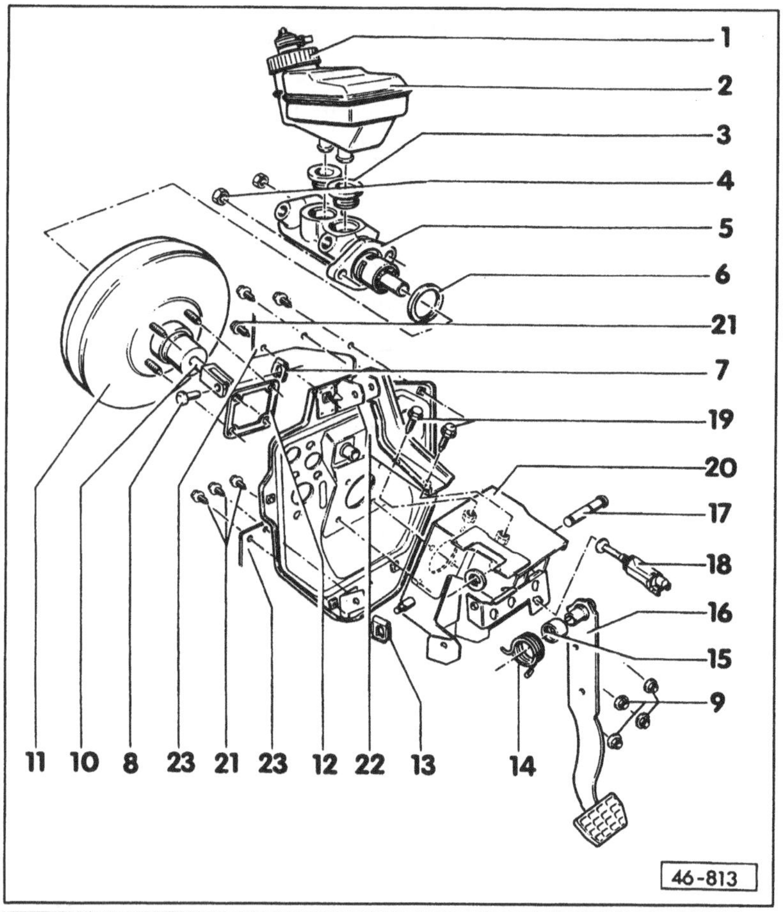

11 - Brake booster

- Before removing, remove Battery and disconnect brake lines.
- Gas engines, intake manifold is vacuum source.
- Diesel engines use vaccum pump for vacuum source.
- Functional check:
  - With engine switched off, depress brake pedal firmly several times (to deplete vacuum in unit).
  - Depress brake pedal with normal foot pressure and hold. Start engine. If booster is working properly, pedal will give slightly under foot (booster assistance becomes effective).
- If faulty, replace complete.

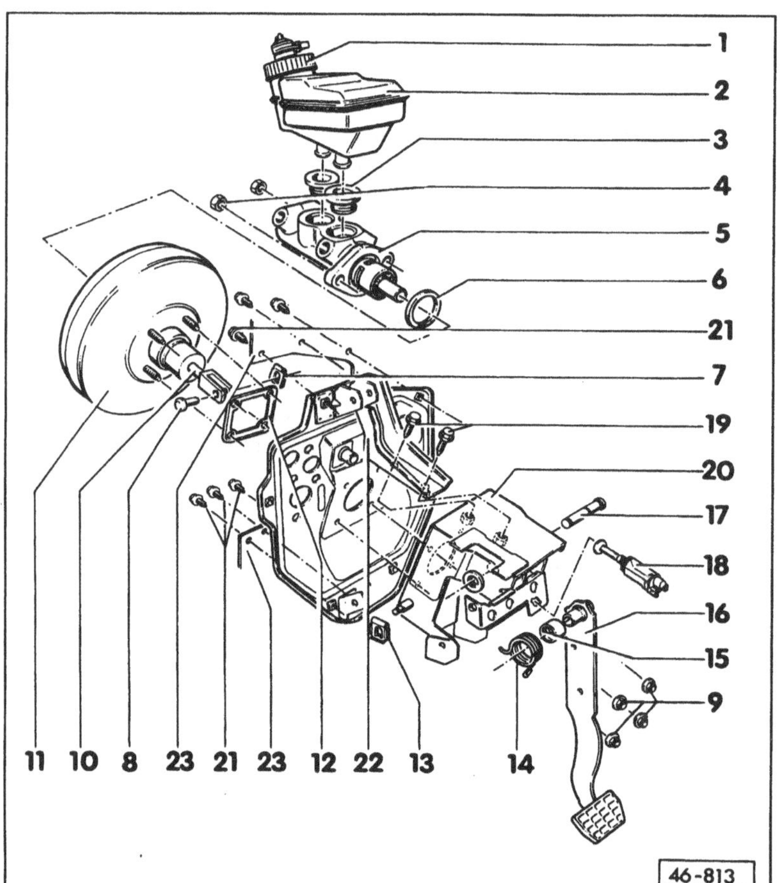

- Non-return valve (in vacuum hose)

  Functional check:
  - must be possible to blow through in direction of arrow.
  - Blowing against arrow must seat valve.

12 - Seal

- For brake booster

13 - Securing clip

14 - Brake pedal return spring

15 - Bushing

16 - Brake pedal

17 - Brake pedal pivot pin

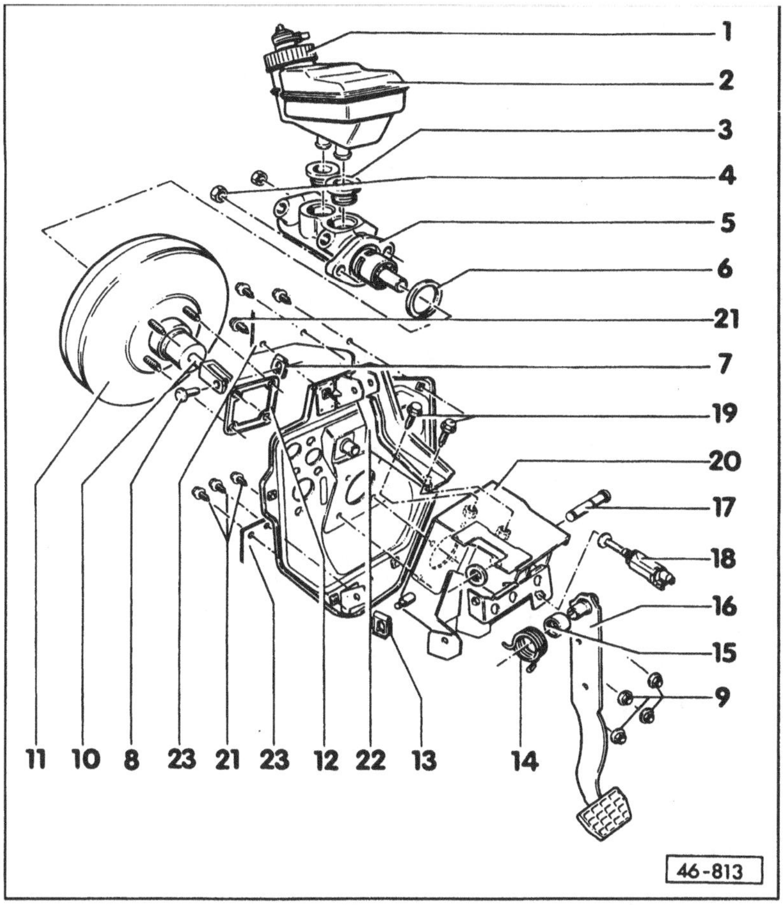

18 - **Brake Light Switch**
- Before installing Brake Light Switch, connect brake pedal to brake booster push rod.
- Before installing, pull plunger out completely.
- Guide switch through mounting hole and secure by turning clockwise 90°.

19 - **Hex bolt, 25 Nm (18 ft lb)**

20 - **Bracket**

21 - **Hex bolt, 25 Nm (18 ft lb)**

22 - **Mounting plate**

23 - **Front body panel**

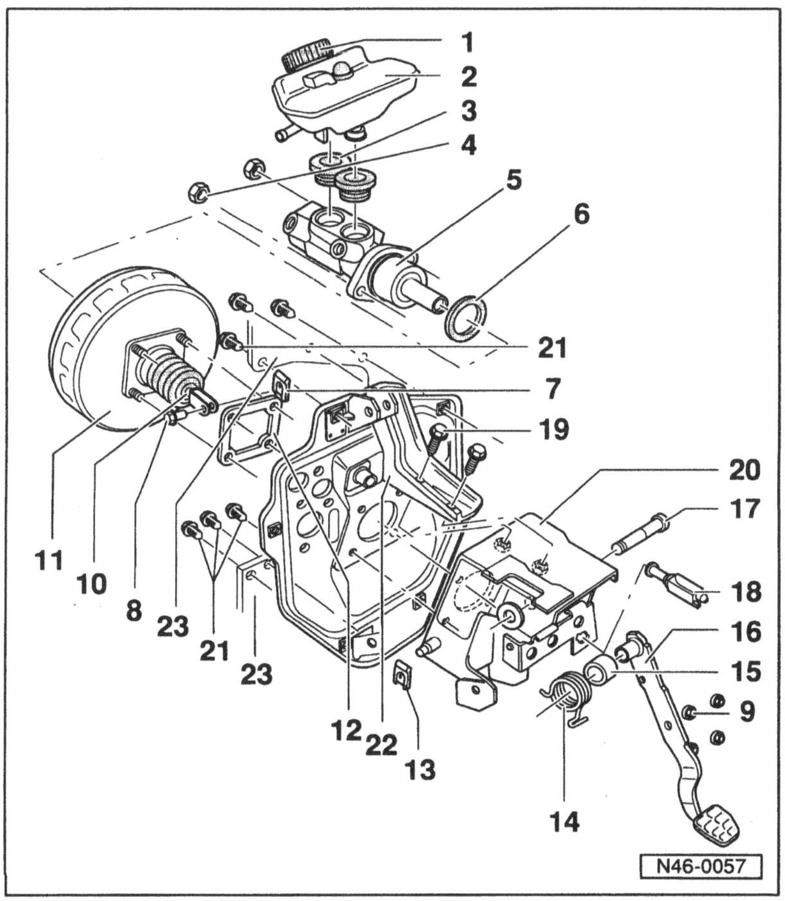

Brake pedal/brake booster/brake master cylinder, removing and installing

Tandem brake booster 6/95 ➤

> ***CAUTION!***
>
> ***Brake pedal travel must not be restricted by floor mats.***

Note:

Grease all mountings before installing, with white lubricating paste, Part No. AOS 126 000 06.

1 - **Cap**

2 - **Brake fluid reservoir**
- With strainer
- For vehicles with brake fluid warning indicator contact pin in reservoir

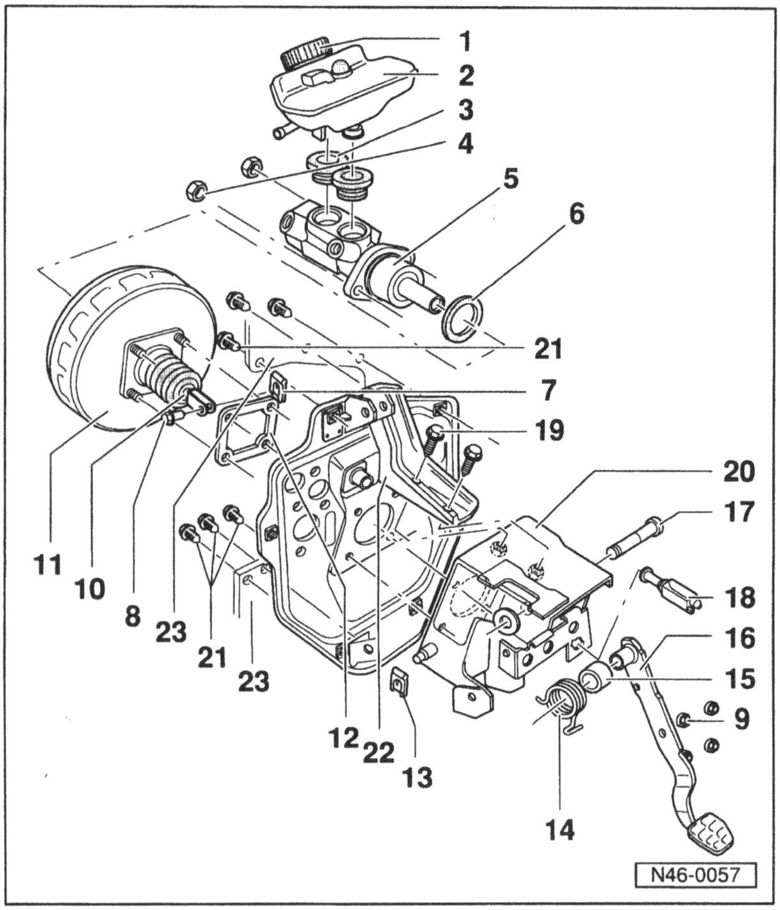

3 - **Rubber grommets**
- Moisten with brake fluid and press into reservoir.

4 - **Hex nut, 20 Nm (15 ft lb)**

5 - **Brake master cylinder, 23.81 mm diameter**
- Cannot be repaired, replace completely if necessary.

6 - **O-ring**
- Replace

7 - **Lock washer**
- Check for secure fit.

8 - **Bolt**

9 - **Self-locking hex nut, 25 Nm (18 ft lb)**
- Replace.

10 - **Brake booster push rod clevis**

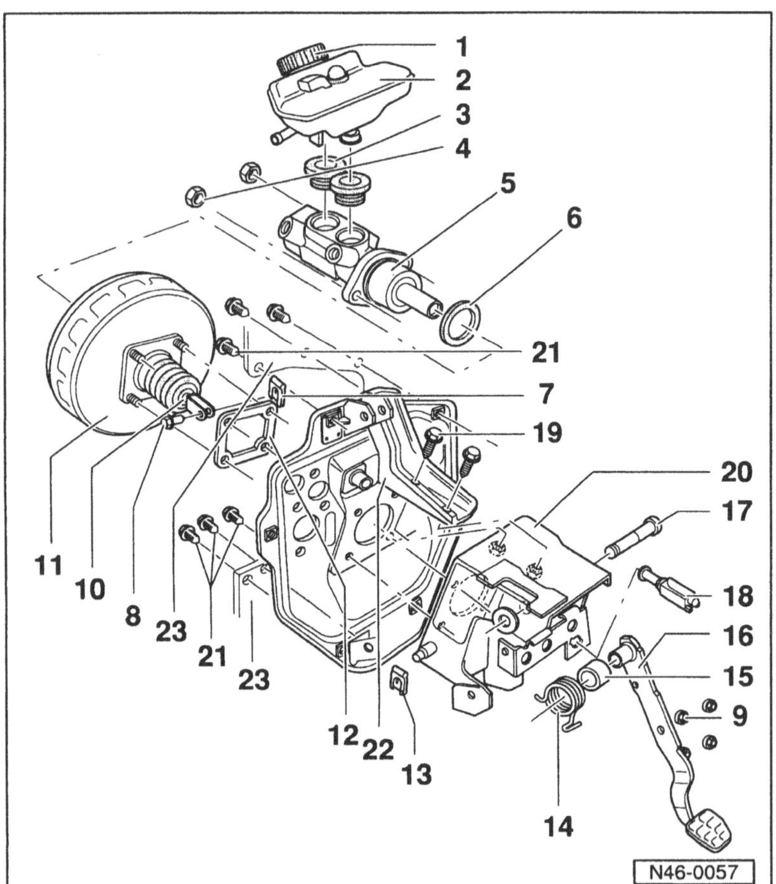

11 - **Brake booster**
- Before removing, remove Battery and disconnect brake lines.
- Gas engine intake manifold is vacuum source.
- Diesel engines use a vacuum pump to create vacuum.
- Function check:
  - With engine switched off, depress brake pedal firmly several times (to deplete vacuum in the unit).
  - Depress brake pedal with normal foot pressure and hold.
  - Start engine. If booster is working properly, pedal will give slightly under foot (booster assistance becomes effective).
- If faulty, replace complete.
- Non-return valve (in vacuum hose)

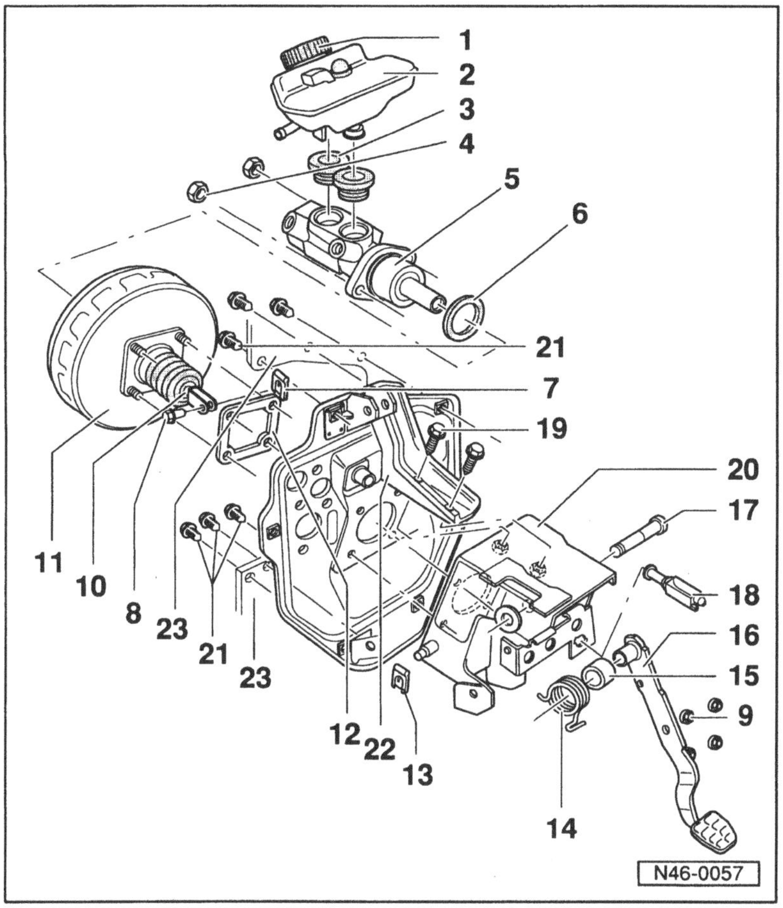

12 - Seal

13 - Securing clip

14 - Brake pedal return spring

15 - Brake pedal bushing

16 - Brake pedal

17 - Brake pedal pivot pin

18 - Brake Light Switch
- ♦ Before installing Brake Light Switch, connect brake pedal to brake booster push rod.
- ♦ Before installing, pull plunger out completely.
- ♦ Guide switch through hole and secure by turning clockwise 90°.

19 - Hex bolt, 25 Nm (18 ft lb)

20 - Bracket

21 - Hex bolt, 25 Nm (18 ft lb)

22 - Bracket

23 - Front body panel

Front brake caliper, assembly

(Mark II caliper)

Notes:

- *When repairing, install all parts supplied in repair kit.*
- *Apply a thin coat of brake fluid to brake cylinder, piston and seals.*

WARNING!
Brake fluid is poisonous. Do not siphon by mouth. Brake fluid is corrosive and must not come in contact with painted parts.

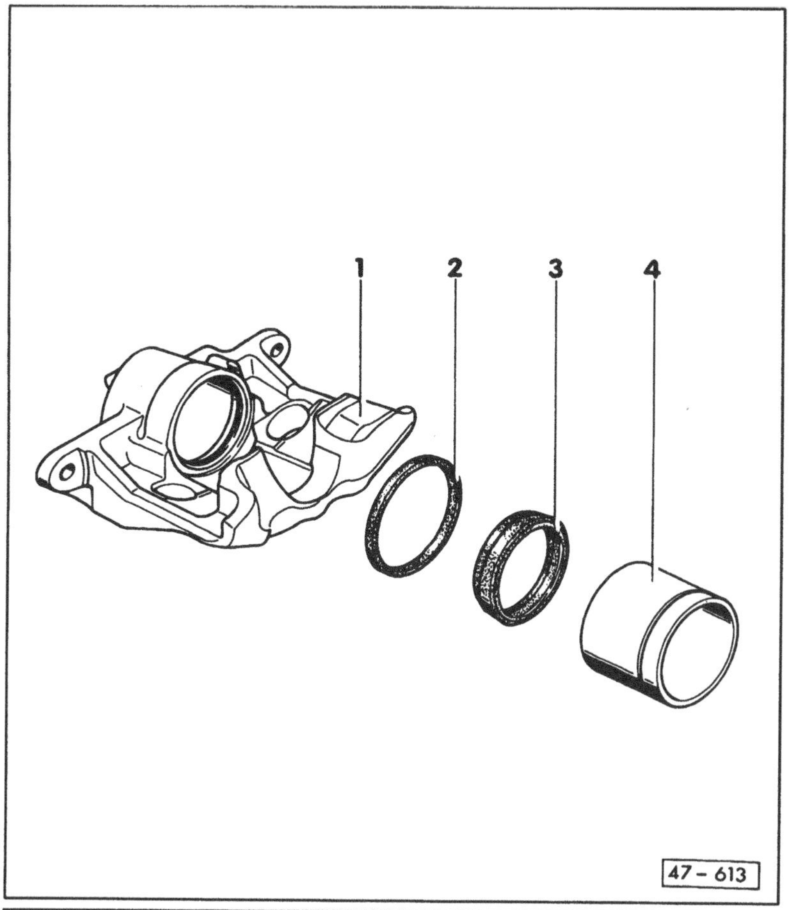

1 - Brake caliper housing

2 - Seal
- Remove using a plastic wedge, ⇒ Page M 47-2.

3 - Dust boot
- Installing, ⇒ Page M 47-3.
- Do not damage when installing piston.

4 - Piston
- Removing and installing, ⇒ Page M 47-2.

Front brake caliper piston, removing and installing

Removing

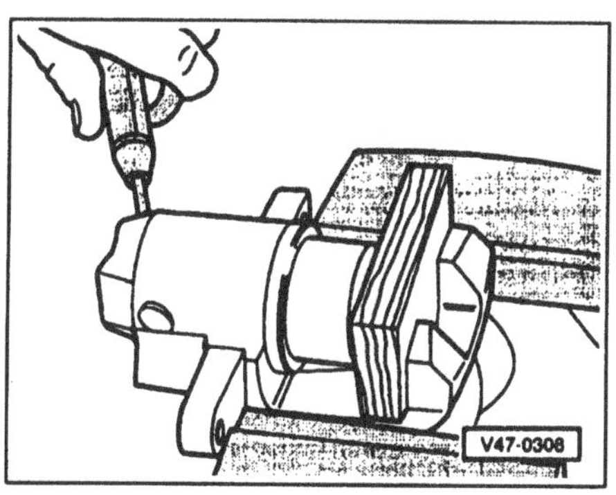

◄ – Force piston out of brake caliper housing, using compressed air.

Note:

Place a piece of wood in the caliper to prevent damaging the piston.

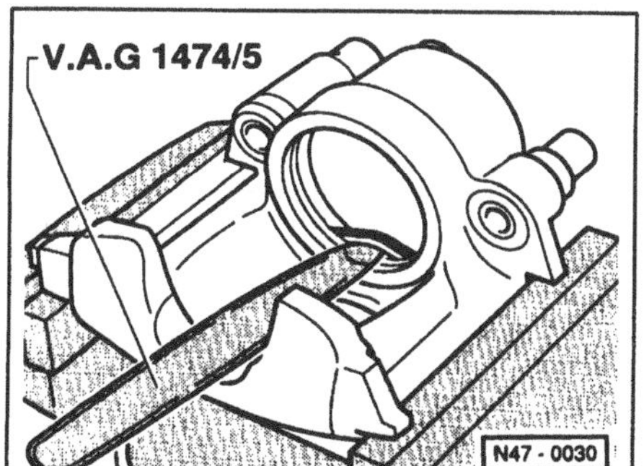

◄ – Remove seal using modified wedge VAG 1474/5, or equivalent.

Note:

When removing, ensure that the surface of the cylinder is not damaged.

Modifying wedge VAG 1474/5, ⇒ Page M 47-3.

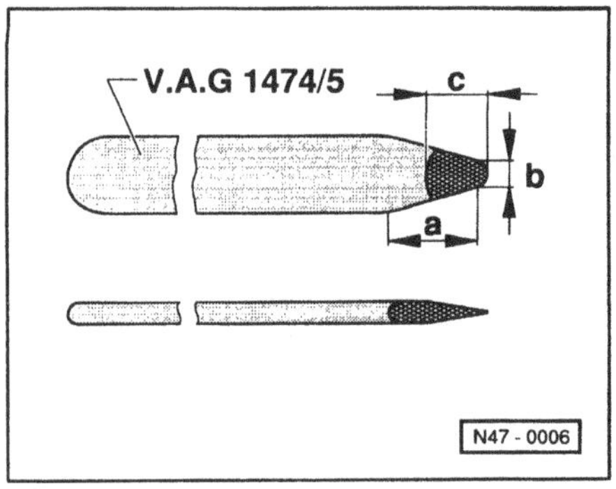

Modifying wedge VAG 1474/5

◄ – Reshape wedge point size in area -a- as per illustration, observing width -b- (6 mm).

– Reshape wedge point to length -c- (10 mm).

– Remove burrs from modified wedge point.

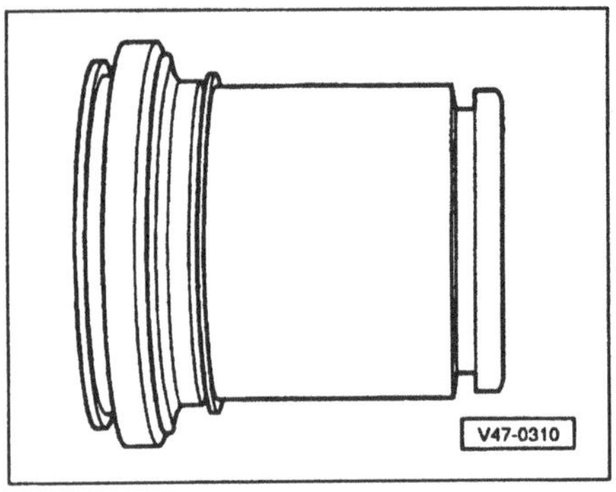

Installing

◄ – Install dust boot with outer sealing lip on piston.

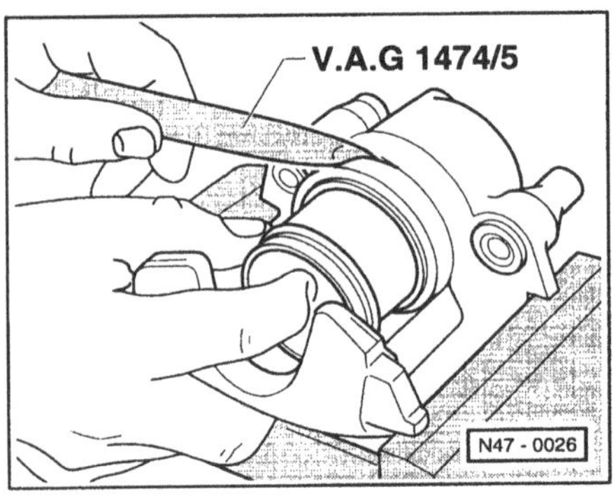

◄ – Using plastic wedge, install inner sealing lip into cylinder groove.

Note:

Hold the piston in front of the caliper housing.

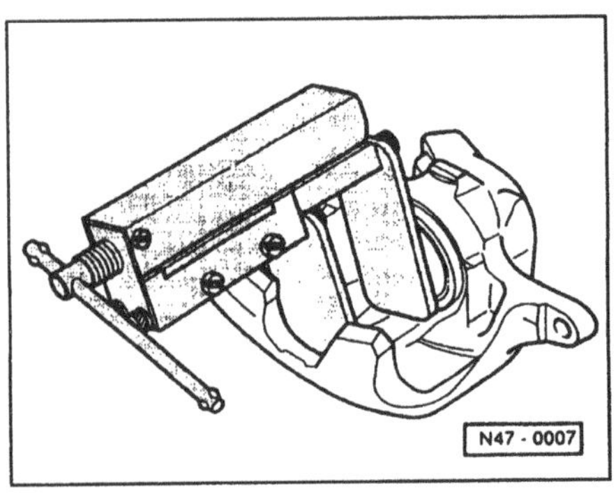

◄ – Press piston into brake caliper housing, using piston resetting tool.

Note:

The dust boot outer sealing lip will then locate in the piston groove.

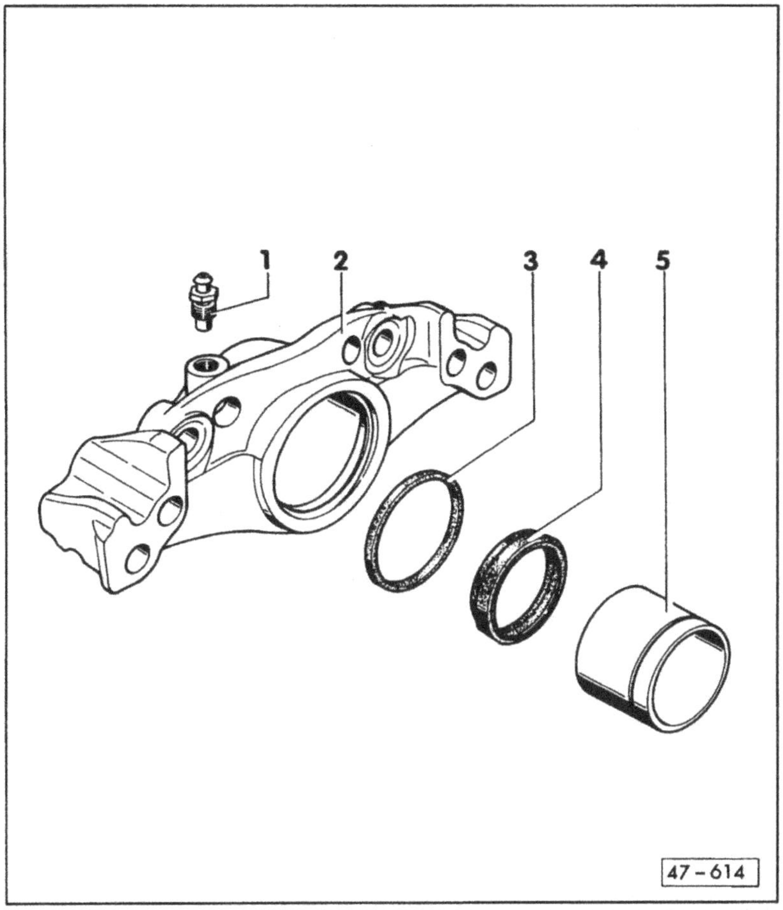

Front brake caliper, assembly

(Floating frame caliper)

Notes:

- *When repairing, install all parts supplied in repair kit.*
- *Apply thin coat of brake fluid to brake cylinder, piston and seals.*

1 - Bleeder screw

2 - Brake caliper housing

3 - Seal
- Remove using a plastic wedge, ⇒ Page M 47-6.

4 - Dust boot
- Installing, ⇒ Page M 47-7.
- Do not damage when installing piston.

5 - Piston
- Removing and installing, ⇒ Page M 47-6.

Front brake caliper piston, removing and installing

(Floating frame caliper)

Removing

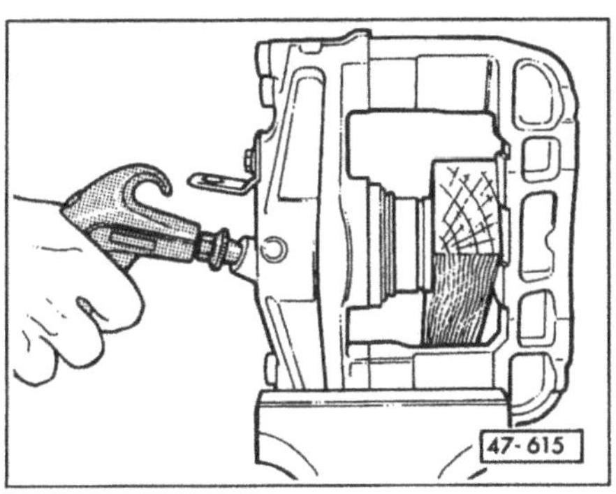

◄ – Force piston out of brake caliper housing, using compressed air.

Note:

Place a piece of wood in the caliper to prevent damaging the piston.

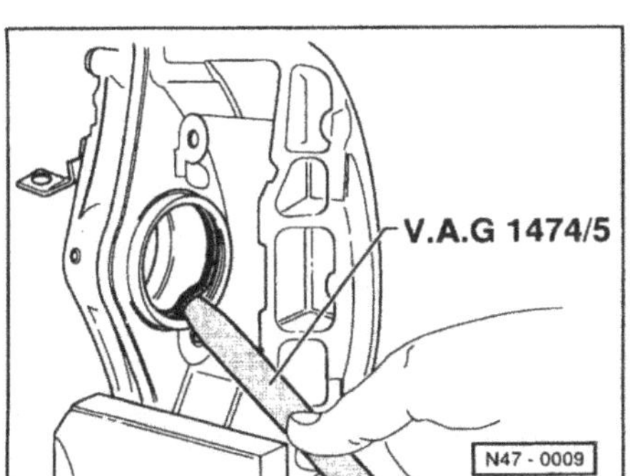

◄ – Remove seal using modified wedge VAG 1474/5, or equivalent.

Note:

When removing ensure that the surface of the cylinder is not damaged.

Modifying wedge VAG 1474/5, ⇒ Page M 47-3.

Installing

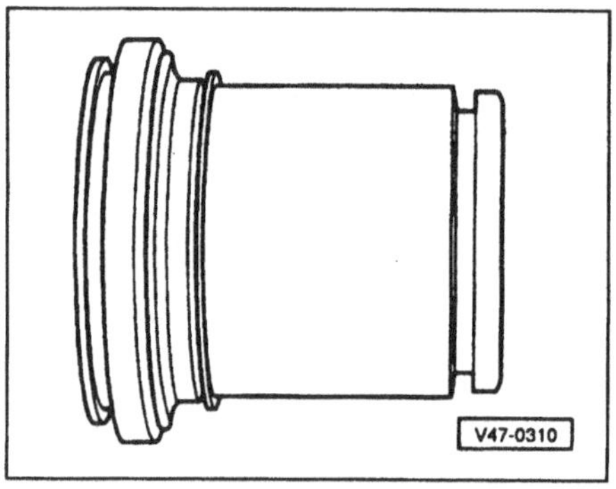

- Place dust boot outer sealing lip on piston.

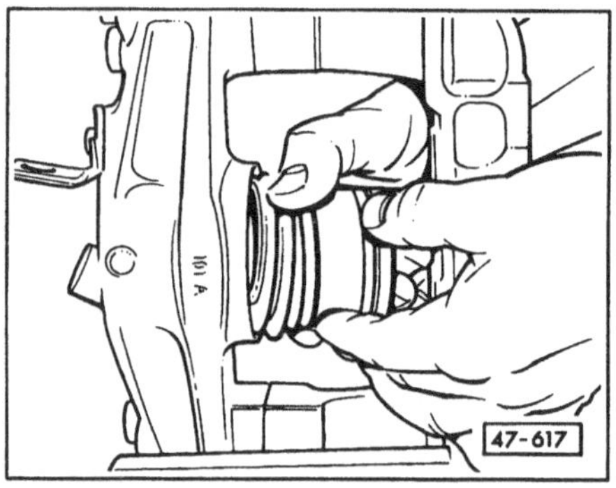

- Install inner sealing lip into cylinder groove.

Note:

Hold piston in front of caliper housing.

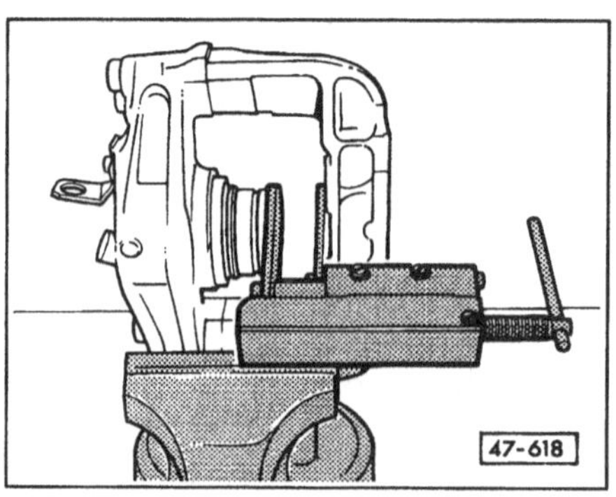

- Press piston into brake caliper housing, using piston resetting tool.

Note:

The dust boot outer sealing lip will then locate in the piston groove.

Brake system, bleeding

(Vehicles with and without ABS)

Changing brake fluid,
⇒ Maintenance manual

Notes:

- *Only use new brake fluid conforming to US standard, FMVSS 116 DOT 4.*
- *Genuine VW/Audi brake fluid conforms to this specification.*
- *Brake fluid is poisonous. Do not siphon by mouth.*
- *Brake fluid is corrosive and must not come into contact with painted parts.*
- *Rinse off spills, using plenty of water.*
- *Brake fluid absorbs moisture from the ambient air and must therefore always be stored in air-tight containers.*

Bleeding

(with brake filling and bleeding tool US 1116)

– Using suction bottle, remove as much brake fluid as possible from brake fluid reservoir.

– Connect US 1116 to brake fluid reservoir and switch on.

– Open bleeder screws and bleed brake calipers and wheel cylinders in following sequence:

1 – Right rear wheel cylinder

2 – Left rear wheel cylinder

3 – Right front brake caliper

4 – Left front brake caliper

Note:

Do not press the horizontal brake pressure regulator lever down when bleeding the brakes on vehicles with a brake pressure regulator.

(without brake filling and bleeding tool US 1116)

- Attach bleeder bottle hose.
- Build-up brake pressure in brake system by pumping brake pedal.
- With brake pedal held down, open and then close bleeder screw.
- Release brake pedal.
- Repeat operation until brake fluid flows with no air present.

Bleeding sequence, ⇒ Page M 47-10.

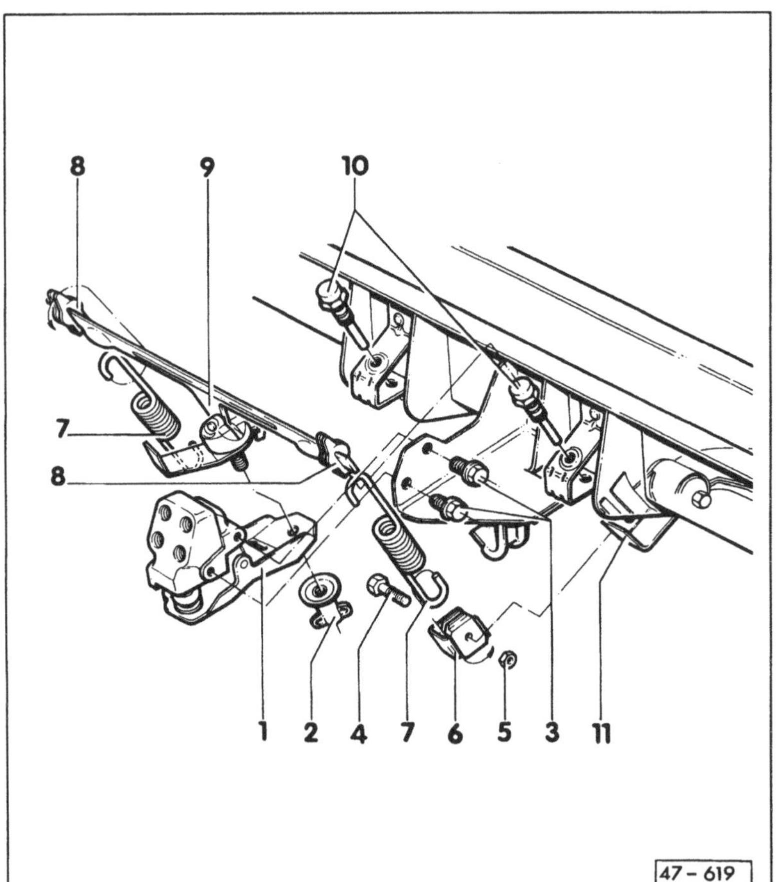

Brake pressure regulator, removing and installing

1 - Brake pressure regulator
- ♦ Adjusting, ⇒ Page M 47-14.

2 - Clip
- ♦ Press onto threaded stud and tighten 2 to 3 turns.

3 - Hex bolt, 20 Nm (15 ft lb)

4 - Hex bolt, 20 Nm (15 ft lb)

5 - Nut

6 - Spring mounting
- ♦ Mark installation position before removing.

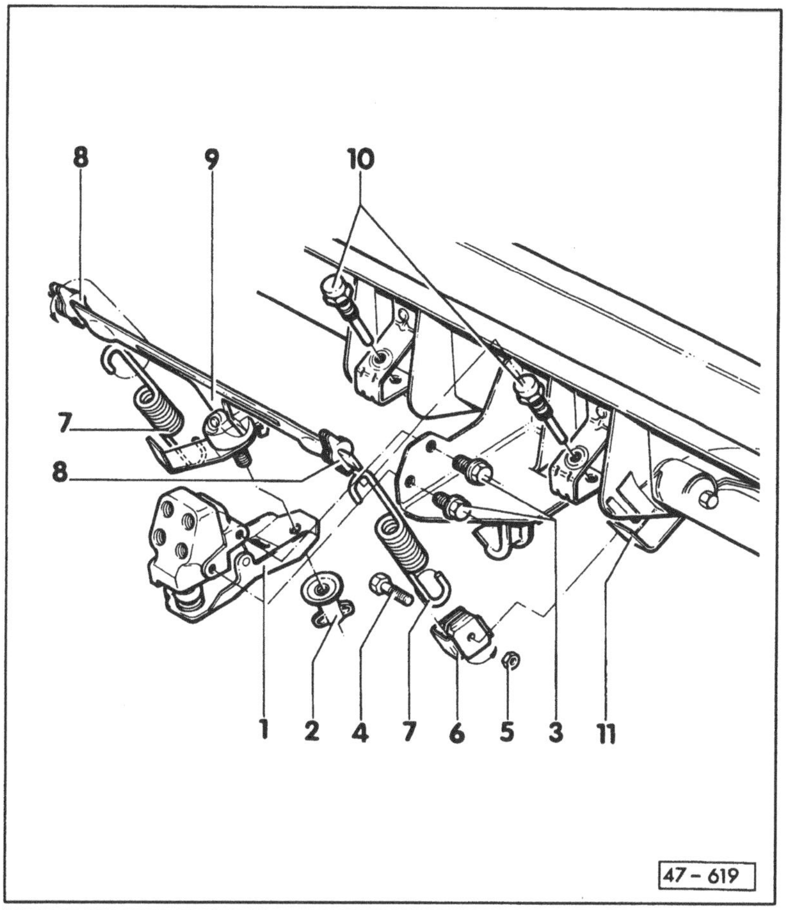

7 - **Spring**
- Note installation position.

8 - **Clip**

9 - **Horizontal beam**
- Slight inclination is permissible, ⇒ Page M 47-19.

10 - **Buffer stop**
- Adjusting, ⇒ Page M 47-20.

11 - **Control arm**

Brake pressure regulator, adjusting

Brake pressure regulator basic setting

Notes:

- *If, for example, one or both control arms are replaced and no markings are given for the installation position of the spring mounting -2- on the control arm, perform the basic setting adjustment as described below.*
- *If basic setting is not necessary, ⇒ Page M 47-15, brake pressure regulator, fine adjustment.*

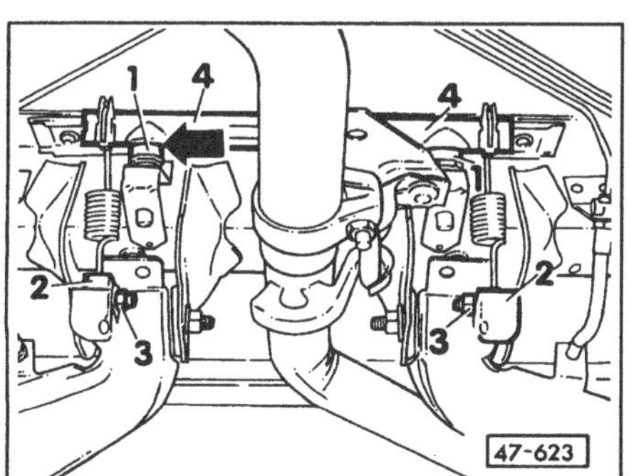

- Remove both shock absorbers, ⇒ Page M 42-16.
- Install special tool 3255, and pre–tension control arm, ⇒ Page M 47-16.
- ◄ Align beam -4- horizontally by turning both buffer stops -1- onto stop on horizontal beam -4- (no preload at buffer stop -1-).

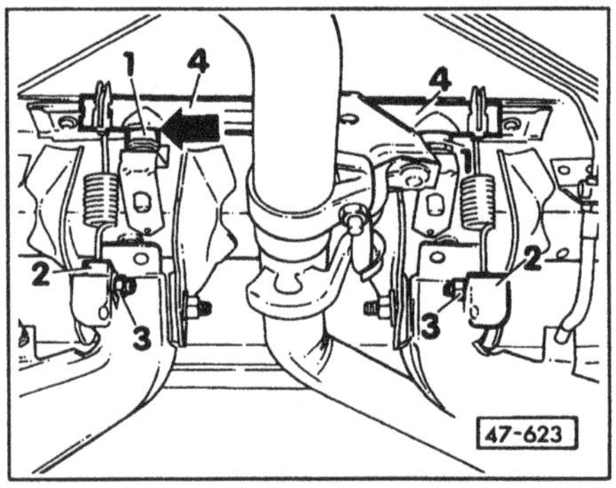

◄ – Move spring mount -2- down until a gap -arrow- of approx. 2 mm exists on opposite side between buffer stop -1- and horizontal beam -4-.

– Move second spring mount -2- down until beam -4- is again horizontal.

– Turn both buffer stops -1- fully back, so largest possible gap exists between horizontal beam -4- and buffer stops.

– Continue with fine adjustment of brake pressure regulator.

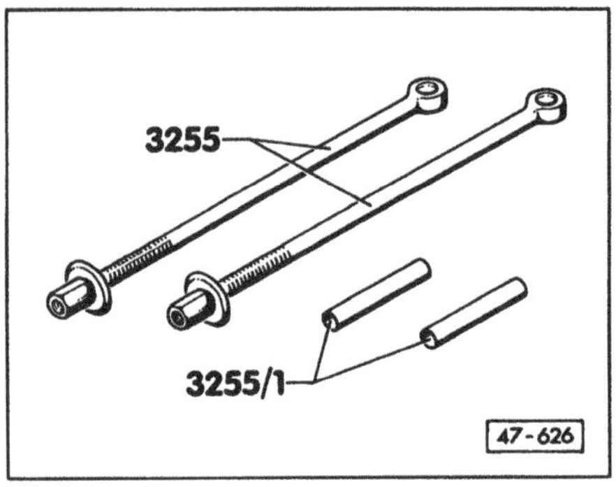

Brake pressure regulator fine adjustment

– Remove both shock absorbers, ⇒ Page M 42-16.

◄ – Slide tubes 3255/1 onto adjuster.

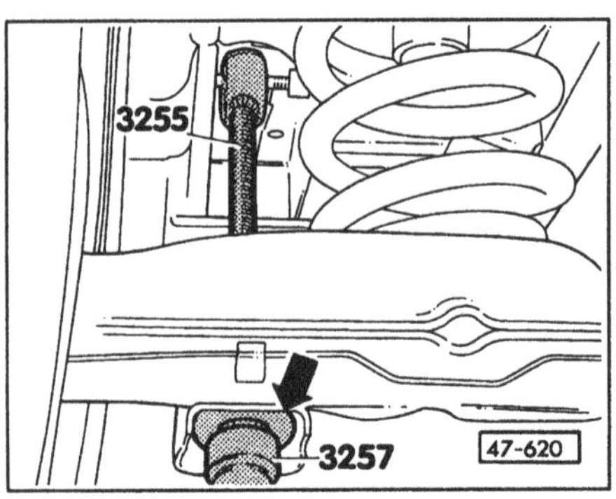

◄ – Install adjuster 3255 with tubes 3255/1 as illustrated.

Notes:

- *Coil spring can be removed to adjust control arm to dimension -x-.*
- *Grease spindle thread, face of nut and washer surface with MoS_2 grease.*
- *Insert a shock absorber rubber bushing between washer -arrow- and control arm to tension control arm.*
- *Do not use an impact screwdriver.*

CAUTION!
Check buffer stop settings, ⇒ Page M 47-20, if poor braking or overbraking of rear axle is noticed.

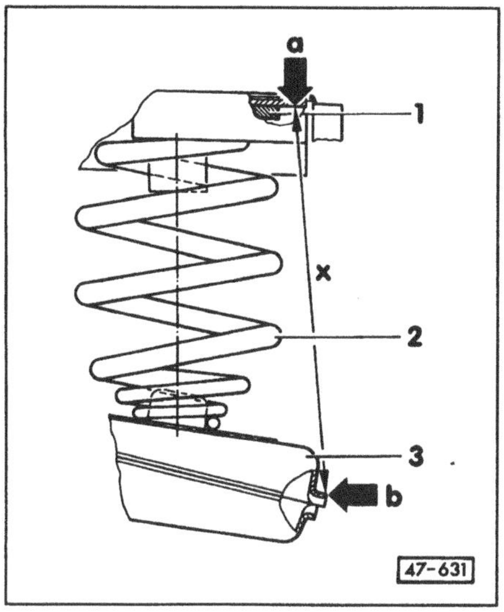

◄ – Pretension control arm to dimension -x- with special tool 3255.

1 - Washer   2 -Spring   3 -Control arm

Notes:

- *Measurement is taken from spring mount steel plate edge -arrow a- to the edge of assembly hole in control arm -arrow b-.*
- *Set dimension -x- according to table below.*

– Continue with brake pressure check, ⇒ Page M 47-18.

Table of brake pressure regulator settings

Rear axle spring Part No.	Color coding	Dimension x ± 1 mm
701 511 105 F	green	281 mm
701 511 105 H/J	orange	270 mm
701 511 105 A/C	white	258 mm
701 511 105	violet	280 mm
701 511 105 B	brown	292 mm
701 511 105 G	blue	246 mm
701 511 105 L	gray	238 mm
701 511 105 K	yellow	284 mm
701 511 105 M	pink	268 mm

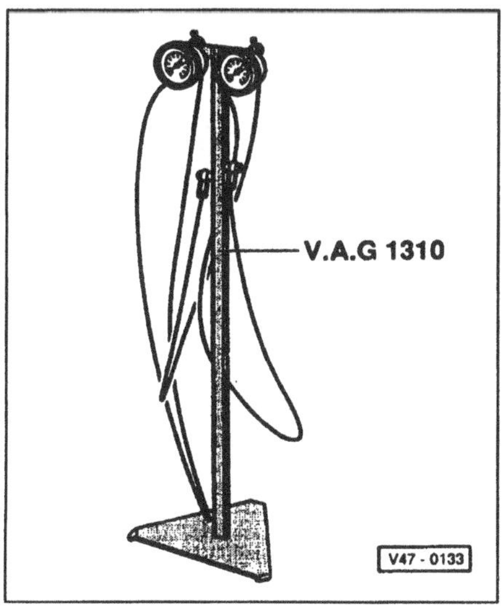

Brake pressure, checking

◄ – Connect pressure gauge VAG 1310 to left front brake caliper and to left rear wheel cylinder.

- **Use VAG 1310/4 adapter for front caliper**
- **Use VAG 1310/3 adapter for rear wheel cylinder**

– Bleed both gauges (bleeding brake system, ⇒ Page M 47-9).

– Depress brake pedal and obtain specified pressure at front axle and read pressure at rear axle.

Front axle: 50 bar (725 psi).

Rear axle: 20 ± 2 bar (290 ± 29 psi).

– Adjust, if necessary, ⇒ Page M 47-19.

– Continue with adjusting buffer stops,⇒ Page M 47-20.

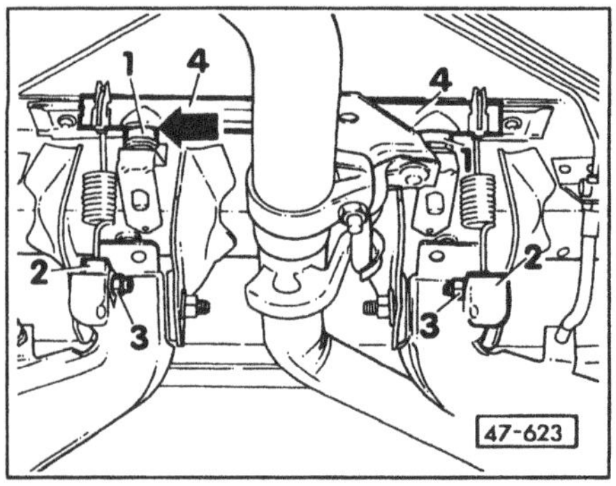

Brake pressure, adjusting

- Turn both buffer stops -1- back onto stop (downward).
- Mark spring mount -2- position with felt tip pen or adhesive tape on control arm.
- Loosen bolt -3- and move mount -2-.

Notes:

- *Moving spring mount -2- down increases pressure.*
- *Moving spring mount -2- up reduces pressure.*

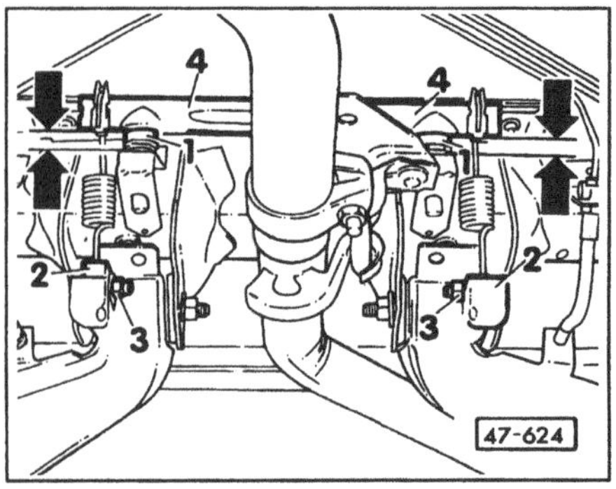

- *It is only necessary to adjust position of second spring mount -2- if horizontal beam -4- is excessively inclined (more than 4 mm, ⇒ arrows).*
- *Moving spring mount -2- 1.0 mm = a pressure change of approximately 4 bar (58 psi) at rear axle.*
- *Slight inclination of horizontal beam -4- is permissible (maximum 4 mm between left and right, ⇒ arrows).*

- Tighten bolt -3-, counter-hold when tightening.
- Correct brake pressure regulator setting at second spring mount -2- if necessary.
- Continue with adjusting buffer stops, ⇒ Page M 47-20.

Buffer stops, adjusting

Note:

Ensure clip, item 2 on Page M 47-12, is securely seated.

- Check brake pressure (if not already checked), ⇒ Page M 47-18.
- If setting is OK, install brake pedal depressor.
- Hold front axle brake pressure to at least 50 bar (725 psi).

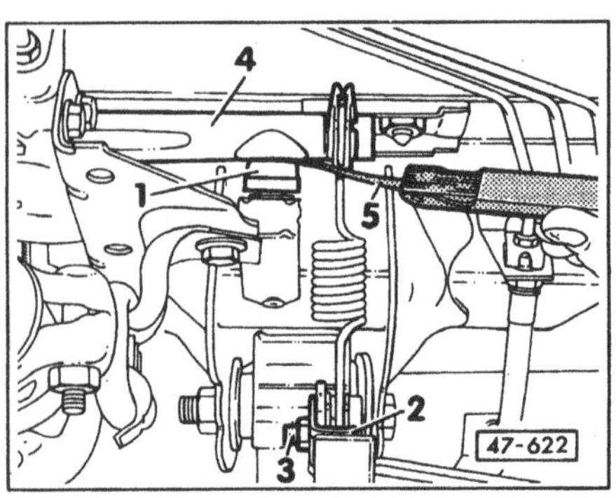

- Turn both buffer stops -1- up, using 6 mm hex key (illustration shows only one buffer stop) until a gap of 0.5 mm can be measured, using a feeler gauge -5- between both buffer stops -1- and horizontal beam -4-.
- Remove VAG 1310 and bleed brake system, ⇒ Page M 47-9.
- Remove special tool 3255.
- Install shock absorbers, ⇒ Page M 42-16.

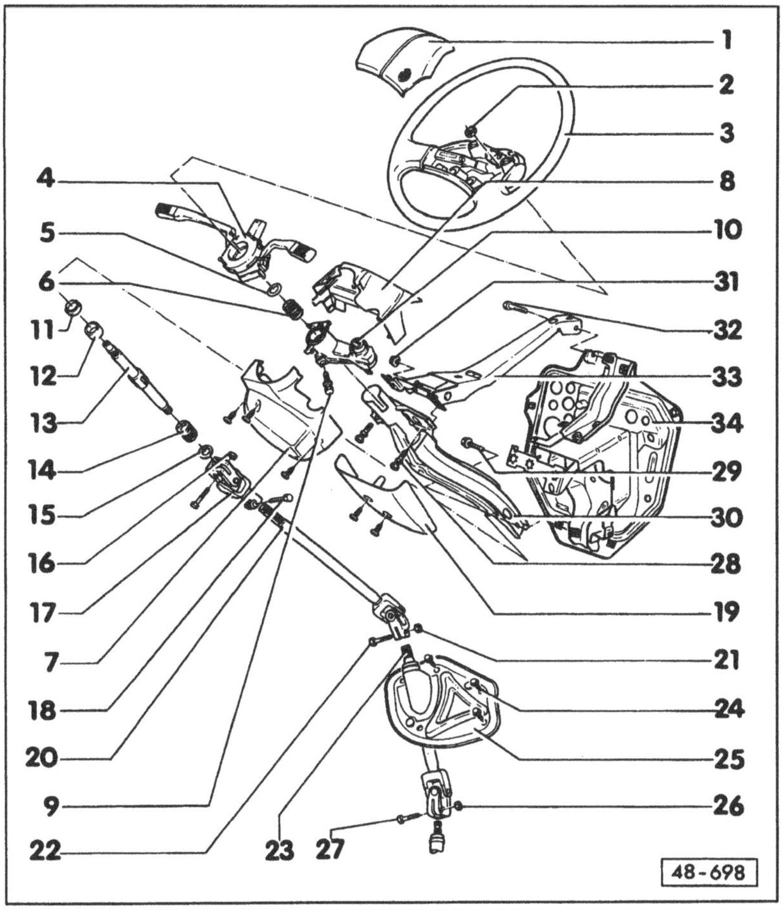

Steering column, assembly

Notes:

- *Always replace self-locking nuts and bolts.*
- *Do not attempt to weld or straighten steering parts.*
- *All components indicated with a 1) should be preassembled away from the vehicle.*

1 - Cover plate
- Unclip.

2 - Hex nut, 70 Nm (52 ft lb)

3 - Steering wheel

4 - Steering column switch 1)

5 - Lock washer 1)
- Installation position, ⇒ Page M 48-7.

6 - Spring 1)

7 - Lower trim

8 - Upper trim

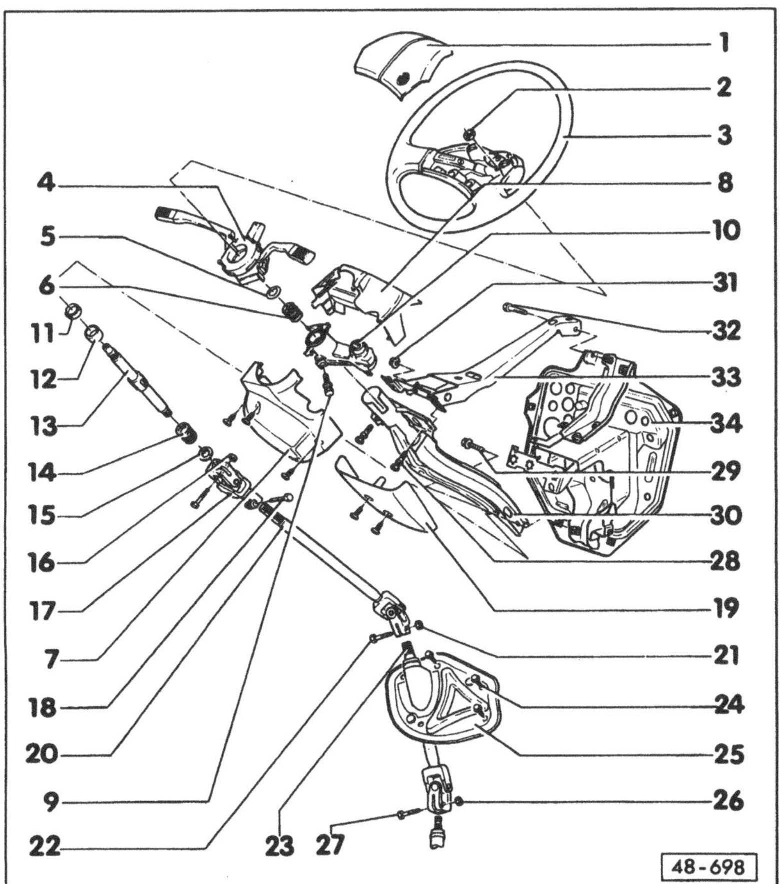

9 - Shear bolt
- Tighten until head shears off.
- Drill out to remove.

10 - Steering lock housing 1)

11 - Support ring 1)

12 - Bushing 1)

13 - Upper steering column tube 1)
- With locking shells for steering lock.

14 - Lower spring 1)

15 - Washer 1)

16 - Self-locking hex nut 1)
- 30 Nm (22 ft lb) and turn 90° further.
- Always replace.

17 - Universal joint 1)

18 - Hex bolt, 30 Nm (22 ft lb) 1)

19 - Trim

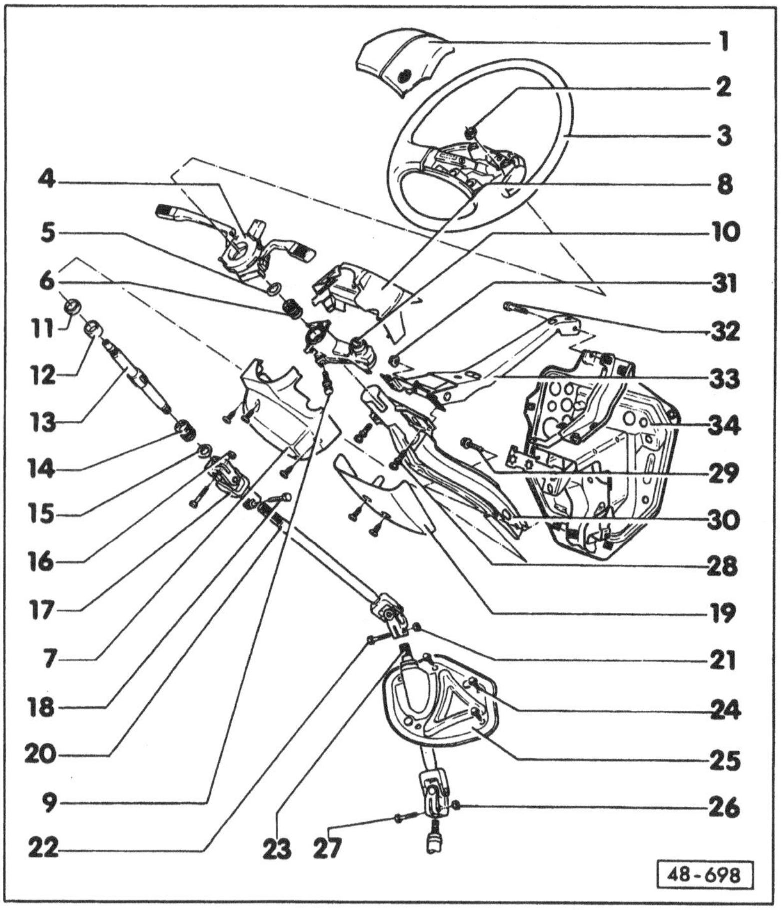

20 - Center steering column tube[1]
- Installation position, ⇒ Page M 48-8.

21 - Self-locking hex nut[1]
- 30 Nm (22 ft lb) and turn 90° further.
- Always replace.

22 - Hex bolt
- Always replace.

23 - Lower steering column tube

24 - Hex bolt, 20 Nm (15 ft lb)

25 - Relay plate with locating pin

26 - Self-locking hex nut[1]
- 30 Nm (22 ft lb) and turn 90° further.
- Always replace.

27 - Hex bolt
- Always replace.

28 - Shear bolt
- Tighten until head shears off.
- Drill out to remove.

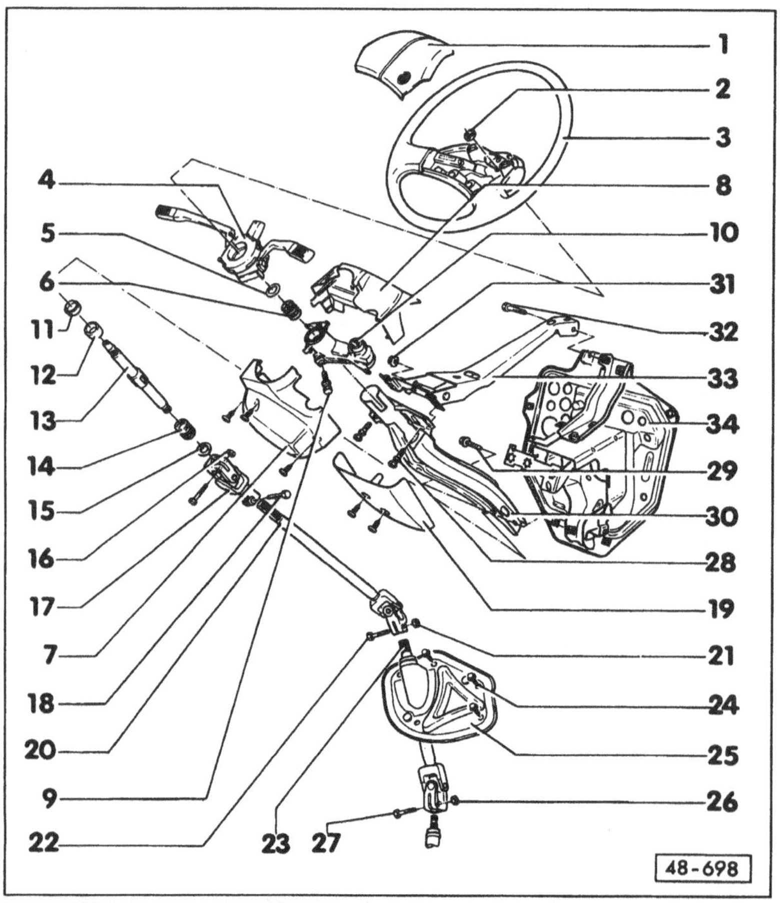

29 - Hex bolt, 20 Nm (15 ft lb)

30 - Column tube with lower mount for steering column tube

31 - Self-locking nut with washer, 16 Nm (12 ft lb)
- Always replace.

32 - Hex bolt, 20 Nm (15 ft lb)

33 - Support

34 - Mounting plate

Steering column, assembling

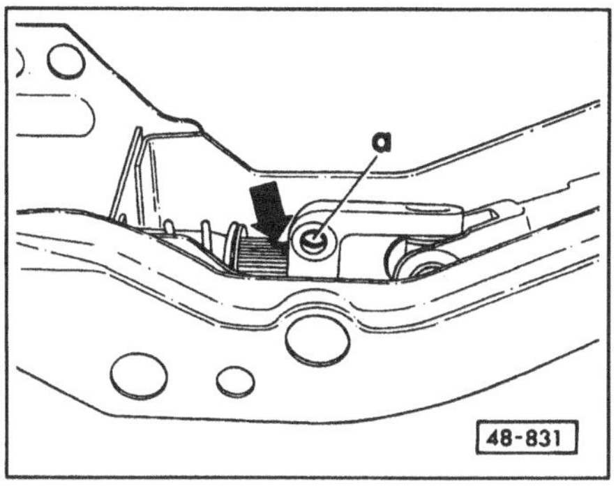

- Slide spacer ring onto upper steering column tube.
- Insert support ring into steering lock housing bearing inner race.
- Guide upper steering column tube into column tube.
- Slide spring and washer onto steering column tube.

◄ – Install universal joint onto steering column tube so notch -arrow- in steering column tube spline aligns with hole -a- in universal joint.

Note:

Ensure components assemble easily when installing universal joint on steering column tube and when installing bolts!

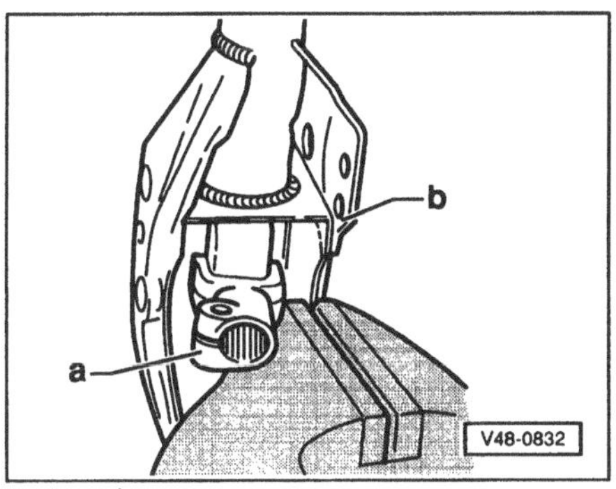

- Bolt universal joint to steering column tube.

◄ – Place column tube in vice ,as illustrated, so universal joint -a- is supported on vice, and push column tube -b- down as far as possible (this compresses the lower steering column tube spring).

- Tighten vice.

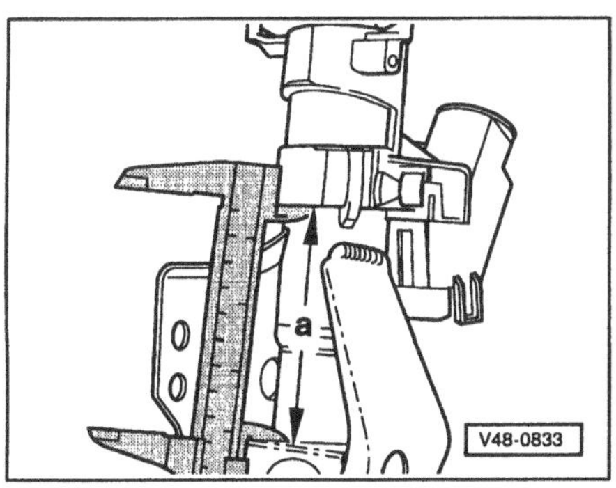

◄ – Push steering lock housing onto column tube until dimension -a- is achieved.

Dimension a: 95 mm ± 1.0 mm (3.740 in. ± 0.039 in.).

- Tighten shear bolt for steering lock until head breaks off.

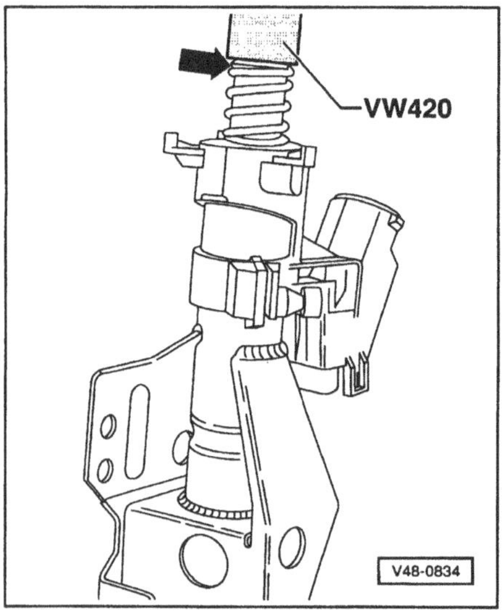

- Install spring with lock washer onto steering lock housing.
- Drive lock washer -arrow- on until correct position is achieved, ⇒ Page M 48-7.

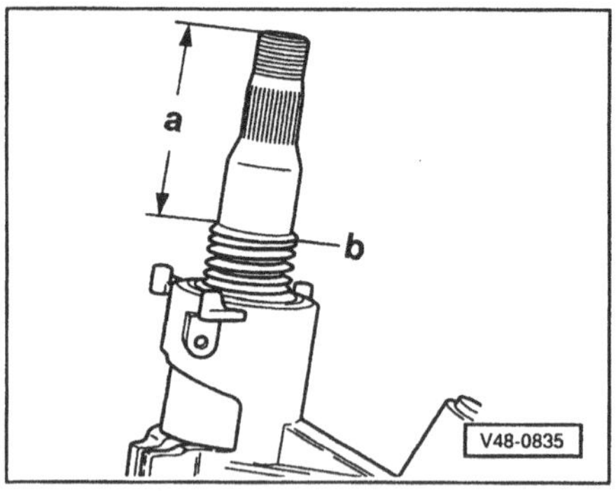

Lock washer installation position

Dimension -a-: 64 mm ± 1.0 mm (2.51 in. ± 0.04 in.).

b - Lock washer.

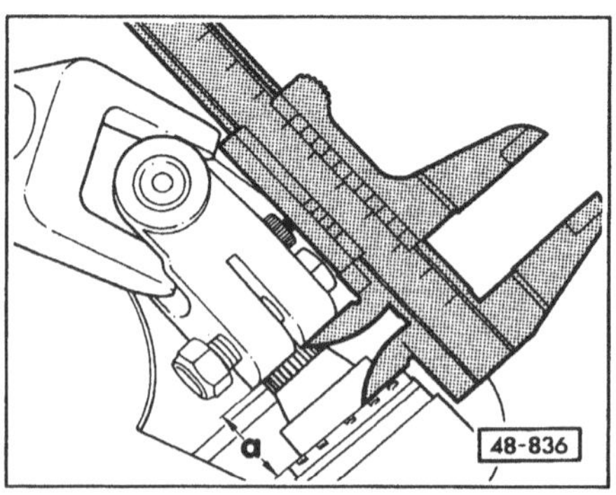

Center steering column tube in installed position

Dimension -a-: 20 mm ± 1.0 mm (0.79 in. ± 0.04 in.).

Assembled steering column, installing

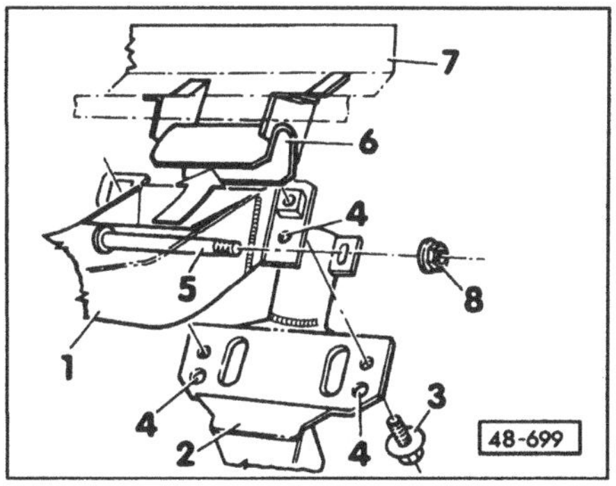

– Bolt strut -1- and column tube -2- loosely to mounting plate.

◄ – Assemble strut -1- and column tube -2- together with shear bolts -3- so holes -4- are aligned and welded bolt -5- locates in bracket -6- of cross member -7-.

– Thread nut -8- to welded bolt -5- (welded to strut).

– Tighten shear bolts -3- on strut mounting plate -1- and column tube -2- (do not yet shear off bolt head).

– Bolt center steering column tube to lower steering column tube.

– Turn steering wheel from lock to lock (with wheels on the ground) and check steering system for uniform ease of movement. Loosen bolted joint and eliminate any stress in system, if necessary.

– Shear off hex head shear bolts.

Power steering pump delivery pressure, checking

Note:

Adapter VAG 1402/6 is required for checking delivery pressure. The tools 1, 2, and 3, shown in illustration V48-0846, are parts of adapter set VAG 1402/6.

Test conditions:

- Power steering belt OK.
- Power steering belt tension OK.
- Power steering system not leaking.
- Hoses/lines are not kinked or restricted.
- Use hydraulic oil, Part No. G 002 000.

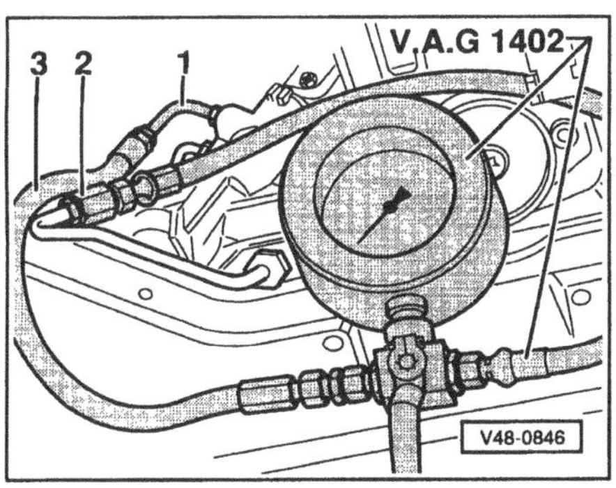

– Remove sound deadener pan.

– Remove pressure line on power steering gear.

◄ – Connect pressure gauge VAG1402, as illustrated (valve on gauge in open position).

- Start engine and top up oil level in reservoir, if necessary.
- Check hose connections for leaks.
- Close valve with engine running at idle speed (leave closed not longer than 5 seconds) and note pressure.

- Specification: 90 –110 bar (1310 – 1595 psi)

Notes:

- *If reading deviates by more than 5 bar (73 psi) above or below specification, replace pump.*
- *Removing and installing vane pump for vehicles with 5-Cylinder gasoline engine from 10/91,* *⇒ Page M 48-32.*
- *If leak is at steering gear pinion, first check line/hose connections and, if necessary, tighten and wipe dry.*
- *Replace steering gear if pinion seal or rack seals in steering gear housing are leaking. Move boot to side to check rack seal.*

Power steering gear, removing and installing

Rules of cleanliness

- *Absolute cleanliness is required when working on power steering system.*
- *Thoroughly clean connections and immediate vicinity before loosening.*
- *Place parts on a clean surface and cover, if not re-installing immediately.*
- *Do not use fluffy cloths.*
- *Carefully cover or seal open components, if repairs cannot be done immediately.*
- *Only remove replacement parts from packaging immediately prior to installation.*
- *Only use genuine packaged parts.*
- *When system is open, do not work with compressed air in immediate area and do not move vehicle.*

Removing

- Install support device 3250, ⇒ Page M 40-3.
- Remove hydraulic fluid with VAG 1358 A , VAG 1307 A, or equivalent.
- Remove sound deadener pan.
- Remove stabilizer bar (only vehicles with reinforced 27 mm diameter stabilizer bar), ⇒ Page M 40-13, Removing and installing front stabilizer bar.

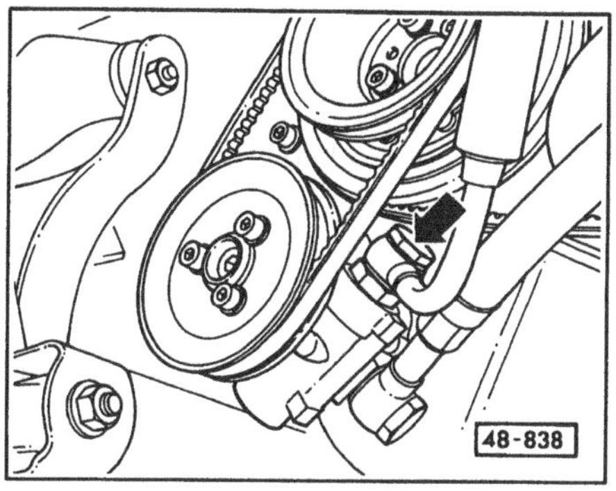

◄ – Open banjo bolt on pump, -arrow-, on all 5-Cylinder gasoline engines from 10/91, and drain hydraulic oil (if not already extracted).

- Disconnect hoses from steering gear (if not already removed), and seal openings with plastic plugs or plastic bags and adhesive tape.
- Seal open ports at pump with plastic plugs.

- Disconnect tie rods at steering rack, ⇒ Page M 48-20.
- Remove stabilizer bar clamps from subframe.
- Remove universal joint bolt at steering gear pinion.
- Disconnect power steering gear from assembly carrier, ⇒ Page M 48-21.
- Push up stabilizer bar (if not already removed).
- Move power steering gear slightly to right and pull steering gear pinion out of universal joint.
- Swing left-hand side of steering gear out to rear.

Installing

Install in reverse order.

Notes:

♦ *If steering gear or tie rod(s) are replaced, check toe setting and adjust if necessary.*

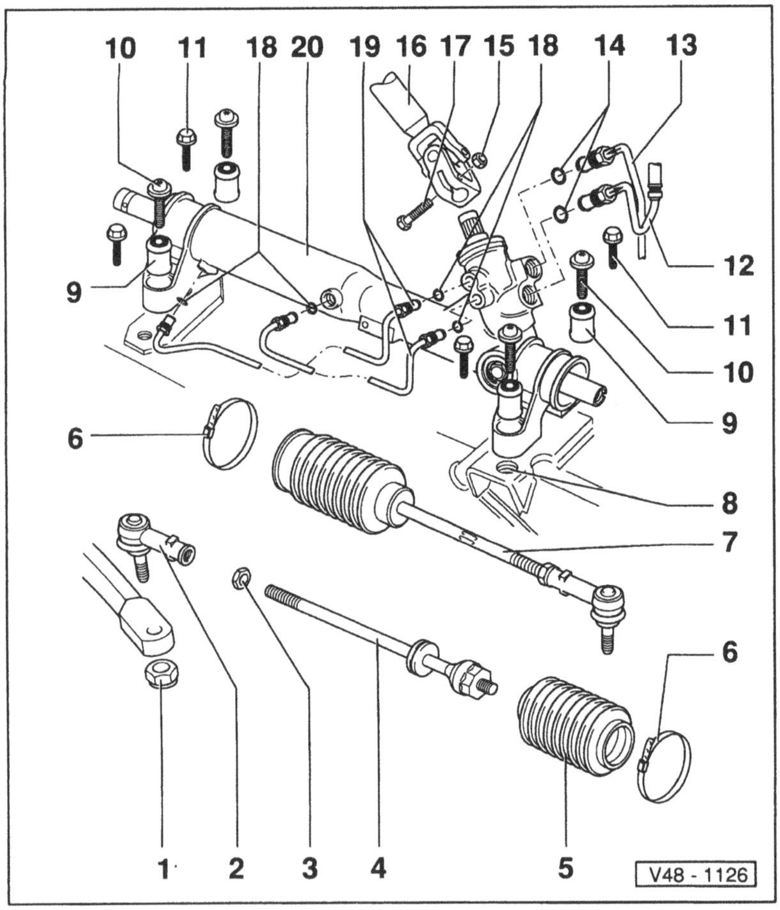

Power steering gear, assembly

No provision is made for servicing power steering gear. Pressure test to determine cause of malfunction.

Notes:

- *Always replace self-locking nuts and bolts.*
- *Do not weld or straighten steering system components.*
- *Only use steering gear grease, Part No. AOF 063 000 04, to grease rack.*
- *Hydraulic fluid Part Number G 002 000.*
- *Quantity of fluid in system, approximately 1 Liter (1.05 qt).*

1 - **Self-locking hex nut, 40 Nm (30 ft lb)**
- Always replace.
- Modification: From VIN 70 SH 002 792, tightening torque changed to 65 Nm (48 ft lb), ⇒ Page M 40-48.

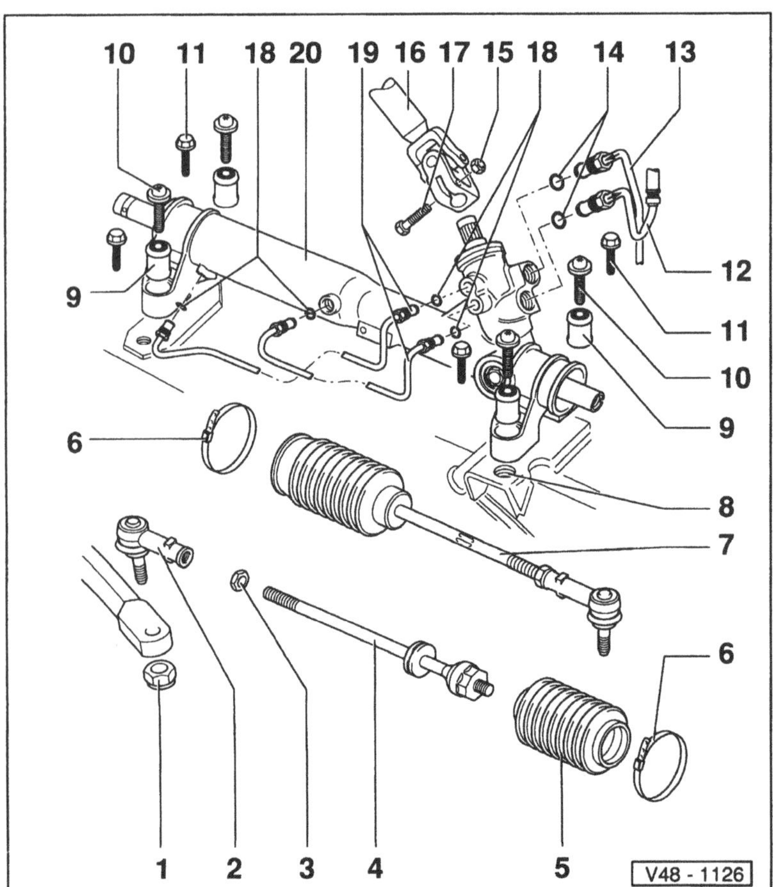

2 - **Tie rod end**
- Ensure dust cap is seated correctly and not damaged.
- Pressing off, ⇒ Fig. 2.
- Modification: From VIN 70 MH 096 449, larger diameter, ⇒ Page M 40-47.
- Modification: From VIN 70 SH 002 792, modified taper and wheel bearing housing dimensions, ⇒ Page M 40-48.

3 - **Hex nut, 55 Nm (41 ft lb)**

4 - **Left tie rod, 70 Nm (52 ft lb)**
- Removing and installing, ⇒ Fig. 1.
- Adjusting toe, ⇒ Page M 44-15.

5 - **Boot**
- Check for damage.
- Can be replaced with steering gear installed.
- Must not be twisted after adjusting tie rod.

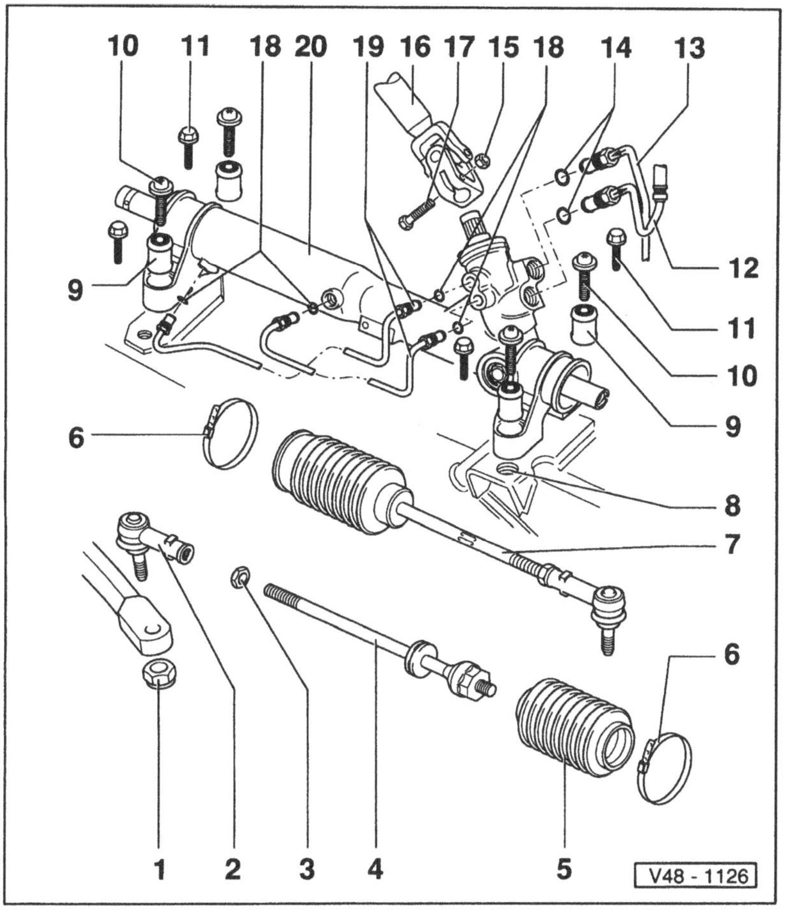

6 - Clamp
- A cable clamp is installed during production.
- For repairs install clamp (⇒ Parts microfiche), and tension with VAG 1275, or equivalent.

7 - Right tie rod, 70 Nm (52 ft lb)
- Removing and installing, ⇒ Fig. 1.
- Adjusting toe, ⇒ Page M 44-15.

8 - Threaded hole in subframe
- Vehicles prior to VIN 70 PH 004 342, M 8 thread.
- Vehicles from VIN 70 PH 004 343, M 10 thread.

9 - Rubber mount
- Pressing out and in, ⇒ Fig. 3.
- Inner bushing is 8 or 10 mm (0.314-0.393 in.) diameter, according to subframe version.

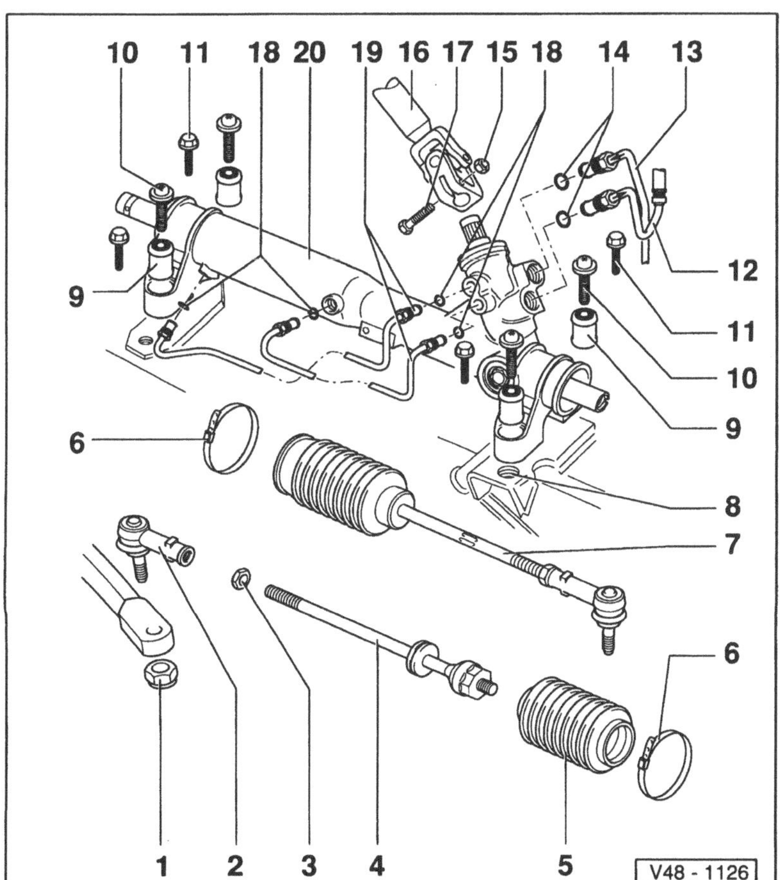

10 - Torx bolt, M 10, 60 Nm (44 ft lb)
- Can be reused.
- 08/92 ➤
- Left bolt behind steering gear is to be replaced by an M10 x 58 hex bolt on vehicles with additional heating.
- Rear torx bolts are replaced by M10 x 58 hex bolts.
- Loosening and tightening, ⇒ Fig. 4, 5 and 6.

11 - Hex bolt, M 8, 33 Nm (24 ft lb)
- Always replace.
- Production ➤ 07/92
- Loosening and tightening, ⇒ Fig. 5.

12 - Return line, 30 Nm (22 ft lb)

13 - Pressure line, 30 Nm (22 ft lb)

14 - O-ring
- Always replace.

15 - Hex bolt, 30 Nm (22 ft lb)

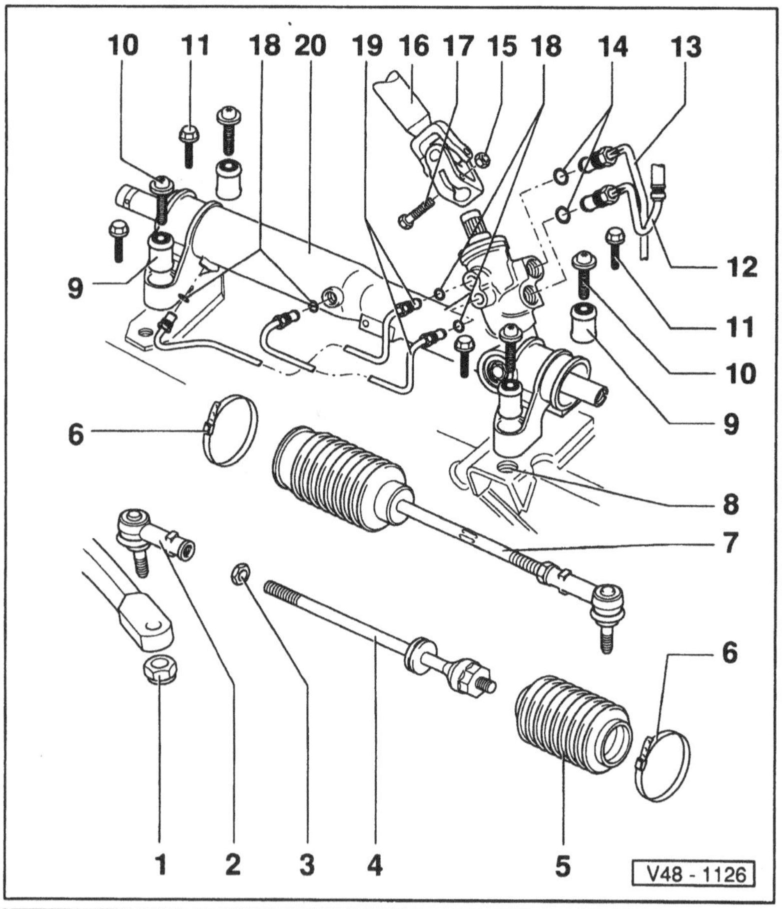

16 - Steering column tube with universal joint

17 - Hex bolt

18 - O-ring
- Always replace.

19 - Control lines
- 14 Nm (10 ft lb).
- When installing, ensure that seals are not damaged.
- Install stress free.
- Do not damage surface; repair paint damage, if necessary.
- Can be removed and installed with steering gear installed.

20 - Power steering gear
- Removing and installing, ⇒ Page M 48-12.

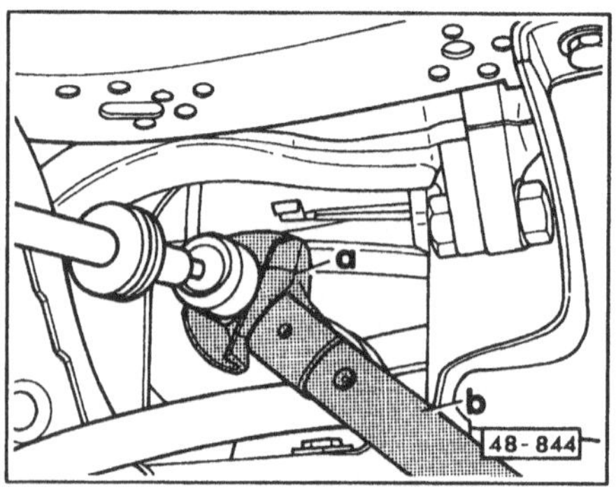

◄ **Fig. 1 Tie rod, removing and installing**

a - VAG 1332/5

b - Torque wrench VAG 1332

Note:
Do not use locking fluid.

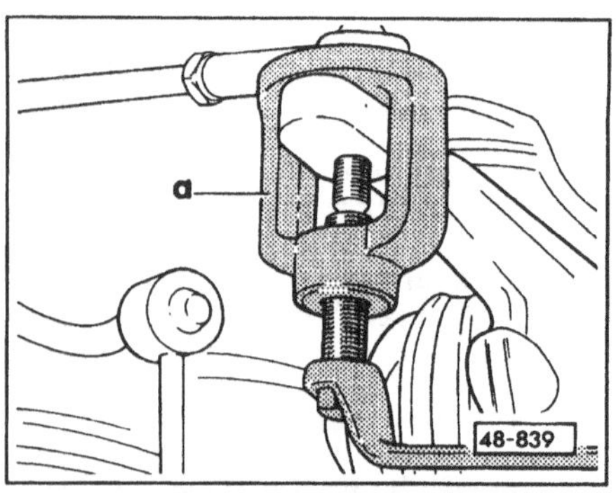

◄ **Fig. 2 Tie rod, pressing off steering arm**

a - Ball joint splitter (commercial type), e.g., Kukko No. 128/2.

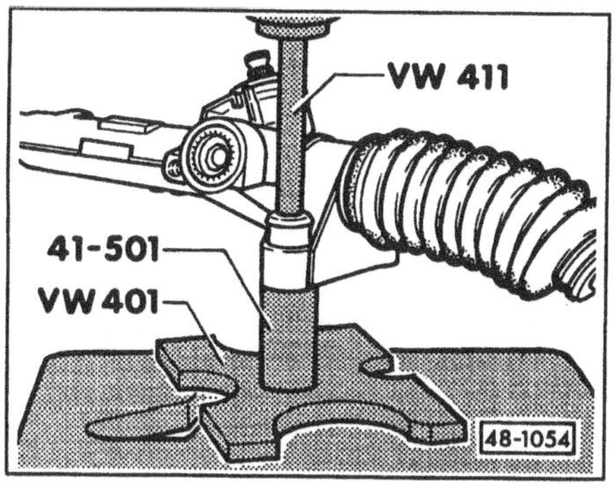

Fig. 3 Rubber bushing, pressing out and in

Note:

Use correct bushing (for bolt M8 or M10) when installing a new steering gear.

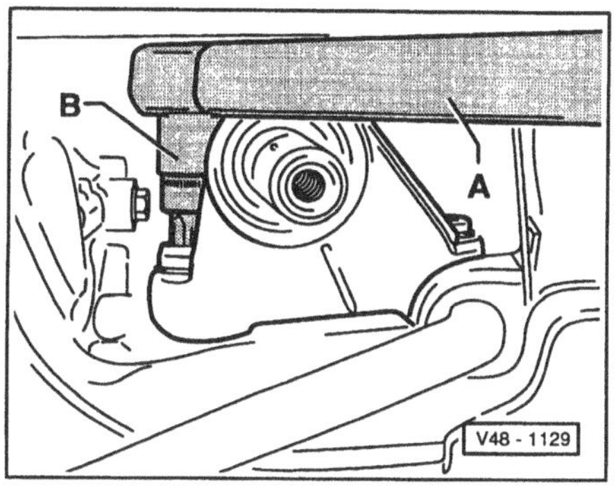

Fig. 4 Bolts for steering gear (Torx bolts), removing and installing

A - Torque wrench VAG 1332

B - Socket insert Torx T50 (commercially available)

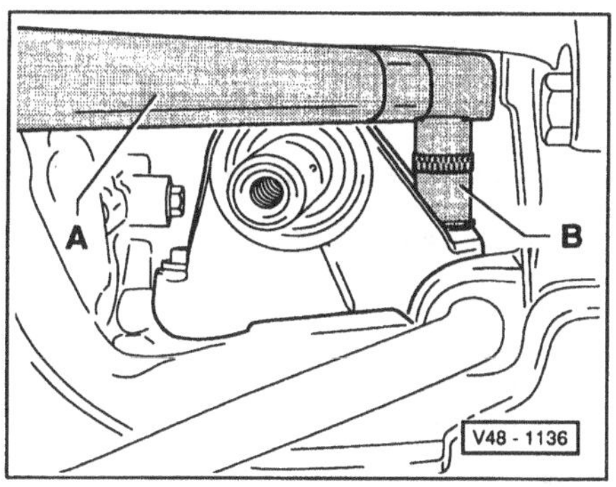

Fig. 5 Bolts for steering box (M8 and M10 hex bolts), removing and installing

A - Torque wrench VAG 1332

B - Socket insert (commercially available)

Note:

Use 13 mm jointed socket, if necessary, for M8 bolts (commercially available).

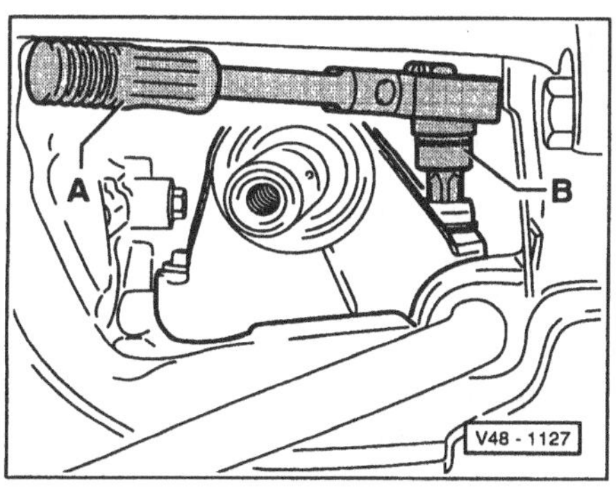

Fig. 6 Left rear torx bolt on vehicles with additional heater, removing

Install an M10 x 58 hex head bolt instead of the torx bolt in this position. Part No. N 906 268 01.

A - Ratchet (commercially available $^3/_8$" drive)

B - Torx T50 socket insert (commercially available)

Note:

The illustration shows a ratchet with a $^3/_8$" drive socket insert. If this tool is not available then push the coolant pipe upward and out of the way as far as possible.

Power steering pump, assembly

There is no provision for servicing the power steering pump. Determine cause of malfunctions by pressure testing. Replace power steering pump, if necessary.

Notes:

- *Pumps supplied as spare parts are not filled with fluid. Therefore, fill pump with hydraulic fluid, Part No. G 002 000, and rotate by hand.*
- *Type of lubricant: hydraulic fluid, Part No. G 002 000*
- *Quantity of fluid in system: approx. 1 Liter (1.05 qt)*
- *Do not reuse drained hydraulic fluid.*
- *Always replace O-rings.*

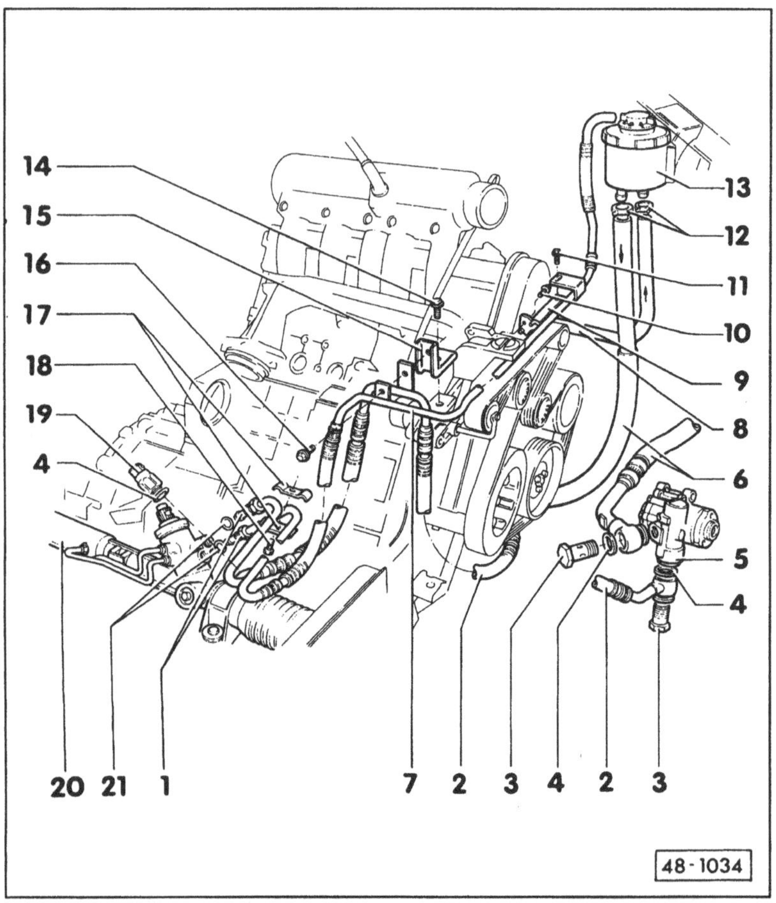

Hydraulic lines, assembly

(5-cylinder gasoline engine 10/91➤)

Notes:* ⇒ *Page M 48-23

1 - 30 Nm (22 ft lb)

2 - Pressure line
- Installing, ⇒ Fig. 1.

3 - Banjo bolt, 35 Nm (26 ft lb)

4 - O-ring

5 - Power steering pump
- Removing and installing, ⇒ Page M 48-32.
- Checking delivery pressure, ⇒ Page M 48-10.
- Before installing fill with oil, ⇒ Note Page M 48-23.

6 - Suction line
- Installing, ⇒ Fig. 2.

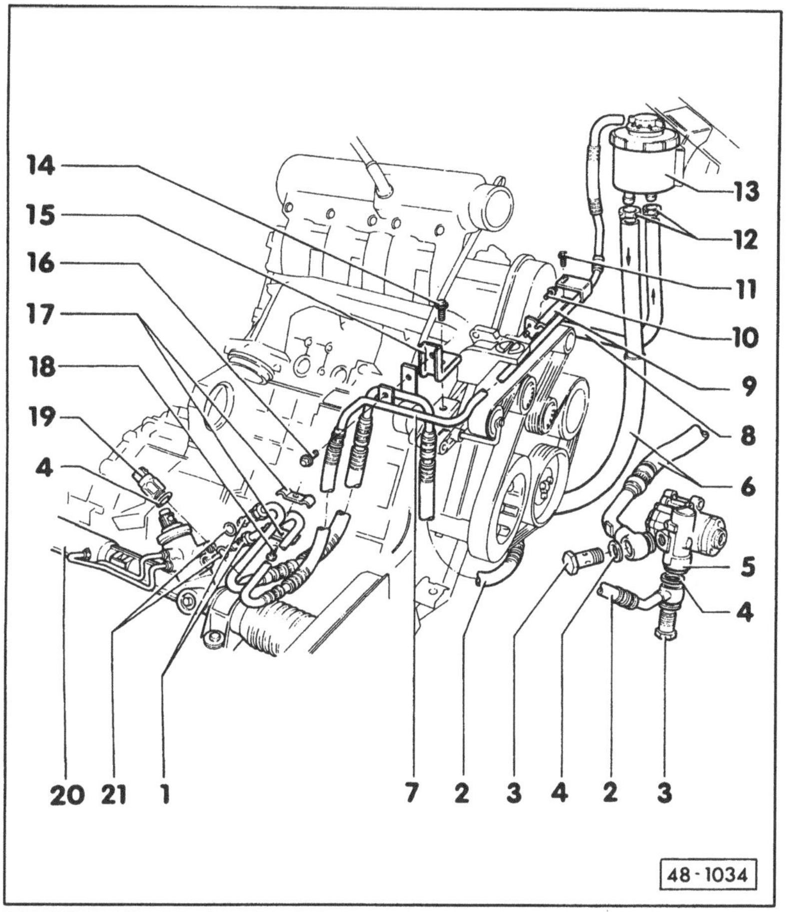

7 - Return line to cooler,
- ⇒ Fig. 3.

8 - Return line,
- ⇒ Fig. 4.

9 - Bracket

10 - 10 Nm (7 ft lb)

11 - 10 Nm (7 ft lb)

12 - Hose clamp

13 - Reservoir
- Oil level: between MIN and MAX marks.

14 - 20 Nm (15 ft lb)

15 - Angle bracket

16 - 20 Nm (15 ft lb)

17 - Clamp
- Installed position, ⇒ Fig. 5.

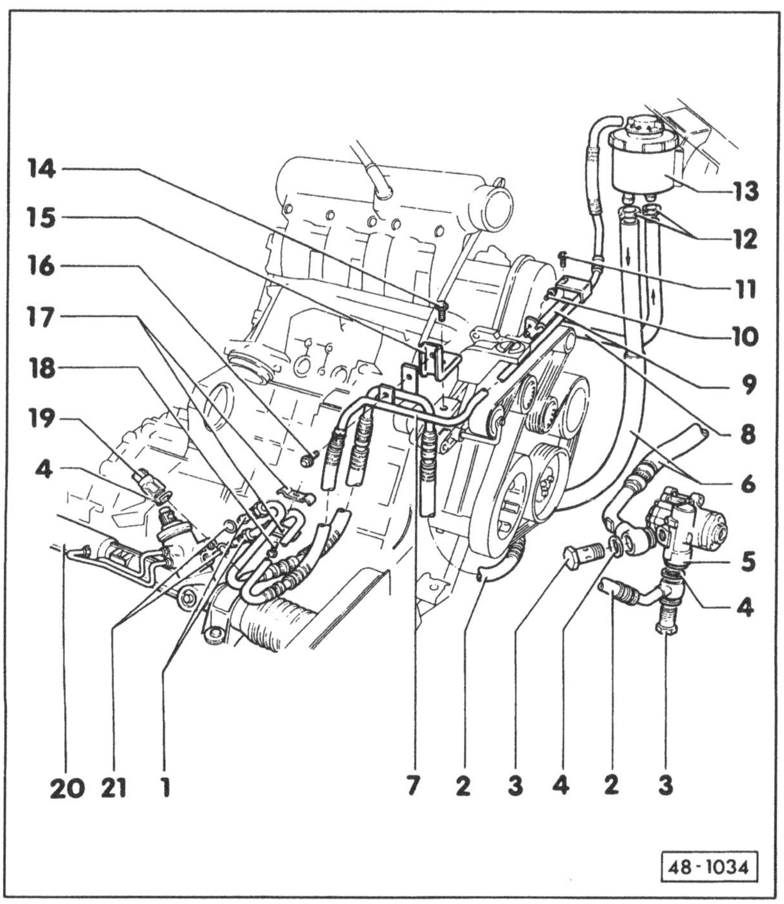

18 - 10 Nm (7 ft lb)
- Installed from above during production; during repairs, install from below

19 - Pressure switch, 15 Nm (11 ft lb)

20 - Power steering gear

21 - Seal

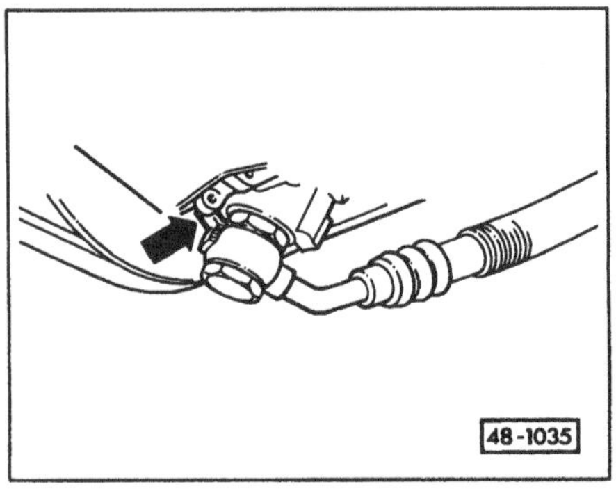

Fig. 1 Pressure line, installing

Note:

Angled stop -arrow- must rest against pump housing when tightening banjo bolt.

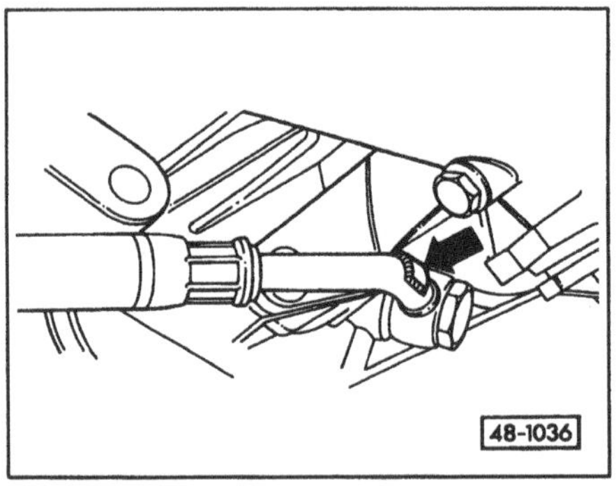

Fig. 2 Suction line, installing

Note:

Angled stop -arrow- must rest against bracket stiffener rib when tightening banjo bolt.

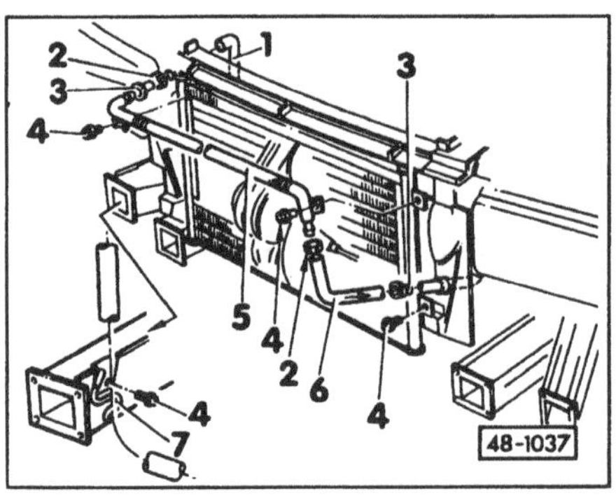

Fig. 3 Return line to cooler, assembly

1 - Return line
2 - Hose clamp
3 - Grommet
4 - Bolt, 8 Nm (70 in.lb)
5 - Coolant line
6 - Return line
7 - Bracket

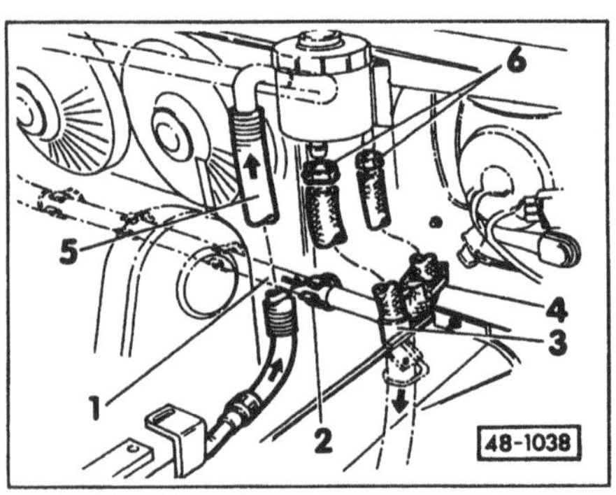

Fig. 4 Return line hose routing

1 - Return line
2 - Hose bracket
3 - Suction hose
4 - Hose bracket
5 - Return line
6 - Hose clamp

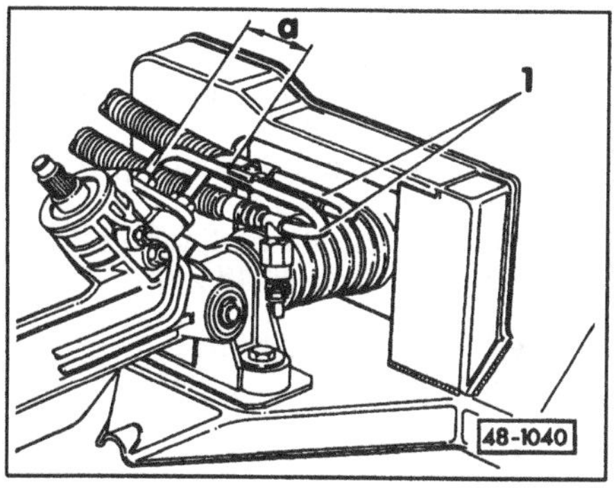

◄ **Fig. 5 Clamp, installed position**

Note:

Install lines parallel to edge of subframe.

Dimension -a-: approx. 60 mm (24 in.)

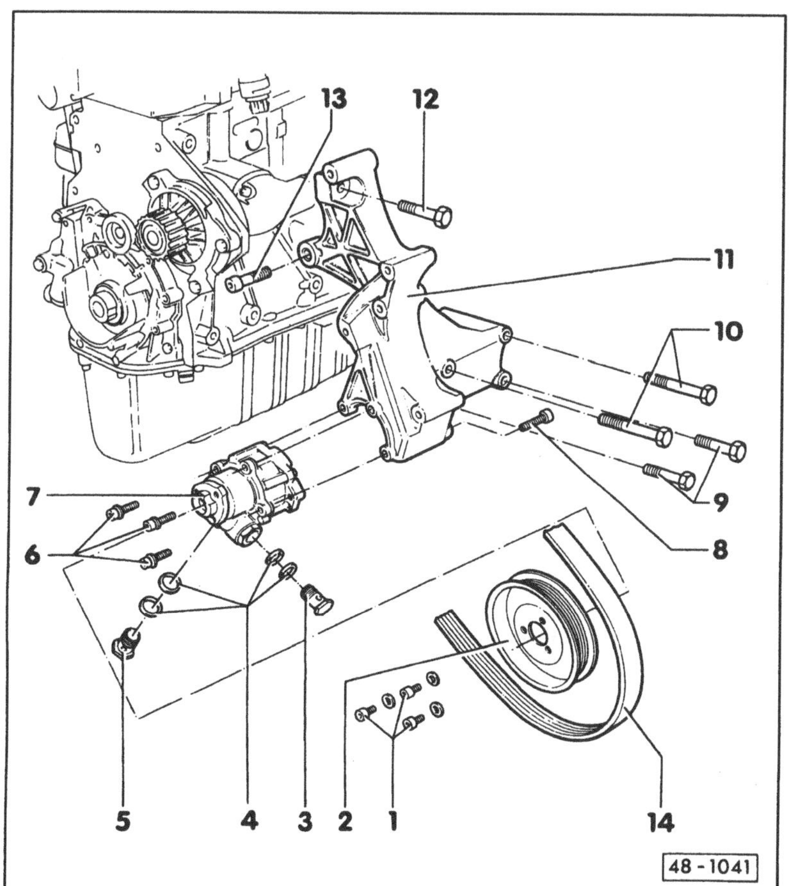

Power steering pump, assembly

(5-cylinder gasoline engine from 10/91)

Notes: ***⇒ Page M 48-23***

1 - **Hex socket-head bolt, 20 Nm (15 ft lb)**
- ♦ Loosening and tightening, ⇒ Page M 48-32.

2 - **Belt pulley**

3 - **Banjo bolt, 30 Nm (22 ft lb)**

4 - **O-ring**
- ♦ Always replace.

5 - **Banjo bolt, 30 Nm (22 ft lb)**

6 - **Hex socket-head bolt, 20 Nm (15 ft lb)**

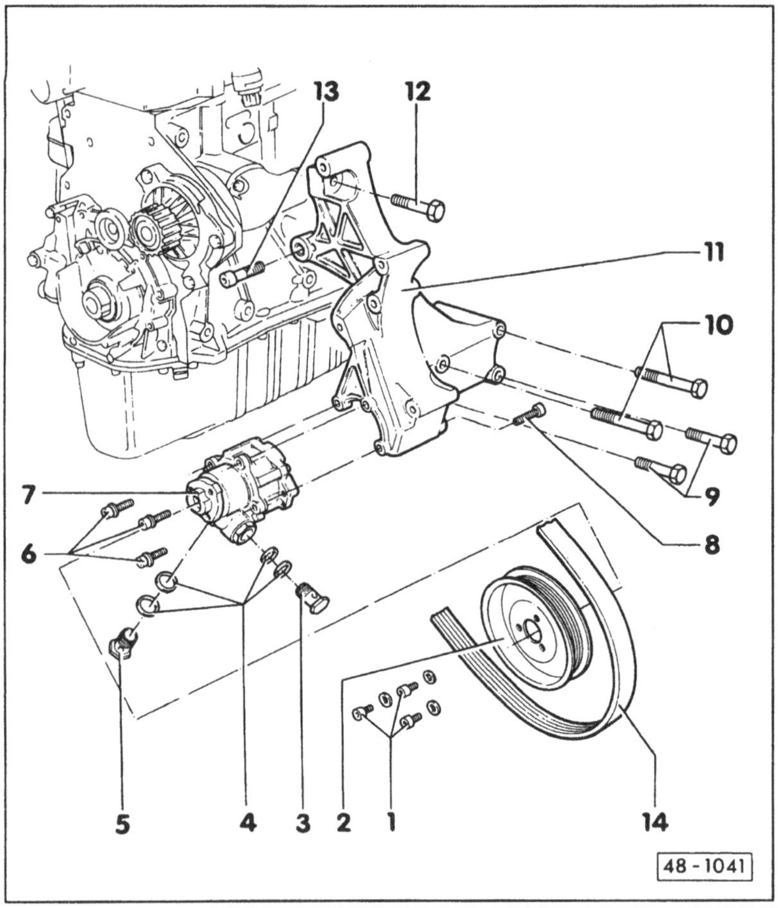

7 - **Power steering pump**
- Checking delivery pressure, ⇒ Page M 48-10.
- Removing and installing, ⇒ Page M 48-32.
- Before installing fill with oil, ⇒ Note Page M 48-23.

8 - **Hex socket-head bolt, 20 Nm (15 ft lb)**

9 - **Hex bolt, 40 Nm (30 ft lb)**

10 - **Hex bolt, 40 Nm (30 ft lb)**

11 - **Bracket**
- Removing and installing:

⇒ 5-Cylinder General – Engine.

12 - **Hex bolt, 40 Nm (30 ft lb)**

13 - **Hex socket-head bolt, 40 Nm (30 ft lb)**

14 - **Ribbed belt**
- Removing and installing:

⇒ 5-Cylinder General – Engine.
- Checking condition,

⇒ Maintenance Manual.

Power steering pump, removing and installing

(5-cylinder gasoline engine 10/91➤)

Removing

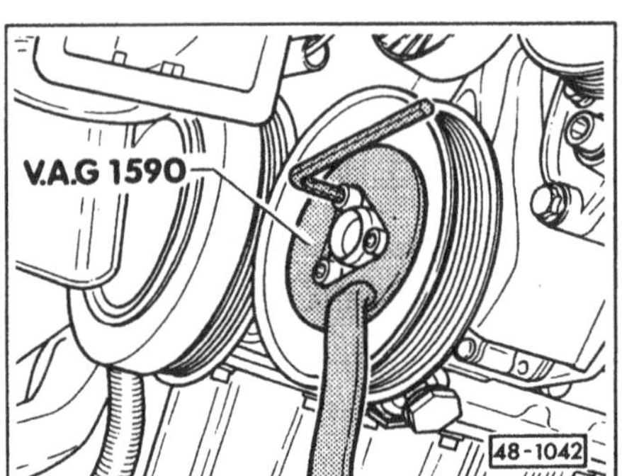

– Remove sound deadener pan.

– Remove ribbed V-belt.
⇒ 5-Cylinder General – Engine.

– Disconnect suction and pressure lines and drain hydraulic fluid.

◄ – Remove belt pulley.

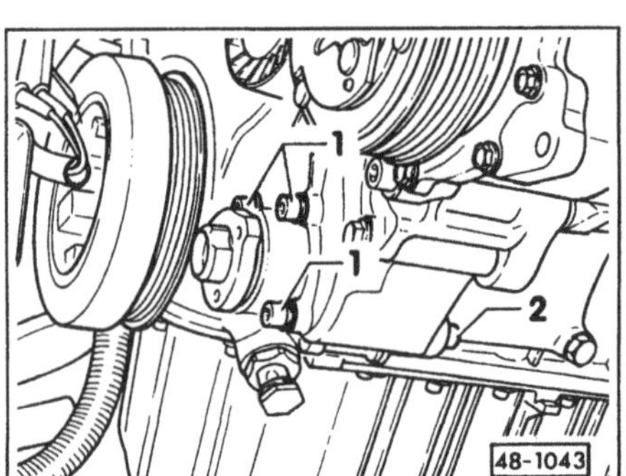

◄ – Remove bolts -1- and -2-.

Installing

Reinstall in reverse sequence.

Suspension, Wheels, Brakes, Steering (1997 ➤)

Section **N**

> ***WARNING!***
> ***Before proceeding with any work, please read the warnings and cautions immediately following this index. Also read any warning and caution that accompanies a procedure in the manual.***

Please read these WARNINGS and CAUTIONS before proceeding with maintenance and repair work.

WARNING—

- Some repairs may be beyond your capability. If you lack the skills, tools and equipment, or a suitable workplace for any procedure described in this manual, we suggest you leave such repairs to an authorized Volkswagen dealer service department, or other qualified shop.

- Volkswagen is constantly improving its cars. Sometimes these changes, both in parts and specifications, are made applicable to earlier models. Therefore, before starting any major jobs or repairs to components on which passenger safety may depend, consult your authorized Volkswagen dealer about the Technical Bulletins that may have been issued since the editorial closing of this manual.

- Do not re-use any fasteners that are worn or deformed in normal use. Many fasteners are designed to be used only once and become unreliable and may fail when used a second time. This includes, but is not limited to, nuts, bolts, washers, self–locking nuts or bolts, circlips and cotter pins. Always replace these fasteners with new parts.

- Never work under a lifted car unless it is solidly supported on stands designed for the purpose. Do not support a car on cinder blocks, hollow tiles or other props that may crumble under continuous load. Never work under a vehicle that is supported solely by a jack. Never work under the car while the engine is running.

- If you are going to work under a car on the ground, make sure that the ground is level. Block the wheels to keep the car from rolling. Disconnect the battery negative terminal (ground strap) to prevent others from starting the vehicle while you are under it.

- Never run the engine unless the work area is well ventilated. Carbon monoxide (CO) kills.

- Do not attempt to work on your vehicle if you do not feel well. You increase the danger of injury to yourself and others if you are tired, upset or have taken medicine or any other substance that may impair you or keep you from being fully alert.

- Disconnect the battery negative (–) terminal (ground strap) whenever you work on the fuel system or the electrical system. Do not smoke or work near heaters or other fire hazards. Keep an approved fire extinguisher handy.

- Batteries produce explosive hydrogen gas. Keep sparks, lighted matches and open flame away from the top of the battery. If hydrogen gas escaping from the cap vents is ignited, it will ignite gas trapped in the cells and cause the battery to explode.

- Connect and disconnect battery cables, jumper cables or a battery charger only with the ignition switched off, to prevent sparks. Do not quick–charge the battery (for boost starting) for longer than one minute, and do not allow charging voltage to exceed 16.5 volts. Wait at least one minute before boosting the battery a second time.

- Illuminate your work area adequately but safely. Use a portable safety light for working inside or under the car. Make sure the bulb is enclosed by a wire cage. The hot filament of an accidentally broken bulb can ignite spilled fuel or oil.

- Tie long hair behind your head. Do not wear a necktie, a scarf, loose clothing, or a necklace when you work near machine tools or running engines. If your hair, clothing, or jewelry were to get caught in the machinery, severe injury could result.

- Finger rings, bracelets and other jewelry should be removed so that they cannot cause electrical shorts, get caught in running machinery, or be crushed by heavy parts.

- Catch draining fuel, oil, or brake fluid in suitable containers. Do not use food or beverage containers that might mislead someone into drinking from them. Store flammable fluids away from fire hazards. Wipe up spills at once, but do not store the oily rags, which can ignite and burn spontaneously.

- Always observe good workshop practices. Wear goggles when you operate machine tools or work with battery acid. Gloves or other protective clothing should be worn whenever the job requires working with harmful substances.

- The Air-Conditioning (A/C) system is filled with a chemical refrigerant, which is hazardous. The A/C system should be serviced only by trained technicians using approved refrigerant recovery/recycling equipment, trained in related safety precautions, and familiar with regulations governing the discharging and disposal of automotive chemical refrigerants.

- Do not expose any part of the A/C system to high temperatures such as an open flame. Excessive heat will increase system pressure and may cause the system to burst.

- Some aerosol tire inflators are highly flammable. Be extremely cautious when repairing a tire that may have been inflated using an aerosol tire inflator. Keep sparks, open flame or other sources of ignition away from the tire repair area. Inflate and deflate the tire at least four times before breaking the bead from the rim. Completely remove the tire from the rim before attempting any repair.

- Friction materials such as brake or clutch discs may contain asbestos fibers. Do not create dust by grinding, sanding, or by cleaning with compressed air. Avoid breathing asbestos fibers and asbestos dust. Breathing asbestos can cause serious diseases such as asbestosis or cancer, and may result in death.

- Greases, lubricants and other automotive chemicals contain toxic substances, many of which are absorbed directly through the skin. Read manufacturer's instructions and warnings carefully. Use hand and eye protection. Avoid direct skin contact.

- An airbag is operated by an explosive device. Handled improperly or without adequate safeguards, it can be accidently activated and cause serious personal injury. To prevent personal injury or airbag system failure, **only trained Volkswagen dealer service technicians** should test, disassemble or service the airbag system.

WARNINGS and CAUTIONS continued on next page ▶

Please read these WARNINGS and CAUTIONS before proceeding with maintenance and repair work.

WARNING (continued) —

- ALWAYS disconnect the airbag power supply before using a computer memory saver. Failure to disconnect the airbag power supply may result in accidental activation of the airbag.
- The airbag unit should not be removed, nor should any work be performed on the airbag system, without first reviewing the warnings and cautions that accompany the airbag system repair information in this manual.
- Never apply stickers or any other coverings on the steering wheel. Do not let chemical cleaners, oil or grease come into contact with the vinyl covering of the airbag unit.
- Never open or otherwise attempt to repair airbag system parts. Always use new parts. Never leave airbag parts or the partially disassembled airbag system unattended.
- Never use a test light or volt meter to conduct electrical tests on the airbag system. The airbag system must only be tested by trained Volkswagen Service technicians using special airbag tester equipment. The airbag system must only be tested while it is installed in the car.
- Do not expose an airbag unit to temperatures above 194°F (90°C) even for a brief period, while handling it during the repair process. Keep the unit clear of heat sources such as hot plates, soldering irons, heat lamps, welding equipment, or the like.
- Always replace airbag system components that have been mechanically damaged (bubbles, cracks, etc.).
- Always set down a removed airbag unit so that the horn pad is facing upward.
- When driving or riding in an airbag-equipped vehicle, never hold test equipment in your hands or lap while the vehicle is in motion. Objects between you and the airbag can increase the risk of injury in an accident.
- The storage, transportation and disposal of airbag units are subject to the laws for flammable solids.

CAUTION—

- If you lack the skills, tools and equipment, or a suitable workshop for any procedure described in this manual, we suggest you leave such repairs to an authorized Volkswagen dealer or other qualified shop. We especially urge you to consult an authorized Volkswagen dealer before beginning repairs on any car that may still be covered wholly or in part by any of the extensive warranties issued by Volkswagen of America, Inc.
- Volkswagen is constantly improving its cars and sometimes these changes, both in parts and specifications, are made applicable to earlier models. Therefore, part numbers listed in this manual are for reference only. Always check with your authorized Volkswagen dealer parts department for the latest information.
- Before starting a job, make certain that you have all the necessary tools and parts on hand. Read all the instructions thoroughly, do not attempt shortcuts. Use tools that are appropriate to the work and use only replacement parts meeting Volkswagen specifications. Makeshift tools, parts and procedures will not make good repairs.
- Use pneumatic and electric tools only to loosen threaded parts and fasteners. Never use these tools to tighten fasteners, especially on light alloy parts. Always use a torque wrench to tighten fasteners to the tightening torque specification listed.
- Be mindful of the environment and ecology. Before you drain the crankcase, find out the proper way to dispose of the oil. Do not pour oil onto the ground, down a drain, or into a stream, pond, or lake. Consult local ordinances that govern the disposal of wastes.
- When connecting cables or ground straps, all connections must be cleaned to bare metal. Be sure mounting bolts are tight.
- Do not replace fuse links with fuse links of other gauge or with standard wire. Always replace the entire length of the fuse link.
- Before doing any electrical welding on vehicles equipped with Anti-lock Brakes (ABS), disconnect the battery negative terminal (ground strap) and the ABS Control Module connector.
- On cars equipped with ABS, the ABS control unit cannot withstand temperatures from a paint–drying booth or a heat lamp in excess of 203°F (95°C) and should not be subjected to temperatures in excess of 185°F (85°C) for more than two hours.
- On cars equipped with anti-theft radios, make sure you know the correct radio activation code before disconnecting the battery or removing the radio. If the wrong code is entered when power is restored, the radio may lock up and be rendered inoperable, even if the correct code is then entered.
- When boost-charging the battery, first remove the fuses for the Engine Control Module (ECM), the Transmission Control Module (TCM), the ABS Control Module, and the On-Board computer. In cases where one or more of these components is not separately fused, disconnect the Control Module connector(s).
- Do not quick-charge the battery (for boost starting) for longer than one minute, and do not exceed 16.5 volts at the battery with the boosting cables attached. Wait at least one minute before boosting the battery a second time.

◀ **WARNINGS and CAUTIONS continued from previous page**

Technical data

Suspension

Wheel base	mm	2920	3320
Track, unladen	mm	1589 front, wheel offset 55 1554 rear, wheel offset 55	1589 front, wheel offset 55 1554 rear, wheel offset 55
		1607 front, wheel offset 46 1572 rear, wheel offset 46	1607 front, wheel offset 46 1572 rear, wheel offset 46
Turning circle	m	11.7	12.9
Wheel lock angle	inner	43°	43°
Ground clearance at permissible gross weight – normal suspension – with lowered suspension	 mm mm	 180 160	 180 160

Steering

Steering gear		Power steering gear
Steering roll radius (negative)	mm	– 5.0
Turns of steering wheel from lock to lock	approx	3.8
Overall steering ratio		17.5
Wheel alignment specifications ⇒ Page N 44-1		

Brakes

Version	Dim.	15″ wheels		
Optional equipment No. (PR-No.)[1]		1LA (floating caliper)	1LE (floating frame)	1LP (floating caliper)
Brake master cylinder diameter	mm	25.4	23.81	23.81
Brake servo unit diameter	inch	9 + 8	9 + 8	9 + 8
Front brakes				
Brake caliper piston diameter	mm	57	54	54
Brake disc diameter	mm	280	280	282
Brake disc thickness	mm	24[2]	24[2]	18
Brake pad thickness	mm	13.5	13	13.5
Rear brakes				
Brake caliper piston diameter	mm	38		
Brake disc diameter	mm	280		
Brake disc thickness	mm	12		
Brake pad thickness	mm	12		

[1] Optional equipment number; Explanation ⇒ Page N 00-3

[2] Ventilated brake disc

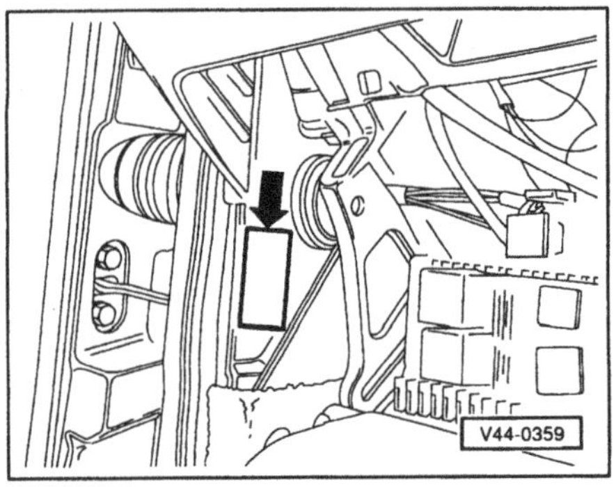

Identification of brake caliper type via optional equipment number (PR No.)

Different PR numbers are printed on vehicle data plate for each brake caliper version.

◄ **Vehicle data plate location -arrow-**

Left-hand A-pillar next to central electric unit.

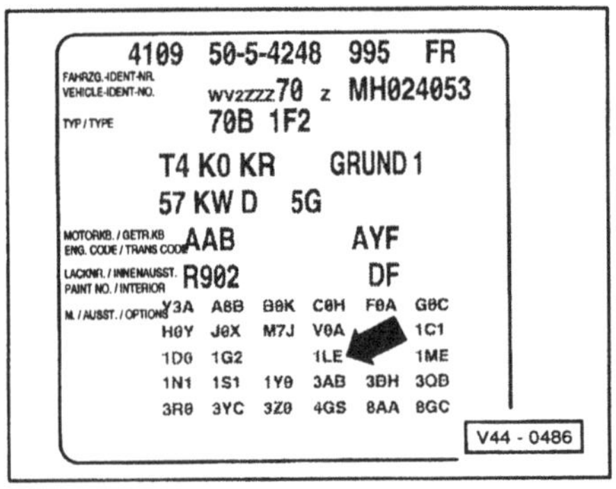

◄ **Location of optional equipment number (PR No.) on vehicle data plate -arrow-**

- 1LU = 15" Floating caliper FN3/57 ITT
- 1LP = 15" Floating caliper Lucas C54
- 1LE = 15" Floating frame caliper Lucas RC 54

Note:

Depending on equipment level, brake caliper code is printed in different locations on vehicle data plate.

Wheels, tires

Tire size[1]	Wheel	Offset
195/70 R15 97 S reinforced	6 J x 15	55 mm
195/70 R15 C 104/102 R	6 J x 15	55 mm
205/65 R15 99 S reinforced	6 J x 15	55 mm
205/65 R15 99 S reinforced	7 J x 15	46 mm
205/65 R15 C 100 T	7 J x 15	46 mm
205/65 R15 C 100 T	6 J x 15	55 mm

[1] Tire pressures ⇒ sticker on vehicle

Caution!
Wheels from vehicles up to 12.95 cannot be used.

Special tools

I - Removing/installing assembly carrier, stabilizer bar, upper control arm / Repair group 40			
Designation	**Tool number**	**Designation**	**Tool number**
Counter-hold	3079	Transmission jack	VAG 1383 A
Front axle carrier	VAG 1712	27 mm AF socket	3257
Torque wrench	VAG 1332	34 mm AF open jaw spanner	VAG 1332/5
II - Removing/installing upper and lower control arms / Repair group 40			
Press tool	VW 412	Press piece	554
Press plate	VW 401/VW 402	Press tool	VW 411
Press tool	VW 409	Centralizing pin	12-551
Guide sleeve	3053	Spindle	VW 771
Press piece	40-105	Press piece	VW 477i
Mounting	3301	Nut from assembly device	3301
Tube	VW 415a	Tube	VW 454
Tube	VW 519		

III - Removing/installing wheel, suspension, drive shaft / Repair group 40			
Designation	**Tool number**	**Designation**	**Tool number**
Press plate	VW 401/VW 402	Press tool	VW 407
Press tool	VW 408a	Thrust piece	40-105
Press tool	VW 412	Thrust piece	3253/6
Press piece	VW 442	Thrust piece	3253/7
Thrust washer	3253/8	Thrust washer	3074
Puller	e.g. Kukko 204-2	7 mm AF multi-tooth sock-et- head wrench	commercial type
Puller	e.g. Kukko 20-10 with hook 2-150	14 mm AF multi-tooth socket-head wrench	commercial type
Centralizing pin	12-551	Ball bearing puller	e.g. Kukko 128/2
Tube	VW 416b	Tube	3253/1
Removing/installing wheel bearing housing / Repair group 40			
Ball bearing puller	e.g. Kukko 128/2		
Servicing drive shaft / Repair group 40			
Press plate	VW 401/VW 402	Press tool	VW 408a
Press tool	VW 407	Torque wrench	VAG 1331
Thrust spindle	3207	Tensioning pliers	1682
Support sleeve	VW 522	Tensioner	40-204A
14 mm multi-tooth socket-head wrench	commercial type		

Servicing rear suspension / Repair group 42			
Designation	**Tool number**	**Designation**	**Tool number**
Thrust piece	VW 442	Puller	3046
Press piece	3128	Tube	30-14
Guide sleeve	3053	Torque wrench	VAG 1331
Pop rivet nut pliers	VAG 1765/VAG 1765A	50 mm AF open jaw wrench	3254
Servicing wheel suspension / Repair group 42			
Tube	3253/1	Bridge	3253/2
Spindle	3253/3	Mounting	3253/4
Disc	3253/5	Disc	3253/6
Press piece	3253/7	Press disc	3253/8
Press disc	3074	Transmission jack	VAG 1383/A
Press piece	40-105	Puller	e.g. Kukko 204-2
27 mm AF socket	3257		
Removing and installing rear drive shaft / Repair group 42			
Press plate	VW 401/VW 402	Press tool	VW 408a
Press tool	VW 407	Tensioner	40-204A
Torque wrench	VAG 1331	Support sleeve	VW 522
Tensioning pliers	1682		

Vehicle alignment / Repair group 44			
Designation	**Tool number**	**Designation**	**Tool number**
Angle gauge	3021	46 mm AF ring wrench	3252
Removing and installing brake pads/linings / Repair group 46			
Designation	**Tool number**		
Piston resetting device	-		
Servicing rear brakes (disc brakes) / Repair group 46			
Re-setting and removal tool	3272		
Servicing front brake caliper / Repair group 47			
Plastic wedge	VAG 1474/5	Piston resetting tool	-
Servicing rear brake caliper			
Plastic wedge	VAG 1474/5	Piston resetting tool	-
Bleeding brake system / Repair group 47			
Brake filler and bleeder unit	VW 1238/1 or US 1116		
Adjusting brake booster / Repair group 47			
Adjusting device	3255	Test unit for brake pressure regulator	VAG 1310

Disassembling and assembling column tube with steering column / Repair group 48			
Designation	**Tool number**	**Designation**	**Tool number**
Separating device	e.g. Kukko 17/0	Puller	e.g. Kukko 18/0
Press plate	VW 401/VW 402	Press tool	VW 408a
Press disc	VW 512	Press plate	30-11
Servicing power steering / Repair group 48			
Adapter set	VAG 1402/6	Test unit for power steering	VAG 1402
Assembly overview: Power steering gear / Repair group 48			
Press plate	VW 401/VW 402	Press tool	VW 411
Torque wrench	VAG 1332	34 mm AF open jaw wrench	VAG 1332/5
Ball joint puller	e.g. Kukko 128/2		
Removing and installing vane pump on vehicles with 5 and 6-cyl. engine / Repair group 48			
Water pump tool	VAG 1590		

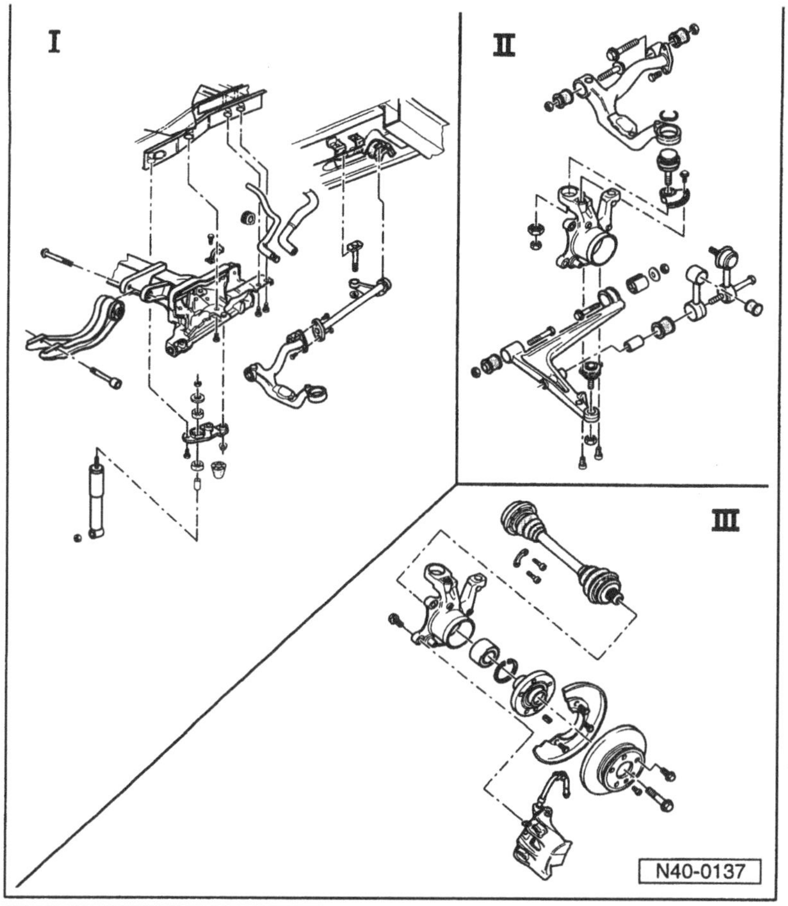

Front suspension, servicing

I - Removing/installing assembly carrier, stabilizer bar, torsion bar and shock absorber,
⇒ Page N 40-3.

II - Removing/installing upper and lower control arms,
⇒ Page N 40-20.

III - Removing/installing wheels and drive shaft,
⇒ Page N 40-38.

Notes:

- *Upper control arm is loaded by torsional force of torsion bar. Therefore, when performing operations listed below, always relieve pressure on torsion bar:*
- *Shock absorber, removing/installing.*
- *Wheel bearing / wheel bearing. housing, removing/installing*
- *Drive shaft, removing/installing.*
- *Lower steering ball joint/control arm, removing/installing.*

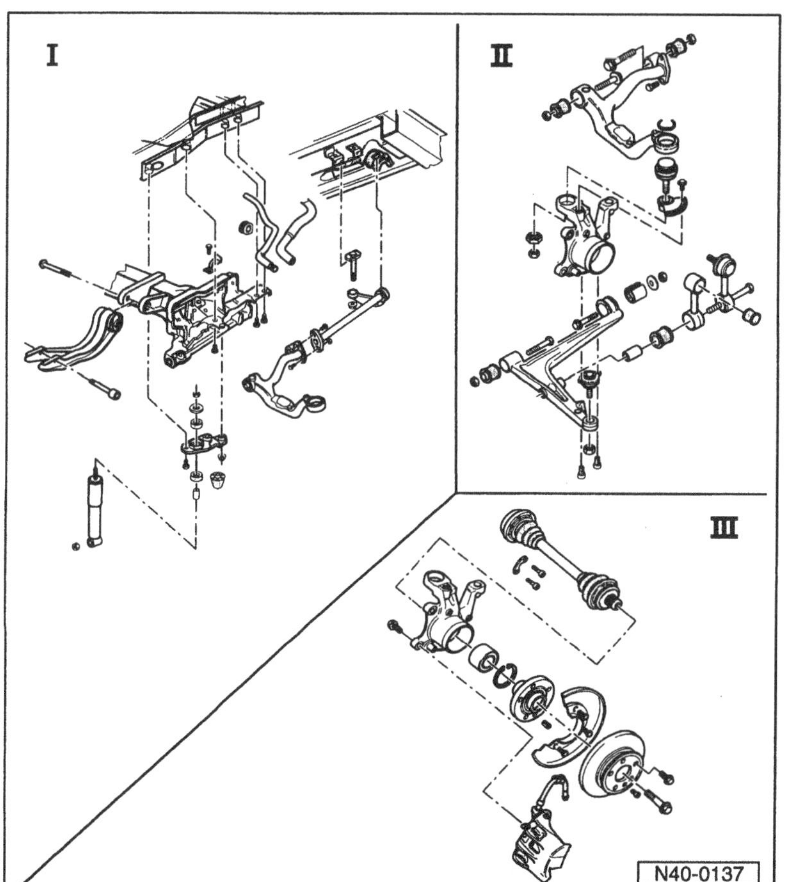

- *Upper control arm, removing/installing.*
- *Eccentric bushing on upper control arm, removing/installing.*
- *Stabilizer bar, removing/installing.*
- *Torsion bar tension, relieving,*
⇒ Page N 40-11.
⇒ Torsion bar, removing/installing/ adjusting.

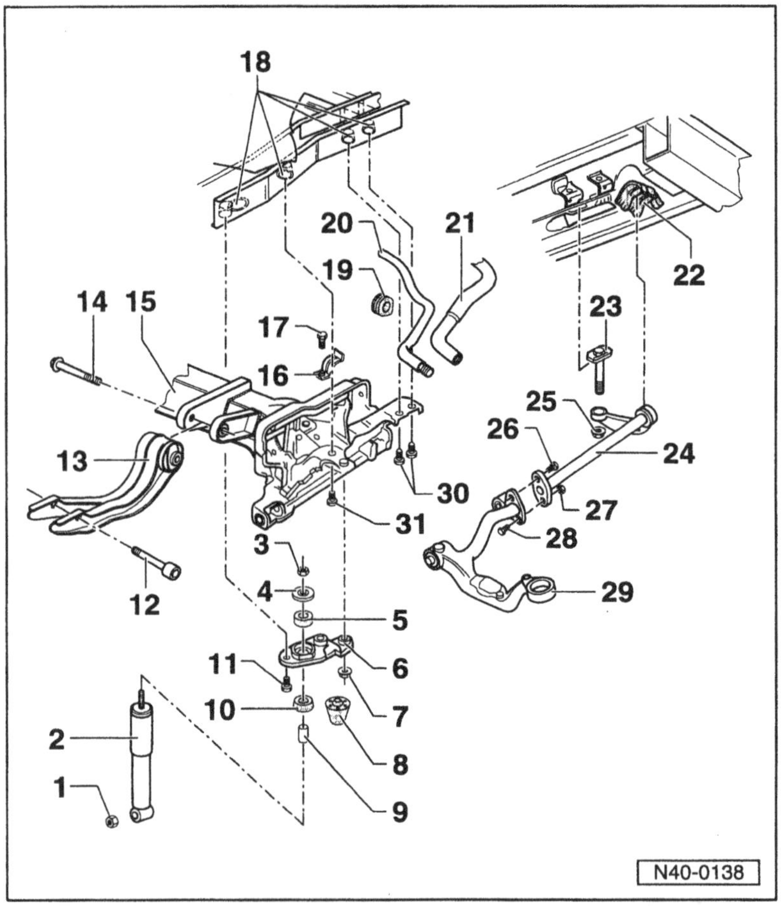

Assembly carrier, stabilizer bar, torsion bar and control arm, removing/installing

Notes:

- *If a vehicle has to be moved after removing drive shaft, first install outer joint and tighten to 50 Nm (37 ft lb), otherwise, wheel bearing damage will result.*
- *Alignment specifications,*

⇒ Page N 44-6.

- *Do not perform welding and straightening operations on load bearing front suspension components.*
- *Always replace self-locking nuts.*
- *Always replace corroded nuts/bolts.*

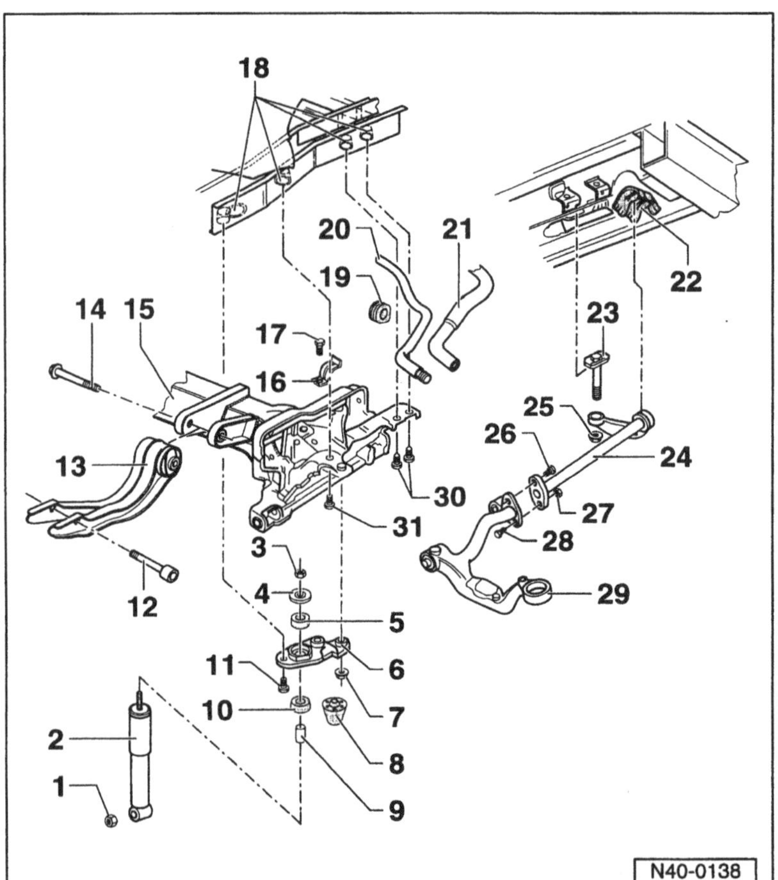

1 - Hex nut, 160 Nm (118 ft lb)

2 - Shock absorber
- Removing/installing,

⇒ Page N 40-10.

- Application:

⇒ Replacement parts catalog.

3 - Hex nut, 25 Nm (18 ft lb)

4 - Washer
- Collar faces down.

5 - Upper bearing bushing

6 - Shock absorber mount

7 - Hex nut, 100 Nm (74 ft lb)

8 - Jounce bumper

9 - Spacer sleve

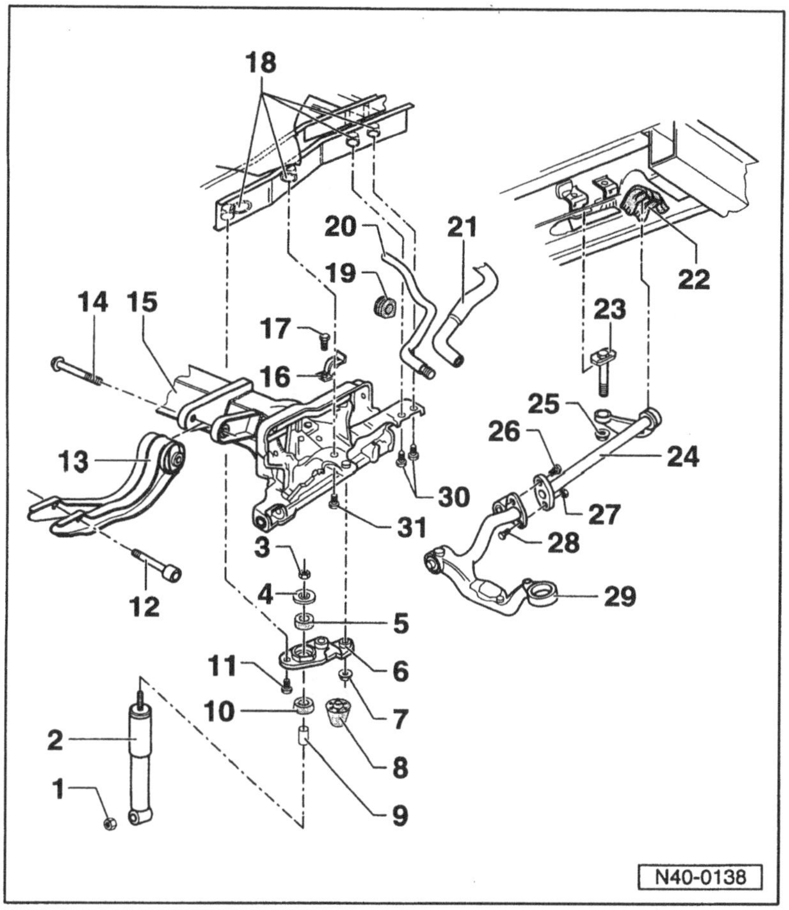

10 - Lower bearing bushing

11 - Hex bolt, 100 Nm (74 ft lb)

12 - Allen bolt
- 80 Nm (59 ft lb) and turn 90° further.

13 - Pendulum support
- Removing and installing,

⇒ from Page N 40-8.

14 - Hex head bolt
- 150 Nm (111 ft lb) and turn 90° further.
- Replace each time after removing.

15 - Sub-frame
- Removing/installing,

⇒ Page N 40-8.

16 - Clamp

17 - Hex bolt, 55 Nm (41 ft lb)

18 - Weld nut
- If damaged, repair,

⇒ Page N 40-19.

19 - Stabilizer bar mount

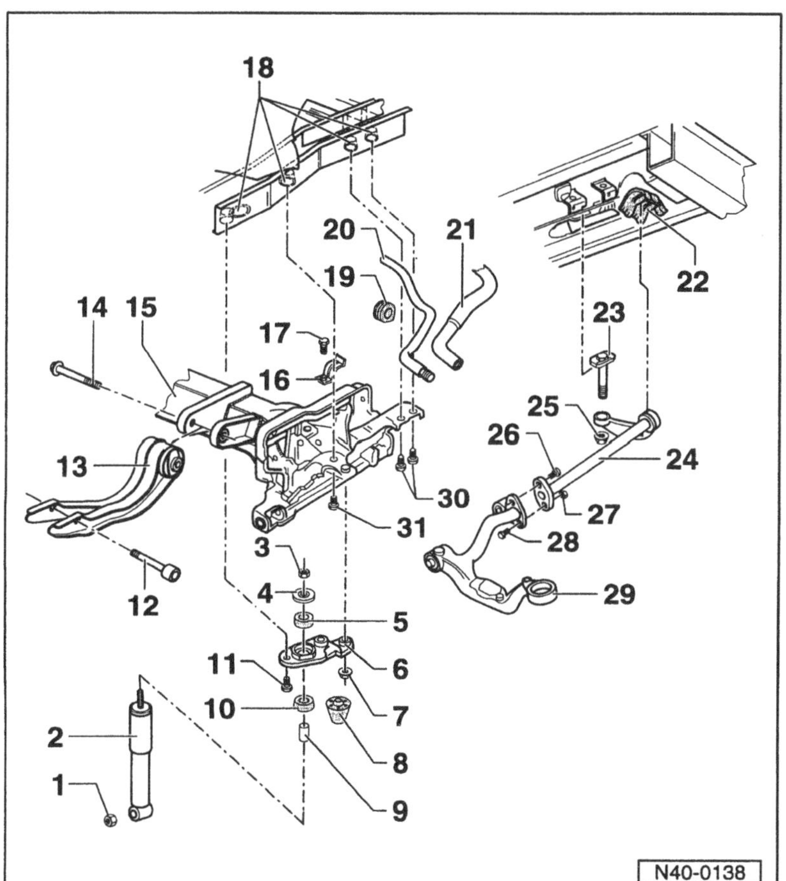

20 - Stabilizer bar
- 23.5 mm dia. (0.93 in.)
- Removing/installing,

⇒ Page N 40-12.

21 - Stabilizer bar
- 27 mm dia. (1.1 in.)
- Removing/installing,

⇒ Page N 40-12.
- Retrofitting,

⇒ Page N 40-14.

22 - Rubber mount

23 - Torsion bar adjustment bolt

24 - Torsion bar
- Right-hand marked with "blue" on end face.
- Left-hand marked with "yellow" on end face.
- Supplied as a spare part, with tensioning lever.
- Removing,

⇒ Page N 40-11.
- Check for paint damage, rectify if necessary.
- Allocation and torsion bar markings,

⇒ Page N 40-15.

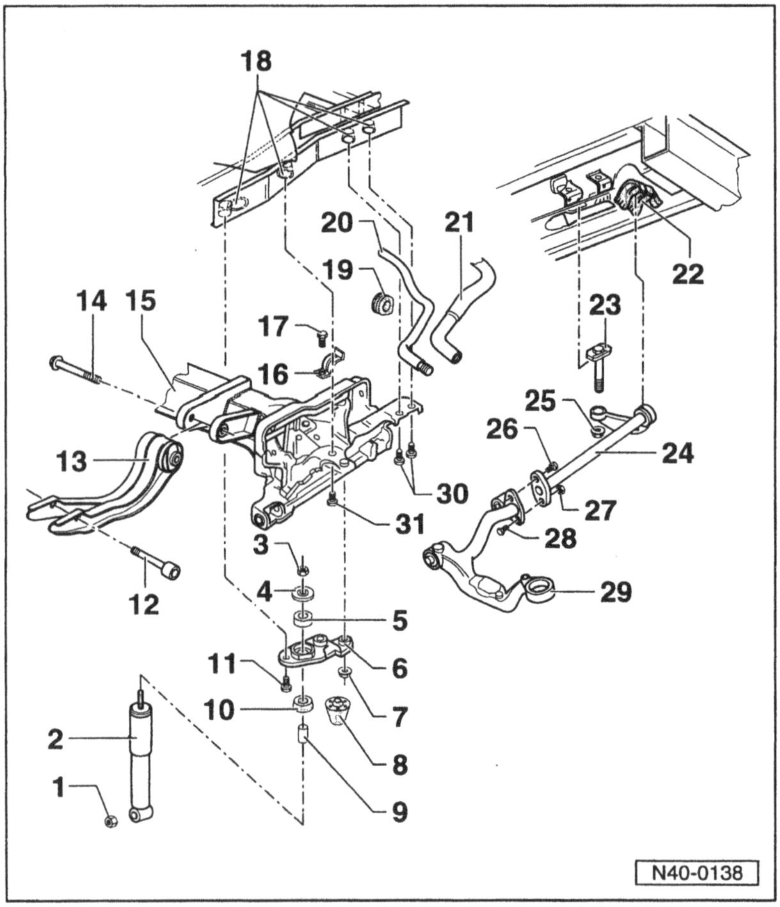

25 - Torsion bar tensioning nut

– Replace.
- ♦ Measure exposed thread length before removing,

⇒ Page N 40-11.
- ♦ Do not use an air tool.

26 - Hex bolt, 160 Nm (118 ft lb)
- ♦ M 14 x 1.5 x 30

27 - Hex nut, 130 Nm (96 ft lb)

28 - Hex bolt
- ♦ M 14 x 1.5 x 42

29 - Upper control arm
- ♦ Can only be removed when assembly carrier has been removed.

30 - Hex bolt, 100 Nm (74 ft lb)
- ♦ M 12 x 1.5 x 32

31 - Hex bolt, 160 Nm (118 ft lb)
- ♦ M 14 x 1.5 x 40

Subframe, removing/installing

– Remove sound dampening pan.

– Remove front wheels.

– Remove brake caliper and secure to body.

– Separate ABS sensor wire at connector.

– Remove fuel tank,
⇒ Engine, Repair group 20.

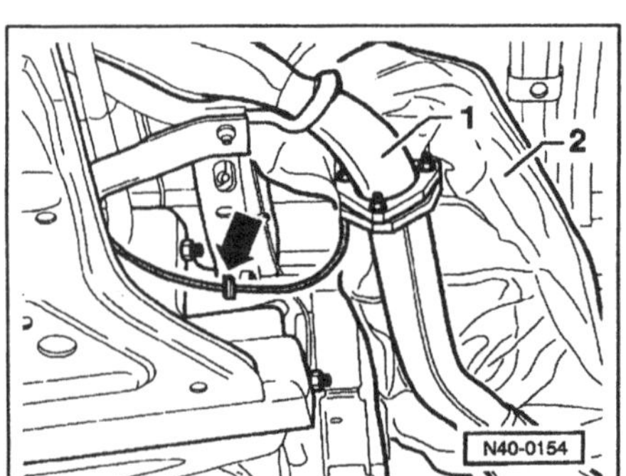

◄ – Unclip wire -arrow- for oxygen sensor at retainer.

– Separate exhaust system behind front exhaust pipe -1-.

– Remove exhaust system heat shield -2- and loosen exhaust system, if necessary.

– Relieve torsion bar tension.
⇒ Page N 40-11, Torsion bar,removing/installing/adjusting.

– Remove universal joint bolt at steering gear and remove lines/hoses from power steering gear, if necessary.

– Disconnect drive shafts at final drive.

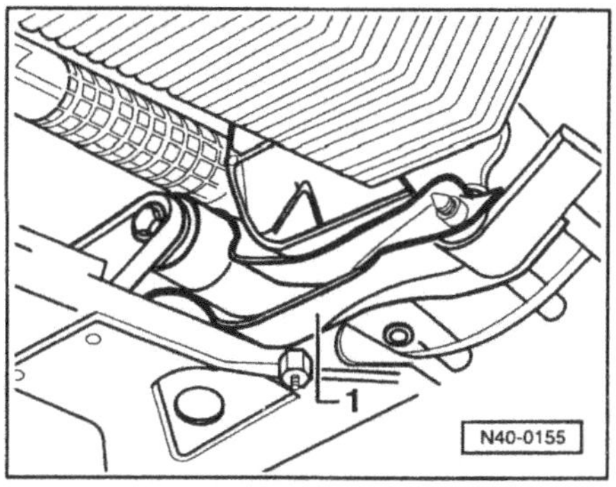

- Remove pendulum support -1- from transmission.
- Place transmission jack VAG 1383 A, with front axle carrier VAG 1712, under assembly carrier.
- Remove assembly carrier bolts.

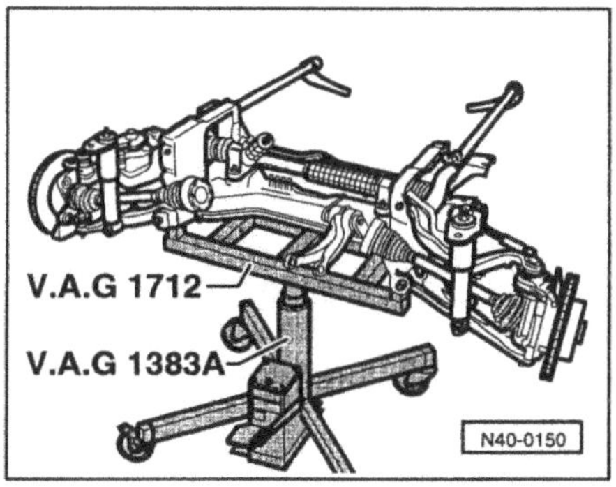

- Lower sub-frame assembly using transmission jack VAG 1383 A.
- At same time guide right torsion bar through front exhaust pipe/catalyst point of separation and guide drive shaft past transmission.

Install in reverse sequence.

Shock absorbers, removing/installing

- Relieve torsion bar tension,

⇒ Page N 40-11, removing/installing torsion bar/adjusting torsion bar.

- Remove shock absorber lower bolt.
- Remove shock absorber mount from body.

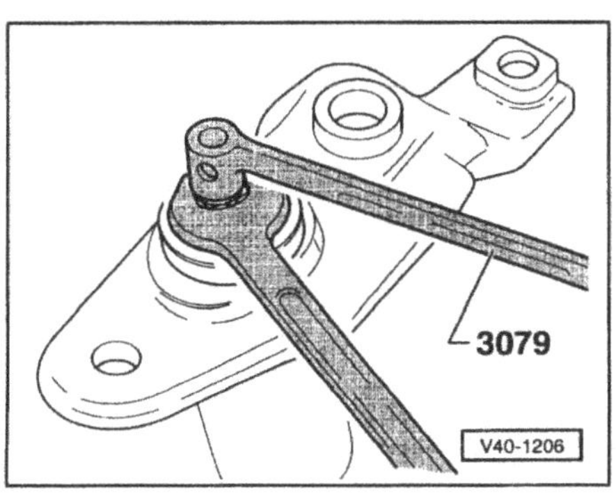

- Remove shock absorber, as illustrated, from mount.

Install in reverse sequence.

Torsion bar, removing/installing/ adjusting

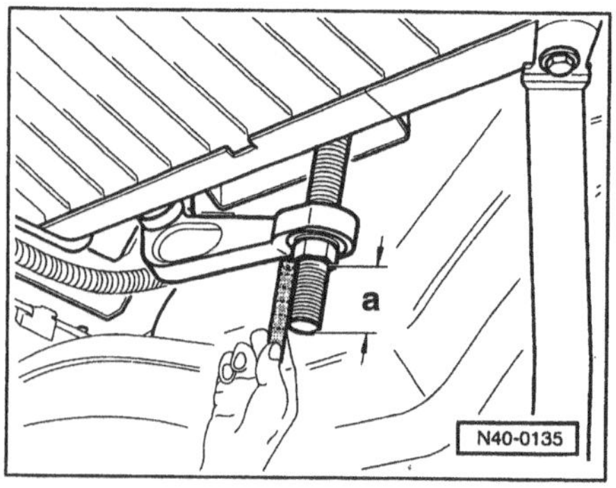

- Remove fuel tank.
⇒ Engine, Repair group 20.
- Remove exhaust system heat shield (if equipped), and loosen exhaust system, if necessary.
- ◄ Measure length of thread projecting from nut to end of threaded rod and note.

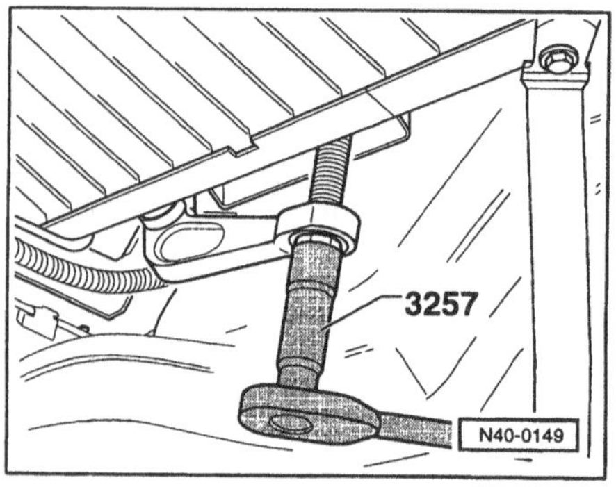

- ◄ Relieve tension on torsion bar with special tool 3257.
- Unbolt torsion bar from control arm and remove.

Notes:

- *Measuring ride height,*
⇒ Page N 44-4.
- *Ride height adjustment data,*
⇒ from Page N 44-6.

Install in reverse order of removal.

Front stabilizer bar, removing/installing

- Remove front wheels.
- Relieve torsion bar tension,
⇒ Page N 40-11, Torsion bar, removing/installing/adjusting.
- Unbolt both shock absorbers from control arms and push completely upward.
- Disconnect connecting link from lower control arm.

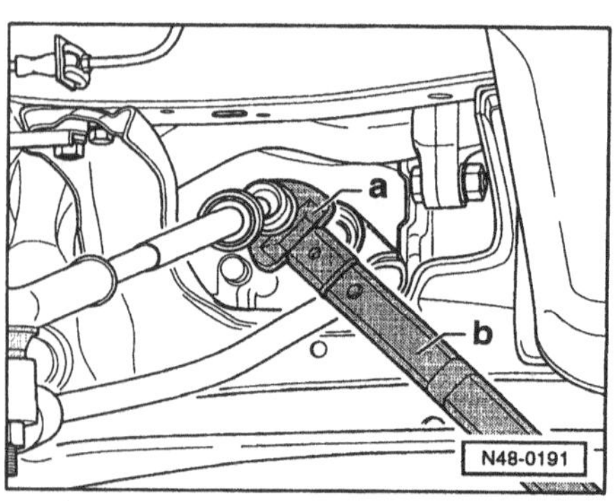

- ◄ Unscrew tie rods from steering gear.

  a - VAG 1332/5

  b - Torque wrench VAG 1332
- Remove front exhaust pipe,
⇒ Repair group 26; Exhaust system, removing/installing.

Vehicles with 27 mm (1.1 in.) dia. heavy duty stabilizer bar

- Remove universal joint bolt on steering pinion and separate steering gear from steering column.
- Remove steering gear from sub-assembly.

Continued for all vehicles

– Remove stabilizer bar clamps from sub-assembly.

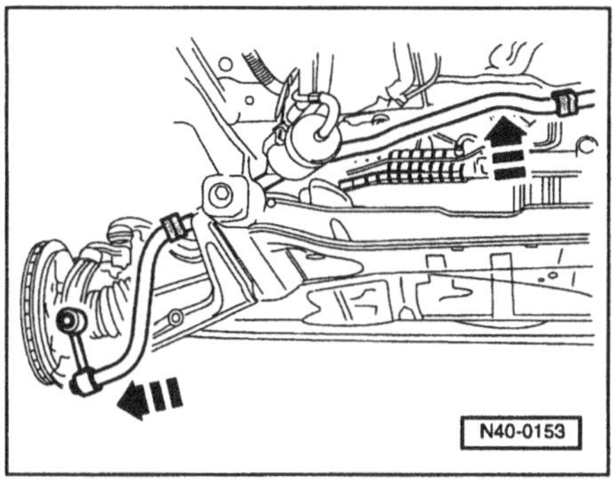

◄ – Turn stabilizer bar upward approximately 90° and remove from left, in one motion. Note right-hand connecting link when removing stabilizer bar.

Install in reverse order of removal.

Notes:

- *Ensure stabilizer bar surface is not damaged during installation. If necessary, repair paint damage and apply underbody sealant.*
- *Connector links remain on stabilizer bar when removing/installing.*

Front stabilizer bar (heavy-duty), retrofitting

Note:

The front heavy duty stabilizer bar can only be installed in conjunction with a rear stabilizer bar.

Parts required (retrofit kit):

- Stabilizer bar, 27 mm dia. (1.1 in.)
- Coupling link (Qty. 2) for 27 mm dia. (1.1 in.) stabilizer bar
- Rubber bushing (Qty. 2) for stabilizer bar

Stabilizer bar, removing/installing,
⇒ Page N 40-12.

Torsion bar, application / marking

Model	Equipment	Torsion bar	color[3)]	dia. mm
Transporter	short wheel base, 800 kg[4)] 1 PO[1)] + 1 BA[2)]	701 411 103 G/104 G	green	25.3
	short wheel base, 1000 kg[4)] 1 PO[1)] + 1 BA[2)]	701 411 103 H/104 H	brown	26.4
	short wheel base, 1200 kg[4)] 1 PO[1)] + 1 BA[2)]	701 411 103 J/104 J	violet	27.9
	long wheel base, 1000 kg[4)] 1 PO[1)] + 1 BA[2)]	701 411 103 H/104 H	brown	26.4
	long wheel base, 1200 kg[4)] 1 PO[1)] + 1 BA[2)]	701 411 103 J/104 J	violet	27.9
	long wheel base, other coachwork builder	701 411 103 H/104 H	brown	26.4
	short wheel base, lowered suspension	701 411 103 J/104 J	violet	27.9
	heavy duty front, irrespective of wheel base, payload, heavy duty rear	701 411 103 J/104 J	violet	27.9

1) Standard suspension

2) Standard shock absorbers

3) Color coding: approx. 75 mm (3 in.) from flange

4) Payload: The payload details are rounded figures. When choosing torsion bars for vehicles with all wheel drive, 100 kg (220.5 lb) must be added to payload (due to increased curb weight).

Model	Equipment	Torsion bar	Color[3)]	dia. mm
Multivan	short wheel base, 800 kg[4)] 1 PO[1)] + 1 BA[2)]	701 411 103 G/104 G	green	25.3
	short wheel base, 800 kg[4)] 6-cyl. engine, 1 PO[1)] + 1 BA[2)]	701 411 103 J/104 J	violet	27.9
	short wheel base, 1000 kg[4)] 1 PO[1)] + 1 BA[2)]	701 411 103 H/104 H	brown	26.4
	short wheel base, 1000 kg[4)] 6-cyl. engine, 1 PO[1)] + 1 BA[2)]	701 411 103 J/104 J	violet	27.9
	short wheel base, 1200 kg[4)] 1 PO[1)] + 1 BA[2)]	701 411 103 J/104 J	violet	27.9
	long wheel base, 1000 kg[4)] 1 PO[1)] + 1 BA[2)]	701 411 103 H/104 H	brown	26.4
	long wheel base, 1200 kg[4)] 1 PO[1)] + 1 BA[2)]	701 411 103 J/104 J	violet	27.9
	short wheel base, lowered suspension	701 411 103 H/104 H	brown	26.4
	H.D. front/rear, regardless of wheel base or payload	701 411 103 J/104 J	violet	27.9

1) Standard suspension

2) Standard shock absorbers

3) Color coding: approx. 75 mm (3 in.) from flange

4) Payload: Payload details are rounded figures. When choosing torsion bars for vehicles with all wheel drive, 100 kg (220.5 lb) must be added to payload (due to increased curb weight).

Model	Equipment	Torsion bar	Color[3]	dia. mm
Transporter	short wheel base, 1000 kg[4] 1 PO[1] + 1 BA[2]	701 411 103 H/104 H	brown	26.4
	short wheel base, 1200 kg[4] 1 PO[1] + 1 BA[2]	701 411 103 J/104 J	violet	27.9
	short wheel base, 2000 kg[4] 1 PO[1]	701 411 103 J/104 J	violet	27.9
	long wheel base, 1000 kg[4] 1 PO[1] + 1 BA[2]	701 411 103 H/104 H	brown	26.4
	long wheel base, 1200 kg[4] 1 PO[1] + 1 BA[2]	701 411 103 J/104 J	violet	27.9
	long wheel base, 2000 kg[4] 1 PO[1]	701 411 103 J/104 J	violet	27.9
	heavy duty front, irrespective of wheel base, payload, heavy duty rear	701 411 103 J/104 J	violet	27.9

[1] Standard suspension

[2] Standard shock absorbers

[3] Color coding: approx. 75 mm (3 in.) from flange

[4] Payload: Payload details are rounded figures. When choosing torsion bars for vehicles with all wheel drive, 100 kg (220.5 lb) must be added to payload (due to increased curb weight).

Model	Equipment	Torsion bar	Color[3]	dia. mm
Ambulance	short wheel base, 1000 kg[4] ambulance suspension	701 411 103 G/104 G	green	25.3
	long wheel base, 1000 kg[4] ambulance suspension	701 411 103 H/104 H	brown	26.4
	long wheel base, 1200 kg[4] other coachwork builder	701 411 103 J/104 J	violet	27.9
High roof ambulance with four wheel drive	long wheel base, 1000 kg[4]	701 411 103 H/104 H	brown	26.4
Camper	short wheel base, 1000 kg[4] 1 PO[1] + 1 BA[2]	701 411 103 H/104 H	brown	26.4
	long wheel base, 1000 kg[4] 1 PO[1] + 1 BA[2]	701 411 103 H/104 H	brown	26.4

[1] Standard suspension

[2] Standard shock absorbers

[3] color coding: approx. 75 mm (3 in.) from flange

[4] Payload: Payload details are rounded figures. When choosing torsion bars for vehicles with all wheel drive, 100 kg (220.5 lb) must be added to payload (due to increased curb weight).

Long member weld nut, repairing

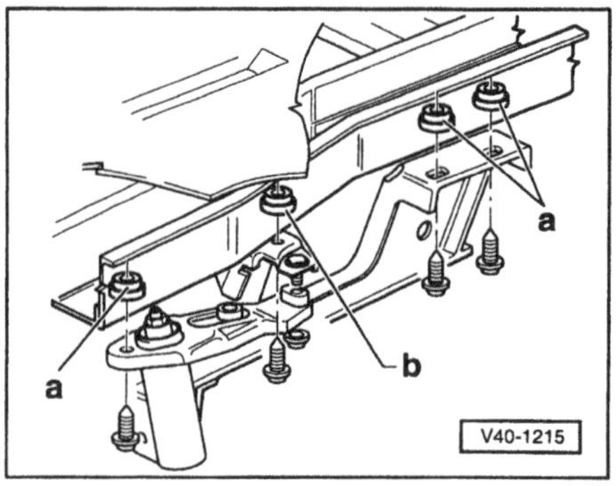

Damaged weld nut threads can be repaired with Heli-coil® thread inserts.

To drill, proceed as follows:

– Remove subframe,
⇒ Page N 40-8.

◄ – Drill through damaged weld nut -a- with a 14.5 mm dia. (0.57 in.) drill and -b- with a 16.5 mm dia. (0.65 in.) drill.

Warning!
Wear protective goggles.

Notes:

- *Do not tilt drill when drilling.*
- *Only use threaded inserts with a pitch of 1.5.*
- *Install Heli-coil® threaded inserts according to manufacturer's instructions.*

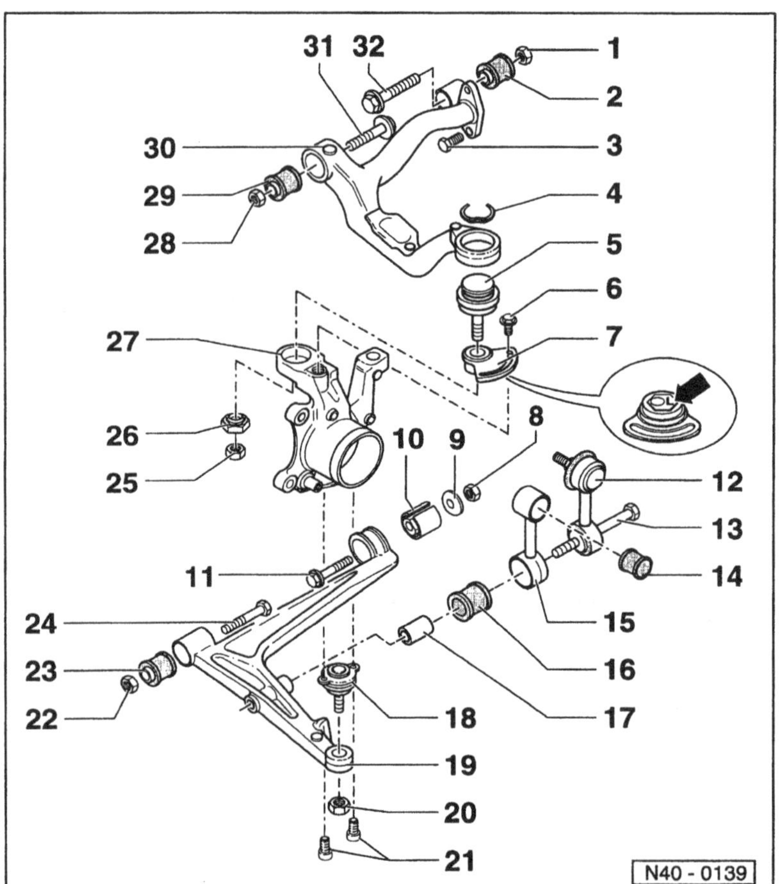

Upper and lower control arms, removing/installing

1 - Hex nut, 100 Nm (74 ft lb)
- Replace hex nut when removed.

2 - Upper control arm rear bushing
- Pressing out/in,

⇒ Fig. 3.

3 - Hex bolt
- M 14 x 1.5 x 42

4 - Circlip
- Pry out with a screwdriver.
- Always replace.

5 - Upper ball joint
- Removing/installing,

⇒ Page N 40-35.

6 - Hex bolt, 60 Nm (44 ft lb)
- M 12 x 1.5 x 22

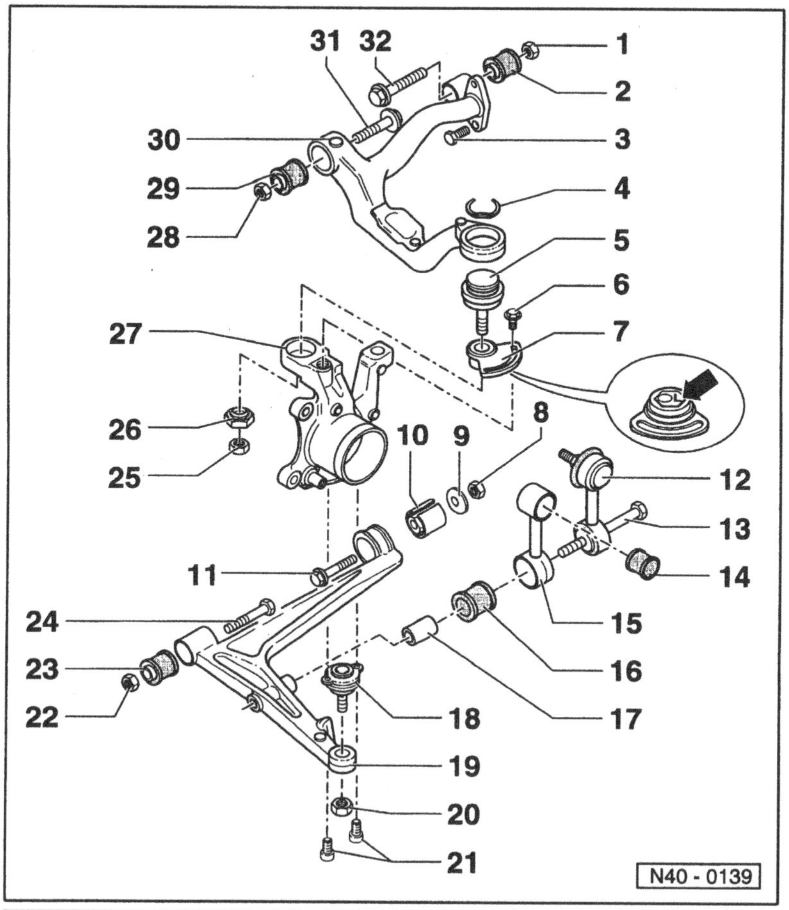

7 - Eccentric bushing
- Right-hand side with "R" marked on underside.
- Left-hand side with "L" marked on underside -arrow-.
- Adjusting camber,

⇒ Page N 44-14.
- Pressing off,

⇒ Page N 40-36.
- Mark installation position before removing.

8 - Hex nut, 130 Nm (96 ft lb)
- Tighten with vehicle on its wheels.

9 - Eccentric washer

10 - Lower control arm rear bushing
- Pressing out,

⇒ Fig. 7.
- Installation position,

⇒ Fig. 8.
- Pre–tensioning,

⇒ Fig. 9.
- Pressing in,

⇒ Fig. 10.

11 - Eccentric bolt
- Adjusting caster,

⇒ Page N 44-17.

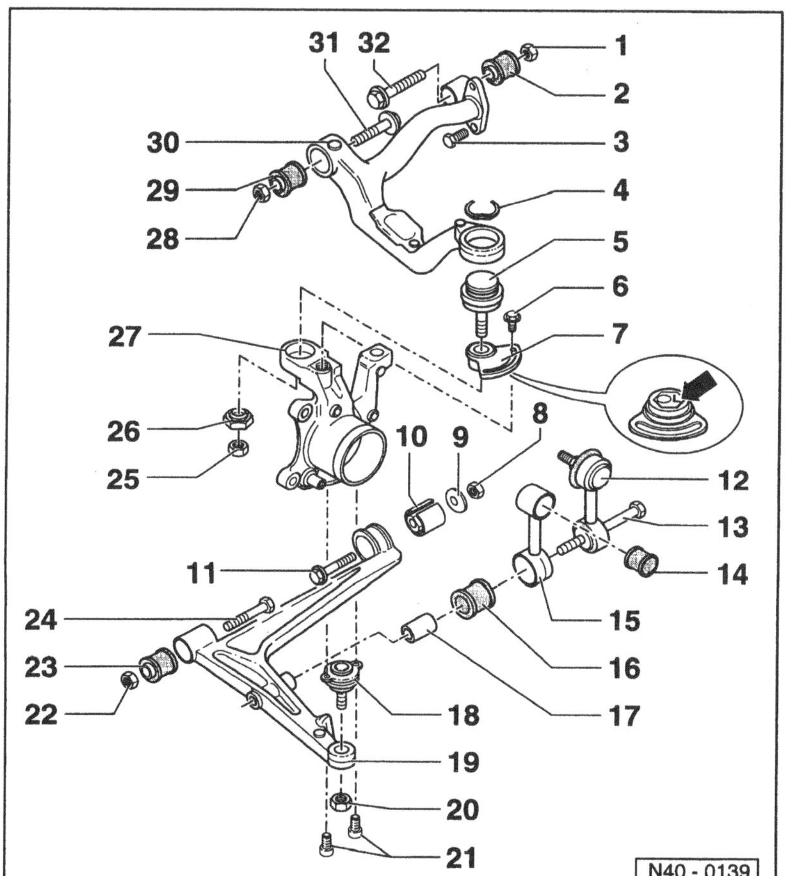

12 - Link, stabilizer bar
- For 27 mm (1.06 in.) stabilizer bar
- Connecting to/from stabilizer bar

⇒ Fig. 13

13 - Hex bolt
- M 14 x 1.5 x 175

14 - Rubber mount
- Only on coupling link for 23.5 mm dia. (0.9 in.) stabilizer bar
- Removing/installing

⇒ Fig. 14

15 - Connecting link
- For 23.5 mm dia. (0.9 in.) stabilizer bar
- Pressing off stabilizer bar

⇒ Fig. 12

16 - Rubber mount
- Removing/installing

⇒ Fig. 15 and 16

17 - Bushing
- Removing/installing

⇒ Fig. 15 and 16

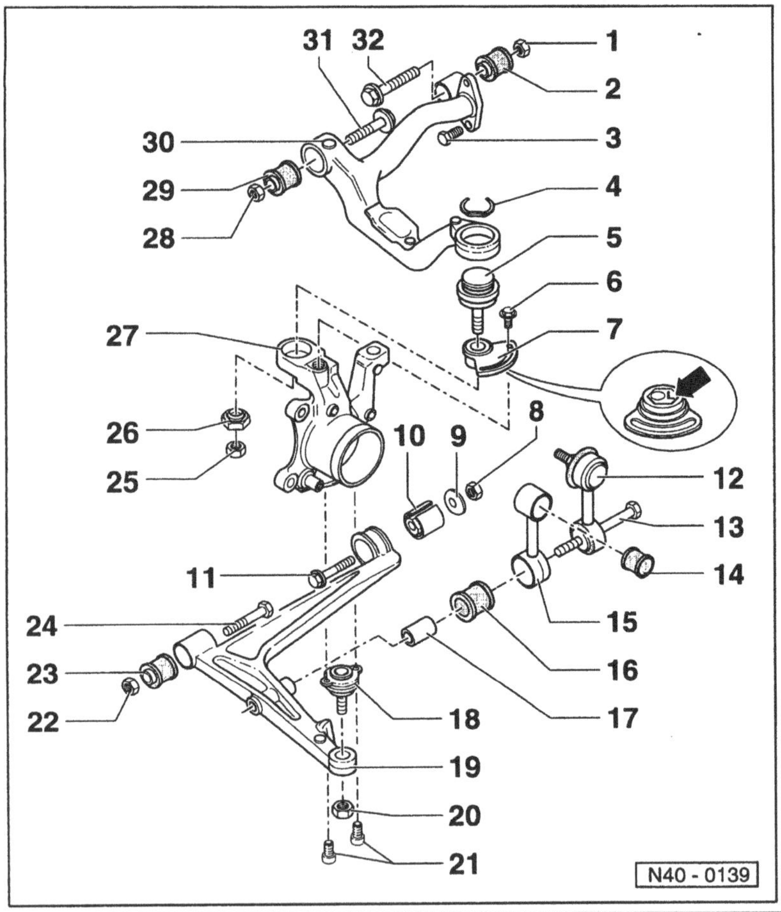

18 - Lower ball joint
- Removing/installing,

⇒ Page N 40-34.

19 - Lower control arm
- When tightening, hold bolts in installation position.

20 - Self-locking hex nut
- 60 Nm (44 ft lb) and turn 90° further.
- Replace after removing.

21 - Multi-point socket head bolt
- Always replace.
- 90 Nm (66 ft lb) and turn 90° further.
- Replace after removing.

22 - Hex nut, 160 Nm (118 ft lb)

23 - Lower control arm front bushing
- Removing,

⇒ Fig. 4.
- Pressing in,

⇒ Fig. 5 and 6.

24 - Hex bolt
- M 14 x 1.5 x 82

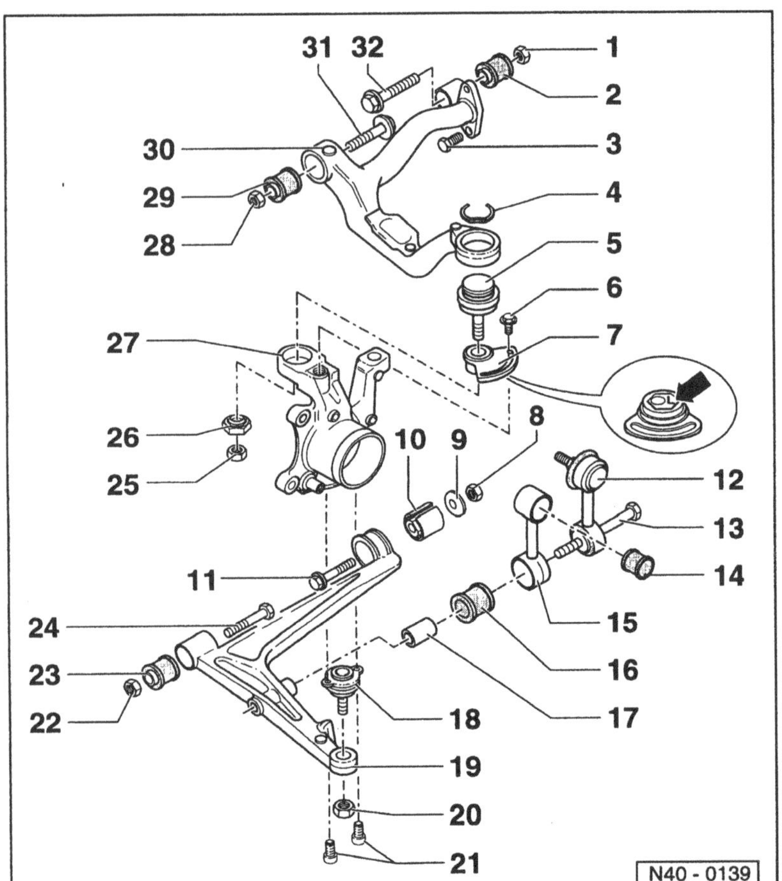

25 - Self-locking hex nut
- 60 Nm (44 ft lb) and turn 90° further.
- Replace after removing.

26 - Eccentric washer

27 - Wheel bearing housing
- Removing/installing,

⇒ Page N 40-34.

28 - Hex nut, 100 Nm (74 ft lb)
- Replace after removing.

29 - Upper control arm front bushing
- Pressing out,

⇒ Fig. 1.
- Pressing in,

⇒ Fig. 2.

30 - Upper control arm
- Can only be removed after subframe has been removed.
- When tightening hold bolts in installation position.

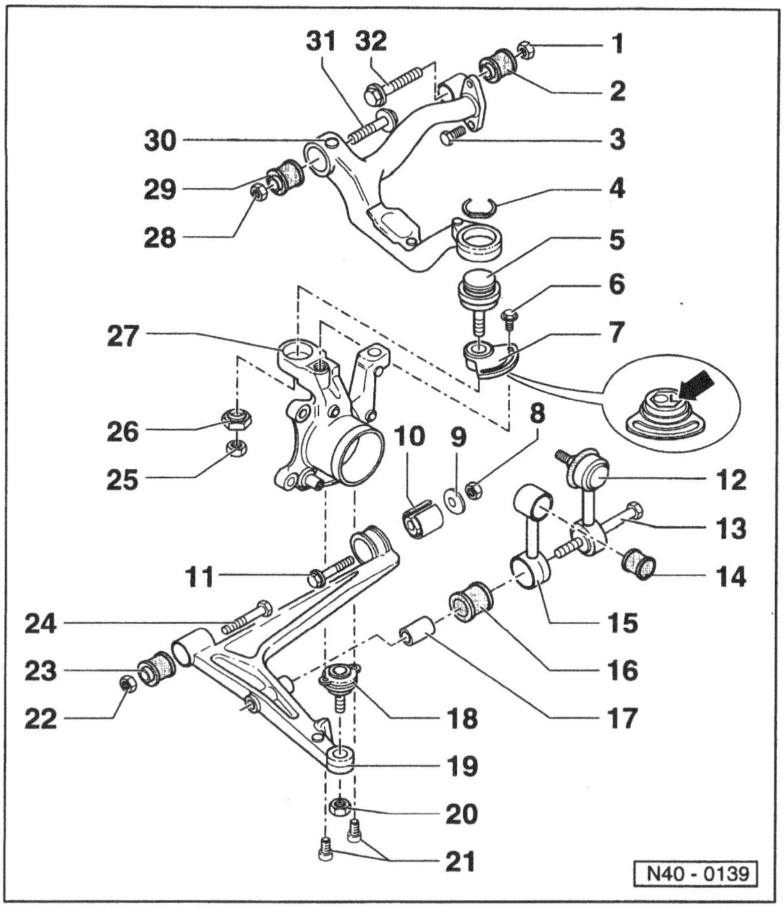

31 - Hex bolt
- M 14 x 1.5 x 80

32 - Hex bolt
- M 14 x 1.5 x 80

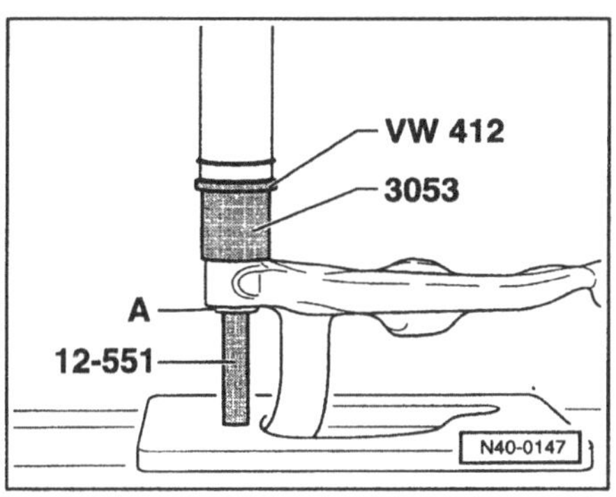

◄ **Fig. 1 Pressing upper control arm front bushing out**

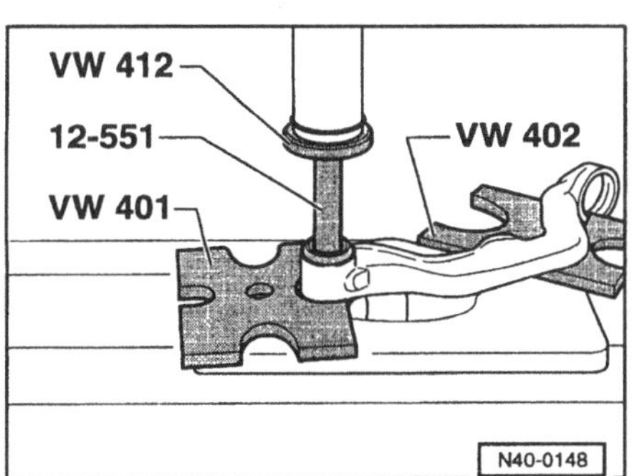

◄ **Fig. 2 Pressing upper control arm front bushing in**

- Press in until shoulders of bushing project evenly on both sides.

Note:

Before pressing in, coat with rubber lubricant (do not use soap).

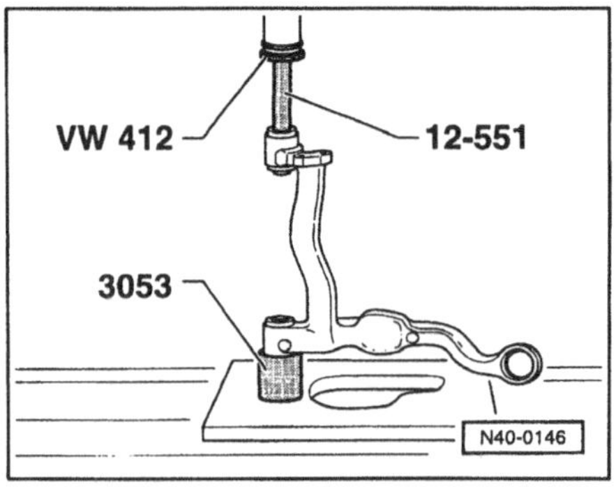

◄ **Fig. 3 Pressing upper control arm rear bushing out and in**

– Press in until shoulders of bushing project evenly on both sides.

Note:

Before pressing in, coat with rubber lubricant (do not use soap).

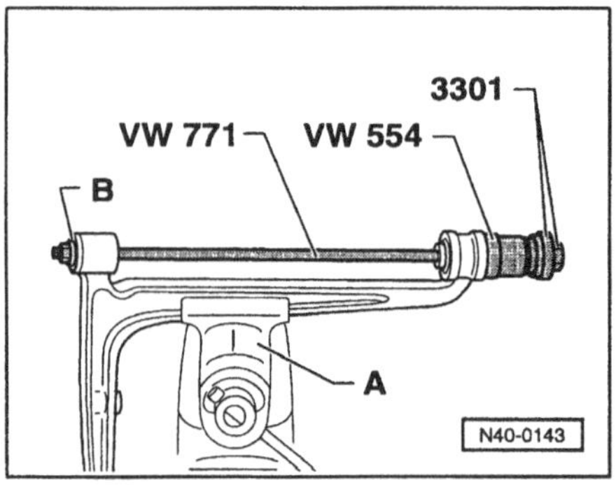

◄ **Fig. 4 Removing lower control arm front bushing**

– Place control arm in vise -A- and pull out front bushing -B- with spindle from multi-purpose tool VW 771.

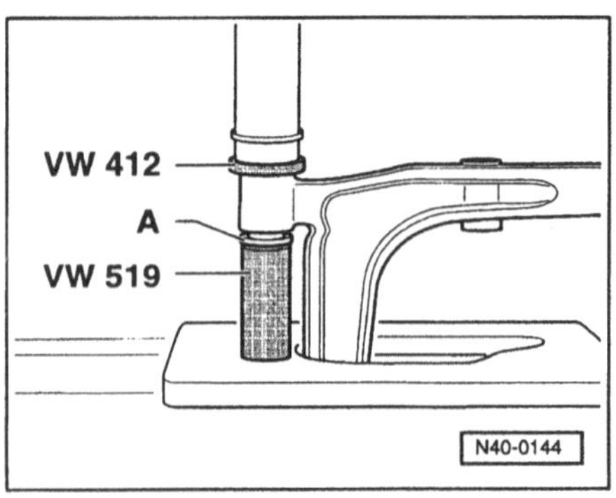

◄ **Fig. 5 Pressing in lower control arm front bushing**

– Press in bushing -A- with pipe VW 519 onto stop in control arm.

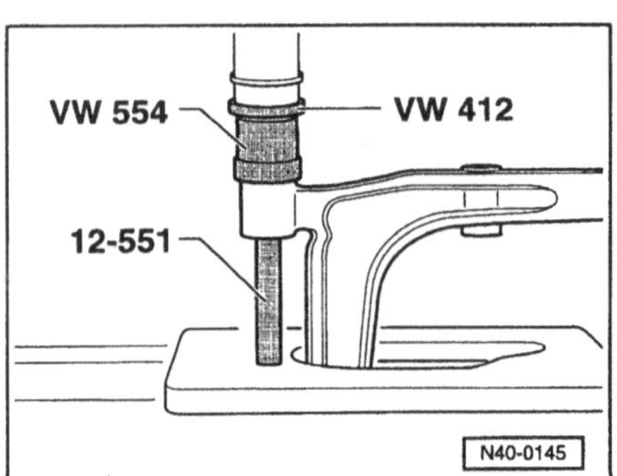

◄ **Fig. 6 Re-pressing lower control arm front bushing**

– Re-press bushing using locating pin 12-551 until outer rubber lip pops-out from control arm.

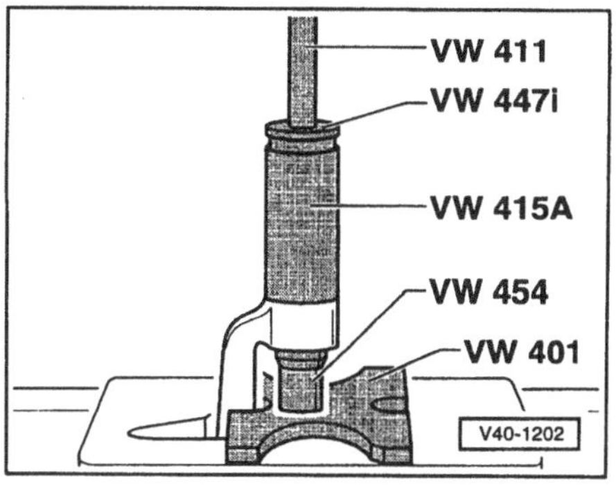

Fig. 7 Pressing lower control arm rear bonded rubber bushing out

- First mark installation position, e.g., by punch mark ⇒ -arrow B- on Fig. 8.

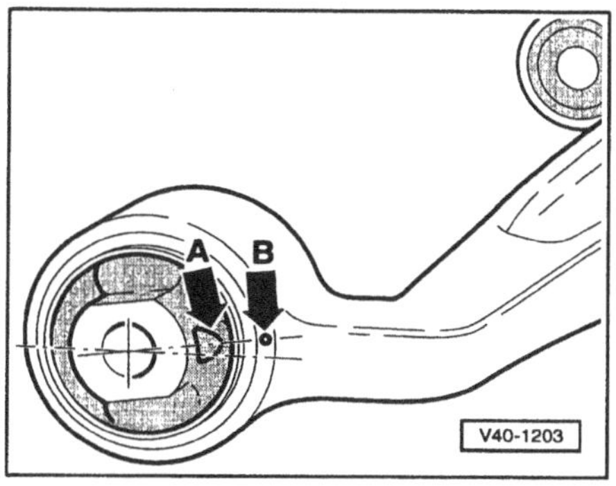

Fig. 8 Rear bonded rubber bushing installation position

Bonded rubber bushing is be pressed in so lug is offset upward, 11° to left and right control arm symmetrical axis.

Note:

Hold control arm in installation position.

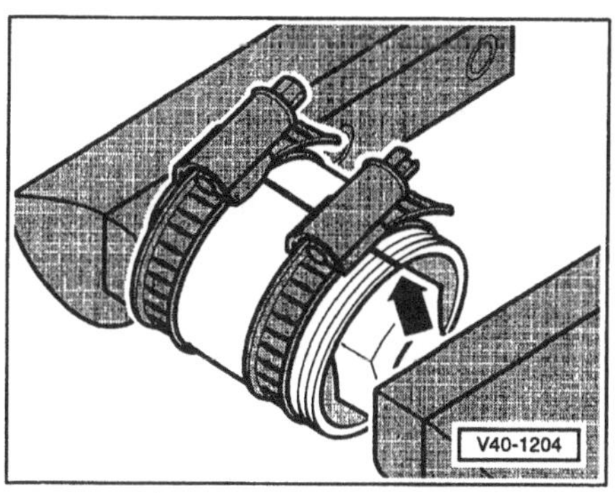

Fig. 9 Pre–tensioning lower control arm rear bonded rubber bushing

- First, pre–tension bonded rubber bushing with hose clips, as shown, so that the two metal shells of bushing are touching -arrow-.

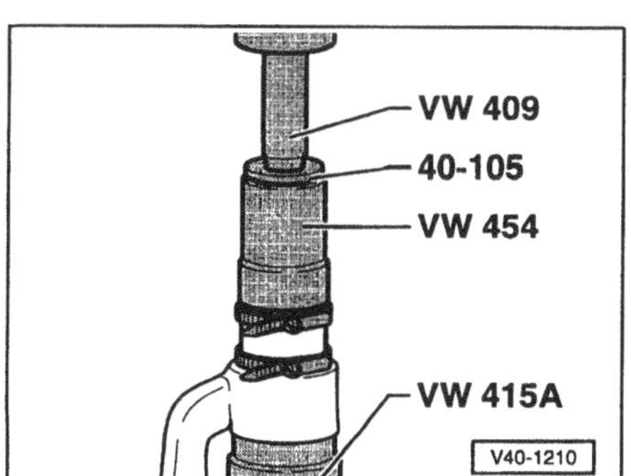

Fig. 10 Pressing rear bonded rubber bushing into lower control arm

- Press pretensioned bonded rubber bushing (⇒ Fig. 9) into installation position (⇒ Fig. 8) far enough so that a clearance of 371 ± 1 mm (14.6 ± 0.04 in.) exists between front rubber bushing inner tube and rear bonded rubber bushing inner tube,

⇒ Fig. 11.

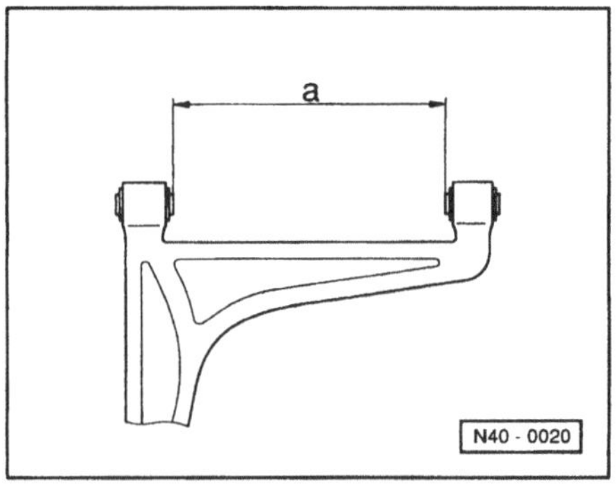

Fig. 11 Lower control arm bonded rubber bushing installation dimension

Dimension a = 371 ± 1 mm (14.6 ± 0.04 in.)

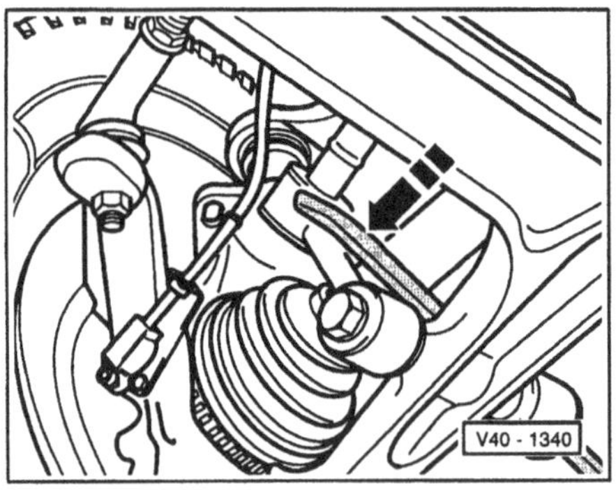

Fig. 12 Pressing coupling link off stabilizer bar, 23.5 mm (0.9 in.) diameter

- Coat inner diameter of coupling link rubber bushing with rubber lubricant, (do not use soap), and press coupling link off stabilizer bar.

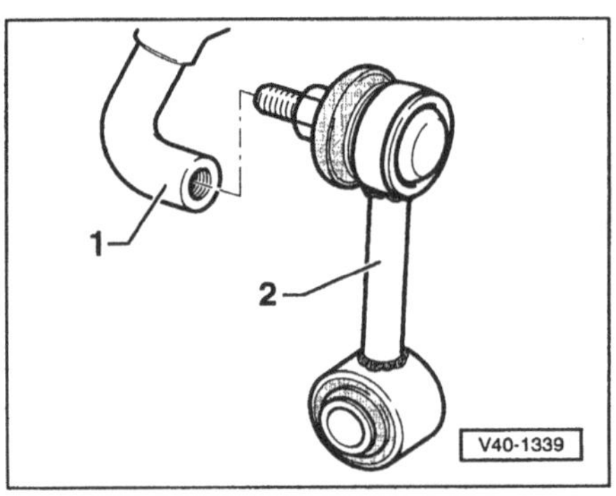

Fig. 13 Fastening coupling link to stabilizer bar 27 mm (1.1 in.) diameter

1 - Heavy-duty stabilizer bar 27 mm (1.1 in.) diameter

2 - Coupling link

Coupling link to stabilizer bar tightening torque:

100 Nm (74 ft lb).

Note:
After installing coupling link, ensure boot is not twisted.

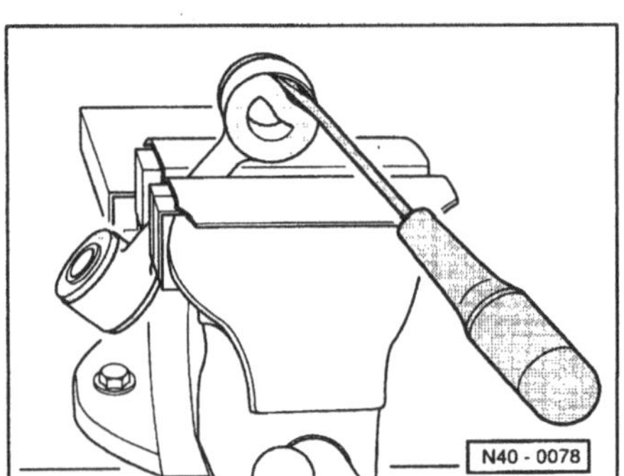

Fig. 14 Removing/installing coupling link rubber bushing on stabilizer bar side

- Pry out rubber bushing with screwdriver, as shown in illustration.
- Press in new rubber bushing.

Note:
Before pressing in, coat with rubber lubricant, (do not use soap).

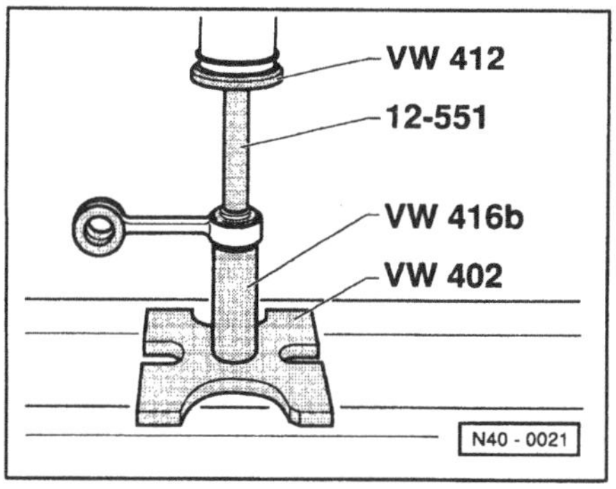

◄ **Fig. 15 Removing coupling link rubber bushing on control arm side**

- Press bushing out as shown in illustration.
- Pry rubber bushing out with a screwdriver.

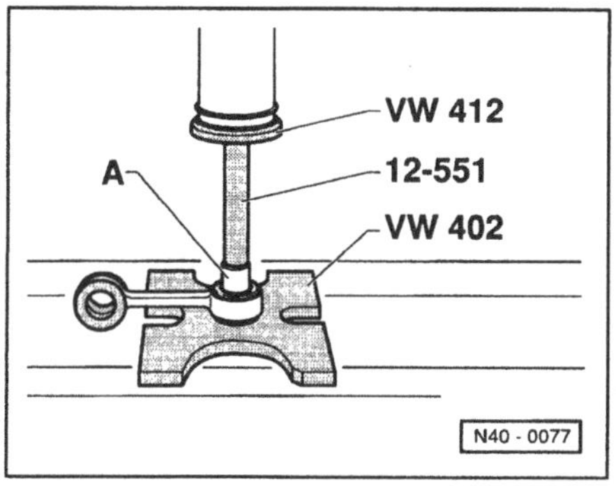

◄ **Fig. 16 Installing coupling link rubber bushing on control arm side**

- Press rubber bushing in coupling link.
- Press bushing -A- in as shown.

Note:

Before pressing in, coat parts with rubber lubricant (do not use soap).

Lower ball joint, removing/installing

- Relieve torsion bar tension,
⇒ Page N 40-11, Torsion bar, Removing/installing/adjusting.
- Remove shock absorber bolt and stabilizer bar coupling link.
- Separate ball joint connection between lower control arm and wheel bearing housing,
⇒ Page N 40-35.

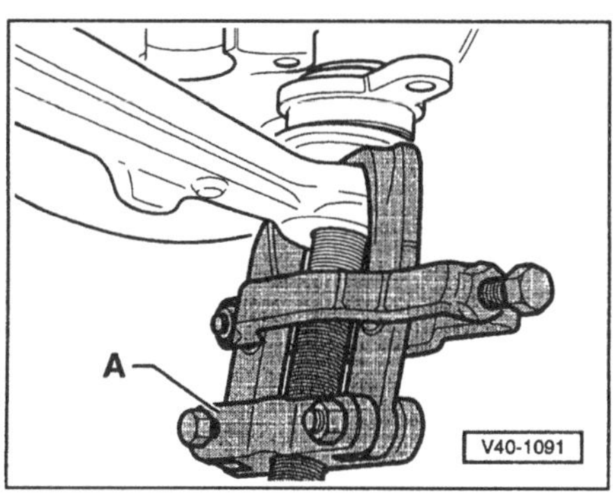

◄ - Press ball joint out.

A - Puller, commercially available, e.g., Kukko 204-2.

Note:

Only use puller with tensioning arm.

Install in reverse order of removal.

Upper ball joint, removing/installing

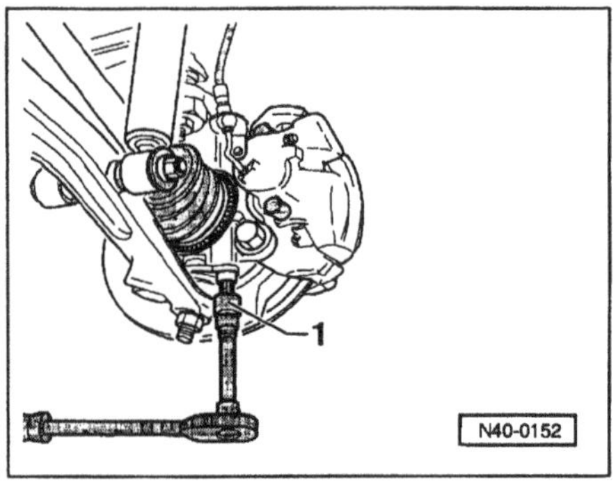

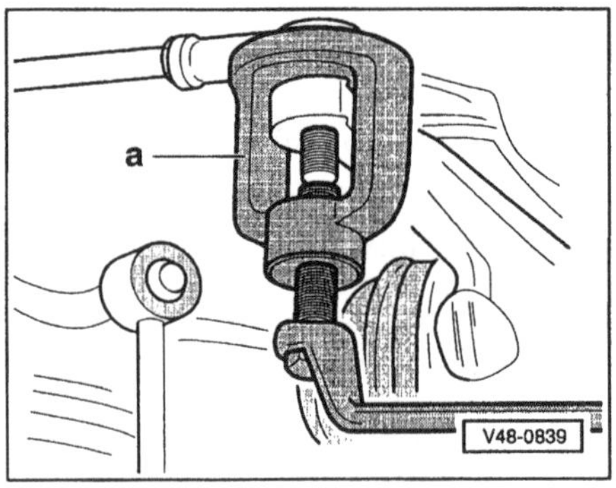

- Relieve torsion bar tension,
⇒ Page N 40-11, Torsion bar, Removing/installing/adjusting.
- Loosen hex bolt for drive shaft/wheel hub, with vehicle on its wheels.
- Remove brake caliper and secure.
- ◄ Separate bolted connection between wheel bearing housing and lower ball joint.

  1 - 14 mm multi-point socket bit
- Remove shock absorber bolt and stabilizer bar coupling link.
- Mark position of eccentric bushing to wheel bearing housing (or reset camber).
- Remove upper ball joint bolt.
- ◄ Press tie rod off steering arm.

  a - Ball joint splitter (commercial type), e.g., Kukko No. 128/2.

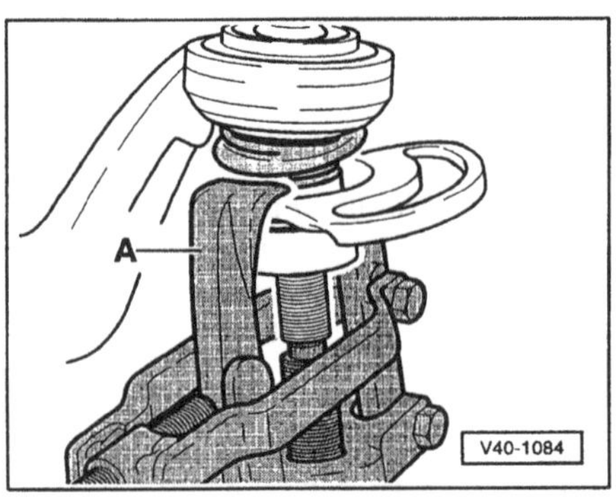

- Remove eccentric disc bolt from wheel bearing housing.
- Remove wheel bearing housing.
- ◄ Remove eccentric bushing as shown in illustration.

  A - Puller, commercially available, e.g., Kukko 204-2.
- Pry out circlip with a screwdriver.

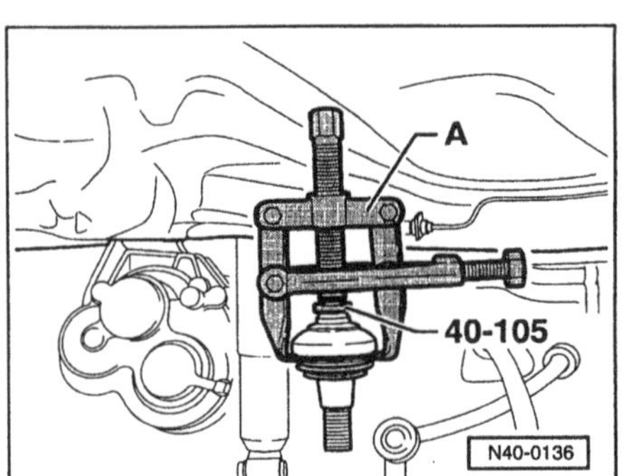

- ◄ Position puller -A- (commercially available, e.g., Kukko 204-2), with tensioning arm, as shown in illustration and press out joint.

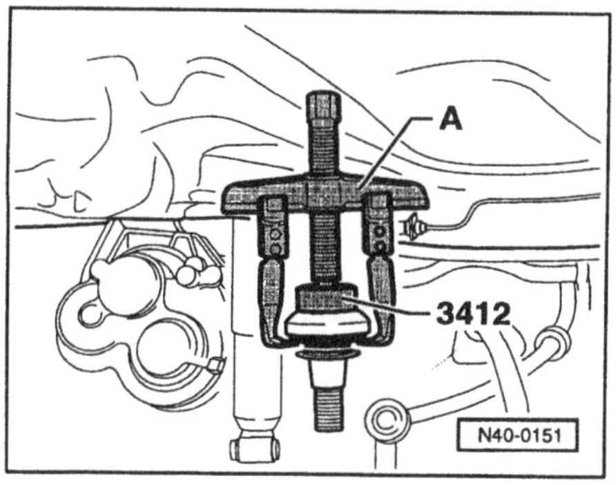

Installing

◄ – Install ball joint in control arm.

A - Two-arm puller, e.g., Kukko 20/2, with hooks 2-150.

– Pull ball joint in onto stop.

– Install new circlip.

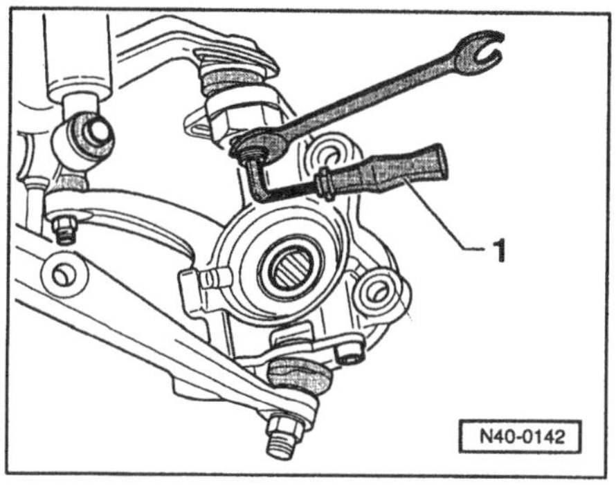

◄ – Bolt upper ball joint to wheel bearing housing.
1 - 7 mm Hex key

Note:

When tightening nut, counter–hold with hex key and simultaneously push up wheel bearing housing with eccentric disc.

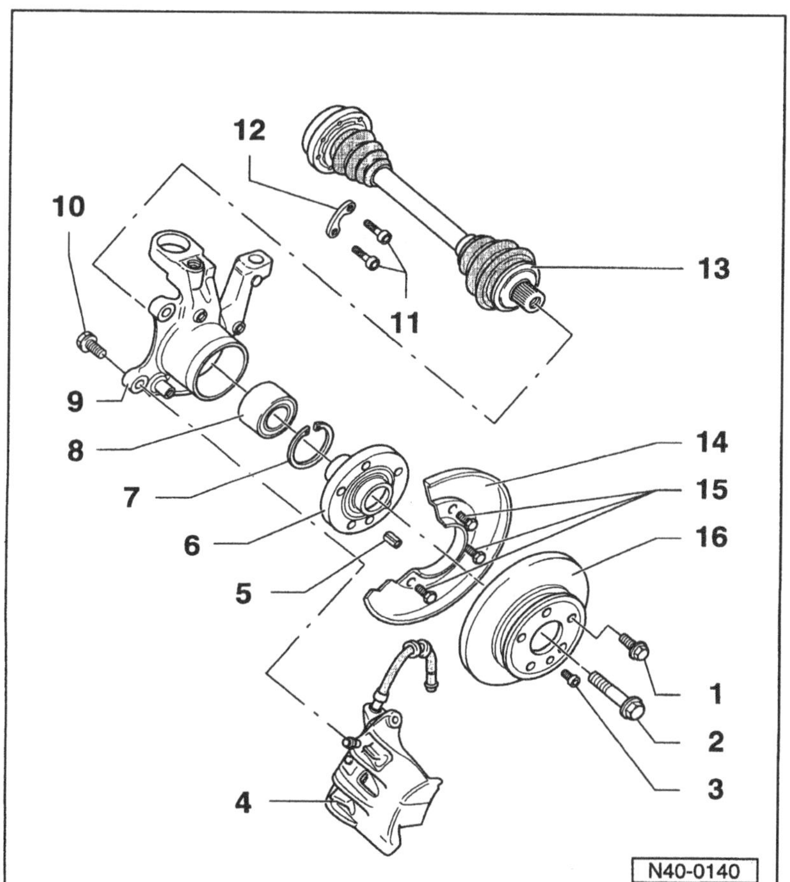

Wheel bearing and drive shaft, removing/installing

Notes:

- *If a vehicle has to be moved after removing drive shaft, install an outer joint instead of a drive shaft. Otherwise, wheel bearing damage will result.*
- *Do not perform welding or straightening operations on load–bearing suspension components, or those components which locate wheels.*
- *Always replace self-locking nuts.*
- *Always replace corroded nuts/bolts.*

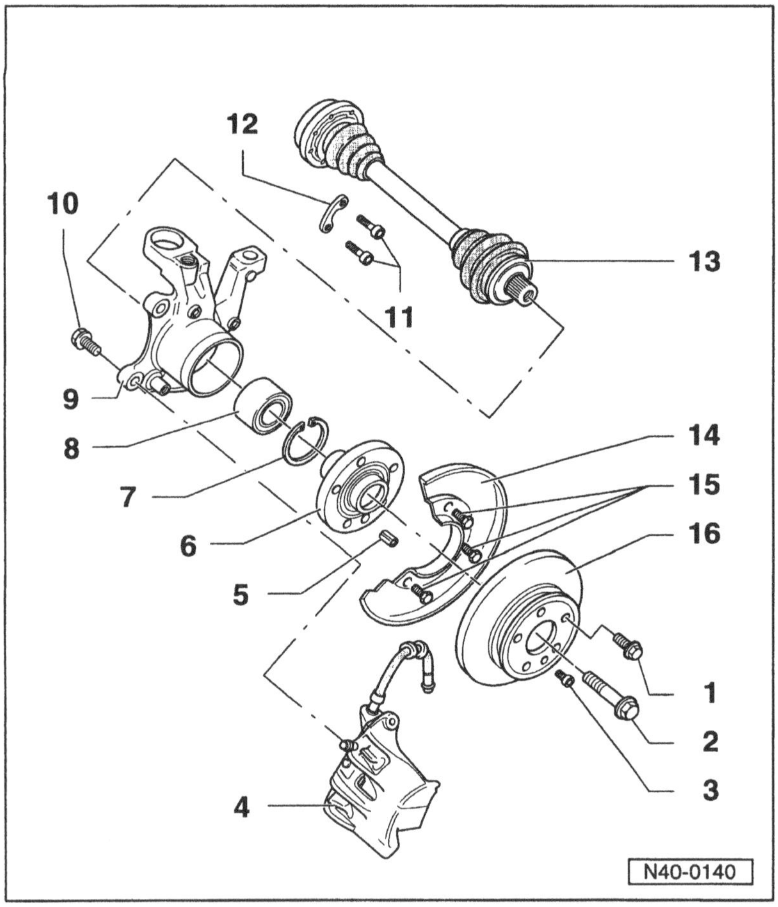

1 - **Wheel bolt, 160 Nm (118 ft lb)**

2 - **Hex bolt**
- ♦ 150 Nm (111 ft lb) and turn 90° further.
- ♦ Replace bolt when removed.

3 - **Hex socket head bolt, 10 Nm (7 ft lb)**

4 - **Brake caliper**
- ♦ Repair instructions,

⇒ Repair group 46 and 47.

5 - **Spring pin**

6 - **Hub**
- ♦ Pressing out,

⇒ Fig. 1.
- ♦ Pressing in,

⇒ Fig. 5.

7 - **Circlip**
- ♦ Ensure seated correctly.

8 - **Wheel bearing**
- ♦ Pressing out,

⇒ Fig. 2.
- ♦ Pressing in,

⇒ Fig. 4.

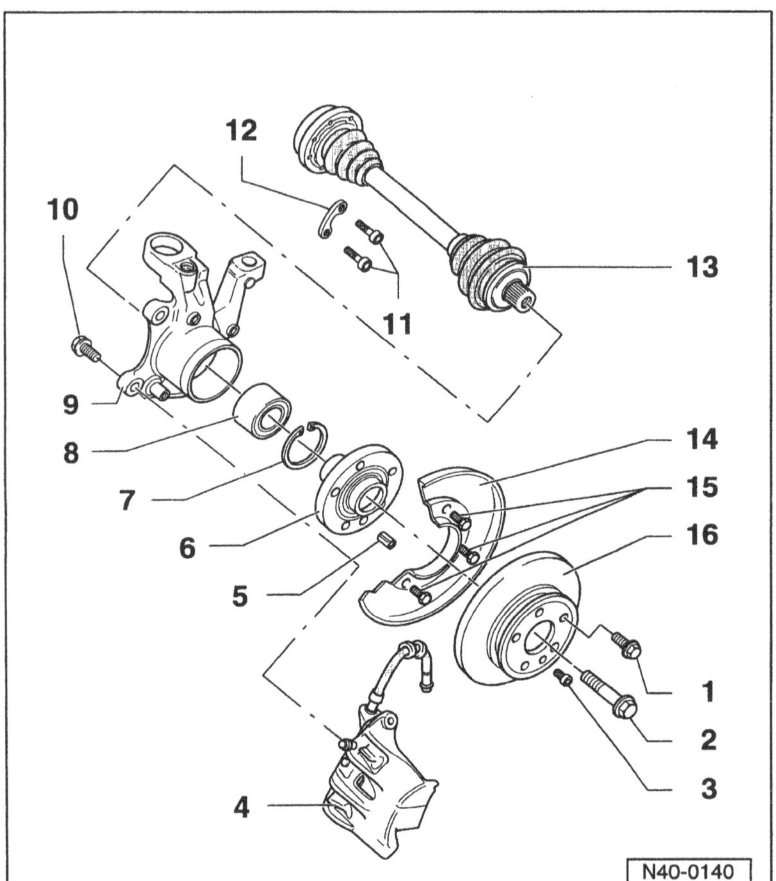

9 - **Wheel bearing housing**
- ♦ Removing/installing,

⇒ Page N 40-44.

10 - **Hex bolt, 280 Nm (207 ft lb)**

11 - **Socket head multi-point bolt, 80 Nm (59 ft lb)**

12 - **Backing plate**

13 - **Drive shaft**
- ♦ Removing/installing,

⇒ Page N 40-46.

14 - **Splash plate**

15 - **Hex bolt, 8 Nm (71 in. lb)**

16 - **Brake disc**
- ♦ Repair instructions,

⇒ Page N 46-1.

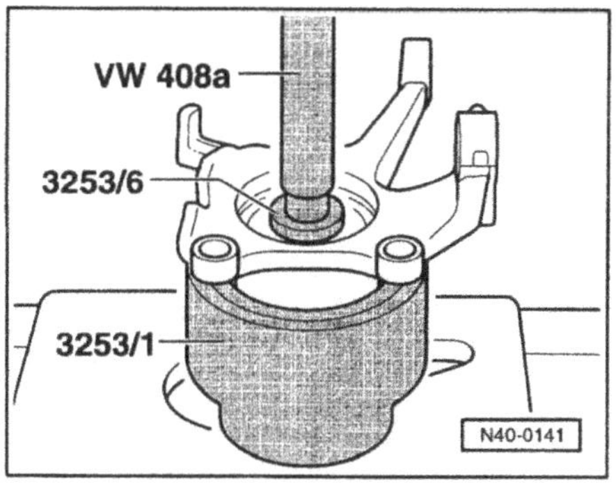

Fig. 1 **Pressing wheel hub out**

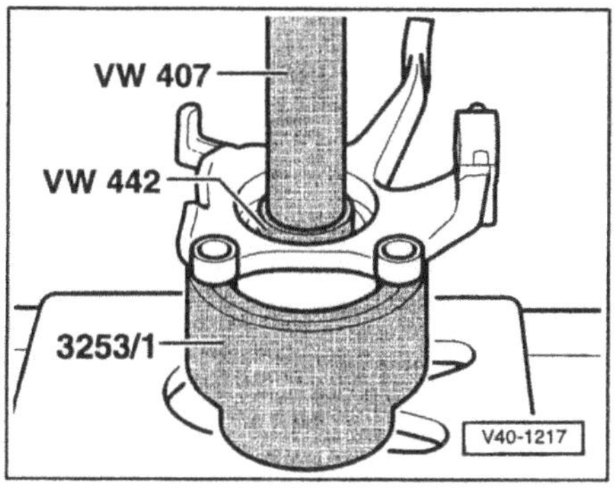

Fig. 2 **Pressing wheel bearing out of wheel bearing housing**

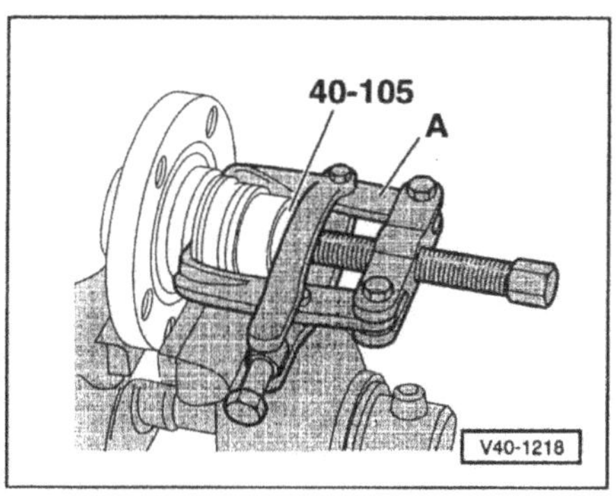

Fig. 3 **Pulling inner race from hub**

Only use puller with leg clamp, e.g., Kukko 204-2 (commercial type).

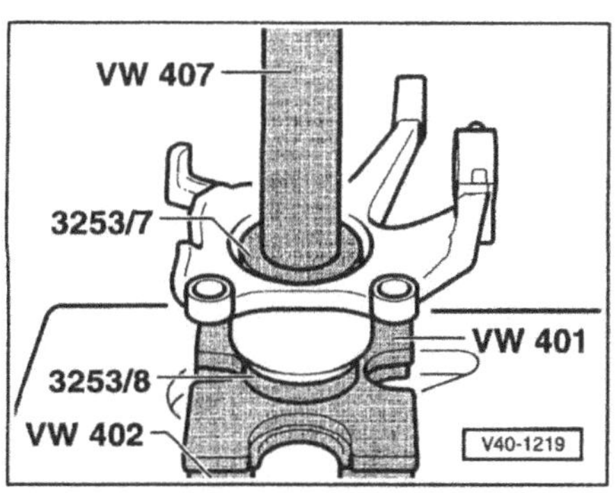

Fig. 4 **Pressing wheel bearing into wheel bearing housing**

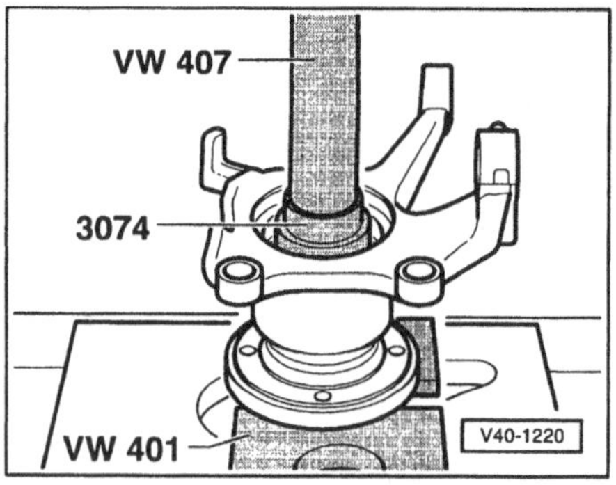

◄ **Fig. 5 Pressing hub into wheel bearing housing**

Wheel bearing housing, removing/installing

– Relieve torsion bar tension.
⇒ Page N 40-11, Torsion bar, removing/installing/adjusting.

> ***Warning!***
>
> ***Loosen drive shaft hex bolt with vehicle on its wheels, to avoid an accident.***

– Remove wheel and raise vehicle.

– Remove stabilizer bar bolt, coupling link/lower shock absorber.

– Remove brake caliper and secure.

– Remove brake disc and cover plate.

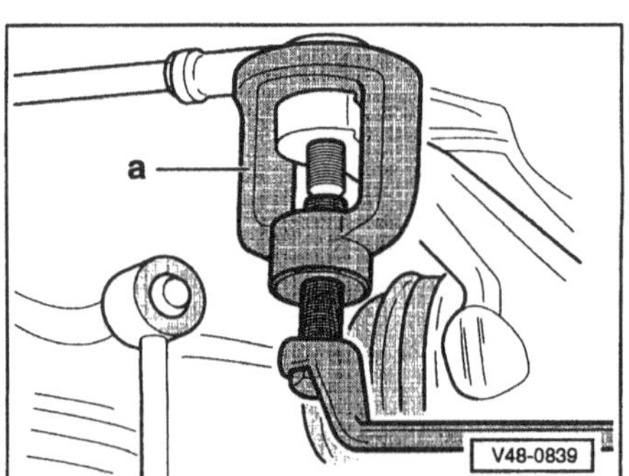

◄ – Press tie rod off steering arm.

a - Ball joint splitter (commercial type), e.g., Kukko No. 128/2.

– Separate wheel bearing housing/lower ball joint connection,
⇒ Page N 40-35.

– Mark eccentric bushing to wheel bearing housing position (e.g., with felt tipped pen), or realign front wheels.

– Separate upper ball joint wheel bearing connection.

– Remove wheel bearing housing.

Install in reverse order of removal.

Drive shafts, removing/installing

Vehicles with manual transmission:

Wheel bearings must not be loaded if a hex bolt is loose.

If a wheel bearing is loaded by vehicles own weight, wheel bearings will be stressed and life expectancy reduced.

– Lift vehicle to relieve front axle load.

– Remove drive shaft hex bolt.

– Remove sound dampening pan.

– Relieve torsion bar tension,
⇒ Page N 40-11, Torsion bar, removing/installing/adjusting.

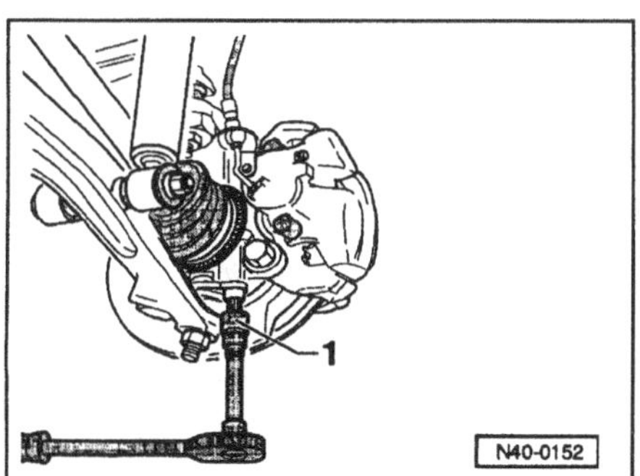

◄ – Separate wheel bearing housing/lower ball joint.

1 - 14 mm multi-point socket bit

– Remove shock absorber/anti-roll bar coupling link bolt.

– Separate drive shaft at final drive flange.

– Push shock absorber completely together.

– Take out drive shaft.

Vehicles with automatic transmission:

Removing

– Lift vehicle to relieve front axle load.

– Remove drive shaft hex bolt.

– Remove noise insulation tray.

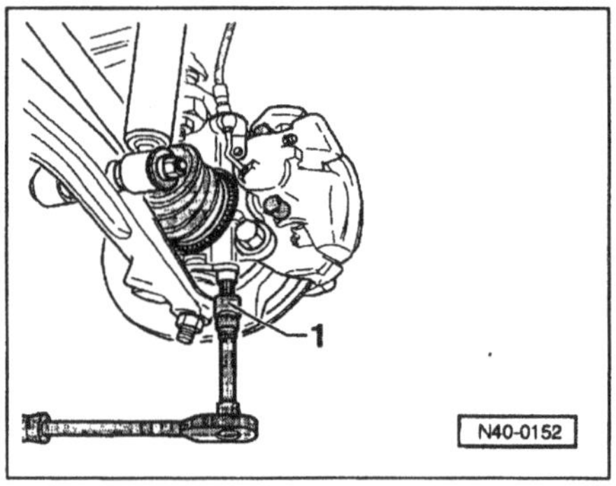

– Relieve torsion bar tension,
⇒ Page N 40-11, Torsion bar, removing/installing/adjusting.

– Detach left shock absorber from lower control arm and push completely together.

◄ – Separate wheel bearing housing/lower ball joint.

1 - 14 mm multi-point socket bit

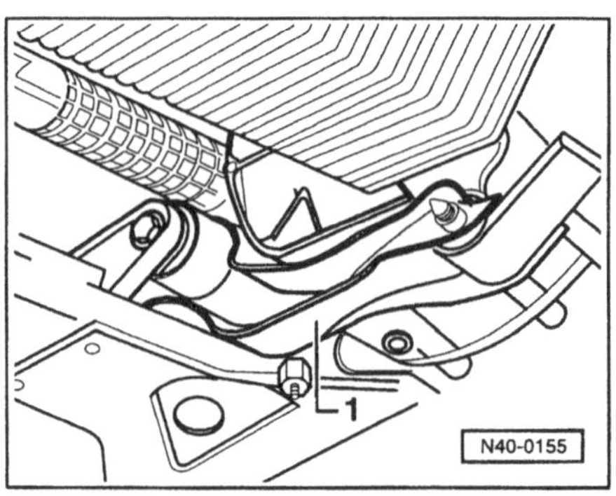

– Remove connector from multi-function switch.

– Separate drive shaft at final drive flange.

◄ – Unbolt pendulum support -1- from transmission.

– Remove drive shaft (push engine/transmission slightly forward).

Install in reverse order of removal.

Tightening torque:

Drive shaft to flange shaft	80 Nm (59 ft lb)
Wheel bearing housing to ball joint	90 Nm (66 ft lb) +90°
Wheel bolts to wheel hub	160 Nm (118 ftlb)
Shock absorber to control arm	160 Nm (118 ftlb)
Drive shaft to wheel hub	150 Nm (111 ft lb) + 90°
Pendulum support to transmission	80 Nm (59 ft lb) + 90°

Drive shaft, assembly overview

Packing outer constant velocity joint with grease:

Re-grease joint, if necessary, when replacing drive shaft boot.

- Press approximately 80 g grease through inner splines of ball hub into outer joint.

- Spread approximately 40 g grease around inside of joint.

Packing inner constant velocity joint with grease:

Re-grease joint, if necessary, when replacing drive shaft boot.

Amount of grease: 120 g.

- Press approximately 60 g G-6.3 grease from each side into joint.

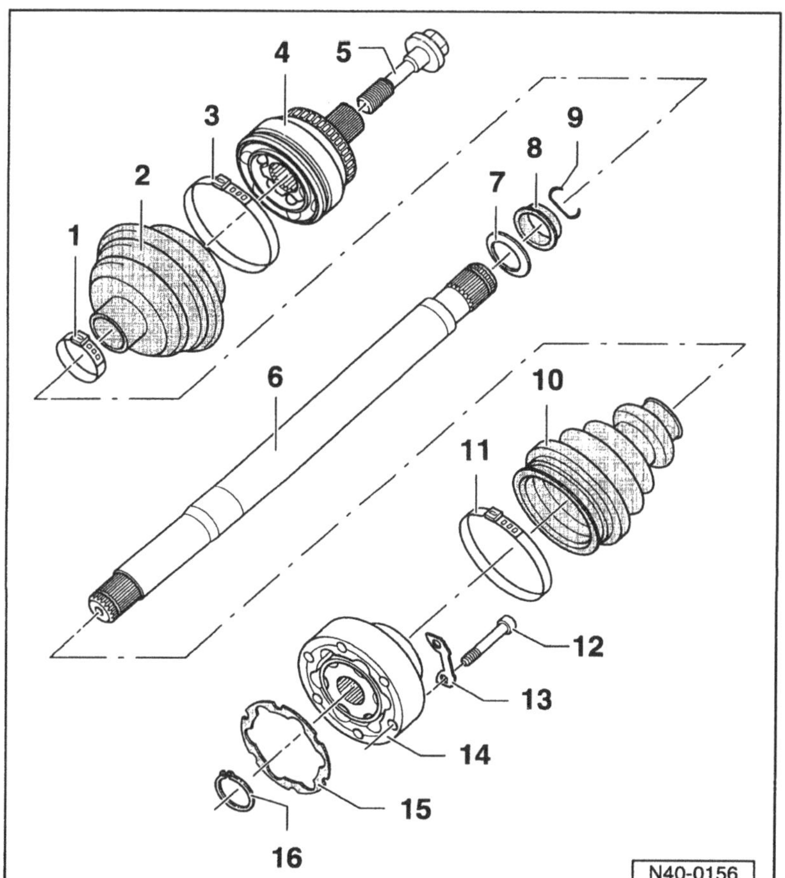

1 - Hose clip
- Replace.
- Tensioning,

⇒ Fig. 3.

2 - Boot, outer constant velocity joint
- Check for tears and chafing, replace if necessary.

3 - Hose clip
- Replace.
- Tensioning,

⇒ Fig. 2.

4 - Outer constant velocity joint
- Replace complete.
- Removing,

⇒ Fig. 1.
- Installing:
- Drive onto shaft with plastic hammer, until compressed circlip seats.
- Greasing,

⇒ Notes, Page N 40-49.

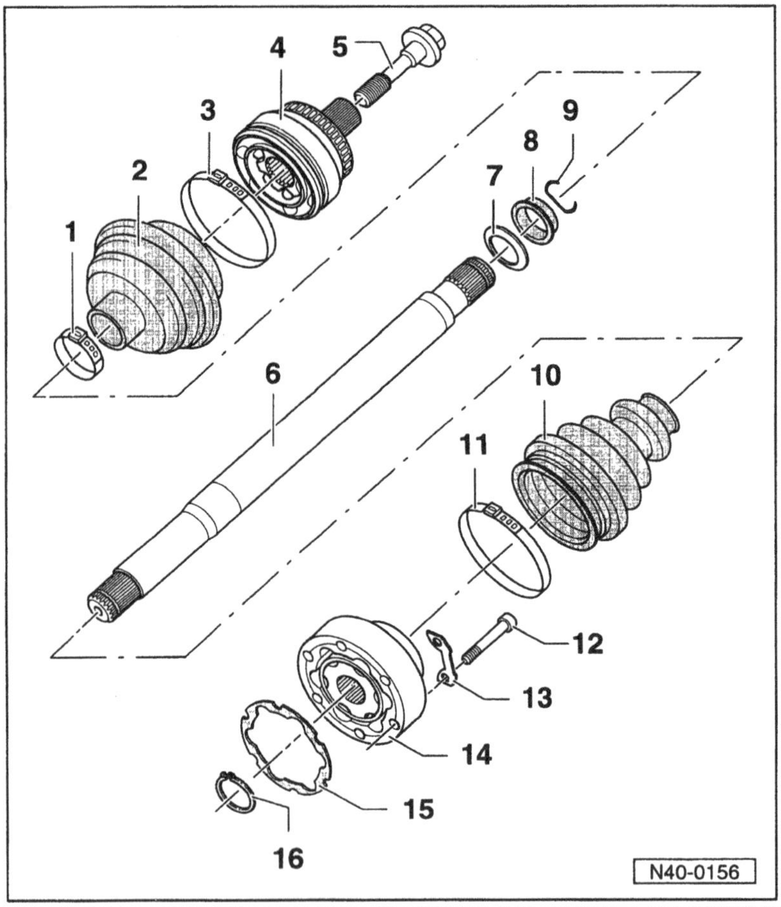

5 - Hex bolt
- 150 Nm (111 ft lb) then turn 90° further.
- Replace during reassembly.

6 - Drive shaft

7 - Dished washer
- Installation position: larger diameter contacts thrust washer.

8 - Thrust washer

9 - Circlip
- Replace.
- Insert in shaft groove.

10 - Inner constant velocity joint boot
- Check for tears and chafing, replace if necessary.
- Clean drive shaft thoroughly before installing new drive shaft boot.

11 - Hose clip
- Replace.
- Tension with VAG 1275 or equivalent.

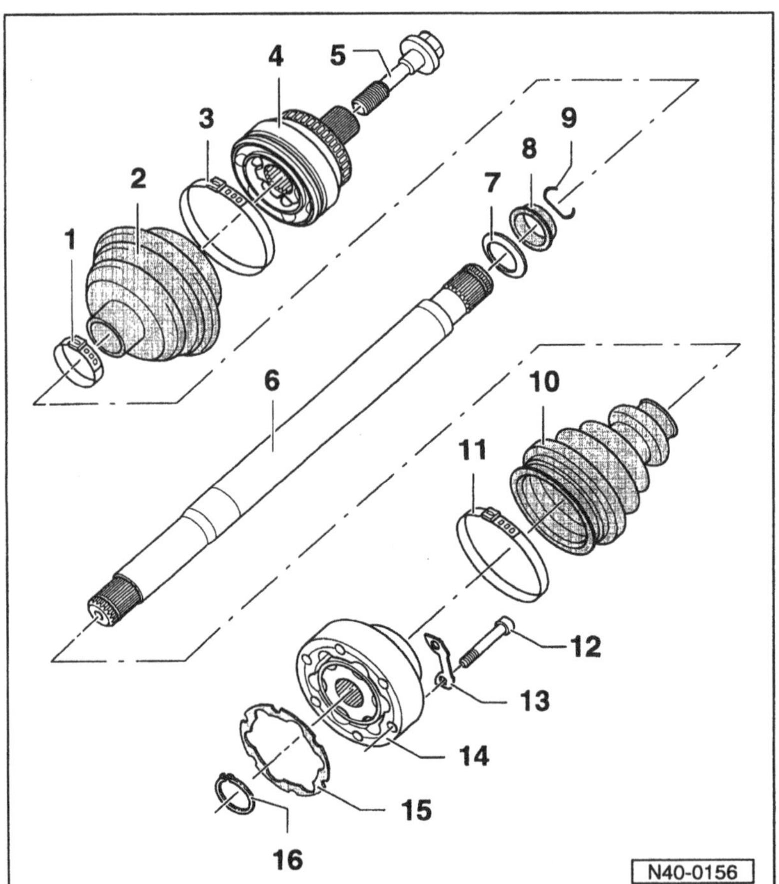

12 - Allen head multi-point bolt, 80 Nm (59 ft lb)

13 - Backing plate

14 - Inner constant velocity joint
- Replace complete.
- Pressing off,

⇒ Fig. 4.
- Pressing on,

⇒ Fig. 5.
- Greasing,

⇒ Notes Page N 40-49.
- Checking:
  - Clean joint.
  - Examine joint, hub, cage, and balls for small indentations (pitting) and signs of wear.
  - Excessive play in joint is noticeable by load change jolts. Replace joint in such cases.
  - Smooth areas and signs of the balls tracking are not a reason for changing joint.

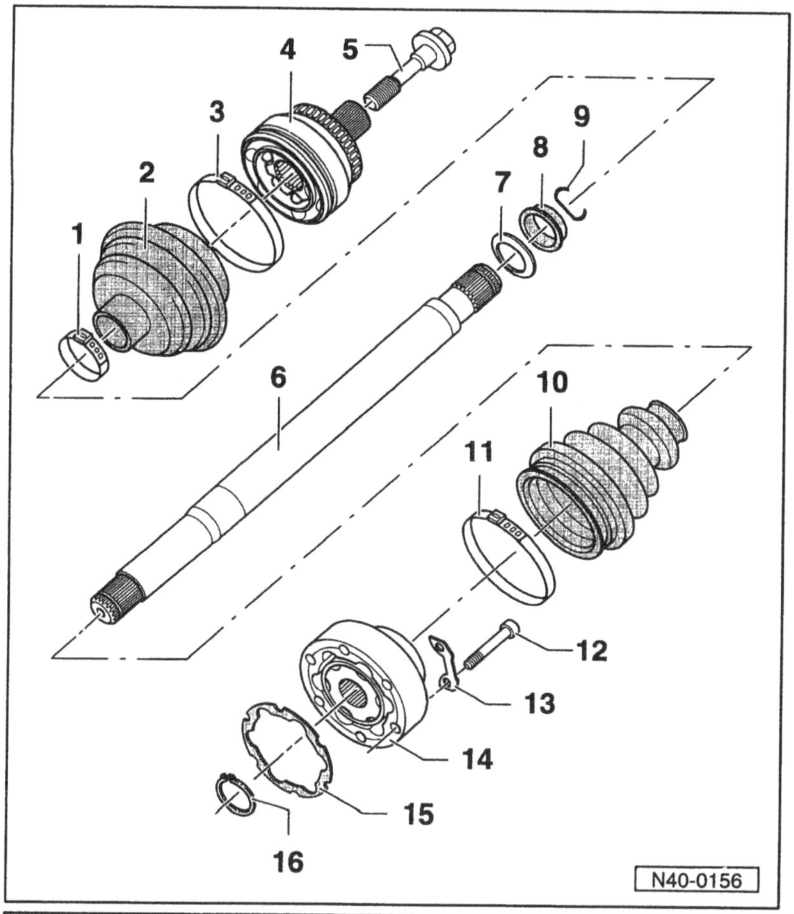

15 - Gasket

- ♦ replacing:
- ♦ Remove protective foil and stick onto joint.

16 - Circlip

- ♦ Replace:
- ♦ Remove/install with long nosed pliers.

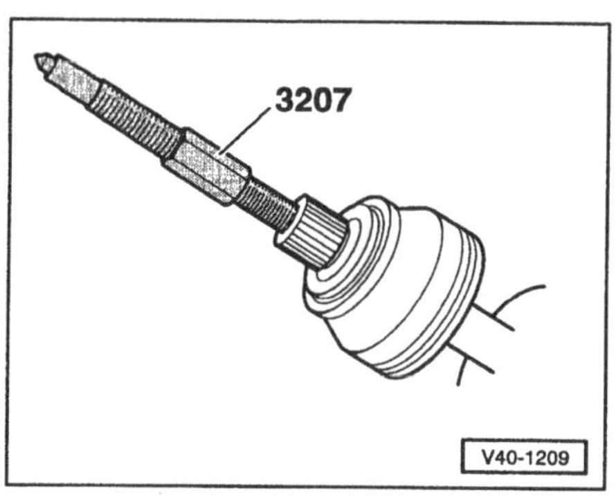

◄ **Fig. 1 Outer constant velocity joint, pressing off**

- Clamp drive shaft in vice, using vice clamps.
- Fold back boot.
- Screw special tool, using the M16 threaded section, into pivot pin far enough to press constant velocity joint off shaft.

Driving joint on

- Drive onto shaft, with plastic hammer, until securing ring engages.

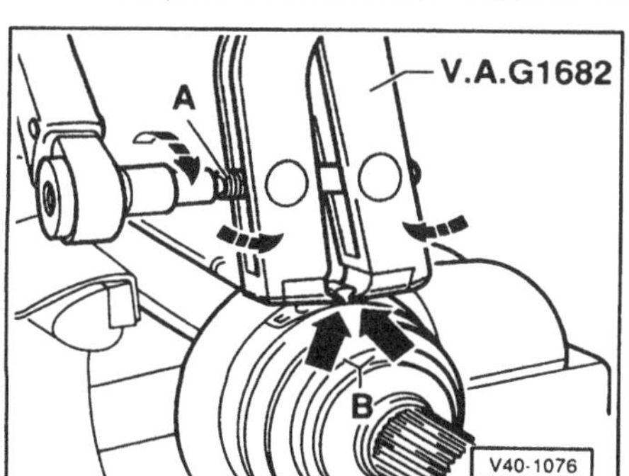

◄ **Fig. 2 Tensioning hose clamp on outer joint**

- Position pliers, VAG 1682, as shown in illustration. Ensure pliers jaws seat in corners -arrow B- of hose clamp.
- Tension hose clamp by turning spindle with a torque wrench (do not tilt pliers).

Notes:

- *Due to hard material of joint boot (compared to rubber), and necessary stainless steel hose clamp, it is only possible to tension hose clamp with pliers VAG 1682 or equivalent.*
- *Tightening torque: 25 Nm (18 ft lb).*
- *Use torque wrench -C- with 5 - 50 Nm (4 - 37 ft lb) range (e.g. VAG 1331).*
- *Ensure spindle thread -A- is not tight. If necessary, lubricate with MOS 2 grease.*
- *If thread is tight, e.g., dirty, the required tensioning force for the hose clamp will not be achieved, inspite of correct tightening torque settings.*

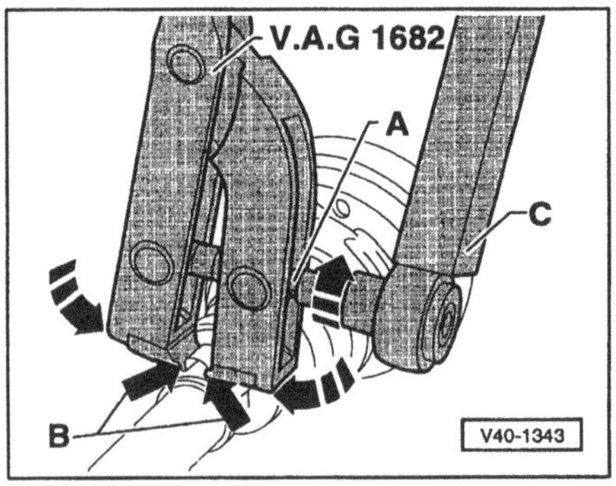

◄ **Fig. 3 Tensioning small hose clamp on outer joint**

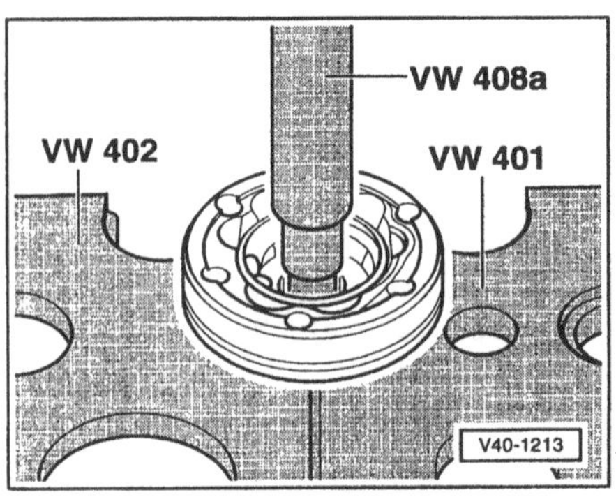

◄ **Fig. 4 Pressing off inner joint**

- Support ball hub when performing this step.
- Remove circlip with long-nosed pliers.

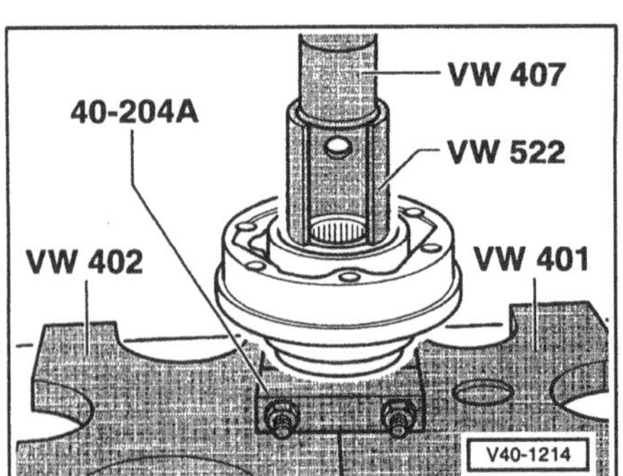

◄ **Fig. 5 Pressing on inner joint**

- Press joint up to stop.
- Insert circlip.

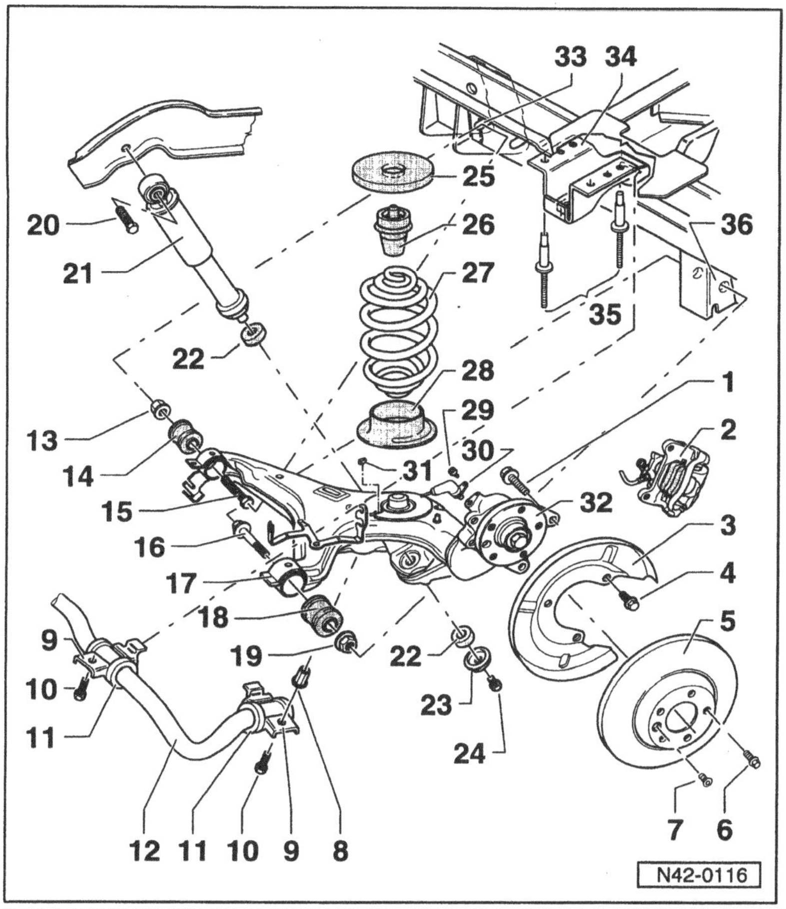

Rear suspension, repairing

Notes:

- *Welding and straightening on wheel locating/supports is not permissible.*
- *Always replace self-locking nuts.*

1 - Hex bolt, 170 Nm (125 ft lb)

2 - Brake caliper

3 - Splash plate

4 - Hex bolt, 8 Nm (71 in. lb)

5 - Brake rotor

6 - Wheel bolt 160 Nm (118 ft lb)

7 - Hex socket head bolt 10 Nm (7 ft lb)

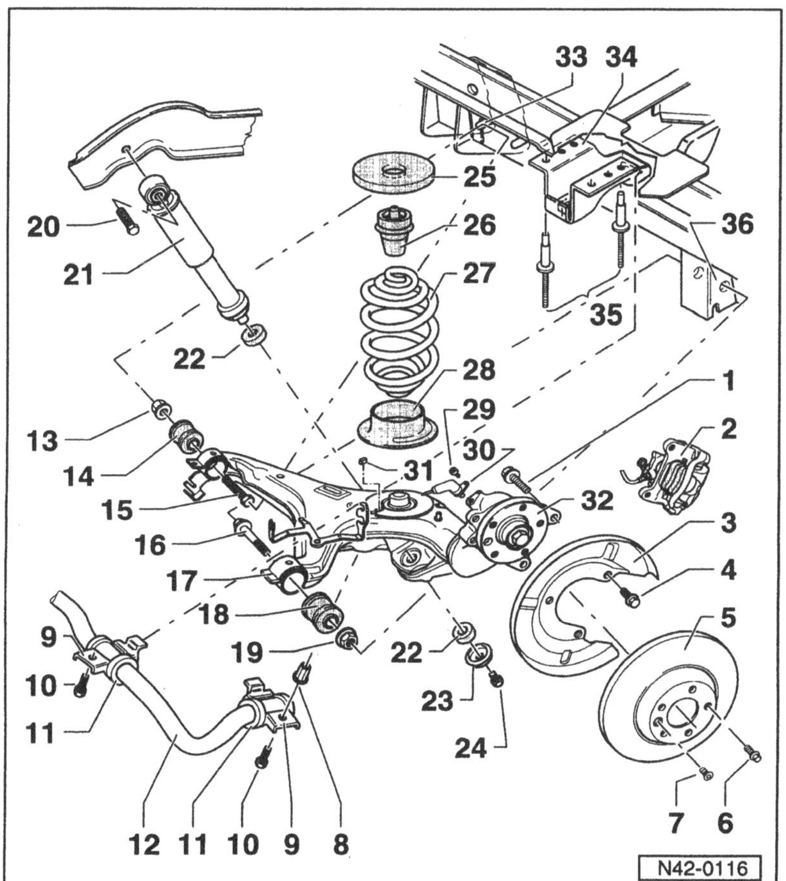

8 - Rivet nut, M8
- Stabilizer bar equipped vehicles only.
- Pulling into control arm,

⇒ Page N 42-11.

9 - Clamp

10 - Hex bolt, 30 Nm (22 ft lb)

11 - Rubber mount

12 - Stabilizer bar
- 26 mm (1 in.) diameter.
- Retrofitting,

⇒ Page N 42-9.

13 - Nut, 100 Nm (74 ft lb)

14 - Inner rubber mount
- Pulling out,

⇒ Fig. 1.
- Pressing in,

⇒ Fig. 2.

15 - Hex bolt
- Mark installation position before removing, or re-align.

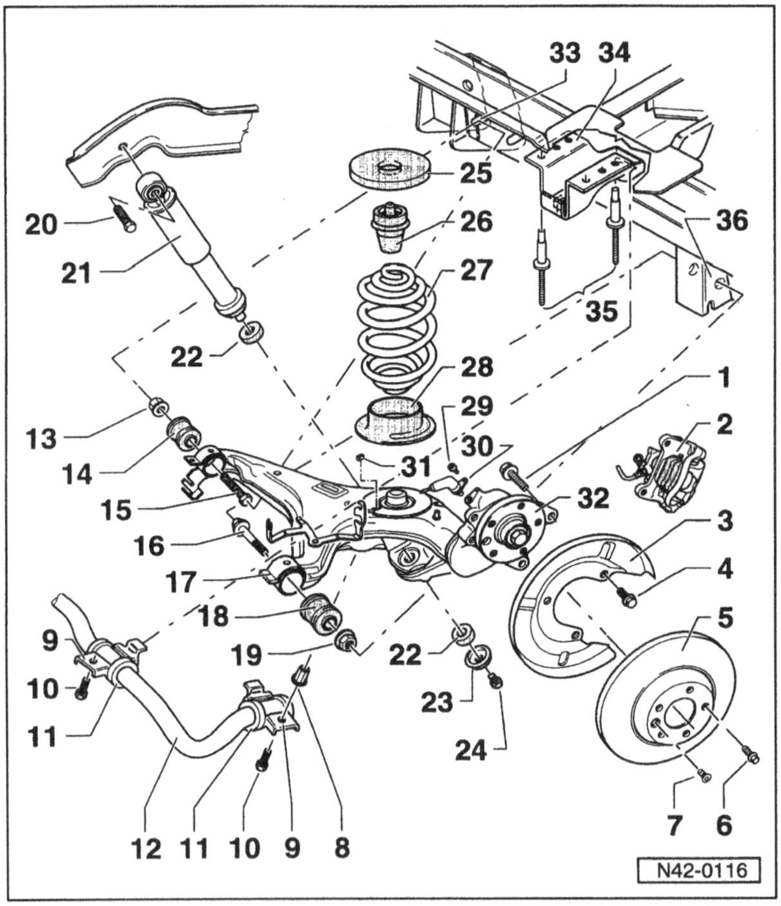

16 - Hex bolt

17 - Control arm
- ♦ Mark installation position on inner mount.
- ♦ When removing, unhook spring for brake pressure regulator at horizontal beam.
- ♦ When replacing, adjust brake pressure regulator,

⇒ Page N 47-26.
- ♦ When tightening bolts, hold in installation position.

18 - Outer rubber mount
- ♦ Pulling out,

⇒ Fig. 1.
- ♦ Pressing in,

⇒ Fig. 2.

19 - Nut, 160 Nm (118 ft lb)

20 - Hex bolt, 100 Nm (74 ft lb)

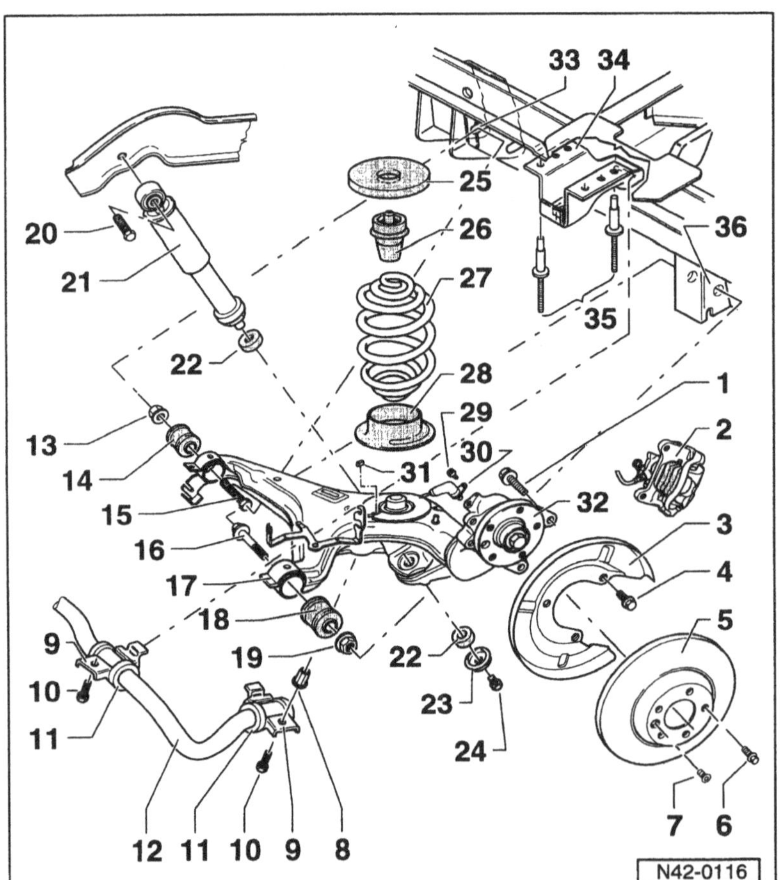

21 - Shock absorber
- ♦ Application,

⇒ Replacement parts catalog.
- ♦ Removing and installing,

⇒ Page N 42-17.
- ♦ Identification:
  black: standard,
  grey: standard,
  green: reinforced.

22 - Rubber mount

23 - Washer
- ♦ Collar points toward shock absorber.

24 - Hex bolt, 45 Nm (33 ft lb)

25 - Upper washer
- ♦ Position spring in recess.

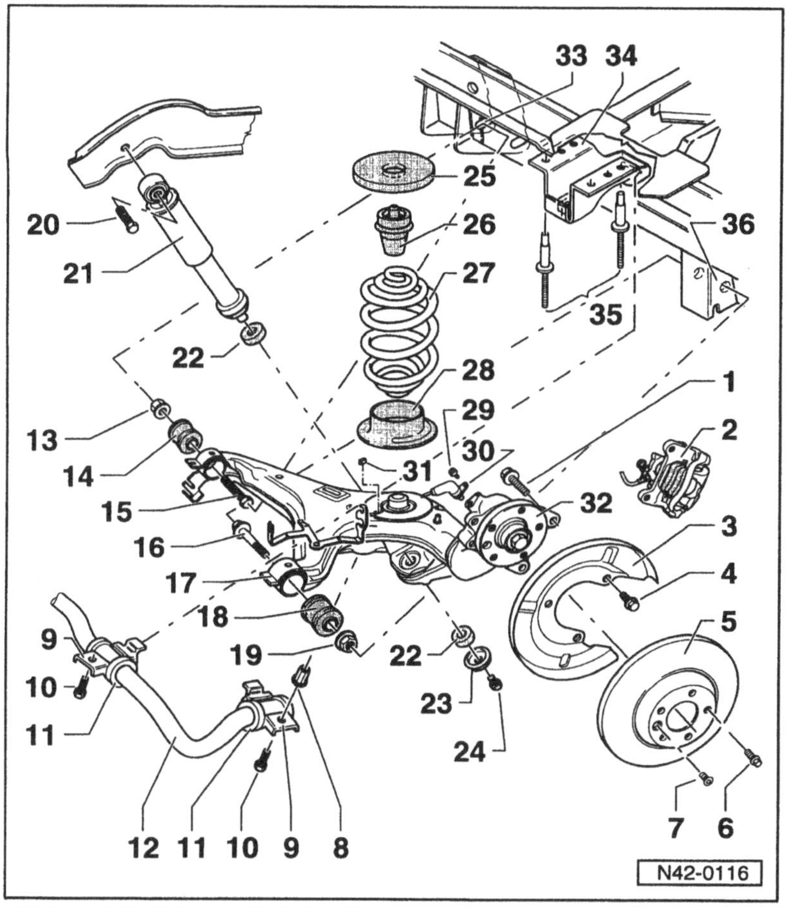

26 - Stop buffer

- Application,
  ⇒ from Page N 42-18.

27 - Coil spring

- Removing: first remove upper bolt from shock absorber,

⇒ Page N 42-17, Shock absorber, removing/installing.

- Disconnect drive shaft on four wheel drive vehicles at final drive flange.
- Installation position,

⇒ Fig. 3.

- Examine for paint damage and correct if necessary.
- Application,
  ⇒ from Page N 42-18.
- Retrofitting reinforced coil springs,

⇒ Page N 42-23.

28 - Cap

- Position spring in recess
- Application,

⇒ from Page N 42-18.

29 - Allen bolt,10 Nm (89 in. lb)

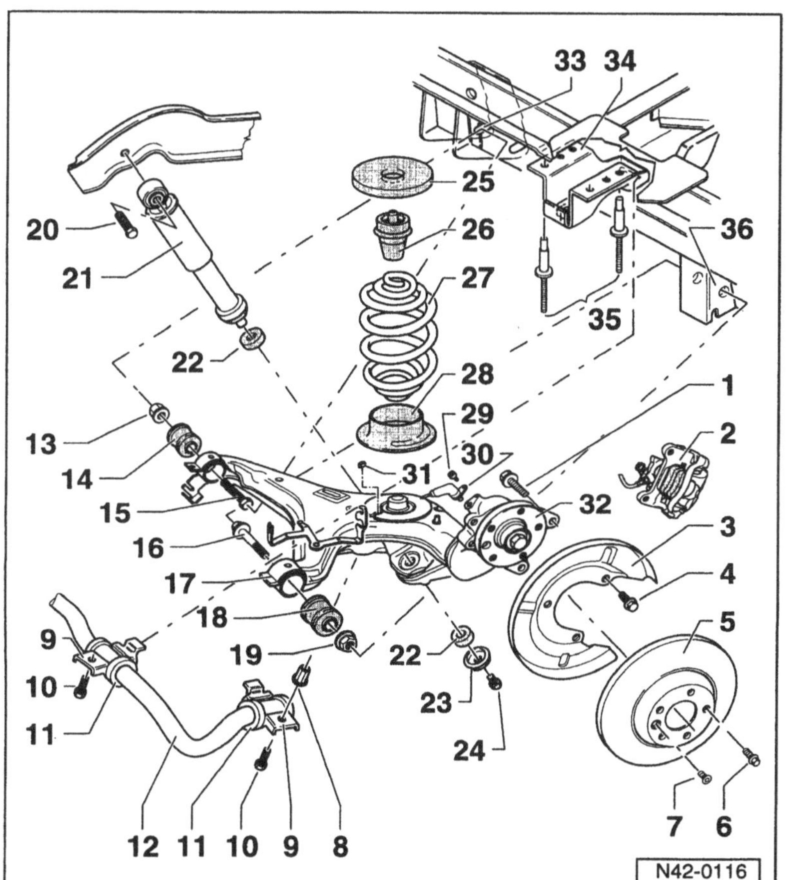

30 - ABS speed sensor

- Before inserting sensor, clean sensor hole inner surface and coat with lubricating paste G 000 650.

31 - Damping buffer

- Determining,

⇒ from Page N 42-18.

- Installing,

⇒ Page N 42-22.

32 - Wheel bearing

- Servicing,

⇒ Page N 42-24.

33 - Inner mount bracket

- With horizontal elongated holes to adjust track, Specification,

⇒ Page N 44-6.

34 - Mounting bracket

- Installing,

⇒ Page N 42-12.

35 - Rivet

36 - Outer mounting bracket

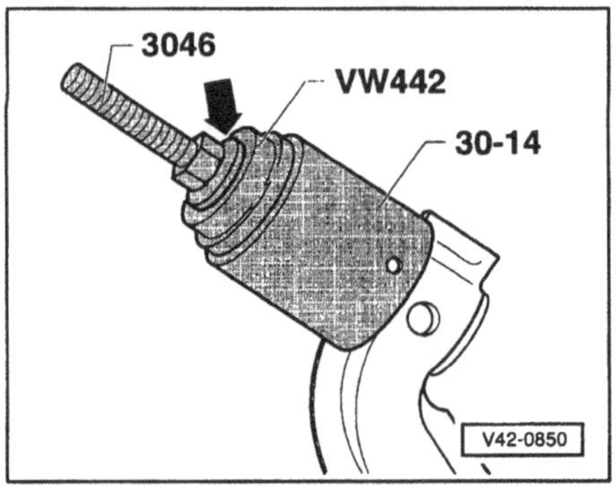

◄ **Fig. 1 Pulling bonded rubber mounting out of control arm**

- Clamp control arm in vice.
- Position washer -arrow- between nut and mount; moisten with oil.
- Pull out bonded rubber mount.

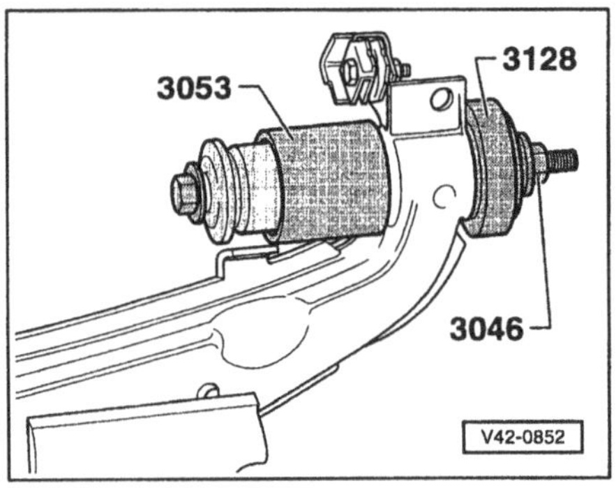

◄ **Fig. 2 Bonded rubber bushing, Installing in control arm**

- Pull bonded rubber bushing in until steel casing is against tool and bushing shoulders are evenly spaced on both sides.

Note:

Coat control arm mounting hole, guide sleeve and bonded rubber mount with rubber lubricant before installing (do not use soap).

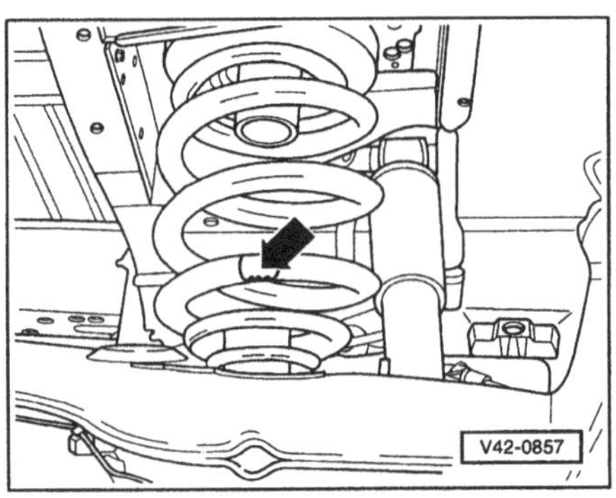

◄ **Fig. 3 Coil spring installation position**

Colored marking -arrow- points rearward.

Rear stabilizer bar, retrofitting

Note:

If a rear stabilizer bar is to be retrofitted, it is also necessary to install a front 27 mm dia. (1.1 in.) stabilizer bar. Retrofit 27 mm dia. (1.1 in.) stabilizer bar to front axle, if necessary, *⇒ Page N 40-14.*

Parts required (retrofitting kit):

- Stabilizer bar 26 mm dia. (1 in.)
- Clamp, stabilizer bar (Qty. 4)
- Rubber bushing (Qty. 4)
- Hex nut M 8 x 28 (Qty. 4)
- Rivet nut M 8 (Qty. 2)
- Mount bracket (Qty. 2)
- Hex bolt M 6 (Qty. 8)
- Rivet nut M 6 (Qty. 8)

Note:

Do not damage painted surface when installing stabilizer bar. Repair paintwork if necessary and apply underbody sealant.

Stabilizer bar, installing to control arm and mounting bracket, ⇒ from Page N 42-1.

Pulling rivet nut M 8 into control arm

– Screw M 8 nut onto M 8 x 45 bolt.

– Screw M 8 x 45 mm bolt far enough into M 8 rivet nut so "nut" is filled out.

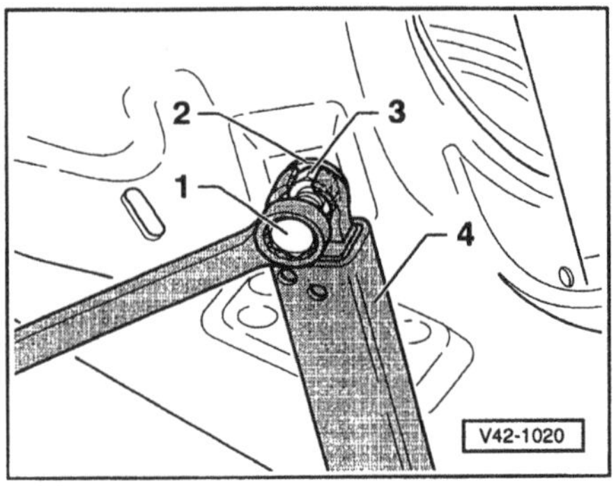

– Turn M 8 nut, by hand, against collar of rivet nut.

– Insert M 8 rivet nut into control arm.

◄ – Pull rivet nut in under pressure by turning M 8 nut and counter-holding at bolt head.

Tightening torque: 25 Nm (18 ft lb)

1 - Bolt M 8 x 45
2 - Rivet nut M 8
3 - Nut M 8
4 - Torque wrench VAG 1331

Note:

When pulling rivet nut in, ensure rivet nut collar is resting, under pressure, against control arm.

Securing mounting bracket in cross member with rivet nut M6

– Drill four of six holes out to 9 mm dia. (0.35 in.).

Warning!
Wear protective goggles.

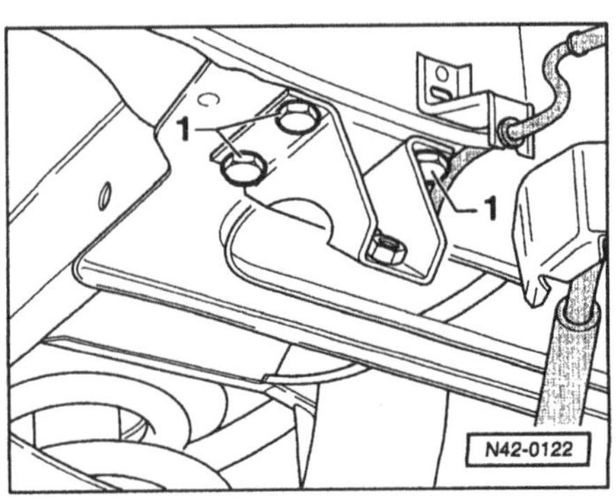

– Pull in rivet nut (use pliers VAG 1765), Page N 42-13; or bolt, nut and washer, ⇒ Page N 42-15.

◄ – Secure mount bracket for bolts,
1 - Bolt, Part No. N 010 218 6.

Tightening torque: 15 Nm (11 ft lb)

– Protect cross member against corrosion.

Pulling in rivet nut M6 with rivet nut pliers VAG 1765/VAG 1765 A

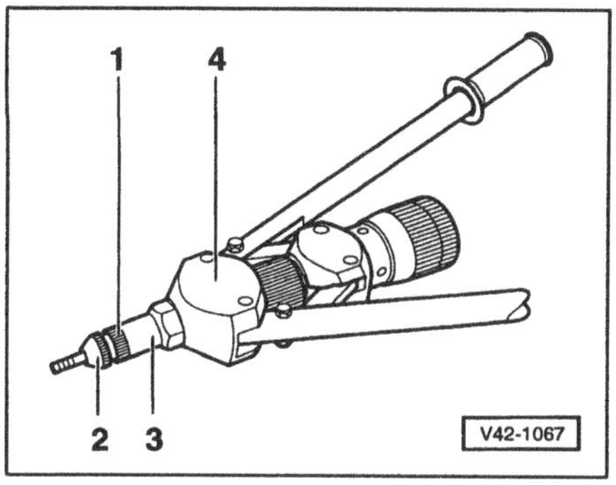

- Position insert required for M6 rivet nut to pliers VAG 1765/VAG 1765 A.

◄ – Loosen lock nut -1-.

– Remove nose piece -2-.

– Remove union nut -3-.

4 - Pliers VAG 1765/VAG 1765 A.

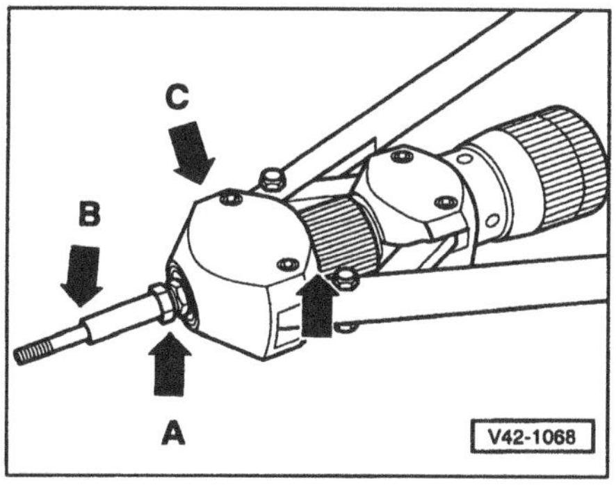

◄ – Loosen lock nut -A-.

– Remove stud -B-.

– Screw on required stud.

– Tighten lock nut -A-.

C - Pliers VAG 1765/VAG 1765 A.

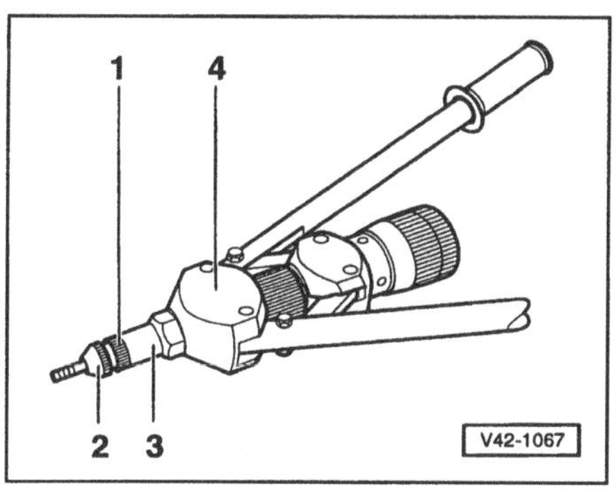

◄ – Thread on union nut -3-.

– Turn knurled ring until it rests tightly against housing -arrow-. (Sets stroke).

– Thread nose piece -2- completely back on. Also, ensure lock nut -1- does not make contact.

– Tighten lock nut -1-.

- Thread on rivet nut. Pliers must be fully open.
- Pull rivet nut in.

Note:

When pulling rivet nut in, ensure rivet collar is resting tightly against surface.

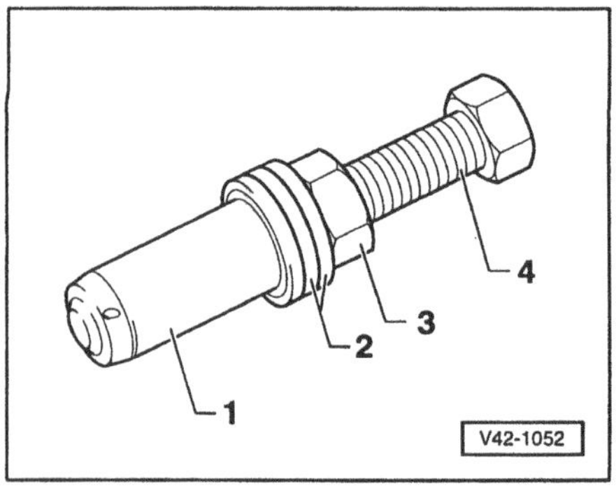

Pulling rivet nut M6 in with bolt, nut and washer

◄ – Screw M 6 x 50 bolt with M 6 nut and two washers far enough into rivet nut for "nut" to be filled out.

1 - Rivet nut M 6, Part No. N 102 952 01

2 - Washers

3 - Nut M 6

4 - Bolt M 6 x 50

– Turn M 6 nut by hand against collar of rivet nut.

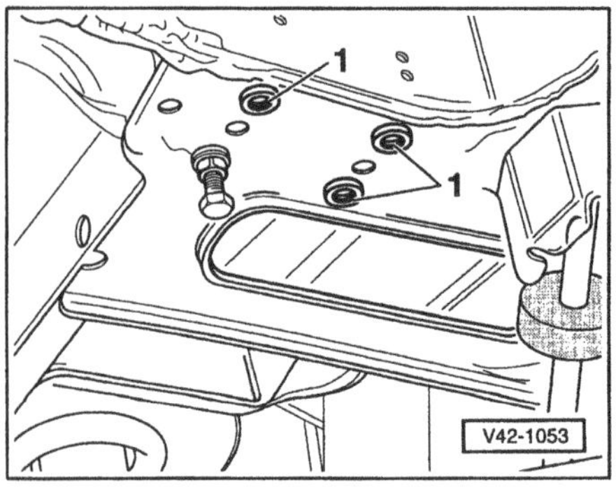

◄ – Insert rivet nut M6 into cross member.

1 - Rivet nuts already pulled in.

– Pull rivet nut in by tightening M6 nut and counter-holding at bolt head.

Notes:

- *Remove protective wax from cross member before pulling rivet nuts in.*
- *When pulling rivet nut in, ensure collar of rivet nut is resting against cross member.*

Tightening torque: 15 Nm (11 ft lb)

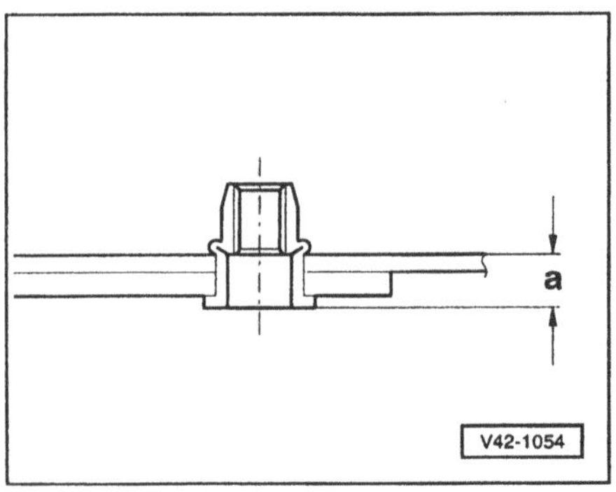

◄ – Remove bolt and check dimension -a-. Dimension -a-: 7 - 8.5 mm (0.28 - 0.33 in.).

Shock absorber, removing/installing

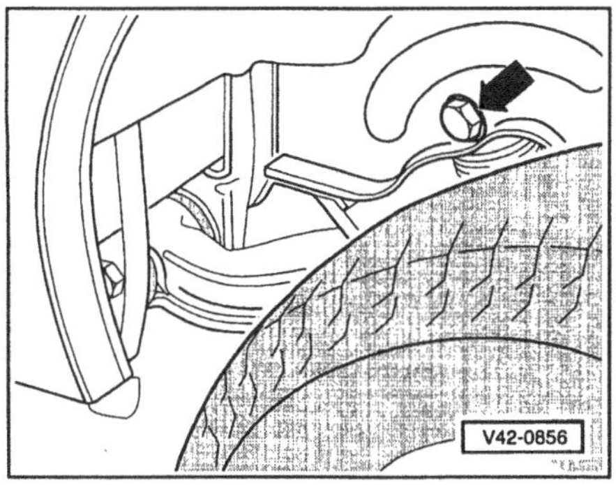

Note:

First disconnect drive shaft at transmission side flange and hang horizontally from bodywork on vehicles with four wheel drive. Otherwise damage may occur to the constant velocity joint.

◄ – Remove bolt -arrow- with vehicle on its wheels (raise vehicle, if necessary, until the bolt is acccessible).

– Raise vehicle to assembly height, this relieves pressure on coil spring.

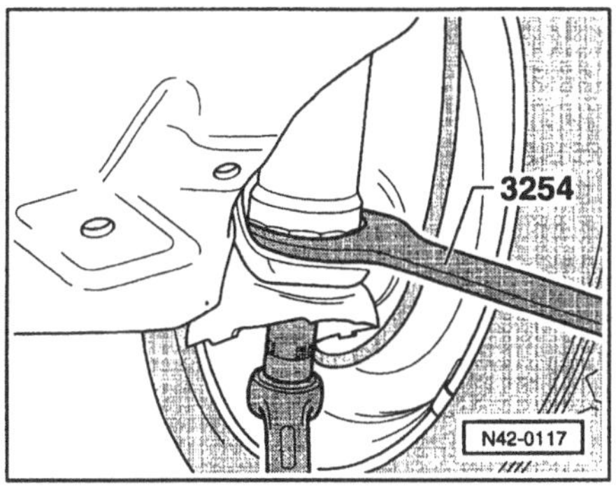

◄ – Remove shock absorber at control arm, (counter-hold shock absorber with special tool 3254).

Install in reverse sequence.

Coil spring and brake pressure regulator spring, determining

Model	Equipment, with payload details[1][2]	Coil spring Part No.	Color	Bump stop	GP. number	Brake pressure regulator spring Part No.	Color
Delivery van,	Short wheel base, 800kg	701 511 105 A/C[3]	white	x	1	701 612 561 B	white
High roof	Short wheebase, 1000kg	701 511 105 A/C[3]	white	x	1	701 612 561 B	white
delivery van,	Short wheelbase, 1200kg	701 511 105 M[3]	pink	x	1	701 612 561 D	black
	Long wheel base, 1000kg	701 511 105 A/C[3]	white	x	1	701 612 561 B	white
	Long wheel base, 1200 kg	701 511 105 M[3]	pink	x	1	701 612 561 D	black
	Long wheel base, other coachwork builder	701 511 105 K[4]	yellow	x	3	701 612 561 B	white
	Short wheel base, lowered suspension	701 511 105 L[3]	grey	x	4	701 612 561 B	white
	Heavy duty front/rear, 1000 kg, regardless of wheel base	701 511 105 M[3]	pink		1	701 612 561 D	black

[1] For vehicles with all wheel drive add 100 kg (220.5 lb) to payload (due to increased curb weight).

[2] Payload details are rounded figures.

[3] Vehicles with these springs are equipped with a damper buffer (Part No. 701 511 137) between spring and control arm,
⇒ Page N 42-22.

[4] Vehicles with these springs are equipped with a cap between spring and control arm,
⇒ Page N 42-5.

Model	Equipment, with payload details[1)2)]	Coil spring Part No.	Color	Bump stop	GP. number	Brake pressure regulator spring Part No.	Color
Caravelle,	short wheel base, 800 kg	701 511 105 P[4)]	blue	x	3	701 612 561	yellow
Multivan,	short wheel base,1000 kg	701 511 105 P[4)]	blue	x	3	701 612 561	yellow
Kombi	short wheel base,1200 kg	701 511 105 M[3)]	pink	x	1	701 612 561 B	white
	short wheel base, 800 kg 6-cyl. gasoline engine, 2.8 liter	701 511 105 L[3)]	grey	x	4	701 612 561 A	green
	short wheel base, 1000 kg 6-cyl. gas engine, 2.8 liter	701 511 105 L[3)]	grey	x	4	701 612 561 A	green
	long wheel base,1000 kg	701 511 105 P[4)]	blue	x	1	701 612 561	yellow
	long wheel base,1200 kg	701 511 105 M[3)]	pink	x	1	701 612 561 B	white
	short wheel base, lowered suspension	701 511 105 L[3)]	grey	x	4	701 612 561 A	green
	short wheel base, rear heavy duty	701 511 105 A/C[3)]	white		3	701 612 561 A	green

1) For vehicles with all wheel drive add 100 kg (220.5 lb) to payload (due to increased curb weight).

2) The payload details are rounded figures.

3) Vehicles with these springs are equipped with a damper buffer (Part No. 701 511 137) between spring and control arm,
⇒ Page N 42-22.

4) Vehicles with these springs are equipped with a cap between spring and control arm,
⇒ Page N 42-5.

Model	Equipment, with payload details[1)2)]	Coil spring Part No.	Color	Bump stop	GP. number	Brake pressure regulator spring Part No.	Color
Caravelle, Multivan,	long wheel base, rear heavy duty	701 511 105 A/C[3)]	white		1	701 612 561 A	green
Kombi	short wheel base, front + rear heavy duty	701 511 105 A/C[3)]	white		1	701 612 561 A	green
	long wheel base, front + rear heavy duty	701 511 105 A/C[3)]	white		1	701 612 561 A	green
Dropside,	short wheel base,1000 kg	701 511 105	violet	x	2	701 612 561 C	brown
Double cab,	short wheel base,1200 kg	701 511 105 B	brown	x	2	701 612 561 C	brown
Chassis with	short wheel base, 2000kg	701 511 105 B	brown	x	2	701 612 561 C	brown
cab,	long wheel base,1000 kg	701 511 105	violet	x	2	701 612 561 C	brown
Chassis with	long wheel base,1200 kg	701 511 105 B	brown	x	2	701 612 561 C	brown
double cab	long wheel base, 2000 kg	701 511 105 B	brown	x	2	701 612 561 C	brown
	heavy duty rear, 1000 kg, irrespective of wheel base, heavy duty front	701 511 105 B	brown		2	701 612 561 C	brown

1) For vehicles with all wheel drive add 100 kg (220.5 lb) to payload (due to increased curb weight).

2) Payload details are rounded figures.

3) Vehicles with these springs are equipped with a damper buffer (Part No. 701 511 137), between spring and control arm,
⇒ Page N 42-22.

Model	Equipment, with payload details[1)2)]	Coil spring Part No.	Color	Bump stop	GP. number	Brake pressure regulator spring Part No.	Color
Ambulance	short wheel base,1000kg	701 511 105 F[3)]	green	x	1	701 612 561	yellow
	long wheel base,1000 kg	701 511 105 F[3)]	green	x	1	701 612 561	yellow
	long wheel base,1200 kg other coachwork builder	701 511 105 B	brown	x	1	701 612 561 C	brown
High roofed ambulance with four wheel drive	long wheel base, 1000kg	701 511 105 P[3)]	blue	x	1	without brake pressure regulator	
Camper	short wheel base,1000kg	701 511 105 K[3)]	yellow	x	3	701 612 561 A	green
	long wheel base, 1000kg	701 511 105 K[3)]	yellow	x	1	701 612 561 A	green

1) For vehicles with all wheel drive add 100 kg (220.5 lb) to payload (due to increased curb weight).

2) Payload details are rounded figures.

3) Vehicles with these springs are equipped with a rubber coated plate between spring and control arm, ⇒ Page N 42-5.

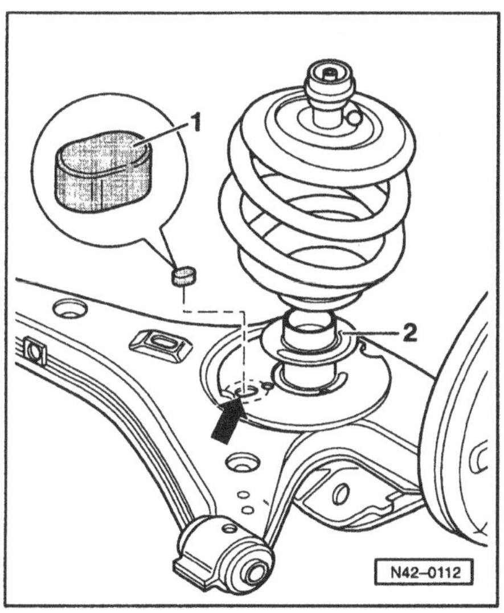

◄ **Damping bumper and cap**

Notes:

- *Damping bumper and cap prevent noise transmission between control arm and coil springs.*
- *Damping bumper is installed in conjunction with coil springs with an index letter of A, C, M, or L, after part number.*

Damping bumper installed in following vehicles:

Van,and Multivan

1 - Damping bumper

- Determining,

⇒ from page N 42-18.

- When replacing damping bumper, clean mount -arrow- on control arm thoroughly.
  Bonding surface must be oil and grease free!
  Remove protective foil before bonding damping bumper.

2 - Cap

- Determination,

⇒ from page N 42-18.

Stronger coil springs, retrofitting

Installing stronger rear axle coil springs has effect of:

- Vehicle springs are not compressed as far when loaded with same payload.

Note:

Vehicle axle loading, as shown on vehicle documents, is not increased by a modification, and must not be exceeded. Exception noted below:

- It is possible to increase maximum payload from 800 kg (1763 lb) to 1000 kg (2205 lb) or 1200 kg (2645 lb) and vehicles with 1000 kg (2205 lb) maximum payload to 1200 kg (2645 lb).

Note:

In this case, changed axle loading must be noted on vehicle documents. Relevant changes, as applicable to suspension group numbers, are to be entered on vehicle data plate by hand (see table),
⇒ Page N 44-3.

> ***Caution!***
>
> ***Brake pressure regulator spring is always determined according to vehicle spring.***

If the brake pressure regulator spring has been changed for another version, adjust the brake pressure regulator,
⇒ Page N 47-26.

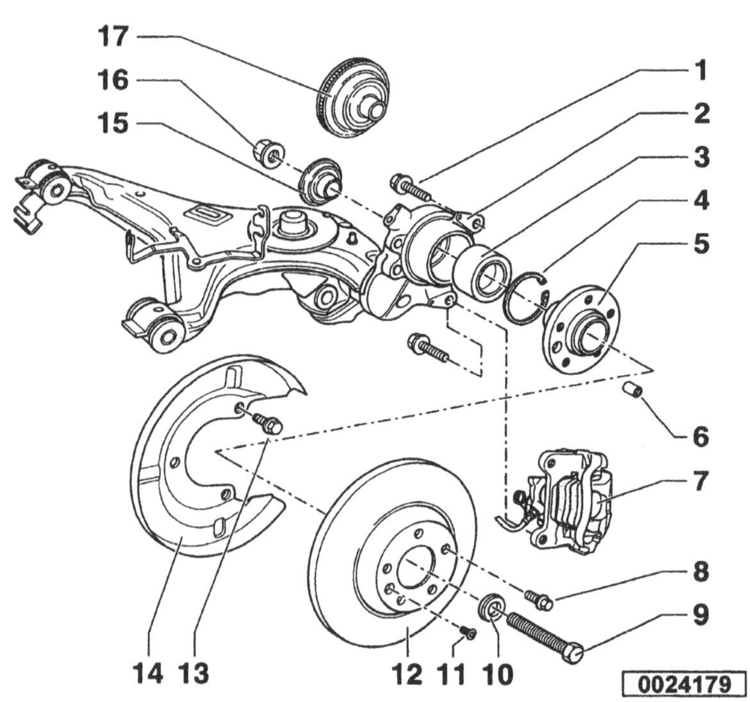

Rear wheel mountings, servicing

1 - Hex bolt, 110 Nm (81 ft lb)

2 - Control arm

3 - Wheel bearing
- Removing,
⇒ Fig. 2.
- Always replace (destroyed during removal).
- Installing (pressed),
⇒ Fig. 4.

4 - Circlip
- Ensure seated correctly.

5 - Hub
- Removing,
⇒ Fig. 1.
- Installing (pressed),
⇒ Fig. 5.
- Removing inner race,
⇒ Fig. 3.

6 - Spring pin

7 - Brake caliper

8 - Wheel bolt 160 Nm (118 ft lb)

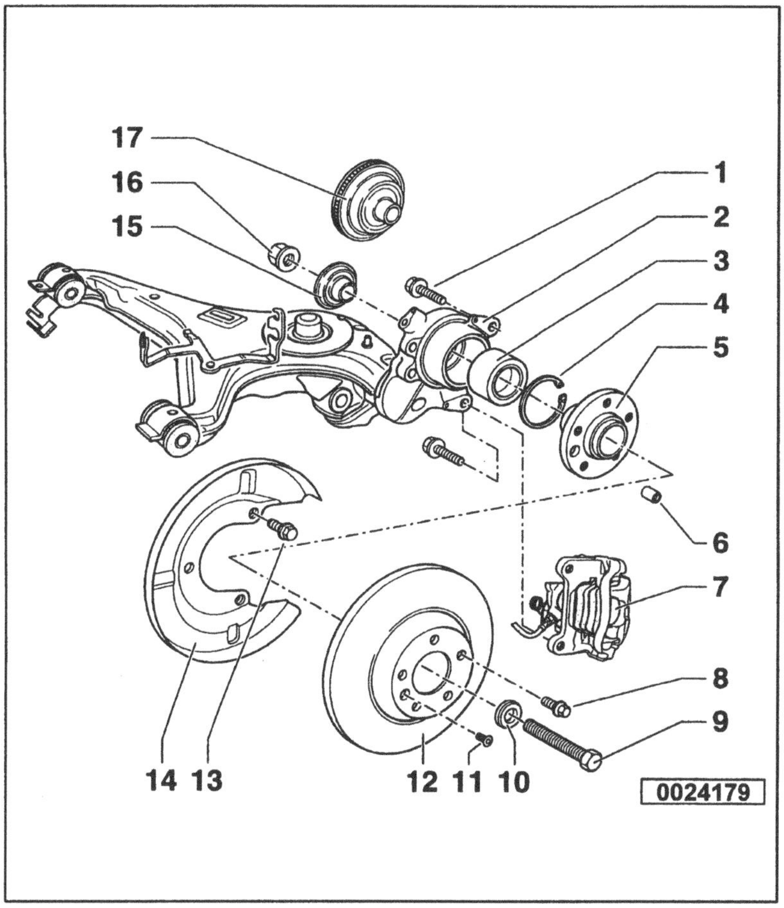

9 - **Hex bolt**
- Replace during reassembly.

10 - **Washer**

11 - **Allen bolt,10 Nm (89 in. lb)**

12 - **Brake rotor**

13 - **Hex bolt, 8 Nm (71 in. lb)**

14 - **Splash shield**

15 - **Tensioning disc**
- Front wheel drive only, without ABS.

16 - **Self-locking Hex nut, 200 Nm (148 ft lb)**
- Always replace.

17 - **Tensioning disc rotor**
- Front wheel drive only, with ABS.

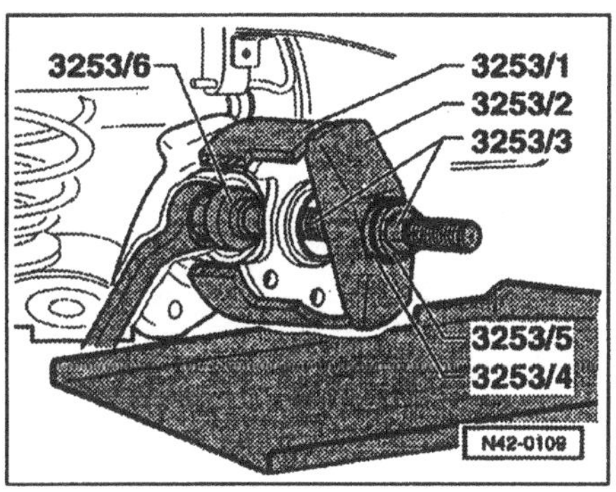

◄ **Fig. 1 Pulling wheel hub out of control arm**

Note:

Place transmission jack VAG 1383/, underneath (danger of accident through falling parts when removing wheel hub).

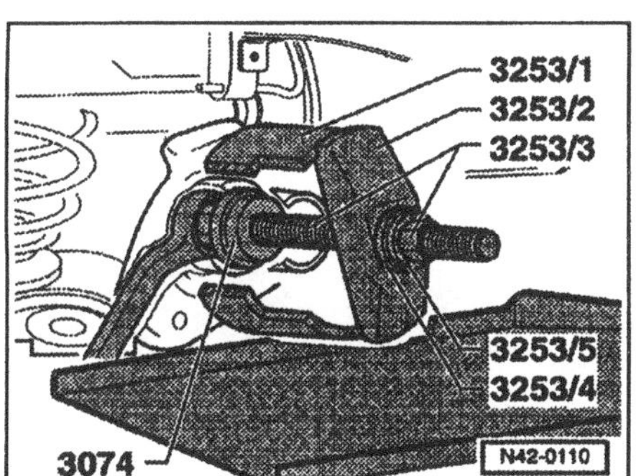

◄ **Fig. 2 Pulling wheel bearing out of control arm**

Note:

- *Place transmission jack VAG 1383/A underneath (danger of accident through falling parts when pulling out wheel bearing).*

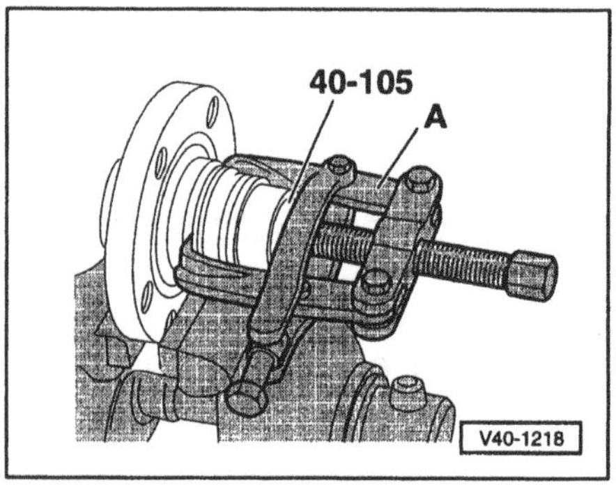

◄ **Fig. 3 Pulling inner race off hub**

Only use puller with leg clamp, e.g., -A- Kukko 204-2 (commercial type).

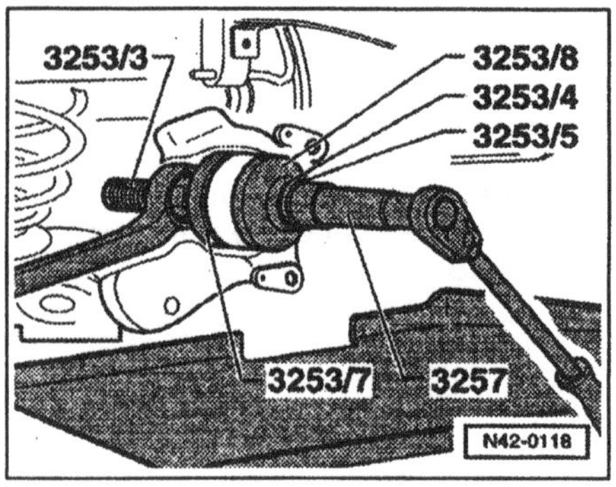

◄ **Fig. 4 Pulling wheel bearing into control arm**

- Pull in wheel bearing to stop.
- Insert circlip with long-nosed pliers.

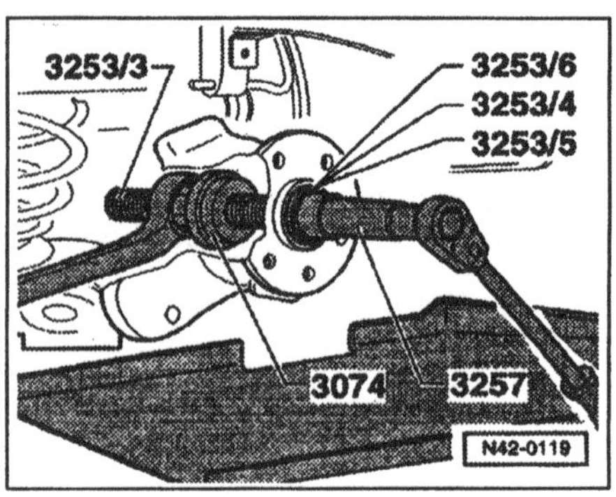

◄ **Fig. 5 Pulling in wheel hub**

- Pull wheel hub onto stop.

Vehicle alignment

Requirements:

- Measuring equipment correctly adjusted.
- Tires inflated to correct pressure.
- Vehicle accurately aligned, suspension bounced and rocked several times.
- Suspension, steering and steering linkage, check for excessive play and damage.
- Vehicle height (ride height) to specification.

Notes:

- *The specifications for vehicle alignment are divided into four groups. The group to which a vehicle is assigned is indicated on the vehicle data plate on the A-pillar at the height of the central electric box,⇒ Page N 44-3.*
- *Vehicle height (ride height) and vehicle longitudinal inclination affect test values. For this reason, measure vehicle height and longitudinal inclination, and adjust vehicle height, if necessary before performing checks/adjustments.*

⇒ Page N 44-5, measuring vehicle inclination.

⇒ Page N 44-4, measuring and, if necessary, adjusting vehicle height.

- *Loaded condition at time of measuring is relative when determining specifications (unloaded, $1/2$ payload, fully loaded).*
- *If alignment machine is approached at an acute angle, move vehicle back and forth several times, to relieve possible stress in bonded rubber mounts.*
- *Align vehicle after accumulating 1000 to 2000 km (600 to 1200 miles); only then will settlings be complete and stable.*

Vehicle data plate, located on A-pillar next to central electric

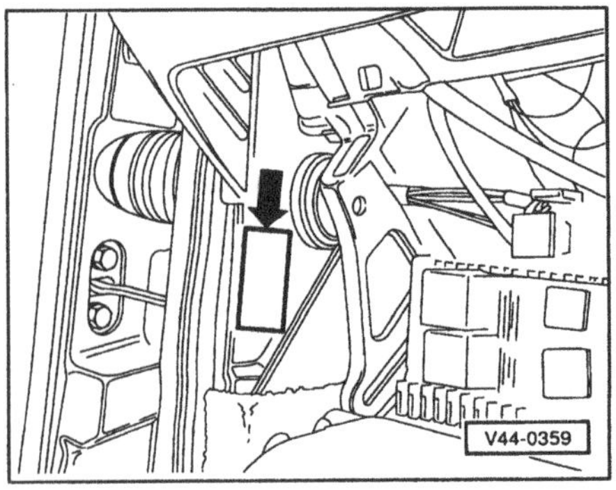

◄ On LHD vehicles - on left side

Vehicles with knee padding, -USA version-, vehicle data plate is located under central electric cover.

Vehicle data plate, group number position

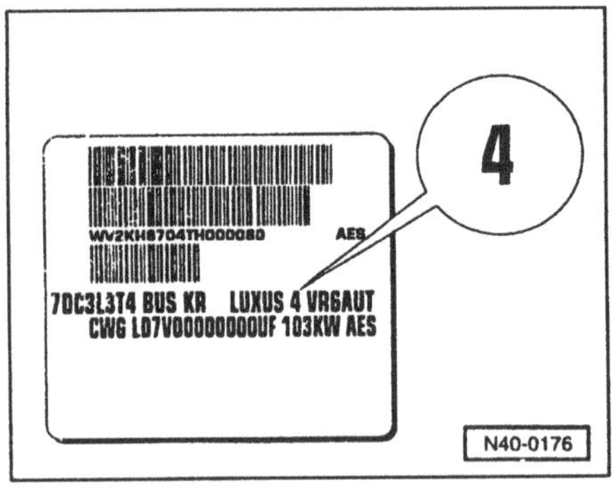

Group numbers 1 - 4 exist at present.

◄ Digit -arrow- indicates vehicle group to which a vehicle is allocated.

The word "LUXUS" on vehicle data plate is given as an example. It is possible that a different word, for example, "Kom." may be printed at this point.

Front axle ride height, measuring/adjusting

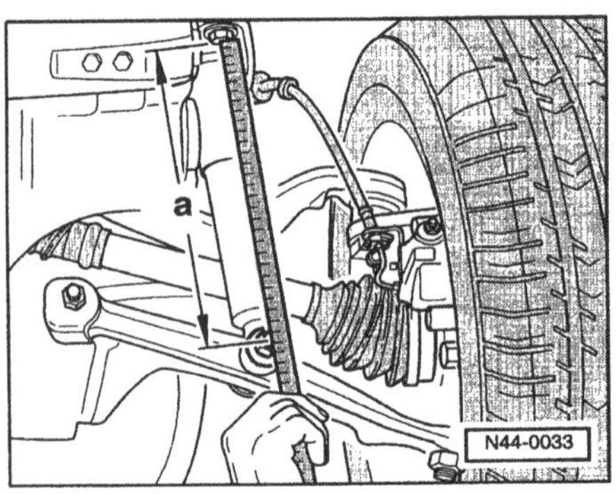

Vehicle must be on wheels.

◄ Dimension -a- is measured from upper shock absorber mount bolt head to center of lower shock absorber bolt. If necessary, adjust to prescribed height by turning nut on torsion bar tensioning lever.

Torsion bar, adjusting,
⇒ Page N 40-11.

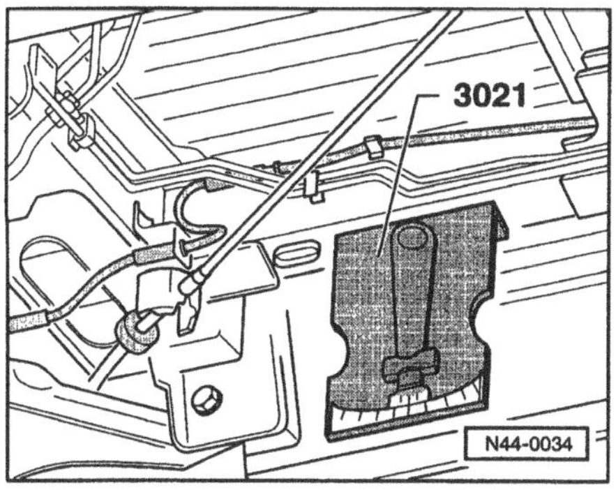

Longitudinal inclination, measuring

◄ – Measure longitudinal inclination at long member.

If vehicle is inclined to front or rear, deduct or add value, measured according to inclination of vehicle.

Figure obtained from alignment instrument		1° 10′
Correction vehicle for vehicle inclination (e.g., 1° forward)	+ 1°	
Caster (for vehicle inclination of 1° forward)		2° 10′

Notes:

- *Vehicle inclined forward: add correction figure.*
- *Vehicle inclined rearward: deduct correction figure.*
- *Ensure long member surface is clean and smooth in measurement area .*

Vehicle alignment specifications

Vehicle group 1

Front axle (double control arm)			
Load state	unloaded[3]	½ payload[3]	fully loaded[3]
Vehicle height[1] front (± 2 mm)	280	267	255
Toe/wheel (not pressed)	+104 + 104	0′ ± 10′	–10′ ± 10′
Total Toe[1] (not pressed)	+20′ ± 20′	0′ ± 20′	–20′ ± 20′
Camber (in straight-ahead position) Maximum permissible difference between sides	–25′ +20′ –40′ max. 30′	–40′ +20′ –40′ max. 30′	–50′ +20′ –40′ max. 30′
Toe angle difference at 20° lock to left and right	2° 10′ ± 30′		
Caster (vehicle inclination zero)[2] Maximum permissible difference between sides	+3° 10′ ± 30′ max. 1°	+3° 20′ ± 30′ max. 1°	+3° 30′ ± 30′ max. 1°

[1] If ride height needs adjustment, complete front axle alignment should be checked. After adjusting vehicle height, push vehicle forward and backward over approx. 2 m (6.5 ft), then re-check height. Checking front height,
⇒ Page N 44-4.

[2] Measuring vehicle longitudinal inclination,
⇒ Page N 44-5.

[3] Load state at time of vehicle alignment measurements.

Rear axle (trailing arm)			
Toe per wheel	+10′ ± 10′	+20′ ± 10′	+30′ ± 10′
Total Toe (at specified camber) Maximum permissible deviation from direction of travel	+20′ ± 20′ max. 30′	+40′ ± 20′ max. 30′	+1° ± 20′ max. 30′
Camber Maximum permissible difference between sides	−30′ ± 30′ max. 30′	−1° 10′ ± 30′ max. 30′	1° 35′ ± 30′ max. 30′

Vehicle group 2

Front axle (double control arm axle)			
Load state	unladen[3]	1/2 payload[3]	fully laden[3]
Vehicle height[1] front (± 2 mm)	280	267	255
Toe per wheel (not pressed)	+10′ ± 10′	0′ ± 10′	−10′ ± 10′
Total Toe[1] (not pressed)	+20′ ± 20′	0′ ± 20′	−20′ ± 20′
Camber (in straight-ahead position) Maximum permissible difference between sides	−25′ +20′ −40′ max. 30′	−40′ +20′ −40′ max. 30′	−50′ +20′ −40′ max. 30′
Toe angle difference at 20° lock to left and right	2° 10′ ± 30′		
Caster (vehicle inclination zero)[2] Maximum permissible difference between sides	+3° 10′ ± 30′ max. 1°	+3° 20′ ± 30′ max. 1°	+3° 30′ ± 30′ max. 1°

[1] If ride height needs adjustment, complete front axle alignment should be checked. After adjusting vehicle height, push vehicle forward and backward over approx. 2 m (6.5 ft), then re-check height. Checking front height,
⇒ Page N 44-4.

[2] Measuring vehicle longitudinal inclination, ⇒ Page N 44-5.

[3] Load state at time of vehicle alignment measurements.

Rear axle (Trailing arm)			
Toe per wheel	+3′ ± 10′	+12′ ± 10′	+20′ ± 10′
Total Toe (at specified camber) Maximum permissible deviation from direction of travel	+6′ ± 20′ max. 30′	+24′ ± 20′ max. 30′	+40′ ± 20′ max. 30′
Camber Maximum permissible difference between sides	0° ± 30′ max. 30′	–40′ ± 30′ max. 30′	–1° 10′ ± 30′ max. 30′

Vehicle group 3

Front axle (double control arm axle)			
Load state	unladen[3]	1/2 payload[3]	fully laden[3]
Vehicle height[1] front (± 2 mm)	273	263	255
Toe per wheel (not pressed)	+5′ ± 10′	0′ ± 10′	–10′ ± 10′
Total Toe[1] (not pressed)	+10′ ± 20′	0′ ± 20′	–20′ ± 20′
Camber (in straight-ahead position) Maximum permissible difference between sides	–35′ +20′ –40′ max. 30′	–45′ +20′ –40′ max. 30′	–50′ +20′ –40′ max. 30′
Toe angle difference at 20° lock to left and right	2° 10′ ± 30′		
Caster (vehicle inclination zero)[1] Maximum permissible difference between sides	+3° 10′ ± 30′ max. 1°	+3° 20′ ± 30′ max. 1°	+3° 30′ ± 30′ max. 1°

1) If ride height needs adjustment, complete front axle alignment should be checked. After adjusting vehicle height, push vehicle forward and backward over approx. 2 m (6.5 ft), then re-check height. Checking front height,
⇒ Page N 44-4.

2) Measuring vehicle longitudinal inclination, ⇒ Page N 44-5.

3) Load state at time of vehicle alignment measurements.

Rear axle (Trailing arm)			
Toe per wheel	+10′ ± 10′	+20′ ± 10′	+30′ ± 10′
Total Toe (at specified camber) Max. permissible deviation from direction of travel	+20′ ± 20′ max. 30′	+40′ ± 20′ max. 30′	+1° ± 20′ max. 30′
Camber Maximum permissible difference between sides	–30′ ± 30′ max. 30′	–1° 10′ ± 30′ max. 30′	–1° 35′ ± 30′ max. 30′

Vehicle group 4

Front axle (double control arm axle)			
Load state	unladen[3]	$^1/_2$ payload[3]	fully laden[3]
Vehicle height[1] front (± 2 mm)	265	260	255
Toe per wheel (not pressed)	0′ ± 10′	–5′ ± 10′	–10′ ± 10′
Total Toe[1] (not pressed)	0′ ± 20′	–10′ ± 20′	–20′ ± 20′
Camber (in straight-ahead position) Maximum permissible difference between sides	–45′ +20′ –40′ max. 30′	–50′ +20′ –40′ max. 30′	–55′ +20′ –40′ max. 30′
Toe angle difference at 20° lock to left and right	2° 10′ ± 30′		
Caster (vehicle inclination zero)[2] Maximum permissible difference between sides	+3° 10′ ± 30′ max. 1°	+3° 20′ ± 20′ max. 1°	+3° 30′ ± 30′ max. 1°

[1] If ride height needs adjustment, complete front axle alignment should be checked. After adjusting vehicle height, push vehicle forward and backward over approx. 2 m (6.5 ft), then re-check height. Checking front height,
⇒ Page N 44-4.

[2] Measuring vehicle longitudinal inclination, ⇒ Page N 44-5.

[3] Load state at time of vehicle alignment measurements.

Rear axle (Trailing arm)			
Toe per wheel	+12′ ± 10′	+22′ ± 10′	+27′ ± 10′
Total Toe (at specified camber) Maximum permissible deviation from direction of travel	+24′ ± 20′ max. 30′	+44′ ± 20′ max. 30′	+54′ ± 20′ max. 30′
Camber Maximum permissible difference between sides	–40′ ± 30′ max. 30′	–1° 10′ ± 30′ max. 30′	–1° 10′ ± 30′ max. 30′

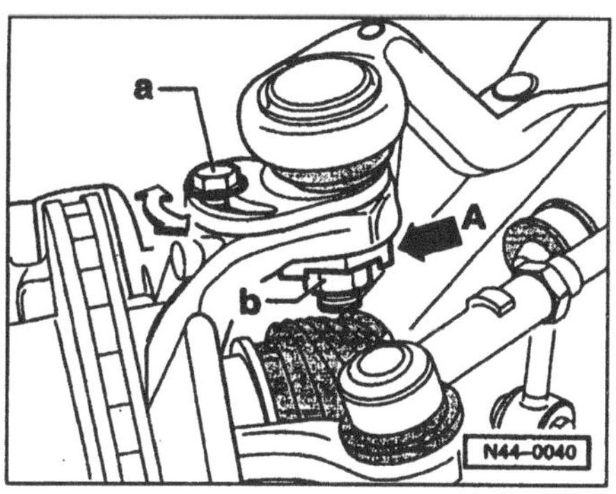

Camber, adjusting

- If necessary, remove right horn cover plate.

◄ – Loosen bolt -a-.

Note:

- *To loosen and tighten with wheel installed, turn steering as necessary.*
- *For ease of illustration, bolted connection is shown without wheel installed.*

– Loosen nut -b-.

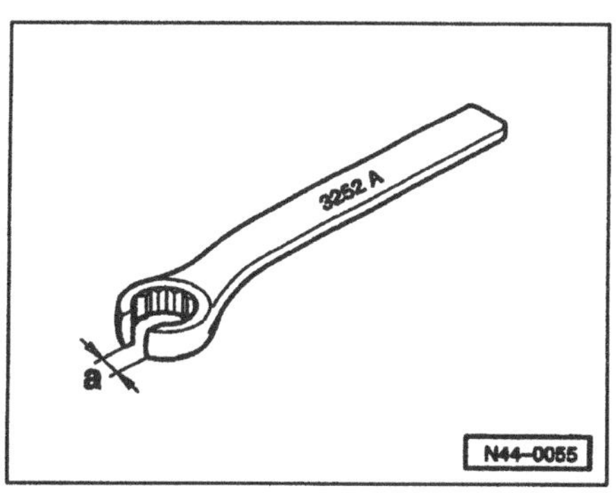

◄ – Obtain special tool 3252A or modify tool 3252 as shown.

- Dimension -a- = 30 mm (1.20 in.)

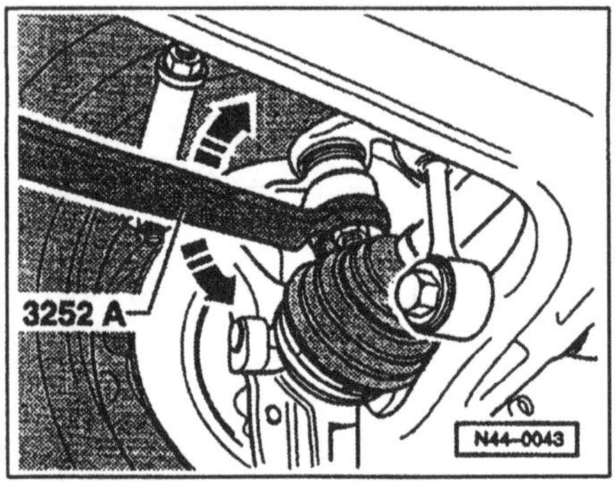

- Turn eccentric bushing with special tool 3252A until specification is achieved.
- Tighten bolt -a- and nut -b- again.

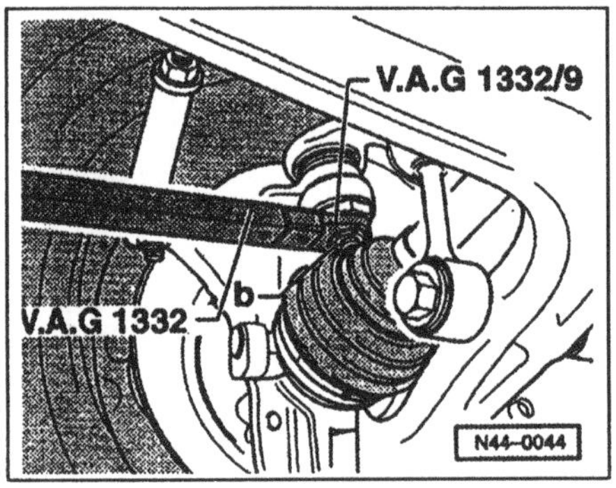

- Tighten nut -b-.
- Tighten bolt -a-; if necessary, remove wheel.

Tightening torque:

Bolt -a-	60 Nm (44 ft lb)
Nut -b-	110 Nm (81 ft lb)

Toe, adjusting

- Turn steering gear to center position.
⇒ Page N 44-16.
- Loosen lock nut.
- Turn both tie rods until specification is achieved.

Note:

- *Steering must remain centered.*
- *After turning tie rod, ensure boots are not twisted.*

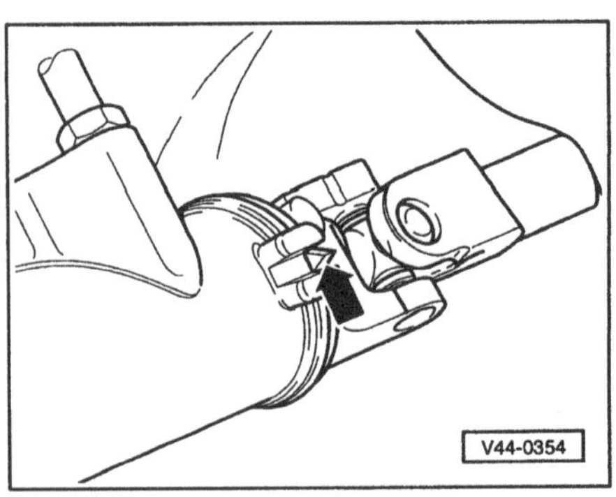

Steering gear, centering

- Turn steering wheel from lock to lock and count number of turns.
- Turn steering wheel back half the number of turns.
- Precisely align steering spindle with aid of point -arrow- on rubber washer and notch in housing. Point and notch must be aligned.

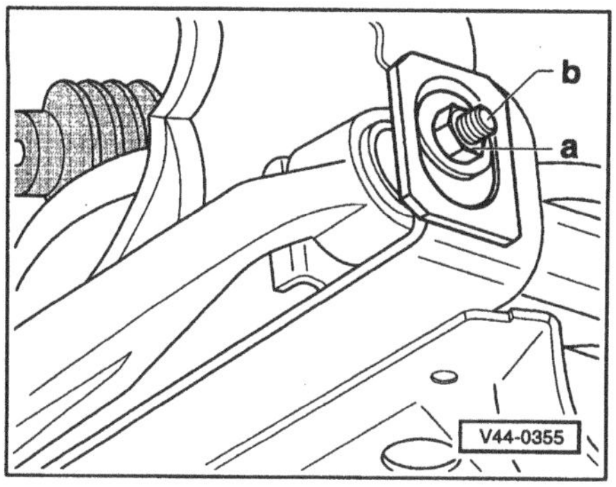

Caster, checking/adjusting

Caster should be checked in the following cases:

- ♦ Steering pulls.
- ♦ Front wheel shimmy.

◄ – Loosen nut -a-.

– Turn eccentric bolt -b- until specification is achieved.

– Tighten nut to 160 Nm (118 ft lb).

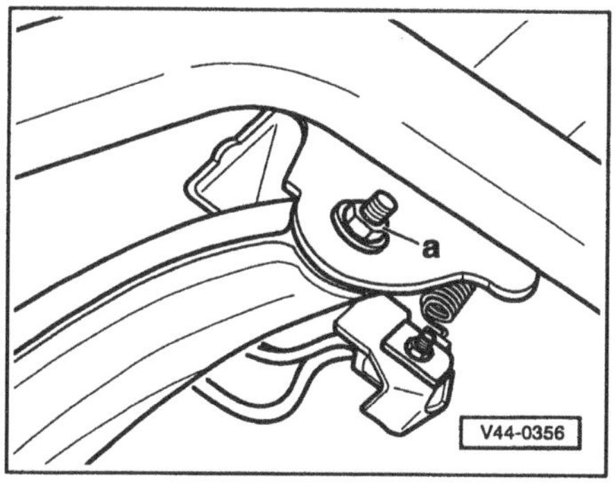

Rear axle, adjusting toe

◄ – Loosen nut -a- on control arm.

– Move control arm forward or backward until specification is achieved (vehicle weight on its wheels).

Note:

If toe setting at rear wheels needs adjusting, also check brake pressure regulator setting and adjust as needed, *⇒ Page N 47-26.*

Direction of travel, calculating:

– If both toe figures have the same preceeding sign (+/+ or –/–), subtract smaller figure from larger figure and divide by 2.

Example:

Left rear wheel Toe		**Right rear wheel Toe**
+ 15′		+ 5′
	15′ - 5′ = 10′ 10′ : 2 = 5′	

Direction of travel, deviation = 5′

– Add figures with different prefix (+/–), divide by 2.

Example:

Left rear wheel Toe		**Right rear wheel Toe**
+ 15′		– 5′
	15′ + 5′ = 20′ 20′ : 2 = 10′	

Direction of travel, deviation = 10′

Individual result is alignment deviation from vehicle longitudinal axis.

Anti-locking brake system (ABS) and anti-locking brake system with electronic differential lock (ABS/EDL) Bosch 5

- The ABS brake system is divided diagonally. Servo-assistance takes place pneumatically through the vacuum servo unit.
- Hydraulic and control modules form an integral component and must be removed and installed together.
- New control modules are not coded and, therefore, require coding after installation.

ABS and ABS/EDL, repairing

Notes:

- *Before repairing anti-locking system, determine cause of malfunction through On Board Diagnostic (OBD).*

⇒ Suspension, Wheels, Brakes, Steering – OBD; Repair group 01.

- *Before working on ABS, ABS/EDL system, disconnect battery ground strap. On vehicles with coded radio, obtain coding.*
- *When working with brake fluid, observe safety precautions,* *⇒ Page N 47-21.*
- *After work for which the brake system had to be opened, bleed the brake system with brake filling and bleeding unit VW 1238/1 or US 1116,* *⇒ Page N 47-21.*
- *During the final road test, perform a controlled brake test at least once (pulsations must be felt at brake pedal).*
- *Absolute cleanliness is required when working on the anti-locking brake system; do not use any products which contain oils, greases etc.*

- *Thoroughly clean all connections and adjacent areas before loosening; do not use aggressive cleaning agents such as brake cleaner, gasoline or thinners.*
- *Place removed parts on a clean surface and cover.*
- *Do not use fluffy cloths (that may have lint).*
- *Cover or seal open components if repairs cannot be completed immediately.*
- *Only remove replacement parts from packaging immediately prior to installation.*
- *Only use genuine packed parts.*
- *When system is open, do not work in the area with compressed air and do not move the vehicle.*
- *Ensure no brake fluid enters harness connectors.*

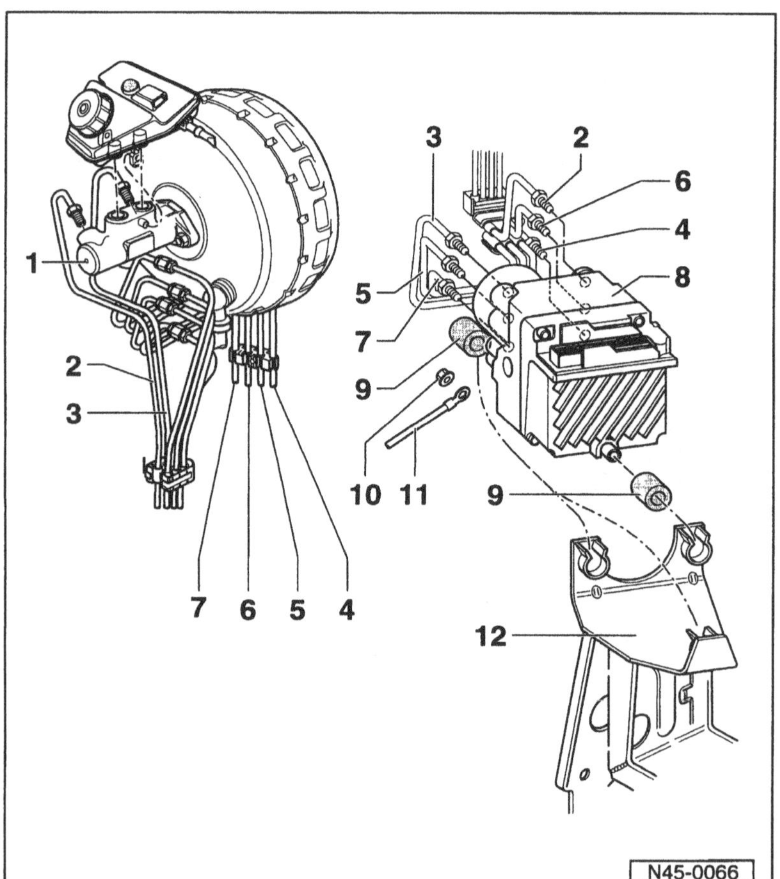

Hydraulic unit, removing/installing

Notes:

- *The hydraulic unit is located on the left in the engine compartment.*
- *When removing or installing brake lines/hoses, ensure lines/hoses are marked and installed as per hydraulic unit markings.*
- *Do not interchange brake lines/ hoses when reconnecting.*

1 - **Tandem brake master cylinder**
- With tandem brake servo unit

2 - **Brake line**
- Secondary piston circuit - hydraulic unit

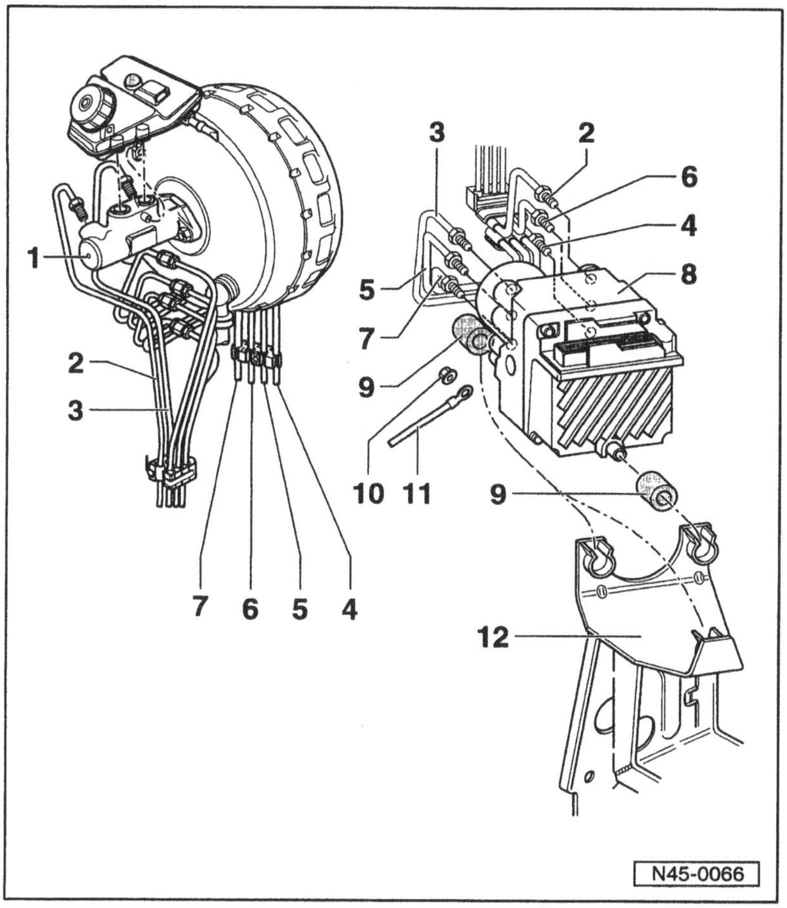

3 - Brake line
- Primary piston circuit - hydraulic unit

4 - Brake line
- Hydraulic unit to front left brake caliper

5 - Brake line
- Hydraulic unit to rear left brake caliper

6 - Brake line
- Hydraulic unit to rear right brake caliper

7 - Brake line
- Hydraulic unit to front right brake caliper

8 - ABS/EDL hydraulic unit
- Fastened to control module
- Replace only as an assembly

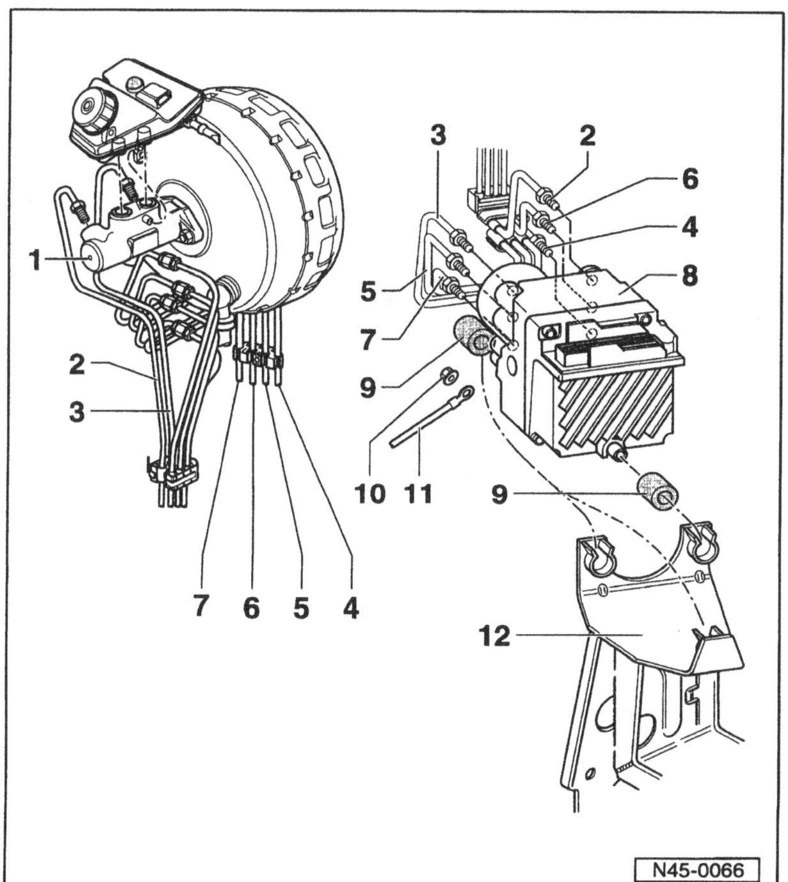

9 - Damper

10 - Ground cable
- Ensure good contact

11 - Hex nut

12 - Mounting bracket

Control module, removing/installing

The control module is bolted to the hydraulic unit and is located on the left side of the engine compartment.

> ***Caution!***
> ***Do not bend the brake lines in the area of the hydraulic unit!***

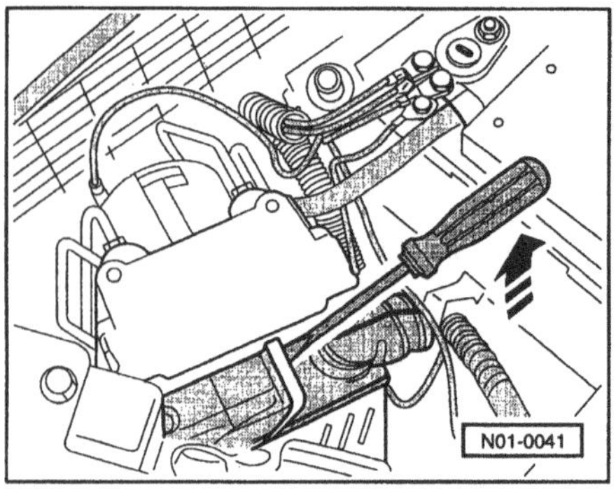

- Obtain radio code, if applicable.
- Disconnect battery ground strap.
- ◄ Remove control module connector.
- Remove as much brake fluid as possible from brake fluid reservoir, using a suction bottle.
- Position brake pedal loading device from VAG 1238 B or equivalent.
- Remove hydraulic unit ground wire.

- Disconnect hydraulic unit brake lines and seal openings with plugs.
- Lift hydraulic unit out upward.

Installing:

Notes:

- ♦ *Only remove hydraulic unit plugs when installing brake line.*
- ♦ *If the plugs are removed too early, brake fluid can escape, no longer guaranteeing the unit is sufficiently filled or adequately bled.*

Further installation is performed in reverse sequence.

Bleeding brake system,
⇒ Page N 47-21.

- Code control module, ⇒ Suspension, Wheels, Brakes, Steering, Repair Group 01.

Tightening torques:

Brake lines on ABS unit:

Thread M 10 x 1	15 Nm (11 ft lb)
Thread M 12 x 1	15 Nm (11 ft lb)

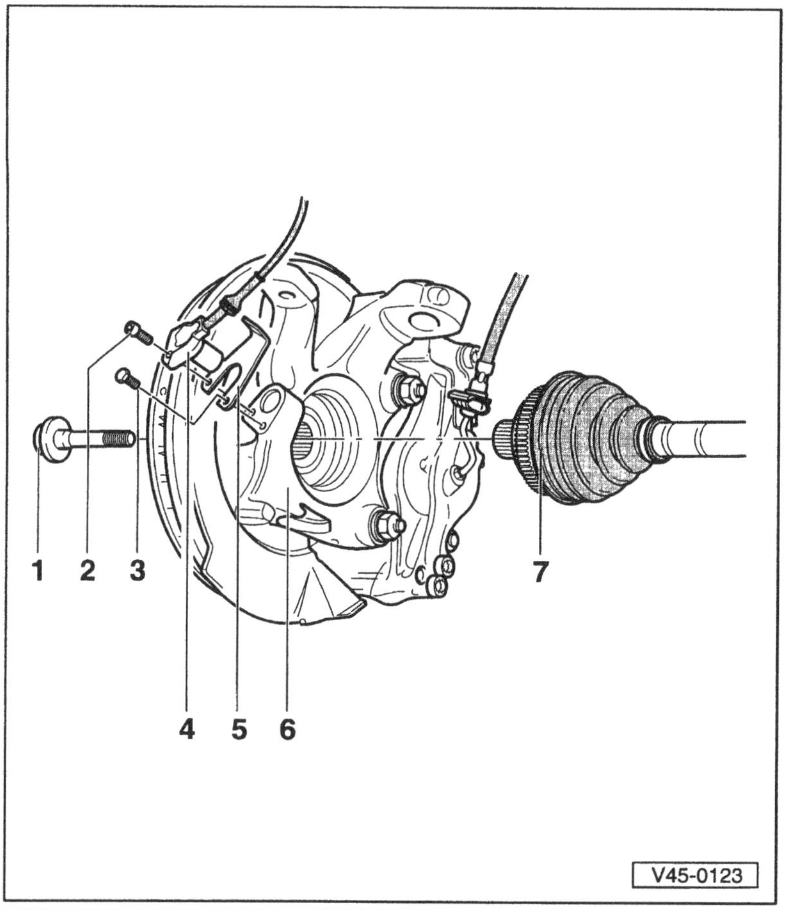

ABS system (front axle), removing/installing

1 - Hex bolt
- 150 Nm (111 ft lb) then turn 90° further.
- Always replace during reassembly.

2 - Allen bolt, 10 Nm (89 in. lb)

3 - Hex bolt

4 - Speed sensor front right/left -G45-/-G47-
- Before inserting sensor, clean sensor mounting hole inner surface and coat with lubricant G 000 650.
- Observe installation sequence: first install speed sensor and then sensor wiring, beginning at sensor and ensuring that rubber grommets are secured in their respective mounts.
- Routing harness,

⇒ Fig. 1.

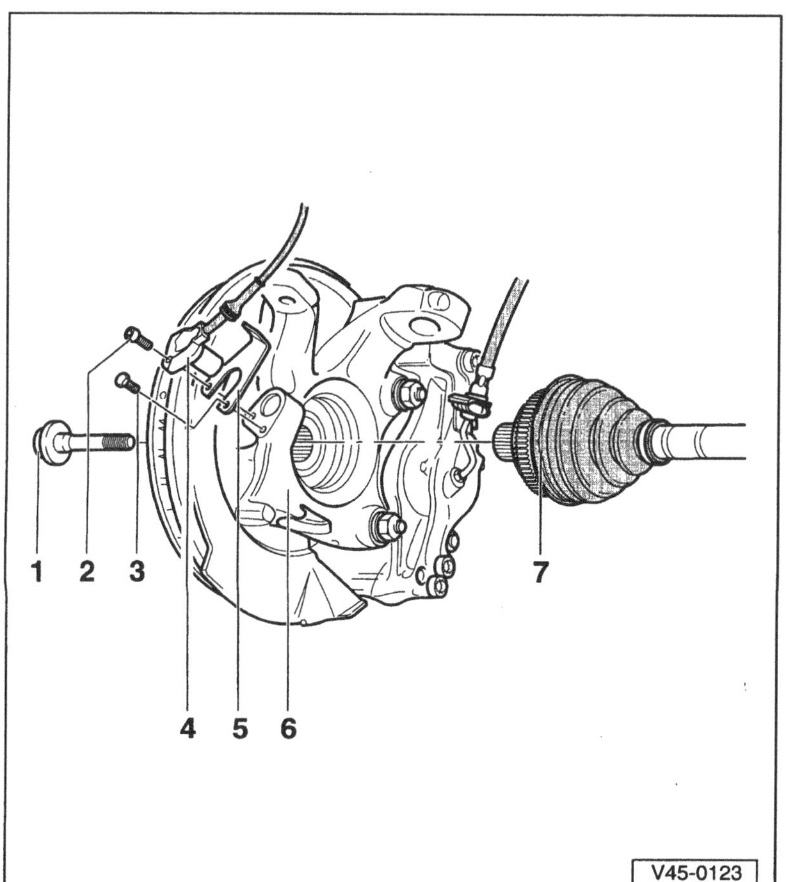

5 - Bracket

6 - Wheel bearing housing

7 - Outer constant velocity joint
- Removing,

⇒ Page N 40-54.

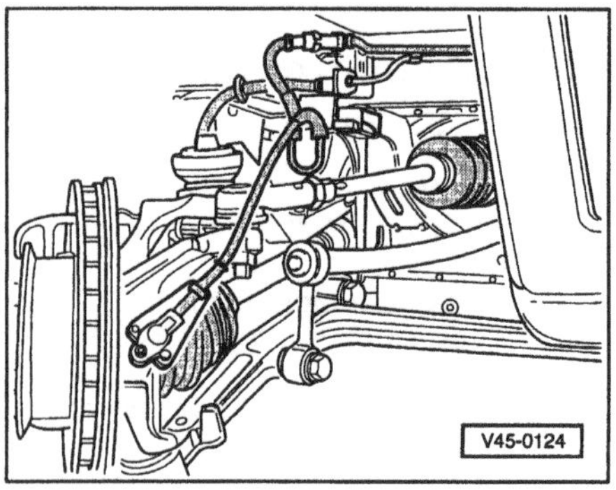

◄ **Fig. 1 Routing speed sensor harness (front axle)**

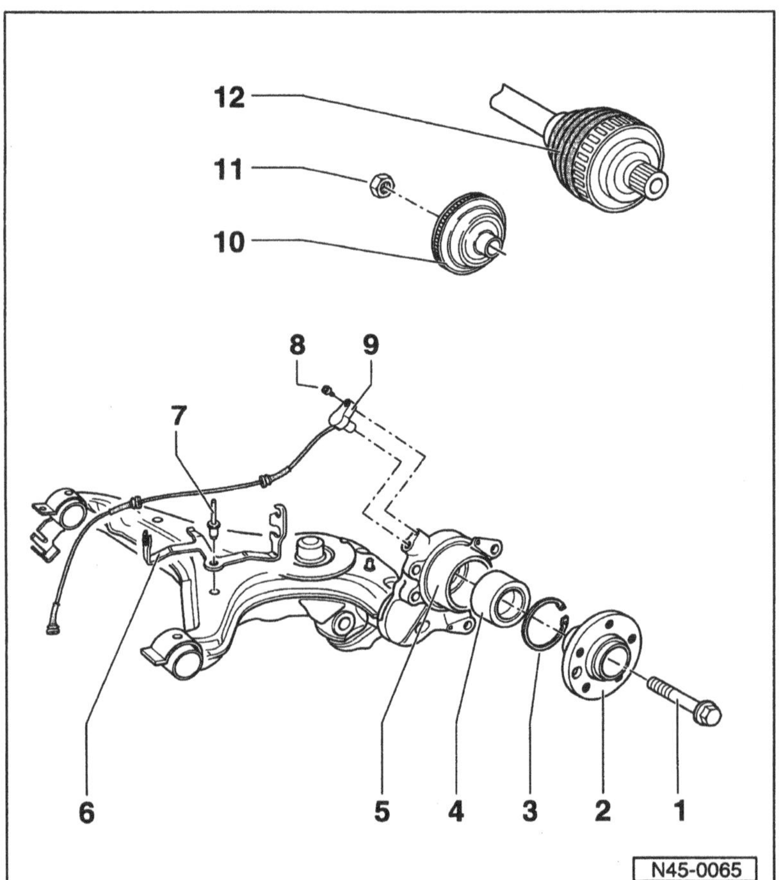

ABS components (rear axle), removing/installing

1 - Hex bolt
- Always replace at reassembly.

Tightening torque
(with all wheel drive):
- 150 Nm (111 ft lb) and turn 90° further.
- All wheel drive vehicles, loosen and tighten only when vehicle is on its wheels (danger of accident).

2 - Hub
- Pulling out and in,

⇒ Page N 42-24.

3 - Circlip
- Ensure circlip is seated correctly.

4 - Wheel bearing
- Pulling out and in,

⇒ Page N 42-24.

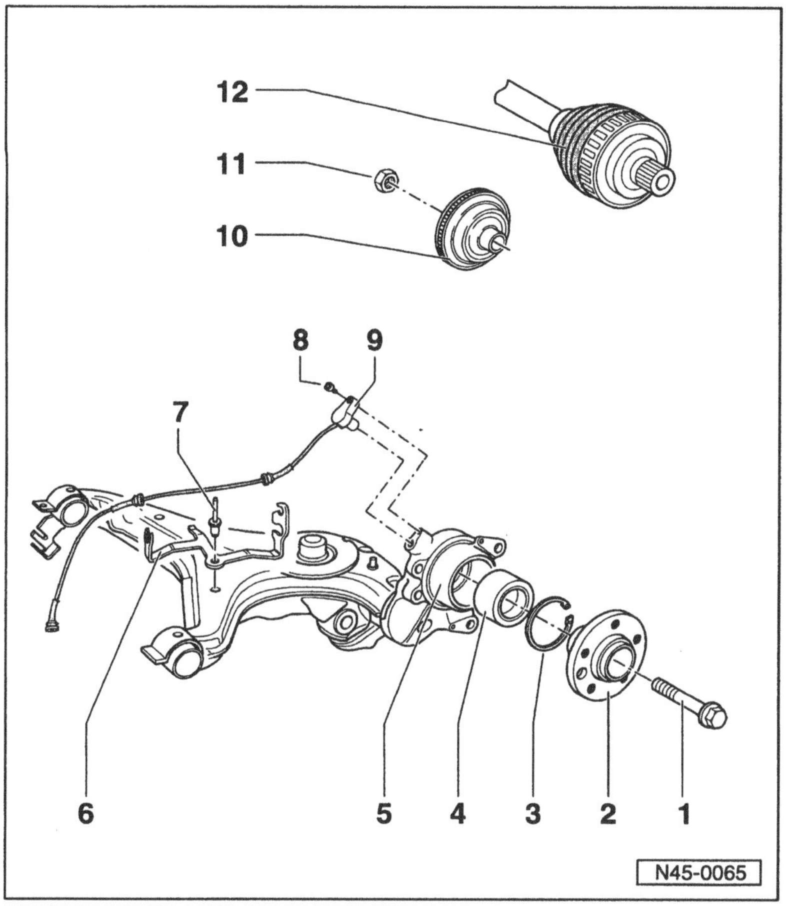

5 - Control arm

6 - Bracket
- When installing, hook into control arm first.

7 - Rivet

8 - Allen bolt, 10 Nm (89 in. lb)

9 - Speed sensor
- ♦ Before inserting sensor, clean sensor mounting hole inner surface and coat with lubricant G 000 650.
- ♦ Observe installation sequence: First install speed sensor and then sensor wiring, beginning at sensor and ensuring that rubber grommets are secured in their respective mounts.

10 - Tensioning disc with rotor
- ♦ Front wheel drive only.

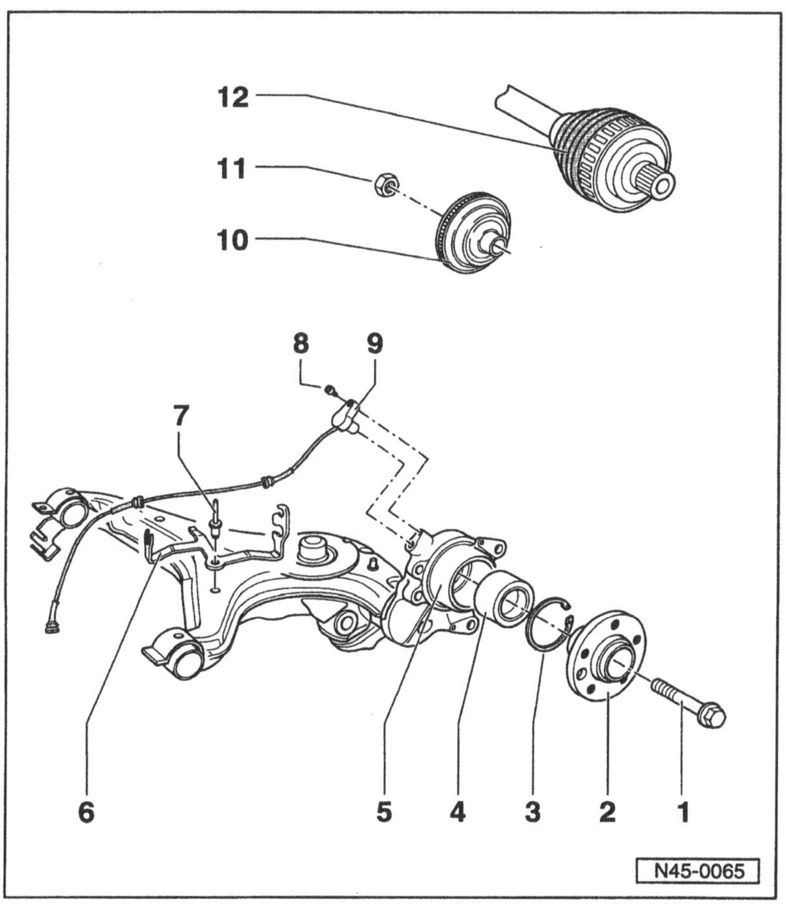

11 - Self-locking hex nut, 200 Nm (148 ft lb)
- ♦ Always replace.
- ♦ Front wheel drive only

12 - Outer constant velocity joint

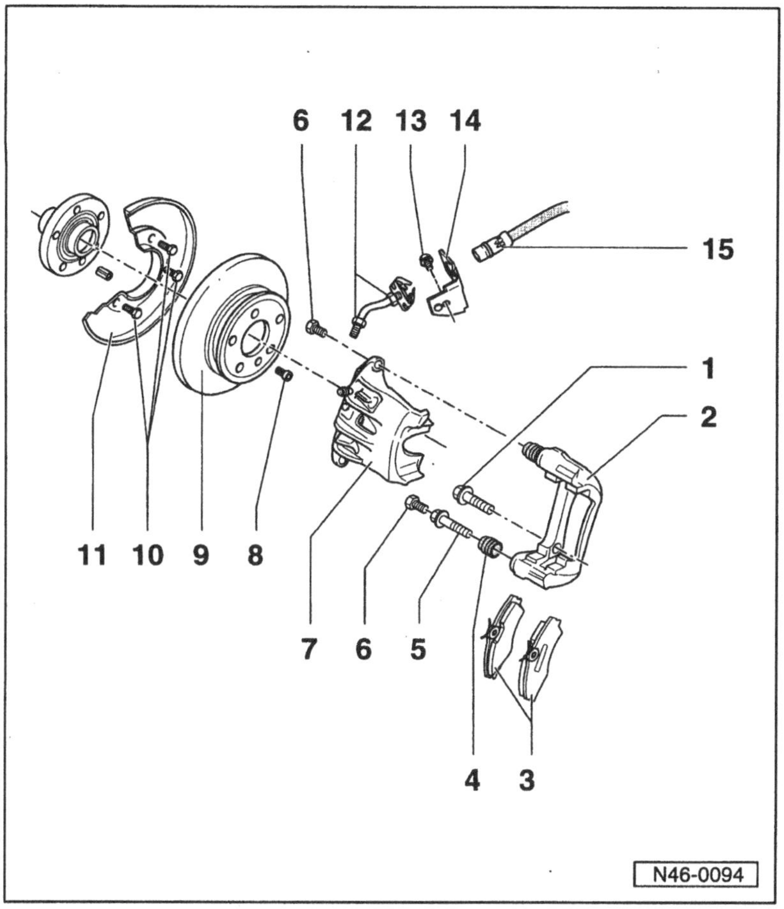

Front brake, assembly (Lucas C54 floating frame caliper)

Notes:

- *After replacing brake pads, depress brake pedal firmly several times, with vehicle stationary, so that the brake pads are properly seated in their normal operating position.*

> ***Warning!***
> ***To draw off brake fluid use a bleeder bottle. Brake fluid is poisonous and must not be siphoned out by mouth through a hose!***

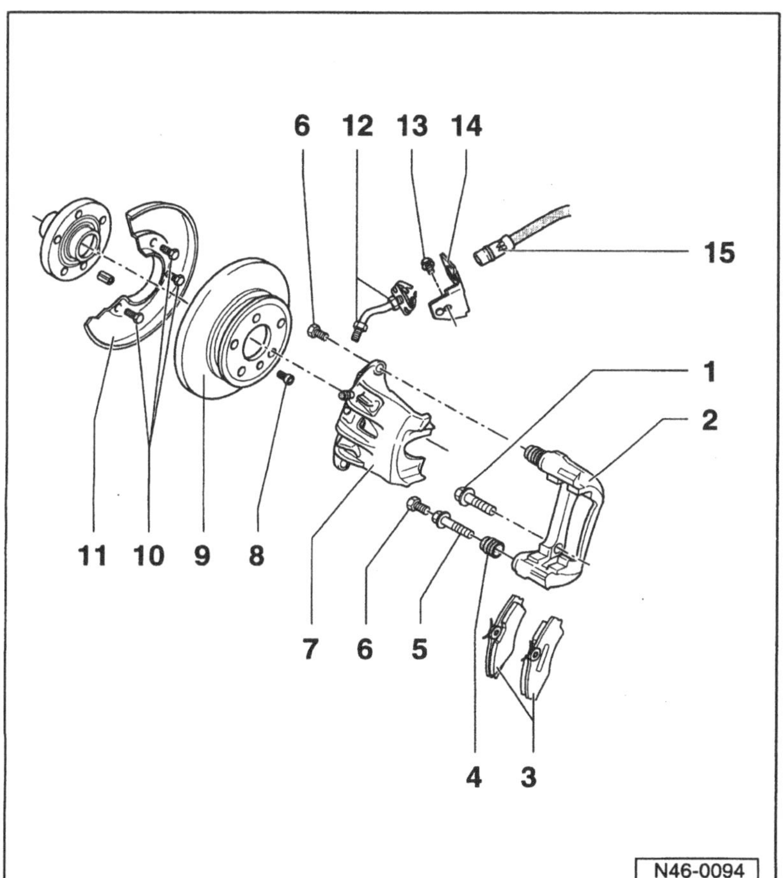

1 - Hex bolt, 280 Nm (207 ft lb)

2 - Brake pad support

3 - Brake pads
- ♦ Always replace both sides.
- ♦ New pad thickness: 13.5 mm.
- ♦ Checking thickness,

⇒ Maintenance manual.
- ♦ Wear limit: 2 mm, excluding backing plate.
- ♦ Removing/installing,

⇒ Page N 46-4.

4 - Protective cap
- – Check for damage.

5 - Guide pin

6 - Self locking bolt, 35 Nm (26 ft lb)
- ♦ Always replace.

7 - Brake caliper
- ♦ Servicing,

⇒ Page N 47-1.
- ♦ Distinguishing features,

⇒ Page N 46-21.

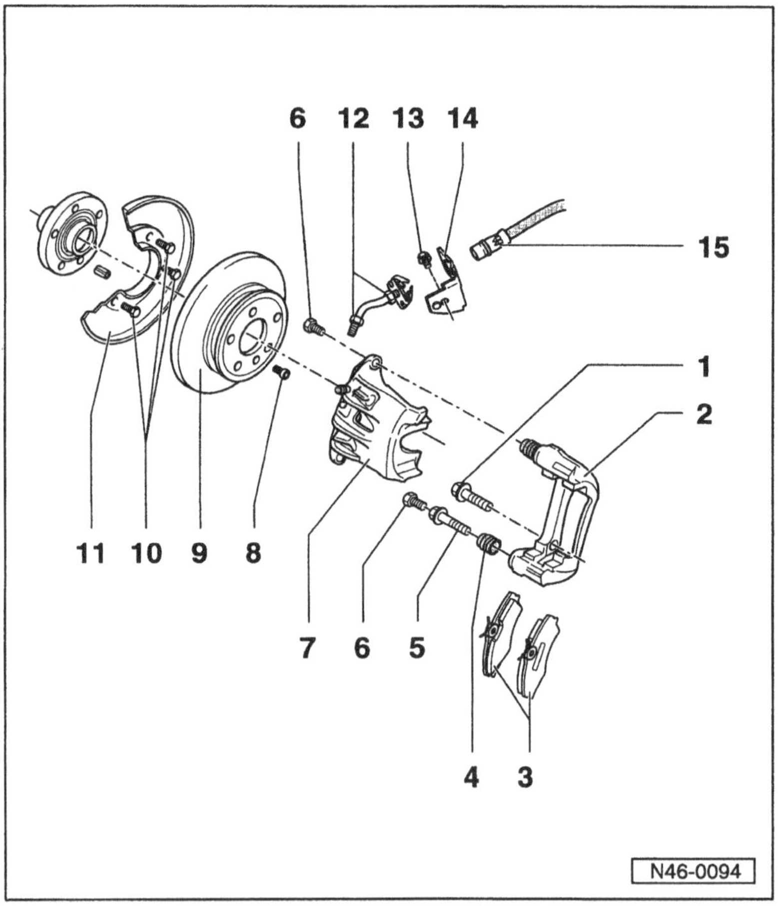

8 - Allen head bolt, 10 Nm (89 in. lb)

9 - Brake rotor
- When worn, always replace both sides.
- To remove, remove brake caliper first.
- Non-ventilated brake rotor diameter: 282 mm.
- Brake rotor thickness: 18 mm.
- Wear limit: 15 mm.

10 - Hex bolt, 8 Nm (71 in. lb)

11 - Splash plate

12 - Brake line with union connection

13 - Hex bolt, 10 Nm (89 in. lb)

14 - Bracket

15 - Brake hose

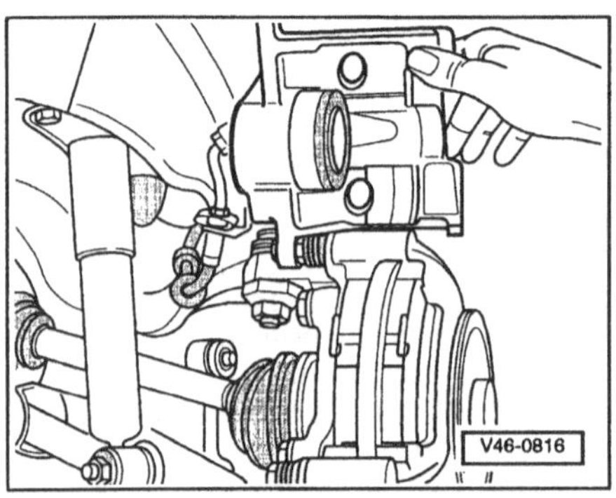

Brake pads, removing/installing

Removing

- Remove lower brake caliper housing bolt. Counter-hold guide pin when loosening.
- ◄ Swing brake caliper housing upward.
- Remove brake pads.
- Secure caliper to prevent it from hanging by brake hose.

Installing

Note:

Before installing new brake pads, press piston back into cylinder with resetting tool. Before pressing piston back, draw off brake fluid from the reservoir with a bleeder bottle. Otherwise, if reservoir has been topped up, brake fluid will overflow and cause damage.

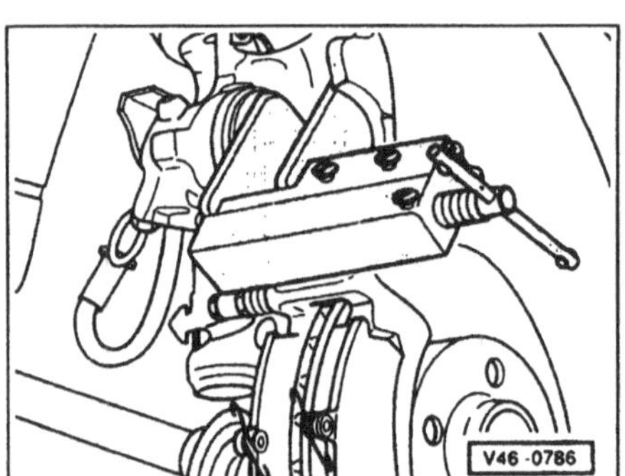

- ◄ Press piston back.
- Install brake pads.

– Swing brake caliper housing down and torque bolts to specification.

Notes:

- *Repair kit contains two self-locking Hex bolts which must be installed.*
- *After replacing brake pads, depress pedal firmly several times with vehicle stationary, so that the brake pads are properly seated in their normal operating position.*

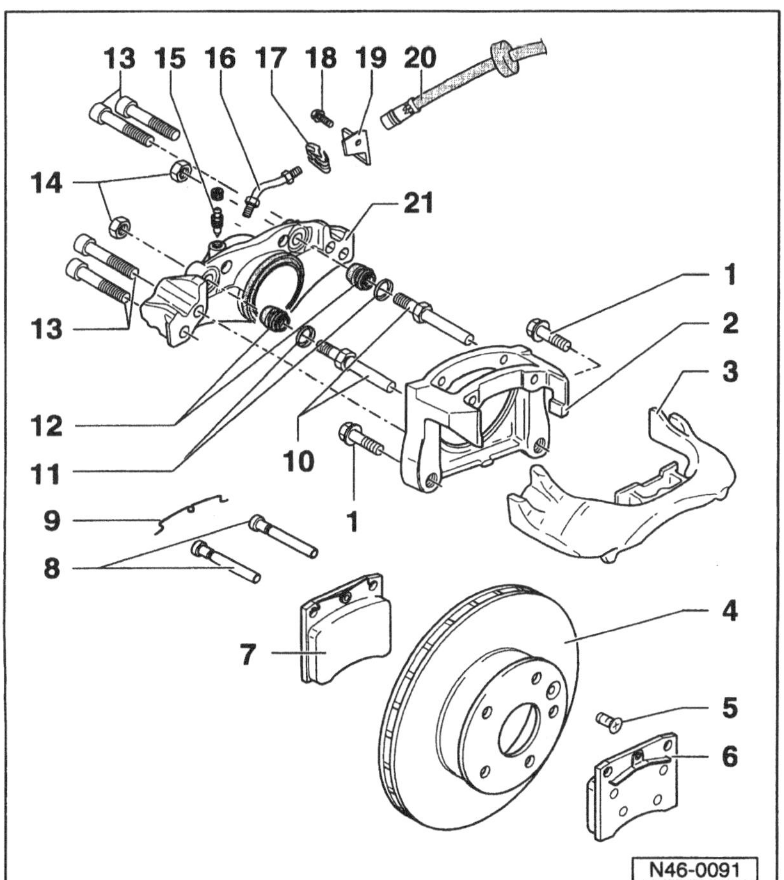

Front brake, assembly (Lucas RC 54 floating frame caliper)

Notes:

- *After replacing brake pads, depress brake pedal firmly several times, with vehicle stationary, to seat brake pads in their normal operating position.*
- *Use bleeder bottle to draw brake fluid from reservoir.*

> ***Warning!***
> ***Brake fluid is poisonous and must not be siphoned out by mouth through a hose!***

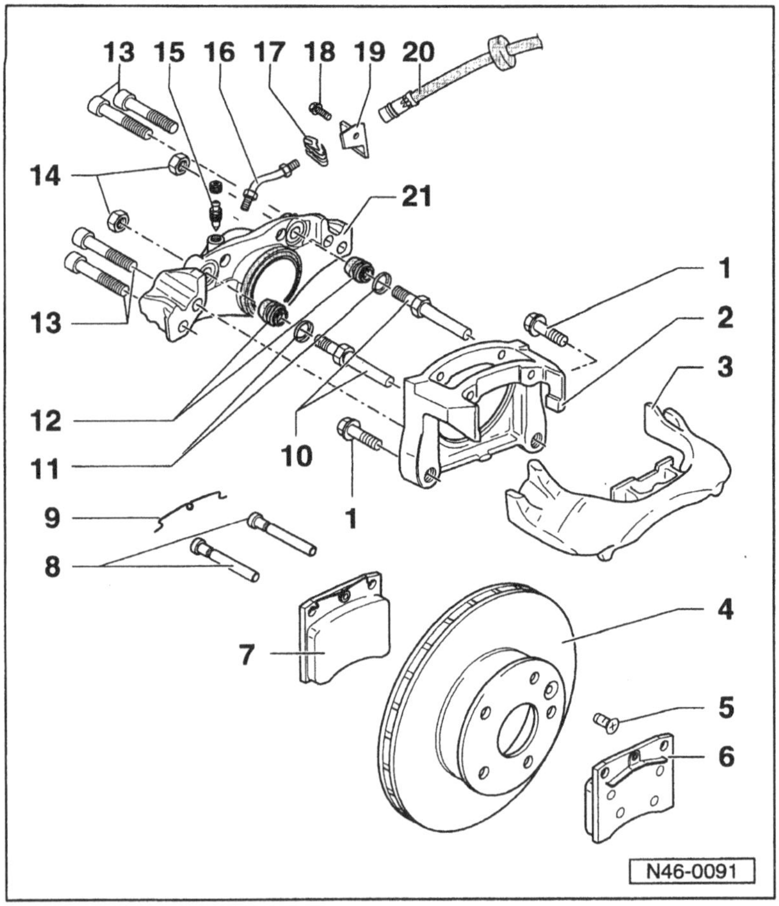

1 - Hex bolt, 280 Nm (207 ft lb)

2 - Brake backplate

3 - Bridge
- Bolt together with brake caliper housing,

⇒ Fig. 1.

4 - Brake disc
- When worn, always replace both sides.
- To remove, first remove lower brake caliper bolt and loosen upper bolt.
- Ventilated brake disc diameter: 280 mm (11 in.).
- Brake disc thickness: 24 mm.
- Wear limit: 20 mm.

5 - Hex socket head bolt, 10 Nm (89 in. lb)

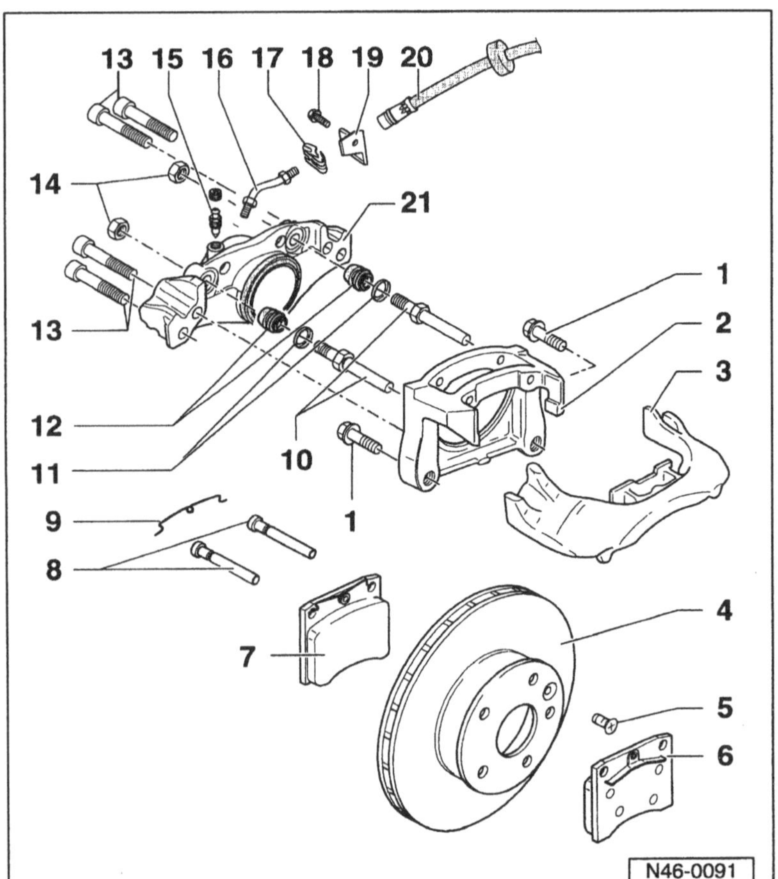

6 - Outer brake pad
- With adhesive foil on backing plate.
- When worn, always replace both sides.
- Removing/installing,

⇒ Page N 46-11.

- Pad thickness: 13 mm.
- Checking thickness,

⇒ Maintenance manual,

- Wear limit: 2 mm, not including backing plate (remaining pad material thickness).

7 - Inner brake pad
- When worn, always replace both sides.
- Removing and installing,

⇒ Page N 46-11.

- Pad thickness: 13 mm.
- Checking thickness,

⇒ Maintenance manual.

- Wear limit: 2 mm, not including backing plate (remaining pad material thickness).

8 - Pad locating pins

9 - Securing spring

10 - Guide pins

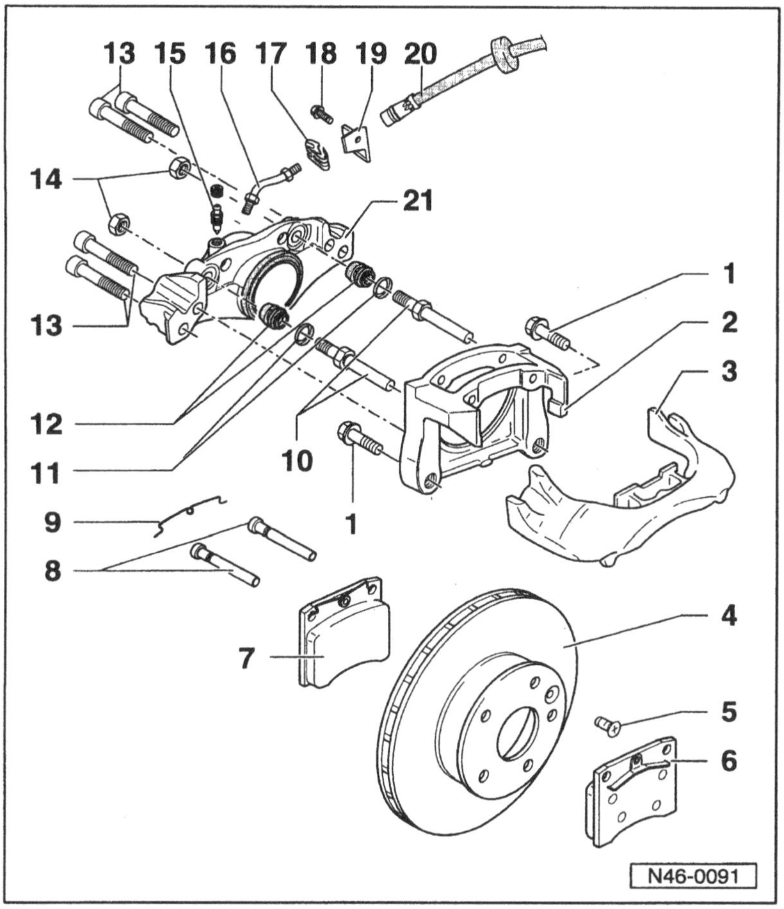

11 - **Metal ring**

12 - **Protective cap**
- Check for damage.

13 - **Multi-point socket bolt,** 90 Nm (66 ft lb).

14 - **Hex nut, 70 Nm (52 ft lb)**

15 - **Bleed screw**

16 - **Brake line with union connection**

17 - **Bracket**

18 - **Hex bolt, 10 Nm (89 in. lb)**

19 - **Bracket for brake hose**

20 - **Brake hose**

21 - **Brake caliper housing**
- Assembly,
⇒ Page N 47-5.
- Bolt together with bridge,
⇒ Fig. 1.
- Distinguishing features,
⇒ Page N 46-21.

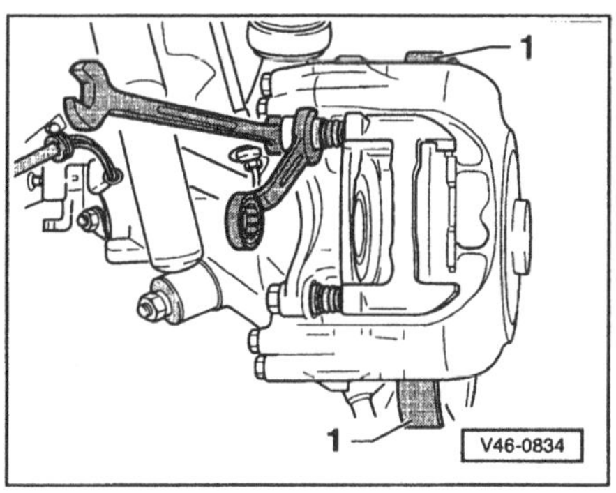

Fig. 1 Brake bridge/caliper housing, bolting together

– Position bridge with multi-point socket head bolts and tighten.

– Insert plate -1- (50 x 24 x 2 mm, locally available) between face of brake disc and lower side of bridge. This ensures sufficient clearance between bridge and brake rotor.

– Press brake caliper housing/bridge against plate and tighten guide pin hex nuts. See note below.

Notes:

- *First tighten leading side guide pins.*
- *Counter-hold at guide pins with an open end wrench, grind to fit, if necessary.*

– Remove plate and ensure brake rotor has sufficient clearance.

Tightening torques

Mult-point socket head bolt	90 Nm (66 ft lb)
Nuts for guide pins	70 Nm (52 ft lb)

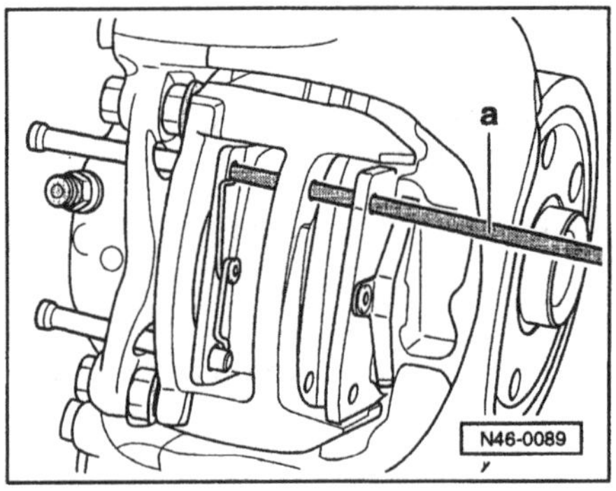

Brake pads, removing and installing

Removing

- Remove securing spring.
- Drive out pad retaining pin with drift -a-.
- Remove inner brake pad.
- Move frame.
- Remove outer brake pad from bridge pad, holding surface with a screwdriver or similar tool.
- Press piston back.

Installing

Note:

Before inserting new brake pads, press the piston back into the cylinder with a piston resetting tool. Before pressing the piston back, draw off brake fluid from the reservoir with a bleeder bottle. Otherwise, particularly if the reservoir has been topped up, fluid will overflow and cause damage.

> ***Warning!***
>
> ***Brake fluid is poisonous and must not be siphoned out by mouth through a hose!***

- Clean brake caliper housing. Bonding surface for brake pad must **be free of adhesive and grease residues**
- Install inner brake pad.
- Install pad retaining pins to hold inner brake pad.
- Pull protective foil off outer brake pad.
- Carefully insert outer brake pad.
- Push pad retaining pins through fully, and install securing spring.

Notes:

- *When installing the outer brake pad, ensure it does not stick to the brake caliper housing before it is in the correct position.*
- *Do not damage adhesive surface.*
- *After installing new brake pads, depress the brake pedal firmly two or three times and hold pedal fully depressed for 5 to 10 seconds, so outer pad is firmly bonded to brake caliper housing. This ensures outer pad noise-free braking.*

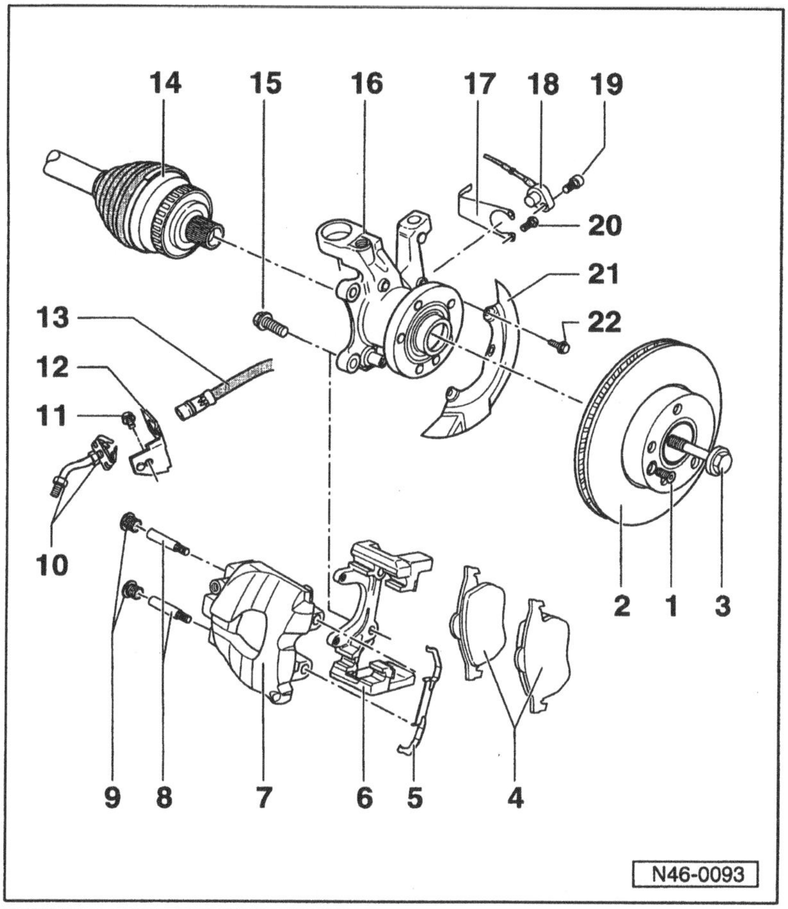

Front brake, assembly (Teves/Ate)

Notes:

- *After replacing brake pads, depress brake pedal firmly, several times,with vehicle stationary, to seat brake pads in their normal operating position.*
- *Use bleeder bottle to draw brake fluid from reservoir.*

> ***Warning!***
> ***Brake fluid is poisonous and must not be siphoned out by mouth through a hose!***

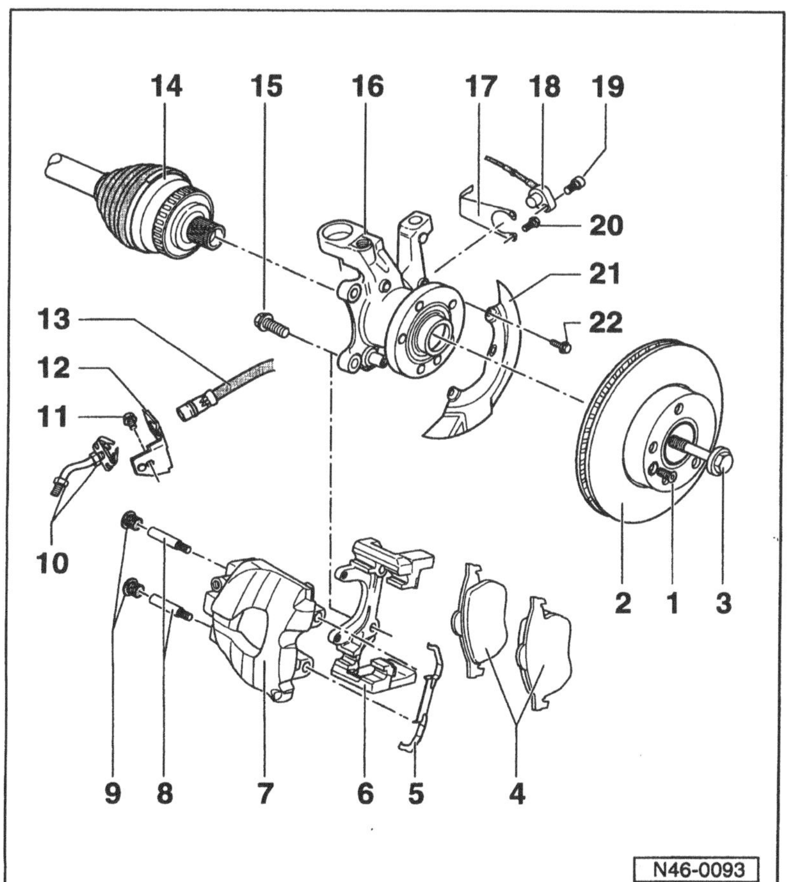

1 - Hex socket head bolt, 10 Nm (89 in. lb)

2 - Brake rotor
- Brake rotor thickness: 24 mm.
- Wear limit: 20 mm.
- When worn, always replace both sides.
- Never remove brake rotors from hub by force. If necessary, use penetrating fluid, otherwise brake discs can be damaged.

3 - Hex bolt
- 150 Nm (111 ft lb) and turn 90° further.
- Replace at reassembly.

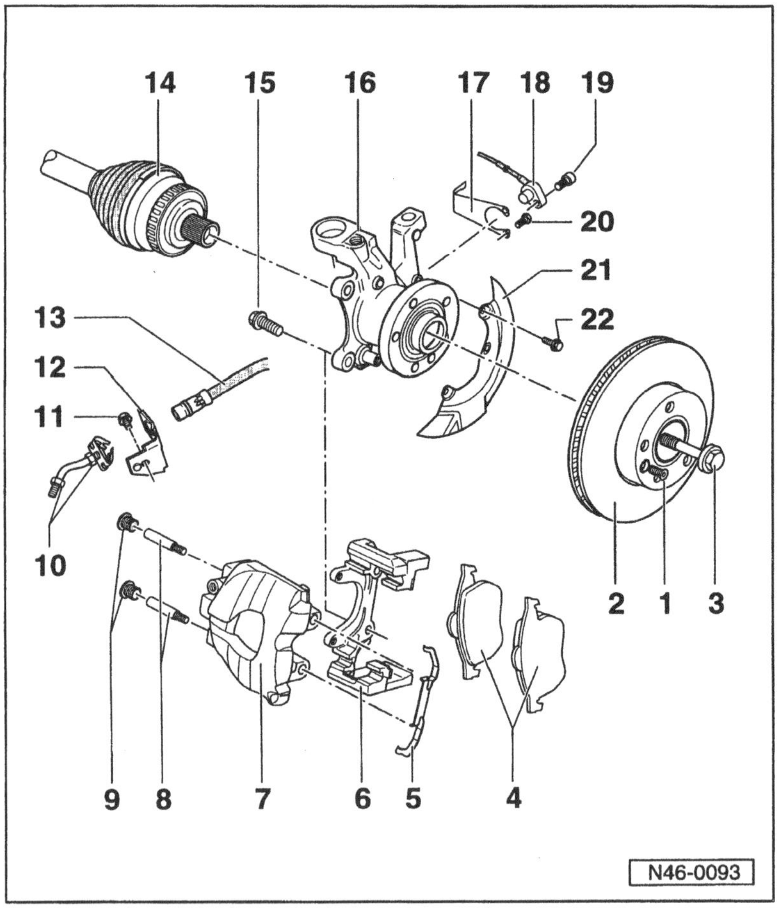

4 - Brake pads
- Pad thickness: 13.5 mm.
- Brake pad, with retainer spring, seats against piston.
- Brake pad with adhesive foil on back plate, outer side.
- Wear limit: 2 mm, not including backing plate (remaining pad material thickness).
- Checking thickness,

⇒ Maintenance manual.
- Always replace both sides.

5 - Retaining spring
- Insert in both brake caliper housing drillings.

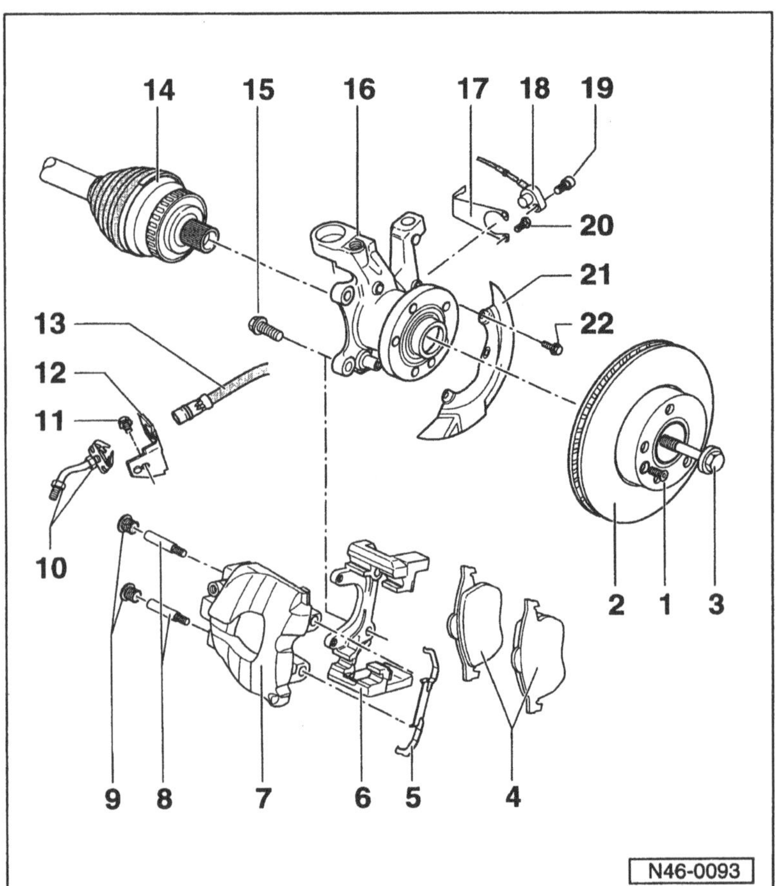

6 - Brake carrier
- Supplied as replacement part, assembled with grease on guide pins.
- If protective caps are damaged, use repair kit.

7 - Brake caliper
- Removing/installing,

⇒ Page N 46-18.

8 - Guide pins, 25 Nm (18 ft lb)
- Removing/installing,

⇒ Page N 46-18.

9 - Protective cap

10 - Brake line with union connection

11 - Hex bolt, 10 Nm (89 in. lb)

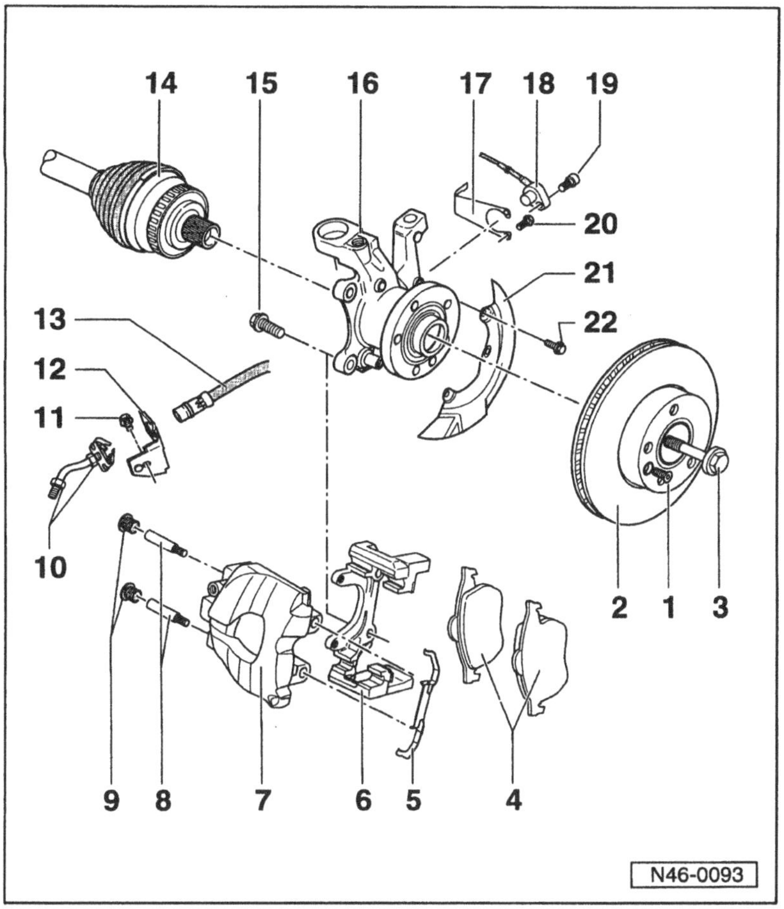

12 - Bracket

13 - Brake hose

14 - Impulse rotor, ABS speed sensor

15 - Hex bolt, 280 Nm (207 ft lb)

16 - Wheel bearing housing

17 - Bracket

18 - ABS speed sensor
- Before installing sensor, clean sensor mounting hole inner surface and coat with lubricant G 000 650.

19 - Allen bolt, 10 Nm (89 in. lb)

20 - Hex bolt

21 - Splash plate

22 - Hex bolt, 8 Nm (71 in. lb)

Brake pads, removing and installing

Removing

Note:

If reusing brake pads mark and return to same location.

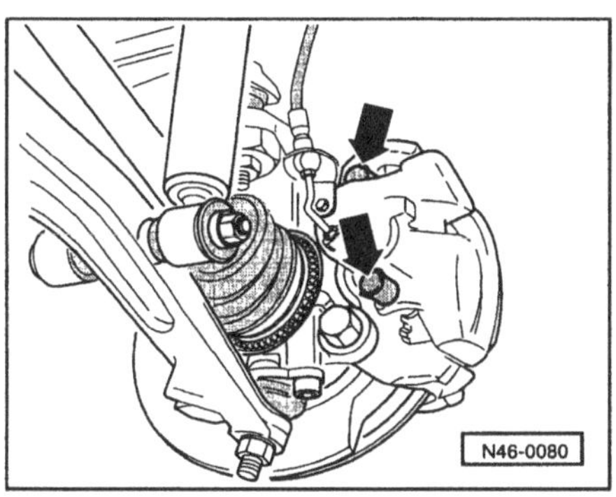

◄ – Remove protective caps.

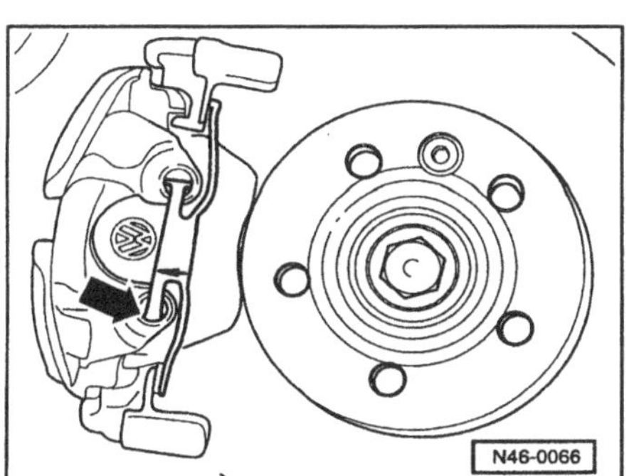

◄ – Unhook brake pad retaining spring from brake caliper housing and remove.

Note:

Release retaining spring from housing -arrow- by pushing in direction of arrow (arrow embossed on retaining spring) and tilt, then remove.

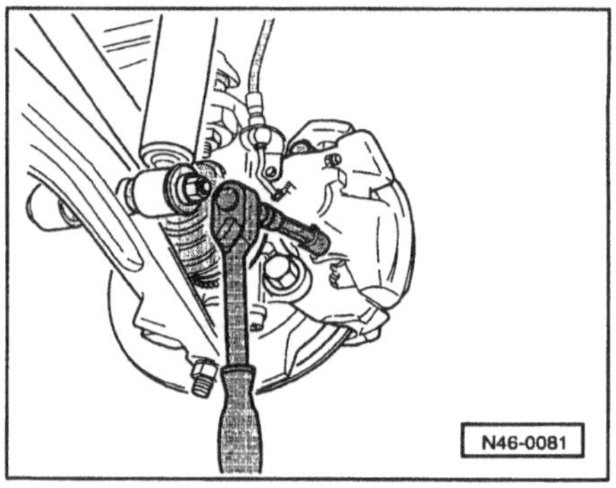

◄ – Remove both guide pins from brake caliper.

– Take off brake caliper housing and lay to one side so that weight of caliper does not stress or damage brake hose.

– Take brake pads out of brake caliper housing or from brake carrier.

– Clean brake caliper housing, in particular bonding surface for brake pad. This area must **be free of adhesive and grease residues.**

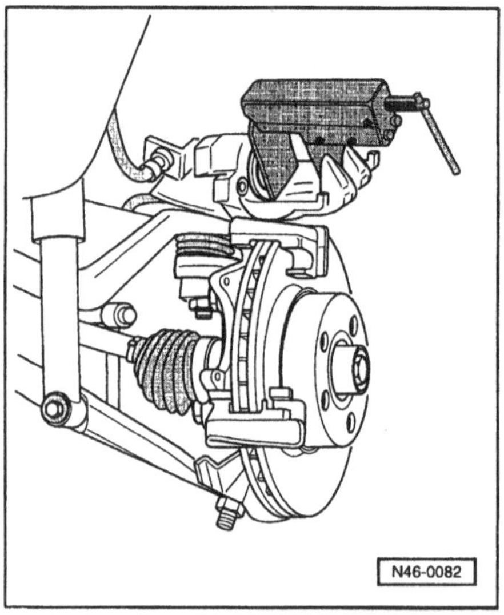

Installing

Note:

Before installing new brake pads, press piston back into the cylinder with the resetting tool. Before pressing the piston back, draw off brake fluid from the reservoir with a bleeder bottle. Otherwise, particularly if reservoir has been topped up, fluid will overflow and cause damage.

◄ – Press piston back.

– Insert brake pad with retaining spring in brake caliper housing (piston).

– Install outer brake pad to brake carrier.

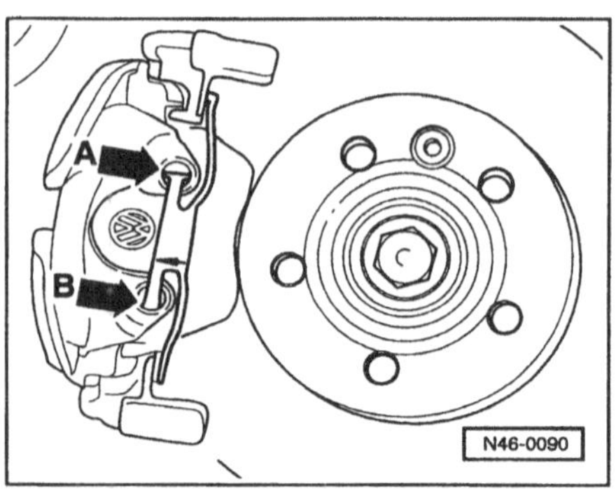

– Bolt brake caliper housing with both guide pins to brake carrier. Torque to 25 Nm (18 ft lb).

– Install both protective caps.

◄ – Install one spring arm under brake carrier and insert one tab in housing retaining hole -arrow A-.

– Tension second spring arm under brake carrier and install second tab into housing retaining hole -arrow B-.

Notes:

- *When inserting the outer brake pad, ensure that it does not stick to the brake caliper housing before it is in the correct installation position.*
- *Do not damage adhesive surface.*
- *After installing new brake pads, depress the brake pedal firmly, two or three times, and hold the pedal fully depressed for 5 ... 10 seconds so that the outer pad is firmly bonded to the brake caliper housing. This ensures that the outer pad is reliably bonded and achieves noise-free braking.*

Front brake caliper distinguishing features

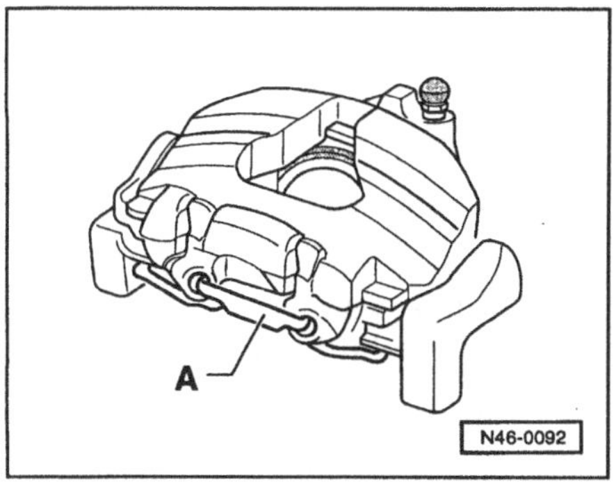

Note:

Position of optional equipment number,⇒ Page N 00-3.

◄ **15" brake caliper (optional equipment number 1LU)**

Distinguishing features:

- Hook retaining spring -A- into brake caliper housing.
- Further technical data, ⇒ Page N 00-2.

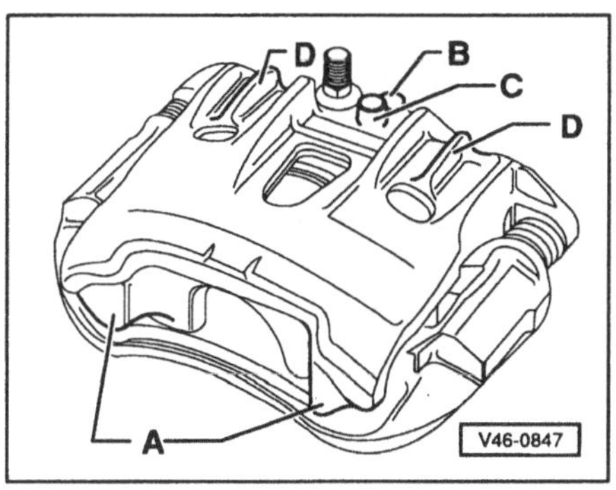

◄ **15" brake caliper (optional equipment number 1LP)**

Distinguishing features:

- Symmetric balance mass -A-.
- VW symbol -B- behind projection -C-.
- Reinforcement rips -D- extend to end.
- Further technical data, ⇒ Page N 00-2.

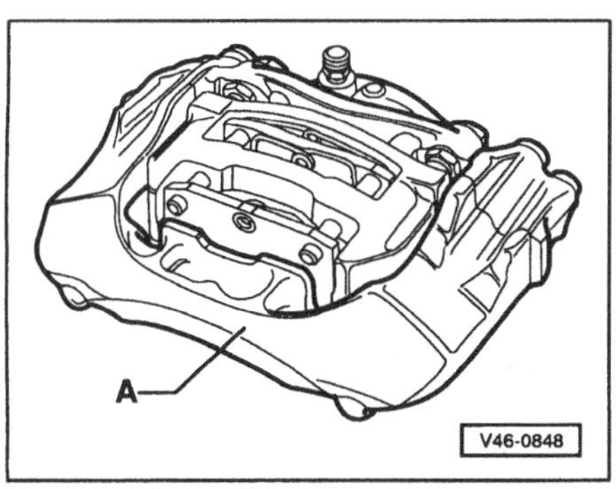

◄ **15" frame brake caliper (optional equipment number 1LE)**

Distinguishing features:

- Frame -A- is closed.
- Brake pads visible from above.
- Further technical data, ⇒ Page N 00-2.

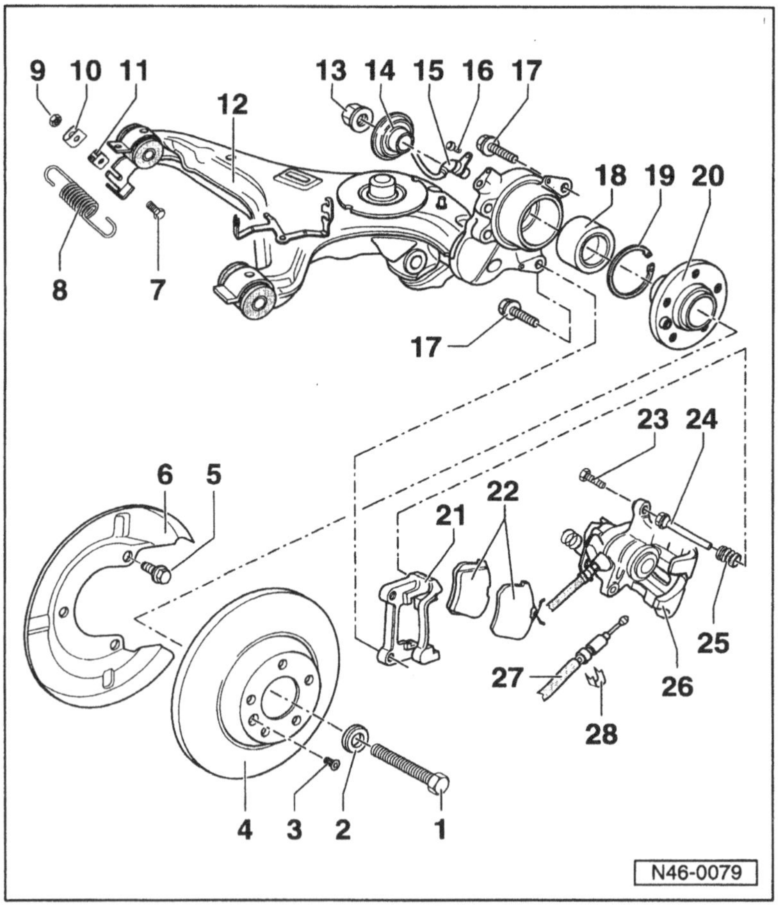

Rear wheel brakes (disc brakes), servicing

1 - **Hex bolt**
  - Replace each time after removing.

  Tightening torque on vehicles with four wheel drive:
  - 150 Nm (111 ft lb) and turn 90° further (1/4 turn).
  - All wheel drive vehicles: loosen and tighten only when vehicle is standing on its wheels (danger of accident).

2 - **Washer**

3 - **Hex socket head bolt, 10 Nm (7 ft lb)**

4 - **Brake disc**
  - Thickness: 12 mm.
  - Wear limit: 10 mm.
  - When worn, always replace both sides.

5 - **Hex bolt, 10 Nm (7 ft lb)**

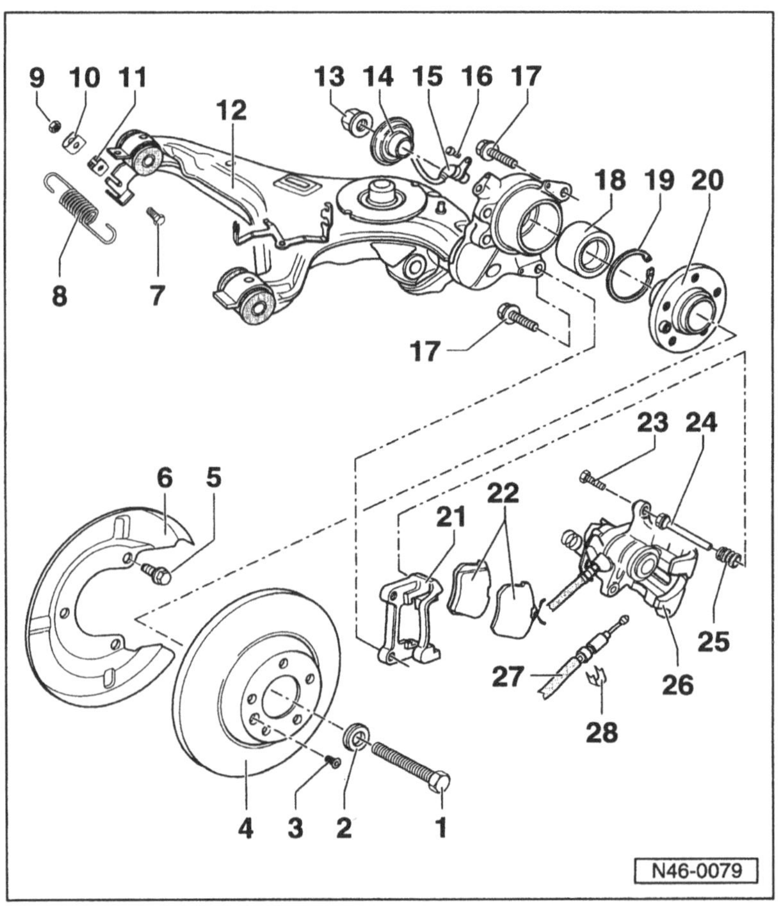

6 - **Splash plate**

7 - **Hex bolt**

8 - **Brake pressure regulator spring**
  - Adjusting brake pressure regulator, ⇒ Page N 47-26.

9 - **Hex nut, 20 Nm (15 ft lb)**

10 - **Retaining plate**
  - Mark installation position before removing.

11 - **Bracket**

12 - **Control arm**
  - Removing and installing, ⇒ Page N 42-1.

13 - **Self-locking hex nut, 200 Nm (148 ft lb)**
  - Always replace.

14 - **Tensioning disc**
  - Vehicles with ABS with impulse wheel for speed sensor.

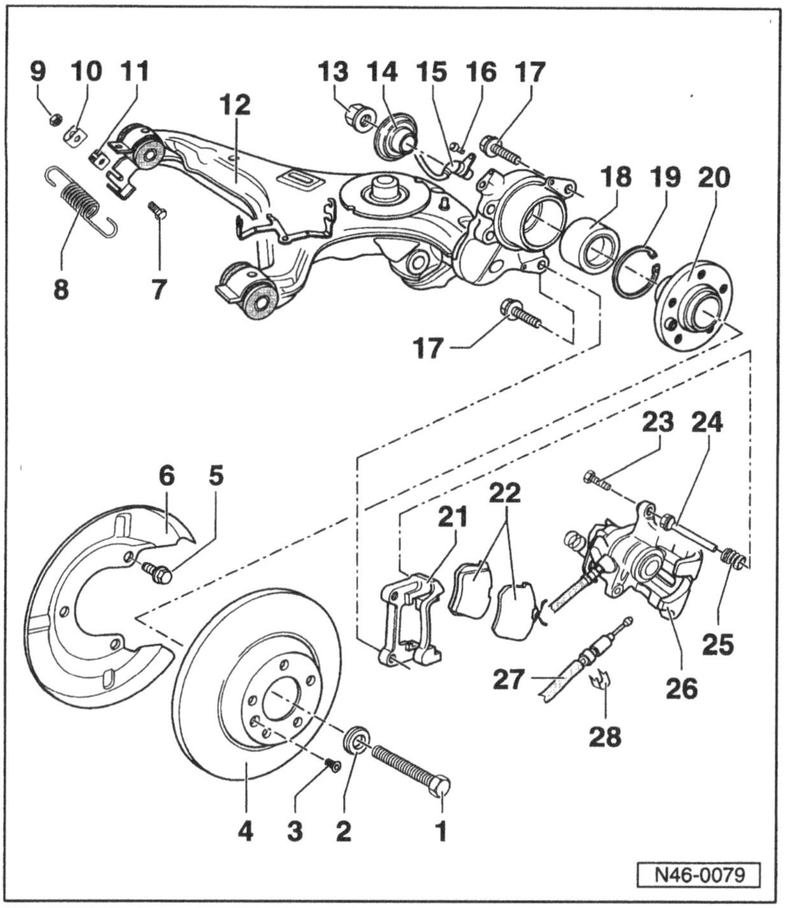

15 - Speed sensor

16 - Hex socket head bolt, 10 Nm (7 ft lb)

17 - Hex bolt, 170 Nm (125 ft lb)

18 - Wheel bearing

19 - Circlip
- ♦ Ensure seated correctly.

20 - Hub

21 - Brake carrier with guide pins and protective cap
- ♦ Supplied as replacement part, assembled with sufficient grease on guide pins.
- ♦ If protective caps or guide pins are damaged, use repair kit. Use grease packet supplied to lubricate guide pins.

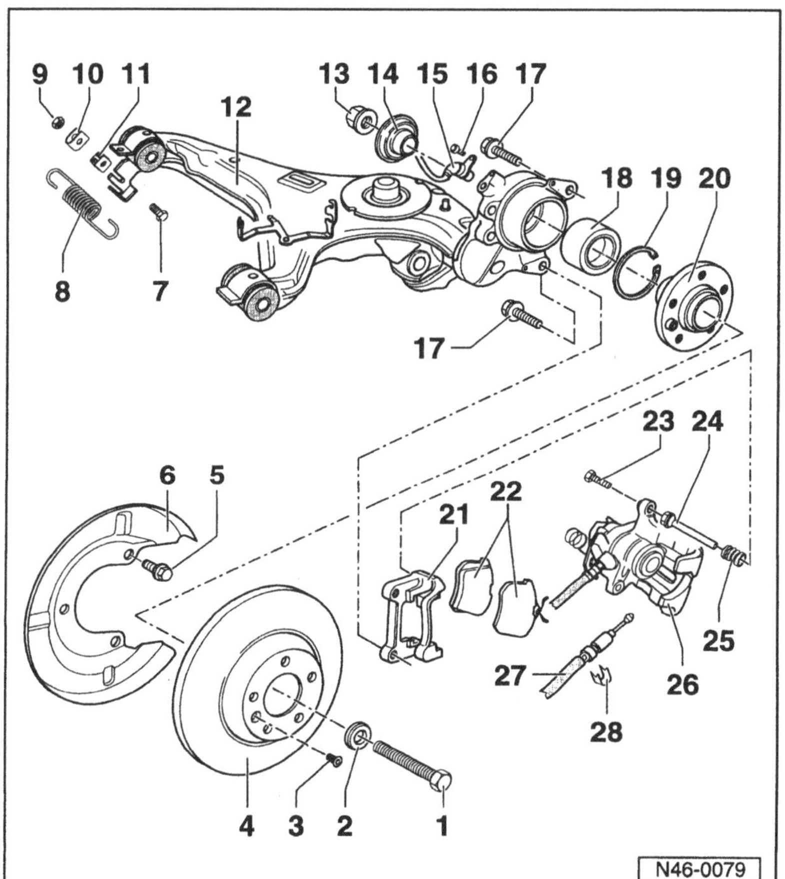

22 - Brake pads
- ♦ Thickness: 12 mm.
- ♦ Checking thickness,

⇒ Maintenance manual.
- ♦ Wear limit: 2 mm, not including backing plate (remaining pad thickness).
- ♦ Always replace both sides.
- ♦ Removing and installing, ⇒ Page N 46-27.

23 - Self-locking hex bolt, 35 Nm (26 ft lb)
- ♦ Always replace.
- ♦ When loosening and tightening, counter–hold on guide pin.

24 - Guide pin

25 - Protective cap

26 - Brake caliper
- ♦ Servicing, ⇒ Page N 47-14.

27 - Parking brake cable
- ♦ Adjusting parking brake, ⇒ Page N 46-30.

28 - Spring clip

Rear brake pads, removing and installing

Note:

Mark brake pads when removing, if they are to be reused. Re-install in their original position to prevent uneven braking!

Removing:

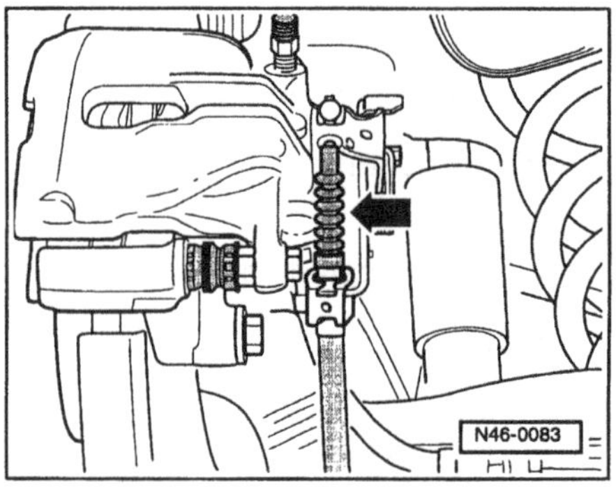

– Remove wheels.

◄ – Remove parking brake cable from brake caliper.

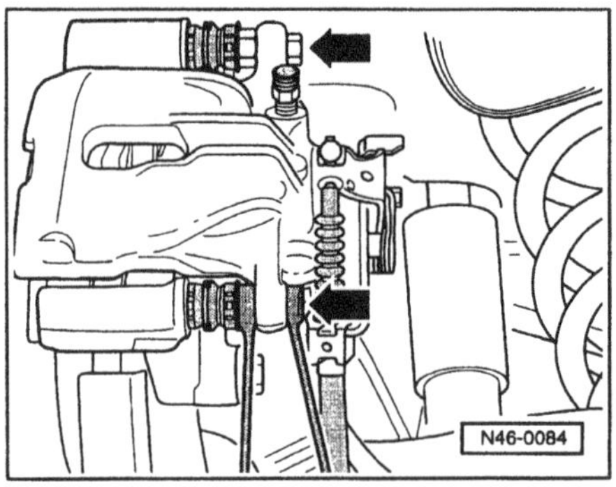

◄ – Remove securing bolts for brake caliper housing -arrows-.

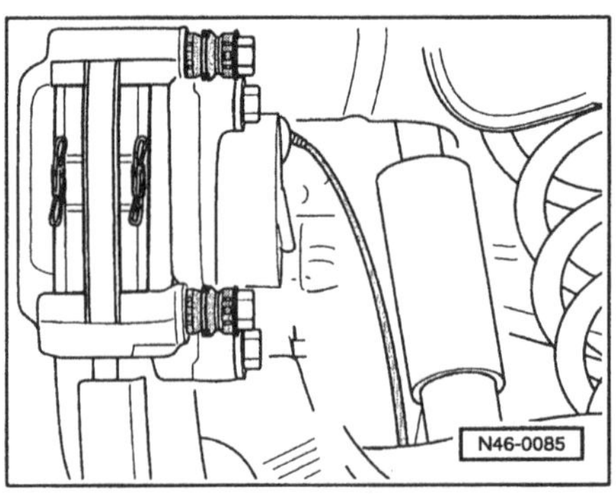

◄ – Remove brake pads.

Installing:

Note:

Before installing new brake pads, turn the piston back into the cylinder. Before turning the pistion back, remove some brake fluid from the reservoir. Use a bleeder bottle or a plastic bottle which is only used for brake fluid. If the reservoir has been topped up, the fluid may overflow and cause damage.

> ***Warning!***
>
> ***Brake fluid is poisonous and must not be siphoned out by mouth through a hose.***

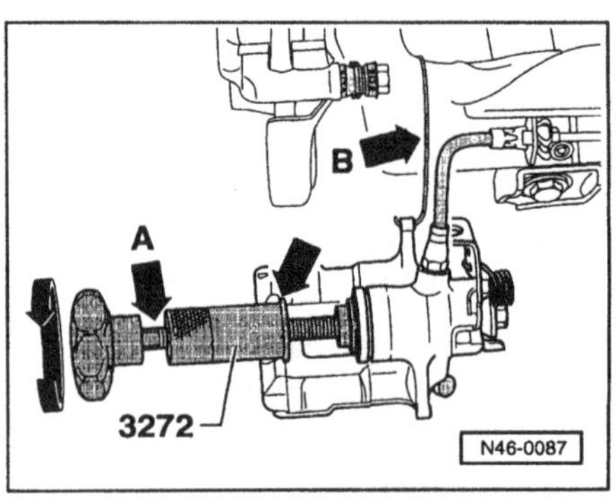

◄ – Move piston in by turning knurled wheel of special tool 3272 clockwise.

Notes:

- *Install special tool 3272 ensuring that the piston protective cap will not be damaged and that the collar -arrow- of the tool is seated against the brake caliper.*
- *If the piston is difficult to move, use a 13 mm open–end wrench on the flats -arrow A- provided for this purpose.*
- *If the piston is pushed back with a piston resetting tool or by operating the foot brake, the automatic adjustment in the brake caliper will be destroyed.*

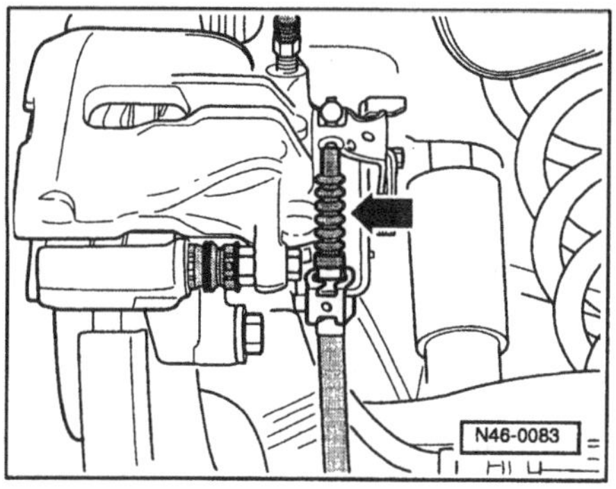

- Install brake pads.
- Secure brake caliper housing with new self-locking bolts.
- ◄ Attach parking brake cable -arrow- to brake caliper housing.
- Adjusting parking brake, ⇒ Page N 46-30.

Note:

- ♦ *After each change of brake pads, firmly depress brake pedal several times, with vehicle stationary, so that the brake pads are properly seated in their normal operating position.*
- ♦ *After changing brake pads, check brake fluid level.*

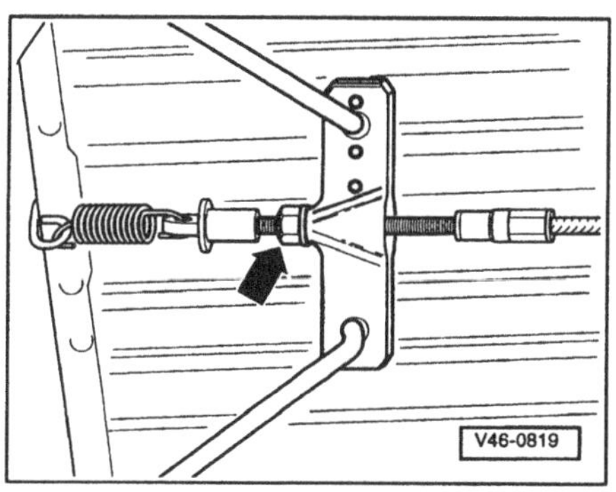

Parking brake, adjusting

Note:

Due to the automatic rear wheel brake adjustment, there is no requirement to adjust the parking brake. Adjustment is only necessary after replacing parking brake cables, brake calipers, brake pads, or brake discs.

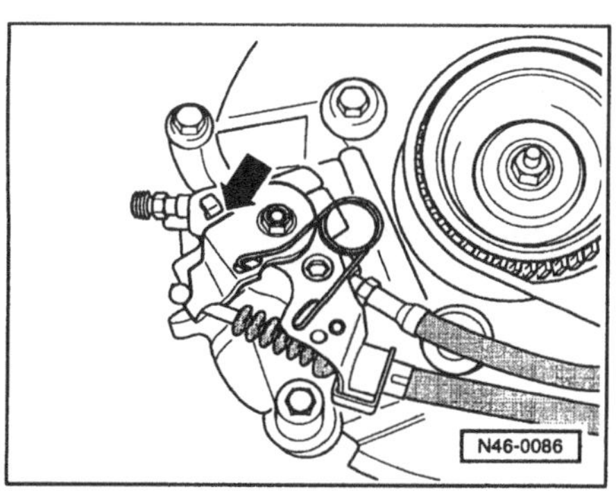

Parking brake lever in rest/off position.

- ◄ Tighten adjusting nut (illustration V46-0819) until lever -arrow- lifts off brake caliper stop.
- ● Maximum 1.0 mm gap to stop is permissible.
- Pull parking brake on and then release again.
- Check that both wheels turn freely.

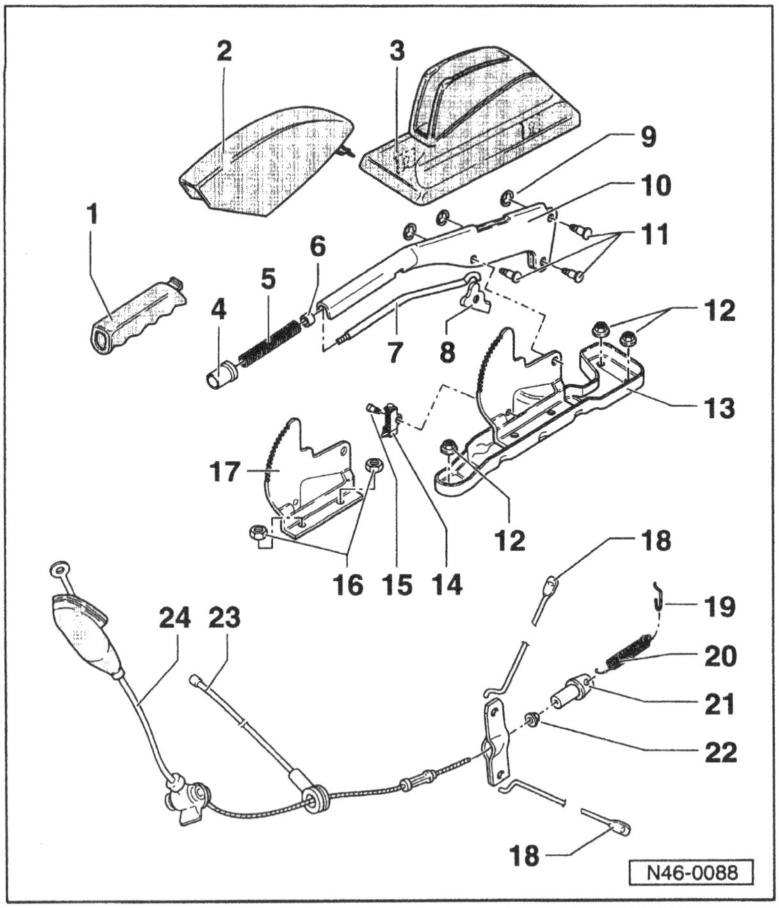

Parking brake lever, removing and installing

1 - **Hand grip**
   - ♦ To remove: press tab below hand grip downward, remove hand grip toward front.

2 - **Parking brake lever trim**

3 - **Trim**

4 - **Push button**

5 - **Spring**
   - ♦ Installation position, ⇒ Fig. 1.

6 - **Spring sleeve**
   - ♦ Installation position, ⇒ Fig. 1.

7 - **Pawl rod**

8 - **Pawl**

9 - **Securing clip**

10 - **Parking brake lever**

11 - **Pin**

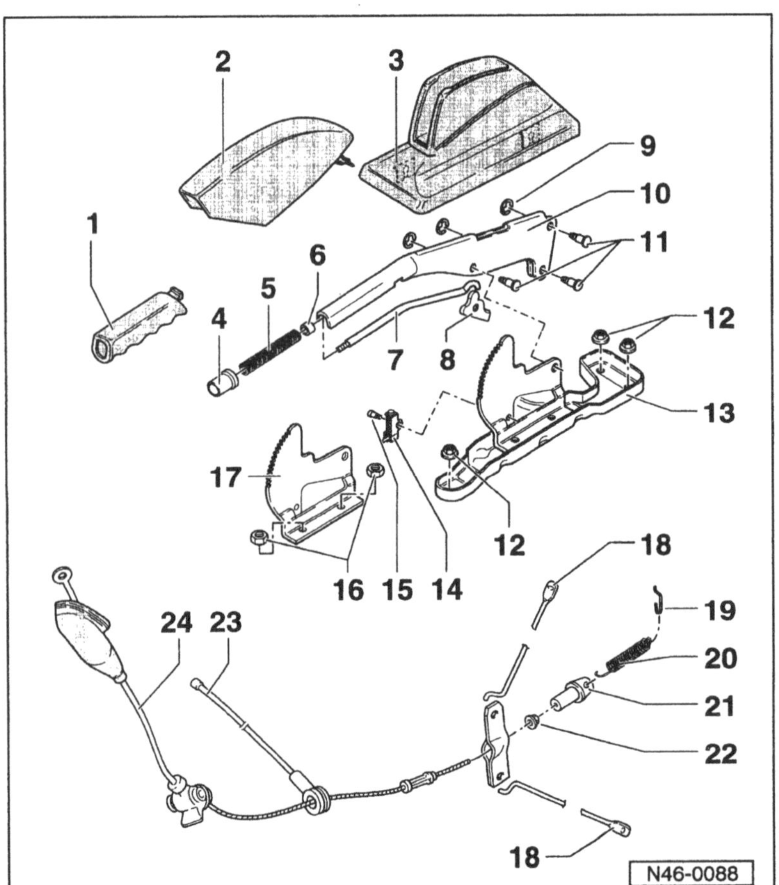

12 - **Self-locking hex nut, 25 Nm (18 ft lb)**
   - ♦ Always replace.

13 - **Parking brake carrier**

14 - **Contact switch**

15 - **Expanding rivet**

16 - **Self-locking hex nut, 25 Nm (18 ft lb)**
   - ♦ Always replace.

17 - **Toothed segment**
   - ♦ Not applicable for US/Canada.

18 - **Pull rod**

19 - **Bracket**

20 - **Return spring**

21 - **Spring holder**

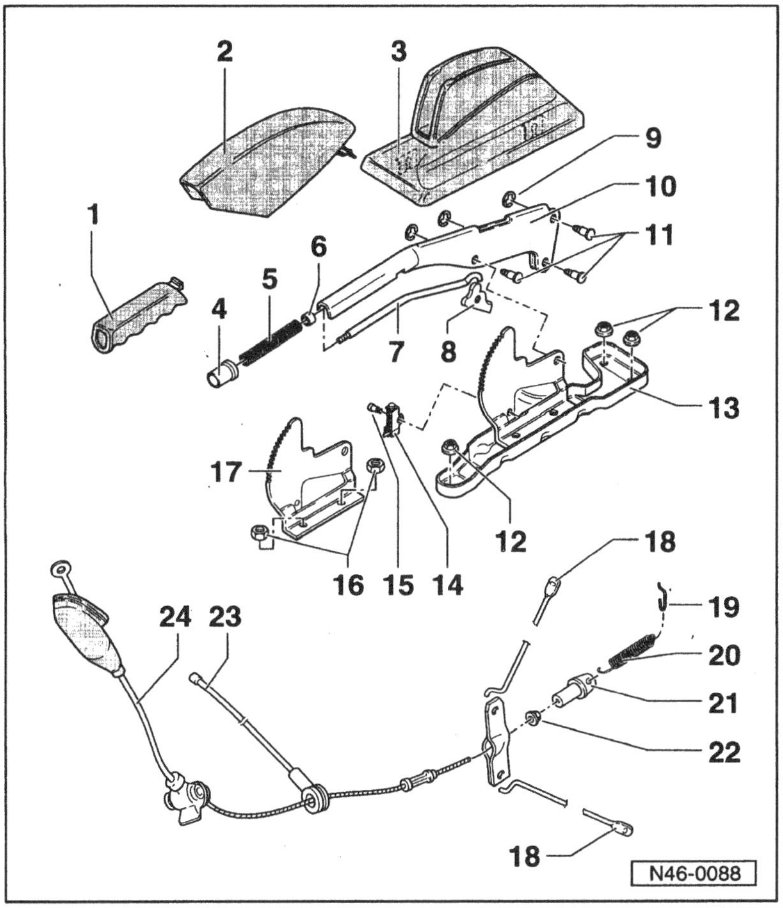

22 - **Adjusting nut**
- Adjusting parking brake, ⇒ Page N 46-30.

23 - **Tensioning cable**
- Only installed on dropside vehicles (pick–ups).

24 - **Parking brake cable**

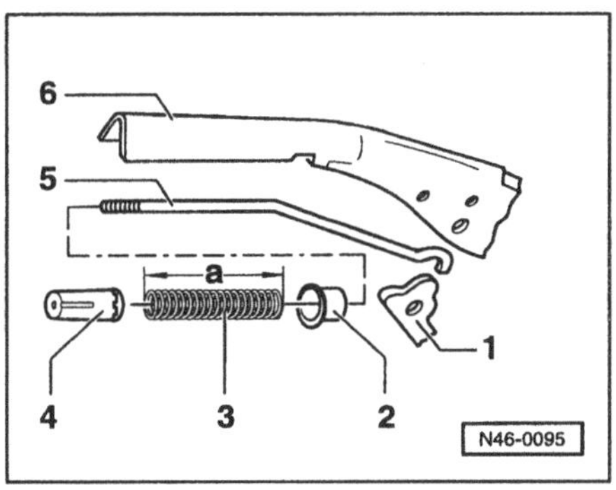

◄ **Fig. 1 Spring sleeve installation position**

1 - Pawl

2 - Spring sleeve

3 - Spring
Dimension -a- approximately 126 mm (5 in.).
When installing, slide spring sleeve -2- onto spring -3-.

4 - Push button
Screw onto stop.

5 - Pawl rod

6 - Parking brake lever

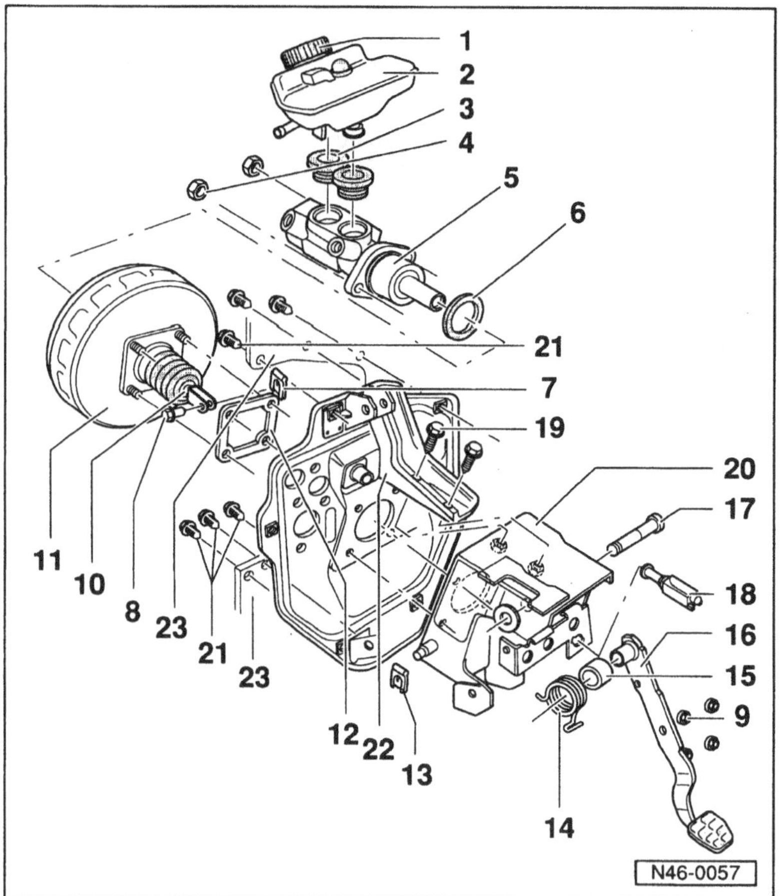

Brake pedal/brake servo/brake master cylinder, removing and installing

> ***Caution!***
>
> ***The brake pedal travel must not be restricted by additional floor coverings.***

Note:

Grease all mountings before installing with white lubricating paste, Part No. AOS 126 000 06.

1 - Cap

2 - Brake fluid reservoir
- ♦ With strainer
- ♦ Contact pin in reservoir on vehicles with a brake fluid level warning device.

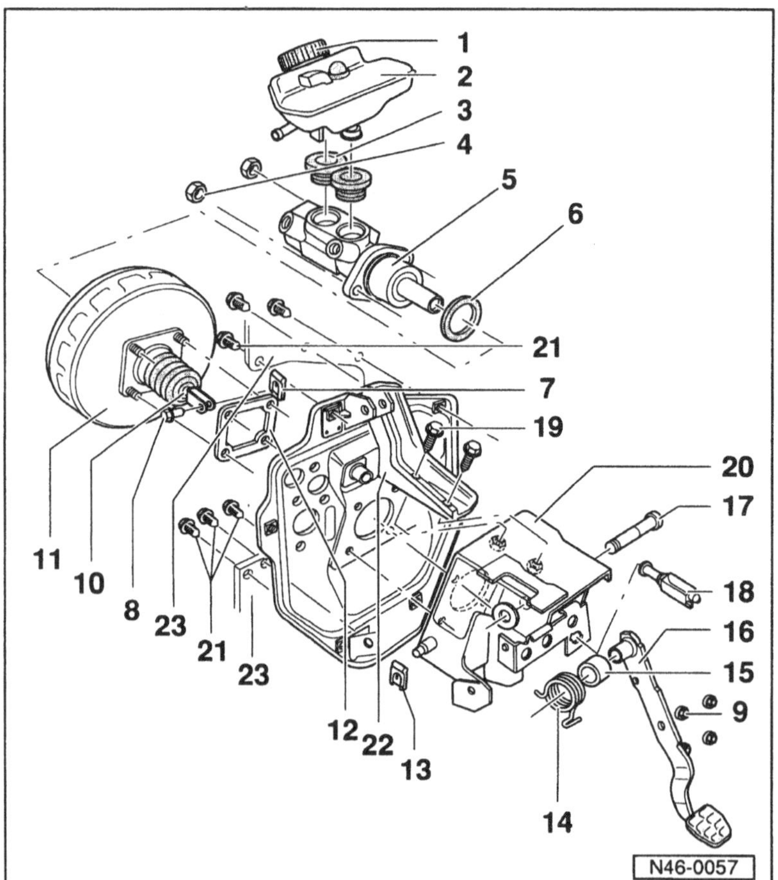

3 - Sealing plug
- ♦ Moisten with brake fluid and press into reservoir.

4 - Hex nut, 20 Nm (15 ft lb)

5 - Brake master cylinder
- ♦ Cannot be repaired. If faulty, replace complete.

6 - Sealing ring
- ♦ Always replace.

7 - Securing clip
- ♦ Ensure secure seating.

8 - Pin

9 - Self-locking hex nut, 25 Nm (18 ft lb)
- ♦ Always replace.

10 - Brake servo push rod clevis

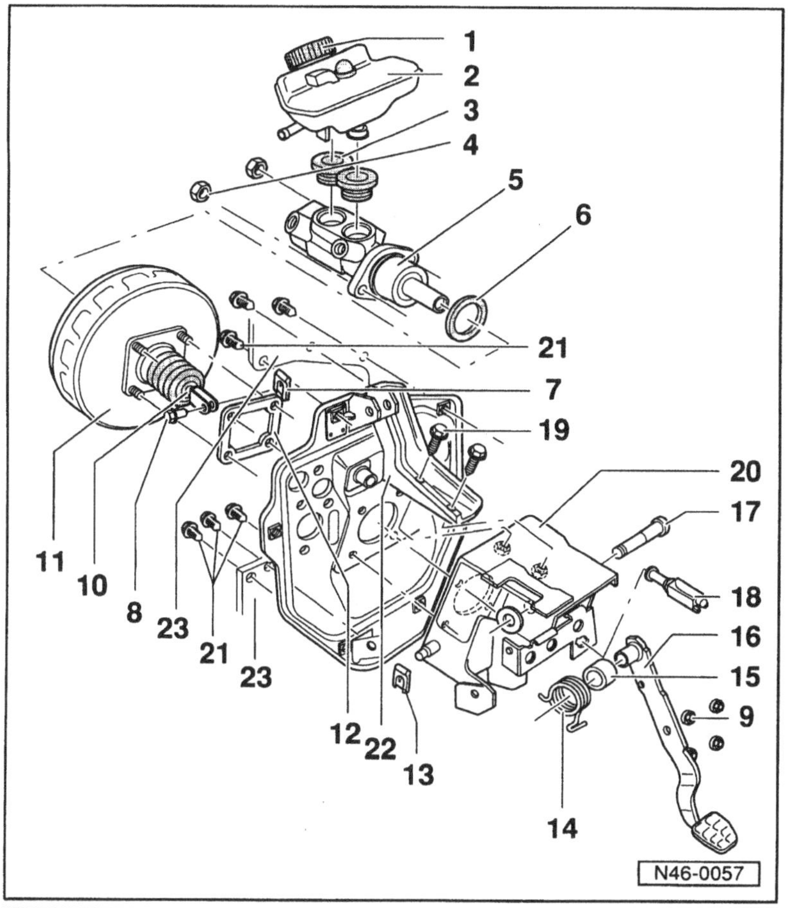

11 - Tandem brake servo

- Before removing tandem brake servo unit, remove battery and disconnect brake lines
- On vehicles with gas engines vacuum required is taken from the intake manifold.
- On Diesel engines an exhauster is installed to create required vacuum.
- Functional check:
  - With engine switched off, depress brake pedal firmly several times (to exhaust vacuum in unit).
  - Now depress brake pedal with average foot pressure, hold and start engine. If servo unit is working properly, pedal will be felt to give slightly under foot (servo assistance becomes effective).
- If faulty replace complete.
- Non-return valve located in vacuum hose

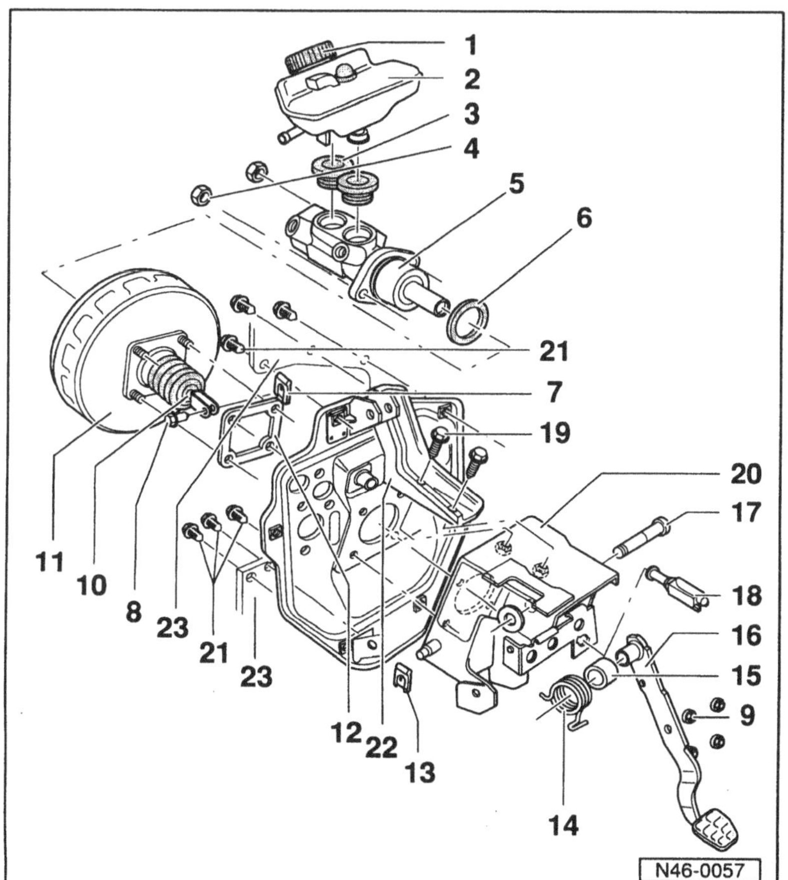

12 - Seal

- For tandem brake servo

13 - Securing clip

14 - Brake pedal return spring

15 - Brake pedal mounting bushing

16 - Brake pedal

17 - Brake pedal pivot pin

18 - Brake Light Switch -F-

- Before installing brake light switch, connect brake pedal to brake servo push rod
- Before installing pull plunger out completely
- Guide switch through mounting hole and secure by turning clockwise 90°

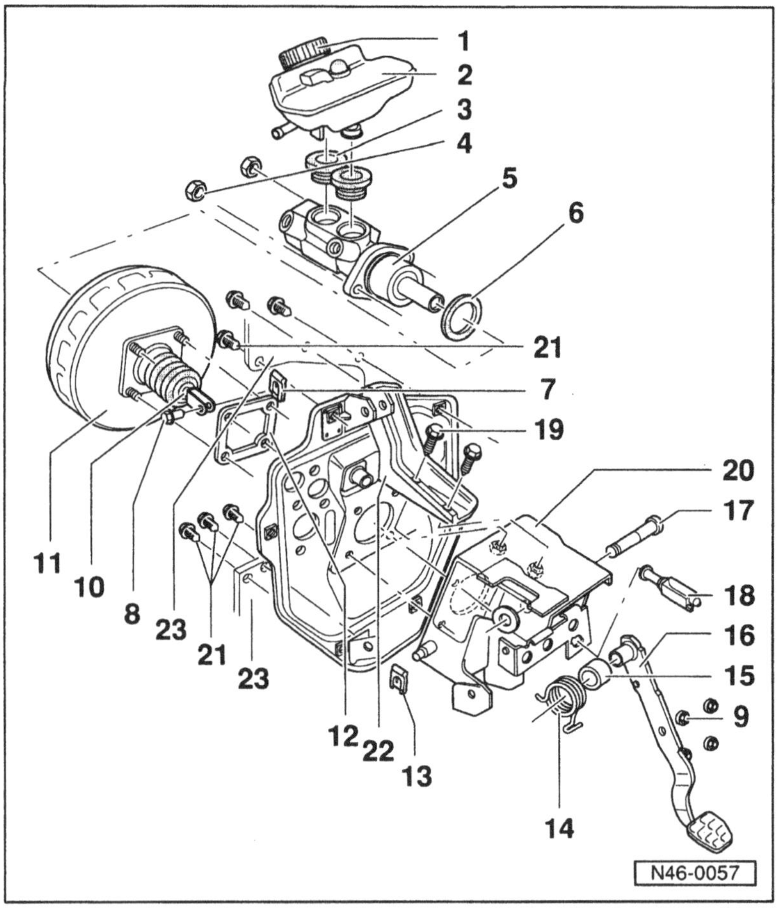

19 - Hex bolt, 25 Nm (18 ft lb)

20 - Mounting bracket

21 - Hex bolt, 25 Nm (18 ft lb)

22 - Mounting plate

23 - Bulkhead

Brake pedal, removing and installing

Note:

- *Grease all mounting and contact surfaces with MoS_2 grease.*
- *To remove brake servo, disconnect brake pedal at clevis only.*

Before removing brake pedal, remove trim:

– Remove brake light switch.

– Remove brake pedal clevis pin connection securing clip.

– Remove pin.

– Remove brake and clutch pedal pivot pin securing clip.

– Remove clutch pedal, ⇒ Repair group 30.

– Remove pivot pin.

– Remove brake pedal.

Installing

– Install in reverse order of removal.

– Adjust brake light switch.

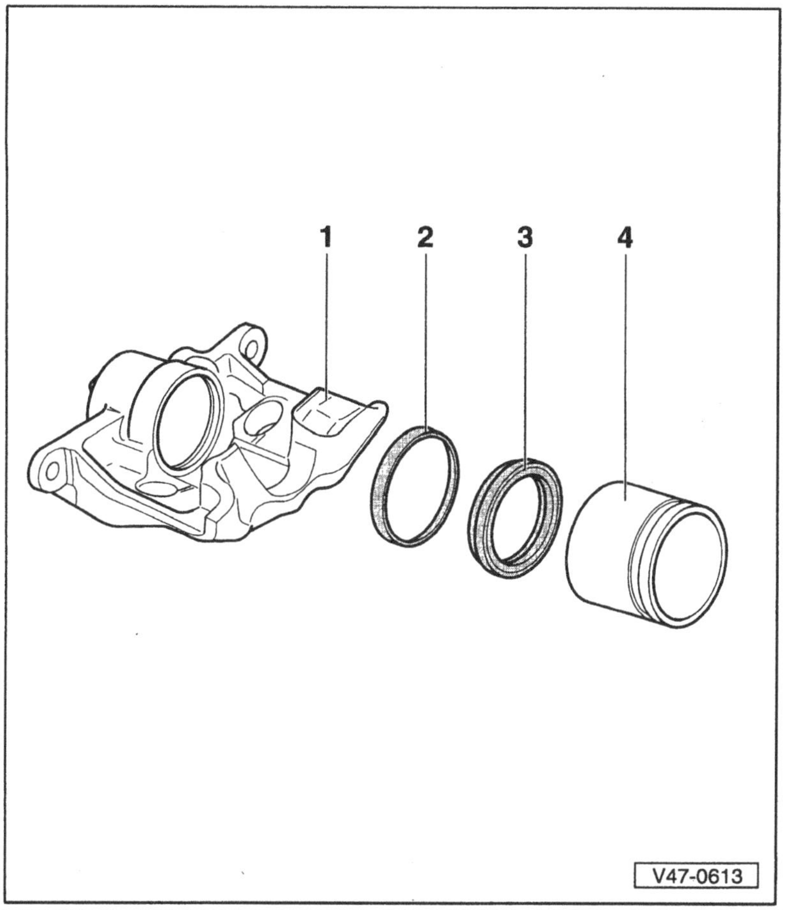

Front brake caliper, servicing

(floating caliper brake)

Notes:

- *When repairing, install all parts supplied in repair kit.*
- *Apply a thin coat of brake fluid to brake cylinder, piston, and seals.*

1 - Brake caliper housing

2 - Seal

- Remove using a plastic wedge, ⇒ Page N 47-2.

3 - Protective seal

- Installing, ⇒ Page N 47-3.
- Do not damage when installing piston.

4 - Piston

- Removing and installing, ⇒ Page N 47-2.

Front brake caliper piston, removing and installing

(floating caliper brake)

Removing

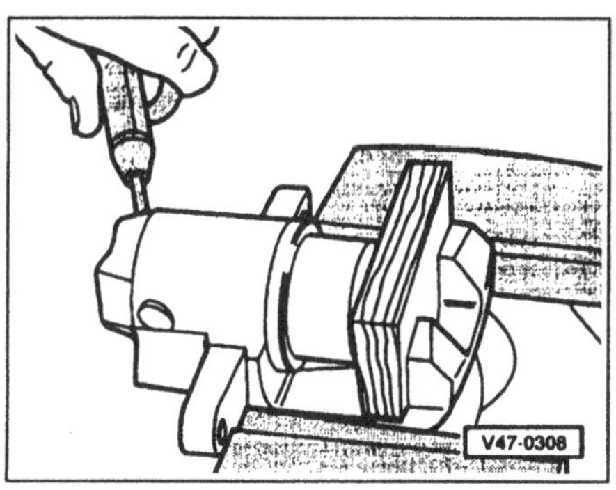

◄ – Force piston out of brake caliper housing, using compressed air.

Note:

Place a piece of wood in the recess to prevent damaging the piston.

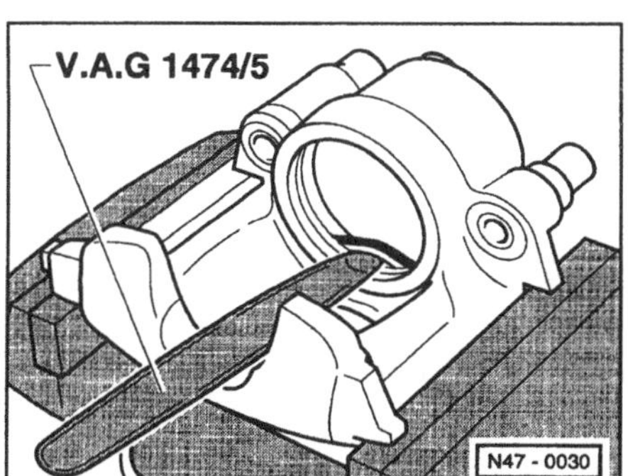

◄ – Remove seal, using modified wedge VAG 1474/5, or equivalent.

Note:

- *When removing, ensure that the surface of the cylinder is not damaged.*
- *Modifying wedge VAG 1474/5, ⇒ Page N 47-3.*

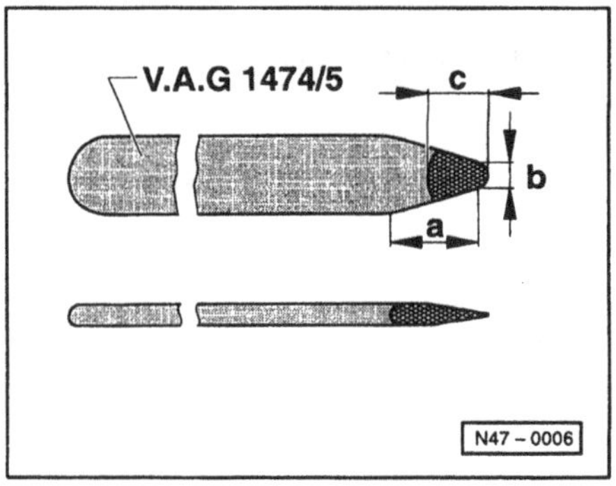

Modifying wedge VAG 1474/5

- Reshape wedge point size in area -a- as per illustration, observing width -b- (6 mm).
- Reshape wedge point to length -c- (10 mm).
- Remove burrs from modified wedge point.

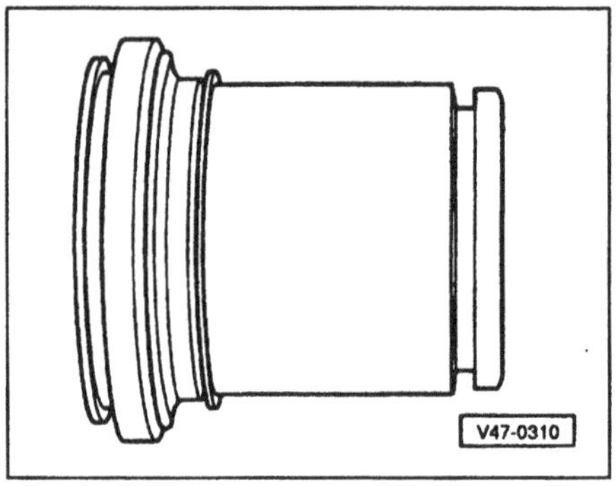

Installing

- Place protective seal with outer sealing lip on piston.

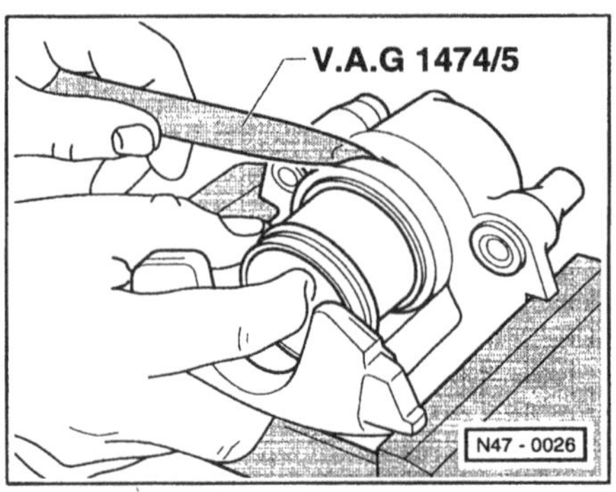

- Using a plastic wedge, insert inner sealing lip into cylinder groove.

Note:

When doing this, hold piston in front of caliper housing.

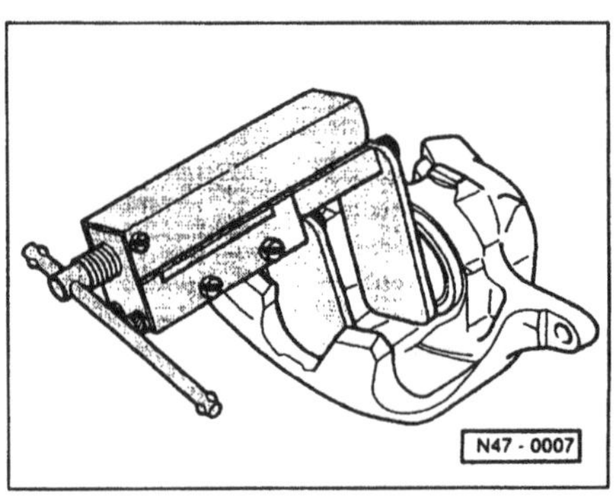

- Press piston into brake caliper housing, using piston resetting tool.

Note:

The protective seal outer sealing lip will then locate in the piston groove.

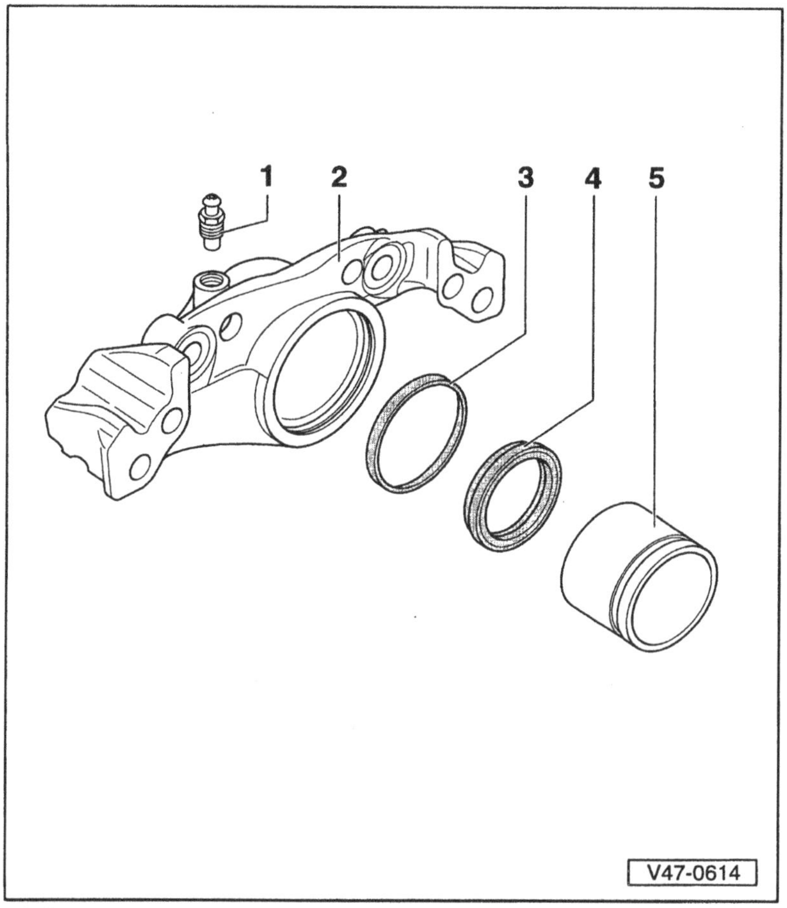

Front brake caliper, servicing (floating frame caliper)

Notes:

- *When repairing, install all parts supplied in repair kit.*
- *Disassembling and assembling brake caliper, ⇒ Page N 46-6.*
- *Apply a thin coat of brake fluid to brake cylinder, piston and seals.*

1 - Bleed screw

2 - Brake caliper housing

3 - Seal
- Remove using a plastic wedge, ⇒ Page N 47-6.

4 - Protective seal
- Installing, ⇒ Page N 47-7.
- Do not damage when installing piston.

5 - Piston
- Removing and installing, ⇒ Page N 47-6.

Front brake caliper piston, removing and installing

(floating frame caliper)

Removing

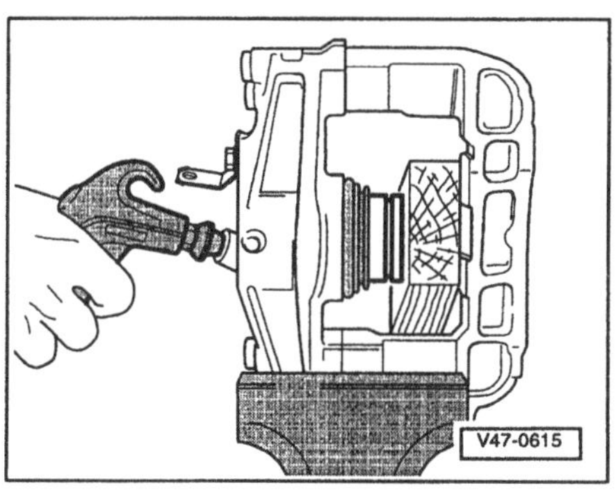

- Force piston out of brake caliper housing using compressed air.

Note:

Place a piece of wood in the recess to prevent damaging the piston.

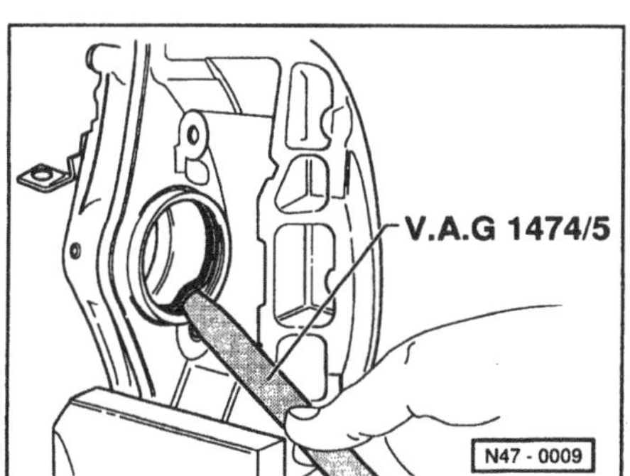

- Remove seal using modified wegde VAG 1474/5 or equivalent.

Note:

- *When removing, ensure that the surface of the cylinder is not damaged.*
- *Modifying wedge VAG 1474/5, ⇒ Page N 47-3.*

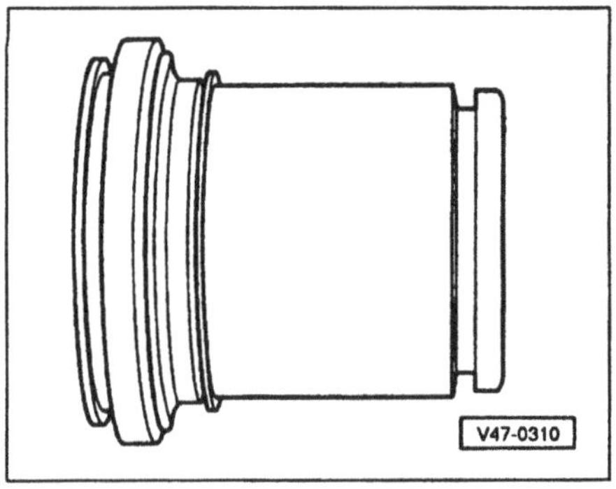

Installing

◄ – Place protective seal with outer sealing lip on piston.

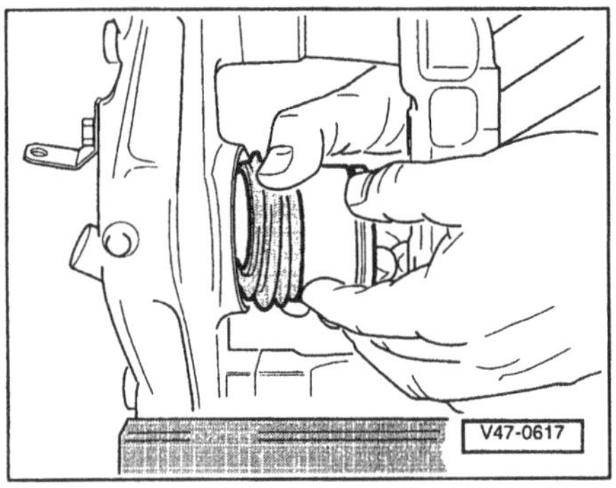

◄ – Insert inner sealing lip into cylinder groove.

Note:

Hold piston in front of caliper housing.

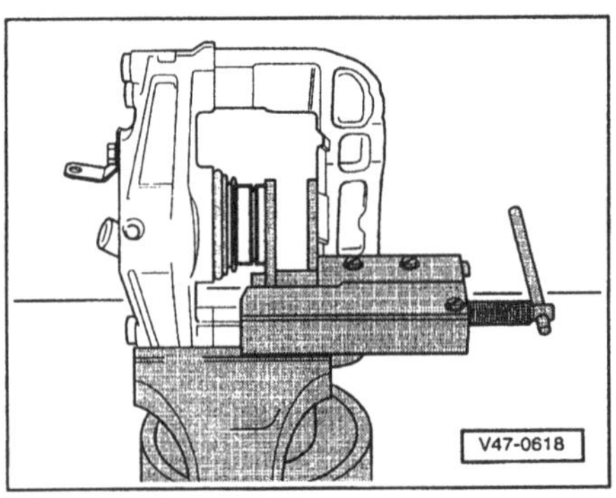

◄ – Press piston into brake caliper housing, using piston resetting tool.

Note:

The protective seal outer sealing lip will then locate in the piston groove.

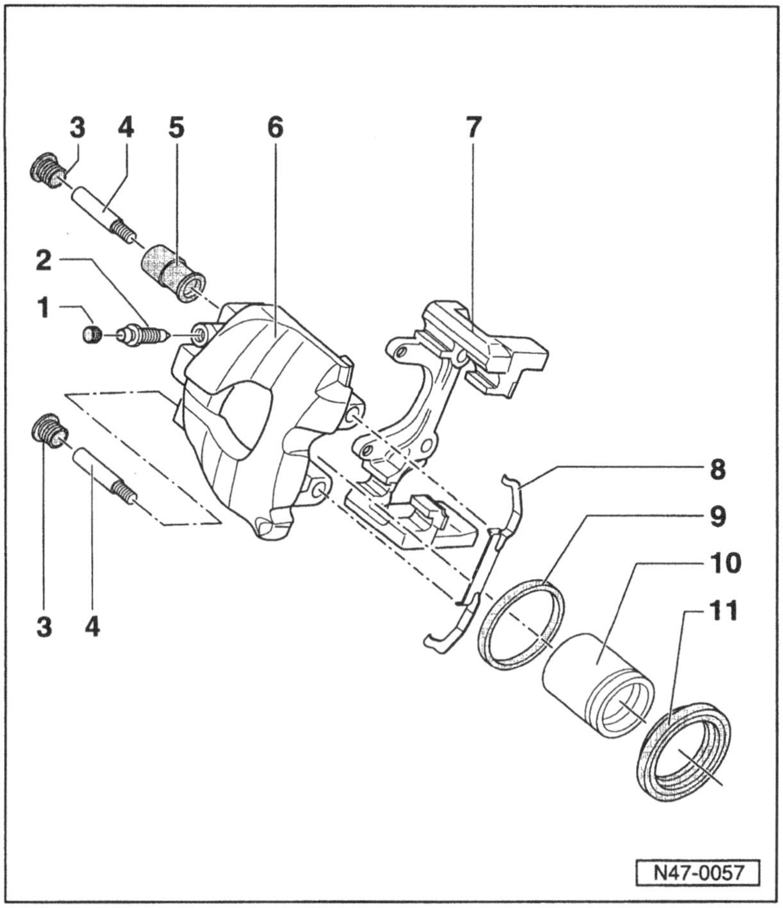

Front brake caliper, servicing (Teves/Ate)

Notes:

- *When repairing, install all parts supplied in repair kit.*
- *Apply a thin coat of brake fluid to brake cylinder, piston, and seals.*

1 - Protective cap
- Push onto bleed valve.

2 - Bleed valve
- Apply thin coat of brake cylinder paste to threads before installing.

3 - Cap
- Insert in mounting bushing.

4 - Guide pin, 25 Nm (18 ft lb)

5 - Mounting bushing
- Insert into brake caliper housing.

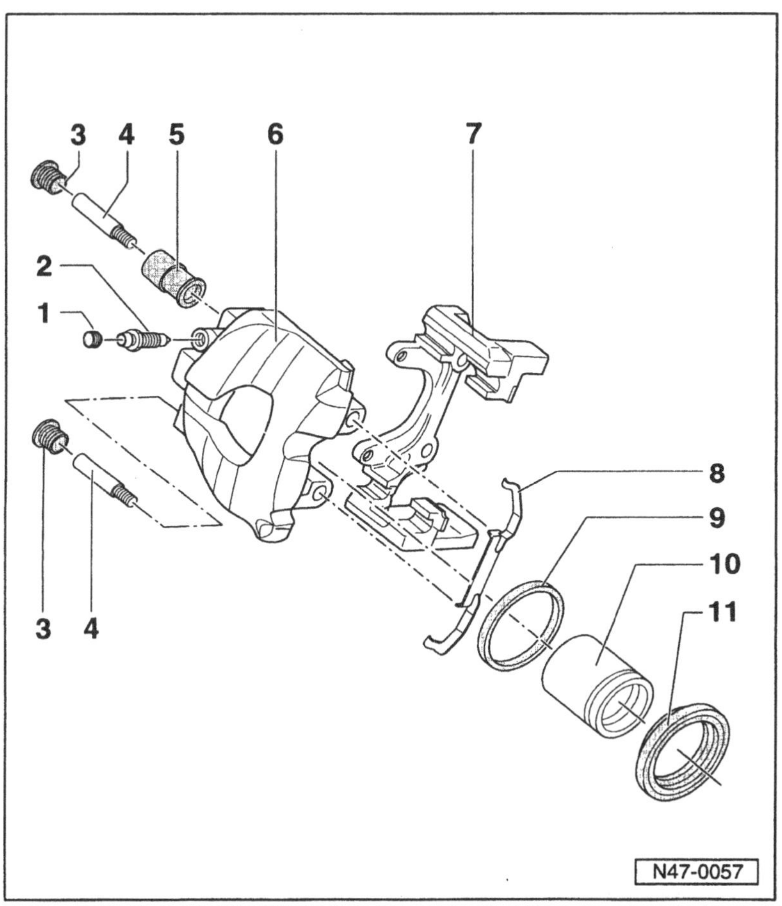

6 - Brake cylinder housing

7 - Brake carrier
- Install on brake caliper housing.

8 - Retaining spring
- Insert with both ends in drillings in brake caliper housing

9 - Seal
- Take out with plastic wedge, ⇒ Fig. 2.

10 - Piston
- Press piston out of brake caliper, using compressed air ⇒ Fig. 1.
- Installing, ⇒ Figs. 3, 4 and 5.

11 - Protective seal
- Do not damage when installing piston.

◄ **Fig. 1 Removing piston**

– Press piston out of brake caliper housing, using compressed air via brake hose connection -arrow-.

Note:

When doing this, place a piece of wood in the shaft so that the piston is not damaged.

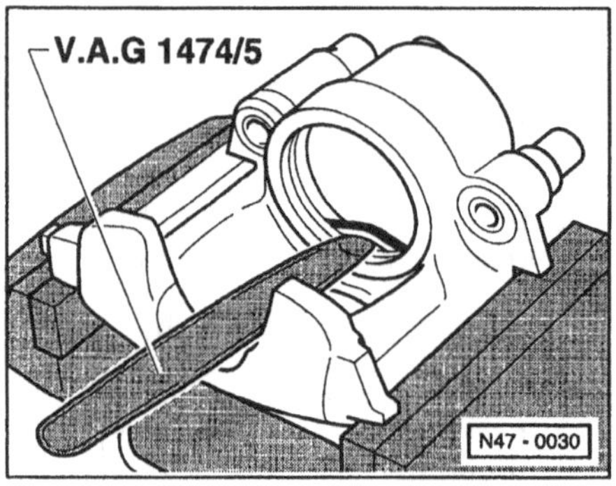

◄ **Fig. 2 Removing seal**

Use a modified VAG 1474/5 plastic wedge, or equivalent.

Note:

Modifying wedge VAG 1474/5, ⇒ Fig. 6.

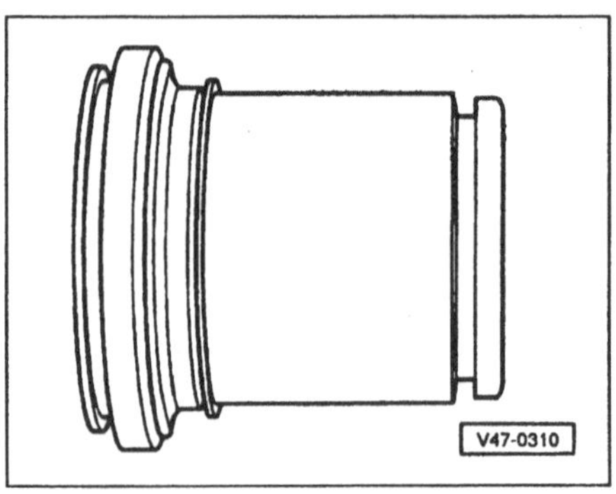

◄ **Fig. 3 Installing protective cap with outer sealing lip on piston**

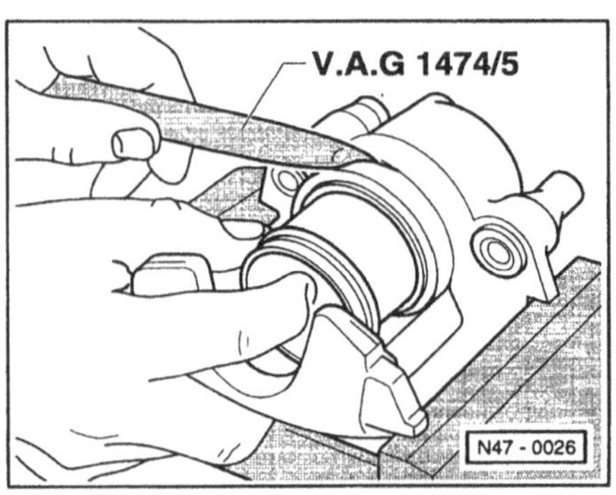

◄ **Fig. 4 Inserting protective cap inner sealing lip into cylinder groove**

Use a modified VAG 1474/5 plastic wedge, or equivalent.

Note:

- *Hold piston before brake caliper housing and ensure that the protective cap is not damaged.*
- *Modifying wedge VAG 1474/5, ⇒ Fig. 6.*

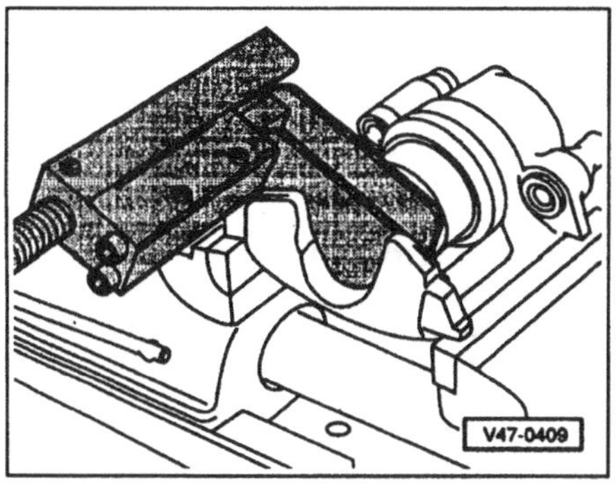

Fig. 5 Pressing piston into brake caliper housing, using piston resetting device

Note:

The outer protective cap sealing lip springs into the groove on the piston.

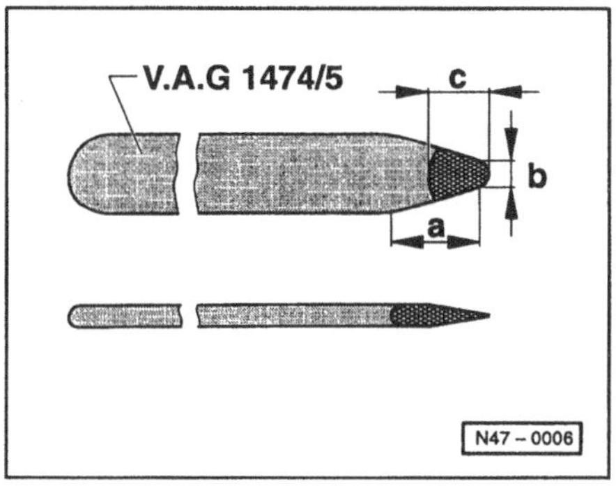

Fig. 6 Modifying VAG 1474/5 wedge

– Reshape wedge point size in area -a- as per illustration observing width -b- (6 mm).

– Reshape wedge point to length -c- (10 mm).

– Remove burrs from modified wedge point.

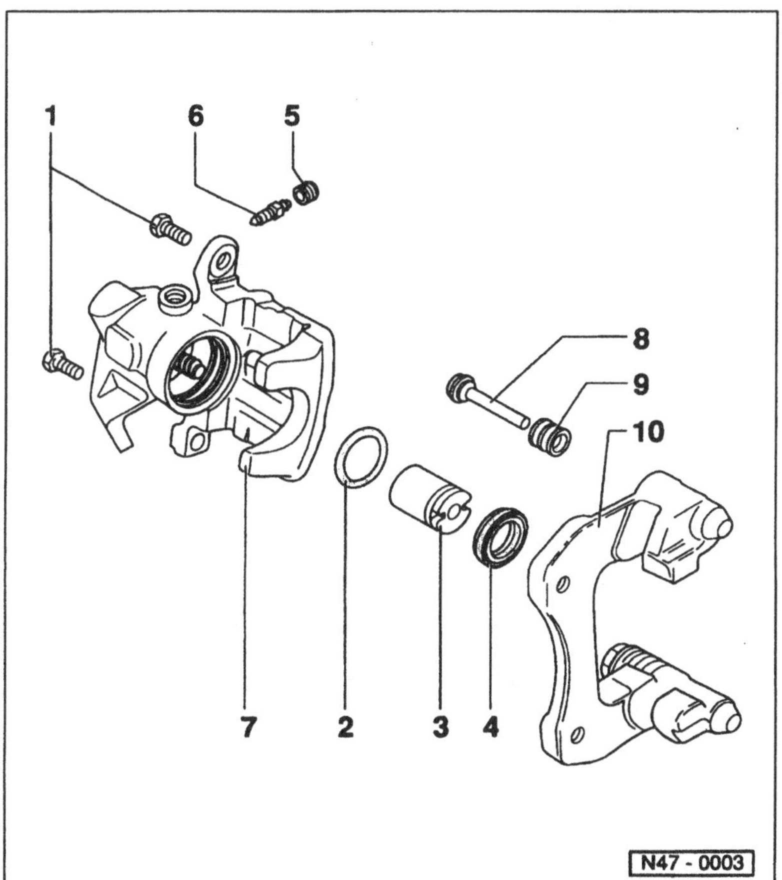

Rear brake caliper, servicing

Notes:

- *Apply a thin coat of brake fluid to brake cylinder, piston, and seals.*
- *When repairing, install all parts supplied in repair kit.*
- *New brake calipers are filled with brake fluid and are pre-bled.*
- *When repairing, it is essential to pre-bleed the brake calipers (without brake pads) before installing in the vehicle,* *⇒ Page N 47-19, Fig. 5.*

1 - Self-locking bolts, 35 Nm (26 ft lb)
- Always replace.
- When loosening and tightening, counter-hold on guide pin.

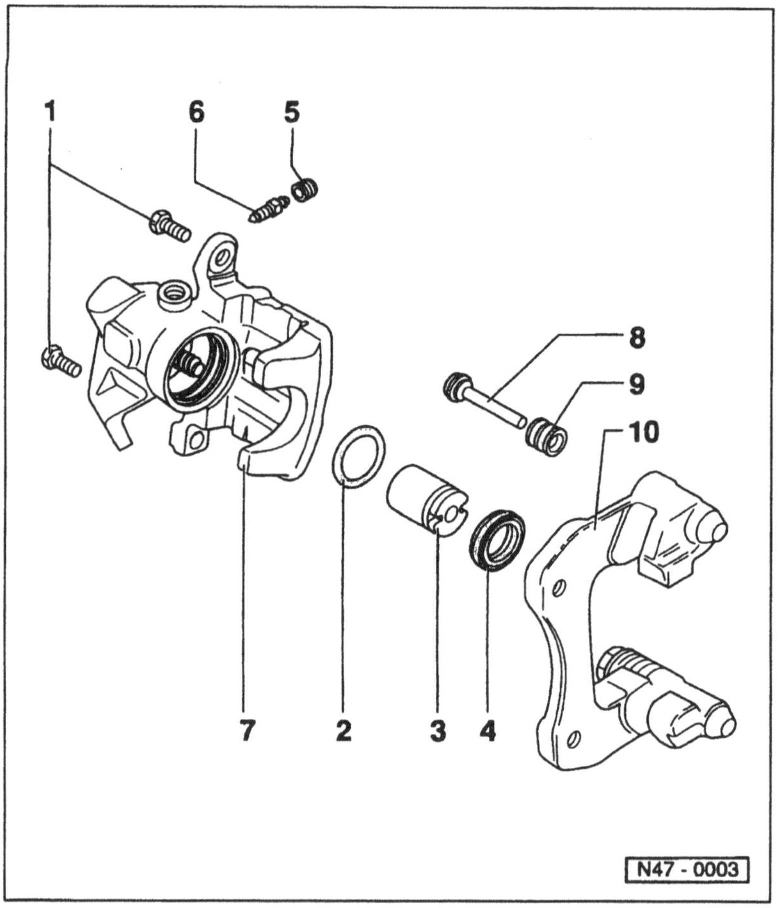

2 - Sealing ring
- Take out using a plastic wedge, ⇒ Fig. 1.

3 - Piston with automatic adjustment
- Removing and installing, ⇒ Page N 47-17.

4 - Protective seal
- Pull outer sealing lip onto piston.
- Fit inner sealing lip into brake caliper housing groove, ⇒ Page N 47-18, Fig. 3.

5 - Dust cap

6 - Bleed valve

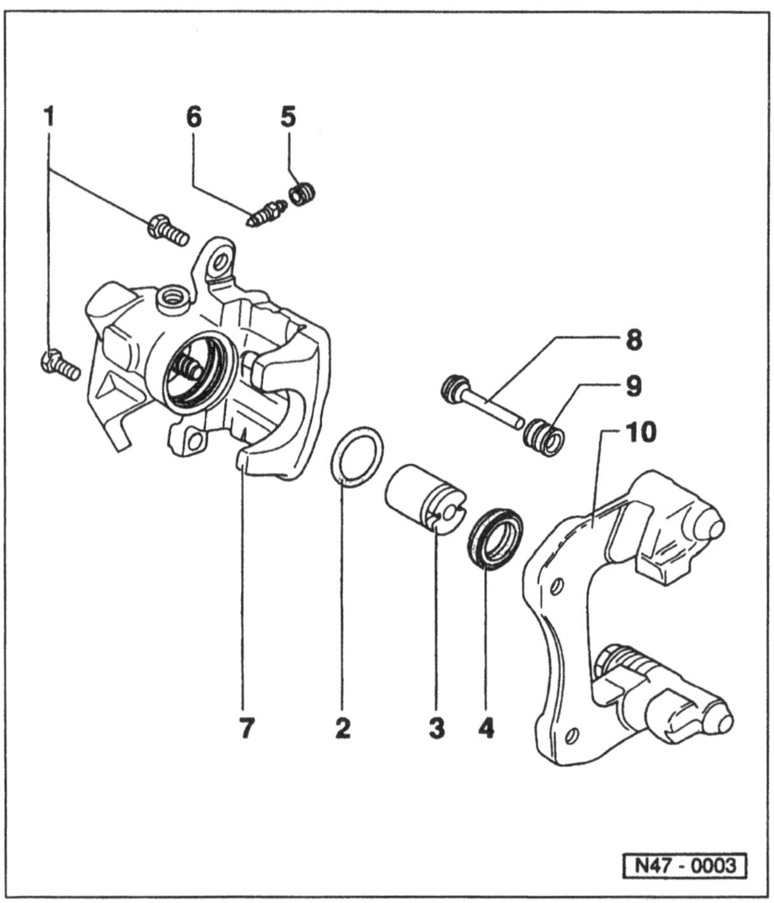

7 - Brake caliper housing with parking brake cable lever
- If a leak exists at parking brake cable lever, replace brake caliper housing.
- After servicing, pre-bleed brake caliper housing, ⇒ Page N 47-19, Fig. 5.

8 - Guide pins
- Grease before pulling on protective cap.

9 - Protective cap
- Pull onto brake carrier and guide pin.

10 - Brake carrier with guide pin and protective cap
- Supplied as replacement part, assembled with sufficient grease on guide pins.
- If protective caps or guide pins are damaged, install repair kit. Use grease packet supplied to lubricate guide pins.

Rear brake caliper piston, removing and installing

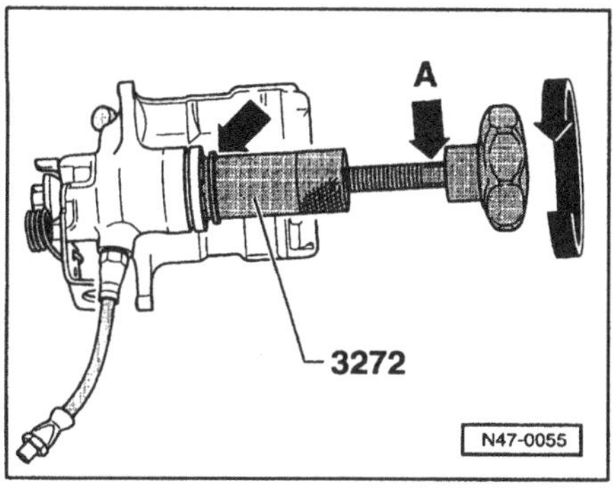

Removing

◄ – Screw piston out of brake caliper housing by turning knurled wheel counter-clockwise.

Notes:

- *If piston is difficult to move, use a 13 mm open end wrench on the flats -arrow A- provided for this purpose.*
- *Install special tool 3272 so that the collar -arrow- is before the piston.*

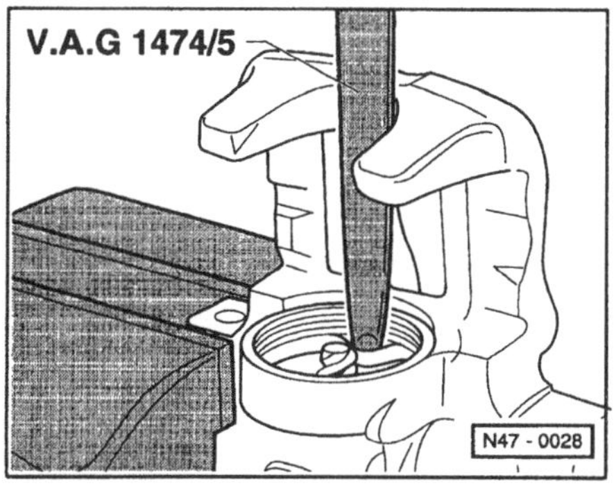

◄ **Fig. 1 Removing seal**

Use modified VAG 1474/5 plastic wedge or equivalent.

Notes:

- *Modifying wedge VAG 1474/5, ⇒ Fig. 6.*
- *Ensure that the surface of the cylinder is not damaged during removal.*

Installing

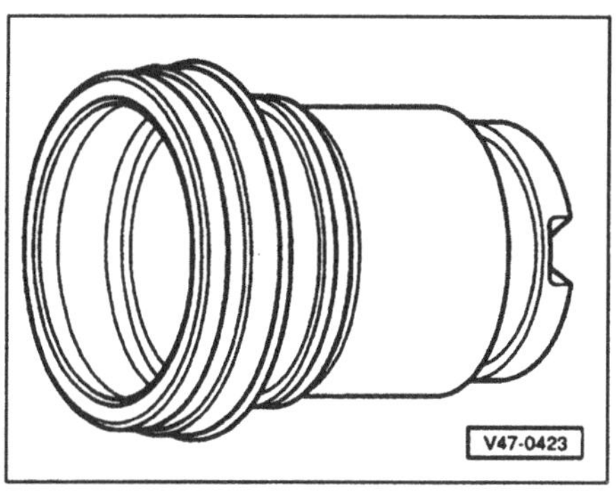

◄ **Fig. 2 Installing protective cap with outer sealing lip on piston**

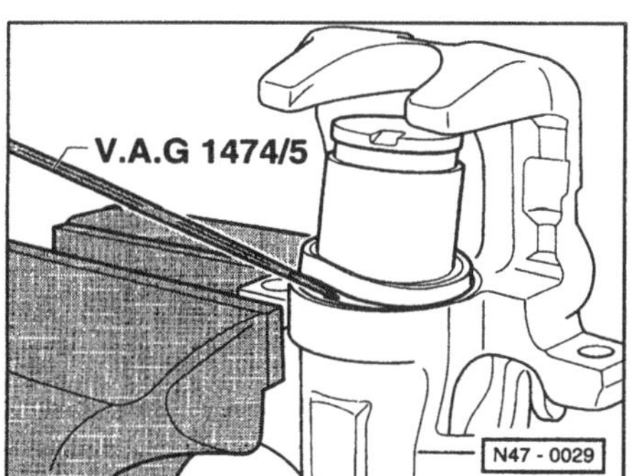

◄ **Fig. 3 Installing protective cap inner sealing lip into brake caliper housing groove**

Note:

- *Use modified plastic wedge VAG 1474/5 or equivalent.*
- *Hold piston in front of brake caliper housing.*
- *Modifying wedge VAG 1474/5, ⇒ Fig. 6.*

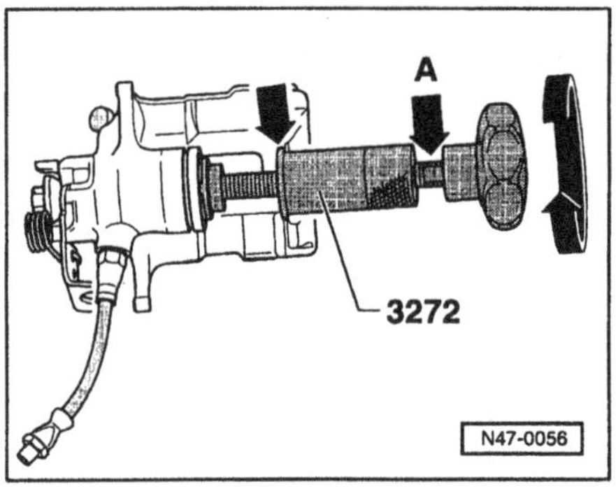

◄ **Fig. 4 Moving piston in by turning wheel clockwise**

Notes:

- *Install special tool 3272 so that the collar -arrow- of the tool is seated against the brake carrier.*
- *If the piston is difficult to move, use a 13 mm open end wrench on the flats -arrow A- provided for this purpose.*
- *If the piston is pushed back with a piston resetting tool or by operating the foot brake, the automatic adjustment in the brake caliper is destroyed.*

– Install brake pads.

◄ **Fig. 5 Pre-bleeding brake caliper**

– Open bleed valve and using a commercially available bleeding unit, fill with brake fluid until fluid comes out of threaded hole (brake hose connection) free of air bubbles.

– Close bleed valve.

Note:

When pre-bleeding, position the brake caliper as shown in the illustration.

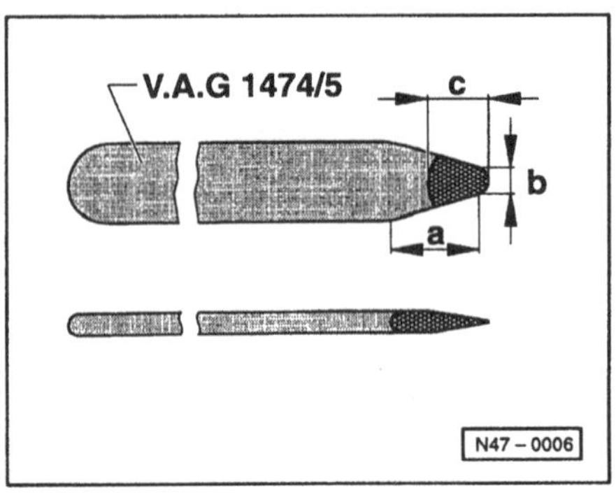

◄ **Fig. 6 Modifying VAG 1474/5 wedge**

– Reshape wedge point size in area -a- as per illustration observing width -b- (6 mm).

– Reshape wedge point to length -c- (10 mm).

– Remove burrs from modified wedge point.

Brake system, bleeding (vehicles with and without ABS)

On vehicles with EDL, the basic setting function of On Board Diagnostic (OBD) is used to bleed the hydraulic unit.

The basic setting is only necessary when at least one chamber in the brake fluid reservoir is completely empty (e.g., leaks in braking system).

Basic setting,
⇒ Booklet: ABS, ABS/EDL On Board Diagnostic (OBD).

Changing brake fluid,
⇒ Maintenance.

Notes:

- *Bleeding the braking system on vehicles with ABS and ABS/EDL is accomplished the same as for vehicles with conventional braking systems.*

Caution!

- ***Only use new brake fluid conforming to US standard FMVSS 116 DOT 4. Genuine VW/Audi brake fluid conforms to this specification.***
- ***Due to its corrosive effect, brake fluid must not come into contact with paintwork.***
- ***Brake fluid is hygroscopic, which means it absorbs moisture from the ambient air and it should therefore always be stored in air tight containers.***
- ***Rinse off spillages using plenty of water.***

Brake system, bleeding with filling and bleeding unit US 1116

Warning!

Brake fluid is poisonous.

- Connect US 1116.
- Install brake pedal loading device.
- Open bleed screws in specified order.

Bleeding sequence

1 – Rear right brake caliper

2 – Rear left brake caliper

3 – Front right brake caliper

4 – Front left brake caliper

Caution!

Do not press the brake pressure regulator horizontal beam down when bleeding the brakes.

Bleeding brake system without brake filling and bleeding appliance

– Build-up brake pressure in brake system by pumping brake pedal.

– Attach bleeder bottle hose and open bleed screw.

– With brake pedal held down, close bleed screw.

– Repeat operation until no air is present.

Bleeding sequence, ⇒ Page N 47-22.

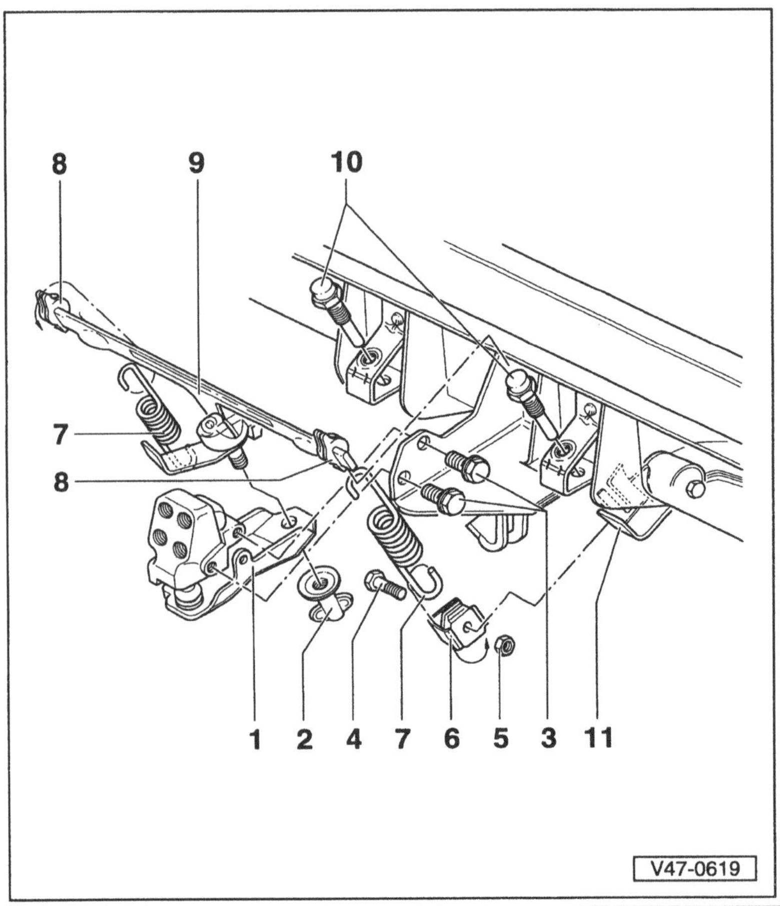

Brake pressure regulator, removing and installing

1 - **Brake pressure regulator**
   - ♦ Adjusting, ⇒ Page N 47-26.

2 - **Securing clip**
   - ♦ Press onto threaded stud and tighten 2 to 3 turns

3 - **Hex bolt, 20 Nm (15 ft lb)**

4 - **Hex bolt, 20 Nm (15 ft lb)**

5 - **Nut**

6 - **Spring mounting**
   - ♦ Mark installation position before removing

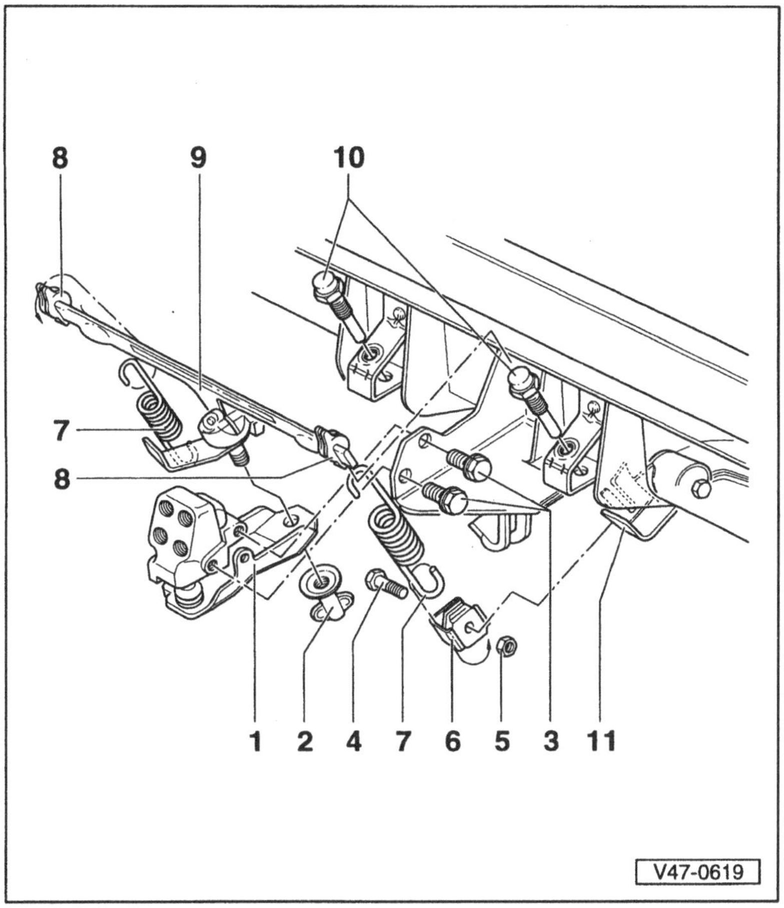

7 - Spring
- Note installation position.
- Observe application, ⇒ Page N 42-18.

8 - Clip

9 - Horizontal beam
- Slight inclination is permissible, ⇒ Page 47–31.

10 - Buffer stop
- Adjusting, ⇒ Page N 47-32.

11 - Control arm

Brake pressure regulator, adjusting

Basic setting

Notes:

- *If, for example, one or both control arms are replaced and no markings are given for the installation position of the spring mounting -2- on the control arm, it is recommended that the basic setting be performed, as described below.*
- *If basic setting is not necessary, ⇒ Page N 47-27, for fine adjustment of the brake pressure regulator.*

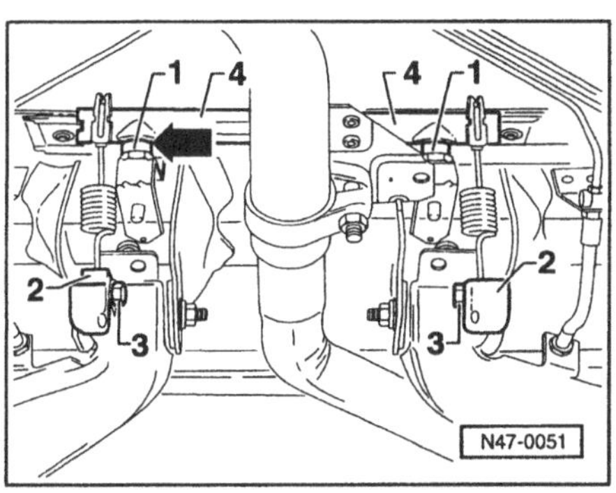

– Remove both shock absorbers, ⇒ Page N 42-17.

– Insert special tool 3255 and pre–tension control arms, ⇒ Page N 47-28.

– Align beam -4- horizontally, by sight, by turning both buffer stops -1- onto stop on horizontal beam -4- (no pre–load at buffer stop -1-).

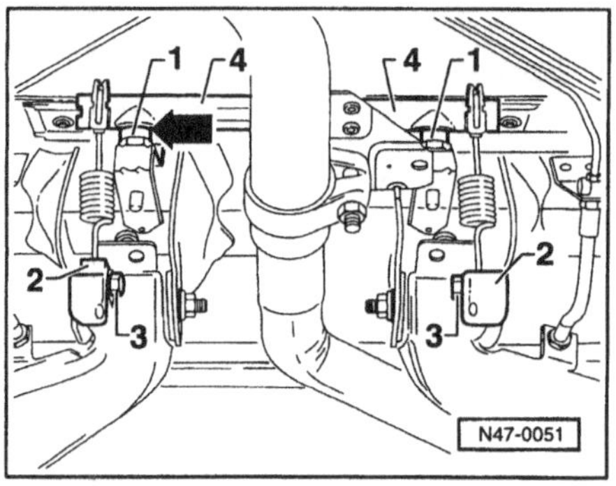

- Move spring mounting -2- down until gap -arrow- of approx. 2 mm exists on opposite side between buffer stop -1- and horizontal beam -4-.
- Move second spring mounting -2- down until beam -4- is again horizontal.
- Turn both buffer stops -1- fully back, so that largest possible gap exists between horizontal beam -4- and buffer stop.
- Continue with fine adjustment of brake pressure regulator, ⇒ Page N 47-27.

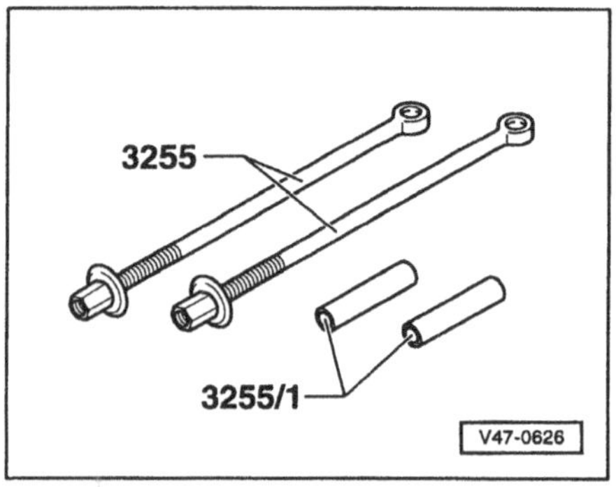

Fine adjustment

- Remove both shock absorbers, ⇒ Page N 42-17.
- Slide tubes 3255/1 onto adjusting device.

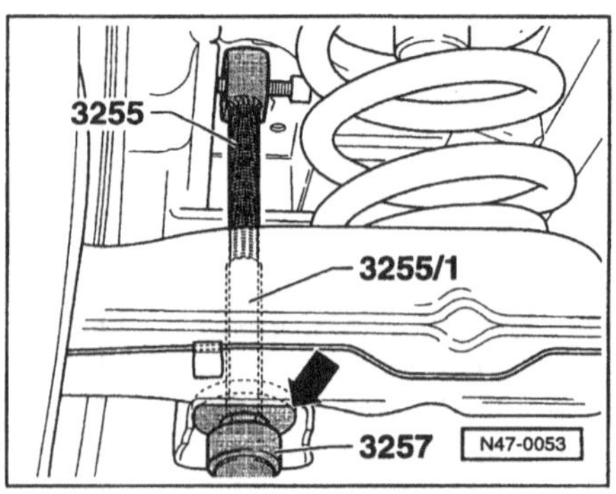

- Install adjusting device 3255 with tubes 3255/1 as illustrated.

Notes:

- *The coil spring can be removed to adjust the control arm to dimension -x-.*
- *Grease the spindle thread, face of nut and washer surface with MoS_2 grease.*
- *Insert a shock absorber rubber bushing between the washer -arrow- and the control arm in order to tension the control arm.*
- *Do not use an impact screwdriver.*
- *First check buffer stop settings, ⇒ Page N 47-32, if complaints of poor braking or over-braking on the rear axle are received.*

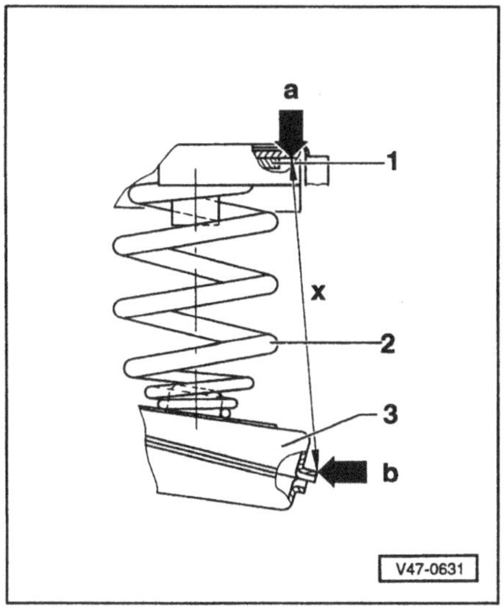

- Pre–tensioning control arm to dimension -x- with special tool 3255.

1 - Washer

2 - Spring

3 - Control arm

Notes:

- *The measurement is taken from the spring mounting steel plate edge -arrow a- to the top edge of the assembly hole in the control arm -arrow b-.*
- *Set dimension -x- according to table on Page N 47-33.*

- Continue with brake pressure check, ⇒ Page N 47-30.

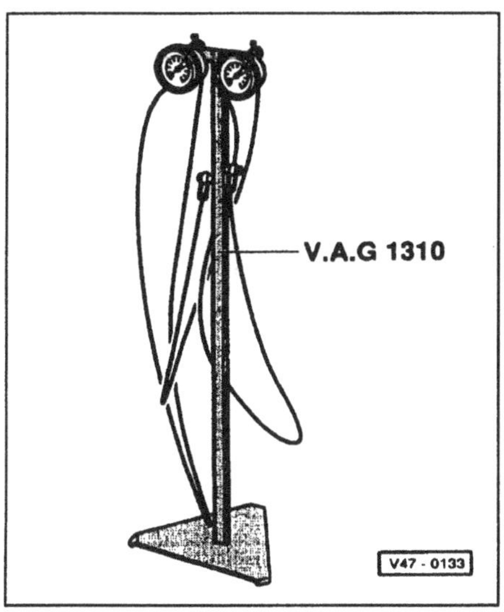

Checking brake pressure

- Connect pressure gauge VAG 1310 to front left brake caliper and to rear left brake caliper.
- Bleed both gauges (bleeding brake system, ⇒ Page N 47-21).
- Depress brake pedal and check pressures.

Front axle	50 bar (725 psi)
Rear axle	20 ± 2 bar (290 ± 29 psi)

- Adjust if necessary, ⇒ Page N 47-31.
- Continue with adjusting buffer stops, ⇒ Page N 47-32.

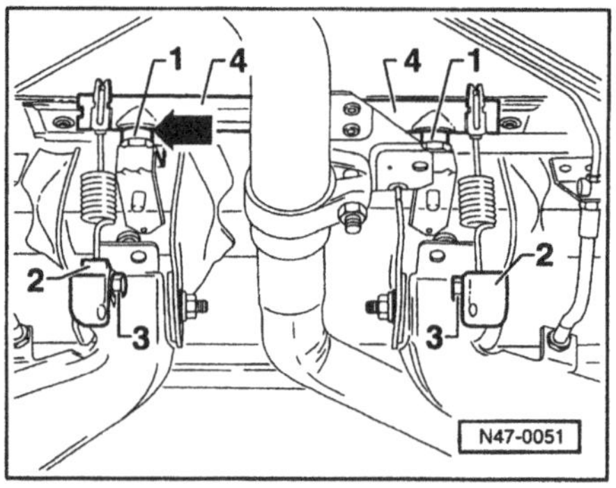

Adjusting brake pressure

- Turn both buffer stops -1- back onto stop (downward).
- Mark spring mounting -2- position with a felt tip pen or adhesive tape on the control arm.
- Loosen bolt -3- and move mounting -2-.

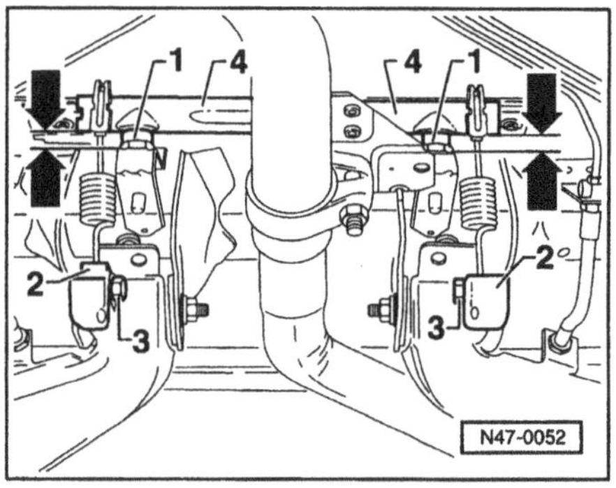

Notes:

- *Moving spring mounting -2- down increases pressure.*
- *Moving spring mounting -2- up reduces pressure.*
- *It is only necessary to adjust the position of the second spring mounting -2- if the horizontal beam -4- is excessively inclined (more than 4 mm, ⇒ arrows).*
- *Moving the spring mounting -2- 1 mm equates to a pressure change of approximately 4 bar at the rear axle.*
- *Slight inclination of the horizontal beam -4- is permissible (maximum 4 mm between left and right, ⇒ arrows).*

- Tighten bolt -3-; counter-hold when tightening.
- Correct brake pressure regulator setting at second spring mounting -2- if necessary.
- Continue with adjusting buffer stops, ⇒ Page N 47-32.

Adjusting buffer stops

Note:

Check whether the securing clip, item 2, on Page N 47-24, is securely seated.

- Check brake pressure (if not already checked), ⇒ Page N 47-30.
- If setting OK, insert pedal depressor.
- Hold front axle brake pressure to at least 50 bar (725 psi).

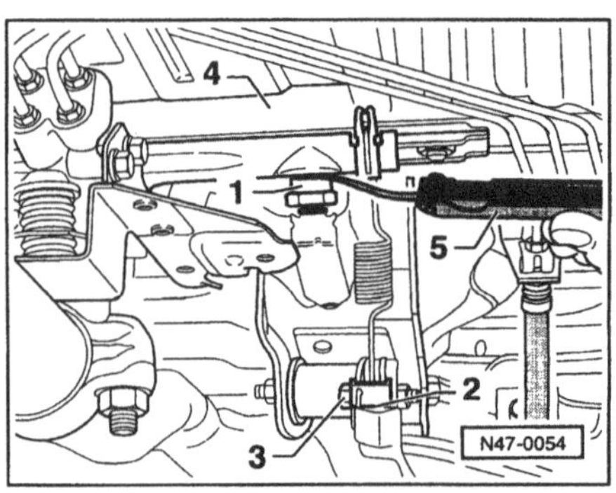

- Turn both buffer stops -1- up, using 6 mm hex key (illustration shows only one buffer stop), until a gap of 0.5 mm can be measured using a feeler gauge -5- between both buffer stops -1- and horizontal beam -4-.
- Remove VAG 1310 and bleed brake system, ⇒ Page N 47-21.
- Remove special tool 3255.
- Install shock absorber, ⇒ Page N 42-17.

Table of brake pressure regulator settings

Rear axle spring	Color coding	Dimension x + 1 mm
701 511 105 F	green	281 mm
701 511 105 H/J	orange	270 mm
701 511 105 A/C	white	258 mm
701 511 105	violet	280 mm
701 511 105 B	brown	292 mm
701 511 105 L	grey	238 mm
701 511 105 K	yellow	284 mm
701 511 105 M	pink	268 mm

Checking brakes on dynamometer rollers

General

Checking brakes on a dynamometer cannot equate 100 % to a braking sequence carried out the road.

Due to the relatively high weight differences between front and rear axles it is normal for the front axle to achieve higher values. Vehicles with low rear axle loading achieve low braking forces at rear axle.

In addition, the influence of weight change, friction, and road surface conditions encountered when braking while driving cannot be reproduced on dynamometer rollers.

The following table shows the minimum braking force at the rear axle dependent on the actual axle loading when the vehicle is stationary.

In the table, the lower limits that must be achieved are shown. A correctly functioning brake system will always achieve or exceed these values.

Test values for the minimum braking force at the rear axle

Rear axle loading (weighed)	**Minimum braking force at rear axle** (both rear wheels)
less than 600 kg	900 N
700 kg	1200 N
800 kg	1500 N
900 kg	1800 N
1000 kg	2200 N
1100 kg	2500 N
1200 kg	2800 N
1300 kg	3100 N
1400 kg	3300 N
1500 kg	3700 N

Column tube with steering column, removing and installing

Removing

- Remove airbag unit,
⇒ Page N 48-13, vehicles with airbag only.

- Remove steering wheel,
⇒ Page N 48-19, steering wheel with airbag.

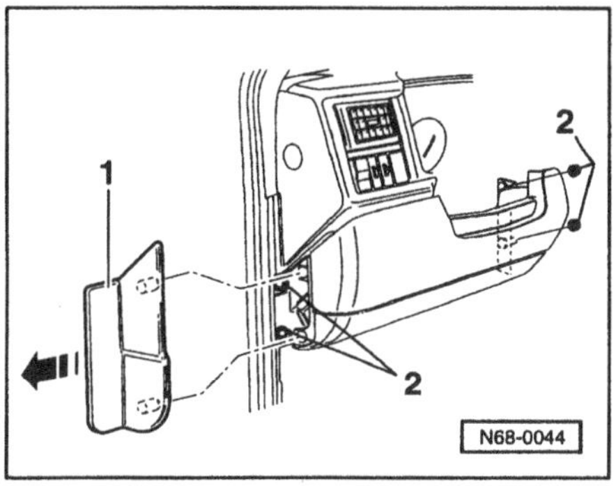

Vehicles with knee padding

◄ - Pry off left-hand trim panel -1-.

- Remove hex nuts and hex bolts -2- (2 each side) and remove driver's side knee padding.

Continued for all vehicles

- Remove trim below steering column switch.

- Remove trim above steering column switch.

- Remove left-hand trim below dash panel.

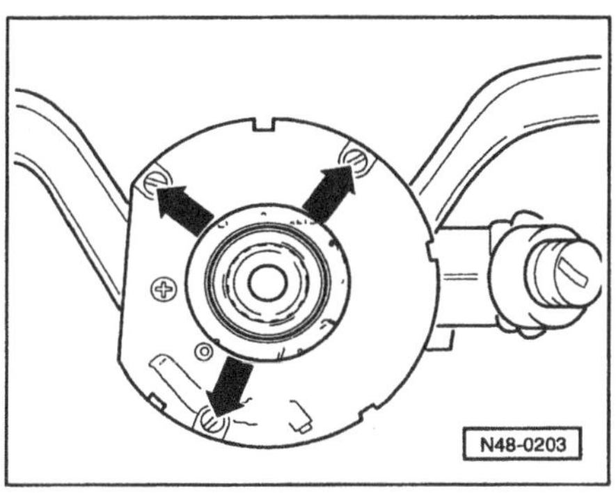

- Separate connector to turn signal and windshield wiper switch.

◄ - Remove screws -arrows-.

- Remove turn signal and windshield wiper switch from steering column.

- Pull connector off steering lock housing.

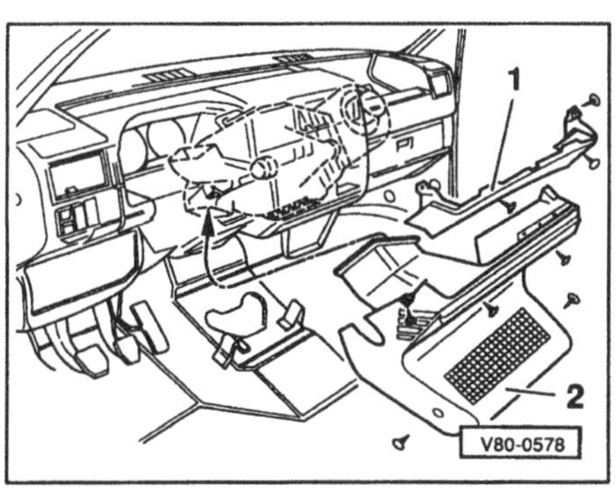

◄ - Remove trim panel -1-.

- Remove footwell vent console -2-.

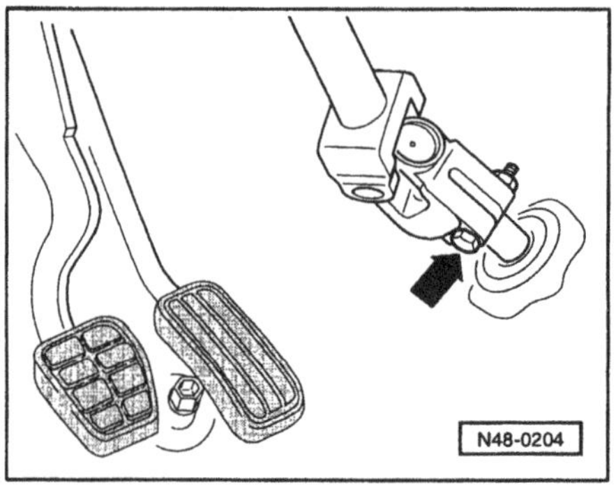

◄ – Remove steering universal joint bolt.

– Pull universal joint off lower steering intermediate shaft.

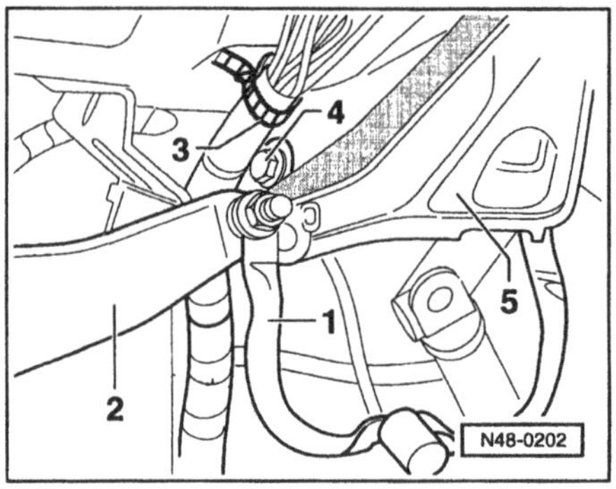

◄ – Remove bracket -1-, if installed, and strut -2- from column tube -5-.

– Unclip cable channel -3-.

– Remove upper securing bolt -4- for column tube.

– Remove lower steering column bolt on pedal cluster mounting bracket.

– Remove column tube with steering column.

Installing

Install in reverse order of removal.

Tightening torques

Steering wheel to steering column	67 Nm (49 ft lb)
Universal joint to steering shaft	30 Nm (22 ft lb) and turn 90° further
Upper securing bolt for column tube	50 Nm (37 ft lb)
Lower securing bolt for column tube	23 Nm (17 ft lb)

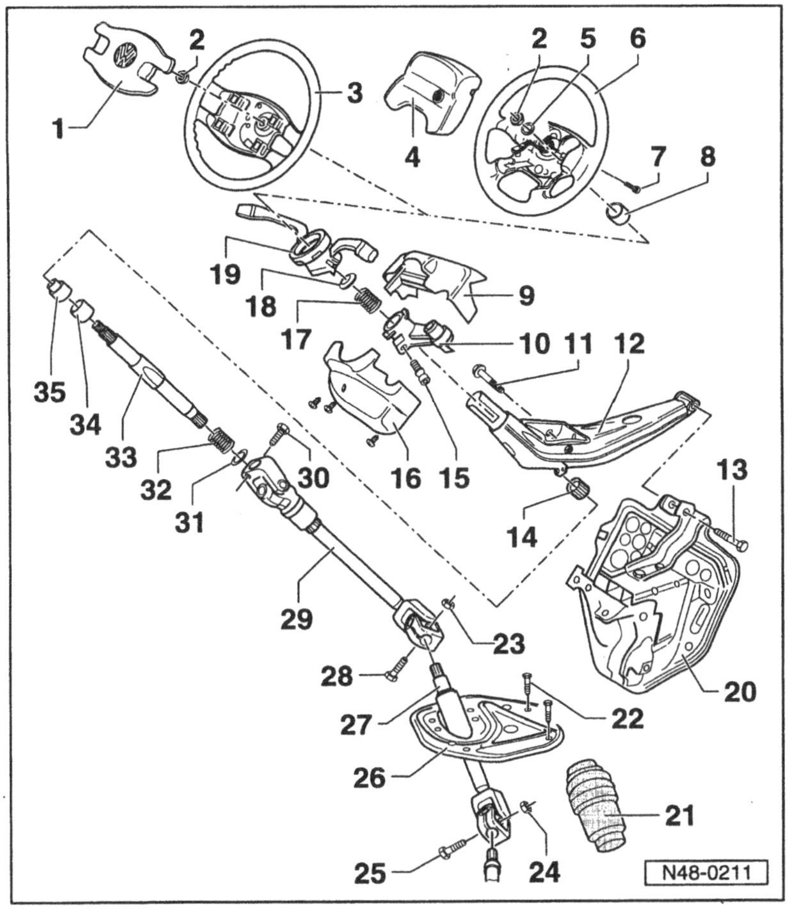

Steering column, column tube and steering wheel, assembly overview

Notes:

- *Always replace self-locking nuts and bolts.*
- *Welding and straightening work on steering components is not permitted.*
- *All components indicated with a [1] should be pre–assembled away from the vehicle.*

1 - Cover plate
- Unclip.

2 - Hex nut, 67 Nm (49 ft lb)

3 - Steering wheel

4 - Airbag unit
⇒ Removing and installing, Page N 48-13.

5 - Washer

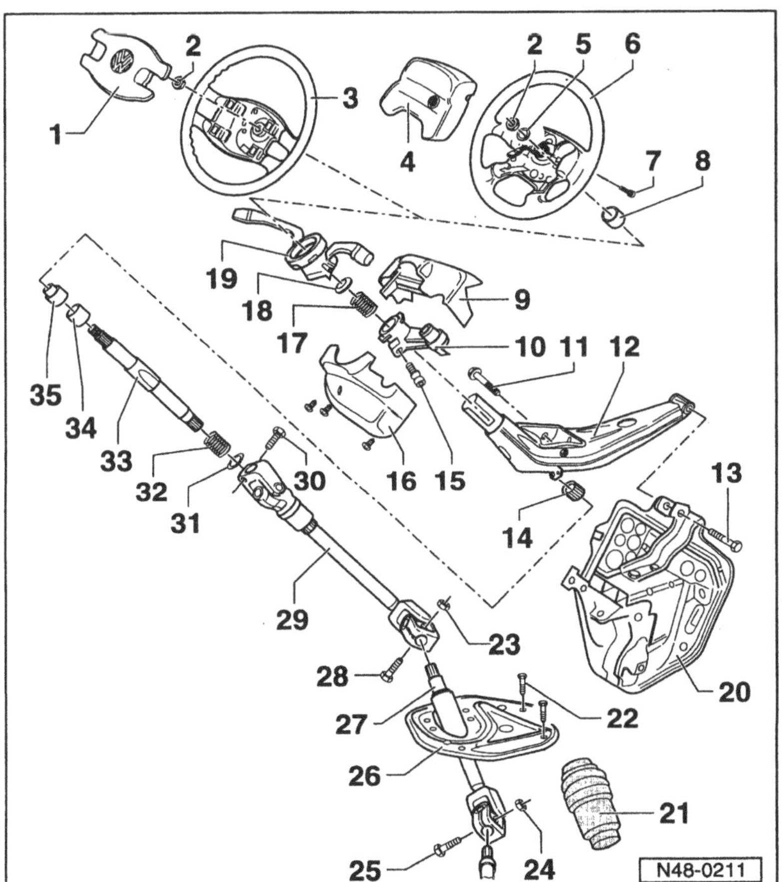

6 - Steering wheel for airbag
⇒ Removing and installing, Page N 48-13.
⇒ Removing and installing spiral spring, Page N 48-21.

7 - Hex socket head bolt, 6.5 Nm (58 in. lb)
- Always replace.

8 - Distance bushing

9 - Upper trim

10 - Steering lock housing[1]

11 - Hex bolt, 50 Nm (37 ft lb)

12 - Column tube

13 - Hex bolt, 23 Nm (17 ft lb)

14 - Ball bearing[1]
- Drive out with VW 295.

⇒ Pressing in, Page N 48-10.

15 - Shear head bolt[1]
- Tighten until head shears off.
- Drill out to remove.

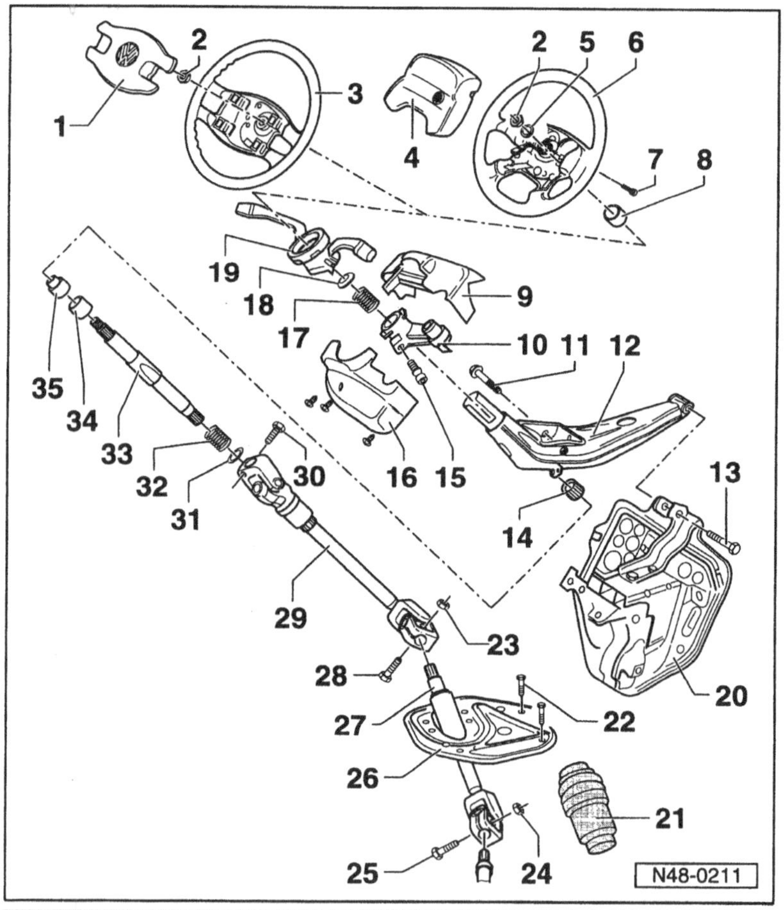

16 - Lower trim

17 - Spring[1)]

18 - Clamping washer[1)]
⇒ Installation position, Page N 48-12.

19 - Steering column switch

20 - Mounting plate

21 - Protective sleeve

22 - Hex bolt, 20 Nm (15 ft lb)

23 - Self-locking hex nut
- 30 Nm (22 ft lb) and turn 90° further.
- Always replace.

24 - Self-locking hex nut
- 30 Nm (22 ft lb) and turn 90° further.
- Always replace.

25 - Hex bolt
- Always replace.

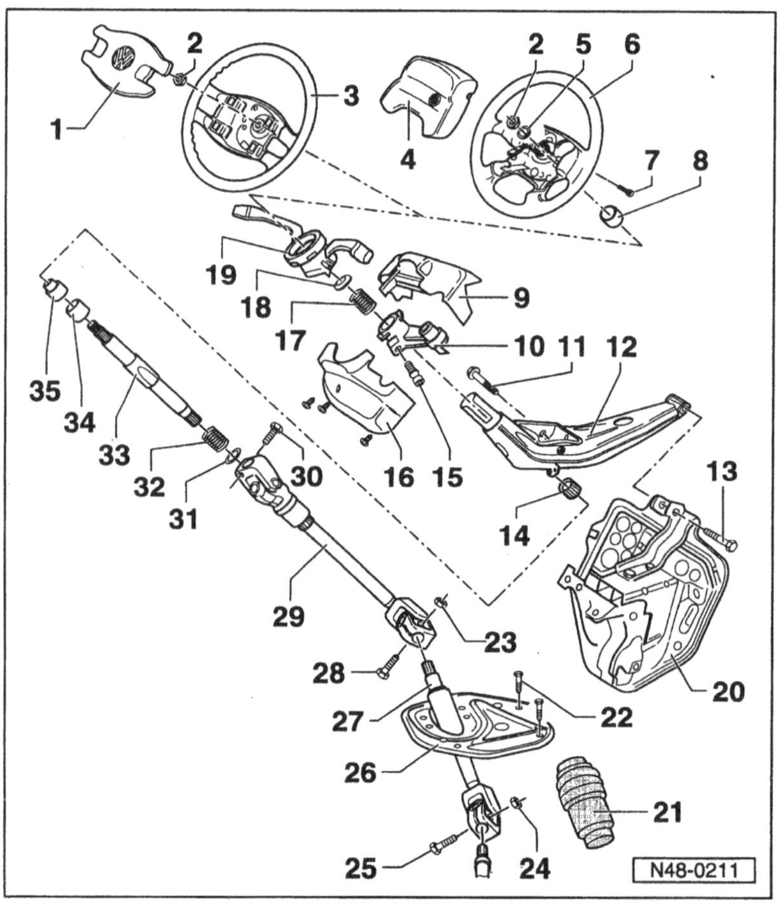

26 - Mounting plate with centralizing pin

27 - Lower steering intermediate shaft

28 - Hex bolt
- Always replace.

29 - Upper steering intermediate shaft[1)]

30 - Hex bolt[1)]
- 30 Nm (22 ft lb) plus 2 turns of 45° each.

31 - Washer[1)]

32 - Lower spring[1)]

33 - Steering column tube[1)]
- With locking shells for steering wheel lock.

34 - Distance bushing[1)]

35 - Support ring[1)]

Column tube with steering column, disassembling and assembling

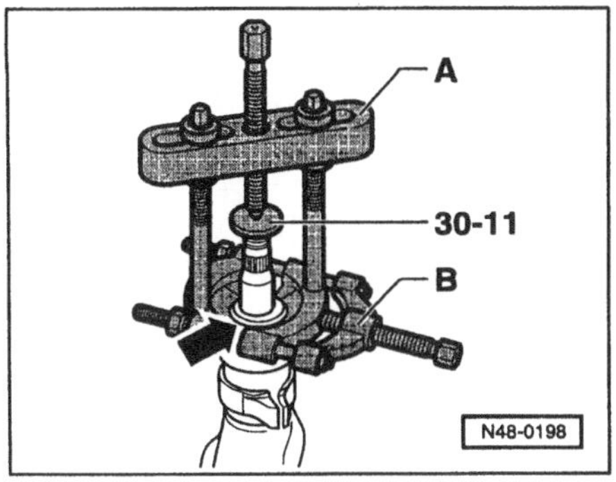

Disassembling

– Drill out steering lock housing securing bolts.

◄ – Pull clamping washer -arrow- and spring off steering column.

A - Puller, e.g., Kukko 18/0

B - Separating device 5 ... 60 mm, e.g., Kukko 17/0

Note:

The steering column is pulled away downward, after pulling off the clamping washer.

– Pull steering lock housing off column tube and steering column.

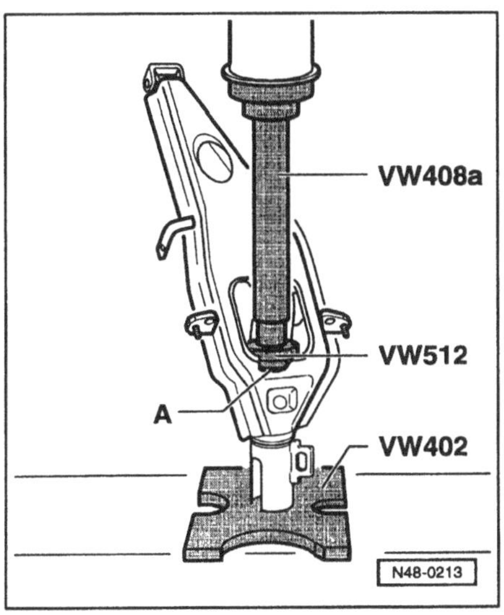

Assembling

◄ – Press ball bearing -A- into column tube.

– Slide spacer ring onto upper steering column tube.

– Install support ring into steering lock housing bearing inner race.

– Guide steering column tube into column tube.

– Slide spring and washer onto steering column tube.

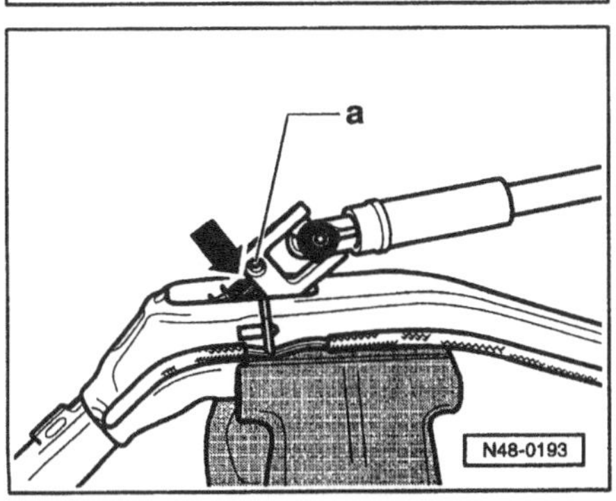

◄ – Install universal joint onto steering column tube, ensuring that notch -arrow- in steering column tube spline aligns with hole -a- in universal joint.

Note:

The components must assemble easily when installing the universal joint on the steering column tube and when inserting the bolts!

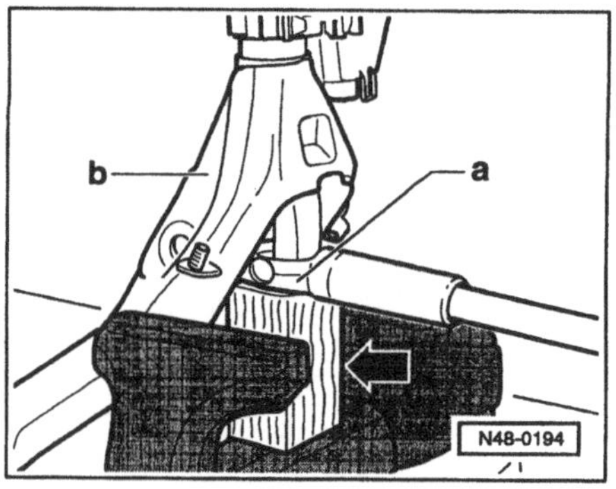

- Bolt universal joint to steering column tube.
- Place column tube in vise as illustrated and support universal joint -a- with wooden block on vise. Push column tube -b- down as far as possible to compresses lower steering column tube spring.

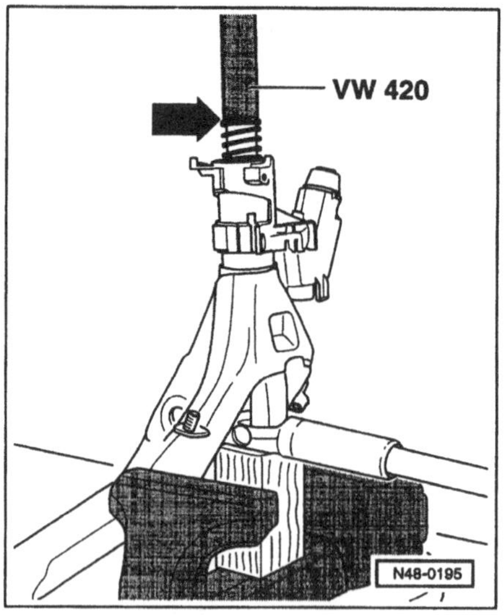

- Tighten vise.
- Install steering lock housing and bolt loosely together with new shear head bolt.
- Install spring with clamping washer onto steering lock housing.
- Drive clamping washer (arrow) on until installation position is achieved,

⇒ Page N 48-12.

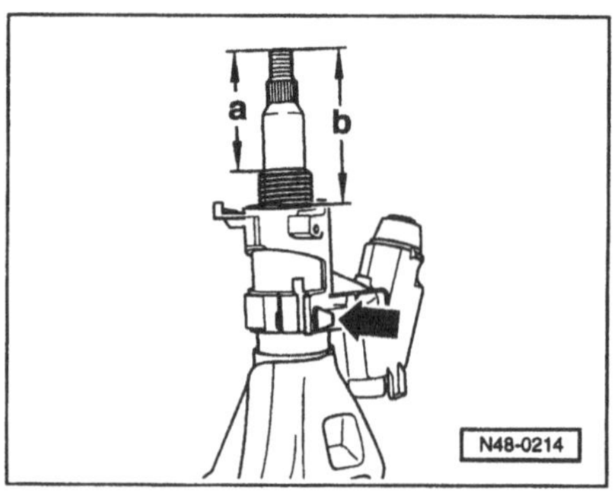

Installation position of clamping washer and steering lock housing

Dimension -a- for clamping washer: minimum 74 mm (2.91 in.).

Dimension -b- for steering lock housing: 92 mm + 1 mm (3.62 in. + 0.039 in.).

- Tighten shear head bolt -arrow- for steering lock until head shears off.

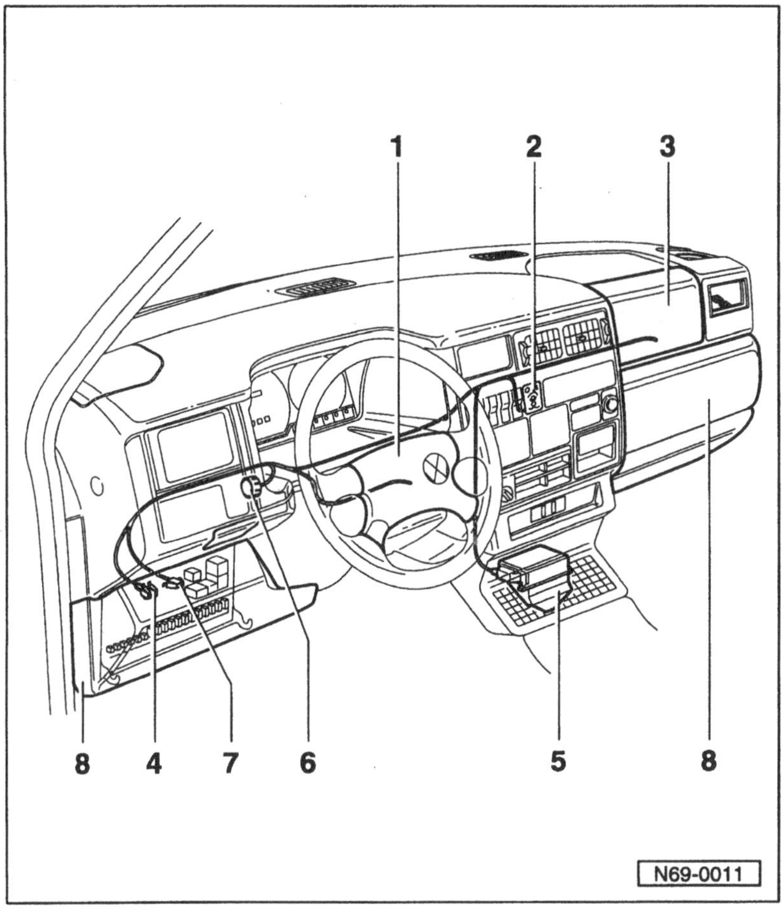

Airbag, removing and installing

1 - **Airbag unit driver's side**
⇒ Removing and installing, Page N 48-18.

2 - **Airbag Malfunction Indicator Lamp -K75-**

3 - **Airbag unit passenger's side**

4 - **Data Link Connectors**

5 - **Airbag Control Module -J 234-**

6 - **Central ground star**

7 - **Relay plate (terminal 15)**

8 - **Knee padding**
⇒ Removing and installing, Page N 48-1.

Airbag, safety precautions

Caution!

- ***Testing, removing, installing and repair work may only be performed by trained Service department personnel.***
- ***Always disconnect the battery ground strap (–) before working on airbag related systems.***
- ***No waiting time is necessary after disconnecting the battery.***
- ***When connecting the airbag system to a voltage source, there must be no person inside the vehicle.***
- ***Any time the battery has been disconnected it will be necessary to reestablish basic settings for the Engine Control Module (ECM) and, on vehicles with automatic transmission, for the Transmission Control Module (TCM), using VAG 1551 Scan Tool (ST).***
- ***On vehicles equipped with anti-theft radios, make sure you know the radio activation code before you disconnect the battery.***
- ***The airbag units should be installed immediately after removing them from their packaging.***
- ***If the work on the airbag system is interrupted, the airbag unit should again be placed in its packaging. Do not leave the airbag unit unattended!***
- ***If the airbag unit or the airbag control module has been dropped from a height of 0.5 meter (19.7 in.) or more, DO NOT install.***

CAUTION!

- ***Airbag units which have been removed should be stored so that the padded side is facing up.***
- ***Always replace any airbag system component that has been mechanically damaged (example: dented or cracked).***
- ***Airbag units which have NOT been activated should be marked and returned to the manufacturer for disposal (use airbag unit transport container).***

Replacing airbag units following an accident

Accident involving activation of airbag

- The following components must always be replaced:
  Driver airbag unit,
  Front passenger airbag unit,
  Airbag control module with holder,
  Coil connector with slip ring,
  Driver seat belt,
  Front passenger seat belt (if used).
- In addition, the following must also be replaced if necessary, (visual check):
  All damaged components.

Accident not involving activation of airbag

- If the Airbag Malfunction Indicator Lamp -K75- does not indicate a malfunction, it is not necessary to replace the airbag components. The seat belts should be checked separately.

After replacing the airbag units, or the control module, affix stickers (tear-off strips) to the registration card and return the registration card to the relevant Importer for registration.

For vehicles produced from 09.98

Removal of driver's side airbag from steering wheel ⇒ Page N 48-51.

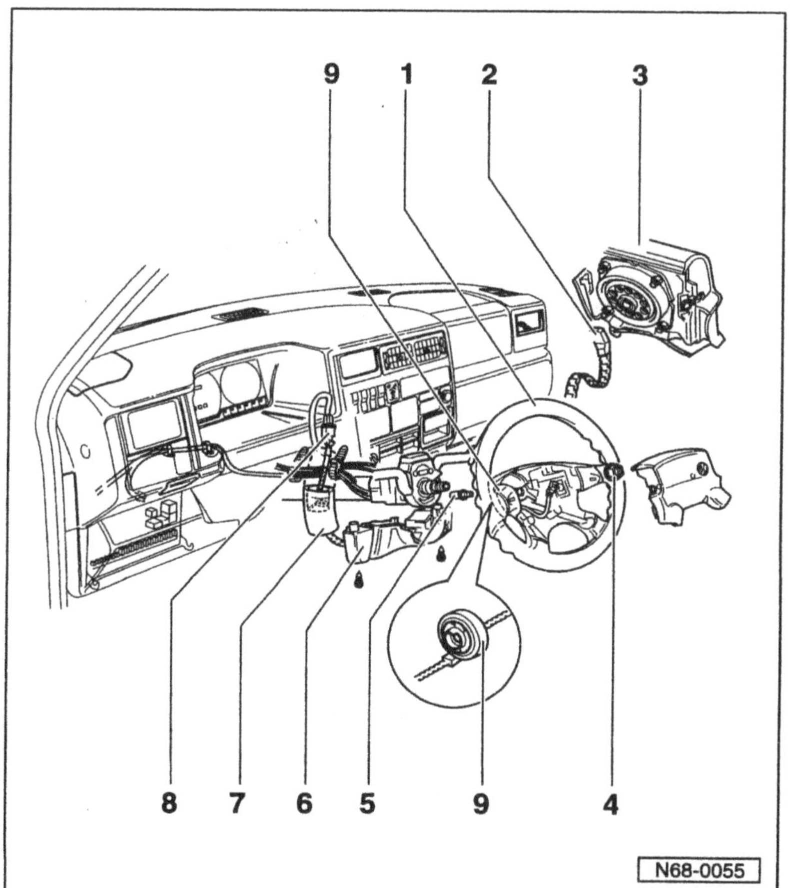

Driver's airbag unit, removing and installing

1 - Steering wheel
- Removing, ⇒ Page N 48-20.

2 - Airbag unit connector

3 - Airbag unit

4 - Hex nut
- **67 Nm (49 ft lb)**

5 - Torx screw (T 30)
- 6.5 Nm (58 in. lb).
- Always replace.

6 - Lower trim

7 - Foam tube

8 - Connector

9 - Airbag spiral spring -F138-
- Removing, ⇒ Page N 48-21.

> ***Caution!***
>
> ***On cars equipped with anti-theft radios, make sure you know the radio activation code before you disconnect the battery.***

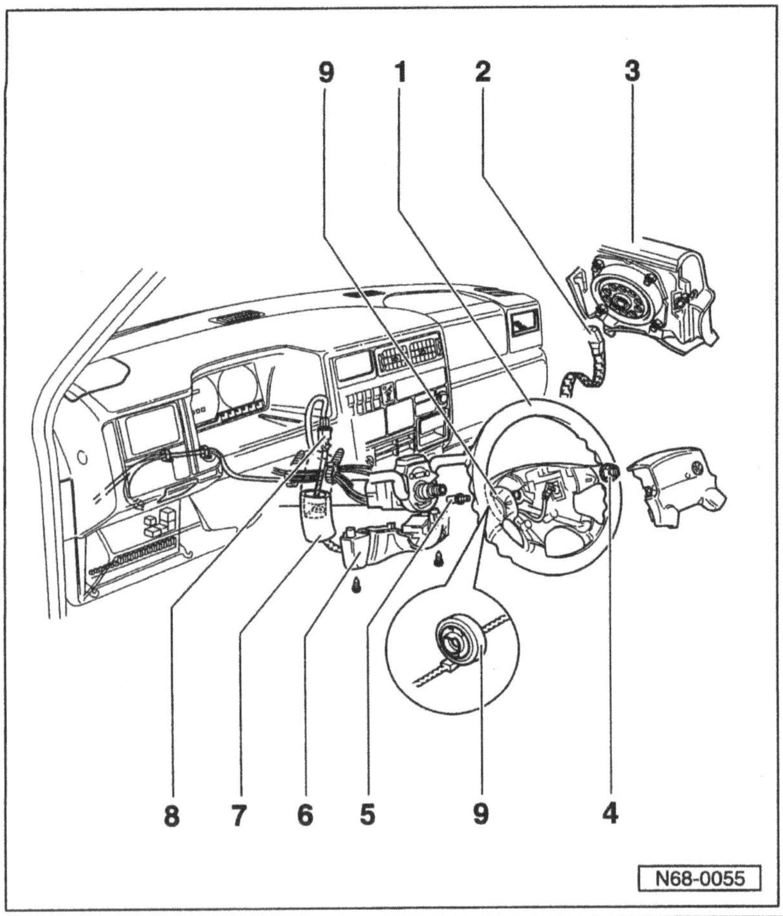

Removing

- Disconnect battery ground strap.
- Turn steering wheel -1- to center position (wheels straight-ahead).
- Remove Torx socket-head screws (T 30) -5-. (through 08.98 production)
- Remove airbag unit -3- from steering wheel and disconnect red connector -2- from airbag unit.

Installing

- Install in reverse order or removal.

Tightening torque (bolts -5-):
6.5 Nm (58 in. lb).

> ***Warning!***
> ***Ensure that there is no person inside the vehicle during this operation.***

After all driver's side airbag components are installed:

- Switch ignition ON.
- Connect battery ground strap (-).

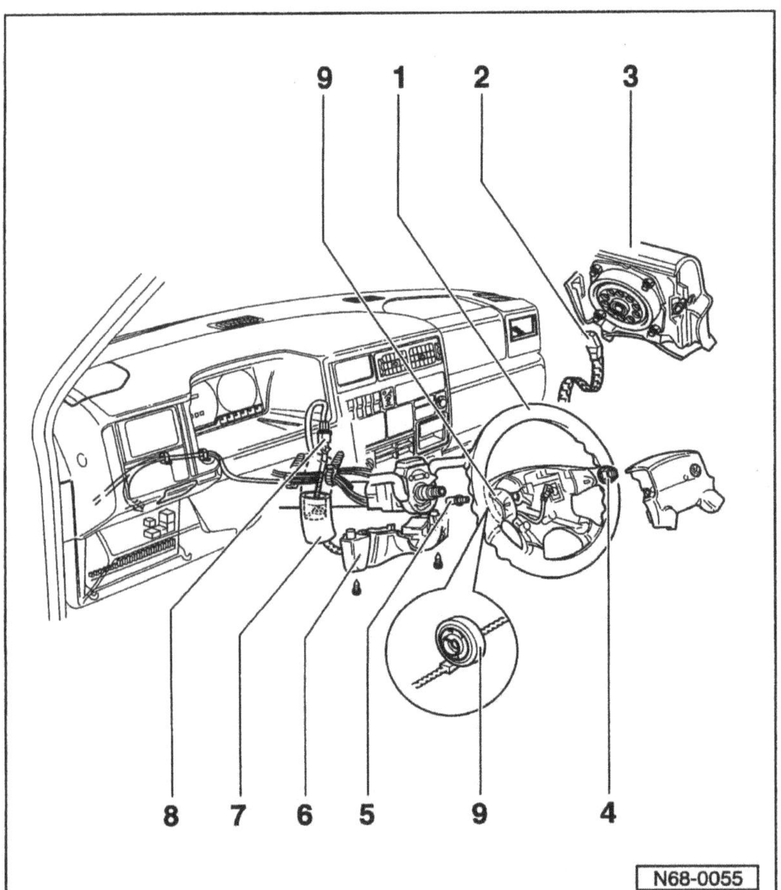

Steering wheel with airbag, removing and installing

Requirement:

♦ Driver's side airbag unit removed,
⇒ Page N 48-18 (through 08.98 prod.)
For vehicles produced from 09.98
⇒ Page N 48-51

- Remove lower trim -6- from steering column switch.
- Pull foam tube -7- off connector -8- and separate connector.
- Remove hex nut -4- and remove steering wheel from steering column.

  Tightening torque (nut -4-):
  67 Nm (49 ft lb).

Note:

Airbag spiral spring -F138- -9- remains on the steering wheel.

Airbag spiral spring -F138-, removing and installing

Requirements:

- Driver's side airbag unit removed, ⇒ Page N 48-18.
- Steering wheel removed, ⇒ Page N 48-20.

– Separate single connector for horn.

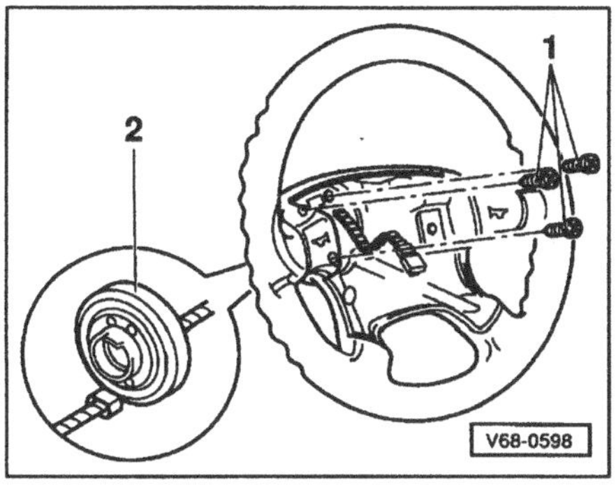

◄ – Remove Phillips screws -1- and remove airbag spiral spring -2- from steering wheel.

Note:

- *The steering must be in the center position (wheels straight-ahead) when the airbag spiral spring -F138- is removed and installed.*
- *Replacement airbag spiral springs are secured in the center position with a cable tie.*

Checking center position

– Turn airbag spiral spring (clockwise) onto stop and then turn back 3.5 turns.

Power steering, servicing

Vane pump delivery pressure, checking

Note:

Adapter VAG 1402/6 is required for checking delivery pressure. The tools 1, 2, and 3, shown in illustration V48-0846, are parts of adapter set VAG 1402/6.

Requirements:

- V-belt/V-belt tension OK.
- System not leaking.
- Hoses/lines are not kinked or restricted.

– Remove noise insulation tray.

– Remove pressure line on power steering gear.

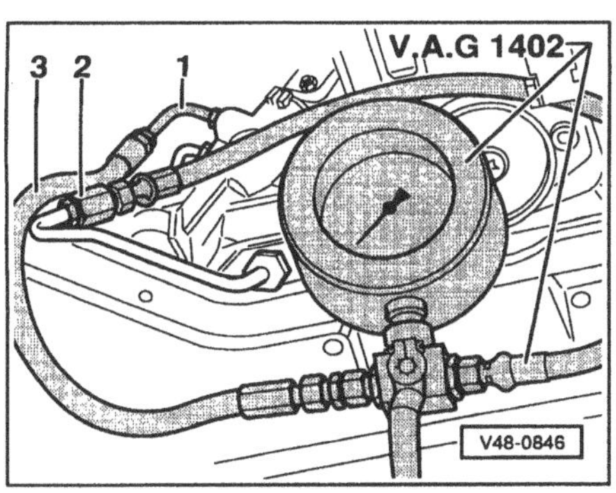

◄ – Connect pressure gauge VAG1402 as illustrated. Cut-off valve on gauge must be open.

– Start engine and top up oil level in reservoir if necessary.

Note:

Use only hydraulic oil G 002 000.

– Check hose connections for leaks.

– With engine running at idling speed, close cut-off valve (not longer than 5 seconds) and read pressure on gauge.

● Specification: 95 ... 110 bar (1377 ... 1595 psi).

Notes:

♦ *If the reading deviates by more than 5 bar above or below specification, replace the pump.*

♦ *Assembly overview: Vane pump for vehicles with 5-Cylinder engine, ⇒ Page N 48-45.*

♦ *Assembly overview: Vane pump for vehicles with VR6 engine, ⇒ Page N 48-48.*

♦ *For leaks at the pinion, first check the lines and connections for leaks and, if necessary, tighten and wipe dry.*

♦ *Replace the steering gear if the pinion seal or the rack seals in the steering gear housing are leaking. Move boot to one side to check rack seal for leaks.*

Power steering gear, removing and installing

Caution!

♦ ***Absolute cleanliness is required when working on the power assisted steering.***

♦ ***Thoroughly clean connections and immediate vicinity before loosening.***

♦ ***Place parts on a clean surface and cover if not re-installing immediately.***

♦ ***Do not use fluffy cloth.***

♦ ***Carefully cover or seal open components, if repairs cannot be carried out immediately.***

♦ ***Only remove replacement parts from packaging immediately prior to installation.***

♦ ***Only use genuine packed parts.***

♦ ***When the system is open, do not work with compressed air in the area and do not move the vehicle.***

Removing

- Relieve torsion bar tension,
⇒ Page N 40-11; Removing and installing torsion bar/adjusting torsion bar.

- Extract hydraulic oil with VAG 1358 A or Fluid Evacuator EZ1 or equivalent.

- Remove noise insulation tray.

- Remove front exhaust pipe.
⇒ Repair group 26; Removing and installing parts of exhaust system.

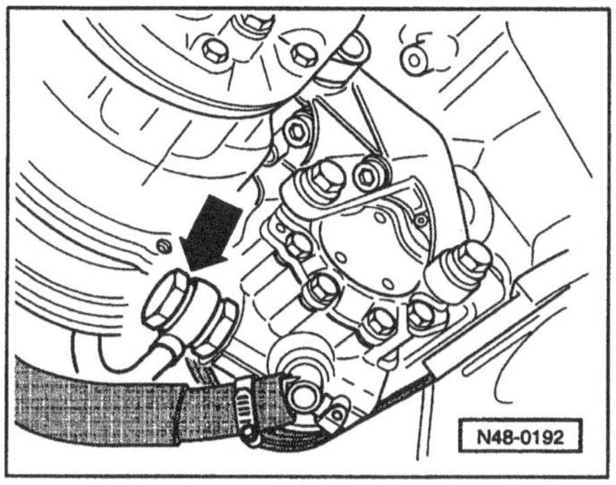

- Remove driveshaft (only vehicles with four wheel drive),
⇒ 5-Speed manual gearbox 02D; Repair group 39; Removing and installing driveshaft.

◄ - Open union bolt on vane pump -arrow- and drain hydraulic oil (if not already extracted).

- Disconnect hoses from steering gear and seal with plastic bag and adhesive tape.

- Seal threaded holes on power steering box with plastic plugs.

- Remove tie rods at steering rack,
⇒ Page N 48-31.

- Remove stabilizer bar clamps from assembly carrier.

- Remove bolt for steering universal joint at steering pinion.

- Unbolt power steering gear from assembly carrier,
⇒ Page N 48-32.

- Push up stabilizer bar.

- Move power steering gear slightly to right to pull power steering gear pinion out of universal joint.

- Swing left-hand side of power steering gear out toward rear of vehicle.

Installing

Install in reverse order of removal.

Notes:

If the steering box or the tie rod(s) are replaced check toe and adjust if necessary.

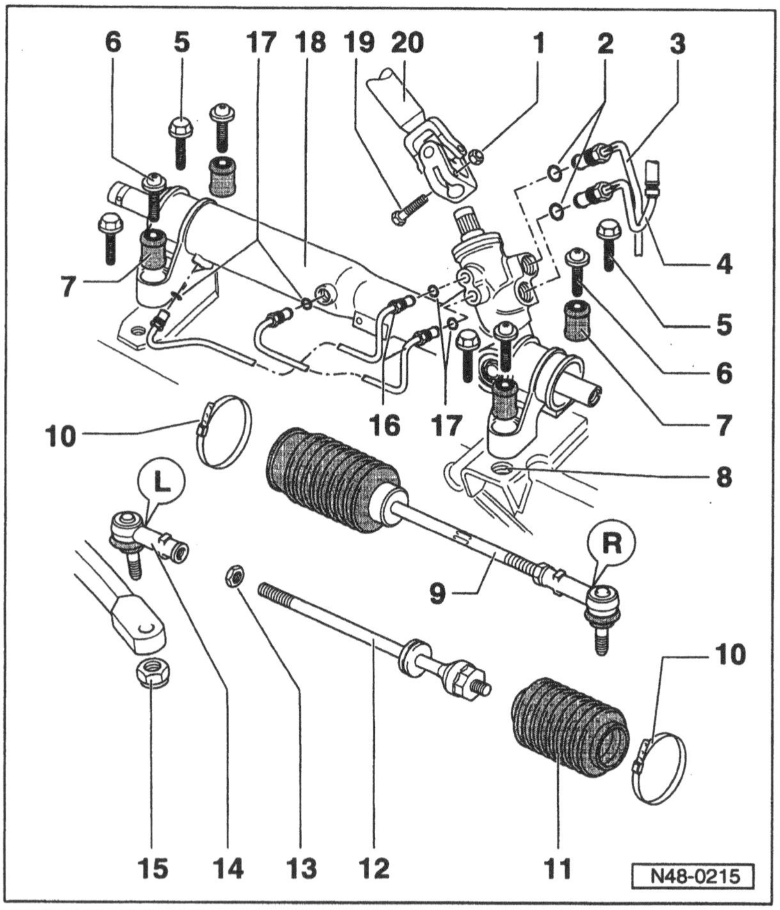

Assembly overview: power steering gear

No provision is made for servicing the power steering gear. For complaints, determine the cause with the help of the pressure test.

Notes:

- *Always replace self-locking nuts and bolts.*
- *Do not weld or straighten steering components.*
- *Only use steering gear grease, Part No. AOF 063 000 04, to grease the steering gear rack.*
- *Use only hydraulic oil, Part No. G 002 000.*
- *Quantity of oil in system: approximately 1 L (1.0 qt).*

1 - Self-locking hex nut
- 30 Nm (22 ft lb) and turn 90° further (1/4 turn).
- Always replace.

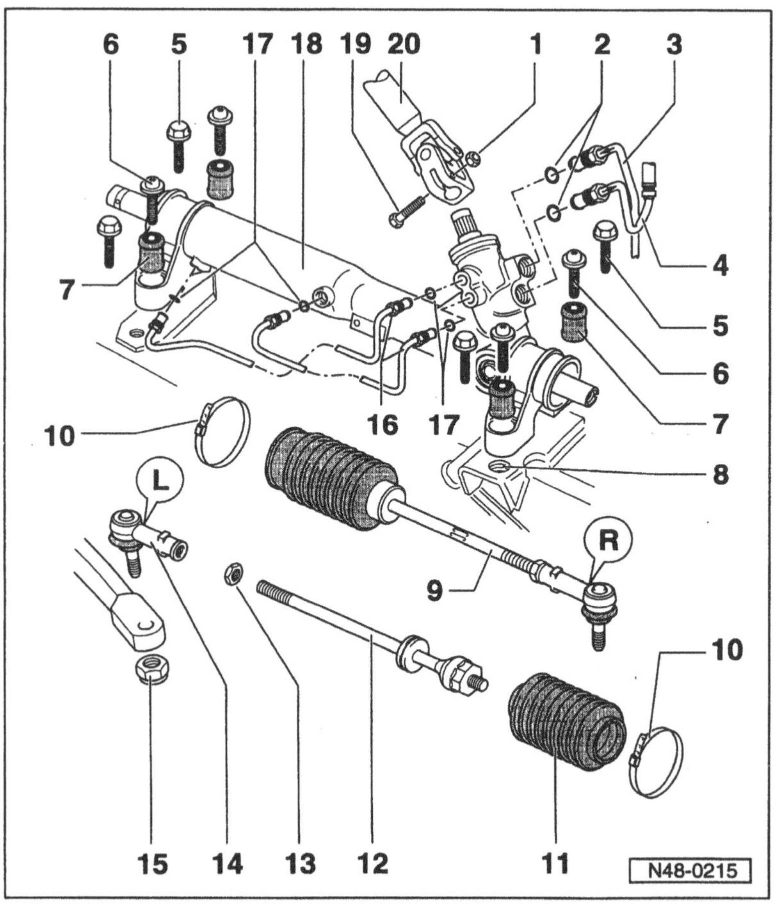

2 - O-ring
- Always replace.

3 - Pressure line, 30 Nm (22 ft lb)

4 - Return line, 30 Nm (22 ft lb)

5 - Hex bolt, M 10
- 60 Nm (44 ft lb),

⇒ Loosening and tightening Fig. 4.

6 - Torx bolt, M 10
- 60 Nm (44 ft lb), can be reused.
- The left-hand bolt behind the steering gear is to be replaced by an M 10 x 58 hex bolt on vehicles with additional heating.

⇒ Loosening and tightening Figs. 4, 5, and 6.

7 - Rubber mounting

⇒ Pressing out and in, Fig. 3.

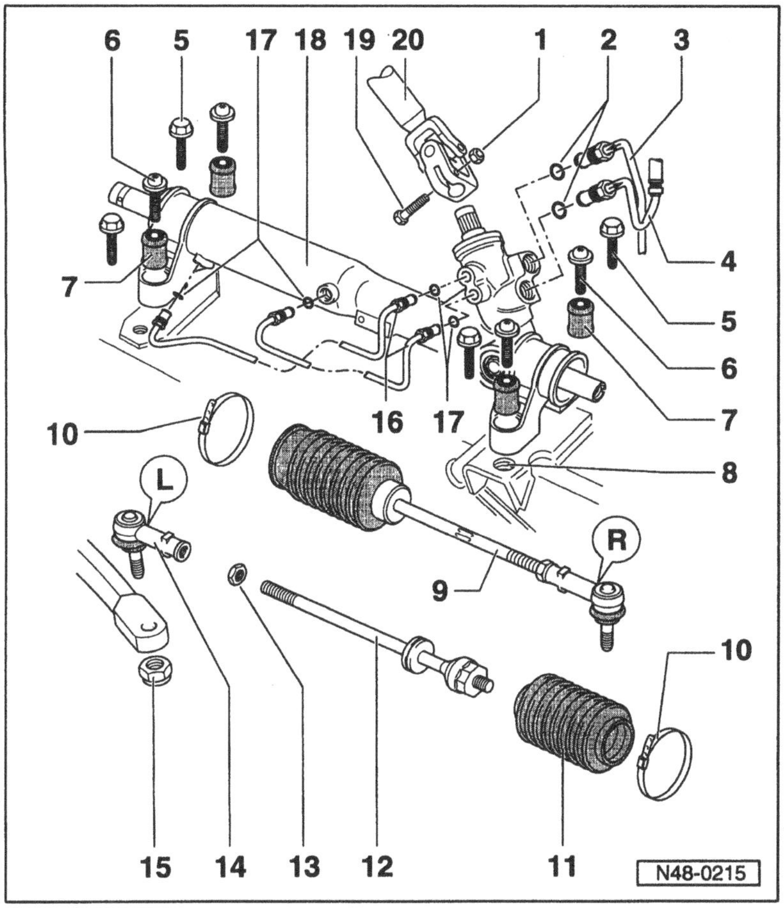

8 - Threaded hole in assembly carrier

9 - Tie rod, right side, 70 Nm (52 ft lb)

⇒ Unbolting and bolting on, Fig. 1.

⇒ Adjusting track, Page N 44-16.

10 - Cable tie

11 - Boot

- Check for damage.
- Can be replaced with steering gear installed.
- Must not be twisted after adjusting toe.

12 - Tie rod, left side, 70 Nm (52 ft lb)

⇒ Unbolting and bolting on, Fig. 1.

⇒ Adjusting track, Page N 44-16.

13 - Hex nut, 55 Nm (41 ft lb)

14 - Tie rod ball joint

- Check that dust caps are seated correctly and are not damaged

⇒ Pressing off, Fig. 2.

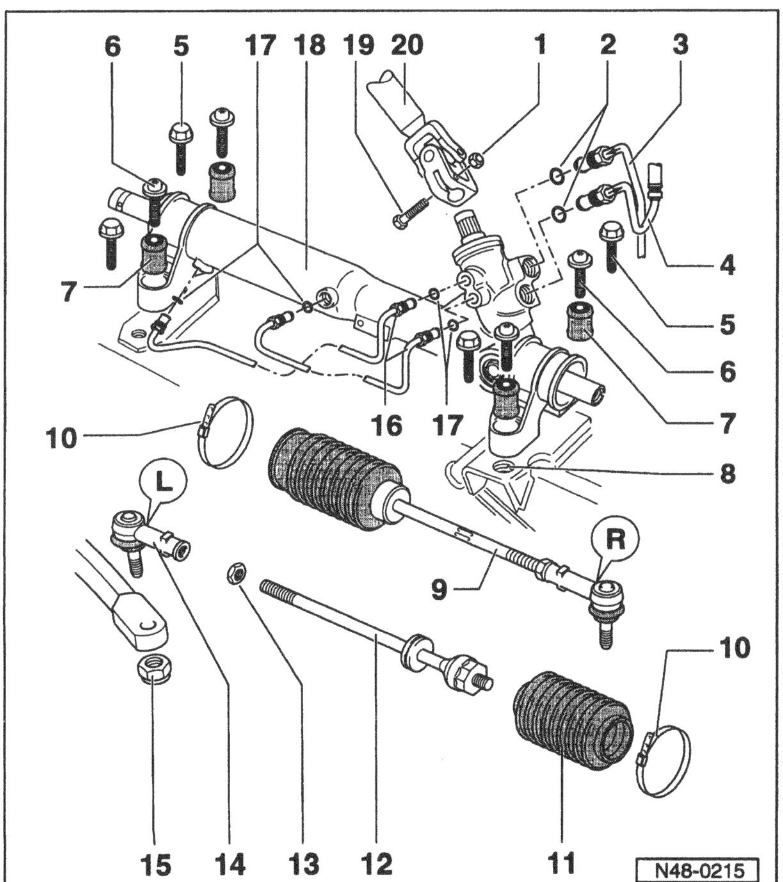

15 - Self-locking hex bolt, 65 Nm (48 ft lb)

- Always replace.

16 - Control lines

- 14 Nm (10 ft lb).
- When installing, ensure that seals are not damaged.
- Install stress–free.
- Do not damage surface; repair paint damage, if necessary.
- Can be removed and installed with steering gear installed.

17 - O-ring

- Always replace.

18 - Power steering gear

⇒ Removing and installing, Page N 48-24.

19 - Hex bolt

20 - Lower steering intermediate shaft with universal joint

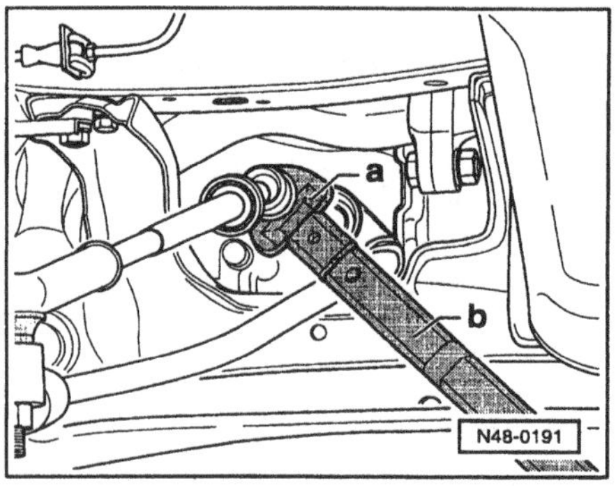

◄ **Fig. 1 Removing and installing tie rod**

a - VAG 1332/5

b - Torque wrench VAG 1332

Note:
Do not use locking fluid.

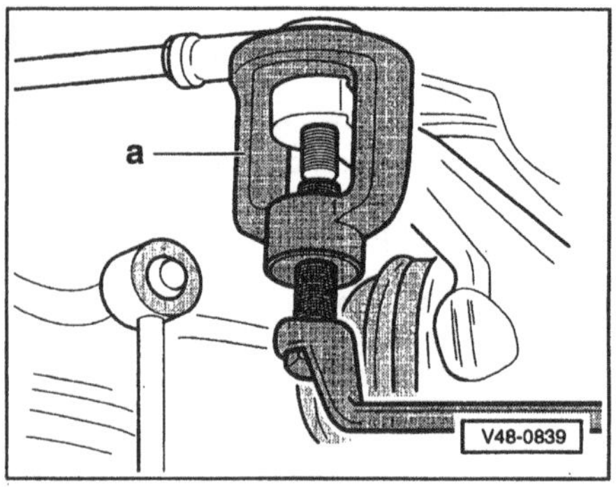

◄ **Fig. 2 Pressing tie rod off steering arm**

a - Ball joint splitter (commercial type), e.g., Kukko No. 128/2.

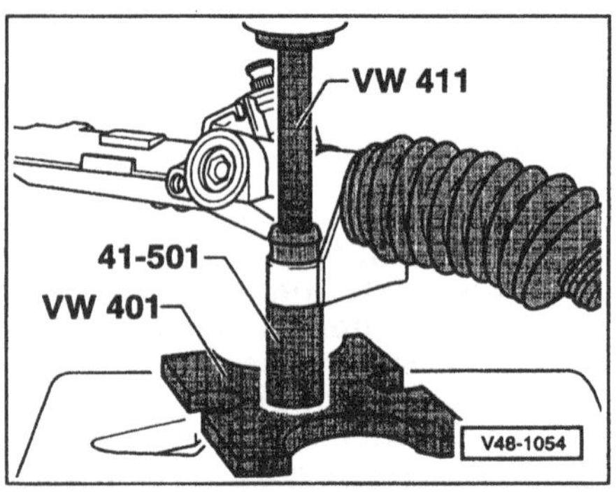

◄ **Fig. 3 Pressing rubber bushing out and in**

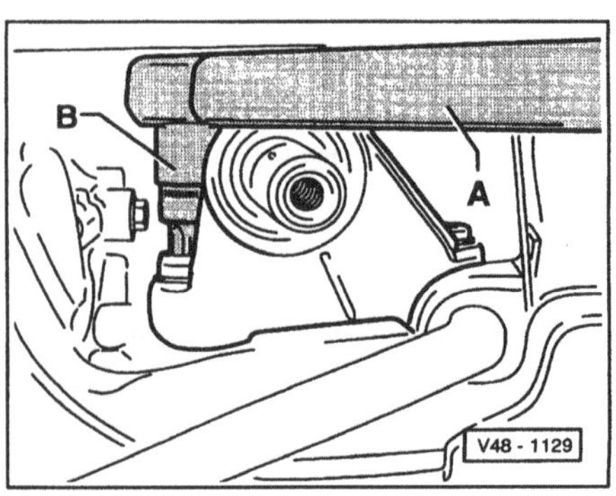

◄ **Fig. 4 Loosening and tightening bolts for steering gear (Torx bolts)**

A - Torque wrench VAG 1332.

B - Socket insert Torx T 50 (commercially available).

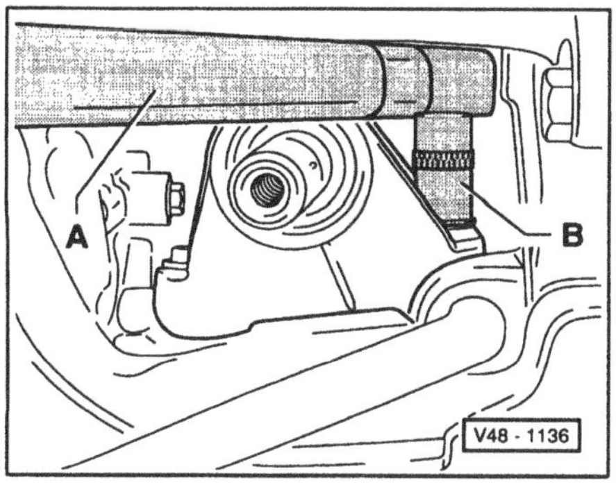

◄ **Fig. 5 Loosening and tightening bolts for steering gear (M 10 hex bolts)**

A - Torque wrench VAG 1332.

B - Socket (commercially available).

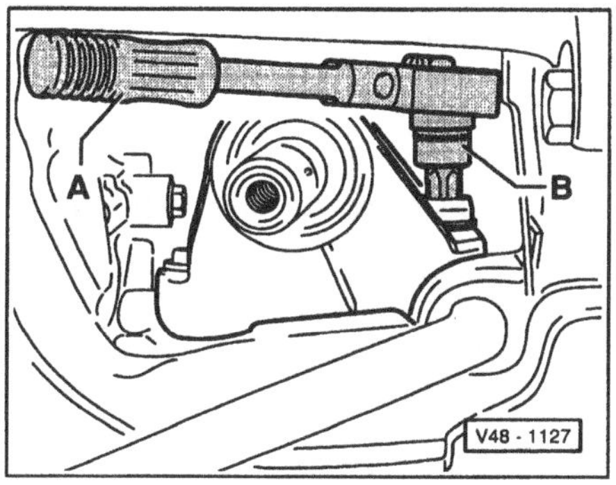

◄ **Fig. 6 Removing left rear torx bolt on vehicles with additional heater**

Use an M10 x 58 hex head bolt instead of the torx bolt in this position, Part No. N 905 886 01.

A - Ratchet (commercially available $^3/_8$" drive).

B - Torx T 50 socket (commercially available).

> ***Caution!***
>
> ***Part numbers are for reference only. Always check with your Parts Department for the latest information.***

Power steering pump, reservoir and hydraulic lines, servicing

No provision is made for servicing the power steering pump. If the pump is malfunctioning determine the cause with the help of a pressure test. If faulty, replace the power steering pump.

Notes:

- *Pumps supplied as spare parts are not filled with oil. Fill the pump with G 002 000 hydraulic oil, and rotate it by hand before installing, or the pump may be damaged or noisy during operation.*
- *Type of oil: hydraulic oil: G 002 000.*
- *Quantity of oil in system: approx. 1 L (1.0 qt).*
- *Do not reuse drained hydraulic oil.*
- *Always replace O-rings.*

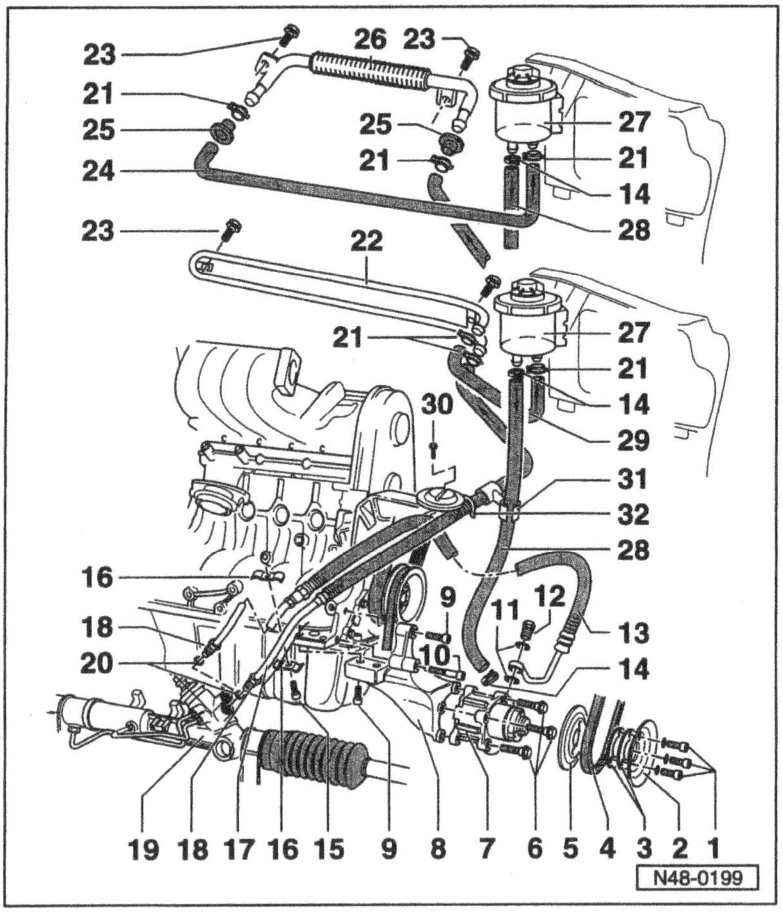

Power steering pump, hydraulic lines/hoses, reservoir (vehicles with VR6 engine), assembly overview

Notes:

- *See notes on page N 48-34.*
- *Items 1 ... 10: Not applicable for US/ Canada vehicles.*
- *For US/Canada vehicles with VR6 engine, ⇒ Page N 48-48, for belt/pulley/ bracket components.*

11 - O-ring
- Always replace.

12 - Banjo bolt, 40 Nm (30 ft lb)

13 - Pressure line
- With non-return valve.
- Ensure hose is routed free of kinks.

14 - Screw type clip
- 4 Nm (35 in. lb).

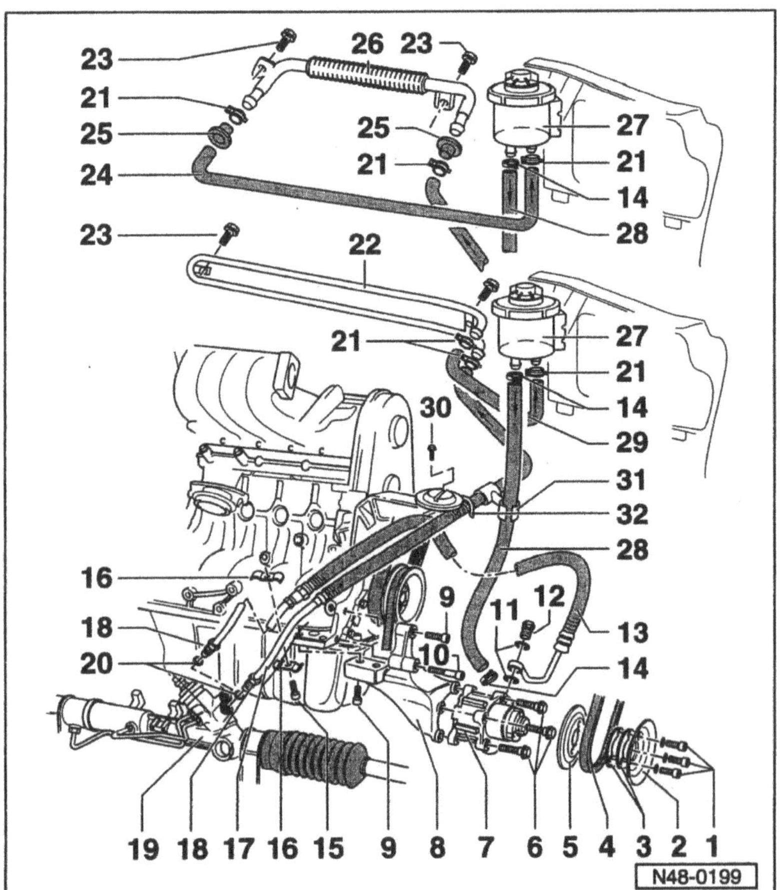

15 - 10 Nm (7 ft lb)
- Installed from above during production, but during repairs, install from below.

16 - Clamp
- Upper and lower parts.
- Install 50 mm (2 in.) from end of hose.

17 - Return line
- Ensure hose is routed free of kinks.
- Route at a distance of 20 mm (3/4 in.) to assembly carrier.

18 - 30 Nm (22 ft lb)

19 - Steering gear

20 - Sealing ring
- Always replace.

21 - Spring-type clamp

22 - Cooling line
- Long front end.

23 - 8 Nm (70 in. lb)

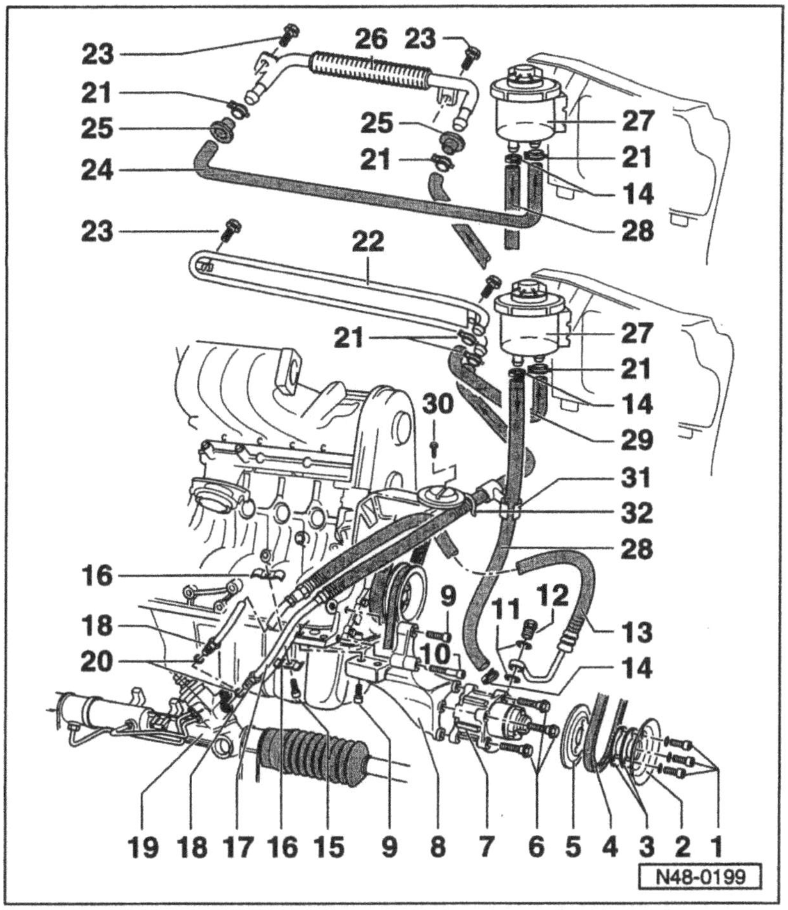

24 - Return hose
- Ensure hose is routed free of kinks.

25 - Bushing

26 - Cooling line
- Short front end.

27 - Reservoir
- Oil level: between MIN and MAX marks.

28 - Suction line
- Ensure hose is routed free of kinks.
- Installation position, ⇒ Fig. 4.

29 - Return hose
- Ensure hose is routed free of kinks.

30 - 10 Nm (7 ft lb)

31 - Line/hose holder
- Install at colored markings.

32 - Bracket
- Installation position, ⇒ Fig. 3.

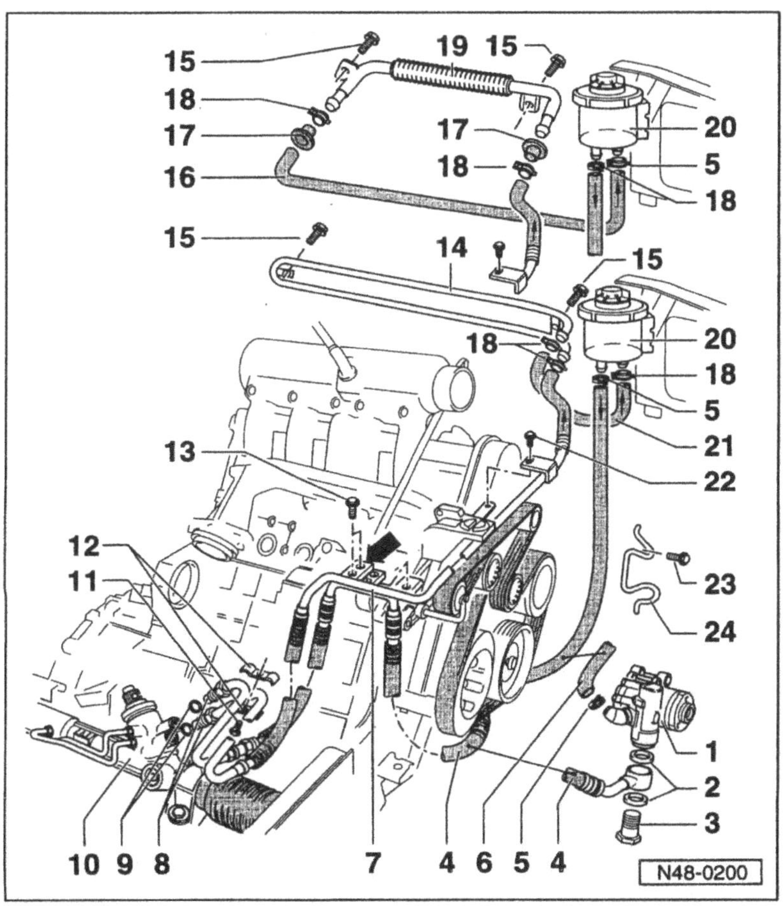

Power steering pump, hydraulic lines, reservoir (vehicles with 5-Cylinder engine), assembly overview

Note:

⇒ Page N 48-34.

1 - Power steering pump
- Removing and installing, ⇒ Page N 48-50.
- Checking delivery pressure, ⇒ Page N 48-22.
- Before installing fill with oil, ⇒ Note Page N 48-34.

2 - O-ring

3 - Banjo bolt, 40 Nm (30 ft lb)

4 - Pressure line
- With non-return valve.
- Installation instructions, ⇒ Fig. 5.

5 - Screw-type clip
- 4 Nm (35 in. lb).

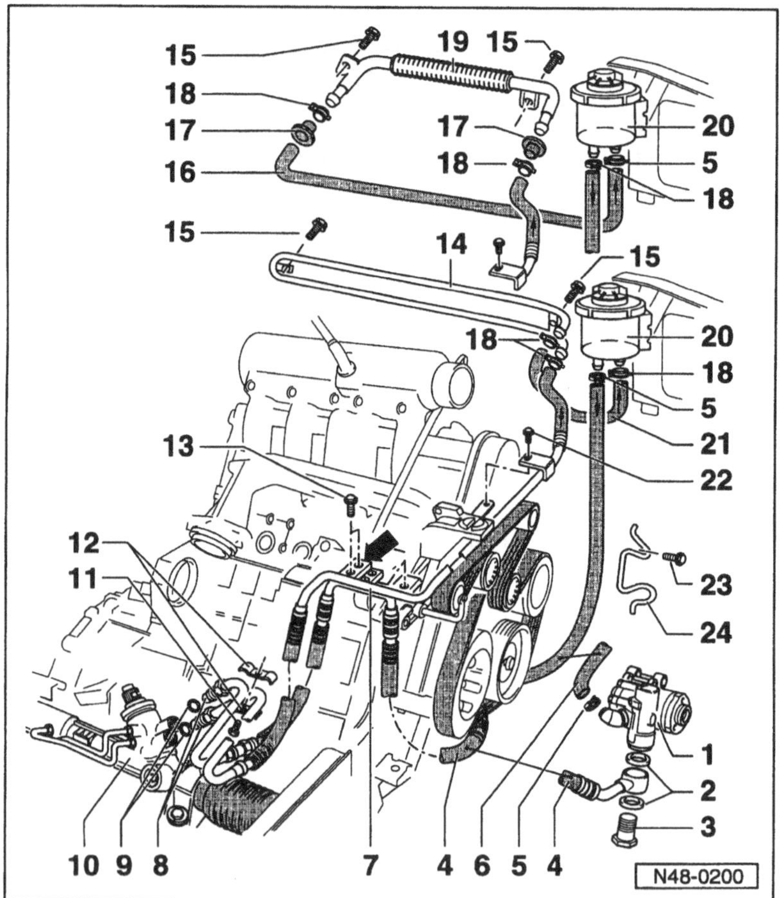

6 - Suction line
- Ensure hose is routed free of kinks.

7 - Return line
- Ensure hose is routed free of kinks.

8 - 30 Nm (22 ft lb)

9 - Sealing ring
- Always replace.

10 - Steering gear

11 - 10 Nm (7 ft lb)
- Installed from above during production but during repairs, install from below.

12 - Clamp
- Upper and lower parts
- Installation position, ⇒ Fig. 6.

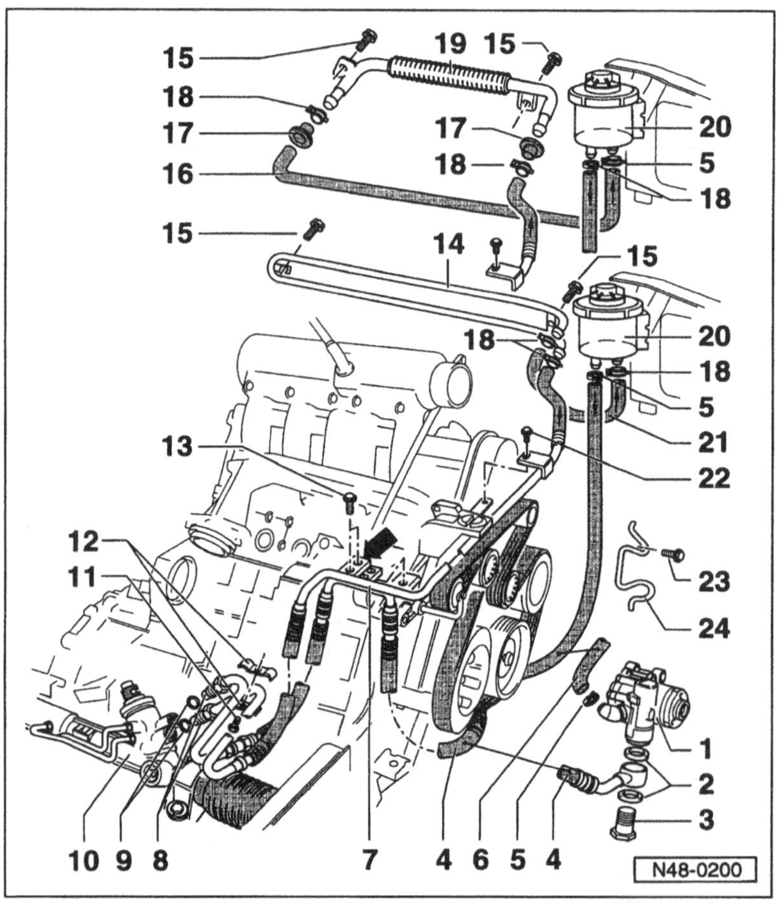

13 - 20 Nm (15 ft lb)
- Installation position with automatic transmission, ⇒ arrow.

14 - Cooling line
- Long front end,
- ⇒ Fig. 1.

15 - 8 Nm (70 in. lb)

16 - Return hose
- Ensure hose is routed free of kinks,
- ⇒ Fig. 2.

17 - Bushing

18 - Spring-type clamp

19 - Cooling line
- Short front end.

20 - Reservoir
- Oil level: between MIN and MAX marks.

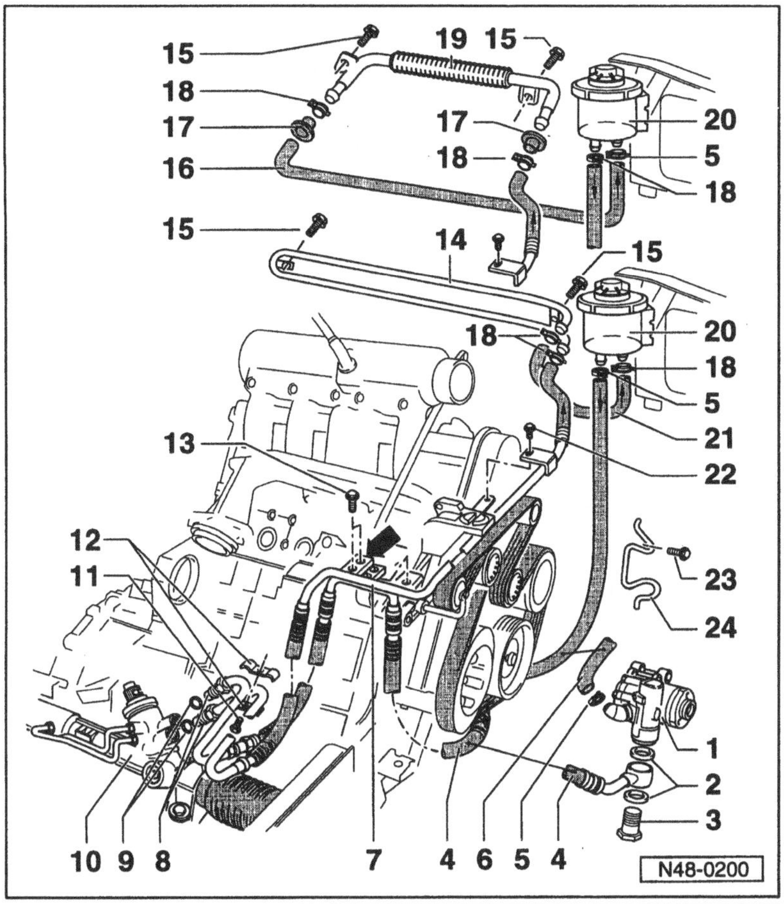

21 - Return hose
- Ensure hose is routed free of kinks.

22 - 10 Nm (7 ft lb)

23 - 8 Nm (71 in. lb)

24 - Bracket
- Installation position, ⇒ Fig. 1.

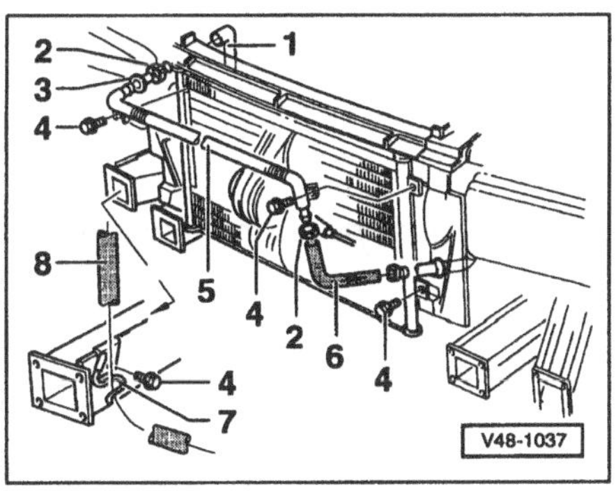

◄ **Fig. 1 Return line to cooling line**

1 - Return line
2 - Spring-type clip
3 - Grommet
4 - Bolt, 8 Nm (71 in. lb)
5 - Cooling line
6 - Return hose
7 - Bracket (only 5 and 6-cyl. engines)
8 - Suction hose

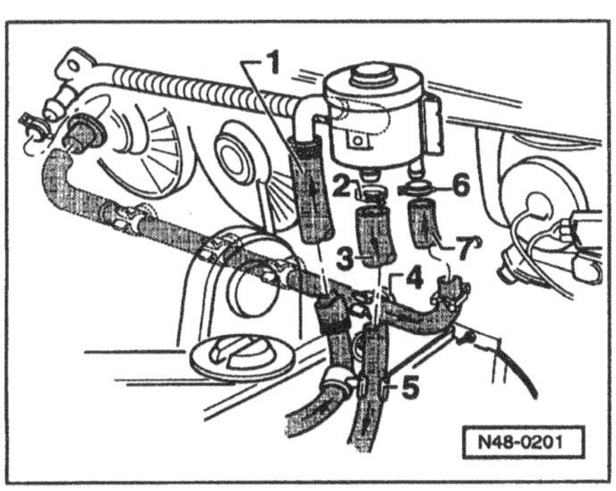

◄ **Fig. 2 Return line hose routing**

1 - Return line
2 - Screw-type clip
3 - Suction hose
4 - Hose retainer
5 - Hose/line holder (6-cyl. engines)
6 - Spring-type clip
7 - Return hose

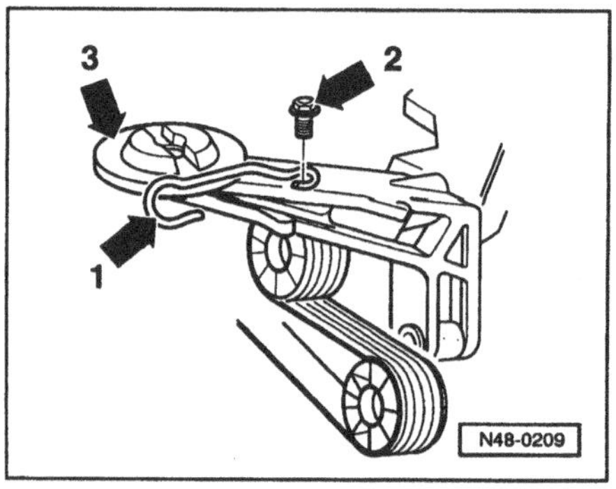

Fig. 3 Retainer installation position

Retainer -arrow 1- must lie against engine support -arrow 3- when tightening hex bolt -arrow 2-.

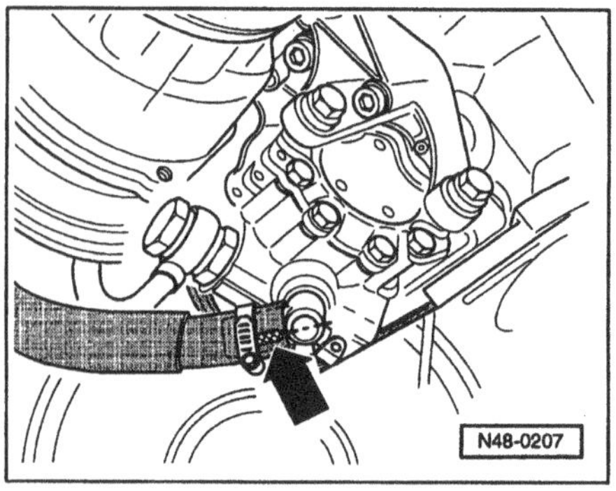

Fig. 4 Suction hose installation position

Colored marks on hose -arrow- must align with molded shape on support.

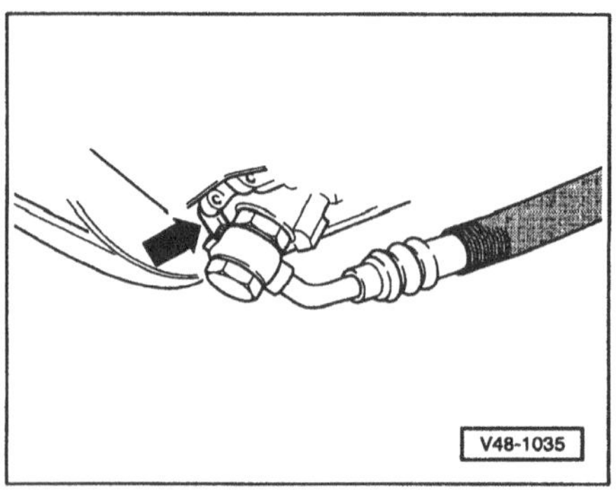

Fig. 5 Pressure line installation instructions

Note:

Angled stop -arrow- must rest against pump housing when tightening the union bolt.

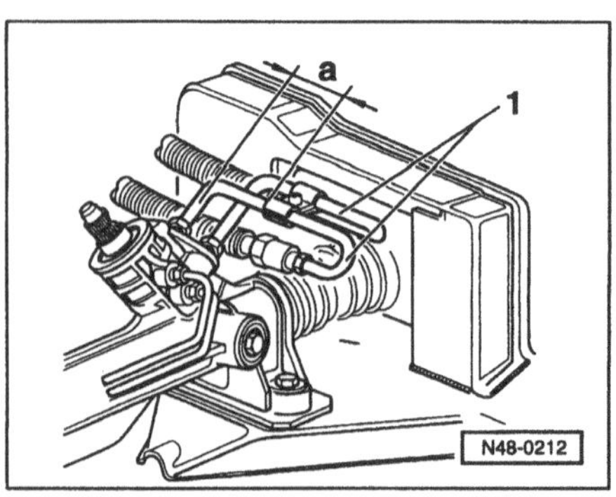

Fig. 6 Clamp installation position

Note:

Install line parallel to edge of assembly carrier.

Dimension -a- approx. 60 mm (2 3/8 in.)

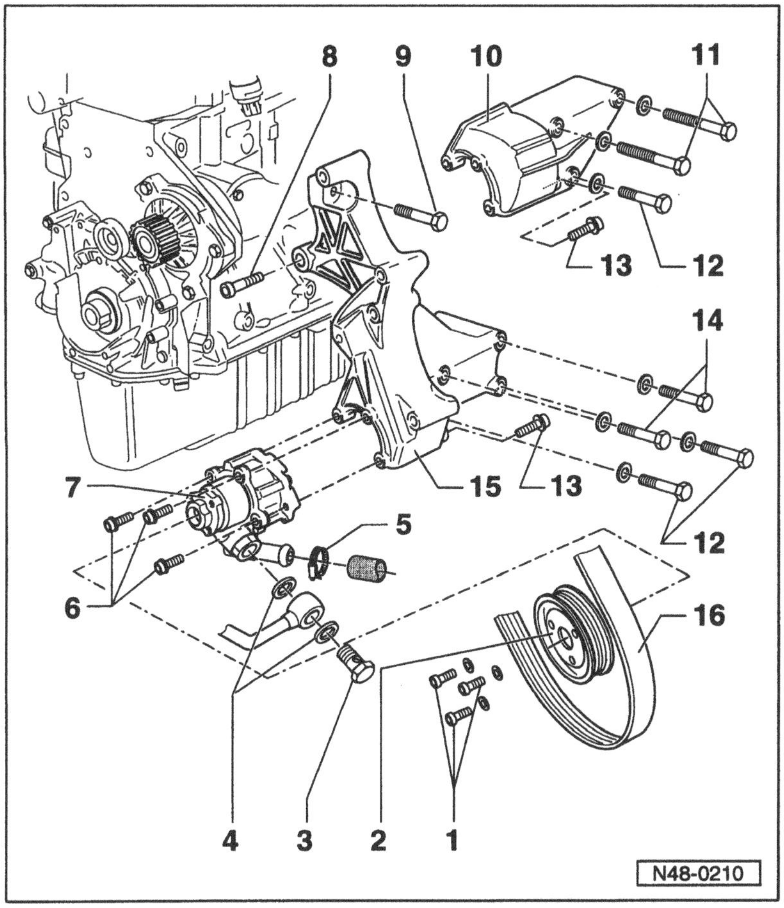

Power steering pump (vehicles with 5-Cylinder engine), assembly overview

Note:

⇒ Page N 48-34.

1 - **Hex socket head bolt, 20 Nm (15 ft lb)**
- ♦ Loosening and tightening, ⇒ Page N 48-50.

2 - **Belt pulley**

3 - **Banjo bolt, 30 Nm (22 ft lb)**

4 - **O-ring**
- ♦ Always replace.

5 - **Screw-type clip**
- ♦ 4 Nm (35 in. lb).

6 - **Hex socket head bolt, 20 Nm (15 ft lb)**

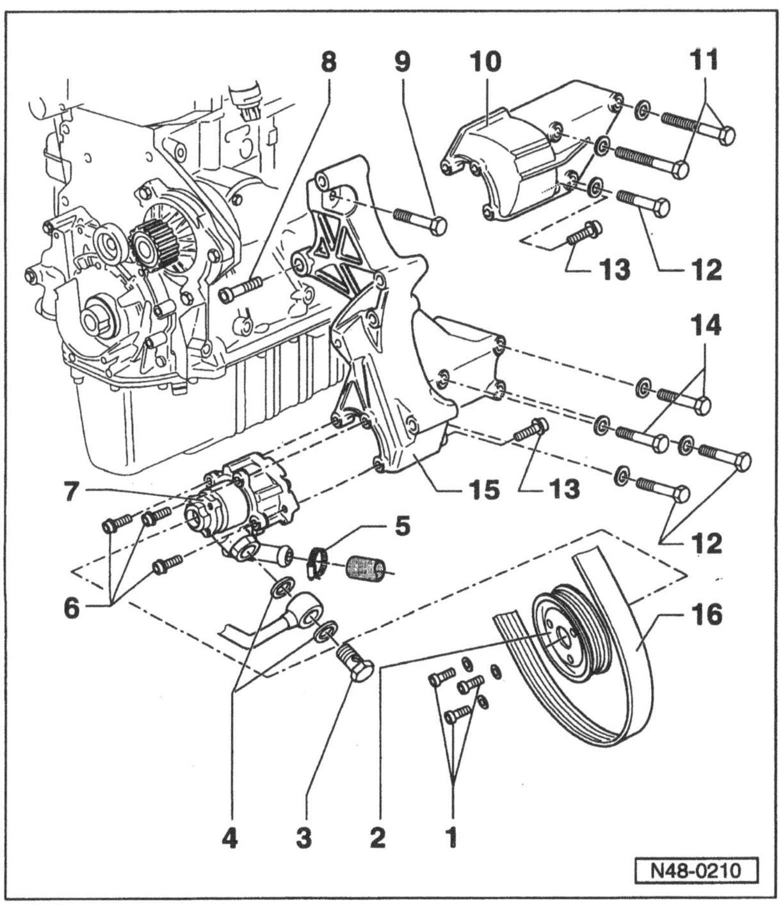

7 - **Power steering pump**
- ♦ Checking delivery pressure, ⇒ Page N 48-22.
- ♦ Removing and installing, ⇒ Page N 48-50.
- ♦ Before installing fill with oil, ⇒ Note Page N 48-34.

8 - **Hex socket head bolt, 40 Nm (30 ft lb)**

9 - **Hex bolt, 40 Nm (30 ft lb)**

10 - **Mounting bracket**
- ♦ Various versions, ⇒ Parts microfiche.

11 - **Hex bolt, 40 Nm (30 ft lb)**

12 - **Hex bolt, 40 Nm (30 ft lb)**
- ♦ Fitted bolt.

13 - **Hex socket head bolt, 20 Nm (15 ft lb)**

14 - **Hex bolt, 40 Nm (30 ft lb)**

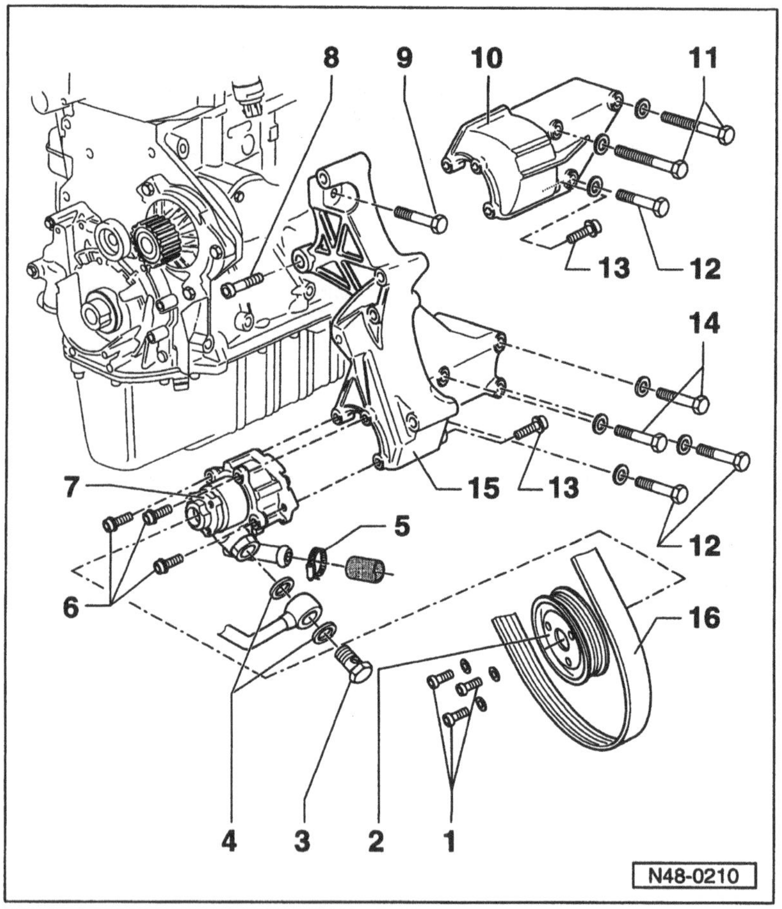

15 - **Mounting bracket**
- ♦ Various versions,
  ⇒ Parts microfiche.

16 - **Ribbed belts**
- ♦ Removing and installing:
⇒ 5-Cylinder engine, Mechanics.
- ♦ Checking condition,
⇒ Maintenance Manual.

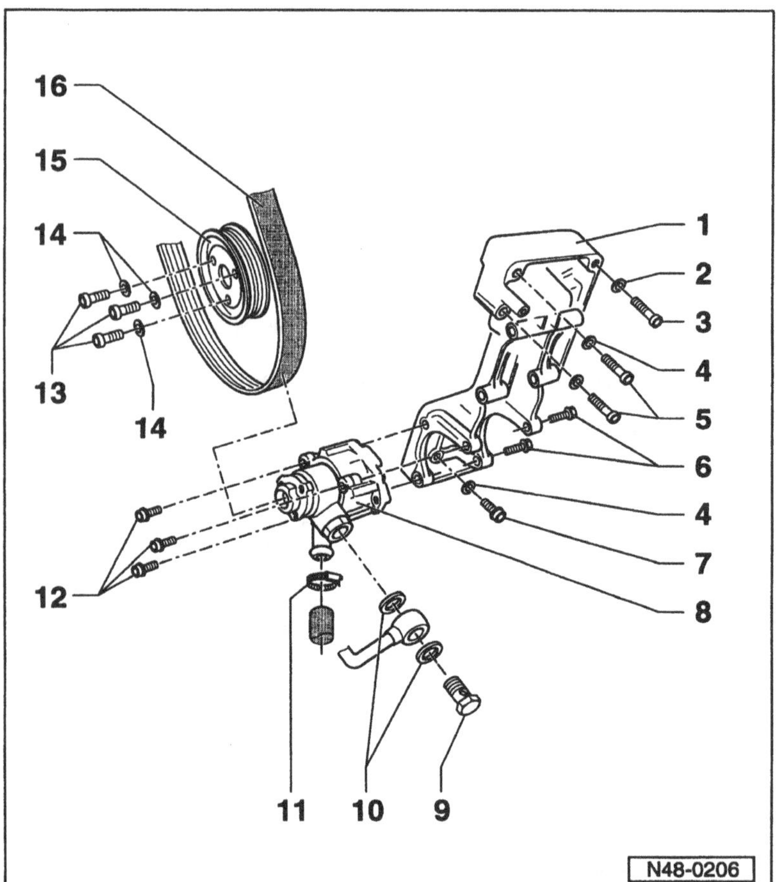

Power steering pump (vehicles with VR6 engine), assembly overview

Note:

⇒ *Page N 48-34.*

1 - **Bracket**

2 - **Washer**

3 - **Hex socket head bolt, M8 x 60, 25 Nm (18 ft lb)**
- ♦ Fitted bolt.
- ♦ Quantity 2.

4 - **Washer**

5 - **Hex socket head bolt, M8 x 60, 25 Nm (18 ft lb)**
- ♦ Quantity 3.

6 - **Hex bolt, 25 Nm (18 ft lb)**

7 - **Hex socket head bolt, M8 x 20, 25 Nm (18 ft lb)**

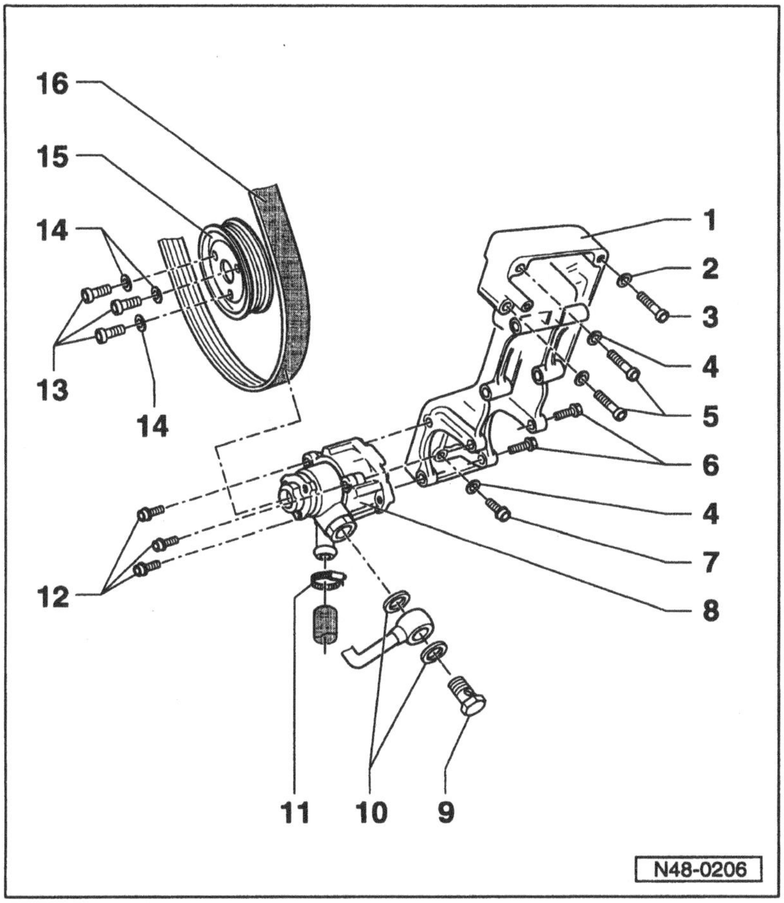

8 - **Power steering pump**
- ♦ Checking delivery pressure, ⇒ Page N 48-22.
- ♦ Removing and installing, ⇒ Page N 48-50.
- ♦ Before installing fill with oil, ⇒ Note Page N 48-34.

9 - **Banjo bolt, 30 Nm (22 ft lb)**

10 - **O-ring**

11 - **Screw-type clip**
- ♦ 4 Nm (35 in. lb).

12 - **Hex socket head bolt, 25 Nm (18 ft lb)**

13 - **Hex socket head bolt, 25 Nm (18 ft lb)**
- ♦ Loosening and tightening, ⇒ Page N 48-50.

14 - **Washer**

15 - **Belt pulley**

16 - **Ribbed belt**

Power steering pump (vehicles with 5- or 6-Cylinder engine), removing and installing

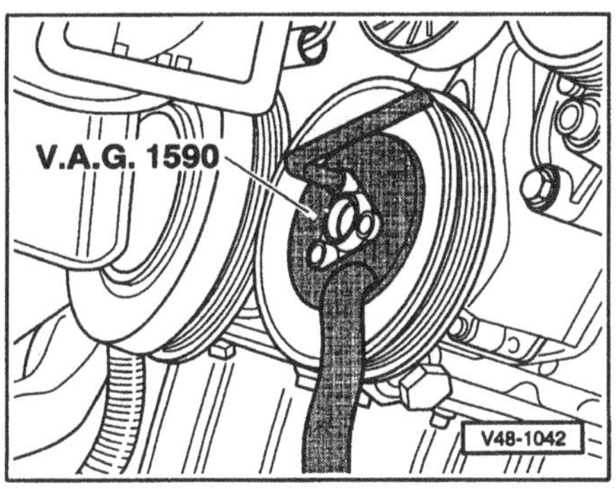

Removing

– Remove sound dampening pan.

– Remove ribbed belt,
⇒ 2.5 Liter 5-Cylinder, General Engine.
⇒ 2.8 Liter VR6, General Engine.

– Remove suction and pressure lines and drain hydraulic oil.

◄ – Remove belt pulley.

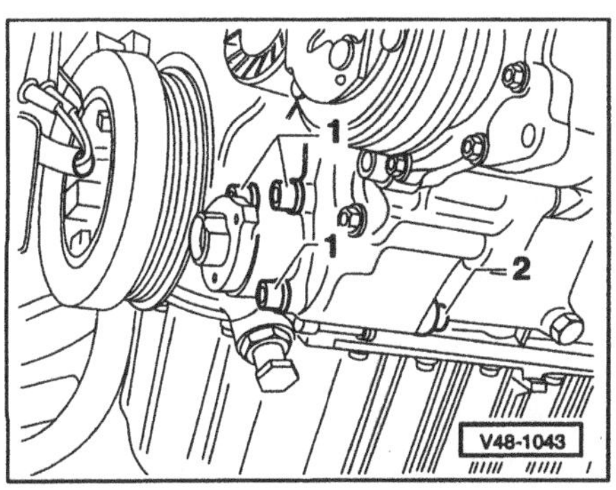

◄ – Remove bolts -1- and -2-.

Installing

Install in reverse order of removal.

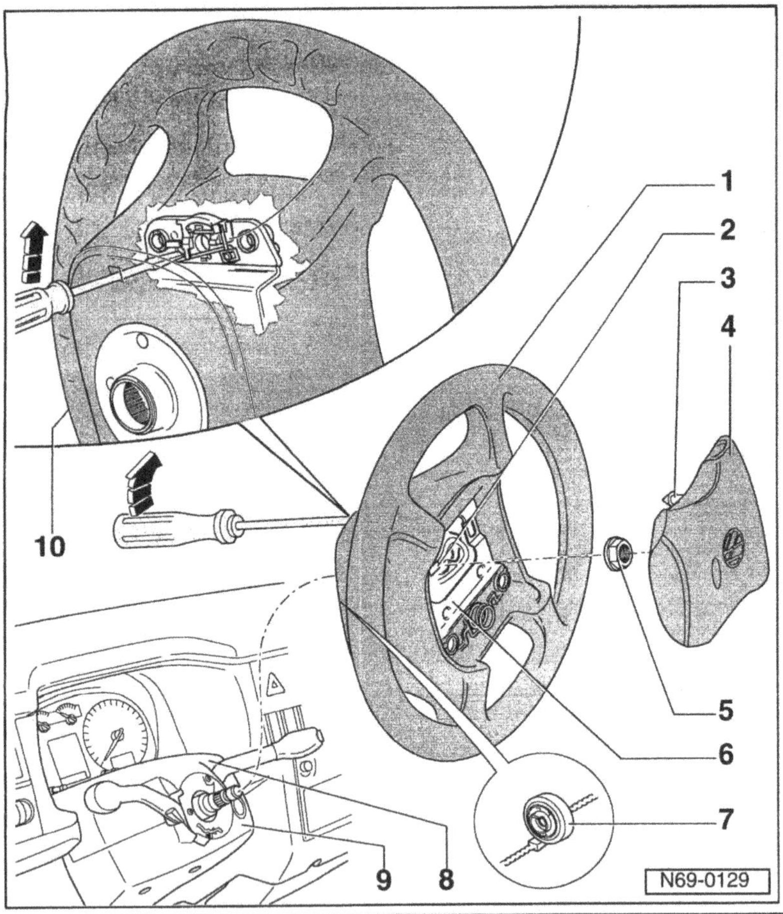

Driver's side airbag unit from 09.98, removing and installing

1 - Steering wheel

2 - Connector

3 - Catches

4 - Airbag unit

5 - 67 Nm (49 ft-lb)

6 - Securing plate

7 - Return spring with slip ring

8 - Steering column trim, upper

9 - Steering column trim, lower

10 - Clip

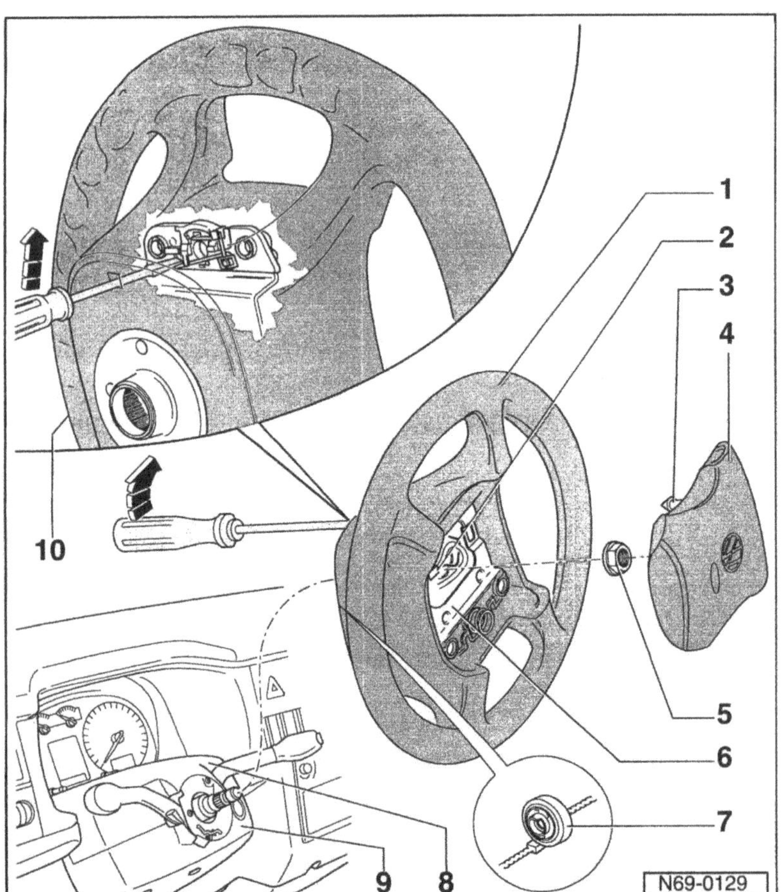

Removing

- Disconnect battery ground (GND).
- Turn steering wheel -1- until the steering wheel spoke is vertical.
- Insert a screwdriver of approx. 7 in. length into bore of steering wheel pan from rear (see inset).
- Press screwdriver in direction of arrow to disengage airbag retaining clip.
- Turn steering wheel 180° and disengage the second catch on the opposite side.
- Turn steering wheel to center position
- Disconnect harness connector from airbag unit.

Installing

Installation is the reverse of removal.

- Switch on ignition.
- Connect battery ground (GND) strap.

Note:

Make sure no one is in the vehicle.

ABS and ABS/EDL On Board Diagnostic

Section P

> ***WARNING!***
> ***Before proceeding with any work, please read the warnings and cautions immediately following this index. Also read any warning and caution that accompanies a procedure in the manual.***

Please read these WARNINGS and CAUTIONS before proceeding with maintenance and repair work.

WARNING—

- Some repairs may be beyond your capability. If you lack the skills, tools and equipment, or a suitable workplace for any procedure described in this manual, we suggest you leave such repairs to an authorized Volkswagen dealer service department, or other qualified shop.

- Volkswagen is constantly improving its cars. Sometimes these changes, both in parts and specifications, are made applicable to earlier models. Therefore, before starting any major jobs or repairs to components on which passenger safety may depend, consult your authorized Volkswagen dealer about the Technical Bulletins that may have been issued since the editorial closing of this manual.

- Do not re-use any fasteners that are worn or deformed in normal use. Many fasteners are designed to be used only once and become unreliable and may fail when used a second time. This includes, but is not limited to, nuts, bolts, washers, self–locking nuts or bolts, circlips and cotter pins. Always replace these fasteners with new parts.

- Never work under a lifted car unless it is solidly supported on stands designed for the purpose. Do not support a car on cinder blocks, hollow tiles or other props that may crumble under continuous load. Never work under a vehicle that is supported solely by a jack. Never work under the car while the engine is running.

- If you are going to work under a car on the ground, make sure that the ground is level. Block the wheels to keep the car from rolling. Disconnect the battery negative terminal (ground strap) to prevent others from starting the vehicle while you are under it.

- Never run the engine unless the work area is well ventilated. Carbon monoxide (CO) kills.

- Do not attempt to work on your vehicle if you do not feel well. You increase the danger of injury to yourself and others if you are tired, upset or have taken medicine or any other substance that may impair you or keep you from being fully alert.

- Disconnect the battery negative (–) terminal (ground strap) whenever you work on the fuel system or the electrical system. Do not smoke or work near heaters or other fire hazards. Keep an approved fire extinguisher handy.

- Batteries produce explosive hydrogen gas. Keep sparks, lighted matches and open flame away from the top of the battery. If hydrogen gas escaping from the cap vents is ignited, it will ignite gas trapped in the cells and cause the battery to explode.

- Connect and disconnect battery cables, jumper cables or a battery charger only with the ignition switched off, to prevent sparks. Do not quick–charge the battery (for boost starting) for longer than one minute, and do not allow charging voltage to exceed 16.5 volts. Wait at least one minute before boosting the battery a second time.

- Illuminate your work area adequately but safely. Use a portable safety light for working inside or under the car. Make sure the bulb is enclosed by a wire cage. The hot filament of an accidentally broken bulb can ignite spilled fuel or oil.

- Tie long hair behind your head. Do not wear a necktie, a scarf, loose clothing, or a necklace when you work near machine tools or running engines. If your hair, clothing, or jewelry were to get caught in the machinery, severe injury could result.

- Finger rings, bracelets and other jewelry should be removed so that they cannot cause electrical shorts, get caught in running machinery, or be crushed by heavy parts.

- Catch draining fuel, oil, or brake fluid in suitable containers. Do not use food or beverage containers that might mislead someone into drinking from them. Store flammable fluids away from fire hazards. Wipe up spills at once, but do not store the oily rags, which can ignite and burn spontaneously.

- Always observe good workshop practices. Wear goggles when you operate machine tools or work with battery acid. Gloves or other protective clothing should be worn whenever the job requires working with harmful substances.

- The Air-Conditioning (A/C) system is filled with a chemical refrigerant, which is hazardous. The A/C system should be serviced only by trained technicians using approved refrigerant recovery/recycling equipment, trained in related safety precautions, and familiar with regulations governing the discharging and disposal of automotive chemical refrigerants.

- Do not expose any part of the A/C system to high temperatures such as an open flame. Excessive heat will increase system pressure and may cause the system to burst.

- Some aerosol tire inflators are highly flammable. Be extremely cautious when repairing a tire that may have been inflated using an aerosol tire inflator. Keep sparks, open flame or other sources of ignition away from the tire repair area. Inflate and deflate the tire at least four times before breaking the bead from the rim. Completely remove the tire from the rim before attempting any repair.

- Friction materials such as brake or clutch discs may contain asbestos fibers. Do not create dust by grinding, sanding, or by cleaning with compressed air. Avoid breathing asbestos fibers and asbestos dust. Breathing asbestos can cause serious diseases such as asbestosis or cancer, and may result in death.

- Greases, lubricants and other automotive chemicals contain toxic substances, many of which are absorbed directly through the skin. Read manufacturer's instructions and warnings carefully. Use hand and eye protection. Avoid direct skin contact.

- An airbag is operated by an explosive device. Handled improperly or without adequate safeguards, it can be accidently activated and cause serious personal injury. To prevent personal injury or airbag system failure, **only trained Volkswagen dealer service technicians** should test, disassemble or service the airbag system.

WARNINGS and CAUTIONS continued on next page ▶

Please read these WARNINGS and CAUTIONS before proceeding with maintenance and repair work.

WARNING (continued) —

- ALWAYS disconnect the airbag power supply before using a computer memory saver. Failure to disconnect the airbag power supply may result in accidental activation of the airbag.
- The airbag unit should not be removed, nor should any work be performed on the airbag system, without first reviewing the warnings and cautions that accompany the airbag system repair information in this manual.
- Never apply stickers or any other coverings on the steering wheel. Do not let chemical cleaners, oil or grease come into contact with the vinyl covering of the airbag unit.
- Never open or otherwise attempt to repair airbag system parts. Always use new parts. Never leave airbag parts or the partially disassembled airbag system unattended.
- Never use a test light or volt meter to conduct electrical tests on the airbag system. The airbag system must only be tested by trained Volkswagen Service technicians using special airbag tester equipment. The airbag system must only be tested while it is installed in the car.
- Do not expose an airbag unit to temperatures above 194°F (90°C) even for a brief period, while handling it during the repair process. Keep the unit clear of heat sources such as hot plates, soldering irons, heat lamps, welding equipment, or the like.
- Always replace airbag system components that have been mechanically damaged (bubbles, cracks, etc.).
- Always set down a removed airbag unit so that the horn pad is facing upward.
- When driving or riding in an airbag-equipped vehicle, never hold test equipment in your hands or lap while the vehicle is in motion. Objects between you and the airbag can increase the risk of injury in an accident.
- The storage, transportation and disposal of airbag units are subject to the laws for flammable solids.

CAUTION—

- If you lack the skills, tools and equipment, or a suitable workshop for any procedure described in this manual, we suggest you leave such repairs to an authorized Volkswagen dealer or other qualified shop. We especially urge you to consult an authorized Volkswagen dealer before beginning repairs on any car that may still be covered wholly or in part by any of the extensive warranties issued by Volkswagen of America, Inc.
- Volkswagen is constantly improving its cars and sometimes these changes, both in parts and specifications, are made applicable to earlier models. Therefore, part numbers listed in this manual are for reference only. Always check with your authorized Volkswagen dealer parts department for the latest information.
- Before starting a job, make certain that you have all the necessary tools and parts on hand. Read all the instructions thoroughly, do not attempt shortcuts. Use tools that are appropriate to the work and use only replacement parts meeting Volkswagen specifications. Makeshift tools, parts and procedures will not make good repairs.
- Use pneumatic and electric tools only to loosen threaded parts and fasteners. Never use these tools to tighten fasteners, especially on light alloy parts. Always use a torque wrench to tighten fasteners to the tightening torque specification listed.
- Be mindful of the environment and ecology. Before you drain the crankcase, find out the proper way to dispose of the oil. Do not pour oil onto the ground, down a drain, or into a stream, pond, or lake. Consult local ordinances that govern the disposal of wastes.
- When connecting cables or ground straps, all connections must be cleaned to bare metal. Be sure mounting bolts are tight.
- Do not replace fuse links with fuse links of other gauge or with standard wire. Always replace the entire length of the fuse link.
- Before doing any electrical welding on vehicles equipped with Anti-lock Brakes (ABS), disconnect the battery negative terminal (ground strap) and the ABS Control Module connector.
- On cars equipped with ABS, the ABS control unit cannot withstand temperatures from a paint–drying booth or a heat lamp in excess of 203°F (95°C) and should not be subjected to temperatures in excess of 185°F (85°C) for more than two hours.
- On cars equipped with anti-theft radios, make sure you know the correct radio activation code before disconnecting the battery or removing the radio. If the wrong code is entered when power is restored, the radio may lock up and be rendered inoperable, even if the correct code is then entered.
- When boost-charging the battery, first remove the fuses for the Engine Control Module (ECM), the Transmission Control Module (TCM), the ABS Control Module, and the On-Board computer. In cases where one or more of these components is not separately fused, disconnect the Control Module connector(s).
- Do not quick-charge the battery (for boost starting) for longer than one minute, and do not exceed 16.5 volts at the battery with the boosting cables attached. Wait at least one minute before boosting the battery a second time.

◀ **WARNINGS and CAUTIONS continued from previous page**

Anti-lock Brakes (ABS) and Anti-lock Brakes with Electronic Differential Lock (ABS/EDL), On Board Diagnostic (OBD)

Function

The ABS control module -J104- is located in the engine compartment near the battery and is equipped with a Diagnostic Trouble Code (DTC) Memory. The On Board Diagnostic (OBD) Data Link Connectors (DLC) are located on the left side of the instrument panel below the heated rear window switch.

Up to 3 DTCs can be stored; in some cases the malfunction type is indicated. ABS control in the control module is then switched off; at the same time the instrument panel ABS warning light illuminates. Conventional brake system remains operational.

The Diagnostic Trouble Code Memory (DTC) must always be checked before beginning OBD. In some cases, the ABS Warning Light -K47- goes out after the ignition is switched off, although the DTC memory is maintained.

An intermittent DTC (e.g. loose connection), is also stored. If the DTC no longer exists after the ignition is switched on again, then the ABS warning light goes out after approx. 2 seconds.

If the DTC is not registered again after the ignition is switched on 50 times, the relevant DTC memory is automatically erased.

Perform "Checking Diagnostic Trouble Code (DTC) memory" ⇒ Page P 01-14 after initiating On Board Diagnostic (OBD) via the VAG 1551 Scan Tool or VAG 1552 Scan Tool, but only if the vehicle is not moving and is in the "rapid data transfer" mode.

OBD is aborted if vehicle is moved at a speed of more than 2.5 km/h (1.56 mph).

ABS is not operational during OBD.

DTC information displayed is used in conjunction with a DTC table ⇒ Page P 01-16 which has information on possible DTC causes.

"Erase DTC memory", eliminating malfunctions ⇒ Page P 01-29.

Diagnostic Trouble Code (DTC), technical data

DTC memory

- Contents of permanent memory are retained after supply voltage is switched off.

Data output

- Rapid data transfer (operating mode 1)
- Blink code output is ignored
- ABS warning light lights up when DTC is detected.

Functions

- Checks control module version
- Checks DTC memory
- Erases DTC memory
- Reads measuring value block

Safety precautions and fundamental points regarding troubleshooting

- *ABS is a vehicle safety system; appropriate knowledge is necessary to work on the system.*
- *Do not drive with the harness connector disconnected from the ABS control module -J104-.*
- *In order to check complaints and perform DTC checking, DTC memory must be checked before starting work on the ABS or ABS/EDL system.*
- *Do not disconnect or re-connector harness connectors when the ignition is switched on.*
- *Do not loosen screws or bolts on the ABS hydraulic unit.*
- *ABS malfunctions are indicated by the ABS warning light coming on; certain malfunctions will be recognized only when exceeding 12 km/h (7.5 mph). A road test is therefore necessary.*

DTC checking requirements

- Tires must be the same (correct) size.
- Conventional brake system including brake light switch and brake lights OK.
- Hydraulic lines and connections not leaking (visual check of hydraulic unit, brake calipers/cylinders).
- Wheel bearings and bearing adjustment OK.
- Ground connection for hydraulic unit, OK.
- Connector on ABS control module -J104- is correctly connected.
- Control module -J104- ground connection OK.
- All fuses OK according to wiring diagram.
- Supply voltage OK (at least 10.5 V).

Necessary technical information

♦ *ABS wiring diagram*
⇒ Electrical Wiring Diagrams

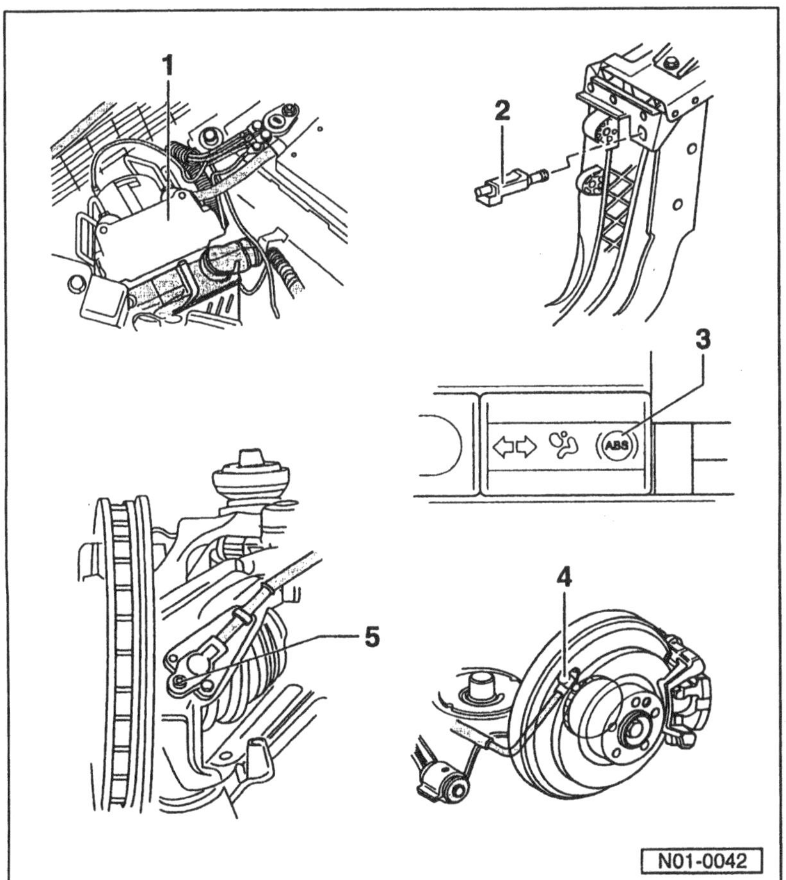

ABS electrical components

All components marked with [1)] are DTC monitored

1 - **ABS hydraulic unit -N55- with ABS control module -J104-[1)]**
   - ♦ Use adapter set VAG 1594 to test electrical connections
   - ♦ Hydraulic unit multi-pin connector

⇒ Page P 01-36

Notes:

- ♦ *Return flow pump and solenoid valves cannot be replaced individually. Replace as a unit.*
- ♦ *Bleed brake system when replacing hydraulic unit.*

⇒ Suspension, Wheels, Brakes, Steering 1997 ➤ ; Repair group 47

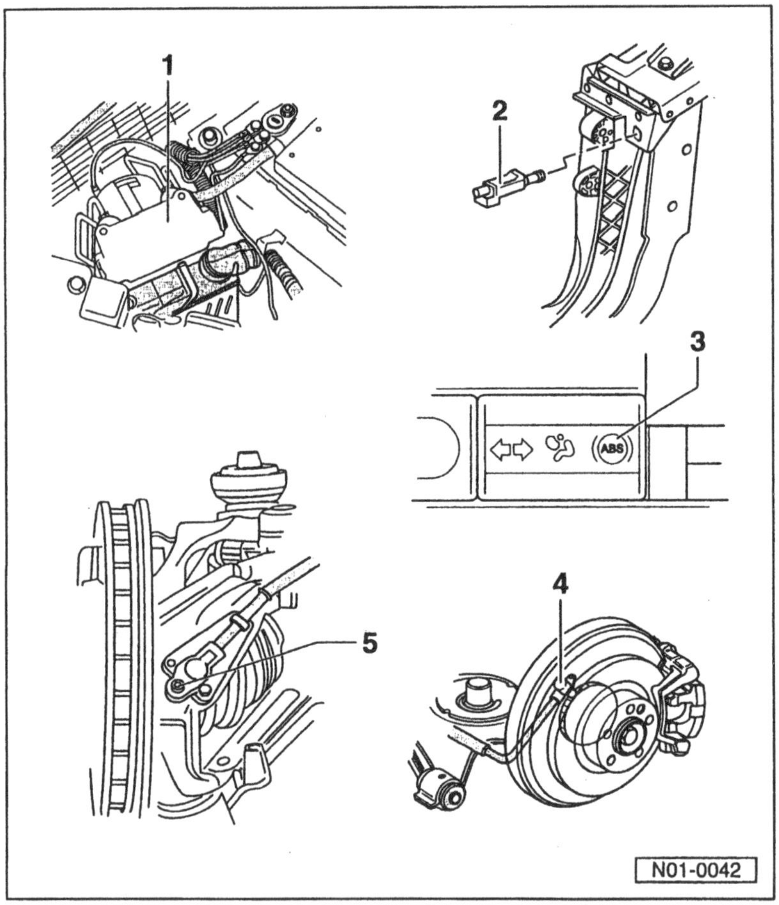

2 - **Brake light switch -F-[1)]**
- Brake light switch plunger must have switched brake lights off when brake pedal is in rest position

3 - **ABS warning light -K47-**
- In center of instrument panel
- Lights up for approx. 2 seconds after ignition is switched on

4 - **Left rear/right rear ABS wheel speed sensor -G44/G46-[1)]**
- Before inserting sensor clean inner surface of sensor mount and lubricate with G 000 650
- Observe assembly sequence: Insert sensor then clip cable into appropriate holder starting from sensor.

5 - **Left front/right front ABS wheel speed sensor -G45/G47-[1)]**
- Installation instructions ⇒ Item 4

VAG 1551 Scan Tool, connecting and selecting function

- Test requirements ⇒ Page P 01-5.

Notes:

- *ABS function is interupted during OBD.*
- *Erase DTC memory after repair and verification are completed.*

– Unclip cover on right side of instrument panel insert.

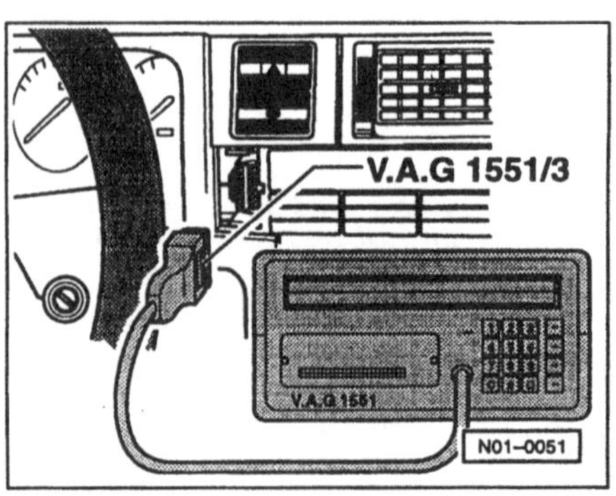

◄ – With ignition switched off connect VAG 1551 Scan Tool with adapter cable VAG 1551/3.

VAG On Board Diagnostic HELP 1-Rapid data transfer [1)] 2-Blink code output[1)]

◄ Indicated on display

[1)] Displayed alternately

Note:

No indication on display, check voltage supply at connector.
⇒ Electrical Wiring Diagrams

Notes:

- ♦ *Depending on program, additional operating information can be printed by pressing VAG 1551 HELP key.*
- ♦ *→ key is used to advance within program*
- ♦ *PRINT key activates printer (warning lamp in key comes on).*

– Switch on ignition.

– Press key -1- for "Rapid data transfer" mode.

Rapid data transfer HELP Enter address word XX

◄ Indicated on display

– Press keys -0- and -3-; 03 inputs address word of system (brake electronics) to be tested.

Rapid data transfer Q 03 Brake electronics

◄ Indicated on display

– Press Q key to confirm input.

703907379A ABS/EDL BOSCH 5 4062 →

◄ Display will apear as shown:

Control module identification number is displayed.

Application of control module
⇒ Parts list

If control module identification number does not appear
⇒ Functions.

– Press → button.

Rapid data transfer HELP Select function XX

◄ Indicated on display

Note:

After pressing HELP key a list of possible functions is printed:

Functions	Page
01 - Check control module version	P 01-10
02 - Check DTC memory	P 01-14
05 - Erase DTC memory	P 01-29
06 - End output	P 01-31
08 - Read measuring value block	P 01-32

Only listed functions can be used for On Board Diagnostic.

After pressing → key, VAG 1551 Scan Tool program returns to start position.

Rapid data transfer	HELP
Select function XX	

◄ Indicated on display (Function selection, e.g. 02 – check DTC memory)

Notes:

Rapid data transfer	HELP
Control module does not answer!	

◄ ♦ *If adjacent DTC information appears in display, possible causes for DTC can be printed by pressing HELP key.*

Control module does not answer!

Possible causes:

1. Incorrect address word entered.

2. Control module not in diagnostic mode.

♦ Ignition not switched on.

♦ Engine speed greater than diagnostic maximum speed.

3. Diagnostic cable inoperable or incorrectly connected.

4. Wiring between control module and data link connector inoperable.

5. Fuse S123 (terminal 15) on additional plate position 5 inoperable.

Rapid data transfer	HELP
K-wire not switching to positive!	

◄ (Ignition must be switched on!)

Rapid data transfer	→
No signal from control module!	

◄ ♦ *DTC have occurred at the start or during the program (external interference?).*

Rapid data transfer	→
Fault in communication build-up	

– Check diagnostic wiring, voltage supply, and ground connection near ABS control module -J104-; electrical test ⇒ Page P 01-35.

– After repairing, enter address code 03 for "Brake electronics" and confirm with Q key.

Rapid data transfer Tester sends address word 03

◄ Indicated on display

703907379A ABS/EDL BOSCH 5 4062

◄ Then the following appears on display:

Control module identification number is displayed.

– Press → button.

Rapid data transfer Select function XX	HELP

◄ Indicated on display

Diagnostic Trouble Code (DTC) memory, checking

– Select function

Rapid data transfer Select function XX	HELP

◄ Indicated on display

– Press keys -0- and -2-; 02 enters "Check DTC memory" function.

Rapid data transfer 02 – Check DTC memory	Q

◄ Indicated on display

– Confirm with Q key.

X DTCs recognized!

◄ Number of DTCs stored or "No DTC recognized" will be displayed.

No DTC recognized!	→

– Press → key.

Stored DTCs are displayed one after the other (print out optional).

– Go to DTC table on Page P 01-16 with DTC information printout.

Program returns to begining after pressing → key:

Rapid data transfer Select function XX	HELP

◄ Indicated on display (Function selection)

Diagnostic Trouble Code (DTC) table, notes

- *All possible DTC recognized by ABS control module -J104- and printed-out on VAG 1551 Scan Tool, are listed on following pages according to 5 digit DTC code.*
- *DTC code appears in "Rapid data transfer" mode only on print-out.*

Example:

DTC code

5 digit

00277

- *DTC type can also appear in DTC table.*
- *Reference is made to individual test steps of electrical test in column "Repairing malfunction".*
- *Before replacing faulty components, check all appropriate connectors, wiring, and ground connections using the Electrical Wiring Diagrams.*
- *After repairs, always check and erase DTC memory with VAG 1551 Scan Tool, also road test (exceeding a speed of 30 km/h (19 mph).*
- *After road test check DTC memory again.*

Diagnostic Trouble Code (DTC) table

VAG 1551 print-out	Possible cause	Malfunction elimination
No DTC recognized	If "No DTC" appears after repairing, On Board Diagnostic (OBD) is ended. If, despite "No DTC recognized" display, ABS system does not operate properly, proceed as follows: 1. Perform road test exceeding 30 km/h (19 mph), 2. Check DTC memory again and repair. 3. If no DTCs are indicated continue checking without OBD and work through complete Electrical checks ⇒ Page P 01-35.	
01201 ABS pump supply voltage	Fuse S53 inoperative	– Check fuses, wiring, connections and voltage supply to control module: ⇒ Electrical check, Page P 01-38, Test step 2
01200 ABS valves supply voltage	Fuse S54 inoperative	– Replace fuse S54 – Check wiring harness wiring ⇒ Electrical check, Page P 01-38, Test step 3

VAG 1551 scan tool print-out	Possible cause	Repairing malfunction
00301 ABS Return Flow Pump -V39-	♦ Open circuit or transfer resistance in ground connections or voltage supply to return flow pump -V39-. ♦ Fuse S53 inoperative ♦ Return flow pump -V39- or hydraulic unit inoperative.	– Locate open circuit or transfer resistance according to "Electrical Wiring Diagrams" and repair. – Locate open circuit or short circuit according to "Electrical Wiring Diagrams" and repair. – Check return flow pump -V39- and ABS hydraulic unit -N55-, Perform Electrical tests ⇒ Page P 01-38, test step 3, "Read measuring value block" ⇒ Page P 01-32, display group number 02 display zone 2.
01130 ABS operation Signal outside tolerance[1]	Electrical interference from external sources (high frequency radiation, e.g. non-insulated ignition cable)	Sequence: – Check all wiring and connections for short to positive or ground – Erase DTC memory – Perform test drive exceeding 30 km/h (19 mph) – Check DTC memory again
65535 Control module	Control module inoperative	– Replace control module

[1] This malfunction can only be recognized above 30 km/h (19 mph).

VAG 1551 scan tool print-out	Possible cause	Repairing malfunction
00526 Brake light switch F Open circuit	♦ Both brake lights -M9-, -M10- or control module -J104- inoperative ♦ Open circuit or short circuit to positive/Ground in wiring between brake light switch and control module -J104-. ♦ Brake light switch inoperative.	– Locate open circuit or short circuit according to ⇒ "Electrical Wiring Diagrams" and correct. – Check brake light switch, "Read measuring value block" ⇒ Page P 01-32, display group number 02 display zone 1
00532 Voltage supply Signal low	♦ Transfer resistance in voltage supply to control module -J104- (terminal 4). ♦ Voltage dip in vehicle electrical system. ♦ Battery inoperative.	– Locate open circuit or transfer resistance according to ⇒ "Electrical Wiring Diagrams" and correct. – Check alternator and voltage regulator ⇒ "Electrical Wiring Diagrams".

VAG 1551 scan tool print-out	Possible cause	Repairing malfunction
00283		
ABS Wheel Speed Sensor, LF -G47-	Open circuit or loose connection in: ♦ ABS wheel speed sensor wire ♦ Connector ♦ ABS wheel speed sensor winding	– Check wiring, connectors, ABS wheel speed sensor and ground to control module:
	ABS wheel speed sensor electrical short circuit	⇒ Electrical check, Page P 01-38,Test steps 9 and 13
Mechanical malfunction[1]	Excessive air gap between speed sensor and impulse rotor (signal not OK)	– Check installation of ABS wheel speed sensor and impulse rotor – Function 08, read measuring value block ⇒ Page P 01-32; checking ABS wheel speed sensor; display group number 01, display zone 1
Signal outside tolerance[1]	Impulse rotor and ABS wheel speed sensor dirty or damaged	

[1] This malfunction can only be recognized above 30 km/h (19 mph).

VAG 1551 scan tool print-out	Possible cause	Repairing malfunction
00285		
ABS Wheel Speed Sensor,RF -G45-	Open circuit or loose connection in: ♦ ABS wheel speed sensor wire ♦ Connector ♦ ABS wheel speed sensor winding	– Check wiring, connectors, ABS wheel speed sensor, and ground to control module:
	Wheel speed sensor electrical short circuit	⇒ Electrical check, Page P 01-38, Test steps 8 and 12
Mechanical malfunction[1]	Excessive air gap between ABS wheel speed sensor and impulse rotor (signal not OK)	– Check installation of ABS wheel speed sensor and impulse rotor – Function 08, read measuring value block ⇒ Page P 01-32; checking ABS wheel speed sensor; display group number 01, display zone 2
Signal outside tolerance[1]	Impulse rotor and ABS wheel speed sensor dirty or damaged	

[1] This malfunction can only be recognized above 30 km/h (19 mph).

VAG 1551 scan tool print-out	Possible cause	Repairing malfunction
00290		
ABS Wheel Speed Sensor,LR -G46-	Open circuit or loose connection in: ♦ ABS wheel speed sensor wire ♦ Connector ♦ ABS wheel speed sensor winding	– Check wiring, connectors, ABS wheel speed sensor, and ground to control module:
	ABS wheel speed sensor electrical short circuit	⇒ Electrical check, Page P 01-38, Test steps 11 and 15
Mechanical malfunction[1]	Excessive air gap between ABS wheel speed sensor and impulse rotor (signal not OK)	– Check installation of ABS wheel speed sensor and impulse rotor – Function 08, read measuring value block ⇒ Page P 01-32; checking ABS wheel speed sensor; display group number 01, display zone 3
Signal outside tolerance[1]	Impulse rotor and ABS wheel speed sensor dirty or damaged	

[1] This malfunction can only be recognized above 30 km/h (19 mph).

VAG 1551 scan tool print-out	Possible cause	Repairing malfunction
00287		
ABS Wheel Speed Sensor,RR -G44-	Open circuit or loose connection in: ♦ ABS wheel speed sensor wire ♦ Connector ♦ ABS wheel speed sensor winding	– Check wiring, connectors, ABS wheel speed sensor, and ground to control module:
	ABS wheel speed sensor electrical short circuit	⇒ Electrical check, Page P 01-38, Test steps 10 and 14
Mechanical fault[1]	Excessive air gap between ABS wheel speed sensor and impulse rotor (signal not OK)	– Check installation of ABS wheel speed sensor and impulse rotor – Function 08, read measuring value block ⇒ Page P 01-32; checking ABS wheel speed sensor; display group number 01, display zone 4
Signal outside tolerance[1]	Impulse rotor or ABS wheel speed sensor dirty or damaged	

[1] This malfunction can only be recognized above 30 km/h (19 mph).

VAG 1551 scan tool print-out	Possible cause	Repairing malfunction
00257 ABS Inlet Valve, LF -N101-	♦ Open circuit, short circuit to positive/Ground. ♦ ABS inlet valve -N101- inoperative.	– Perform "Electrical test" ⇒ Page P 01-38, test steps 1 + 3. – If after performing electrical tests no cause can be found, check wiring and connectors for loose connections. – If none of the mentioned measures correct the malfunction, replace control module.
00259 ABS Inlet Valve, RF -N99-	♦ Open circuit, short circuit to positive/Ground. ♦ ABS inlet valve -N99- inoperative.	– Perform "Electrical test" ⇒ Page P 01-38, test steps 1 + 3. – If after performing electrical tests no malfunctions can be found, check wiring and connectors for loose connections. – If these measures do not correct malfunction, replace control module.

VAG 1551 scan tool print-out	Possible cause	Repairing malfunction
00274 ABS Inlet Valve, LR -N134-	♦ Open circuit, short circuit to positive/Ground. ♦ ABS inlet valve -N134- inoperative.	– Perform "Electrical test" ⇒ Page P 01-38, test steps 1 + 3. – If after performing electrical tests no cause can be found, check wiring and connectors for loose connections. – If these measures do not correct malfunction, replace control module.
00273 ABS Inlet Valve, RR -N133-	♦ Open circuit, short circuit to positive/Ground. ♦ ABS inlet valve -N133- inoperative.	– Perform "Electrical test" ⇒ Page P 01-38, test steps 1 + 3. – If after performing electrical tests no cause can be found, check wiring and connectors for loose connections. – If these measures do not correct malfunction, replace control module.

VAG 1551 scan tool print-out	Possible cause	Repairing malfunction
00265 ABS Outlet Valve, LR -N102-	♦ Open circuit, short circuit to positive/Ground. ♦ ABS outlet valve -N102- inoperative.	– Perform "Electrical test" ⇒ Page P 01-38, Test steps 1 + 3. – If after performing electrical tests no cause can be found, check wiring and connectors for loose connections. – If these measures do not correct malfunction, replace control module.
00267 ABS Outlet Valve, RF -N100-	♦ Open circuit, short circuit to positive/Ground. ♦ ABS outlet valve -N100- inoperative.	– Perform "Electrical test" ⇒ Page P 01-38, test steps 1 + 3. – If after performing electrical tests no cause can be found, check wiring and connectors for loose connections. – If these measures do not correct malfunction, replace control module.

VAG 1551 scan tool print-out	Possible cause	Repairing malfunction
00276 ABS Outlet Valve, LR -N136-	♦ Open circuit, short circuit to positive/Ground. ♦ ABS outlet valve -N136- inoperative.	– Perform "Electrical test" ⇒ Page P 01-38, test steps 1 + 3. – If after performing electrical tests no cause can be found, check wiring and connectors for loose connections. – If these measures do not correct malfunction, replace control module.
00275 ABS Outlet Valve, RR -N135-	♦ Open circuit, short circuit to positive/Ground. ♦ ABS outlet valve -N135- inoperative.	– Perform "Electrical test" ⇒ Page P 01-38, test steps 1 + 3. – If after performing electrical tests no cause can be found, check wiring and connectors for loose connections. – If these measures do not correct malfunction, replace control module.

VAG 1551 scan tool print-out	Possible cause	Repairing malfunction
00644 EDL Switch-Over Valve, LF -N168-	♦ Open circuit, short circuit to positive/Ground. ♦ EDL Switch-Over Valve -N168- inoperative. -USV1-	– Perform "Electrical test" ⇒ Page P 01-38 test steps 1 + 3. – If after performing electrical tests no cause can be found, check wiring and connectors for loose connections. – If these measures do not correct malfunction, replace control module.
00642 EDL Switch-Over Valve, RF -N166-	♦ Open circuit, short circuit to positive/Ground. ♦ EDL Switch-over Valve -N166- inoperative. -USV2-	– Perform "Electrical test" ⇒ Page P 01-38 test steps 1 + 3. – If after performing electrical tests no cause can be found, check wiring and connectors for loose connections. – If these measures do not correct malfunction, replace control module.

VAG 1551 scan tool print-out	Possible cause	Repairing malfunction
00645 EDL Outlet Valve, LF -N169-	♦ Open circuit, short circuit to positive/Ground. ♦ EDL outlet valve -N169- inoperative. EDS -ASV1-	– Perform "Electrical test" ⇒ Page P 01-38 test steps 1 + 3. – If after performing electrical tests no cause can be found, check wiring and connectors for loose connections. – If these measures do not correct malfunction, replace control module.
00643 EDL Outlet Valve, RF -N167-	♦ Open circuit, short circuit to positive/Ground. ♦ EDL outlet valve -N167- inoperative. -ASV2-	– Perform "Electrical test" ⇒ Page P 01-38 test steps 1 + 3. – If after performing electrical tests no cause can be found, check wiring and connectors for loose connections. – If these measures do not correct malfunction, replace control module.

Diagnostic Trouble Code (DTC) memory, erasing and ending output

Requirement:

- DTC memory has been checked

Erasing DTC memory

– Press → key,

Rapid data transfer HELP Select function XX	Indicated on display (Clear for function selection)

– Press keys -0- and -5-; 05 enters "Erase DTC memory" function.

Rapid data transfer Q 05 Erase DTC memory	Indicated on display

– Confirm with Q key

Rapid data transfer → DTC memory is erased!	Indicated on display

– Press → key

Rapid data transfer HELP Select function XX	Indicated on display (Clear for function selection)

Attention! DTC Memory is not interrogated	

Notes:

- *If "DTC Memory is not interrogated" is displayed, the OBD test sequence is faulty.*
- *Follow the program sequence exactly: first check DTC memory, then erase.*

DTC memory is not erased if:

- Ignition was off between "Check DTC memory" and "Erase DTC memory"
- Vehicle has been moved at more than 2.5 km/h (1.56 mph).

Ending output

Rapid data transfer HELP Select function XX

Indicated on display (clear for function selection)

- Press keys -0- and -6-, to select "End output" function 06.

Rapid data transfer Q 06 End output

Indicated on display

- Confirm with Q key

Rapid data transfer Help Enter address word XX

Indicated on display

- Switch ignition off.
- Disconnect VAG 1551 Scan Tool.
- Switch ignition on.

ABS warning light -K47- must go out after approx. 2 seconds.

- Perform road test briefly exceeding 30 km/h (19 mph).

- ABS warning light must not light up again.

Measuring value block, reading

- Connect VAG 1551 Scan Tool, enter address code "03 Brake electronics" and advance until "Select function XX" is indicated on display ⇒ from page P 01-8.

Rapid data transfer HELP Select function XX

Indicated on display:

- Press keys 0 and 8 to select "Read measuring value block" function 08.

Rapid data transfer Q 08 – Read Measuring Value Block

Indicated on display:

- Confirm with Q key.

Read measuring value block Enter display group number XX

Indicated on display:

- Enter display group number ⇒ List of selectable display group numbers, Page P 01-33.
- Confirm with Q key.

Read measuring value block 1 → →1 →2 →3 →4

- There are always 4 display zones -arrows-. Decoding individual values in display zones 1 to 4 ⇒ Test table page P 01-34.

List of selectable display group numbers

Momentary wheel speed is displayed and is used to check ABS wheel speed sensor application to wheel (vehicle must be raised and wheels turned by hand).

Read measuring value block			1 →
1 km/h	1 km/h	1 km/h	1 km/h

Read measuring value block		2 →
0	0	1

Display group No.	Display field	Designation
01	1	Wheel speed, left front (km/h)
	2	Wheel speed, right front (km/h)
	3	Wheel speed, left rear (km/h)
	4	Wheel speed right rear (km/h)
02	1	Brake light switch
	2	Voltage at return flow pump
	3	Voltage at solenoid valve
	4	Vacant

Test table (Measuring value block)

Display group No.	Display field	Designation	• Requirements	Indicated on VAG 1551
01	1	Wheel speed, left front (km/h)		1 ... 19 [1)]
	2	Wheel speed, right front (km/h)		1 ... 19 [1)]
	3	Wheel speed, rear left (km/h)		1 ... 19 [1)]
	4	Wheel speed, rear right (km/h)		1 ... 19 [1)]
02	1	Brake light switch	• Brake light switch not operated	0
			• Brake light switch operated	1
	2	Voltage at return flow pump	• No voltage (OK)	0
			• Voltage present (faulty)	1
	3	Voltage at solenoid valve	• Voltage present (OK)	1
			• No voltage present (faulty)	0
	4	Vacant		

[1)] Control module On Board Diagnostic (OBD) is terminated above 19 km/h (11.5 mph).

Electrical test

Test equipment

- Digital multimeter Fluke 83 or equivalent
- VW 1594 connector test kit
- Test box VAG 1598 and adapter VAG 1598/24

Notes:

- *If readings obtained deviate from specifications, locate cause according to wiring diagram and correct.*
- *Test box VAG 1598 socket designations are identical to terminal designations on the ABS control module -J104-.*
- *To prevent electrical component damage select relevant multimeter scale before connecting test cables/leads.*

Test prerequisites:

- Battery voltage OK.
- Fuses OK.
- Ground connections OK.

Multi-pin connector terminal assignment

Note:

All terminals not listed are currently not assigned and must never be connected to other components!

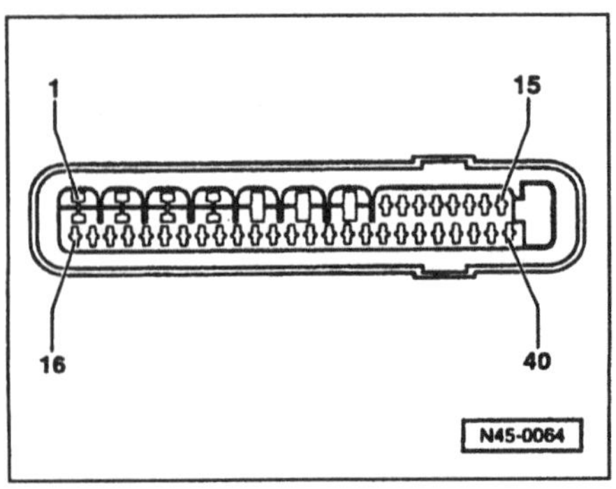

◄ **T40 wiring harness/ABS control module -J104- connector assignment**

Terminal	Connected to component ...
1	⇒ Ground terminal 31
2	⇒ Battery + via S53
3	⇒ Battery + via S54
4	⇒ Terminal 15 voltage supply via S123
16	⇒ Left front ABSwheel speed sensor -G47-
17	⇒ Left front ABS wheel speed sensor -G47-
18	⇒ Left rear ABS wheel speed sensor -G46-
19	⇒ Left rear ABS wheel speed sensor -G46-
20	⇒ Right rear ABS wheel speed sensor -G44-
21	⇒ Right rear ABS wheel speed sensor -G44-

Terminal	Connected to component ...	Terminal	Connected to component ...
22	⇒ Right front ABS wheel speed sensor -G45-	29	⇒ K wire connection T16/7
24	⇒ Right front ABS speed sensor -G45-	40	⇒ ABS warning light activation
28	⇒ Brake light switch -F-		

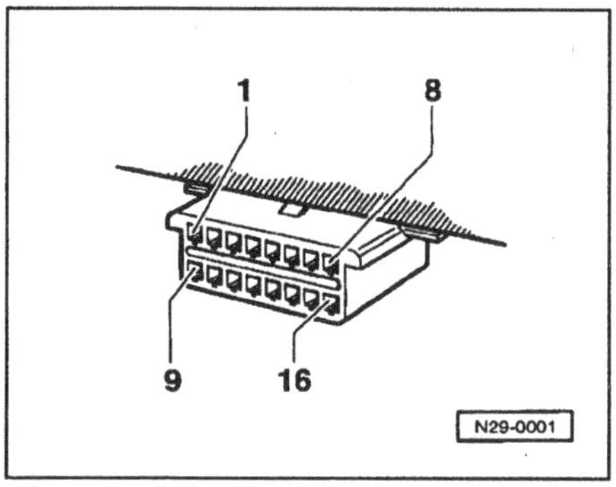

◄ **Data Link Connector (DLC) terminal assignment**

Terminal 4	=	Ground (terminal 31)
Terminal 16	=	Positive (terminal 30) via S21
Terminal 7	=	K-wire via junction TV 14 (above relay plate), connector T1V to ABS control module -J104- contact 29

Note:

DLC located vertically in vehicle.

Test table

Switch to measuring range: Voltage measurement (20 V =)					
Test step	**VAG 1598/24 socket**	**Item tested**	**• Test conditions – Additional operations**	**Specification**	**Repairing malfunction**
1	4 + 1	Voltage supply (terminal 15) on ABS control module -J104-	• Ignition switched on	10.0-14.5 V	– Check wire from terminal 4 via fuse S123 to relay plate W/3 – Check wire from terminal 1 to Ground, near -J104-
2	2 + 1	Voltage supply for pump motor (terminal 30) on control module -J104-	• Ignition switched off	10.0 - 14.5 V	– Check wire from terminal 1 to Ground. – Check wire from terminal 2 via fuse S53 to battery +.
3	3 + 1	Voltage supply for inlet and outlet solenoid valves (terminal 30) on ABS control module -J104-	• Ignition switched off	10.0 - 14.5 V	– Check wire from terminal 1 to Ground. – Check wire from terminal 3 via fuse S54 to battery +.

Switch to measuring range:
Voltage measurement (20 V =)

Test step	VAG 1598/24 socket	Item tested	• Test conditions – Additional operations	Specification	Repairing malfunction
4	1 + 28	Function of brake light switch -F-	• Ignition switched off – Brake pedal not operated	0.0 - 0.5 V	– Check fuse S20 and brake light switch -F-. – Check wire from terminal 1 to Ground.
			– Operate brake pedal	10.0 - 14.5 V	– Check wire from terminal 28 to terminal W/4 (relay plate).
5	-	ABS warning light -K47-, function	• Ignition switched off	-K47- does not light up	
			– Switch ignition on	-K47- lights up	– Check wiring from -K47- to relay plate X/4 and X/8 – Check -K47- – Check wire from relay plate W/2 to ABS control module -J104-, terminal 40 – Check Ground wire to return flow pump -V39-

Switch to measuring range:
Voltage measurement (20 V =) in test step 6, Resistance measurement (200 Ω) in test step 7

Test step	VAG 1598/24 socket	Item tested	• Test conditions – Additional operations	Specification	Repairing malfunction
6	–	Voltage supply for VAG 1551 Scan Tool, connector T16	Connect digital multimeter Fluke 83 with aux. cables from adapter set VW 1594 to T16 [1]:		– Check wire from T16/4 to Ground. – Check wire from T16/16 via S21 to terminal 30.
			terminal 4 - Ground terminal 16 - positive	10.0 -14.5 V	
7	–	K-wire resistance for OBD, connector T16	• Ignition switched off – Using adapter set VW 1594 connect socket 29 of VAG 1598 and terminal 7 of connector T16 to the digital multimeter Fluke 83		– Check wire from TV 14 to terminal 29.
				max. 1.5 Ω	– Check wire from T16/7 via TV14 to terminal 29.

[1] ⇒ page P 01-37, data link connector terminal assignment

Switch to measuring range: Resistance measurement (2 kΩ)					
Test step	**VAG 1598/24 socket**	**Item tested**	**• Test conditions – Additional operations**	**Specification**	**Repairing malfunction**
8	22 + 24	Resistance of ABS wheel speed sensor, RF -G45-	• Ignition switched off	1.2...2.0 kΩ	– Check connector T_{2d}. – Check resistance of ABS wheel speed sensor (1.2...2.0 kΩ). – Check wire to ABS wheel speed sensor, move wire while checking (loose connection).
9	16 + 17	Resistance of ABS wheel speed sensor, LF -G47-	• Ignition switched off	1.2...2.0 kΩ	– Check connector T_{2f} – Check resistance of ABS wheel speed sensor (1.2...2.0 kΩ) – Check wire to ABS wheel speed sensor, move wire while checking (loose connection).

Switch to measuring range: Resistance measurement (2 kΩ)					
Test step	**VAG 1598/24 socket**	**Item tested**	**• Test conditions – Additional operations**	**Specification**	**Repairing malfunction**
10	20 + 21	Resistance of ABS wheel speed sensor, RR -G44-	• Ignition switched off	1.2...2.0 kΩ	– Check connector T_{2c} – Check resistance of ABS wheel speed sensor (1.2...2.0 kΩ) – Check wire to ABS wheel speed sensor, move wire while checking (loose connection).
11	18 + 19	Resistance of ABS wheel speed sensor, LR -G46-	• Ignition switched off	1.2...2.0 kΩ	– Check connector T_{2e} – Check resistance of ABS wheel speed sensor (1.2...2.0 kΩ) – Check wire to ABS wheel speed sensor, move wire while checking (loose connection).

Switch to measuring range: Voltage measurement (2 V ≈)					
Test step	**VAG 1598/24 socket**	**Item tested**	**• Test conditions – Additional operations**	**Specification**	**Repairing malfunction**
12	22 + 24	ABS wheel speed sensor, RF -G45- voltage signal	• Vehicle raised		– Check installation of ABS wheel speed sensor and impulse rotor.
			• Ignition switched off		
			– Turn right front wheel at approx. 1 rev./sec.	min. 65 mV alternating voltage	– Check if ABS wheel speed sensor has been interchanged. ⇒ Read measuring value block P 01-32
13	16 + 17	ABS wheel speed sensor, LF -G47- voltage signal	• Vehicle raised		– Check installation of ABS wheel speed sensor and impulse rotor.
			• Ignition switched off		
			– Turn left front wheel at approx. 1 rev./sec.	min. 65 mV alternating voltage	– Check if ABS wheel speed sensor has been interchanged.⇒ Read measuring value block P 01-32

Switch to measuring range: Voltage measurement (2 V ≈)					
Test step	**VAG 1598/24 socket**	**Item tested**	**• Test conditions – Additional operations**	**Specification**	**Repairing malfunction**
14	20 + 21	ABS wheel speed sensor, RR -G44- voltage signal	• Vehicle raised		– Check installation of ABS wheel speed sensor and impulse rotor.
			• Ignition switched off		
			– Turn right rear wheel at approx. 1 rev./sec.	190 mV to1140 mV	– Check if ABS wheel speed sensor has been interchanged. ⇒ Read measuring value block P 01-32
15	18 + 19	ABS wheel speed sensor, LR -G46- voltage signal	• Vehicle raised		– Check installation of ABS wheel speed sensor and impulse rotor.
			• Ignition switched off		
			– Turn left front wheel at approx. 1 rev./sec.	190 mV to1140 mV	– Check if ABS wheel speed sensor has been interchanged. ⇒ Read measuring value block P 01-32